## Earth, Moon, and Sun

| | |
|---|---|
| **Earth** | |
| Mass | $M_E = 5.98 \times 10^{24}$ kg |
| Equatorial radius | $R_E = 6.378 \times 10^6$ m |
| Polar radius | $R'_E = 6.357 \times 10^6$ m |
| Mean density | 5520 kg/m$^3$ |
| Surface gravity | $g = 9.81$ m/s$^2$ = 32.2 ft/s$^2$ |
| Period of rotation | 1 sidereal day = 23 h 56 min 4 s = $8.616 \times 10^4$ s |
| Moment of inertia: | |
| about polar axis | $I = 0.331\ M_E R_E^2$ |
| about equatorial axis | $I' = 0.329 M_E R_E^2$ |
| Mean distance from Sun | $1.50 \times 10^{11}$ m |
| Period of revolution (period of orbit) | 1 year = 365 days 6 h = $3.16 \times 10^7$ s |
| Orbital speed | 29.8 km/s |
| **Moon** | |
| Mass | $7.35 \times 10^{22}$ kg |
| Radius | $1.74 \times 10^6$ m |
| Mean density | 3340 kg/m$^3$ |
| Surface gravity | 1.62 m/s$^2$ |
| Period of rotation | 27.3 days |
| Mean distance from Earth | $3.84 \times 10^8$ m |
| Period of revolution | 1 sidereal month = 27.3 days |
| **Sun** | |
| Mass | $M_s = 1.99 \times 10^{30}$ kg |
| Radius | $6.96 \times 10^8$ m |
| Mean density | 1410 kg/m$^3$ |
| Surface gravity | 274 m/s$^2$ |
| Period of rotation | ~ 26 days |
| Luminosity | $3.9 \times 10^{26}$ W |

## Greek Alphabet

| | | | | | |
|---|---|---|---|---|---|
| A | $\alpha$ | alpha | N | $\nu$ | nu |
| B | $\beta$ | beta | Ξ | $\xi$ | xi |
| Γ | $\gamma$ | gamma | O | $o$ | omicron |
| Δ | $\delta$ | delta | Π | $\pi$ | pi |
| E | $\varepsilon$ | epsilon | P | $\rho$ | rho |
| Z | $\zeta$ | zeta | Σ | $\sigma$ | sigma |
| H | $\eta$ | eta | T | $\tau$ | tau |
| Θ | $\theta$ | theta | Y | $\upsilon$ | upsilon |
| I | $\iota$ | iota | Φ | $\phi$ | phi |
| K | $\kappa$ | kappa | X | $\chi$ | chi |
| Λ | $\lambda$ | lambda | Ψ | $\psi$ | psi |
| M | $\mu$ | mu | Ω | $\omega$ | omega |

# Contents

* This section is optional.

* This section is optional.

* This section is optional.

INTERLUDE III. AUTOMOBILE COLLISIONS AND AUTOMOBILE STRUCTURE



* This section is optional.

* This section is optional.

## INTERLUDE V. ENERGY, ENTROPY, AND ENVIRONMENT

# Preface

In this Second Edition of *Physics,* I have incorporated divers alterations and additions recommended by the many thoughtful users of the First Edition. My objectives in the book remain the same: to present a contemporary, modern view of classical mechanics and electromagnetism, and to offer the student a glimpse of what is going on in physics today. Thus, throughout the book, I encourage students to keep in mind the atomic structure of matter and to think of the material world as a multitude of restless electrons, protons, and neutrons. For instance, in the mechanics chapters, I emphasize that all macroscopic bodies are systems of particles; and in the electricity chapters, I introduce the concepts of positive and negative charge by referring to protons and electrons, not by referring to the antiquated procedure of rubbing glass rods with silk rags (which, according to experts on triboelectricity, can give the wrong sign if the silk has been thoroughly cleaned). I try to make sure that students are always aware of the limitations of the nineteenth-century fiction that matter and electric charge are continua. Blind reliance on this fiction has often been justified by the claim that engineering students need physics as a tool, and that the atomic structure of matter is of little concern to them. But if physics is a tool, it is also a work of art, and its style cannot be dissociated from its function. In this book I give a physicist's view of physics, because I believe that it is fitting that all students should gain some appreciation for the artistic style of the toolmaker.

## Core Chapters and Interludes

The book contains two kinds of chapters: *core* chapters and *interlude* chapters. The 41 core chapters cover the essential topics of introductory physics: mechanics of particles, rigid bodies, and fluids; oscillations; wave motion; heat and thermodynamics; electricity and magnetism; optics; and special relativity. An expanded version of the book includes another five core chapters covering quantum physics, nuclei, and elementary particles.

The organization of the core chapters is fairly traditional, with some innovations. For instance, I start the study of magnetism (Chapter 30) with the law for the magnetic force between two moving point charges; this magnetic force is no more complicated than the force between two current elements, and the crucial advantage is that magnetism can be developed from the magnetic-force law in much the same way as electricity is developed from Coulomb's Law. This approach is consistent with the underlying philosophy of the book: particles are primary entities and should always be treated first, whereas macroscopic bodies and currents are composite entities which should be treated later. As another innovation, I include a simple derivation of the electric radiation field of an accelerated charge (Chapter 35); this calculation relies on Richtmeyer and Kennard's clever analysis of the kinks in the electric field lines of an accelerated charge (a set of computer-generated film loops available from the Educational Development Center shows how such kinks propagate along the field lines; these film loops tie in very well with the calculations of Chapter 35).

The interlude chapters (the expanded version contains 11) present some of the fascinating discoveries and applications of physics today: crystal structure and symmetry, the expansion of the universe, automobile safety, ionizing radiation, energy resources, atmospheric electricity, plasmas, superconductivity, general relativity, lasers, and fission. All of these interludes are optional—they rely on the core chapters, but the core chapters do not rely on them. Users of the First Edition will perceive that Chapter 46, on elementary particles, was an interlude in the First Edition. It has now been recast as a core chapter to round out the set of core chapters covering modern physics. However, it remains an interlude in spirit, and can be treated as such.

The inspiration for the interludes grew out of my unhappiness over a paradox afflicting the typical undergraduate physics curriculum: liberal-arts students in a nonmathematical physics course often get to see more of the beauty and excitement of today's physics than do science and engineering students in a calculus-based physics course. While liberal-arts students get a glimpse of gluons, black holes, or the Big Bang, science and engineering students are expected to calculate the motion of blocks on top of other blocks sliding down an inclined plane, or the motion of a baseball thrown in some direction or another by a man (or woman) riding in an elevator. To some extent this is unavoidable—science and engineering students need to learn and practice classical mechanics and electromagnetism, and they have little time left for dabbling in the arcane mysteries of contemporary physics. Nevertheless, most teachers will occasionally find an hour or two to tell their students a little of what is going on in physics today. I wrote the interludes to lend encouragement and support to such excursions to the frontiers of physics.

My choice of topics for the interludes reflect the interests expressed by my students. Over the years, I have often been asked: When will we get to quarks? or Are you going to tell us about gravitational collapse? and I came to feel that such curiosity must not be allowed to wither away. Obviously, in the typical introductory course it will be impossible to cover all of the interludes (I have usually covered two per term), but the broad range of topics will permit teachers to select according to their own tastes. The interludes are mainly descriptive rather than analytic. In them, I try to avoid formulas and instead give students a qualitative feeling for the underlying physics, keeping the discussion simple so that students can read them on their own. Thus, the interludes could be used for supplementary reading, not necessarily accompanied by lectures. For the inquisitive student, each interlude includes a collection of qualitative questions and an extensive annotated list of further readings.

## Optional Sections and Chapters

Optional sections and chapters have been indicated by a large asterisk (*). All such optional sections or chapters and all the interludes can be omitted without loss of continuity. Some other sections and chapters could possibly be omitted. For the guidance of teachers, the outline of quintessentials attached to this Preface lists all the sections that are indispensable for the logical coherence of the text.

## Mathematical Prerequisites

In order to accommodate students who are taking an introductory calculus course concurrently, derivatives are used slowly and hesitantly at first (Chapter 2), and routinely later on. Likewise, the use of integrals is postponed as far as possible (Chapter 7), and they come into heavy use only in the second volume (after Chapter 21). For students who need a review of calculus, Appendix 5 contains a concise primer on derivatives and integrals.

## Examples, Problems, Questions, and Summaries

The core chapters include generous collections of solved examples (about 300 altogether), of qualitative questions (about 850 altogether), and of problems (about 2150 altogether). Answers to the even-numbered problems are given in Appendix 11. The problems are grouped by sections, with the most difficult problems at the end of each section. The levels of difficulty of the problems are roughly indicated by no star, one star (*), or two stars (**). No-star problems are easy and straightforward; they are mostly of the plug-in type. One-star problems are of medium difficulty; they contain a few complications requiring the combination of several concepts or the manipulation of several formulas. Two-star problems are difficult and challenging; they demand considerable thought and perhaps some insight, and they occasionally demand substantial mathematical skills.

I have tried to make the problems interesting to the student by drawing on realistic examples from technology, sports, and everyday life. Many of the problems are based on data extracted from engineering handbooks, car-repair manuals, *Jane's Book of Aircraft, The Guinness Book of World Records,* newspaper reports, etc. Many other problems deal with atoms and subatomic particles; these are intended to reinforce the atomistic view of the material world. In some cases, cognoscenti will perhaps consider the use of classical physics somewhat objectionable in a problem that really ought to be handled by quantum mechanics. But I believe that the advantages of familiarization with atomic quantities and magnitudes outweigh the disadvantages of a naïve use of classical mechanics.

Each chapter also includes a collection of qualitative questions intended to stimulate thought and to test the grasp of basic concepts (some of these questions are discussion questions that do not have a unique answer). Moreover, each chapter contains a brief summary of the main physical quantities and laws introduced in it. The virtue of these summaries lies in their brevity. They include essential definitions and equations, because the statements in the body of each chapter are adequate.

## Units

The SI system of units is used exclusively. In the abbreviations for the units, I follow the dictates of the Conférence Générale des Poids et Mesures of 1971, although I deplore the majestic stupidity of the decision to replace the old, self-explanatory abbreviations amp, coul, nt, sec, °K by an alphabet soup of cryptic symbols A, C, N, s, K, etc. For the sake of clarity, I spell out the names of units in full whenever the abbreviations are likely to lead to ambiguity and confusion.

For reference purposes, the definitions of the British units have been retained. But these units are not used in examples or in problems, with the exception of a handful of problems in the first chapter. In the definitions of the British units, the pound (lb) is taken to be the unit of mass, and the pound-force (lbf) is taken to be the unit of force. This is in accord with the practice approved by the American National Standards Institute (ANSI), the Institute of Electrical and Electronic Engineers (IEEE), and the United States Department of Defense.

## Changes from the First Edition

In response to comments from users of the First Edition, I have reorganized some chapters and added some new chapters and sections to provide more thorough and better-balanced coverage. The discussion of angular momentum, which used to be dispersed over two chapters, now appears in one place, in Chapter 12. The discussion of the magnetic force between moving charges in Chapter 30 has been made clearer and simpler. Concise, but self-contained, introductions to dot and cross products of vectors have been included where they are first needed (Chapters 7 and 12, respectively); this makes the book more flexible, since it is now possible to skip the sections on these products in Chapter 3. The chapter on gravitation has been moved forward, so it now precedes the discussion of systems of particles; this move seems sensible since all the simple problems in gravitation assume a fixed center of force, and they can therefore be regarded as single-particle problems. The chapter on special relativity and the chapter on elementary particles have been moved toward the end of the book, in accord with their historical position. However, the prerequisites for these two chapters have not been changed appreciably, and students could read them much sooner, during the first half of the course.

Several new chapters have been added: (14) "Statics and Elasticity"; (38) "Mirrors, Lenses, and Optical Instruments"; (44) "Quantum Structure of Atoms, Molecules, and Solids"; (45) "Nuclei"; and a new Interlude III, "Automobile Collisions and Automobile Structure." The sections that have been added are: 1.7 Significant Figures, Conversion of Units, and Consistency of Units; 19.3 Kinetic Pressure and the Maxwell Distribution; 19.5 The Mean Free Path; 20.3 Thermometers and Thermal Equilibrium; 29.6 Electrical Measurements; 29.7 The RC Circuit; 32.7 The RL Circuit; and 37.4 Polarization.

Besides these additions, the new edition incorporates a multitude of stylistic and pedagogical improvements. For instance, *Comments and Suggestions* have been appended to many of the solved examples in the text; these comments point out generalizations and/or special features of the results, and they provide the student with helpful hints on how to apply (and how not to apply) the methods illustrated in the examples to the solution of problems.

The number of solved examples and of problems has been increased by about 25%, and more attention has been paid to achieving a balance among examples and among problems of diverse levels of difficulty.

## Study Guide

An excellent study guide for this book has been written by Professors Van E. Neie (Purdue University) and Peter J. Riley (University of Texas, Austin). This guide includes for every chapter a brief introduction laying out the objectives; a list of key terms for review; detailed commentaries on each of the main ideas; and a large collection of interesting sample problems, which alternate between worked problems (with full solutions) and guided problems (which provide step-by-step schemes that lead students to the solutions). Those with Macintosh computers should also be aware of an exciting new project by Eric Mazur (Harvard University). His software gives access to any of its three component parts: interactively solved problems, demonstrations, and summary notes.

## Acknowledgments

I have greatly benefited from comments by reviewers and users of the book. Many of the alterations in this Second Edition originated from users who, to my pleasant surprise, took the trouble to send me their recommendations for improvements, and sometimes even sent me data for the construction of additional realistic and interesting problems. For very detailed, comprehensive reviews, I am indebted to John R. Boccio (Swarthmore College), Roger W. Clapp, Jr. (University of South Florida), A. Douglas Davis (Eastern Illinois University), Anthony P. French (Massachusetts Institute of Technology), J. David Gavenda (University of Texas, Austin), Roger D. Kirby (University of Nebraska), Roland M. Lichtenstein (Rensselaer Polytechnic Institute), Richard T. Mara (Gettysburg College), John T. Marshall (Louisiana State University), Delo E. Mook (Dartmouth College), Harvey S. Picker (Trinity College), Peter J. Riley (University of Texas, Austin), and Peter L. Scott (University of California, Santa Cruz). For briefer reviews, I am indebted to John R. Albright (Florida State University), I. H. Bailey (Western Australia Institute of Technology), Frank Crawford (University of California, Berkeley), Sumner Davis (University of California, Berkeley), John F. Devlin (University of Michigan, Dearborn), Phil Eastman (University of Waterloo, Canada), Hugh Evans (University of Waterloo, Canada), Frank A. Ferrone (Drexel University), James R. Gaines (Ohio State University), Stephen Gasiorowicz (University of Minnesota), Michael A. Guillen (Harvard University), Lyle Hoffman (Lafayette College), Walter Knight (University of California, Berkeley), Jean P. Krisch (University of Michigan, Ann Arbor), L. B. Meyer (Duke University), Frank Moscatelli (Swarthmore College), Hermann Nann (Indiana University), Mette Owner-Peterson (The Technical University of Denmark), Norman Pearlman (Purdue University), P. Bruce Pipes (Dartmouth College), Jack Prince (Bronx Community College), Malvin Ruderman (Columbia University), Kenneth Schick (Union College), Mark P. Silverman (Trinity College), Gerald A. Smith (Michigan State University), Julia A. Thompson and David Kraus (University of Pittsburgh), Som Tiagi (Drexel University), and Gary A. Williams (University of California, Los Angeles). I am also indebted to the experts who reviewed the interludes: Edmond Brown (Rensselaer Polytechnic Institute; "The Architecture of Crystals"), Priscilla W. Laws (Dickinson College; "Radiation and Life"), Alan H. Guth (Massachusetts Institute of Technology; "The Big Bang and the Expansion of the Universe"), Frank Richardson (National Highway Traffic Safety Administration; "Automobile Collisions and Automobile Structure"), Donald F. Kirwan (University of Rhode Island; "Energy, Entropy, and the Environment"), Irving Kaplan (Massachusetts Institute of Technology; "Nuclear Fission"), Bernard Vonnegut (State University of New York at Albany; "Atmospheric Electricity"), Sam Cohen (Princeton University Plasma Physics Laboratory; "Plasma"), and Margaret L. A. MacVicar (Massachusetts Institute of Technology; "Superconductivity").

I thank Richard D. Deslattes and Barry N. Taylor of the National Bureau of Standards for information on units, standards, and precision measurements. I thank some of my colleagues at Union College: C. C. Jones prepared the beautiful photographs of interference and diffraction by light (Chapters 39 and 40); Barbara C. Boyer gave valuable advice and assistance on photographs dealing with biological material; and the staff of the library helped me find many a tidbit of information needed for an example or a problem.

I thank the editorial staff of W. W. Norton & Co. for their zeal. Drake McFeely, editor, gave this Second Edition the same meticulous attention he gave the First, and he was indefatigable in collecting comments from reviewers and users. Avery Hudson supervised all aspects of the production process; he also improved my sentences, and helped me to attain a high level of stylistic consistency in the text and in the diagrams. And Ruth Mandel provided the detective work needed to find the many splendid new photographs that have been added to the book.

H. C. O.
February 1989

## Quintessentials

The sections in the following list are indispensable for the logical coherence of the text. Teachers pressed for time can take this list of quintessentials as a base line, and add sections to suit their personal preferences.

Chapter 2: all
Chapter 3: 3.1, 3.2, 3.3
Chapter 4: 4.1, 4.2, 4.4
Chapter 5: all
Chapter 6: 6.2, 6.3, 6.5, 6.7
Chapter 7: all
Chapter 8: 8.1, 8.2, 8.5, 8.7
Chapter 10: 10.1, 10.2, 10.3
Chapter 11: 11.1, 11.2
Chapter 12: all
Chapter 13: 13.1, 13.2, 13.4
Chapter 15: 15.1, 15.2, 15.3
Chapter 16: 16.1, 16.2, 16.5, 16.6
Chapter 19: 19.1, 19.2, 19.3
Chapter 20: 20.1

# THE WORLD OF PHYSICS

Physics is the study of matter. In a very literal sense, physics is the greatest of all natural sciences: it encompasses the smallest particles, such as electrons and quarks; and it also encompasses the largest bodies, such as galaxies and the entire universe. The smallest particles and the largest bodies differ in size by a factor of more than $10^{40}$! In the following pictures we will survey the world of physics and attempt to develop some rough feeling for the sizes of things in this world. This preliminary survey sets the stage for our explanations of the mechanisms that make things behave in the way they do. Such explanations are at the heart of physics, and they are the concern of the later chapters of this book.

The pictures fall into two sequences. In the first sequence we zoom out: we begin with a picture of a woman's face and proceed step by step to pictures of the entire Earth, the Solar System, the Galaxy, and the universe. This ascending sequence contains 27 pictures, with the scale decreasing in steps of factors of 10.

In the second sequence we zoom in: we again begin with a picture of the face and close in on the eye, the retina, the rod cells, the molecules, the atoms, and the subatomic particles. This descending sequence contains 15 pictures, with the scale increasing in steps of factors of 10.

Most of our pictures are photographs. Many of these have only become available in recent years; they were taken by high-flying U-2 aircraft, Landsat satellites, astronauts on the Moon, or sophisticated electron microscopes. For some of our pictures no photographs are available and we have to rely, instead, on carefully prepared drawings.

## PART I: THE LARGE-SCALE WORLD

**SCALE 1:1.5** This is Charlotte, an intelligent biped of the planet Earth, Solar System, Orion Spiral Arm, Milky Way Galaxy, Local Group, Local Supercluster. She is made of $5.1 \times 10^{27}$ atoms, with $1.8 \times 10^{28}$ electrons, the same number of protons, and $1.4 \times 10^{28}$ neutrons.

**SCALE 1:1.5 × 10** Charlotte has a height of 1.7 meters and a mass of 55 kilograms. Her chemical composition (by mass) is 65% oxygen, 18.5% carbon, 9.5% hydrogen, 3.3% nitrogen, 1.5% calcium, 1% phosphorus, and 0.35% of other elements.

The matter in Charlotte's body and the matter in her immediate environment occur in three states of aggregation: *solid, liquid,* and *gas.* All these forms of matter are made of atoms and molecules, but solid, liquid, and gas are qualitatively different because the arrangements of the atomic and molecular building blocks are different.

In a solid, each building block occupies a definite place. When a solid is assembled out of molecular or atomic building blocks, these blocks are locked in place once and for all, and they cannot move or drift about except with great difficulty. This rigidity of the arrangement is what makes the aggregate hard — it makes the solid "solid." In a liquid, the molecular or atomic building blocks are not rigidly connected. They are thrown together at random and they move about fairly freely, but there is enough adhesion between neighboring blocks to prevent the liquid from dispersing. Finally, in a gas, the molecules or atoms are almost completely independent of one another. They are distributed at random over the volume of the gas and are separated by appreciable distances, coming in touch only occasionally during collisions. A gas will disperse spontaneously if it is not held in confinement by a container or by some restraining force, such as gravity.

The molecules of a gas are forever moving around at high speed. For example, at a temperature of 0°C the average speed of a molecule of nitrogen in air is about 450 meters/second, faster than the speed of sound. The speed of the molecules is directly related to the temperature: the speed increases if the air is heated and decreases if the air is cooled. The molecules of a liquid also move; their speeds are not very different from those of the molecules of a gas. However, since there is little space between the molecules in a liquid, the motion is continually interrupted by collisions. Because of these frequent collisions, the path of a molecule consists of a series of random zigzags, and the molecule takes a long time to wander from one part of the liquid to another. Even in a solid there is some motion of the building blocks. But the motion of each atom or molecule is merely a vibration around its assigned position — the atom behaves as if kept on a short leash and although it moves back and forth at a high speed, it never strays beyond some tight limits.

If we regard the motion of the atoms of a gas in a container as analogous to the bouncing of a few dice in a shaker, then the motion of atoms in a liquid is analogous to the random wandering of the dice in a shaker that has been loosely but completely filled with dice; and the motion of atoms in a solid is analogous to the impotent rattling of the dice in a shaker that has been tightly and regularly packed full of dice.

The eternal, dancing motion of the atoms is the key to the transformations of state from solid to liquid to gas and vice versa. If we heat a solid, the vibrational motion becomes more violent and the atoms or molecules finally shake themselves out of place — the solid softens and melts, turning into a liquid. If we heat this resulting liquid further, the motion ultimately becomes so violent that the adhesion between the atoms or molecules cannot prevent some of them from escaping from the surface of the liquid. Gradually more and more escape — the liquid evaporates, turning into a gas.

**SCALE 1:1.5 × 10²** The building behind Charlotte is the New York Public Library, one of the largest libraries on Earth. This library holds 9,300,000 volumes, containing roughly 10% of the total accumulated knowledge of our terrestrial civilization.

**SCALE 1:1.5 × 10³** The New York Public Library is located at the corner of Fifth Avenue and 42nd Street, in the middle of New York City.

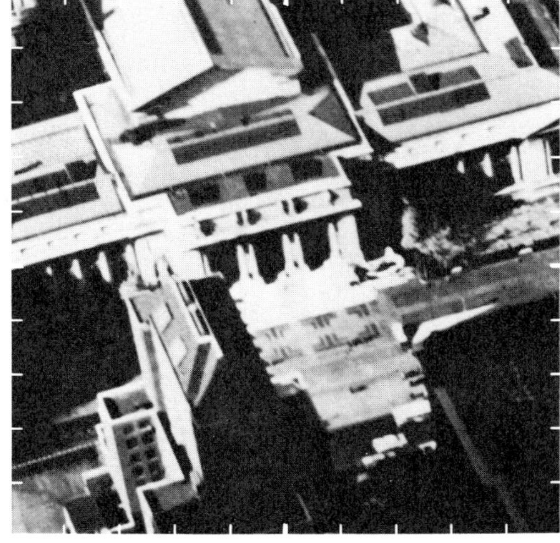


**SCALE 1:1.5 × 10⁴** This aerial photograph shows an area of 1 kilometer × 1 kilometer in the vicinity of the New York Public Library. The streets in this part of the city are laid out in a regular rectangular pattern. The library is the building in the park in the upper middle of the picture. The Empire State Building shows up in the lower left of the picture. This skyscraper is 381 meters high and it casts a long shadow. Completed in 1930, it was for many years the tallest building in the world. Although the concrete and steel used in its construction are usually thought of as rigid materials, they have some elasticity — in a strong wind the top of the Empire State Building sways as much as 20 centimeters to a side.

This photograph was taken from an airplane flying at an altitude of 3700 meters. North is at the top of the photograph.

**SCALE 1:1.5 × 10⁵** This photograph shows a large portion of New York City. We can recognize the library and its park as a small rectangular patch slightly above the center of the picture. The central mass of land is the island of Manhattan, with the Hudson River on the left and the East River on the right. Three bridges over the East River connect Manhattan with other parts of the city. The hundred-year-old Brooklyn Bridge is the southernmost of these bridges. It was the first steel-wire suspension bridge ever built and it stands as a spectacular and graceful achievement of nineteenth-century engineering.

This photograph was taken from a U-2 aircraft flying at an altitude of about 20,000 meters.

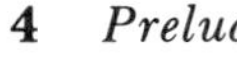


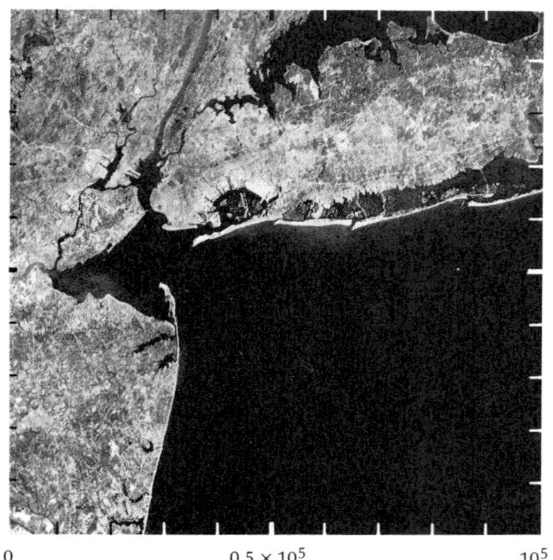

**SCALE 1:1.5 × 10⁶** In this photograph, Manhattan is in the upper left quadrant. On this scale, we can no longer distinguish the pattern of streets in the city. However, we can distinguish many highways, bridges, and causeways. For example, the thin white line crossing the dark water south of the tip of Manhattan is the Verrazano-Narrows Bridge. With a center span of 1300 m, it is one of the longest suspension bridges in the world. The vast expanse of water in the lower right of the picture is part of the Atlantic Ocean. The mass of land in the upper right is Long Island, with Long Island Sound to the north of it. Parallel to the south shore of Long Island we can see a string of very narrow islands; they almost look man-made. These are barrier islands; they are heaps of sand piled up by ocean waves in the course of thousands of years.

This photograph was taken by a Landsat satellite orbiting the Earth at an altitude of 920 kilometers.

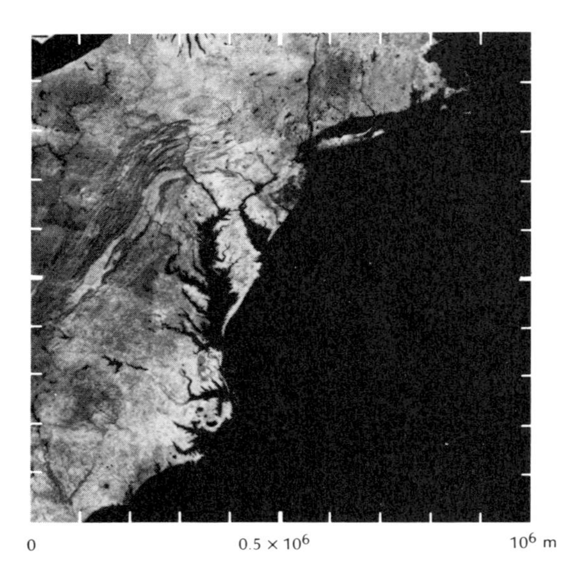

0 $\quad 0.5 \times 10^6 \quad 10^6$ m

**SCALE 1:1.5 × 10⁷** Here we see the eastern coast of the United States, from Cape Cod to Cape Fear. Cape Cod is the hook near the northern end of the coastline, and Cape Fear is the promontory near the southern end of the coastline. If we move along the coast starting at the north, we first come to Long Island; then to Delaware Bay and Chesapeake Bay, two deep indentations in the coastline; and then to Cape Hatteras, at the extreme end of the large bulge of land thrusting eastward into the Atlantic. Several rivers show up as thin, dark lines: the Hudson running due south from the northern edge of the photograph, the Delaware flowing into Delaware Bay, and the Susquehanna and Potomac flowing into Chesapeake Bay. The wrinkles in the land west of Chesapeake Bay are the Appalachian Mountains. Note that on this scale no signs of human habitation are visible. However, at night the lights of large cities would stand out clearly.

This picture is a mosaic, assembled by joining together many Landsat photographs such as the one above.

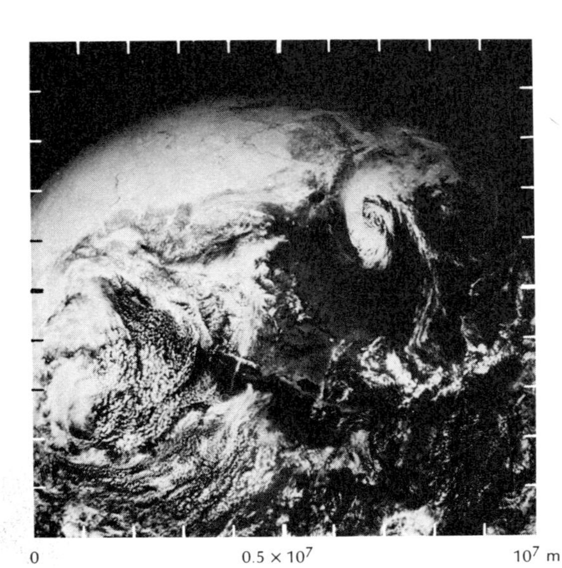

0 $\quad 0.5 \times 10^7 \quad 10^7$ m

**SCALE 1:1.5 × 10⁸** In this photograph, taken by the Apollo 16 astronauts during their trip to the Moon, we see a large part of the Earth. Through the gap in the clouds in the lower middle of the picture, we can see the coast of California and Mexico. We can recognize the peninsula of Baja California and the Gulf of California. In the middle right of the photograph we can recognize the Gulf of Mexico. Charlotte's location, the East Coast of the United States, is covered by a big system of swirling clouds in the upper right of the photograph.

Note that a large part of the area visible in this photograph is ocean. About 71% of the surface of the Earth is ocean; only 29% is land. The atmosphere covering this surface is about 100 kilometers thick; on the scale of this photograph, its thickness is about 0.7 millimeter. Seen from a large distance, the predominant colors of the planet Earth are blue (oceans) and white (clouds).

**SCALE 1:1.5 × 10⁹** This photograph of the Earth was taken by the Apollo 16 astronauts standing on the surface of the Moon. Sunlight is striking the Earth from the top of the picture.

As is obvious from this and from the preceding photograph, the Earth is a sphere. Its radius is $6.38 \times 10^6$ meters and its mass is $5.98 \times 10^{24}$ kilograms. Its chemical composition (by mass) is 38.6% iron, 28.6% oxygen, 14.4% silicon, 11.1% magnesium, 3.0% nickel, 1.6% sulfur, 1.3% aluminum, and 1.3% of other elements.

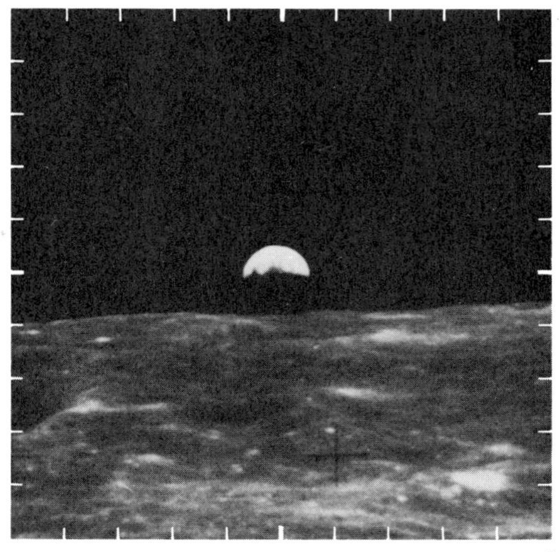


**SCALE 1:1.5 × 10¹⁰** In this picture, we see the Earth, the Moon, and its orbit. This picture is a drawing, not a photograph; none of our manned spacecraft has traveled sufficiently far away to take a photograph of such a panoramic view. (Many of the pictures on the following pages are also drawings.) As in the preceding picture, the Sun is far below the bottom of the picture. The position of the Moon is that of January 1, 1980. On this day, the Moon was almost full.

The orbit of the Moon around the Earth is an *ellipse*, but an ellipse that is very close to a circle. The solid curve in the picture is the orbit of the Moon and the dashed curve is a circle; by comparing these two curves we can see how little the ellipse deviates from a circle. The point on the ellipse closest to the Earth is called the *perigee* and the point farthest from the Earth is called the *apogee*. The distance between the Moon and the Earth is roughly 30 times the diameter of the Earth. The Moon takes 27.3 days to travel once around the Earth.

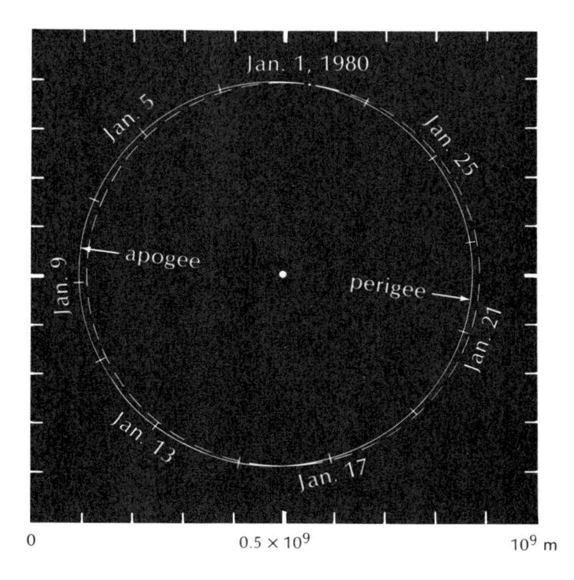


**SCALE 1:1.5 × 10¹¹** This picture shows the Earth, the Moon, and portions of their orbits around the Sun. On this scale, both the Earth and the Moon look like small dots. Again, the Sun is far below the bottom of the picture. In the middle, we see the Earth and the Moon in their positions for January 1, 1980. On the right and on the left we see, respectively, their positions for 1 day before and 1 day after this date.

Note that the net motion of the Moon consists of the combination of two simultaneous motions: the Moon orbits around the Earth, which in turn orbits around the Sun. The net orbit of the Moon around the Sun is, roughly, a twelve-sided polygon with rounded sides and corners.

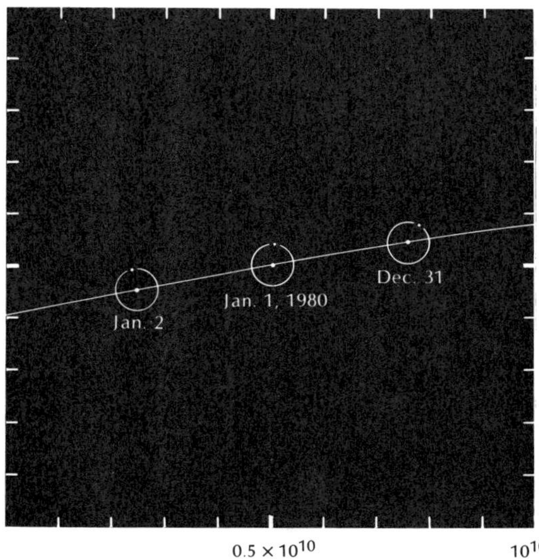

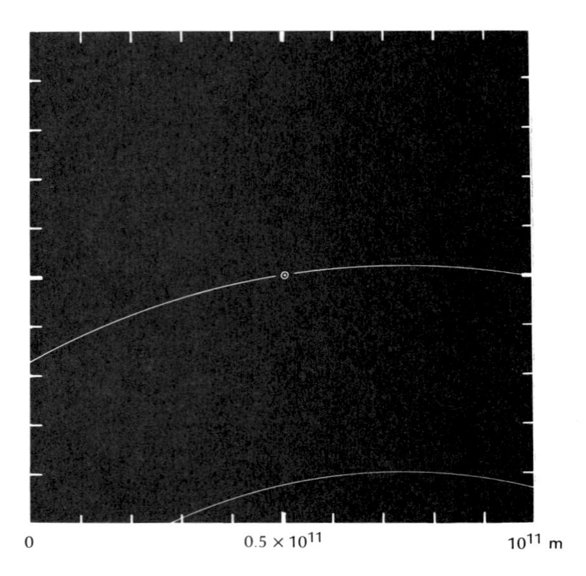


**SCALE 1:1.5 × 10¹²** Here we see the orbits of the Earth and of Venus. However, Venus itself is beyond the edge of the picture. The small circle is the orbit of the Moon. The dot representing the Earth is much larger than what it should be, although the draftsman has drawn it as minuscule as possible. On this scale, even the Sun is quite small; if it were included in this picture, it would be only 1 millimeter across.

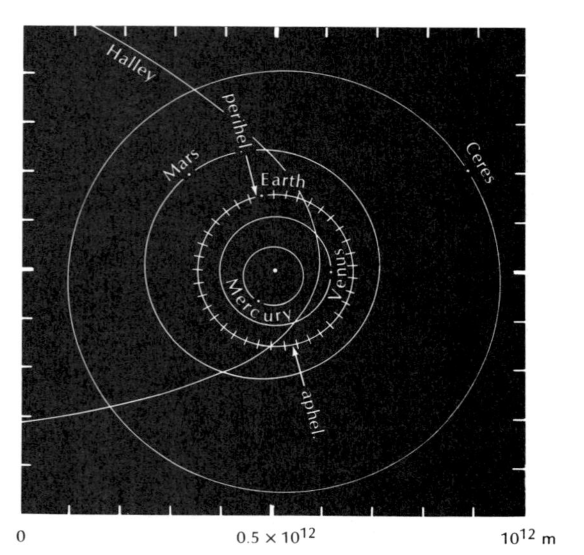


**SCALE 1:1.5 × 10¹³** This picture shows the positions of the Sun and the inner planets: Mercury, Venus, Earth, and Mars. The positions of the planets are those of January 1, 1980. The orbits of all these planets are ellipses, but they are close to circles. The point of the orbit nearest to the Sun is called the *perihelion* and the point farthest from the Sun is called the *aphelion*. The Earth reaches perihelion about January 3 and aphelion about July 6 of each year.

All the planets travel around their orbits in the same direction: counterclockwise in our picture. The marks along the orbit of the Earth indicate the successive positions at intervals of 10 days. The orbits of the planets are not quite in the same plane. In the picture, we see the orbit of the Earth exactly face on; the orbits of the other planets are slightly tilted, but this tilt is not shown in the picture.

Beyond the orbit of Mars, a large number of asteroids orbit around the Sun. The four largest of these asteroids are Ceres, Pallas, Juno, and Vesta; the first of these has been included in our picture, but the others have been omitted to prevent excessive clutter. Furthermore, a large number of comets orbit around the Sun. Most of these have pronounced elliptical orbits. The comet Halley has been included in our picture.

The Sun is a sphere of radius $6.96 \times 10^8$ meters. On the scale of the picture, the Sun looks like a very small dot, even smaller than the dot drawn here. The mass of the Sun is $1.99 \times 10^{30}$ kilograms. Its chemical composition (by mass) is 75% hydrogen, 23% helium, 0.8% oxygen, 0.4% carbon, 0.2% nitrogen, and about 0.6% of other elements. Since the mass of the Sun is much larger than that of the planets, hydrogen and helium are by far the most abundant atoms in the Solar System. All the other atoms taken together account for less than 2% of the mass in the Solar System and in our Galaxy. Thus, the chemical elements found on the Earth and in our bodies must be regarded as mere traces, mere impurities — we are made of very rare stuff.

The matter in the Sun is in the *plasma* state, sometimes called the fourth state of matter. Plasma is a very hot gas in which violent collisions between the atoms in their random thermal motion have fragmented the atoms, ripping electrons off them. An atom that has lost one or more electrons is called an *ion*. Thus, plasma consists of a mixture of electrons and ions, all milling about at high speed and engaging in frequent collisions. These collisions are accompanied by the emission of light, making the plasma luminous.

**SCALE 1:1.5 × 10¹⁴** This picture shows the positions of the outer planets of the Solar System: Jupiter, Saturn, Uranus, Neptune, and Pluto. On this scale, the orbits of the inner planets are barely visible. As in our other pictures, the positions of the planets are those of January 1, 1980.

The outer planets move slowly and their orbits are very large; thus they take a long time to go once around their orbit. The extreme case is that of Pluto, which takes 248 years to complete one orbit.

Uranus, Neptune, and Pluto are so far away and so faint that their discovery only became possible through the use of telescopes. Uranus was discovered in 1781, Neptune in 1846, and the tiny Pluto in 1930.

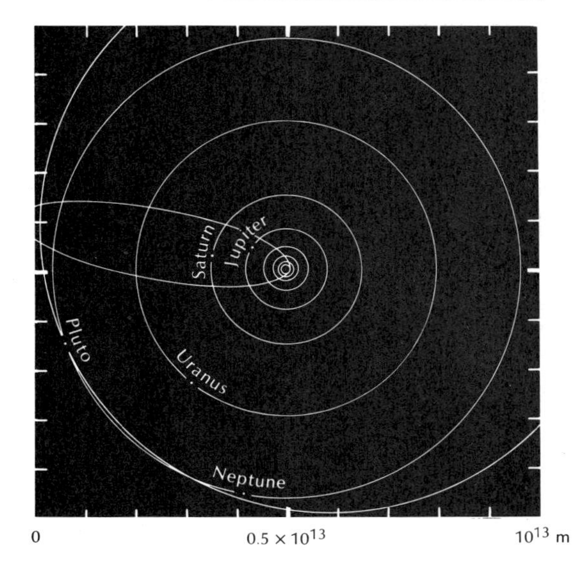


**SCALE 1:1.5 × 10¹⁵** We now see that the Solar System is surrounded by a vast expanse of empty space. Actually, this space is not quite empty. The Solar System is encircled by a large cloud of millions of comets whose orbits crisscross the sky in all directions. Furthermore, the interstellar space in this picture and in the succeeding pictures contains traces of gas and of dust. The interstellar gas is mainly hydrogen; its density is typically 1 atom per cubic centimeter.

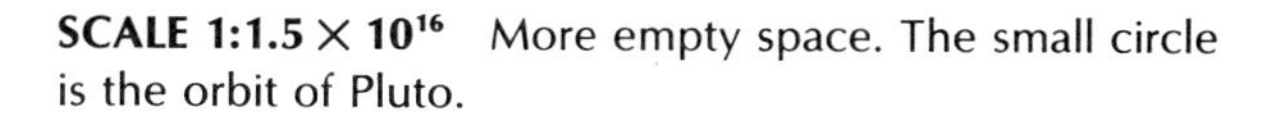
**SCALE 1:1.5 × 10¹⁶** More empty space. The small circle is the orbit of Pluto.

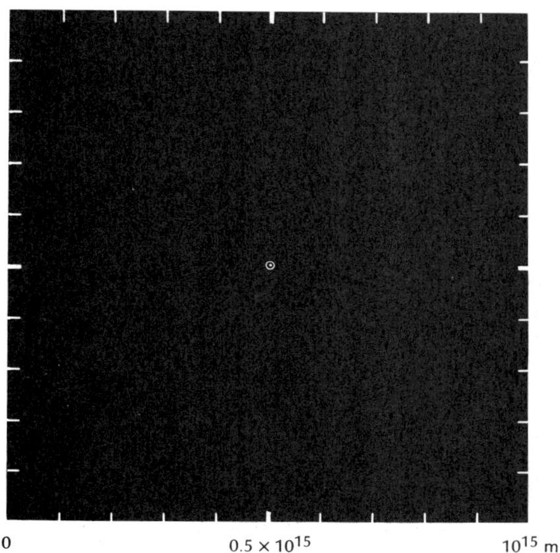

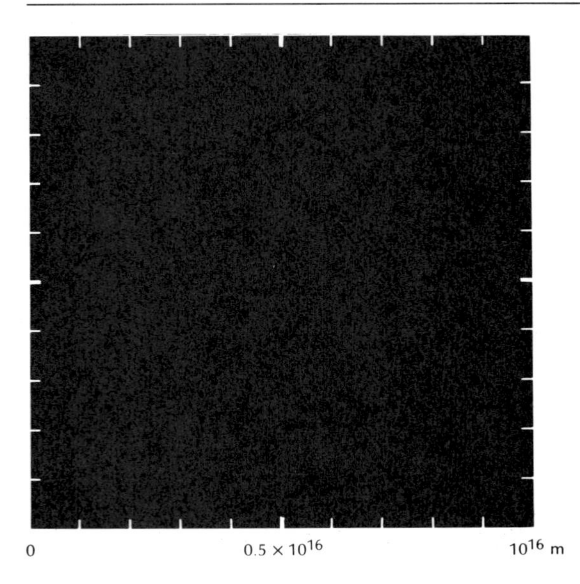


**SCALE 1:1.5 × 10^17** And more empty space. On this scale, the Solar System looks like a minuscule dot, 0.1 millimeter across.

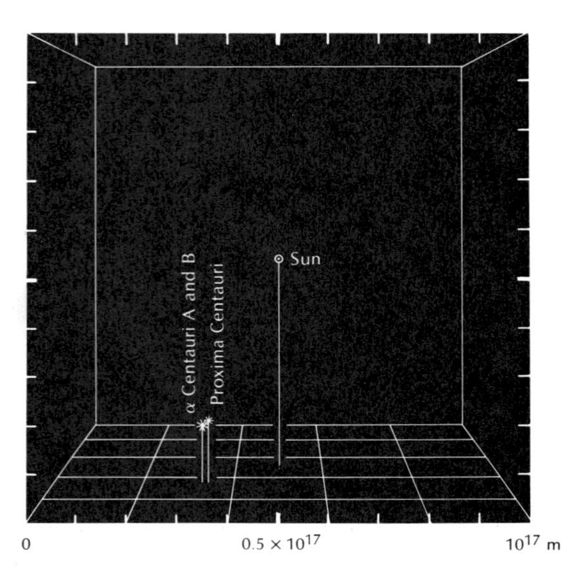


**SCALE 1:1.5 × 10^18** Here, at last, we see the stars nearest to the Sun. The picture shows all the stars within a cubical box $10^{17}$ meters × $10^{17}$ meters × $10^{17}$ meters centered on the Sun: Alpha Centauri A, Alpha Centauri B, and Proxima Centauri. All three are in the constellation Centaurus, in the southern sky.

The star closest to the Sun is Proxima Centauri. This is a very faint, reddish star (a "red dwarf"), at a distance of $4.0 \times 10^{16}$ meters from the Sun. Astronomers like to express stellar distances in light-years: Proxima Centauri is 4.2 light-years from the Sun, which means light takes 4.2 years to travel from this star to the Sun.

Proxima Centauri is too faint to be seen by the naked eye. The nearest stars that can be seen by the naked eye are Alpha Centauri A and Alpha Centauri B. The former is a bright star quite similar to our Sun; the latter is a fainter, orange star. These two stars are so close together that we need to use a telescope to distinguish between them. They form a double star, continually orbiting around each other. This double star is at a distance of 4.3 light-years from the Sun.

Like the Sun, all the stars are giant balls of luminous plasma. In all the stars, the temperature is so high that almost all the atoms are ionized; only near the surface of the stars can some atoms survive intact.

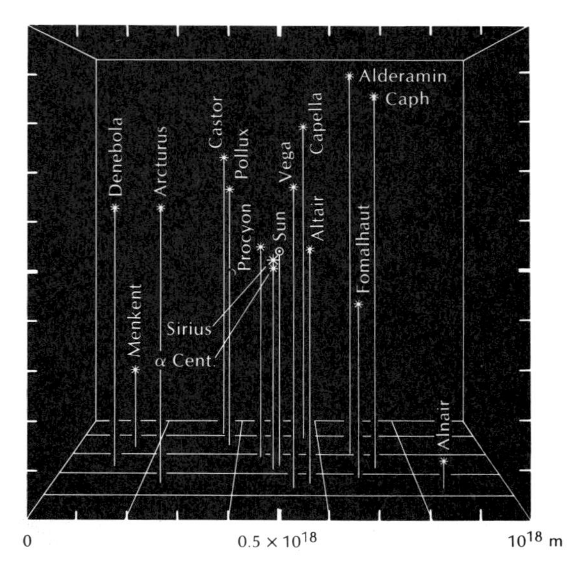


**SCALE 1:1.5 × 10^19** This picture displays the brightest stars within a cubical box $10^{18}$ meters × $10^{18}$ meters × $10^{18}$ meters centered on the Sun. There are many more stars in this box besides those shown — the total number of stars in this box is about 2000.

Sirius is the brightest of all the stars in the night sky. If it were at the same distance from the Earth as the Sun, it would be 28 times brighter than the Sun. Sirius has a much fainter companion very close to it.

**SCALE 1:1.5 × 10²⁰** Here we expand our box to $10^{19}$ meters × $10^{19}$ meters × $10^{19}$ meters, again showing only the brightest stars and omitting many others. The total number of stars within this box is about 2 million. We recognize several clusters of stars in this picture: the Pleiades Cluster, the Hyades Cluster, the Coma Berenices Cluster, and the Perseus Cluster. Each of these has hundreds of stars crowded into a fairly small patch of sky.

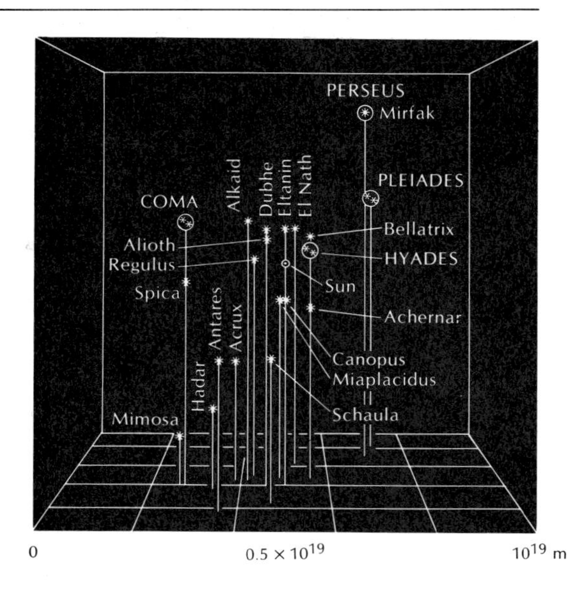


**SCALE 1:1.5 × 10²¹** Now there are so many stars in our field of view that they appear to form clouds of stars. There are about a million stars in this photograph, and there are many more stars too faint to show up distinctly. Although this photograph is not centered on the Sun, it simulates what we would see if we could look toward the Solar System from very far away. The photograph shows a view of the Milky Way in the direction of the constellation Sagittarius. When we look from our Solar System in this direction, we see the clouds of stars in the neighboring spiral arm of our Galaxy (the next picture shows a galaxy and its spiral arms). This neighboring arm is the Sagittarius Spiral Arm; the cloud of stars in which our Sun is located belongs to the *Orion Spiral Arm*.

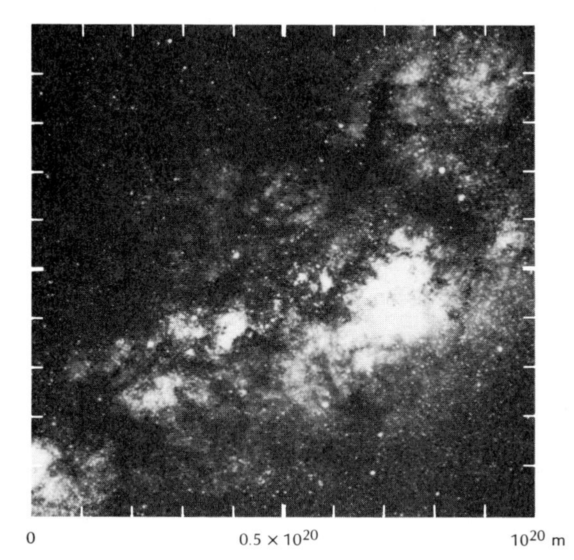


**SCALE 1:1.5 × 10²²** This is the spiral galaxy NGC 5457. Its clouds of stars are arranged in spiral arms wound around a central bulge. The bright central bulge is the nucleus of the galaxy; it has a more or less spherical shape. The surrounding region, with the spiral arms, is the disk of the galaxy. This disk is quite thin; it has a thickness of only about 3% of its diameter. The stars making up the disk circle around the galactic center in a counterclockwise direction.

Our Sun is in a spiral galaxy of roughly similar shape and size: the *Galaxy of the Milky Way*. The total number of stars in this galaxy is about $10^{11}$. The Sun is in one of the spiral arms, roughly one-third inward from the edge of the disk toward the center. The Sun takes about 250 million years to complete one orbit around the galactic center. Recent observations of the galactic center with radiotelescopes suggest that there is a large black hole at the center, a black hole several million times as massive as the Sun.

This photograph, like the remaining photographs of this section, was made with the great 5-meter telescope on Palomar Mountain.

**SCALE 1:1.5 × $10^{23}$** Galaxies are often found in clusters of several galaxies. Some of these clusters consist of just a few galaxies, others of hundreds or even thousands. The photograph shows a modest cluster, or group, of four galaxies beyond the constellation Hercules. The group contains an elliptical galaxy like a luminous egg (lower left), two spiral galaxies close together (top), and a spiral with a bar (bottom).

Our Galaxy is part of another modest cluster, the *Local Group,* consisting of our own Galaxy, the great Andromeda galaxy, the Triangulum galaxy, the Large Magellanic Cloud, plus 16 other small galaxies.

According to recent investigations, the dark, apparently empty, space near galaxies contains some form of distributed matter, with a total mass several times as large as the mass in the luminous, visible galaxies. But the composition of this dark, invisible, extragalactic matter is not known.

In this and the following photograph, the stars in the foreground have been erased to prevent their confusion with galaxies.

**SCALE 1:1.5 × $10^{24}$** The Local Group lies on the fringes of a very large cluster of galaxies, called the *Local Supercluster.* This is a cluster of clusters of galaxies. At the center of the Local Supercluster is the Virgo Cluster with several thousand galaxies. Seen from a large distance, our supercluster would present a view comparable to this photograph, which shows a rich cluster beyond the constellation Hydra, a cluster that is at a large distance from us.

All the distant galaxies are moving away from us and away from each other. For instance, the Hydra Cluster is moving away from our Galaxy at the rate of 6030 kilometers per second. This motion of recession of the galaxies is analogous to the outward motion of, say, the fragments of a grenade after its explosion. The motion of the galaxies suggests that the universe began with a big explosion, the *Big Bang,* that launched the galaxies away from each other.

**SCALE 1:1.5 × $10^{25}$** On this scale a galaxy equal in size to our own Galaxy would look like a fuzzy dot, 0.1 millimeter across. Thus, the galaxies are too small to show up clearly on a photograph. Instead we must rely on a plot of the positions of the galaxies. The plot shows the positions of about 200 of the brightest galaxies in an angular patch of the sky, 45° by 45°.

Since we are looking into a volume of space, some of the galaxies are in the foreground, some are in the background; but our plot takes no account of perspective. The distance scale has been computed for those galaxies that are at a middle distance, at about $1.4 \times 10^{8}$ light-years from the Earth; the scale is not valid for galaxies in the foreground or in the background.

The dense cluster of galaxies in the lower half of the picture is the Virgo Cluster. The loose cluster of galaxies in the upper half of the picture is the Coma Cluster. This cluster has roughly the same number of galaxies as the Virgo Cluster but is much farther away; hence only the very brightest galaxies of the Coma Cluster have been included in the picture.

**SCALE 1:1.5 × $10^{26}$** Now there are so many galaxies in our field of view that it is impractical to plot the positions of individual galaxies. In this picture the sky has been divided into small squares, and the brightness of each square has been adjusted so it is proportional to the number of galaxies in that square. The picture includes about $2.3 \times 10^5$ galaxies in an angular patch of the sky, 40° by 40°. The distance scale has been computed for those galaxies that are at a middle distance, at about $1.4 \times 10^9$ light-years from the Earth.

The galaxies in the picture are distributed more or less at random. If we look closely, we can recognize some chains or "filaments" of galaxies and some voids in between. But there are no conspicuous features. On a large scale, the universe is fairly uniform.

This is the last of our pictures in the ascending series. We have reached the limits of zoom-out. If we wanted to draw another diagram, 10 times larger than this, we would need to know the shape and the size of the entire universe. We do not yet know that.

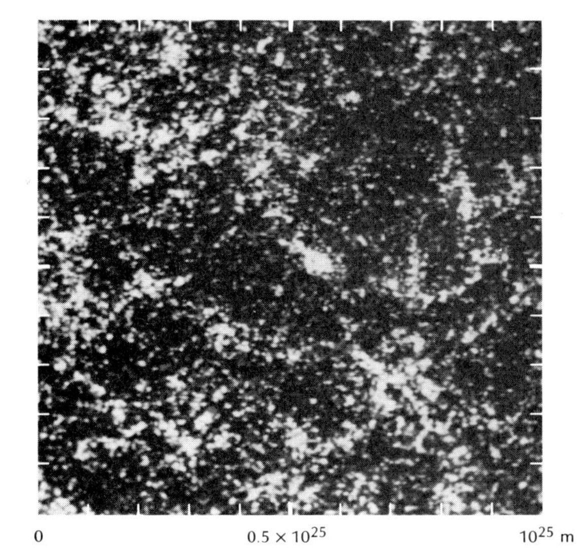


## PART II: THE SMALL-SCALE WORLD

**SCALE 1:1.5** We now return to Charlotte and zoom in on her eye. The surface of her skin appears smooth and firm. But this is an illusion. Matter appears continuous because the number of atoms in each cubic centimeter is extremely large. In a cubic centimeter of human tissue there are about $10^{23}$ atoms. This large number creates the illusion that matter is continuously distributed — we only see the forest and not the individual trees. The solidity of matter is also an illusion. The atoms in our bodies are mostly vacuum. As we will discover in the following pictures, within each atom the volume actually occupied by subatomic particles amounts to only about 1 part in $10^{13}$.

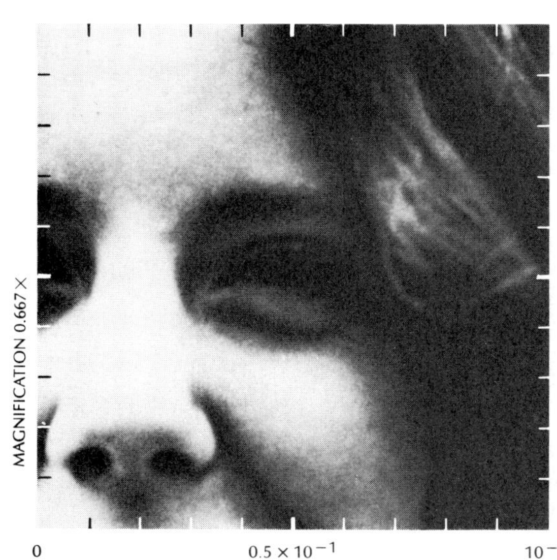

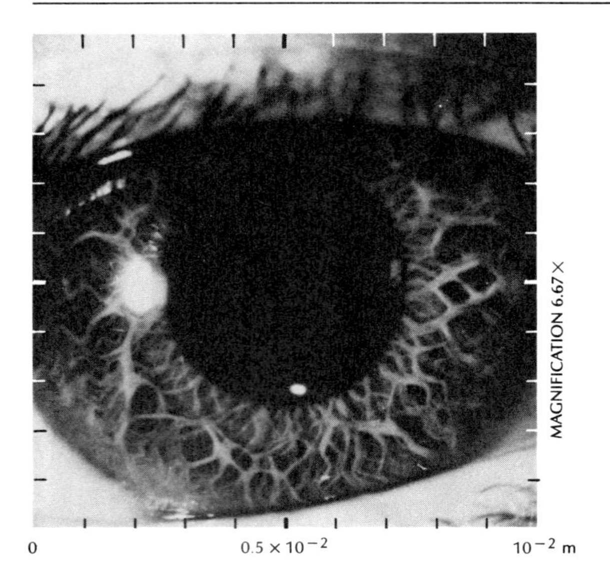


**SCALE 1:1.5 × $10^{-1}$** Our eyes are very sophisticated sense organs; they collect more information than all our other sense organs taken together. The photograph shows the pupil and the iris of Charlotte's eye. Annular muscles in the iris change the size of the pupil and thereby control the amount of light that enters the eye. In strong light the pupil automatically shrinks to about 2 millimeters; in very weak light it expands to as much as 7 millimeters.

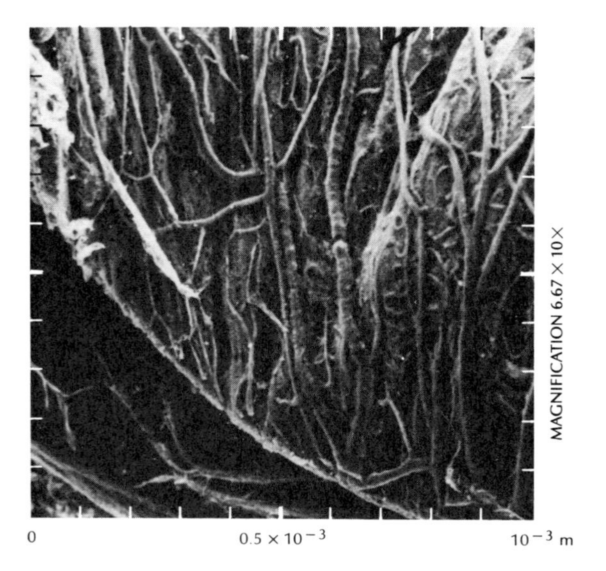


**SCALE 1:1.5 × $10^{-2}$** This photograph shows the delicate network of blood vessels and nerve fiber bundles on the front surface of the retina, the light-sensitive membrane lining the interior of the eyeball. The rear surface of the retina is densely packed with two kinds of cells that sense light: cone cells and rod cells. In a human retina there are about 6 million cone cells and 120 million rod cells. The cone cells distinguish colors; the rod cells distinguish only brightness and darkness, but they are more sensitive than the cone cells and therefore give us vision in faint light ("night vision").

This and the following photographs were made with *electron microscopes.* An ordinary microscope uses a beam of light to illuminate the object; an electron microscope uses a beam of electrons. Electron microscopes can achieve much sharper contrast and much higher magnification than ordinary microscopes.

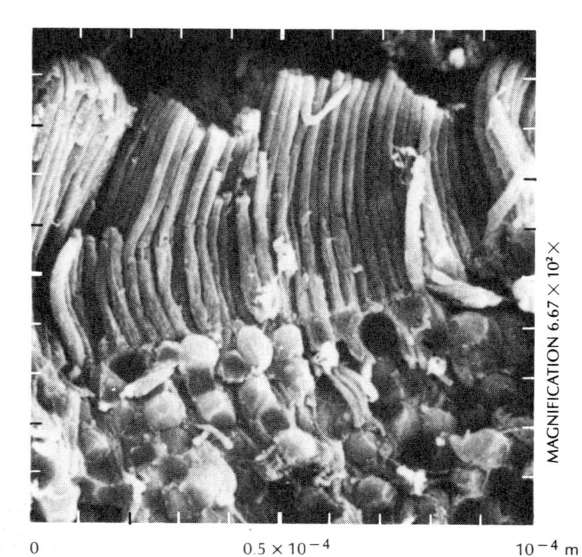


**SCALE 1:1.5 × $10^{-3}$** Here we have a clear photograph of rod cells. To make this photograph, the retina was cut apart and the microscope was aimed at the edge of the cut. In the top half of the picture we see tightly packed rods. Each rod is connected to the main body of a cell. In the bottom part of the picture we can distinguish tightly packed cell bodies. The round balls lying about in the middle are the nuclei of some cells. The cutting of the retina has broken some cells apart and has exposed their nuclei.

**SCALE 1:1.5 × 10⁻⁴** This is a close-up view of a few rods, showing the scarfed joints between the upper and the lower portions. The upper portions of the rods contain a special pigment — visual purple — that is very sensitive to light. The absorption of light by this pigment initiates a chain of chemical reactions that finally trigger nerve pulses from the eye to the brain.

**SCALE 1:1.5 × 10⁻⁵** These are strands of DNA, or deoxyribonucleic acid, as seen with an electron microscope at very high magnification. DNA is found in the nuclei of cells. It is a long molecule made by stringing together a large number of nitrogenous base molecules on a backbone of sugar and phosphate molecules. The base molecules are of four kinds, the same in all living organisms. But the sequence in which they are strung together varies from one organism to another. This sequence spells out a message — the base molecules are the "letters" in the "words" of this message. The message contains all the genetic instructions governing the metabolism, growth, and reproduction of the cell.

The strands of DNA in the photograph are encrusted with a variety of small protein molecules. At intervals, the strands of DNA are wrapped around larger protein molecules that form lumps looking like the beads of a necklace.

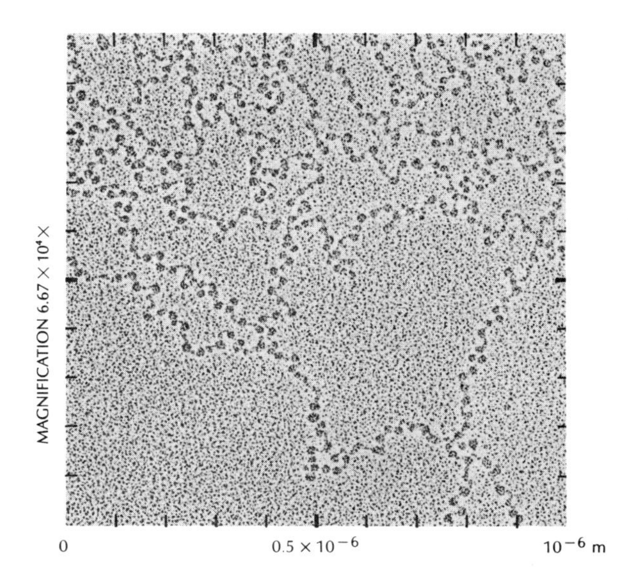


**SCALE 1:1.5 × 10⁻⁶** The magnification in the preceding photograph is very difficult to exceed. Only in the case of a few special samples of materials and of atoms has it become possible to attain higher magnifications. Here we have a photograph of the tip of a fine platinum needle as viewed with an *ion microscope*. This kind of microscope uses helium ions instead of light to "illuminate" the object.

Each bright spot in the photograph is the image of a platinum atom. The needle is made of row upon row of atoms, arranged in a beautiful symmetric pattern. Materials with a regular arrangement of atoms are called *crystals*. Here we have direct visual evidence that platinum is a crystal.

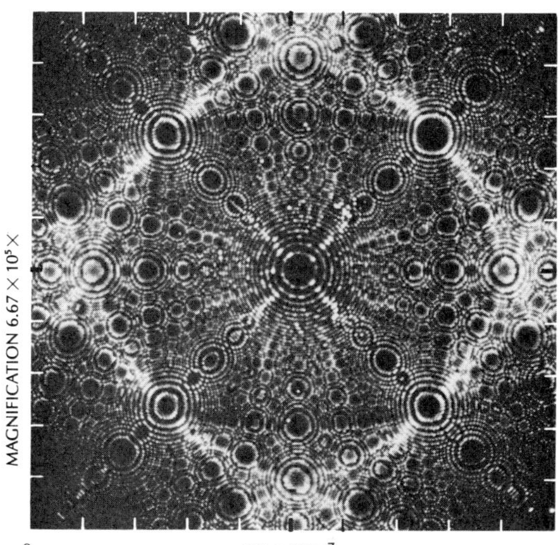

0 0.5 × 10⁻⁸ 10⁻⁸ m

MAGNIFICATION 6.67 × 10⁶ ×

**SCALE 1:1.5 × $10^{-7}$** This is a photograph of uranium atoms, taken with an electron microscope of extremely high power. The small white dots are individual uranium atoms; the large white blobs are clusters of uranium atoms. The atoms are stuck on a very thin film of carbon.

At present we know of more than 100 kinds of atoms or chemical elements. The *Periodic Table* P.1 is a list of these different kinds of atoms. The first entry in this table is the lightest atom — hydrogen (H) — with a mass of $1.67 \times 10^{-27}$ kilogram; the last entry is one of the heaviest known atoms — hahnium (Ha) — with a mass 260 times as large. Some of the heavy atoms near the end of the table do not occur in nature; they can only be manufactured artificially in nuclear reactors or accelerators by "alchemy" or transmutation of other elements. Recently, nuclear physicists have claimed the discovery of three or four elements beyond hahnium. But these elements have not yet been officially baptized.

In Table P.1 two numbers are included with each kind of atom. The first number indicates the position of the atom in the table. This is called the *atomic number.* As we will see later, it represents the number of electrons belonging to the atom. The second number gives the *mass* of the atom in atomic mass units (1 atomic mass unit = 1 u = $1.66 \times 10^{-27}$ kilogram). For example, the oxygen atom has numbers 8 and 15.9994; i.e., it is listed as the eighth entry in the table and it has a mass of 15.9994 u, or $15.9994 \times 1.66 \times 10^{-27}$ kilogram. With only a few exceptions, the masses of atoms in Table P.1 increase monotonically along the table.

The classical method for distinguishing between different atoms is, of course, chemical analysis — we identify the atoms by the reactions in which they engage. But a modern method that has become increasingly important is spectroscopic analysis — we identify the atoms by the light that they give off when stimulated by heat or by an electric current. In Table P.1, the atoms have been arranged in columns according to their chemical and spectroscopic properties. Different columns of this table contain atoms with similar properties; these columns are called *groups* and they are labeled IA, IIA, etc. For instance, the group IA contains hydrogen (H), lithium (Li), sodium (Na), potassium (K), etc.; these are the alkalis. The group 0 contains helium (He), neon (Ne), argon (Ar), krypton (Kr), etc.; these are the noble gases. All atoms in a given group are chemically

**Table P.1** THE PERIODIC TABLE OF CHEMICAL ELEMENTS[a]

| IA | IIA | IIIB | IVB | VB | VIB | VIIB | VIII | VIII | VIII | IB | IIB | IIIA | IVA | VA | VIA | VIIA | 0 |
|---|---|---|---|---|---|---|---|---|---|---|---|---|---|---|---|---|---|
| 1 H 1.00794 | | | | | | | | | | | | | | | | | 2 He 4.00260 |
| 3 Li 6.941 | 4 Be 9.01218 | | | | | | | | | | | 5 B 10.81 | 6 C 12.011 | 7 N 14.0067 | 8 O 15.9994 | 9 F 18.998403 | 10 Ne 20.179 |
| 11 Na 22.98977 | 12 Mg 24.305 | | | | | | | | | | | 13 Al 26.98154 | 14 Si 28.0855 | 15 P 30.97376 | 16 S 32.06 | 17 Cl 35.453 | 18 Ar 39.948 |
| 19 K 39.0983 | 20 Ca 40.08 | 21 Sc 44.9559 | 22 Ti 47.88 | 23 V 50.9415 | 24 Cr 51.996 | 25 Mn 54.9380 | 26 Fe 55.847 | 27 Co 58.9332 | 28 Ni 58.69 | 29 Cu 63.546 | 30 Zn 65.38 | 31 Ga 69.72 | 32 Ge 72.59 | 33 As 74.9216 | 34 Se 78.96 | 35 Br 79.904 | 36 Kr 83.80 |
| 37 Rb 85.4678 | 38 Sr 87.62 | 39 Y 88.9059 | 40 Zr 91.22 | 41 Nb 92.9064 | 42 Mo 95.94 | 43 Tc (98) | 44 Ru 101.07 | 45 Rh 102.9055 | 46 Pd 106.42 | 47 Ag 107.8682 | 48 Cd 112.41 | 49 In 114.82 | 50 Sn 118.69 | 51 Sb 121.75 | 52 Te 127.60 | 53 I 126.9045 | 54 Xe 131.29 |
| 55 Cs 132.9054 | 56 Ba 137.33 | 57–71 Rare Earths | 72 Hf 178.49 | 73 Ta 180.9479 | 74 W 183.85 | 75 Re 186.207 | 76 Os 190.2 | 77 Ir 192.22 | 78 Pt 195.08 | 79 Au 196.9665 | 80 Hg 200.59 | 81 Tl 204.383 | 82 Pb 207.2 | 83 Bi 208.9804 | 84 Po (209) | 85 At (210) | 86 Rn (222) |
| 87 Fr (223) | 88 Ra 226.0254 | 89-103 Actinides | 104 Rf (261) | 105 Ha (260) | 106 (263) | 107 (262) | 108 (265) | 109 (266) | | | | | | | | | |

| Rare Earths (Lanthanides) | 57 La 138.9055 | 58 Ce 140.12 | 59 Pr 140.9077 | 60 Nd 144.24 | 61 Pm (145) | 62 Sm 150.36 | 63 Eu 151.96 | 64 Gd 157.25 | 65 Tb 158.9254 | 66 Dy 162.50 | 67 Ho 164.9304 | 68 Er 167.26 | 69 Tm 168.9342 | 70 Yb 173.04 | 71 Lu 174.967 |
|---|---|---|---|---|---|---|---|---|---|---|---|---|---|---|---|
| Actinides | 89 Ac 227.0278 | 90 Th 232.0381 | 91 Pa 231.0359 | 92 U 238.0289 | 93 Np 237.0482 | 94 Pu (244) | 95 Am (243) | 96 Cm (247) | 97 Bk (247) | 98 Cf (251) | 99 Es (252) | 100 Fm (257) | 101 Md (258) | 102 No (259) | 103 Lr (260) |

[a] In each box, the upper number is the *atomic number.* The lower number is the *atomic mass.* Numbers in parentheses denote the atomic mass of the most stable or best-known isotope of the element; all other numbers represent the average mass of a mixture of several isotopes as found in naturally occurring samples of the element.

similar, that is, they have the same valence and they engage in similar reactions. These atoms are also spectroscopically similar, that is, they emit light with a similar pattern of colors; thus, the colors emitted by sodium display a pattern that has the same general features as the colors emitted by hydrogen.

Atoms are the building blocks of molecules and these, in turn, are the building blocks of everything surrounding us. Among the simplest, and most familiar, of all molecules are the water molecule and the oxygen molecule. The water molecule consists of two hydrogen atoms and one oxygen atom; that of oxygen of two oxygen atoms. But many other molecules are much more complex and much larger. For example, the molecule of DNA shown in the photograph on page 13 consists of about 100,000,000 atoms of carbon, hydrogen, oxygen, nitrogen, phosphorus, etc., assembled in a long strand which is coiled into a helical structure. The molecules of synthetic polymers, such as nylon, polyethylene, synthetic rubber, etc., consist of an even larger number of atoms assembled in extremely long chains within which identical molecular units are strung together one after another.

**SCALE 1:1.5 × $10^{-8}$** This is a photograph of a neon atom. The atom is a sphere, with a somewhat fuzzy surface. The atoms of the noble-gas elements (helium, neon, argon, krypton, xenon, and radon) all have this same spherical shape, although their sizes are slightly different. The atoms of other elements have other shapes; for example, atoms of carbon (in compounds) have an approximately tetrahedral shape, with round edges.

This picture represents the highest magnification that has been attained to date. The picture was obtained with an *electron-holography microscope* by a two-stage process. First, the atom was illuminated with a beam of electrons, and the intensity pattern of the emerging electrons was recorded on a photographic plate. Then this photographic plate was illuminated with a beam of laser light. This gives the highly magnified image of the atom seen in our picture.

All the pictures on the remaining pages of this section are drawings. At these higher magnifications no photographs are available.

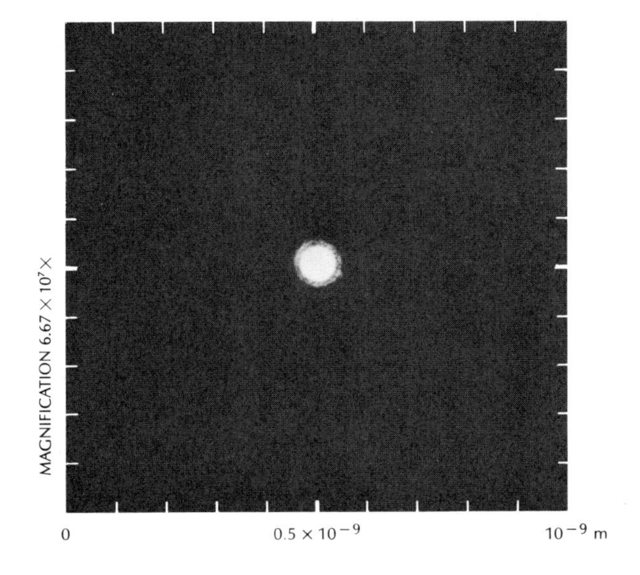


**SCALE 1:1.5 × $10^{-9}$** The drawing shows the interior of an atom of neon. This atom consists of 10 electrons orbiting around a nucleus. In the drawing, the electrons have been indicated by small dots, and the nucleus by a slightly larger dot at the center of the picture. These dots have been drawn as small as possible, but even so the size of these dots does not give a correct impression of the actual size of the electrons and of the nucleus. The electron is smaller than any other particle we know; maybe the electron is truly pointlike and has no size at all. The nucleus has a finite size, but this size is much too small to show up on the drawing. Note that the electrons tend to cluster near the center of the atom. However, the overall size of the atom depends on the distance to the outermost electron; this electron defines the outer edge of the atom.

The electrons move around the nucleus in a very complicated motion. The drawing shows the electrons as they would be seen at one instant of time with a *Heisenberg microscope*. This is a hypothetical microscope that employs gamma rays instead of light rays to illuminate an object; no such microscope has yet been built.

The mass of each electron is 9.1 × $10^{-31}$ kilogram, but most of the mass of the atom is in the nucleus; the 10 electrons of the neon atom have only 0.03% of the total mass of the atom. The number of electrons in the neon atom equals the atomic number of neon. This equality of the number of electrons and the atomic number holds for all kinds of atoms. Thus, the hydrogen atom has one electron, the helium atom has two electrons, and so on.

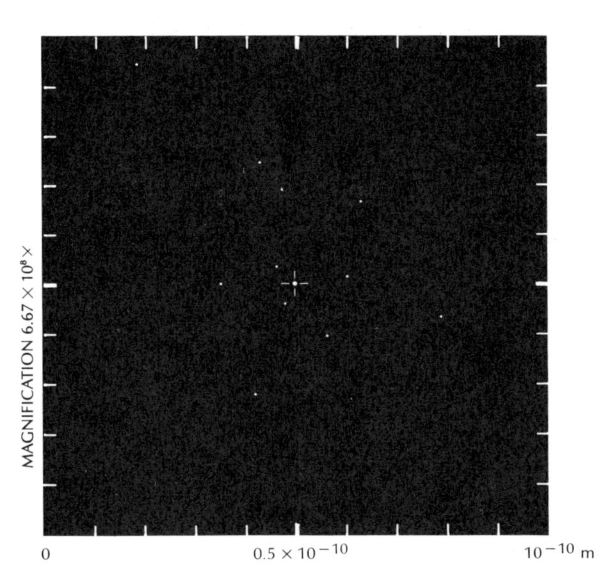

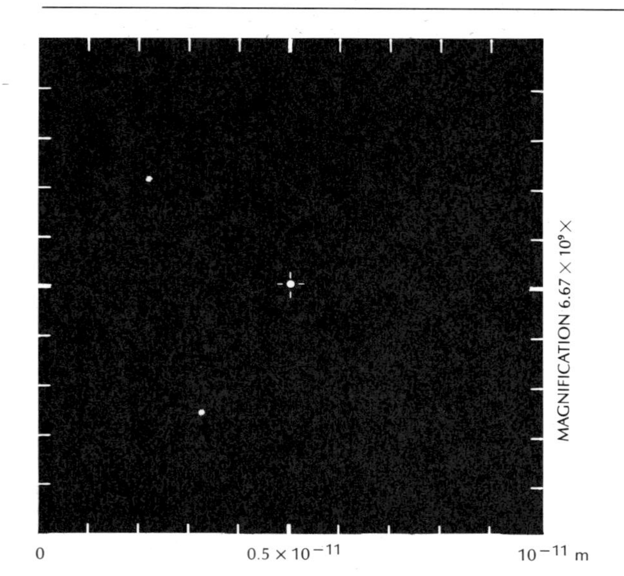


**SCALE 1:1.5 × $10^{-10}$** Here we are closing in on the nucleus. We are seeing the central part of the atom. Only two electrons are in our field of view; the others are beyond the margin of the drawing. The size of the nucleus is still much smaller than the size of the dot at the center of the picture.

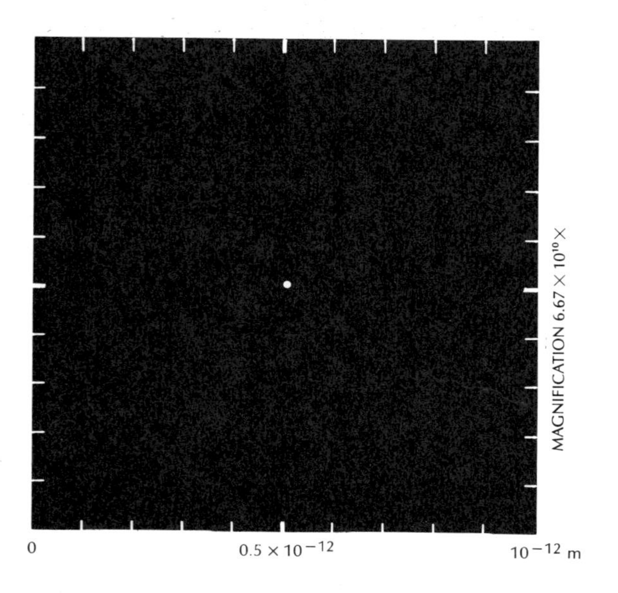


**SCALE 1:1.5 × $10^{-11}$** In this drawing we finally see the nucleus in its true size. At this magnification, the nucleus of the neon atom looks like a small dot, 0.5 millimeter in diameter. Since the nucleus is extremely small and yet contains most of the mass of the atom, the density of the nuclear material is enormous. If we could assemble a drop of pure nuclear material of a volume of 1 cubic centimeter, it would have a mass of 2.3 × $10^{11}$ kilograms, or 230 million metric tons!

Our drawings show clearly that most of the volume within the atom is empty space. The nucleus only occupies a very small fraction of this volume.

**SCALE 1:1.5 × $10^{-12}$** We can now begin to distinguish the nuclear structure. The nucleus has a nearly spherical shape, but it is slightly lumpy.

**SCALE 1:1.5 × 10$^{-13}$** At this extreme magnification we can see the details of the nuclear structure. The nucleus of the neon atom is made up of 10 protons (white balls) and 10 neutrons (gray balls). Each proton and each neutron is a sphere with a diameter of about 2.1 × 10$^{-15}$ meter, and a mass of 1.67 × 10$^{-27}$ kilogram. In the nucleus, these protons and neutrons are tightly packed together, so tightly that they almost touch. The protons and neutrons move around the volume of the nucleus at high speed in a complicated motion.

Note that the number of protons matches the number of electrons, or the atomic number. Neon atoms always have 10 protons in their nucleus, but they do not always have 10 neutrons. In naturally occurring samples of neon gas, about 91% of the atoms have 10 neutrons, but 8.8% have 12 neutrons, and 0.2% have 11 neutrons. Atoms with the same numbers of protons but different numbers of neutrons are called *isotopes.* The naturally occurring isotopes of neon are designated $^{20}$Ne, $^{22}$Ne, and $^{21}$Ne. The superscript on the chemical symbol indicates the sum of the number of protons and the number of neutrons; this superscript is called the *mass number,* since it is approximately equal to the mass of the isotope in atomic mass units.

Besides the isotopes $^{20}$Ne, $^{21}$Ne, and $^{22}$Ne, neon also has other isotopes ranging from $^{17}$Ne to $^{27}$Ne (see Table P.2). However, these other isotopes do not occur naturally; they can only be produced artificially by transmutation of elements in a nuclear reactor or accelerator. These artificial isotopes are extremely unstable, lasting for only a few minutes or seconds before spontaneously decaying into fluorine or sodium.

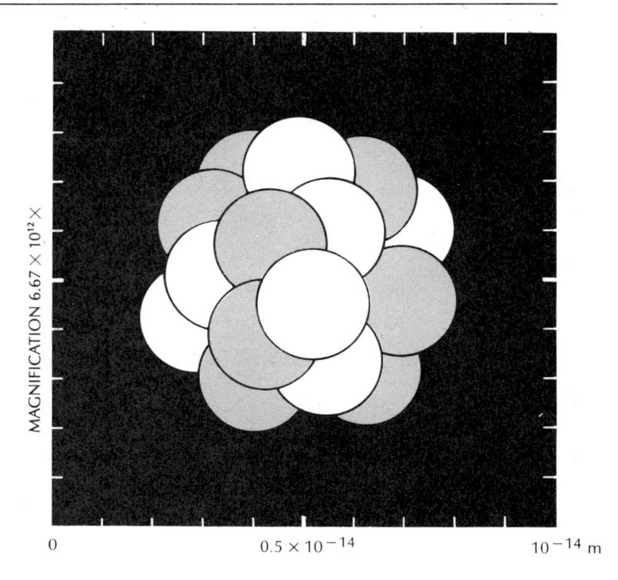


All chemical elements have several isotopes. Hydrogen has three isotopes ($^{1}$H, or ordinary hydrogen; $^{2}$H, or deuterium; $^{3}$H, or tritium), helium has five isotopes ($^{3}$He, $^{4}$He, $^{5}$He, $^{6}$He, $^{8}$He), etc. Gold has 33 known isotopes, more than any other chemical element. Some of these isotopes occur in nature, others are unstable and can only be produced by artificial means. Table P.2 lists all the known isotopes of the first few elements.

**Table P.2** EXCERPT FROM THE CHART OF ISOTOPES[a]

| Z | Element | N = 0 | 1 | 2 | 3 | 4 | 5 | 6 | 7 | 8 | 9 | 10 | 11 | 12 | 13 | 14 | 15 | 16 | 17 |
|---|---|---|---|---|---|---|---|---|---|---|---|---|---|---|---|---|---|---|---|
| 10 | Ne<br>20.179 | | | | | | | | 17 Ne<br>0.109 s<br>17.0177 | 18 Ne<br>1.67 s | 19 Ne<br>17.4 s | 20 Ne<br>90.5%<br>19.992439 | 21 Ne<br>0.27%<br>20.993847 | 22 Ne<br>9.22%<br>21.991384 | 23 Ne<br>37.6 s | 24 Ne<br>3.38 m | 25 Ne<br>0.61 s | 26 Ne<br>26.0005 | 27 Ne<br>27.0072 |
| 9 | F<br>18.9984 | | | | | | | 15 F<br>15.0180 | 16 F<br>~10$^{-19}$s<br>16.011 | 17 F<br>66.0 s | 18 F<br>109.8 m | 19 F<br>100%<br>18.998403 | 20 F<br>11.1 s | 21 F<br>4.36 s | 22 F<br>4.0 s | 23 F<br>2.2 s | 24 F<br>24.0093 | 25 F<br>25.0138 | |
| 8 | O<br>15.9994 | | | | | | 13 O<br>0.0089 s | 14 O<br>70.5 s | 15 O<br>122 s<br>15.003065 | 16 O<br>99.756%<br>15.994915 | 17 O<br>0.037%<br>16.99913 | 18 O<br>0.204%<br>17.999159 | 19 O<br>26.9 s | 20 O<br>13.6 s | 21 O<br>3.4 s | 22 O<br>23.0101 | 23 O<br>23.0193 | | |
| 7 | N<br>14.0067 | | | | | 11 N<br>11.0267 | 12 N<br>0.011 s | 13 N<br>9.97 m<br>13.005739 | 14 N<br>99.63%<br>14.003074 | 15 N<br>0.37%<br>15.000109 | 16 N<br>7.11 s | 17 N<br>4.16 s | 18 N<br>0.63 s | 19 N<br>0.42 s<br>19.0176 | 20 N<br>20.0238 | 21 N<br>21.0289 | | | |
| 6 | C<br>12.011 | | | | 9 C<br>0.127 s | 10 C<br>19.4 s | 11 C<br>20.4 m<br>11.011433 | 12 C<br>98.89%<br>12.00000 | 13 C<br>1.11%<br>13.003355 | 14 C<br>5730 y<br>14.003242 | 15 C<br>2.45 s | 16 C<br>0.74 s | 17 C<br>17.0226 | 18 C<br>18.0267 | 19 C<br>19.0370 | | | | |
| 5 | B<br>10.811 | | | | 8 B<br>0.774 s | 9 B<br>~8×10$^{-19}$s<br>9.01333 | 10 B<br>19.8%<br>10.012938 | 11 B<br>80.2%<br>11.009305 | 12 B<br>0.020 s | 13 B<br>0.017 s | 14 B<br>0.016 s | | | 17 B<br>17.0986 | | | | | |
| 4 | Be<br>9.01218 | | | 6 Be<br>≥3×10$^{-21}$s<br>6.01973 | 7 Be<br>53.3 d<br>7.016930 | 8 Be<br>~1×10$^{-16}$s<br>8.005305 | 9 Be<br>100%<br>9.012183 | 10 Be<br>1.6×10$^{6}$ y | 11 Be<br>13.8 s | 12 Be<br>0.011 s | | 14 Be<br>14.0440 | | | | | | | |
| 3 | Li<br>6.941 | | | 5 Li<br>~10$^{-21}$s<br>5.0125 | 6 Li<br>7.5%<br>6.015123 | 7 Li<br>92.5%<br>7.016005 | 8 Li<br>0.85 s | 9 Li<br>0.17 s | | 11 Li<br>0.009 s | | | | | | | | | |
| 2 | He<br>4.00260 | | 3 He<br>0.00013%<br>3.016029 | 4 He<br>~100%<br>4.002603 | 5 He<br>2×10$^{-21}$s<br>5.0122 | 6 He<br>0.802 s | | 8 He<br>0.122 s | | | | | | | | | | | |
| 1 | H<br>1.0079 | 1 H<br>99.985%<br>1.007825 | 2 H<br>0.015%<br>2.014102 | 3 H<br>12.33 y<br>3.01649 | | | | | | | | | | | | | | | |
| 0 | | | 1 n<br>10.6m<br>1.008665 | | | | | | | | | | | | | | | | |

[a] The number Z, increasing vertically along the chart, is the number of protons in the isotope; it coincides with the atomic number. The number N, increasing horizontally, is the number of neutrons. In each box, the number directly below the symbol for the isotope gives the abundance in percent for naturally occurring isotopes, or else the half-life for unstable, artificially produced isotopes (the half-life is the time required for one-half of a sample of unstable isotope to decay). The number at the bottom gives the mass of the neutral atom (nucleus plus Z electrons) in atomic mass units.

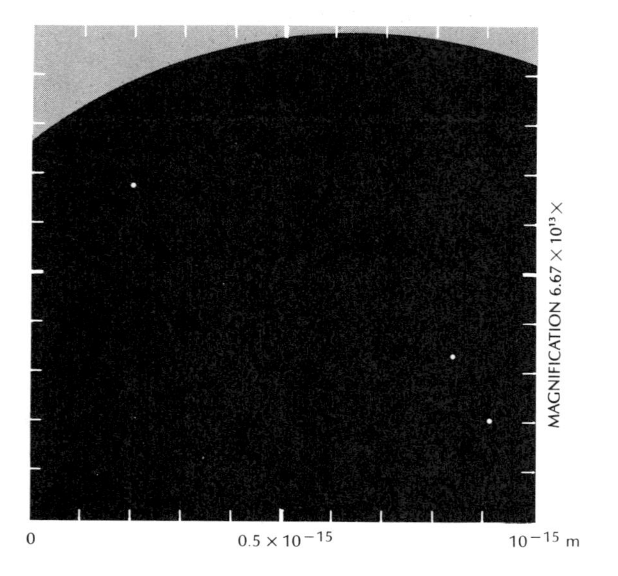


**SCALE 1:1.5 × 10⁻¹⁴** This final picture shows three pointlike bodies within a proton. These pointlike bodies are *quarks* — each proton and each neutron is made of three quarks. Recent experiments have told us that the quarks are much smaller than protons or neutrons, but we do not yet know their precise size. Hence the dots in the drawing probably do not give a fair description of the size of quarks. The quarks within protons and neutrons are of two kinds, called *up* and *down*. The proton consists of two *up* quarks and one *down* quark bound together; the neutron consists of one *up* quark and two *down* quarks bound together.

Besides these two kinds of quarks, physicists have discovered four other kinds, called by the quaint names *strange, charmed, top,* and *bottom*. These quaint quarks do not occur in pieces of ordinary matter; they can only be manufactured under exceptional circumstances in high-energy collisions between subatomic particles.

This final picture takes us to the limits of our knowledge of the subatomic world. As a next step we would like to zoom in on the quarks and show what they are made of. But we do not yet know whether they are made of anything else.

CHAPTER 1

# Measurement of Space, Time, and Mass

The investigator of any phenomenon — an earthquake, a flash of lightning, a collision between two ships — must begin by asking the question: Where and when did it happen? Phenomena happen at points in space and at points in time. A complicated phenomenon — such as a collision between two ships — is spread out over many points of space and time. But no matter how complicated, any phenomenon can be fully described by stating what happened at diverse points of space at successive instants of time. A happening at one point of space and one point of time is called an **event.** For example, the first contact between the bows of two colliding ships is an event. The entire complicated phenomenon is a succession of such individual events. Since space and time form the arena for all physical phenomena, we will commence our study of physics with an examination of the properties of space and time.

Ships and other macroscopic bodies are made of atoms. Although atoms are themselves made of smaller constituents — electrons and nuclei — the atoms do not suffer any internal changes in ordinary mechanical phenomena, and we can therefore regard atoms as indivisible, unchanging entities for most practical purposes. Since the sizes of the atoms are extremely small compared to the sizes of macroscopic bodies, we can regard atoms as almost pointlike masses. A pointlike mass of no discernible size or internal structure is called an **ideal particle.** Such a particle may be thought of as an infinitesimal grain of mass, a grain so small that its size can be ignored for all purposes. At any given instant of time, the ideal particle occupies a single point of space. Furthermore, the particle has a mass. And that is all: if we know the position of the particle as a function of time, and we know its mass, then we know everything that can be known about the particle. Position, time, and mass give a *complete* description of the behavior and the

*Ideal particle*

attributes of an ideal particle.[1] Since every macroscopic body consists of particles, we can — in principle — describe the behavior and the attributes of such a body by describing the particles within the body. Thus, measurements of position, time, and mass are of fundamental significance in physics.

## 1.1 Space and Time

To determine the position of an event or of a particle, we first take some convenient point of space as **origin** and then measure the position relative to this origin. For this purpose, we imagine a grid of lines around the origin and check the location of the event within this grid; that is, we imagine that space is filled with three-dimensional "graph paper" and specify the position by means of coordinates read off this graph paper. The most common kind of coordinates are **rectangular coordinates** $x$, $y$, $z$, which rely on a rectangular grid (Figure 1.1). The three mutually perpendicular lines through the origin are called the $x$, $y$, and $z$ axes; the rectangular grid is erected on these axes. The coordinates $x$, $y$, $z$ of a point $P$ simply indicate how far we must move parallel to the corresponding axis in order to go from the origin $O$ to the point $P$. For example, the point $P$ shown in Figure 1.1 has coordinates $x = 3$ units, $y = 5$ units, $z = 3$ units.

*Systems of coordinates*

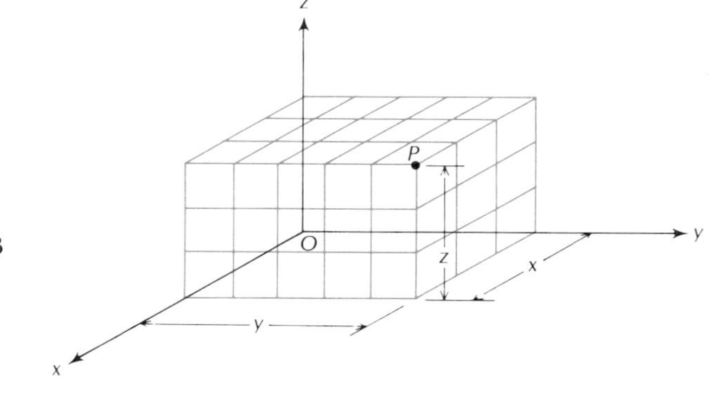


**Fig. 1.1** Rectangular coordinates $x$, $y$, $z$. The point $P$ has coordinates $x = 3$, $y = 5$, $z = 3$ units.

Instead of rectangular coordinates, we can just as well use **spherical coordinates** or **cylindrical coordinates.** These rely on a spherical or cylindrical grid, respectively. The spherical coordinates are essentially the "latitude" angle, the "longitude" angle, and the radial distance from the origin (Figure 1.2); the cylindrical coordinates are the "longitude" angle (sometimes called the "bearing angle"), the height above the $x$–$y$ plane, and the distance along the $x$–$y$ plane (Figure 1.3). Which kind of coordinates is chosen is a matter of convenience. For example, an airport controller will usually describe the position of an approaching aircraft by means of a cylindrical grid with origin at the airport, i.e., he will give the bearing angle of the aircraft, its altitude above the horizontal plane, and its horizontal distance from the airport. On the other hand, the pilot of an aircraft on a transatlantic flight will find it

[1] We will ignore for now the possibility that the particle also has an electric charge. Electricity is the subject of Chapters 22–36.

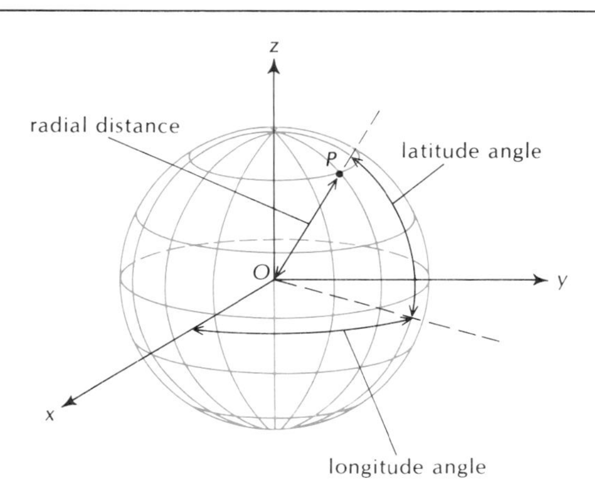


**Fig. 1.2** Spherical coordinates: longitude, latitude, and radial distance.

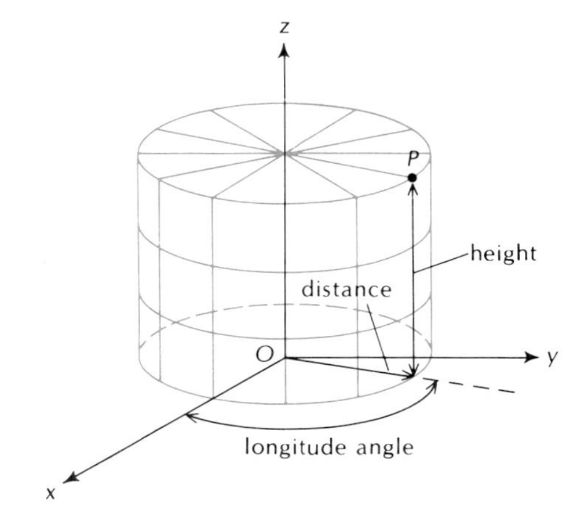


**Fig. 1.3** Cylindrical coordinates: longitude, horizontal distance, and height.

more convenient to describe position by means of a spherical grid with origin at the center of the Earth, that is, he will give the (geographic) latitude, longitude, and altitude (the latter is in obvious correspondence with the distance to the Earth's center).

*Three dimensions of space*

It is a fact of experience that, regardless of the kind of coordinate grid we choose, exactly *three* coordinates are necessary and sufficient to locate a point relative to an origin. This is the rigorous meaning of the assertion that the space in which we live is three dimensional. Using language adapted to a rectangular grid, we can say that space has length, width, and height (or depth).

*Euclidean geometry*

That our space is three dimensional is its most elementary and important property. Next in importance is the property that the geometry of our space is **Euclidean,** at least to a very good approximation. This means that the postulates and theorems of Euclid are valid for the real world. For example, in any arbitrary plane triangle laid out in space, the sum of the interior angles is 180° (Figure 1.4).[2]

It must be emphasized that the three-dimensional character and the Euclidean geometry of our space rest on direct **empirical evidence,** that is, evidence directly derived from observation and experimentation. We can perceive the three-dimensional character simply by feeling around space with our hands or looking about with our binocular vision. The Euclidean geometry requires observations of a more intricate kind. One way to check it is by precisely measuring the sum of the interior angles of a plane triangle. For a plane triangle defined by three points in the space in the vicinity of Earth, such an experiment can be done with surveying instruments; the measured result for the sum of interior angles agrees with 180° to within the experimental error, a few tenths of a second of arc. Thus, the direct experimental evidence shows that on or near the Earth, the Euclidean geometry is good to within better than a few parts in $10^6$. (Some indirect evidence, based on the theory of relativity, indicates that the Euclidean geometry is probably good to within a few parts in $10^{10}$.)

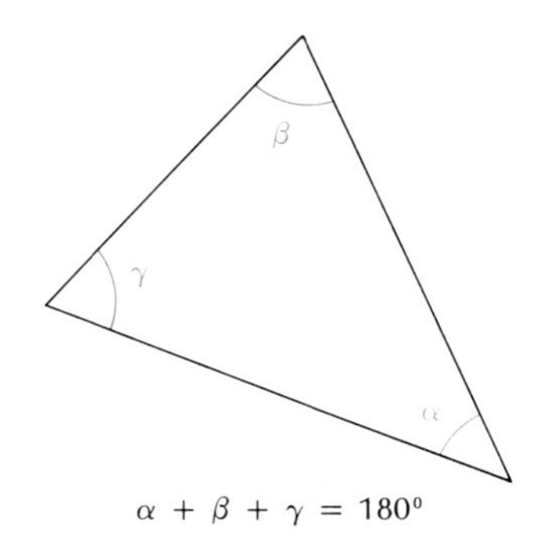


$\alpha + \beta + \gamma = 180°$

**Fig. 1.4** A plane triangle with interior angles $\alpha$, $\beta$, and $\gamma$.

[2] Appendix 4 gives a review of trigonometry, including the definitions of degrees and radians. The Greek alphabet appears on the endpapers.

Equal in importance to the property that space is Euclidean ranks the property that time is absolute. The meaning of this is as follows: Suppose we have two identical clocks, initially at rest side by side, synchronized and running at exactly the same rate. We put one of the clocks in motion and transport it along some path, perhaps with varying speed. Then we bring the traveling clock back to the location of its companion clock and compare them. If the two clocks are still synchronized, regardless of where the traveling clock was transported and with what speed it was moved, then time is absolute; that is, the rate at which time elapses is independent of position and independent of speed.[3]

Experiments indicate that time is absolute to a very high precision. Physicists take advantage of this fact in order to synchronize clocks in laboratories located at different places on Earth; they use special portable atomic clocks (see Section 1.4), which they first synchronize with the clock in one laboratory, and then transport to the other laboratory in order to transfer the synchronization.

The theory based on the assumptions of absolute time and Euclidean geometry of space is called **classical physics,** or **Newtonian physics.** The foundations of this theory were laid down by Isaac Newton in the seventeenth century. Newton asserted in his *Principia Mathematica* that

> absolute, true, and mathematical time, of itself, and from its own nature, flows equably without relation to anything external.

He took for granted that the geometry of space is Euclidean, and never even bothered to state this assumption explicitly.

For more than 200 years, Newton's theory stood unchallenged. But in the early part of this century, Albert Einstein proposed his theories of Special Relativity and General Relativity and brought about a revolutionary revision of our concepts of space and time. According to Einstein's theories, time is not absolute, and the geometry of space in the vicinity of a large massive body is not Euclidean. The failure of absolute time was explicitly verified in delicate experiments with very accurate atomic and nuclear clocks. Physicists found that when these clocks are moved with high speeds or lifted to high altitude above the surface of the Earth, they lose or gain time relative to clocks at rest on the surface of the Earth. For instance, experiments with an extremely accurate atomic clock flying in an aircraft at 900 kilometers per hour have shown that the rate of this flying clock is slowed by about 1 part in $10^{12}$, as compared to the rate of an identical clock that is kept stationary on the ground. And astronomers investigating the propagation of light and radio waves near the Sun ("bending" of light rays near the Sun) found that the sum of the interior angles of a triangle constructed with light rays deviates from 180° by 1 or 2 seconds of arc. Such deviations from Euclidean geometry are also believed to occur near the Earth, but they are much smaller than the deviations near the Sun.[4]

---

[3] This, of course, relies on the tacit assumption that the clocks are designed in such a way that the shocks (accelerations) they receive during their motion do not directly (mechanically) affect their rate. For example, a good wristwatch is designed so that shocks do not affect it. In contrast, a pendulum clock is very sensitive to accelerations and would be quite unsuitable for this experiment.

[4] The question arises whether the deviation from Euclidean geometry is to be blamed on the light rays. But (indirect) experimental and theoretical evidence suggests that a deviation of about the same magnitude would persist even if the triangle were laid out with tightly stretched strings.

Since the deviations from absolute time and from Euclidean geometry are so small as to be barely detectable with the most accurate clocks and the most sensitive telescopes and radiotelescopes, we will ignore Einstein's theories for the purposes of most of the following chapters and rely, instead, on Newton's theory with its absolute time and Euclidean geometry. Although Newton's theory is only an approximation to the real world, it is an eminently satisfactory one, quite adequate for the description of all the phenomena we encounter in everyday life and (almost) all the phenomena we encounter in the realm of engineering. Physics is full of such approximations made in the interest of simplicity. Nevertheless, physics is regarded as an "exact" science, because the physicist usually knows what approximations can be made in the description of a given phenomenon without introducing excessive errors.

## 1.2 Frames of Reference

The measurement of the position of an event by means of a coordinate grid erected around an origin is a *relative measurement,* which is to say the values of the coordinates depend on the choice of origin and on the choice of grid. Obviously, if two rectangular grids are based on origins that are displaced from one another, then the two sets of coordinates will have a corresponding difference (Figure 1.5a). It is also possible for two grids to differ in orientation (1.5b), and this can lead to extra differences in the sets of coordinates. Furthermore, the origin of one grid may be in motion relative to another grid (1.5c), in which case not only will the two sets of coordinates of a given point be different, but the difference will change in time. Position measurements are always relative; they only become meaningful when the origin and the coordinate grid used for the measurements is carefully specified.

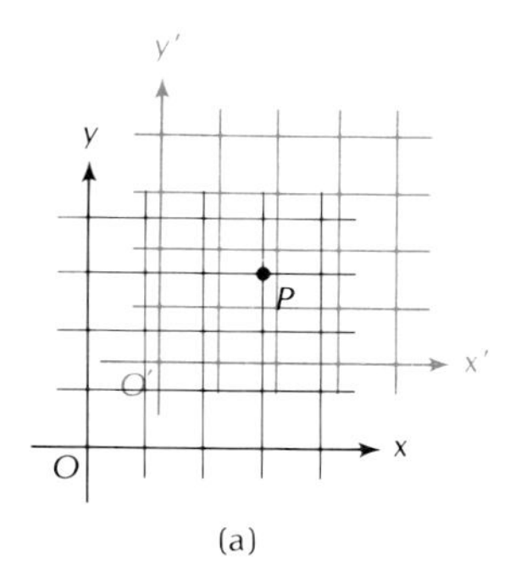


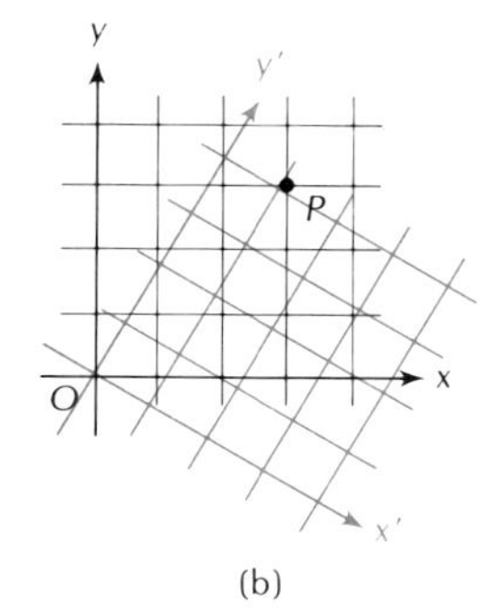


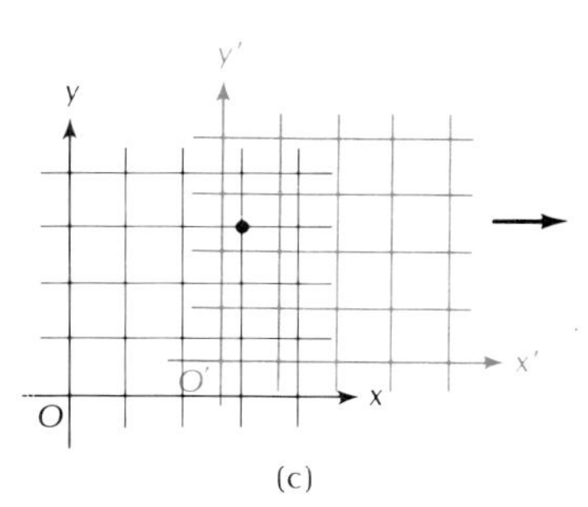


**Fig. 1.5** Rectangular grids $x$–$y$ and $x'$–$y'$. (a) The grid $x'$–$y'$ is displaced relative to the grid $x$–$y$ by a fixed amount.
(b) The grid $x'$–$y'$ is rotated relative to the grid $x$–$y$.
(c) The grid $x'$–$y'$ is in motion relative to the grid $x$–$y$.

The choice of origin of coordinates and the choice of its motion are matters of convenience. For example, the navigator of a ship often finds it convenient to take the midpoint of his ship as origin and to imagine a grid around this origin; the grid then moves with the ship.[5] If the navigator plots the track of a second ship on this grid, he can tell at a glance what the distance of closest approach will be, and whether the other ship is on a collision course (crosses the origin).

*Reference frame*

A coordinate grid together with a set of synchronized clocks is called a **reference frame.** The coordinate grid and the clocks are used to determine the space and time coordinates of events. If the motion of a reference frame coincides with the motion of some given body — a particle, a ship, the Earth, or whatever — so that the body remains permanently at rest in this reference frame, then this reference frame is called the **rest frame** of that body. We can think of such a rest frame as rigidly attached to the body. A rest frame is often labeled with the name of the body to which it is attached; thus, we speak of the rest frame of the Earth (moving with the Earth), the rest frame of a ship (moving with the ship), and so on.

The clocks associated with a given reference frame will be assumed to move with the reference frame. It is convenient to suppose that the

[5] In the U.S. Navy, the coordinates based on this moving grid are called "relative coordinates."

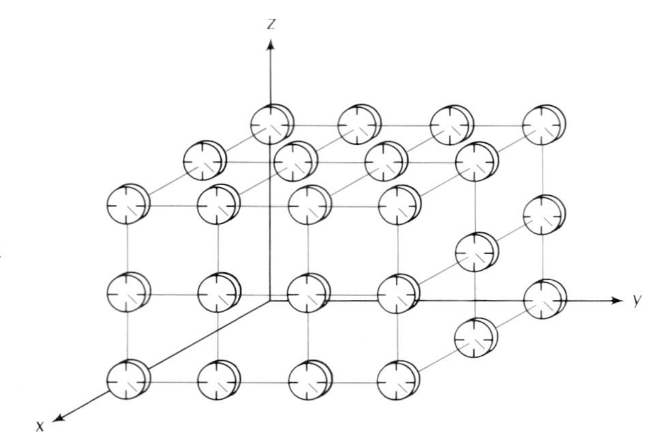


**Fig. 1.6** A reference frame consisting of a coordinate grid and synchronized clocks.

clocks are arranged at regular intervals along the coordinate grid (Figure 1.6) so that whenever an event occurs anywhere, one of the clocks will be in the immediate vicinity and can be used to record the time.

As long as we deal with phenomena for which the notion of an absolute time is a satisfactory approximation, we do not really need to insist that each reference frame have its own clocks moving with it. If time is absolute, then all clocks in all reference frames agree, and it is just as well to use the clocks of one chosen reference frame for all other reference frames. But when we wish to deal with relativistic phenomena involving speeds near the speed of light (as in Chapter 41), then the lack of an absolute time forces us to insist that each reference frame have its own set of synchronized clocks. In the relativistic domain, both the rates of clocks and their synchronization are properties peculiar to a given reference frame; that is, there is no absolute time, only a "relative" time associated with each reference frame. Although we will ignore this problem of the relativity of time until Chapter 41, for the sake of uniformity we will suppose that each reference frame always has its own clocks, whether the approximation of an absolute time applies or not.

## 1.3 The Unit of Length

For the construction of the coordinate grid to be used for a measurement of position, we need a **unit of length** that tells us where the points $x = 1$, $y = 1$, $z = 1$ are located along the respective axes. The unit of length is a basic amount of length against which any other amount of length can be compared.

*SI, or metric system*

The **International System of Units,** or **SI,**[6] uses the **meter** as the unit of length, the **kilogram** as the unit of mass, and the **second** as the unit of time. As we will see in Section 1.6, units of length, time, and mass are necessary and sufficient for the measurement of any physical quantity. Because the SI unit of length is the meter, this system of units is often called the **metric system.**

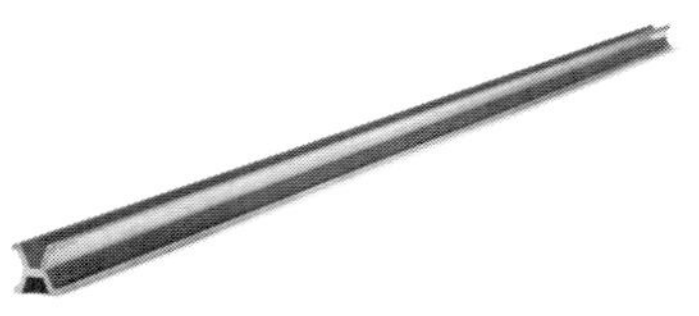

**Fig. 1.7** International standard meter bar.

For many years, the standard of length that told us the amount of length in one meter (1 m) was the standard meter bar kept at the Bureau International des Poids et Mesures (BIPM) at Sèvres, France. This is a bar made of platinum–iridium alloy with a fine scratch mark near each end (Figure 1.7). By definition, the distance between these

[6] *Système International.*

scratches was taken to be exactly one meter. The length of the meter was originally chosen so as to make the polar circumference of the Earth exactly 40 million meters (Figure 1.8); however, modern determinations of this circumference show it to be about 0.02% more than 40 million meters.

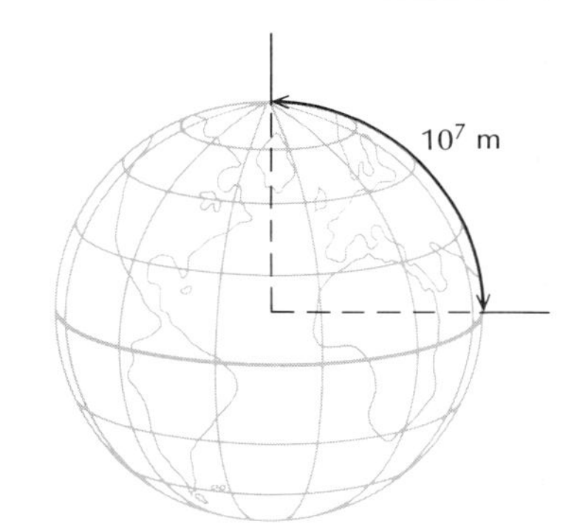


**Fig. 1.8** One-quarter of the polar circumference of the Earth equals $10^7$ m.

Copies of the prototype standard meter were manufactured in France and distributed to other countries to serve as secondary standards. The length standards used in industry and engineering have been derived from these secondary standards. For example, Figure 1.9 shows a set of gauge blocks commonly used as length standards in machine shops.

The precision of the standard meter is limited by the coarseness of the scratch marks at its ends. Although these scratches were made with great care, a microscope reveals them as irregular bands whose midpoints are somewhat uncertain. For the sake of higher precision, a new standard of length was adopted in 1960. This standard was the wavelength of the orange light emitted by krypton atoms. Just as a flute stimulated by a stream of air emits a sound wave consisting of a regular succession of wave crests and wave troughs, where the air is, respectively, compressed and rarified, a krypton atom stimulated by an electric current emits a light wave consisting of a regular succession of wave crests and wave troughs of light. The distance from one wave crest to the next is called the **wavelength.** The wavelength of the light emitted by krypton atoms (under suitable conditions) serves as the standard of length. Since all krypton atoms are exactly alike, they all emit exactly the same wavelength, that is, all krypton atoms are exactly in tune. This makes the atomic standard of length universally accessible to all laboratories — krypton is readily available everywhere and it is not necessary to keep a "prototype krypton atom" at the BIPM. The meter was defined as 1,650,763.73 times the wavelength of the light emitted by krypton. We can measure any unknown length by comparing it with the krypton wavelength. Figure 1.10 shows a high-precision interferometer used for such a comparison; essentially, this interferometer counts the wavelengths that fit into the unknown length.

**Fig. 1.9** Some gauge blocks. The numbers on the blocks give their length or thickness in inches.

**Fig. 1.10** Special interferometer used at the BIPM for the comparison of lengths. The interferometer automatically counts the wavelengths when the carriage C moves through a given length.

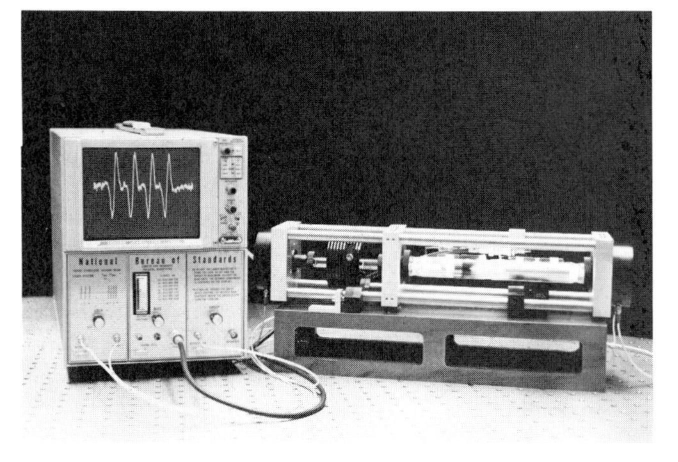


**Fig. 1.11** Stabilized laser at the National Bureau of Standards.

The wavelength of the light emitted by a krypton atom suffers from a slight uncertainty that arises from processes occurring within the atom during the act of emission. This lack of sharpness of the wavelength limits the precision of the krypton standard. Recently, stabilized lasers have been developed which can attain much higher precision. The wavelength of the light generated by these lasers has an uncertainty of less than 2 parts in $10^{12}$. This led to the adoption of a new (and final) definition of the meter in 1983: *The meter is the length of path traveled by a light wave in vacuum in a time interval of* 1/299,792,458 *second.* This new definition hinges on the definition of the unit of time, which we will give in the next section. And it also hinges on a standard value of the speed of light — since the meter is adjusted so that light travels one meter in 1/299,792,458 second, the speed of light is exactly

*Meter,* m

*Speed of light*

$$c = 2.99792458 \times 10^8 \text{ meters per second}$$

The assignment of a fixed value to the speed of light makes good sense because, according to Einstein's theory of relativity, the speed of light is a fundamental, immutable constant, ideally suited to play the role of a standard of speed. Note that the definition of the meter does not mention stabilized lasers. However, the practical implementation of this definition relies on these lasers and on the uniformity of the light waves that they emit. Because of the uniformity of the light waves, the time interval between the emission of one wave crest and the next can be determined with extreme precision and, upon multiplying this time interval by the speed of light, we obtain the wavelength of the laser in meters. We can then use this laser wavelength as a standard for the measurement of an unknown length, in much the same way as we used the krypton wavelength.

By using a stabilized laser in conjunction with an interferometer (such as shown in Figure 1.11), we can measure a change of position (that is, a displacement) of one or several millimeters with an error of only $\pm 10^{-12}$ m. For a change of position of the order of one or several meters, the errors tend to be somewhat larger, about $\pm 10^{-8}$ m. However, if we attempt to measure the length of a body (distance between the two ends), then the irregularity of the body's surface and the deformation of this surface by contact with the measuring device engender ambiguities in just what is meant by "length." These ambiguities prevent measurements of the length of a body to within better than $\pm 10^{-9}$ m.

Table 1.1 lists a few distances and sizes in meters, from the largest to the smallest. Many of these distances and sizes have already been mentioned in the Prelude. Some of the numbers are given with two significant figures (a digit and one decimal), others with one significant figure (a digit without decimals). Of course, we can measure most of these lengths with much higher precision than that. For instance, we can measure wavelengths of light to within 9 or 10 significant figures; but to do this we would first have to decide just what kind of light to measure. Since Table 1.1 is only intended to present some typical examples of lengths, such high precision is not called for. Table 1.2 lists some multiples and submultiples of the meter and their abbreviations.

*British system*

In the **British system of units**, the unit of length is the foot. By definition, one **foot** (1 ft) is exactly 0.3048 m. Table 1.3 gives multiples and submultiples of the foot; these multiples and submultiples are not decimal, a great computational nuisance which will lead to the com-

**Table 1.1** SOME DISTANCES AND SIZES

| | |
|---|---|
| Distance to boundary of observable universe | $\sim 1 \times 10^{26}$ m |
| Distance to Andromeda galaxy | $2.1 \times 10^{22}$ m |
| Diameter of our Galaxy | $7.6 \times 10^{20}$ m |
| Distance to nearest star (Proxima Centauri) | $4.0 \times 10^{16}$ m |
| Earth–Sun distance | $1.5 \times 10^{11}$ m |
| Radius of Earth | $6.4 \times 10^{6}$ m |
| Wavelength of radio wave (AM band) | $\sim 3 \times 10^{2}$ m |
| Length of ship *Queen Elizabeth* | $3.1 \times 10^{2}$ m |
| Height of man (average male) | 1.8 m |
| Length of one pace (U.S. Army) | 0.76 m |
| Diameter of 5¢ coin | $2.1 \times 10^{-2}$ m |
| Diameter of red blood cell (human) | $7.5 \times 10^{-6}$ m |
| Wavelength of visible light | $\sim 5 \times 10^{-7}$ m |
| Diameter of smallest virus (potato spindle) | $2 \times 10^{-8}$ m |
| Diameter of atom | $\sim 1 \times 10^{-10}$ m |
| Diameter of atomic nucleus (iron) | $8 \times 10^{-15}$ m |
| Diameter of proton | $2 \times 10^{-15}$ m |

**Table 1.2** MULTIPLES AND SUBMULTIPLES OF THE METER

| | |
|---|---|
| kilometer | 1 km = $10^{3}$ m |
| meter | 1 m |
| centimeter | 1 cm = $10^{-2}$ m |
| millimeter | 1 mm = $10^{-3}$ m |
| micron | 1 $\mu$m = $10^{-6}$ m |
| nanometer | 1 nm = $10^{-9}$ m |
| Ångstrom | 1 Å = $10^{-10}$ m |
| fermi | 1 f = $10^{-15}$ m |

**Table 1.3** MULTIPLES AND SUBMULTIPLES OF THE FOOT

| | |
|---|---|
| mile | 1 mi = 5280 ft = 1609.38 m |
| yard | 1 yd = 3 ft = 0.9144 m |
| foot | 1 ft = 0.3048 m |
| inch | 1 in. = $\frac{1}{12}$ ft = 2.540 cm |

plete abolition of this system of units in the near future. At present, only two countries, Burma and the United States, are still dawdling with the British system.

## 1.4 The Unit of Time

The unit of time in both the metric and British systems is the **second.** Originally one second (1 s) was defined as $1/(60 \times 60 \times 24)$, or 1/86,400, of a mean solar day. The solar day is the time interval between two successive passages of the Sun over a given meridian on the Earth, say, over the meridian of Greenwich.

The length of the solar day depends on the rate of rotation of the Earth, which, unfortunately, is not quite uniform. There are not only periodic seasonal variations in the rotation rate, but also long-term variations. The seasonal variations are caused by changes in the motions of air, water, and ice on the Earth; such seasonal variations can be averaged out by recourse to the *mean* solar day rather than an actual day. The long-term variations are mainly caused by friction that

**Fig. 1.12** Cesium atomic clock at the National Bureau of Standards.

the ocean tides exert on the Earth, which leads to a gradual slowing down of the rotation rate. Thus, between 1900 and now, the time that the Earth takes to complete 365 rotations has increased by about 1 second.

*Second, s*

To avoid such a lack of uniformity in the time scale, we now use an atomic standard of time. This standard is the period of one vibration of atoms of cesium. The second is defined as the time needed for 9,192,631,770 vibrations of cesium atoms. This coincides with the mean solar second as it was in the year 1900.

Figure 1.12 shows one of the atomic clocks at the National Bureau of Standards in Boulder, Colorado. The vibrations of the cesium atoms regulate the rate of this atomic clock just as the vibrations of a balance wheel regulate an ordinary wristwatch or the vibrations of a small quartz crystal regulate a quartz wristwatch. The roomful of complicated equipment shown in Figure 1.12 is needed to amplify the feeble vibrations of cesium atoms to a level that permits them to control the clock. The best cesium clocks are good to 1 part in $10^{13}$, that is, they lose or gain no more than 1 second in 300,000 years.

The time kept by the cesium clocks at the National Bureau of Standards is called **Coordinated Universal Time** (abbreviated UTC). Precise time signals keyed to the UTC time scale are continuously transmitted by radio station WWV, Fort Collins, Colorado. These time signals can be picked up worldwide on shortwave receivers tuned to 2.5, 5, 10, 15, or 20 megahertz. The time announced by WWV is Greenwich Mean Time, which is exactly 5 hours ahead of Eastern Standard Time.[7] Precise time signals are also announced continuously by telephone — the telephone number is (303) 499-7111. The radio signal is accurate to within $1 \times 10^{-3}$ s,[8] and the telephone signal is accurate to within $30 \times 10^{-3}$ s.

Incidentally: Because the rotation of the Earth is slower than it used to be, Coordinated Universal Time gradually drifts ahead of mean solar time. To avoid any excessive divergence between these two time scales, the National Bureau of Standards inserts a leap second once a

---

[7] Radio station CHU, Ottawa, Ontario, transmits similar time signals at 3.33, 7.33, and 14.67 megahertz.

[8] In order to take advantage of this high accuracy, the receiving station must make an allowance for the travel time of the radio signal. This travel time can amount to as much as $\frac{1}{10}$ s if the receiving station is at a remote location on the Earth's surface.

year, usually on December 31 — the last minute of the year then has 61 seconds rather than 60 seconds.

For extremely high-precision comparisons between clocks located in different laboratories, radio signals are inadequate, because irregularities in their propagation through the atmosphere introduce uncertainties in their travel time from one laboratory to the other. To deal with this difficulty, portable atomic clocks have been built. These can be synchronized with a clock in one laboratory and then carried to another laboratory, transferring the synchronization. Figure 1.13 shows one such portable atomic clock; the clock and its batteries fit in a suitcase. This particular model makes use of the vibrations of rubidium atoms and, under typical transport conditions, it is accurate to 1 part in $10^{12}$. The portable clock has been used for comparisons among the clocks of the National Bureau of Standards, the Naval Observatory (Washington, D.C.), and the Bureau International de l'Heure (Paris).

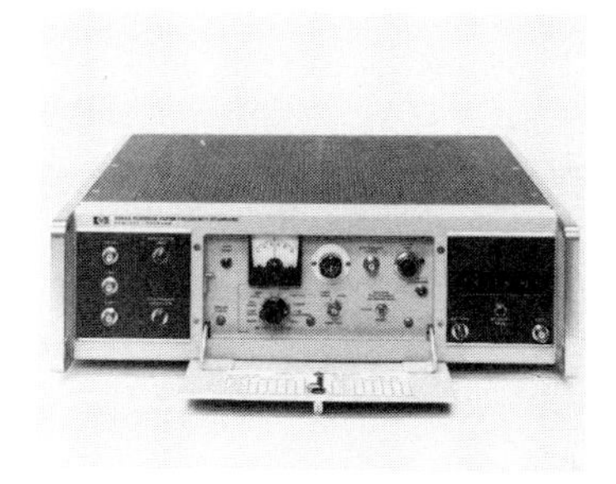

**Fig. 1.13** Portable rubidium atomic clock built by Hewlett-Packard.

Table 1.4 lists some typical time intervals. There is a curious coincidence between the numbers appearing in Tables 1.1 and 1.4: the ratio of the longest and shortest lengths of objects in our universe is about $10^{40}$, and the ratio of the longest and shortest times associated with objects in our universe is also about $10^{40}$. As we will see in Section 1.5, the number $10^{40}$ recurs in Table 1.6, which lists typical masses: the ratio of the largest and smallest masses of objects in our universe is about $(10^{40})^2$. Is this coincidence between these large numbers purely accidental or is it a clue pointing to a hidden connection between cosmology and elementary-particle physics? We do not know . . . yet.

Table 1.5 gives multiples and submultiples of the second.

**Table 1.4** SOME TIME INTERVALS

| | |
|---|---|
| Age of the universe | $\sim 4 \times 10^{17}$ s |
| Age of the Solar System | $1.4 \times 10^{17}$ s |
| Age of the oldest written records (Sumerian) | $1.6 \times 10^{11}$ s |
| Life-span of man (average) | $2.2 \times 10^{9}$ s |
| Travel time for light from nearest star | $1.4 \times 10^{8}$ s |
| Revolution of Earth (1 year) | $3.2 \times 10^{7}$ s |
| Rotation of Earth (1 day) | $8.6 \times 10^{4}$ s |
| Life-span of free neutron (average) | $9.2 \times 10^{2}$ s |
| Travel time for light from Sun | $5 \times 10^{2}$ s |
| Travel time for light from Moon | 1.3 s |
| Period of heartbeat (human) | $\sim 0.9$ s |
| Period of sound wave (middle C) | $3.8 \times 10^{-3}$ s |
| Period of radio wave (AM band) | $\sim 1 \times 10^{-6}$ s |
| Life-span of $\pi^+$ particle (average) | $2.6 \times 10^{-8}$ s |
| Period of light wave | $\sim 2 \times 10^{-15}$ s |
| Period of X ray | $\sim 3 \times 10^{-19}$ s |
| Life-span of shortest lived, unstable particle | $\sim 10^{-24}$ s |

**Table 1.5** MULTIPLES AND SUBMULTIPLES OF THE SECOND

| | |
|---|---|
| century | 1 century = 100 yr = $3.156 \times 10^{9}$ s |
| year | 1 year = $3.156 \times 10^{7}$ s = 365.25 days |
| day | 1 day = 86,400 s |
| hour | 1 h = 3600 s |
| minute | 1 min = 60 s |
| millisecond | 1 ms = $10^{-3}$ s |
| microsecond | 1 $\mu$s = $10^{-6}$ s |
| nanosecond | 1 ns = $10^{-9}$ s |
| picosecond | 1 ps = $10^{-12}$ s |

## 1.5 The Unit of Mass

*Kilogram*, kg

In the metric system, the unit of mass is the **kilogram.** The standard of mass is a cylinder of platinum–iridium alloy kept at the Bureau International des Poids et Mesures (Figure 1.14). By definition, the mass of this cylinder is exactly one kilogram (1 kg). Mass is the only fundamental unit for which we do not, as yet, have an atomic standard. It would obviously be logical to take the mass of, say, a hydrogen or krypton atom as the standard of mass. But it has so far been impossible to compare a macroscopic mass with such an atomic standard with satisfactory accuracy. Comparison of an unknown macroscopic mass with the platinum–iridium kilogram is much easier than comparison with the mass of a single atom.

**Fig. 1.14** International standard kilogram in its protective bell jar.

Mass is measured with a balance, an instrument that compares the **weight** of an unknown mass with the weight of a standard mass. Weight is directly proportional to mass, and hence equal weights imply equal masses (the precise distinction between mass and weight will be spelled out in Chapter 6). Figure 1.15 shows an extremely accurate balance especially designed by the National Bureau of Standards to handle the standard kilogram and similar masses. This balance has a pan at one end of the balance beam, and a permanently fixed counterweight at the other. To compare the weights of two masses, each of them in turn is placed in the pan and compared with the counterweight; this yields an indirect comparison of the two masses. This balance can compare two masses of about 1 kg to within a few parts in $10^9$.

The masses of individual atoms cannot be measured with such a high precision. The most recent and best determination of the mass of an individual atom (in terms of kilograms) is accurate to within 1 part in $10^6$. Essentially, such a determination is equivalent to a determination of **Avogadro's number** $N_A$, or the number of atoms per mole.[9] For instance, since the mass of one mole of carbon is exactly 12 grams, the mass of one carbon atom is

**Fig. 1.15** One-kilogram balance of the National Bureau of Standards.

$$[\text{mass of carbon atom}] = \frac{12 \text{ grams}}{N_A} \tag{1}$$

To find $N_A$, it is necessary to count the number of atoms in one mole; this is most conveniently done in a crystal where the regular arrangement of the atoms makes the counting fairly straightforward. The best available data lead to

$$N_A = 6.02214 \times 10^{23} \tag{2}$$

with an uncertainty of about $\pm 4 \times 10^{17}$, or about 6 parts in $10^7$.

Because of the uncertainty in $N_A$, an atomic standard of mass is at present not competitive with the standard kilogram. However, the

*Mole*

[9] One **mole** of any chemical element (or chemical compound) is defined as that amount of matter containing exactly as many atoms (or molecules) as there are atoms in 12 grams of carbon. The "atomic mass" of a chemical element (or the "molecular mass" of a compound) is the mass of one mole. Thus, according to the table of atomic masses (see Appendix 9), one mole of C has a mass of 12.0 grams, one mole of $O_2$ has a mass of 32.0 grams, one mole of $H_2O$ has a mass of 18.0 grams, etc.

masses of atoms can be measured *relative to one another* with considerably higher accuracy than relative to the kilogram. For this reason, the masses of atoms are often stated in terms of **atomic mass units** (u). In these units, the mass of the carbon atom is assigned the standard value

*Atomic mass unit,* u

$$[\text{mass of carbon atom}] = 12 \text{ atomic mass units} = 12 \text{ u}$$

and other masses of atoms are referred to this standard. In terms of kilograms, the value of 1 u is approximately $1.660540 \times 10^{-27}$ kg.

Table 1.6 lists some examples of masses expressed in kilograms.

Table 1.7 gives multiples and submultiples of the kilogram.

In the British system, the unit of mass is the **pound** (lb), which equals 0.453592 kg.

**Table 1.6** SOME MASSES

| | |
|---|---|
| Observable universe | $\sim 10^{55}$ kg |
| Galaxy | $4 \times 10^{41}$ kg |
| Sun | $2.0 \times 10^{30}$ kg |
| Earth | $6.0 \times 10^{24}$ kg |
| Ship *Queen Elizabeth* | $7.6 \times 10^{7}$ kg |
| Jet airliner (Boeing 747, empty) | $1.6 \times 10^{5}$ kg |
| Skylab | $7.0 \times 10^{4}$ kg |
| Automobile | $1.5 \times 10^{3}$ kg |
| Man (average male) | 73 kg |
| 5¢ coin | $5.2 \times 10^{-3}$ kg |
| Raindrop | $2 \times 10^{-6}$ kg |
| Red blood cell | $9 \times 10^{-14}$ kg |
| Smallest virus (potato spindle) | $4 \times 10^{-21}$ kg |
| Atom (iron) | $9.5 \times 10^{-26}$ kg |
| Proton | $1.7 \times 10^{-27}$ kg |
| Electron | $9.1 \times 10^{-31}$ kg |

**Table 1.7** MULTIPLES AND SUBMULTIPLES OF THE KILOGRAM

| | |
|---|---|
| metric ton (tonne) | $1 \text{ t} = 10^{3}$ kg |
| kilogram | 1 kg |
| gram | $1 \text{ g} = 10^{-3}$ kg |
| milligram | $1 \text{ mg} = 10^{-6}$ kg |
| atomic mass unit | $1 \text{ u} = 1.660 \times 10^{-27}$ kg |
| pound | $1 \text{ lb} = 0.4536$ kg |

EXAMPLE 1. How many atoms are there in a 5¢ coin? The coin is made of nickel and has a mass of $5.2 \times 10^{-3}$ kg, or 5.2 grams.

SOLUTION: According to the periodic table of chemical elements in the Prelude, the atomic mass of nickel is 58.71. Thus, one mole of nickel is 58.71 grams, and the number of moles in 5.2 grams is

$$\frac{5.2\ g}{58.71\ g} = 0.089$$

Multiplying this by Avogadro's number, we find that the number of atoms in the coin is

$$0.089 \times 6.022 \times 10^{23} = 5.3 \times 10^{22}$$

Alternatively, we can obtain this result by recalling that the atomic mass is the mass of one atom expressed in u. Hence the mass of one nickel atom is $58.71 \times 1.66 \times 10^{-27}$ kg $= 9.75 \times 10^{-26}$ kg, and the number of atoms in $5.2 \times 10^{-3}$ kg is

$$\frac{5.2 \times 10^{-3}\ \text{kg}}{9.75 \times 10^{-26}\ \text{kg}} = 5.3 \times 10^{22}$$

## 1.6 Derived Units

The meter, second, and kilogram are the fundamental units, or **base units,** of the metric system. Any other physical quantity can be measured by introducing a **derived unit** constructed by some combination of the fundamental units. For example, **volume** can be measured with a derived unit that is the cube of the unit of length. Thus, in the metric system the unit of volume is the cubic meter ($m^3$); Table 1.8 gives the submultiples of this unit.

*Volume*

**Table 1.8** SUBMULTIPLES OF THE CUBIC METER

| | |
|---|---|
| cubic meter | 1 $m^3$ |
| liter | 1 liter $= 10^{-3}\ m^3 = 10^3\ cm^3$ |
| cubic centimeter | 1 $cm^3 = 10^{-6}\ m^3$ |
| cubic millimeter | 1 $mm^3 = 10^{-9}\ m^3$ |
| cubic foot | 1 $ft^3 = 0.02832\ m^3$ |
| U.S. gallon | 1 gal. $= 0.003785\ m^3$ |

*Density*

Similarly, **density** can be measured with a derived unit that is the ratio of a unit of mass and a unit of volume. In the metric system, the unit of density is the kilogram per cubic meter ($kg/m^3$). We will see that speed, acceleration, force, etc., are also measured with derived units. Can *all* physical quantities be given units that are some combinations of meters, seconds, and kilograms? Yes, they all can be given such derived units, but sometimes it is very awkward to do so. For example, electric charge could, in principle, be measured in $(kg)^{1/2} \cdot (m)^{3/2}/s$, but in practice this unit does not lend itself to precise experimental determination, and it is convenient to introduce an entirely different unit of charge, independent of meters, seconds, and kilograms.

In essence, three fundamental units are always sufficient because physics, in a way, reduces all the complicated properties of the world to particle properties. When faced with a complicated physical quantity, such as electric charge, the physicist always asks how this quantity would affect particles. And particles can be described totally by position as a function of time and by mass — requiring three, and only three, units. Note that there is no harm, and there may be some convenience, in using a system with more than three "fundamental" units; however, in such a system one or more of the "fundamental" units will, in principle, be redundant.

A system of units consists of a complete set of fundamental units and derived units for the measurement of all physical quantities. The International System of Units, or SI, employed in this book, is the most widely accepted system of units for mechanical, thermodynamic, electric, and optical measurements. It is based on six "fundamental" units: the meter for length, the second for time, the kilogram for mass, the kelvin for temperature, the ampere for electric current, and the candela for luminosity of a light source.[10] The last three of these units are, strictly, redundant: temperature, electric current, and luminosity can be measured with suitable combinations of meters, seconds, and kilograms, but for practical reasons it happens to be convenient to make up independent units for these quantities.

We will deal with the definitions of the units of temperature and electric charge and current in later parts of the book, when we come to the study of heat and electricity. For the study of mechanics, in the first part of the book, we need only the units of length, time, and mass and some derived units constructed from these.

## 1.7 Significant Figures; Conversion of Units and Consistency of Units

The numbers in Tables 1.1, 1.4, and 1.6 are written in scientific notation, with powers of ten. This not only has the advantage that very large or very small numbers can be written compactly, but it also serves to indicate the precision of the numbers. For instance, a scientist observing the 1985 marathon at which Carlos Lopes set his new world record of 2 h, 7 min, 11.0 s, would have reported the running time as $7.6310 \times 10^3$ s, or $7.631 \times 10^3$ s, or $7.63 \times 10^3$ s, or $7.6 \times 10^3$ s depending on whether the measurement of time was made with a stopwatch, or a wristwatch with a seconds hand but no stop button, or a wristwatch without a seconds hand, or a "designer" watch with one of those daft blank faces without any numbers at all. The first of these watches permits measurements to within $\frac{1}{10}$ s, the second to within about 1 s, the third to within 10 or 20 s, and the fourth to within 1 or 2 minutes (if the scientist is good at guessing the position of the hand on the blank face). We will adopt the rule that only as many digits and decimals, or **significant figures,** are to be written down as are known fairly reliably. In accord with this rule, the number $7.6310 \times 10^3$ s comprises five significant figures, of which the last (0) represents tenths of a second; the number $7.631 \times 10^3$ s comprises four significant figures, of which the last (1) represents seconds; and so on. The scientific notation therefore gives us an immediate indication of the precision to within which the number is known.

*Significant figures*

When numbers in scientific notation are added, subtracted, multiplied, or divided, the final result can be no more accurate than the original numbers on which it is based. Hence, the final result should always be rounded off so that its precision is consistent with the least precise of the original numbers that enters the calculation. In a multiplication or division, the final result should only retain as many significant figures as found in the original number with the least significant figures. For example, the result of multiplying $7.63 \times 10^3$ by $7.6 \times 10^3$ is $5.7988 \times 10^7$, which should be rounded off to $5.8 \times 10^7$, since one

[10] Officially, the list of fundamental units of the SI system also includes the mole as a unit of "amount of substance."

of the original numbers has only two significant figures. And in an addition or subtraction, the final result should retain only as many decimals as found in the number with the least decimals (assuming all the numbers in the addition or subtraction have been written with the same power of ten). For example, the sum of $7.631 \times 10^3$ and $7.6 \times 10^3$ is $15.231 \times 10^3$, which should be rounded off to $15.2 \times 10^3$. When rounding off a number ending in 5, always round upward, because this minimizes the errors introduced by rounding (if you round off 1.5 to 2, the deviation between the original number and the round number is 25%, reckoned relative to the round number; but if you round off 1.5 to 1, the deviation is 50%).

Sometimes a number known to within many significant figures is rounded off to fewer significant figures for the sake of convenience, when high accuracy is not required. For instance, the exact value of the speed of light is $2.99792458 \times 10^8$ m/s, but for most purposes, it is adequate to round this off to $3.00 \times 10^8$ m/s, and we will often employ this approximate value of the speed of light in our calculations.

---

EXAMPLE 2. The light-year is a unit of length commonly used by astronomers to measure the distance between stars. One light-year is the distance that light travels in one year. Given that the speed of light has the (approximate) value $3.00 \times 10^8$ m/s, express one light-year in meters.

SOLUTION: According to Table 1.5, there are $3.156 \times 10^7$ seconds in one year. The distance that light travels in one year is therefore

$$[\text{distance}] = [\text{speed}] \times [\text{time}] \tag{3}$$

$$= 3.00 \times 10^8 \frac{\text{m}}{\text{s}} \times 3.156 \times 10^7 \text{ s}$$

$$= 9.47 \times 10^{15} \text{ m}$$

COMMENTS AND SUGGESTIONS: Note that the final result of this calculation has been rounded off to three significant figures, the same as the number of significant figures specified in the given data. Any additional significant figures in the final result would be unreliable and misleading. In fact, even the third significant figure in the answer is not quite reliable — a calculation starting with the exact value of the speed of light shows that, to within three significant figures, one light-year equals $9.46 \times 10^{15}$ m, and not $9.47 \times 10^{15}$ m. It is always wise to doubt the accuracy of the last significant figure, in the final result and (sometimes) also in the initial data.

Furthermore, note that in the second line of the calculation, the units of time cancel, and the final result has the units of length, as it should. We will adopt the general rule that in any calculation with the equations of physics, *the units are to be included in the calculation and they are to be multiplied and divided as though they were algebraic quantities.* These algebraic operations with the units should automatically yield the correct units for the final result — any inconsistency in the expected cancellations is a sure sign of trouble. Thus, the requirement of consistency of units provides a useful check on the calculation and on the equations, and gives some protection against costly mistakes.

---

*Dimensions*

The requirement of consistency of units in the equations of physics can be reformulated in a more general way as a requirement of consistency of **dimensions.** In this context, the dimensions of a physical quantity are said to be length, time, mass, or some product or ratio of these if the units of this quantity are those of length, time, or mass, or some product or ratio of these. Thus, volume has the dimensions of $(\text{length})^3$, density has the dimensions of $(\text{mass})/(\text{length})^3$, speed has the dimensions of (length)/(time), and so on. In any equation of physics, *the dimensions of the two sides of the equation must be the same.* For instance, we can test the consistency of Eq. (3) by examining the cancellations of the dimensions of the quantities appearing in this equation,

$$(\text{length}) = \frac{(\text{length})}{(\text{time})} \times (\text{time})$$

Dimensions are customarily used in preliminary tests of the consistency of equations, when there is some suspicion of a mistake in the equation. A test of the consistency of dimensions tells us no more and no less than a test of the consistency of units, but has the advantages that we need not commit ourselves to a particular choice of units, and we need not worry about conversions among multiples and submultiples of the units. Bear in mind that if an equation fails this consistency test, it is proved wrong; but if it passes, it is not proved right.

*Conversions of units*

In many calculations, it will be necessary to convert quantities expressed in one set of units to another set of units. As illustrated in the following example, such conversions involve no more than simple substitutions of the equivalent amounts in the two sets of units (a comprehensive list of equivalent amounts in different units will be found in Appendix 7).

---

EXAMPLE 3. The density of water is $1.00 \times 10^3$ kg/m$^3$. Express this in g/cm$^3$ and in lb/ft$^3$.

SOLUTION: Since 1 kg = $10^3$ g and 1 m = $10^2$ cm, we find

$$1.00 \times 10^3 \frac{\text{kg}}{\text{m}^3} = 1.00 \times 10^3 \times \frac{10^3\ \text{g}}{(10^2\ \text{cm})^3} = 1.00 \times 10^3 \times \frac{10^3\ \text{g}}{10^6\ \text{cm}^3}$$

$$= 1.00 \frac{\text{g}}{\text{cm}^3}$$

Similarly, since 1 kg = 1/0.454 lb and 1 m = 1/0.305 ft,

$$1.00 \times 10^3 \frac{\text{kg}}{\text{m}^3} = 1.00 \times 10^3 \times \frac{1/0.454\ \text{lb}}{(1/0.305\ \text{ft})^3}$$

$$= 1.00 \times 10^3 \times \frac{(0.305)^3\ \text{lb}}{0.454\ \text{ft}^3} = 62.5 \frac{\text{lb}}{\text{ft}^3}$$

COMMENTS AND SUGGESTIONS: An alternative method for the conversion of units from one set of units to another involves the following: Since 1 kg = $10^3$ g, we have the identity

$$1 = \frac{10^3\ \text{g}}{1\ \text{kg}}$$

and, similarly,

$$1 = \frac{1\ \text{m}}{10^2\ \text{cm}}$$

This means that any quantity can be multiplied by $10^3$ g/1 kg or 1 m/$10^2$ cm without changing its value. Thus, starting with $1.0 \times 10^3$ kg/m³, we obtain

$$1.00 \times 10^3 \frac{\text{kg}}{\text{m}^3} = 1.00 \times 10^3 \frac{\text{kg}}{\text{m}^3} \times \frac{10^3\ \text{g}}{1\ \text{kg}} \times \left(\frac{1\ \text{m}}{10^2\ \text{cm}}\right)^3$$

$$= 1.00 \times 10^3 \times 10^{-3} \frac{\text{kg}}{\text{m}^3} \frac{\text{g}}{\text{kg}} \frac{\text{m}^3}{\text{cm}^3}$$

$$= 1.00 \frac{\text{g}}{\text{cm}^3}$$

Note that the kg and m cancel and only $\text{g/cm}^3$ is left, as desired.

*Conversion factors*

Factors such as $10^3$ g/1 kg and 1 m/$10^2$ cm, whose numerical values are 1, are called **conversion factors.** To change the units of a quantity, simply multiply the quantity by whatever conversion factors will bring about the cancellation of the old units.

---

EXAMPLE 4. We can obtain a rough estimate of the size of a molecule by means of the following simple experiment. Take a droplet of oil and let it spread out on a smooth surface of water. When the oil slick attains its maximum area, it consists of a monomolecular layer, that is, it consists of a single layer of oil molecules which stand on the water surface side by side. Given that an oil droplet of mass $8.4 \times 10^{-7}$ kg and of density 920 kg/m³ spreads out into an oil slick of maximum area 0.55 m², calculate the length of an oil molecule.

SOLUTION: The volume of the oil droplet is

$$[\text{volume}] = [\text{mass}]/[\text{density}] \tag{4}$$

$$= \frac{8.4 \times 10^{-7}\ \text{kg}}{920\ \text{kg/m}^3} = 9.1 \times 10^{-10}\ \text{m}^3$$

The volume of the oil slick must be exactly the same. This latter volume can be expressed in terms of the thickness and the area of the oil slick:

$$[\text{volume}] = [\text{thickness}] \times [\text{area}]$$

Consequently,

$$[\text{thickness}] = [\text{volume}]/[\text{area}]$$

$$= \frac{9.1 \times 10^{-10}\ \text{m}^3}{0.55\ \text{m}^2} = 1.7 \times 10^{-9}\ \text{m}$$

The length of a molecule is the same as this thickness, $1.7 \times 10^{-9}$ m.

---

EXAMPLE 5. Some engineers have proposed that for long-distance travel between cities we should dig perfectly straight connecting tunnels through the Earth (Figure 1.16). A train running along such a tunnel would initially pick up speed in the first half of the tunnel as if running downhill; it would reach maximum speed at the midpoint of the tunnel; and it would gradually slow down in the second half of the tunnel, as if running uphill. Suppose such a tunnel were dug between San Francisco and Washington, D.C. The distance between these cities, measured along the Earth's surface, is 3900 km.

(a) What is the distance along the straight tunnel?
(b) What is the depth of the tunnel at its midpoint, somewhere below Kansas?
(c) What is the downward slope of the tunnel relative to the horizontal direction at San Francisco?

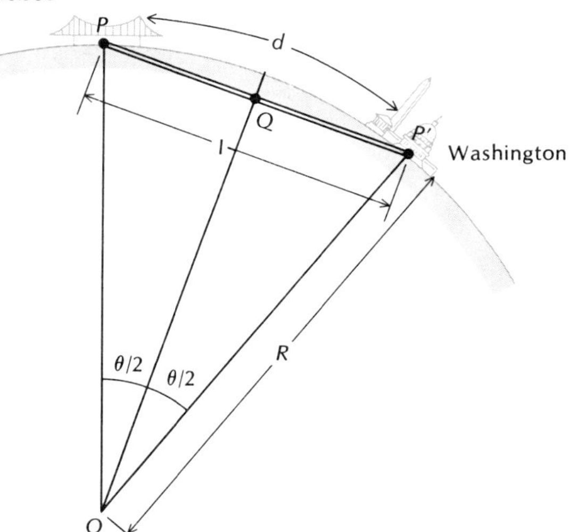


**Fig. 1.16** Cross section through the Earth showing San Francisco and Washington and the straight tunnel connecting them. The distance $d$ is measured along the surface of the Earth; the distance $l$ is the length of the tunnel. The distance $R$ is the radius of the Earth.

SOLUTION: (a) Figure 1.16 is a cross section through the Earth in the plane that contains the center of the Earth ($O$) and the two cities ($P$ and $P'$). The circle represents the surface of the Earth. Figure 1.16 shows the tunnel connecting the two cities and the radial lines from the center of the Earth to the two cities. The angle, in radians, between these two radial lines is[11]

$$\theta = \frac{[\text{distance along circle}]}{[\text{radius}]} = \frac{d}{R} \tag{5}$$

With $d = 3.9 \times 10^3$ km and $R = 6.4 \times 10^3$ km, this yields

$$\theta = \frac{3.9 \times 10^3 \text{ km}}{6.4 \times 10^3 \text{ km}} = 0.609 \text{ radian}$$

To convert radians to degrees, we note that $2\pi$ radians = 360°, so the conversion factor is $360°/2\pi$ radians:

$$\theta = 0.609 \text{ radian} \times 360°/2\pi \text{ radian} = 34.9°$$

The distance $QP$ is one side of the triangle $OQP$. Trigonometry tells us that in this triangle

$$QP = OP \sin(\theta/2)$$

or

$$QP = R \sin(\theta/2)$$

The length of the tunnel is twice $QP$:

$$l = 2QP = 2R \sin(\theta/2)$$

With $R = 6.4 \times 10^3$ km and $\theta = 34.9°$,

$$l = 2 \times 6.4 \times 10^3 \text{ km} \times \sin(34.9°/2)$$

$$= 3.8 \times 10^3 \text{ km}$$

[11] A review of trigonometry and the definitions of angles in degrees and radians will be found in Appendix 4.

(b) The distance $OQ$ is another side of the right triangle $OQP$. Trigonometry tells us that

$$OQ = OP \cos(\theta/2) = R \cos(\theta/2)$$

The depth $h$ of the tunnel at its midpoint equals the difference between $R$ and $OQ$,

$$h = R - OQ = R - R\cos(\theta/2)$$

$$= R[1 - \cos(\theta/2)]$$

which gives

$$h = 6.4 \times 10^3 \text{ km} \times [1 - \cos(34.9°/2)]$$

$$= 2.9 \times 10^2 \text{ km}$$

*Slope*

(c) The **slope** of a line relative to the horizontal direction is defined as the ratio of the vertical increment to the horizontal increment or decrement (Figure 1.17):

$$[\text{slope}] = \frac{\Delta y}{\Delta x} \qquad (6)$$

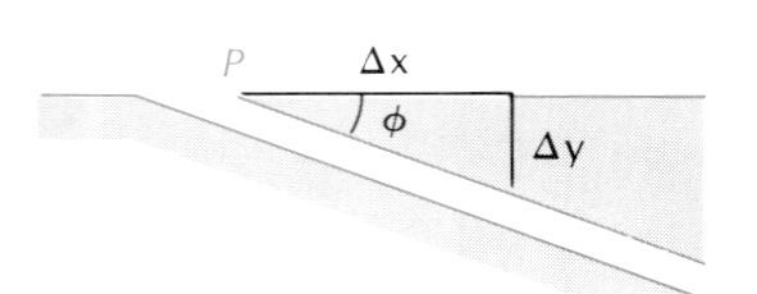


**Fig. 1.17** At San Francisco, the tunnel goes down a vertical distance $\Delta y$ while it advances a horizontal distance $\Delta x$. The ratio $\Delta y/\Delta x$ is the slope of the tunnel.

By trigonometry, this is simply the tangent of the angle between the line and the horizontal,

$$[\text{slope}] = \tan\phi \qquad (7)$$

Since the horizontal at San Francisco is perpendicular to $OP$, and the line of the tunnel is perpendicular to $OQ$, the angle $\phi$ equals the angle $\theta/2$ (see Figure 1.16). Hence

$$[\text{slope}] = \tan\phi = \tan(\theta/2) = \tan(34.9°/2) = 0.31$$

Engineers usually write slopes as ratios; in this notation, the slope is 31:100. Thus, at San Francisco the tunnel has to go down 31 m vertically for every 100 m it goes horizontally.

## SUMMARY

**Fundamental assumptions of classical physics:**

The motion of a particle can be described by giving the position as a function of time.

Space is three dimensional.

Space is Euclidean.

Time is absolute.

**Reference frame:** A coordinate grid with a set of synchronized clocks.

**Units of length, time, and mass:** meter, second, kilogram.

**Standards of length, time, and mass:** speed of light implemented by stabilized laser, cesium atomic clock, cylinder of platinum–iridium.

## QUESTIONS

1. Try to estimate by eye the lengths, in centimeters or meters, of a few objects in your immediate environment. Then measure them with a ruler or meter stick. How good were your estimates?

2. How close is your watch to standard time right now? Roughly how many minutes does your watch gain or lose per month?

3. What is meant by the phrase *a point in time*?

4. Mechanical clocks (with pendulums) were not invented until the tenth century A.D. What clocks were used by the ancient Greeks and Romans?

5. By counting aloud "One, two, three, . . . , ten, one, . . ." at a reasonable rate you can measure seconds fairly accurately. Try to measure 30 seconds in this way. How good a timekeeper are you?

6. Pendulum clocks are affected by the temperature and pressure of air. Why?

7. During the debate over the "densepack" configuration of buried silos for American intercontinental ballistic missiles (ICBMs), the proponents argued that Soviet ICBMs would be ineffective against the "densepack" because the explosion of the first few incoming Soviet ICBMs would destroy most of the other incoming ICBMs ("fratricide"). A physicist was quick to point out (the *New York Times,* December 1, 1982) that the Soviets could avoid "fratricide" by making all their ICBMs explode at almost exactly the same instant. What method could be used to bring about simultaneous explosions within, say, 1 microsecond?

8. In 1761 an accurate chronometer built by John Harrison was tested aboard H.M.S. *Deptford* during a voyage at sea for 5 months. During this voyage, the chronometer accumulated an error of less than 2 minutes. For this achievement, Harrison was ultimately awarded a prize of £20,000 that the British government had offered for the discovery of an accurate method for the determination of geographical longitude at sea. Explain how the navigator of a ship uses a chronometer and observation of the position of the Sun in the sky to find longitude.

9. Captain Lecky's *Wrinkles in Practical Navigation,* a famous nineteenth-century textbook of celestial navigation, recommends that each ship carry three chronometers for accurate timekeeping. What can the navigator do with three chronometers that he cannot do with two?

10. Suppose that by an "act of God" (or by the act of a thief) the standard kilogram at Sèvres were destroyed. Would this destroy the metric system?

11. Estimate the masses, in grams or kilograms, of a few bodies in your environment. Check the masses with a balance if you have one available.

12. Consider the piece of paper on which this sentence is printed. If you had available suitable instruments, what physical quantities could you measure about this piece of paper? Make the longest list you can and give the units. Are all these units derived from the meter, second, and kilogram?

13. Could we take length, time, and density as the three fundamental units? What could we use as a standard of density?

14. Could we take length, mass, and density as the three fundamental units? Length, mass, and speed?

## PROBLEMS

### Section 1.3

1. What is your height in feet? In meters?

2. With a ruler, measure the thickness of this book, excluding the cover. Deduce the thickness of each of the sheets of paper making up the book.

3. A football field measures 100 yd × $53\frac{1}{3}$ yd. Express this in meters.

4. Express the last four entries in Table 1.1 in inches.

5. Express the following fractions of an inch in millimeters: $\frac{1}{2}$, $\frac{1}{4}$, $\frac{1}{8}$, $\frac{1}{16}$, $\frac{1}{32}$, and $\frac{1}{64}$ in.

6. As seen from the Earth, the Sun has an angular diameter of 0.53°. The distance between the Earth and the Sun is $1.5 \times 10^{11}$ m. From this, calculate the radius of the Sun.

7. Analogies can often help us to imagine the very large or very small distances that occur in astronomy or in atomic physics.
   (a) If the Sun were the size of a grapefruit, how large would the Earth be? How far away would the nearest star be?
   (b) If your head were the size of the Earth, how large would an atom be? How large would a red blood cell be?

8. One of the most distant objects yet observed by astronomers is the quasar Q1208+1011, at a distance of 12.4 billion light-years from the Earth. If you wanted to plot the position of this quasar on the same scale as the diagram at the top of page 9 of the Prelude, how far from the center of the diagram would you have to place this quasar?

9. On the scale of the second diagram on page 6 of the Prelude, what would have to be the size of the central dot if it is to represent the size of the Sun faithfully?

*10. The Earth is approximately a sphere of radius $6.37 \times 10^6$ m. Calculate the distance from the pole to the equator, measured along the surface of the Earth. Calculate the distance from the pole to the equator, measured along a straight line passing through the Earth.

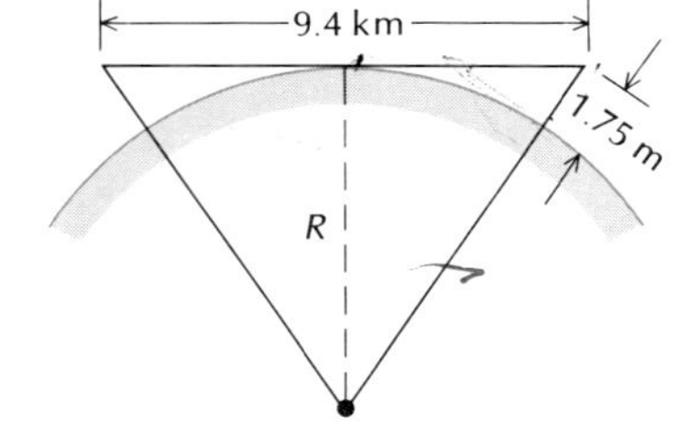


**Fig. 1.18** The distance between the physicist and the pole is 9.4 km.

*11. A nautical mile (nmi) equals 1.151 mi, or 1852 m. Show that a distance of 1 nmi along a meridian of the Earth corresponds to a change of latitude of 1 minute of arc.

**12. A physicist plants a vertical pole at the waterline on the shore of a calm lake. When she stands next to the pole, its top is at eye level, 175 cm above the waterline. She then rows across the lake and walks along the waterline on the opposite shore until she is so far away from the pole that her entire view of it is blocked by the curvature of the surface of the lake, that is, the entire pole is below the horizon (Figure 1.18). She finds that this happens when her distance from the pole is 9.4 km. From this information, deduce the radius of the Earth.

## Section 1.4

13. What is your age in days? In seconds?

14. The age of the Earth is $4.5 \times 10^9$ years. Express this in seconds.

*15. Each day at noon a mechanical wristwatch was compared with WWV time signals. The watch was not reset. It consistently ran late, as follows: June 24, late 4 s; June 25, late 20 s; June 26, late 34 s; June 27, late 51 s.
   (a) For each of the three 24-hour intervals, calculate the rate at which the wristwatch lost time. Express your answer in seconds lost per hour.
   (b) What is the average of the rates of loss found in part (a)?
   (c) When the wristwatch shows $10^h30^m$ on June 30, what is the correct WWV time? Do this calculation with the average rate of loss of part (b) and also with the largest of the rates of loss found in part (a). Estimate to within how many seconds the wristwatch can be trusted on June 30 after the correction for rate of loss has been made.

*The asterisks indicate the levels of difficulty of the problems (see the Preface).

*16. The navigator of a sailing ship seeks to determine his longitude by observing at what time (Greenwich Mean Time) the Sun reaches the zenith at his position (local noon). Suppose that the navigator's chronometer is in error and is late by 1 s compared to Greenwich Mean Time. What will be the consequent error of longitude (in minutes of arc)? What will be the error in position (in kilometers) if the ship is on the equator?

### Section 1.5

17. What is your mass in pounds? In kilograms? In atomic mass units?

18. What percentage of the mass of the Solar System is in the planets? What percentage is in the Sun? Use the data given in the table printed on the endpapers of the book.

19. The atomic mass of fissionable uranium is 235.0 g. What is the mass of a single uranium atom? Express your answer in kilograms and in atomic mass units.

20. The atom of uranium consists of 92 electrons, each of mass $9.1 \times 10^{-31}$ kg, and a nucleus. What percentage of the total mass is in the electrons and what percentage is in the nucleus of the atom?

21. How many water molecules are there in 1 gallon (3.79 liters) of water? How many oxygen atoms? Hydrogen atoms?

*22. (a) How many molecules of water are there in one cup of water? A cup is about 250 $cm^3$.
(b) How many molecules of water are there in the ocean? The total volume of the ocean is $1.3 \times 10^{18}$ $m^3$.
(c) Suppose you pour a cup of water into the ocean, allow it to become thoroughly mixed, and then take a cup of water out of the ocean. On the average, how many of the molecules originally in the cup will again be in the cup?

23. How many molecules are there in one cubic centimeter of air? Assume that the density of air is 1.3 $kg/m^3$ and that it consists entirely of nitrogen molecules ($N_2$).

*24. How many atoms are there in the Sun? The mass of the Sun is $1.99 \times 10^{30}$ kg and its chemical composition (by mass) is approximately 70% hydrogen and 30% helium.

*25. The chemical composition of air is (by mass) 75.5% $N_2$, 23.2% $O_2$, and 1.3% Ar. What is the average "molecular mass" of air; that is, what is the mass of $6.02 \times 10^{23}$ molecules of air?

*26. How many atoms are there in a human body of 73 kg? The chemical composition (by mass) of a human body is 65% oxygen, 18.5% carbon, 9.5% hydrogen, 3.3% nitrogen, 1.5% calcium, 1% phosphorus, and 0.35% other elements (ignore the "other elements" in your calculation).

### Section 1.7

27. The distance from our Galaxy to the Andromeda galaxy is $2.2 \times 10^6$ light-years. Express this distance in meters.

28. In analogy with the light-year, we can define the light-second as the distance light travels in one second and the light-minute as the distance light travels in one minute. Express the Earth–Sun distance in light-minutes. Express the Earth–Moon distance in light-seconds.

29. Astronomers often use the astronomical unit (AU), the parsec (pc), and the light-year. The AU is the distance from the Earth to the Sun;[12] $1\text{ AU} = 1.496 \times 10^{11}$ m. The pc is the distance at which 1 AU subtends an

[12] Strictly, it is the semimajor axis of the Earth's orbit.

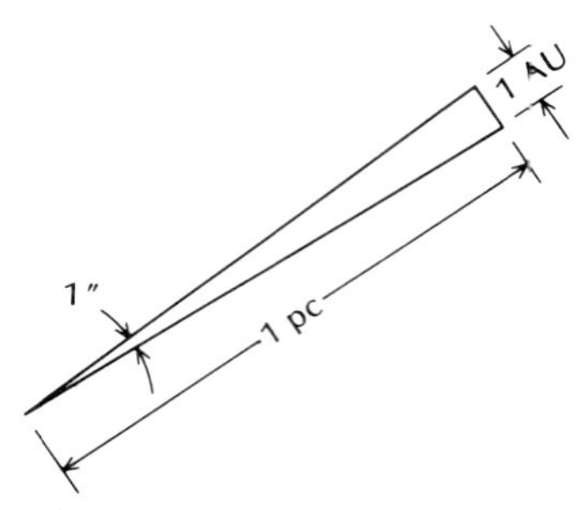


**Fig. 1.19**

angle of exactly 1 second of arc (Figure 1.19). The light-year is the distance that light travels in 1 year.
(a) Express the pc in AU.
(b) Express the pc in light-years.
(c) Express the pc and the light-year in meters.

30. The density of copper is 8.9 g/cm$^3$. Express this in kg/m$^3$, lb/ft$^3$, and lb/in.$^3$.

31. The federal highway speed limit was 55 mi/h. Express this in kilometers per hour, feet per second, and meters per second.

32. What is the volume of an average human body? (Hint: The density of the body is about the same as that of water.)

33. Meteorologists usually report the amount of rain in terms of the depth in inches to which the water would accumulate on a flat surface if it did not run off. Suppose that 1 in. of rain falls during a storm. Express this in cubic meters of water per square meter of surface. How many kilograms of water per square meter of surface does this amount to?

34. A fire hose delivers 300 liters of water per minute. Express this in m$^3$/s. How many kilograms of water per second does this amount to?

35. The total volume of the oceans of Earth is $1.3 \times 10^{18}$ m$^3$. What percentage of the mass of the Earth is in the oceans?

36. The nucleus of an iron atom is spherical and has a radius of $4.6 \times 10^{-15}$ m; the mass of the nucleus is $9.5 \times 10^{-26}$ kg. What is the density of the nuclear material? Express your answer in metric tons per cubic centimeter.

37. Our Sun has a radius of $7.0 \times 10^8$ m and a mass of $2.0 \times 10^{30}$ kg. What is its average density? Express your answer in grams per cubic centimeter.

38. Pulsars, or neutron stars, typically have a radius of 20 km and a mass equal to that of our Sun ($2.0 \times 10^{30}$ kg). What is the average density of such a pulsar? Express your answer in metric tons per cubic centimeter.

39. The nuclei of all atoms have approximately the same density of mass. The nucleus of a copper atom has a mass of $1.06 \times 10^{-25}$ kg and a radius of $4.8 \times 10^{-15}$ m. The nucleus of a lead atom has a mass of $3.5 \times 10^{-25}$ kg; what is its radius? The nucleus of an oxygen atom has a mass of $2.7 \times 10^{-26}$ kg; what is its radius? Assume that the nuclei are spherical.

40. The table printed on the endpapers gives the masses and radii of the major planets. Calculate the average density of each planet and make a list of the planets in order of decreasing densities. Is there a correlation between the density of a planet and its distance from the Sun?

41. A small single-engine plane is flying at a height of 5000 m at a (horizontal) distance of 18 km from the San Francisco airport when the engine quits. The pilot knows that, without the engine, the plane will glide downward at an angle of 15°. Can she reach San Francisco?

*42. You are crossing the Atlantic in a sailboat and hoping to make a landfall in the Azores. The highest peak on the Azores has a height of 2300 m. From what distance can you see this peak just emerging over the horizon?

*43. In the Galapagos (on the equator) the small island of Marchena is 60 km west of the small island Genovesa. If the sun sets at 8:00 P.M. at Genovesa, when will it set at Marchena?

*44. For tall trees, the diameter at the base (or the diameter at any given point of the trunk, such as the midpoint) is roughly proportional to the 3/2 power of the length. The tallest sequoia in Sequoia National Park in California has a length of 81 m, a diameter of 7.6 m at the base, and a mass of 6100 metric tons. A petrified sequoia found in Nevada has a length of 90 m. Estimate its diameter at the base, and estimate the mass it had when it was still alive.

CHAPTER 2

# Kinematics in One Dimension

Kinematics is the study of the geometry of motion: it deals with the mathematical description of motion in terms of position, velocity, and acceleration. Kinematics serves as a prelude to dynamics, which studies force as the cause of changes in motion. In this chapter, and in the next nine chapters, we will be concerned only with **translational motion** of a particle, which is defined as change of position of the particle as a function of time. In the case of an ideal particle — a body with no size and no internal structure — position as a function of time gives a complete description. In the case of a more complicated body — a man, a ship, or a planet — position as a function of time does not give a complete description. Such a complicated body has many internal parts which can rotate and move in relation to one another. Nevertheless, insofar as we are not interested in the size, orientation, and internal structure of a body, we may find it useful to concentrate on its translational motion and ignore all other motions. Under these circumstances we may pretend that the motion of the complicated body is particle motion. The position of the body is then to be specified by the position of some fiducial point[1] marked on it, such as its center, or its leading edge. For example, we can describe the motion of a ship steaming out of New York harbor as particle motion — for most purposes it will be sufficient to know the position of the center of the ship as a function of time. However, if the ship were to suffer a collision, we could not describe the motion of the ship during the collision as particle motion — in this case it would be essential to know the changes in internal structure and in orientation as a function of time.

*Translational motion*

[1] A fiducial point is a reference point, which we imagine painted on the body.

## 2.1 Average Speed

Consider a particle in motion, that is, a particle whose position changes as a function of time. The particle travels along a path, straight or curved. In some given time interval the particle will travel a certain distance measured along the path. The **average speed** of the particle is defined as the ratio of this distance to the magnitude of the time interval,

*Average speed*

$$[\text{average speed}] = \frac{[\text{distance traveled}]}{[\text{time taken}]} \tag{1}$$

Thus, speed is the rate of change of the distance, or the change of distance per unit time. We see from Eq. (1) that the unit of speed is the unit of length divided by the unit of time. In the metric system, the unit of speed is the meter per second (m/s). In the British system, the unit of speed is the foot per second (ft/s). Table 2.1 gives some examples of typical speeds.

**Table 2.1** SOME SPEEDS

| | |
|---|---|
| Light | $3.0 \times 10^8$ m/s |
| Recession of fastest known quasar | $2.7 \times 10^8$ m/s |
| Electron around nucleus (hydrogen) | $2.2 \times 10^6$ m/s |
| Earth around Sun | $3.0 \times 10^4$ m/s |
| SST airliner (Tu-144, maximum airspeed) | $7.1 \times 10^2$ m/s |
| Rifle bullet (muzzle velocity) | $\sim 7 \times 10^2$ m/s |
| Rotation of Earth (at equator) | $4.6 \times 10^2$ m/s |
| Random motion of molecules in air (average) | $4.5 \times 10^2$ m/s |
| Sound | $3.3 \times 10^2$ m/s |
| Jet airliner (Boeing 747, maximum airspeed) | $2.7 \times 10^2$ m/s |
| Cheetah (maximum) | 28 m/s |
| Federal highway speed limit (55 mi/h) | 25 m/s |
| Man (maximum) | 12 m/s |
| Man (walking briskly) | 1.3 m/s |
| Snail | $\sim 10^{-3}$ m/s |
| Glacier | $\sim 10^{-6}$ m/s |
| Rate of growth of hair (human) | $3 \times 10^{-9}$ m/s |
| Continental drift | $\sim 10^{-9}$ m/s |
| Rate of subsidence of Southeast England | $\sim 10^{-10}$ m/s |

EXAMPLE 1. A runner takes 11 seconds to run a 100-meter dash. What is his average speed relative to the Earth?

SOLUTION: According to Eq. (1),

$$[\text{average speed}] = \frac{[\text{distance}]}{[\text{time taken}]} = \frac{100\text{ m}}{11\text{ s}} = 9.1\text{ m/s}$$

EXAMPLE 2. The Earth moves around the Sun in a circular orbit of radius $1.50 \times 10^8$ km. What is the average speed of the Earth relative to the Sun?

SOLUTION: The Earth moves once around its orbit in 1 year, that is, in $3.16 \times 10^7$ s. The distance covered in this time is $2\pi \times 1.50 \times 10^8$ km. Hence

$$[\text{average speed}] = \frac{2\pi \times 1.50 \times 10^8 \text{ km}}{3.16 \times 10^7 \text{ s}} = 29.8 \text{ km/s}$$

Motion and speed are *relative:* the value of the speed depends on the frame of reference with respect to which it is calculated. Example 1 gives the speed of the runner *relative to the surface of the Earth.* However, since the Earth is moving around the Sun, his speed *relative to the Sun* is approximately 29.8 km/s. Thus, questions regarding speed are meaningless unless the frame of reference is first specified. In everyday language "speed" usually means speed relative to the Earth's surface. The reference frame relative to which the speed is reckoned will usually be clear from the context. For example, in Table 2.1 the speed of the rifle bullet is reckoned relative to the rifle (which may be in motion relative to the Earth). We will be careful to specify the frame of reference whenever it is not clear from the context.

## 2.2 Average Velocity in One Dimension

For the rest of this chapter, we will consider the special case of motion along a straight line, that is, motion in one dimension. For convenience, we will assume that the straight line coincides with the $x$ axis. We can then give a mathematical description of the motion of the particle by specifying the $x$ coordinate as a function of time. Graphically, we can represent the motion by means of a plot of the $x$ coordinate vs. the time coordinate. For example, Figure 2.1 shows such a plot of the position coordinate $x$ vs. the time coordinate $t$ for an automobile that starts from rest, accelerates along a straight road for 10 seconds, and then brakes and comes to a full stop 4.3 seconds later (the plot is based on data from an acceleration test of a Maserati sports car). The position is measured from the starting point on the road to a fiducial point which we imagine marked on the automobile. The translational motion of the automobile can be represented by the motion of the fiducial point, and therefore this motion of the automobile can be regarded as particle motion. For a particle, the plot of the position vs. time provides a complete description of the motion. In the terminology of modern physics, the curve in such a plot is called the **worldline** of the particle (this terminology is borrowed from the theory of relativity). Thus, the curve plotted in Figure 2.1 is the worldline of the automobile.

*Worldline*

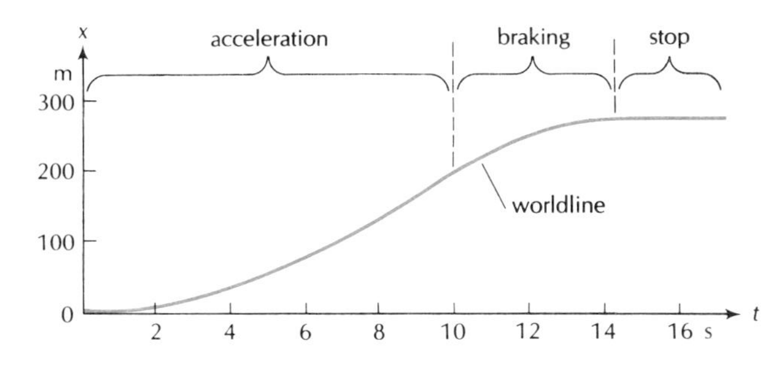


**Fig. 2.1** Worldline of an automobile that accelerates ($0 \text{ s} < t < 10 \text{ s}$), then brakes ($10 \text{ s} < t < 14.3 \text{ s}$), and then remains stopped ($t > 14.3 \text{ s}$) (based on data from a road test of a Maserati Bora by *Road & Track* magazine).

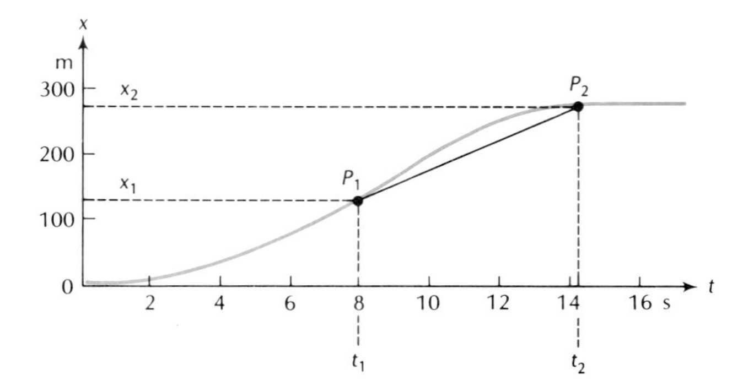


**Fig. 2.2** The straight line connecting the two points $P_1$ and $P_2$ on the worldline has a slope $(x_2 - x_1)/(t_2 - t_1)$; this slope is the average velocity.

Suppose that at time $t_1$ the automobile is at position $x_1$ and at a subsequent time $t_2$ the automobile is at position $x_2$ (see Figure 2.2). Then $x_2 - x_1$ is the change of position that occurred in the time interval $t_2 - t_1$. The change of position divided by the time interval is called the **average velocity,**

*Average velocity*

$$\bar{v} = \frac{x_2 - x_1}{t_2 - t_1} \tag{2}$$

Here, the overbar on the symbol for velocity is a standard notation used in physics to indicate an average quantity. The expression (2) can also be written as

$$\boxed{\bar{v} = \frac{\Delta x}{\Delta t}} \tag{3}$$

with $\Delta x = x_2 - x_1$ and $\Delta t = t_2 - t_1$ (here, the Greek letter $\Delta$ is a standard notation used to indicate a change in a quantity). Thus, the average velocity is the average rate of change of the position.

Graphically, in the plot of position vs. time, the average velocity is the vertical separation between the points $P_1$ and $P_2$ on the worldline divided by the horizontal separation. Hence, the average velocity is the slope of the line connecting the points $P_1$ and $P_2$ (see Figure 2.2). For instance, if $t_1 = 8.0$ s and $t_2 = 14.0$ s, then the average velocity or, alternatively, the slope of the straight line connecting the points $P_1$ and $P_2$ in Figure 2.2 is

$$\bar{v} = \frac{x_2 - x_1}{t_2 - t_1} = \frac{270\text{ m} - 130\text{ m}}{14.0\text{ s} - 8.0\text{ s}} = \frac{140\text{ m}}{6.0\text{ s}} = 23\text{ m/s}$$

Note that, according to the general formula (2), the velocity is positive or negative depending on whether $x_2$ is larger or smaller than $x_1$, that is, depending on whether the $x$ coordinate increases or decreases in the time interval $t_2 - t_1$. This means that the sign of the velocity depends on the direction of motion. If the motion is in the positive $x$ direction — as in the example plotted in Figure 2.2 —, the velocity is positive; if the motion is in the negative $x$ direction, the velocity is negative. In contrast, the speed [defined by Eq. (1)] is always positive.

The average velocity takes into account only the net change in position; it does not take into account how this change in position is accomplished in detail. For example, for any back and forth motion (round trip) in which the particle returns to its initial position, the average velocity is zero, just as though the particle had not moved at all. The average speed takes into account only the distance traveled; it does not take into account the direction of travel or whether the travel produced any net change of position. The following example will help to make this distinction clear.

EXAMPLE 3. A runner runs 100 m on a straight track in 11 s and then walks back in 80 s. What are the average velocity and the average speed for each part of this motion and for the complete motion?

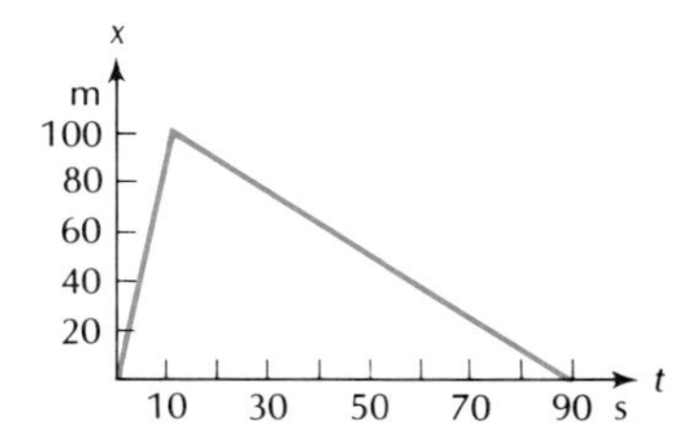


**Fig. 2.3** Worldline of a runner.

SOLUTION: The worldline of the runner is shown in Figure 2.3. The motion has two parts: the run ($0 \text{ s} \leq t \leq 11 \text{ s}$) and the walk ($11 \text{ s} \leq t \leq 91 \text{ s}$). The average velocity for the run is

$$\bar{v} = \frac{\Delta x}{\Delta t} = \frac{+100 \text{ m}}{11 \text{ s}} = +9.1 \text{ m/s}$$

The average velocity for the walk is

$$\bar{v} = \frac{-100 \text{ m}}{80 \text{ s}} = -1.2 \text{ m/s}$$

The average velocity for the entire motion is

$$\bar{v} = \frac{0 \text{ m}}{91 \text{ s}} = 0 \text{ m/s}$$

This is zero because the net change of position is zero.

The average speeds for the run and walk are, respectively, 9.1 m/s and 1.2 m/s. The average speed for the entire motion is

$$\frac{200 \text{ m}}{91 \text{ s}} = 2.2 \text{ m/s}$$

COMMENTS AND SUGGESTIONS: The magnitude of the average velocity is always smaller than the average speed, except for the case of a motion that proceeds in one direction and never reverses — then the magnitude of the average velocity coincides with the average speed.

## 2.3 Instantaneous Velocity

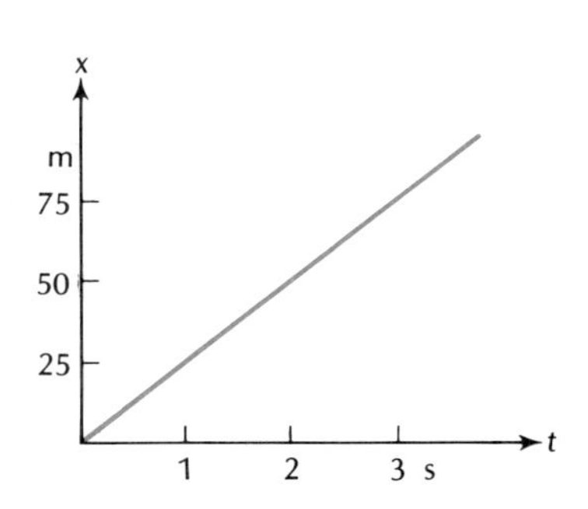


**Fig. 2.4** Worldline of an automobile moving at constant velocity.

If a particle moves at constant velocity (uniform motion), the worldline is a straight line, with a slope equal to the velocity. For example, Figure 2.4 shows a plot of position vs. time for an automobile moving along a straight road at a constant velocity of 25 m/s. This plot is a straight line of constant slope — the slope in any time interval is equal to the slope in any other time interval. Thus, the average velocity is the same for all time intervals — it is always 25 m/s. Since the velocity is always the same, we may regard the *instantaneous* velocity for this motion as identical to the average velocity.

If a particle moves with a varying velocity (accelerated motion), the worldline is a curve. The worldline of the accelerating automobile shown in Figure 2.1 gives us an example of this: the automobile first accelerates and then decelerates, and the worldline is a curve of varying slope. How can we construct a definition of the instantaneous velocity of the automobile on the basis of the worldline plotted in Figure 2.1? The instantaneous velocity is, of course, what is displayed on the speedometer of the automobile (except that the speedometer does not make a distinction between positive velocity and negative velocity — it does not care whether you are driving eastward or westward along a street, and only displays the *magnitude* of the instantaneous velocity; that is, it displays the instantaneous speed).

To arrive at a definition of the instantaneous velocity, consider the instant $t = 4$ s. We can find an *approximate* value for the velocity at this instant by taking a small time interval of, say, 0.001 s centered on 4 s, that is, a time interval from 3.9995 s to 4.0005 s. In this time interval the automobile moves some small distance $\Delta x$, and we can approximate the actual curved worldline by a straight line segment connecting the endpoints of the interval (see Figure 2.5a). According to the discussion at the beginning of this section, the instantaneous velocity associated with a straight worldline is simply the slope of the worldline; hence the instantaneous velocity at $t = 4$ s can be evaluated approximately as the slope of the short line segment shown in Figure 2.5a. Whether this is a good approximation depends on how closely the straight line segment coincides with the actual curved worldline. Obviously, the approximation can be improved by taking a shorter time interval, 0.0001 s or

**Fig. 2.5** (a) Over the small time interval $\Delta t$, we can approximate the worldline by a straight line segment connecting the endpoints of the interval. Within the accuracy of this plot, this straight line segment coincides with the tangent that touches the worldline at $t = 4$ s. (b) The tangent to the worldline at $t = 4$ s has a slope of 34.0 m/2.0 s = 17.0 m/s.

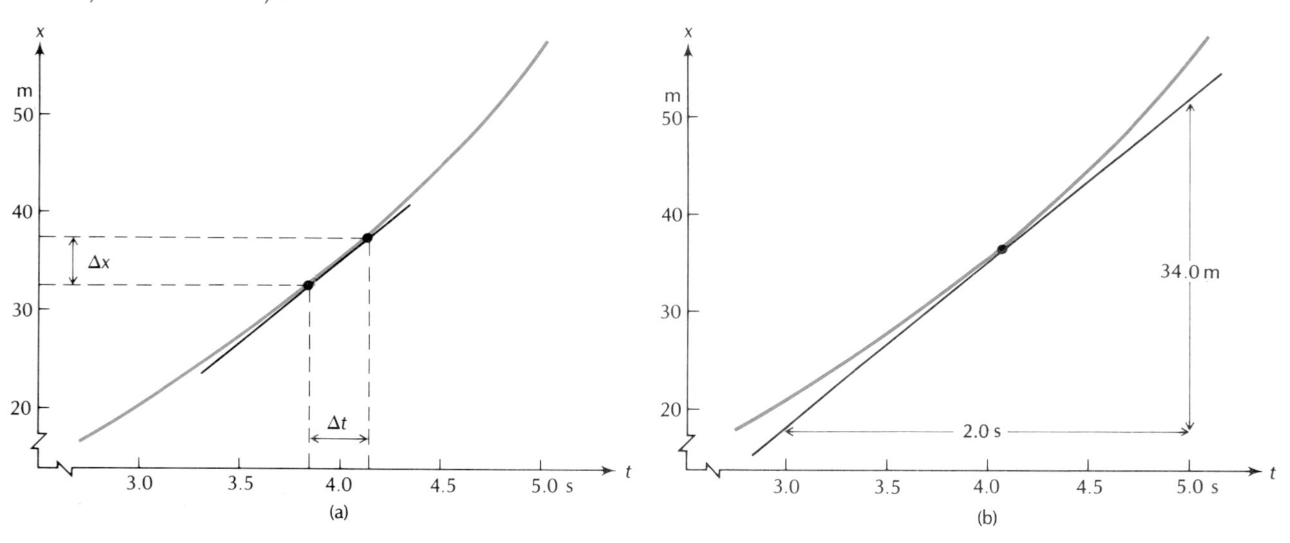


*Instantaneous velocity as slope*

even less. In the limiting case of an extremely small time interval (infinitesimal time interval), the straight line segment coincides with the tangent to the worldline at $t = 4$ s. Hence, the **instantaneous velocity** at a given time *equals the slope of the tangent to the worldline at that time.* For example, drawing the tangent that touches the worldline at $t = 4$ s (Figure 2.5b) and measuring its slope on the graph, we readily find that this slope is 17 m/s; hence the instantaneous velocity at $t = 4$ s is 17 m/s.

By drawing tangents at other points of the worldline and measuring their slopes, we can obtain a complete table of instantaneous velocities as a function of time. Figure 2.6 is a plot of the results of such a determination of the instantaneous velocities. The velocity is initially zero

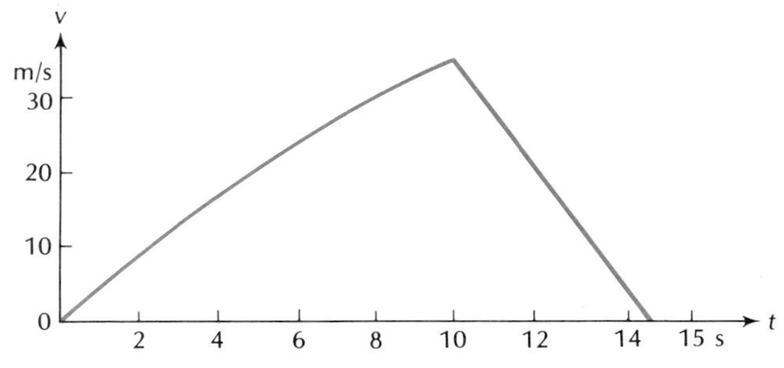


**Fig. 2.6** Instantaneous velocity as a function of time.

(zero slope in Figure 2.1), then increases, reaching a maximum of 34.9 m/s at $t = 10$ s (maximum slope in Figure 2.1), and finally decreases to zero at $t = 14.3$ s (zero slope in Figure 2.1).

The above is a graphical method for the determination of the instantaneous velocity; this method is convenient if the data have been given as a plot of position vs. time (as in Figure 2.1). However, if the data have been given analytically, as formulas, then an analytic method, based on formulas, is more convenient. Suppose that the worldline is described by an equation that gives $x$ as a function of $t$. For example, the position of the automobile plotted in Figure 2.1 is described by the formulas

$$x = 2.376t^2 - 0.042t^3 \qquad \text{for } 0 \text{ s} \leq t \leq 10 \text{ s} \tag{4}$$

and

$$x = -555.8 + 115.4t - 4.022t^2 \qquad \text{for } 10 \text{ s} \leq t \leq 14.3 \text{ s} \tag{5}$$

where the distance is measured in meters and the time in seconds.[2] As we will see in Eqs. (9) and (10), the instantaneous velocity can then be calculated directly from these expressions. We recall that as a first approximation to the instantaneous velocity, we took the average velocity

$$\overline{v} = \frac{\Delta x}{\Delta t} \tag{6}$$

for a small time interval $t = 0.001$ s centered around $t = 4$ s. To improve on this approximation, we took an even smaller time interval and finally an infinitesimal time interval. Thus the exact instantaneous velocity is obtained by evaluating Eq. (6) in the limiting case of vanishing $\Delta t$, that is, by taking the limit as $\Delta t$ tends to zero,

$$v = \lim_{\Delta t \to 0} \frac{\Delta x}{\Delta t} \tag{7}$$

In the notation of differential calculus this limit is written as

*Instantaneous velocity as derivative*

$$\boxed{v = \frac{dx}{dt}} \tag{8}$$

[2] In the formulas (4) and (5), $x$ and $t$ are treated as pure numbers, unaccompanied by units. Since we are not including the units in the equations, we have to specify them by a separate explicit statement. We will occasionally adopt this procedure for the sake of calculational convenience.

and is called the *derivative of x with respect to t.* Thus, the **instantaneous velocity** is the derivative of the position with respect to time.

According to the rules of calculus, the derivative of the function $ct^n$ with respect to time is $cnt^{n-1}$ (where $c$ is some arbitrary constant), that is,

$$\frac{d(ct^n)}{dt} = cnt^{n-1}$$

Furthermore, the derivative of the sum of two such functions is the sum of their derivatives.[3] Hence, the derivative of $2.376t^2 - 0.042t^3$ with respect to time is $2 \times 2.376t - 3 \times 0.042t^2$; and the derivative of $-555.8 + 115.4t - 4.022t^2$ with respect to time is $115.4 - 2 \times 4.022t$. The instantaneous velocities implied by the formulas (4) and (5) are then

$$v = \frac{dx}{dt} = 4.752t - 0.126t^2 \qquad \text{for } 0\text{ s} \le t \le 10\text{ s} \tag{9}$$

$$v = \frac{dx}{dt} = 115.4 - 8.044t \qquad \text{for } 10\text{ s} \le t \le 14.3\text{ s} \tag{10}$$

where velocity is measured in meters per second. At, say, $t = 4$ s, Eq. (9) gives $v = 4.752 \times 4 - 0.126 \times 4^2 = 17.0$ m/s, which agrees with the graphical determination.

## 2.4 Acceleration

Any motion with a change of velocity is accelerated motion. If a particle has velocity $v_1$ at time $t_1$ and velocity $v_2$ at time $t_2$, then the **average acceleration** for this time interval is defined as the change of velocity divided by the change of time,

*Average acceleration*

$$\bar{a} = \frac{v_2 - v_1}{t_2 - t_1} = \frac{\Delta v}{\Delta t} \tag{11}$$

where $\Delta v = v_2 - v_1$ and $\Delta t = t_2 - t_1$. Accordingly, the average acceleration is the average rate of change of the velocity.

On a plot of velocity vs. time, the average acceleration is the slope of the straight line connecting the points $P_1$ and $P_2$ that correspond to $t = t_1$ and $t = t_2$ on the velocity curve (Figure 2.7).

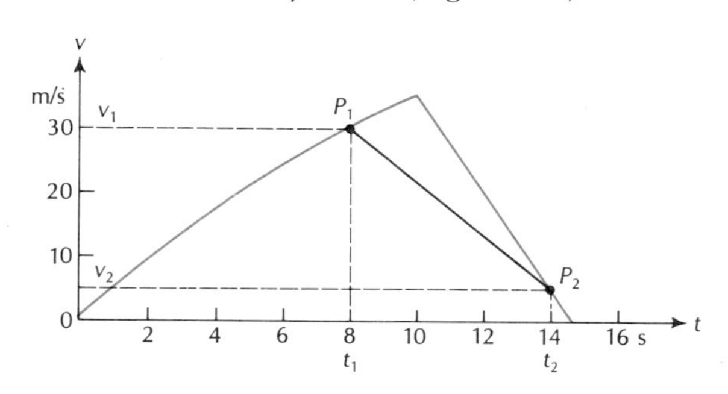


**Fig. 2.7** The colored curve shows the instantaneous velocity as a function of time. The straight black line connecting the points $P_1$ and $P_2$ has a slope $(v_2 - v_1)/(t_2 - t_1)$; this slope is the average acceleration.

[3] Appendix 5 contains a brief review of the rules for differentiation.

The acceleration can be positive or negative, depending on the sign of the velocity change $v_2 - v_1$. If the velocity is positive and increasing in magnitude, the acceleration is positive; if the velocity is positive and decreasing in magnitude, the acceleration is negative. However, note that if the velocity is *negative* (motion in the negative $x$ direction) and increasing in magnitude, that is, becoming more negative, the acceleration is *negative.* Thus, an automobile speeding up while moving in the negative $x$ direction has a negative acceleration; conversely, an automobile braking while moving in the negative $x$ direction has positive acceleration.

The unit of acceleration is the unit of velocity divided by the unit of time. In the metric system the unit of acceleration is the (m/s)/s, or m/s$^2$; in the British system it is the ft/s$^2$. Table 2.2 gives some examples of accelerations.

**Table 2.2** SOME ACCELERATIONS

| | |
|---|---|
| Protons in Fermilab accelerator | $9 \times 10^{13}$ m/s$^2$ |
| Ultracentrifuge | $3 \times 10^{6}$ m/s$^2$ |
| Baseball struck by bat | $3 \times 10^{4}$ m/s$^2$ |
| Soccer ball struck by foot | $3 \times 10^{3}$ m/s$^2$ |
| Automobile crash (100 km/h into fixed barrier) | $1 \times 10^{3}$ m/s$^2$ |
| Parachutist during opening of parachute (extreme) | $3.2 \times 10^{2}$ m/s$^2$ |
| Gravity on surface of Sun | $2.7 \times 10^{2}$ m/s$^2$ |
| Explosive seat ejection from aircraft (extreme) | $1.5 \times 10^{2}$ m/s$^2$ |
| F16 aircraft pulling out of dive | 80 m/s$^2$ |
| Loss of consciousness in man ("blackout") | 70 m/s$^2$ |
| Gravity on surface of Earth | 9.8 m/s$^2$ |
| Braking of automobile | ~8 m/s$^2$ |
| Gravity on surface of Moon | 1.7 m/s$^2$ |
| Rotation of Earth (at equator) | $3.4 \times 10^{-2}$ m/s$^2$ |

The **instantaneous acceleration** is the slope of the tangent to the velocity curve in a plot of velocity vs. time. For example, at $t = 4$ s, we can draw the tangent to the velocity curve of Figure 2.6 and find that the slope, or instantaneous acceleration, is 3.74 m/s$^2$. By drawing many such tangents at different times, we can prepare a table of instantaneous accelerations as a function of time for the automobile whose velocity is given by Figure 2.6. This acceleration is plotted in Figure 2.8. At the initial instant, $t = 0$, the acceleration is large (large slope in Figure 2.6); as the automobile gains velocity, the acceleration gradually drops (decreasing slope in Figure 2.6); at $t = 10$ s, the brakes

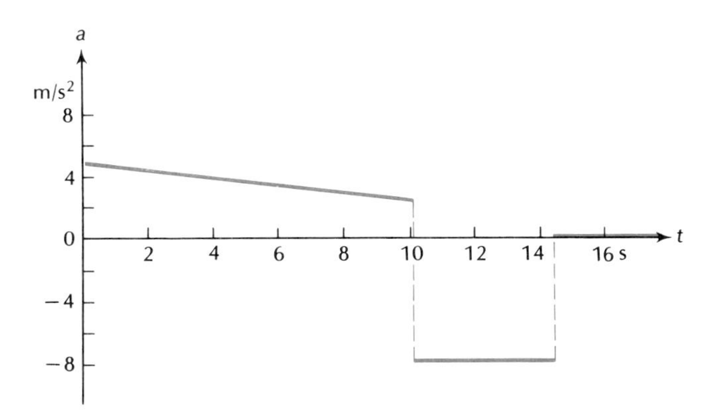


**Fig. 2.8** Instantaneous acceleration as a function of time.

are applied, leading to a large negative value of the acceleration (negative slope in Figure 2.6); this negative acceleration, or **deceleration,** is maintained until the automobile comes to a halt. Note that the plot of Figure 2.8 has discontinuities at $t = 10$ s and $t = 14.3$ s; at $t = 10$ s the acceleration suddenly jumps from a positive value to a negative value and at $t = 14.3$ s it suddenly jumps to zero.

In the language of calculus, the instantaneous acceleration is the derivative of the instantaneous velocity with respect to time,

$$a = \lim_{\Delta t \to 0} \frac{\Delta v}{\Delta t} \tag{12}$$

or

*Instantaneous acceleration*

$$\boxed{a = \frac{dv}{dt}} \tag{13}$$

Equivalently, we can say that the acceleration is the second derivative of $x$ with respect to time,

$$a = \frac{d}{dt}\left(\frac{dx}{dt}\right) \quad \text{or} \quad a = \frac{d^2x}{dt^2} \tag{14}$$

For example, the accelerations calculated by differentiation of the formulas (9) and (10) are

$$a = 4.752 - 0.252t \qquad \text{for } 0 \text{ s} \leq t \leq 10 \text{ s} \tag{15}$$

$$a = -8.044 \qquad \text{for } 10 \text{ s} \leq t \leq 14.3 \text{ s} \tag{16}$$

where the acceleration is measured in $m/s^2$. According to Eq. (16), in the time interval between 10 and 14.3 s, the acceleration is constant; this agrees with the result obtained from the graphical method (see Figure 2.8).

## 2.5 Motion with Constant Acceleration

Constant acceleration implies a constant slope in the plot of velocity vs. time; thus the plot is a straight line. In this case the velocity increases (or decreases) by equal amounts in each 1-second interval. For example, in the interval between 10 and 14.3 s, the velocity plotted in Figure 2.6 decreases by 8.044 m/s in each second.

In the case of constant acceleration, there are some simple relations between acceleration, velocity, position, and time. Suppose that the initial velocity (at time $t = 0$) is $v_0$ and that the velocity increases at a constant rate given by the constant acceleration $a$. After a time $t$ has elapsed, the velocity will have increased from the initial value $v_0$ to the value

*Constant acceleration: $v$, $a$, and $t$*

$$\boxed{v = v_0 + at} \tag{17}$$

Suppose that the initial position is $x_0$. After a time $t$ has elapsed, the position will have changed by an amount equal to the product of average velocity multiplied by time, that is, the position will have changed from the initial value $x_0$ to

$$x = x_0 + \bar{v}t \tag{18}$$

Since the velocity increases uniformly with time, the average value of the velocity is simply the average of the initial value and the final value,

$$\bar{v} = \tfrac{1}{2}(v_0 + v) = \tfrac{1}{2}(v_0 + v_0 + at)$$

$$= v_0 + \tfrac{1}{2}at \tag{19}$$

Substituting this into Eq. (18), we find

$$x = x_0 + (v_0 + \tfrac{1}{2}at)t \tag{20}$$

that is,

$$x = x_0 + v_0t + \tfrac{1}{2}at^2 \tag{21}$$

Thus, with constant acceleration, the change in position is

$$\boxed{x - x_0 = v_0t + \tfrac{1}{2}at^2} \tag{22}$$

*Constant acceleration: x, a, and t*

The right side consists of two terms: the term $v_0t$ represents the change in position that the particle would suffer if moving at constant velocity, and the term $\frac{1}{2}at^2$ represents the effect of the acceleration.

---

EXAMPLE 4. Use the rules of calculus to evaluate the second derivative of $x$ from Eq. (22), and verify that this second derivative equals the acceleration $a$.

SOLUTION: From Eq. (22),

$$x = x_0 + v_0t + \tfrac{1}{2}at^2$$

In the differentiation, $x_0$, $v_0$, and $a$ are constants; hence

$$\frac{dx}{dt} = \frac{d}{dt}\left(x_0 + v_0t + \tfrac{1}{2}at^2\right) = v_0 + \tfrac{1}{2}a \times 2t = v_0 + at$$

and

$$\frac{d^2x}{dt^2} = \frac{d}{dt}(v_0 + at) = a$$

COMMENTS AND SUGGESTIONS: This calculation amounts to an alternative proof of Eq. (22), since, if we were to change the powers of $t$ in Eq. (22) or the constants multiplying them, then the first derivative would not agree with the initial velocity $v_0$ at time $t = 0$, and the second derivative would not agree with the constant acceleration $a$.

---

Equations (17) and (22) express velocity and position in terms of time. By eliminating the time $t$ between these two equations, we obtain a direct relation between position and velocity, which is sometimes useful. According to Eq. (17),

$$t = \frac{v - v_0}{a} \tag{23}$$

and if we substitute this into Eq. (22), we obtain

$$x - x_0 = v_0\left(\frac{v - v_0}{a}\right) + \tfrac{1}{2}a\left(\frac{v - v_0}{a}\right)^2 \tag{24}$$

which can be rearranged to give

*Constant acceleration: $x$, $a$, and $v$*

$$\boxed{a(x - x_0) = \tfrac{1}{2}(v^2 - v_0^2)} \tag{25}$$

Figure 2.9 shows plots of position, velocity, and acceleration for motion with constant acceleration. In these plots, the motion for negative values of $t$ (instants before $t = 0$) has also been included with the assumption that $a$ always has the same constant value. Note that the worldline in Figure 2.9a has the shape of a parabola; this shape of the worldline is a distinctive characteristic of motion with constant acceleration.

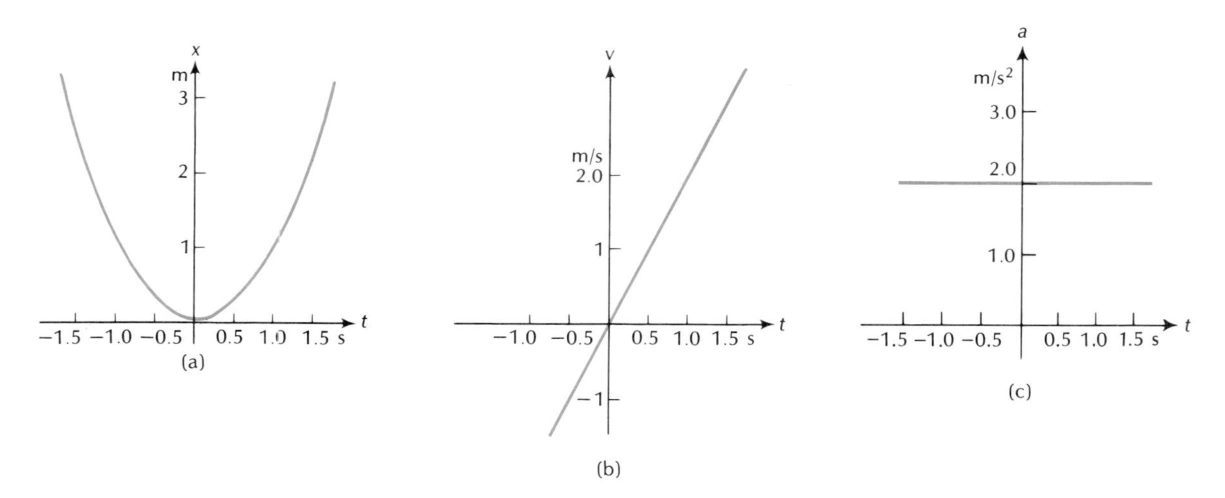


**Fig. 2.9** (a) Position, (b) velocity, and (c) acceleration for an example of motion with constant acceleration, $a = 2.0\ \text{m/s}^2$.

EXAMPLE 5. An automobile initially traveling at 50 km/h crashes into a stationary, rigid barrier. The automobile comes to rest after its front end crumples by 0.40 m. Assuming constant deceleration of the passenger compartment during the collision, what is the value of the deceleration? How long does it take the passenger compartment to stop?

SOLUTION: To solve this problem, we must first decide what equations to use. For motion with constant acceleration, the relevant equations are Eqs. (17), (22), and (25). The initial velocity, the final velocity, and the change in position are known quantities. To calculate the acceleration, it will be best to use Eq. (25), since there the acceleration appears as the *only* unknown quantity.

If we solve Eq. (25) for $a$, we find

$$a = \frac{v^2 - v_0^2}{2(x - x_0)}$$

The initial velocity is $v_0 = 50$ km/h $= 50 \times 10^3$ m/3600 s $= 13.9$ m/s. The final velocity is $v = 0$. The change in position corresponding to this change of velocity is $x - x_0 = 0.40$ m. Consequently,

$$a = \frac{v^2 - v_0^2}{2(x - x_0)} = \frac{-(13.9 \text{ m/s})^2}{2 \times 0.40 \text{ m}} = -2.4 \times 10^2 \text{ m/s}^2$$

We can then calculate the time the passenger compartment takes to stop from Eq. (17):

$$t = \frac{v - v_0}{a} = \frac{-13.9 \text{ m/s}}{-2.4 \times 10^2 \text{ m/s}^2} = 5.8 \times 10^{-2} \text{ s}$$

COMMENTS AND SUGGESTIONS: In this example, our equations automatically lead to a negative sign for the acceleration, as is appropriate for deceleration. And our equations automatically lead to the correct units for the final results — the units in the equation for the acceleration and in the equation for the time cancel in such a way as to leave us with m/s$^2$ and with s, respectively.

Note that if we had attempted to use Eqs. (17) and (22) instead of Eq. (25), we would have had to deal simultaneously with the two unknowns $a$ and $t$, both of which appear in Eqs. (17) and (22). This means we would have had to solve a system of two equations in two unknowns, which would have been slightly more tedious than the method of solution we adopted.

## 2.6 The Acceleration of Gravity

A body released near the surface of the Earth will accelerate downward under the influence of the pull of gravity. If the frictional resistance of the air has been eliminated (by placing the body in an evacuated container), then the body is in **free fall** and the downward motion proceeds with constant acceleration. It is a remarkable fact that the value of this acceleration is exactly the same for all bodies released at the same location; that is, the value of the acceleration is completely independent of the speed, mass, size, shape, chemical composition, etc., of the bodies. Figure 2.10 shows a simple experimental demonstration of the equality of the accelerations of two bodies in free fall. The universality of the rate of free fall is one of the most precisely and rigorously tested laws of nature; a long series of careful experiments have tested the equality of the rates of free fall of different bodies to within 1 part in $10^{10}$, and in some special cases even to within 1 part in $10^{12}$. [4]

**Galileo Galilei,** *1564–1642, Italian mathematician, astronomer, and physicist. He was professor at Pisa and at Padua, and became chief mathematician to Cosimo II de'Medici, grand duke of Tuscany. Galileo demonstrated experimentally that all bodies fall with the same acceleration, and he deduced that the trajectory of a projectile is a parabola. With a telescope of his own design, he discovered the satellites of Jupiter and sun spots. He vociferously defended the heliocentric system of Copernicus, for which he was sentenced by the Inquisition.*

The acceleration produced by gravity near the surface of the Earth is usually denoted by $g$. The numerical value of $g$ is approximately

$$g \cong 9.81 \text{ m/s}^2 \cong 32.2 \text{ ft/s}^2 \qquad (26)$$

*Acceleration of gravity, g*

[4] The tests to within 1 part in $10^{12}$ involved bodies falling toward the Sun rather than toward the Earth.

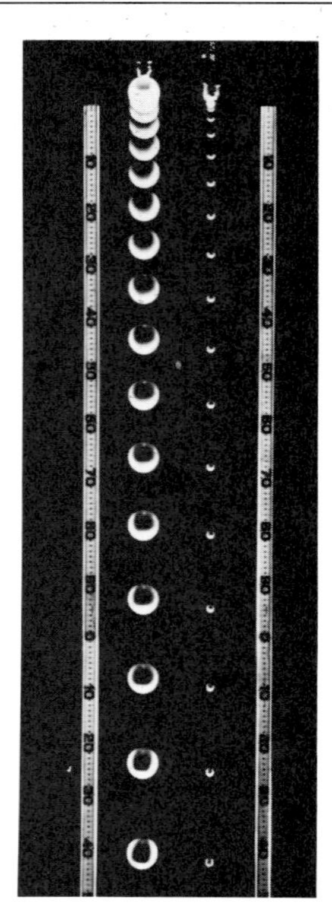

**Fig. 2.10** Stroboscopic photograph of two free-falling balls of unequal masses released simultaneously. This photograph was made by leaving the shutter of the camera open, and triggering a flash at regular intervals. The balls fall with equal accelerations.

The exact value of the acceleration varies from location to location on the Earth. For instance, the value of $g$ increases with geographical latitude and decreases with altitude. Table 2.3 gives the values of $g$ for a few locations. The variation with latitude shows up clearly in this table. However, 9.81 m/s$^2$ is accurate to within $\pm 0.02$ m/s$^2$ for all of North America and for all altitudes up to about 6 km.

**Table 2.3** VARIATION OF g WITH LATITUDE

| *Station* | *Latitude* | *g* |
|---|---|---|
| Quito, Ecuador | 0° N | 9.780 m/s$^2$ |
| Madras, India | 13° N | 9.783 |
| Hong Kong | 22° N | 9.788 |
| Cairo, Egypt | 30° N | 9.793 |
| New York, U.S.A. | 41° N | 9.803 |
| London, England | 51° N | 9.811 |
| Oslo, Norway | 60° N | 9.819 |
| Murmansk, U.S.S.R. | 69° N | 9.825 |
| Spitsbergen | 80° N | 9.831 |
| North Pole | 90° N | 9.832 |

To describe vertical free-fall motion, take the $x$ axis in the upward direction. Then $a = -g$ and Eqs. (17), (21), and (25) become

$$\boxed{v = v_0 - gt} \tag{27}$$

$$\boxed{x = x_0 + v_0 t - \tfrac{1}{2}gt^2} \tag{28}$$

$$\boxed{-g(x - x_0) = \tfrac{1}{2}(v^2 - v_0^2)} \tag{29}$$

Although these equations are strictly valid only for bodies falling in a vacuum, they are good approximations for dense and compact bodies, such as pieces of metal or stone, released in air. For such bodies the frictional resistance is unimportant as long as the speed is low (the exact restriction to be imposed on the speed depends on the mass and shape of the body and on the desired accuracy). In the following calculations we will usually ignore the resistance of air, even when the speeds are not all that low.

EXAMPLE 6. At Acapulco, professional divers amuse tourists by jumping from a 36-m-high cliff into the sea. How long do they fall? What is their impact velocity?

SOLUTION: For this problem, the relevant equations are Eqs. (27), (28), and (29). The known quantities are the change of position and the initial velocity; the unknown quantities are the time of fall and the final velocity. To calculate the time from the known quantities, we will use Eq. (28), in which the time is the only unknown. With $x - x_0 = -36$ m and $v_0 = 0$, Eq. (28) yields

$$-36 \text{ m} = -\tfrac{1}{2}(9.8 \text{ m/s}^2)t^2$$

which can be solved for $t$,

$$t = \sqrt{\frac{2 \times 36}{9.8} \text{ s}^2} = 2.7 \text{ s}$$

The impact velocity is then

$$v = -gt = -9.8 \text{ m/s}^2 \times 2.7 \text{ s}$$
$$= -26 \text{ m/s}$$

This is about 94 km/h!

COMMENTS AND SUGGESTIONS: The data for many of the examples and problems in this book have been taken from newspapers, magazines, manuals, and other everyday sources which do not express numbers in scientific notation. This means that we must make a guess at how many significant figures are in the data. In the present example, it seems reasonable to guess that the height of the cliff has been reported with a precision of two significant figures, and not three or more, that is, we guess that 36 m means $3.6 \times 10$ m, and not $3.60 \times 10$ m. Accordingly, all the numbers in the calculation have been rounded to two significant figures.

When the precision of the data given by some everyday source is not explicitly specified, it is usually a good guess that the data has two significant figures, sometimes three, but hardly ever more than three.

---

EXAMPLE 7. A powerful bow, such as used to establish world records in archery, can launch an arrow at a velocity of 90 m/s. How high will such an arrow rise if aimed vertically upward? How long will it take to return to the ground? How fast will it be going when it hits the ground? Ignore friction.

SOLUTION: At the highest point of the motion, the instantaneous velocity is $v = 0$ (the motion reverses at this point, and stops for an instant). Hence, for the upward motion, we can regard the initial and final velocities as known. The height reached and the time are unknown. Equation (29) gives us the height in terms of the known quantities,

$$x - x_0 = \frac{-(v^2 - v_0^2)}{2g} = \frac{-0 + (90 \text{ m/s})^2}{2 \times 9.8 \text{ m/s}^2} = 4.1 \times 10^2 \text{ m}$$

Equation (27) gives us the time for the upward motion:

$$t = \frac{v_0 - v}{g} = \frac{90 \text{ m/s} - 0}{9.8 \text{ m/s}^2} = 9.2 \text{ s}$$

The downward motion takes exactly as long as the upward motion; hence the time for the arrow to return to its starting point is $2 \times 9.2 \text{ s} = 18.4$ s.

Since we are ignoring friction, the downward motion is simply the reverse of the upward motion. The velocity of the arrow when it hits the ground must therefore be the same as its initial velocity, except for sign; i.e., it must be $-90$ m/s.

COMMENTS AND SUGGESTIONS: Instead of obtaining the final velocity from this physical argument that compares the upward and downward motions, we can obtain the final velocity directly from the mathematical properties of our equations. With $x = x_0$, Eq. (29) yields $v^2 = v_0^2$, which has a positive and a negative square root, $v = \pm v_0 = \pm 90$ m/s; the plus sign is for the upward motion, and the minus sign is for the downward motion. Thus, as in other examples, we see that the equations automatically provide information about the direction of motion.

---

EXAMPLE 8. Suppose you throw a stone straight up with an initial velocity of 15 m/s and, 2.0 s later, you throw a second stone straight up with the same initial velocity. The first stone going down will meet the second stone going up. At what height do the two stones meet?

SOLUTION: According to Eq. (28), the equation for the worldline of the first stone is

$$x_1 = x_0 + v_0 t - \tfrac{1}{2}gt^2 \tag{30}$$

The second stone is thrown 2.0 s later; hence the equation for its worldline is

$$x_2 = x_0 + v_0(t - 2.0\ \text{s}) - \tfrac{1}{2}g(t - 2.0\ \text{s})^2$$

We can write this as

$$x_2 = x_0 + v_0(t - t_2) - \tfrac{1}{2}g(t - t_2)^2 \tag{31}$$

where $t_2 = 2.0$ s. In these equations both the position and the time are unknown quantities — we have too many unknowns.

To proceed with the solution, we need an extra relationship between the unknowns. The extra relationship we have available is the condition that the two stones meet; expressed mathematically, $x_1 = x_2$ or

$$v_0 t - \tfrac{1}{2}gt^2 = v_0(t - t_2) - \tfrac{1}{2}g(t - t_2)^2 \tag{32}$$

This equation does not directly tell us the height of the meeting, but it does tell us the time of the meeting and, once we know this time, we can return to Eq. (30) to find the height. Simplifying Eq. (32), we obtain

$$v_0 t - \tfrac{1}{2}gt^2 = v_0 t - v_0 t_2 - \tfrac{1}{2}gt^2 + gtt_2 - \tfrac{1}{2}gt_2^2$$

or

$$0 = -v_0 t_2 + gtt_2 - \tfrac{1}{2}gt_2^2$$

Canceling a factor $t_2$ in this equation, we can solve for $t$,

$$t = \frac{v_0}{g} + \tfrac{1}{2}t_2 = \frac{15\ \text{m/s}}{9.8\ \text{m/s}^2} + \tfrac{1}{2} \times 2.0\ \text{s} = 2.53\ \text{s}$$

According to Eq. (30), the height of the meeting is then

$$x_1 - x_0 = 15\ \text{m/s} \times 2.53\ \text{s} - \tfrac{1}{2} \times 9.8\ \text{m/s}^2 \times (2.53\ \text{s})^2$$

$$= 6.6\ \text{m}$$

COMMENTS AND SUGGESTIONS: This is a fairly complicated calculation, in which the equations given in this chapter do not, by themselves, determine the unknowns. The crucial extra ingredient needed for the solution is the relation $x_1 = x_2$ [Eq. (32)] between the position coordinates of the two stones. This example illustrates that sometimes the conditions of a problem imply special mathematical relations among the unknown quantities. Try to exploit such special mathematical relations whenever you are faced with a problem that seems to have too many unknowns and too few equations.

*1 G = 9.81 m/s²*

Acceleration is occasionally measured in multiples of a "standard" acceleration of gravity; we will call this "standard" acceleration 1 G, where

$$1\ \text{G} = 9.80665\ \text{m/s}^2 \cong 9.81\ \text{m/s}^2 \qquad (33)$$

Note the distinction between $g$ and G: $g$ is the acceleration of gravity on or near the surface of the Earth — its value is approximately 9.81 m/s$^2$, but its exact value depends on location (see Table 2.3). G is a unit of acceleration — its value is exactly 9.80665 m/s$^2$ by definition. The acceleration of gravity at the Earth's surface is approximately 1 G, the acceleration of gravity on the Moon's surface (see Table 2.2) is 0.17 G, etc.

Measurements of the acceleration of gravity at different locations on the surface of the Earth are commonly performed by a pendulum method (we will describe this method in Chapter 15). However, in recent years the Bureau International des Poids et Mesures has developed a very sophisticated device which permits an extremely accurate determination of the value of $g$ by directly observing the motion of a small projectile in free fall in an evacuated chamber. Figure 2.11 shows this device and the projectile used in these measurements. The projectile is launched vertically upward by a rubber band and the time that it takes to rise a given distance and fall back is measured; the value of $g$ can then be determined from this distance and time. For the sake of high precision, the distance the projectile travels is measured with an interferometer to within $\pm 10^{-9}$ m. This permits the determination of $g$ to within 3 parts in $10^9$, or $\pm 3 \times 10^{-8}$ m/s$^2$. One strange finding emerging from these high-precision experiments is that the value of $g$ fluctuates slightly with time. From one year to the next $g$ increases or decreases by as much as $4 \times 10^{-7}$ m/s$^2$. The cause of these fluctuations is not known, but they are presumably due to changes in the distribution of mass within the Earth.

In all the above calculations we have ignored the effects of the frictional resistance of air on the falling bodies. By holding a hand out of the window of a speeding automobile, you can readily feel that air offers a substantial frictional resistance to motion at speeds of a few tens of kilometers per hour; this frictional resistance increases with speed (roughly in proportion to the square of the speed). Hence a fall-

**Fig. 2.11** (a) Jaeger gravimeter for the accurate measurement of $g$. (b) Projectile used in the measurement.

(a)

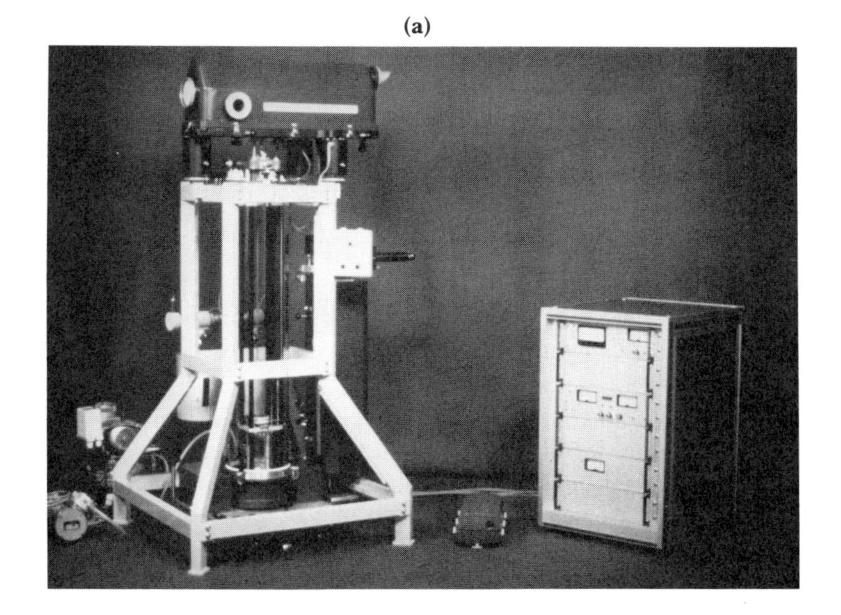

(b)

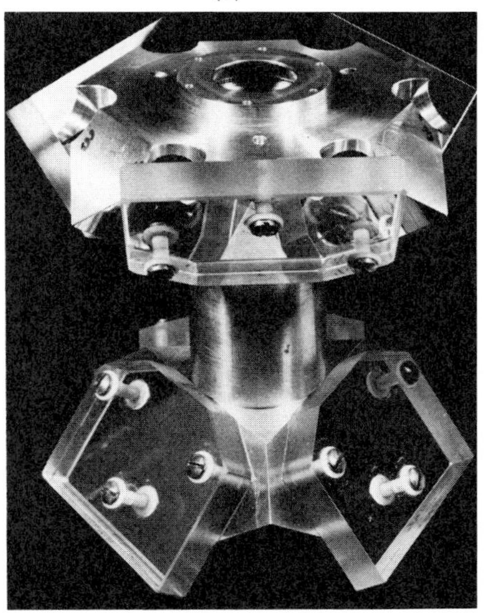

ing, accelerating body will experience a larger and larger frictional resistance as its speed increases. Ultimately, this resistance becomes so large that it counterbalances the pull of gravity — the body ceases to accelerate and attains a constant speed. This ultimate speed is called the **terminal speed.** The precise value of the terminal speed depends on the mass of the body, the size, and the shape; for instance, a sky diver with a closed parachute attains a terminal speed of about 200 km/h, whereas a sky diver with an open parachute attains a terminal speed of only about 18 km/h.

*Terminal speed*

## Supplement: Advice on Solving Problems

The solving of problems is an art; there is no simple recipe for obtaining the solutions. Most of the problems in this and the following chapters are applications of the concepts and principles developed in the text. The examples scattered throughout each chapter illustrate typical cases of problem solving. Sometimes, you will be able to solve a problem by imitating one of these examples. But if you can't see how to begin the solution, try the following steps:

1. Carefully read and reread the problem and prepare a complete list of given (known) and sought (unknown) quantities. Make sure you do not overlook some given bit of information, and make sure you do not overlook any of the questions you are expected to answer. Keep in mind that the problem may involve some implicit assumptions; for instance, an implicit assumption about the reference frame relative to which the velocities are reckoned or an implicit assumption about negligible friction.
2. Draw a sketch of the situation described by the problem and label the known and unknown quantities.
3. If the problem deals with some kind of motion, try to visualize the progress of the motion in time, as though you were watching a movie.
4. Ask yourself what physical conditions characterize the situation sketched and visualized in the preceding steps, and ask yourself what physical principles are applicable. For instance, does the motion proceed with constant velocity? With constant acceleration? Does the principle of universality of the rate of free fall apply?
5. Having decided what physical conditions and principles are applicable, examine the mathematical relations that are valid under the given conditions. Then try to select an equation that expresses the unknowns in terms of the known quantities (see Examples 5, 6, and 7). Be discriminating in your selection — sometimes an equation will tempt you because it displays all the desired quantities, but it will be an invalid equation if the assumptions that went into its derivation are not satisfied in your problem.
6. You will often find that you seem to have too many unknowns and too few equations. Then ask yourself: Are there any extra mathematical relations that the conditions of the problem impose on the unknowns? Can you combine several equations to eliminate some of the unknowns? Are there any quantities that you can calculate from the known quantities? Do these calculated quantities bring you nearer to the answer (see Example 8)?

7. It is good practice to solve all the equations by algebraic manipulations, and substitute numbers only at the very end; this makes it easier to spot and correct mistakes.
8. When you substitute numbers into your equations, also include the units of these numbers. The units in your equations should then combine or cancel in such a way as to yield the correct units for the final result. If the units do not combine or cancel in the expected way, something has gone wrong with your algebra.
9. After you have finished your calculations, always check whether the answer is plausible. For instance, if your calculation yields the result that a diver jumping off a cliff hits the water at 3000 km/h, then somebody has made a mistake somewhere!
10. Lastly, remember to round off your final answer to the same number of significant figures as given in the data for the problem.

## SUMMARY

**Average speed:** $\dfrac{[\text{distance traveled}]}{[\text{time taken}]}$

**Average velocity:** $\overline{v} = \dfrac{\Delta x}{\Delta t}$

**Instantaneous velocity:** $v = \dfrac{dx}{dt}$

**Average acceleration:** $\overline{a} = \dfrac{\Delta v}{\Delta t}$

**Instantaneous acceleration:** $a = \dfrac{dv}{dt} = \dfrac{d^2x}{dt^2}$

**Motion with constant acceleration:** $v = v_0 + at$

$$x = x_0 + v_0 t + \tfrac{1}{2}at^2$$

$$a(x - x_0) = \tfrac{1}{2}(v^2 - v_0^2)$$

**Acceleration of gravity:** $g \cong 9.81\ \text{m/s}^2$

$$\cong 32.2\ \text{ft/s}^2$$

## QUESTIONS

1. The motion of a runner can be regarded as particle motion, but the motion of a gymnast cannot. Explain.

2. According to newspaper reports, the world record for speed skiing is 203.160 km/h. This speed was measured on a 100-meter "speed-trap." The skier took about 1.7 s to cross this trap. In order to calculate speed to six significant figures, we need to measure distance and time to six significant figures. What accuracy in distance and time does this require?

3. Do our sense organs permit us to feel velocity? Acceleration?

4. What is your velocity at this instant? Is this a well-defined question? What is your acceleration at this instant?

5. Does the speedometer of your car give speed or velocity? Does the speedometer care whether you drive eastward or westward along a road?

6. Suppose that at one instant of time the velocity of a body is zero. Can this body have a nonzero acceleration at this instant? Give an example.

7. Give an example of a body in motion with instantaneous velocity and acceleration of the same sign. Give an example of a body in motion with instantaneous velocity and acceleration of opposite signs.

8. Experienced drivers recommend that when driving in traffic you should stay at least 2 s behind the car in front of you. This is equivalent to a distance of about two car lengths for every 10 mi/h. Why is it necessary to leave a larger distance between the cars when the speed is larger?

9. Is the average speed equal to the average magnitude of the velocity? Is the average speed equal to the magnitude of the average velocity?

10. In the seventeenth century Galileo Galilei measured the acceleration of gravity by rolling balls down an inclined plane. Why did he not measure the acceleration directly by dropping a stone from a tower?

11. Why did astronauts find it easy to jump on the Moon? If an astronaut can jump to a height of 20 cm on the Earth (with his spacesuit), how high can he jump on the Moon?

12. According to Table 2.3, the value of $g$ at the North Pole is 9.832 m/s$^2$. What do you expect for the value of $g$ at the South Pole? Why?

13. Interpolate Table 2.3 to find the value of $g$ at your latitude.

14. A particle is initially at rest at some height. If the particle is allowed to fall freely, what distance does it cover in the time from $t = 0$ s to $t = 1$ s? From $t = 1$ s to $t = 2$ s? From $t = 2$ s to $t = 3$ s? Show that these successive distances are in the ratios 1:3:5:7. . . .

15. An elevator is moving upward with a constant velocity of 5 m/s. If a passenger standing in this elevator drops an apple, what will be the acceleration of the apple relative to the elevator?

16. Some people are fond of firing guns into the air when under the influence of drink or patriotic fervor. What happens to the bullets? Is this practice dangerous?

17. A woman riding upward in an elevator drops a penny in the elevator shaft when she is passing by the third floor. At the same instant, a man standing at the elevator door at the third floor also drops a penny in the elevator shaft. Which coin hits the bottom first? Which coin hits with the higher speed? Neglect friction.

18. If you take friction into account, how does this change your answers to Question 17?

19. Suppose that you drop a $\frac{1}{2}$-kg packet of sugar and a $\frac{1}{2}$-kg ball of lead from the top of a building. Taking air friction into account, which will take the shorter time to reach the ground? Suppose that you place the sugar and the lead in identical sealed glass jars before dropping them. Which will now take the shorter time?

20. Is air friction important in the falling motion of a raindrop? If a raindrop were to fall without friction from a height of 300 m and hit you, what would it do to you?

21. Galileo claimed in his *Dialogues* that "the variation of speed in air between balls of gold, lead, copper, porphyry, and other heavy materials is so slight that in a fall of 100 cubits a ball of gold would surely not outstrip one of copper by as much as four fingers. Having observed this, I came to the conclusion that in a medium totally void of resistance all bodies would fall with the same speed." One cubit is about 46 cm and four fingers are about 10 cm. According to Galileo's data, what is the maximum percent difference between the accelerations of the balls of gold and of copper?

22. An archer shoots an arrow straight up. If you consider the effects of the frictional resistance of air, would you expect the arrow to take a longer time to rise or to fall?

## PROBLEMS

### Section 2.1

1. The speed of nerve impulses in mammals is typically $10^2$ m/s. If a shark bites the tail of a 30-m-long whale, roughly how long will it take before the whale knows of this?

2. The world record for the 100-yard run is 9.0 seconds. What is the corresponding average speed in miles per hour?

3. In 1958 the nuclear-powered submarine *Nautilus* took 6 days and 12 hours to travel submerged 5068 km across the Atlantic from Portland, England, to New York City. What was the average speed (in km/h) for this trip?

4. A galaxy beyond the constellation Corona Borealis is moving directly away from our Galaxy at the rate of 21,600 km/s. This galaxy is now at a distance of $1.4 \times 10^9$ light-years from our Galaxy. Assuming that the galaxy has always been moving at a constant speed, how many years ago was it right on top of our Galaxy?

5. On one occasion a tidal wave (tsunami) originating near Java was detected in the English Channel 32 h later. Roughly measure the distance from Java to England by sea (round the Cape of Good Hope) on a map of the world and calculate the average speed of the tidal wave.

6. A hunter shoots an arrow at a deer running directly away from him. When the arrow leaves the bow, the deer is at a distance of 40 m. When the arrow strikes, the deer is at a distance of 50 m. The speed of the arrow is 65 m/s. What must have been the speed of the deer? How long did the arrow take to travel to the deer?

7. In 1971, Francis Chichester, in the yacht *Gypsy Moth V,* attempted to sail the 4000 nautical miles (nmi) from Portuguese Guinea to Nicaragua in no more than 20 days.
   (a) What minimum average speed (in nautical miles per hour) does this require?
   (b) After sailing 13 days, he still had 1720 nmi to go. What minimum average speed did he require to reach his goal in the remaining 7 days? Knowing that his yacht could at best achieve a maximum speed of 10 nmi/h, what could he conclude at this point?

*8. The fastest land animal is the cheetah, which runs at a speed of up to 101 km/h. The second fastest is the antelope, which runs at a speed of up to 88 km/h.
   (a) Suppose that a cheetah begins to chase an antelope. If the antelope has a head start of 50 m, how long does it take the cheetah to catch the antelope? How far will the cheetah have traveled by this time?
   (b) The cheetah can only maintain its top speed for about 20 s (and then has to rest), whereas the antelope can do it for a considerably longer time. What is the maximum head start the cheetah can allow the antelope?

*9. The table printed on the endpapers gives the radii of the orbits of the planets ("mean distance from the Sun") and the times required for moving around the orbit ("period of revolution").
   (a) Calculate the speed of motion of each of the nine planets in its orbit around the Sun. Assume that the orbits are circular.
   (b) In a logarithmic graph of speed vs. radius, plot the logarithm of the

speed of each planet and the logarithm of its radial distance from the Sun as a point. Draw a curve through the nine points. Can you represent this curve by a simple equation?

### Section 2.2

10. Suppose you throw a baseball straight up so that it reaches a maximum height of 8.00 m and returns to you 2.55 s after you throw it. What is the average speed for this motion of the ball? What is the average velocity?

11. Consider the automobile with the worldline plotted in Figure 2.1. What is the average velocity for the interval from $t = 0$ to $t = 10$ s? From $t = 10$ s to $t = 14.3$ s?

### Sections 2.3 and 2.4

12. In an experiment with a water-braked rocket sled, an Air Force volunteer (?) was subjected to an acceleration of 82.6 G for 0.04 s. What was his change of speed in this time interval?

13. A Porsche racing car takes 2.2 s to accelerate from 0 to 96 km/h (60 mi/h). What is the average acceleration?

14. A soccer player kicks a stationary football and sends it flying. Slow-motion photography shows that the ball is in contact with the foot for $8.2 \times 10^{-3}$ s and leaves with a speed of 25 m/s. What is the average acceleration of the ball while in contact with the foot?

15. (a) The solid curve in Figure 2.12 is a plot of velocity vs. time for a Triumph sports car undergoing an acceleration test. By drawing tangents to the velocity curve, find the accelerations at time $t = 0$, 10, 20, 30, and 40 s; express these accelerations in G.
(b) The dashed curve in Figure 2.12 is a plot of velocity vs. time for the same car when coasting with its gears in neutral. Find the accelerations at time $t = 0$, 10, 20, 30, and 40 s.

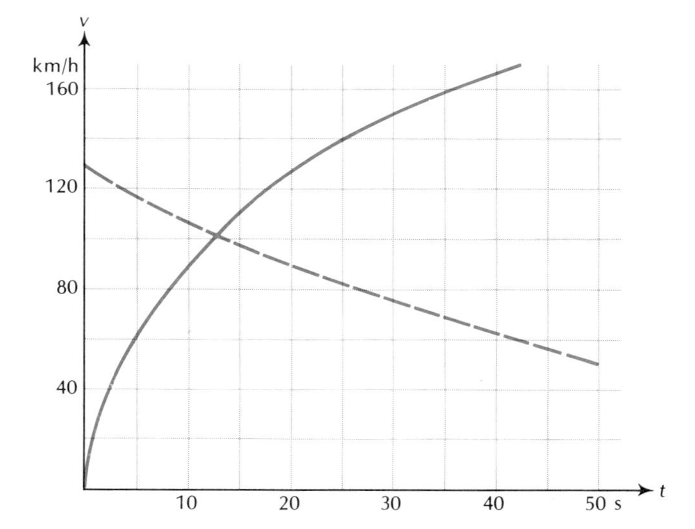


**Fig. 2.12** Instantaneous velocity for a Triumph sports car undergoing an acceleration test.

16. A particle moves along the $x$ axis with an equation for the worldline as follows:

$$x = 2.0 + 6.0t - 3.0t^2$$

where $x$ is measured in meters and $t$ is measured in seconds.
(a) What is the position of the particle at $t = 0.50$ s? What is the velocity? What is the acceleration?

(b) What is the position at $t = 2.0$ s? What is the velocity? What is the acceleration?

(c) What is the average velocity for the time interval $t = 0.50$ s to $t = 2.0$ s? What is the average acceleration?

17. Consider the solid curve in Figure 2.12 showing velocity vs. time for an accelerating sports car.

(a) From this curve, estimate the average velocity for the interval $t = 0$ s to $t = 5$ s. (Hint: The average velocity for a small time interval is roughly the average of the initial and final velocities; alternatively, it is roughly the velocity at the midpoint of the time interval.) Estimate how far the car travels in this time interval.

(b) Repeat the calculation of part (a) for every 5-s time interval between $t = 5$ s and $t = 45$ s.

(c) What is the *total* distance that the car will have traveled at the end of 45 s?

*18. Table 2.4 gives the horizontal velocity as a function of time for a projectile of 100 lb fired horizontally from a naval 6-inch gun. The velocity decreases with time because of the frictional drag of the air.

**Table 2.4** EFFECT OF AIR RESISTANCE ON A PROJECTILE

| *Time* | *Velocity* | *Time* | *Velocity* |
|---|---|---|---|
| 0 s | 657 m/s | 1.80 s | 557 m/s |
| 0.30 | 638 | 2.10 | 542 |
| 0.60 | 619 | 2.40 | 528 |
| 0.90 | 604 | 2.70 | 514 |
| 1.20 | 588 | 3.00 | 502 |
| 1.50 | 571 | | |

(a) On a piece of graph paper, make a plot of velocity vs. time and draw a smooth curve through the points of the plot.

(b) Estimate the average velocity for each time interval [the estimated average velocity for the time interval $t = 0$ s to $t = 0.30$ s is $\frac{1}{2}(657 + 638)$ m/s, etc.]. From the average velocities calculate the distances that the projectile travels in each time interval. What is the total distance that the projectile travels in 3.0 s?

(c) By directly counting the squares on your graph paper, estimate the area (in units of s · m/s) under the curve plotted in part (a) and compare this area with the result of part (b).

19. The instantaneous velocity of the projectile described in Problem 18 can be approximately represented by the formula (valid for $0 \text{ s} \leq t \leq 3 \text{ s}$)

$$v = 655.9 - 61.14t + 3.26t^2$$

where $v$ is measured in meters per second and $t$ in seconds. Calculate the instantaneous acceleration of the projectile at $t = 0$ s, at $t = 1.50$ s, and at $t = 3.00$ s.

*20. The equation of the worldline of a particle is given by

$$x = A \cos (bt)$$

where $A$ and $b$ are constants. Assume that $A = 2.0$ m and $b = 1.0$ radian/s.

(a) Roughly plot the worldline of this particle for the time interval $0 \text{ s} \leq t \leq 7.0 \text{ s}$.

(b) At what time does the particle pass the origin ($x = 0$)? What are its velocity and acceleration at this instant?

(c) At what time does the particle reach maximum distance from the origin? What are its velocity and acceleration at this instant?

*21. The motion of a rocket burning its fuel at a constant rate while moving through empty interstellar space can be described by

$$x = u_{ex}t + u_{ex}(1/b - t)\ \ln(1 - bt)$$

where $u_{ex}$ and $b$ are constants ($u_{ex}$ is the exhaust velocity of the gases at the tail of the rocket and $b$ is proportional to the rate of fuel consumption).

(a) Find a formula for the instantaneous velocity of the rocket.
(b) Find a formula for the instantaneous acceleration.
(c) Suppose that a rocket with $u_{ex} = 3.0 \times 10^3$ m/s and $b = 7.5 \times 10^{-3}$/s takes 120 s to burn all its fuel. What is the instantaneous velocity at $t = 0$ s? At $t = 120$ s?
(d) What is the instantaneous acceleration at $t = 0$ s? At $t = 120$ s?

Section 2.5

22. The takeoff speed of a jetliner is 360 km/h. If the jetliner is to take off from a runway of length 2100 m, what must be its acceleration along the runway (assumed constant)?

23. A British 6-inch naval gun has a barrel 6.63 m long. The muzzle speed of a projectile fired from this gun is 657 m/s. Assuming that upon detonation of the explosive charge the projectile moves along the barrel with constant acceleration, what is the magnitude of this acceleration? How long does it take the projectile to travel the full length of the barrel?

24. The nearest star is Proxima Centauri, at a distance of 4.2 light-years from the Sun. Suppose we wanted to send a spaceship to explore this star. To keep the astronauts comfortable, we want the spaceship to travel with a constant acceleration of 1.0 G at all times (this will simulate ordinary gravity within the spaceship). If the spaceship accelerates at 1.0 G until it reaches the midpoint of its trip and then decelerates at 1.0 G until it reaches Proxima Centauri, how long will the one-way trip take? What will be the speed of the spaceship at the midpoint? Do your calculations according to Newtonian physics (actually, the speed is so large that the calculation should be done according to relativistic physics; see Chapter 41).

25. In an accident on motorway M.1 in England, a Jaguar sports car made skid marks 290 m long while braking. Assuming that the deceleration was 1 G during this skid (this is approximately the maximum deceleration that a car with rubber wheels can attain on ordinary pavements), calculate the initial speed of the car before braking.

26. The front end of a German automobile has been designed so that upon impact it progressively crumples by as much as 0.7 m. Suppose that the automobile crashes into a solid brick wall at 80 km/h. During the collision the passenger compartment decelerates over a distance of 0.7 m. Assume that the deceleration is constant. What is the magnitude of the deceleration? If the passenger is held by a safety harness, is he likely to survive? (Hint: Compare the deceleration with the acceleration listed for a parachutist in Table 2.2.)

27. A jet-powered car racing on the Salt Flats in Utah went out of control and made skid marks 9.6 km long. Assuming that the deceleration during the skid was about 0.5 G, what must have been the initial speed of the car? How long did the car take to come to a stop?

28. The operation manual of a passenger automobile states that the stopping distance is 50 m when the brakes are fully applied at 96 km/h. What is the deceleration? What is the stopping time?

29. With an initial speed of 260 km/h, the French TGV (*tres grande vitesse*) train takes 1500 m to stop on a level track. Assume that the deceleration is constant. What is the magnitude of the deceleration? What is the time taken for stopping?

*30. In a "drag" race a car starts at rest and attempts to cover 440 yd in the shortest possible time. The world record for a piston-engined car is 5.637 s; while setting this record, the car reached a final speed of 250.69 mi/h at the 440-yd mark.

(a) What was the average acceleration for the run?
(b) Prove that the car did not move with constant acceleration.
(c) What would have been the final speed if the car had moved with constant acceleration so as to reach 440 yd in 5.637 s?

*31. In a large hotel, a fast elevator takes you from the ground floor to the 21st floor. The elevator takes 17 s for this trip: 5 s at constant acceleration, 7 s at constant velocity, and 5 s at constant deceleration. Each floor in the hotel has a height of 2.5 m. Calculate the values of the acceleration and deceleration (assume they are equal). Calculate the maximum speed of the elevator.

*32. (a) At the World Trade Center in New York City, the elevator takes 55 s to descend from the 107th floor to ground level, a distance of 400 m. What is the average speed of the elevator for this trip?
(b) The elevator is at rest at the beginning and at the end of the trip. If you wanted to program the elevator so that it completes the trip in the specified time with a minimum acceleration and a minimum deceleration, how would you have to accelerate and decelerate the elevator? What would be these minimum values of the acceleration and deceleration? What would be the maximum speed during the trip?

*33. The driver of an automobile traveling at 96 km/h perceives an obstacle on the road and slams on the brakes.

(a) Calculate the total stopping distance (in meters). Assume that the reaction time of the driver is 0.75 s (so that there is a time interval of 0.75 s during which the automobile continues at constant speed while the driver gets ready to apply the brakes) and that the deceleration of the automobile is 0.80 G when the brakes are applied.
(b) Repeat the calculation of part (a) for initial speeds of 15, 30, 45, 60, and 75 km/h. Make a plot of stopping distance vs. initial speed.

*34. An automobile is traveling at 90 km/h on a country road when the driver suddenly notices a cow in the road 30 m ahead. The driver attempts to brake the automobile, but the distance is too short. With what velocity does the automobile hit the cow? Assume that, as in Problem 33, the reaction time of the driver is 0.75 s and that the deceleration of the automobile is 0.80 G when the brakes are applied.

*35. In a collision, an automobile initially traveling at 50 km/h decelerates at a constant rate of 200 m/s$^2$. A passenger not wearing a seat belt crashes against the dashboard. Before the collision, the distance between the passenger and the dashboard was 0.60 m. With what speed, relative to the automobile, does the passenger crash into the dashboard? Assume that the passenger has no deceleration before contact with the dashboard.

**36. Figure 2.13 (copied from the operation manual of an automobile) describes the passing ability of the automobile at low speed. From the data supplied in this figure, calculate the acceleration of the automobile during the pass and the time required for the pass. Assume constant acceleration.

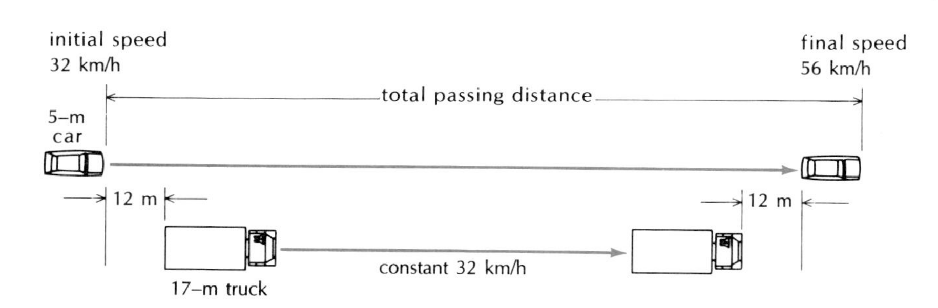


**Fig. 2.13** Diagram from the operation manual of an automobile.

*37. For a body released from rest falling through a viscous medium (for instance, an iron pellet falling in a jar full of oil), the speed is given by the formula

$$v = -g\tau + g\tau\, e^{-t/\tau}$$

where $\tau$ is a constant that depends on the size and shape of the body and on the viscosity of the medium, and e = 2.718 . . . is the basis of the natural logarithms.

(a) Find the acceleration as a function of time.
(b) Show that for $t \to \infty$, the speed approaches the terminal value $-g\tau$.
(c) By differentiation, verify that the equation for the worldline consistent with the above expression for the speed is

$$x = -g\tau t - g\tau^2\, e^{-t/\tau} + g\tau^2 + x_0$$

(d) Show that for small values of $t$ ($t << \tau$), the equation for the worldline is approximately $x \simeq -\frac{1}{2}gt^2 + x_0$.

Section 2.6

38. An apple drops from the top of the Empire State Building, 380 m above street level. How long does the apple take to fall? What is its impact velocity on the street? Ignore air resistance.

39. Peregrine falcons dive on their prey with speeds of up to 130 km/h. From what height must a falcon fall freely to achieve this speed? Ignore air resistance.

Fig. 2.14 The volcano Loki.

40. The muzzle speed of a 22-caliber bullet fired from a rifle is 366 m/s. If there were no air resistance, how high would this bullet rise when fired straight up?

41. An engineer standing on a bridge drops a penny toward the water and sees the penny splashing into the water 3.0 s later. How high is the bridge?

42. The volcano Loki on Io, one of the moons of Jupiter, ejects debris to a height of 200 km (Figure 2.14). What must be the initial ejection velocity of the debris? The acceleration of gravity on Io is 1.80 m/s$^2$. There is no atmosphere on Io, hence no air resistance.

*43. The nozzle of a fire hose discharges water at the rate of 280 liters/min at a speed of 26 m/s. How high will the stream of water rise if the nozzle is aimed straight up? How many liters of water will be in the air at any given instant?

*44. According to an estimate, a man who survived a fall from a 56-m cliff took 0.015 s to stop upon impact on the ground. What was his speed just before impact? What was his average deceleration during impact?

*45. Queeche gorge in Vermont has a depth of 45 m. If you want to measure this depth to within 10% by timing the fall of a stone dropped from the bridge across the gorge, how accurately must you measure the time? Is an ordinary watch with a second hand adequate for this task or do you need a stopwatch?

*46. A golf ball released from a height of 1.5 m above a concrete floor bounces back to a height of 1.1 m. If the ball is in contact with the floor for $6.2 \times 10^{-4}$ s, what is the average acceleration of the ball while in contact with the floor?

*47. In 1978 the stuntman A. J. Bakunas died when he jumped from the 23rd floor of a skyscraper and hit the pavement. The air bag that was supposed to cushion his impact ripped.

(a) The height of his jump was 96 m. What was his impact speed?

(b) The air bag was 3.7 m thick. What would have been the man's deceleration had the air bag not ripped? Assume that his deceleration would have been uniform over the 3.7-m interval.

*48. The HARP (High-Altitude Research Project) gun can fire an 84-kg projectile containing scientific instruments straight up to an altitude of 180 km. If we pretend there is no air resistance, what muzzle speed is required to attain this altitude? How long does the projectile remain at a height in excess of 100 km, the height of interest for high-altitude research?

*49 The International Geodesy Association has adopted the following formula for the acceleration of gravity as a function of the latitude $\Theta$ (at sea level):

$$g = 978.0318 \text{ cm}/s^2 \times (1 + 53.024 \times 10^{-4} \sin^2 \Theta - 5.9 \times 10^{-6} \sin^2 2\Theta)$$

(a) According to this formula, what is the acceleration of gravity at the equator? At a latitude of 45°? At the pole?
(b) Show that according to this formula, $g$ has a minimum at the equator, a maximum at the pole, and no minima or maxima at intermediate latitudes.

*50. At a height of 1500 m, a dive bomber in a vertical dive at 300 km/h shoots a cannon at a target on the ground. Relative to the bomber the initial speed of the projectile is 700 m/s. What will be the impact speed of the projectile on the ground? How long will it take to get there? Ignore air friction in your calculation.

**51. Suppose you throw a stone straight up with an initial speed of 15.0 m/s.
(a) If you throw a second stone straight up 1.00 s after the first, with what speed must you throw this second stone if it is to hit the first at a height of 11.0 m? (There are two answers. Are both plausible?)
(b) If you throw the second stone 1.30 s after the first, with what speed must you throw this second stone if it is to hit the first at a height of 11.0 m?

***52. Raindrops drip from a spout at the edge of a roof and fall to the ground. Assume that the drops drip at a steady rate of $n$ drops per second (where $n$ is large) and that the height of the roof is $h$.
(a) How many drops are in the air at one instant?
(b) What is the median height of these drops (i.e., the height above and below which an equal number of drops are found)?
(c) What is the average of the heights of these drops?

CHAPTER 3

# Vectors

The mathematical concept of vector turns out to be very useful for the description of the position, velocity, and acceleration in two- or three-dimensional motion. We will see that a vector description of the motion permits us to give precise meaning to the *direction* of the velocity and acceleration in two or three dimensions. Furthermore, we will see in later chapters that the vector concept is also useful for the description of many other physical quantities, such as force and momentum, which have both a magnitude and a direction. The present chapter is an introduction to vectors and to the mathematics of vectors — their addition, subtraction, and multiplication. This chapter contains no physics; instead, it develops mathematical tools that we will need for handling the physics in subsequent chapters.

## 3.1 The Displacement Vector and the General Definition of a Vector

*Displacement vector*

We begin with the concepts of displacement and the displacement vector. The displacement of a particle is simply a change of its position. If a particle moves from a point $P_1$ to a point $P_2$, we can represent the change of position graphically by an arrow, or directed line segment, from $P_1$ to $P_2$. This directed line segment is the **displacement vector** of the particle. For example, if a ship moves from Liberty Island to the Battery in New York harbor, then the displacement vector is as shown in Figure 3.1. Note that the displacement vector only tells us where the final position ($P_2$) is in relation to the initial position ($P_1$); it does not tell us what path the ship followed between the two positions. Thus

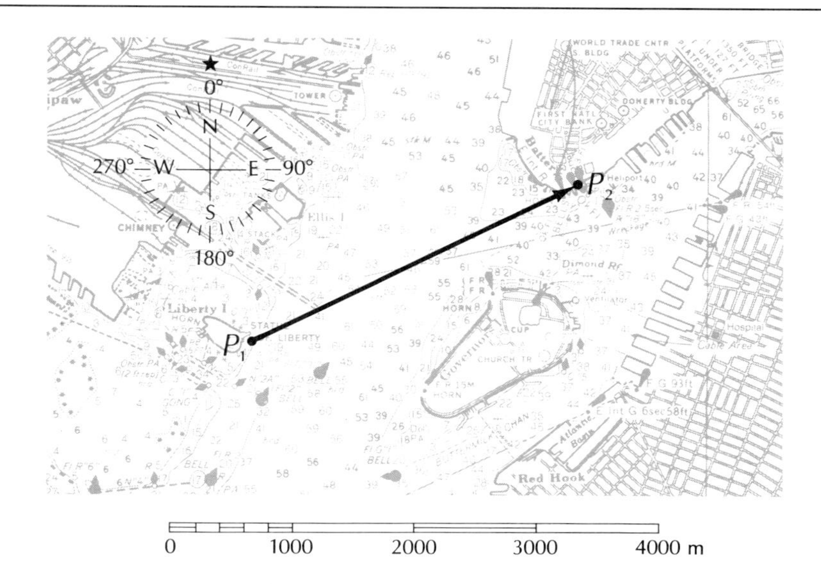


**Fig. 3.1** Displacement vector for a ship moving from Liberty Island to the Battery in New York harbor. (Excerpt from a National Ocean Survey Chart.)

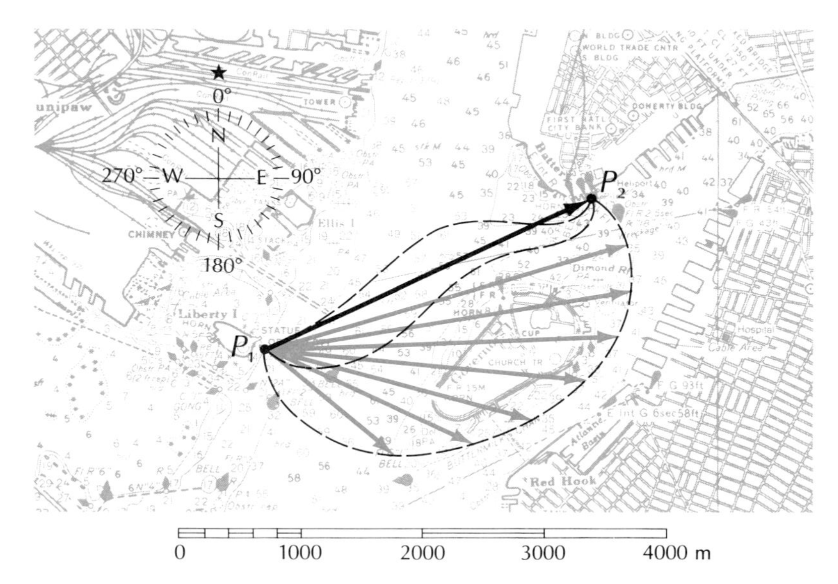


**Fig. 3.2** Three alternative paths from Liberty Island to the Battery. All of these result in the same final displacement. Also shown are displacement vectors at successive times for a ship moving along one of these paths.

any of the paths shown by the dashed lines in Figure 3.2 results in the same final displacement vector. If we wanted to describe the complete path vectorially, then we would have to draw a sequence of vectors representing the displacement of the ship at different times.

The displacement vector in Figure 3.1 has a *length* and a *direction.* Instead of describing the vector graphically by drawing a picture, we can describe it numerically by giving the numerical value of its length (in, say, meters) and the numerical value of the angle (in, say, degrees) it makes with some fiducial direction. For example, we can specify the displacement vector in Figure 3.1 by stating that it is 2890 m long and points at an angle of 65° east of north.

Since the displacement vector describes a *change* in position, any two directed line segments of identical length and direction represent equal vectors, regardless of whether the endpoints of the line segments are the same. Thus, the two parallel directed line segments shown in

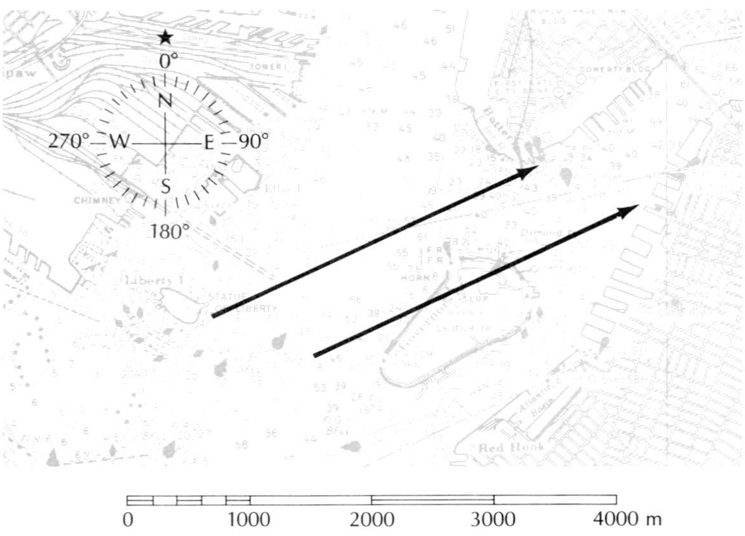

**Fig. 3.3** These two displacement vectors are equal.

Figure 3.3 do not represent different vectors; both involve the same *change* of position (same distance and same direction), and both represent equal displacement vectors.

In printed books, vectors are usually indicated by boldface letters, such as **A,** and we will follow this convention. In handwritten calculations, an alternative notation consisting of either a small arrow, such as $\vec{A}$, or a wavy underline, such as $\underset{\sim}{A}$, is usually more convenient. We will denote the length of a displacement vector by absolute-value signs, such as $|\mathbf{A}|$, or simply by an italic letter, such as $A$.

The displacement vector serves as prototype for all other vectors. To decide whether some mathematical quantity endowed with magnitude and direction is a vector, we compare its mathematical properties with those of the displacement vector. *Any quantity that has magnitude and direction and that behaves mathematically like*[1] *the displacement vector is a* **vector.** For example, velocity, acceleration, and force are vectors; they can be represented graphically by directed line segments of a length equal to the magnitude of the velocity, acceleration, or force (in some suitable units) and a corresponding direction.

*Vector*

By contrast, any quantity that has a magnitude but *no* direction is called a **scalar.** For example, length, time, mass, area, volume, density, and energy are scalars; they can be completely specified by their numerical magnitude. Note that the length of a displacement vector, such as the length 2890 m of the displacement vector in Figure 3.1, is a quantity that has magnitude but no direction, i.e., the length of a vector is a scalar.

*Scalar*

## 3.2 Vector Addition and Subtraction

Since by definition all vectors have the mathematical properties of displacement vectors, we can investigate all the mathematical operations with vectors by looking at displacement vectors. The most important of these mathematical operations is **vector addition.**

*Vector addition*

[1] The exact meaning of "behaves mathematically like" will be spelled out in Section 3.5.

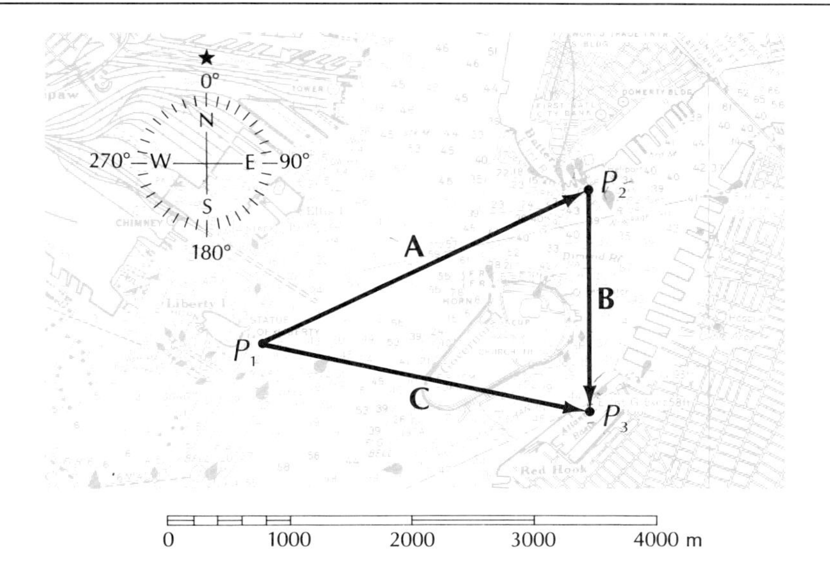


**Fig. 3.4** The displacement **A** is followed by the displacement **B**. The net displacement is **C**.

Two displacements carried out in succession result in a net displacement, which can be regarded as the sum of the two individual displacements. For example, Figure 3.4 shows a displacement vector **A** (from $P_1$ to $P_2$) and a displacement vector **B** (from $P_2$ to $P_3$). The net displacement vector is the directed line segment from $P_1$ to $P_3$; this net displacement vector is denoted by **C** in Figure 3.4. This vector **C** can be regarded as the sum of the individual displacements,

*Resultant*

$$\mathbf{C} = \mathbf{A} + \mathbf{B} \tag{1}$$

The sum of two vectors is often called the **resultant** of these vectors. Thus, **C** is called the resultant of **A** and **B**.

EXAMPLE 1. A ship moves from Liberty Island in New York harbor to the Battery and from there to the Atlantic Basin (Figure 3.4). The first displacement is 2890 m at 65° east of north; the second is 1830 m due south. What is the resultant?

SOLUTION: The resultant of the two displacement vectors **A** and **B** is the vector **C**, from the tail of **A** to the tip of **B**. For a graphical determination of **C**, we can measure the length of **C** directly on the chart using the scale of length marked on the chart, and we can measure the direction of **C** with a protractor (the way the navigator of the ship would solve the problem).

For a more precise numerical determination of **C**, we note that **A**, **B**, and **C** form a triangle. We can therefore find **C** by using standard trigonometric methods.[2] The lengths of the known sides are $A = 2890$ m and $B = 1830$ m; the angle between these sides is 65° (Figure 3.5). By the law of cosines

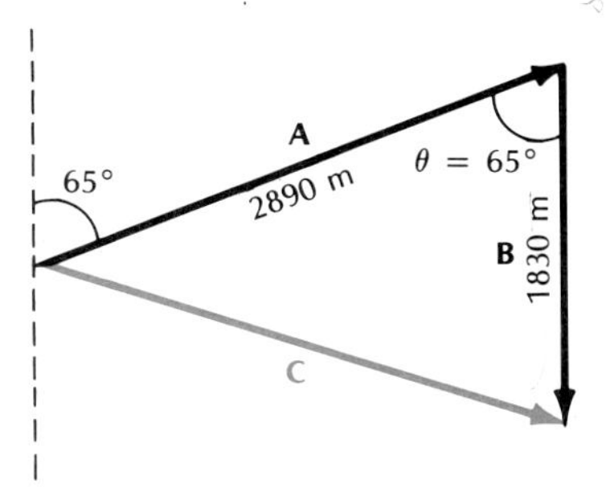


**Fig. 3.5** Vector triangle.

$$C^2 = A^2 + B^2 - 2AB \cos 65° \tag{2}$$

from which

$$C = \sqrt{(2890)^2 + (1830)^2 - 2 \times 2890 \times 1830 \times \cos 65°} \text{ m}$$

$$= 2690 \text{ m}$$

[2] Appendix 4 gives a review of trigonometry.

By the law of sines the angle $\theta$ between $C$ and $A$ is given by

$$\frac{\sin\theta}{1830} = \frac{\sin 65^\circ}{2690}$$

and

$$\theta = 38.1^\circ$$

The angle between the northerly direction and **C** is then $38.1^\circ + 65^\circ = 103.1^\circ$. Thus the resultant displacement is 2690 m at $13.1^\circ$ south of east.

The procedure for the addition of any arbitrary vectors — such as velocity, acceleration, and force vectors — mimics that for displacement vectors. If **A** and **B** are two arbitrary vectors (Figure 3.6a), then their resultant can be obtained by placing the tail of **B** on the head of **A;** the directed line segment connecting the tail of **A** to the head of **B** is the resultant (Figure 3.6b). Alternatively, the resultant can be obtained by placing the tail of **B** on the tail of **A** and drawing a parallelogram with **A** and **B** as two of the sides; the diagonal of the parallelogram is then the resultant (Figure 3.6c).

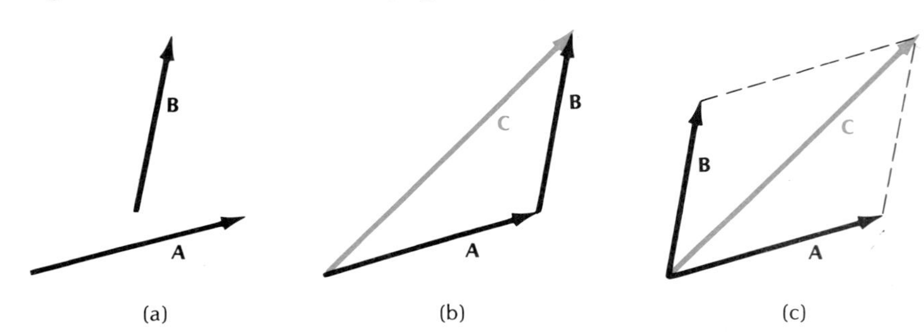


**Fig. 3.6** The vector sum **A + B;** the resultant is **C.** (a) The two vectors **A** and **B.** (b) Addition of **A** and **B** by the tail-to-head method. (c) Addition of **A** and **B** by the parallelogram method.

Note that the order in which the two vectors are added makes no difference to the final result. Whether we place the tail of **A** on the head of **B** or the tail of **B** on the head of **A,** the resultant is the same (Figure 3.7). Hence

*Commutative law*

$$\mathbf{A} + \mathbf{B} = \mathbf{B} + \mathbf{A} \tag{3}$$

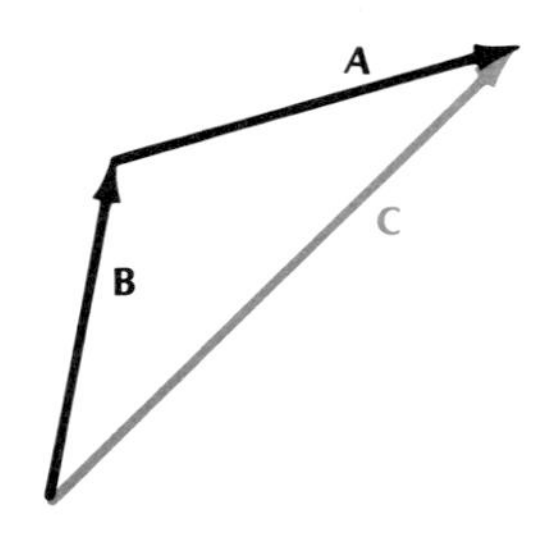


**Fig. 3.7** The vector sum **B + A.** The resultant **C** is the same as in Figure 3.6b.

This identity is called the **commutative law** for vector addition; it indicates that, just as in ordinary addition of numbers, the order of the terms is irrelevant.

Three or more vectors can be added in succession. For example, the resultant of three vectors **A, B,** and **D** can be obtained by first adding **A** and **B** and then adding **D** (Figure 3.8a). Alternatively, this resultant can be obtained by first adding **B** and **D** and then adding this to **A** (Figure 3.8b). Comparison of these figures shows that the grouping of terms makes no difference; in both cases we obtain the same final resultant. Hence

*Associative law*

$$(\mathbf{A} + \mathbf{B}) + \mathbf{D} = \mathbf{A} + (\mathbf{B} + \mathbf{D}) \tag{4}$$

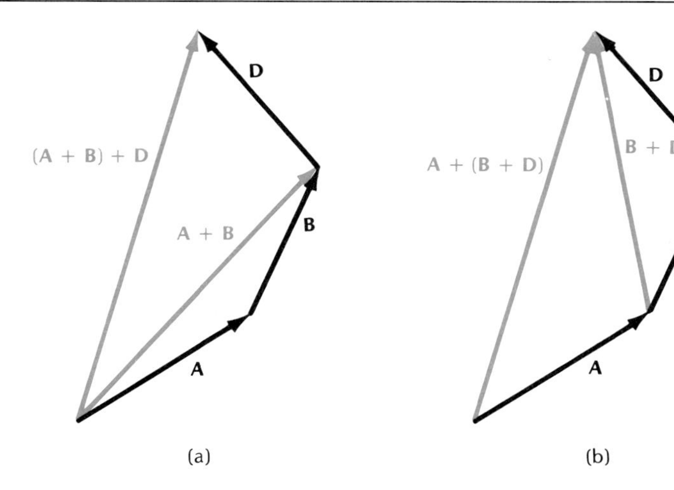


**Fig. 3.8** (a) First add **A** and **B;** then add **D.** (b) First add **B** and **D;** then add this to **A.**

This is called the **associative law** of vector addition. Taken together, Eqs. (3) and (4) show that the resultant of a given number of vectors can be evaluated by summing the vectors in any convenient order.

The magnitude of the resultant of two (or more) vectors is usually less than the sum of the magnitudes of the vectors. Thus, if

$$\mathbf{C} = \mathbf{A} + \mathbf{B} \tag{5}$$

then

$$C \leq A + B \tag{6}$$

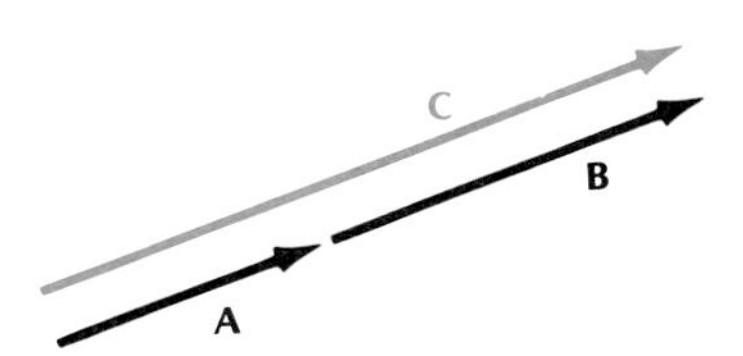


**Fig. 3.9** Parallel vectors **A** and **B** and their resultant **C.**

This inequality simply expresses the fact that in a triangle (see Figure 3.6b) the length of any one side is less than the sum of the lengths of the other two sides. Only in the special case where **A** and **B** are parallel (see Figure 3.9) will the magnitude of **C** equal the sum of the magnitudes of **A** and **B;** it can never exceed this sum.

The negative of a given vector **A** is a vector of the same magnitude, but opposite direction; this new vector is denoted by **−A** (Figure 3.10). Obviously,

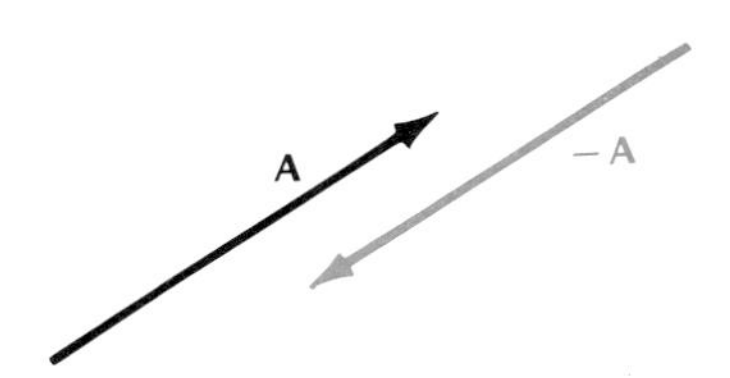


**Fig. 3.10** The negative of the vector **A.**

$$\boxed{\mathbf{A} + (-\mathbf{A}) = \mathbf{0}} \tag{7}$$

which says that the sum of a vector and its negative gives a vector of zero magnitude.

The **subtraction of two vectors A** and **B** is defined as the sum of **A** and **−B,**

*Vector subtraction*

$$\boxed{\mathbf{A} - \mathbf{B} = \mathbf{A} + (-\mathbf{B})} \tag{8}$$

Figure 3.11a shows the vector sum of **A** and **−B.** Figure 3.11b shows that the same result can be obtained by placing the tail of **B** on the tail of **A** and drawing the same vector parallelogram as for addition: one of the diagonals of the parallelogram is then the sum **A + B,** and the other diagonal is the difference **A − B.**

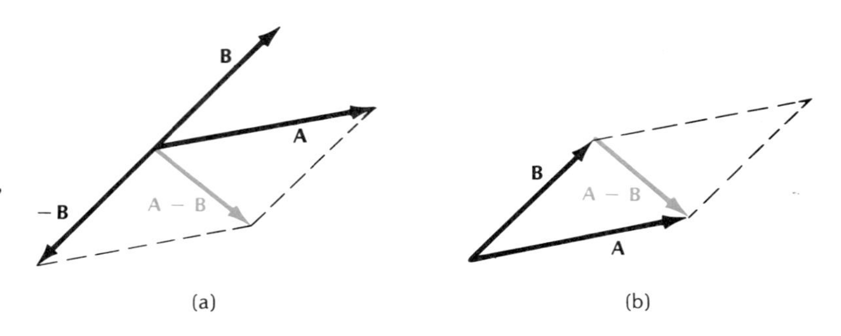


**Fig. 3.11** The vector difference **A** − **B**. (a) Construction of **A** − **B** by means of the addition of **A** and −**B**. (b) Construction of **A** − **B** by means of the "other diagonal" of the vector parallelogram, from the head of **B** to the head of **A**.

A vector can be multiplied by a positive or negative number. For instance, if **A** is a given vector, then 3**A** is a vector of the same direction and of a magnitude three times as large (Figure 3.12a); and −3**A** is a vector of the opposite direction and, again, of a magnitude three times as large (Figure 3.12b). In particular, if we multiply a vector by −1, we obtain the negative of that vector,

$$(-1)\mathbf{A} = -\mathbf{A}$$

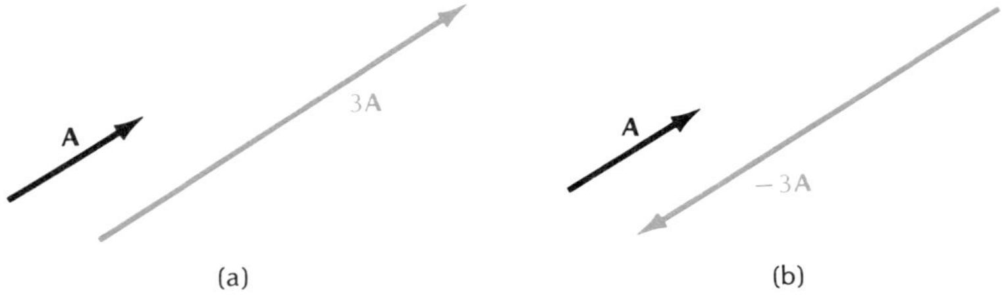

**Fig. 3.12** (a) Vector **A** multiplied by 3. (b) Vector **A** multiplied by −3.

## 3.3 The Position Vector; Components of Vectors

To describe the position of a point $P$ in three dimensions, we must choose an origin and construct a coordinate grid. If the grid is rectangular, then the position of a point will be given by the three rectangular coordinates, $x$, $y$, $z$. Alternatively, we can describe the position of a point by means of the displacement vector from the origin to the point (Figure 3.13). This displacement vector is called the **position vector,** and is usually denoted by **r**.

*Position vector*

The explicit connection between **r** and $x$, $y$, $z$ can be expressed mathematically as follows. We can define a vector $\hat{\mathbf{x}}$ (pronounced "$x$ hat") that has a magnitude $|\hat{\mathbf{x}}| = 1$ and points in the positive $x$ direction; likewise, we can define vectors $\hat{\mathbf{y}}$ and $\hat{\mathbf{z}}$ that have magnitudes $|\hat{\mathbf{y}}| = 1$ and $|\hat{\mathbf{z}}| = 1$ and point in the positive $y$ and $z$ directions, respec-

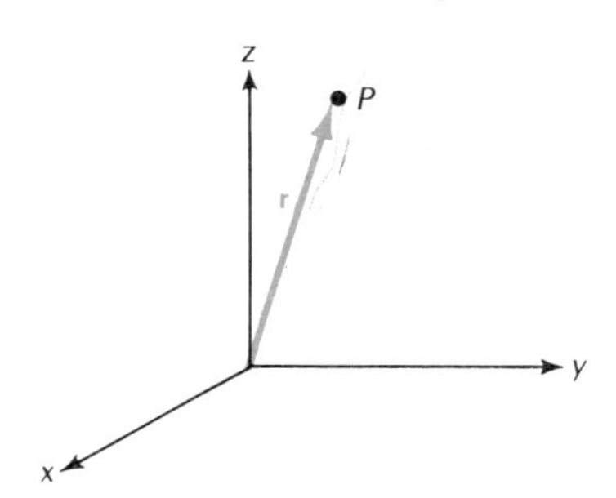


**Fig. 3.13** The position vector **r** of the point $P$.

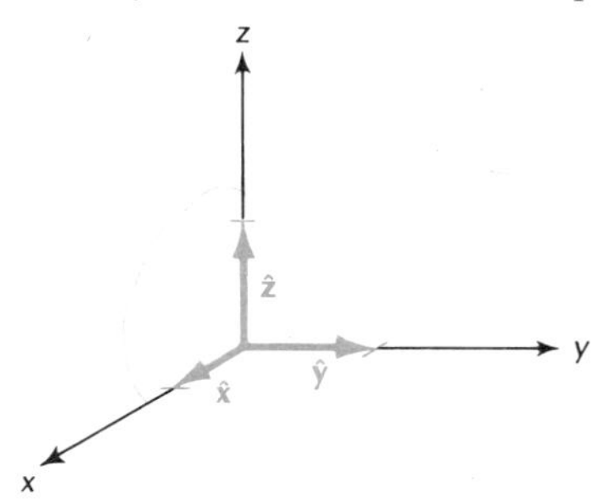


**Fig. 3.14** The unit vectors $\hat{\mathbf{x}}$, $\hat{\mathbf{y}}$, $\hat{\mathbf{z}}$.

tively (Figure 3.14).[3] These vectors are called **unit vectors.** Now consider the vector sum $x\hat{\mathbf{x}} + y\hat{\mathbf{y}} + z\hat{\mathbf{z}}$. This sum consists of a displacement of magnitude $x$ in the $x$ direction, followed by a displacement of magnitude $y$ in the $y$ direction, followed by a displacement of magnitude $z$ in the $z$ direction. This vector sum brings us from the origin to the point $x, y, z$ (Figure 3.15). Hence this vector sum coincides with the position vector,

*Unit vectors $\hat{x}$, $\hat{y}$, $\hat{z}$*

$$\mathbf{r} = x\hat{\mathbf{x}} + y\hat{\mathbf{y}} + z\hat{\mathbf{z}} \tag{9}$$

The three numbers $x$, $y$, and $z$ are called the **components** of the vector **r**. For example, if the point $P$ has coordinates $x = 2$ m, $y = 3$ m, $z = 5$ m, then the position vector of $P$ is

*Components of vectors*

$$\mathbf{r} = (2\ \mathrm{m})\hat{\mathbf{x}} + (3\ \mathrm{m})\hat{\mathbf{y}} + (5\ \mathrm{m})\hat{\mathbf{z}} \tag{10}$$

and the $x$ component of this vector is 2 m, the $y$ component is 3 m, and the $z$ component is 5 m.

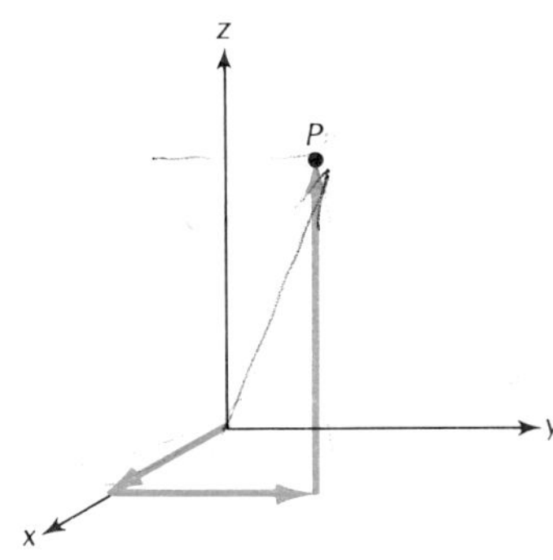


**Fig. 3.15** The sum of three perpendicular displacements brings us from the origin to the point $x$, $y$, $z$.

Graphically, the components of **r** can be obtained by dropping perpendiculars from the point $P$ to the three coordinate axes $x$, $y$, $z$. The intercepts of these perpendiculars with the axes give the components. Figure 3.16 shows the vector **r** of Eq. (10) and its components.

For any arbitrary vector **A**, the $x$, $y$, and $z$ components can be defined by analogy with those of the position vector. First the tail of the vector is placed at the origin of coordinates and perpendiculars are dropped from the tip of the vector to the coordinate axes. Once again the intercepts of these perpendiculars with the axes give the components (Figure 3.17). If we designate the numerical values of these components as $A_x$, $A_y$, and $A_z$, then the vector **A** can be expressed as

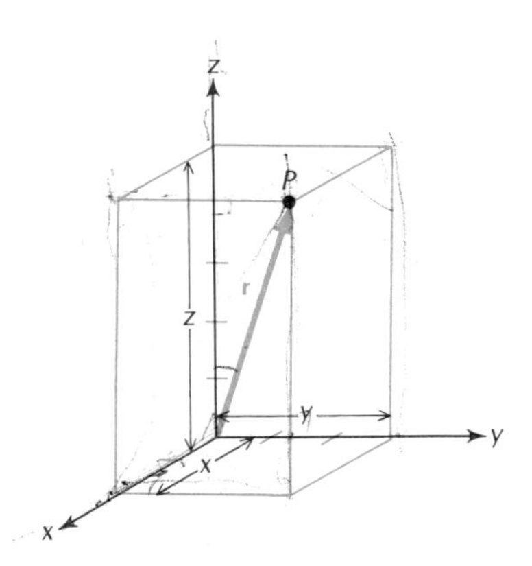


**Fig. 3.16** The components $x$, $y$, $z$ of the position vector **r**.

$$\mathbf{A} = A_x\hat{\mathbf{x}} + A_y\hat{\mathbf{y}} + A_z\hat{\mathbf{z}} \tag{11}$$

Figure 3.18 shows the special case of a vector **A** in two dimensions. We see from this figure that the $x$ and $y$ components of the vector can

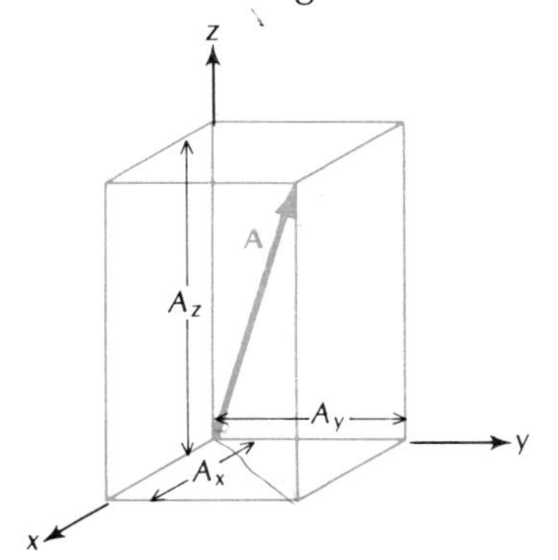


**Fig. 3.17** The components $A_x$, $A_y$, $A_z$ of an arbitrary vector **A**.

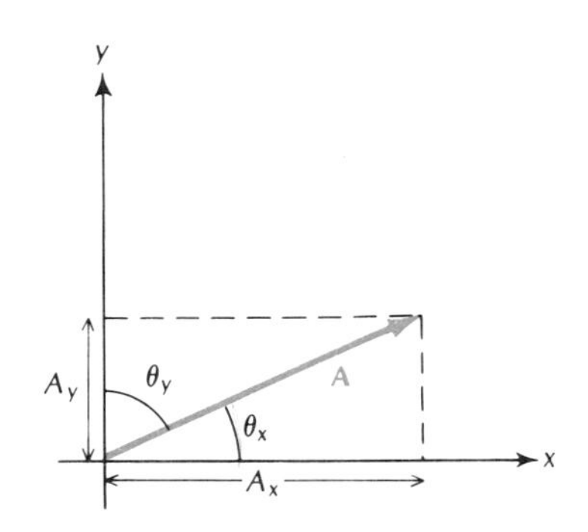


**Fig. 3.18** Vector **A** in two dimensions and its components.

[3] Note that the magnitude of each unit vector is a pure number, without any meters, seconds, or kilograms. Nevertheless, when drawing the unit vectors (as in Figure 3.14), we will adopt the convention of showing these vectors as though they had a length of 1 meter; thus, $\hat{\mathbf{x}}$ starts at the point $x = 0$ meter and ends at $x = 1$ meter.

be expressed in terms of the magnitude of the vector and the angles that it makes with the $x$ and $y$ axes,

$$A_x = A \cos \theta_x \qquad A_y = A \cos \theta_y \tag{12}$$

Here $\theta_x$ is reckoned from the positive $x$ axis toward the vector; values of $\theta_x$ larger than 90° give a negative value for $A_x$. The same is true for $\theta_y$.

Figure 3.18 also shows that, by the Pythagorean theorem, the magnitude of **A** can be expressed in terms of $A_x$ and $A_y$,

$$A = \sqrt{A_x^2 + A_y^2} \tag{13}$$

Upon generalization to three dimensions, Eqs. (12) and (13) become

$$\boxed{A_x = A \cos \theta_x \quad A_y = A \cos \theta_y \quad A_z = A \cos \theta_z} \tag{14}$$

$$\boxed{A = \sqrt{A_x^2 + A_y^2 + A_z^2}} \tag{15}$$

Note that in the special case of the position vector, Eqs. (14) and (15) take the form

$$x = r \cos \theta_x \quad y = r \cos \theta_y \quad z = r \cos \theta_z \tag{16}$$

$$r = \sqrt{x^2 + y^2 + z^2} \tag{17}$$

The last of these equations is a familiar expression for the distance between the point $x, y, z$ and the origin of coordinates.

Expressed in terms of components, vector addition (or subtraction) turns out to be merely addition (or subtraction) of components. Thus, given two vectors

$$\mathbf{A} = A_x\hat{\mathbf{x}} + A_y\hat{\mathbf{y}} + A_z\hat{\mathbf{z}} \tag{18}$$

and

$$\mathbf{B} = B_x\hat{\mathbf{x}} + B_y\hat{\mathbf{y}} + B_z\hat{\mathbf{z}} \tag{19}$$

Their sum is

$$\begin{aligned}\mathbf{A} + \mathbf{B} &= A_x\hat{\mathbf{x}} + A_y\hat{\mathbf{y}} + A_z\hat{\mathbf{z}} + B_x\hat{\mathbf{x}} + B_y\hat{\mathbf{y}} + B_z\hat{\mathbf{z}} \\ &= (A_x + B_x)\hat{\mathbf{x}} + (A_y + B_y)\hat{\mathbf{y}} + (A_z + B_z)\hat{\mathbf{z}}\end{aligned} \tag{20}$$

Hence the $x$ component of the resultant is the sum of the $x$ components of **A** and **B**, and so on.

---

EXAMPLE 2. The eye of the hurricane is 200 km from Miami on a bearing of 30° south of east. A reconnaissance airplane is 100 km due north of Miami. What displacement vector will bring the plane to the eye of the hurricane?

SOLUTION: In Example 1 we saw how to use a graphical method and a trigonometric method for finding an unknown vector. Here we will see how to use the component method.

For this calculation we need to make a choice of coordinate system, that is, a choice of origin and of axes. We can make this choice in any way that happens to be convenient. In Figure 3.19 we have placed the origin on Miami, with the $x$ axis eastward and the $y$ axis northward. In this coordinate system, the airplane has a position vector **A** with components

$$A_x = 0 \text{ km} \qquad A_y = 100 \text{ km}$$

The hurricane has a position vector **B** with components

$$B_x = 200 \text{ km} \times \cos 30° = 173 \text{ km}$$

$$B_y = 200 \text{ km} \times \cos 120° = -100 \text{ km}$$

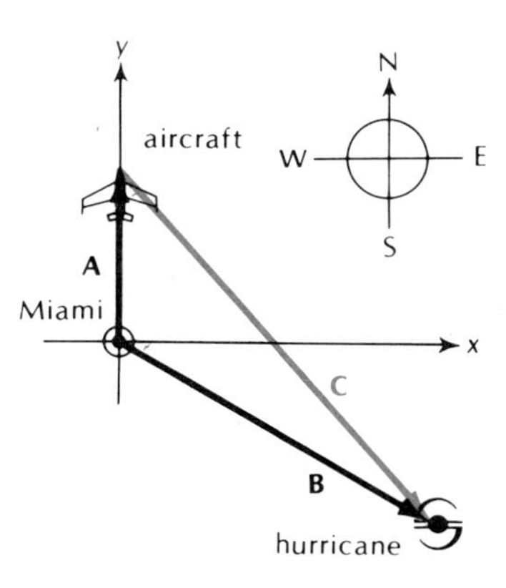


**Fig. 3.19** Displacement vector **C** from the airplane to the eye of the hurricane.

As we can see from Figure 3.19, the displacement **C** from the airplane to the hurricane is the *difference* between these position vectors, that is, **C** = **B** − **A.** This vector **C** has components

$$C_x = B_x - A_x = 173 \text{ km} - 0 \text{ km} = 173 \text{ km}$$

$$C_y = B_y - A_y = -100 \text{ km} - 100 \text{ km} = -200 \text{ km}$$

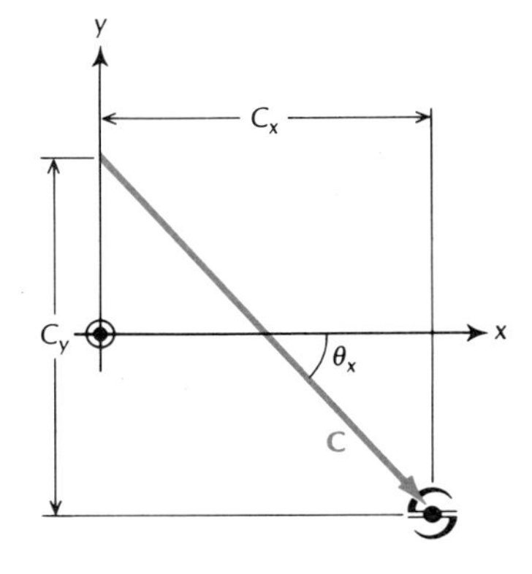


**Fig. 3.20** Components of the displacement vector **C.**

Figure 3.20 shows the components of the vector **C**. The magnitude of **C** is

$$C = \sqrt{C_x^2 + C_y^2} = \sqrt{(173)^2 + (200)^2} \text{ km} = 264 \text{ km}$$

and the angle between **C** and the $x$ axis is given by

$$\tan \theta_x = \frac{C_y}{C_x} = \frac{-200}{173} \tag{21}$$

$$\theta_x = -49.1°$$

Hence a displacement of 264 km at 49.1° south of east will bring the airplane to the hurricane.

Comments and Suggestions: We have become acquainted with three methods for vector addition: the graphical method (drawing of vector parallelograms or triangles), the trigonometric method (solution of triangles; see Example 1), and the component method. Of these the first is the quickest, but it suffers from limited accuracy. Which of the other two methods is most convenient depends on how the vectors are specified. For vectors specified in terms of lengths and angles, the trigonometric method is usually best; for vectors specified in terms of components, the component method is best. However, if more than two vectors are to be added, then the component method is almost always the best choice, regardless how the vectors are specified.

## 3.4* Vector Multiplication

There are several ways of multiplying vectors. The reason for this diversity is that in forming the "product" of two vectors, we must take into account both their magnitudes *and* their directions. Depending on how we combine these quantities, we obtain different kinds of vector products. The two most important vector products are the dot product and the cross product.

* This section is optional. The dot product will be used in Section 7.2 and the cross product will be used in Section 12.5. Those sections include brief, self-contained expositions of the vector products.

DOT PRODUCT The **dot product** (also called the **scalar product** or the **inner product**) of two vectors **A** and **B** is denoted by $\mathbf{A} \cdot \mathbf{B}$. This quantity is simply the product of the magnitudes of the two vectors and the cosine of the angle $\phi$ between them (Figure 3.21),

*Dot product*

$$\boxed{\mathbf{A} \cdot \mathbf{B} = AB \cos \phi} \tag{22}$$

**Fig. 3.21** The vectors **A** and **B** and the angle between them.

Thus, the dot product of two vectors simply gives a number, that is, a scalar rather than a vector. The number will be positive if $\phi < 90°$ and negative if $\phi > 90°$. If the two vectors are perpendicular, then their dot product is zero. Note that the dot product is commutative; as in ordinary multiplication, the order of the factors is irrelevant:

$$\boxed{\mathbf{A} \cdot \mathbf{B} = \mathbf{B} \cdot \mathbf{A}}$$

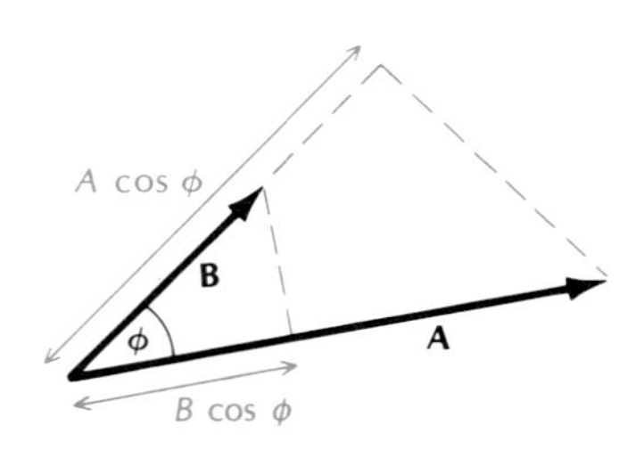


**Fig. 3.22** The component of **A** along **B** is $A \cos \phi$; the component of **B** along **A** is $B \cos \phi$.

We see from Figure 3.22 that the dot product $\mathbf{A} \cdot \mathbf{B}$ can be regarded as $B$ times the component of **A** along **B**, or as $A$ times the component of **B** along **A**. The special case of the dot product of a vector with itself gives the square of the magnitude of the vector:

$$\boxed{\mathbf{A} \cdot \mathbf{A} = AA \cos 0 = A^2} \tag{23}$$

The dot products of the unit vectors with themselves and with each other are

$$\begin{array}{lll} \hat{\mathbf{x}} \cdot \hat{\mathbf{x}} = 1 & \hat{\mathbf{y}} \cdot \hat{\mathbf{y}} = 1 & \hat{\mathbf{z}} \cdot \hat{\mathbf{z}} = 1 \\ \hat{\mathbf{x}} \cdot \hat{\mathbf{y}} = \hat{\mathbf{y}} \cdot \hat{\mathbf{x}} = 0 & \hat{\mathbf{x}} \cdot \hat{\mathbf{z}} = \hat{\mathbf{z}} \cdot \hat{\mathbf{x}} = 0 & \hat{\mathbf{y}} \cdot \hat{\mathbf{z}} = \hat{\mathbf{z}} \cdot \hat{\mathbf{y}} = 0 \end{array} \tag{24}$$

By means of these equations it is possible to express the dot product of two arbitrary vectors in terms of their components:[4]

$$\begin{aligned} \mathbf{A} \cdot \mathbf{B} &= (A_x\hat{\mathbf{x}} + A_y\hat{\mathbf{y}} + A_z\hat{\mathbf{z}}) \cdot (B_x\hat{\mathbf{x}} + B_y\hat{\mathbf{y}} + B_z\hat{\mathbf{z}}) \\ &= A_xB_x\hat{\mathbf{x}} \cdot \hat{\mathbf{x}} + A_xB_y\hat{\mathbf{x}} \cdot \hat{\mathbf{y}} + A_xB_z\hat{\mathbf{x}} \cdot \hat{\mathbf{z}} \\ &\quad + A_yB_x\hat{\mathbf{y}} \cdot \hat{\mathbf{x}} + A_yB_y\hat{\mathbf{y}} \cdot \hat{\mathbf{y}} + A_yB_z\hat{\mathbf{y}} \cdot \hat{\mathbf{z}} \\ &\quad + A_zB_x\hat{\mathbf{z}} \cdot \hat{\mathbf{x}} + A_zB_y\hat{\mathbf{z}} \cdot \hat{\mathbf{y}} + A_zB_z\hat{\mathbf{z}} \cdot \hat{\mathbf{z}} \end{aligned} \tag{25}$$

With the values given by Eq. (24), this reduces to

*Dot-product formula*

$$\boxed{\mathbf{A} \cdot \mathbf{B} = A_xB_x + A_yB_y + A_zB_z} \tag{26}$$

[4] In working out this product, we make use of the fact that vector multiplication obeys the **distributive law:** $(\mathbf{C} + \mathbf{D}) \cdot \mathbf{E} = \mathbf{C} \cdot \mathbf{E} + \mathbf{D} \cdot \mathbf{E}$. To prove this law, we need only note that this product is the magnitude of **E** times the component of $\mathbf{C} + \mathbf{D}$ along **E**; and that the component of $\mathbf{C} + \mathbf{D}$ along any direction is equal to the sum of the component of **C** plus the component of **D**.

Thus, the dot product is simply the sum of the products of the $x$, $y$, and $z$ components of the two vectors.

Finally, note that the components of a vector are equal to the dot product of the vector and the corresponding unit vectors. For instance,

$$\hat{\mathbf{x}} \cdot \mathbf{A} = \hat{\mathbf{x}} \cdot (A_x\hat{\mathbf{x}} + A_y\hat{\mathbf{y}} + A_z\hat{\mathbf{z}})$$

$$= A_x\hat{\mathbf{x}} \cdot \hat{\mathbf{x}} + A_y\hat{\mathbf{x}} \cdot \hat{\mathbf{y}} + A_z\hat{\mathbf{x}} \cdot \hat{\mathbf{z}} = A_x \tag{27}$$

---

EXAMPLE 3. Find the dot product of the vectors **A** and **B** of Example 2.

SOLUTION: The vector **A** has a magnitude $A = 100$ km and **B** has a magnitude $B = 200$ km; the angle between the vectors is $\phi = 120°$. Hence

$$\mathbf{A} \cdot \mathbf{B} = AB \cos \phi = 100 \text{ km} \times 200 \text{ km} \times \cos 120°$$

$$= -10{,}000 \text{ km}^2 \tag{28}$$

Alternatively, the calculation can be done by components:

$$\mathbf{A} \cdot \mathbf{B} = A_xB_x + A_yB_y + A_zB_z$$

$$= 0 \text{ km} \times 173 \text{ km} + 100 \text{ km} \times (-100 \text{ km}) + 0 \text{ km} \times 0 \text{ km}$$

$$= -10{,}000 \text{ km}^2 \tag{29}$$

This agrees with Eq. (28).

---

EXAMPLE 4. Find the angle between the vectors $\mathbf{A} = 2\hat{\mathbf{x}} + \hat{\mathbf{y}}$ and $\mathbf{B} = -\hat{\mathbf{x}} + 3\hat{\mathbf{y}} + 2\hat{\mathbf{z}}$.

SOLUTION: Since

$$\mathbf{A} \cdot \mathbf{B} = AB \cos \phi$$

the cosine of the angle between the vectors is

$$\cos \phi = \frac{\mathbf{A} \cdot \mathbf{B}}{AB} = \frac{A_xB_x + A_yB_y + A_zB_z}{\sqrt{A_x^2 + A_y^2 + A_z^2}\,\sqrt{B_x^2 + B_y^2 + B_z^2}}$$

$$= \frac{(2)(-1) + (1)(3) + (0)(2)}{\sqrt{5}\,\sqrt{14}} = 0.120$$

This implies an angle of 83.1°.

COMMENTS AND SUGGESTIONS: This trick for calculating the unknown angle between two vectors is also useful for calculating the angle between two lines in space; simply take two vectors pointing along the lines and calculate the angle between them.

---

CROSS PRODUCT The **cross product** (also called the **vector product**) of two vectors **A** and **B** is denoted by **A** × **B.** This quantity is a *vector* with a magnitude equal to the product of the magnitude of the two vectors and the sine of the angle between them. Thus if we write the resulting vector as

$$\mathbf{C} = \mathbf{A} \times \mathbf{B} \tag{30}$$

then the magnitude of this vector is

*Cross product*

$$C = AB \sin \phi \tag{31}$$

The direction of **C** is defined to be along the perpendicular to the plane formed by **A** and **B** (Figure 3.23). The direction of **C** along this perpendicular is given by the **right-hand rule:** *put the fingers of your right hand along* **A** *(Figure 3.24a) and curl them toward* **B** *in the direction of the smaller angle from* **A** *to* **B** *(Figure 3.24b); the thumb then points along* **C.** Note that the fingers must be curled from the first vector in the product toward the second. Thus, **A** × **B** is not the same as **B** × **A.** For the latter product the fingers must be curled from **B** toward **A** (rather than vice versa); hence, the direction of the vector **B** × **A** is opposite to that of **A** × **B,**

*Right-hand rule*

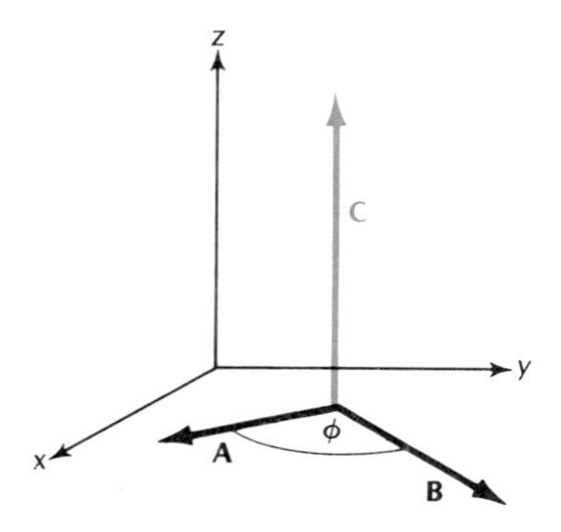


**Fig. 3.23** The vectors **A** and **B** and their cross product **C** = **A** × **B.**

$$\mathbf{B} \times \mathbf{A} = -\mathbf{A} \times \mathbf{B} \tag{32}$$

Accordingly, the cross product of two vectors is *not* commutative; in contrast to ordinary multiplication, the result does depend on the order of the factors.

**Fig. 3.24** The right-hand rule. If the fingers curl from **A** toward **B,** the thumb points along **C.**

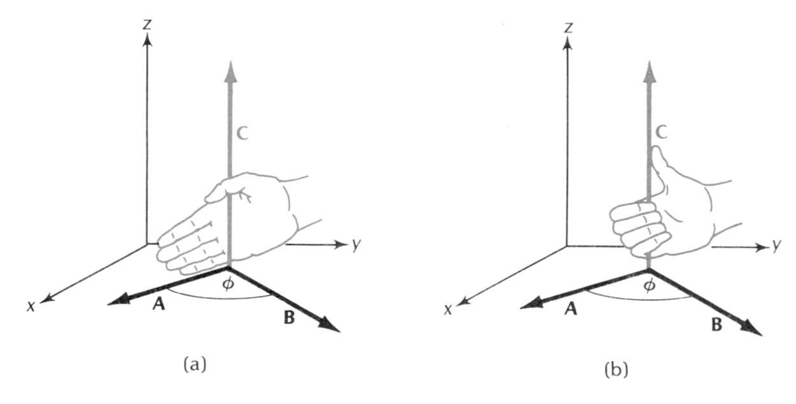


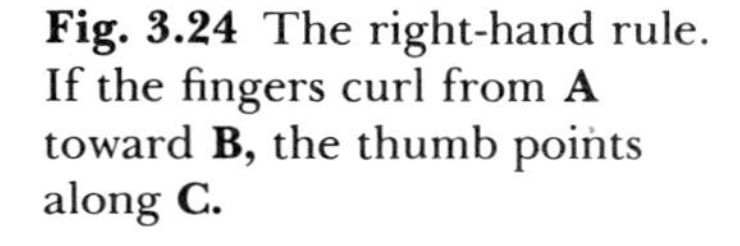


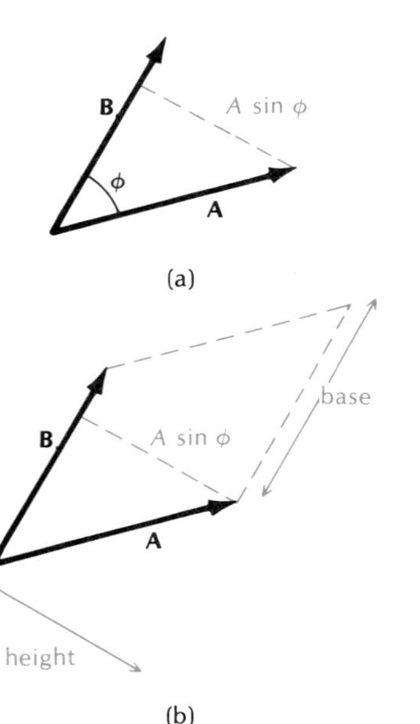


**Fig. 3.25** (a) $A \sin \phi$ is the component of **A** perpendicular to **B.** (b) $AB \sin \phi$ is the area of the parallelogram.

As we see from Figure 3.25a, the magnitude of **A** × **B** is equal to the product of the magnitude of **B** times the component of **A** perpendicular to **B** (or the magnitude of **A** times the component of **B** perpendicular to **A**). Furthermore, from Figure 3.25b, we see that the magnitude of **A** × **B** is equal to the area of the parallelogram formed out of the vectors **A** and **B.**

If the vectors **A** and **B** are parallel, then their cross product is zero; in particular, the cross product of any vector with itself is zero,

$$\mathbf{A} \times \mathbf{A} = 0 \tag{33}$$

The cross products of the unit vectors are

$$\begin{array}{lll} \hat{\mathbf{x}} \times \hat{\mathbf{x}} = 0 & \hat{\mathbf{y}} \times \hat{\mathbf{y}} = 0 & \hat{\mathbf{z}} \times \hat{\mathbf{z}} = 0 \\ \hat{\mathbf{x}} \times \hat{\mathbf{y}} = -\hat{\mathbf{y}} \times \hat{\mathbf{x}} = \hat{\mathbf{z}} & \hat{\mathbf{z}} \times \hat{\mathbf{x}} = -\hat{\mathbf{x}} \times \hat{\mathbf{z}} = \hat{\mathbf{y}} & \hat{\mathbf{y}} \times \hat{\mathbf{z}} = -\hat{\mathbf{z}} \times \hat{\mathbf{y}} = \hat{\mathbf{x}} \end{array} \tag{34}$$

(This assumes that the coordinate system is *right-handed,* that is, the coordinate axes $x$, $y$, and $z$ have the relative directions shown in Figure 3.24. If one of these axes were reversed, the coordinates would become *left-handed,* and all the signs in Eq. (34) would have to be reversed. We will always use right-handed coordinates in this book.)

Using these equations, we can express the cross product of two arbitrary vectors in terms of their components.

$$\begin{aligned}\mathbf{A}\times\mathbf{B} &= (A_x\hat{\mathbf{x}} + A_y\hat{\mathbf{y}} + A_z\hat{\mathbf{z}})\times(B_x\hat{\mathbf{x}} + B_y\hat{\mathbf{y}} + B_z\hat{\mathbf{z}})\\ &= A_xB_x\hat{\mathbf{x}}\times\hat{\mathbf{x}} + A_xB_y\hat{\mathbf{x}}\times\hat{\mathbf{y}} + A_xB_z\hat{\mathbf{x}}\times\hat{\mathbf{z}}\\ &\quad + A_yB_x\hat{\mathbf{y}}\times\hat{\mathbf{x}} + A_yB_y\hat{\mathbf{y}}\times\hat{\mathbf{y}} + A_yB_z\hat{\mathbf{y}}\times\hat{\mathbf{z}}\\ &\quad + A_zB_x\hat{\mathbf{z}}\times\hat{\mathbf{x}} + A_zB_y\hat{\mathbf{z}}\times\hat{\mathbf{y}} + A_zB_z\hat{\mathbf{z}}\times\hat{\mathbf{z}}\\ &= 0 + A_xB_y\hat{\mathbf{z}} - A_xB_z\hat{\mathbf{y}}\\ &\quad - A_yB_x\hat{\mathbf{z}} + 0 + A_yB_z\hat{\mathbf{x}}\\ &\quad + A_zB_x\hat{\mathbf{y}} - A_zB_y\hat{\mathbf{x}} + 0 \end{aligned} \tag{35}$$

and, collecting terms, we have

$$\boxed{\begin{aligned}\mathbf{A}\times\mathbf{B} &= (A_yB_z - A_zB_y)\hat{\mathbf{x}} + (A_zB_x - A_xB_z)\hat{\mathbf{y}}\\ &\quad + (A_xB_y - A_yB_x)\hat{\mathbf{z}}\end{aligned}} \tag{36}$$

*Cross-product formula*

[This messy result can be compactly written as a determinant,

$$\mathbf{A}\times\mathbf{B} = \begin{vmatrix}\hat{\mathbf{x}} & \hat{\mathbf{y}} & \hat{\mathbf{z}}\\ A_x & A_y & A_z\\ B_x & B_y & B_z\end{vmatrix} \tag{37}$$

By multiplying out this determinant according to the rules for $3\times3$ determinants, you can check that this is the same as Eq. (36).]

EXAMPLE 5. What is the cross product of the vectors **A** and **B** of Example 2?

SOLUTION: Figure 3.26 shows the vectors **A** and **B.** The magnitudes of the vectors are 100 km and 200 km, respectively; the angle between them is 120°. Hence the magnitude of the cross product is

$$\begin{aligned}C = |\mathbf{A}\times\mathbf{B}| &= AB\sin\Phi\\ &= 100\text{ km}\times 200\text{ km}\times\sin 120^\circ\\ &= 17{,}300\text{ km}^2\end{aligned}$$

The direction of **C** is perpendicular and into the plane of Figure 3.26.

We can also do this calculation by means of the cross-product formula of Eq. (36). With $A_x = 0$ km, $A_y = 100$ km, $A_z = 0$ km, $B_x = 173$ km, $B_y = -100$ km, $B_z = 0$ km, the first two terms in Eq. (36) drop out, since they involve factors of $A_z$ or $B_z$ which are zero. The last term gives

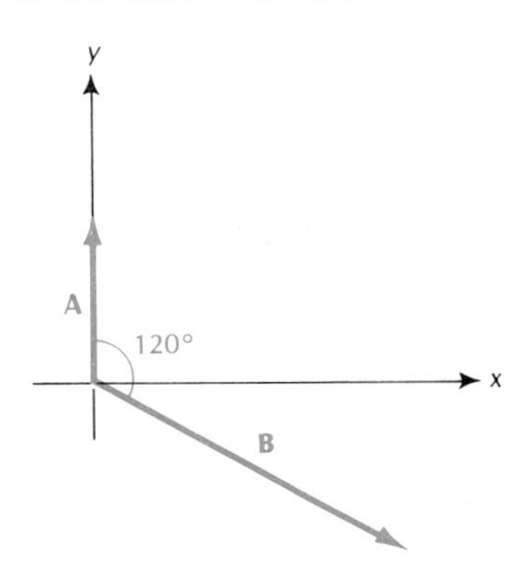


**Fig. 3.26** The vectors **A** and **B** of Example 2.

$$\mathbf{A} \times \mathbf{B} = 0 + 0 + (A_xB_y - A_yB_x)\hat{\mathbf{z}}$$

$$= [0 \text{ km} \times (-100 \text{ km}) - 100 \text{ km} \times 173 \text{ km}]\hat{\mathbf{z}}$$

$$= -17{,}300\hat{\mathbf{z}} \text{ km}^2$$

This vector points along the negative $z$ axis, that is, again into the plane of Figure 3.26 (note that if the $x$ axis is toward the right and the $y$ axis is upward, then the $z$ axis must be *out* of the plane of Figure 3.26).

COMMENTS AND SUGGESTIONS: The cross-product formula (36) automatically yields the direction of the cross product (the negative $z$ direction in the present example). However, to interpret the result, we need to know the direction of the $z$ axis relative to the directions of the $x$ and $y$ axes in Figure 3.26; this means we must look back at Figure 3.24, which shows the relative directions of the axes.

## 3.5* Vectors and Coordinate Rotations

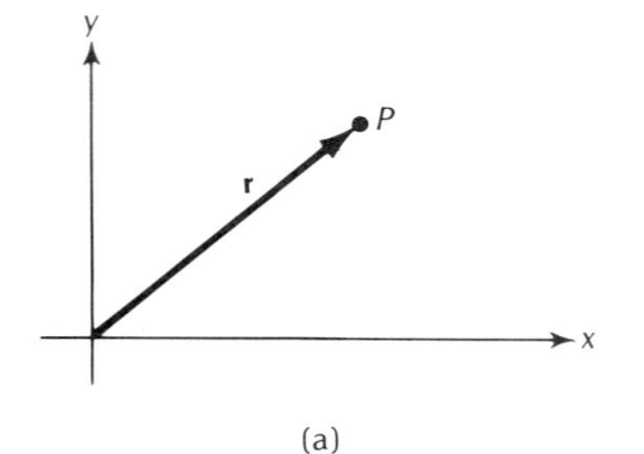


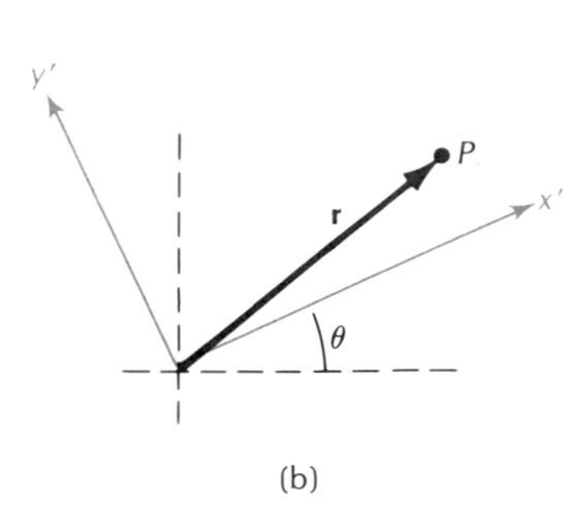


**Fig. 3.27** Two sets of rectangular coordinate axes: (a) $x$, $y$ and (b) $x'$, $y'$.

According to the definition we gave in Section 3.1, a vector is a quantity that behaves mathematically like a displacement vector. This definition was somewhat vague, since we did not spell out precisely what is meant by the mathematical "behavior" of a displacement vector. A branch of advanced mathematics called tensor analysis tells us that the crucial behavior is that under coordinate rotations. In this section we will briefly discuss coordinate rotations. For the sake of simplicity we will focus on rotations in two dimensions; of course, a similar discussion can be given in three dimensions. Figure 3.27 shows two sets of rectangular coordinate axes; the axes $x'$, $y'$ are rotated by an angle $\theta$ relative to the axes $x$ and $y$. Figure 3.27 also shows a point $P$ and its position vector. The position vector $\mathbf{r}$ in Figures 3.27a and 3.27b is exactly the same (the displacement from the origin is the same), but the values of the components of the position vector depend on the choice of coordinates, and these values are different. In the first coordinate system we have

$$\mathbf{r} = x\hat{\mathbf{x}} + y\hat{\mathbf{y}} \tag{38}$$

and in the second coordinate system

$$\mathbf{r} = x'\hat{\mathbf{x}}' + y'\hat{\mathbf{y}}' \tag{39}$$

The relationship between the quantities $x$, $y$ and $x'$, $y'$ is

*Coordinate rotation*

$$x' = x \cos\theta + y \sin\theta \tag{40}$$

$$y' = -x \sin\theta + y \cos\theta \tag{41}$$

* This section is optional. It provides some useful background for the Galilean transformations and the Lorentz transformations.

To derive these two equations, we use Figure 3.28. In terms of the angles $\theta$ and $\phi$ shown in this figure, we have

$$x = r\cos\phi \qquad y = r\sin\phi \tag{42}$$

$$x' = r\cos(\phi - \theta) \qquad y' = r\sin(\phi - \theta) \tag{43}$$

Using the trigonometric identity $\cos(\phi - \theta) = \cos\phi\cos\theta + \sin\phi\sin\theta$, we then obtain

$$\begin{aligned} x' &= r\cos\phi\cos\theta + r\sin\phi\sin\theta \\ &= x\cos\theta + y\sin\theta \end{aligned}$$

which is Eq. (40). The derivation of Eq. (41) is similar.

Taken together, Eqs. (40) and (41) are **transformation equations**; they show how the rotation from one coordinate system to another changes the components of the position vector, that is, the coordinates of a point.

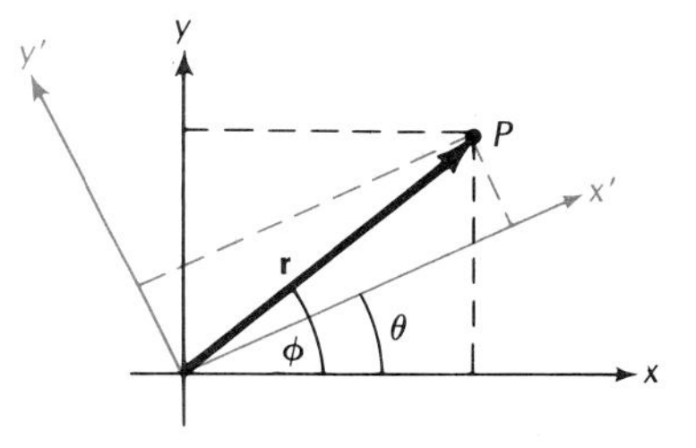


**Fig. 3.28** The angles $\theta$ and $\phi$.

We are now ready for a rigorous definition of a vector: *in two dimensions, a* **vector** *is an object with two components* $A_x$ *and* $A_y$ *that transform under a coordinate rotation in exactly the same way as the components of the position vector.* Thus, if the vector **A** has the form

*Rigorous vector definition*

$$\mathbf{A} = A_x\hat{\mathbf{x}} + A_y\hat{\mathbf{y}} \tag{44}$$

in the first coordinate system and the form

$$\mathbf{A} = A'_x\hat{\mathbf{x}}' + A'_y\hat{\mathbf{y}}' \tag{45}$$

in the second coordinate system, then the respective components are related as follows:

$$A'_x = A_x\cos\theta + A_y\sin\theta \tag{46}$$

$$A'_y = -A_x\sin\theta + A_y\cos\theta \tag{47}$$

According to the rigorous statement given above, vectors are defined by their transformation properties. In three dimensions the situation is similar, although the equations for three-dimensional rotations of coordinates are rather more messy; the three components of the vector must obey transformation equations identical to those for the three components of the position vector.

The magnitude of any vector is unchanged by a coordinate rotation. This is obvious from the fact that the rotation does not change the vector; it only changes its components. Mathematically, the preservation of the magnitude can be expressed as

$$\sqrt{A_x'^2 + A_y'^2} = \sqrt{A_x^2 + A_y^2} \tag{48}$$

Alternatively, this identity can be verified by explicit calculation: simply substitute the expressions (46) and (47) into the left side of Eq. (48); the sines and cosines will cancel.

The dot product of any two vectors is also unchanged by a coordinate rotation. This is a simple consequence of the fact that a rotation changes neither the magnitudes of the two vectors, nor the angle between them. The preservation of the dot product can be expressed as

$$A_x'B_x' + A_y'B_y' + A_z'B_z' = A_xB_x + A_yB_y + A_zB_z \tag{49}$$

Again, this identity can be verified by explicit calculation.

According to the rigorous definitions, a scalar is a numerical quantity that is unchanged by coordinate rotations. Thus, the length of any vector is a scalar and the dot product of any two vectors is a scalar. Any numerical constant such as $\pi$, or $\sqrt{2}$, or 666, or whatever, is of course also a scalar.

## SUMMARY

**Vector:** Quantity with magnitude and direction; it behaves like a displacement vector.

**Addition of vectors:** Use the parallelogram method or the tail-to-head method; alternatively, add the components.

**Unit vectors:** $\hat{\mathbf{x}}$, $\hat{\mathbf{y}}$, $\hat{\mathbf{z}}$

**Position vector:** $\mathbf{r} = x\hat{\mathbf{x}} + y\hat{\mathbf{y}} + z\hat{\mathbf{z}}$

**Components of a vector:**

$$\mathbf{A} = A_x\hat{\mathbf{x}} + A_y\hat{\mathbf{y}} + A_z\hat{\mathbf{z}}$$
$$A_x = A\cos\theta_x$$
$$A_y = A\cos\theta_y$$
$$A_z = A\cos\theta_z$$
$$A = \sqrt{A_x^2 + A_y^2 + A_z^2}$$

**Dot product:** $\mathbf{A}\cdot\mathbf{B} = AB\cos\phi$
$= A_xB_x + A_yB_y + A_zB_z$

**Cross product:** magnitude of $\mathbf{A}\times\mathbf{B}$ is $AB\sin\phi$
direction is given by right-hand rule

**Behavior of a vector under coordinate rotation:**

$$A_x' = A_x\cos\theta + A_y\sin\theta$$
$$A_y' = -A_x\sin\theta + A_y\cos\theta$$

## QUESTIONS

1. A large oil tanker proceeds from Kharg Island (Persian Gulf) to Rotterdam via the Cape of Good Hope. A small oil tanker proceeds from Kharg Island to Rotterdam via the Suez Canal. Are the displacement vectors of the two tankers equal? Are the distances covered equal?

2. An airplane flies from Boston to Houston and back to Boston. Is the displacement zero in the reference frame of the Earth? In the reference frame of the Sun?

3. Does a vector of zero magnitude have a direction? Does it matter?

4. Can the magnitude of a vector be negative? Zero?

5. Two vectors have nonzero magnitude. Under what conditions will their sum be zero? Their difference?

6. Is it possible for the sum of two vectors to have the same magnitude as the difference of the two vectors?

7. Three vectors have the same magnitude. Under what conditions will their sum be zero?

8. The magnitude of a vector is never smaller than the magnitude of any one component of the vector. Explain.

9. Two vectors have nonzero magnitude. Under what conditions will their dot product be zero? Their cross product?

10. Suppose $\mathbf{A} \cdot \mathbf{B} > 0$. What can you conclude about the angle between **A** and **B?**

11. If **A** and **B** are any arbitrary vectors, then $\mathbf{A} \cdot (\mathbf{B} \times \mathbf{A}) = 0$. Explain.

12. Why is there no vector division? (Hint: If **A** and **B** are given and if $A = \mathbf{B} \cdot \mathbf{C}$, then there exist several vectors **C** that satisfy this equation, and similarly for $\mathbf{A} = \mathbf{B} \times \mathbf{C}$.)

13. Assume that **A** is some nonzero vector. If $\mathbf{A} \cdot \mathbf{B} = \mathbf{A} \cdot \mathbf{C}$, can we conclude that $\mathbf{B} = \mathbf{C}$? If $\mathbf{A} \times \mathbf{B} = \mathbf{A} \times \mathbf{C}$, can we conclude that $\mathbf{B} = \mathbf{C}$? What if *both* $\mathbf{A} \cdot \mathbf{B} = \mathbf{A} \cdot \mathbf{C}$ and $\mathbf{A} \times \mathbf{B} = \mathbf{A} \times \mathbf{C}$?

## PROBLEMS

### Section 3.2

1. A ship moves from the Golden Gate bridge in San Francisco Bay to Alcatraz Island and from there to Point Blunt. The first displacement vector is 10.2 km due east, and the second is 5.9 km due north. What is the resultant displacement vector?

2. In midtown Manhattan, the street blocks have a uniform size of 80 m × 280 m, with the short side oriented at 29° east of north ("uptown") and the long side oriented at 29° north of west. Suppose you walk three blocks uptown and then two blocks to the left. What is the magnitude and direction of your displacement vector?

3. The displacement vector **A** has a length of 350 m in the direction 45° west of north; the displacement vector **B** has a length of 120 m in the direction 20° east of north. Find the magnitude and direction of the resultant of these vectors.

4. Figure 3.29 shows the successive displacements of an aircraft flying a search pattern. The initial position of the aircraft is $P$ and the final position is $P'$. What is the net displacement (magnitude and direction) between $P$ and $P'$? Find the answer both graphically (by carefully drawing a page-size diagram with protractor and ruler and measuring the resultant) and trigonometrically (by solving triangles).

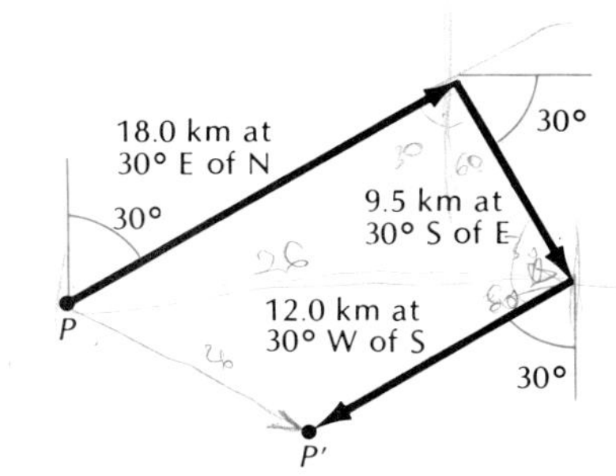


**Fig. 3.29** Successive displacement vectors of an aircraft.

5. A sailboat tacking against the wind moves as follows: 3.2 km at 45° east of north, 4.5 km at 50° west of north, 2.6 km at 45° east of north. What is the net displacement for the entire motion?

6. The resultant of two displacement vectors has a length of 5.0 m and a direction due north. One of the displacement vectors has a length of 2.2 m and a direction 35° east of north. What is the other displacement vector?

7. The vector **A** has a length of 6.2 units in a direction 30° south of east. The

vector **B** has length of 9.6 units in a direction due south. What is the sum **A** + **B** of these vectors? What is the difference **A** − **B**?

8. Three displacement vectors **A, B,** and **C** are, respectively, 4 cm at 30° west of north, 8 cm at 30° east of north, and 3 cm due north. Carefully draw these vectors on a sheet of paper. Find **A** + **B** + **C** graphically. Find **A** + **B** − **C** graphically.

9. During the maneuvers preceding the Battle of Jutland, the British battle cruiser *Lion* moved as follows (distances are in nautical miles): 1.2 nmi due north, 6.1 nmi at 38° east of south, 2.9 nmi at 59° east of south, 4.0 nmi at 89° east of north, and 6.5 nmi at 31° east of north.
   (a) Draw each of these displacement vectors and draw the net displacement vector.
   (b) Graphically or algebraically find the distance between the initial position and the final position.

*10. The Earth moves around the Sun in a circle of radius $1.50 \times 10^{11}$ m at (approximately) constant speed.
   (a) Taking today's position of the Earth as origin, draw a diagram showing the position vector 3 months, 6 months, 9 months, and 12 months later.
   (b) Draw the displacement vector between the 0-month and the 3-month position; the 3-month and the 6-month position, etc. Calculate the magnitude of the displacement vector for one of these 3-month intervals.

*11. Both Singapore and Quito are (nearly) on the Earth's equator; the longitude of Singapore is 104° East and that of Quito is 78° West. What is the magnitude of the displacement vector between these cities? What is the distance between them measured along the equator?

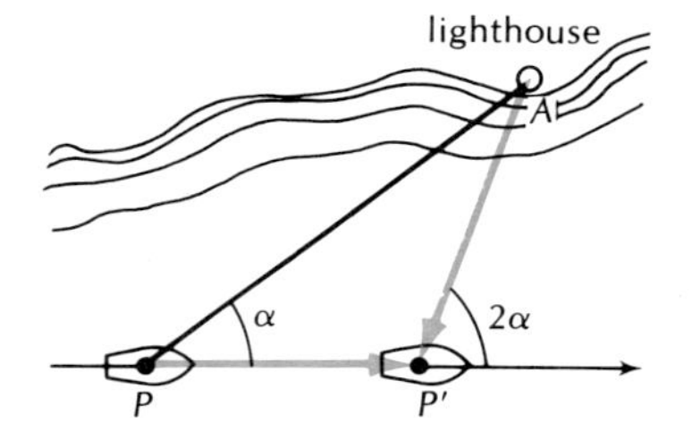


**Fig. 3.30**

*12. By a method known as "doubling the angle on the bow," the navigator of a ship can determine his position relative to a fixed point, such as a lighthouse. Figure 3.30 shows the (straight) track of a ship passing by a lighthouse. At the point *P,* the navigator measures the angle $\alpha$ between the line of sight to the lighthouse and the direction of motion of the ship. He then measures how far the ship advances through the water until the angle between the line of sight and the direction of motion is twice as large as it was initially. Prove that the magnitude of the displacement vector *PP′* equals the magnitude of the position vector *AP′* of the ship relative to the lighthouse.

*13. The radar operator of a stationary Coast Guard cutter observes that at $10^h30^m$ an unidentified ship is at a distance of 9.5 km on a bearing of 60° east of north and at $11^h10^m$ the unidentified ship is at a distance of 4.2 km on a bearing of 33° east of north. What is the displacement vector of the unidentified ship at $11^h10^m$? Assuming that the unidentified ship continues on the same course at the same speed, what will be its displacement vector at $11^h30^m$? What will be its distance and bearing from the cutter?

*14. A room measures 4 m in the $x$ direction, 5 m in the $y$ direction, and 3 m in the $z$ direction. A lizard crawls along the walls from one corner of the room to the diametrically opposite corner. If the starting point is the origin of coordinates, what is the displacement vector? What is the length of the displacement vector? If the lizard chooses the shortest path along the walls, what is the length of its path?

*15. Suppose that two ships proceeding at constant speeds are on converging straight tracks. Prove that the ships will collide if and only if the bearing of each remains constant as seen from the other. This constant-bearing rule is routinely used by mariners to check whether there is danger of collision. (Hint: A convenient method of proof is to draw the displacement vector from one ship to the other at several successive times.)

*16. The fastest crossing of the Atlantic by sail was achieved in 1916 by the four-masted ship *Lancing,* which sailed from New York (latitude 40°48′ north, longitude 73°58′ west) to Cape Wrath, Scotland (latitude 58°36′ north, longitude 5°1′ west), in $6\frac{3}{4}$ days. What was the magnitude of the displacement vector for this trip?

### Section 3.3

17. A vector length of 5.0 m is in the $x$–$y$ plane at an angle of 30° with the $x$ axis. What is the $x$ component of this vector? The $y$ component?

18. A displacement vector has a magnitude of 12.0 km in the direction 40° west of north. What is the north component of this vector? The west component?

19. Air traffic controllers usually describe the position of an aircraft relative to the airport by altitude, horizontal distance, and bearing. Suppose an aircraft is at altitude 500 m, distance 15 km, and bearing 35° east of north. What are the $x$, $y$, and $z$ components (in meters) of the position vector? The $x$ axis is east, the $y$ axis is north, and the $z$ axis is vertical.

20. The displacement vectors **A** and **B** are in the $x$–$y$ plane. Their components are $A_x = 3$ cm, $A_y = 2$ cm, $B_x = -1$ cm, $B_y = 3$ cm.
   (a) Draw a diagram showing these vectors.
   (b) Calculate the resultant of **A** and **B.** Draw the resultant in your diagram.

21. A vector in the $x$–$y$ plane has a magnitude of 8.0 units. The angle between the vector and the $x$ axis is 52°. What are the $x$, $y$, and $z$ components of this vector?

22. Given that a vector has a magnitude of 6.0 units and makes angles of 45° and 85° with the $x$ and $y$ axes, respectively, find the $x$ and $y$ components of this vector. Does the given information determine the $z$ component? What can you say about the $z$ component?

23. An air traffic controller notices that one aircraft approaching the airport is at altitude 2500 m, (horizontal) distance 120 km, and bearing 20° south of east. A second aircraft is at altitude 3500 m, distance 110 km, and bearing 25° south of east. What is the displacement vector from the first aircraft to the second? Express your answer in terms of altitude, (horizontal) distance, and bearing.

24. What is the magnitude of the vector $3\hat{\mathbf{x}} + 2\hat{\mathbf{y}} - \hat{\mathbf{z}}$?

25. Suppose that $\mathbf{A} = -5\hat{\mathbf{x}} - 3\hat{\mathbf{y}} + \hat{\mathbf{z}}$ and $\mathbf{B} = 2\hat{\mathbf{x}} + \hat{\mathbf{y}} - 3\hat{\mathbf{z}}$. Calculate the following.
   (a) $\mathbf{A} + \mathbf{B}$
   (b) $\mathbf{A} - \mathbf{B}$
   (c) $2\mathbf{A} - 3\mathbf{B}$

26. A vector has components $A_x = 5.0$, $A_y = -3.0$, $A_z = 1.0$. What is the magnitude of this vector? What is the angle between this vector and the $x$ axis? The $y$ axis? The $z$ axis?

*27. Find a vector that has the same direction as $3\hat{\mathbf{x}} - 6\hat{\mathbf{y}} + 2\hat{\mathbf{z}}$ but a magnitude of 2 units.

*28. Given that $\mathbf{A} = 6\hat{\mathbf{x}} - 2\hat{\mathbf{y}}$ and $\mathbf{B} = -4\hat{\mathbf{x}} - 3\hat{\mathbf{y}} + 8\hat{\mathbf{z}}$, find a vector **C** such that $3\mathbf{A} - 2\mathbf{C} = 4\mathbf{B}$.

### Section 3.4

29. Calculate the dot product of the vectors $5\hat{\mathbf{x}} - 2\hat{\mathbf{y}} + \hat{\mathbf{z}}$ and $2\hat{\mathbf{x}} - \hat{\mathbf{z}}$.

30. Calculate the dot product of the vectors **A** and **B** described in Example 1.

31. Find the magnitude of the vector $-2\hat{\mathbf{x}} + \hat{\mathbf{y}} + 2\hat{\mathbf{z}}$. Find the magnitude of the vector $3\hat{\mathbf{x}} - 6\hat{\mathbf{y}} + 2\hat{\mathbf{z}}$. Find the angle between these two vectors.

32. The displacement vector **A** has a length of 50 m and a direction of 30° east of north; the displacement vector **B** has a length of 35 m and a direction 70° west of north. What is the dot product of these vectors?

*33. Suppose that

$$\mathbf{A} = \hat{\mathbf{x}} \cos \omega t + \hat{\mathbf{y}} \sin \omega t$$

where $\omega$ is a constant. Find $d\mathbf{A}/dt$ (note that $\hat{\mathbf{x}}$ and $\hat{\mathbf{y}}$ behave as constants in differentiation). Show that $d\mathbf{A}/dt$ is perpendicular to **A**.

*34. Suppose that $\mathbf{A} = 3\hat{\mathbf{x}} + 4\hat{\mathbf{y}}$ and $\mathbf{B} = -\hat{\mathbf{x}} + 3\hat{\mathbf{y}} - 2\hat{\mathbf{z}}$. Find the component of **A** along the direction of **B**. Find the component of **B** along the direction of **A**.

*35. A vector **A** has components $A_x = 2$, $A_y = -1$, $A_z = -4$. Find a vector (give its components) that has the same direction as **A** but a magnitude of 1 unit.

**36. Find a unit vector that bisects the angle between the vectors $\hat{\mathbf{y}} + 2\hat{\mathbf{z}}$ and $3\hat{\mathbf{x}} - \hat{\mathbf{y}} + \hat{\mathbf{z}}$.

**37. Find a unit vector that points toward a position halfway between the two position vectors $4\hat{\mathbf{x}} + 2\hat{\mathbf{y}}$ and $-\hat{\mathbf{x}} + 3\hat{\mathbf{y}} + 2\hat{\mathbf{z}}$.

**38. Find the angle between the diagonal of a cube and one of its edges. (Hint: Suppose that the edges of the cube are parallel to the vectors $\hat{\mathbf{x}}$, $\hat{\mathbf{y}}$, and $\hat{\mathbf{z}}$. What vector is then parallel to the diagonal?)

39. The displacement vector **A** has a length of 50 m and a direction of 30° east of north; the displacement vector **B** has a length of 35 m and a direction 70° west of north. What is the magnitude and direction of the cross product $\mathbf{A} \times \mathbf{B}$? The cross product $\mathbf{B} \times \mathbf{A}$?

40. The displacement vector **A** has a length of 6.0 m in the direction 30° east of north; the displacement vector **B** has a length of 8.0 m in the direction 40° south of east. Find the magnitude and the direction of $\mathbf{A} \times \mathrm{B}$.

41. Calculate the cross product of the vectors **A** and **B** described in Example 1.

42. Given that $\mathbf{A} = 2\hat{\mathbf{x}} - 3\hat{\mathbf{y}} + 2\hat{\mathbf{z}}$ and $\mathbf{B} = -3\hat{\mathbf{x}} + 4\hat{\mathbf{z}}$, calculate the cross product $\mathbf{A} \times \mathbf{B}$.

43. The vectors **A, B,** and **C** have components $A_x = 3$, $A_y = -2$, $A_z = 2$, $B_x = 0$, $B_y = 0$, $B_z = 4$, $C_x = 2$, $C_y = -3$, $C_z = 0$. Calculate the following.

(a) $\mathbf{A} \cdot (\mathbf{B} + \mathbf{C})$
(b) $\mathbf{A} \times (\mathbf{B} + \mathbf{C})$
(c) $\mathbf{A} \cdot (\mathbf{B} \times \mathbf{C})$
(d) $\mathbf{A} \times (\mathbf{B} \times \mathbf{C})$

*44. Find a unit vector perpendicular to both $4\hat{\mathbf{x}} + 3\hat{\mathbf{y}}$ and $-\hat{\mathbf{x}} - 3\hat{\mathbf{y}} + 2\hat{\mathbf{z}}$.

**45. Show that the magnitude of $\mathbf{A} \cdot (\mathbf{B} \times \mathbf{C})$ is the volume of the parallelepiped determined by **A, B,** and **C.**

**46. Show that $\mathbf{A} \times (\mathbf{B} \times \mathbf{C}) = \mathbf{B}(\mathbf{A} \cdot \mathbf{C}) - \mathbf{C}(\mathbf{A} \cdot \mathbf{B})$. (Hint: Choose the orientation of your coordinate axes in such a way that **B** is along the $x$ axis and that **C** is in the $x$–$y$ plane.)

### Section 3.5

47. With respect to a given coordinate system, a vector has components $A_x = 5$, $A_y = -3$, $A_z = 0$.

(a) What are the components $A_x'$ and $A_y'$ of this vector in a new coordinate system whose $x'$ and $y'$ axes make angles of 30° with the old $x$ and $y$ axes?
(b) Calculate the length of the vector from its $A_x$ and $A_y$ components. Calculate the length of the vector from its $A_x'$ and $A_y'$ components.

48. In the vicinity of New York City, the direction of magnetic north is 11°55′ west of true north (that is, a magnetic compass needle points 11°55′ west of north). Suppose that an aircraft flies 5.0 km on a bearing of 56° east of magnetic north.

(a) What are the north and east components of this displacement in a coordinate system based on the direction of magnetic north?
(b) What are the north and east components of the displacement in a coordinate system based on the direction of true north?

49. Verify Eq. (48) by explicit substitution of Eqs. (46) and (47).

50. Verify Eq. (49) by explicit substitution of Eqs. (46) and (47).

*51. In one rectangular coordinate system, a vector has components $A_x = 5.00$, $A_y = 3.00$, $A_z = 0$. In another coordinate system, it has components $A'_x = 0.10$, $A'_y = -5.83$, $A'_z = 0$. Show that the two coordinate systems are related by a rotation. What is the angle of rotation?

*52. A vector has components $A_x = 6$, $A_y = -3$, $A_z = 0$ in a given rectangular coordinate system. Find a new coordinate system such that the only nonzero component of the vector is $A'_x$.

*53. Show that if the axes $x'$, $y'$ are rotated by an angle $\theta$ relative to the axes $x$, $y$, then the corresponding unit vectors are related as follows:

$$\hat{\mathbf{x}}' = \hat{\mathbf{x}} \cos\theta + \hat{\mathbf{y}} \sin\theta$$

$$\hat{\mathbf{y}}' = \hat{\mathbf{y}} \cos\theta - \hat{\mathbf{x}} \sin\theta$$

**54. Suppose that the coordinates $x'$, $y'$, $z'$ are related to the coordinates $x$, $y$, $z$ by a rotation through an angle $\theta$ about the $z$ axis [as in Eqs. (40) and (41)]. Suppose that the coordinates $x''$, $y''$, $z''$ are related to $x'$, $y'$, $z'$ by a rotation through an angle $\phi$ about the $x'$ axis.
(a) What is the equation that relates the $x''$, $y''$, $z''$ coordinates to the $x'$, $y'$, $z'$ coordinates?
(b) What is the equation that relates the $x''$, $y''$, $z''$ coordinates to the $x$, $y$, $z$, coordinates?

# THE ARCHITECTURE OF CRYSTALS*

Atoms and molecules are the building blocks within all the pieces of matter in our immediate environment — they are the building blocks within sticks, stones, water, air, our own bodies, and the entire Earth. The number of atoms in ordinary pieces of matter is extremely large. For example, the number of copper atoms in a penny coin is about $3 \times 10^{22}$. This large number of atoms creates the illusion that matter is continuously distributed. Yet the physical and chemical properties of a substance, such as copper, are entirely determined by the properties of its atoms. The density of copper, its elasticity and strength, its color, its melting point and boiling point, its thermal and electrical characteristics, etc., all hinge on the properties of the atomic building blocks and on the manner in which these building blocks are assembled into a large-scale structure. Unfortunately, because the number of atoms in even a small piece of matter is so extremely large, the calculation of the macroscopic properties of a piece of matter from the microscopic properties of its constituent atoms is a formidable mathematical task. Physicists have devoted much effort to the study of metals and minerals, solids with a regular structure, which makes them more amenable to mathematical analysis. The building blocks in metals and minerals are arranged in an orderly, repetitive pattern, reminiscent of the orderly pattern of soldiers standing on parade. Solids with such an orderly, repetitive arrangement of atoms or molecules are called **crystals.** The study of all the conceivable geometrical arrangements of the atomic or molecular building blocks in a crystal is called **crystallography;** alternatively it might be called the architecture of crystals.

Regularity in the arrangement of the building blocks implies **symmetry.** In its precise mathematical meaning, a symmetry of a body is any geometric operation that leaves the body unaltered. For instance, the human body has bilateral symmetry, or reflection symmetry: we can exchange the right side of the body point by point with the left side, leaving the body unaltered (Figure I.1). In this chapter we will become acquainted with a variety of other geometric symmetries

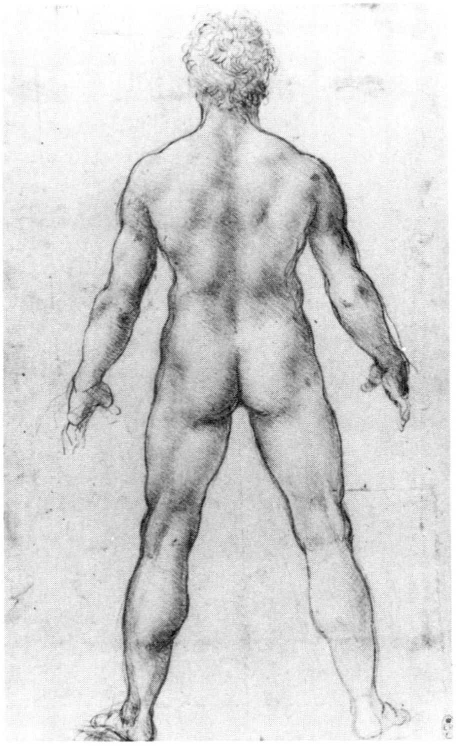

**Fig. I.1** *Male nude.* Red chalk drawing by Leonardo da Vinci. (Courtesy the Royal Library, Windsor Castle.)

and we will uncover some of the relationships between these symmetries and the physical properties of crystals. But the significance of the concept of symmetry in physics runs much deeper than is apparent from the concrete geometric examples in this chapter. For instance, the repetitive pattern of behavior of the atoms listed in the Periodic Table (see Table P.1) can be traced to an underlying symmetry of the equation governing the motion of the electrons in the atoms. And the regular pattern of behavior of the elementary particles in "families" of particles (see Chapter 46) can be traced to a symmetry of the parameters describing the internal structure of these particles. Such abstract mathematical symmetries play a key role in physics. In fact, much of modern theoretical physics can be described as a search for symmetry.

* This chapter is optional.

## I.1 THE STRUCTURE OF CRYSTALS

In all solids the building blocks, whether molecules or atoms, are permanently locked into their positions. What distinguishes a crystalline solid, such as a metal, from a noncrystalline solid, such as wood or glass, is the regularity of the arrangement of the building blocks — they form an orderly, repetitive lattice, somewhat like a three-dimensional, rectangular coordinate grid.[1] Thus, metals and other crystalline solids can be regarded as giant supermolecules in which the atoms have a precisely organized arrangement, just as they have in an ordinary molecule. A crystalline solid can be regarded as a three-dimensional polymer rather than a stringlike, one-dimensional polymer.

The regularity of the arrangement of the building blocks of crystals can be seen very clearly in Figures I.2–I.5. The first of these shows a "crystal" made of tobacco necrosis virus particles. Each individual virus particle is a giant molecule, nearly spherical in shape with a diameter of about 250 Å; because of the relatively large size of the molecule, the structure of this crystal can be seen with an **electron microscope.** Figure I.3 shows the regular arrangement of the molecules in a barium titanate crystal; this picture was taken with a new, very powerful electron microscope. More complicated microscopes are needed to see the structure of other crystals. Figure I.4 shows the arrangement of iron disulfide molecules in a marcasite crystal. This picture was produced with a **two-wavelength microscope,** which employs both X rays and visible light. Figure I.5 shows the atoms on the surface of the spherical tip of a needle of platinum. The orderly arrangement of the layers of platinum atoms manifests itself in the beautiful symmetry of this picture; the large-scale pattern of circles and lines is due to the protruding ridges of the atomic layers, ridges that correspond to the intersection of the curved surface of the needle with different atomic layers. Figure I.5 was prepared with an **ion microscope,** which "illuminates" the needle with ions rather than with ordinary light.

**Fig. I.2** Electron-microscope photograph of a crystal of tobacco necrosis virus particles. The magnification is about 50,000×.

[1] If a small sample of molten metal is very suddenly cooled (quenched), its atoms will sometimes freeze in a disordered (amorphous) configuration, rather like a glass. In the following we will ignore such exceptional, freakish states of metals.

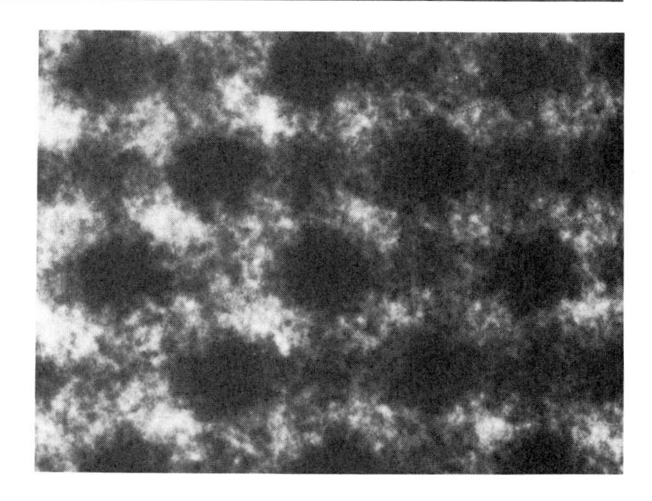

**Fig. I.3** This electron-microscope photograph shows the arrangement of $BaTiO_3$ molecules in a crystal of barium titanate. The dark spheres are titanate ions and the lighter spheres are barium ions; the distance between them is about 2 Å. (Courtesy R. Gronsky, Lawrence Berkeley Laboratories.)

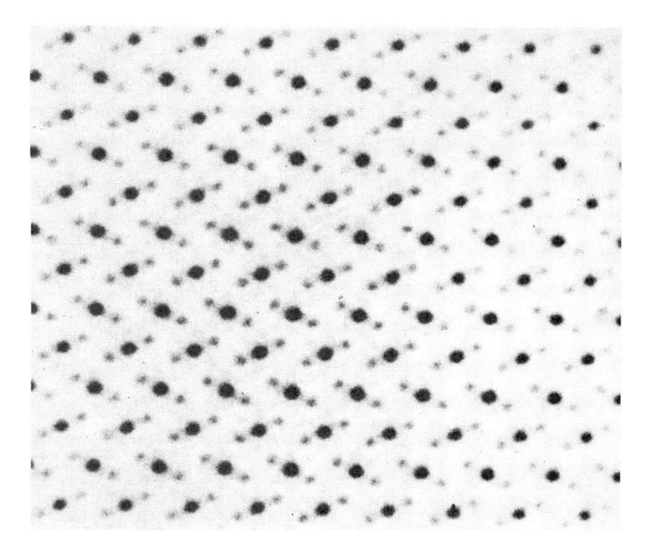

**Fig. I.4** Arrangement of $FeS_2$ molecules in a crystal of marcasite. The large dots are iron atoms and the smaller dots are sulfur atoms; the distance between an iron atom and the nearest sulfur atom is about 1 Å.

**Fig. I.5** Image of the tip of a platinum needle viewed with an ion microscope. The radius of the tip is approximately 1000 Å and the magnification is approximately $10^6$×. The positions of the dots show the positions of individual atoms, but the size of the dots is larger than the size of the atoms. (Courtesy T. T. Tsong, Pennsylvania State University.)

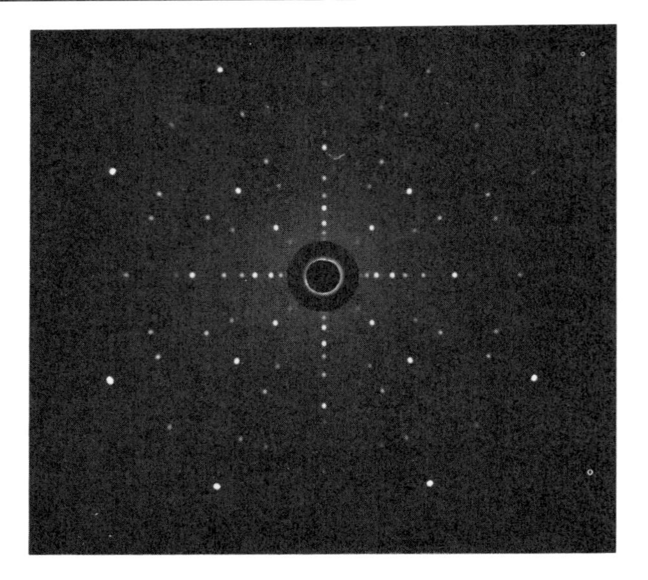

**Fig. I.6** Pattern of Laue spots produced by a silicon crystal. (Courtesy R. P. Goehner, General Electric.)

**Fig. I.7** Pattern of Laue spots produced by a $SrTiO_3$ crystal. (Courtesy J. M. Karasinski, IBM Watson Research Center.)

Electron and ion microscopes provide impressive and beautiful pictures of the arrangement of the atoms on the surfaces of crystals. However, the layer of atoms on the surface of a crystal is often different from and less regular than a typical internal layer. The preferred method for the investigation of the internal regularity of crystals is **X-ray diffraction**; this method does not produce spectacular visual displays such as in the above figures, but it does give us the most detailed information about the structure of crystals. The method is based on the following: if a crystal is illuminated with a beam of X rays, these rays do not all pass through in a straight line; rather, some are deflected and emerge at different angles. This deflection of X rays is called diffraction. The deflection of balls in a pinball machine is a crude analog of the diffraction of X rays by a crystal. The atoms are represented by the pins, and the X rays are the balls that make their way along the "alleys" between the pins. The incident balls are deflected by the pins one way or another and they finally emerge in certain directions; similarly, the X rays are deflected by the atoms and they finally emerge from the crystals in a few select directions. A photographic plate is used to register the emerging X rays. Figures I.6 and I.7 show some of the patterns of impact points of X rays on a photographic plate placed behind a crystal exposed to an X-ray beam. The large dark spot in the center of these pictures is a hole cut in the plate for the passage of the undeflected beam. The small bright spots, or **Laue spots,** were made by deflected rays. The regularity of the pattern of Laue spots corresponds to the regularity of the arrangement of atoms within the crystal. From a careful study of the pattern of spots, crystallographers can deduce the arrangement of the atoms within the crystal.

Occasionally, the underlying regularity of the microscopic structure manifests itself in the gross macroscopic shape of the crystal. For example, the rectangular shape of grains of ordinary salt (NaCl) is a direct manifestation of the rectangular (or, more precisely, cubic) arrangement of the atoms in this crystal. Figure I.8

shows grains of salt magnified a hundred times; if we crush these small rectangular crystallites, the resulting fragments will again be rectangular. Figure I.9 shows a single, large crystal of topaz; the regularity of the angles between the faces of this crystal is a manifestation of the regularity of the arrangement of the atoms in this crystal. But perhaps the most beautiful example is found in the shape of ice crystals in snowflakes (Figure I.10); the hexagonal pattern in these crystals is a direct manifestation of the underlying hexagonal arrangement of the water molecules. Note that although the center of the ice crystal is a very precise hexagon, the six "trees," or dendrites, that grow outward from the corners of the hexagon are not identical. These ice trees are about as similar as one might expect of, say, neighboring pine trees growing in some particular spot of the woods. The ice trees on a given crystal tend to be similar for the same reason that the pine trees are similar — they grew up together in the same environment.

**Fig. I.8** Grains of salt; magnification 100×.

**Fig. I.9** Crystal of topaz.

**Fig. I.10** Snowflakes. All these snowflakes have a hexagonal pattern.

Metals, such as iron, tin, or brass, usually do not display any obvious geometric features that reveal their underlying regularity. These metals are **polycrystalline;** that is, they are not made of a single crystal but, rather, of many small crystallites, or microcrystals, jumbled together. Such crystallites form the angular, flaky pattern that can be seen on the surface of the zinc plating commonly used on "galvanized" sheet metal (Figure I.11). In other metals the edges of the crystallites can be made visible by etching the surface of the metal with acid. The acid penetrates more deeply along the edges of the crystallites, where the atoms do not fit together very well; hence the acid outlines these edges in relief (Figure I.12).

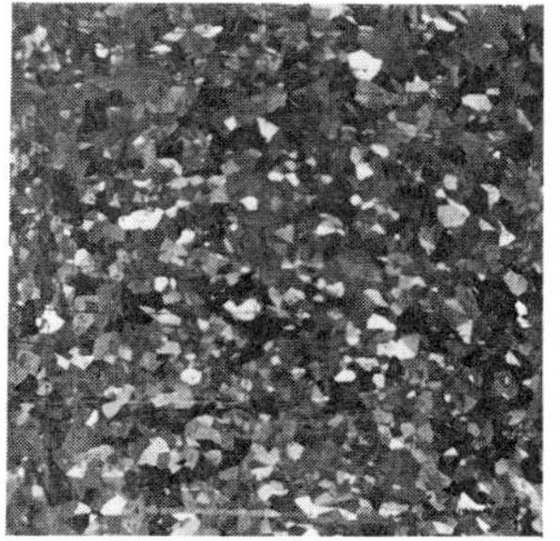

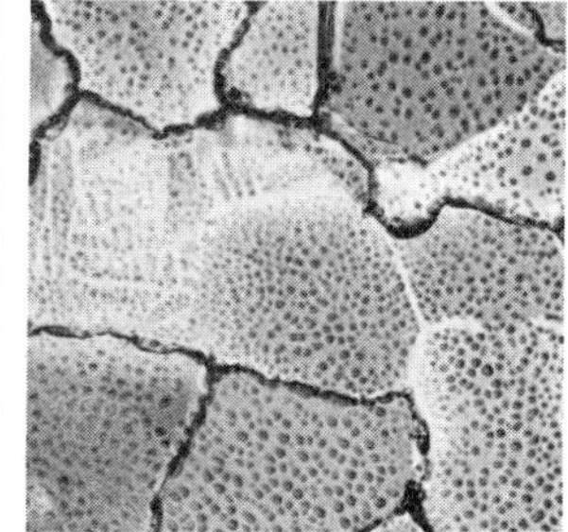

**Fig. I.11** (left) Crystals on the surface of galvanized sheet metal. Approximately natural size.

**Fig. I.12** (right) Boundaries of crystals in a lead–tin alloy, made visible by etching the surface with acid. Magnification 250×. (Courtesy H. B. Huntington, Rensselaer Polytechnic Institute.)

## I.2 SYMMETRY

The essential features in the structure of a crystal are *regularity* and *repetition*. The fundamental building blocks — whether atoms or molecules — are stacked together in a three-dimensional array of regular geometric shapes that repeat and repeat in all directions almost forever. Of course, any real crystal has a beginning and an end; however, the number of repeated building blocks is many thousands or many millions in any one direction and, for the purposes of mathematical description, it is useful to pretend that the crystal is infinite.

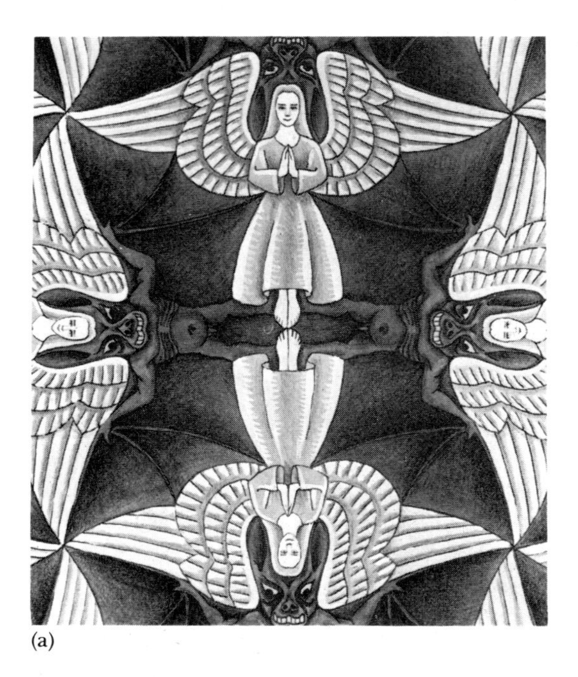

(a)

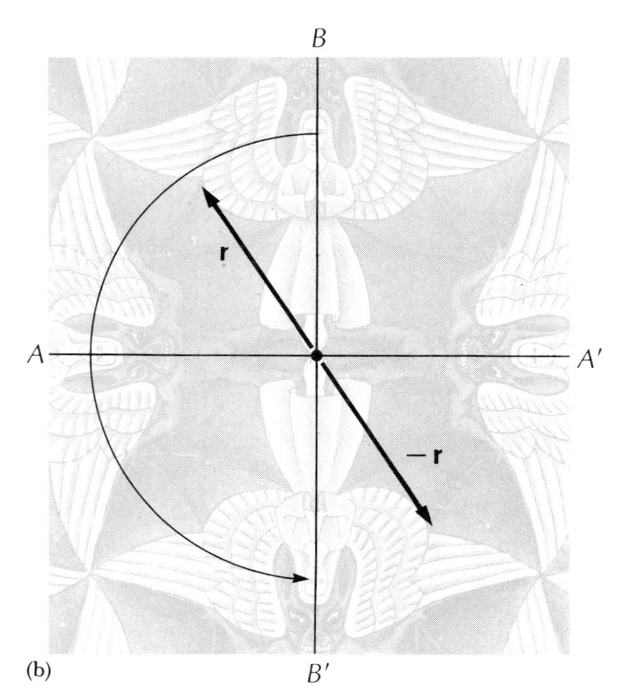


(b)

**Fig. I.13** (a) A drawing by M. C. Escher. (b) The symmetries of the drawing are reflection about the horizontal midline *AA′*, reflection about the vertical midline *BB′*, rotation by 180° about the center, and inversion about the center.

To study the geometric properties of the arrangement of building blocks in a crystal, we need to become familiar with different kinds of symmetries. It will be best to begin with some simple examples of symmetries in two dimensions.

Figure I.13a shows a drawing by the artist M. C. Escher. The pattern of this drawing relies on regularity and repetition; for instance, the design of each of the four angels is essentially the same — the pattern has symmetries. To arrive at a precise description of the symmetries, let us ask what operations we can perform with this pattern that leave it in a condition indistinguishable from its initial condition. First of all, we can imagine exchanging each point above the horizontal midline with a corresponding point below the midline (Figure I.13b). In this operation the midline acts like a mirror — the operation is a **reflection** about the midline. The reflected pattern is indistinguishable from the original pattern. Hence reflection about the horizontal midline is symmetry of the pattern. Obviously, reflection about the vertical midline is also a symmetry.

The pattern has some other symmetries. For instance, we can perform a **rotation** of this pattern by 180° about an axis through its center perpendicular to its face (Figure I.13b). The rotated pattern is indistinguishable from the original pattern. The axis of this rotation is a symmetry axis; it is called a twofold symmetry axis because it involves two symmetric orientations (0° and 180°).

Finally, we can imagine exchanging each point with a corresponding point on the opposite side of the center (Figure I.13b). This operation can be described more rigorously as a replacement of the point of position vector $\mathbf{r}$ (relative to the center) by the point of position vector $-\mathbf{r}$. This operation is called an **inversion** through the center. Although it is another symmetry of the pattern, it is not an independent symmetry: under inversion each point suffers exactly the same net displacement as under a rotation of 180°. However, this identity between inversion and rotation only holds true for a two-dimensional object. For a three-dimensional object, inversion through a point and rotation are independent operations.

Figure I.14a shows another drawing by Escher. This pattern relies on repetition in an obvious way. Let us pretend that the pattern repeats forever in both directions. Then the operation of translation by one step in

(a)

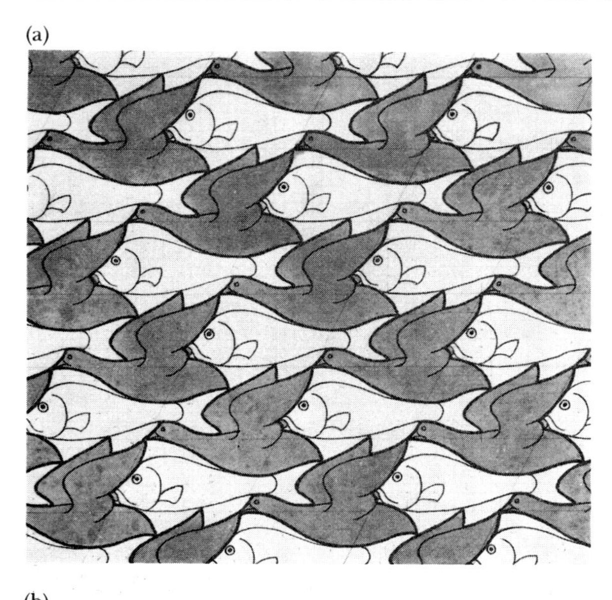

(b)

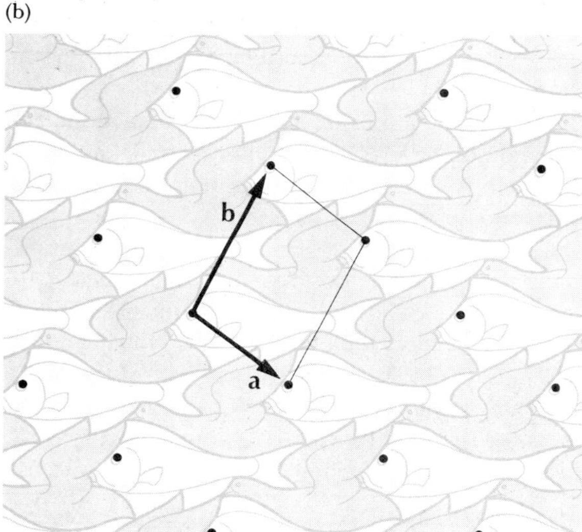


**Fig. I.14** (a) Another drawing by M. C. Escher. (b) The primitive cell is a parallelogram. The vectors **a** and **b** are primitive vectors, from a lattice point to the two nearest, distinct lattice points.

an oblique downward direction or by one step in an oblique upward direction leaves the pattern unchanged. Hence **translation** is a symmetry of this pattern.

The pattern of Figure I.14a can be regarded as a two-dimensional "crystal" and it is therefore worthwhile to investigate its geometric properties in some detail. The entire pattern consists of identical basic building blocks arranged in a regular lattice. Figure I.14b shows one such basic building block. In the language of crystallography, the basic building block is called a **primitive cell;** we will adopt this terminology. To describe the positions of the primitive cells, we take one corner of each cell as a fiducial point, called

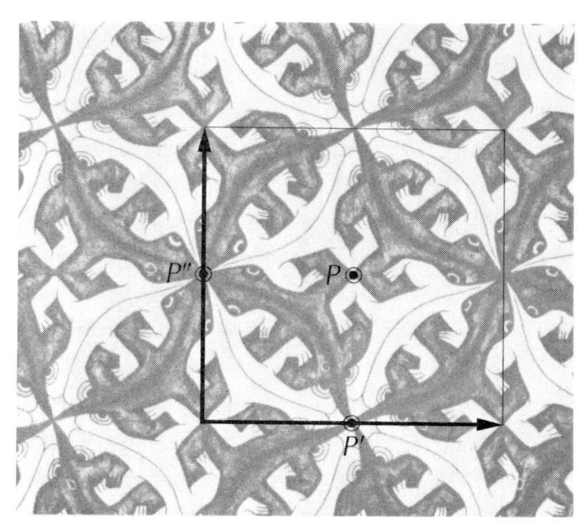


**Fig. I.15** Another drawing by M. C. Escher. The primitive cell is a square.

a lattice point. The positions of the cells within the complete pattern can then be indicated by the lattice points shown in Figure I.14b. The displacement vectors from one lattice point to the two nearest, distinct lattice points are called the **primitive vectors** of the lattice. We will designate these vectors by **a** and **b** (Figure I.14b). Any displacement vector from one lattice point to another lattice point is then a (positive or negative) multiple of one of the primitive vectors or a sum of such multiples; that is, any such displacement vector is of the form

$$\Delta \mathbf{r} = m\mathbf{a} + n\mathbf{b}$$

where $m$ and $n$ are positive or negative integers or zero. The primitive vectors completely characterize the translation symmetry of the lattice.

A two-dimensional pattern can simultaneously have translation, rotation, and reflection symmetries. Figure I.15 shows yet another drawing by Escher, a pattern with several simultaneous symmetries. The basic building block, or primitive cell, is a square. The translation symmetry is characterized by primitive vectors of equal length. Obviously, rotation is an extra symmetry of the pattern. A rotation by 180° about a perpendicular axis through the point $P$ is a symmetry; this axis is a twofold symmetry axis since it involves two symmetric orientations of the pattern (0° and 180°). A rotation by 90°, or by a multiple of 90°, about a perpendicular axis through either of the points $P'$ or $P''$ is also a symmetry; each of these axes is a fourfold symmetry axis since they both involve four symmetric orientations (0°, 90°, 180°, and 270°).

Besides translation, rotation, and reflection symmetries, a two-dimensional pattern can have yet another

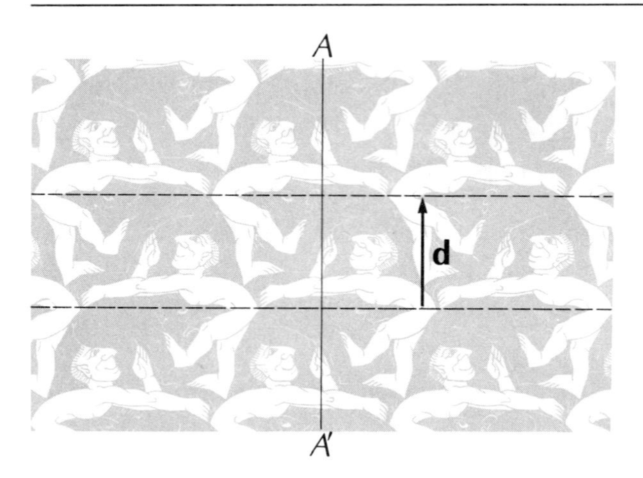


**Fig. I.16** A woodcut by M. C. Escher. The line *AA'* is a glide line.

symmetry that involves a joint reflection and translation, neither of which is a symmetry by itself. Figure I.16, a woodcut by Escher, illustrates this other symmetry. If we reflect the pattern about the vertical line *AA'* and then shift the pattern through the displacement indicated by the vector **d**, the pattern is left unchanged. The line *AA'* is called a **glide line.**

The most general symmetry of a two-dimensional "crystal" is some combination of translation symmetry, rotation symmetry, reflection symmetry, and glide-line symmetry. By taking all these symmetries into account and by investigating all the requirements for their compatibility, one can show that a two-dimensional "crystal" structure must have one or another of 17 distinct kinds of symmetry combinations.

In three dimensions symmetry combinations are more complicated. The basic building blocks are three dimensional and there are three independent primitive vectors. Furthermore, whereas in the two-dimensional case the inversion through a point could be ignored (it coincides with at 180° rotation), in the three-dimensional case this symmetry must be taken into consideration separately. And there is yet another possible symmetry that involves a joint rotation and translation, neither of which is a symmetry by itself (the corresponding rotation and translation axis is called a **screw axis**). Thus, the most general symmetry of a three-dimensional crystal is some combination of translation symmetry, rotation symmetry, reflection symmetry, inversion symmetry, glide-plane symmetry, and screw-axis symmetry. It turns out that a three-dimensional crystal must have one or another of 230 distinct kinds of symmetry combinations.[2]

[2] These kinds of symmetry combinations are called **space groups.** Symmetry combinations involving all the above operations *except* translation are called **point groups.**

## I.3 CLOSE-PACKED STRUCTURES

The crystals formed by the atoms or molecules of a given substance may involve cubic, rectangular, hexagonal, rhombohedral, or other lattices. Just what configuration the building blocks will adopt depends on the details of the interatomic forces. These forces are essentially electric attractions and repulsions among the electrons and nucleus of one atom and those of another. But differences in the arrangement of electrons within the atoms of different elements lead to differences in the corresponding interatomic forces. Crudely, we might say that the atoms of different chemical elements have different shapes and hence fit together in different ways when packed together in a crystal.

Some atoms have a perfect spherical shape — the force that such an atom exerts on another atom is completely independent of the relative orientation of the atoms. Helium, neon, argon, and krypton are atoms of this kind. At normal temperatures they form gases, but at very low temperatures they will aggregate into crystals. The atoms of, say, helium attract one another and so they will settle into a configuration in which each is as close to its neighbors as possible, a configuration called **close packed.**

The geometric problem of close packing such atoms is similar to the problem of close packing oranges into a crate or stacking cannonballs in a mound. The recipe for close packing is simple: Begin by assembling a first layer (Figure I.17); the atoms of each row of this layer fit into the notches of the adjacent rows. Note that inspection of Figure I.17 reveals a hexagonal pattern — each atom is surrounded by six neighboring atoms.

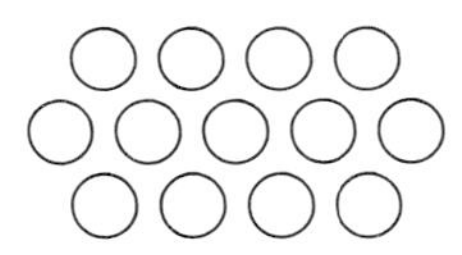

**Fig. I.17** A layer of close-packed spherical atoms.

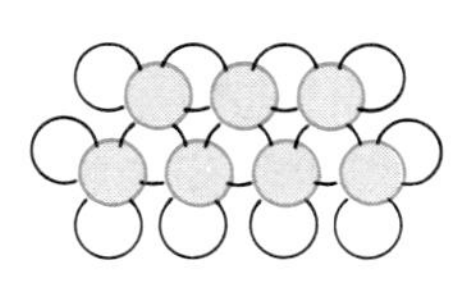

**Fig. I.18** Second layer of close-packed atoms (colored).

Next, construct a second layer on top of the first; to achieve close packing, it is obviously necessary to place each atom of this second layer in a hollow of the first (Figure I.18). Likewise, place another layer on top of the second. There are two alternative ways of placing this third layer. Either the atoms of the third layer are placed directly above those of the first layer (the third layer then duplicates the first layer; Figure I.19a), or else the atoms of the third layer are placed with a lateral offset relative to those of the first layer (the

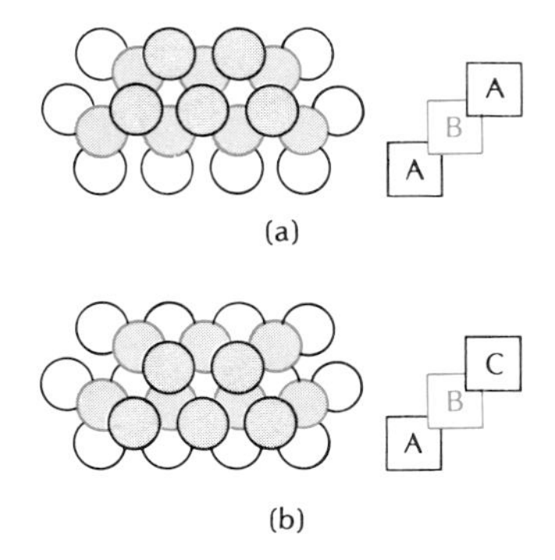


**Fig. I.19** Third layer of close-packed atoms (gray): (a) configuration *ABA* and (b) configuration *ABC*.

third layer is then shifted relative to the first; Figure I.19b), but when a fourth layer is next added, it duplicates either the first or the second layer. Thus there are at most three distinct layers, which we can designate by the letters *A, B,* and *C.* Any general close-packed configuration can then be described by a sequence of these letters, such as *ABCBA . . . ,* in which adjacent letters are different.

In crystals, close packing is usually accomplished by regular repetition of two distinct layers *AB,* or else by regular repetition of three distinct layers *ABC.* The configuration is then *ABAB . . .* or else *ABCABC . . . ;* in the former the third layer is a duplicate of the first, and in the latter the fourth layer is a duplicate of the first. Both of these configurations are equally close packed. And both have a high degree of symmetry.

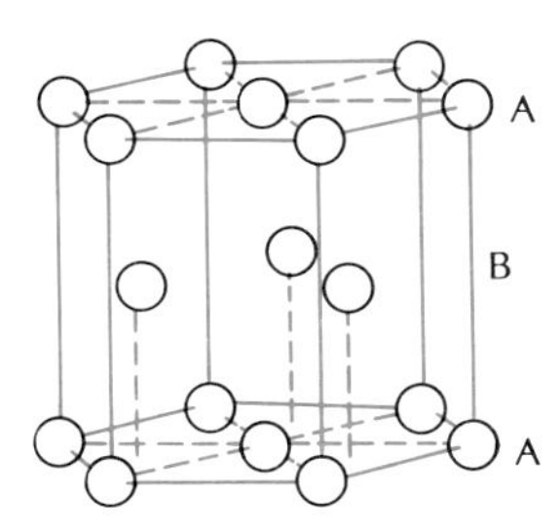


**Fig. I.20** Close-packed configuration with layers *ABA.*

Figure I.20 shows atoms in the *ABAB . . .* configuration; this figure gives an "exploded" view in which the interatomic distances have been exaggerated for the sake of clarity. Obviously, there is a threefold rotation axis passing vertically through the center of Figure I.20. There are also three twofold rotation axes passing horizontally through the center. Besides, there are several vertical and horizontal planes of reflection.

Figure I.21a shows atoms in the *ABCABC . . .* configuration. To discover the symmetries of this configuration, it is best to begin with the observation that the atoms are actually arranged in a cubic structure (Figure I.21b). This kind of cubic structure is called a **face-centered cube** because there is an atom at the center of each face. The *ABCABC . . .* configuration therefore has all the symmetries of a cube. There is a fourfold rotation axis passing through the center of each face, a threefold rotation axis through each vertex, a twofold rotation axis through the middle of each edge, and there are several planes of reflection (to visualize these symmetries, it is best to manipulate a cube).

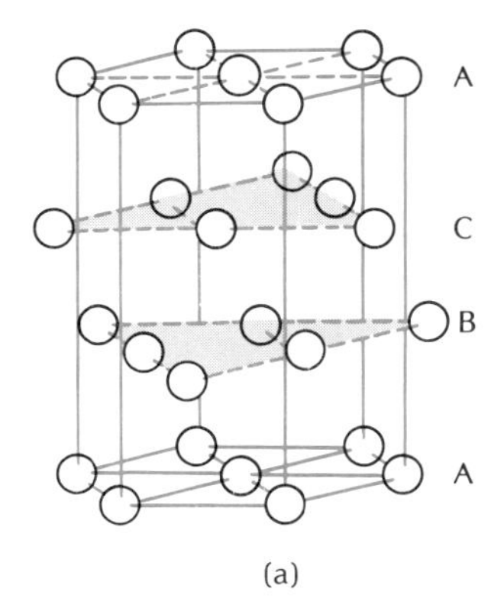


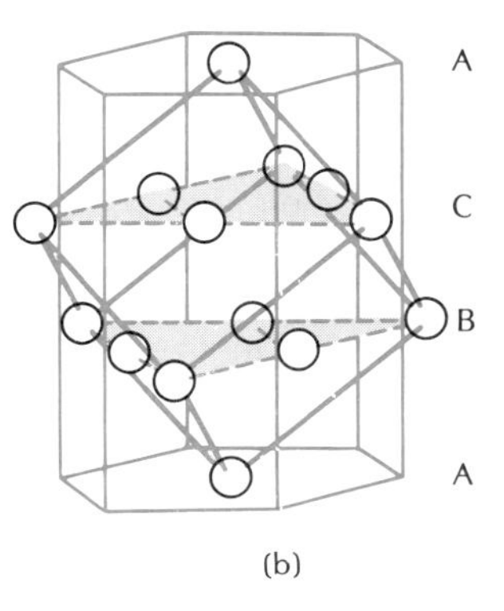


**Fig. I.21** (a) Close-packed configuration with layers *ABCA.* (b) The atoms of this close-packed configuration are arranged in a face-centered cube.

Which of the two close-packed configurations will be adopted by the atoms of a particular element depends on what forces an atom in one layer exerts on an atom two or more layers away — these forces between remote atoms are weak but they decide the issue. Helium crystallizes in the *ABAB . . .* configuration; neon, argon, and krypton crystallize in the *ABCABC . . .* configuration.

Many metals, such as magnesium, zinc, aluminum, and copper, also adopt one or the other of these simple close-packed structures. This is perhaps somewhat unexpected. The atoms of metals do not have the spherical symmetry of the atoms of the noble gases; so why do they adopt close packing, which, as we saw, is characteristic of spheres? The answer to this puzzle is found in the electronic structure of these atoms. The atoms of metal are nonspherical because the outermost one or two electrons of each atom stick out in asymmetric ways; but except for these outermost electrons, the atoms are spherical — the atomic cores are spherical. When such atoms aggregate in a crystal, they release the outermost electrons and the residual spherical cores then settle very nicely in a close-packed arrangement.

What happens to the released electrons? They remain free to wander all over the crystal, forming a gas of electrons. Thus, a metal consists of two interpenetrating components: a lattice made of atomic cores and a gas of electrons. Note that the atomic cores are ions, that is, atoms with missing electrons. This reminds one of a plasma where ions and electrons coexist in an intimate mixture. The difference between a metal and a plasma lies in the arrangement of the ions: in a metal the ions form a lattice, whereas in a plasma the ions form a gas.

The presence of a gas of free electrons inside a metal accounts for many of the characteristic properties of metals. For instance, metals have a remarkable ability to conduct heat and electricity, because the electron gas carries heat and electricity very efficiently from one end of the metal to the other.

Besides the close-packed crystal lattice described above, there are many other kinds of crystal lattices. For example, Figure I.22 shows the crystal structure of sodium; this structure is called a **body-centered cube** because there is an atom at the center of each cube.

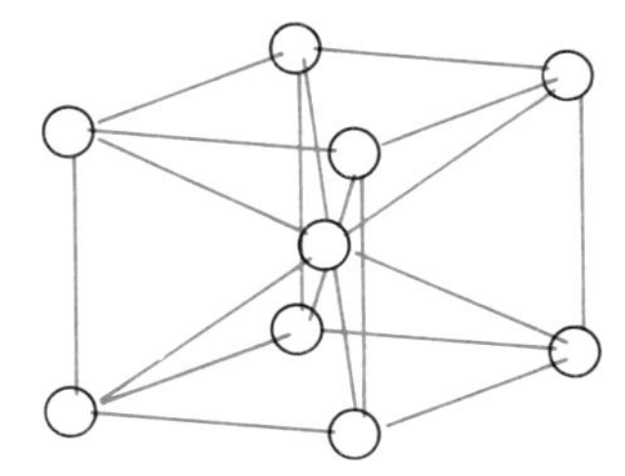

**Fig. I.22** Structure of sodium crystal.

Figures I.23–I.25 show the crystal structures of diamond, graphite, and sodium chloride (ordinary salt). These figures give a glimpse of the great variety of regular symmetric arrangements of atoms found in crystals.

Many of the physical properties of solids can be readily explained in terms of their crystal structure. Diamond is hard and graphite is soft, yet both consist of pure carbon (diamond burns like coal if ignited at high temperature). The qualitative explanation of this

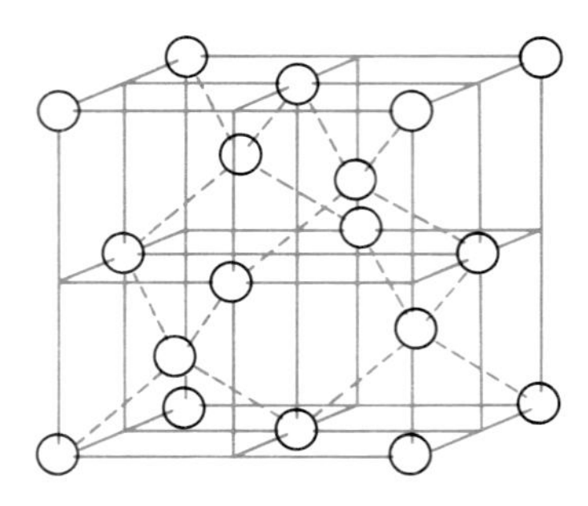

**Fig. I.23** Structure of diamond crystal.

difference is as follows. In diamond, each carbon atom is surrounded by four other carbon atoms; each atom is at the center of a tetrahedron with four surrounding atoms at the vertices of this tetrahedron (Figure I.23). The central carbon atom forms bonds with these surrounding atoms (it is known from chemistry that carbon atoms tend to form four tetrahedral bonds). These strong interlocking bonds give the diamond crystal its extreme hardness and high melting point. In graphite the carbon atoms are arranged in parallel sheets; within each sheet the atoms form adjoining hexagonal rings (Figure I.24). The bonds within each sheet are

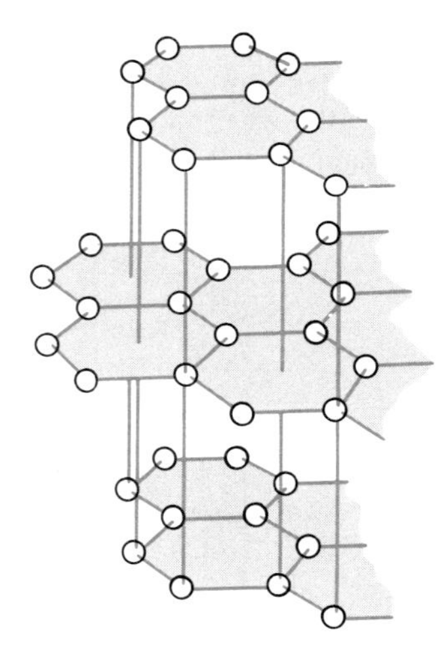

**Fig. I.24** Structure of graphite crystal.

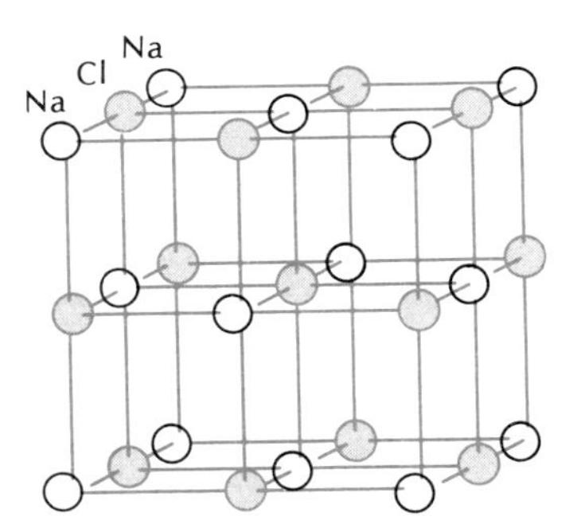


**Fig. I.25** Structure of NaCl crystal. The white balls are sodium atoms and the colored balls are chlorine atoms.

fairly strong, but the bonds between one sheet and the next are fairly weak. Hence the sheets easily slide over one another. This sliding motion gives graphite its softness and its lubricating quality, properties that find practical application in graphite ("lead") pencils and graphite grease.

The ability of many metals to form alloys with hydrogen, carbon, nitrogen, and boron can also be readily explained in terms of their crystal structure. These alloys are solid solutions; the atoms of solute penetrate into the lattice of the metal and occupy the remaining empty spaces, or interstices, between the atoms of metal. Surprisingly, the close-packed structure of metals is more favorable to the formation of such a solid solution than other structures. How can a close-packed lattice accommodate extra atoms? The answer to this puzzle is that although the net volume of the interstitial spaces in a crystal of close-packed spheres is a small fraction of the total volume of the crystal, the shape of each interstitial space is such that it can easily accommodate an extra sphere, provided that the extra sphere is somewhat smaller. A simple calculation shows that an extra sphere will fit into an interstitial space if its radius is smaller than 0.59 times the radius of the close-packed spheres. Atoms of hydrogen, carbon, nitrogen, and boron have sizes that are not far from this limit and they therefore fit quite nicely into the interstitial spaces of metals.

Incidentally: Alloys of hydrogen with metals are of great interest in the technology of fuels. Pure hydrogen is an excellent fuel — when burned with oxygen, it releases a large amount of energy without producing any pollutants (the product of the combustion is pure water). Hydrogen could be used as a substitute for gasoline in the propulsion of automobiles. However, a tank full of hydrogen in the form of a liquid or a compressed gas poses a frightful explosive hazard. It would be much safer to store the hydrogen within a chunk of metal. For instance, magnesium readily absorbs hydrogen, forming an alloy or solid solution. Magnesium can store up to 67 kg of hydrogen per cubic meter of (porous) magnesium, so the storage of an amount of hydrogen equivalent to 80 liters of gasoline would require about 320 liters of hydrogen–magnesium alloy. Although this is an inconveniently large volume (and weight), it does eliminate the explosion hazard. To supply the engine, the hydrogen can be gradually released from the magnesium by means of a small amount of heat applied to the alloy.

## I.4 DEFECTS IN CRYSTALS

Crystals found in nature are never perfect. A large crystal is likely to contain within it a number of microcrystals jumbled together at different angles (Figure I.26); the mismatch at the boundaries of the microcrystals is a conspicuous defect of the crystal. But even within a single microcrystal, one often finds a variety of defects. Sometimes atoms are missing from the lattice, leaving empty sites (Figure I.27). Sometimes extra atoms are squeezed between the regular rows and columns of the lattice (Figure I.28). Furthermore, sometimes a few lattice sites are occupied by atoms of a different element (Figure I.29); the presence of such a different atom tends to distort the crystal lattice in the immediate vicinity. These three kinds of isolated

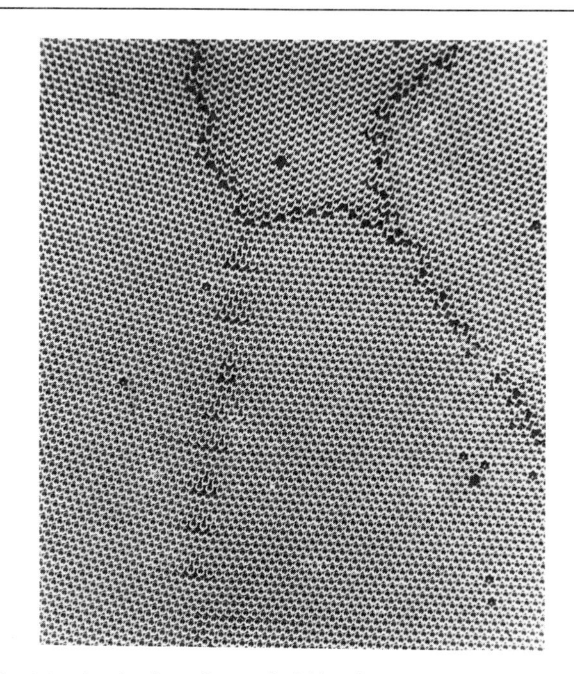

**Fig. I.26** A raft of small soap bubbles floating on water. The raft consists of several "crystals" with different orientations. The boundaries between these crystals are very conspicuous. (Courtesy C. S. Smith, MIT.)

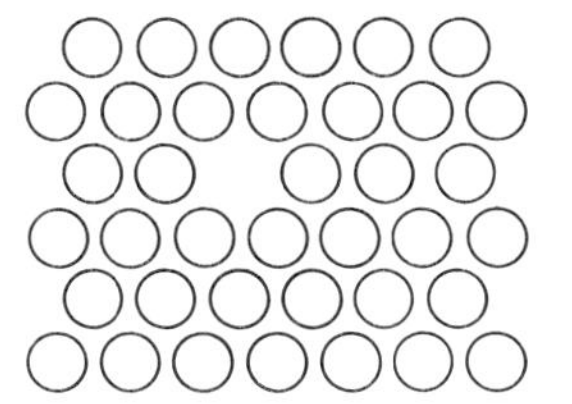

**Fig. I.27** Vacancy in a crystal lattice.

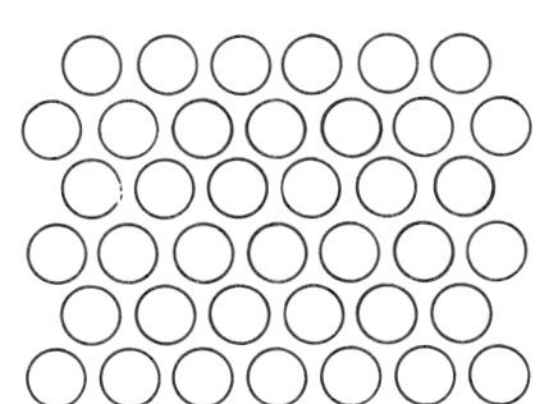

**Fig. I.28** Interstitial in a crystal lattice.

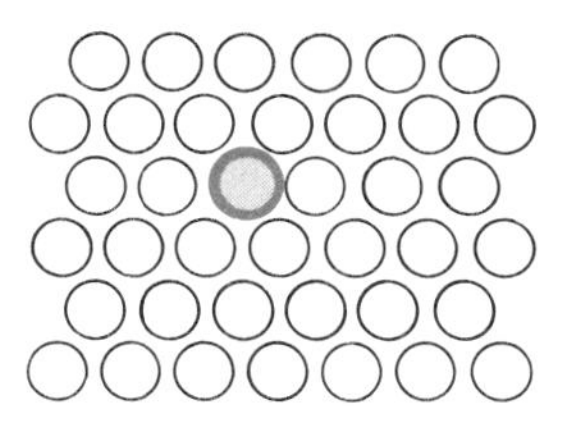

**Fig. I.29** Substitutional impurities in a crystal lattice.

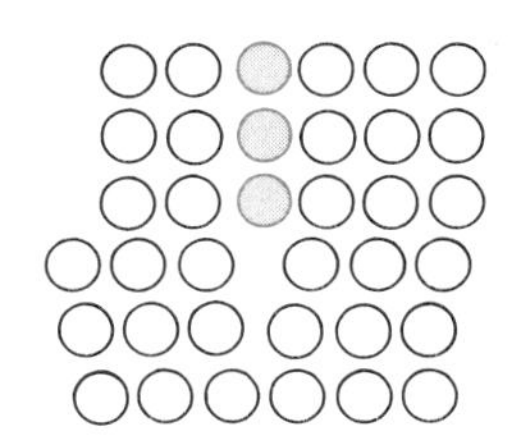

**Fig. I.30** Dislocation in a crystal lattice.

defects are called, respectively, **vacancies, interstitials,** and **substitutional impurities.**

Figure I.30 shows a lattice with a large defect: the crystal has an extra atomic plane slipped between the regular layers. This kind of defect is called a **dislocation.** The crystals in ordinary pieces of cast metal have an abundance of such defects — anywhere between $10^5$ and $10^9$ dislocations per square centimeter.

Dislocations have a crucial effect on the strength of the metal. If a bar of metal is subjected to a large bending force, it will become permanently deformed, that is, its crystal planes will suffer a permanent displacement. This displacement does not occur by the simultaneous slippage of an entire atomic layer over the adjacent layer — such a motion would involve the simultaneous disruption of all the atomic bonds joining one layer to the next and that would require a prohibitively large force. Rather, the layer moves step by step, by means of propagating dislocation. Figure I.31 shows how the dislocation propagates. Note that only one atom moves at a time, and hence no very large force is required to generate this motion. The displacement of crystal layers by the propagation of a dislocation is analogous to the displacement of a rug by the passage of a wrinkle: to move a heavy rug a few inches, it is much easier to run a wrinkle down the rug than to pull the whole rug along all at once by brute force.

Dislocations move through a crystal so easily that metals would be rather soft were it not for a snag: when a moving dislocation collides with another dislocation or when it reaches the boundary between one microcrystal and the adjacent one, it finds it hard to keep going — the sudden change in the orientation of the crystal planes at such places acts as an obstacle. Hence a crystal with many defects will be hard. The hardening of iron and steel by cold-working takes advantage of this effect. The treatment generates dislocations and breaks up the microcrystals into smaller microcrystals; the corresponding increase in the number of obstacles increases the resistance that the metal offers to deformation.

The presence of substitutional atoms also tends to harden a metal. The distortion of the lattice in the vicinity of the substitutional atoms acts as an obstacle to any moving dislocation. This effect is largely responsible for the fact that the alloy of two metals (for example, copper and tin) is harder than either of the pure metals.

The industrial production of steel is a prominent practical example of how the mechanical properties of a material can be controlled by deliberate modification of its crystal structure. Ordinary steel is an alloy of iron and carbon (up to 1.7% carbon). At high temperature (above 1000°C), the atoms in iron crystals adopt a close-packed structure, a face-centered cube (see Figure I.21). As we remarked in the preceding section, the carbon atoms fit into the interstitial spaces of the close-packed iron lattice. However, at room temperature, the atoms in iron crystals adopt a new cubic structure, a body-centered cube (see Figure I.22). The transformation from the face-centered to the body-centered structure occurs at a temperature between 700°C and 900°C (depending on the amount of carbon). Thus, drastic changes occur in a crystal of steel as it cools from an initially high temperature.

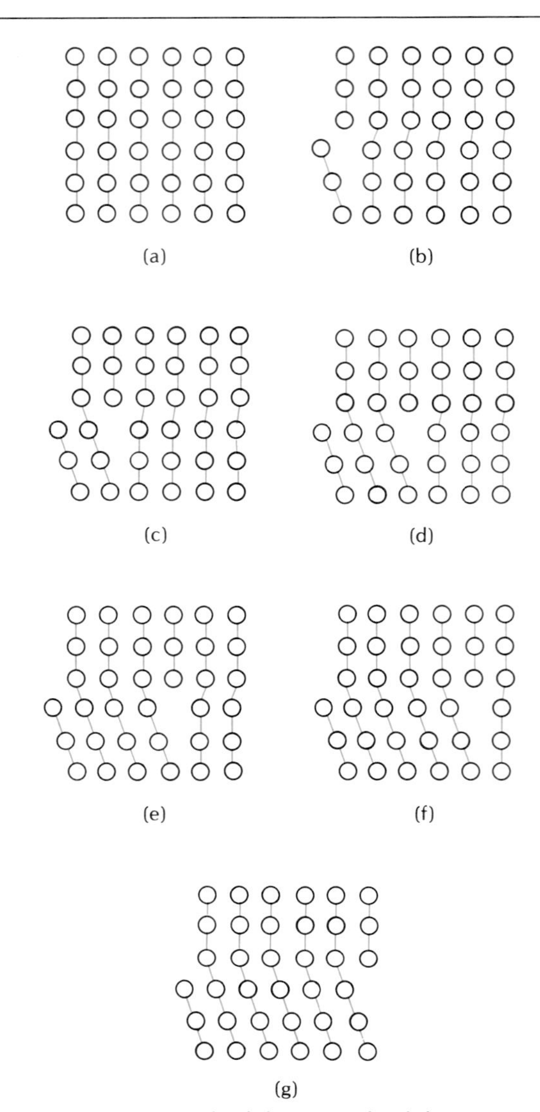


**Fig. I.31** Propagation of a dislocation. The dislocation moves from left to right.

What happens to the carbon atoms during this drastic change of the structure of the iron lattice depends crucially on the rate of cooling. If the steel is subjected to very sudden cooling (quenching) by immersion in water or oil, the carbon atoms do not have time to move from their sites; they remain fixed in their positions as the iron lattice changes. But since the carbon atoms do not fit into the interstitial spaces of the body-centered lattice, this lattice will be strongly distorted in the vicinity of each carbon atom. As we have seen, such defects in the lattice will make the material hard — the quenched steel is very hard. Unfortunately, this steel is also quite brittle — a sudden blow can crack and shatter it. If the steel is cooled slowly, the carbon atoms have time to segregate and the steel develops crystallites of two different kinds: one kind consists of almost pure iron and the other consists of molecules of $Fe_3C$ (iron carbide). These two kinds of crystallites form alternating plates (or lamellae) of irregular size and shape. The resulting steel is fairly soft.

The two varieties of steel described above are extremes. For most applications the steel should be neither excessively hard (and brittle) nor excessively soft. One method for producing steel of intermediate qualities relies on careful heat treatment. By slowly heating (tempering) hard steel, one can encourage its carbon atoms to segregate and to form crystallites of iron compounds. Skillful control of the temperature and the duration of this heat treatment allows metallurgists to achieve a wide range of grades of hardness and shock resistance. It is even possible to achieve different mechanical qualities in different parts of one single piece of steel. For instance, in the manufacture of armor plate, the outer face of the plate is made very hard and the inner part somewhat softer; when struck by a projectile, the outer part resists penetration while the inner part absorbs the shock and prevents the outer part from shattering.

The sword blades forged by the old Japanese swordsmiths achieve their remarkable strength by a similar combination of a very hard cutting edge backed by a soft, elastic core. The intermingling of the two kinds of steel gives the surface of these blades their rich, distinctive texture (Figures I.32 and I.33).

Vacancies and impurities in a crystal lattice also have drastic effects on the optical and electric properties of the crystal. For instance, the deep red color of ruby is due to chromium impurities and the deep blue color of sapphire is due to chromium, iron, and titanium impurities; in both of these precious stones, the bulk of the crystal is aluminum oxide, but the impurities determine the color.

All of solid-state electronics depends on impurities. The diodes and transistors in solid-state circuitry are made of semiconductors, such as silicon or germanium. A pure semiconductor is a fairly poor conductor of electricity — it has some electrons within it, but not enough to carry a large electric current. However, selected impurities introduced into the crystal of semiconductor — such as indium impurities in a crystal of silicon — will yield up some of their electrons to the crystal, that is, these impurities act as sources of free electrons. This dramatically improves the semiconductor's ability to conduct electricity. The manufacture of solid-state devices relies on such manipulation of the electronic properties of materials by artificially "doping" the crystals with impurities. As in other practical applications of crystals, we again see that the imperfections put the crystal at our service. Perfect crystals are beautiful, but imperfect crystals are useful.

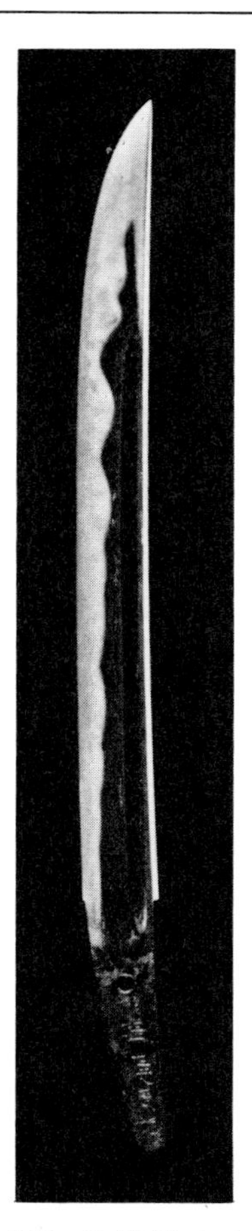

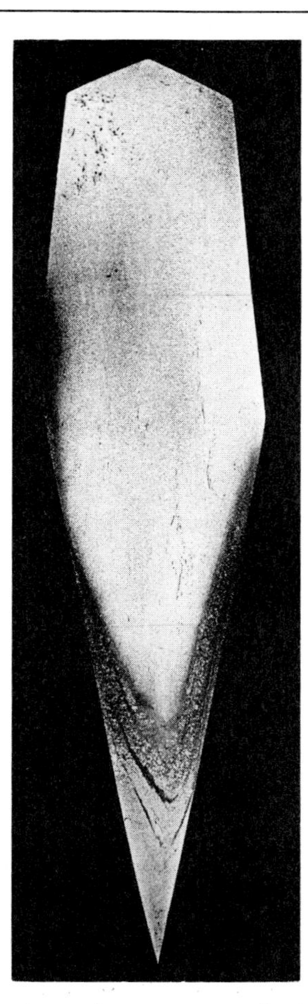

**Fig. I.32** (left) Blade by the sixteenth-century swordsmith Hiromitsu. Note the fine texture in the bright zone along the cutting edge.

**Fig. I.33** (right) Cross section of a blade attributed to the thirteenth-century swordsmith Nagamitsu. The core is soft steel, and the cutting edge hard steel. The zone in between consists of different kinds of steel in layers produced by the forging process and heat treatment. (Courtesy C. S. Smith, MIT.)

## Further Reading

*Snow Crystals* by W. A. Bentley and W. J. Humphreys (Dover, New York, 1962) contains a spectacular collection of more than 2000 photographs of snowflakes. *On Growth and Form* by D. W. Thompson (Cambridge University Press, Cambridge, 1961) gives many beautiful examples of symmetry and regularity in plants and animals. P. W. Medawar described this book as "beyond comparison the finest work of literature in all the annals of science that have been recorded in the English tongue."

*Symmetry* by H. Weyl (Princeton University Press, Princeton, N.J., 1952) is a classic study by a renowned mathematical physicist of the symmetries found in plants, animals, art, and architecture. *Symmetry Discovered* by J. Rosen (Cambridge University Press, Cambridge, 1975) gives a careful elementary introduction to the concept of symmetry and includes some applications of symmetry in the solution of problems in physics. *The Ambidextrous Universe* by M. Gardner (Basic Books, New York, 1964) is a charming exploration of the role of reflection symmetry in biology, chemistry, and physics.

*Symmetry Aspects of M. C. Escher's Periodic Drawings* by C. H. MacGillavry (Abrams, New York, 1976) presents a mathematical analysis of the symmetries of Escher's drawings. *The Mathematics of Islamic Art* (Metropolitan Museum of Art, New York, 1979) is a teaching packet with 20 beautiful slides that provide an introduction to symmetry in Islamic art.

*A Search for Structure* by C. S. Smith (MIT Press, Cambridge, 1982) is a collection of elegant and perceptive essays by an expert metallurgist that describe how the microscopic structure of materials has influenced their utilization in art and technology. In his investigations of history and art, Smith establishes that the first discovery of useful materials and processes almost always occurred in the decorative arts, and the practical applications were only recognized later.

*Crystals and Crystal Growing* by A. H. Holden and P. Singer (Doubleday, New York, 1960) provides a nice, elementary introduction to the general properties of crystals. This book includes recipes for growing crystals; it also includes a list of further books and articles. *Crystals and Light* by E. A. Wood (Van Nostrand, New York, 1964) is a concise introduction to the symmetry of crystals, with a minimum of mathematics. It contains a thorough discussion of how the internal structure affects the external shape. The second half of the book deals with the propagation of light in crystals and with the relationship between optical properties and symmetry. *Crystals and X Rays* by H. S. Lipson (Wykeham, London, 1970) describes the diffraction of X rays by crystals and how we can extract information about the structure of crystals and molecules from X-ray diffraction experiments. The book includes an account of the first discovery of a crystal structure (NaCl) by W. H. Bragg and W. L. Bragg.

The following is a list of some recent articles dealing with crystals and symmetry:

"Snow Crystals," C. Knight and N. Knight, *Scientific American*, January 1973

"Crystallography," A. Guinier, *Physics Today*, February 1975

"Geometry of Surface Layers," P. J. Estrup, *Physics Today*, April 1975

"Electron Microscopy of Atoms in Crystals," J. M. Cowley and S. Iijima, *Physics Today*, March 1977

"Disclinations," W. F. Harris, *Scientific American*, December 1977

"Hydrogen Storage in Metal Hydrides," J. J. Reilly and G. D. Sandrock, *Scientific American*, February 1980

"Ion Implantation of Surfaces," S. T. Picraux and P. S. Peercy, *Scientific American*, March 1985

"The Scanning Tunneling Microscope," G. Binning and H. Rohrer, *Scientific American*, August 1985

## Questions

1. Can you think of any animals whose external shape has a symmetry other than bilateral symmetry?

2. List the symmetries of (a) a circle, (b) a triangle, (c) a rectangle, and (d) a square. In what sense does a circle have more symmetry than a square, and a square more symmetry than a rectangle?

3. List the symmetries of (a) a sphere, (b) a cylinder, (c) a cube, (d) a parallelepiped, and (e) an ellipsoid of revolution.

4. What are the symmetries of this book? Assume that the book is closed and ignore the print on the pages and on the cover.

5. What are the rotation and the reflection symmetries of the pattern in Figure I.5? In Figure I.6?

6. Try to make a drawing of interlocking figures, in the style of Escher. Why is this difficult?

7. Find another glide line in Figure I.16.

8. Identify the primitive cell and the primitive vectors for the sodium chloride crystal of Figure I.25.

9. Find a primitive cell and the primitive vectors for one of the "crystals" shown in Figure I.26.

10. Figure I.34 shows an old Iranian wall panel. Identify the primitive cell and the primitive vectors. List the rotation and the reflection symmetries of this pattern.

**Fig. I.34** Glazed tile panel; Nishapur, thirteenth to fourteenth century.

11. For shipment to stores, round cans of canned goods are packed in cardboard boxes, usually 24 cans to a box. Are these cans close packed? Why does the manufacturer not take advantage of close packing?

12. If you buy apples by the bushel (a unit of volume), do you get more weight if you select small apples or large apples? Assume that the apples are close packed.

13. Consider a plane array of close-packed circles of equal size. What fraction of the area is between the circles?

14. What defects can you see in the "crystals" shown in Figure I.26?

CHAPTER 4

# Kinematics in Three Dimensions

We now take up the description of translational motion in three dimensions. This is a straightforward generalization of the one-dimensional case we studied in Chapter 2; in essence, three-dimensional motion consists of three one-dimensional motions occurring simultaneously. We will see that the vector formalism developed in the preceding chapter permits us to describe the motion in a very concise manner.

## 4.1 The Velocity and Acceleration Vectors

In the case of one-dimensional motion along a straight line we defined the average velocity by Eq. (2.3):

$$\bar{v} = \frac{\Delta x}{\Delta t} \tag{1}$$

where $\Delta x$ is the change in position in the time interval $\Delta t$.

In three-dimensional motion, the change in position, or the displacement, will involve simultaneous changes in $x$, $y$, and $z$; correspondingly, the **average velocity** has three components:

$$\bar{v}_x = \frac{\Delta x}{\Delta t} \qquad \bar{v}_y = \frac{\Delta y}{\Delta t} \qquad \bar{v}_z = \frac{\Delta z}{\Delta t} \tag{2}$$

In vector notation this can be written compactly as

*Average velocity*

$$\boxed{\bar{\mathbf{v}} = \frac{\Delta \mathbf{r}}{\Delta t}} \tag{3}$$

where **r** is the position vector [Eq. (3.9)] and $\Delta\mathbf{r}$ is the change in position, i.e, it is the displacement

$$\Delta\mathbf{r} = \Delta x\, \hat{\mathbf{x}} + \Delta y\, \hat{\mathbf{y}} + \Delta z\, \hat{\mathbf{z}} \tag{4}$$

The average velocity given by Eq. (3) is the displacement (a vector) multiplied by $1/\Delta t$ (a number or scalar); thus, velocity is a *vector* quantity.

As in the case of one-dimensional motion (see Section 2.2), the average velocity takes into account only the net displacement in the time interval $\Delta t$. Hence the average velocity ignores the details of the motion; it gives no credits for back and forth motion or for the length of the path, so that two motions between points $P_1$ and $P_2$ completed in the same time have the same average velocity, regardless of the exact path taken (Figure 4.1).

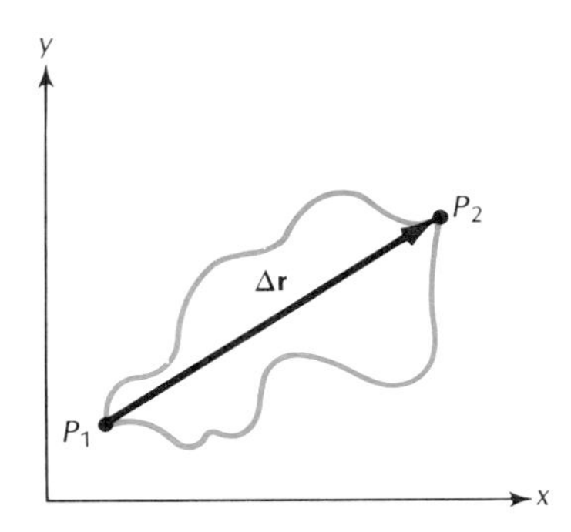


**Fig. 4.1** Two alternative paths connecting the points $P_1$ and $P_2$. If the motions along these paths take the same time, then they have the same average velocity, regardless of the difference in the lengths of the paths.

The **instantaneous velocity** gives a more precise description of the motion. As in the one-dimensional case, the instantaneous velocity is obtained by evaluating the average velocity in the limit of vanishing $\Delta t$:

$$v_x = \frac{dx}{dt} \qquad v_y = \frac{dy}{dt} \qquad v_z = \frac{dz}{dt} \tag{5}$$

In vector notation these equations become

$$\mathbf{v} = v_x\hat{\mathbf{x}} + v_y\hat{\mathbf{y}} + v_z\hat{\mathbf{z}} \tag{6}$$

or

$$\mathbf{v} = \frac{dx}{dt}\,\hat{\mathbf{x}} + \frac{dy}{dt}\,\hat{\mathbf{y}} + \frac{dz}{dt}\,\hat{\mathbf{z}} \tag{7}$$

Since the unit vectors $\hat{\mathbf{x}}$, $\hat{\mathbf{y}}$, and $\hat{\mathbf{z}}$ are time independent (constant), they have zero time derivative; hence $\hat{\mathbf{x}}\; dx/dt = d(\hat{\mathbf{x}}x)/dt$, etc., and Eq. (7) may also be written as

$$\mathbf{v} = \frac{d}{dt}\,(\hat{\mathbf{x}}x + \hat{\mathbf{y}}y + \hat{\mathbf{z}}z) \tag{8}$$

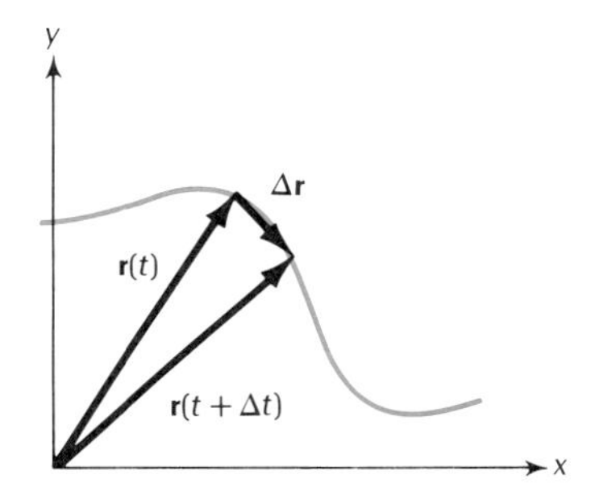


**Fig. 4.2** Position vectors of a particle at times $t$ and $t + \Delta t$. Their difference is $\Delta\mathbf{r}$.

or

$$\boxed{\mathbf{v} = \frac{d\mathbf{r}}{dt}} \tag{9}$$

*Instantaneous velocity*

Thus, the **instantaneous velocity vector** is the time derivative of the position vector.

The velocity vector of Eq. (9) has the direction of the tangent to the path of the particle. This can be seen from Figure 4.2, which shows the position vectors of a particle at the two times $t$ and $t + \Delta t$. In the limiting case $\Delta t \to 0$, the vector $\Delta\mathbf{r}$ will be tangent to the path and hence the velocity vector will also be tangent to the path.

EXAMPLE 1. A golfer launches a ball in the eastward direction with an initial speed of 30.0 m/s at an upward angle of 34° with the horizontal (see Fig-

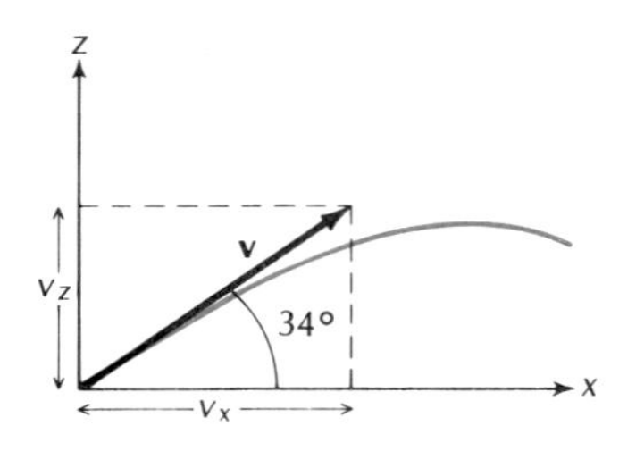


**Fig. 4.3** Velocity vector of a golf ball.

ure 4.3). What are the components of the instantaneous velocity of the ball in the reference frame of the ground? The $x$, $y$, and $z$ axes point south, east, and up, respectively.

SOLUTION: According to Figure 4.3, the velocity vector is in the $x$–$z$ plane; it makes an angle of 34° with the $x$ axis, an angle of 56° with the $z$ axis, and an angle of 90° with the $y$ axis. Hence

$$v_x = v \cos 34° = 30.0 \text{ m/s} \times \cos 34° = 24.9 \text{ m/s}$$

$$v_y = v \cos 90° = 0 \text{ m/s}$$

$$v_z = v \cos 56° = 30.0 \text{ m/s} \times \cos 56° = 16.8 \text{ m/s}$$

In everyday language, the words *velocity* and *speed* are synonymous. However, we will adopt the convention that the *speed* is the magnitude of the velocity vector:

$$[\text{speed}] = v = \left|\frac{d\mathbf{r}}{dt}\right| \tag{10}$$

or, in rectangular coordinates,

$$[\text{speed}] = v = \sqrt{v_x^2 + v_y^2 + v_z^2} \tag{11}$$

According to this rigorous definition, velocity is a vector and speed a scalar; velocity has magnitude and direction, whereas speed has only magnitude. Note that Eqs. (10) and (11) give the *instantaneous* speed; since $|d\mathbf{r}/dt| = |d\mathbf{r}|/dt$, the instantaneous speed is the increment of path length divided by the increment of time. In contrast, the *average* speed is the time average of this instantaneous speed, i.e., it is the total path length divided by the time taken [see Eq. (2.1)].

The definitions of average and instantaneous acceleration in three dimensions are straightforward generalizations of those in one dimension. For the **average acceleration** we have

$$\bar{a}_x = \frac{\Delta v_x}{\Delta t} \qquad \bar{a}_y = \frac{\Delta v_y}{\Delta t} \qquad \bar{a}_z = \frac{\Delta v_z}{\Delta t} \tag{12}$$

or

*Average acceleration*

$$\boxed{\bar{\mathbf{a}} = \frac{\Delta \mathbf{v}}{\Delta t}} \tag{13}$$

and for the **instantaneous acceleration** we have

$$a_x = \frac{dv_x}{dt} = \frac{d^2x}{dt^2} \qquad a_y = \frac{dv_y}{dt} = \frac{d^2y}{dt^2} \qquad a_z = \frac{dv_z}{dt} = \frac{d^2z}{dt^2} \tag{14}$$

or

*Instantaneous acceleration*

$$\boxed{\mathbf{a} = \frac{d\mathbf{v}}{dt} = \frac{d^2\mathbf{r}}{dt^2}} \tag{15}$$

It is an important consequence of this definition that there is an acceleration whenever the velocity changes either in magnitude or in *direction.* Thus, a motion at constant speed is an accelerated motion whenever its direction changes.

EXAMPLE 2. An automobile enters a 90° curve at a constant speed of 25 m/s and emerges from this curve 6.0 s later. What is the average acceleration for this time interval?

SOLUTION: Figure 4.4a shows the path of the automobile with the initial velocity $\mathbf{v}_1$ and the final velocity $\mathbf{v}_2$. Figure 4.4b shows the change $\Delta\mathbf{v} = \mathbf{v}_2 - \mathbf{v}_1$. The vector triangle is a right isosceles triangle; its hypotenuse is

$$|\Delta\mathbf{v}| = |v_1|/\cos 45°$$

$$= (25 \text{ m/s})/\cos 45°$$

$$= 35.4 \text{ m/s}$$

Hence the average acceleration has a magnitude

$$|\bar{\mathbf{a}}| = \left|\frac{\Delta\mathbf{v}}{\Delta t}\right| = \frac{35.4 \text{ m/s}}{6.0 \text{ s}} = 5.9 \text{ m/s}^2$$

The direction of this acceleration is along the vector $\mathbf{v}_2 - \mathbf{v}_1$.

COMMENTS AND SUGGESTIONS: This example illustrates how a mere change of direction of the velocity, without any change of speed, leads to an average acceleration. Of course, the automobile also has an instantaneous acceleration, but this is somewhat more difficult to calculate (in Section 4.4 we will examine the instantaneous acceleration of an automobile or a particle moving with uniform speed around a circular curve, and we will see that the instantaneous acceleration is at right angles to the instantaneous velocity). Note that if the automobile changes its speed while moving around the curve, then there are two simultaneous contributions to the acceleration: one from the change of the magnitude of the velocity, and one from the change of direction of the velocity. The resulting (average) acceleration can always be calculated from Eq. (13), since the vector quantity $\Delta\mathbf{v}$ automatically includes all changes in magnitude and in direction of the velocity.

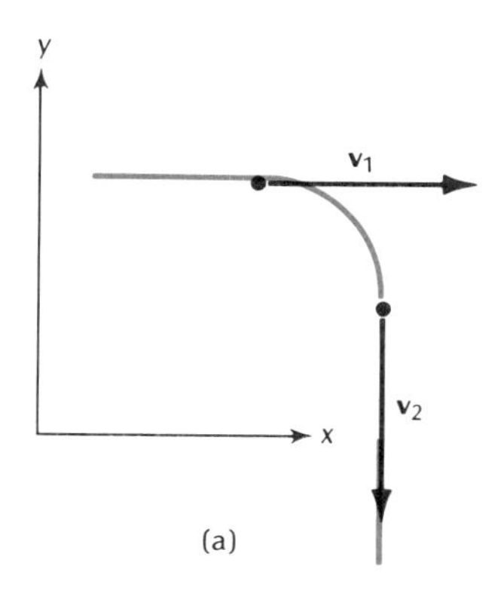


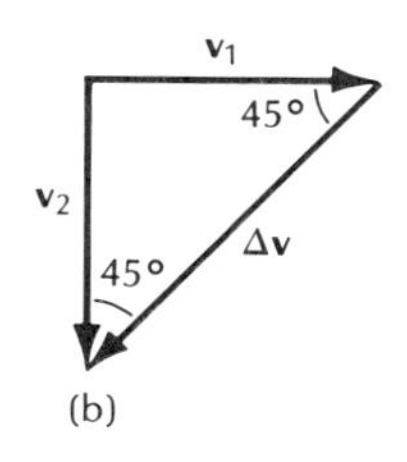


**Fig. 4.4** (a) Path of an automobile through a 90° curve. (b) The initial and the final velocity vectors $\mathbf{v}_1$ and $\mathbf{v}_2$.

## 4.2 Motion with Constant Acceleration

For a particle moving in three dimensions with constant acceleration, we can derive some equations relating acceleration, velocity, position, and time analogous to the equations that hold for motion in one dimension (see Section 2.5). We begin with Eq. (13), which tells us that a constant acceleration $\mathbf{a}$ produces a velocity change $\mathbf{a}\,\Delta t$ in a time interval $\Delta t$. Suppose that the initial velocity vector is $\mathbf{v}_0$; then the velocity vector after a time $t$ has elapsed will be

$$\boxed{\mathbf{v} = \mathbf{v}_0 + \mathbf{a}t} \tag{16}$$

*Constant acceleration:* **v**, **a**, *and t*

Furthermore, a mathematical argument similar to that used in the one-dimensional case [see Eqs. (2.18)–(2.22)] leads us to the following expression for the change of the position vector:

*Constant acceleration:*
**r**, **a**, *and t*

$$\mathbf{r} - \mathbf{r}_0 = \mathbf{v}_0 t + \tfrac{1}{2}\mathbf{a}t^2 \tag{17}$$

The $x$, $y$, and $z$ components of Eqs. (16) and (17) are, respectively,

$$v_x = v_{0x} + a_x t \tag{18}$$

$$v_y = v_{0y} + a_y t \tag{19}$$

$$v_z = v_{0z} + a_z t \tag{20}$$

and

$$x - x_0 = v_{0x}t + \tfrac{1}{2}a_x t^2 \tag{21}$$

$$y - y_0 = v_{0y}t + \tfrac{1}{2}a_y t^2 \tag{22}$$

$$z - z_0 = v_{0z}t + \tfrac{1}{2}a_z t^2 \tag{23}$$

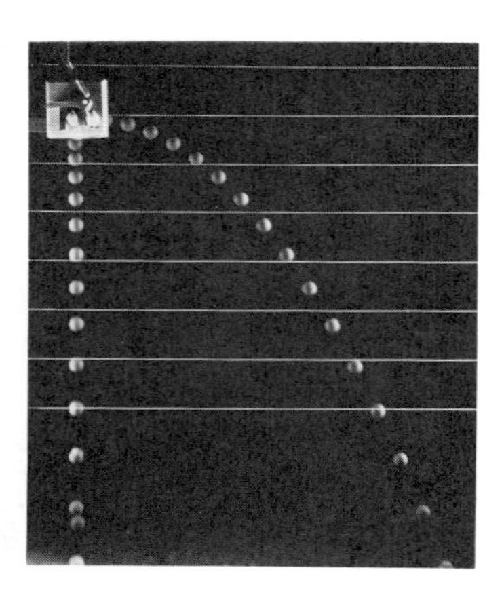

**Fig. 4.5** Stroboscopic photograph of two balls that have been released simultaneously from a platform; one ball has a horizontal velocity, the other does not.

Equations (18)–(23) state that the $x$, $y$, and $z$ components of the motion proceed completely independently of one another. Thus, the $x$ acceleration only affects the $x$ velocity, and the change in the $x$ position is entirely determined by the $x$ acceleration and the initial $x$ velocity. Figure 4.5 shows an experimental demonstration of this independence between different components of the motion. Two balls were released simultaneously from a platform above the surface of the Earth; one was merely dropped from rest, the other was launched with an initial horizontal velocity. Since the vertical accelerations are the same, Eqs. (20) and (23) predict that the vertical components of the motions are exactly the same for both balls; only the horizontal components of the motions should differ. The stroboscopic images of the two balls at successive instants of time show that they do in fact fall downward in unison, reaching equal heights at the same time. Thus, the vertical displacements are the same for both balls; only the horizontal displacements differ.

## 4.3 The Motion of Projectiles

Near the surface of the Earth the pull of gravity gives a freely falling body a downward acceleration of about 9.81 m/s$^2$. If we ignore friction with the air, this is the only acceleration that the body experiences when launched from some initial position with some initial velocity. The motion of such a launched body, or **projectile,** is therefore motion with constant vertical acceleration and zero horizontal acceleration. This kind of motion is called projectile motion, or ballistic motion.

The initial velocity of the projectile will in general have both a vertical and a horizontal component. If we take the $z$ axis in the upward vertical direction and the $x$ axis in the direction of the initial horizontal velocity, we have $a_x = 0$, $a_y = 0$, $a_z = -g = -9.81$ m/s$^2$, and $v_{0y} = 0$. (Note that $a_z$ is *negative* because the $z$ axis points upward and the acceleration of gravity is downward.) Furthermore, for the sake of simplicity, let us assume that the origin of coordinates coincides with the initial position of the projectile, so that $x_0 = 0$, $y_0 = 0$, and $z_0 = 0$. The components of the velocity and position will then be, according to Eqs. (18)–(23),

$$v_x = v_{0x} \tag{24}$$

$$v_y = 0 \tag{25}$$

$$v_z = v_{0z} - gt \tag{26}$$

$$x = v_{0x}t \tag{27}$$

$$y = 0 \tag{28}$$

$$z = v_{0z}t - \tfrac{1}{2}gt^2 \tag{29}$$

*Equations for projectile motion*

With our choice of coordinates, the value of $y$ is initially zero and always remains zero. Thus the motion of the projectile is confined to the $x$–$z$ plane — the motion is *two dimensional.* We will from now on ignore the $y$ coordinate entirely in our study of projectile motion.

---

EXAMPLE 3. Consider the golf ball launched as described in Example 1. At what time does this projectile reach its maximum height? What is this maximum height? What is the horizontal distance at this instant?

SOLUTION: The relevant equations are Eqs. (24)–(29). The known quantities are the components of the initial velocity; according to Example 1, these components are $v_{0x} = 24.9$ m/s and $v_{0z} = 16.8$ m/s. Furthermore, we know that the projectile reaches maximum height when $v_z = 0$. In terms of these values of $v_{0z}$ and $v_z$, Eq. (26) determines the time of maximum height,

$$0 = 16.8 \text{ m/s} - gt$$

or

$$t = \frac{16.8 \text{ m/s}}{g} = \frac{16.8 \text{ m/s}}{9.81 \text{ m/s}^2} = 1.71 \text{ s}$$

With this value of the time, we can calculate the maximum height,

$$\begin{aligned} z &= v_{0z}t - \tfrac{1}{2}gt^2 \\ &= 16.8 \text{ m/s} \times 1.71 \text{ s} - \tfrac{1}{2} \times 9.81 \text{ m/s}^2 \times (1.71 \text{ s})^2 \\ &= 14.4 \text{ m} \end{aligned}$$

and the horizontal distance,

$$x = v_{0x}t = 24.9 \text{ m/s} \times 1.71 \text{ s} = 42.6 \text{ m}$$

In our calculation we have ignored the effects of air resistance. For a heavy and compact projectile with such a fairly low velocity, this is a good approximation.

COMMENTS AND SUGGESTIONS: In this example, and in many other examples of projectile motion, we proceed in two steps: we first perform a complete determination of the vertical motion, and only then deal with the horizontal motion. We can separate the problem into these two parts because the vertical motion and the horizontal motion are independent. Note that the determination of the vertical motion is essentially the same as for one-dimensional free-fall motion, as in the examples of Chapter 2.

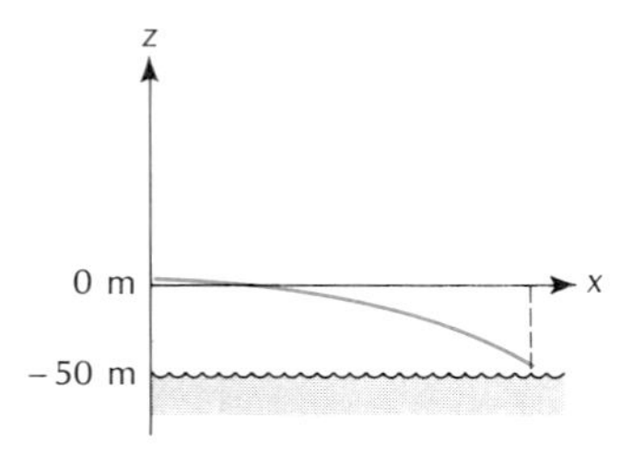


**Fig. 4.6** Trajectory of a bomb.

EXAMPLE 4. In low-level bombing (at "smokestack level"), a World War II bomber releases a bomb at a height of 50 m above the surface of the sea while in horizontal flight at a constant speed of 320 km/h. How long does the bomb take to fall to the surface? How far ahead (horizontally) of the point of release is the point of impact?

SOLUTION: It is convenient to place the origin of coordinates at the point of release, 50 m above the level of the sea, with the $x$ axis along the horizontal path of the bomber (Figure 4.6). The initial velocity of the bomb is the same as that of the bomber:

$$v_{0x} = 320 \text{ km/h} = 88.9 \text{ m/s}$$

$$v_{0z} = 0$$

When the bomb reaches the level of the sea, its vertical position is $z = -50$ m (relative to the origin!). With these known quantities, Eq. (29) gives us the time of impact:

$$-50 \text{ m} = -\tfrac{1}{2}gt^2$$

or

$$t = \sqrt{2 \times 50 \text{ m}/g} = \sqrt{2 \times 50 \text{ m}/(9.81 \text{ m/s}^2)} = 3.19 \text{ s}$$

At this time, the horizontal position of the bomb is

$$x = v_{0x}t = 88.9 \text{ m/s} \times 3.19 \text{ s} = 284 \text{ m}$$

**Fig. 4.7** A "string" of bombs released from a bomber. The bombs continue to move forward with the same horizontal velocity as that of the bomber, and their horizontal position therefore remains fixed relative to the bomber.

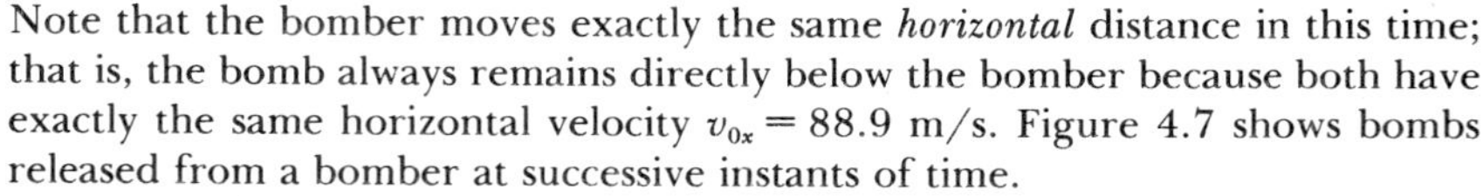

Note that the bomber moves exactly the same *horizontal* distance in this time; that is, the bomb always remains directly below the bomber because both have exactly the same horizontal velocity $v_{0x} = 88.9$ m/s. Figure 4.7 shows bombs released from a bomber at successive instants of time.

In our calculation we again ignored air resistance. For low-level bombing, this is an acceptable approximation, but in high-level bombing the resistive corrections are very important.

COMMENTS AND SUGGESTIONS: Instead of placing the origin of coordinates at the height of the bomber, we could have placed the origin at sea level. But this would have required a modification of Eq. (29), since the origin would not have coincided with the launch point of the projectile. If the initial values $x_0$, $y_0$, and $z_0$ of the components of the position of the projectile are different from zero, then these values must be included in Eqs. (24)–(29), as they were in Eqs. (18)–(23).

The trajectory of a bomb, or any other projectile, is a parabola. The mathematical proof of this statement rests on Eqs. (27) and (29). Suppose that time is reckoned from the instant at which the projectile reaches its maximum height (so that $t < 0$ before the instant of maximum height, and $t > 0$ after). Then $v_{0z} = 0$ and

$$x = v_{0x}t \tag{30}$$

$$z = -\tfrac{1}{2}gt^2 \tag{31}$$

Combining these equations, we find

$$z = -\tfrac{1}{2}g(x/v_{0x})^2 \tag{32}$$

This equation has the form $z = [\text{constant}] \times x^2$, which is the equation of a **parabola** in the $x$–$z$ plane (Figure 4.8).

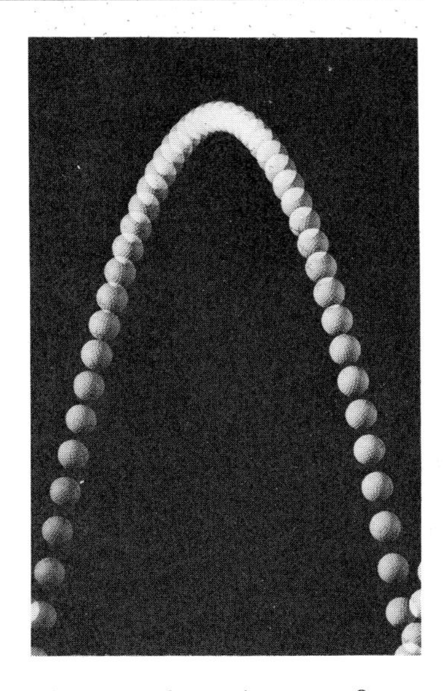

**Fig. 4.8** The trajectory of a projectile (golf ball) in free fall is a parabola. (Courtesy H. E. Edgerton, MIT.)

If air resistance is negligible, the trajectory of a projectile launched with some horizontal and vertical velocity is always some portion of a parabola. In the case of a bomb dropped by an airplane, the relevant portion starts at the apex and descends to the target. In the case of a shot or bullet fired from a gun or a ball or stone thrown by hand, the relevant portion begins at ground level, rises to the apex, and descends to the target. In the latter case, it is often important to calculate the **maximum height** reached, the **time of flight** (time between the instants of launch and impact), and the **range** (distance between the points of launch and impact). For the following calculation we will assume that the launch point and the target are at the same level.

We again assume that the origin of coordinates coincides with the launch point. The projectile is launched with some initial horizontal ($v_{0x}$) and vertical ($v_{0z}$) velocity. As explained in Example 3, it reaches maximum height at the time $t_{max}$, when

$$0 = v_{0z} - gt_{max}$$

Consequently,

$$t_{max} = v_{0z}/g \tag{33}$$

At this time the height reached is

$$z_{max} = v_{0z}t_{max} - \tfrac{1}{2}gt_{max}^2 = v_{0z}(v_{0z}/g) - \tfrac{1}{2}g(v_{0z}/g)^2 \tag{34}$$

that is,

$$z_{max} = \tfrac{1}{2}v_{0z}^2/g \tag{35}$$

To find the time of impact, we must evaluate Eq. (29) with $z = 0$,

$$0 = v_{0z}t - \tfrac{1}{2}gt^2 \tag{36}$$

This equation has two solutions: $t = 0$ and $t = 2v_{0z}/g$. The latter solution gives the time of flight,

$$t_{flight} = 2v_{0z}/g \tag{37}$$

Comparison of Eqs. (33) and (37) shows that the time of flight is exactly twice the time required to reach maximum height; this is so because the motion is symmetric about the apex of the parabola.

The range is simply the horizontal distance attained at the time $t = t_{flight}$,

$$x_{max} = v_{0x}t_{flight} = 2v_{0x}v_{0z}/g \tag{38}$$

The formulas for $z_{max}$, $t_{flight}$, and $x_{max}$ can be expressed in terms of the magnitude of the initial velocity (the launch speed or the muzzle

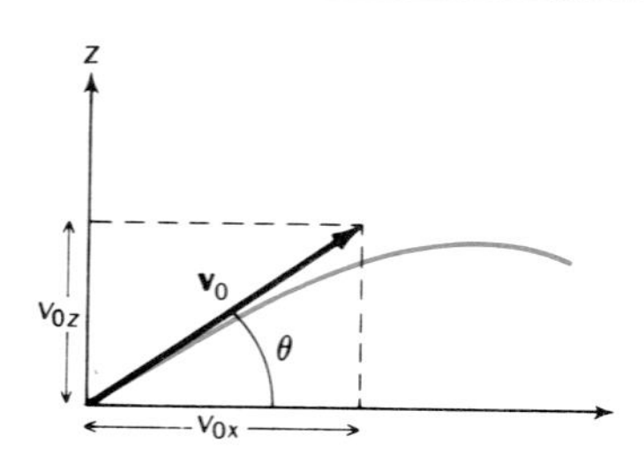


**Fig. 4.9** The elevation angle $\theta$ gives the initial direction of motion of the projectile.

speed) and the angle between the horizontal and the direction of the initial velocity (the elevation angle). Figure 4.9 shows the initial velocity vector $\mathbf{v}_0$ and the elevation angle $\theta$; we have

$$v_{0x} = v_0 \cos\theta \tag{39}$$

$$v_{0z} = v_0 \sin\theta \tag{40}$$

Hence

*Maximum height*

$$\boxed{z_{\text{max}} = \frac{v_0^2 \sin^2\theta}{2g}} \tag{41}$$

*Time of flight*

$$\boxed{t_{\text{flight}} = \frac{2v_0 \sin\theta}{g}} \tag{42}$$

$$x_{\text{max}} = \frac{2v_0^2 \sin\theta \cos\theta}{g}$$

We can also write this last formula as

*Range*

$$\boxed{x_{\text{max}} = \frac{v_0^2 \sin 2\theta}{g}} \tag{43}$$

Keep in mind that the calculation leading to Eqs. (41)–(43) assumed that the launch point and the impact point are at the same level ($z = 0$). If the target is higher or lower, the projectile may reach impact before or after the time given by Eq. (42) and the range will be shorter or longer than that given by Eq. (43). Furthermore, keep in mind that although Eqs. (41)–(43) are often handy for the solution of problems of projectile motion, they do not give us a complete description of the motion. The complete description of the motion is contained in Eqs. (24)–(29), and we can always extract anything we want to know about the motion from these equations.

Figure 4.10 shows the trajectories of projectiles of equal launch speeds as a function of elevation angle. Note that the range is maximum for $\theta = 45°$ and that angles that are equal amounts above or below 45° yield the same range; for instance, 60° and 30° yield the same range.

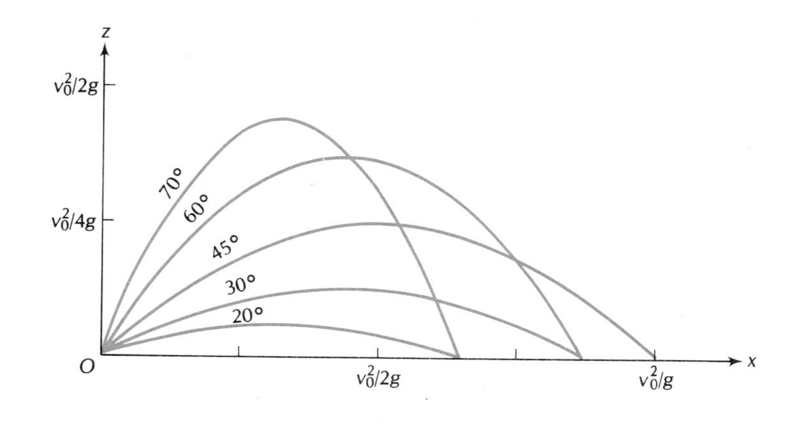


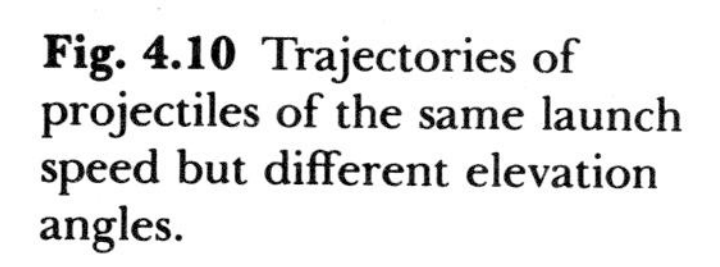

**Fig. 4.10** Trajectories of projectiles of the same launch speed but different elevation angles.

EXAMPLE 5. The world record for the discus throw set by M. Wilkins in 1976 was 70.87 m. What is the least initial speed required to achieve this range? Ignore the height of the hand.

SOLUTION: To achieve the largest range with the least initial speed, the thrower must launch the discus with an elevation angle of 45°. Equation (43) then gives

$$v_0^2 = gx_{\text{max}}/\sin 2\theta = 9.81\ \text{m/s}^2 \times 70.87\ \text{m}/\sin 90°$$

$$= 6.95 \times 10^2\ \text{m}^2/\text{s}^2$$

and

$$v_0 = 26.4\ \text{m/s}$$

The curve labeled 45° in Figure 4.10 shows the trajectory of the discus.

EXAMPLE 6. A fireman standing at a horizontal distance $x$ from a burning building aims the stream of water from a fire hose up at an angle $\theta$ (Figure 4.11). The water leaves the nozzle of the fire hose with a speed $v_0$. At what height $z$ will the stream of water strike the building?

SOLUTION: Each water particle moves like a projectile. However, we cannot use Eqs. (41)–(43) because the launch point and the impact point are *not* at the same height. We must begin with Eqs. (27) and (29):

$$x = v_{0x}t$$

$$z = v_{0z}t - \tfrac{1}{2}gt^2$$

From the first of these equations we obtain $t = x/v_{0x}$, which, when substituted into the second equation, gives us the vertical height of the impact point,

$$z = \frac{v_{0z}}{v_{0x}}x - \frac{1}{2}\frac{gx^2}{v_{0x}{}^2}$$

With $v_{0x} = v_0 \cos\theta$ and $v_{0z} = v_0 \sin\theta$, this becomes

$$z = x\tan\theta - \frac{1}{2}\frac{gx^2}{v_0^2\cos^2\theta} \tag{44}$$

(a)

(b)

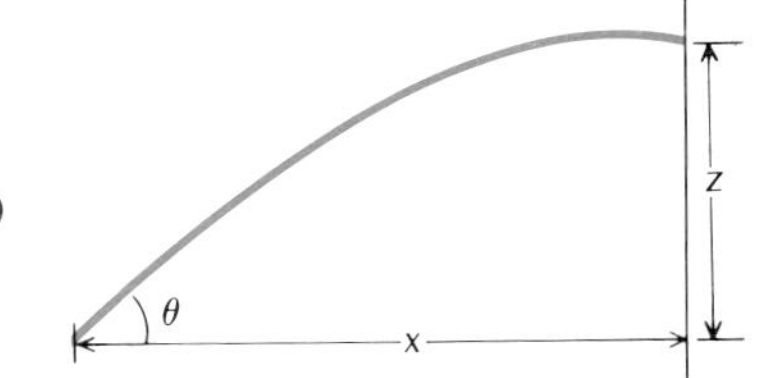


**Fig. 4.11** (a) Stream of water from a fire hose follows the trajectory of a projectile. (b) Coordinates of the launch point and of the impact point.

In Figure 4.10, as in all calculations of this section, air resistance has been neglected. This is an adequate approximation for fairly heavy and compact projectiles moving at fairly low speeds, but it is not adequate for projectiles moving at high speeds. Figure 4.12 shows the trajectory of a high-speed projectile, for which air resistance is large. The trajectory is a **ballistic curve;** it cannot be described by a simple mathematical formula, but it can be calculated numerically by taking into account the empirically measured magnitude of the air resistance. Obviously, the trajectory for such a projectile is very different from the trajectory without air resistance predicted by the simple formulas of this section. If we use these simple formulas in the calculation of the motion of a high-speed projectile, our results will bear only a very vague resemblance to reality.

*Ballistic curve*

The motion of a ballistic missile above the atmosphere of the Earth is an instance of projectile motion without any air resistance, and we might be tempted to apply the simple equations of this section to the

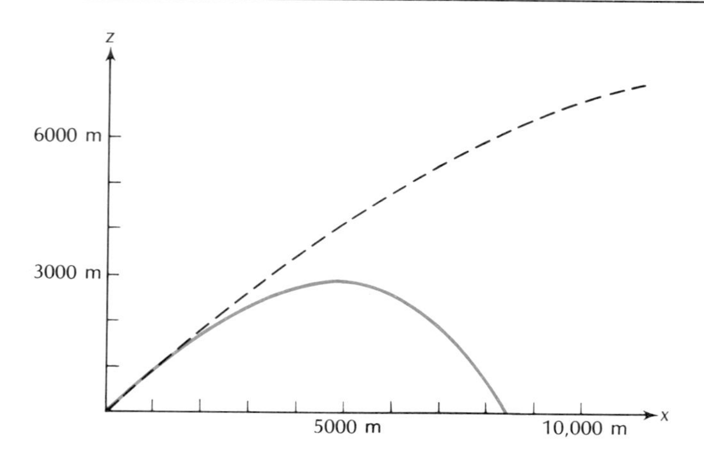


**Fig. 4.12** Trajectory of a projectile with large air resistance. The projectile is a pointed cylinder of diameter 7.7 cm, mass 6.9 kg, launched with $v_0 = 550$ m/s, $\theta = 45°$. The dashed line shows the theoretical trajectory without air resistance. (Based on C. Cranz and K. Becker, *Exterior Ballistics*.)

motion of such a missile. However, for long-range missiles, such as intercontinental ballistic missiles (ICBMs), the curvature of the surface of the Earth, the rotation of the Earth, and the decrease of gravity with height lead to complications in the motion, and our simple equations are not valid. For a short-range missile or for a short portion of the trajectory of a long-range missile, these extra complications are less significant, and our equations are approximately valid.

## 4.4 Uniform Circular Motion

*Uniform circular motion*

**Uniform circular motion** is motion with constant speed along a circular path, such as the motion of an automobile around a circular curve or the motion of a planet in a circular orbit around the Sun. Figure 4.13 shows the positions at different times of a particle in uniform circular motion. All the velocity vectors shown have the same magnitude (same speed), but they differ in direction. Because of this change of direction, uniform circular motion is *accelerated motion.*

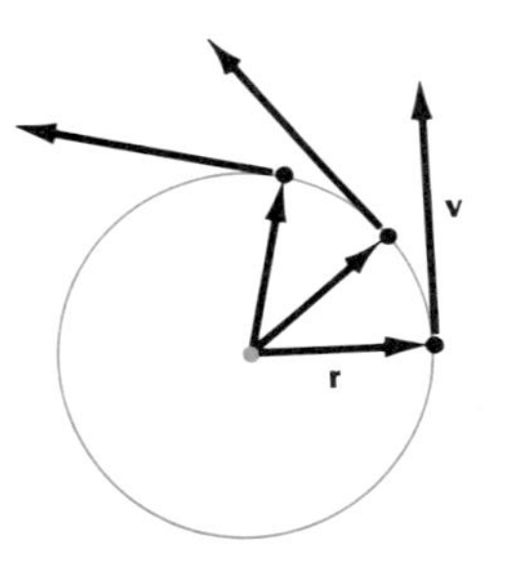


**Fig. 4.13** Position and velocity vectors for a particle in uniform circular motion.

Suppose that the radius of the circle is $r$ and the constant speed of the particle is $v$. To find the value of the instantaneous acceleration, we must look at the velocity change in a very short time interval $\Delta t$. Figure 4.14a shows the particle at two positions $\mathbf{r}_1$ and $\mathbf{r}_2$ a short time apart. The figure also shows the two velocity vectors $\mathbf{v}_1$ and $\mathbf{v}_2$; both velocity vectors have the same magnitude $|\mathbf{v}_1| = |\mathbf{v}_2| = v$. The position vectors make a small angle $\Delta\theta$ with one another, and so do the velocity vectors. Figure 4.14b shows the two velocity vectors tail to tail and their difference $\Delta\mathbf{v} = \mathbf{v}_2 - \mathbf{v}_1$. If $\Delta t$ is very small, then $\Delta\theta$ (measured in radians) will also be small, and the magnitude of $\Delta\mathbf{v}$ will be approximately

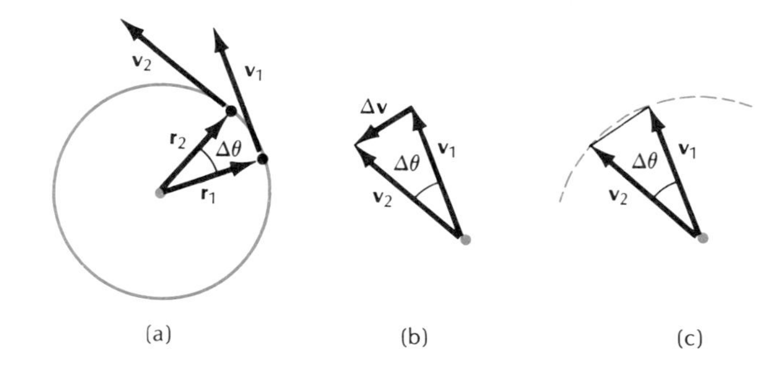


**Fig. 4.14** (a) The position vectors at times $t$ and $t + \Delta t$. (b) The velocity vectors at times $t$ and $t + \Delta t$. The difference between the velocity vectors is $\Delta\mathbf{v}$. (c) The circular arc (colored) nearly coincides with the straight line segment connecting the tips of the vectors $\mathbf{v}_1$ and $\mathbf{v}_2$.

$$|\Delta \mathbf{v}| \cong v\, \Delta\theta \tag{45}$$

To justify this approximation, note that in Figure 4.14c the length of the circular arc of radius $v$ connecting the tips of $\mathbf{v}_1$ and $\mathbf{v}_2$ is nearly equal to the length of the straight line segment connecting these tips, an approximation which becomes exact in the limit $\Delta\theta \to 0$. Since the straight line segment has a length $|\Delta \mathbf{v}|$ and the circular arc has a length [radius] × [angle in radians] $= v \times \Delta\theta$, their approximate equality implies Eq. (45). The magnitude of the acceleration is therefore

$$a = \frac{|\Delta \mathbf{v}|}{\Delta t} \cong v \frac{\Delta\theta}{\Delta t} \tag{46}$$

To evaluate the ratio $\Delta\theta/\Delta t$, we note that the time interval between the two positions shown in Figure 4.14a is

$$\Delta t = \frac{[\text{path length}]}{[\text{speed}]} = \frac{[\text{radius}] \times [\text{angle in radians}]}{[\text{speed}]}$$

$$= r\, \Delta\theta / v$$

Hence Eq. (46) becomes

$$a \cong v \frac{\Delta\theta}{r\, \Delta\theta / v} = \frac{v^2}{r} \tag{47}$$

In the limit $\Delta t \to 0$, this relation becomes exact, so that the acceleration for uniform circular motion is

*Centripetal acceleration*

$$\boxed{a = v^2/r} \tag{48}$$

The instantaneous *direction* of this acceleration remains to be determined. From Figure 4.14b it is clear that in the limit $\Delta\theta \to 0$, the direction of $\Delta \mathbf{v}$ will be perpendicular to the velocity vectors $\mathbf{v}_1$ and $\mathbf{v}_2$ (which will be parallel in this limit); hence the instantaneous acceleration vector is perpendicular to the instantaneous velocity vector. Since the velocity vector corresponding to circular motion is tangential to the circle, the acceleration points along the radius, toward the center of the circle. Figure 4.15 shows the velocity and acceleration vectors at several positions along the circular path. The acceleration of uniform circular motion, with a magnitude given by Eq. (48) and a direction toward the center of the circle, is called **centripetal acceleration.**

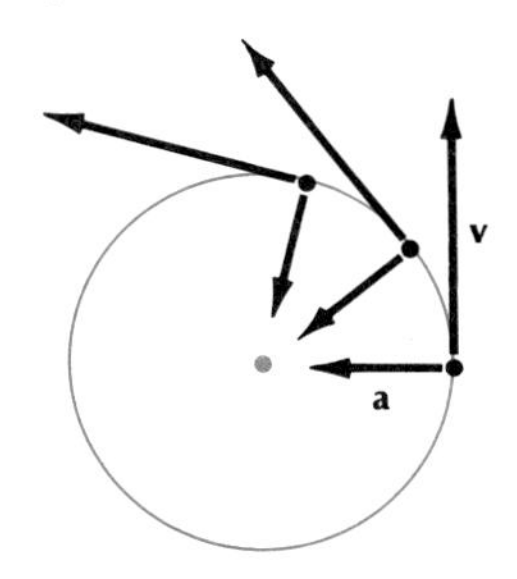


**Fig. 4.15** Acceleration vectors for a particle in uniform circular motion.

EXAMPLE 7. In tests of the effects of high acceleration on the human body, astronauts at the National Aeronautics and Space Administration (NASA) Manned Spacecraft Center in Houston are placed in a capsule that is whirled around a path of radius 15 m at the end of a revolving girder (Figure 4.16). If the girder makes 24 revolutions per minute, what is the acceleration of the capsule? How many G does this amount to?

SOLUTION: The circumference of the circular path is $2\pi \times$ [radius] $= 2\pi \times$ 15 m. Since the capsule takes (1/24) min, or (60/24) s, to go around this circumference, the speed is

**Fig. 4.16** Centrifuge at the Manned Spacecraft Center in Houston.

$$v = \frac{2\pi \times 15 \text{ m}}{60/24 \text{ s}} = 38 \text{ m/s}$$

From Eq. (48), the centripetal acceleration is then

$$a = \frac{v^2}{r} = \frac{(38 \text{ m/s})^2}{15 \text{ m}} = 95 \text{ m/s}^2$$

This acceleration is 9.7 G.

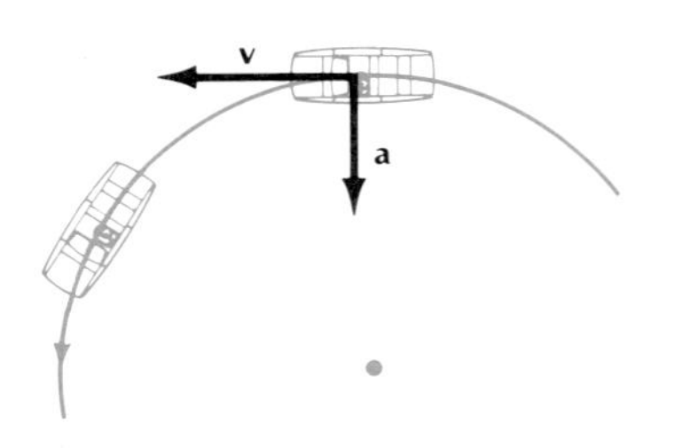


**Fig. 4.17** Automobile in uniform circular motion.

EXAMPLE 8. An automobile enters a circular curve of radius 30 m (Figure 4.17). If the wheels of the automobile can tolerate a maximum transverse acceleration of 8.0 m/s² without skidding, what is the maximum permissible speed?

SOLUTION: While traveling around the curve, the automobile is in uniform circular motion. Equation (48) gives

$$v^2 = ra = 30 \text{ m} \times 8.0 \text{ m/s}^2 = 2.4 \times 10^2 \text{ m}^2/\text{s}^2$$

or

$$v = 15 \text{ m/s} = 56 \text{ km/h}$$

## 4.5 The Relativity of Motion and the Galilean Transformations

Motion is relative — the values of the velocity and acceleration of a particle depend on the frame of reference in which these quantities are measured. For example, consider one reference frame attached to the shore and a second reference frame attached to a ship moving due east at a constant speed of 10 m/s. Suppose that observers in both reference frames measure and plot the coordinates of a sailboat passing by (Figure 4.18). The coordinates and the time measured in the first reference frame will be denoted by $x$, $y$, $z$, and $t$; those measured in the second reference frame will be denoted by $x'$, $y'$, $z'$, and $t'$. The values

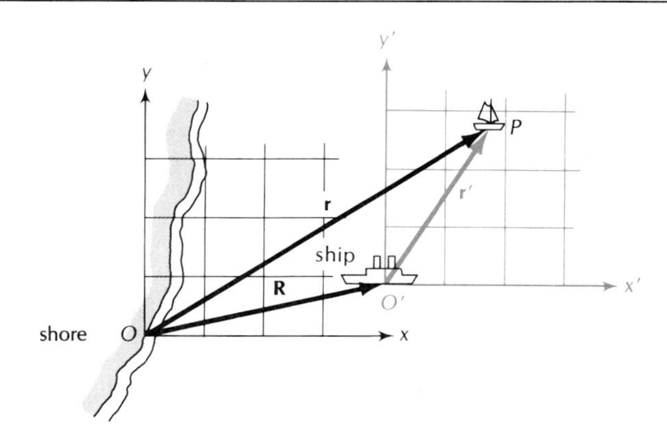


**Fig. 4.18** The coordinate grid $x'$–$y'$ of the ship moves relative to the coordinate grid $x$–$y$ of the shore.

of these two sets of coordinates measured in the two reference frames will then be different. However, the coordinates obtained in one reference frame are related to those obtained in the other reference frame by a simple transformation formula which can be deduced as follows.

We begin with the time coordinate. Since time is absolute, at least to a very good approximation (see Section 1.1), we can assume without loss of generality that the clocks in both reference frames are permanently synchronized; hence the measured values of the time coincide,

$$\boxed{t' = t} \tag{49}$$

Next we consider the space coordinates. Figure 4.18 shows the coordinate grids associated with both the shore and the ship; these coordinate grids have the same orientation (the axes are parallel), but they move past one another at a speed of 10 m/s. The position vector of the sailboat is designated by $\mathbf{r}$ in the coordinate grid of the shore, and by $\mathbf{r}'$ in the coordinate grid of the ship. To establish the mathematical relations between $\mathbf{r}$ and $\mathbf{r}'$, we will make the simplifying (but unnecessary) assumption that the origins $O$ and $O'$ of the two coordinate grids coincide at the time $t = 0$. The displacement $\mathbf{R}$ from the origin $O$ to the origin $O'$ at any later time is then

$$\mathbf{R} = \mathbf{V}_O t \tag{50}$$

where $\mathbf{V}_O$ is the velocity of the second coordinate grid with respect to the first, i.e., $\mathbf{V}_O$ is the velocity of the ship with respect to the shore. Inspection of Figure 4.18 then indicates that the position vector $\mathbf{r}$ is simply the sum of the position vector $\mathbf{r}'$ and the displacement $\mathbf{R}$,

$$\mathbf{r} = \mathbf{r}' + \mathbf{R} \tag{51}$$

or

$$\mathbf{r} = \mathbf{r}' + \mathbf{V}_O t \tag{52}$$

However, there is a catch in this argument: $\mathbf{r}'$ is a vector defined in the reference frame of the ship, whereas $\mathbf{R}$ is a vector defined in the reference frame of the shore; and before we add these vectors according to the usual rules of vector addition, we ought to ask whether these rules remain valid for vectors defined in *different* reference frames. The justification for the vector addition in Eq. (51) rests on the fact that length is absolute — at least to a very good approximation (see Section 1.1). As a consequence, the observers in the reference frame of the shore and those in the reference frame of the ship agree on the measured length (and direction) of the line segment from the ship to the sailboat (from $O'$ to $P$), that is, the value of the displacement vector $\mathbf{r}'$ is the same regardless of whether it is measured in the reference frame of the ship or in that of the shore. It is therefore quite legitimate to add this vector $\mathbf{r}'$ to the vector $\mathbf{R}$, as in Eq. (51), since we may pretend that both vectors have been measured in the reference frame of the shore.

Equation (52) can be rewritten as

$$\boxed{\mathbf{r}' = \mathbf{r} - \mathbf{V}_O t} \tag{53}$$

If the velocity $\mathbf{V}_O$ is along the $x$ axis, as in Figure 4.19, then the components of this equation are

$$x' = x - V_O t \tag{54}$$

$$y' = y \tag{55}$$

$$z' = z \tag{56}$$

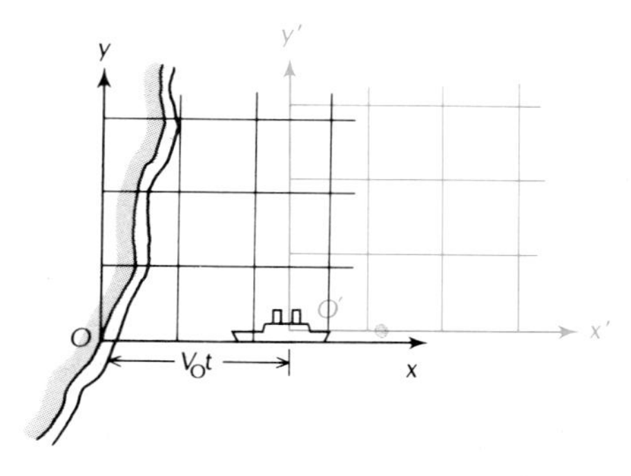


**Fig. 4.19** The coordinate grid $x'$–$y'$ moves along the $x$ axis of the coordinate grid $x$–$y$.

Taken together, Eqs. (49), (54), (55), and (56) constitute the transformation equations that change the time and the space coordinates of one reference frame into those of another. These transformation equations for reference frames in uniform motion relative to one another are called the **Galilean transformations.** It is worth emphasizing that these equations hinge on the assumption that length and time are absolute. As we saw in Chapter 1, these assumptions lie at the foundations of Newton's theory of space and time. The Galilean transformation equations are therefore valid as long as Newton's theory is valid, that is, as long as we are dealing with the velocities of everyday experience, which are small compared to the velocity of light. New transformation equations are required when the velocities are comparable to the velocity of light, and they will be the subject of Chapter 41.

*Galilean transformations*

The Galilean transformation equations for velocity and acceleration can be obtained directly from Eq. (53) by differentiation. Thus, the velocity of the sailboat relative to the ship of Figure 4.18 is

$$\mathbf{v}' = \frac{d\mathbf{r}'}{dt'} = \frac{d\mathbf{r}'}{dt} = \frac{d}{dt}(\mathbf{r} - \mathbf{V}_O t) = \frac{d\mathbf{r}}{dt} - \mathbf{V}_O \tag{57}$$

that is,

$$\boxed{\mathbf{v}' = \mathbf{v} - \mathbf{V}_O} \tag{58}$$

This shows that the velocity of the sailboat relative to the ship is merely the difference between the velocities of sailboat and ship relative to the shore. If this relation seems intuitively obvious, it is only because the absolute character of time and length are so deeply engrained in our intuition.

Likewise, the acceleration of the sailboat relative to the ship is

$$\mathbf{a}' = \frac{d\mathbf{v}'}{dt'} = \frac{d\mathbf{v}'}{dt} = \frac{d}{dt}(\mathbf{v} - \mathbf{V}_O) = \frac{d\mathbf{v}}{dt} - 0 \tag{59}$$

that is,

$$\mathbf{a}' = \mathbf{a} \tag{60}$$

Hence the accelerations of the sailboat measured in the two reference frames are exactly the same. This means that acceleration is absolute — the acceleration does not depend on the reference frame. But note that the absolute character of acceleration hinges on the assumption that $\mathbf{V}_O$ is constant — if not, then the two accelerations $\mathbf{a}'$ and $\mathbf{a}$ would differ by $d\mathbf{V}_O/dt$.

EXAMPLE 9. Off the coast of Miami, the Gulf Stream current has a velocity of 4.8 km/h in a direction due north. The captain of a motorboat wants to travel on a straight course from Miami to Bimini Island, due east of Miami. His boat has a speed of 18 km/h relative to the water. In what direction must he head his boat? What will be its speed relative to the shore?

SOLUTION: Obviously, the captain must head his boat somewhat to the south to compensate for the drift produced by the Gulf Stream current. Figure 4.20a shows the path of the boat relative to the shore, and the heading of the boat. Figure 4.20b shows the velocity vector $\mathbf{V}_O$ of the water relative to the shore and the velocity vector $\mathbf{v}'$ of the boat relative to the water. According to Eq. (58), the velocity vector $\mathbf{v}$ of the boat relative to the shore is the sum of $\mathbf{v}'$ and $\mathbf{V}_O$,

$$\mathbf{v} = \mathbf{v}' + \mathbf{V}_O \tag{61}$$

By hypothesis, this vector $\mathbf{v}$ points due east (Figure 4.20b). Since the vector triangle is a right triangle, the angle between $\mathbf{v}'$ and the eastward direction is given by

$$\sin\theta = V_O/v'$$

so that

$$\theta = \sin^{-1}\left(\frac{V_O}{v'}\right) = \sin^{-1}\left(\frac{4.8\ \text{km/h}}{18\ \text{km/h}}\right) = 15°$$

Thus, the boat must head 15° south of east. The speed of the boat relative to the shore is

$$v = v'\cos\theta = (18\ \text{km/h}) \times \cos 15° = 17\ \text{km/h}$$

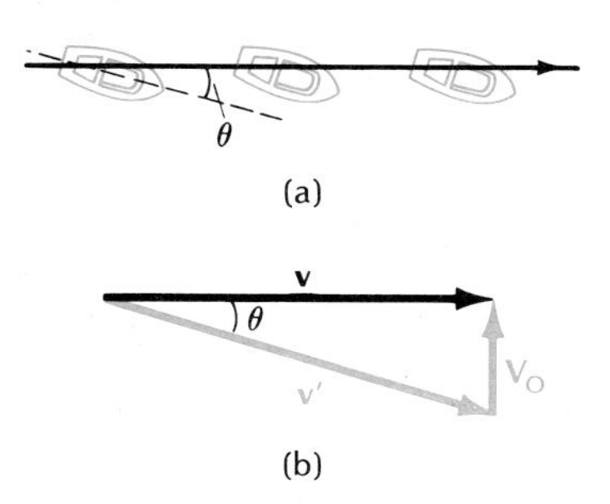


**Fig. 4.20** (a) The path of the boat is toward the east, but the heading of the boat is south of east. (b) The velocity vector $\mathbf{V}_O$ of the water relative to the shore, and the velocity vector $\mathbf{v}'$ of the boat relative to the water. The sum of these vectors equals the velocity $\mathbf{v}$ of the boat relative to the shore.

COMMENTS AND SUGGESTIONS: As a check on such a calculation, it is a good idea to draw a careful scale diagram of the vectors **v** and $\mathbf{V}_O$ and measure the heading angle directly on the diagram (navigators of boats often prefer such a graphical solution of current and drift problems). The calculation in this example is also useful for problems involving the drift of an aircraft produced by a crosswind.

## SUMMARY

**Average velocity:** $\bar{\mathbf{v}} = \dfrac{\Delta \mathbf{r}}{\Delta t}$

**Instantaneous velocity:** $\mathbf{v} = \dfrac{d\mathbf{r}}{dt}$

**Average acceleration:** $\bar{\mathbf{a}} = \dfrac{\Delta \mathbf{v}}{\Delta t}$

**Instantaneous acceleration:** $\mathbf{a} = \dfrac{d\mathbf{v}}{dt} = \dfrac{d^2\mathbf{r}}{dt^2}$

**Motion with constant acceleration:** $\mathbf{v} = \mathbf{v}_0 + \mathbf{a}t$

$$\mathbf{r} - \mathbf{r}_0 = \mathbf{v}_0 t + \tfrac{1}{2}\mathbf{a}t^2$$

**Motion of a projectile:** $v_x = v_{0x} = v_0 \cos\theta$

$$v_z = v_{0z} - gt = v_0 \sin\theta - gt$$

$$x = v_{0x}t$$

$$z = v_{0z}t - \tfrac{1}{2}gt^2$$

**Range, maximum height, and time of flight (over flat ground):** $x_{\text{max}} = \dfrac{v_0^2 \sin 2\theta}{g}$

$$z_{\text{max}} = \frac{v_0^2 \sin^2\theta}{2g}$$

$$t_{\text{flight}} = \frac{2v_0 \sin\theta}{g}$$

**Centripetal acceleration in uniform circular motion:** $a = v^2/r$

**Galilean transformations:** $t' = t$

$$\mathbf{r}' = \mathbf{r} - \mathbf{V}_O t$$

$$\mathbf{v}' = \mathbf{v} - \mathbf{V}_O$$

## QUESTIONS

1. The speedometer of your automobile shows that you are proceeding at a steady 80 km/h. Is it nevertheless possible that your automobile is in accelerated motion?

2. Can an automobile have eastward instantaneous velocity and northward instantaneous acceleration? Give an example.

3. Consider an automobile that is rounding a curve and braking at the same

time. Draw a diagram showing the relative directions of the instantaneous velocity and acceleration.

4. A projectile is launched over level ground. Its initial velocity has a horizontal component $v_{0x}$ and a vertical component $v_{0y}$. What is the average velocity of the projectile between the instants of launch and of impact?

5. What is the acceleration of a projectile when it reaches the top of its trajectory?

6. Are the velocity and the acceleration of a projectile ever perpendicular? Parallel?

7. If you throw a crumpled piece of paper, its trajectory is not a parabola. How does it differ from a parabola and why?

8. If a projectile is subject to air resistance, then the elevation angle for maximum range is not 45°. Do you expect the angle to be larger or smaller than 45°?

9. Figure 4.12 shows the trajectory of a high-speed projectile subject to strong air friction. The trajectory is not symmetric about its highest point. Why not? Would you expect the ascending or the descending portion of the trajectory to take more time?

10. Baseball pitchers are fond of throwing curve balls. How does the trajectory of such a ball differ from the simple parabolic trajectory we studied in this chapter? What accounts for the difference?

11. A particle travels once around a circle with uniform circular motion. What is the average velocity and the average acceleration?

12. A pendulum is swinging back and forth. Is this uniform circular motion? Draw a diagram showing the directions of the velocity and the acceleration at the top of the swing. Draw a similar diagram at the bottom of the swing.

13. Suppose that at the top of its parabolic trajectory a projectile has a horizontal speed $v_{0x}$. The segment at the top of the parabola can be approximated by a circle, called the osculating circle (Figure 4.21). What is the radius of this circle? (Hint: The projectile is instantaneously in uniform circular motion at the top of the parabola.)

**Fig. 4.21** The osculating circle.

14. Why do raindrops fall down at a pronounced angle with the vertical when seen from the window of a speeding train? Is this angle necessarily the same as that of the path of a water drop sliding down along the outside surface of the window?

15. When a sailboat is sailing to windward ("beating"), the wind feels much stronger than when the sailboat is sailing downwind ("running"). Why?

16. Rain is falling vertically. If you run through the rain, at what angle should you hold your umbrella? If you don't have an umbrella, should you bend forward while running?

17. In the reference frame of the ground, the path of a sailboat beating to windward makes an angle of 45° with the direction of the wind. In the reference frame of the sailboat, the angle is somewhat smaller. Explain.

18. Suggest a method for measuring the speed of falling raindrops.

19. According to a theory proposed by Galileo, the tides on the oceans are caused by the Earth's rotational motion about its axis combined with its translational motion around the Sun. At midnight these motions are in the same direction; at noon they are in opposite directions (Figure 4.22). Thus, any point on the Earth alternately speeds up and slows down. Galileo was of the opinion that the speeding up and slowing down of the ocean basins would make the water slosh back and forth, thus giving rise to tides. What is wrong with this theory? (Hint: What is the acceleration of a point on the Earth? Does this acceleration depend on the translational motion?)

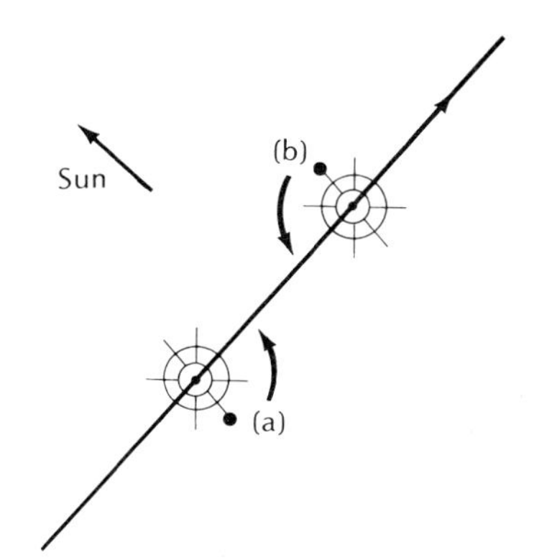


**Fig. 4.22** Rotational and translational motions of the Earth: (a) at midnight and (b) at noon.

## PROBLEMS

### Section 4.1

1. A sailboat tacking against the wind moves as follows: 3.2 km at 45° east of north, 4.5 km at 50° west of north, and 2.6 km at 45° east of north. The entire motion takes 1 h 15 min.
   (a) What is the total displacement for this motion?
   (b) What is the average velocity for this motion?
   (c) What is the speed if it is assumed to be constant?

2. In one-half year, the Earth moves halfway around its orbit, a circle of radius $1.50 \times 10^{11}$ m centered on the Sun. What is the average speed and what is the magnitude of the average velocity for this time interval?

3. The fastest bird is the spine-tailed swift, which reaches speeds of 171 km/h. Suppose that you wish to shoot such a bird with a 22-caliber rifle that fires a bullet with a speed of 366 m/s. If you fire at the instant when the bird is 30 m directly overhead, how many meters ahead of the bird must you aim the rifle? Ignore gravity in this problem.

4. Suppose that the position vector of a particle is given by the following function of time:

$$\mathbf{r} = (6 + 2t^2)\hat{\mathbf{x}} + (3 - 2t + 3t^2)\hat{\mathbf{y}}$$

where distance is measured in meters and time in seconds.
   (a) What is the instantaneous velocity vector at $t = 2$ s? What is the magnitude of this vector?
   (b) What is the instantaneous acceleration vector at $t = 2$ s? What is the magnitude of this vector?

5. Suppose that the acceleration vector of a particle moving in the $x$–$y$ plane is

$$\mathbf{a} = 3\hat{\mathbf{x}} + 2\hat{\mathbf{y}}$$

where the acceleration is measured in m/s$^2$. The velocity vector and the position vector are zero at $t = 0$.
(a) What is the velocity vector of this particle as a function of time?
(b) What is the position vector as a function of time?

6. Suppose that a particle moving in three dimensions has a position vector

$$\mathbf{r} = (4 + 2t)\hat{\mathbf{x}} + (3 + 5t + 4t^2)\hat{\mathbf{y}} + (2 - 2t - 3t^2)\hat{\mathbf{z}}$$

where distance is measured in meters and time in seconds.
(a) Find the instantaneous velocity vector.
(b) Find the instantaneous acceleration vector. What is the magnitude and the direction of the acceleration?

7. A particle is moving in the $x$–$y$ plane; the components of its position vector are

$$x = A \cos bt \qquad y = A \sin bt$$

where $A$ and $b$ are constants.
   (a) What are the components of the instantaneous velocity vector? The instantaneous acceleration vector?
   (b) What is the magnitude of the instantaneous velocity? The instantaneous acceleration?

### Sections 4.2 and 4.3

8. The fastest recorded speed of a baseball thrown by a pitcher is 162.3 km/h

(100.9 mi/h), achieved by Nolan Ryan in 1974 at Anaheim Stadium. If the baseball leaves the pitcher's hand with a horizontal velocity of this magnitude, how far will the ball have fallen vertically by the time it has traveled 20 m horizontally?

9. At Acapulco, professional divers jump from a 36-m-high cliff into the sea (compare Example 2.6). At the base of the cliff, a rocky ledge sticks out for a horizontal distance of 6.4 m. With what minimum horizontal velocity must the divers jump off if they are to clear this ledge?

10. A stunt driver wants to make his car jump over 10 cars parked side by side below a horizontal ramp (Figure 4.23). With what minimum speed must he drive off the ramp? The vertical height of the ramp is 2.0 m and the horizontal distance he must clear is 24 m.

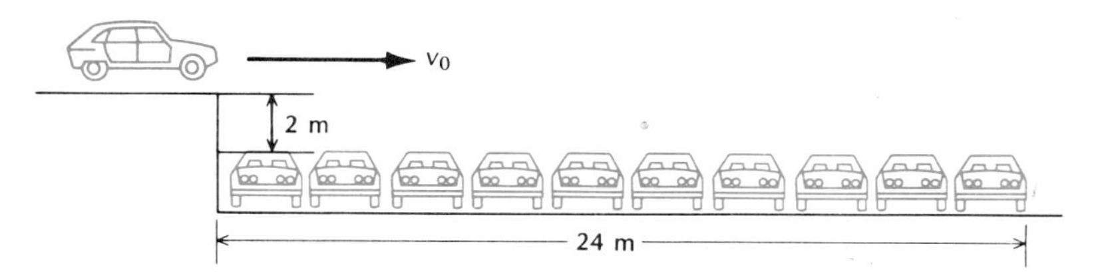


**Fig. 4.23**

11. A particle has an initial position vector $\mathbf{r} = 0$ and an initial velocity $\mathbf{v}_0 = 3\hat{\mathbf{x}} + 2\hat{\mathbf{y}}$ (where distance is measured in meters and velocity in meters per second). The particle moves with a constant acceleration $\mathbf{a} = \hat{\mathbf{x}} - 4\hat{\mathbf{y}}$ (measured in m/s$^2$). At what time does the particle reach a maximum $y$ coordinate? What is the position vector of the particle at that time?

12. According to a reliable report, in 1795 a member of the Turkish embassy in England shot an arrow to a distance of 441 m. According to a less reliable report, a few years later the Turkish Sultan Selim shot an arrow to 889 m. In each of these cases calculate what must have been the minimum initial speed of the arrow.

13. A golfer claims that a golf ball launched with an elevation angle of 12° can reach a horizontal range of 250 m. Ignoring air friction, what would the initial speed of such a golf ball have to be? What maximum height would it reach?

14. A gunner wants to fire a gun at a target at a horizontal distance of 12,500 m from his position.

(a) If his gun fires with a muzzle velocity of 700 m/s and if $g = 9.81$ m/s$^2$, what is the correct elevation angle? Pretend that there is no air resistance.

(b) If the gunner mistakenly assumes $g = 9.80$ m/s$^2$, by how many meters will he miss the target?

15. According to the *Guinness Book of World Records,* during a catastrophic explosion at Halifax on December 6, 1917, William Becker was thrown through the air for some 1500 m and was found, still alive, in a tree. Assume that Becker left the ground and returned to the ground (ignore the height of the tree) at an angle of 45°. With what speed did he leave the ground? How high did he rise? How long did he stay in flight?

16. Volcanoes on the Earth eject rocks at speeds of up to 700 m/s. Assume that the rocks are ejected in all directions; ignore the height of the volcano and ignore air friction.

(a) What is the maximum height reached by rocks?

(b) What is the maximum horizontal distance reached by rocks?

(c) Is it reasonable to ignore air friction in these calculations?

17. In a circus act at the Ringling Bros. and Barnum & Bailey Circus, a "human cannonball" was fired from a large cannon with a muzzle velocity of

87 km/h. Assume that the firing angle was 45° from the horizontal. How many seconds did the human cannonball take to reach maximum height? How high did he rise? How far from the cannon did he land?

18. The world record for the javelin throw by a woman established in 1976 by Ruth Fuchs in Berlin was 69.11 m (226 ft 9 in.). If Fuchs had thrown her javelin with the same initial velocity in Buenos Aires rather than in Berlin, how much farther would it have gone? The acceleration of gravity is 9.8128 $m/s^2$ in Berlin and 9.7967 $m/s^2$ in Buenos Aires. Pretend that air resistance plays no role in this problem.

19. The motion of a ballistic missile can be regarded as the motion of a projectile, because along the greatest part of its trajectory the missile is in free fall, outside of the atmosphere. Suppose that a missile is to strike a target 1000 km away. What minimum speed must the missile have at the beginning of its trajectory? What maximum height does it reach when launched with this minimum speed? How long does it take to reach its target? For these calculations assume that $g = 9.8$ $m/s^2$ everywhere along the trajectory and ignore the (short) portions of the trajectory inside the atmosphere.

**Fig. 4.24**

20. The natives of the South American Andes throw stones by means of slings which they whirl over their heads (see Figure 4.24). They can accurately throw a $\frac{1}{5}$-kg stone to a distance of 50 m.

(a) What is the minimum speed with which the stone must leave the sling to reach this distance?

(b) Just before release, the stone is being whirled around a circle of radius 1.0 m with the speed calculated in part (a). How many revolutions per second does the stone make?

21. A large stone-throwing engine designed by Archimedes could throw a 77-kg stone over a range of 180 m. What must have been the initial speed of the stone if thrown at an initial angle of 45° with the horizontal?

22. The nozzle of a fire hose ejects 280 liters of water per minute at a speed of 26 m/s. How far away will the stream of water land if the nozzle is aimed at an angle of 35° with the horizontal? How many liters of water are in the air at any given instant?

23. According to an ancient Greek source, a stone-throwing machine on one occasion achieved a range of 730 m. If this is true, what must have been the minimum initial speed of the stone as it was ejected from the engine? When thrown with this speed, how long would the stone have taken to reach its target?

*24. (a) A golfer wants to drive a ball to a distance of 240 m. If he launches the ball with an elevation angle of 14°, what is the appropriate initial speed? Ignore air resistance.

(b) If the speed is too great by 0.6 m/s, how much farther will the ball travel when launched at the same angle?

(c) If the elevation angle is 0.5° larger than 14°, how much farther will the ball travel if launched with the speed calculated in part (a)?

*25. Show that for a projectile launched with an elevation angle of 45°, the maximum height reached is one-quarter of the range.

*26. During a famous jump in Richmond, Virginia, in 1903, the horse Heatherbloom with its rider jumped over an obstacle 8 ft 8 in. high while covering a horizontal distance of 37 ft. At what angle and with what speed did the horse leave the ground? Make the (somewhat doubtful) assumption that the motion of the horse is particle motion.

*27. With what elevation angle must you launch a projectile if its range is to equal twice its maximum height?

*28. In a baseball game, the batter hits the ball and launches it upward at an angle of 52° with a speed of 38 m/s. At the same instant, the center fielder starts to run toward the (expected) point of impact of the ball from a distance of 45 m. If he runs at 8.0 m/s, can he reach the point of impact before the ball?

*29. The gun of a coastal battery is emplaced on a hill 50 m above the water level. It fires a shot with a muzzle speed of 600 m/s at a ship at a horizontal distance of 12,000 m. What elevation angle must the gun have if the shot is to hit the ship? Pretend that there is no air resistance.

*30. In a flying ski jump, the skier acquires a speed of 110 km/h by racing down a steep hill and then lifts off into the air from a horizontal ramp. Beyond this ramp, the ground slopes downward at an angle of 45°.

(a) Assuming that the skier is in free-fall motion after he leaves the ramp, at what distance down the slope will he land?

(b) In actual jumps, skiers reach distances of up to 165 m. Why does this not agree with the result you obtained in part (a)?

*31. Olympic target archers shoot arrows at a bull's-eye 12 cm across from a distance of 90 m. If the initial speed of the arrow is 70 m/s, what must be the elevation angle? If the archer misaims the arrow by 0.03° in the vertical direction, will it hit the bull's-eye? If the archer misaims the arrow by 0.03° in the horizontal direction, will it hit the bull's-eye? Assume that the height of the bull's-eye above the ground is the same as the height of the bow, and ignore air resistance.

*32. The muzzle speed for a Lee-Enfield rifle is 630 m/s. Suppose you fire this rifle at a target 700 m away and at the same level as the rifle.

(a) In order to hit the target, you must aim the barrel at a point above the target. How many meters above the target must you aim? Pretend that there is no air resistance.

(b) What will be the maximum height that the bullet reaches along its trajectory?

(c) How long does the bullet take to reach the target?

*33. In artillery, it is standard practice to fire a sequence of trial shots at a target before commencing to fire "for effect." The artillerist first fires a shot short of the target, then a shot beyond the target, and then makes the necessary adjustment in elevation so that the third shot is exactly on target. Suppose that the first shot fired from a gun aimed with an elevation angle of 7°20′ lands 180 m short of the target; the second shot fired with an elevation of 7°35′ lands 120 m beyond the target. What is the correct elevation angle to hit the target?

*34. A hay-baling machine throws each finished bundle of hay 2.5 m up in the air so it can land on a trailer waiting 5 m behind the machine. What must be the speed with which the bundles are launched? What must be the angle of launch?

*35. A battleship steaming at 45 km/h fires a gun at right angles to the longitudinal axis of the ship. The elevation angle of the gun is 30° and the muzzle velocity of the shot is 720 m/s; the gravitational acceleration is 9.80 m/s$^2$. What is the range of this shot in the reference frame of the ground? Pretend that there is no air resistance.

**36. The maximum speed with which you can throw a stone is about 25 m/s (a professional baseball pitcher can do much better than this). Can you hit a window 50 m away and 13 m up from the point where the stone leaves your hand? What is the maximum height of a window you can hit at this distance?

**37. A gun standing on sloping ground (see Figure 4.25) fires up the slope. Show that the range of the gun (measured along the slope) is

$$l = \frac{2v_0^2 \cos^2\theta}{g \cos \alpha} (\tan \theta - \tan \alpha)$$

where $\alpha$ is the angle of the slope and the other symbols have their usual meaning. For what value of $\theta$ is this range maximum?

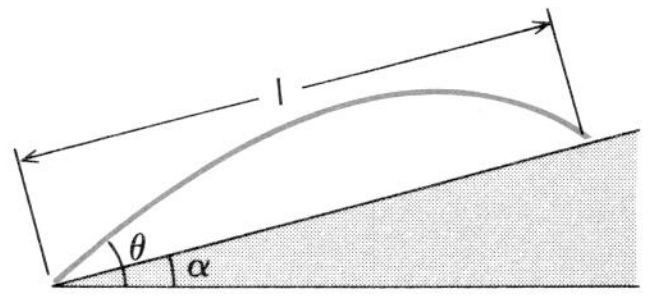

**Fig. 4.25**

**Fig. 4.26**

**38. When a tractor leaves a muddy field and drives on the highway, clumps of mud will sometimes come off the rear wheels and be launched into the air (see Figure 4.26). In terms of the speed $u$ of the tractor and the radius $R$ of the wheel, find the maximum height that a clump of dirt can reach. In your calculation be careful to take into account both the initial velocity of the clump and the initial height at which it comes off the wheel. Evaluate numerically for $u = 30$ km/h and $R = 0.80$ m. (Hint: Solve this problem in the reference frame of the tractor.)

**39. A gun on the shore (at sea level) fires a shot at a ship which is heading directly toward the gun at a speed of 40 km/h. At the instant of firing, the distance to the ship is 15,000 m. The muzzle velocity of the shot is 700 m/s. Pretend that there is no air resistance.

(a) What is the required elevation angle for the gun? Assume $g = 9.80$ m/s$^2$.
(b) What is the time interval between firing and impact?

**40. A ship is steaming at 30 km/h on a course parallel to a straight shore at a distance of 17,000 m. A gun emplaced on the shore (at sea level) fires a shot with a muzzle velocity of 700 m/s when the ship is at the point of closest approach. If the shot is to hit the ship, what must be the elevation angle of the gun? How far ahead of the ship must the gun be aimed? Give the answer to the latter question both in meters and in minutes of arc. Pretend that there is no air resistance. (Hint: Solve this problem by the following method of successive approximations. First calculate the time of flight of the shot, neglecting the motion of the ship; then calculate how far the ship moves in this time; and then calculate the elevation angle and the aiming angle required to hit the ship at this new position.)

### Section 4.4

41. An ultracentrifuge spins a small test tube in a circle of radius 10 cm at 1000 revolutions per second. What is the centripetal acceleration of the test tube? How many G does this amount to?

42. The blade of a circular saw has a diameter of 20 cm. If this blade rotates at 7000 revolutions per minute (its maximum safe speed), what is the speed and what is the centripetal acceleration of a point on the rim?

43. At the Fermilab accelerator (one of the world's largest atom smashers), protons are forced to travel in an evacuated tube in a circular orbit of diameter 2.0 km. The protons have a speed nearly equal to the speed of light (99.99995% of the speed of light). What is the centripetal acceleration of these protons? Express your answer in m/s$^2$ and in G.

44. An automobile travels at a steady 90 km/h along a road leading over a small hill. The top of the hill is rounded so that, in the vertical plane, the road approximately follows an arc of circle of radius 70 m. What is the centripetal acceleration of the automobile at the top of the hill?

45. The Earth moves around the Sun in a circular path of radius $1.50 \times 10^{11}$ m at uniform speed. What is the magnitude of the centripetal acceleration of the Earth toward the Sun?

46. An automobile has wheels of diameter 64 cm. What is the centripetal acceleration of a point on the rim of this wheel when the automobile is traveling at 95 km/h?

47. The Earth rotates about its axis once in one sidereal day of 23 h 56 min. Calculate the centripetal acceleration of a point located on the equator. Calculate the centripetal acceleration of a point located at a latitude of 45°.

*48. When looping the loop, the Blue Angels stunt pilots of the U.S. Navy fly their jet aircraft along a vertical circle of radius 1000 m (Figure 4.27). At the top of the circle, the speed is 350 km/h; at the bottom of the circle, the speed is 620 km/h. What is the centripetal acceleration at the top? At the bottom? In the reference frame of one of these aircraft, what is the acceleration that the pilot feels at the top and at the bottom, i.e., what is the acceleration relative to the aircraft of a small body, such as a coin, released by the pilot?

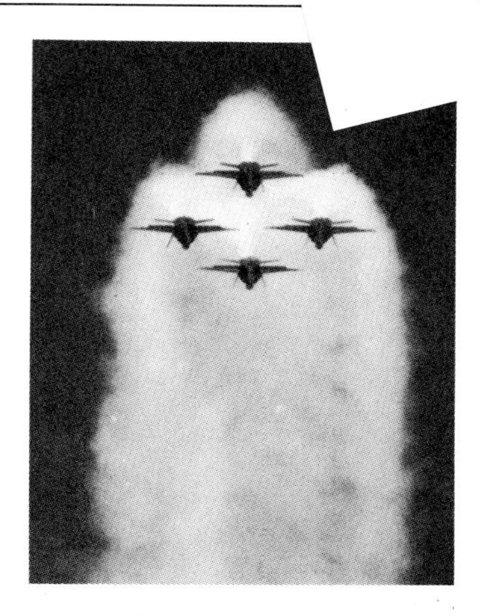

Fig. 4.27

**49. The table printed on the endpapers lists the radii of the orbits of the planets ("mean distance from the Sun") and the times required for moving around the orbit ("period of revolution"). For the following problem assume that the planets move along circular orbits at constant speed.

(a) Calculate the magnitude of the centripetal acceleration of each planet.

(b) In a logarithmic graph of centripetal acceleration vs. radius, plot the log of the acceleration vs. the log of the radius and draw a smooth curve through the points. Can you discover a simple equation that describes the acceleration as a function of radius?

### Section 4.5

50. On a rainy day, a steady wind is blowing at 30 km/h. In the reference frame of the *air,* the raindrops are falling vertically with a speed of 10 m/s. What is the magnitude and direction of the velocity of the raindrops in the reference frame of the ground?

51. A lump of concrete falls off a crumbling overpass and strikes an automobile traveling on a highway below. The lump of concrete falls 5 m before impact and the automobile has a speed of 90 km/h.

(a) What is the speed of impact of the lump in the reference frame of the automobile?

(b) What is the angle of impact?

52. On a rainy day, raindrops are falling with a vertical velocity of 10 m/s. If an automobile drives through the rain at 25 m/s, what is the velocity (magnitude and direction) of the raindrops relative to the automobile?

53. A battleship steaming at 13 m/s toward the shore fires a shot in the forward direction. The elevation angle of the gun is 20° and the muzzle speed of the shot is 660 m/s. What is the velocity vector of the shot relative to the shore?

54. A wind of 30 m/s is blowing from the west. What will be the speed, relative to the ground, of a sound signal traveling due north? The speed of sound, relative to air, is 330 m/s.

55. On a windy day, a hot-air balloon is ascending at a rate of 0.5 m/s relative to the air. Simultaneously, the air is moving with a horizontal velocity of 12 m/s. What is the velocity (magnitude and direction) of the balloon relative to the ground?

*56. A blimp is motoring at constant altitude. The airspeed indicator on the blimp shows that its speed relative to the air is 20 km/h and the compass shows that the heading of the blimp is 10° east of north. If the air is moving over the ground with a velocity of 15 km/h due east, what is the velocity (magnitude and direction) of the blimp relative to the ground? For an observer on the ground, what is the angle between the longitudinal axis of the blimp and the direction of motion?

*57. A sailboat is moving in a direction 50° east of north at a speed of 14 km/h. The wind measured by an instrument aboard the sailboat has an apparent (relative to the sailboat) speed of 32 km/h coming from an apparent direction of 10° east of north. Find the true (relative to the ground) speed and direction of the wind.

*58. (a) In still air, a high-performance sailplane has a rate of descent (or sinking rate) of 0.50 m/s at a forward speed (or airspeed) of 60 km/h. Suppose the plane is at an initial altitude of 1500 m. How far can it travel horizontally in still air before it reaches the ground?

(b) Suppose the plane is in a (horizontal) wind of 20 km/h. With the same initial conditions, how far can it travel in the downwind direction? In the upwind direction?

*59. A wind is blowing at 50 km/h from a direction 45° west of north. The pilot of an airplane wishes to fly on a route due north from an airport. The airspeed of the airplane is 250 km/h.

(a) In what direction must the pilot point the nose of the airplane?
(b) What will be the airplane's speed relative to the ground?

*60. At the entrance of Ambrose Channel at New York Harbor, the tidal current at one time of the day has a velocity of 4.2 km/h in a direction 20° south of east. Consider a ship in this current; suppose that the ship has a speed of 16 km/h relative to the water. If the helmsman keeps the bow of the ship aimed due north, what will be the actual velocity (magnitude and direction) of the ship relative to the ground?

*61. A white automobile is traveling at a constant speed of 90 km/h on a highway. The driver notices a red automobile 1.0 km behind, traveling in the same direction. Two minutes later, the red automobile passes the white automobile.

(a) What is the average speed of the red automobile relative to the white?
(b) What is the speed of the red automobile relative to the ground?

*62. Two automobiles travel at equal speeds in opposite directions on two separate lanes of a highway. The automobiles move at constant speed $v_0$ on straight parallel tracks separated by a distance $h$. Find a formula for the rate of change of the distance between the automobiles as a function of time; take the instant of closest approach as $t = 0$. Plot $v$ vs. $t$ for $v_0 = 60$ km/h, $h = 50$ m.

*63. A ferryboat on a river has a speed $v$ relative to the water. The water of the river flows with a speed $V$ relative to the ground. The width of the river is $d$.

(a) Show that the ferryboat takes a time $2d/\sqrt{v^2 - V^2}$ to travel across the river and back.
(b) Show that the ferryboat takes a time $2dv/(v^2 - V^2)$ to travel a distance $d$ up the river and back. Which trip takes a shorter time?

**64. An AWACS aircraft is flying at high altitude in a wind of 150 km/h from due west. Relative to the air, the heading of the aircraft is due north and its speed is 750 km/h. A radar operator on the aircraft spots an unidentified target approaching from northeast; relative to the AWACS aircraft, the bearing of the target is 45° east of north, and its speed is 950 km/h. What is the speed of the unidentified target relative to the ground?

CHAPTER 5

# Dynamics — Newton's Laws

**Dynamics** is the study of forces and their effects on the motion of bodies. So far we have dealt only with the mathematical description of motion — the definitions of position, velocity, and acceleration, and the relationships among these quantities. We did not inquire what causes a body to accelerate. In this chapter we will see that the cause of acceleration is a force exerted by some external agent. The fundamental properties of force and the relationship between force and acceleration are given by Newton's three laws of motion. The first of these laws describes the natural state of motion of a free body on which no external forces are acting, whereas the other two laws deal with the behavior of bodies under the influence of forces. The first law was actually discovered by Galileo Galilei early in the seventeenth century, but it remained for Isaac Newton, in the second half of the seventeenth century, to formulate a coherent theory of forces and to lay down a complete set of equations from which the motion of bodies under the influence of arbitrary forces can be calculated.

**Sir Isaac Newton,** *1642–1727, English mathematician and physicist, widely regarded as the greatest scientist of all time. He was professor at Cambridge, president of the Royal Society, and later became master of the Mint. His brilliant discoveries in mechanics were published in 1687 in his book* Principia Mathematica, *one of the glories of the Age of Reason. In this book, Newton laid down the laws of motion and the law of universal gravitation, and he demonstrated that planets in the sky as well as bodies on the Earth obey the same mathematical equations. For over 200 years, Newton's laws stood as the unchallenged basis of all our attempts at a scientific explanation of the physical world; even today, these laws still remain as the essential ingredient of engineering physics. Newton was also the foremost mathematician of his time, and he shares with the philosopher Gottfried Leibniz the credit for discovering the calculus.*

## 5.1 Newton's First Law

Everyday experience suggests that a force — a push or pull — is needed to keep a body moving at constant velocity. For example, if the wind pushing a sailboat suddenly ceases, the boat will coast along for some distance, but it will gradually slow down, stop, and remain stopped until a new gust of wind comes along. However, what actually slows down the sailboat is not the *absence* of a propulsive force but, rather, the *presence* of friction forces. Under ideal frictionless conditions a body in motion would continue to move forever. Experiments with pucks or gliders riding on a cushion of air on a low-friction air table or air track give some hint of the persistence of motion; but in order to eliminate friction entirely, it is best to use bodies moving in a vacuum, without even air against which to rub. Observations of particles moving in evacuated tubes as well as observations of celestial bodies — planets, satellites, and comets — moving in the emptiness of interplanetary space show that a body left to itself persists indefinitely in its state of uniform motion.[1] A body on which no external force is acting is called a **free body,** and the experiments and observations concerning such bodies are summarized in **Newton's First Law:**

*Newton's First Law*

> *A body at rest remains at rest and a body in motion continues to move at constant velocity unless acted upon by an external force.*

This law is also called the **law of inertia,** since it expresses the tendency of bodies to maintain their original state of motion.

One crucial restriction on Newton's First Law concerns the choice of reference frame: the law is not valid in all reference frames, but only in certain special frames. It is obvious that if this law is valid in one given reference frame, then it cannot be valid in a second reference frame that has an accelerated motion relative to the first. For example, in the reference frame of the ground, a ball at rest on the floor of a train station remains at rest, but in the reference frame of an accelerating train leaving the station, a ball initially at rest on the floor of a car has a "spontaneous" acceleration toward the rear of the train, in contradiction to Newton's First Law. Those special reference frames in which the law is valid are called **inertial reference frames.** Thus the reference frame of the ground is an inertial reference frame, but that of the accelerating train is not.

*Inertial reference frame*

How do we know whether or not a given reference frame is inertial? We can only tell by making a test: take a free body, that is, a body isolated from all external forces, and observe its motion; if the body persists in its state of uniform motion, then the reference frame is inertial. This test procedure suggests that Newton's First Law is nothing but a rule for finding the "right" reference frames. But the First Law is much more than that. The decisive fact is that after *one* body has been used to test that the reference frame is inertial, all *other* free bodies will obey the First Law in that reference frame without any further quibble. With this in mind, we can reformulate the essence of the First Law as follows: *If one free body remains at rest or in uniform motion in a given reference frame, then so will all other free bodies.* This is a law of nature, i.e., it is a statement about the behavior of the physical world that can be verified by experiments. Thus, the First Law plays a dual role: it is a law of nature as well as a definition of what is meant by an inertial reference frame. Several other laws of physics, including Newton's Second Law (see below), also play such dual roles.

Note that if some given reference frame is inertial, any other refer-

---

[1] The motion of celestial bodies is not motion with uniform velocity — planets and comets have centripetal accelerations toward the Sun, which are produced by the gravitational pull of the Sun. Nevertheless, observation of the celestial bodies provides evidence for the persistence of uniform motion because the centripetal acceleration produced by gravity is found to be the *only* acceleration experienced by these bodies — the components of the motion perpendicular to the centripetal direction are unaccelerated, and thereby testify to the persistence of uniform motion in the absence of external forces.

ence frame in uniform translational motion relative to the first will also be inertial, and any other reference frame in accelerated motion relative to the first will not be inertial. Thus, any two inertial reference frames can differ only by some constant relative velocity;[2] they cannot differ by an acceleration. This implies that, as measured with respect to inertial reference frames, *acceleration is absolute:* when a particle has some acceleration in one inertial reference frame, then it will have exactly the same acceleration in any other. By contrast, the velocity of the particle is relative; the coordinates in one inertial reference frame are related to those in another by the Galilean transformations of Eq. (4.53), and the velocities are related by the simple additive law of Eq. (4.58).

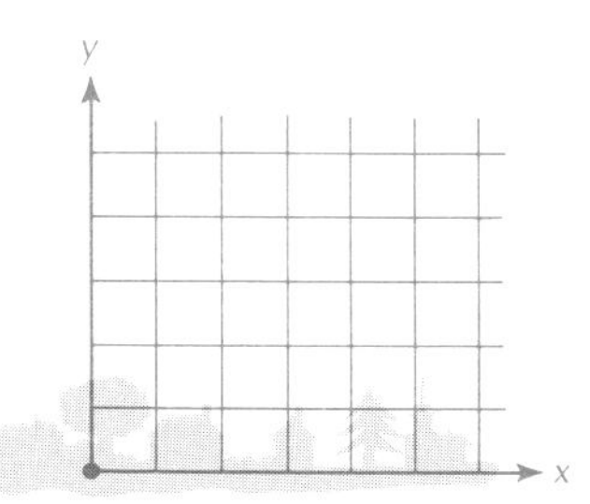


**Fig. 5.1** A coordinate system with the origin of coordinates fixed at the surface of the Earth.

Finally, we must answer an important question: Which of the reference frames in practical use for everyday measurements are inertial? For the description of everyday phenomena, the most commonly used reference frame is one attached to the ground, with the origin of coordinates fixed at some point on the surface of the Earth (Figure 5.1). Although crude experiments indicate that this reference frame is inertial (for example, an isolated railroad car at rest on a level track remains at rest), more precise experiments show that this reference frame is not inertial. A famous experiment of this kind is Foucault's pendulum experiment (Figure 5.2). When a long pendulum on a twist-free suspension is allowed to oscillate back and forth, observation shows that the pendulum gradually drifts sideways, that is, the orientation of the plane of oscillation gradually rotates relative to the Earth. Of course, the pendulum bob is not a free body, since gravity and the pendulum string both pull on it; but neither of these forces can cause the transverse rotational motion that is observed. Hence this motion indicates that Newton's First Law is not valid in the reference frame of the Earth. The explanation is that the Earth rotates about its axis — it rotates relative to an inertial reference frame. Since rotational motion has a centripetal acceleration, a reference frame attached to the Earth has an acceleration and is not an inertial reference frame. Thus, the rotational motion of the plane of oscillation of the Foucault pendulum arises from the rotation of the Earth: the pendulum tends to swing in a fixed plane, but the Earth rotates under it.

**Fig. 5.2** Foucault pendulum at the Smithsonian Institution, Washington, D. C.

Although a reference frame attached to the ground is not exactly inertial, the numerical value of the centripetal acceleration of points on the surface of the Earth is fairly small. Figure 5.1 shows the origin of coordinates at a point on the surface of the Earth; this point is in uniform circular motion around the Earth's axis. If the origin is at a geographical latitude $\theta$, then the radius of the circular motion is $r = R_E \cos\theta$ (where $R_E$ is the radius of the Earth; Figure 5.3) and the speed of the motion is $v = 2\pi r/T$ (where $T = 23$ h 56 min, the time for one rotation of the Earth).[3] The centripetal acceleration is then

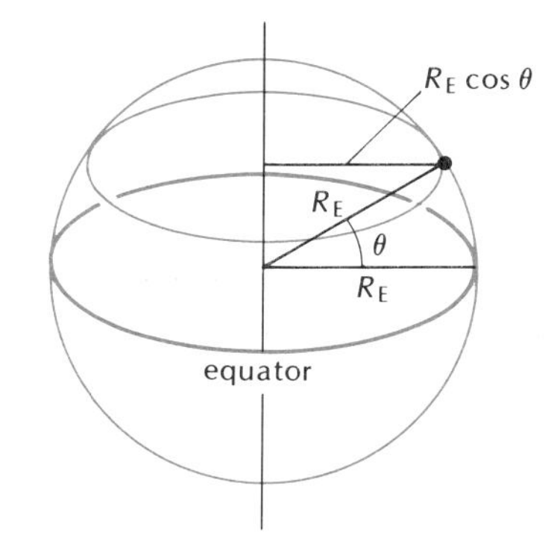


**Fig. 5.3** A point on the surface of the Earth at latitude $\theta$ moves along a circle of radius $R_E \cos\theta$ as the Earth rotates.

[2] They can also differ in the orientation of the axes. But, in the present context, such a difference is of no interest.

[3] The period of rotation of the Earth with respect to an inertial reference frame is 23 h 56 min rather than 24 h. The reason is that, as seen from the Earth, the Sun advances with respect to the background of "fixed" stars. Consequently, in order to catch up with the Sun, the Greenwich meridian must turn through a bit more than one revolution; this makes the time between successive meridian passages of the Sun (solar day, 24 h) slightly longer than the time between successive meridian passages of a fixed star (sidereal day, 23 h 56 min). The difference is small and can often be neglected.

$$a = \frac{v^2}{r} = \left(\frac{2\pi r}{T}\right)^2 \frac{1}{r} = 4\pi^2 \frac{R_E \cos\theta}{T^2} \tag{1}$$

At the equator ($\cos\theta = 1$), this gives the result $a = 0.034$ m/s$^2$; at other latitudes, this result is reduced by a factor of $\cos\theta$. For most practical purposes, we can neglect this small centripetal acceleration and we can regard a reference frame attached to the ground as inertial to a good approximation. What contributes to the success of this approximation is that, to some extent, the effects of the noninertial acceleration of the reference frame can be hidden among the effects of gravity. For example, at the equator (say, in Quito, Ecuador), the centripetal acceleration of the reference frame gives a free body a relative centrifugal acceleration of 0.034 m/s$^2$ upward; simultaneously, gravity gives it an acceleration of 9.814 m/s$^2$ downward; the result is a net acceleration of 9.780 m/s$^2$ downward. Tables of values of $g$ (e.g., Table 2.3) always list this net acceleration, that is, these tables already take into account the acceleration of the reference frame. When using such tables one can therefore pretend that the reference frame attached to the ground is inertial. However, there are limits to this pretense — the Foucault experiment shows that not all the effects of the acceleration of the reference frame can be hidden in the value of $g$. Incidentally, most of the variation of $g$ with geographical latitude (see Table 2.3) arises from the variation of the centripetal acceleration with latitude [see Eq. (1)]. Some extra variation arises from the ellipsoidal shape of the Earth, which makes the pull of gravity somewhat stronger at the poles than at the equator.

Hereafter, unless otherwise stated, we will take it for granted that the reference frames in which we express the laws of physics are inertial reference frames, either exactly inertial or at least so nearly inertial that no appreciable deviation from Newton's First Law occurs within the region of space and time on which we wish to concentrate.

**Ernst Mach** (makh), *1838–1916, Austrian philosopher and physicist, professor at Prague and at Vienna. He performed a profound analysis of the logical foundations of physics. His book* The Science of Mechanics *is a brilliant critical examination of the historical development of Newtonian mechanics.*

## 5.2 Newton's Second Law

Newton's Second Law establishes the relation between the force acting on a body and the acceleration caused by this force. The mass of the body enters this relation as a constant of proportionality. Thus, before we can deal with the Second Law, we need a precise definition of mass. The standard of mass (standard kilogram) has already been described in Section 1.5. It remains to set up a procedure for comparing an unknown mass with this standard. The common procedure makes use of a balance (see Figure 1.15). But this is not quite satisfactory — a balance will not function in interstellar space, far from the pull of gravity of the Earth or any other celestial body. It is therefore desirable to contrive a measurement procedure that can be used anywhere in the universe. This is a question of principle, but it also is a practical question. During the Skylab mission, three astronauts were kept for about 2 months under zero-gravity conditions in a capsule orbiting the Earth, and scientists who wanted a daily record of the astronauts' masses had to invent some measurement device not subject to the limitations of an ordinary balance.

A perfectly general procedure, originally proposed by Ernst Mach, for the measurement of an unknown mass is the following. Take the body of unknown mass and take a standard mass (the standard kilogram or some other known mass) and let them interact, that is, let them exert forces on each other. The mechanism involved in the generation of these forces is irrelevant. For example, the bodies may exert forces on each other by means of springs, rubber bands, strings, etc. The only restriction is that the device used to communicate the forces from one body to the other must itself have a mass that is negligible compared to both the unknown mass and the standard mass (as we will see in Section 5.4, the forces acting on the two bodies are then of equal magnitudes and opposite directions). Under the influence of their mutual pushes or pulls, the body of unknown mass and the standard mass will accelerate away from or toward one another. Designate the magnitudes of the accelerations of these bodies (in an inertial reference frame) by $a$ and $a_s$, respectively, and denote the values of their masses by $m$ and $m_s$, respectively. The unknown mass is then defined by the relation

$$\boxed{\frac{m}{m_s} = \frac{a_s}{a}} \tag{2}$$

*Definition of mass*

According to this relation, the ratio of the masses is the inverse ratio of the accelerations; thus, the unknown mass is large if its acceleration is small. This is of course quite reasonable — a large mass has large inertia and is therefore hard to accelerate. The rigorous definition given by Eq. (2) expresses the intuitive notion that mass is a measure of the resistance that a body offers to changes in its velocity.

For the practical application of Eq. (2), instead of letting the body of unknown mass and the standard mass exert forces directly and simultaneously on each other, it is sometimes more convenient to subject these two bodies separately and successively to the same force (provided by a spring or a rubber band attached to a fixed support). Under these circumstances, the ratio of the accelerations of the two bodies will be the same as when the bodies interact directly, and the ratio of masses can, again, be calculated from Eq. (2).

---

EXAMPLE 1. The mass measurement device aboard Skylab consisted of a small chair that could be accelerated back and forth by a spring attached to it (see Figure 5.4). Instruments connected to the chair measured the accelera-

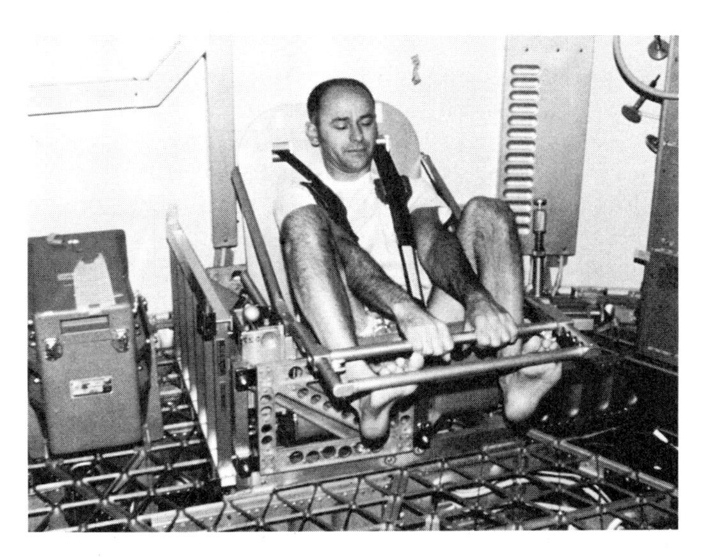

**Fig. 5.4** Body-mass measurement device on Skylab.

tion.[4] Suppose that with a known standard mass of 66.9 kg placed in the chair, the bent spring produced an acceleration of 0.0262 $m/s^2$. With the standard mass removed and with astronaut J. R. Lousma sitting in the chair, the bent spring (with the same amount of bending) produced an acceleration of 0.0204 $m/s^2$. Deduce the mass of Lousma. Ignore the mass of the chair.

SOLUTION: Since the bent spring provides the same force when the standard mass is placed in the chair and when the astronaut is placed in the chair, Eq. (2) is valid for the operation of the Skylab device. Substituting the measured values of the accelerations, we find

$$m = \frac{a_s}{a} m_s = \frac{0.0262 \text{ m/s}^2}{0.0204 \text{ m/s}^2} \times 66.9 \text{ kg} = 85.9 \text{ kg} \tag{3}$$

for the mass of Lousma.

---

Now that we have available a precise definition of mass, we can state Newton's Second Law, which tells us what acceleration an external force produces when acting on a body of a given mass. This law summarizes the results of experiments and observations on bodies moving under the action of external forces. Qualitatively, a force is any push or pull exerted on a body. It is intuitively obvious that such a push or pull has a direction as well as a magnitude — in fact, force is a vector quantity and therefore can be represented graphically by an arrow. The precise quantitative definition of force is included in the Second Law, and we will discuss this definition later on. For the sake of simplicity, we assume that only one force is acting on the body, but we will eliminate this assumption in the next section. **Newton's Second Law** asserts:

*Newton's Second Law*

*A force acting on a body causes an acceleration which is in the direction of the force and has a magnitude inversely proportional to the mass of the body,*

$$\mathbf{a} = \mathbf{F}/m \tag{4}$$

or

$$\boxed{m\mathbf{a} = \mathbf{F}} \tag{5}$$

The Second Law is subject to the same restrictions as the First Law: it is valid only in inertial reference frames. Furthermore, the Second Law, just like the First Law, plays a dual role: it is a law of nature and it also serves as a precise definition of force. To measure a given force — say, the force generated by a spring that has been stretched a certain amount — we apply this force to the standard mass. If the resulting acceleration of the standard mass is $a_s$, then the force has a magnitude

$$F = m_s a_s$$

After the standard mass has been used to measure the force, any other masses to which this same force is applied will be found to obey the

[4] The instruments actually measured the period of the back-and-forth oscillations of the chair; this amounts to an indirect measurement of the acceleration.

Second Law. In regard to these other masses, the Second Law is a law of nature — it is an assertion about the physical world that can be verified by experiments.

The most precise empirical test of the Second Law is supplied by the study of the motion of celestial bodies. The observed motions of the planets, satellites, and comets have been found to agree with this law with extreme precision. Another severe empirical test of the law is supplied by experiments with electrons, protons, and ions (atoms with missing electrons or with excess electrons) moving in electric and magnetic fields. The motion of such particles is in accord with the Second Law as long as the speed of the particles is low compared with the speed of light, but the motion displays deviations from the Second Law when the speed is comparable to the speed of light. These deviations from the Second Law at high speeds indicate a failure of the physics of Newton. As we will see in Chapter 41, high-speed motion belongs to the realm of the theory of relativity, where the physics of Einstein supersedes the physics of Newton. However, the motion of ordinary bodies with ordinary speeds — such as the motion of baseballs, automobiles, ships, aircraft, or space shuttles — is in complete agreement with Newton's Second Law, and we can always trust this law when dealing with such bodies.

*Newton,* N

In the metric system, the unit of force is the **newton** (N); this is the force that will give the standard mass an acceleration of 1 $m/s^2$,

$$1 \text{ newton} = 1 \text{ N} = 1 \text{ kg} \cdot \text{m/s}^2 \tag{6}$$

Table 5.1 lists the magnitudes of some typical forces.

**Table 5.1** SOME FORCES

| | |
|---|---|
| Gravitational pull of Sun on Earth | $3.5 \times 10^{22}$ N |
| Gravitational pull of Earth on Moon | $2.0 \times 10^{20}$ N |
| Thrust of Saturn V rocket engines | $3.3 \times 10^{7}$ N |
| Pull of large tugboat | $1 \times 10^{6}$ N |
| Thrust of jet engines (Boeing 747) | $7.7 \times 10^{5}$ N |
| Pull of large locomotive | $5 \times 10^{5}$ N |
| Gravitational pull of Earth on automobile | $1.5 \times 10^{4}$ N |
| Decelerating force on automobile during braking | $1 \times 10^{4}$ N |
| Force between two protons in a nucleus | $\sim 10^{4}$ N |
| Accelerating force on automobile | $7 \times 10^{3}$ N |
| Gravitational pull of Earth on man | $7.2 \times 10^{2}$ N |
| Maximum upward force exerted by forearm (isometric) | $2.7 \times 10^{2}$ N |
| Gravitational pull of Earth on 1 kg | 9.8 N |
| Gravitational pull of Earth on apple | 2 N |
| Gravitational pull of Earth on 5¢ coin | $5.1 \times 10^{-2}$ N |
| Force between electron and nucleus of atom (hydrogen) | $8 \times 10^{-8}$ N |

*Pound-force*

In the British system, the unit of force is the **pound-force** (lbf); this unit can be defined in terms of newtons:

$$1 \text{ pound-force} = 1 \text{ lbf} = 4.44822 \text{ N} \tag{7}$$

By means of the conversion factors for kg to lb and for m to ft, the pound-force can be expressed in the alternative form

$$1 \text{ lbf} = 32.174 \text{ lb ft/s}^2 \tag{8}$$

The conversion factor for lb ft/s² to lbf has the same numerical value as the standard gravitational acceleration, 32.174 ft/s². The reason for this coincidence is that 1 lbf equals the downward gravitational pull that the Earth exerts on 1 lb under standard conditions. Note that in the British system, the product of the unit of mass and the unit of acceleration does not give the unit of force — instead, it gives 1/32.174 times the unit of force [see Eq. (8)].[5] This must be kept in mind whenever mass and force are substituted into Newton's Second Law. For instance, if a force of 40 lbf acts on a body of 20 lb, the acceleration is

$$\begin{aligned} a &= F/m = 40 \text{ lbf}/20 \text{ lb} = 2 \text{ lbf/lb} \\ &= 2 \times (32.174 \text{ lb ft/s}^2)/\text{lb} = 64.348 \text{ ft/s}^2 \end{aligned} \tag{9}$$

**Fig. 5.5** *Spirit of America.*

EXAMPLE 2. The jet-engined car *Spirit of America* (see Figure 5.5), which set a world record for speed on the Salt Flats of Utah, had a mass of 4100 kg and its engine could develop up to 68,000 N of thrust. What acceleration could this car achieve?

SOLUTION: According to Newton's Second Law, the horizontal acceleration is

$$a_x = \frac{F_x}{m} = \frac{68{,}000 \text{ N}}{4100 \text{ kg}} = 17 \text{ m/s}^2 \tag{10}$$

This is about 1.7 G.

EXAMPLE 3. A body in free fall has a downward acceleration produced by the gravitational pull of the Earth. What must be this gravitational pull on a body of 1 kg if it is to give this body the standard gravitational acceleration, 9.80665 m/s²?

SOLUTION: With $m = 1$ kg, Newton's Second Law yields

$$F_z = ma_z = 1 \text{ kg} \times (-9.80665 \text{ m/s}^2) = -9.80665 \text{ N}$$

This brings us to the question of the practical measurement of mass and force. In laboratories on the Earth, the most common and most precise mass measurements are carried out with beam balances that compare the downward gravitational pull that the Earth exerts on these masses (see Figure 5.6). This gravitational pull is called the weight. The precise meaning of weight will be further discussed in Section 6.2, and we will see that measurements of mass via weight give results consistent with those obtained by the primary procedure based on Eq. (2). Likewise, measurements of force are often carried out by comparing the unknown force with a known weight. Alternatively, force can be measured with a spring balance (see Figure 5.7) that matches the unknown force with a known force supplied by a stretched spring. The spring can be calibrated by hanging known weights on it and checking how far it stretches.

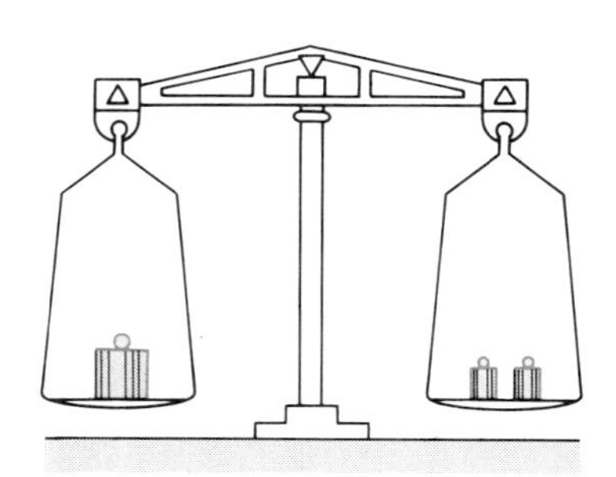

**Fig. 5.6** Beam balance.

*Slug*

[5] It is possible to define a new unit of mass such that the product of this unit of mass and the unit of acceleration gives the British unit of force. This unit of mass is the **slug,**

$$\begin{aligned} 1 \text{ slug} &= 1 \text{ lbf}/(\text{ft/s}^2) \\ &= 32.174 \text{ lb} = 14.594 \text{ kg} \end{aligned}$$

However, this unit of mass is hardly ever used by practicing engineers or by anybody else.

The masses of electrons and protons and the masses of ions are too small to be measured by their weight. Instead, they are measured with a procedure based on Eq. (5). The procedure involves applying a known force to the particle, measuring the resulting acceleration, and then calculating the mass from Eq. (5). The device used for such mass determinations is called a mass spectrometer. It consists of an evacuated vessel into which is shot a steady succession of particles, in the form of a beam. In the absence of forces, this beam would proceed along a straight path. But the spectrometer applies known electric and magnetic forces to the particles, causing a deflection of the beam. The measured value of the deflection indicates the acceleration, and Eq. (5) then permits the calculation of the masses of the particles. Table 5.2 lists the masses of the electron, the proton, and the neutron (the latter particle does not respond to electric and magnetic forces, and hence its mass cannot be determined directly by the mass-spectrometer method; but its mass can be deduced from the masses of ions containing known numbers of neutrons in their atomic nuclei).

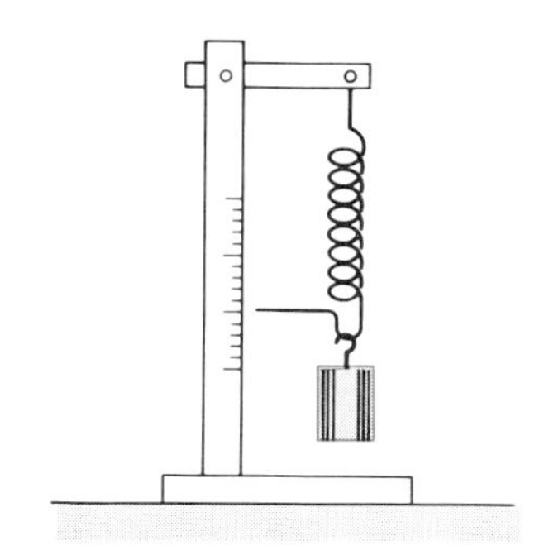

**Fig. 5.7** Spring balance.

*Masses of electron, proton, and neutron*

**Table 5.2** THE MASSES OF THE ELECTRON, PROTON, AND NEUTRON

| *Particle* | *Mass*[a] |
|---|---|
| Electron | $9.110 \times 10^{-31}$ kg |
| Proton | $1.673 \times 10^{-27}$ |
| Neutron | $1.675 \times 10^{-27}$ |

[a] More precise values of these masses will be found in Appendix 8.

## 5.3 The Superposition of Forces

More often than not, a body will be subjected to the simultaneous action of several forces. For example, Figure 5.8 shows a barge under tow by two tugboats. The forces acting on the barge are the pull of the first tow rope, the pull of the second tow rope, and the frictional resistance of the water; these forces are indicated by the arrows in Figure 5.8.[6] Newton's Second Law tells us what each of these forces would do

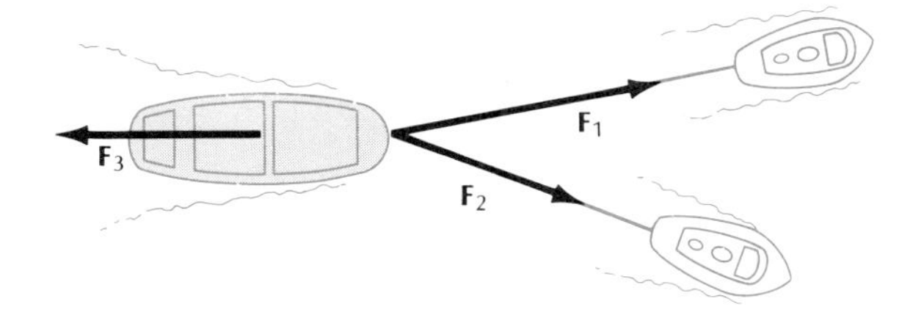


**Fig. 5.8** A barge under tow by two tugboats. The frictional resistance of the water opposes the motion.

if acting by itself. The question now is: How can we calculate the simultaneous effect of two or more forces? The answer is supplied by the **principle of superposition** for forces:

[6] These are the horizontal forces. There are also vertical forces: the downward pull of gravity (weight) and the upward pressure of the water (buoyancy). These vertical forces will be ignored (they cancel each other).

*If several forces* $\mathbf{F}_1$, $\mathbf{F}_2$, $\mathbf{F}_3$, . . . , *act simultaneously on a body, then the acceleration is the same as that produced by a single force given by*

*Principle of superposition*

$$\mathbf{F}_{\text{net}} = \mathbf{F}_1 + \mathbf{F}_2 + \mathbf{F}_3 + \cdots \tag{11}$$

*Net force*

The single force that has the same effect as the combination of the individual forces is called the **net force** or the **resultant force.** According to Eq. (11), the net force is merely the vector sum of the individual forces. In terms of the net force, Newton's Second Law becomes

$$m\mathbf{a} = \mathbf{F}_{\text{net}} \tag{12}$$

Equation (12) can be interpreted as follows: Each force ($\mathbf{F}_1$, $\mathbf{F}_2$, $\mathbf{F}_3$, . . .) acting by itself produces its own acceleration ($\mathbf{a}_1 = \mathbf{F}_1/m$, $\mathbf{a}_2 = \mathbf{F}_2/m$, $\mathbf{a}_3 = \mathbf{F}_3/m$, . . .); and all the forces acting together produce a net acceleration $\mathbf{a}$, which, according to Eq. (12), is simply the sum of these individual accelerations,

$$\mathbf{a} = \frac{1}{m}(\mathbf{F}_1 + \mathbf{F}_2 + \mathbf{F}_3 + \cdots) \tag{13}$$

$$= \mathbf{a}_1 + \mathbf{a}_2 + \mathbf{a}_3 + \cdots \tag{14}$$

Thus, the principle of superposition of forces is equivalent to the assertion that each force produces an acceleration independently of the presence or absence of other forces.

We must emphasize that this principle is a law of nature which has the same status as Newton's laws. The most precise empirical test of this principle emerges from the study of planetary motion. There, one finds that the net force on a planet is indeed the vector sum of all the gravitational pulls exerted by the Sun and by the other planets. Somewhat less precise tests of this principle can be performed in a laboratory experiment by pulling on a body with known forces in known directions.

---

EXAMPLE 4. Suppose that the two towropes in Figure 5.8 pull with horizontal forces of $2 \times 10^5$ and $1.5 \times 10^5$ N, and that these forces make angles of 10° and 20° with the long axis of the barge (Figure 5.9). Suppose that the friction force is zero. What is the net horizontal force on the barge?

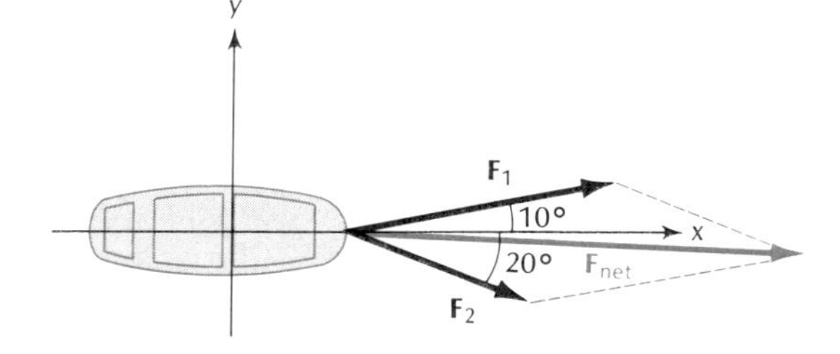


**Fig. 5.9** The forces $\mathbf{F}_1$ and $\mathbf{F}_2$ pull on the barge. The friction force is assumed to be zero. The net force is the vector sum of $\mathbf{F}_1$ and $\mathbf{F}_2$.

SOLUTION: The net force is

$$\mathbf{F}_{\text{net}} = \mathbf{F}_1 + \mathbf{F}_2$$

where $\mathbf{F}_1$ is the force of the first towrope and $\mathbf{F}_2$ that of the second. With the $x$ and $y$ axes as shown in Figure 5.9, the net force has components

$$F_{\text{net},x} = 2.0 \times 10^5 \text{ N} \times \cos 10° + 1.5 \times 10^5 \text{ N} \times \cos 20°$$

$$= 3.38 \times 10^5 \text{ N}$$

$$F_{\text{net},y} = 2.0 \times 10^5 \text{ N} \times \sin 10° - 1.5 \times 10^5 \text{ N} \times \sin 20°$$

$$= -1.66 \times 10^4 \text{ N}$$

This net force is shown in Figure 5.9.

Comments and Suggestions: Here we have used the component method to add the two force vectors. We could equally well have used the trigonometric method, and found the net force by solving the vector triangle. What method we adopt is a matter of preference. However, we must always bear in mind that if a problem asks for a vector quantity, such as force, the final answer must list all the components of the vector, or else the magnitude *and* the direction of the vector.

## 5.4 Newton's Third Law

Consider a tugboat pushing on a barge (Figure 5.10a) with a force of, say, $2 \times 10^5$ N. This force of $2 \times 10^5$ N describes the action of the tugboat on the barge. However, there is also a reciprocal action (or reaction) of the barge on the tugboat; the presence of the barge slows down the motion of the tugboat — the barge pushes back on the tugboat. Thus the mutual interaction of the tugboat and the barge involves two forces: the "action" force of the tugboat on the barge and the "reaction" force of the barge on the tugboat. These forces are said to form an **action–reaction pair.** Which of the forces is regarded as "action" and which as "reaction" is irrelevant. It may seem reasonable to regard the push of the tugboat as an action; then the push of the barge back on the tugboat is a reaction. However, it is equally valid to regard the push the barge exerts on the tugboat as an action and then the push of the tugboat is a reaction. The important point is that forces always occur in pairs; each of them cannot exist without the other.

*Action and reaction*

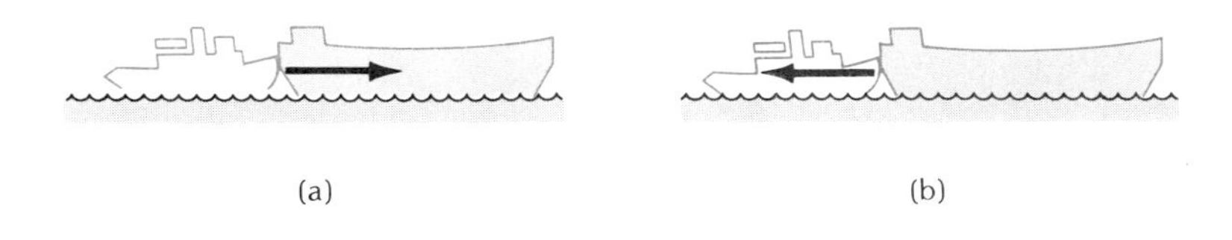


**Fig. 5.10** (a) Tugboat pushes on barge (action). (b) Barge pushes on tugboat (reaction).

**Newton's Third Law** gives the quantitative relationship between the action force and the corresponding reaction force:

*Newton's Third Law*

> *Whenever a body exerts a force on another body, the latter exerts a force of equal magnitude and opposite direction on the former.*

To return to the example given above, the tugboat exerts a force of $2 \times 10^5$ N on the barge; hence by the Third Law the barge exerts an

**Fig. 5.11** Horizontal forces acting on the tugboat.

**Fig. 5.12** Automobile pushes on ground; ground pushes on automobile. (This diagram shows only the horizontal forces.)

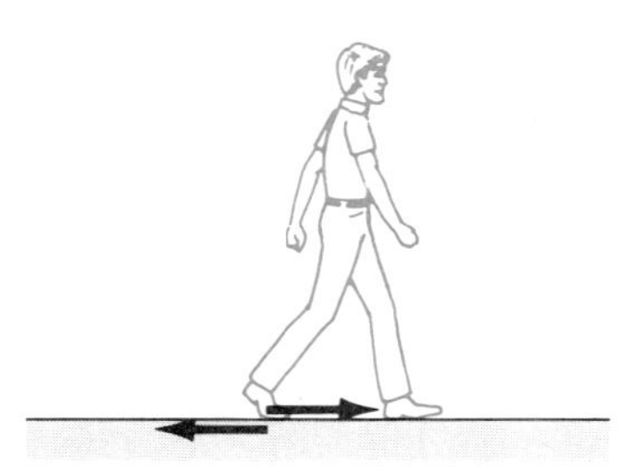

**Fig. 5.13** Man pushes on ground; ground pushes on man. (This diagram shows only the horizontal forces.)

opposite force of $2 \times 10^5$ N on the tugboat (Figure 5.10b). Note that although these forces are of equal magnitudes, they act on different bodies and their effects are quite different: the first force gives an acceleration to the barge (if there is no other force acting on the barge), whereas the second force merely slows the tugboat and prevents it from accelerating as much as it would if the barge were not there. Thus, although action and reaction are forces of equal magnitudes and of opposite directions, their effects do not cancel because *they are applied to different bodies.*

Note that the force that propels the tugboat forward is another example of a reaction force: the propeller of the tugboat presses against the water (action) and the water consequently presses against the propeller (reaction) and pushes the tugboat along. Figure 5.11 shows the two horizontal forces[7] exerted on the tugboat; the reaction of the barge is in the backward direction and the reaction of the water is in the forward direction.

Reaction forces play a crucial role in all machines that produce locomotion by pushing against the ground, water, or air. For example, an automobile moves by pushing backward on the ground with its wheels; the reaction of the ground then pushes the automobile forward (Figure 5.12). A man walks by pushing backward on the ground; the reaction of the ground then pushes the man forward (Figure 5.13), and so on.

Reaction forces exist even if the two interacting bodies are not in direct contact, so the forces between them must bridge the intervening empty space. For example, consider an apple in free fall at some height above the ground. The Earth pulls on the apple by means of gravity. If this pull has a magnitude of, say, 2 N, then the Third Law requires that the apple pull on the Earth with an opposite force of 2 N (Figure 5.14). This reaction force is also a form of gravity — it is the gravity that the apple exerts on the Earth. However, the effect of the apple on the motion of the Earth is insignificant because the mass of the Earth is so large (about $6 \times 10^{24}$ kg) that a force of only 2 N can hardly move it.

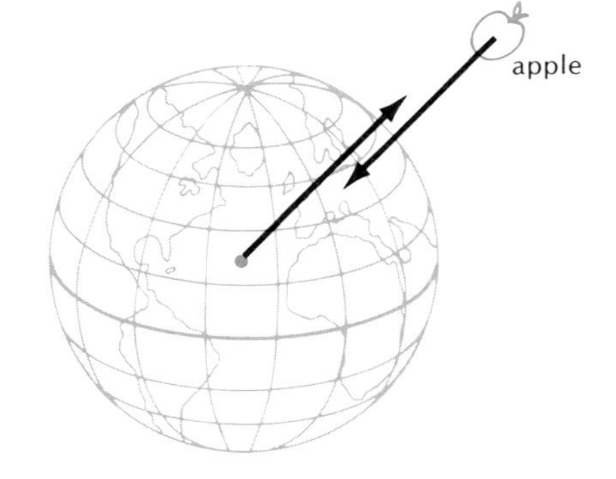


**Fig. 5.14** Earth pulls on apple; apple pulls on Earth.

EXAMPLE 5. A tugboat tows an empty barge of mass 25,000 kg by means of a strong steel cable of mass 200 kg (Figure 5.15). If the tugboat exerts a pull of 3000 N on the cable, what is the acceleration of the barge? What is the tension in the cable at its forward end? At its rearward end? At its midpoint?

[7] There are also vertical forces (weight and buoyant force), but they cancel each other. Friction against the water has been ignored.

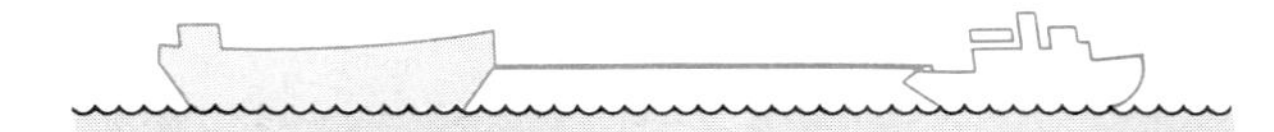

**Fig. 5.15** A tugboat tows a barge by means of a steel cable.

Assume that the cable is horizontal and does not sag, and ignore the friction of the water on the barge.

SOLUTION: The force of 3000 N exerted by the tugboat must accelerate both the cable and the barge; that is, it must accelerate a total mass of $m_{\text{barge}} + m_{\text{cable}}$. Hence the resulting acceleration is

$$a = \frac{F}{m_{\text{barge}} + m_{\text{cable}}} = \frac{3000\text{ N}}{25{,}000\text{ kg} + 200\text{ kg}} = 0.119\text{ m/s}^2$$

By **tension** at the end of the cable is meant the force with which the cable pulls on what is attached to it. Since the tugboat pulls on the cable with a force of 3000 N, Newton's Third Law requires that the cable pull on the tugboat with an equally large force; thus, the tension at the forward end is

$$T_1 = 3000\text{ N}$$

To find the tension at the rearward end, we note that in order to accelerate the barge at the rate of 0.119 m/s$^2$, the cable must exert a pull of $m_{\text{barge}}a$. Hence the tension at the rearward end is

$$T_2 = m_{\text{barge}}a = 25{,}000\text{ kg} \times 0.119\text{ m/s}^2 = 2976\text{ N}$$

By tension at the midpoint of the cable is meant the force with which the forward half of the cable pulls on the rearward half (or vice versa; Figure 5.16). To find this force, we note that it must accelerate both the rearward half of the cable and the barge, that is, it must accelerate a total mass of

$$m_{\text{barge}} + \tfrac{1}{2}m_{\text{cable}}$$

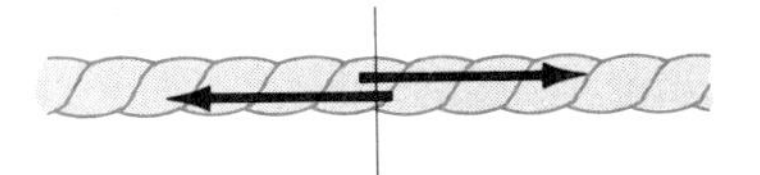

**Fig. 5.16** Forward portion of cable pulls on rearward portion, and vice versa.

Hence the required tension is

$$T_3 = (m_{\text{barge}} + \tfrac{1}{2}m_{\text{cable}})a$$

$$= (25{,}000\text{ kg} + 100\text{ kg}) \times 0.119\text{ m/s}^2 = 2988\text{ N}$$

From this calculation it is evident that the tension in the cable steadily decreases along its length from a value of 3000 N at the forward end to a value of 2976 N at the rearward end. The difference between the tensions at the ends is 24 N; this net force of 24 N acting on the cable is of course exactly what is required to accelerate the cable at the rate of 0.119 m/s$^2$.

COMMENTS AND SUGGESTIONS: In this example, the difference of 24 N between the tensions at the two ends of the cable is only a small fraction of the tension — the difference is about 1% of the tension. For practical purposes one can usually pretend that a freely hanging cable, rope, string, or chain has the same tension everywhere along its entire length, that is, the cable transmits the tension without change in magnitude. This assumption is justified whenever the mass of the cable is negligible compared with the mass pulled by the cable. In subsequent problems we will always neglect the mass of the cable unless it is explicitly stated otherwise.

Note that the definition of mass by means of Eq. (2) hinges on Newton's Third Law and on the assumption that the mass of the device that transmits the force from the unknown mass to the standard mass is negligible. If so, the device will transmit a push or a pull from the unknown mass to the standard mass without change — the two bodies will experience forces of equal magnitude and their accelerations will be inversely proportional to their masses.

EXAMPLE 6. In order to pull an automobile out of the mud in which it is stuck, a man stretches a rope tautly from the front end of the automobile to a stout tree. He then pushes sideways against the rope at the midpoint (Figure 5.17). When he pushes with a force of 900 N, the angle between the two halves of the rope on his right and left is 170°. What is the tension in the rope under these conditions?

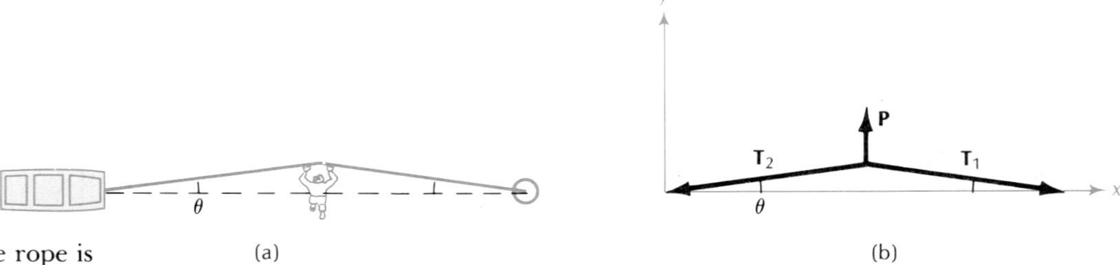


**Fig. 5.17** (a) The rope is stretched between the automobile and the tree. The man pushes at the midpoint. (b) Forces acting on the point of the rope where the man pushes. The angle $\theta$ is 5°.

SOLUTION: Figure 5.17b shows the three forces acting on the rope at the point where the man pushes. These forces are the push **P** and the tensions $\mathbf{T}_1$ and $\mathbf{T}_2$ toward the right and left; the magnitudes of these tensions are equal, $|\mathbf{T}_1| = |\mathbf{T}_2| = T$. Since this point of the rope has next to no mass ($m = 0$), Newton's Second Law tells us that the net force acting on this point is zero ($\mathbf{F}_{\text{net}} = m\mathbf{a} = 0$). Thus, the three forces **P**, $\mathbf{T}_1$, and $\mathbf{T}_2$ acting on this point cancel, that is, they are in equilibrium. With the $x$ and $y$ axes as shown in Figure 5.17b, the components of the forces are

$$P_x = 0 \qquad P_y = P$$

$$T_{1,x} = T\cos\theta \qquad T_{1,y} = -T\sin\theta$$

$$T_{2,x} = -T\cos\theta \qquad T_{2,y} = -T\sin\theta$$

The net force has an $x$ component

$$F_{\text{net},\,x} = 0 + T\cos\theta - T\cos\theta$$

and a $y$ component

$$F_{\text{net},\,y} = P - T\sin\theta - T\sin\theta$$

The $x$ component of the net force is identically zero, regardless of the value of $T$. To make the $y$ component zero, we must satisfy the equation

$$P - 2T\sin\theta = 0$$

This tells us that the tension is

$$T = \frac{P}{2\sin\theta} = \frac{900\text{ N}}{2\sin 5°} = 5.2 \times 10^3\text{ N}$$

Thus, with this rope trick, the push of 900 N generates a tension that is almost six times as large. Of course, once the automobile moves forward, the angle $\theta$ will increase and the tension will decrease. To take full advantage of the rope trick the man must then shorten the rope before pushing again.

COMMENTS AND SUGGESTIONS: This is a typical equilibrium problem: we find an unknown force from the requirement that all the forces acting at some point must cancel. The crucial step in the solution of such an equilibrium problem lies in a suitable choice of the point at which the cancellation occurs. For instance, we could have chosen a point of the rope away from the midpoint, say, a point near the tree; then the equilibrium condition would merely have told us that the tensions pulling this point to the right and to the left are of equal magnitudes — which is true, but not helpful.

## 5.5 The Momentum of a Particle

Newton's laws can be expressed very neatly in terms of momentum, a vector quantity of great importance in physics. The **momentum** of a particle is defined as the product of its mass and velocity:

$$\boxed{\mathbf{p} = m\mathbf{v}} \tag{15}$$

*Momentum*

Thus, the momentum vector **p** has the same direction as the velocity vector, but a magnitude that is $m$ times as large.

The units of momentum are kg · m/s in the metric system and lb · ft/s in the British system.

Expressed in terms of momentum, Newton's First Law merely states that, in the absence of external forces, the momentum of a particle remains constant,

$$\boxed{\mathbf{p} = [\text{constant}]} \tag{16}$$

*Newton's First Law in terms of momentum*

In physics, it is customary to say that a quantity that remains constant during motion is **conserved.** Thus Eq. (16) states that the momentum of a free particle is conserved. Of course, we could equally well say that the velocity of a free particle is conserved; but the deeper significance of momentum will emerge when we study the motion of a system of many particles exerting forces on one another (see Chapter 10). We will find that the total momentum of such a system is conserved — any momentum lost by one particle is compensated by a momentum gain of some other particle or particles. Conserved quantities are useful in physics because they permit us to make some predictions about the motion of a particle or a system of particles without any need for a tedious calculation of the full details of the motion. In later sections we will become acquainted with other quantities that are conserved under suitable conditions, such as angular momentum and energy.

To express the Second Law in terms of momentum, we make use of the fact that the mass of the particle is constant, and we transform $m\mathbf{a}$ into the following:

$$m\mathbf{a} = m\frac{d\mathbf{v}}{dt} = \frac{d}{dt}(m\mathbf{v}) = \frac{d}{dt}(\mathbf{p})$$

The Second Law can then be written

$$\boxed{\frac{d\mathbf{p}}{dt} = \mathbf{F}} \tag{17}$$

*Newton's Second Law in terms of momentum*

This says that the rate of change of momentum equals the force.

Finally, note that we can also express Newton's Third Law in terms of momentum. Since the action force is exactly opposite to the reaction force, the rate of change of momentum generated by the action force on one body is exactly opposite to the rate of change of momentum generated by the reaction force on the other body. Hence we can state the Third Law as follows:

*Whenever two bodies exert forces on one another, the resulting changes of momentum are of equal magnitudes and opposite directions.*

This fact will lead us to a general law of conservation of momentum for a system of particles (see Section 10.1).

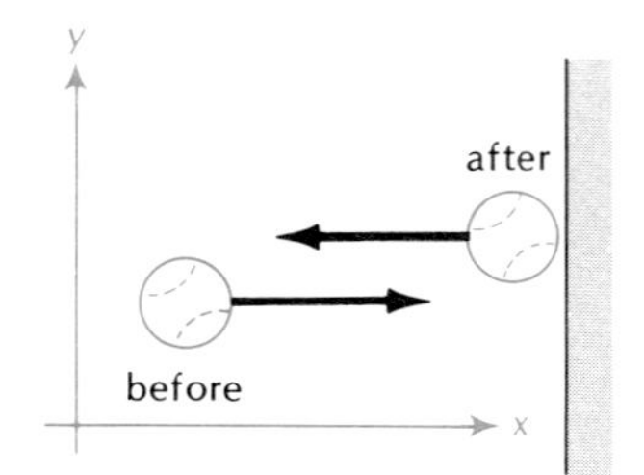


**Fig. 5.18** A tennis ball bounces off a wall.

EXAMPLE 7. A tennis player smashes a ball of mass 60 g at a vertical wall. The ball hits the wall at right angles with a speed of 40 m/s and bounces straight back with the same speed. What is the change of the momentum of the ball during the impact?

SOLUTION: Take the positive $x$ axis along the direction of the initial motion of the ball. The momentum of the ball before impact is then in the positive direction:

$$p_x = mv_x = 0.060 \text{ kg} \times 40 \text{ m/s} = 2.4 \text{ kg} \cdot \text{m/s}$$

The momentum of the ball after impact has the same magnitude but opposite direction (Figure 5.18):

$$p'_x = -2.4 \text{ kg} \cdot \text{m/s}$$

The change of momentum is

$$\Delta p_x = p'_x - p_x = -2.4 \text{ kg} \cdot \text{m/s} - 2.4 \text{ kg} \cdot \text{m/s}$$

$$= -4.8 \text{ kg} \cdot \text{m/s}$$

This change of momentum is produced by the (large) force that acts on the ball during impact on the wall.

COMMENTS AND SUGGESTIONS: In this calculation, the directions of the momentum before and after the impact or, equivalently, the signs of the components $p_x$ and $p'_x$, are crucial. If we had paid no attention to the directions, or the signs, we would have reached the wrong conclusion that the momentum is the same before and after the impact, and that there is no change of momentum.

## 5.6 Newtonian Relativity

As has been emphasized in the early part of this chapter, Newton's First and Second Laws are valid only in inertial reference frames. These laws therefore distinguish from the set of all conceivable reference frames a subset of preferred, inertial reference frames. Different inertial reference frames are in uniform translational motion relative to one another. If a given event is observed in two such reference frames, then the measured coordinates are related by the Galilean transformations [Eqs. (4.54)–(4.56)].

Newton's laws do not make any intrinsic distinction between two inertial reference frames. For instance, Newton's laws in the reference frame of a ship steaming away from the shore at constant velocity are exactly the same as in the reference frame of the shore;[8] these laws de-

[8] For the purposes of this example, we will pretend that the reference frame of the Earth is inertial.

pend only on the *acceleration,* which is the same in both reference frames. Hence the equations governing the motion of billiard balls on a pool table aboard the ship are exactly the same as on shore. No experiment with billiard balls, or any other mechanical experiment, aboard the ship will reveal its uniform motion with respect to the shore. Only *changes* in the uniform motion of the ship can be detected. For example, if the ship rolls, or pitches, or strikes an iceberg and suddenly decelerates, then the billiard balls will misbehave, developing "spontaneous" accelerations relative to the ship.

Thus, uniform motion is relative, whereas accelerated motion is absolute; that is, the uniform motion of a reference frame can only be detected relative to another reference frame, whereas the accelerated motion of a reference frame can be detected by experiments within that reference frame. The inertial guidance systems used aboard ships, aircraft, and missiles take advantage of the absolute character of acceleration to keep track of the motion of the reference frame. These guidance systems rely on **accelerometers** to measure absolute accelerations. Essentially, an accelerometer consists of a spring with one end attached to the reference frame and the other end attached to a mass (a spring balance; Figure 5.19). When the reference frame accelerates, the inertia of the mass stretches the spring; the amount of stretch is directly proportional to the acceleration. High-precision accelerometers

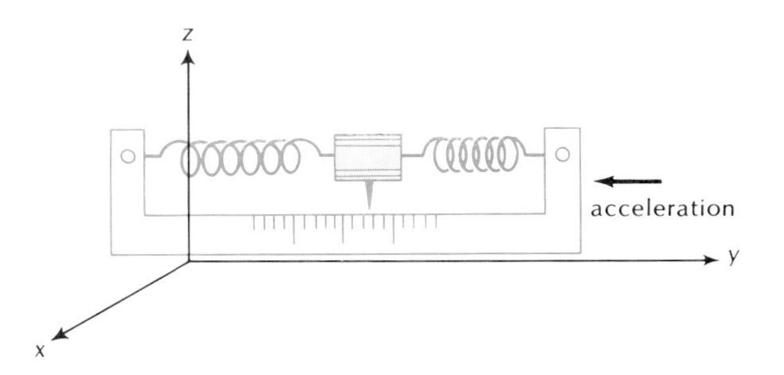


**Fig. 5.19** Accelerometer.

(see Figure 5.20) used in inertial guidance systems are capable of measuring accelerations to within better than $\pm 10^{-6}$ G. A complete inertial guidance system contains three accelerometers to measure the acceleration along three perpendicular axes; the system also contains gyroscopes which maintain the axes in a fixed orientation (see Chapter 13). From a knowledge of the acceleration (both the magnitude and direction) as a function of time and a knowledge of the initial position and velocity, a computer can automatically calculate the position and velocity of the ship at any later time.

**Fig. 5.20** Three-axis accelerometer, consisting of three separate accelerometers mounted at right angles to each other.

The impossibility of detecting uniform translational motion by experiments within an inertial reference frame was recognized by Galileo, who argued that it is impossible to distinguish between a "stationary" Earth and a "moving" Earth by mechanical experiments performed on the Earth's surface.[9] The assertion that the laws of mechanics are the same in all inertial reference frames is called the **Galilean** or **Newtonian Principle of Relativity.**

[9] However, Galileo suffered from some confusion: he thought (erroneously) that the ocean tides result from a combination of the Earth's translational motion and its rotational motion about its axis. If this were so, it would make an absolute distinction between a stationary and a moving Earth.

## SUMMARY

**Newton's First Law:** In an inertial reference frame, a body at rest remains at rest and a body in motion continues to move at constant velocity unless acted upon by an external force.

**Definition of mass:** $m/m_s = a_s/a$

**Newton's Second Law:** $m\mathbf{a} = \mathbf{F}$

**Superposition of forces:** $\mathbf{F}_{net} = \mathbf{F}_1 + \mathbf{F}_2 + \mathbf{F}_3 + \cdots$

**Newton's Third Law:** Whenever a body exerts a force on another body, the latter exerts a force of equal magnitude and opposite direction on the former.

**Momentum:** $\mathbf{p} = m\mathbf{v}$

$$\frac{d\mathbf{p}}{dt} = \mathbf{F}$$

## QUESTIONS

1. If a glass stands on a table on top of a sheet of paper, you can remove the paper without touching the glass by jerking the paper away very sharply. Explain why the glass more or less stays put.

2. Tribes of natives in the Amazon jungle use extremely long and heavy arrows (3 m or more). Why? (Hint: What is likely to happen to an arrow flying through dense jungle?)

3. Make a critical assessment of the following statement: An automobile is a device for pushing the air out of the way of the passenger so that his body can continue to its destination in its natural state of motion at uniform velocity.

4. When rounding a curve in your automobile, you get the impression that a force tries to pull you toward the outside of the curve. Is there such a force?

5. If the Earth were to stop spinning (other things remaining equal), the value of $g$ at all points of the surface except the poles would become slightly larger. Why?

6. Does the mass of a body depend on the frame of reference from which we observe the body? Answer this by appealing to the definition of mass.

7. Suppose that a (strange) body has negative mass. Suppose you tie this body to a body of positive mass of the same magnitude by means of a stretched rubber band. Describe the motion of the two bodies.

8. Does the magnitude or the direction of a force depend on the frame of reference?

9. A fisherman wants to reel in a large dead shark hooked on a thin fishing line. If he jerks the line, it will break; but if he reels it in very gradually and smoothly, it will hold. Explain.

10. The following statements appeared in *Tennis Trade* magazine:

> The racquet is an instrument of work and a transmitter of energy. The amount of energy you transfer to the ball can best be defined as a Force ($F$). The amount of this Force can then be equated to its mass times its acceleration.
>
> $$F = ma$$
>
> $F$ = Force is the amount of energy you transfer to the ball by the racquet.
> $m$ = Mass is the weight of the racquet divided by gravity, or $m = w/g$, gravity is generally 32 ft/sec and for all practical purposes, simply refer to it as weight.

$a$ = Acceleration is the speed with which you swing the racquet.

From the above equation ($F = ma$), you can see that the faster one swings the racquet the greater the force will be transmitted to the ball.

Which of these statements are false? (If a statement makes no sense, regard it as false.) Would any of these statements help you to become a better tennis player?

11. If a body crashes into a water surface at high speed, the impact is almost as hard as on a solid surface. Explain.

12. Moving downwind, a sailboat can go no faster than the wind. Moving across the wind, a sailboat can go faster than the wind. How is this possible? (Hint: What are the horizontal forces on the sail and on the keel of a sailboat?)

13. In the situation described in Example 6, why is it best to push at the mid-point of the rope?

14. A boy and girl are engaged in a tug-of-war (Figure 5.21). (a) Draw a diagram showing the horizontal forces on the boy, (b) on the girl, and (c) on the rope. Which of these forces are action–reaction pairs?

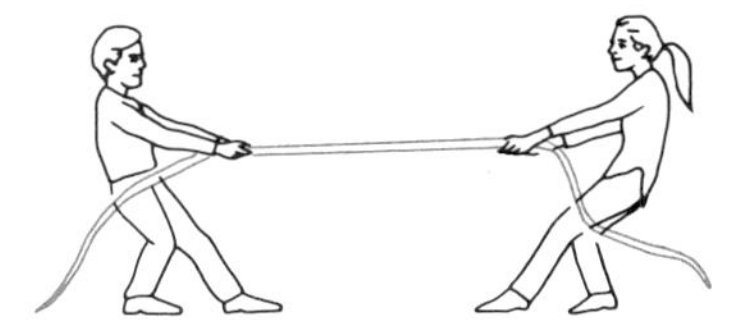

**Fig. 5.21** A boy and a girl in a tug-of-war.

15. In an experiment performed in 1654, Otto von Guericke, mayor of Magdeburg and inventor of the air pump, gave a demonstration of air pressure before Emperor Ferdinand. He had two teams of 15 horses each pull in opposite directions on two evacuated hemispheres held together by nothing but air pressure. The horses failed to pull these hemispheres apart (Figure 5.22). If each horse exerted a pull of 3000 N, what was the tension in the harness attached to each hemisphere? If the harness attached to one of the hemispheres had simply been tied to a stout tree, what would have been the tension exerted by a single team of horses hitched to the other harness? What would have been the tension exerted by the two teams of horses hitched in series to the other harness? Can you guess why von Guericke hitched up his horses in the way he did?

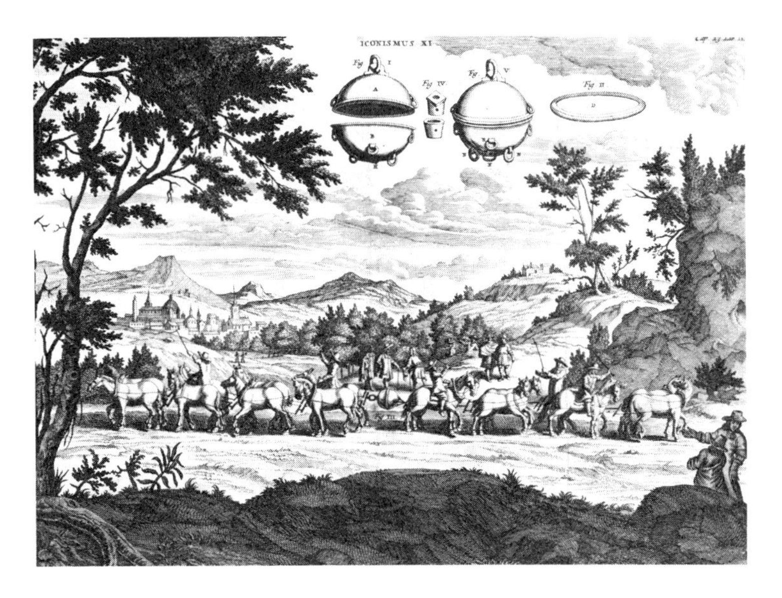


**Fig. 5.22** The Magdeburg hemispheres.

16. In a tug-of-war, two teams of children pull on a rope (Figure 5.23). Is the tension constant along the entire length of the rope? Along what portion of the rope is it constant?

**Fig. 5.23** Two teams of children in a tug-of-war.

17. When you are standing on the Earth, your feet exert a force (push) against the surface. Why does the Earth not accelerate away from you?

18. When an automobile accelerates on a level road, the force that produces this acceleration is the push of the road on the wheels. If so, why does the automobile need an engine?

19. You are in a small boat in the middle of a calm lake. You have no oars, and you cannot put your hands in the water because the lake is full of piranhas. The boat carries a large load of coconuts. How can you get to the shore?

20. A tennis ball in horizontal flight bounces off a vertical wall. Is the momentum conserved?

21. You are inside a ship that is trying to make headway against the strong current of a river. Without looking at the shore or other outside markers, is there any way you can tell whether the ship is making any progress?

22. A submarine uses its inertial guidance system to determine position and velocity at any instant without the need of visual observation of shore points, stars, or other outside markers. Does this conflict with the Newtonian Principle of Relativity?

## PROBLEMS

### Section 5.2

1. On a flat road, a Maserati sports car can accelerate from 0 to 80 km/h (0 to 50 mi/h) in 5.8 s. The mass of the car is 1620 kg. What are the average acceleration and the average force on the car?

2. The Grumman F-14B fighter plane has a mass of 16,000 kg and its engines develop a thrust of $2.7 \times 10^5$ N when at full power. What is the maximum horizontal acceleration that this plane can achieve? Ignore friction.

3. Pushing with both hands, a sailor standing on a pier exerts a horizontal force of 270 N on a destroyer of 3400 metric tons. Assuming that the mooring ropes do not interfere and that the water offers no resistance, what is the acceleration of the ship? How far does the ship move in 60 s?

4. A woman of 57 kg is held firmly in the seat of her automobile by a lap-and-shoulder seat belt. During a collision, the automobile decelerates from 50 to 0 km/h in 0.12 s. What is the average horizontal force that the seat belt exerts on the woman? Compare the force with the weight of the woman.

5. A heavy freight train has a total mass of 16,000 metric tons. The locomotive exerts a pull of 670,000 N on this train. What is the acceleration? How long does it take to increase the speed from 0 to 50 km/h?

6. With brakes fully applied, a 1500-kg automobile decelerates at the rate of 8.0 m/s² on a flat road. What is the braking force acting on the automobile? Draw a diagram showing the direction of motion of the automobile and the direction of the braking force.

7. Consider the impact of the automobile on a barrier described in Example 2.5. If the mass of the automobile is 1400 kg, what is the average force acting on the automobile during the deceleration?

8. In a crash at the Silverstone circuit in England, a race-car driver suffered more than thirty fractures and dislocations and several heart stoppages after a deceleration from 174 km/h to 0 km/h within a distance of about 66 cm. If the deceleration was constant during the crash and the mass of the driver was 75 kg, what was the deceleration and the force on the driver?

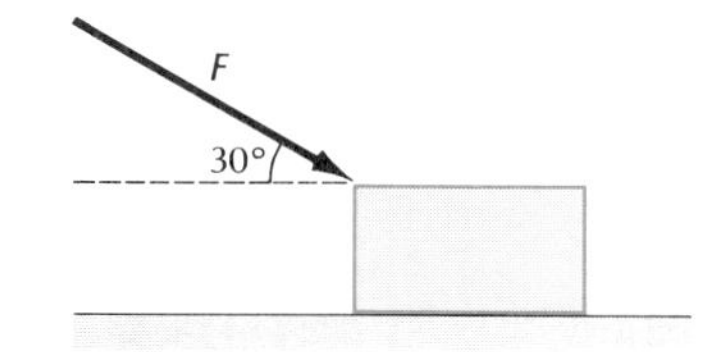


**Fig. 5.24**

9. A box of 25 kg sits on a smooth, frictionless table. If you push down on the box at an angle of 30° with a force of 80 N (see Figure 5.24), what is the acceleration of the box?

10. The projectile fired by the gun described in Problem 2.23 has a mass of 45 kg. What is the force on this projectile as it moves along the barrel?

*11. Figure 2.12 shows the plot of velocity vs. time for a Triumph sports car coasting along with its gears in neutral. The mass of the car is 1160 kg. From the values of the deceleration at the times $t = 0$, 10, 20, 30, and 40 s (see Problem 2.15b), calculate the friction force that the car experiences at these times. Make a plot of friction force vs. velocity.

*12. Table 2.4 gives the velocity of a projectile as a function of time. The projectile slows down because of the friction force exerted by the air. For the first 0.30-s time interval and for the last 0.30-s time interval, calculate the average friction force.

*13. A proton moving in an electric field has an equation of motion

$$\mathbf{r} = (5.0 \times 10^4 t)\hat{\mathbf{x}} + (2.0 \times 10^4 t - 2.0 \times 10^5 t^2)\hat{\mathbf{y}} - (4.0 \times 10^5 t^2)\hat{\mathbf{z}}$$

where distance is measured in meters and time in seconds. The proton has a mass of $1.7 \times 10^{-27}$ kg. What are the components of the force acting on this proton? What is the magnitude of the force?

*14. The speed of a projectile traveling horizontally and slowing down under the influence of air friction can be approximately represented by

$$v = 655.9 - 61.14t + 3.26t^2$$

where $v$ is measured in meters per second and $t$ in seconds; the mass of the projectile is 45.36 kg (see Problems 2.18 and 2.19). Find a formula for the force of air friction as a function of time.

## Section 5.3

15. While braking, an automobile of mass 1200 kg decelerates along a level road at 0.80 G. What is the horizontal force that the road exerts on each wheel of the automobile? Assume all wheels contribute equally to the braking. Ignore the friction of the air.

16. In 1978, in an accident at a school in Harrisburg, Pennsylvania, several children lost parts of their fingers when a nylon rope suddenly snapped during a giant tug-of-war among 2300 children. The rope was known to have a breaking tension of 58,000 N. Each child can exert a pull of approximately 130 N. Was it safe to employ this rope in this tug-of-war?

*17. Two forces $\mathbf{F}_1$ and $\mathbf{F}_2$ act on a particle of 6.0 kg. The forces are

$$\mathbf{F}_1 = 2\hat{\mathbf{x}} - 5\hat{\mathbf{y}} + 3\hat{\mathbf{z}}$$
$$\mathbf{F}_2 = -4\hat{\mathbf{x}} + 8\hat{\mathbf{y}} + \hat{\mathbf{z}}$$

where the force is measured in newtons.

(a) What is the net force vector?

(b) What is the acceleration vector of the particle and what is the magnitude of the acceleration?

*18. The Earth exerts a gravitational pull of $2.0 \times 10^{20}$ N on the Moon; the Sun exerts a gravitational pull of $4.3 \times 10^{20}$ N on the Moon. What is the net force on the Moon when the angular separation between the Earth and the Sun is 90° as seen from the Moon?

*19. A sailboat is propelled through the water by the combined action of two forces: the push ("lift") of the wind on the sail and the push of the water on the keel. Figure 5.25 shows the magnitudes and the directions of these forces acting on a medium-sized sailboat (this oversimplified diagram does not include the drag of wind and water). What is the resultant of the forces in Figure 5.25?

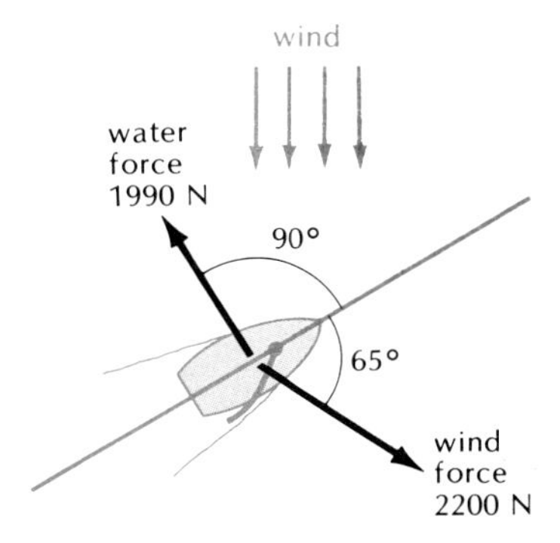


**Fig. 5.25** Forces on a sailboat. The angles are measured relative to the line of motion.

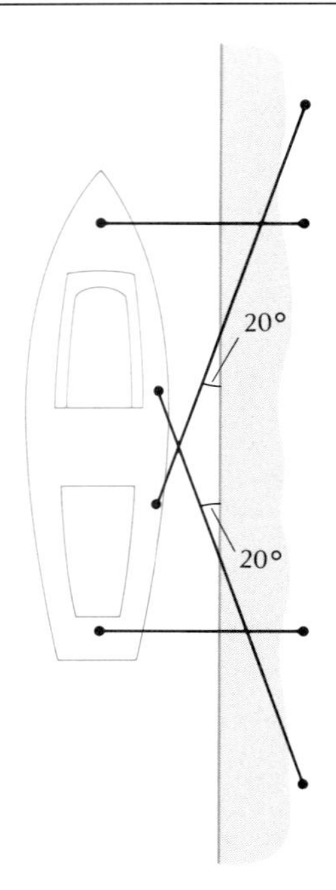

**Fig. 5.26** Ropes holding a boat at a dock.

*20. A boat is tied to a dock by four (horizontal) ropes. Two ropes, with a tension of 260 N each, are at right angles to the dock. Two other ropes, with a tension of 360 N each, are at an angle of 20° with the dock (Figure 5.26). What is the resultant of these forces?

**21. In a tug-of-war, a jeep of mass 1400 kg and a tractor of mass 2000 kg pull on a horizontal rope in opposite directions. At one instant, the tractor pulls on the rope with a force of $1.50 \times 10^4$ N while its wheels push horizontally against the ground with a force of $1.60 \times 10^4$ N. Calculate the instantaneous accelerations of the tractor and of the jeep; calculate the horizontal push of the wheels of the jeep. Assume the rope does not stretch or break.

## Section 5.4

22. A small truck of 2800 kg collides with an initially stationary automobile of 1200 kg. The acceleration of the truck during the collision is $-500$ m/s$^2$ and it lasts for 0.02 s. What is the acceleration of the automobile? What is the speed of the automobile after the collision? Assume that the frictional force due to the road can be neglected during this collision.

23. A diver of mass 75 kg is in free fall after jumping off a high platform.

(a) What is the force that the Earth exerts on the diver? What is the force that the diver exerts on the Earth?
(b) What is the acceleration of the diver? What is the acceleration of the Earth?

*24. A long freight train consists of 250 cars each of mass 64 metric tons. The pull of the locomotive accelerates this train at the rate of 0.043 m/s$^2$ along a level track. What is the tension in the coupling that holds the first car to the locomotive? What is the tension in the coupling that holds the last car to the next to last car? Ignore friction.

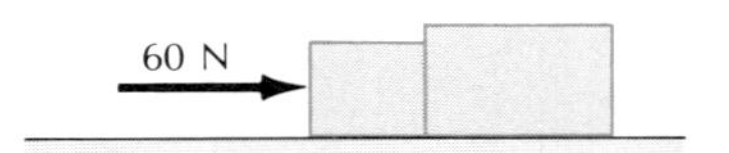

**Fig. 5.27** Two boxes in contact.

*25. Two heavy boxes of masses 20 kg and 30 kg sit on a smooth, frictionless surface. The boxes are in contact and a horizontal force of 60 N pushes horizontally against the smaller box (Figure 5.27). What is the acceleration of the two boxes? What is the force that the smaller box exerts on the larger box? What is the force that the larger box exerts on the smaller box?

*26. An archer pulls the string of her bow back with her hand with a force of 180 N. If the two halves of the string above and below her hand make an angle of 120° with each other, what is the tension in each half of the string?

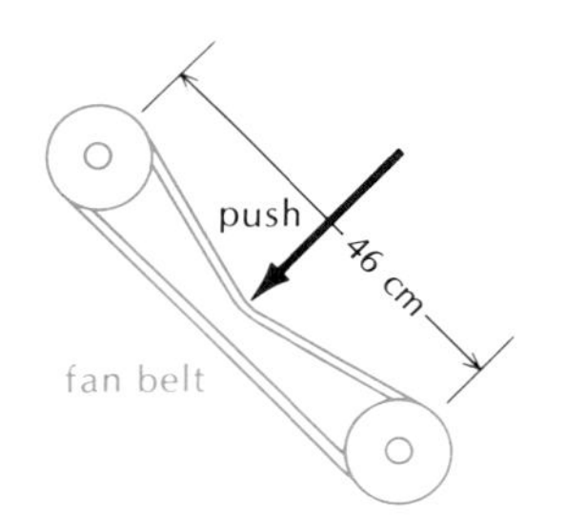

**Fig. 5.28** Push on a fan belt.

*27. A mechanic tests the tension in a fan belt by pushing against it with his thumb (Figure 5.28). The force of the push is 130 N and it is applied to the midpoint of a segment of belt 46 cm long. The lateral displacement of the belt is 2.5 cm. What is the tension in the belt (while the mechanic is pushing)?

*28. On a sailboat, a rope holding the foresail passes through a block (a pulley) and is made fast on the other side to a cleat (Figure 5.29). The two parts of the rope make an angle of 140° with each other. The sail pulls on the rope with a force of $1.2 \times 10^4$N. What is the force that the rope exerts on the block?

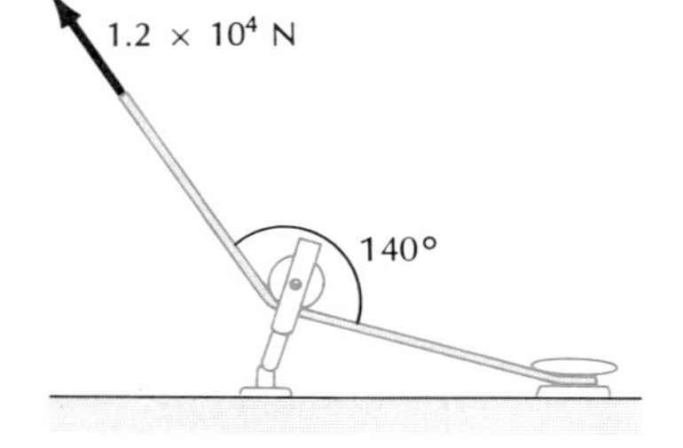

**Fig. 5.29** The left end of the rope is attached to the sail, the right end is attached to a cleat.

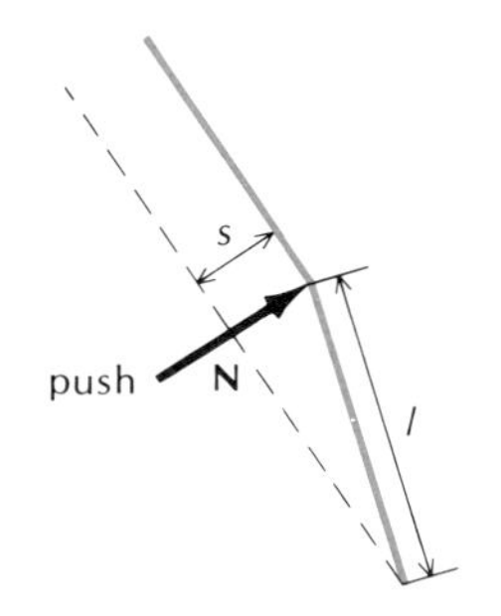

Fig. 5.30 Push on a wire rope.

**29. A sailor tests the tension in a wire rope holding up a mast by pushing against the rope with his hand at a distance $l$ from the lower end of the rope. When he exerts a transverse push $N$, the wire rope suffers a transverse displacement $s$ (Figure 5.30).

(a) Show that for $s \ll l$ the tension in the wire rope is given approximately by the formula

$$T = Nl/s$$

In your calculation, assume that the distance to the upper end of the rope is effectively infinite, i.e., the total length of the rope is much larger than $l$.

(b) What is the tension in a rope that suffers a transverse displacement of 2.0 cm under a force of 150 N applied at a distance of 1.5 m from the lower end?

*30. On a windy day, a small tethered balloon is held by a long string making an angle of 70° with the ground. The vertical buoyant force on the balloon (exerted by the air) is 67 N . During a sudden gust of wind, the (horizontal) force of the wind is 200 N; the tension in the string is 130 N. What are the magnitude and direction of the force on the balloon?

*31. A crate of mass 2000 kg is hanging from a crane at the end of a cable 12 m long. If we attach a horizontal rope to this crate and gradually apply a pull of 1800 N, what angle will the cable finally make with the vertical?

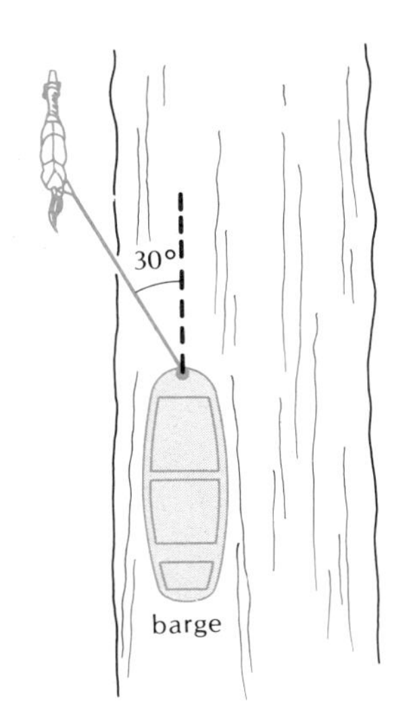

Fig. 5.31 Horse pulling barge.

*32. A horse, walking along the bank of a canal, pulls a barge. The horse exerts a pull of 300 N on the barge at an angle of 30° (Figure 5.31). The bargeman relies on the rudder to steer the barge on a straight course parallel to the bank. What transverse force (perpendicular to the bank) must the rudder exert on the barge?

**33. A flexible massless rope is placed over a cylinder of radius $R$. A tension $T$ is applied to each end of the rope, which remains stationary (see Figure 5.32). Show that each small segment $d\theta$ of the rope in contact with the cylinder pushes against the cylinder with a force $Td\theta$ in the radial direction. By integration of the forces exerted by all the small segments, show that the net vertical force on the cylinder is $2T$ and the net horizontal force is zero.

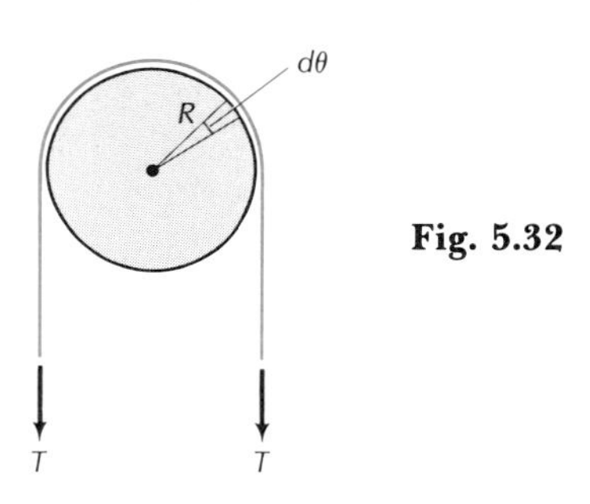

Fig. 5.32

Section 5.5

34. What is the momentum of an automobile of 900 kg moving at 65 km/h? If a truck of 7200 kg is to have the same momentum as the automobile, what must be its speed?

35. Using the entries listed in Tables 1.6 and 2.1, find the magnitude of the momentum for each of the following: Earth moving around Sun, jet airliner at maximum airspeed, automobile at federal speed limit (55 mi/h), man walking, electron moving around a nucleus.

36. The push that a bullet exerts during impact on a target depends on the momentum of the bullet. A Remington .244 rifle, used for hunting deer, fires

a bullet of 90 grains (1 grain is $\frac{1}{7000}$ lb) with a speed of 975 m/s. A Remington .35 rifle fires a bullet of 200 grains with a speed of 674 m/s. What is the momentum of each bullet?

37. An electron of mass $9.1 \times 10^{-31}$ kg is moving in the $x$–$y$ plane; its speed is $2.0 \times 10^5$ m/s and its direction of motion makes an angle of 25° with the $x$ axis. What are the components of the momentum of the electron?

38. A sky diver of mass 75 kg is in free fall. What is the rate of change of his momentum? Ignore friction.

*39. A soccer player kicks a ball and sends it flying with an initial speed of 26 m/s at an upward angle of 30°. The mass of the ball is 0.43 kg. Ignore friction.

(a) What is the initial momentum of the ball?
(b) What is the momentum when the ball reaches maximum height on its trajectory?
(c) What is the momentum when the ball returns to the ground? Is this final momentum the same as the initial momentum?

*40. Consider the proton with the equation of motion given in Problem 13. What are the components of the momentum of this proton at time $t = 0$?

*41. The Earth moves around the Sun in a circle of radius $1.5 \times 10^{11}$ m at a speed of $3.0 \times 10^4$ m/s. The mass of the Earth is $6.0 \times 10^{24}$ kg. Calculate the magnitude of the rate of change of the momentum of the Earth from these data. (Hint: The magnitude of the momentum does not change, but the direction does.)

CHAPTER 6

# Dynamics — Forces and the Solution of the Equation of Motion

Newton's Second Law is usually called the **equation of motion.** If the force on a particle is known, then the Second Law determines the acceleration, and from this the complete motion of the particle can be calculated. Thus, in principle, the motion of the particle is completely predictable. As the great French mathematician and astronomer Pierre Simon de Laplace expressed it,

> If an intellect were to know, for a given instant, all the forces that animate nature and the condition of all the objects that compose her, and were also capable of subjecting these data to analysis, then this intellect would encompass in a single formula the motions of the largest bodies in the universe as well as those of the smallest atom; nothing would be uncertain for this intellect, and the future as well as the past would be present before its eyes.

**Pierre Simon, marquis de Laplace,** *1749–1827, professor at the École Militaire, Paris. He investigated the theory of the mutual disturbances that the planets exert on one another, and used Newton's equations to establish that the Solar System is stable — in the long run the mutual disturbances average to zero. For his great contributions to celestial mechanics he was called the "Newton of France."*

To find a solution of the equation of motion means to find a force $\mathbf{F}$ and a position vector $\mathbf{r}(t)$ which is a function of time such that the equation $m\mathbf{a} = \mathbf{F}$ is satisfied. For a physicist, the typical problem involves a known force and an unknown motion; for example, the physicist knows the force between the planets and the Sun and he seeks to calculate the motion of these bodies. But for an engineer, the reverse problem with a known motion and an unknown force is often of practical importance; for example, the engineer knows that a train is to round a given curve at 100 km/h and he seeks to calculate the forces that the track and the wheels must withstand. A special problem with known motion is the problem of statics; here we know that the body is at rest ($\mathbf{v} = 0$ and $\mathbf{a} = 0$) and we wish to compute the forces that will maintain this condition of equilibrium. Thus, depending on the circumstances, we can regard either the right side or the left side of the equation $m\mathbf{a} = \mathbf{F}$ as an unknown that is to be calculated from what we know about the other side.

Before we look at some examples of solutions of the equation of motion, we will briefly discuss some general properties of forces.

## 6.1 The Four Fundamental Forces

In everyday experience we encounter an enormous variety of forces: the gravity of the Earth which pulls all bodies downward, contact forces between rigid bodies that resist their interpenetration, elastic forces that oppose the deformation of springs and beams, pressure forces exerted by air or water on bodies immersed in them, adhesive forces exerted by a layer of glue bonding two surfaces, friction forces that resist the motion of a surface sliding over another, electrostatic forces between two electrified bodies, magnetic forces between the poles of magnets, and so on.

Besides these forces that act in the macroscopic world of everyday experience, there are many others that act in the microscopic world of atomic and nuclear physics. There are intermolecular forces that attract or repel molecules to or from each other, interatomic forces that bind atoms into molecules or repel them if they come too close to each other, atomic forces within the atom that hold its parts together, nuclear forces that act on the parts of the nucleus, and even more esoteric forces which only act for a brief instant when subnuclear particles are made to suffer violent collisions in high-energy experiments performed in accelerator laboratories.

Yet, at a fundamental level, this bewildering variety of forces involves only four different kinds of force. The four fundamental forces are the gravitational force, the electromagnetic force, the "strong" force, and the "weak" force (see Figure 6.1a–d).

The **gravitational force** is a mutual attraction between all masses. If we reckon the strength of the forces in terms of their effect on elementary particles, then gravitation is the weakest of the four forces. The

**Fig. 6.1** (a) The gravitational force is responsible for the free-fall motion of this diver. (b) Electric forces between the atoms give rise to the contact forces and elastic forces that act in this impact of a club on a golf ball. (c) The "strong" force brings about the thermonuclear fusion reactions in this explosion of a hydrogen bomb. (d) The "weak" force causes the decays of several elementary particles and the creation of new particles, which made these tracks in a bubble chamber at an accelerator laboratory (the created particles made the distinctive **V**-shaped tracks at the upper left and lower right).

(a)

(b)

(c)

(d)

gravitational attraction between two neighboring protons in a nucleus is only about $10^{-34}$ N, which is completely insignificant. On the surface of the Earth, we feel the force of gravity only because the mass of the Earth is so large; the forces between the individual particles in our bodies and in the Earth are insignificant, but our bodies and the Earth contain a very large number of particles, and their forces add up.

The **electromagnetic force** is an attraction or repulsion between electric charges. The electric and the magnetic forces, once considered to be separate, are now grouped together because they are closely related: the magnetic force is nothing but an extra electric force that acts whenever charges are in motion. The electric force is of medium strength; between two neighboring protons, it is about 90 N. Of all the forces, the electric force plays the most pervasive role in our lives. With the exception of the Earth's gravity, every force in our immediate macroscopic environment is electric. Contact forces between rigid bodies, elastic forces, pressure forces, adhesive forces, friction forces, and so on, are nothing but electric forces between the charged particles in the atoms of one body and those in the atoms of another.

The **"strong" force** acts mainly within the nuclei of atoms. It serves as the nuclear glue that prevents the pieces of the nucleus from flying apart. This nuclear force is called "strong" because it is the strongest of the four forces. Between two neighboring protons in a nucleus, this force is typically 10,000 N. It can be either attractive or repulsive: the strong force will push the protons apart if they come too near to each other, and it will pull them together if they begin to drift too far apart.

Finally, the **"weak" force** only manifests itself in certain reactions among elementary particles. Most of the reactions caused by the weak force are radioactive decay reactions; they involve the spontaneous breakup of a particle into several other particles. This force is called "weak" because it is very weak; between two neighboring protons in a nucleus, its strength is estimated at only about $10^{-2}$ N.

Table 6.1 lists the four fundamental forces in order of increasing strength.

**Table 6.1** THE FOUR FUNDAMENTAL FORCES

| *Force* | *Acts on* | *Strength*[a] | *Range* |
|---|---|---|---|
| Gravitational | All masses | $10^{-34}$ N | Infinite |
| Weak | Most elementary particles | $10^{-2}$ | Less than $10^{-17}$ m |
| Electromagnetic | Electric charges | $10^{2}$ | Infinite |
| Strong | Nuclear particles | $10^{4}$ | $10^{-15}$ m |

[a]The strength listed here is the force (in newtons) between two protons separated by a distance equal to their diameter, $2 \times 10^{-15}$ m.

Gravitation and electromagnetism both bridge empty space and reach from one particle to another, even if the intervening distance is very large. A spectacular instance of such **action-at-a-distance** is the gravitation of the Sun: the Earth feels the gravitational pull of the Sun even though there is a gap of 100 million miles between these bodies. Another instance is the magnetism of the Earth: a compass needle feels the pull of the magnetic north pole even when it is thousands of miles away. In Chapter 23 we will inquire into the mechanism involved in

*Action-at-a-distance*

the transmission of force through empty space; we will see that this involves a **field,** or a disturbance that spreads through space and carries the interaction from one body to another. But for now, we will ignore the mechanism and simply think of forces as bridging space and acting at a distance. Note that even so-called contact forces, such as friction or the force that opposes the interpenetration of solid bodies, rely on action at some finite distance. When your finger pushes against a page of this book, the subatomic particles making up two neighboring atoms in "contact" are not really touching, and the forces must bridge the gap of empty space between these particles.

The strong force also bridges space, reaching from one particle to another. However, its reach is very short, not much more than $10^{-15}$ m. The reach of the weak force is even shorter; it acts only if the elementary particles are just about interpenetrating one another. The distance over which a force acts is called its **range.** Table 6.1 lists the ranges of the four forces.

In the next few sections we will look at the practical aspects of a few forces of great importance in the macroscopic world: the gravity of the Earth (or the weight), the friction force, and the restoring force of a spring.

## 6.2 Weight

**Weight** is the pull of the Earth's gravity, that is, weight is a *force.* Consequently, weight is a vector quantity — it has a direction (downward) as well as a magnitude. The units of weight are the units of force, that is, newtons or pounds.

Consider a body of mass $m$ in free fall near the surface of the Earth. Under the influence of gravity, this body will accelerate downward with an acceleration $g$. According to Newton's Second Law, the magnitude of the gravitational force that causes this acceleration must be

$$F = ma = mg \tag{1}$$

This gravitational force is what we call the weight. Usually we will denote the weight by the vector symbol **w.** The magnitude of the weight **w** is

*Weight*

$$\boxed{w = mg} \tag{2}$$

If the body is not in free fall but is held in a stationary position by some supports, then the weight is of course still the same as given by Eq. (2); however, the supports prevent this weight from producing the downward motion.

EXAMPLE 1. What is the weight of a 74-kg man? Assume $g = 9.81\ \text{m/s}^2$.

SOLUTION: By Eq. (2), the weight is

$$w = mg = 74\ \text{kg} \times 9.81\ \text{m/s}^2$$

$$= 726\ \text{N}$$

(Alternatively, in British units the mass is 74 kg × 1 lb/0.454 kg = 163 lb, and the weight is

$$w = mg = 163 \text{ lb} \times 32.2 \text{ ft/s}^2$$
$$= 163 \text{ lbf})$$

Since the value of $g$ depends on location, the weight of a body also depends on its location. For example, if a 74-kg man travels from London ($g = 9.81$ m/s$^2$) to Hong Kong ($g = 9.79$ m/s$^2$), his weight will decrease from 726 N to about 724 N, a difference of about 2 N. And if this man were to travel to a point 6400 km above the surface of the Earth ($g = 2.45$ m/s$^2$), his weight would decrease to 181 N! This example illustrates an essential distinction between mass and weight: mass is an *intrinsic* property of a body, measuring the resistance (inertia) with which the body opposes changes in its motion; the definition of mass is designed in such a way that a given body has the same mass regardless of its position in the universe. Weight is an *extrinsic* property of a body, measuring the pull of gravity on the body; it depends on the (gravitational) environment in which the body is located and is therefore a function of position.

*Mass vs. weight*

In practice, the mass of a body is usually measured by "weighing" the body on a balance, either a beam balance or a spring balance. A beam balance with equal arms will be in equilibrium if the weight (force) on each balance pan is the same (Figure 6.2a). Since, to a very high accuracy, the value of $g$ is the same at both balance pans, the equality of weights implies the equality of masses; thus, a beam balance can be used to compare masses via their weight. A spring balance can likewise be used to compare masses via their weight (Figure 6.2b). The spring balance can be directly calibrated to read in mass units by suspending known masses from it (of course, this calibration will only remain accurate as long as the balance is kept in a fixed location where $g$ has some fixed value).

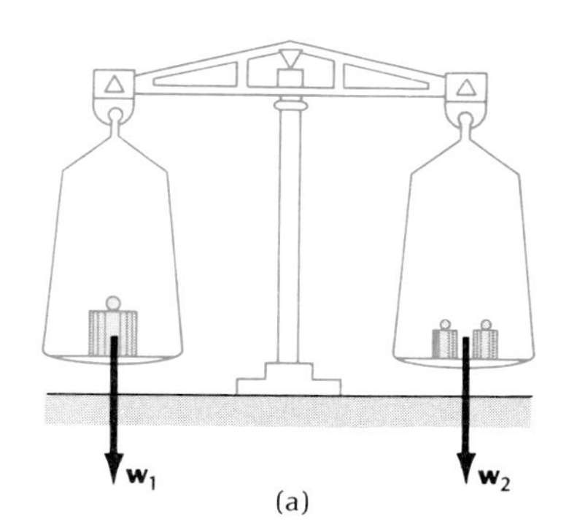


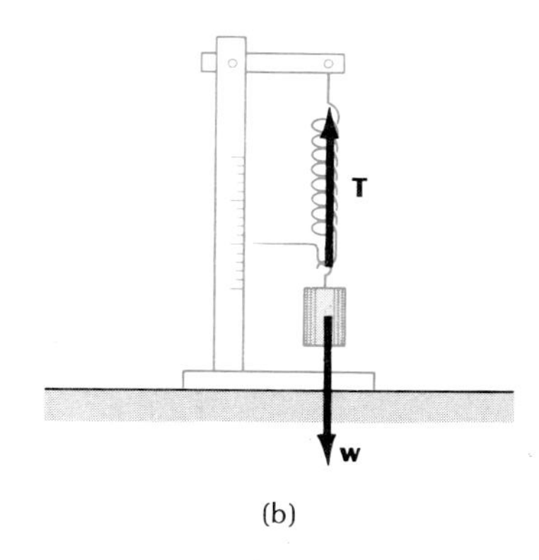


**Fig. 6.2** (a) A beam balance is in equilibrium when the weights in the two pans are equal. (b) A spring balance is in equilibrium when the weight matches the force exerted by the spring.

In everyday usage, *mass* and *weight* are often confused. The typical labels on packages of commercial goods display this confusion. Such labels state, "This package has a weight of 1 lb," whereas they should state, "This package has a mass of 1 lb." The balances used to measure commercial goods compare an unknown mass with a set of known, standard masses; thus these balances are calibrated in mass, not in weight. Because of the common misuse of the word *weight*, it is often necessary to guess its intended meaning from the context in which it appears.[1]

A body deep in intergalactic space, far from any star or planet, will not experience any gravitational pull at all — the weight of the body will be zero. This condition of weightlessness can be simulated in the vicinity of the Earth by means of a freely falling reference frame. Consider an observer in free fall, say, a diver who has jumped off a springboard and is accelerating downward with the acceleration of gravity. If the diver releases an apple from his hand, the apple accelerates downward at the same rate as the diver; that is, the apple remains at rest relative to the diver. Thus, in the rest frame of the diver (a freely falling reference frame accelerating downward with the acceleration of grav-

[1] The Soviet Union seems to be the only country free of this confusion. Russian labels on packages give the mass (MACCA).

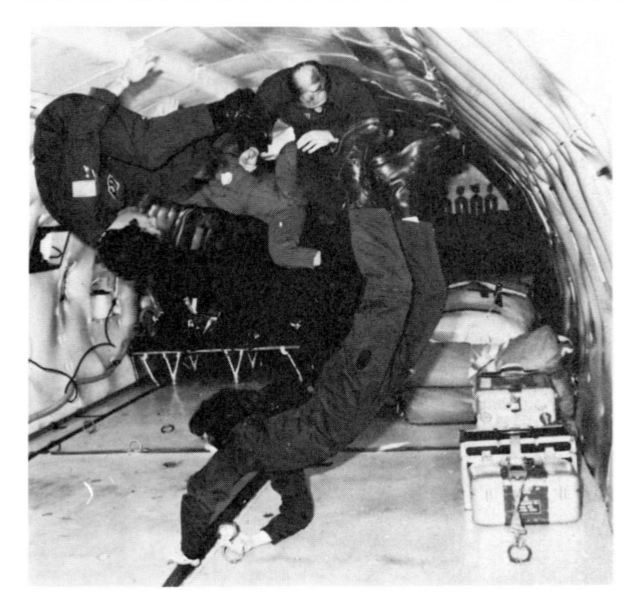

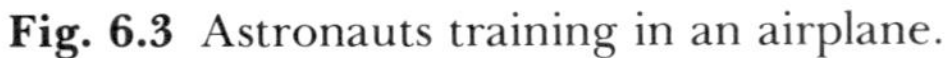

**Fig. 6.3** Astronauts training in an airplane.

**Fig. 6.4** Astronauts floating in the cargo bay of the Space Shuttle.

ity!) a body at rest remains at rest and, more generally, a body in motion continues to move at constant velocity. This means that in such a freely falling reference frame the gravitational pull is *apparently* zero; the weight is *apparently* zero. Of course, this simulated weightlessness arises from the accelerated motion of the reference frame — in the unaccelerated, inertial reference frame of the ground, the weight of the apple is certainly not zero. Nevertheless, if the diver insists on looking at things from his own reference frame, he will judge the weight of the apple, and also the weight of his own body, as zero. This sensation of weightlessness is also simulated within an airplane flying along a parabola, imitating the motion of a (frictionless) projectile (Figure 6.3); and it is also simulated in a spacecraft orbiting the Earth (Figure 6.4). Both of these motions are free-fall motions.

---

EXAMPLE 2. In preparation for walking in low-gravity conditions on the surface of the Moon, astronauts at NASA trained on a Moon simulator consisting of an inclined plane across which they could walk while suspended in a harness (Figure 6.5a). Suppose that the pull of the harness is parallel to the inclined plane. What must be the angle of inclination if the forces perpendicular to the plane are to simulate the conditions on the Moon? The acceleration of gravity on the surface of the Moon is $g_{\text{Moon}} = 1.6\ \text{m/s}^2$.

SOLUTION: Suppose that the astronaut takes a small jump so that his feet momentarily lose contact with the plane. Then the only forces acting on the astronaut are his weight **w** and the tension **T** supplied by the harness. Figure 6.5b shows these forces (we will pretend that the astronaut can be regarded as a particle so that all the forces act at one point). Taking the $x$ axis parallel to the inclined plane and the $y$ axis perpendicular, we can resolve the weight **w** into two components: a component $w_x$ parallel to the plane and a component $w_y$ perpendicular to the plane. These components have the values

$$w_x = -mg \sin \theta \qquad w_y = -mg \cos \theta$$

Since the harness restrains the astronaut from moving in the $x$ direction, the

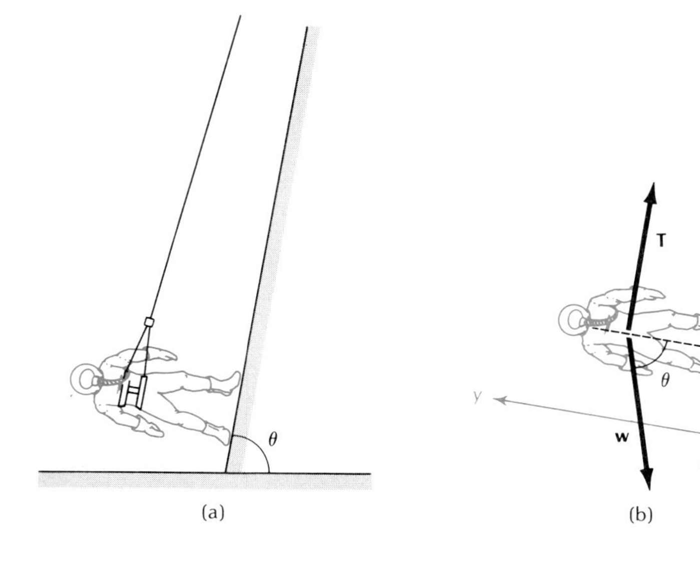


**Fig. 6.5** (a) Astronaut on inclined plane. The astronaut can walk forward and backward, and he can jump perpendicularly to the plane. (b) Forces acting on the astronaut during a small jump.

component $w_x$ is of no interest in the present problem (the force $w_x$ is actually compensated by the tension **T**). The component $w_y$ produces an acceleration in the $y$ direction,

$$a_y = \frac{w_y}{m} = -g \cos \theta$$

In order to simulate Moon conditions, we want this acceleration to match the acceleration of gravity on the Moon, so that

$$-g \cos \theta = -g_{\text{Moon}}$$

From this we find

$$\cos \theta = \frac{g_{\text{Moon}}}{g} = \frac{1.6 \text{ m/s}^2}{9.8 \text{ m/s}^2} = 0.16$$

and $\theta = \cos^{-1}(0.16) = 81°$.

## 6.3 Motion with a Constant Force

If the force acting on a particle is constant, then the acceleration is also constant. The motion is then given by Eqs. (4.18)–(4.23), with $a_x = F_x/m$, $a_y = F_y/m$, and $a_z = F_z/m$.

The motion of projectiles is an obvious example of motion with constant force. If we neglect air resistance, the only force on the projectile is the weight, $F = w = mg$. The weight is constant provided that the acceleration of gravity is constant along the path of the projectile, a provision that is well satisfied for most projectiles except ballistic missiles, which reach very large heights and very great distances, where the acceleration of gravity changes in magnitude and in direction. With the $z$ axis in the vertical upward direction, the acceleration of the projectile is $a_x = 0$, $a_y = 0$, and $a_z = -g$. We have already discussed the resulting motion in Section 4.3.

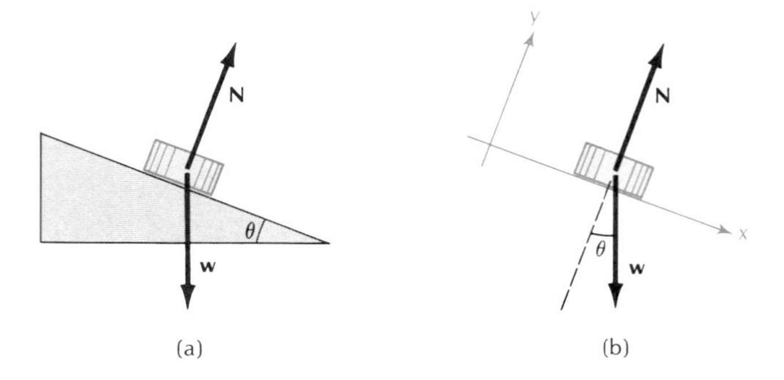


**Fig. 6.6** (a) Block sliding down an inclined plane. (b) "Free-body" diagram showing all the forces on the block.

*Normal force*

Another instance of motion with constant force is the motion of a body sliding on a ramp, or an inclined plane. Figure 6.6a shows a block of mass $m$ sliding down a smooth, frictionless plane inclined at an angle $\theta$ to the horizontal. There are two forces acting on the block: the weight **w** pointing vertically downward and the **normal force N** pointing in a direction perpendicular to the inclined plane. The normal force **N** is a contact force that arises from the repulsion between the atoms of the block and the atoms of the plane. This repulsion prevents the block from penetrating the plane. The contact force acts uniformly all over the bottom surface of the block, but in Figure 6.6a the force is shown as though acting only at the center of the surface. The contact force is perpendicular, or "normal," to the plane because we are assuming that there is no friction. If there were friction, then the contact force would have both a component perpendicular to the plane and a component parallel to the plane; the latter would be a friction force in the opposite direction of the sliding motion of the block along the plane.

*"Free-body" diagram*

Figure 6.6b shows the block (regarded as a particle) and all the forces acting on it: these forces are the weight **w** (vertically downward) and the normal force **N** (perpendicular to the plane). The inclined plane is not shown in this diagram — the effect of the plane is entirely contained in the force **N.** Such a diagram, showing the body and all the forces acting on it, but not showing the surrounding bodies that exert these forces, is called a **"free-body" diagram.** (In this context *free* does not mean free of force; it means that the surroundings are represented by the exerted forces.) Such a diagram eliminates clutter, and helps us to focus on the forces that we need to formulate the equation of motion of the body. Looking at Figure 6.6b, we see that the net force on our block is the vector sum of **N** and **w.** Taking the $x$ axis parallel to the plane and the $y$ axis perpendicular, we find that the components of these two forces are

$$N_x = 0 \qquad N_y = N$$

$$w_x = mg \sin\theta \qquad w_y = -mg\cos\theta$$

and the components of the net force are

$$F_x = N_x + w_x = mg\sin\theta$$

$$F_y = N_y + w_y = N - mg\cos\theta \tag{3}$$

Newton's equation of motion then gives the corresponding components of the acceleration of the block, regarded as a particle:

$$a_x = F_x/m = g \sin\theta \tag{4}$$

$$a_y = F_y/m = N/m - g\cos\theta \tag{5}$$

Equation (4) tells us that the acceleration of the block along the plane is $g \sin\theta$. In the case of a horizontal plane ($\theta = 0$), there is no acceleration; and in the case of a vertical plane ($\theta = 90°$), the acceleration is that of free fall. Both of these extreme cases are as expected.

Equation (5) can be used to evaluate $N$. Since the motion is necessarily parallel to the plane, the acceleration $a_y$ perpendicular to the plane is identically zero; hence

$$0 = N/m - g\cos\theta \tag{6}$$

or

$$N = mg\cos\theta \tag{7}$$

In the case of a horizontal plane ($\theta = 0$), the normal force has a magnitude $mg$, that is, the magnitude matches the weight; and in the case of a vertical plane ($\theta = 90°$), the normal force vanishes.

---

EXAMPLE 3. The world's steepest railroad track, found in Guatemala, has a slope of 1:11 (Figure 6.7a). What force is required to move a boxcar of 20 metric tons at constant speed along this track? Ignore friction and treat the motion of the boxcar as a particle motion.

SOLUTION: The forces on the boxcar are the weight **w**, the normal force **N**, and the force **T** that pulls the boxcar along the track. Figure 6.7b shows these forces in a "free-body" diagram. For the sake of convenience, all the forces are drawn as though acting at the center of the boxcar (the translational motion of the boxcar is particle motion, and for this motion, it makes no difference where the forces act).

The net force **F** is the vector sum **w** + **N** + **T**. With the coordinate axes arranged as in Figure 6.7, the $x$ and $y$ components of the net force are

$$F_x = mg\sin\theta - T$$

$$F_y = -mg\cos\theta + N \tag{8}$$

Since the acceleration is supposed to vanish, $F_x = 0$ and therefore

$$T = mg\sin\theta \tag{9}$$

For a slope of 1:11, the angle $\theta$ is $\theta = \tan^{-1}(1/11) = 5.19°$. Hence

$$T = 20 \times 10^3 \text{ kg} \times 9.81 \text{ m/s}^2 \times \sin 5.19°$$

$$= 1.78 \times 10^4 \text{ N}$$

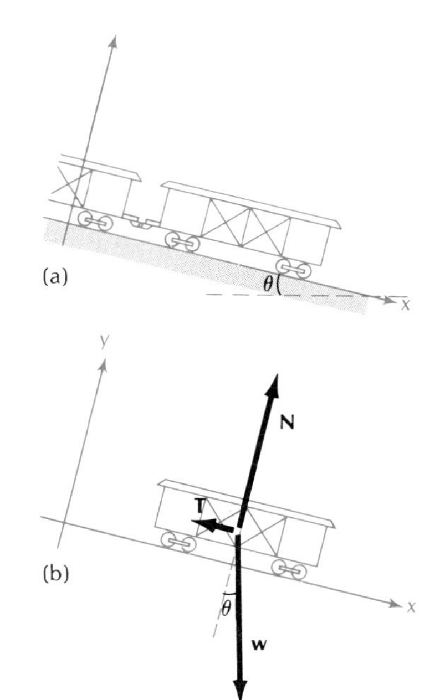


**Fig. 6.7** (a) Boxcar on a steep railroad track. (b) "Free-body" diagram for the boxcar.

COMMENTS AND SUGGESTIONS: The value of the tension is the same whether the boxcar moves up the track, down the track, or remains at rest. Whenever the acceleration is zero (and the friction is zero), all these motions require the same tension. Contrary to what our intuition might suggest, no more force is required to pull the boxcar upward at constant speed than to restrain it so it rolls downward at constant speed.

Note that the dependence of Eq. (9) on the angle $\theta$ is reasonable: the tension increases with the angle $\theta$, from a smallest value $T = 0$ at an angle of 0° to a largest value $T = mg$ at an angle of 90°. These two extreme cases of the tension agree exactly with what we would expect for a car sitting on a flat track and for a car hanging on a vertical cliff, and this agreement gives us some confidence that Eq. (9) is correct. Upon completion of the solution of a complicated problem, it is good practice to test whether the final formulas have the expected behavior in extreme cases. If the final formulas have an unreasonable behavior, there is probably a mistake in the solution.

Yet another instance of motion with constant force is illustrated in Figure 6.8a. Two masses, $m_1$ and $m_2$, hang on the two ends of a string which runs over a frictionless pulley. Since the two masses are linked by the string, it is necessary to solve their equations of motion simultaneously. We will assume that the string and the pulley are massless and that the pulley runs freely, without offering any resistance to the motion of the string. Under these conditions, the string merely transmits the tension from one mass to the other; consequently, the upward tension forces exerted by the ends of the string on each mass are exactly equal. (Keep in mind that this equality of tensions is a consequence of the special conditions of this example — if the pulley had a mass or if a brake or a motor were coupled to the shaft of the pulley, preventing it from running freely, then the tensions would *not* be equal.)

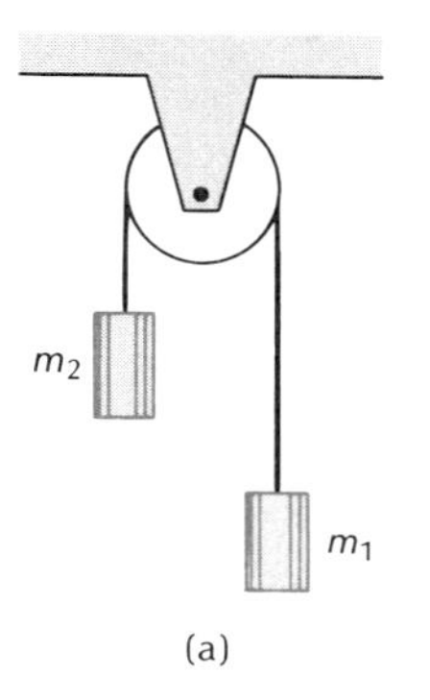


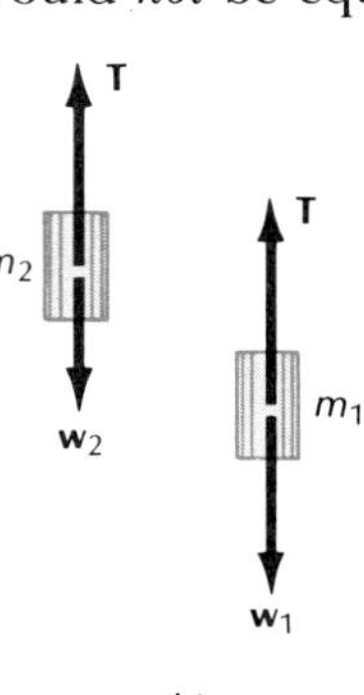


**Fig. 6.8** (a) Two bodies linked by a string. (b) "Free-body" diagrams for the masses $m_1$ and $m_2$.

Besides the tension, the only force acting on each mass is its weight. Figure 6.8b shows the separate "free-body" diagrams for the masses $m_1$ and $m_2$. For a system of several bodies, such as the system of two bodies we are dealing with here, the "free-body" diagrams are especially helpful, because they permit us to view each body by itself and give us a clear picture of what happens to each separate body. The vectors $\mathbf{T}$ in Figure 6.8b represent the tensions, and $\mathbf{w}_1$ and $\mathbf{w}_2$ represent the weights. The vertical force on $m_1$ is $T - w_1$, or $T - m_1 g$; the vertical force on $m_2$ is $T - w_2$, or $T - m_2 g$. Consequently, the equations for the vertical motion of the two masses are

$$m_1 a_1 = T - m_1 g \tag{10}$$

$$m_2 a_2 = T - m_2 g \tag{11}$$

where the forces and accelerations are regarded as positive when directed upward. Equations (10) and (11) contain three unknowns: $T$, $a_1$, and $a_2$. To solve for these three unknowns, we need one more relation between the unknowns. Since the two masses are tied together by a fixed length of string, their accelerations are always of the same magnitudes and in opposite directions, that is,

$$a_1 = -a_2 \tag{12}$$

With this extra relation, Eqs. (10) and (11) become

$$m_1a_1 = T - m_1g \tag{13}$$

$$-m_2a_1 = T - m_2g \tag{14}$$

These are two simultaneous equations for the two unknowns $T$ and $a_1$. To solve these equations, we simply subtract each side of the second equation from each side of the first equation; this gives

$$m_1a_1 - (-m_2a_1) = T - m_1g - (T - m_2g)$$

Here the unknown $T$ cancels out and leaves us with a simple equation for $a_1$,

$$m_1a_1 + m_2a_1 = -m_1g + m_2g$$

We can then immediately solve for $a_1$:

$$a_1 = \frac{m_2 - m_1}{m_1 + m_2}\,g \tag{15}$$

Substituting this result into Eq. (13), we obtain a simple equation for $T$:

$$T - m_1g = m_1\,\frac{m_2 - m_1}{m_1 + m_2}\,g$$

which gives

$$T = m_1\,\frac{m_2 - m_1}{m_1 + m_2}\,g + m_1g$$

or, combining the two terms on the right side,

$$T = \frac{2gm_1m_2}{m_1 + m_2} \tag{16}$$

As we might have expected, Eq. (15) shows that the acceleration is zero if the masses are equal — the two masses are then in equilibrium.

EXAMPLE 4. A passenger elevator consists of an elevator cage of 900 kg (empty) and a counterweight of 990 kg connected by a cable running over a pair of pulleys (Figure 6.9). Neglect the masses of the cable and of the pulleys. (a) What is the upward acceleration of the elevator cage if the pulleys are permitted to run freely? What is the tension in the cable? (b) What are the tensions in the cable if the pulleys are locked (by means of a brake) so that the elevator remains stationary?

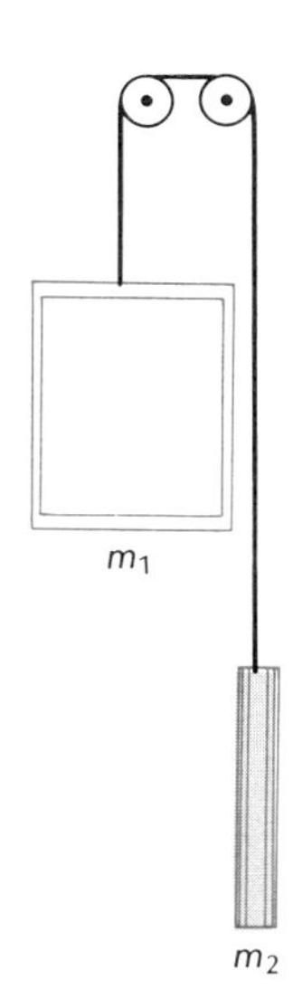

**Fig. 6.9** Elevator with counterweight.

SOLUTION: (a) Such an elevator with a counterweight is equivalent to a system of two masses $m_1$ and $m_2$ connected by a string running over a pulley, as discussed above. With $m_1 = 900$ kg and $m_2 = 990$ kg, Eq. (15) gives

$$a_1 = \frac{990\text{ kg} - 900\text{ kg}}{990\text{ kg} + 900\text{ kg}}\,g = \frac{90}{1890}\,g$$

$$= 0.048g = 0.47\text{ m/s}^2$$

and Eq. (16) gives

$$T = \frac{2g \times 900\text{ kg} \times 990\text{ kg}}{900\text{ kg} + 990\text{ kg}} = 9.2 \times 10^3\text{ N}$$

(b) If the pulleys are locked, the tension in the cable on either side of the pulleys must match the weight hanging from that side. Thus

$$T_1 = m_1 g = 8.8 \times 10^3 \text{ N} \quad \text{and} \quad T_2 = m_2 g = 9.7 \times 10^3 \text{ N}$$

COMMENTS AND SUGGESTIONS: When the pulleys are locked, the tensions in the right and the left parts of the cable are *not* equal because the locked pulleys grip each part of the cable and make the tensions in the right and the left parts independent, as though there were two separate cables rigidly attached to the locked pulleys.

---

From the examples given in this section, we see that the solution of a problem of motion with forces always proceeds in a sequence of steps. The first step is always a careful enumeration of all the forces. To keep track of these forces, it is advisable to display them on a "free-body" diagram, with an identifying label for each force. We must keep in mind that the only forces to be included in the "free-body" diagram are the forces that act *on* the body, not the forces exerted *by* the body. Then we must choose some inertial reference frame and draw coordinate axes on the "free-body" diagram, preferably placing one of the axes along the direction of motion. If the motion proceeds along a sloping ramp, as in the examples of motion of the block or boxcar on an inclined plane, it is convenient to use tilted coordinate axes, with one axis along the slope. Next, we must examine the components of the individual forces and the components of the net force. And, as a final step, we must apply Newton's Second Law.

If the problem involves several moving bodies interacting by strings or by contact forces, as in the example of the two masses and the pulley, then a separate "free-body" diagram is required for each moving body, and Newton's Second Law is to be applied to each body.

## 6.4 Friction

Friction forces, which we have ignored up to now, play an important role in our environment and provide us with many interesting examples of motion with constant force. Suppose that a block of steel, in the shape of a brick, slides on a tabletop of steel. If the block has some initial velocity, friction will gradually slow it down and ultimately stop it. If the steel surfaces are clean and unlubricated, the block will decelerate at the rate of about 6 m/s², or 0.6 G. Figure 6.10 shows the forces acting on the block. The weight **w** acts downward with a magnitude $mg$. The normal force **N** exerted by the table on the block acts upward. The magnitude of the normal force must be $mg$ so that it exactly balances the weight. The friction force $\mathbf{f}_k$ acts horizontally, parallel to the tabletop, in a direction opposite to the motion. This force, just like the normal force, is a contact force which acts over the entire bottom surface of the block; however, for the sake of simplicity, in Figure 6.10 it is shown as though acting at the center of the surface.

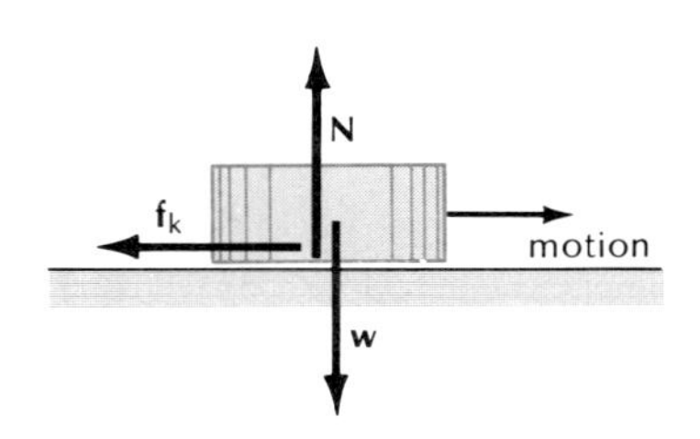


**Fig. 6.10** Forces on a block sliding on a plate.

The friction force arises from adhesion between the two pieces of metal: the atoms in the block form bonds with the atoms in the tabletop. The bonds between the atoms in the two metals can be so strong that occasionally small fragments of steel are plucked out from their surface and they adhere to the opposing surface (Figure 6.11). However, the adhesion does not occur uniformly over the entire bottom surface of the block. On a microscopic scale, the apparently smooth

**Fig. 6.11** Fragment of soft steel adhering to the surface of hard steel of a ball bearing. The fragment was plucked out of a slider of soft steel that was pushed strongly against the surface. Magnification 1600×. (Courtesy S. J. Calabrese, Rensselaer Polytechnic Institute.)

surface of a machined and polished metal contains a great many irregular protuberances (Figure 6.12). When two such surfaces are placed one over the other, the microscopic area of contact is much smaller than the apparent macroscopic area. It is rather like turning Switzerland upside down and placing it on top of Austria — only the tips of the mountains will touch. Thus, for two metal surfaces, intimate contact will occur only at isolated spots at the tips of the protuberances — and only at these spots will the atoms of one surface adhere to those of the other. During sliding, the surfaces are instantaneously "spot-welded" together; then the welds rupture and new welds form, and so on.

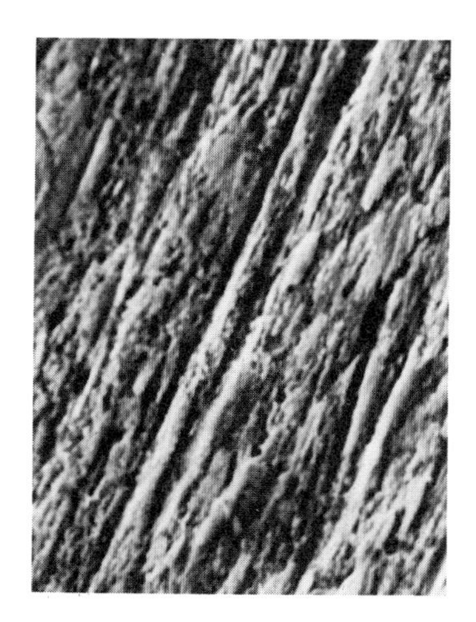

**Fig. 6.12** Surface of a finely polished, mirror-smooth ball bearing of steel, magnified 3300×. The ridges were produced by the machining of the ball bearing. (Courtesy S. J. Calabrese, Rensselaer Polytechnic Institute.)

Although at the microscopic level the phenomenon of friction is very complicated, at the macroscopic level the resulting friction force can often be described adequately by a simple empirical law, first enunciated by Leonardo da Vinci:

> *The magnitude of the force of friction between unlubricated, dry surfaces sliding one over the other is proportional to the normal force pressing the surfaces together and is independent of the (macroscopic) area of contact and of the relative speed.*

Friction involving surfaces in relative motion is called **kinetic friction** (or sliding friction). According to the above law, the force of kinetic friction can be written mathematically as

$$f_k = \mu_k N \tag{17}$$

*Kinetic friction*

where $\mu_k$ is the **coefficient of kinetic friction,** a constant characteristic of the materials involved.

Note that Eq. (17) states that the *magnitudes* of the forces $\mathbf{f}_k$ and $\mathbf{N}$ are proportional; the *directions* of these forces are, however, very different: $\mathbf{N}$ is perpendicular to the surface of contact, and $\mathbf{f}_k$ is parallel to this surface, in a direction opposite to that of the motion.

The above simple "law" of friction lacks the general validity of, say, Newton's laws. It is only approximately valid and it is phenomenological, i.e., it is merely a descriptive summary of empirical observations which does not rest on any detailed theoretical understanding of the mechanism that causes friction. Deviations from this simple law occur at extremely high and at extremely low speeds; in the former case the friction tends to be smaller and in the latter case larger. However, we can ignore these deviations in many everyday engineering problems in which the speeds are not at high or low extremes. The simple law is then quite a good approximation for a wide range of materials, and is at its best for metals.

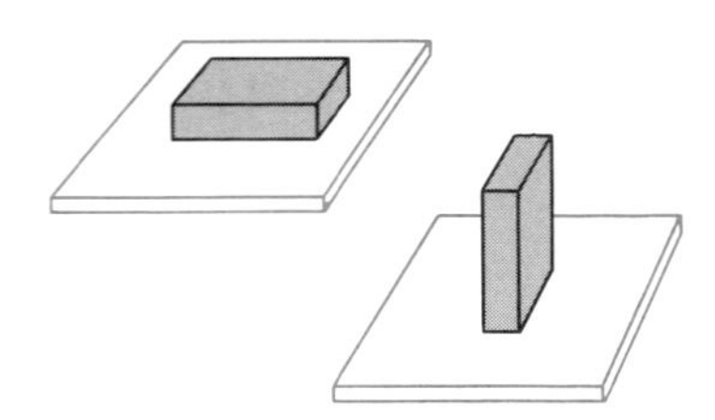

**Fig. 6.13** Steel block sliding on a steel plate.

The fact that the friction force is independent of the (macroscopic) area of contact means that the friction force of the block sliding on the steel tabletop is the same whether the block slides on a large face or on one of the small faces (Figure 6.13). This may seem surprising at first: since friction results from adhesion, we might expect the friction force to be larger when the block slides on its large face, because the contact area is larger. However, what determines the amount of adhesion is not the macroscopic contact area, but the microscopic contact area, and the latter is pretty much independent of whether the block rests on a large face or on a small face. The steel of the block only makes contact with the steel of the tabletop at the tips of the irregular protuberances projecting from its surface; the normal force squeezes the protuberances of one surface against those of the other and causes these protuberances to deform to some extent so that they mate more closely with each other. When the block rests on a large face, the number of contacting protuberances is larger than when it rests on a small face, but since the normal force is distributed over a larger number of such protuberances, the deformation of each is less than when the block rests on a small face. Thus, in one case there is a large number of contacting protuberances, each involving a small area; and in the other case there is a somewhat smaller number of contacting protuberances, each involving a somewhat larger area. The net result is that in both cases the sum of all the microscopic contact areas is the same and, consequently, the friction force is the same.

**Leonardo da Vinci,** *1452–1519, Italian artist, engineer, and scientist. Famous for his brilliant achievements in painting, sculpture, and architecture, Leonardo also made pioneering contributions to science. His insatiable curiosity led him to investigate problems in mathematics, astronomy, mechanics, hydraulics, geology, botany, and anatomy. Leonardo's investigations of friction were forgotten, and the laws of friction were rediscovered two hundred years later by Guillaume Amontons, a French physicist.*

Table 6.2 lists typical friction coefficients $\mu_k$ for a variety of materials. The values given are only approximate. The friction depends to some extent on the condition of the surfaces and also on the temperature. For example, if two surfaces of the same metal are cleaned while in a vacuum so as to prevent any contamination, they will then adhere very strongly and the friction will be very large — the two surfaces "weld" together, forming a single chunk of metal. The coefficients listed in Table 6.2 involve surfaces exposed to air; consequently the metal surfaces are actually covered by films of oxide and there will be very few contacts between the pure metals.

The low friction of ice and snow is due to the formation of a lubricating film of water between the ice surface and the other surface. The

**Table 6.2** KINETIC AND STATIC FRICTION COEFFICIENTS[a]

| *Materials* | $\mu_k$ | $\mu_s$ |
|---|---|---|
| Steel on steel | 0.6 | 0.7 |
| Steel on lead | 0.9 | 0.9 |
| Steel on copper | 0.4 | 0.5 |
| Copper on cast iron | 0.3 | 1.1 |
| Copper on glass | 0.5 | 0.7 |
| Waxed ski on snow | | |
| at −10°C | 0.2 | 0.2 |
| at 0°C | 0.05 | 0.1 |
| Rubber on concrete | ~1 | ~1 |

[a] The friction coefficient depends on the condition of the surfaces. The values in this table are typical for unlubricated, dry surfaces, but should not be trusted blindly.

sliding process heats the top layer of the ice and melts it; the body skidding on the ice actually floats on this film of water. High speeds and a temperature near the melting point of ice favor this lubrication mechanism.

Although the independence of the friction force of the area of (macroscopic) contact is an excellent approximation for metals, it fails for some other materials, such as plastics and rubber. For these materials, the coefficient of friction depends on the shape of the sliding surfaces. But in the following examples we will deal only with materials free of such complications.

EXAMPLE 5. A ship is launched toward the water on a slipway making an angle of 5° with the horizontal (see Figure 6.14a). The coefficient of kinetic friction between the bottom of the ship and the slipway is $\mu_k = 0.08$. What is the acceleration of the ship along the slipway?

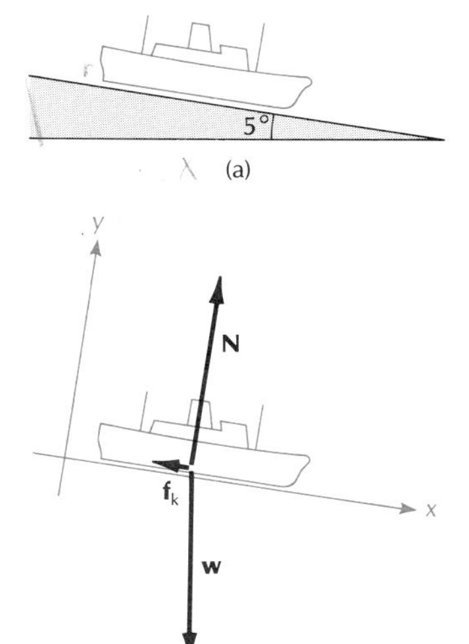

**Fig. 6.14** (a) Ship on slipway. (b) "Free-body" diagram for the ship.

SOLUTION: Figure 6.14b is a "free-body" diagram for the ship. The forces shown are the weight **w**, the normal force **N**, and the friction force $\mathbf{f}_k$. The magnitude of the normal force is $N = mg \cos \theta$ [see Eq. (7)], and the magnitude of the friction force is

$$f_k = \mu_k N = \mu_k \, mg \cos \theta \tag{18}$$

With the $x$ axis parallel to the plane of the slipway, the net force has an $x$ component $F_x = w_x - f_k$, or $F_x = mg \sin \theta - \mu_k mg \cos \theta$. Hence,

$$ma_x = mg \sin \theta - \mu_k \, mg \cos \theta$$

or

$$a_x = (\sin \theta - \mu_k \cos \theta)g$$

$$= (\sin 5° - 0.08 \cos 5°)g$$

$$= 0.0075g = 0.073 \text{ m/s}^2$$

EXAMPLE 6. A man pushes a heavy crate over a smooth floor. The man pushes downward and forward, so that his push makes an angle of 30° with the horizontal (Figure 6.15a). The mass of the crate is 60 kg and the coefficient of kinetic friction is $\mu_k = 0.50$. What force must the man exert to keep the crate moving at uniform velocity?

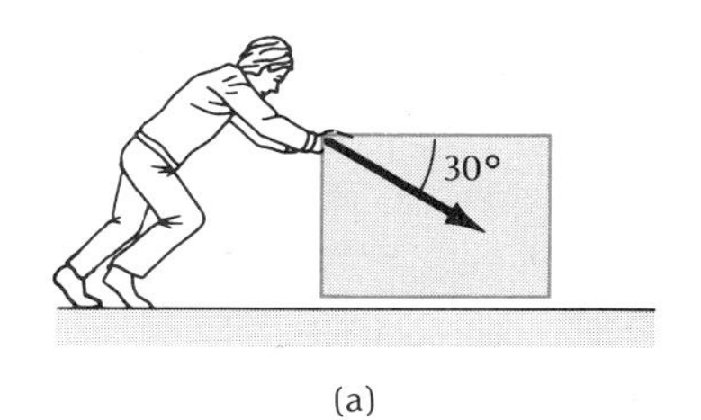


SOLUTION: Figure 6.15b shows a "free-body" diagram for the crate. The forces on the crate are the push **P** of the man, the weight **w,** the normal force **N,** and the friction force $\mathbf{f}_k$. Note that because the man pushes the crate down against the floor, the magnitude of the normal force is not equal to $mg$; we will have to treat the magnitude of the normal force as an unknown. The horizontal and vertical components of the forces are, respectively,

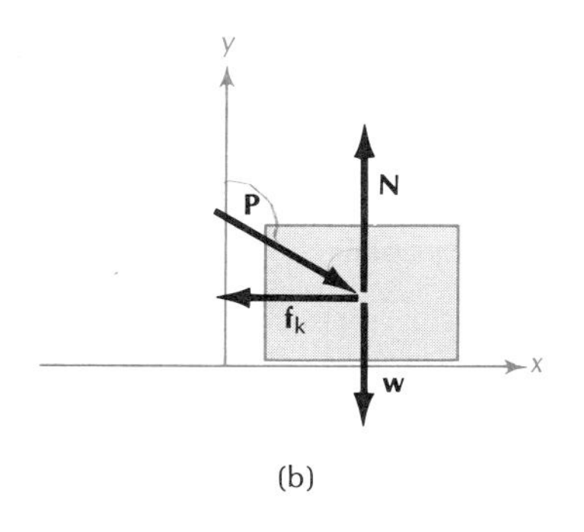

**Fig. 6.15** (a) Man pushing a crate. (b) "Free-body" diagram for the crate.

$$P_x = P \cos 30° \qquad P_y = -P \cos 60°$$

$$w_x = 0 \qquad w_y = -mg$$

$$N_x = 0 \qquad N_y = N$$

$$f_{k,x} = -\mu_k N \qquad f_{k,y} = 0$$

Since the acceleration of the crate is zero in both the $x$ and $y$ directions, the net force in each of these directions must be zero,

$$-P\cos 30° + 0 + 0 - \mu_k N = 0$$

$$-P\cos 60° - mg + N + 0 = 0$$

These are two equations for the two unknowns $P$ and $N$. By adding $\mu_k$ times the second of these equations to the first, we find

$$P(\cos 30° - \mu_k \cos 60°) - \mu_k mg = 0$$

from which

$$P = \frac{\mu_k mg}{\cos 30° - \mu_k \cos 60°}$$

$$= \frac{0.50 \times 60\ \text{kg} \times 9.8\ \text{m/s}^2}{\cos 30° - 0.50 \times \cos 60°} = 4.8 \times 10^2\ \text{N}$$

Friction forces also act between two surfaces at rest. If a force is exerted against the side of, say, a steel block initially at rest on a steel tabletop, the block will not move unless the force is sufficiently large to overcome the friction that holds it in place. For example, on a 1-kg steel block resting on steel, we must apply a lateral push of about 7 N to start the motion. Friction between surfaces at rest is called **static friction.** The maximum magnitude of the static friction force, that is, the magnitude that this force attains when the lateral push is just about to start the motion, can be described by an empirical law quite similar to that for the kinetic friction force:

> *The magnitude of the maximum force of friction between unlubricated, dry surfaces at rest with respect to each other is proportional to the normal force and independent of the (macroscopic) area of contact.*

Mathematically,

$$\boxed{f_{s,\max} = \mu_s N} \tag{19}$$

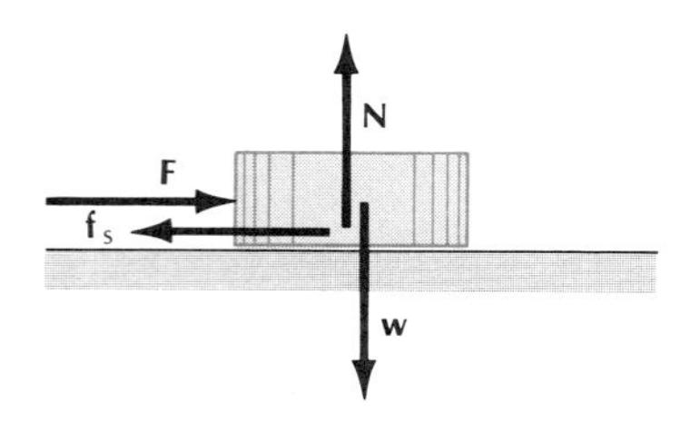


**Fig. 6.16** Forces on a steel block held at rest by friction on a steel plate.

where $\mu_s$ is the **coefficient of static friction,** which depends on the materials. The direction of $\mathbf{f}_{s,\max}$ is parallel to the surface, so as to oppose the lateral push that tries to move the body (Figure 6.16).

The force in Eq. (19) has been labeled with the subscript *max* because it represents the largest static friction force that the surfaces can support without beginning to slide, i.e., $f_{s,\max}$ is the friction force at the "breakaway" point, when the lateral push is just about to start the motion. Of course, if the lateral push is less than this critical value, then the static friction force $f_s$ is also less than $f_{s,\max}$ and exactly matches the magnitude of the lateral push. Thus, in general,

*Static friction*

$$\boxed{f_s \leq \mu_s N} \tag{20}$$

For most materials $\mu_s > \mu_k$, and therefore the maximum static friction force is larger than the kinetic friction force. This implies that if the lateral push applied to the block is large enough to overcome the static friction and to start the block moving, it will more than compensate for the subsequent kinetic friction, and it will therefore accelerate the block continuously.

The larger value of the static friction is due to several effects that increase the adhesion between surfaces at rest. Prolonged contact between the two surfaces produces a gradual plastic deformation ("creep") of the protuberances, permitting the surfaces to mate more closely, with more intimate contact. Furthermore, atoms migrate from one surface into the other, and the surfaces actually grow into each other. Finally, the films of contaminants covering the surfaces tend to split apart when subjected to a steady tangential push, such as the lateral push supported by the surfaces before "breakaway"; and gaps in the films then permit stronger adhesion between the bare materials.

Table 6.2 includes some typical values of the coefficient of static friction $\mu_s$.

EXAMPLE 7. The coefficient of static friction of hard rubber on a street surface is $\mu_s = 0.8$. What is the steepest slope of a street on which an automobile with rubber tires (and locked wheels) can rest without slipping?

SOLUTION: The "free-body" diagram is shown in Figure 6.17. The angle $\theta$ is assumed to be at its maximum value, that is, the friction force has its maximum value $f_{s,max} = \mu_s N = \mu_s mg \cos\theta$. The component of the net force along the inclined plane is

$$F_x = w_x - f_{s,max}$$

$$= mg \sin\theta - \mu_s mg \cos\theta \tag{21}$$

If the automobile is to remain stationary, $F_x = 0$ and

$$0 = mg \sin\theta - \mu_s mg \cos\theta$$

or

$$\mu_s = \tan\theta \tag{22}$$

With $\mu_s = 0.8$, this gives $\theta = 39°$, or a slope of 4:5.

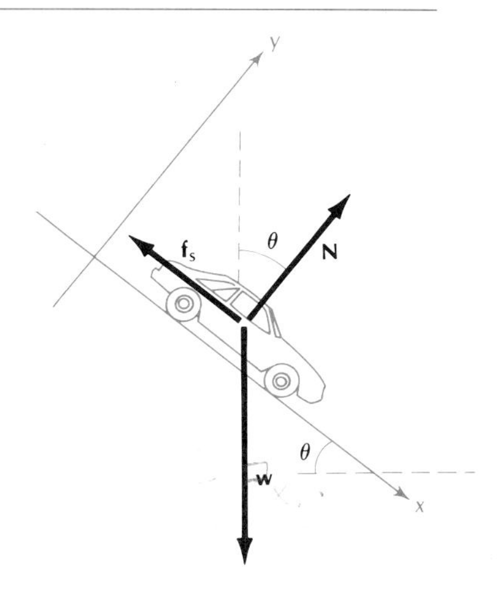


**Fig. 6.17** "Free-body" diagram for an automobile parked on a very steep street.

EXAMPLE 8. An automobile is braking on a level road. What is the maximum deceleration that the automobile can achieve without skidding? As in the preceding example, assume that the tires of the automobile have a coefficient of static friction $\mu_s = 0.8$.

SOLUTION: If the wheels are rolling without skidding, their rubber surface does *not* slide on the street surface (the point of contact between the wheel and the street is instantaneously at rest on the street — we will further explore this feature of rolling motion in Section 13.5). Hence, the relevant friction force is the *static* friction force. The maximum value of this force is

$$f_{s,\,max} = \mu_s N$$

where the magnitude of the normal force $N$ is simply $mg$, since the normal force must balance the weight of the automobile (see Figure 6.18). The deceleration is then given by

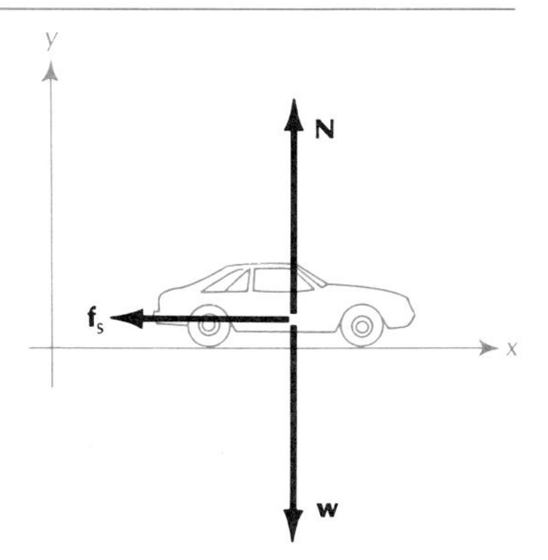


**Fig. 6.18** "Free-body" diagram for braking automobile.

$$ma_x = -f_{s,\,max} = -\mu_s mg$$

which yields

$$a_x = -\mu_s g = -0.8 \times 9.8\ \text{m/s}^2 = -7.8\ \text{m/s}^2$$

## 6.5 Restoring Force of a Spring; Hooke's Law

**Robert Hooke,** *1635–1703, English experimental physicist, curator and secretary of the Royal Society. He was regarded as the foremost mechanic of his time, making many improvements in telescopes and other astronomical instruments, clocks, and watches. Hooke recognized that the motion of the planets must be considered as a mechanical problem, and he proposed the inverse-square law for the gravitational force.*

A body is said to be **elastic** if it suffers a deformation when a stretching or compressing force is applied to it and returns to its original shape when the force is removed. For example, suitable forces can stretch a coil spring or a rubber band and they can bend a flexible rod or a beam of metal or wood. Even bodies normally regarded as rigid, such as the balls of a ball bearing made of hardened steel, are somewhat elastic — they will deform if a sufficiently large force is applied to them.

The force with which a body resists deformation is called its **restoring force.** If we stretch a spring by pulling with our hand, we can feel the restoring force opposing our pull. The restoring force and the force that produces the deformation are of equal magnitudes; they are an action–reaction pair.

Under static conditions, the restoring force with which an elastic body opposes whatever pulls on it often obeys a simple empirical law known as **Hooke's Law:**

*Hooke's Law*

> *The magnitude of the restoring force is directly proportional to the deformation.*

This is not a general law of physics — the *exact* restoring force produced by the deformation of an elastic body depends in a complicated way on the shape of the body and on the detailed properties of the material of the body. Hooke's Law is only an approximate, phenomenological description of the restoring force. However, it is often a quite good approximation, provided that the deformation is small.

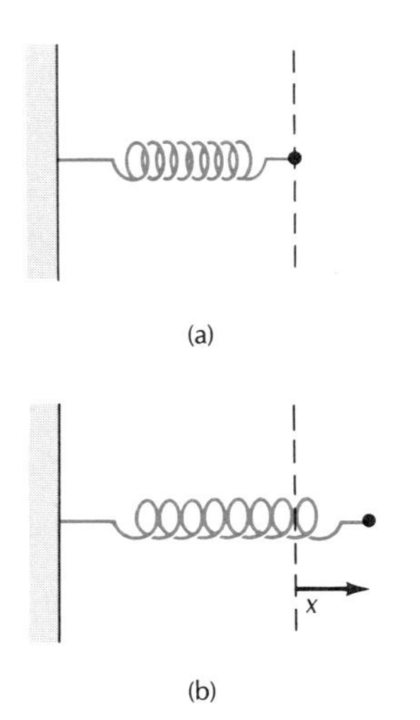


**Fig. 6.19** (a) Spring, relaxed. (b) Spring, stretched by a length $x$.

As a special case, consider a coil spring. Figure 6.19a shows such a spring in its relaxed state; it is loosely coiled so that it can be compressed as well as stretched. Suppose that the left end of the spring is attached to a rigid support (wall), and we apply a stretching or compressing force to the right end. Under the influence of this force, the spring will settle into a new equilibrium configuration such that the restoring force exactly balances the externally applied force. We can measure the deformation of the spring by the displacement that the right end undergoes relative to its initial position. In Figure 6.19b this displacement is denoted by $x$. A positive value of $x$ corresponds to an elongation of the spring, and a negative value corresponds to a compression. Clearly, $x$ is nothing but the change in the length of the spring.

Hooke's Law for the value of the restoring force, then, is that the force is directly proportional to the change in the length of the spring,

*Restoring force of a spring*

$$\boxed{F = -kx} \tag{23}$$

*Spring constant*

The constant of proportionality $k$ is the **spring constant**; it is a positive number characteristic of the spring. The spring constant is a measure of the stiffness of the spring — a stiff spring has a high value of $k$, and a soft spring a low value of $k$. The unit for the spring constant is the newton per meter (N/m) or the pound-force per foot (lbf/ft). The negative sign in Eq. (23) indicates that the restoring force opposes the deformation: if the spring is elongated (positive $x$), then the restoring force is negative and opposes the external stretching force (Figure 6.20a); if the spring is compressed (negative $x$), then the restoring force is positive and opposes the external compressing force (Figure 6.20b).

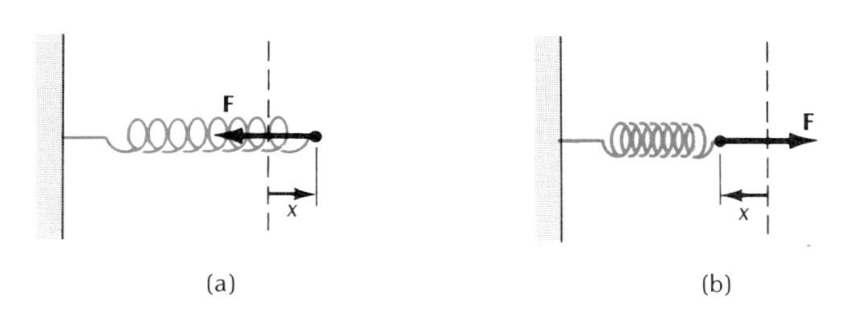


**Fig. 6.20** (a) If the change $x$ in the length of the spring is positive, then the restoring force $F$ is negative. (b) If $x$ is negative, then $F$ is positive.

Coil springs usually obey Hooke's Law quite closely unless the deformation is excessively large. If a spring is stretched beyond its elastic limit, it will suffer a permanent deformation and not snap back to its original shape when released. Such damage to the spring destroys the simple proportionality given by Eq. (23). Furthermore, if a spring is stretched much beyond its elastic limit, it will break and then, of course, the restoring force disappears altogether. Figure 6.21 is a plot of the restoring force versus length for a small steel spring stretched beyond its elastic limit.

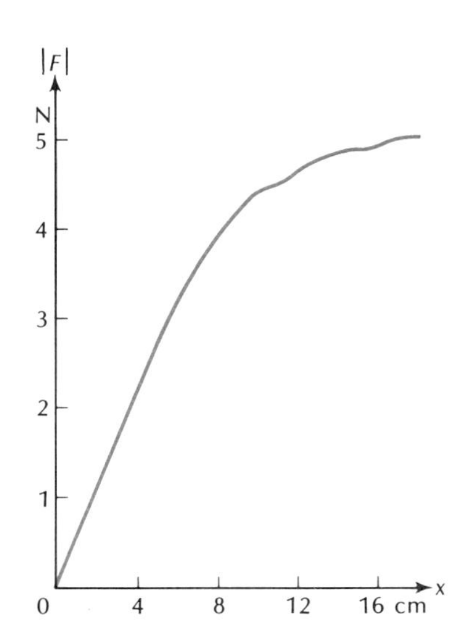


**Fig. 6.21** Restoring force vs. length for a small steel spring. When stretched beyond 7 cm, the spring suffered a permanent deformation.

EXAMPLE 9. The manufacturer's specifications for the coil spring for the front suspension of a Triumph sports car call for a spring of 10 coils with a relaxed length of 0.316 m, and a length of 0.205 m when under a load of 399 kg. What is the spring constant?

SOLUTION: The restoring force that will balance the weight of 399 kg is $F = 399\text{ kg} \times 9.81\text{ m/s}^2 = 3.91 \times 10^3\text{ N}$. The corresponding change of length is $x = 0.205\text{ m} - 0.316\text{ m} = -0.111\text{ m}$. Hence,

$$k = -\frac{F}{x} = -\frac{3.91 \times 10^3\text{ N}}{-0.111\text{ m}} = 3.52 \times 10^4\text{ N/m}$$

EXAMPLE 10. If the spring described in the preceding example is cut into two equal pieces, what will be the spring constant of each piece?

SOLUTION: If the spring is cut into two equal pieces, the spring constant of each piece will be twice as large, i.e., $k' = 7.04 \times 10^4$ N/m. This can be best understood by noting that if a 5-coil spring is to be compressed by, say, 10 cm, each coil must be deformed by 2 cm; whereas if a 10-coil spring is to be compressed by 10 cm, each coil must be deformed by only 1 cm. The force required to deform one coil by 2 cm is twice as large as the force required to deform one coil by 1 cm. Consequently, the spring constant of the 5-coil spring is twice as large as the spring constant of the 10-coil spring.

COMMENTS AND SUGGESTIONS: This example leads to a general rule: other things being equal, the spring constant is inversely proportional to the number of coils. With this rule we can understand how the suspension springs of an automobile can be stiffened by means of clamps that rigidly hold two adjacent coils so as to prevent their relative motion. Such a clamp makes the two coils act as one, that is, it effectively reduces the total number of coils.

## 6.6* Motion with a Variable Force

If the force acting on a body is a known function of position or of time, then Newton's Second Law determines the acceleration of the body at each instant, and from this acceleration the complete motion can be calculated. The examples of the preceding sections illustrate how such calculations are done in the simple case of a constant force. However, this case is an exception — most forces that occur in nature are *not* constant. In this section we will take a brief look at an example involving a motion with a variable force. Mathematically, the solution of the equation of motion with a variable force is often quite difficult. Consider one-dimensional motion (along the $x$ axis) with a force that is a known function of position, $F = F(x)$. The equation of motion is

$$m\frac{d^2x}{dt^2} = F(x) \tag{24}$$

*Differential equation*

This is a **differential equation** for the function $x(t)$, that is, an equation that contains the function $x(t)$ and its derivatives. The function $x(t)$ is regarded as the unknown; finding a solution of the differential equation means finding a function $x(t)$ such that Eq. (24) is satisfied. The solution will be mathematically simple only if $F(x)$ is some very special function. For example, if the force is that of a spring obeying Hooke's Law, $F(x) = -kx$, then Eq. (24) can be solved fairly easily in terms of simple mathematical functions (see Section 14.2). But, in general, the

* This section is optional.

solution of Eq. (24) cannot be expressed in "closed" form, that is, it cannot be expressed in terms of well-known, standard mathematical functions such as sines, cosines, or exponentials.

If no convenient exact solution is available, one may have recourse to approximate solutions obtained by numerical methods. To illustrate such a numerical solution, we will look at the example of an inverse-square force,

$$F(x) = -\frac{A}{x^2} \tag{25}$$

where $A$ is a constant. At points close to $x = 0$, this force has a very large value; the point $x = 0$ is called the **center of force.** The direction of the force is toward this center,[2] and the magnitude decreases as the distance $x$ from this center increases. According to Eq. (25), whenever $x$ increases by a factor of two, the force decreases by a factor of four, etc. Figure 6.22 is a plot of $F(x)$.

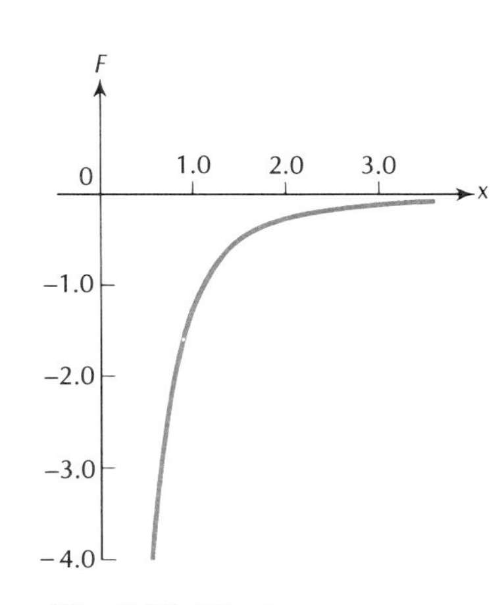


**Fig. 6.22** The inverse-square force as a function of $x$, with $A = 1$.

The equation of motion of the particle exposed to this force is then

$$m\frac{d^2x}{dt^2} = -\frac{A}{x^2}$$

or

$$a = -\frac{1}{x^2}\frac{A}{m} \tag{26}$$

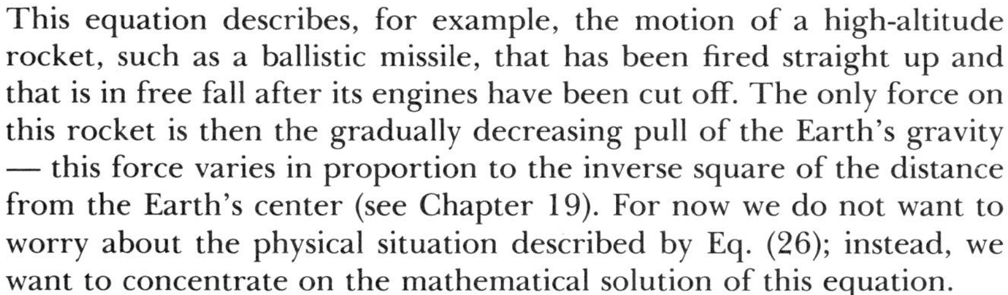

This equation describes, for example, the motion of a high-altitude rocket, such as a ballistic missile, that has been fired straight up and that is in free fall after its engines have been cut off. The only force on this rocket is then the gradually decreasing pull of the Earth's gravity — this force varies in proportion to the inverse square of the distance from the Earth's center (see Chapter 19). For now we do not want to worry about the physical situation described by Eq. (26); instead, we want to concentrate on the mathematical solution of this equation.

To solve the equation numerically, we regard the time variable as made up of discrete intervals, each of which lasts a time $\Delta t$. The instants of time that characterize the beginning and the end of each of these small intervals are

$$\begin{aligned} t_0 &= 0 \\ t_1 &= \Delta t \\ t_2 &= 2\ \Delta t \\ t_3 &= 3\ \Delta t, \qquad \text{etc.} \end{aligned} \tag{27}$$

and the values of $x$ at these times are

[2] The minus sign in Eq. (25) indicates that the force is in the negative direction; i.e., the force is toward the origin whenever $x$ is on the positive $x$ axis. Throughout the following calculation we will assume $x > 0$.

$$\begin{aligned} &x_0, && \text{position at } t = t_0 \\ &x_1, && \text{position at } t = t_1 \\ &x_2, && \text{position at } t = t_2 \\ &x_3, && \text{position at } t = t_3, \quad \text{etc.} \end{aligned} \tag{28}$$

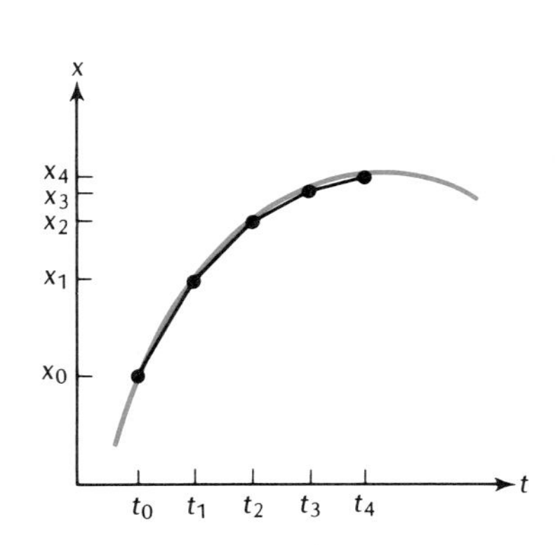


**Fig. 6.23** The worldline (colored) of a particle can be approximated by short straight segments (black).

We now approximate the exact worldline of the particle by a sequence of straight worldline segments (Figure 6.23). This means that the velocities within each time interval are regarded as constant,

$$\begin{aligned} &v_0, && \text{velocity for } t_0 \leq t < t_1 \\ &v_1, && \text{velocity for } t_1 \leq t < t_2 \\ &v_2, && \text{velocity for } t_2 \leq t < t_3, \quad \text{etc.} \end{aligned} \tag{29}$$

Hence, the values of $x$ at succeeding times $t_i$ and $t_{i+1}$ are related by

$$x_{i+1} = x_i + v_i \, \Delta t \tag{30}$$

where $i = 0, 1, 2, 3$, etc.

The acceleration [Eq. (26)] will gradually change the velocity of the particle. In the context of our approximation, all of this change is supposed to occur at the ends of the time intervals. The values of the velocities at succeeding times $t_i$ and $t_{i+1}$ differ by about $a \, \Delta t$, where we take the acceleration $a$ at $x_i$, that is, $a = -(1/x_i^2)(A/m)$. Thus

$$v_{i+1} = v_i - \frac{1}{x_i^2} \frac{A}{m} \Delta t \tag{31}$$

The two sets of equations for position and velocity [Eqs. (30) and (31), respectively] can be solved recursively, beginning with the initial values $x_0$ and $v_0$, which are assumed known. The segmented worldline obtained in this way is an approximation to the precise curved worldline; the approximation becomes exact in the limit $\Delta t \to 0$. Of course, we cannot achieve this limit in our numerical work — we will have to be satisfied with some small, but finite, value at $\Delta t$.

For the following calculation we will assume that $A = 1.0$ and that the mass of the particle is $m = 1.0$ (in some suitable units). Equations (30) and (31) then become

$$x_{i+1} = x_i + v_i \, \Delta t \tag{32}$$

$$v_{i+1} = v_i - \Delta t / x_i^2 \tag{33}$$

For the initial values we will assume $x_0 = 1.0$ and $v_0 = 1.0$. For the time interval we take $\Delta t = 0.1$; this choice of $\Delta t$ is small enough to give reasonable accuracy without excessive labor. The calculation then proceeds step by step:

$$t_0 = 0$$
$$x_0 = 1.0$$
$$v_0 = 1.0$$

$$t_1 = 0.1$$
$$x_1 = x_0 + v_0\,\Delta t = 1.0 + 1.0 \times 0.1 = 1.100$$
$$v_1 = v_0 - \frac{\Delta t}{x_0^2} = 1.0 - \frac{0.1}{(1.0)^2} = 0.900$$

$$t_2 = 0.2$$
$$x_2 = x_1 + v_1\,\Delta t = 1.100 + 0.90 \times 0.1 = 1.190$$
$$v_2 = v_1 - \frac{\Delta t}{x_1^2} = 0.900 - \frac{0.1}{(1.10)^2} = 0.817$$

and so on.

Table 6.3 gives the result of the complete calculation (done on a pocket calculator) and Figure 6.24 gives a plot of position as a function of time. At the time $t = 2.5$, the particle reaches a maximum distance $x \simeq 1.99$ from the center of force; at this time the velocity changes from positive to negative and the particle begins to move back. Clearly, what has happened is that the attractive force has halted the outward motion of the particle and begun to pull it back in. The subsequent inward motion is nothing but the outward motion in reverse.

**Table 6.3** NUMERICAL SOLUTION OF EQS. (32) AND (33)

| $t_i$ | $x_i$ | $v_i$ |
|---|---|---|
| 0.0 | 1.000 | 1.000 |
| 0.1 | 1.100 | 0.900 |
| 0.2 | 1.190 | 0.817 |
| 0.3 | 1.272 | 0.747 |
| 0.4 | 1.346 | 0.685 |
| 0.5 | 1.415 | 0.630 |
| 0.6 | 1.478 | 0.580 |
| 0.7 | 1.536 | 0.534 |
| 0.8 | 1.589 | 0.492 |
| 0.9 | 1.638 | 0.452 |
| 1.0 | 1.684 | 0.415 |
| 1.1 | 1.726 | 0.380 |
| 1.2 | 1.763 | 0.346 |
| 1.3 | 1.798 | 0.314 |
| 1.4 | 1.829 | 0.283 |
| 1.5 | 1.857 | 0.253 |
| 1.6 | 1.883 | 0.224 |
| 1.7 | 1.905 | 0.196 |
| 1.8 | 1.924 | 0.168 |
| 1.9 | 1.941 | 0.141 |
| 2.0 | 1.956 | 0.115 |
| 2.1 | 1.967 | 0.088 |
| 2.2 | 1.976 | 0.063 |
| 2.3 | 1.982 | 0.037 |
| 2.4 | 1.986 | 0.012 |
| 2.5 | 1.987 | −0.014 |

The dots in Figure 6.24 show the approximate results of Table 6.3 and the smooth curve shows the exact solution of the equation of motion. Our approximation is quite good, and it can be made even better

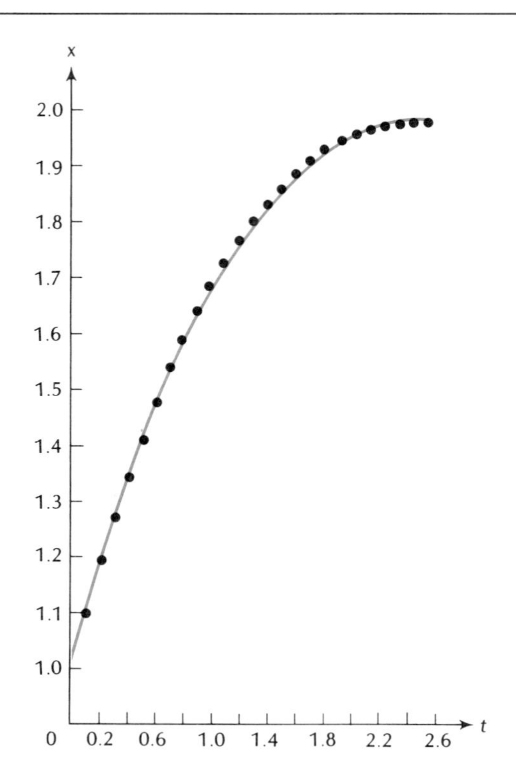


**Fig. 6.24** The dots represent the results listed in Table 6.3. The smooth curve is the exact worldline.

by repeating the calculation using a shorter time interval, say, $\Delta t = 0.05$. Such a reduction of the time interval by, say, a factor of 2 can also serve as a self-consistent test of the approximation: if the numerical calculations with $\Delta t = 0.1$ and $\Delta t = 0.05$ give just about the same results, then we can be reasonably confident that the results are pretty close to the exact solution.

Incidentally: The method of numerical solution adopted in the above example is not the most efficient. For example, we can improve the precision of the calculation by inserting in place of the velocity $v_i$ in Eq. (32) the average velocity $\frac{1}{2}(v_i + v_{i+1})$. When programming numerical calculations on an electronic digital computer, it usually pays to take advantage of such tricks. But the main reason for the high precision attainable with modern electronic computers is their high speed of computation, which makes it feasible to proceed by a very large number of steps, each involving only a very small value of $\Delta t$. For example, for the accurate numerical solution of the equation of motion of a planet, astronomers trace out the orbit by taking several hundred thousand steps around the Sun.

However, these refinements must not be allowed to obscure the important principle involved in Eq. (30) and (31). These equations very explicitly display the essential *deterministic* feature of the equation of motion: given the initial values of position and velocity, a step-by-step calculation leads us inevitably to the values of position and velocity at any later time.

## 6.7 Dynamics of Uniform Circular Motion

As we saw in Section 4.4, uniform circular motion is accelerated motion with a centripetal acceleration. If the motion proceeds with speed

$v$ along a circle of radius $r$, the magnitude of the centripetal acceleration is

$$a = v^2/r \tag{34}$$

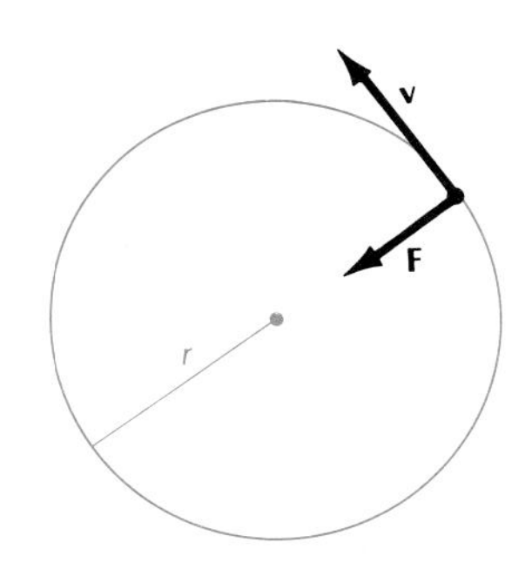


**Fig. 6.25** Centripetal force for a particle in uniform circular motion.

According to Newton's Second Law, this acceleration must be caused by a force having the same centripetal direction as the acceleration, that is, the direction of the force must be toward the center of the circle (Figure 6.25). For instance, the centripetal acceleration of a stone being whirled around a circle at the end of a string is caused by the pull of the string toward the center of the circle, and the centripetal acceleration of the Earth moving in its (nearly) circular orbit around the Sun is caused by the gravitational pull of the Sun.

The magnitude of the centripetal force required to maintain uniform circular motion is

*Centripetal force for uniform circular motion*

$$\boxed{F = ma = mv^2/r} \tag{35}$$

This equation can be used to calculate the magnitude of the force if the speed of the motion is known, or it can be used to calculate the speed if the force is known. The following are several examples of such calculations involving different kinds of forces.

---

EXAMPLE 11. A stone of mass 0.2 kg is tied to a string of length 0.8 m. If you hold the end of this string in your hand and whirl the stone around a circle at the rate of 2 revolutions per second, what is the tension in the string? Ignore gravity in this example.

SOLUTION: The speed of the stone is the distance around the circle divided by the time per revolution.

$$v = \frac{[\text{distance}]}{[\text{time}]} = \frac{2\pi \times 0.8\ \text{m}}{0.5\ \text{s}} = 10\ \text{m/s}$$

The tension in the string plays the role of centripetal force on the stone. According to Eq. (35), the magnitude of this centripetal tension force on the stone is then

$$F = ma = \frac{mv^2}{r} = 0.2\ \text{kg} \times \frac{(10\ \text{m/s})^2}{0.8\ \text{m}} = 25\ \text{N}$$

---

EXAMPLE 12. What is the maximum speed with which an automobile can round a curve of radius 100 m without skidding? Assume that the road is flat and that the coefficient of static friction between the tires and the road surface is $\mu_s = 0.8$.

SOLUTION: The "free-body" diagram for the automobile is given in Figure 6.26. The forces on the automobile are the weight **w**, the normal force **N**, and the friction force $\mathbf{f}_s$. The weight balances the normal force, that is, $N = mg$. The horizontal friction force plays the role of centripetal force; hence the magnitude of the friction force must be

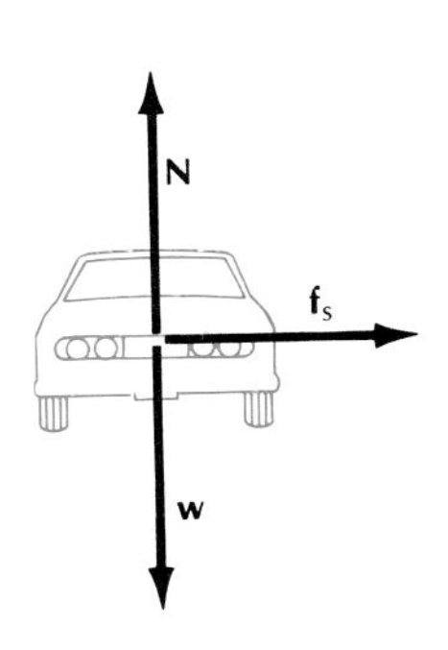


**Fig. 6.26** "Free-body" diagram for an automobile rounding a curve.

$$f_s = mv^2/r$$

The friction is *static* because, by assumption, there is no lateral slippage. At the maximum speed the friction force has its maximum value $f_s = \mu_s mg$ so that

$$\mu_s mg = mv^2/r$$

This yields

$$v = \sqrt{\mu_s g r}$$

$$= \sqrt{0.8 \times 9.81 \text{ m/s}^2 \times 100 \text{ m}}$$

$$= 28.0 \text{ m/s} \qquad (36)$$

EXAMPLE 13. At a speedway in Texas, a curve of radius 500 m is banked at an angle of 22° (Figure 6.27a). If the driver of a racing car does not wish to rely on friction, at what speed should he take this curve?

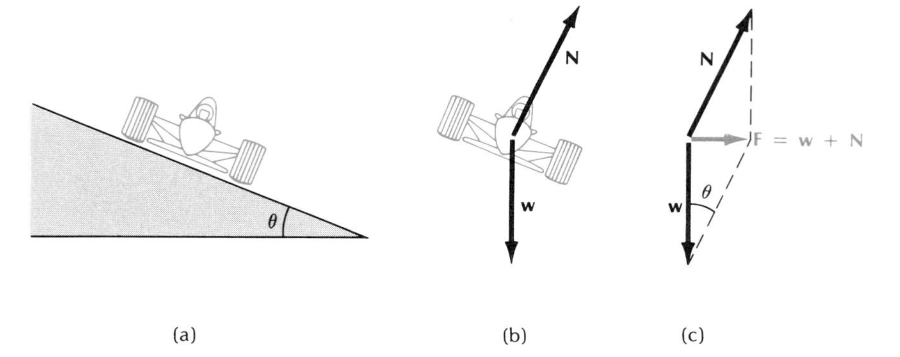


**Fig. 6.27** (a) Car on a banked curve. (b) "Free-body" diagram for a car rounding a banked curve. (c) Resultant of the forces **N** and **w.**

SOLUTION: The "free-body" diagram of the car is shown in Figure 6.27b. Friction is assumed to be absent and hence the normal force **N** and the weight **w** are the only forces acting on the racing car.[3] The resultant of these forces must play the role of centripetal force. Hence the resultant must be horizontal, as in Figure 6.27c. From this figure we see that the magnitude of the resultant is $F = w \tan \theta$, which must coincide with the magnitude of the centripetal force,

$$w \tan \theta = mv^2/r$$

or

$$mg \tan \theta = mv^2/r$$

which yields

$$v = \sqrt{rg \tan \theta}$$

$$= \sqrt{500 \text{ m} \times 9.81 \text{ m/s}^2 \times \tan 22^\circ}$$

$$= 44.5 \text{ m/s} \qquad (37)$$

This is 160 km/h. If the car goes faster than this, it will tend to skid up the embankment; if it goes slower it will tend to skid down the embankment unless friction holds it there.

COMMENTS AND SUGGESTIONS: Note that the centripetal force is *not* included in the "free-body" diagram, and it should never be. The only forces (pushes or pulls) acting on the car are the normal force and the weight. Neither of these forces is centripetal, but their *resultant* is centripetal and provides the centripetal acceleration required to keep the car in the curve.

[3] Air resistance also acts on the car, but is compensated by the propulsive force that the wheels produce by reaction on the ground.

To prevent confusion, do not include resultants in the "free-body" diagram; instead, draw the resultant on a separate diagram, as in Fig. 6.27c. And *never* include the quantity $mv^2/r$ in a "free-body" diagram or in a force diagram. This quantity is not a force. It is merely the product of the mass and the centripetal acceleration — this acceleration is caused by a force or by the resultant of several forces already included among the pushes and pulls displayed in the "free-body" diagram.

*Centrifugal force*

It is commonly said that circular motion generates **centrifugal forces,** that is, forces away from the center. But this statement must be treated with circumspection. The force exerted *on* the body that causes the body to move with circular motion is always a centripetal force; the only centrifugal force in uniform circular motion is the *reaction force* exerted *by* the body. In the above examples, the force exerted on the stone by the string and the force exerted on the automobile by the road are centripetal, whereas the reaction force exerted by the stone on the string and by the automobile on the road are centrifugal. We must always bear in mind that such centrifugal forces are merely reaction forces which do not act on the body in circular motion, but only on the string, the road, or other device that holds the body in its circular motion. And we must bear in mind that our intuition can easily mislead us about centrifugal forces. For instance, if you are riding in an automobile rounding a curve at high speed, you have to hold on to the edge of the seat to keep from sliding outward, and this gives you the sensation that something is pulling you to the outside of the curve, as though your weight had acquired an extra, centrifugal component. However, you are suffering from an illusion. There is actually no such centrifugal force pulling you outward — the only force on your body is the centripetal force which the seat exerts on you. The situation is similar to what happens when the automobile is accelerating along a straight road; the backrest then exerts a forward force on you, and this gives you the illusion that something is pressing you against the backrest, as though your weight had acquired an extra, backward horizontal component. In general, the sensations you experience when your body is subjected to an acceleration are equivalent to those produced by an extra, apparent weight, of the same magnitude as the force causing the acceleration, but of opposite direction. The sensation of weightlessness you experience when in free fall is an extreme case of such an apparent alteration of your weight — the extra weight appears to cancel your normal weight, leaving you weightless.

EXAMPLE 14. A pilot in a fast jet aircraft loops the loop (Figure 6.28). Assume that the aircraft maintains a constant speed of 200 m/s and that the radius of the loop is 1.5 km. (a) At the bottom of the loop, what is the apparent weight that the pilot feels? Express the answer as a multiple of his normal weight. (b) What is his apparent weight at the top of the loop?

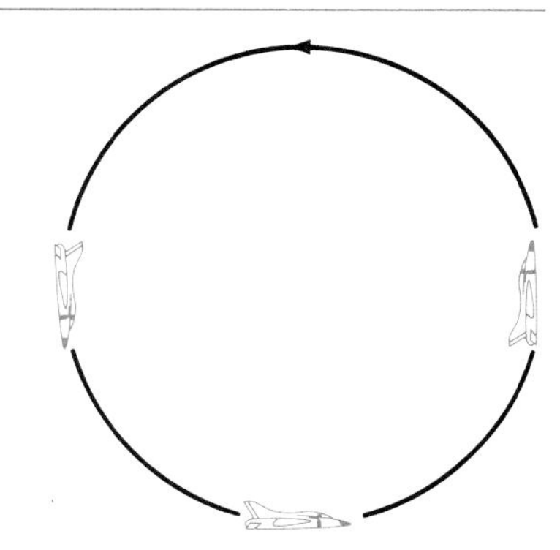

**Fig. 6.28** Jet looping the loop.

SOLUTION: (a) Figure 6.29a shows a "free-body" diagram for the pilot at the bottom of the loop. The forces acting on him are the (true) weight **w** and the normal force **N** exerted by the chair. The net vertical upward force is $N - mg$, and this must provide the centripetal acceleration:

$$N - mg = \frac{mv^2}{r}$$

Hence

$$N = mg + \frac{mv^2}{r} = mg\left(1 + \frac{v^2}{gr}\right)$$

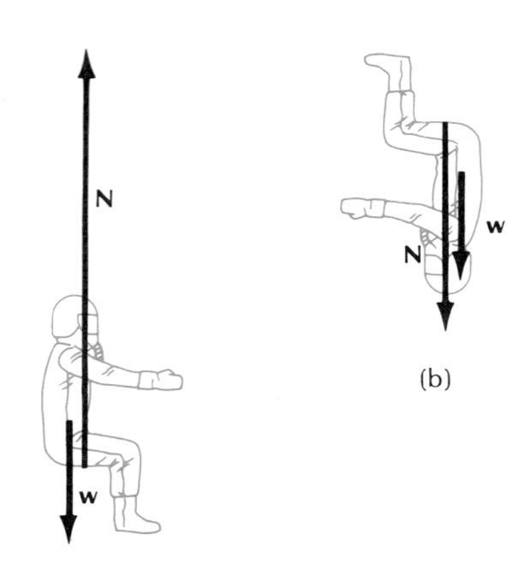


**Fig. 6.29** "Free-body" diagrams of jet pilot at (a) the bottom and (b) the top of the loop.

With $v = 200$ m/s and $r = 1.5 \times 10^3$ m, this gives

$$N = mg\left(1 + \frac{(200 \text{ m/s})^2}{9.8 \text{ m/s}^2 \times 1.5 \times 10^3 \text{ m}}\right)$$

$$= mg \times 3.7$$

This is the force the pilot feels pressing on the seat of his pants. Since this force is 3.7 times larger than the force the pilot would feel when sitting in a stationary chair on the ground, the pilot will say he is experiencing an apparent weight 3.7 times his normal weight, or experiencing 3.7 G (in pilot jargon, he will say he is "pulling 3.7 G").

(b) Figure 6.29b shows the "free-body" diagram at the top of the loop. The net vertical downward force is $N + mg$ and

$$N + mg = \frac{mv^2}{r}$$

Hence

$$N = -mg + \frac{mv^2}{r} = mg\left(-1 + \frac{v_2}{gr}\right)$$

This gives

$$N = mg\left(-1 + \frac{(200 \text{ m/s})^2}{9.8 \text{ m/s}^2 \times 1.5 \times 10^3 \text{ m}}\right)$$

$$= mg \times 1.7$$

Thus, the apparent weight is 1.7 times the normal weight. Note that the pilot presses *up* against the chair. Thus, he will have the impression that the direction of gravity is reversed — the sky is *below* his feet and the Earth is *above* his head. This sensation can be very disorienting for inexperienced pilots.

## SUMMARY

**The four fundamental forces:** gravitational, "weak," electromagnetic, "strong"

**Weight:** magnitude: $w = mg$
direction: Downward

**Kinetic friction:** $f_k = \mu_k N$

**Static friction:** $f_s \leqslant \mu_s N$

**Restoring force of a spring (Hooke's Law):** $F = -kx$

**Force required for uniform circular motion:** magnitude: $mv^2/r$
direction: Centripetal

## QUESTIONS

1. According to the adherents of parapsychology, some people are endowed with the supernormal power of psychokinesis, i.e., spoon-bending-at-a-distance

via mysterious psychic forces emanating from the brain. Physicists are confident that the only forces acting between pieces of matter are those listed in Table 6.1, none of which is implicated in psychokinesis. Given that the brain is nothing but a (very complicated) piece of matter, what conclusions can a physicist draw about psychokinesis?

2. If you carry a spring balance from London to Hong Kong, do you have to recalibrate it? If you carry a beam balance?

3. When you stretch a rope horizontally between two fixed points, it always sags a little, no matter how great the tension. Why?

4. What are the forces on a soaring bird? How can the bird gain altitude without flapping its wings?

5. An automobile is parked on a street. (a) Draw a "free-body" diagram showing the forces acting on the automobile. What is the net force? (b) Draw a "free-body" diagram showing the forces that the automobile exerts on the Earth. Which of the forces in diagrams (a) and (b) are action–reaction pairs?

6. A ship sits in calm water. What are the forces acting on the ship? Draw a "free-body" diagram for the ship.

7. Some old-time roofers claim that when walking on a rotten roof, it is important to "walk with a light step so that your full weight doesn't rest on the roof." Can you walk on a roof with less than your full weight? What is the advantage of a light step?

8. In a tug-of-war on sloping ground the party on the low side has the advantage. Why?

9. The label on a package of sugar claims that the contents are "1 lb or 454 g." What is wrong with this statement?

10. A physicist stands on a bathroom scale in an elevator. When the elevator is stationary, the scale reads 73 kg. Describe qualitatively how the reading of the scale will fluctuate while the elevator makes a trip to a higher floor.

11. How could you use a pendulum suspended from the roof of your automobile to measure its acceleration?

12. When an airplane flies along a parabolic path similar to that of a projectile, the passengers experience a sensation of weightlessness. How would the airplane have to fly to give the passengers a sensation of enhanced weight?

13. A frictionless chain hangs over two adjoining inclined planes (Figure 6.30a). Prove the chain is in equilibrium, i.e., the chain will not slip to the left or to the right. [Hint: One method of proof, due to the seventeenth-century engineer and mathematician Simon Stevin, asks you to pretend that an extra piece of chain is hung from the ends of the original chain (Figure 6.30b). This makes it possible to conclude that the original chain cannot slip.]

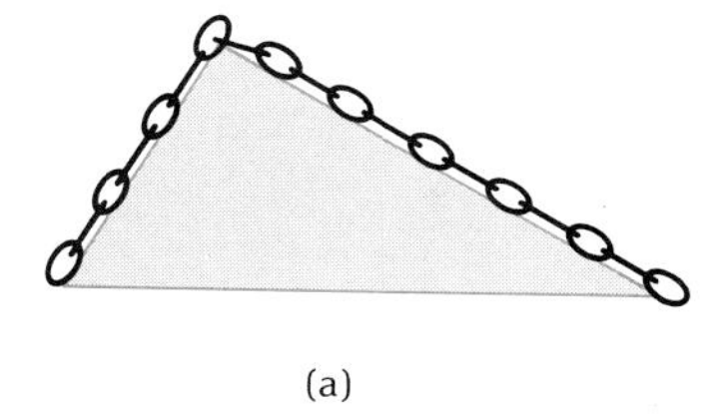

(a)

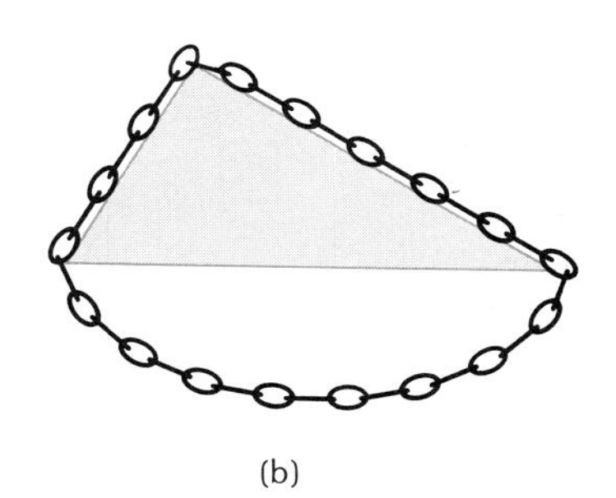

(b)

**Fig. 6.30**

14. Seen from a reference frame moving with the wave, the motion of a surfer is analogous to the motion of a skier down a mountain.[4] If the wave were to last forever, could the surfer ride it forever? In order to stay on the wave as long as possible, in what direction should the surfer ski the wave?

15. Excessive polishing of the surfaces of a block of metal increases its friction. Explain.

16. Some drivers like to spin the wheels of their automobiles for a quick start. Does this give them greater acceleration? (Hint: $\mu_s > \mu_k$.)

17. Cross-country skiers like to use a ski wax that gives their skis a large coefficient of static friction, but a low coefficient of kinetic friction. Why is this useful? How do "waxless" skis achieve the same effect?

[4] There is, however, one complication: surf waves grow higher as they approach the beach. Ignore this complication.

18. Designers of locomotives usually reckon that the maximum force available for moving the train ("tractive force") is $\frac{1}{4}$ or $\frac{1}{5}$ of the weight resting on the drive wheels of the locomotive. What value of the friction coefficient between the wheels and the track does this implicitly assume?

19. When an automobile with rear-wheel drive accelerates from rest, the maximum acceleration that it can attain is less than the maximum deceleration that it can attain while braking. Why? (Hint: Which wheels of the automobile are involved in acceleration? In braking?)

20. Can you think of some materials with $\mu_k > 1$?

21. For a given initial speed, the stopping distance of a train is much longer than that of a truck. Why?

22. Why does the traction on snow or ice of an automobile with rear-wheel drive improve when you place extra weight over the rear wheels?

23. Why are wet streets slippery?

24. In order to stop an automobile on a slippery street in the shortest distance, it is best to brake as hard as possible without initiating a skid. Why does skidding lengthen the stopping distance? (Hint: $\mu_s > \mu_k$.)

25. Suppose that in a panic stop, a driver locks the wheels of his automobile and leaves skid marks on the pavement. How can you deduce his initial speed from the length of the skid marks?

**Fig. 6.31** Drag racer at the start of a race.

26. Hot-rod drivers in drag races find it advantageous to spin their wheels very fast at the start so as to burn and melt the rubber of their tires (Figure 6.31). How does this help them to attain a larger acceleration than expected from the static coefficient of friction?

27. A curve on a highway consists of a quarter of a circle connecting two straight segments. If this curve is banked perfectly for motion at some given speed, can it be joined to the straight segments without a bump? How could you design a curve that is banked perfectly along its entire length and merges smoothly into straight segments without any bump?

28. Automobiles with rear engines (such as the old VW "Beetle") tend to oversteer, that is, in a curve their rear end tends to swing toward the outside of the curve, turning the car excessively into the curve. Explain.

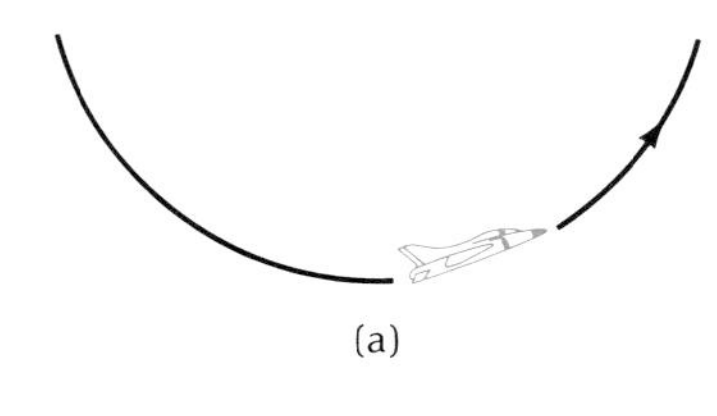

(a)

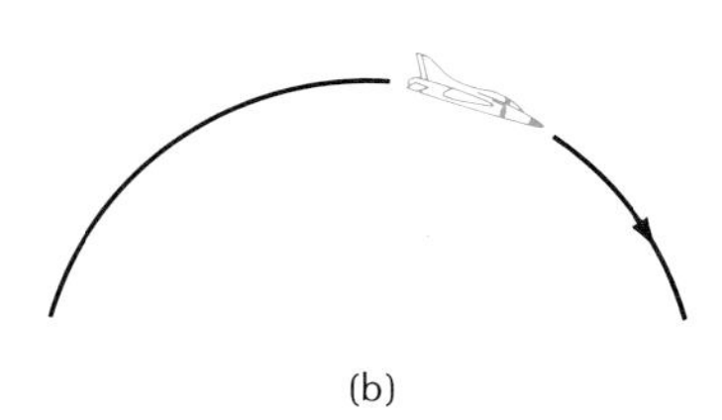

(b)

**Fig. 6.32** Aircraft (a) pulling out of a dive and (b) beginning a dive.

29. (a) If a pilot in a fast aircraft very suddenly pulls out of a dive (Figure 6.32a), he will suffer "blackout" caused by loss of blood pressure in the brain. If he suddenly begins a dive while climbing (Figure 6.32b), he will suffer "redout" caused by excessive blood pressure in the brain. Explain.

(b) A pilot wearing a gee suit — a tightly fitting garment that squeezes the tissues of the legs and abdomen — can tolerate 5 G while pulling out of a dive. How does the gee suit prevent blackout? A pilot can tolerate no more than −2 G while beginning a dive. Why does the gee suit not help against redout?

30. While rounding a curve at high speed, a motorcycle rider leans the motorcycle toward the center of the curve. Why?

## PROBLEMS

### Section 6.2

1. What is the weight (in pounds-force) of a 1-lb bag of sugar in New York? In Hong Kong? In Quito? Use the data given in Table 2.3.

2. A bar of gold of mass 500 g is transported from Paris ($g = 9.8094$ m/s$^2$) to San Francisco ($g = 9.7996$ m/s$^2$).

(a) What is the decrease of the weight of the gold? Express your answer as a fraction of the initial weight.
(b) Does the decrease of weight mean that the bar of gold is worth less in San Francisco?

3. A woman stands on a chair. Her mass is 60 kg and the mass of the chair is 20 kg. What is the force that the chair exerts on the woman? What is the force that the floor exerts on the chair?

4. A chandelier of 10 kg hangs from a cord attached to the ceiling and a second chandelier of 3 kg hangs from a cord below the first (see Figure 6.33). Draw the "free-body" diagram for the first chandelier and the "free-body" diagram for the second chandelier. Find the tension in each of the cords.

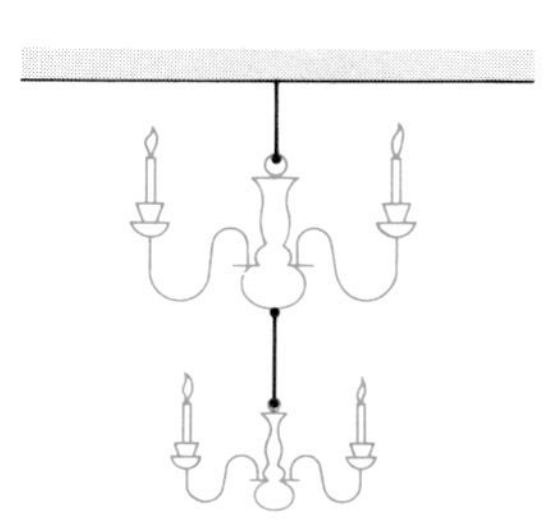

**Fig. 6.33**

### Section 6.3

5. At lift-off, the Saturn V rocket used for the Apollo missions has a mass of $2.45 \times 10^6$ kg.
(a) What is the minimum thrust that the rocket engines must develop to achieve lift-off?
(b) The actual thrust that the engines develop is $3.3 \times 10^7$ N. What is the vertical acceleration of the rocket at lift-off?
(c) At burnout, the rocket has spent its fuel and its remaining mass is $0.75 \times 10^6$ kg. What is the acceleration just before burnout? Assume that the motion is still vertical and that the strength of gravity is the same as when the rocket is on the ground.

6. An elevator accelerates upward at 1.8 m/s². What is the normal force on the feet of an 80-kg passenger standing in the elevator? By how much does this force exceed his weight?

7. A parachutist of mass 80 kg approaches the ground at 5 m/s. Suppose that when he hits the ground, he decelerates at a constant rate (while his legs buckle under him) over a distance of 1 m. What is the force the ground exerts on his feet during the deceleration?

8. If the elevator described in Example 4 carries four passengers of 70 kg each, what speed will the elevator attain running down freely from a height of 10 m, starting from rest?

9. A boy on a skateboard rolls down a hill of slope 1:5. What is his acceleration? What speed will he reach after rolling for 50 m? Ignore friction.

10. A skier of mass 75 kg is sliding down a frictionless hillside inclined at 35° to the horizontal.
(a) Draw a "free-body" diagram showing all the forces acting on the skier (regarded as a particle); draw a separate diagram showing the resultant of these forces.
(b) What is the magnitude of each force? What is the magnitude of the resultant?
(c) What is the acceleration of the skier?

11. Suppose that the last car of a train becomes uncoupled while the train is moving upward on a slope of 1:6 at a speed of 48 km/h. How far will the car coast up the slope before it stops? Ignore friction.

12. A bobsled slides down an icy track making an angle of 30° with the horizontal. How far must the bobsled slide in order to attain a speed of 90 km/h if initially at rest? When will it attain this speed? Assume that the motion is frictionless.

13. Figure 6.34 shows a spherical ball hanging on a string on a smooth, frictionless wall. The mass of the ball is $m$, its radius is $R$, and the length of the string is $l$. Draw a "free-body" diagram with all the forces acting on the ball. Find the normal force between the ball and the wall. Show that $N \to 0$ as $l \to \infty$.

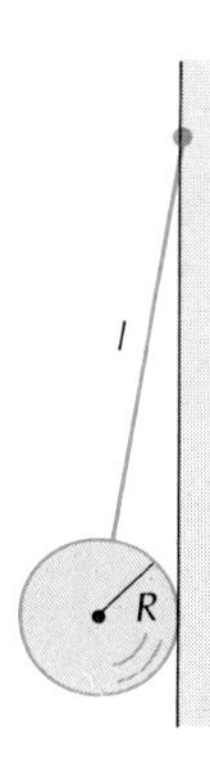


**Fig. 6.34**

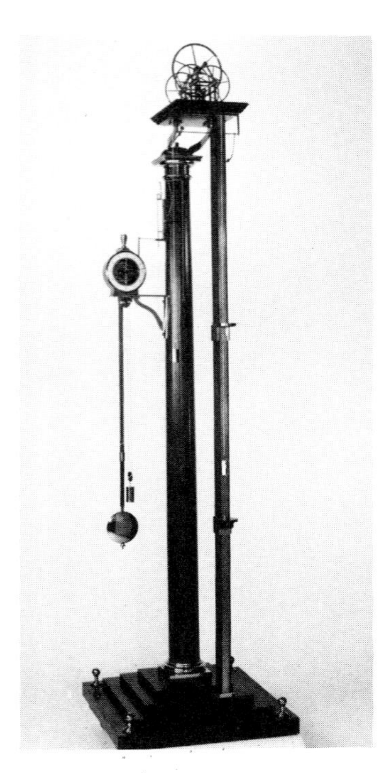

**Fig. 6.35** Atwood's machine.

14. Figure 6.35 shows two masses hanging from a string running over a pulley (see also Figure 6.8a). Such a device can be used to measure the acceleration of gravity; it is then called **Atwood's machine**. If the masses are nearly equal, then the acceleration $a$ of the masses will be much smaller than $g$; this makes it convenient to measure $a$ and then to calculate $g$ by means of Eq. (15). Suppose that an experimenter using masses $m_1 = 400.0$ g and $m_2 = 402.0$ g finds that the masses move a distance of 0.50 m in 6.4 s starting from rest. What value of $g$ does this imply? Assume the pulley is massless.

*15. A woman pushes horizontally on a cardboard box of 60 kg sitting on a frictionless ramp inclined at an angle of 30° (see Figure 6.36).

(a) Draw the "free-body" diagram for the box.

(b) Calculate the magnitudes of all the forces acting on the box under the assumption that the box is at rest or in uniform motion along the ramp.

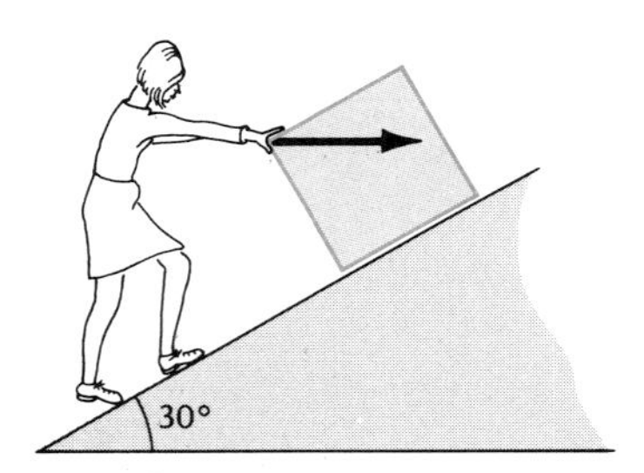


**Fig. 6.36**

*16. During takeoff, a jetliner is accelerating along the runway at 1.2 m/s$^2$. In the cabin, a passenger holds a pocket watch by a chain (a plumb). Draw a "free-body" diagram with the forces acting on the watch. What angle will the chain make with the vertical during this acceleration?

*17. In a closed subway car, a girl holds a helium-filled balloon by a string. While the car is traveling at constant velocity, the string of the balloon is exactly vertical.

(a) While the subway car is braking, will the string be inclined forward or backward relative to the car?

(b) Suppose that the string is inclined at an angle of 20° with the vertical and remains there. What is the acceleration of the car?

**18. A string passes over a frictionless, massless pulley attached to the ceiling (see Figure 6.37). A mass $m_1$ hangs from one end of this string, and a second massless, frictionless pulley hangs from the other end. A second string passes over the second pulley, and a mass $m_2$ hangs from one end of this string, whereas the other end is attached firmly to the ground. Draw separate "free-body" diagrams for the mass $m_1$, the second pulley, and the mass $m_2$. Find the accelerations of the mass $m_1$, the second pulley, and the mass $m_2$.

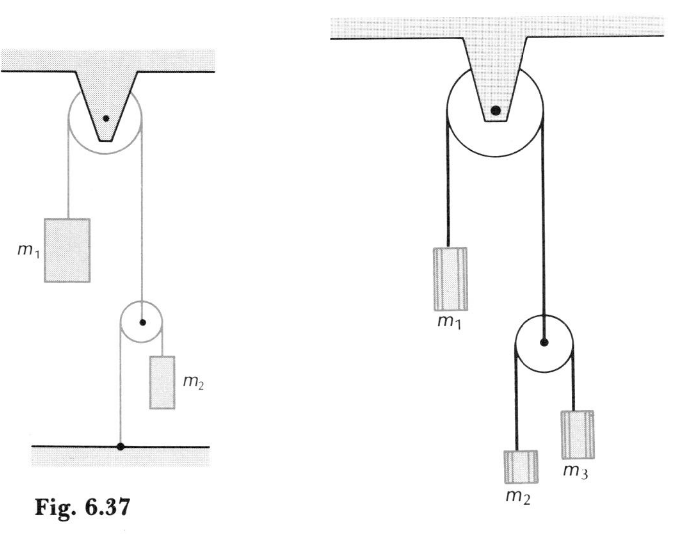


**Fig. 6.37**

**Fig. 6.38**

**19. A particle sliding down a frictionless ramp is to attain a given *horizontal* displacement $\Delta x$ in a minimum amount of time. What is the best angle for the ramp? What is the minimum time?

***20. A mass $m_1$ hangs from one end of a string passing over a frictionless, massless pulley. A second frictionless, massless pulley hangs from the other end of the string (Figure 6.38). Masses $m_2$ and $m_3$ hang from a second string passing over this second pulley. Find the acceleration of the three masses, and find the tensions in the two strings.

Section 6.4

21. The ancient Egyptians moved large stones by dragging them across the sand in sleds. How many Egyptians were needed to drag an obelisk of 700 metric tons? Assume that $\mu_k = 0.3$ for the sled on sand and that each Egyptian exerted a horizontal force of 360 N.

22. The base of a winch is bolted to a mounting plate with four bolts. The base and the mounting plate are flat surfaces made of steel; the friction coefficient of these surfaces in contact is $\mu_s = 0.4$. The bolts provide a normal force of 2700 N each. What maximum static friction force will act between the steel surfaces and help to oppose lateral slippage of the winch on its base?

23. According to tests performed by the manufacturer, an automobile with an initial speed of 65 km/h has a stopping distance of 20 m on a level road. Assuming that no skidding occurs during braking, what is the value of $\mu_s$ between the wheels and the road required to achieve this stopping distance?

24. A crate sits on the load platform of a truck. The coefficient of friction between the crate and the platform is $\mu_s = 0.4$. If the truck stops suddenly, the crate will slide forward and crash into the cab of the truck. What is the maximum braking deceleration that the truck may have if the crate is to stay put?

25. When braking (without skidding) on a dry road, the stopping distance of a sports car with a high initial speed is 38 m. What would have been the stopping distance of the same car with the same initial speed on an icy road? Assume that $\mu_s = 0.95$ for the dry road and $\mu_s = 0.20$ for the icy road.

26. If the coefficient of static friction between the tires of an automobile and the road is $\mu_s = 0.8$, what is the minimum distance the automobile needs in order to stop without skidding from an initial speed of 90 km/h?

27. In a remarkable accident on motorway M.1 (in England), a Jaguar car initially speeding "in excess of 100 m.p.h." skidded 290 m before coming to rest. Assuming that the wheels were completely locked during the skid, and that the coefficient of kinetic friction between the wheels and the road was 0.8, find the initial speed.

28. Because of a failure of its landing gear, an airplane has to make a belly landing on the runway of an airport. The landing speed of the airplane is 90 km/h and the coefficient of kinetic friction between the belly of the airplane and the runway is $\mu_k = 0.6$. How far will the airplane slide along the runway?

*29. A girl pulls a sled along a level dirt road by means of a rope attached to the front of the sled (Figure 6.39). The mass of the sled is 40 kg, the coefficient of kinetic friction is $\mu_k = 0.60$, and the angle between the rope and the road is 30°. What pull must the girl exert to move the sled at constant velocity?

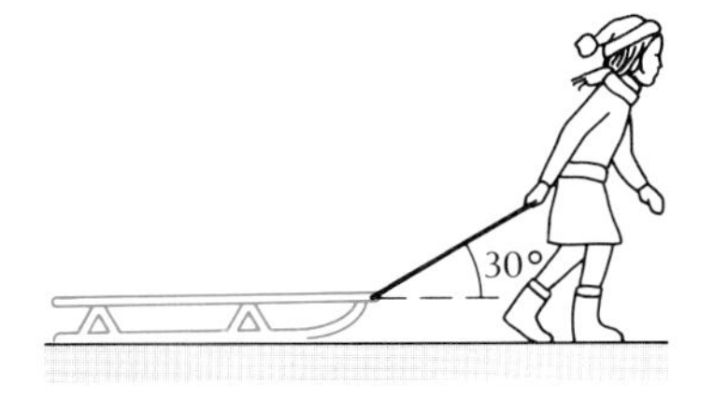


**Fig. 6.39** Girl pulling a sled.

*30. The "Texas" locomotives of the old T&P railway had a mass of 200,000 kg, of which 136,000 kg rested on the driving wheels. What maximum acceleration could such a locomotive attain (without slipping) when pulling a train of 100 boxcars of mass 18,000 kg each on a level track? Assume that the coefficient of static friction between the driving wheels and the track is 0.25.

*31. During braking, a truck has a steady deceleration of 7.0 m/s². A box sits on the platform of this truck. The box begins to slide when the braking begins and, after sliding a distance of 2.0 m (relative to the truck), it hits the cab of the truck. With what speed (relative to the truck) does the box hit? The coefficient of kinetic friction for the box is $\mu_k = 0.50$.

*32. The Schleicher ASW-22 is a high-performance sailplane of a wingspan of 24 m and a mass of 750 kg (including the pilot). At a forward speed (airspeed) of 35 knots, the sink rate, or the rate of descent, of this sailplane is 0.46 m/s.

Draw a "free-body" diagram showing the forces on the plane. What is the friction force (antiparallel to the direction of motion) exerted on the plane by air resistance under these conditions? What is the lift force (perpendicular to the direction of motion) generated by air streaming past the wings?

*33. The friction force (including air friction and rolling friction) acting on an automobile traveling at 65 km/h amounts to 500 N. What slope must a road have if the automobile is to roll down this road at a constant speed of 65 km/h (with its gears in neutral)? The mass of the automobile is $1.5 \times 10^3$ kg.

*34. In a downhill race, a skier slides down a 40° slope. Starting from rest, how far must he slide down the slope in order to reach a speed of 130 km/h? How many seconds does it take him to reach this speed? The friction coefficient between his skis and the snow is $\mu_k = 0.1$. Ignore the resistance offered by the air.

*35. To measure the coefficient of static friction of a block of plastic on a plate of steel, an experimenter places the block on the plate and then gradually tilts the plate. The block suddenly begins to slide when the plate makes an angle of 38° with the horizontal. What is the value of $\mu_s$?

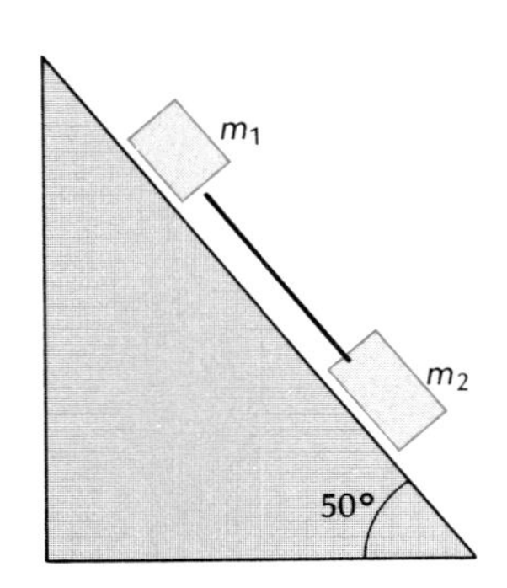

**Fig. 6.40** Two masses connected by a string sliding down a ramp.

*36. A block of wood rests on a sheet of paper lying on a table. The coefficient of static friction between the block and the paper is $\mu_s = 0.7$ and that between the paper and the table is $\mu_s = 0.5$. If you tilt the table, at what angle will the block begin to move?

**37. On a level road, the stopping distance for an automobile is 35 m for an initial speed of 90 km/h. What is the stopping distance of the same automobile on a similar road with a downward slope of 1:10?

**38. Two masses, of 2.0 kg each, connected by a string slide down a ramp making an angle of 50° with the horizontal (Figure 6.40). The mass $m_1$ has a coefficient of kinetic friction 0.60 and the mass $m_2$ has a coefficient of kinetic friction 0.40. Find the acceleration of the masses and the tension in the string.

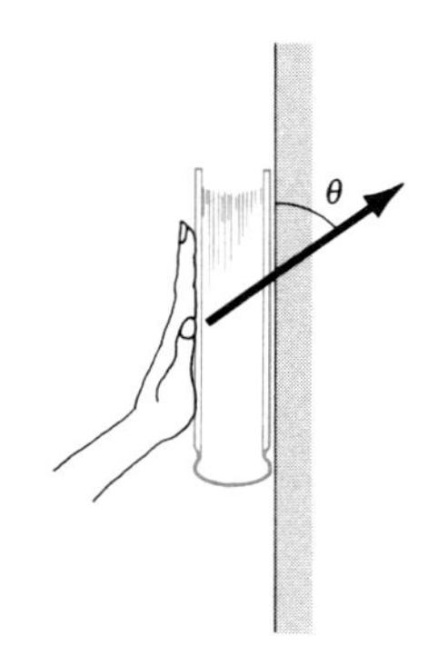

**Fig. 6.41**

**39. You are holding a book against a wall by pushing with your hand. Your push makes an angle of $\theta$ with the wall (see Figure 6.41). The mass of the book is $m$ and coefficient of static friction between the book and the wall is $\mu_s$.
(a) Draw the "free-body" diagram for the book.
(b) Calculate the magnitude of the push you must exert to (barely) hold the book stationary.
(c) For what value of the angle $\theta$ is the magnitude of the required push as small as possible? What is the magnitude of this smallest possible push?
(d) If you push at an angle larger than 90°, you must push very hard to hold the book in place. For what value of the angle will it become impossible to hold the book in place?

**40. A box is being pulled along a level floor at constant velocity by means of a rope attached to the front end of the box. The rope makes an angle $\theta$ with the horizontal. Show that for a given mass $m$ of the box and a given coefficient of kinetic friction $\mu_k$, the tension required in the rope is minimum if $\tan\theta = \mu_k$. What is the tension in the rope when at this optimum angle?

**41. Consider the man pushing the crate described in Example 6. Assume that instead of pushing down at an angle of 30°, he pushes down at an angle $\theta$. Show that he will not be able to keep the crate moving if $\theta$ is larger than $\tan^{-1}(1/\mu_k)$.

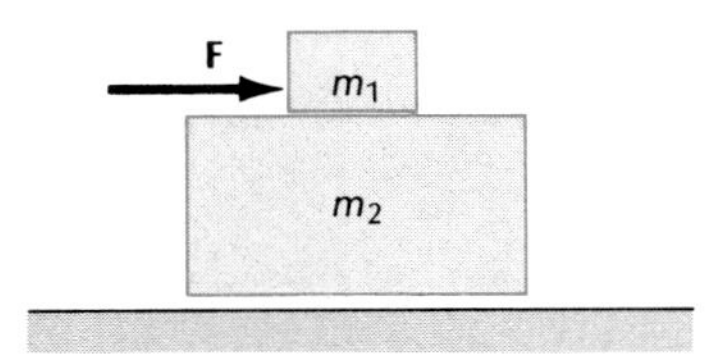

**Fig. 6.42**

**42. A block of mass $m_1$ sits on top of a larger block of mass $m_2$ which sits on a flat surface (Figure 6.42). The coefficient of kinetic friction between the upper and lower blocks is $\mu_1$, and that between the lower block and the flat surface is $\mu_2$. A horizontal force **F** pushes against the upper block, causing it to slide; the friction force between the blocks then causes the lower block to slide also. Find the acceleration of the upper block and the acceleration of the lower block.

**43. Two masses $m_1 = 1.5$ kg and $m_2 = 3.0$ kg are connected by a thin string running over a massless pulley. One of the masses hangs from the string; the other mass slides on a 35° ramp with a coefficient of kinetic friction $\mu_k = 0.40$ (Figure 6.43). What is the acceleration of the masses?

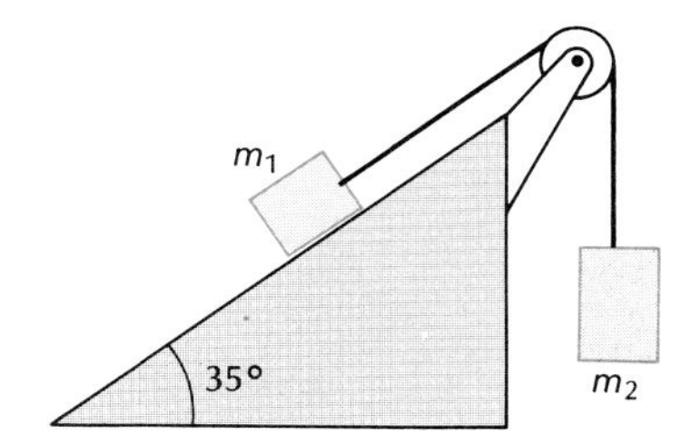


**Fig. 6.43**

**44. A man pulls a sled up a ramp by means of a rope attached to the front of the sled (Figure 6.44). The mass of the sled is 80 kg, the coefficient of kinetic friction between the sled and the ramp is $\mu_k = 0.70$, the angle between the ramp and the horizontal is 25°, and the angle between the rope and the ramp is 35°. What pull must the man exert to keep the sled moving at constant velocity?

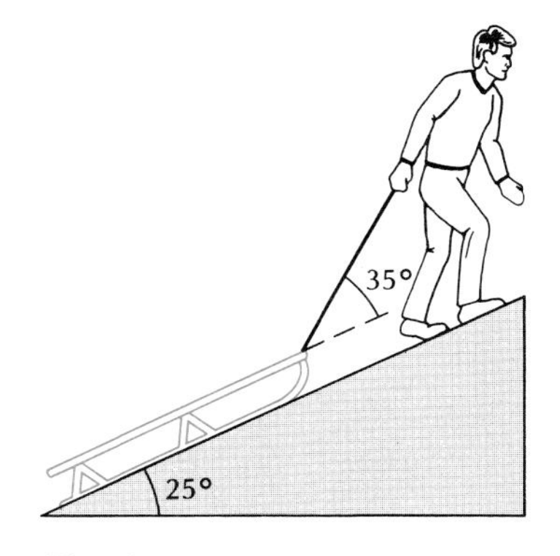


**Fig. 6.44**

**45. Two blocks of masses $m_1$ and $m_2$ are sliding down an inclined plane making an angle $\theta$ with the horizontal. The leading block has a coefficient of kinetic friction $\mu_k$; the trailing block has a coefficient of kinetic friction $2\mu_k$. A string connects the two blocks; this string makes an angle $\phi$ with the ramp (Figure 6.45). Find the acceleration of the blocks and the tension in the string.

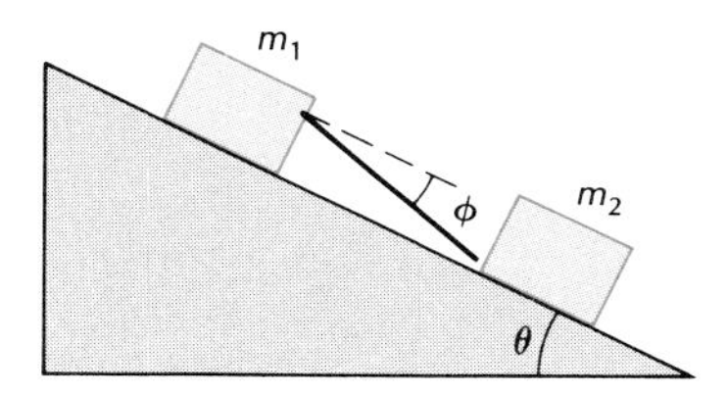


**Fig. 6.45** Two blocks connected by a string sliding down an inclined plane.

Section 6.5

46. Attempting to measure the force constant of a spring, an experimenter clamps the upper end of the spring in a vise and suspends a mass of 1.5 kg from the lower end. This stretches the spring by 0.20 m. What is the force constant of the spring?

47. A spring with a force constant $k = 150$ N/m has a relaxed length of 0.15 m. What force must you exert to stretch this spring to twice its length? What force must you exert to compress this spring to one-half its length?

48. The body of an automobile is held above the axles of the wheels by means of four springs, one near each wheel. Assume that the springs are vertical and that the forces on all the springs are the same. The mass of the automobile is 1200 kg and the spring constant of each spring is $2.0 \times 10^4$ N/m. When the automobile is stationary on a level road, how far are the springs compressed from their relaxed length?

49. A rubber band of relaxed length 6.3 cm stretches to 10.2 cm under a force of 1.0 N, and to 16.5 cm under a force of 2.0 N. Does this rubber band obey Hooke's Law?

*50. Suppose that a uniform spring with a constant $k = 120$ N/m is cut into two pieces, one twice as long as the other. What are the spring constants of the two pieces?

*51. Show that if two springs, of constants $k_1$ and $k_2$, are connected in series (Figure 6.46), the net spring constant $k$ of the combination is given by

$$\frac{1}{k} = \frac{1}{k_1} + \frac{1}{k_2}$$

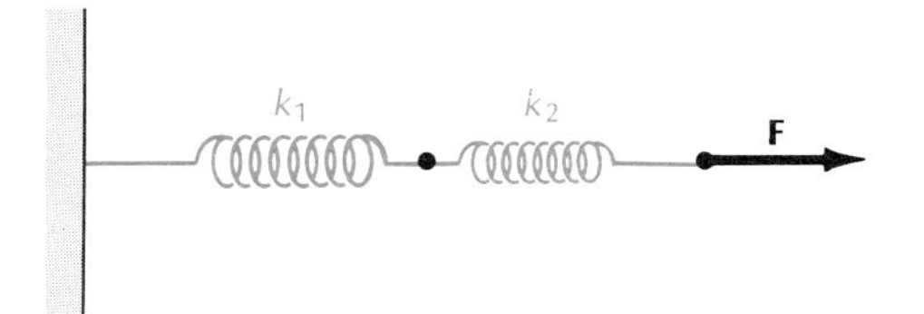


**Fig. 6.46** Springs in series.

*52. Show that if two springs, of constants $k_1$ and $k_2$, are connected in parallel (Figure 6.47), the net spring constant $k$ of the combination is given by

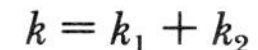

$$k = k_1 + k_2$$

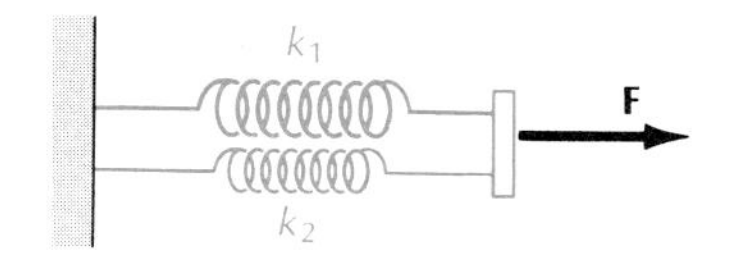


**Fig. 6.47** Springs in parallel.

### Section 6.7

**53. A particle of mass 2.0 kg constrained to move along the $x$ axis experiences a force $F = -40x$, where $F$ is measured in newtons and $x$ in meters. The initial position of the particle is $x = 0.2$ m and the initial velocity is zero. By numerical solution of the equation of motion, find how long it takes the particle to reach the origin $x = 0$. (Hint: Take time intervals of about 0.05 s.)

**54. Under the influence of a friction force proportional to the velocity, a particle has an acceleration $a = -0.2v$, where $a$ is measured in $m/s^2$ and $v$ in m/s. The initial velocity of the particle is 10 m/s. By numerical solution of the above equation, find how long it takes for the velocity to decrease by a factor of 2. (Hint: Use time intervals of 0.20 s.)

**55. Equation (26) describes the motion of a ballistic missile when its engines have cut off and the missile is coasting upward under the influence of the gradually decreasing gravity of the Earth. For a missile of m = 1000 kg, the value of the force constant is $A = 3.99 \times 10^{17}$ N $\cdot$ m$^2$. Suppose that the initial conditions are $x_0 = 6.5 \times 10^6$ m and $v_0 = 6.0 \times 10^3$ m/s. Solve the equation of motion numerically and find the maximum value of $x$ that the missile reaches.

### Section 6.6

56. A swing consists of a seat supported by a pair of ropes 5 m long. A 60-kg woman sits in the swing. Suppose that the speed of the woman is 5.0 m/s at the instant the swing goes through its lowest point. What is the tension in each of the two ropes? Ignore the masses of the seat and of the ropes.

57. The Moon moves around the Earth in a circular orbit of radius $3.8 \times 10^8$ m in 27 days. The mass of the Moon is $7.3 \times 10^{22}$ kg. From these data, calculate the magnitude of the force required to keep the Moon in its orbit.

58. A man of 80 kg is standing in the cabin of a Ferris wheel of radius 30 m rotating at 1 rev/min. What is the force that the feet of the man exert on the floor of the cabin when he reaches the highest point? The lowest point?

59. An automobile enters a curve of radius 45 m at 70 km/h. Will the automobile skid? The curve is not banked and the coefficient of static friction between the wheels and the road is 0.8.

60. A few copper coins are lying on the (flat) dashboard of an automobile. The coefficient of static friction between the copper and the dashboard is 0.5. Suppose the automobile rounds a curve of radius 90 m. At what speed of the automobile will the coins begin to slide? The curve is *not* banked.

61. A curve of radius 400 m has been designed with a banking angle such that an automobile moving at 75 km/h does not have to rely on friction to stay in the curve. What is the banking angle?

*62. Two identical automobiles enter a curve side-by-side, one traveling on the inside lane, the other on the outside. The curve is an arc of circle, and it is unbanked. Each automobile travels through the curve at the maximum speed tolerated without skidding. Which automobile has the higher speed? Which automobile emerges from the curve first? Prove your answer.

*63. The highest part of a road over the top of a hill follows an arc of a vertical circle of radius 50 m. With what minimum speed must you drive an automobile along this road if its wheels are to lose contact with the road at the top of the hill?

*64. A woman holds a pail full of water by the handle and whirls it around a vertical circle at constant speed. The radius of this circle is 0.9 m. What is the minimum speed that the pail must have at the top of its circular motion if the water is not to spill out of the upside-down pail?

*65. A stone of 0.90 kg attached to a string is being whirled around a vertical circle of radius 0.92 m. Assume that during this motion the speed of the stone is constant. If at the top of circle, the tension in the string is (just about) zero, what is the tension ir string at the bottom of the circle?

*66. In ice speedway r motorcycles run at a high speed on an ice-covered track and are kept fr dding by long spikes on their wheels. Suppose that a motorcycle runs a curve of radius 30 m at a speed of 96 km/h. What is the angle of in of the force exerted by the track on the wheels?

*67. An auto aveling at speed $v$ on a level surface approaches a brick wall (Figure hen the automobile is at a distance $d$ from the wall, the driver su izes that he must either brake or turn. If the coefficient of static fr een the tires and the surface is $\mu_s$, what is the minimum distance ver needs to stop (without turning)? What is the minimum dist driver needs to complete a 90° turn (without braking)? What tic for the driver?

is attached to the lower end of a string of length $l$; the upper end g is held fixed. Suppose that the string initially makes an angle $\theta$ ertical. With what horizontal velocity must we launch the mass so ntinues to travel at constant speed along a horizontal circular path he influence of the combined forces of the tension of the string and ? This device is called a **conical pendulum** (Figure 6.49).

. An automobile of mass 1200 kg rounds a curve at a speed of 25 m/s. e radius of the curve is 400 m and its banking angle is 6°. What is the magnitude of the normal force on the automobile? The friction force?

*70. An airplane flies in a horizontal circular path of radius 1500 m at 320 km/h. At what angle should the wings be banked? [Hint: The force exerted by the air on the wings (lift) is perpendicular to the wings.]

**71. A flexible drive belt runs over a flywheel turning freely on a frictionless axle (see Figure 6.50). The mass per unit length of the drive belt is $\sigma$ and the tension in the drive belt is $T$. The speed of the drive belt is $v$. Show that each small segment $d\theta$ of the drive belt exerts a radial force $(T - \sigma v^2)d\theta$ on the flywheel. For what value of $v$ is this force zero?

**72. A circle of rope of mass $m$ and radius $r$ is spinning about its center so that each point of the rope has a speed $v$. Calculate the tension in the rope.

**73. The rotor of a helicopter consists of two blades 180° apart. Each blade has a mass of 140 kg and a length of 3.6 m. What is the tension in each blade at the hub when rotating at 320 rev/min? Pretend that each blade is a uniform thin rod.

**74. Assume that the Earth is a sphere and that the force of gravity ($mg$) points precisely toward the center of the Earth. Taking into account the rotation of the Earth about its axis, calculate the angle between the direction of a plumb line and the direction of the Earth's radius as a function of latitude. What is this deviation angle at a latitude of 45°?

**75. A curve of radius 120 m is banked at an angle of 10°. If an automobile with wheels with $\mu_s = 0.9$ is to round this curve without skidding, what is the maximum permissible speed?

**76. Figure 6.51 shows a pendulum hanging from the edge of a horizontal disk which rotates around its axis at a constant rate. The angle $\alpha$ that the rotating pendulum makes with the vertical increases with the speed of rotation, and can therefore be used as an indicator of this speed. Find a formula for the speed $v_0$ of the edge of the disk in terms of the angle $\alpha$, the radius $R$ of the disk, and the length $l$ of the pendulum. If $R = 0.20$ m and $l = 0.30$ m, what is the speed when $\alpha = 45°$?

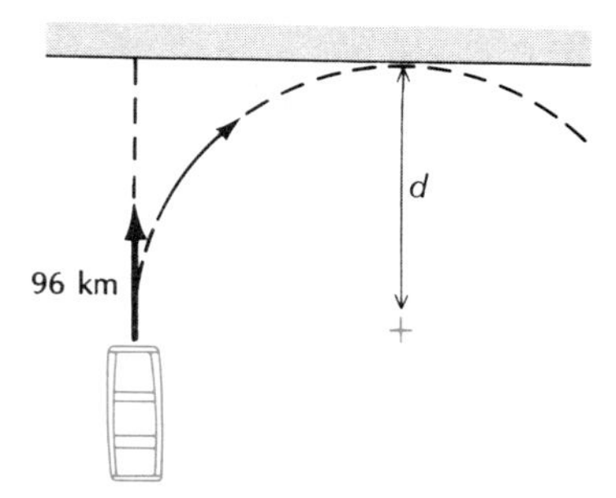

**Fig. 6.48** Automobile approaching a brick wall.

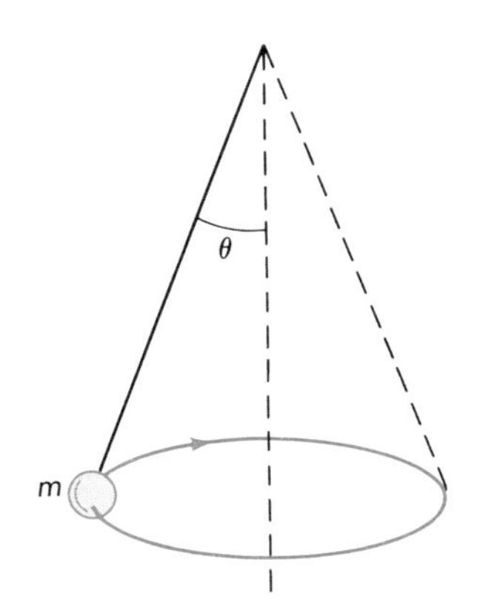

**Fig. 6.49** Mass suspended from a string swinging around in a circle (conical pendulum).

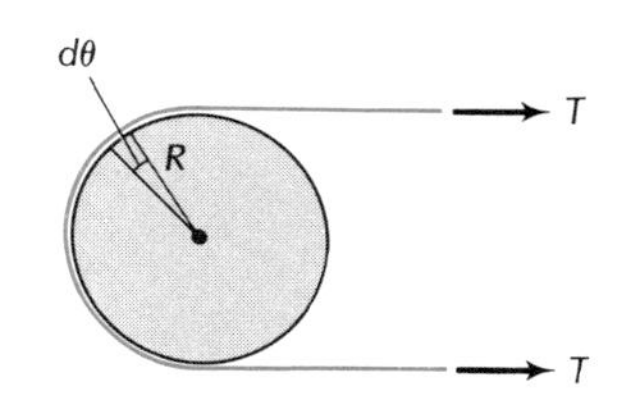

**Fig. 6.50**

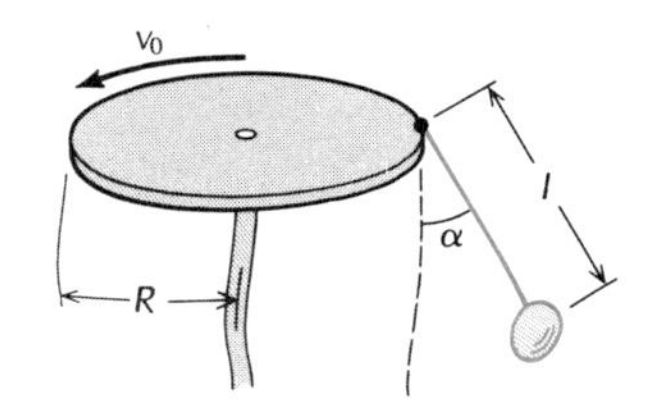

**Fig. 6.51**

CHAPTER 7

# Work and Energy

*Conservation laws*

**Conservation laws** play an important role in the world of matter. Such laws assert that some quantity is conserved, i.e., some quantity remains constant even when matter suffers drastic changes involving motions, collisions, and reactions. One familiar example of a conservation law is the conservation of mass. Expressed in its simplest form, this law asserts that the mass of a given atom, or a given subatomic particle, remains constant. In the two preceding chapters we took this conservation law for granted, and we treated the particle mass appearing in Newton's Second Law as a constant, time-independent quantity. In everyday life and in commercial and industrial operations, we always rely implicitly on the conservation of mass. For instance, in the chemical plants that reprocess the uranium fuel for nuclear reactors, the batches of uranium compounds are carefully weighed at several checkpoints to make sure that no uranium is diverted for nefarious purposes. Obviously, this procedure would make no sense if the particle masses were changeable, so that the net mass of a batch could increase or decrease spontaneously. In a more general form, the law of conservation of mass asserts that the sum of the masses of the particles and other kinds of matter participating in any reaction or interaction remains constant, even when some of the particles are destroyed and other are created in the reaction. This conservation law of mass is a basic law of nature, which has been exhaustively tested and confirmed by experiment.

Other conservation laws involving mechanical quantities are the conservation of energy, momentum, and angular momentum. Furthermore, elementary-particle physicists have discovered several other conservation laws involving esoteric quantities such as baryon number, lepton number, etc.

Physicists take advantage of conservation laws in several ways. In mechanics, they use the conservation laws for energy, momentum, and angular momentum to make predictions about some aspect of the motion of a particle or of a system of particles when it is undesirable or impossible to calculate the full details of the motion from Newton's Second Law. This is especially helpful in those cases where the force law is not known exactly (some examples of this kind will be treated in Chapter 10). In the study of atoms, nuclei, and elementary particles, physicists use the conservation laws to decide what reactions are possible — if a hypothetical reaction violates a conservation law, then it is impossible. But, besides these more or less practical aspects, conservation laws also have a deep theoretical significance. There is an intimate connection between conservation laws and symmetry: the existence of a quantity that remains constant during the motions, collisions, and reactions of matter indicates the presence of a mathematical symmetry in the equations for the fundamental forces governing these phenomena. Such a mathematical symmetry means that certain terms in the equation can be interchanged without affecting its validity. Although the details of this connection are beyond the scope of our discussion, it is obvious that the laws of force must have some special features if they are to avoid violations of the conservation laws. In the search for the fundamental laws of force, the conservation laws therefore give the crucial clues.

**James Prescott Joule** (jool), *1818–1889, English physicist. He established experimentally that heat is a form of mechanical energy, and he made the first direct measurement of the mechanical equivalent of heat. By a series of meticulous mechanical, thermal, and electrical experiments, Joule provided empirical proof of the general law of conservation of energy.*

Both the present chapter and the next deal with the conservation of energy. This conservation law is one of the basic laws of nature. Although we will derive this law from Newton's laws, it is actually much more general and remains valid even when Newton's laws fail. No violation of the law of conservation of energy has ever been discovered.

## 7.1 Work in One Dimension

In spelling out the definition of work, we will begin with the simple case of one-dimensional motion and then proceed to the general case of three-dimensional motion. Consider a particle moving in one dimension, say, along the $x$ axis. If a constant force $F_x$, also along the $x$ axis, acts on the particle, then the **work** done by this force on the particle as it moves some given distance is defined as the product of the force and the displacement,

$$W = F_x \, \Delta x \tag{1}$$

*Work done by constant force*

This rigorous definition of work is motivated by our intuitive notion of what constitutes "work" — if we push a heavy crate over a floor, the work that we perform is proportional to the magnitude of the force we have to exert and to the distance we move the crate. But the ultimate justification of our definition lies in the consequences of this definition, to be discussed in the following sections.

Note that $F_x$ is positive if the force is in the positive $x$ direction and negative if in the negative $x$ direction, that is, $F_x$ is the component of the force along the $x$ axis. According to Eq. (1), the work is positive if the force and the displacement are in the same direction, and negative if they are in opposite directions. The meaning of negative work is that the particle, by its reaction force, does positive work on the agent that exerts the force — negative work *on* the particle means positive work *by* the particle, and positive work *on* the particle means negative work *by* the particle.

Equation (1) gives the work done by one of the forces acting on the particle. If several forces act, then Eq. (1) can be used to calculate the work done by each force.

In the metric system the unit of work is the **joule** (J), which is the work done by a force of 1 N during a displacement of 1 m:

*Joule, J*

$$1 \text{ joule} = 1 \text{ J} = 1 \text{ N} \cdot \text{m} = 1 \text{ kg} \cdot \text{m}^2/\text{s}^2 \tag{2}$$

In the British system the corresponding unit is the **foot-pound-force** (ft · lbf), commonly called foot-pound, which is the work done by a force of 1 lbf during a displacement of 1 ft:

$$1 \text{ ft} \cdot \text{lbf} = 1 \text{ ft} \times 1 \text{ lbf} = 1.356 \text{ J} \tag{3}$$

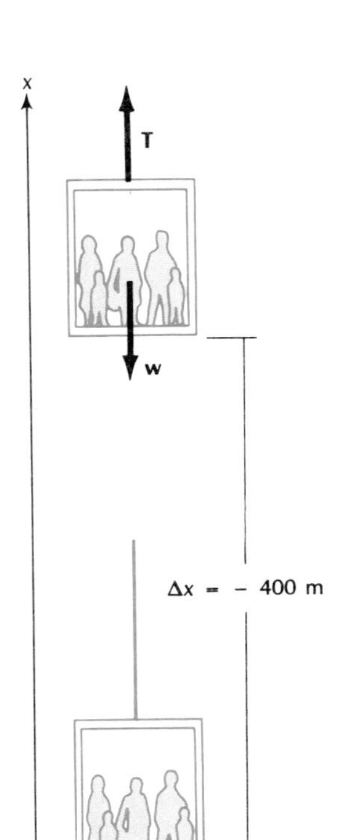


**Fig. 7.1** A descending elevator.

EXAMPLE 1. A 900-kg elevator cage descends 400 m within a skyscraper. What is the work done by gravity on the elevator cage during this displacement?

SOLUTION: With the $x$ axis arranged vertically upward (Figure 7.1), the displacement is negative, $\Delta x = -400$ m, and the weight is also negative, $F_x = -mg = -8.8 \times 10^3$ N. Hence the work is

$$W = F_x \Delta x = (-8.8 \times 10^3 \text{ N}) \times (-400 \text{ m})$$

$$= 3.5 \times 10^6 \text{ J} \tag{4}$$

Note that the work done by gravity is completely independent of the details of the motion; the work depends only on the total vertical displacement and not on the velocity or acceleration of the motion.

EXAMPLE 2. Assuming that the elevator cage of Example 1 descends at constant velocity, what is the work done by the tension of the suspension cable?

SOLUTION: For motion at constant velocity, the tension force must exactly balance the weight. Therefore the tension force has the same magnitude as the weight but the opposite direction, $F_x = T = +8.8 \times 10^3$ N. The work done by this force is then

$$W = F_x \Delta x = (8.8 \times 10^3 \text{ N}) \times (-400 \text{ m})$$

$$= -3.5 \times 10^6 \text{ J} \tag{5}$$

COMMENTS AND SUGGESTIONS: In this example, the work done by the tension is exactly the opposite of the work done by gravity — the total work done by both forces together is zero. However, the result (5) depends implicitly on the assumptions made about the motion. Only for unaccelerated motion does the tension remain constant at $8.8 \times 10^3$ N. For instance, if the elevator cage were allowed to move downward with the acceleration $g$, then the tension would be zero; the work done by the tension would then also be zero, while the work done by gravity would still be $3.5 \times 10^6$ J.

Although the rigorous definition of work contained in Eq. (1) agrees to some extent with our intuition of what constitutes "work," the rigorous definition clashes with our intuition in some instances. For example, consider a man holding a bowling ball in a fixed position in his outstretched hand (Figure 7.2). Our intuition suggests that the man does work — yet Eq. (1) indicates that no work is done on the ball, since $\Delta x = 0$! The resolution of this conflict hinges on the observation that, although the man does no work *on the ball,* he does work *within his own muscles* and, consequently, grows tired of holding the ball. A contracted muscle is never in a state of complete rest: within it, atoms, cells, and muscle fibers engage in complex chemical and mechanical processes which involve motion and work. These internal motions go on continuously while the external shape of the muscle remains rigid. This means that work is done, and wasted, internally within the muscle while no work is done externally on the bone to which the muscle is attached or on the bowling ball supported by the bone. The man does work, but no useful, external work. Obviously, it is very inefficient to hold up a weight by means of rigidly contracted muscles. The human machine would be much more efficient if it had some mechanism for locking the joints (with a pin or a clamp) whenever rigidity was desired; then no muscular effort would be required to hold a weight and no internal work would be wasted.

**Fig. 7.2** Man holding a bowling ball.

Another conflict between our intuition and the rigorous definition of work arises when we contemplate a body in motion. Suppose that the man of Figure 7.2 with the bowling ball in his hand rides in an elevator moving steadily upward. In this case, the displacement is not zero and the force (push) exerted by the hand on the ball does work — the displacement and the force are both in the same direction and consequently the hand continuously does positive work on the ball. Nevertheless, to the man the ball feels no different when riding in the elevator than when standing on the ground. The man plays a passive role: the elevator floor does work on his feet, and the man transmits a suitable fraction of this work to the bowling ball. (The elevator and its cable also play a passive role: they merely transmit the work generated by the motor.) This example illustrates that the amount of work done on a body depends on the frame of reference. In the reference frame of the ground, the ball is moving upward and work is done on it; in the reference frame of the elevator, the ball is at rest and no work is done on it. The lesson we learn from this is that before proceeding with a calculation of work, we must be careful to specify the frame of reference.

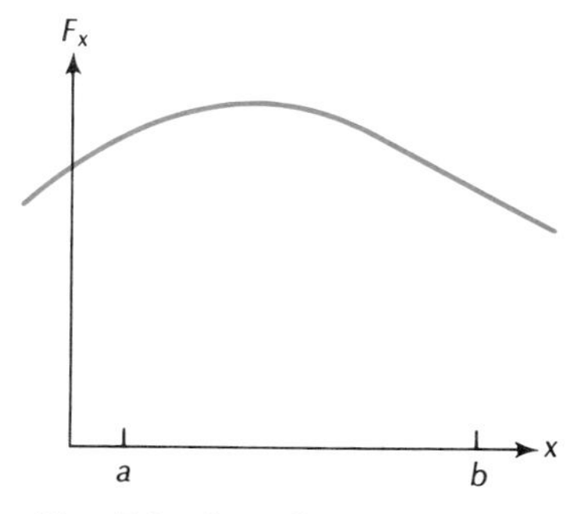


**Fig. 7.3** Plot of $F_x$ vs. $x$.

Next we will have to consider the definition of work for a force that is not constant. Suppose that the force is some function of position,

$$F_x = F_x(x) \tag{6}$$

Figure 7.3 shows a plot of typical function $F_x(x)$. To evaluate the work done by this force on a particle during a displacement from $x = a$ to $x = b$, we divide the total displacement into a large number of small intervals, each of length $\Delta x$ (Figure 7.4). The beginnings and ends of these intervals are located at $x_0, x_1, x_2, \ldots, x_n$, where $x_0 = a$ and $x_n = b$. Within each interval the force can be regarded as approximately constant — within the interval $x_i$ to $x_{i+1}$ (where $i = 0, 1, 2, 3, \ldots$), the force is approximately $F_x(x_i)$. This approximation is obviously at its best if $\Delta x$ is very small. The work done by this force as the particle moves from $x_i$ to $x_{i+1}$ is then

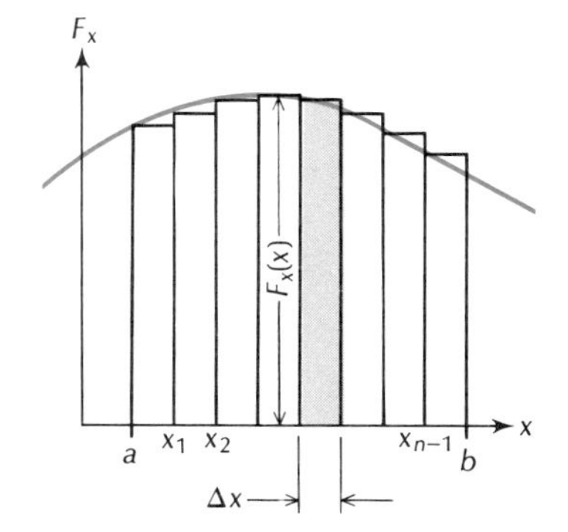


**Fig. 7.4** The total displacement from $a$ to $b$ has been divided into small intervals of length $\Delta x$.

$$\Delta W_i = F_x(x_i)\,\Delta x \tag{7}$$

and the total work done as the particle moves from $a$ to $b$ is simply the sum of all the small amounts of work associated with the small intervals:

$$W = \sum_{i=0}^{i=n-1} \Delta W_i = \sum_{i=0}^{i=n-1} F_x(x_i)\, \Delta x \tag{8}$$

Note that $F_x(x_i)\, \Delta x$ is the area of the rectangle of height $F_x(x_i)$ and base $\Delta x$ (see Figure 7.4). Thus, Eq. (8) gives the sum of all the rectangular areas shown in Figure 7.4.

Equation (8) is of course only an approximation for the work. In order to improve this approximation, we must use a smaller interval $\Delta x$. In the limiting case $\Delta x \to 0$ (and $n \to \infty$), we obtain an exact expression for the work. Thus, the exact definition for the work done by a variable force is

$$W = \lim_{\Delta x \to 0} \sum_i F_x(x_i)\, \Delta x \tag{9}$$

This expression is called the **integral** of the function $F_x(x)$ between the limits $a$ and $b$. The usual notation for this integral is

*Work done by variable force*

$$W = \int_a^b F_x(x)\, dx \tag{10}$$

where $\int$ is called the integral sign and $F_x(x)$ is called the integrand. The quantity in Eq. (9) or (10) is exactly equal to the area bounded by the curve representing $F_x(x)$, the $x$ axis, and the vertical lines $x = a$ and $x = b$ in Figure 7.3; areas above the $x$ axis are reckoned positive and areas below the $x$ axis are reckoned negative.[1]

EXAMPLE 3. A spring exerts a restoring force of $F_x(x) = -kx$ on a particle attached to it (compare Section 6.5). What is the work done by the spring on the particle when it moves from $x = a$ to $x = b$?

SOLUTION: By Eq. (10),

$$W = \int_a^b F_x(x)\, dx = \int_a^b (-kx)\, dx \tag{11}$$

To evaluate this integral we rely on a theorem of calculus which states that the (definite) integral of the function $x^n$ is

$$\int_a^b x^n\, dx = \left[\frac{x^{n+1}}{n+1}\right]_a^b = \frac{b^{n+1}}{n+1} - \frac{a^{n+1}}{n+1}$$

Therefore

$$W = \int_a^b (-\,kx)\, dx = -k\left[\tfrac{1}{2}x^2\right]_a^b = -\tfrac{1}{2}k(b^2 - a^2) \tag{12}$$

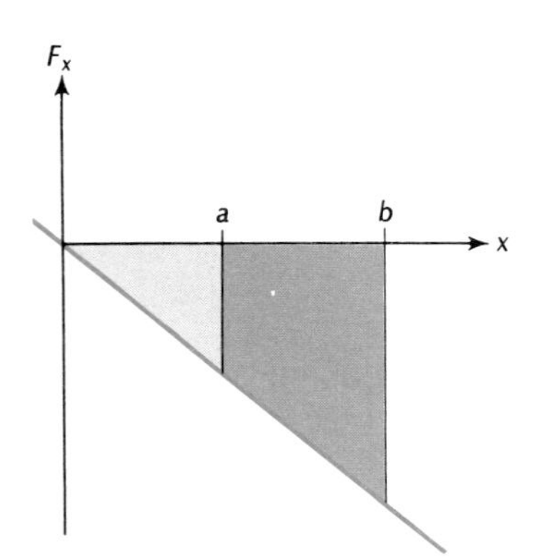


**Fig. 7.5** The straight line is a plot of the function $F_x(x) = -kx$. The quadrilateral area to the right of $a$ and the left of $b$ represents the work.

This result can also be obtained by calculating the area in a plot of force v. position. Figure 7.5 shows the force $F_x(x) = -kx$ as a function of $x$. The triangular area between the origin and $x = b$ is $\frac{1}{2}$[base] × [height] $= \frac{1}{2}b \times kb$, or $\frac{1}{2}kb^2$. Likewise, the triangular area between the origin and $x = a$ is $\frac{1}{2}ka^2$. The quadrilateral area that represents the work is the difference between the two triangu-

[1] A brief review of integrals is given in Appendix 5.

lar areas, that is, $\frac{1}{2}kb^2 - \frac{1}{2}ka^2$. Taking into account that areas below the $x$ axis must be reckoned as negative, we see that this area calculation agrees with Eq. (12).

## 7.2 Work in Three Dimensions; the Dot Product

In the case of three-dimensional motion, the definition of work must take into account the different directions of force and displacement. If a constant force **F** acts on a particle, then the work done by this force on the particle during a displacement $\Delta\mathbf{r}$ is defined as the product of the force, the length of the displacement, and the cosine of the angle between the force and the displacement,

$$W = F\,\Delta r\cos\theta \tag{13}$$

This expression can be regarded as the product of the length of the displacement and the component of the force along the displacement (see Figure 7.6). Thus, only the component of the force along the displacement produces work. The work is positive or negative, depending on the angle $\theta$. If the force is parallel to the direction of motion ($\theta = 0$) or antiparallel to the direction of motion ($\theta = 180°$), then the work is simply $F\,\Delta r$ or $-F\,\Delta r$, respectively; this is just as in the one-dimensional case. If the force is perpendicular to the direction of motion ($\theta = 90°$), then the work vanishes; for instance, this means that neither the normal force **N** on a body sliding on a surface nor the centripetal force on a body in circular motion does any work — both of these forces are always perpendicular to the (instantaneous) motion.

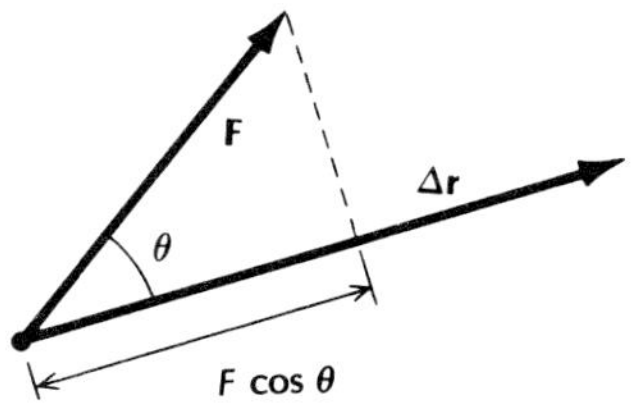

**Fig. 7.6** A constant force **F** acts during a displacement $\Delta\mathbf{r}$. $F\cos\theta$ is the component of **F** along $\Delta\mathbf{r}$.

For any two arbitrary vectors **A** and **B**, the product of their magnitudes and the cosine of the angle between them is called the **dot product** of the vectors. The standard notation for the dot product consists of the two vector symbols separated by a dot,

*Dot product*

$$\mathbf{A}\cdot\mathbf{B} = A\,B\cos\theta \tag{14}$$

In terms of the dot product of the two vectors **F** and $\Delta\mathbf{r}$, the expression (13) for the work takes the concise form

$$W = \mathbf{F}\cdot\Delta\mathbf{r} \tag{15}$$

It would seem that the difference between Eqs. (13) and (15) is merely a matter of appearances, since, by definition, the dot product $\mathbf{F}\cdot\Delta\mathbf{r}$ is equal to $F\,\Delta r\cos\theta$. However, the advantage of Eq. (15) lies in that the dot product can be shown to have the usual distributive properties of ordinary multiplication.[2] This means that if the force in Eq. (15) is the sum of several forces, or if the displacement is the sum of several displacements, then the product can be evaluated by term-by-term multiplication. For instance, suppose that the force is the sum of

[2] For more details on the dot product of vectors, see Section 3.4.

a force in the $x$ direction and a force in the $y$ direction, and likewise for the displacement,

$$\mathbf{F} = F_x\hat{\mathbf{x}} + F_y\hat{\mathbf{y}} \tag{16}$$

and

$$\Delta\mathbf{r} = \Delta x\,\hat{\mathbf{x}} + \Delta y\,\hat{\mathbf{y}} \tag{17}$$

Then

$$\begin{aligned} W = \mathbf{F}\cdot\Delta\mathbf{r} &= (F_x\hat{\mathbf{x}} + F_y\hat{\mathbf{y}})\cdot(\Delta x\,\hat{\mathbf{x}} + \Delta y\,\hat{\mathbf{y}}) \\ &= F_x\Delta x\,\hat{\mathbf{x}}\cdot\hat{\mathbf{x}} + F_x\Delta y\,\hat{\mathbf{x}}\cdot\hat{\mathbf{y}} + F_y\,\Delta x\,\hat{\mathbf{y}}\cdot\hat{\mathbf{x}} + F_y\Delta y\,\hat{\mathbf{y}}\cdot\hat{\mathbf{y}} \end{aligned} \tag{18}$$

The product of any unit vector with itself equals 1 (since the cosine of the angle between the vector and an identical vector is $\cos\theta = 1$), and the product of any unit vector with a different unit vector at right right angles is zero (since the cosine of the angle between such vectors is $\cos 90° = 0$). Hence

$$\hat{\mathbf{x}}\cdot\hat{\mathbf{x}} = 1 \quad \hat{\mathbf{y}}\cdot\hat{\mathbf{y}} = 1 \quad \hat{\mathbf{x}}\cdot\hat{\mathbf{y}} = \hat{\mathbf{y}}\cdot\hat{\mathbf{x}} = 0 \tag{19}$$

and Eq. (18) becomes

$$W = \mathbf{F}\cdot\Delta\mathbf{r} = F_x\Delta x + F_y\Delta y \tag{20}$$

This equation shows that each separate component of the force does work as though the motion were one dimensional [compare Eq. (1)]. Note, however, that Eq. (20) does *not* mean that the work has several components. The work is a scalar quantity, with only one component, and Eq. (20) merely shows how this one-component quantity is to be calculated from the several components of the vectors **F** and $\Delta$**x.**

Equations (19) and (20) can be readily generalized to include the $z$ components of the force and the displacement,

$$\hat{\mathbf{z}}\cdot\hat{\mathbf{z}} = 1 \quad \hat{\mathbf{x}}\cdot\hat{\mathbf{z}} = \hat{\mathbf{z}}\cdot\hat{\mathbf{x}} = 0 \quad \hat{\mathbf{y}}\cdot\hat{\mathbf{z}} = \hat{\mathbf{z}}\cdot\hat{\mathbf{y}} = 0 \tag{21}$$

and

*Work done by constant force (three dimensions)*

$$\boxed{W = \mathbf{F}\cdot\Delta\mathbf{r} = F_x\Delta x + F_y\Delta y + F_z\Delta z} \tag{22}$$

Obviously, the mathematical argument leading to Eq. (22) is valid for any two arbitrary vectors **A** and **B;** thus, expressed in components, the dot product of two arbitrary vectors is

$$\mathbf{A}\cdot\mathbf{B} = A_xB_x + A_yB_y + A_zB_z \tag{23}$$

From this equation [and also from Eq. (14)] we see that the order of the factors does not affect the value of the dot product, that is,

$$\mathbf{A}\cdot\mathbf{B} = \mathbf{B}\cdot\mathbf{A} \tag{24}$$

This commutative rule for the dot product of two vectors is the same as for ordinary multiplication of two numbers.

EXAMPLE 4: Suppose you push a block of mass $m$ up an inclined plane to a height $\Delta z$. What is the work done by gravity on a block during this displacement?

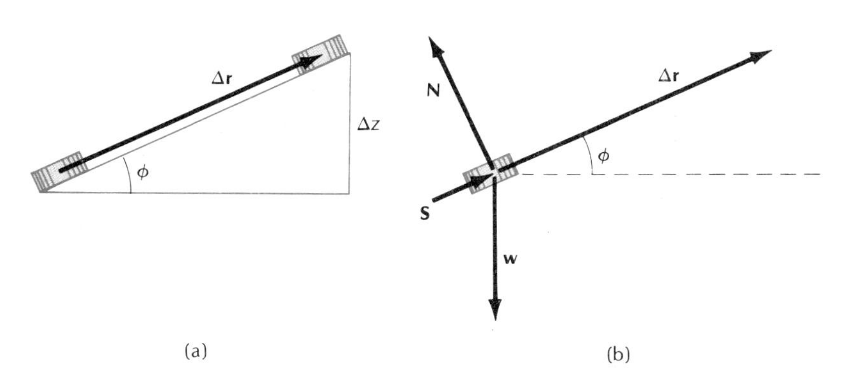


**Fig. 7.7** (a) A block moving along an inclined plane. (b) "Free-body" diagram showing the weight **w**, the normal force **N**, the push **S** and the displacement of the block.

SOLUTION: Figure 7.7a shows the inclined plane; the block moves up the full length of this plane, so the displacement has a magnitude

$$\Delta r = \frac{\Delta z}{\sin \phi}$$

Fig. 7.7b is a "free-body" diagram for the block. The angle between the weight and the displacement vector is $\pi/2 + \phi$. According to Eq. (13), the work done by gravity is

$$W = mg\ \Delta r \cos\left(\frac{\pi}{2} + \phi\right) = mg \frac{\Delta z}{\sin \phi}(-\sin \phi) = -mg\ \Delta z$$

Alternatively, we can use components to evaluate the work. The $z$ component of the weight is $-mg$, and the other components are zero. According to Eq. (22), the work done by gravity is then

$$W = 0 \times \Delta x + 0 \times \Delta y + (-mg) \times \Delta z = -mg\ \Delta z$$

COMMENTS AND SUGGESTIONS: Which of the two expressions (13) and (22) we use for the calculation of the work is a matter of convenience. If the force and the displacement vectors are specified by magnitudes and angles, then Eq. (13) is the most convenient. If the vectors are specified by components, then Eq. (22) is the most convenient.

Note that the work done by gravity on the block moving along an inclined plane depends only on the change of height $\Delta z$, and not on the angle of the plane. More generally, consider a block or a particle that moves from a point $P_1$ to a point $P_2$ along some arbitrary three-dimensional path. Figure 7.8 shows these two points and their coordinates; as always, the $z$ axis is directed vertically upward. The components of the weight are $F_x = 0$, $F_y = 0$, $F_z = -mg$. Hence, as in Example 4, Eq. (22) yields

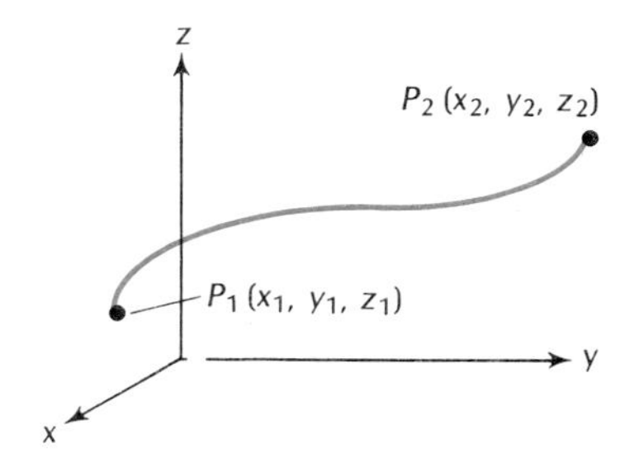


**Fig. 7.8** Path of a particle from $P_1$ to $P_2$. The $z$ axis is directed vertically upward.

$$\boxed{W = -mg\ \Delta z} \tag{25}$$

This establishes that the work done by gravity depends *only* on the vertical separation between the points $P_1$ and $P_2$ — neither the horizontal separation between the points nor the exact shape of the path is of any relevance. The result (25) can also be written as

$$W = -mg(z_2 - z_1) \tag{26}$$

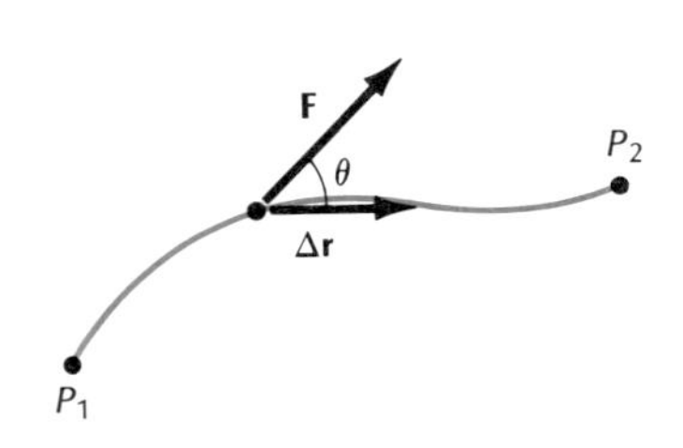


**Fig. 7.9** A force **F** acts during a small displacement $\Delta\mathbf{r}$.

where $z_1$ and $z_2$ are the $z$ coordinates of the two points.

The definition of work in three dimensions in the case of a force that is not constant involves an integral, just as in the analogous case in one dimension. Suppose that a particle moves along some given path from a point $P_1$ to a point $P_2$ (Figure 7.9). We can then divide the path into short segments, each of length $\Delta\mathbf{r}$. Within each such segment the force and the direction of motion are approximately constant and the work done on the particle is

$$\Delta W = F\,\Delta r \cos\theta$$

The total work is approximately the sum of all such small amounts of work. In the limit $\Delta r \to 0$, this sum becomes the integral

$$W = \int_{P_1}^{P_2} F\cos\theta\, dr \tag{27}$$

where $dr$ is the element of length along the path followed by the particle. Equation (27) can also be written more compactly as

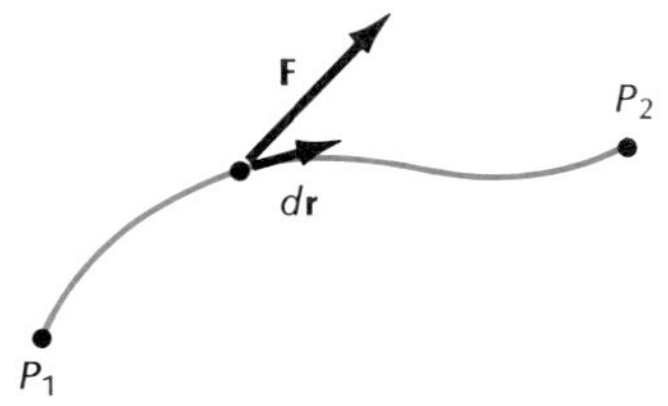

$$W = \int_{P_1}^{P_2} \mathbf{F}\cdot d\mathbf{r} \tag{28}$$

where $d\mathbf{r}$ is a vector of magnitude $dr$ and of direction tangent to the path (Figure 7.10). Alternatively, this can be written

**Fig. 7.10** The infinitesimal displacement $d\mathbf{r}$ is tangent to the path.

$$W = \int_{P_1}^{P_2} (F_x\, dx + F_y\, dy + F_z\, dz)$$

or

*Work done by variable force (three dimensions)*

$$\boxed{W = \int_{P_1}^{P_2} F_x\, dx + \int_{P_1}^{P_2} F_y\, dy + \int_{P_1}^{P_2} F_z\, dz} \tag{29}$$

As in the calculation of the work by a constant force, which of the two alternative expressions (27) and (29) we use is a question of convenience. In the following example, we will find it convenient to express the force in components and to use Eq. (29).

EXAMPLE 5. At a construction site a laborer pushes horizontally against a large bucket full of concrete of total mass 600 kg suspended from a crane by a 20-m cable (Figure 7.11a). How much work must the laborer do on the bucket to push it slowly 2 m away from the vertical? How much work does gravity do on the bucket?

SOLUTION: Figure 7.11b is a "free-body" diagram of the bucket, regarded as a particle, displaced sideways by an angle $\phi$. The forces on the bucket are the weight **w**, the tension **T** of the cable, and the push **S** of the laborer. The motion is slow, i.e., the acceleration is just about zero, and therefore the forces must balance. For the $x$ components and the $z$ components of the forces, this implies

$$-T \sin \phi + S = 0$$

$$T \cos \phi - mg = 0$$

Eliminating $T$ between these two equations, we find

$$S = mg \tan \phi \tag{30}$$

Since the push is entirely in the $x$ direction, we can use Eq. (29) with $F_x = S$, $F_y = 0$, and $F_z = 0$. The point $P_1$ corresponds to $x = 0$ and the point $P_2$ to $x = 2\text{m}$. Hence

$$W = \int_{P_1}^{P_2} F_x \, dx = \int_0^{2\text{ m}} mg \tan \phi \, dx \tag{31}$$

Here $x$ is the variable of integration and we therefore ought to express $\tan \phi$ as a function of $x$. Instead of doing this, let us adopt $\phi$ as the variable of integration and express $dx$ in terms of $d\phi$. From Figure 7.11a,

$$\sin \phi = x/l$$

where $l$ is the length of the cable. Hence

$$dx = l \cos \phi \, d\phi$$

Furthermore, when $x = 2$ m, $\sin \phi = \frac{2}{20}$ and $\phi = 0.10$ radian $= 5.74°$. Our expression for the work then becomes

$$W = \int_0^{0.10} mg \tan \phi \, (l \cos \phi \, d\phi) = \int_0^{0.10} mgl \sin \phi \, d\phi$$

$$= mgl\big[-\cos \phi\big]_0^{0.10} = mgl(1 - \cos 0.10) = mgl(1 - \cos 5.74°)$$

$$= 600 \text{ kg} \times 9.8 \text{ m/s}^2 \times 20 \text{ m} \times 0.0050 = 590 \text{ J} \tag{32}$$

The work done by gravity can be calculated directly from Eq. (26). The initial value of $z$ is $z_1 = -l$ and the final value is $z_2 = -l(\cos 5.74°)$. Hence the work is

$$W = -mg(-l \cos 5.74° + l)$$

$$= -mgl(1 - \cos 5.74°) \tag{33}$$

Comparing this with Eq. (32), we see that the work done by gravity is exactly the opposite of the work done by the push, i.e., it is $-590$ J. Thus, the net work done by both forces together is zero.

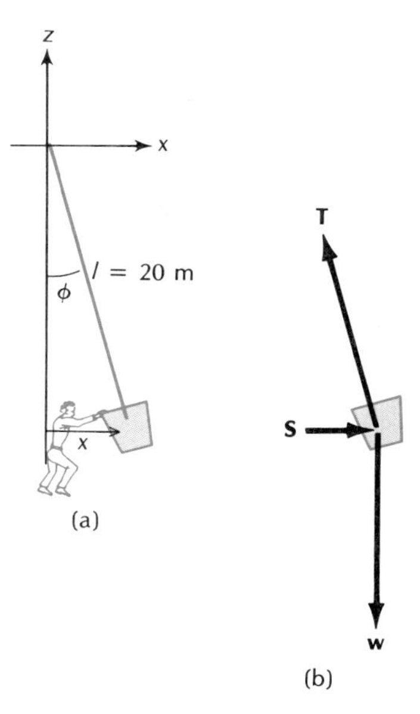


**Fig. 7.11** (a) Bucket hanging on a cable. (b) "Free-body" diagram showing the forces on the bucket.

## 7.3 Kinetic Energy

We will now derive an important identity between the work done on a particle and the change of speed of the particle. Suppose that the particle moves from a point $P_1$ to a point $P_2$ and that during this motion the *net* force acting on it is $\mathbf{F}$. This net force may be a function of position and of time, but in any case the work done by $\mathbf{F}$ on the particle is

$$W = \int_{P_1}^{P_2} \mathbf{F} \cdot d\mathbf{r} \tag{34}$$

or

$$W = \int_{P_1}^{P_2} (F_x\, dx + F_y\, dy + F_z\, dz) \tag{35}$$

Consider the first term in this integral. Since the net force equals the mass times the acceleration, the first term equals

$$\int_{P_1}^{P_2} F_x\, dx = \int_{P_1}^{P_2} m \frac{dv_x}{dt}\, dx \tag{36}$$

The velocity $v_x$ is a function of time; but for the purpose of the integration in Eq. (36), it is better to regard the velocity as a function of position (along the worldline of the particle, time and position are in one-to-one correspondence and therefore it is always possible to change from one of these variables to the other). Then

$$\frac{dv_x}{dt} = \frac{dv_x}{dx}\frac{dx}{dt} = \frac{dv_x}{dx} v_x = v_x \frac{dv_x}{dx} \tag{37}$$

and, consequently,

$$\begin{aligned} \int_{P_1}^{P_2} F_x\, dx &= \int_{P_1}^{P_2} m \frac{dv_x}{dt}\, dx = \int_{P_1}^{P_2} m v_x \frac{dv_x}{dx}\, dx \\ &= \int_{v_{1,x}}^{v_{2,x}} m v_x\, dv_x = m\Big[\tfrac{1}{2} v_x^2\Big]_{v_{1,x}}^{v_{2,x}} \\ &= \tfrac{1}{2} m v_{2,x}^2 - \tfrac{1}{2} m v_{1,x}^2 \end{aligned} \tag{38}$$

where $v_{1,x}$ is the $x$ velocity at $P_1$ and $v_{2,x}$ is the $x$ velocity at $P_2$.

The other two terms in Eq. (35) can be transformed similarly, with the final result

$$\begin{aligned} W &= \int_{P_1}^{P_2} (F_x\, dx + F_y\, dy + F_z\, dz) \\ &= \tfrac{1}{2} m v_{2,x}^2 - \tfrac{1}{2} m v_{1,x}^2 + \tfrac{1}{2} m v_{2,y}^2 - \tfrac{1}{2} m v_{1,y}^2 + \tfrac{1}{2} m v_{2,z}^2 - \tfrac{1}{2} m v_{1,z}^2 \end{aligned}$$

or

$$W = \tfrac{1}{2} m v_2^2 - \tfrac{1}{2} m v_1^2 \tag{39}$$

where, according to the usual notation, $v^2 = v_x^2 + v_y^2 + v_z^2$.

Equation (39) shows that the change in the square of the velocity is proportional to the work done by the net force between $P_1$ and $P_2$. In Examples 2 and 5 we looked at motions in which the speed was constant; that is, the square of the velocity was constant. According to Eq.

(39), the net work should then be zero and, of course, this is exactly what we found by the explicit calculations contained in these examples.

The quantity

$$\boxed{K = \tfrac{1}{2}mv^2} \qquad (40)$$

***Kinetic energy***

is called the **kinetic energy** of the particle. With this terminology, Eq. (39) says that *the change in the kinetic energy equals the work done on the particle by the net force,*

$$K_2 - K_1 = W \qquad (41)$$

or

$$\boxed{\Delta K = W} \qquad (42)$$

***Work–energy theorem***

This result is called the **work–energy theorem.** Keep in mind that the work in Eqs. (39), (41), and (42) must be evaluated with the *net* force, that is, all the forces that do work on the particle must be included.

When a force does positive work on a particle initially at rest, the kinetic energy increases. The acquired kinetic energy gives the particle a capacity to do work. If the moving particle is allowed to push against some obstacle, then the particle does positive work on the obstacle and the kinetic energy decreases. The total amount of work the particle can perform on the obstacle is equal to its kinetic energy. Thus the kinetic energy represents accumulated work, or latent work, which under suitable conditions can become actual work.

The acquisition of kinetic energy through work and the subsequent production of work by this kinetic energy are neatly illustrated in the operation of a waterwheel driven by falling water. In a flour mill of an old Spanish Colonial design, the water runs down from a reservoir in a steep, open channel (Figure 7.12). The motion of a water particle is essentially that of a particle sliding down an inclined plane. If we ignore friction, then the only force that does work on the water particle is gravity. This work is positive; consequently, the kinetic energy of the water increases and attains a maximum value at the lower end of the channel. The stream of water emerges from this channel with high kinetic energy and hits the blades of the waterwheel; the water pushes on the wheel, turns it, and gives up its kinetic energy while doing work — and the wheel turns the millstones and does useful work on them. Thus, the work that gravity does on the descending water is ultimately converted into useful work, with the kinetic energy playing an intermediate role in this process. The operation of turbines in a hydroelectric power station involves the same principle; however, the situation is somewhat more complicated because the water flows in a closed pipe rather than in an open channel and pressure plays an important role.

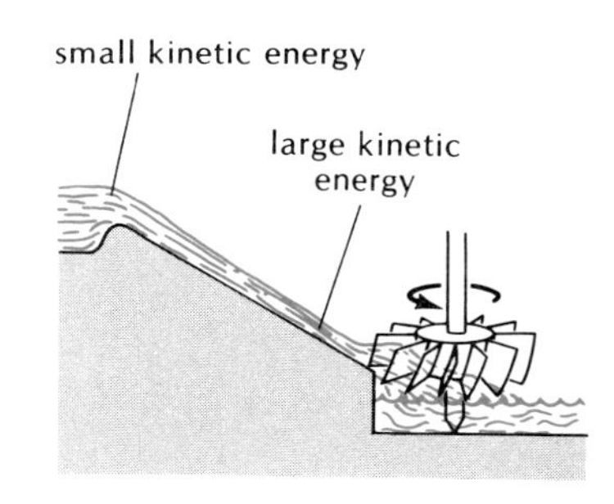


**Fig. 7.12** Horizontal waterwheel.

The units of kinetic energy are the same as the units of work: the joule and the foot-pound. Table 7.1 lists some typical kinetic energies.

It is sometimes convenient to express the kinetic energy directly in terms of momentum by means of the formula

**Table 7.1** SOME KINETIC ENERGIES

| | |
|---|---|
| Orbital motion of Earth | $2.6 \times 10^{33}$ J |
| Rotational motion of Earth | $2.1 \times 10^{29}$ J |
| Ship *Queen Elizabeth* (at cruising speed) | $9 \times 10^{9}$ J |
| Jet airliner (Boeing 747 at maximum speed) | $7 \times 10^{9}$ J |
| Automobile (at 90 km/h) | $5 \times 10^{5}$ J |
| Rifle bullet | $4 \times 10^{3}$ J |
| Man walking | 60 J |
| Highest energy found in a single cosmic ray | 50 J |
| Falling raindrop | $4 \times 10^{-5}$ J |
| Proton from large accelerator (Fermilab) | $1.6 \times 10^{-7}$ J |
| Fission fragments of one uranium-235 nucleus | $2.6 \times 10^{-11}$ J |
| Proton in nucleus | $3 \times 10^{-12}$ J |
| Electron in atom (hydrogen) | $2.2 \times 10^{-18}$ J |
| Air molecule (at room temperature) | $6.2 \times 10^{-21}$ J |

*Kinetic energy as a function of momentum*

$$\boxed{K = \frac{p^2}{2m}} \tag{43}$$

By substituting $p = mv$ into this, we immediately recognize that Eq. (43) agrees with Eq. (40).

## 7.4 Gravitational Potential Energy

*Potential energy*

As we have seen, the kinetic energy represents the capacity of a particle to do work by virtue of its speed. We will now become acquainted with another form of energy that represents the capacity of the particle to do work by virtue of its position in space. This is the **potential energy.** In this section, we will examine the special case of gravitational potential energy for a particle moving under the influence of the constant force of gravity, and we will formulate a law of conservation of energy for such a particle. In the next chapter we will examine other cases of potential energy and formulate the general law of conservation of energy.

When the constant force of gravity, $F_z = -mg$, acts on a particle undergoing a displacement from $(x_1, y_1, z_1)$ to $(x_2, y_2, z_2)$, it does an amount of work [see Eq. (26)]

$$W = -mg(z_2 - z_1)$$

We can write this as

$$W = -mgz_2 + mgz_1 \tag{44}$$

where the function $mgz$ is called the **gravitational potential energy** of the particle,

*Gravitational potential energy*

$$\boxed{[\text{gravitational potential energy}] = mgz} \tag{45}$$

According to Eq. (44), the change in the gravitational potential energy between the points $z_1$ and $z_2$ equals the negative of the work done by gravity on the particle moving between these points.

Note that the construction of the potential-energy function hinges on the path independence of the work. If different paths from the point $z_1$ to the point $z_2$ required different amounts of work, then the work could not be expressed as a function of $z_1$ and $z_2$, and we could not construct a potential energy which is a unique, well-defined function of position. Also note that Eq. (45) assumes that the $z$ axis is directed vertically upward, so $z$ represents the height measured upward from the surface of the Earth or from some other given reference level.

The gravitational potential energy represents the capacity of the particle to do work by virtue of its height above the surface of the Earth. A particle high above the surface of the Earth is endowed with a large amount of latent work which can be exploited and converted into actual work by allowing the particle to push against some obstacle as it descends. The total amount of work that can be extracted during the descent is equal to the change in potential energy. Of course, the work extracted in this way really arises from the Earth's gravity — the particle can do work on the obstacle because gravity is doing work on the particle. Hence the gravitational potential energy is really a joint property of both the particle and the Earth; it is a property of the configuration of the particle–Earth system.

If the only force acting on a particle is gravity, then by combining Eqs. (41) and (44) we can obtain a relation between potential energy and kinetic energy,

$$-mgz_2 + mgz_1 = K_2 - K_1 \tag{46}$$

which we can rewrite as follows:

$$K_1 + mgz_1 = K_2 + mgz_2 \tag{47}$$

This equality indicates that the quantity $K + mgz$ is a constant of the motion, that is, it has the same value at the point $P_2$ as at the point $P_1$. Thus

$$K + mgz = [\text{constant}] \tag{48}$$

The sum of the kinetic and potential energies is called the **mechanical energy** of the particle. It is usually designated by the symbol $E$:

$$\boxed{E = K + mgz} \tag{49}$$

*Mechanical energy*

This energy represents the total capacity of the particle to do work by virtue of both its speed and its position.

Equation (48) shows that if the only force acting on the particle is gravity, then the mechanical energy remains constant:

$$\boxed{E = K + mgz = [\text{constant}]} \tag{50}$$

*Law of conservation of mechanical energy*

This is the **law of conservation of mechanical energy.**

**Christiaan Huygens** (hoigens), *1629–1695, Dutch mathematician and physicist. He invented the pendulum clock, made improvements in the manufacture of telescope lenses, and discovered the rings of Saturn. Huygens investigated the theory of collisions of elastic bodies and the theory of oscillations of the pendulum, and he stated the law of conservation of mechanical energy for motion under the influence of gravity.*

Apart from its practical significance in terms of work, the mechanical energy is very helpful in the study of the motion of the particle. Since the sum of the potential and kinetic energies must remain constant during the motion, an increase in one must be compensated by a decrease in the other; this means that during the motion, kinetic energy is converted into potential energy and vice versa. If we make use of the formula for $K$, Eq. (50) becomes

$$E = \tfrac{1}{2}mv^2 + mgz = [\text{constant}] \tag{51}$$

This shows explicitly how the particle trades speed for height during the motion: whenever $z$ increases, $v$ must decrease (and conversely) so as to keep the sum of the two terms on the left side of Eq. (51) constant. Note that Eq. (2.29) is essentially Eq. (51) written in a slightly different form.

An important facet of Eq. (51) is that it is valid not only for a particle in free fall (a projectile), but also for a particle sliding on a surface or a track of arbitrary shape, provided that there is no friction. Of course, under these conditions besides the gravitational force there also acts the normal force; but this force does no work and hence does not affect the left side of Eq. (46) or any of the subsequent Eqs. (47)–(51). This gives us a glimpse of the elegance and power of the law of conservation of mechanical energy. Whereas the derivation of Eq. (2.29) is valid only for a freely falling particle, the derivation of Eq. (51) is valid under much more general conditions.

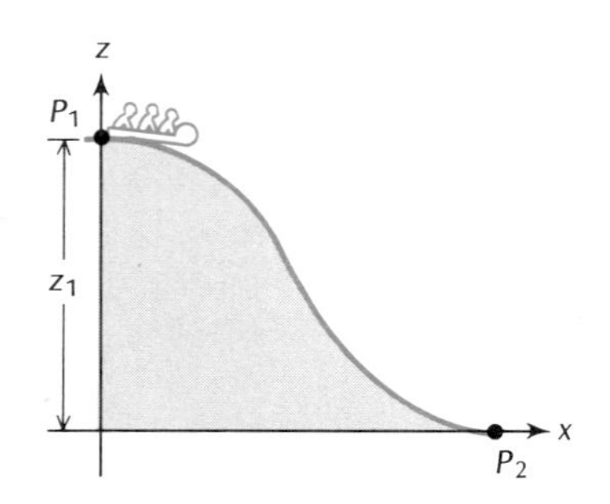


**Fig. 7.13** A bobsled run.

EXAMPLE 6. A bobsled run leading down a hill at Lake Placid (the site of the 1980 Winter Olympics) descends 148 m from its highest point to its lowest. Suppose that a bobsled, initially at rest at the highest point, slides down this run without friction. What speed will the bobsled attain at the lowest point?

SOLUTION: If the $z$ coordinate is measured upward from the lowest point, the coordinates of the highest and the lowest points are $z_1 = 148$ m and $z_2 = 0$ m, respectively (Figure 7.13). According to Eq. (49), the energy at the start of the motion is

$$E = \tfrac{1}{2}mv_1^2 + mgz_1 = 0 + mgz_1 \tag{52}$$

and the energy at the end of the motion is

$$E = \tfrac{1}{2}mv_2^2 + mgz_2 = \tfrac{1}{2}mv_2^2 + 0 \tag{53}$$

The conservation of energy implies that the right sides of Eqs. (52) and (53) are equal,

$$\tfrac{1}{2}mv_2^2 = mgz_1 \tag{54}$$

or

$$v_2 = \sqrt{2gz_1} = \sqrt{2 \times 9.8\ \text{m/s}^2 \times 148\ \text{m}} = 54\ \text{m/s}$$

COMMENTS AND SUGGESTIONS: This example illustrates how energy conservation can be exploited to answer a question about motion. To obtain the final speed by direct computation of forces and accelerations would have been extremely difficult — it would have required detailed knowledge of the shape of the path. With energy conservation we can bypass these complications.

The use of energy conservation in a problem of motion typically involves three steps: First write an expression for the energy at one point of the motion [Eq. (52)], then write an expression for the energy at another point [Eq. (53)], and then rely on conservation of energy to equate the two expressions [Eq. (54)]. This yields one equation, which can be solved for one unknown coordinate or one unknown speed. (If the problem involves more than one unknown, then the one equation supplied by energy conservation is not sufficient and must be supplemented by some extra information.)

Note that the value of the gravitational potential energy $mgz$ depends on the level from which we measure the $z$ coordinate. However, the *change* in $mgz$ does not depend on the choice of this level, and therefore any choice will lead to the same result for the change in the kinetic energy. It is usually most convenient to place the zero level for the $z$ coordinate either at the ending point of the motion (as in this example) or at the starting point (as in the next example). And always remember that the formula $mgz$ for the gravitational potential energy assumes that the $z$ axis is directed vertically *upward.*

---

EXAMPLE 7. A pendulum consists of a mass $m$ tied to one end of a string of length $l$. The other end of the string is attached to a fixed point of support (Figure 7.14a). Suppose that the pendulum is initially held at an angle of 45° with the vertical. If the pendulum is released from this position, what will be the tension in the string at the instant the mass passes through its lowest position?

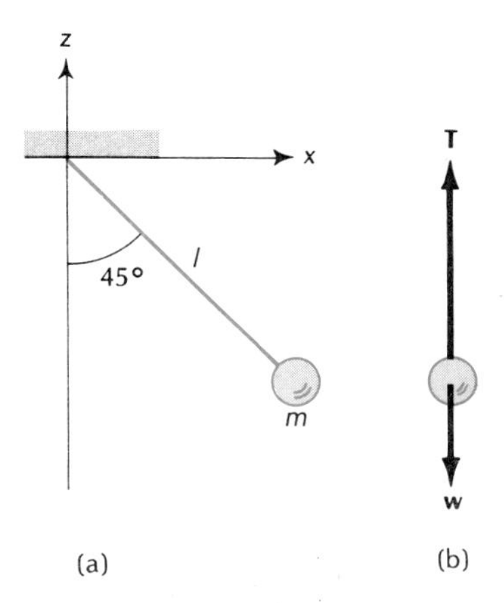


**Fig. 7.14** (a) A pendulum. Initially the pendulum makes an angle of 45° with the vertical. (b) "Free-body" diagram for the mass $m$ at the instant it reaches the lowest point.

SOLUTION: First, we must calculate the speed attained by the mass. If the $z$ coordinate is measured upward from the point of support, the initial coordinate of the mass is $z_1 = -l \cos \theta = -l \cos 45°$ and the lowest coordinate is $z_2 = -l$. The energy at the initial position is

$$E = \tfrac{1}{2}mv_1^2 + mgz_1 = 0 - mgl \cos 45° \qquad (55)$$

and the energy at the lowest position is

$$E = \tfrac{1}{2}mv_2^2 + mgz_2 = \tfrac{1}{2}mv_2^2 - mgl \qquad (56)$$

The conservation of energy then implies

$$\tfrac{1}{2}mv_2^2 - mgl = -mgl \cos 45°$$

from which we find the speed, or the square the speed, attained by the mass at its lowest point:

$$v_2^2 = 2(gl - gl \cos 45°) = 2gl(1 - \cos 45°)$$

Accordingly, the centripetal acceleration is

$$v_2^2/l = 2g(1 - \cos 45°) \qquad (57)$$

The mass multiplied by the centripetal acceleration gives us the magnitude of the centripetal force. This force has two contributions: a positive (upward) contribution due to the tension $T$ in the string and a negative (downward) contribution due to the weight $mg$ (Figure 7.14b). Thus, the net centripetal force is $T - mg$, and

$$mv_2^2/l = T - mg$$

or

$$2mg(1 - \cos 45°) = T - mg$$

Consequently,

$$T = mg + 2mg(1 - \cos 45°)$$
$$= (3 - \sqrt{2}/2)mg \tag{58}$$

## SUMMARY

**Dot product:** $\mathbf{A} \cdot \mathbf{B} = AB \cos \theta$

**Work done by a constant force:** $W = \mathbf{F} \cdot \Delta\mathbf{r} = F_x\,\Delta x + F_y\,\Delta y + F_z\,\Delta z$

**Work done by a variable force:**

$$W = \int_{P_1}^{P_2} \mathbf{F} \cdot d\mathbf{r} = \int_{P_1}^{P_2} F_x\,dx + \int_{P_1}^{P_2} F_y\,dy + \int_{P_1}^{P_2} F_z\,dz$$

**Work done by gravity:** $W = -mg\,\Delta z$

**Kinetic energy:** $K = \frac{1}{2}mv^2$

$$= \frac{p^2}{2m}$$

**Work–energy theorem:** $\Delta K = W$

**Gravitational potential energy:** $mgz$

**Mechanical energy:** $E = K + mgz$

**Conservation of mechanical energy:** $E = K + mgz = [\text{constant}]$

## QUESTIONS

1. Does the work of a force on a body depend on the frame of reference in which it is calculated? Give some examples.

2. Does your body do work (external or internal) when standing at rest? When walking steadily along a level road?

3. Consider a pendulum swinging back and forth. During what part of the motion does the weight do positive work? Negative work?

4. According to Eq. (40), $K = \frac{1}{2}mv_x^2 + \frac{1}{2}mv_y^2 + \frac{1}{2}mv_z^2$. Does this mean that the kinetic energy has $x$, $y$, and $z$ components?

5. Consider a woman steadily climbing a flight of stairs. The external forces on the woman are her weight and the normal force of the stairs against her feet. During the climb, the weight does negative work, while the normal force does no work. Under these conditions how can the kinetic energy of the woman remain constant? (Hint: The entire woman cannot be regarded as a particle, since her legs are not rigid; but the upper part of her body can be regarded as a particle, since it is rigid. What is the force of her legs against the upper part of her body? Does this force do work?)

6. An automobile increases its speed from 80 to 95 km/h. What is the percent increase of kinetic energy? What is the percent reduction of travel time for a given distance?

7. Two blocks in contact slide past one another and exert friction forces on

one another. Can the friction force *increase* the kinetic energy of one block? Of both? Does there exist a reference frame in which the friction force decreases the kinetic energy of both blocks?

8. When an automobile with rear-wheel drive is accelerating on, say, a level road, the horizontal force of the road on the rear wheels gives the automobile momentum, but it does not give the automobile any energy because the point of application of this force (point of contact of wheel on ground) is instantaneously at rest if the wheel is not slipping. What force gives the body of the automobile energy? Where does the energy come from? (Hint: Consider the force that the rear axle exerts against its bearings.)

9. A dictionary gives the following definition of *zoom:* "to climb in an airplane suddenly and sharply at an angle greater than normal, using the energy of momentum." Does this make any sense?

10. Why do elevators have counterweights? (See Figure 6.9.)

11. A parachutist jumps out of an airplane, opens a parachute, and lands safely on the ground. Is the mechanical energy for this motion conserved?

12. If you release a tennis ball at some height above a hard floor, it will bounce up and down several times, with a gradually decreasing amplitude. Where does the ball suffer a loss of mechanical energy?

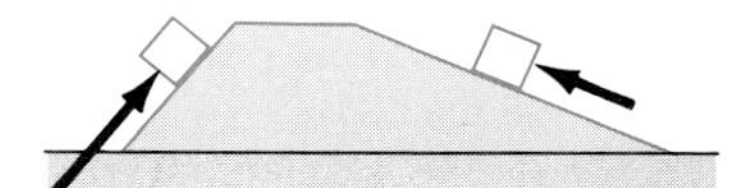

**Fig. 7.15** Two ramps of different steepness.

13. Two ramps, one steeper than the other, lead from the floor to a loading platform (Figure 7.15). It takes more force to push a (frictionless) box up the steeper ramp. Does this mean it takes more work to raise the box from the floor to the platform?

14. Consider the two ramps described in the preceding question. Taking friction into account, which ramp requires less work for raising a box from the floor to the platform?

15. A stone is tied to a string. Can you whirl this stone in a vertical circle with constant speed? Can you whirl this stone with constant energy? For each of these two cases, describe how you must move your hand.

## PROBLEMS

### Section 7.1

1. If it takes a horizontal force of 300 N to push a stalled automobile along a level road at constant speed, how much work must you do to push this automobile a distance of 5.0 m?

2. In an overhead lift, a champion weight lifter raises 254 kg from the floor to a height of 1.98 m. How much work does he do?

3. Suppose that the force required to push a saw back and forth through a piece of wood is 35 N. If you push this saw back and forth 30 times, moving it forward 12 cm and back 12 cm each time, how much work do you do?

4. A man pushes a crate along a flat concrete floor. The mass of the crate is 120 kg and the coefficient of friction between the crate and the floor is $\mu_k = 0.5$. How much work does the man do if, pushing horizontally, he moves the crate 15 m at constant speed?

5. A woman slowly lifts a 20-kg package of books from the floor to a height of 1.8 m, and then slowly returns it to the floor. How much work does she do on the package while lifting? How much work does she do on the package while lowering? What is the total work she does on the package? From the information given, can you deduce how much work she expends internally in her muscles, that is, how many calories she expends?

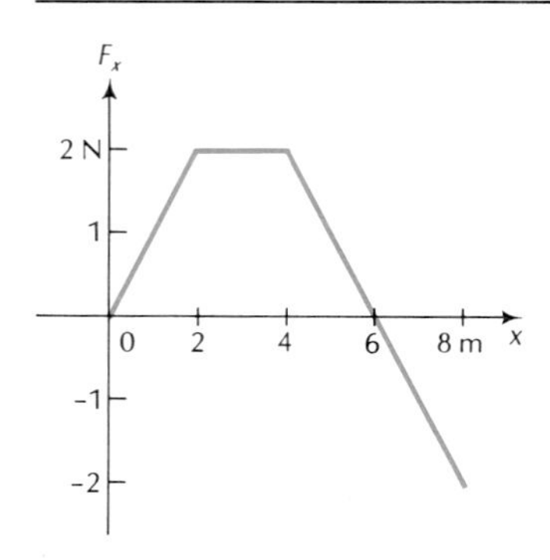


**Fig. 7.16**

6. A record for stair climbing was achieved by a man who raced up the 1600 steps of the Empire State Building to a height of 320 m in 10 min 59 s. If his mass was 75 kg, how much work did he do against gravity? At what average rate (in J/s) did he do this work?

7. A spring has a force constant of $3.5 \times 10^4$ N/m. The spring is initially relaxed. How much work must you do to compress the spring by 0.10 m? How much work must you do to compress the spring a further 0.10 m?

8. A particle moving along the $x$ axis is subjected to a force $F_x$ that depends on position as shown in the plot in Figure 7.16. From this plot, find the work done by the force as the particle moves from $x = 0$ to $x = 8$ m.

*9. The ends of a relaxed spring of length $l$ and force constant $k$ are attached to two points on two walls.
   (a) How much work must you do to push the midpoint of the spring up or down a distance $y$ (see Figure 7.17)?
   (b) How much force must you exert to hold the spring in this configuration? Verify that the magnitude of the force coincides with the negative of the derivative of the work with respect to $y$.

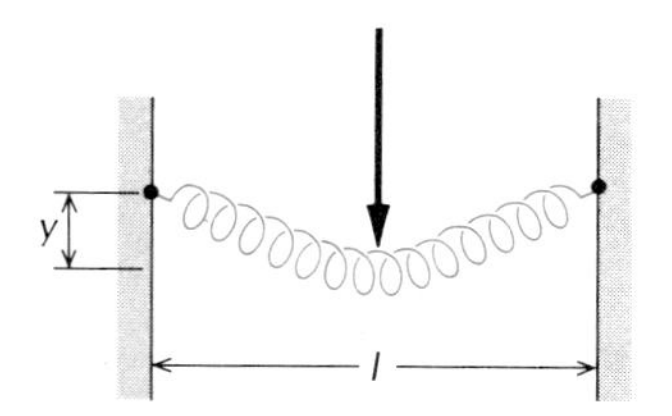


**Fig. 7.17** The midpoint of the spring has been pushed down a distance $y$. When the spring is relaxed, its length matches the distance between the points of attachment.

*10. A man pushes a heavy box up an inclined ramp making an angle of 30° with the horizontal. The mass of the box is 60 kg and the coefficient of kinetic friction between the box and the ramp is 0.45. How much work must the man do to push the box to a height of 2.5 m at constant speed? Assume that the man pushes on the box in a direction parallel to the surface of the ramp.

*11. An elevator consists of an elevator cage and a counterweight attached to the ends of a cable which runs over a pulley (Figure 7.18). The mass of the cage (with its load) is 1200 kg and the mass of the counterweight is 1000 kg. The elevator is driven by an electric motor attached to the pulley. Suppose that the elevator is initially at rest at the first floor of the building, and the motor makes the elevator accelerate upward at the rate of 1.5 m/s$^2$.
   (a) What is the tension in the part of the cable attached to the elevator cage? What is the tension in the part of the cable attached to the counterweight?
   (b) The acceleration lasts exactly 1.0 s. How much work has the electric motor done in this interval? Ignore friction forces and ignore the mass of the pulley.
   (c) After the acceleration interval of 1.0 s, the motor pulls the elevator upward at constant speed until it reaches the third floor, exactly 10.0 m above the first floor. What is the total amount of work that the motor has done up to this point?

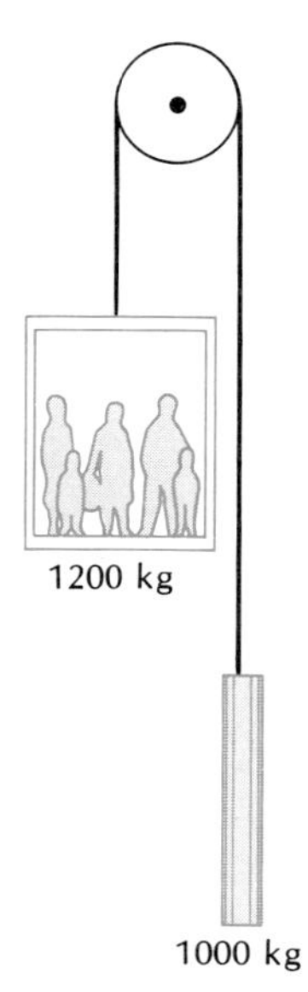


**Fig. 7.18** Elevator cage and counterweight.

*12. A particle moves along the $x$ axis from $x = 0$ to $x = 2$. A force $F_x(x) = 2x^3 + 8x$ acts on the particle (the distance $x$ is measured in meters, and the force in newtons). Calculate the work done by the force $F_x(x)$ during this motion.

### Section 7.2

13. A particle moves in the $x$–$y$ plane from the origin $x = 0$, $y = 0$ to the point $x = 2$, $y = -1$ while under the influence of a force $\mathbf{F} = 3\hat{\mathbf{x}} + 2\hat{\mathbf{y}}$. How much work does this force do on the particle during this motion? The distances are measured in meters and the force in newtons.

14. A block sliding on a table is subjected to a push of 50 N making a downward angle of 60° with the direction of motion. What is the work done by this force when the block moves a distance of 1.6 m?

15. A 2-kg stone thrown upward reaches a height of 4 m at a horizontal distance of 6 m from the point of launch. What is the work done by gravity during this displacement?

16. A man pulls a cart along a level road by means of a short rope stretched

over his shoulder and attached to the front end of the cart. The friction force that opposes the motion of the cart is 250 N.

(a) If the rope is attached to the cart at shoulder height, how much work must the man do to pull the cart 50 m at constant speed?

(b) If the rope is attached to the cart below shoulder height so it makes an angle of 30° with the horizontal, what is the tension in the rope? How much work must the man now do to pull the cart 50 m? Assume that the friction force is unchanged.

17. The driver of a 1200-kg automobile notices that, with its gears in neutral, it will roll downhill at a constant speed of 110 km/h on a road of slope 1:20. Draw a "free-body" diagram for the automobile, showing the force of gravity, the normal force (exerted by the road), and the friction force (exerted by the road and by air resistance). What is the magnitude of the friction force on the automobile under these conditions? What is the work done by the friction force while the automobile travels 1.0 km down the road?

*18. During a storm, a sailboat is anchored in a 10-m-deep harbor. The wind pushes against the boat with a steady horizontal force of 7000 N.

(a) The anchor rope that holds the boat in place is 50 m long and is stretched straight between the anchor and the boat (Figure 7.19a). What is the tension in the rope?

(b) How much work must the crew of the sailboat do to pull in 30 m of the anchor rope, bringing the boat nearer to the anchor (Figure 7.19b)? What is the tension in the rope when the boat is in this new position?

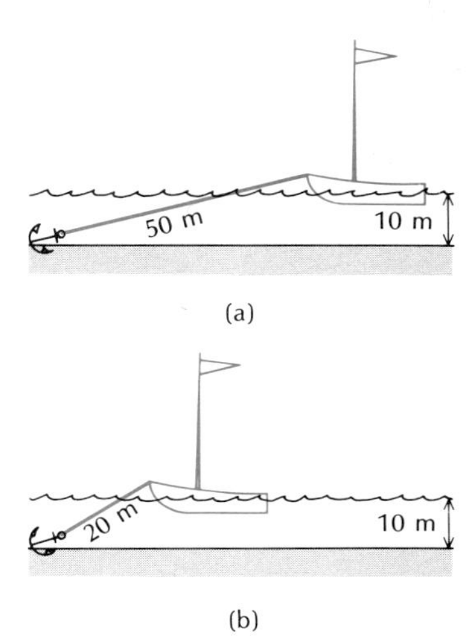


**Fig. 7.19** A sailboat at anchor.

*19. By means of a towrope, a girl pulls a sled loaded with firewood along a level, icy road. The coefficient of friction between the sled and the road is $\mu_k = 0.10$ and the mass of the sled plus its load is 150 kg. The towrope is attached to the front of the sled and makes an angle of 30° with the horizontal. How much work must the girl do on the sled to pull it 1.0 km at constant speed?

*20. A horse pulls a sled along a snow-covered curved ramp. Seen from the side, the surface of the ramp follows an arc of a circle of radius $R$ (Figure 7.20). The pull of the horse is always parallel to this surface. The mass of the sled is $m$ and the coefficient of sliding friction between the sled and the surface is $\mu_k$. How much work must the horse do on the sled to pull it to a height $(1 - \sqrt{2}/2)R$, corresponding to an angle of 45° along the circle (Figure 7.20)? How does this compare with the amount of work required to pull the sled from the same starting point to the same height along a straight ramp inclined at 22.5°?

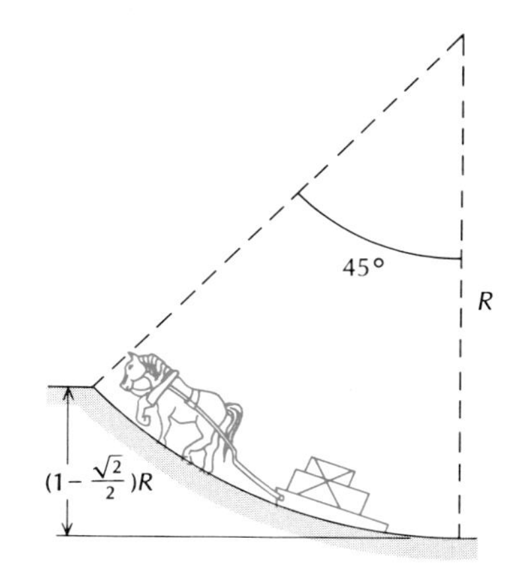


**Fig. 7.20** A horse pulling a sled along a curved ramp.

*21. Suppose that the force acting on a particle is a function of position; the force has components $F_x = 4x^2 + 1$, $F_y = 2x$, $F_z = 0$, where force is measured in newtons and distance in meters. What is the work done by this force if the particle moves on a straight line from $x = 0, y = 0, z = 0$ to $x = 2, y = 2, z = 0$?

### Section 7.3

22. Calculate the kinetic energy that the Earth has owing to its motion around the Sun.

23. The electron in a hydrogen atom has a speed of $2.2 \times 10^6$ m/s. What is the kinetic energy of this electron?

24. The fastest skier is Graham Wilkie, who attained 212.52 km/h on a steep slope at Les Arcs, France. The fastest runner is Robert Hayes, who briefly attained 44.88 km/h on a level track. Assume that the skier and the runner each have a mass of 75 kg. What is the kinetic energy of each? By what factor is the kinetic energy of the skier larger than that of the runner?

25. The Skylab satellite disintegrated when it reentered the atmosphere. Among the pieces that crashed down on the surface of the Earth, one of the heaviest was a lead-lined film vault of 1770 kg which had an estimated impact

speed of 120 m/s on the surface. What was its kinetic energy? How many kilograms of TNT would we have to explode to release the same amount of energy? (One kilogram of TNT releases $4.6 \times 10^6$ J.)

26. An automobile of mass 1600 kg is traveling along a straight road at 80 km/h.
   (a) What is the kinetic energy of this automobile in the reference frame of the ground?
   (b) What is the kinetic energy in the reference frame of a motorcycle traveling in the same direction at 60 km/h?
   (c) What is the kinetic energy in the reference frame of a truck traveling in the opposite direction at 60 km/h?

27. According to statistical data, the probability that an occupant of an automobile suffers a lethal injury when involved in a crash is proportional to the square of the speed of the automobile.
   (a) At a speed of 80 km/h, the probability is approximately 3%. What are the probabilities at 95 km/h, 110 km/h, and 125 km/h?
   (b) Do the data suggest that the injuries are related to the kinetic energy of the automobile or to the momentum?

28. For the projectile described in Problem 2.18, calculate the initial kinetic energy ($t = 0$) and calculate the final kinetic energy ($t = 3.0$ s). How much energy does the projectile lose to friction in 3.0 s?

29. The mass of the bullet and the muzzle velocity for several rifles manufactured around 1900 are as follows: Lee-Enfield, 13.9 g, 628 m/s; Lebel, 15.0 g, 632 m/s; Mauser, 14.7 g, 638 m/s; Springfield, 9.7 g, 792 m/s.
   (a) For each of these rifles compute the momentum and the kinetic energy of the bullet.
   (b) Roughly, the momentum determines the push that the bullet can exert upon impact on the target and the kinetic energy determines the damage (breakage) that the bullet can do within the target. Which of these bullets can exert the largest push? Which can do the most damage?

30. Suppose you throw a stone straight up so it reaches a maximum height $h$. At what height does the stone have one-half of its initial kinetic energy? At what height does the stone have one-half of its initial momentum?

31. The velocity of small bullets can be roughly measured with ballistic putty. When the bullet strikes a slab of putty, it penetrates a distance that is roughly proportional to the kinetic energy. Suppose that a bullet of velocity 160 m/s penetrates 0.8 cm into the putty and a second, identical bullet fired from a more powerful gun penetrates 1.2 cm. What is the velocity of the second bullet?

32. A particle moving along the $x$ axis is subject to a force

$$F_x = -ax + bx^3$$

where $a$ and $b$ are constants.
   (a) How much work does this force do as the particle moves from $x_1$ to $x_2$?
   (b) If this is the only force acting on the particle, what is the change of kinetic energy during this motion?

*33. A large stone-throwing engine designed by Archimedes could throw a 77-kg stone over a range of 180 m. Assume the stone is thrown at an initial angle of 45° with the vertical.
   (a) Calculate the initial kinetic energy of this stone.
   (b) Calculate the kinetic energy of the stone at the highest point of the trajectory.

*34. With the brakes fully applied, a 1500-kg automobile decelerates at the rate of 8.0 m/s$^2$.
   (a) What is the braking force acting on the automobile?

(b) If the initial speed is 90 km/h, what is the stopping distance?
(c) What is the work done by the braking force in bringing the automobile to a stop from 90 km/h?
(d) What is the change in the kinetic energy of the automobile?

*35. A box of mass 40 kg is initially at rest on a flat floor. The coefficient of kinetic friction between the box and the floor is $\mu_k = 0.60$. A woman pushes horizontally against the box with a force of 250 N until the box attains a speed of 2.0 m/s.
(a) What is the change of kinetic energy of the box?
(b) What is the work done by the friction force on the box?
(c) What is the work done by the woman on the box?

### Section 7.4

36. It has been reported that at Cherbourg, France, waves smashing on the coast lifted a boulder of 3200 kg over a 6-m wall. What minimum energy must the waves have given to the boulder?

37. A 75-kg man walks up the stairs from the first to the third floor of a building, a height of 10 m. How much work does he do against gravity? Compare your answer with the food energy he acquires by eating an apple (see Table 8.1).

38. What is the kinetic energy and what is the gravitational potential energy (relative to the ground) of a jetliner of mass 73,000 kg cruising at 880 km/h at an altitude of 9000 m?

39. A golf ball of mass 50 g released from a height of 1.5 m above a concrete floor bounces back to a height of 1.0 m. What is the amount of energy lost during the impact?

40. Surplus energy from an electric power plant can be temporarily stored as gravitational energy by using this surplus energy to pump water from a river into a reservoir at some altitude above the level of the river. If the reservoir is 250 m above the level of the river, how much water (in cubic meters) must we pump in order to store $2 \times 10^{13}$ J?

41. The track of a cable car on Telegraph Hill in San Francisco rises more than 60 m from its lowest point. Suppose that a car is ascending at 13 km/h along the track when it breaks away from its cable at a height of exactly 60 m. It will then coast up the hill some extra distance, stop, and begin to race down the hill. What speed does the car attain at the lowest point of the track? Ignore friction.

42. In pole vaulting, the jumper achieves great height by converting his kinetic energy of running into gravitational potential energy (Figure 7.21). The pole plays an intermediate role in this process. When the jumper leaves the ground, part of his translational kinetic energy has been converted into kinetic energy of rotation (with the foot of the pole as the center of rotation) and part has been converted into elastic potential energy of deformation of the pole. When the jumper reaches his highest point, all of this energy has been converted into gravitational potential energy. Suppose that a jumper runs at a speed of 10 m/s. If the jumper converts all of the corresponding kinetic energy into gravitational potential energy, how high will his center of mass rise? The actual height reached by pole vaulters is 5.7 m (measured from the ground). Is this consistent with your calculation?

**Fig. 7.21**

43. Because of a brake failure, a bicycle with its rider careens down a steep hill 45-m high. If the bicycle starts from rest and if there is no friction, what is the final speed attained at the bottom of the hill?

44. Under favorable conditions, an avalanche can reach extremely great speeds because the snow rides down the mountain on a cushion of trapped air that makes the sliding motion nearly frictionless. Suppose that a mass of

$2 \times 10^7$ kg of snow breaks loose from a mountain and slides down into a valley 500 m below the starting point. What is the speed of the snow when it hits the valley? What is its kinetic energy? The explosion of 1 short ton (2000 lb) of TNT releases $4.2 \times 10^9$ J. How many tons of TNT release the same energy as the avalanche?

45. A parachutist of mass 60 kg jumps out of an airplane at an altitude of 800 m. Her parachute opens and she lands on the ground with a speed of 5.0 m/s. How much energy has been lost to air friction in this jump?

*46. In some barge canals built in the nineteenth century, barges were lifted from a low level of the canal to a higher level by means of wheeled carriages. In a French canal, barges of 70 metric tons were placed on a carriage of 35 tons which was pulled, by a wire rope, to a height of 12 m along an inclined track 500 m long.

(a) What was the tension in the wire rope?
(b) How much work was done to lift the barge and carriage?
(c) If the cable had broken just as the carriage reached the top, what would have been the final speed of the carriage when it crashed at the bottom?

*47. A roller coaster near St. Louis is 34 m high at its highest point.

(a) What is the maximum speed that a car can attain by rolling down from the highest point if initially at rest? Ignore friction.
(b) Some people claim that cars reach a maximum speed of 100 km/h. If this is true, what must be the initial speed of a car at the highest point?

*48. A center fielder throws a baseball of mass 0.17 kg with an initial speed of 28 m/s and elevation angle of 30°. What is the kinetic energy and what is the potential energy of the baseball when it reaches the highest point of its trajectory? Ignore friction.

*49. A jet aircraft looping the loop (see Problem 4.48) flies along a vertical circle of radius 1000 m with a speed of 620 km/h at the bottom of the circle and a speed of 350 km/h at the top of the circle. The change of speed is mainly due to the downward pull of gravity. For the given speed at the bottom of the circle, what speed would you expect at the top of the circle if the thrust of the aircraft's engine were exactly balancing the friction force of air (as in the case for level flight)?

*50. A pendulum consists of a mass tied to a string of length 1.0 m. Suppose that this pendulum is initially held at an angle of 30° with the vertical (see Figure 7.22) and then released. What is the speed with which the mass swings through its lowest point? At what angle will the mass have one half of this speed?

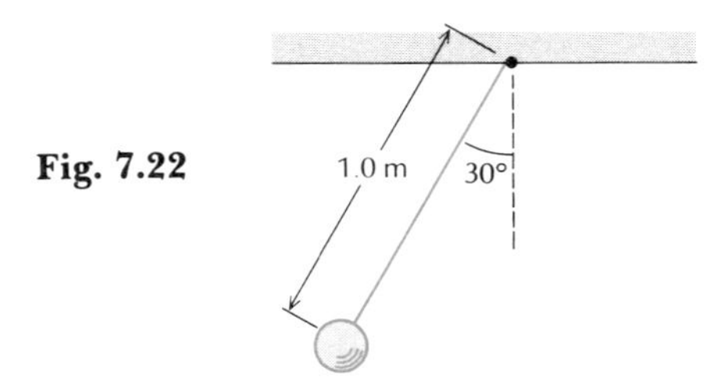

**Fig. 7.22**

**Fig. 7.23** A roller coaster with a full loop.

**51. A stone is tied to a string of length $R$. A man whirls this stone in a vertical circle. Assume that the energy of the stone remains constant as it moves around the circle. Show that if the string is to remain taut at the top of the circle, the speed of the stone at the bottom of the circle must be at least $\sqrt{5gR}$.

**52. In a loop coaster at an amusement park, cars roll along a track that is bent in a full vertical loop (Figure 7.23). If the upper portion of the track is an arc of a circle of radius $R = 10$ m, what is the minimum speed that a car must have at the top of the loop if it is not to fall off? If the highest point of the loop has a height $h = 40$ m, what is the minimum speed with which the car must enter the loop at its bottom? Ignore friction.

**53. You are to design a roller coaster in which cars start from rest at a height $h = 30$ m, roll down into a valley, and then up a mountain (Figure 7.24).

(a) What is the speed of the cars at the bottom of the valley?
(b) If the passengers are to feel 8 G at the bottom of the valley, what must be the radius $R$ of the arc of circle that fits the bottom of the valley?
(c) The top of the next mountain is an arc of circle of the same radius $R$. If the passengers are to feel 0 G at the top of this mountain, what must be its height $h'$?

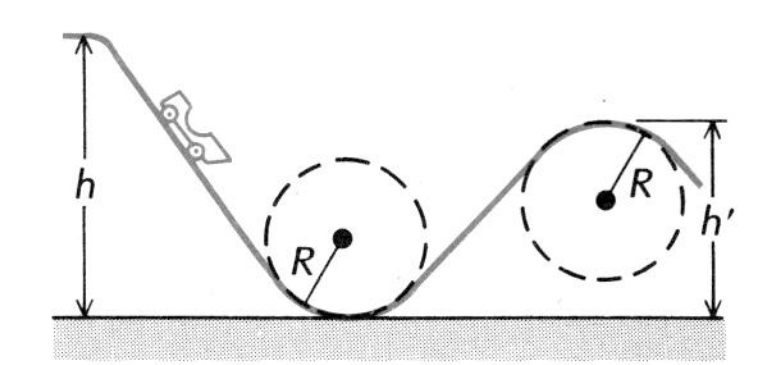


**Fig. 7.24** A roller coaster.

**54. One portion of the track of a toy roller coaster is bent into a full vertical circle of radius $R$. A small cart rolling on the track enters the bottom of the circle with a speed $2\sqrt{gR}$. Show that this cart will fall off the track before it reaches the top of the circle and find the (angular) position at which the cart loses contact with the track.

**55. A particle initially sits on top of a large smooth sphere of radius $R$ (Figure 7.25). The particle begins to slide down the sphere, without friction. At what angular position will the particle lose contact with the surface of the sphere? Where will the particle land on the ground?

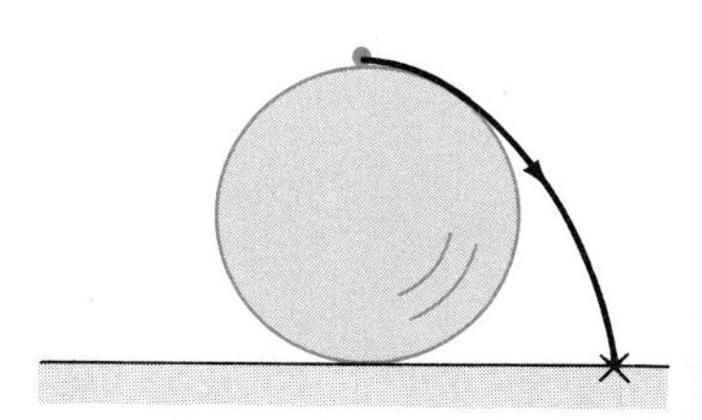

**Fig. 7.25** Particle sliding down a sphere.

CHAPTER 8

# Conservation of Energy

In the preceding chapter we found how to formulate a law of conservation of mechanical energy for a particle moving under the influence of the force of gravity. Now we will seek to formulate a law of conservation of mechanical energy when other forces act on the particle, and we will state the general law of conservation of energy.

## 8.1 Conservative Forces

The example of a particle moving under the influence of the force of gravity taught us how to obtain a constant energy by taking the sum of the kinetic and potential energies. In order to formulate a similar conservation law for a particle moving under the influence of some other force, we have to begin by asking what conditions must be met by the force if a potential energy is to exist. The answer to this question is that the force must be conservative.

*Conservative force*

The definition of a **conservative force** is as follows. Consider a particle that moves from a point $P_1$ to a point $P_2$ along some path. Suppose that the force $\mathbf{F}$ that acts on the particle is a function of position only, so that the force does not depend on the velocity of the particle and does not depend (explicitly[1]) on time. The work done by the force on the particle can be calculated from Eq. (7.28),

$$W = \int_{P_1}^{P_2} \mathbf{F} \cdot d\mathbf{r} \tag{1}$$

[1] Since the position of the particle is a function of time, the force at the position of the particle is also a function of time. However, we will assume that the force at any given position is the same regardless of the time at which the particle arrives at that position.

*The force* **F** *is conservative if this work depends only on the position of the points* $P_1$ *and* $P_2$ *and not on the shape of the path between* $P_1$ *and* $P_2$. Thus, all the paths shown in Figure 8.1 give exactly the same result for the integral in Eq. (1).

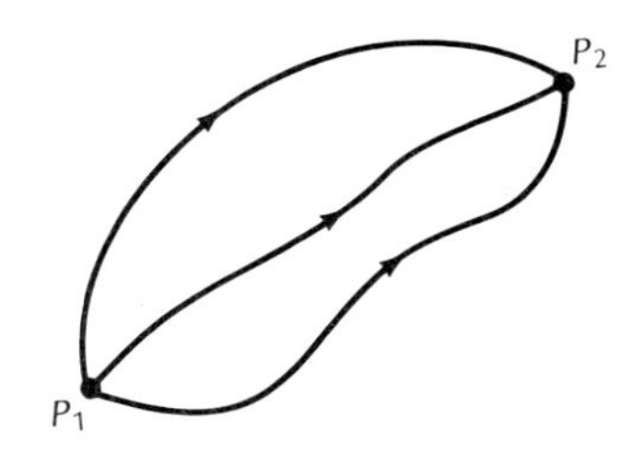


**Fig. 8.1** Several alternative paths for a particle moving from $P_1$ to $P_2$.

The characteristic property of a conservative force can also be expressed another way: the force is conservative if the work is exactly zero for any round trip along a closed path. Figure 8.2 shows such a closed path which starts and ends at the point $P_1$. If the work for this complete path is zero, then the work done between $P_1$ and some intermediate pont $P_2$ must exactly cancel the work done between $P_2$ and $P_1$, that is,

$$0 = \underset{\text{(path I)}}{\int_{P_1}^{P_2} \mathbf{F} \cdot d\mathbf{r}} + \underset{\text{(path II)}}{\int_{P_2}^{P_1} \mathbf{F} \cdot d\mathbf{r}} \tag{2}$$

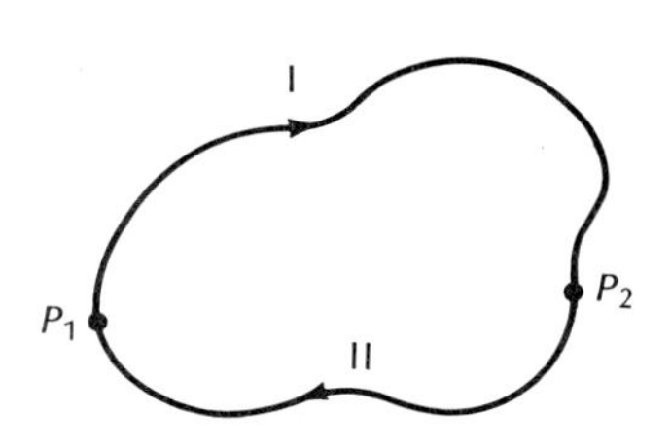


**Fig. 8.2** Round trip along a closed path.

If we exchange the upper and lower limits on the second integral, we obtain

$$0 = \underset{\text{(path I)}}{\int_{P_1}^{P_2} \mathbf{F} \cdot d\mathbf{r}} - \underset{\text{(path II)}}{\int_{P_1}^{P_2} \mathbf{F} \cdot d\mathbf{r}} \tag{3}$$

But this simply states that the work done in going from $P_1$ to $P_2$ along path I is the same as the work done along path II, that is, the work is path independent. Thus the criterion of path independence is equivalent to the criterion of zero work for any round trip.

Gravity is an obvious example of a conservative force. According to Eq. (7.30), the work done by gravity between the points $P_1$ and $P_2$ with coordinates $(x_1, y_1, z_1)$ and $(x_2, y_2, z_2)$ is a function of only $z_1$ and $z_2$, that is, the work depends only on the location of the points, and not on the path between them.

The force exerted by a spring on a particle in one-dimensional motion is another example of a conservative force. According to Eq. (7.12), the work is a function of only $x_1$ and $x_2$, that is, it again depends only on the location of the points and not on how much the particle zigzagged back and forth while proceeding from one point to the other.

The force of kinetic friction is an example of a force that is *not* conservative. Obviously, the longer the path, the greater the work done by friction. Thus the criterion of path independence fails — the work depends on the length of the path and not just on the location of the endpoints. The criterion of zero work for a round trip also fails — the work done by friction is negative for all portions of the path and no cancellation is possible.

However, the nonconservative character of friction is to some extent an illusion which arises from our failure to keep careful track of the motion of all the particles involved. According to the criterion of zero work for any round trip, we ought to examine motions in which all the particles return to their starting points. If we slide a metal block back and forth on a tabletop, this condition is never quite met because the atoms in the block and in the tabletop are disturbed in an uncontrollable manner. When we reverse the motion of the block, the motion of the atoms will not be reversed — during the return trip the atoms do not return to their original place; instead they are even more disturbed. Hence, from a microscopic point of view, we recognize that the work done by the friction force on the block during a round trip is not a fair test of the conservative character of the adhesion forces involved in the friction mechanism. If all the atomic motions are taken into account, then it actually turns out that these adhesion forces *are* conservative. Of course, for practical purposes, we adopt a macroscopic point of view and ignore the atomic motions — friction forces must then be regarded as nonconservative.

## 8.2 Potential Energy of a Conservative Force

Whenever the force acting on a particle is conservative, it is possible to construct a corresponding potential energy by the following recipe: Take a reference point $P_0$; this point may be the origin of coordinates, or some point at a large distance from the region where the force acts, or any other convenient point. At the point $P_0$, assign to the potential energy some value $U(P_0)$; this value may be any convenient number, for instance, $U(P_0) = 0$. At any other point $P$, assign to the potential energy the value

*Potential energy*

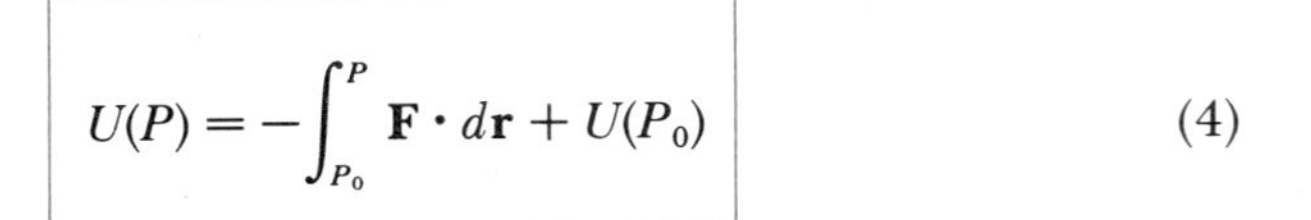

$$U(P) = -\int_{P_0}^{P} \mathbf{F} \cdot d\mathbf{r} + U(P_0) \tag{4}$$

**Joseph Louis, Comte Lagrange** (lagranj) *1736–1813, French mathematician and theoretical astronomer, director of the Berlin Academy and professor at the École Polytechnique, Paris. He was the most eminent mathematician of the eighteenth century, barring only the great Leonhard Euler. In the elegant mathematical treatise* Analytical Mechanics — *which is probably the only book on physics completely devoid of diagrams — Lagrange formulated Newtonian mechanics in the language of advanced calculus and introduced the general definition of the potential-energy function. Lagrange is also known for his calculations of the motion of planets and for his influential role in securing the adoption of the metric system of units.*

The integral appearing here is to be evaluated along any path connecting $P_0$ and $P$ — in view of the conservative character of $\mathbf{F}$, any choice of path will give the same result. The integral may therefore be regarded as a function of the position of $P$ (it is also a function of the position of $P_0$, but since the latter point is kept fixed, this dependence can be ignored). Thus, $U(P)$ is some well-defined function of position.

We can now readily verify that the potential-energy function constructed according to the recipe of the preceding paragraph has all the properties we wish. First of all, the change in the potential energy between two points $P_1$ and $P_2$ equals the negative of the work done by the force between these two points, as shown by the following equalities:

$$\begin{aligned} U(P_2) - U(P_1) &= -\int_{P_0}^{P_2} \mathbf{F} \cdot d\mathbf{r} + U(P_0) + \int_{P_0}^{P_1} \mathbf{F} \cdot d\mathbf{r} - U(P_0) \\ &= -\int_{P_0}^{P_2} \mathbf{F} \cdot d\mathbf{r} - \int_{P_1}^{P_0} \mathbf{F} \cdot d\mathbf{r} \\ &= -\int_{P_1}^{P_2} \mathbf{F} \cdot d\mathbf{r} \end{aligned} \tag{5}$$

Note that $U(P_0)$ cancels in this relation. Thus, the choice of $P_0$ and the choice of $U(P_0)$ in no way affect the calculation of the work from the potential energy. This justifies the somewhat cavalier attitude adopted in the preceding paragraph: the constant $U(P_0)$ can be given any arbitrary value because the quantity of physical significance is the *change* in the potential energy rather than the potential energy itself.

Next, we can show that the total mechanical energy is conserved. Equation (7.41) states that the change in kinetic energy equals the work, that is,

$$K_2 - K_1 = W = \int_{P_1}^{P_2} \mathbf{F} \cdot d\mathbf{r} \tag{6}$$

But, according to Eq. (5), the integral appearing in Eq. (6) is simply $U(P_1) - U(P_2)$. Hence

$$K_2 - K_1 = U(P_1) - U(P_2)$$

or

$$K_2 + U(P_2) = K_1 + U(P_1)$$

This proves that $K + U$ is a constant of the motion,

$$K + U = [\text{constant}] \tag{7}$$

As in the gravitational case, the **mechanical energy** is defined as the sum of the kinetic and potential energies,

$$\boxed{E = K + U} \tag{8}$$

*Mechanical energy*

and Eq. (7) shows that this energy is conserved,

$$\boxed{E = K + U = [\text{constant}]} \tag{9}$$

*Law of conservation of mechanical energy*

This is the law of **conservation of mechanical energy.**

As a concrete illustration of these general mathematical results, consider the motion of a particle under the influence of the force of gravity, $F_z = -mg$. For the reference point $P_0$, take the origin $x = 0$, $y = 0$, $z = 0$, and for the potential energy at this point, take $U(P_0) = 0$. Then the general recipe (4) gives[2]

$$U(P) = -\int_0^x F_x\, dx' - \int_0^y F_y\, dy' - \int_0^z F_z\, dz'$$

$$= -\int_0^z F_z\, dz' = -\int_0^z (-mg)\, dz' = mgz$$

Hence the potential energy is a function of $z$ only,

$$U(P) = U(z) = mgz$$

and the conserved mechanical energy is

$$E = K + U = \tfrac{1}{2}mv^2 + mgz$$

[2] The variables of integration in this integral have been written as $x'$, $y'$, $z'$ in order to distinguish them from the limits of integration $x$, $y$, $z$.

This, of course, agrees with Eq. (7.51). Thus, our general recipe for the potential energy reproduces the formula for gravitational potential energy that we derived in Section 7.4.

As another illustration, consider the motion of a particle in one dimension (along the $x$ axis) under the influence of a force $F_x(x) = -kx$ produced by an elastic spring. The corresponding potential energy can be constructed as follows. For the reference point $P_0$ take $x = 0$, and for the potential energy at this point take $U(P_0) = 0$. Then[3]

$$U(x) = -\int_0^x (-kx')\, dx' = k\left[\tfrac{1}{2}x'^2\right]_0^x$$

or

*Potential energy of a spring*

$$\boxed{U(x) = \tfrac{1}{2}kx^2} \tag{10}$$

Hence the conserved mechanical energy is

$$E = K + U = \tfrac{1}{2}mv^2 + \tfrac{1}{2}kx^2 \tag{11}$$

This expression for the energy gives us some information about the general features of the motion; it shows how, say, an increase of $x$ requires a decrease of $v$ so as to keep $E$ constant.

Note that if instead of the arbitrary choice $U(P_0) = 0$ we had made the equally good (and equally arbitrary) choice $U(P_0) = 3.0$ J, then the energy in Eqs. (10) and (11) would have had an extra 3.0 J added to it. This only changes the base value from which the energy is reckoned; it does not alter the fact that the energy is conserved — and that is what is really important.

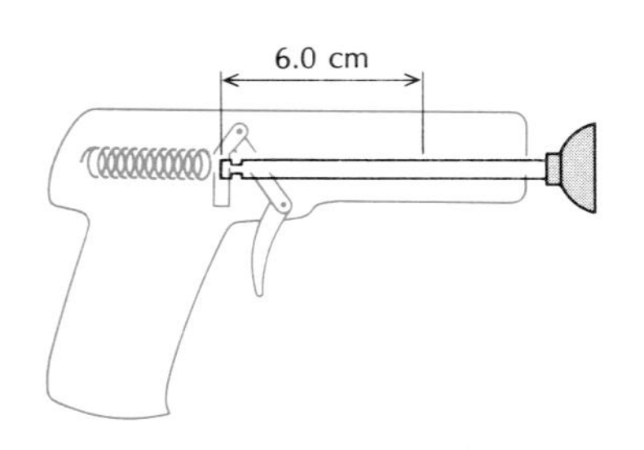


**Fig. 8.3** A toy gun. The spring is compressed 6.0 cm.

EXAMPLE 1. A child's toy gun shoots a dart by means of a compressed spring. The constant of the spring is $3.2 \times 10^2$ N/m and the mass of the dart is 8.0 g. Before shooting, the spring is compressed by 6.0 cm and the dart is placed in contact with the spring (Figure 8.3); the spring is then released. What will be the speed of the dart when the spring reaches its equilibrium length?

SOLUTION: The dart can be regarded as a particle moving under the influence of a force $F = -kx$. According to Eq. (11), the initial energy at $x_1 = -6.0$ cm is

$$E = \tfrac{1}{2}mv_1^2 + \tfrac{1}{2}kx_1^2 = 0 + \tfrac{1}{2}kx_1^2 \tag{12}$$

When the spring reaches its equilibrium length (at $x_2 = 0$), the energy will be

$$E = \tfrac{1}{2}mv_2^2 + \tfrac{1}{2}kx_2^2 = \tfrac{1}{2}mv_2^2 + 0 \tag{13}$$

Conservation of energy demands that the right sides of Eqs. (12) and (13) be equal,

$$\tfrac{1}{2}mv_2^2 = \tfrac{1}{2}kx_1^2 \tag{14}$$

Hence the speed of the dart will be

[3] The variable of integration in this integral has been written as $x'$ in order to distinguish it from the limit of integration $x$.

$$v_2 = \sqrt{k/m}\, x_1$$

$$= \sqrt{(3.2 \times 10^2 \text{ N/m})8.0 \times 10^{-3} \text{ kg}} \times 6.0 \times 10^{-2} \text{ m}$$

$$= 12 \text{ m/s}$$

COMMENTS AND SUGGESTIONS: This is another typical instance of the use of the law of conservation of energy in the solution of a problem of motion. As in the examples of motion under the influence of gravity at the end of Chapter 7, the solution involves three steps: write an expression for the energy at one point of the motion [Eq. (12)], write an expression for the energy at another point [Eq. (13)], and rely on conservation of energy to equate the two expressions [Eq. (14)]. Such a three-step procedure is almost always helpful in dealing with a problem involving a conservation law.

As yet another illustration, consider the motion of a particle in one dimension with the force $F(x) = -A/x^2$, that is, the inverse-square force mentioned in Section 6.6. For the reference point $P_0$ take $x = \infty$, and for the potential energy at this point take $U(P_0) = 0$. Then

$$U(x) = -\int_{\infty}^{x} \frac{-A}{x'^2}\, dx' = A\left[-\frac{1}{x'}\right]_{\infty}^{x} = -\frac{A}{x}$$

The corresponding conserved energy is

$$E = K + U = \tfrac{1}{2}mv^2 - \frac{A}{x} \qquad (15)$$

This expression again shows how changes in $x$ and $v$ must be related.

EXAMPLE 2. A particle of mass $m$ is moving along the $x$ axis under the influence of the force $F(x) = -A/x^2$; assume that (in some suitable units) $m = 1.0$ and $A = 1.0$. The initial position and velocity of the particle are $x = 1.0$ and $v = 1.0$. What is the energy of the particle? What is the maximum distance that the particle can reach along the (positive) $x$ axis?

SOLUTION: We have already analyzed this motion in detail by means of numerical approximation methods in Section 6.6. It will be instructive to compare our earlier approximate result for the maximum distance with the exact result obtained from energy conservation.

The energy is given by Eq. (15) with $A = 1.0$, $m = 1.0$, $x = 1.0$, and $v = 1.0$:

$$E = \tfrac{1}{2}mv^2 - \frac{A}{x} = \tfrac{1}{2} \times 1.0 \times (1.0)^2 - \frac{1.0}{1.0} = -\tfrac{1}{2}$$

The units here are the same as in Section 6.6 and will not be further specified.

The energy at any later time must be the same:

$$-\tfrac{1}{2} = \tfrac{1}{2}mv^2 - \frac{A}{x}$$

When the particle reaches maximum distance along the $x$ axis, the velocity will be zero, that is,

$$-\tfrac{1}{2} = -\frac{A}{x}$$

which gives $x = 2.0$. This exact result agrees fairly well with the approximate result, $x \cong 1.986$, that we obtained by numerical methods (see Table 6.3).

If several conservative forces simultaneously act on a particle, then the net potential energy is the sum of all the potential energies of all these forces. The total mechanical energy is the sum of the kinetic energy and the net potential energy; the total mechanical energy is, of course, conserved.

Finally, what if both a conservative and a nonconservative force, such as friction, act on a particle? Then the change in the kinetic energy plus the change in the potential energy of the conservative force will be equal to the work done by the nonconservative force,

$$\Delta K + \Delta U = W_{\text{nonconserv}}$$

that is,

*Loss of mechanical energy by nonconservative force*

$$\boxed{\Delta E = W_{\text{nonconserv}}} \tag{16}$$

This equation for the change of energy in the presence of a nonconservative force is not nearly as useful in the investigation of the motion of a particle as is the equation of energy conservation that would hold if this force were absent. The latter equation tells us how arbitrary changes in position and in speed are related (see Examples 1 and 2). The former equation does no such thing because the value of $W_{\text{nonconserv}}$ depends not only on the net change of position, but also on the details of the motion between the initial and the final positions; thus, $W_{\text{nonconserv}}$ cannot be evaluated explicitly until the details of the motion are known. The main use of Eq. (16) is in the evaluation of $W_{\text{nonconserv}}$ *after* position and speed have been calculated by other means.

**Fig. 8.4** Crash test of an automobile.

EXAMPLE 3. In a crash test, an automobile of mass 1800 kg is dropped from a height of 9.50 m onto an instrumental target (see Figure 8.4). The automobile rebounds 0.10 m after this impact. How much work was done by nonconservative forces during the impact?

SOLUTION: The initial energy of the automobile is purely potential,

$$U_1 = mgz_1 = 1800 \text{ kg} \times 9.81 \text{ m/s}^2 \times 9.50 \text{ m} = 1.68 \times 10^5 \text{ J}$$

The final energy, at the top of the rebound, is again purely potential,

$$U_2 = mgz_2 = 1800 \text{ kg} \times 9.81 \text{ m/s}^2 \times 0.10 \text{ m} = 0.018 \times 10^5 \text{ J}$$

Hence the change of mechanical energy during the impact is

$$\Delta E = 1.68 \times 10^5 \text{ J} - 0.018 \times 10^5 \text{ J} = 1.66 \times 10^5 \text{ J}$$

and, according to Eq. (16), this must be the work done by the nonconservative forces. The work done by these forces produces the crushing of the front end of the automobile, and it also produces some amount of heat by internal friction within the deforming pieces of metal.

## 8.3* Calculation of the Force from the Potential Energy

It is sometimes desirable to calculate the force from the potential energy function. Such a calculation may seem irrelevant, since, according to Eq. (4), the potential energy is defined in terms of the force. However, in modern theoretical physics potential energies play a very important role — in many instances interactions are primarily described in terms of the corresponding potential energy and the force is only calculated afterward.

Suppose that the points $P$ and $P_0$ in Eq. (4) are separated only by an infinitesimal displacement $d\mathbf{r}$; then $U(P)$ will differ from $U(P_0)$ only by an infinitesimal quantity

$$dU = U(P) - U(P_0) = -\mathbf{F} \cdot d\mathbf{r}$$

We can also write this as

$$dU = -F_x\,dx - F_y\,dy - F_z\,dz \tag{17}$$

Now assume that the infinitesimal displacement is entirely in the $x$ direction. Then $dy = 0$ and $dz = 0$, so that

$$dU = -F_x\,dx \tag{18}$$

or

$$F_x = -\frac{dU}{dx} \tag{19}$$

Thus, the $x$ component of the force is the derivative of $U$ with respect to $x$, *with $y$ and $z$ being held constant,* that is, when we differentiate the function $U$ with respect to $x$ in Eq. (19), both $y$ and $z$ are to be treated just as any other constants. Since the differentiation only applies to one of the variables, the derivative appearing in Eq. (19) is called a **partial derivative** and is usually written with the following special notation:

$$\boxed{F_x = -\frac{\partial U}{\partial x}} \tag{20}$$

*Force as the derivative of potential energy*

Likewise,

$$F_y = -\frac{\partial U}{\partial y} \tag{21}$$

$$F_z = -\frac{\partial U}{\partial z} \tag{22}$$

* This section is optional.

For instance, if $U$ is the potential energy of an elastic spring, $U(x) = \frac{1}{2}kx^2$, then Eq. (20) yields a force

$$F_x = -\frac{\partial U}{\partial x} = -\frac{\partial}{\partial x}(\tfrac{1}{2}kx^2) = -kx$$

which is as expected. Note that in this instance $U$ does not depend on $y$ or $z$ and therefore Eqs. (21) and (22) yield $F_y = 0$ and $F_z = 0$, which means there is no force in the $y$ and $z$ directions.

EXAMPLE 4. In a diatomic molecule the atoms can move relative to one another within certain limits. According to a simple theory of interatomic forces, the potential energy for the motion of each atom is

$$U(x) = U_0(e^{-2(x-x_0)/b} - 2e^{-(x-x_0)/b})$$

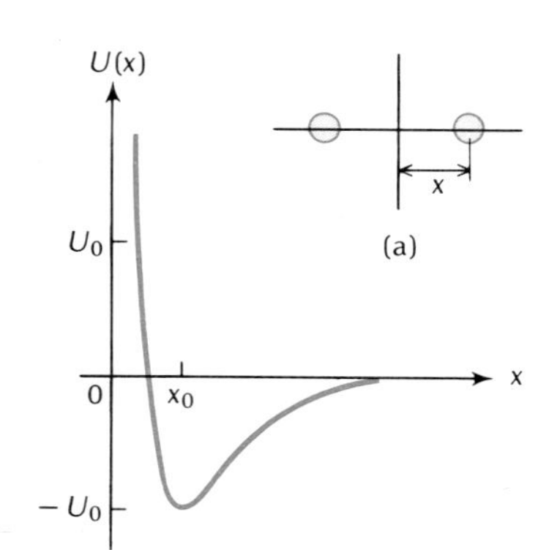


**Fig. 8.5** (a) A molecule consisting of two atoms. (b) Plot of the potential energy as a function of $x$.

Here e = 2.718 . . . ; $U_0$, $x_0$, and $b$ are constants; and $x$ is the distance from the one atom to the midpoint of the molecule (Figure 8.5). What is the force on the atom? For what value of $x$ is this force zero?

SOLUTION: By Eq. (20), the force is the negative of the derivative of $U$ with respect to $x$:

$$F_x = -\frac{\partial U}{\partial x} = -U_0\left(-\frac{2}{b}e^{-2(x-x_0)/b} + \frac{2}{b}e^{-(x-x_0)/b}\right)$$

The force is zero if

$$-\frac{2}{b}e^{-2(x-x_0)/b} + \frac{2}{b}e^{-(x-x_0)/b} = 0$$

which happens when $x = x_0$.

## 8.4* The Curve of Potential Energy

If a particle of some given energy is moving in one dimension under the influence of a conservative force, then Eq. (9) permits us to calculate the speed of the particle as a function of position. Suppose that the potential energy is $U = U(x)$; then Eq. (9) states

$$E = \tfrac{1}{2}mv^2 + U(x)$$

or

$$v^2 = \frac{2}{m}[E - U(x)] \qquad (23)$$

Since the left side of this equation is never negative, we can immediately conclude that the particle must always remain within a range of values of $x$ for which $U(x) \leq E$. If the particle reaches a point at which

* This section is optional.

$U(x) = E$, then $v = 0$, that is, the particle will stop at this point and its motion will reverse. Such a point is called a **turning point** of the motion.

*Turning point*

According to Eq. (23), $v^2$ is directly proportional to $E - U(x)$; thus, $v^2$ is large wherever the difference between $E$ and $U(x)$ is large. We can therefore gain some insights into the qualitative features of the motion by drawing a graph of potential energy on which it is possible to display the difference between $E$ and $U(x)$. For example, Figure 8.6 shows the curve of potential energy for an atom in a diatomic molecule (compare Example 4). The equation of this curve is

$$U(x) = U_0(e^{-2(x - x_0)/b} - 2e^{-(x - x_0)/b}) \tag{24}$$

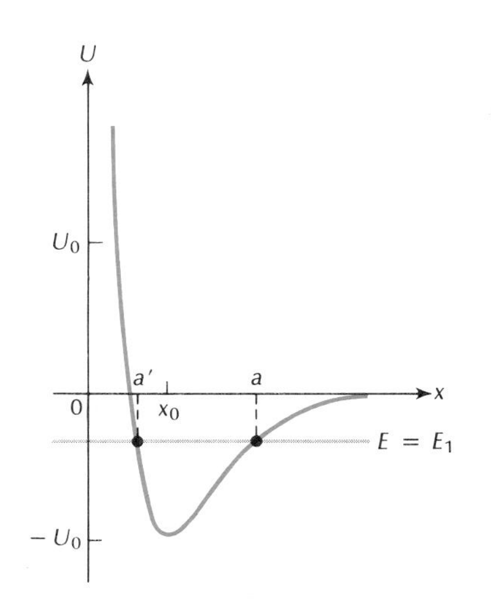


**Fig. 8.6** Potential-energy curve for an atom in a diatomic molecule. The horizontal line is the energy level of the particle. The turning points are at $x = a$ and at $x = a'$.

On this graph, we indicate the value of the energy of the particle by a horizontal line. We will call this horizontal line the **energy level** of the particle. At any point $x$, we can then see the difference between $E$ and $U(x)$ at a glance; according to Eq. (23), this tells us $v^2$. For instance, suppose that a particle has an energy $E = E_1$. Figure 8.6 shows this energy level. Obviously, the particle has maximum speed at the point $x = x_0$, where the separation between the energy level and the potential-energy curve is maximum. The speed gradually decreases as the particle moves toward the right. The potential-energy curve intersects the energy level at $x = a$; at this point the speed of the particle will reach zero, so this point is a turning point of the motion. The particle then moves toward the left, again attaining its greatest speed at $x = x_0$. The speed gradually decreases as the particle continues to move toward the left, and the speed reaches zero at $x = a'$, the second turning point of the motion. Here the particle begins to move toward the right, and so on. Thus the particle continues to move back and forth between the two turning points — the particle is confined between the two turning points. The regions $x > a$ and $x < a'$ are forbidden regions; the region $a' \leq x \leq a$ is permitted. The particle is said to be in a **bound orbit.** The motion is periodic, i.e., it repeats again and again whenever the particle returns to its starting point.

*Energy level*

*Bound orbit*

The location of the turning points depends on the energy. For a particle with a lower energy level the turning points are closer together. The lowest possible energy level intersects the potential-energy

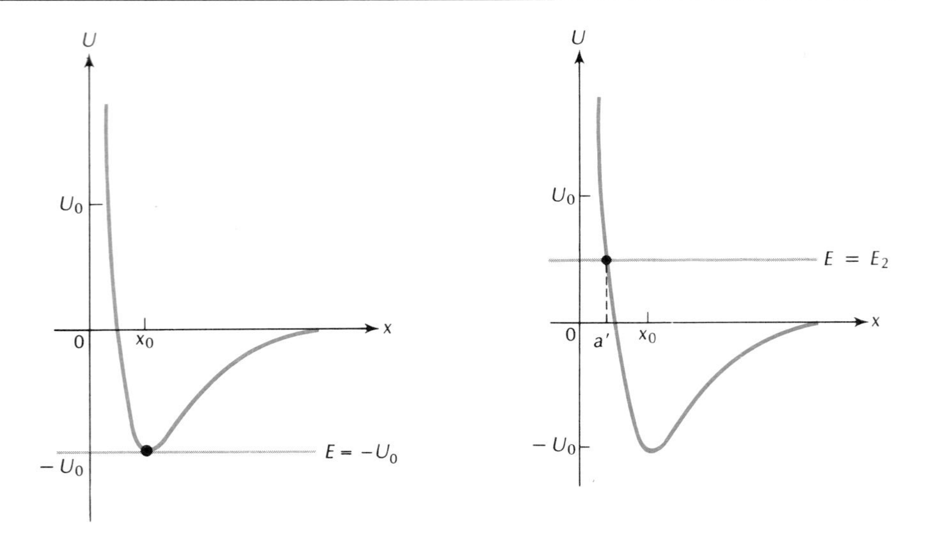


**Fig. 8.7** (left) Energy level of a stationary particle.

**Fig. 8.8** (right) Energy level of a particle in an unbound orbit. There is a single turning point at $x = a'$.

curve at its minimum (see $E = -U_0$ in Figure 8.7); the two turning points then merge into the single point $x = x_0$. A particle with this lowest possible energy cannot move at all — it remains stationary at $x = x_0$. The point $x = x_0$ is an **equilibrium point.** Note that the potential-energy curve has zero slope at $x = x_0$; this corresponds to zero force, $F_x = -\partial U/\partial x = 0$ (see Example 4).

*Equilibrium point*

In Figure 8.6, the right side of the potential-energy curve never rises above $U = 0$. Consequently, if the energy level is above this value (for instance $E = E_2$; see Figure 8.8), then there is only one single turning point on the left, and no turning point on the right. A particle with energy $E_2$ will continue to move toward the right forever; it is not confined. Such a particle is said to be in an **unbound orbit.**

*Unbound orbit*

Of course, the above qualitative analysis based on the curve of potential energy cannot tell us the details of the motion such as, say, the travel time from one point to another. But the qualitative analysis is useful because it gives us a quick survey of the types of orbits that are possible for different values of the energy. If we need more details, we must solve Eq. (23). In view of $v = dx/dt$, Eq. (23) is a differential equation for $x$; this differential equation can be solved by integration, but the integrals are usually rather difficult to evaluate, especially if the potential is a complicated function such as $U(x)$ of Eq. (24).

## 8.5 Other Forms of Energy

As we saw in Section 8.2, if the forces acting on a particle are conservative, then the mechanical energy of the particle is conserved. But if some of the forces acting on the particle are not conservative, then the mechanical energy of the particle will not always remain constant. For instance, if friction forces are acting, they do negative work and decrease the mechanical energy.

However, it is a remarkable fact about our physical universe that whenever mechanical energy is lost by a body, this energy never disappears — it is merely transmuted into other forms of energy. Thus, in the case of friction, the mechanical energy lost by the body is transformed into kinetic and potential energy of the atoms in the body and in the surface against which it is rubbing. The energy that the atoms acquire in the rubbing process is disorderly kinetic and potential energy — it is spread among the atoms in an irregular, random fashion. The disorderly kinetic and potential energy of the atoms of a body is what is meant by **heat.** Hence friction produces heat.

*Heat*

Heat is a form of energy, but whether it is to be regarded as a new form of energy or not depends on the point of view one wishes to adopt. Taking a macroscopic point of view, we ignore the atomic motions; then heat is to be regarded as distinct from mechanical energy. Taking a microscopic point of view, we recognize heat as kinetic and potential energy of the atoms; then heat is to be regarded as mechanical energy. (We will further discuss heat in Chapter 20.)

Chemical energy and nuclear energy are two other forms of energy. The former is kinetic and potential energy of the electrons within the atoms; the latter is kinetic and potential energy of the protons and neutrons within the nuclei of atoms. As in the case of heat, whether these are to be regarded as new forms of energy depends on the point of view.

Electric and magnetic energy are forms of energy associated with electric charges and with light and radio waves. (We will examine these forms of energy in Chapters 26 and 32.)

Table 8.1 lists some examples of different forms of energy. All the energies in Table 8.1 have been expressed in joules. However, for reasons of tradition and convenience, some other energy units are often used in specialized areas of physics and engineering. The energy of atomic and subatomic particles is usually measured in **electron-volts** (eV), where

*Alternative energy units*

$$1 \text{ electron-volt} = 1 \text{ eV} = 1.602 \times 10^{-19} \text{ J}$$

The multiples of this unit are $1 \text{ keV} = 10^3 \text{ eV}$, $1 \text{ MeV} = 10^6 \text{ eV}$, and $1 \text{ GeV} = 10^9 \text{ eV}$.

**Table 8.1** SOME ENERGIES

| | |
|---|---|
| Nuclear fuel in Sun | $1 \times 10^{45}$ J |
| Explosion of a supernova | $1 \times 10^{44}$ J |
| Fossil fuel available on Earth | $2.0 \times 10^{23}$ J |
| Yearly energy expenditure of the United States | $8 \times 10^{19}$ J |
| Volcanic explosion (Krakatoa) | $6 \times 10^{18}$ J |
| Annihilation of 1 kg of matter-antimatter | $9.0 \times 10^{16}$ J |
| Nuclear fuel in nuclear reactor | $1 \times 10^{16}$ J |
| Explosion of thermonuclear bomb (1 megaton) | $4.2 \times 10^{15}$ J |
| Fission of 1 kg of uranium | $8.2 \times 10^{13}$ J |
| Gravitational potential energy of jet airliner (Boeing 747 at 9000 m) | $2 \times 10^{10}$ J |
| Lightning flash | $1 \times 10^{9}$ J |
| Combustion of 1 gal. of gasoline | $1.3 \times 10^{8}$ J |
| Daily food intake of man (3000 kcal) | $1.3 \times 10^{7}$ J |
| Explosion of 1 kg of TNT | $4.6 \times 10^{6}$ J |
| Metabolization of one apple (110 kcal) | $4.6 \times 10^{5}$ J |
| Kinetic energy of running man | $4 \times 10^{3}$ J |
| One push-up | $3 \times 10^{2}$ J |
| Fission of one uranium nucleus | $3.2 \times 10^{-11}$ J |
| Annihilation of electron-positron pair | $1.6 \times 10^{-13}$ J |
| Energy of ionization of hydrogen atom | $2.2 \times 10^{-18}$ J |

**Hermann von Helmholtz,** *1821–1894, Prussian surgeon, biologist, mathematician, and physicist, professor at Berlin. His scientific contributions ranged from the invention of the ophthalmoscope and studies of the physiology and physics of vision and hearing, to the measurement of the speed of light and studies in theoretical mechanics. Helmholtz formulated the general law of conservation of energy, treating it as a consequence of the basic laws of mechanics and electricity.*

The energy supplied by electric power plants is usually expressed in **kilowatt-hours** (kW · h):

$$1 \text{ kilowatt-hour} = 1 \text{ kW} \cdot \text{h} = 3.600 \times 10^6 \text{ J}$$

And the thermal energy supplied by the combustion of fuels is expressed in **kilocalories** (kcal):

$$1 \text{ kilocalorie} = 1 \text{ kcal} = 4.187 \times 10^3 \text{ J}$$

or in **British thermal units** (Btu):

$$1 \text{ Btu} = 1.055 \times 10^3 \text{ J}$$

We will learn more about the definitions of these units in later chapters.

*General law of conservation of energy*

All these forms of energy can be transformed into one another. For example, in an internal combustion engine, chemical energy of the fuel is transformed into heat and kinetic energy; in a hydroelectric power station, gravitational potential energy of the water is transformed into electric energy; in a nuclear reactor, nuclear energy is transformed into heat, light, kinetic energy, etc. However, in any such transformation process the sum of all the energies of all the pieces of matter involved in the process remains constant: *the form of the energy changes, but the total amount of energy does not change.* This is the **general law of conservation of energy.**

## 8.6* Mass and Energy

One of the great discoveries made by Einstein early in this century is that energy can be transformed into mass and mass can be transformed into energy. Thus, *mass is a form of energy.* The amount of energy contained in an amount $m$ of mass is

*Mass is energy*

$$\boxed{E = mc^2} \tag{25}$$

where $c$ is the speed of light ($c = 3.00 \times 10^8$ m/s). This formula is a consequence of Einstein's theory of relativity. The most spectacular experimental demonstration of this formula is found in the annihilation of matter and antimatter. If a proton collides with an antiproton, or an electron with an antielectron, they react violently and annihilate each other in an explosion that generates an intense flash of very energetic light. In this reaction the mass of the particles is entirely converted into energy of light. According to Eq. (25), the annihilation of just 1 metric ton of matter and antimatter (500 kg of each) would release an amount of energy

$$E = mc^2 = 10^3 \text{ kg} \times (3 \times 10^8 \text{ m/s})^2 = 9 \times 10^{19} \text{ J} \tag{26}$$

* This section is optional.

This is enough energy to satisfy the needs of the United States for a full year. Unfortunately, antimatter is not readily available in large amounts. On Earth, antiparticles can only be obtained from reactions induced by the impact of beams of high-energy particles on a target. These collisions occasionally result in the creation of a particle–antiparticle pair. Such pair creation is the reverse of pair annihilation: the creation process transforms some of the kinetic energy of the collision into mass, and a subsequent annihilation merely gives back the original energy.

But the relationship between energy and mass of Eq. (25) also comprises another aspect. *Energy has mass.* Whenever the energy of a body is changed, its mass (and weight) is changed. The change in mass that accompanies a given change of energy is

$$\Delta m = \Delta E / c^2 \tag{27}$$

*Energy has mass*

For instance, if the kinetic energy of a body increases, its mass (and weight) increases. At speeds small compared to the speed of light, the mass increment is not noticeable. But, when a body approaches the speed of light, the mass increment becomes very large. The electrons produced by the Stanford Linear Accelerator provide an extreme example of this effect: these electrons have a speed of 99.99999997% of the speed of light and their mass is 44,000 times the mass of electrons at rest!

The mass that a particle has when at rest is sometimes called its **rest mass** and the corresponding energy [Eq. (25)] is called the **rest-mass energy.** The masses listed in tables of particles (see, e.g., Table 5.2) are always the rest masses. In Chapter 41 we will study the theory of Special Relativity and obtain a formula for the increase of mass with velocity. However, in all other chapters we will neglect the dependence of mass on velocity because the effect is insignificant at the velocities that we encounter in everyday experience.

*Rest mass*

The fact that energy has mass indicates that energy is a form of mass. Conversely, as we have seen above, mass is a form of energy. Hence mass and energy must be regarded as essentially the same thing. The laws of conservation of mass and conservation of energy are therefore not two independent laws — each implies the other. For example, consider the fission reaction of uranium inside the reactor vessel of a nuclear power plant. The complete fission of 1 kg of uranium yields an energy of $8.2 \times 10^{13}$ J. The reaction conserves energy — it merely transforms nuclear energy into heat, light, and kinetic energy, but does not change the total amount of energy. The reaction also conserves mass — if the reactor vessel is hermetically sealed and thermally insulated from its environment, then the reaction does not change the mass of the contents of the vessel. However, if we open the vessel during or after the reaction, and let some of the heat and light escape, then the mass of the residues will not match the mass of the original amount of uranium. The mass of the residues will be about 0.1% smaller than the mass of the original uranium. This mass defect represents the mass carried away by the energy that has escaped. Thus the often repeated statement that nuclear reactions convert mass into energy is misleading. True, the mass of the residues is less than the mass of the original uranium, but the escaped energy carries with it just the right amount of mass to balance the accounts: the net mass re-

**James Watt,** *1736–1819, Scottish inventor and engineer. He modified and improved an earlier steam engine designed by Thomas Newcomen, adding a separate condenser. Watt introduced the* horsepower *as a unit of mechanical power.*

mains constant. A nuclear reaction merely transforms energy into new forms of energy and mass into new forms of mass. In this regard a nuclear reaction is not essentially different from a chemical reaction. The net mass remains constant in any chemical reaction, but the mass of the residues of an exothermic chemical reaction is slightly less than the original mass. The heat released in such a chemical reaction carries away some mass, but, in contrast to a nuclear reaction, this amount of mass is so small as to be quite immeasurable.

## 8.7 Power

The power delivered by a force to a body is the rate at which the force does work on that body. If the force does an amount of work $\Delta W$ in an interval of time $\Delta t$, then the **average power** is

*Average power*

$$\bar{P} = \frac{\Delta W}{\Delta t} \tag{28}$$

The **instantaneous power** is the time derivative of the work,

*Instantaneous power*

$$P = \frac{dW}{dt} \tag{29}$$

In the metric system, the unit of power is the **watt** (W), which is a rate of work of 1 joule per second,

*Watt,* W

$$1 \text{ watt} = 1 \text{ W} = 1 \text{ J/s} \tag{30}$$

In the British system, the unit of power is the ft · lbf/s. In practice, engineers often prefer to measure power in **horsepower** (hp) units, where

*Horsepower,* hp

$$1 \text{ horsepower} = 1 \text{ hp} = 550 \text{ ft} \cdot \text{lbf/s} = 745.7 \text{ W} \tag{31}$$

This is roughly the rate at which a (very strong) horse can do work.

---

EXAMPLE 5. An elevator has a mass of 900 kg. How many horsepower must the motor deliver to the elevator if it is to raise the elevator at the rate of 1.8 m/s? The elevator has no counterweight.

SOLUTION: By means of the elevator cable, the motor must exert an upward force $F = mg$ to raise the elevator. If the elevator is raised a distance $\Delta z$, the work done by this force is

$$\Delta W = F\,\Delta z$$

and the rate at which work is done is

$$P = \frac{\Delta W}{\Delta t} = F\frac{\Delta z}{\Delta t} = Fv \tag{32}$$

This gives

$$P = mgv = 900 \text{ kg} \times 9.8 \text{ m/s}^2 \times 1.8 \text{ m/s}$$

$$= 1.6 \times 10^4 \ W$$

$$= 1.6 \times 10^4 \ W \times \frac{1 \text{ hp}}{746 \ W} = 21 \text{ hp}$$

---

Equation (32) is a special instance of a general formula, which expresses the instantaneous power in terms of force and velocity. During an infinitesimal displacement $d\mathbf{r}$, a force $\mathbf{F}$ will perform an amount of work

$$dW = \mathbf{F} \cdot d\mathbf{r}$$

The instantaneous power delivered by this force is then

$$P = \frac{dW}{dt} = \mathbf{F} \cdot \frac{d\mathbf{r}}{dt} \qquad (33)$$

which leads to the general formula

$$\boxed{P = \mathbf{F} \cdot \mathbf{v}} \qquad (34)$$

*Power delivered by a force*

Note that if this equation for the power is employed for an extended body, the velocity $\mathbf{v}$ must be taken to be the velocity of the specific point of the body at which the force acts. For instance, consider a woman walking up a hill. She pushes against the ground, and the reaction force exerted by the ground on her feet pushes her up the hill. However, when her foot is in contact with the ground, it is instantaneously at rest ($\mathbf{v} = 0$) and therefore the reaction force exerted by the ground on the foot does no work and delivers no energy to the foot. This, of course, is in accord with common sense — the energy required for the walk up the hill is not supplied by the ground, but by the woman's leg muscles.

---

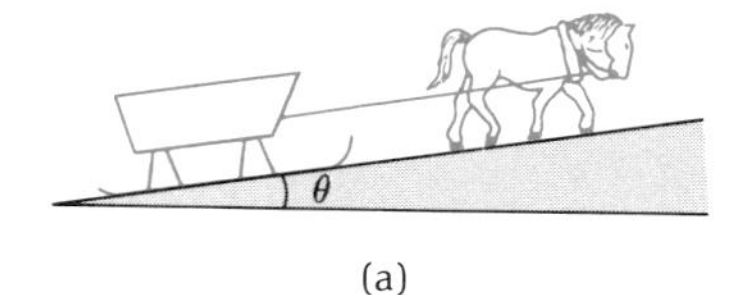


(a)

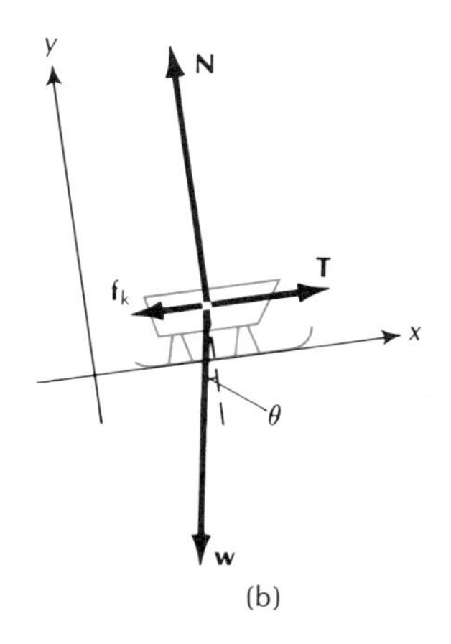


(b)

**Fig. 8.9** (a) Horse dragging a sled and (b) the "free-body" diagram for the sled.

EXAMPLE 6. A horse drags a sled up a steep snow-covered street of slope 1:7 (Figure 8.9a). The sled has a mass of 300 kg and the coefficient of sliding friction between the sled and the snow is 0.12. If the horse pulls parallel to the surface of the street and delivers a power of 1.0 hp, what is the maximum (constant) speed with which the horse can drag the sled? What fraction of the horse's power is expended against friction? What fraction against gravity?

SOLUTION: Figure 8.9b is a "free-body" diagram for the sled, showing the weight ($w = mg$), the normal force ($N = mg \cos \theta$), the friction force ($f_k = \mu_k N$), and the pull of the horse ($T$). Since the acceleration along the street is zero,

$$0 = T + w_x - f_k = T - mg \sin \theta - f_k$$

so that

$$T = mg \sin \theta + f_k = mg \sin \theta + \mu_k mg \cos \theta$$

The power delivered by the horse to the sled is

$$P = \mathbf{T} \cdot \mathbf{v} = (mg \sin \theta + \mu_k mg \cos \theta)v$$

from which

$$v = \frac{P}{mg(\sin\theta + \mu_k \cos\theta)}$$

With $\theta = \tan^{-1}\frac{1}{7} = 8.13°$, this yields

$$v = \frac{746 \text{ W}}{300 \text{ kg} \times 9.8 \text{ m/s}^2 \times (\sin 8.13° + 0.12 \cos 8.13°)}$$

$$= 0.98 \text{ m/s}$$

The power exerted by the friction force is

$$P_{\text{friction}} = \mathbf{f}_k \cdot \mathbf{v} = -f_k v = -\mu_k mg \cos\theta \times v$$

$$= -0.12 \times 300 \text{ kg} \times 9.8 \text{ m/s}^2 \times \cos 8.13° \times 0.98 \text{ m/s}$$

$$= -3.4 \times 10^2 \text{ W} = -0.46 \text{ hp}$$

and the power exerted by gravity is

$$P_{\text{gravity}} = \mathbf{w} \cdot \mathbf{v} = w_x v = -mg \sin\theta \times v$$

$$= -300 \text{ kg} \times 9.8 \text{ m/s}^2 \times \sin 8.13° \times 0.98 \text{ m/s}$$

$$= -4.1 \times 10^2 \text{ W} = -0.54 \text{ hp}$$

Thus 46% of the horse's power is expended against friction and 54% against gravity.

---

Since the velocity depends on the frame of reference, the power also depends on the frame of reference. For instance, consider a man holding a bowling ball in his outstretched hand while riding in an elevator moving steadily upward. In the reference frame of the elevator, his hand delivers no power to the ball; but in the reference frame of the ground, his hand delivers positive power to the ball — and the weight of the ball delivers an equal amount of negative power to the ball.

The above equations all refer to *mechanical* power. In general, power is the rate at which energy is transformed from one form to another or transported from one place to another. Table 8.2 gives some examples of different kinds of power.

**Table 8.2** SOME POWERS

| | |
|---|---|
| Light and heat emitted by the Sun | $3.9 \times 10^{26}$ W |
| Solar light and heat incident on the Earth | $1.7 \times 10^{17}$ W |
| Mechanical power generated by hurricane | $2 \times 10^{13}$ W |
| Total power used in United States (average) | $2 \times 10^{12}$ W |
| Large electric power plant | $\sim 10^{9}$ W |
| Jet airliner engines (Boeing 747) | $2.1 \times 10^{8}$ W |
| Automobile engine | $1.5 \times 10^{5}$ W |
| Radio emission by large radio transmitter | $1 \times 10^{5}$ W |
| Solar light and heat per square meter at Earth | $1.4 \times 10^{3}$ W |
| Electricity used by toaster | $1 \times 10^{3}$ W |
| Work output of man (athlete at maximum) | $2 \times 10^{2}$ W |
| Electricity used by light bulb | $1 \times 10^{2}$ W |
| Heat output of man (average) | $1 \times 10^{2}$ W |
| Heat and work output of bumblebee (in flight) | $2 \times 10^{-2}$ W |
| Atom radiating light | $\sim 10^{-10}$ W |

Incidentally: Multiplication of a unit of power by a unit of time gives a unit of energy. An example of this is the kilowatt-hour (kW · h), already mentioned in Section 8.5:

$$1 \text{ kilowatt-hour} = 1 \text{ kW} \cdot \text{h} = 10^3 \text{ W} \cdot 3600 \text{ s}$$
$$= 3.6 \times 10^6 \text{ J}$$

This is the unit commonly used to measure electric energy delivered by power plants.

## SUMMARY

**Conservative force:** The work done by the force does not depend on the path; it depends only on the positions of the endpoints of the path.

**Potential energy of a conservative force:**

$$U(P) = -\int_{P_0}^{P} \mathbf{F} \cdot d\mathbf{r} + U(P_0)$$

**Mechanical energy:** $E = K + U$

**Conservation of energy:** $E = K + U = [\text{constant}]$

**Potential energy of a spring:** $U(x) = \frac{1}{2}kx^2$

**Loss of mechanical energy by nonconservative force:** $\Delta E = W_{\text{nonconserv}}$

**Force as derivative of potential energy:**

$$F_x = -\frac{\partial U}{\partial x} \qquad F_y = -\frac{\partial U}{\partial y} \qquad F_z = -\frac{\partial U}{\partial z}$$

**Mass is a form of energy:** $E = mc^2$

**Energy has mass:** $\Delta m = \Delta E/c^2$

**Average power:** $\overline{P} = \dfrac{\Delta W}{\Delta t}$

**Instantaneous power:** $P = \dfrac{dW}{dt}$

**Mechanical power delivered by a force:** $P = \mathbf{F} \cdot \mathbf{v}$

## QUESTIONS

1. A body slides on a smooth horizontal plane. Is the normal force of the plane on the body a conservative force? Can we define a potential energy for this force according to the recipe in Section 8.2?

2. If you stretch a spring so far that it suffers a permanent deformation, is the force exerted by the spring during this operation conservative?

3. Is there any frictional dissipation of mechanical energy in the motion of the planets of the Solar System or in the motion of their satellites? (Hint: Consider the tides.)

4. What happens to the kinetic energy of an automobile during braking without skidding? With skidding?

5. Consider a stone thrown vertically upward. If we take friction against the air into account, we see that $\frac{1}{2}mv^2 + mgz$ must *decrease* as a function of time. From this, prove that the stone will take longer for the downward motion than for the upward motion.

6. An automobile travels down a road leading from a mountain peak to a valley. What happens to the gravitational potential energy of the automobile? How is it dissipated?

7. Suppose you wind up a watch and then place it into a beaker full of nitric acid and let it dissolve. What happens to the potential energy stored in the spring of the watch?

**Fig. 8.10** Thermonuclear explosion.

8. News reporters commonly speak of "energy consumption." Is it accurate to say that energy is *consumed*? Would it be more accurate to say that energy is *dissipated*?

9. The explosive yield of thermonuclear bombs (Figure 8.10) is usually reported in kilotons or megatons of TNT. Would the explosion of a 1-megaton hydrogen bomb really produce the same effects as the explosion of 1 megaton of TNT (a mountain of TNT more than a hundred meters high)?

10. When you heat a potful of water, does its mass increase?

11. Since mass is a form of energy, why don't we measure mass in the same units as energy? How could we do this?

12. In the annihilation of matter and antimatter, a particle and an antiparticle — such as a proton and an antiproton, or an electron and an antielectron — disappear explosively upon contact, giving rise to an intense flash of light. Is energy conserved in this reaction? Is mass conserved?

13. It takes about 5000 hp to keep a 26-m motor yacht moving at its top speed of 88 km/h (50 knots). What happens to this power?

14. In order to travel at 130 km/h, an automobile of average size needs an engine delivering about 40 hp to overcome the effects of air friction, road friction, and internal friction (in the transmission and drive train). Why do most drivers think they need an engine of 150 or 200 hp?

## PROBLEMS

### Section 8.1

1. A particle moves along the $x$ axis under the influence of a variable force $F_x = 2x^3 + 1$ (where force is measured in newtons and distance in meters).
   (a) Show that this force is conservative; that is, show that for any back-and-forth motion that starts and ends at the same place (round trip), the work done by the force is zero.
   (b) Show that the same is true for any force $F_x = F_x(x)$ that is an arbitrary function of position.

2. Consider a force that is a function of the velocity of the particle (and is not perpendicular to the velocity). Show that the work for a round trip along a closed path can then be different from zero.

*3. A particle moves along a circle $x^2 + y^2 = R^2$ in the $x$–$y$ plane. Suppose that a force with components $F_x = -y$ and $F_y = x$ (where force is measured in newtons and distance in meters) acts on the particle. Show that this force is not conservative.

### Section 8.2

4. The force acting on a particle moving along the $x$ axis is given by the formula $F_x = K/x^4$, where $K$ is a constant. Find the corresponding potential energy. Assume that $U(x) = 0$ for $x = \infty$.

5. A particle moves along the $x$ axis under the influence of a variable force $F_x = 5x^2 + 3x$ (where force is measured in newtons and distance in meters). What is the potential energy associated with this force? Assume that $U(x) = 0$ at $x = 0$.

6. The spring from an automobile suspension has a spring constant $3.53 \times 10^4$ N/m (see Example 6.9). How much work must you do to compress this spring from its relaxed length of 0.316 m to 0.205 m?

*7. A bow may be regarded mathematically as a spring. The archer stretches this "spring" and then suddenly releases it so that the bowstring pushes against the arrow. Suppose that when the archer stretches the "spring" 0.52 m, he must exert a force of 160 N to hold the arrow in this position. If he now releases the arrow, what will be the speed of the arrow when the "spring" reaches its equilibrium position? The mass of the arrow is 0.020 kg. Pretend that the "spring" is massless.

*8. A particle is subjected to a force that depends on position as follows:

$$\mathbf{F} = 4\hat{\mathbf{x}} + 2x\hat{\mathbf{y}}$$

where the force is measured in newtons and the distance in meters.

(a) Calculate the work done by this force as the particle moves from the origin to the point $x = 1$ m, $y = 1$ m along the straight-line path I shown in Figure 8.11.

(b) Calculate the work done by this force if the particle moves from the point $x = 1$ m, $y = 1$ m to the origin along the path II consisting of a horizontal and a vertical segment (see Figure 8.11). Is the force conservative?

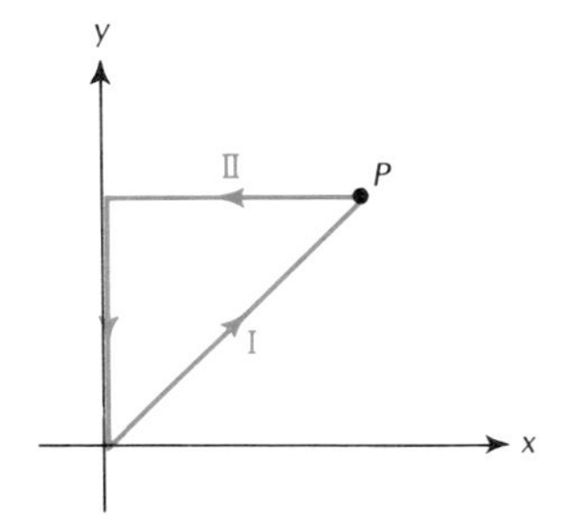


**Fig. 8.11**

*9. A mass $m$ hangs on a vertical spring of spring constant $k$.

(a) How far will this hanging mass have stretched the spring from its relaxed length?

(b) If you now push up on the mass and lift it until the spring reaches its relaxed length, how much work will you have done against gravity? Against the spring?

*10. A 3.0-kg block sliding on a horizontal surface is accelerated by a compressed spring. At first, the block slides without friction. But after leaving the spring, the block travels over a new portion of the surface, with a coefficient of friction 0.20, for a distance of 8.0 m before coming to rest (see Figure 8.12). The force constant of the spring is 120 N/m.

(a) What was the maximum kinetic energy of the block?

(b) How far was the spring compressed before being released?

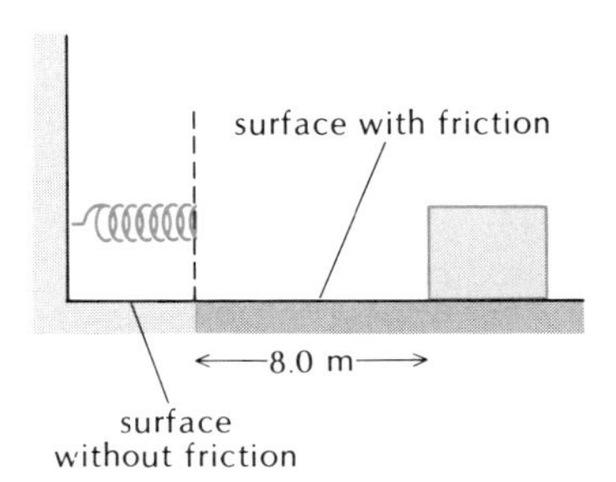


**Fig. 8.12** Block released from a spring.

*11. A particle moving in the $x$–$y$ plane experiences a conservative force

$$\mathbf{F} = by\hat{\mathbf{x}} + bx\hat{\mathbf{y}}$$

where $b$ is a constant.

(a) What is the work done by this force as the particle moves from $x_1 = 0$, $y_1 = 0$ to $x_2 = x$, $y_2 = y$? (Hint: Use a path from the origin to the point $x_2$, $y_2$ consisting of a segment parallel to the $x$ axis and a segment parallel to the $y$ axis.)

(b) What is the potential energy associated with this force? Assume that the potential energy is zero when the particle is at the origin.

*12. The four wheels of an automobile of mass 1200 kg are suspended below the body by vertical springs of constant $k = 7.0 \times 10^4$ N/m. If the forces on all wheels are the same, what will be the maximum instantaneous deformation of the springs if the automobile is lifted by a crane and dropped on the street from a height of 0.8 m?

*13. A rope can be regarded as a long spring; when under tension, it stretches and stores elastic potential energy. Consider a nylon rope similar to that which snapped during a giant tug-of-war at a school in Harrisburg (see Problem 5.16). Under a tension of 58,000 N (applied at its ends), the rope of initial length 300 m stretches to 390 m. What is the elastic energy stored in the rope at this tension? What happens to this energy when the rope breaks?

*14. Among the safety features on elevator cages are spring-loaded brake pads which grip the guide rail if the elevator cable should break. Suppose that an elevator cage of 2000 kg has two such brake pads, arranged to press against opposite sides of the guide rail, each with a force of $1.0 \times 10^5$ N. The friction coefficient for the brake pads sliding on the guide rail is 0.15. Assume that the elevator cage is falling freely with an initial speed of 10 m/s when the brake pads come into action. How long will the elevator cage take to stop? How far will it travel? How much energy is dissipated by friction?

**15. Mountain climbers use nylon safety rope whose elasticity plays an important role in cushioning the sharp jerk if a climber falls and is suddenly stopped by the rope.

(a) Suppose that a climber of 80 kg attached to a 10-m rope falls freely from a height of 10 m above to a height of 10 m below the point at which the rope is anchored to a vertical wall of rock. Treating the rope as a spring with $k = 4.9 \times 10^3$ N/m (which is the appropriate value for a braided nylon rope of 9.2 mm diameter), calculate how much the rope stretches while stopping the fall of the climber. Calculate the maximum force that the rope exerts on the climber during stopping.

(b) Repeat the calculations for a rope of 5 m and an initial height of 5 m. Assume that this second rope is made of the same material as the first, and remember to take into account the change in the spring constant due to the change in length. Compare your results for (a) and (b) and comment on the advantages and disadvantages of long ropes vs. short ropes.

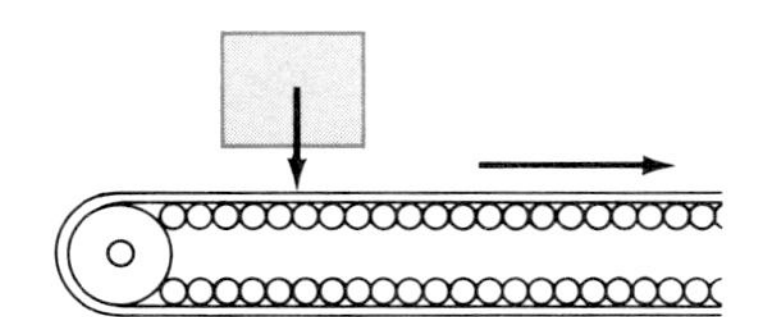

**Fig. 8.13** Package dropped onto a conveyer belt.

**16. A package is dropped on a horizontal conveyor belt (Figure 8.13). The mass of the package is $m$, the speed of the conveyor belt is $v$, and the coefficient of kinetic friction for the package on the belt is $\mu_k$. For what length of time will the package slide on the belt? How far will it move in this time? How much energy is dissipated by friction? How much energy does the belt supply to the package (including the energy dissipated by friction)?

### Section 8.3

17. The potential energy of a particle moving in the $x$–$y$ plane is $U = a/(x^2 + y^2)^{1/2}$, where $a$ is a constant. What is the force on the particle? Draw a diagram showing the particle at the position $x,y$ and the force vector.

18. The potential energy of a particle moving along the $x$ axis is $U(x) = K/x^2$, where $K$ is a constant. What is the corresponding force acting on the particle?

19. According to theoretical calculations, the potential energy of two quarks (see the Prelude) separated by a distance $r$ is $U = \eta r$, where $\eta = 1.18 \times 10^{24}$ eV/m. What is the force between the two quarks? Express your answer in newtons.

### Section 8.4

20. The potential energy of one of the atoms in the hydrogen molecule may be taken to be (see Example 4)

$$U(x) = U_0(e^{-2(x - x_0)/b} - 2e^{-(x - x_0)/b})$$

with $U_0 = 2.36$ eV, $x_0 = 0.37$ Å, and $b = 0.34$ Å.[4] Under the influence of the force corresponding to this potential, the atom moves back and forth along

[4] These values of $U_0$, $x_0$, and $b$ are half as large as those usually quoted, because we are looking at the motion of only *one* atom relative to the center of the molecule.

the $x$ axis within certain limits. If the energy of the atom is $E = -1.15$ eV, what will be the turning points of the motion, i.e., at what positions $x$ will the kinetic energy be zero? [Hint: Solve this problem graphically by making a careful plot of $U(x)$; from your plot find the values of $x$ that yield $U(x) = -1.15$ eV.]

21. Suppose that the potential energy of a particle moving along the $x$ axis is

$$U(x) = \frac{b}{x^2} - \frac{2c}{x}$$

where $b$ and $c$ are positive constants.

(a) Plot $U(x)$ as a function of $x$; assume $b = c = 1$ for this purpose. Where is the equilibrium point?

(b) Suppose the energy of the particle is $E = -\frac{1}{2}c^2/b$. Find the turning points of the motion.

(c) Suppose that the energy of the particle is $E = \frac{1}{2}c^2b$. Find the turning points of the motion. How many turning points are there in this case?

22. A particle moves along the $x$ axis under the influence of a conservative force with a potential energy $U(x)$. Figure 8.14 shows the plot of $U(x)$ vs. $x$. Figure 8.14 shows several alternative energy levels for the particle: $E = E_1$, $E = E_2$, and $E = E_3$. Assume that the particle is initially at $x = x_0$. For each of the three alternative energies describe the motion qualitatively, answering the following questions.

(a) Roughly, where are the turning points (right and left)?

(b) Where is the speed of the particle maximum? Where is the speed minimum?

(c) Is the orbit bound or unbound?

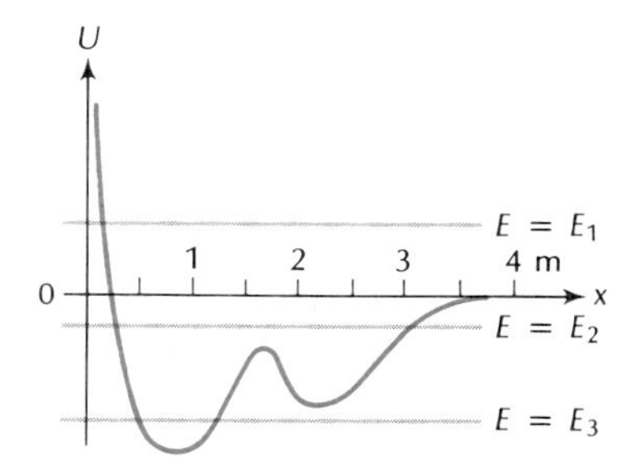


**Fig. 8.14** Plot of $U(x)$ vs. $x$.

Section 8.5

23. Express the last three entries in Table 8.1 in electron-volts.

24. The chemical formula for TNT is $CH_3C_6H_2(NO_2)_3$. The explosion of 1 kg of TNT releases $4.6 \times 10^6$ J. Calculate the energy released per molecule of TNT. Express your answer in electron-volts.

25. Using the data of Table 8.1, calculate the amount of gasoline that would be required if all the energy requirements of the United States were to be met by the consumption of gasoline. How many gallons per day would have to be consumed?

26. The following table lists the fuel consumption and the passenger capacity of several vehicles. Assume that the energy content of the fuel is that of gasoline (see Table 8.1). Calculate the amount of energy used by each vehicle per passenger per mile. Which is the most energy-efficient vehicle? The least energy efficient?

| Vehicle | Passenger capacity | Fuel consumption |
|---|---|---|
| Motorcycle | 1 | 60 mi/gal. |
| Snowmobile | 1 | 12 |
| Automobile | 4 | 12 |
| Intercity bus | 45 | 5 |
| Concorde SST | 110 | 0.12 |
| Jetliner | 360 | 0.1 |

*27. When a humpback whale breaches, or jumps out of the water (see Figure 8.15), it typically leaves the water at an angle of about 70° at high speed and sometimes attains a height of 3 m, measured from the water surface to the center of the whale. For a rough estimate of the energy requirements for such a breach, we can treat the translational motion of the whale as that of a particle moving from the surface of the water upward to a height of 3 m (for a

**Fig. 8.15**

more accurate calculation, we would have to take into account the buoyancy of the whale, which assists it in getting out of the water, but let us ignore this). What is the initial speed of the whale when it emerges from the water? Express the speed in knots. What is the initial kinetic energy of a whale of 33 metric tons? Express the energy in kilocalories.

*28. The following table gives the rate of energy dissipation by a man engaged in diverse activities; the energies are given per kilogram of body mass:

RATE OF ENERGY DISSIPATION OF MAN (MALE, PER KG OF BODY MASS)

| | |
|---|---|
| Standing | 1.3 kcal/(kg · h) |
| Walking (5 km/h) | 3.3 |
| Running (8 km/h) | 8.2 |
| Running (16 km/h) | 15.2 |

Suppose the man wants to travel a distance of 2.5 km in one-half hour. He can walk this distance in exactly half an hour, or run slow and then stand still until the half hour is up, or run fast and then stand still until the half hour is up. What is the energy per kg of body mass dissipated in each case? Which program uses the most energy? Which the least?

### Section 8.6

29. The atomic bomb dropped on Hiroshima had an explosive energy equivalent to that of 20,000 tons of TNT, or $8.4 \times 10^{13}$ J. How many kilograms of rest mass must have been converted into energy in this explosion?

30. How much energy will be released by the annihilation of one electron and one antielectron (both initially at rest)? Express your answer in electron-volts.

31. The mass of the Sun is $2 \times 10^{30}$ kg. The thermal energy in the Sun is about $2 \times 10^{41}$ J. How much does the thermal energy contribute to the mass of the Sun?

*32. In a high-speed collision between an electron and an antielectron, the two particles can annihilate and create a proton and an antiproton. The reaction

$$e + \bar{e} \rightarrow p + \bar{p}$$

converts the rest-mass energy and kinetic energy of the electron and antielectron into the rest-mass energy of the proton and antiproton. Assume that the electron and the antielectron collide head on with opposite velocities of equal magnitudes and that the proton and antiproton are at rest immediately after the reaction. Calculate the kinetic energy of the electron required for this reaction; express your answer in electron-volts.

### Section 8.7

33. For an automobile traveling at a steady speed of 65 km/h, the friction of the air and the rolling friction of the ground on the wheels provide a total external friction force of 500 N.

(a) At what rate does this force remove momentum from the automobile? At what rate does it remove energy?

(b) To keep the automobile going at constant velocity, the momentum loss and the energy loss must be compensated for. What body supplies the necessary momentum? What body supplies the necessary energy?

34. In 1979, B. Allen flew a very lightweight propeller airplane across the English Channel. His legs, pushing bicycle pedals, supplied the power to turn the propeller. To keep the airplane flying, he had to supply about 0.30 hp. How much energy did he supply for the full flight lasting 2 h 49 min? Express your answer in kilocalories.

35. The ancient Egyptians and Romans relied on slaves as a source of mechanical power. One slave, working desperately by turning a crank, can deliver about 200 W of mechanical power (at this power the slave would not last long). How many slaves would be needed to match the output of a modern automobile engine (150 hp)? How many slaves would an ancient Egyptian have to own in order to command the same amount of power as the average per capita power used by residents of the United States (14 kW)?

36. An electric clock uses 2 W of electric power. How much electric energy (in kilowatt-hours) does this clock use in 1 year? What happens to this electric energy?

37. Nineteenth-century engineers reckoned that a laborer turning a crank can do steady work at the rate of 5000 ft · lbf/min. Suppose that four laborers working a manual crane attempt to lift a load of 9 short tons (1 short ton = 2000 lb). If there is no friction, what is the rate at which they can lift this load? How long will it take them to lift the load 15 ft?

38. The driver of an automobile traveling on a straight road at 80 km/h pushes forward with his hands on the steering wheel with a force of 50 N. What is the rate at which his hands do work on the steering wheel in the reference frame of the ground? In the reference frame of the automobile?

39. A crane is powered by an electric motor delivering 60 hp. What is the maximum speed with which this crane can raise a load of 10 metric tons? Assume that 28% of the power of the motor is lost to friction within the crane.

40. A horse walks along the bank of a canal and pulls a barge by means of a long horizontal towrope making an angle of 35° with the bank. The horse walks at the rate of 5 km/h and the tension in the rope is 400 N. What horsepower does the horse deliver?

41. A 900-kg automobile accelerates from 0 to 80 km/h in 7.5 s. What are the initial and the final translational kinetic energies of the automobile? What is the average power delivered by the engine in this time interval? Express your answer in horsepower.

42. A six-cylinder internal combustion engine, such as used in an automobile, delivers an average power of 150 hp while running at 3000 rev/min. Each of the cylinders fires once every two revolutions. How much energy does each cylinder deliver each time it fires?

*43. The ancient Egyptians moved large stones by dragging them across the sand in sleds. Suppose that 6000 Egyptians are dragging a sled with a coefficient of sliding friction $\mu_k = 0.3$ along a level surface of sand.

(a) If each Egyptian exerts a force of 360 N, what is the maximum weight they can move at constant speed?

(b) If each Egyptian delivers a mechanical power of 0.20 hp, what is the maximum speed with which they can move this weight?

44. An automobile engine typically has an efficiency of about 25%, i.e., it converts about 25% of the chemical energy available in gasoline into mechanical energy. Suppose that an automobile engine has a mechanical output of 110 hp. At what rate (in gallons per hour) will this engine consume gasoline? See Table 8.1 for the energy content of gasoline.

45. In a braking test, a 990-kg automobile takes 2.1 s to come to a full stop from an initial speed of 60 km/h. What is the amount of energy dissipated in the brakes? What is the average power dissipated in the brakes? Ignore external friction in your calculation and express the power in horsepowers.

46. The takeoff speed of a DC-3 airplane is 100 km/h. Starting from rest, the airplane takes 10 s to reach this speed. The mass of the (loaded) airplane is 11,000 kg. What is the average power delivered by the engines to the airplane during takeoff?

47. The Sun emits energy in the form of radiant heat and light at the rate of

$3.9 \times 10^{26}$ W. At what rate does this energy carry away mass from the Sun? How much mass does this amount to in 1 year?

48. The energy of sunlight arriving at the surface of the Earth amounts to about 1 kW per square meter of surface (facing the Sun). If all of the energy incident on a collector of sunlight could be converted into useful energy, how many square meters of collector area would we need to satisfy all of the energy demands in the United States? See Table 8.1 for the energy expenditure of the United States.

49. Equations (2.9) and (2.15) give the velocity and the acceleration of an accelerating Maserati sports car as a function of time. The mass of this automobile is 1770 kg. What is the instantaneous power delivered by the engine to the automobile? Plot the instantaneous power as a function of time in the time interval from 0 to 10 s. At what time is the power maximum?

50. The ship *Globtik Tokyo,* a supertanker, has a mass of 650,000 metric tons when fully loaded.
  (a) What is the kinetic energy of the ship when her speed is 26 km/h?
  (b) The engines of the ship deliver a power of 44,000 hp. According to the energy requirements, how long a time does it take the ship to reach a speed of 26 km/h, starting from rest? Make the assumption that 50% of the engine power goes into friction or into stirring up the water and 50% remains available for the translational motion of the ship.
  (c) How long a time does it take the ship to stop from an initial speed of 26 km/h if her engines are put in reverse? Estimate roughly how far she will travel during this time.

51. At Niagara Falls, 6200 $m^3$ per second of water fall down a height of 49 m.
  (a) What is the rate (in watts) at which gravitational potential energy is wasted by the falling water?
  (b) What is the amount of energy (in kilowatt-hours) wasted in 1 year?
  (c) Power companies get paid about 5 cents per kilowatt-hour of electric energy. If all of the gravitational potential energy wasted at Niagara Falls could be converted into electric energy, how much money would this be worth?

52. The movement of a grandfather clock is driven by a 5-kg weight which drops a distance of 1.5 m in the course of a week. What is the power delivered by the weight to the movement?

53. In a waterfall on the Alto Paraná river (between Brazil and Paraguay), the height of fall is 33 m and the average rate of flow is 13,000 $m^3$ of water per second. What is the power wasted by this waterfall?

54. A 27,000-kg truck has a 550-hp engine. What is the maximum speed with which this truck can move up a 10° slope?

*55. In order to overcome air friction and other mechanical friction, an automobile of mass 1500 kg requires a power of 20 hp from its engine to travel at 64 km/h on a level road. Assuming the friction remains the same, what power does the same automobile require to travel uphill on an incline of slope 1:10 at the same speed? Downhill on the same incline at the same speed?

*56. With the gears in neutral, an automobile rolling down a long incline of slope 1:10 reaches a terminal speed of 95 km/h. At this speed the rate of decrease of the gravitational potential energy matches the power required to overcome air friction and other mechanical friction. What power (in horsepower) must the engine of this automobile deliver to drive it at 95 km/h on a level road? The mass of the automobile is 1500 kg.

*57. When jogging at 12 km/h on a level road, a 70-kg man uses 750 kcal/h. How many kilocalories per hour does he require when jogging up a 1:10 incline at the same speed? Assume that the frictional losses are independent of the value of the slope.

*58. Each of the two Wright "Cyclone" engines on a DC-3 airplane generates a power of 850 hp. The mass of the loaded plane is 10,900 kg. The plane can climb at the rate of 260 m/min. When the plane is climbing at this rate, what percentage of the engine power is used to do work against gravity?

*59. A fountain sends a stream of water 10 m up in the air. The base of the stream is 10 cm across. What power is expended to send the water to this height?

*60. The record of 203.1 km/h for speed skiing set by Franz Weber at Velocity Peak, Colorado, was achieved on a mountain slope inclined downward at 51°. At this speed, the force of friction (air and sliding friction) balances the pull of gravity along the slope, so that the motion proceeds at constant velocity.

(a) What is the rate at which gravity does work on the skier? Assume that the mass of the skier is 75 kg.
(b) What is the rate at which sliding friction does work? Assume that the coefficient of friction is $\mu_k = 0.03$.
(c) What is the rate at which air friction does work?

*61. A windmill for the generation of electric power has a propeller of diameter 1.8 m. In a wind of 40 km/h, this windmill delivers 200 W of electric power.

(a) At this wind speed, what is the rate at which the air carries kinetic energy through the circular area swept out by the propeller? The density of air is 1.29 kg/m$^3$.
(b) What percentage of the kinetic energy of the air passing through this area is converted into electric energy?

*62. Consider a projectile traveling horizontally and slowing down under the influence of air resistance, as described in Problems 2.18 and 2.19. The mass of this projectile is 45.36 kg and the speed as a function of time is

$$v = 655.9 - 61.14t + 3.26t^2$$

where speed is measured in m/s and time in seconds.

(a) What is the instantaneous power removed from the projectile by the air resistance?
(b) What is the kinetic energy at time $t = 0$? At time $t = 3.0$ s?
(c) What is the average power for the time interval from 0 to 3.0 s?

*63. A small electric kitchen fan blows 8.5 m$^3$/min of air at a speed of 5.0 m/s out of the kitchen. The density of air is 1.3 kg/m$^3$. What electric power must the fan consume to give the ejected air the required kinetic energy?

*64. The final portion of the Tennessee River has a downward slope of 0.074 m per kilometer. The rate of flow of water in the river is 280 m$^3$/s. Assume that the speed of the water is constant along the river. How much power is wasted by friction of the water against the riverbed per kilometer?

*65. Off the coast of Florida, the Gulf Stream has a speed of 4.6 km/h and a rate of flow of $2.2 \times 10^3$ km$^3$/day. At what rate is kinetic energy flowing past the coast? If all this kinetic energy could be converted into electric power, how many kilowatts would it amount to?

*66. (a) With its engines switched off, a small two-engine airplane of mass 1100 kg glides downward at an angle of 13° at a speed of 90 knots. Under these conditions, the weight of the plane, the lift force (perpendicular to the direction of motion) generated by air flowing over the wings, and the frictional force (opposite to the direction of motion) exerted by air are in balance. Draw a "free-body" diagram for these forces, and calculate their magnitudes.
(b) Suppose that with its engine switched on, the plane climbs at an upward angle of 13° at a speed of 90 knots. Draw a "free-body" dia-

gram for the forces acting on the airplane under these conditions; include the push that the air exerts on the propeller. Calculate the magnitudes of all the forces.

(c) Calculate the power that the engine must deliver to compensate for the rate of increase of the potential energy of the plane and the power lost to friction. For a typical small plane of 1100 kg, the actual engine power required for such a climb of 13° is about 400 hp. Explain the discrepancy between your result and the actual engine power. (Hint: What does the propeller do to the air?)

*67. The reaction that supplies the Sun with energy is

$$\mathrm{H + H + H + H \rightarrow He} + [\text{energy}]$$

(The reaction involves several intermediate steps, but this need not concern us now.) The mass of the hydrogen (H) atom is 1.00813 u and that of the helium (He) atom is 4.00388 u.

(a) How much energy is released in the reaction of four hydrogen atoms (by the conversion of rest mass into energy)?
(b) How much energy is released in the reaction of 1 kg of hydrogen atoms?
(c) The Sun releases energy at the rate of $3.9 \times 10^{26}$ W. At what rate (in kg/s) does the Sun consume hydrogen?
(d) The Sun contains about $1.5 \times 10^{30}$ kg of hydrogen. If it continues to consume hydrogen at the same rate, how long will the hydrogen last?

CHAPTER 9

# Gravitation

The high accuracy of celestial mechanics is legendary. The theoretical calculations of celestial mechanics are based on Newton's laws of motion and on Newton's law of universal gravitation, according to which every mass exerts an attractive gravitational force on every other mass. These mutual gravitational attractions govern the motions of all the celestial bodies. The theoretical calculations of the motions yield long-range predictions for the positions of the planets, satellites, and comets; these predicted positions agree very precisely with astronomical observations. For example, the predicted planetary angular positions agree with the observed positions to within a few seconds of arc, even after a lapse of tens of years. Most of the limitations in the accuracy of celestial mechanics arise not from any defect in the theory, but from the approximations that must be made to simplify the lengthy computations that take into account all the mutual attractions of all the planets.

By the nineteenth century Newton's theory of gravitation had proved itself so trustworthy that when astronomers noticed an irregularity in the motion of Uranus, they could not bring themselves to believe that the theory was at fault. Instead, they suspected that a new unknown planet caused these irregularities by its gravitational pull on Uranus. J. C. Adams and U. J. J. Leverrier proceeded to calculate the expected position of this hypothetical planet — and the new planet, later named Neptune, was immediately found at just about the expected position. This discovery of Neptune was a spectacular success of the theory of gravitation.

Newton's theory of gravitation has had other great successes, but it has also had a (minor) failure. Leverrier discovered an enigmatic discrepancy in the motion of Mercury; roughly, Mercury wanders ahead of its predicted position by 43 seconds of arc in each century. This (small) deviation cannot be explained by Newton's theory. It was finally explained by Einstein's theory of gravitation, or General Relativity, a new theory based on a radical revision of our fundamental concepts of space and time (see Interlude IX).

Apart from the barely noticeable defect in the motion of Mercury, Newton's theory of gravitation has stood up remarkably well to the test of centuries; and it remains one of the most accurate and successful theories in all of physics, that is to say, in all of science.

## 9.1 Newton's Law of Universal Gravitation

Within the Solar System, planets orbit around the Sun and satellites orbit around planets. These circular, or nearly circular, motions require a centripetal force pulling the planets toward the Sun and the satellites toward the planets. It was Newton's great discovery that this interplanetary force holding celestial bodies in their orbits is of the same kind as the force of gravity that causes apples, and other things, to fall downward near the surface of the Earth. The **law of universal gravitation** formulated by Newton states:

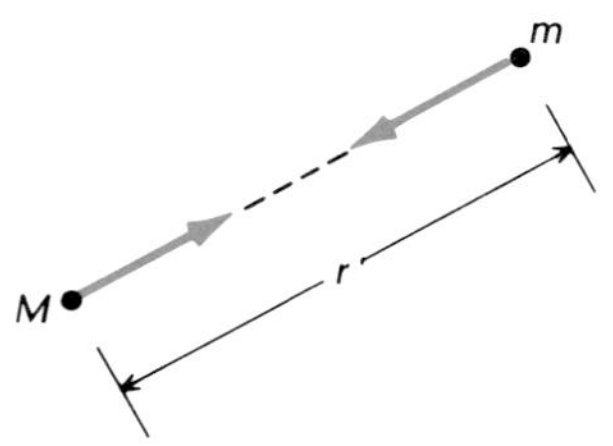

> *Every particle attracts every other particle with a force directly proportional to the product of their masses and inversely proportional to the square of the distance between them.*

**Fig. 9.1** Two particles attract each other gravitationally. The forces are of equal magnitudes and opposite directions.

Expressed mathematically, the magnitude of the gravitational force that two particles of masses $M$ and $m$ separated by a distance $r$ exert on each other is

*Law of universal gravitation*

$$F = \frac{GMm}{r^2} \tag{1}$$

where $G$ is a universal constant. The direction of the force on each particle is toward the other particle.

Figure 9.1 shows the direction of the force on each particle. Note that the two forces are of equal magnitude and opposite direction; they form an action–reaction pair.

The constant $G$ is known as the **gravitational constant.** In metric units its value is

*Gravitational constant, G*

$$G = 6.67 \times 10^{-11}\ \text{N} \cdot \text{m}^2/\text{kg}^2 \tag{2}$$

EXAMPLE 1. What is the magnitude of gravitational force between a 70-kg man and a 70-kg woman separated by a distance of 10 m? Treat both masses as particles.

SOLUTION: From Eq. (1),

$$F = \frac{GMm}{r^2}$$

$$= \frac{6.67 \times 10^{-11}\ \text{N} \cdot \text{m}^2/\text{kg}^2 \times 70\ \text{kg} \times 70\ \text{kg}}{(10\ \text{m})^2}$$

$$= 3.3 \times 10^{-9}\ \text{N}$$

This is a very small force, but as we will see in Section 9.2, the measurement of such small forces is not beyond the reach of sensitive instruments.

The gravitational force of Eq. (1) is an inverse-square force: it decreases by a factor of 4 when the distance increases by a factor of 2; it decreases by a factor of 9 when the distance increases by a factor of 3; etc. Figure 9.2 is a plot of the magnitude of the gravitational force as a function of the distance. Although the force decreases with distance, it never quite reaches zero. Thus every particle in the universe continually attracts every other particle at least a little bit, even if the distance between them is very, very large.

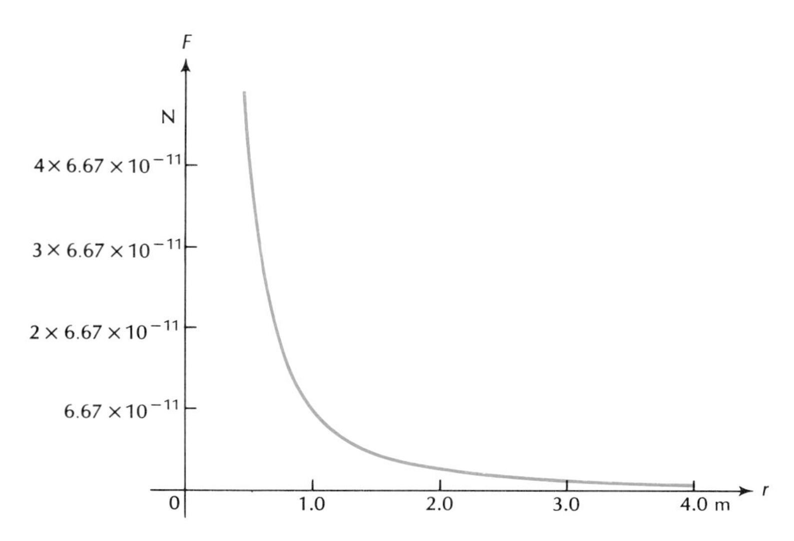


**Fig. 9.2** Magnitude of the gravitational force exerted by one particle of 1 kg on another particle of 1 kg.

The gravitational force does not require any contact between the interacting particles. In reaching from one remote particle to another, the gravitational force somehow bridges the empty space between the particles. This is called *action-at-a-distance*. For now we will not inquire into the mechanism that transmits the force over the empty space, but we will explore this question later in Chapters 23 and 46.

It is also quite remarkable that the gravitational force between two particles is unaffected by the presence of intervening masses. For example, a particle in Washington attracts a particle in Peking with exactly the force given by Eq. (1), even though all of the bulk of the Earth lies between Washington and Peking. This means that it is impossible to shield a particle from the gravitational attraction of another particle.

Since the gravitational attraction between two particles is completely independent of the presence of other particles, it follows that the net gravitational force between two bodies (for example, the Earth and the Moon, or the Earth and an apple) is merely the vector sum of the individual forces between all the particles making up the bodies, that is, the gravitational force obeys the principle of superposition. We will prove in Section 9.6 that this implies that the net gravitational force between two spherical bodies acts just as though the mass of each body were concentrated at the center of its respective sphere. This important result is called **Newton's theorem.** Since the Sun, the planets, and most of their satellites are almost exactly spherical, we can treat all these celestial bodies as pointlike particles in all calculations concerning their gravitational attractions. For instance, the magnitude of gravitational force exerted by the Earth on a particle above its surface is

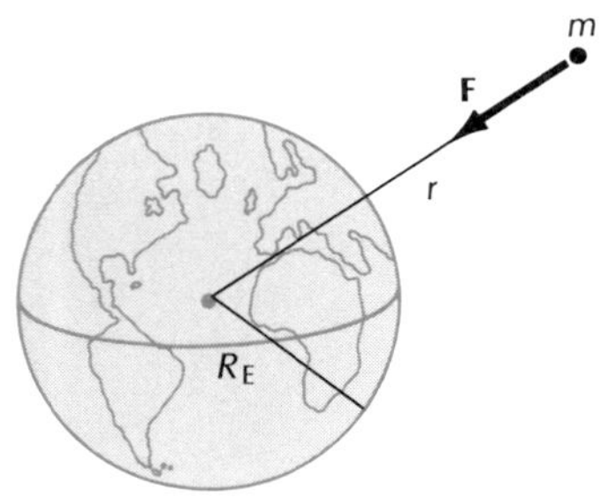


**Fig. 9.3** The gravitational force exerted by the Earth on a particle is directed toward the center of the Earth.

$$F = \frac{GM_E m}{r^2} \tag{3}$$

where $m$ is the mass of the particle, $M_E$ the mass of the Earth, and $r$ the distance from the *center* of the Earth (Figure 9.3). This expression, however, is only valid if the particle is outside the Earth. If the particle is inside the Earth (for instance, in a mine shaft), then the force is smaller (see Section 9.6).

If the particle is at the surface of the Earth, at a radius $r = R_E$, then Eq. (3) gives a force

$$F = \frac{GM_E m}{R_E^2} \tag{4}$$

The corresponding acceleration of the mass $m$ is

$$a = \frac{F}{m} = \frac{GM_E}{R_E^2} \tag{5}$$

But this acceleration is what we usually call the acceleration of gravity:

*Acceleration of gravity, g*

$$\boxed{g = \frac{GM_E}{R_E^2}} \tag{6}$$

This equation establishes the connection between the ordinary gravity we experience at the surface of the Earth and Newton's law of universal gravitation.

If the particle is at high altitude above the surface of the Earth, then the acceleration of gravity is less than that given by Eq. (6). At a distance $r$ from the center, Eq. (3) gives an acceleration

$$a = \frac{GM_E}{r^2} \tag{7}$$

which can also be written

$$a = \frac{R_E^2}{r^2} g \tag{8}$$

where, as in Eq. (6), $g$ is the acceleration at the surface.

## 9.2 The Measurement of $G$

The gravitational constant $G$ is rather difficult to measure with precision. The trouble is that the gravitational forces between masses of lab-

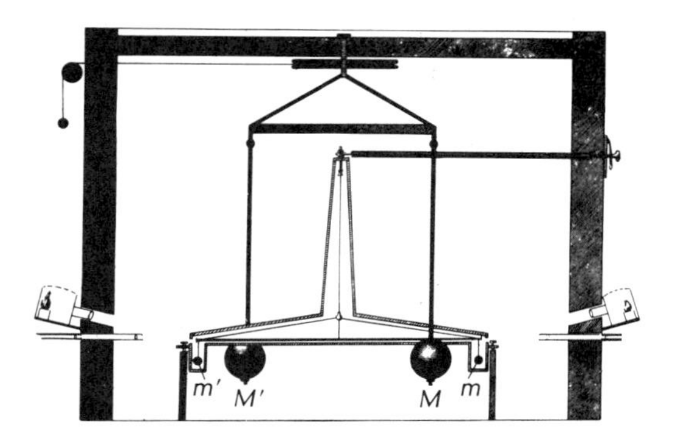


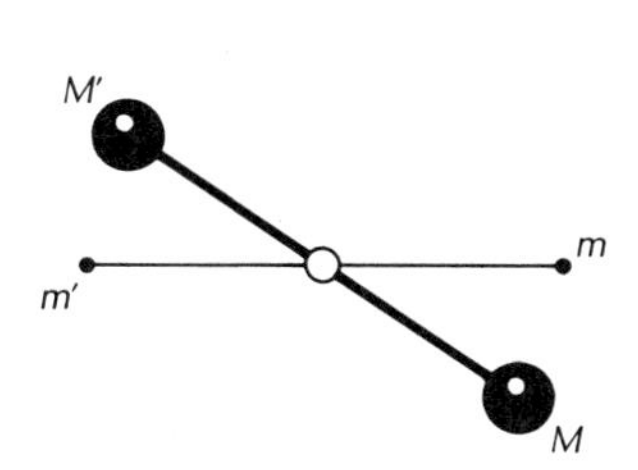


**Fig. 9.4** (left) The torsion balance used by Cavendish.

**Fig. 9.5** (right) Top view of torsion balance.

oratory size are extremely small, and thus a very delicate apparatus is needed to detect these forces. Measurements of $G$ are usually done with Cavendish's **torsion balance** (Figures 9.4 and 9.5). Two equal, small spherical masses $m$, $m'$ are attached to a lightweight horizontal beam which is suspended at its middle by a thin vertical fiber. When the beam is left undisturbed, it will settle into an equilibrium position such that the fiber is completely untwisted. If two equal, large masses $M$, $M'$ are brought near the small masses $m$, $m'$, the gravitational attraction between each small mass and its neighboring large mass tends to rotate the beam counterclockwise (as seen from above). The twist of the fiber opposes this rotation, and the net result is that the beam settles into a new equilibrium position in which the forces on the beam generated by the gravitational attraction between the masses is exactly balanced by the force generated by the twisted fiber. The gravitational constant can then be calculated from the measured values of the angular displacement between the two equilibrium positions, the values of the masses and their distances, and the value of the force constant[1] of the fiber. [A preliminary measurement of the force constant (torque constant) of the fiber is required; in Example 15.6 we will describe a simple procedure for doing this.]

**Henry Cavendish,** *1731–1810, English experimental physicist and chemist. Cavendish's main work was on the chemistry of gases; he isolated hydrogen and identified it as a separate chemical element. His torsion balance for the absolute measurement of the gravitational force was based on an earlier design used by Coulomb for the measurement of the electric force.*

Several modern methods for the measurement of $G$ rely on clever modifications of the basic Cavendish balance described above. The best available data lead to $G = 6.673 \times 10^{-11}\ \text{N} \cdot \text{m}^2/\text{kg}^2$ with an uncertainty of about 0.01%. Thus $G$ is only known to within four significant figures, whereas most other fundamental constants of physics are known to within six or eight significant figures. Fortunately, the lack of precision in the measurements of $G$ does not affect the high accuracy of celestial mechanics: the gravitational force exerted by, say, the Sun only depends on the product $G \times$ [mass of the Sun] and this *product* can be determined very precisely from planetary observations even though the individual factors cannot be determined very precisely.

Incidentally: The mass of the Earth can be calculated from Eq. (6) using the known values of $G$, $R_E$, and $g$:

$$M_E = \frac{R_E^2 g}{G} = \frac{(6.38 \times 10^6\ \text{m})^2 \times 9.81\ \text{m/s}^2}{6.67 \times 10^{-11}\ \text{N} \cdot \text{m}^2/\text{kg}}$$

$$= 5.98 \times 10^{24}\ \text{kg} \qquad (9)$$

For a more precise calculation it is necessary to take into account that the Earth is an ellipsoid rather than a sphere.

[1] Just as a spring has a spring constant, a fiber has a force constant that characterizes its torque.

This calculation would seem to be a rather roundabout way to arrive at the mass of the Earth, but there is no direct route, since we cannot place the Earth on a balance. Because this calculation requires a prior measurement of the value of $G$, the Cavendish experiment has often been described figuratively as "weighing the Earth."

## 9.3 Circular Orbits

Although the mutual gravitational forces of the Sun on a planet and of the planet on the Sun are equal in magnitude, the mass of the Sun is much larger than the mass of a planet and hence its acceleration is much smaller. It is therefore an excellent approximation to regard the Sun as fixed and immovable, and it then remains only to investigate the motion of the planet. If we designate the masses of the Sun and planet by $M_S$ and $m$, and their separation by $r$, then the magnitude of the gravitational force on the planet is

$$F = \frac{GM_S m}{r^2} \tag{10}$$

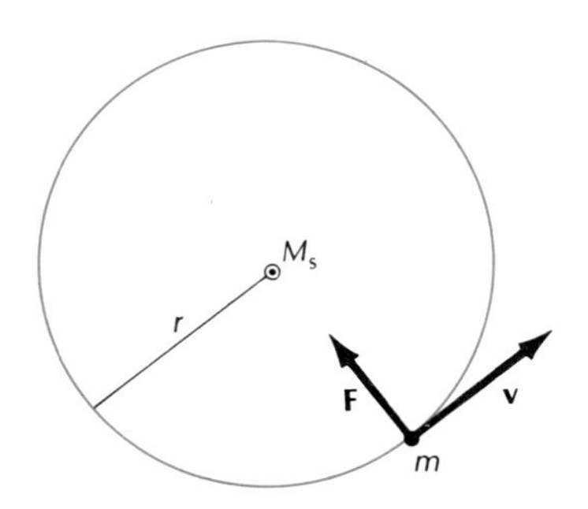


**Fig. 9.6** Circular orbit of a planet around the Sun.

This force points toward the center of the Sun, that is, the center of the Sun is the center of force (Figure 9.6). For a particle moving under the influence of such a central force, the simplest conceivable motion is uniform circular motion, with the gravitational force acting as centripetal force. The motion of the planets in our Solar System is somewhat more complicated than that — as we will see in the next section, the planets move along ellipses, instead of circles. However, none of these planetary ellipses deviates very much from a circle, and as a first approximation we can pretend that the planetary orbits are circles.

If the speed of the planet is $v$, the centripetal acceleration is $v^2/r$, and the equation of motion is

$$\frac{mv^2}{r} = \frac{GM_S m}{r^2} \tag{11}$$

or

$$\boxed{v^2 = \frac{GM_S}{r}} \tag{12}$$

**Nicholas Copernicus,** *1473–1543, Polish astronomer. In his book* De Revolutionibus Orbium Coelestium *he formulated the heliocentric system for the description of the motion of the planets, according to which the Sun is immovable and the planets orbit around it. This new system — ardently defended by Galileo — gradually supplanted the old Ptolemaic system, according to which the Earth is immobile at the center of the universe.*

The speed of the planet is related to the circumference of the orbit and to the time $T$ for one revolution:

$$v = \frac{2\pi r}{T} \tag{13}$$

The time $T$ is called the **period** of the orbit. With this, Eq. (12) becomes

$$\frac{4\pi^2 r^2}{T^2} = \frac{GM_S}{r} \tag{14}$$

or

$$T^2 = \frac{4\pi^2}{GM_S} r^3 \qquad (15)$$

*Period for circular orbit*

This says that the square of the period is proportional to the cube of the radius of the orbit, with a constant of proportionality depending on the mass of the central body.

---

EXAMPLE 2. Both Venus and the Earth have approximately circular orbits around the Sun. The period of the orbit of Venus is 0.615 year and the period of the orbit of the Earth is 1 year. According to Eq. (15), by what factor do the sizes of the two orbits differ?

SOLUTION: From Eq. (15), the orbital radius is proportional to the $\frac{2}{3}$ power of the period. Thus,

$$\frac{r_E}{r_V} = \frac{T_E^{2/3}}{T_V^{2/3}} = \frac{(1 \text{ year})^{2/3}}{(0.615 \text{ year})^{2/3}} = 1.38$$

---

EXAMPLE 3. Equation (15) can be used to find the mass of the Sun from the observed values of the orbital radius and period of a planet. Given that the (mean) orbital radius of the Earth is $1.496 \times 10^{11}$ m, find the mass of the Sun.

SOLUTION: The period of the orbital motion of the Earth is 1 year $= 3.156 \times 10^7$ s. Consequently, the mass of the Sun must be

$$M_S = \frac{4\pi^2 r^3}{GT^2}$$

$$= \frac{4\pi^2 \times (1.496 \times 10^{11} \text{ m})^3}{6.673 \times 10^{-11} \text{ N}\cdot\text{m}^2/\text{kg}^2 \times (3.156 \times 10^7 \text{ s})^2}$$

$$= 1.989 \times 10^{30} \text{ kg}$$

COMMENTS AND SUGGESTIONS: A similar method can be used to find the mass of a planet from the observed values of the orbital radius and the period of one of its satellites.

---

The equation of motion for a moon or an artificial satellite in a circular orbit around a planet is analogous to Eq. (15). The planet now plays the role of the central body, and, in Eq. (15), its mass appears in place of the mass of the Sun.

---

EXAMPLE 4. The Early Bird communications satellite is in a circular equatorial orbit around the Earth. The period of the orbit is exactly 1 day so that the satellite always holds a fixed station relative to the rotating Earth. What must be the radius of such a "synchronous" or "geostationary" orbit?

SOLUTION: Since the central body is the Earth, the equation analogous to Eq. (15) is

$$T^2 = \frac{4\pi^2}{GM_E} r^3 \qquad (16)$$

Fig. 9.7 Orbit of a "geostationary" satellite around Earth.

and

$$r = \left(\frac{GM_E T^2}{4\pi^2}\right)^{1/3}$$

$$= \left(\frac{6.67 \times 10^{-11}\ \text{N}\cdot\text{m}^2/\text{kg}^2 \times 5.98 \times 10^{24}\ \text{kg} \times (24 \times 60 \times 60\ \text{s})^2}{4\pi^2}\right)^{1/3}$$

$$= 4.23 \times 10^7\ \text{m}$$

The orbit is shown in Figure 9.7, which is drawn to scale. A number of communication satellites have been placed in this geostationary orbit. These satellites routinely relay radio and TV signals from one continent to another.

## 9.4 Elliptical Orbits; Kepler's Laws

*Kepler's First, Second, and Third Laws*

Although the orbits of the planets around the Sun are approximately circular, none of these orbits are *exactly* circular. The deviations from circularity are most pronounced in the orbits of Mercury, Mars, and Pluto. We will not attempt the general solution of the equation of motion for such noncircular orbits. A complete calculation establishes that with the inverse-square force of Eq. (10), the planetary orbits are ellipses. This is **Kepler's First Law:**

*The orbits of the planets are ellipses with the Sun at one focus.*

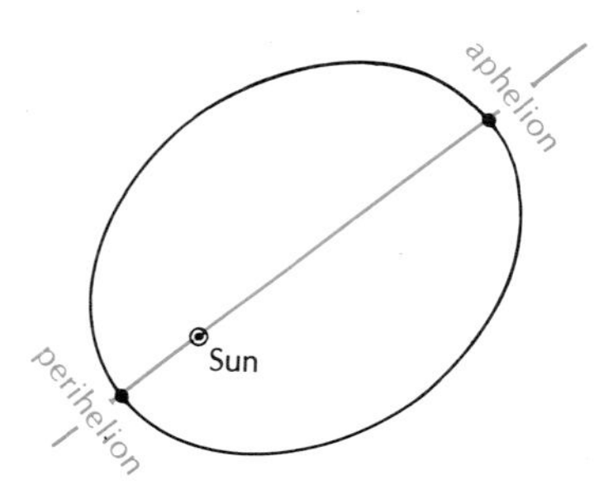


**Fig. 9.8** Elliptical orbit. The Sun is at one focus of the ellipse.

Figure 9.8 shows an elliptical planetary orbit (for the sake of clarity, the elongation of this ellipse has been exaggerated; actual planetary orbits have only very small elongations, or eccentricities). The point closest to the Sun is called the **perihelion;** the point farthest from the Sun is called the **aphelion.** The sum of the perihelion and the aphelion distances is the major axis of the ellipse. The distance from the center of the ellipse to the perihelion (or aphelion) is the semimajor axis; this distance equals the average of the perihelion and aphelion distances.

Kepler originally discovered his First Law and his other two laws (see below) early in the seventeenth century, by direct analysis of the available observational data on planetary motions. Thus, Kepler's laws were originally purely phenomenological statements, that is, they described the phenomenon of planetary motion but did not explain its causes. The explanation only came later, when Newton laid down his laws of motion and his law of universal gravitation and deduced the features of planetary motion from these fundamental laws.

**Kepler's Second Law** describes the variation in the speed of the motion of a planet:

*The radial line segment from the Sun to the planet sweeps out equal areas in equal times.*

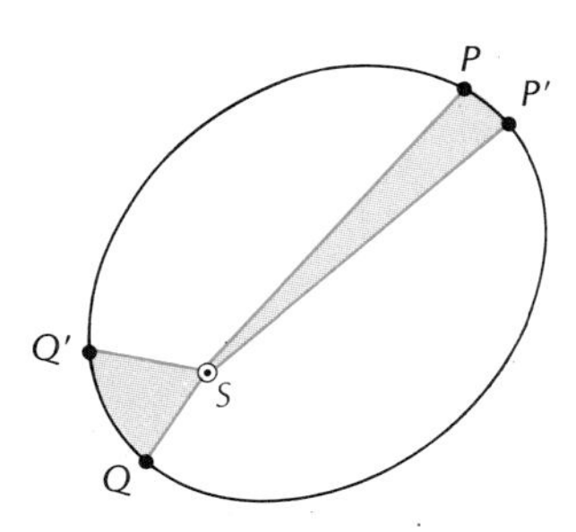


**Fig. 9.9** For equal time intervals, the areas $QQ'S$ and $PP'S$ are equal.

Figure 9.9 illustrates this law. The two colored areas are equal and the planet takes equal times to move from $P$ to $P'$ and from $Q$ to $Q'$. According to Figure 9.9, the speed of the planet is larger when it is near the Sun (at $Q$) than when it is far from the Sun (at $P$).

This Second Law, also called the law of areas, is a direct consequence of the centripetal direction of the gravitational force. We can prove this law by a simple geometrical argument. Consider three successive positions $P$, $Q$, $R$ of the planet along the orbit, separated by relatively small distances. Suppose that the time intervals between $P$, $Q$ and between $Q$, $R$ are equal, say, each interval is 1 second. Fig. 9.10 shows the positions $P$, $Q$, $R$. Between these positions the curved orbit can be approximated by straight line segments $PQ$ and $QR$. Since the time intervals are one unit of time (1 second), the straight line segments $PQ$ and $QR$ represent the average velocities in the two time intervals. The velocities differ because the gravitational force causes an acceleration. However, since the direction of the force is toward the center, parallel to the radius, the component of the velocity perpendicular to the radius cannot change. The component of the velocity perpendicular to the radius is represented by the line segment $PP'$ for the first time interval, and it is represented by $RR'$ for the second time interval. These line segments perpendicular to the radius are, respectively, the heights of the triangles $SQP$ and $SQR$ (see Figure 9.10). Since these heights are equal and since both triangles have the same base $SQ$, their areas must be equal. Thus, the areas swept out by the radial line in the two time intervals must be equal, as asserted by Kepler's Second Law. Note that this geometrical argument depends only on the fact that the force is directed toward a center; it does not depend on the magnitude of the force. This means that Kepler's Second law is valid not only for planetary motion, but also for motion with any kind of central force.

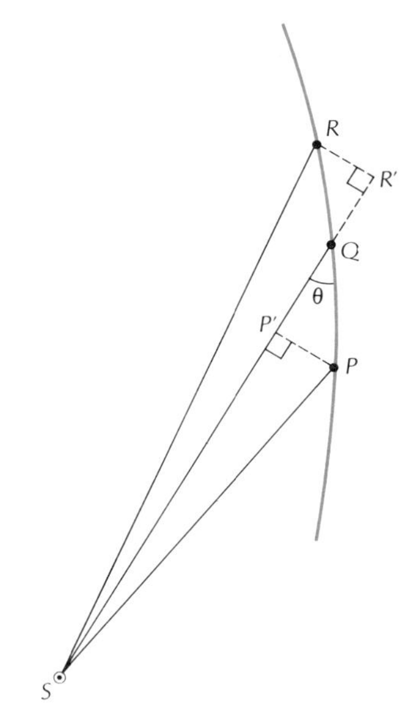


**Fig. 9.10** In one second the planet travels from $P$ to $Q$ and in the next second from $Q$ to $R$. The radial line segment sweeps out the triangular area $SPQ$ in the first second and the triangular area $SQR$ in the next second.

Let us express Kepler's Second Law in another way. The triangle $SPQ$ swept out in 1 second has a base $r$ and a height $PP' = PQ \times \sin\theta$, where $\theta$ is the angle between the velocity and the radial line (see Figure 9.10b). Thus, the area of this triangle is $\frac{1}{2} \times r \times PQ \times \sin\theta$. The line segment $PQ$ represents the magnitude of the velocity, that is, $PQ = v$. The area of the triangle swept out per unit time, or the rate of sweeping of area, is then $\frac{1}{2} \times rv \sin\theta$, and Kepler's Second Law asserts that this rate is constant,

$$\frac{1}{2} \times rv \sin\theta = [\text{constant}] \tag{17}$$

In a later chapter we will become acquainted with the **angular momentum,** which is defined as[2]

$$L = mrv \sin\theta \tag{18}$$

Apart from an irrelevant factor of $\frac{1}{2}$ and an irrelevant factor of $m$, Eq. (17) therefore asserts that the angular momentum of the planet is constant. Thus, Kepler's Second Law is equivalent to a conservation law for angular momentum.

**Johannes Kepler,** *1571–1630, German astronomer, professor at Graz and at Linz, and court mathematician to the Emperor Rudolph II. Kepler relied on the theoretical framework of the Copernican system and he extracted his three laws by a meticulous analysis of the observational data on planetary motions collected by the great Danish astronomer Tycho Brahe.*

---

EXAMPLE 5. The orbit of the Earth around the Sun is an ellipse of very small elongation (small eccentricity). At perihelion, the Earth–Sun distance is $1.47 \times 10^{11}$ m; at aphelion the Earth-Sun distance is $1.52 \times 10^{11}$ m. By what factor is the speed of the Earth at perihelion greater than the speed at aphelion?

[2] To be precise, the angular momentum is a vector, and $mrv \sin\theta$ is the magnitude of this vector.

SOLUTION: At aphelion and at perihelion, the velocity is perpendicular to the radial line, that is, $\theta = 90°$. The conservation of the quantity $\frac{1}{2} \times rv \sin\theta$ therefore implies that

$$r_1 v_1 = r_2 v_2 \tag{19}$$

where the subscripts 1 and 2 refer to perihelion and aphelion, respectively. The ratio of the speeds is

$$\frac{v_1}{v_2} = \frac{r_2}{r_1} = \frac{1.52 \times 10^{11}\ \text{m}}{1.47 \times 10^{11}\ \text{m}} = 1.03$$

which means the speed at perihelion is 3% larger than the speed at aphelion.

**Kepler's Third Law** relates the period of the orbit to the size of the orbit:

> *The square of the period is proportional to the cube of the semimajor axis of the planetary orbit.*

This Third Law, or law of periods, is nothing but the generalization of Eq. (15) to elliptical orbits.

Table 9.1 lists the orbital data on the planets of the Solar System. The mean distance listed in this table is the average of the perihelion and aphelion distances, i.e., it is the semimajor axis of the ellipse. The difference between the perihelion and aphelion distances gives an indication of the elongation of the ellipse. Figure 9.11 shows the orbits of the planets Mercury, Venus, Earth, Mars, Jupiter, and Saturn (the orbits of Uranus, Neptune, and Pluto are considerably larger and would have to be plotted on a separate diagram; see the Prelude).

**Table 9.1** THE PLANETS

| *Planet* | *Mass* | *Mean distance from Sun (semimajor axis)* | *Perihelion distance* | *Aphelion distance* | *Period* |
|---|---|---|---|---|---|
| Mercury | $3.30 \times 10^{23}$ kg | $57.9 \times 10^6$ km | $45.9 \times 10^6$ km | $69.8 \times 10^6$ km | 0.241 year |
| Venus | $4.87 \times 10^{24}$ | 108 | 107 | 109 | 0.615 |
| Earth | $5.98 \times 10^{24}$ | 150 | 147 | 152 | 1.00 |
| Mars | $6.42 \times 10^{23}$ | 228 | 207 | 249 | 1.88 |
| Jupiter | $1.90 \times 10^{27}$ | 778 | 740 | 816 | 11.9 |
| Saturn | $5.67 \times 10^{26}$ | 1430 | 1350 | 1510 | 29.5 |
| Uranus | $8.70 \times 10^{25}$ | 2870 | 2730 | 3010 | 84.0 |
| Neptune | $1.03 \times 10^{26}$ | 4500 | 4460 | 4540 | 165 |
| Pluto | $1.5 \times 10^{22}$ | 5890 | 4410 | 7360 | 248 |

Kepler's three laws apply not only to planets but also to satellites and to comets.[3] For example, Figure 9.12 shows the orbits of the main moons of Jupiter and Figure 9.13 shows the orbits of a few of the many artificial satellites of the Earth. All these orbits are ellipses. Table 9.2 gives the orbital data for the main moons of Jupiter; besides these four moons, Jupiter has another eight, but these other moons are much smaller. Table 9.3 gives the orbital data for the first six artificial

[3] Kepler's laws only apply to *periodic* comets, that is, those that return at regular intervals. Some comets only make a single pass by the Sun and never return; their orbits are parabolas or hyperbolas and neither Kepler's First nor Third Law applies to them.

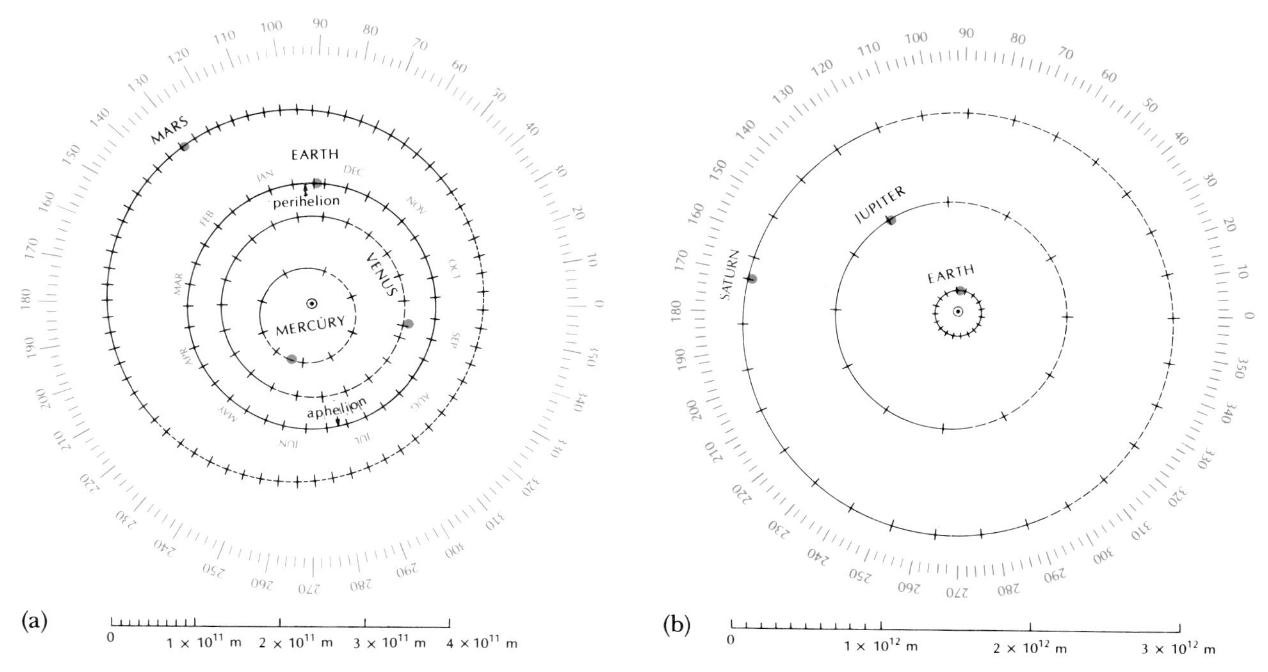


**Fig. 9.11** (a) Orbits of Mercury, Venus, Earth, and Mars. The colored dots indicate the positions of the planets on January 1, 1980. The tick marks indicate the positions at intervals of 10 days. The orbit of the Earth is shown in the plane of the page. The orbits of the other planets are slightly tilted relative to this plane; the portions of the orbits above or below this plane are represented by solid or dashed lines, respectively. (b) Orbits of Jupiter and Saturn. The tick marks indicate the positions at intervals of 1 year.

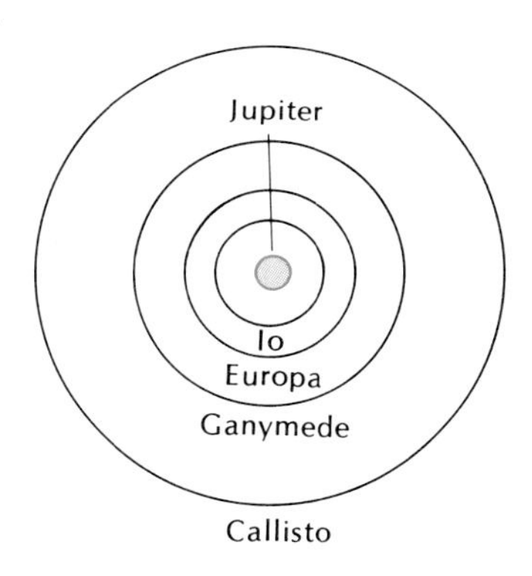


**Fig. 9.12** Orbits of the main moons of Jupiter.

**Table 9.2** THE MAIN MOONS OF JUPITER

| *Moon* | *Mass* | *Mean distance from Jupiter (semimajor axis)* | *Perijove distance* | *Apjove distance* | *Period* |
|---|---|---|---|---|---|
| Io | $8.9 \times 10^{22}$ kg | $422 \times 10^3$ km | $422 \times 10^3$ km | $422 \times 10^3$ km | 1.77 days |
| Europa | $4.8 \times 10^{22}$ | 671 | 671 | 671 | 3.55 |
| Ganymede | $1.5 \times 10^{23}$ | 1070 | 1068 | 1071 | 7.16 |
| Callisto | $1.1 \times 10^{23}$ | 1883 | 1870 | 1896 | 16.69 |

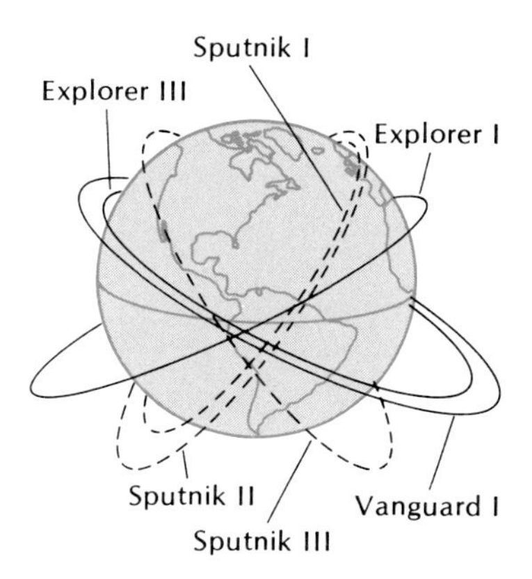


**Fig. 9.13** Orbits of the first artificial satellites of the Earth.

**Table 9.3** THE FIRST ARTIFICIAL SATELLITES OF THE EARTH

| *Satellite* | *Mass* | *Mean distance from center of Earth (semimajor axis)* | *Perigee distance* | *Apogee distance* | *Period* |
|---|---|---|---|---|---|
| Sputnik I | 83 kg | $6.97 \times 10^3$ km | $6.60 \times 10^3$ km | $7.33 \times 10^3$ km | 96.2 min |
| Sputnik II | 3000 | 7.33 | 6.61 | 8.05 | 104 |
| Explorer I | 14 | 7.83 | 6.74 | 8.91 | 115 |
| Vanguard I | 1.5 | 8.68 | 7.02 | 10.3 | 134 |
| Explorer III | 14 | 7.91 | 6.65 | 9.17 | 116 |
| Sputnik III | 1320 | 7.42 | 6.59 | 8.25 | 106 |

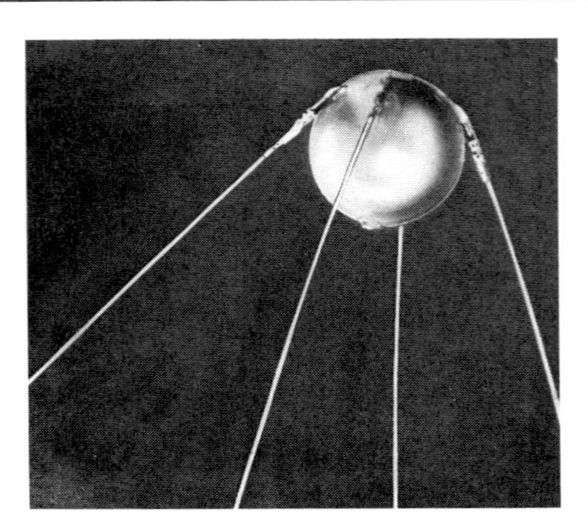

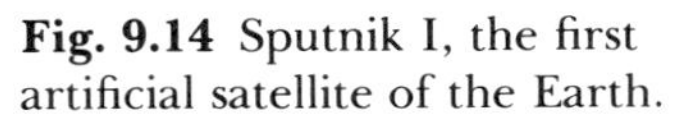

**Fig. 9.14** Sputnik I, the first artificial satellite of the Earth.

**Fig. 9.15** Vanguard I.

satellites of the Earth (Figures 9.14 and 9.15); these satellites were launched in 1957 and 1958 and, except for Vanguard I, they all burned up in the atmosphere after a few months or a few years because they were not sufficiently far from the Earth to avoid the effects of residual atmospheric friction.

---

EXAMPLE 6. An astronaut in a spacecraft is in a circular orbit of radius $9.6 \times 10^3$ km around the Earth. At one point of the orbit, he briefly fires the thrusters of his spacecraft in the forward direction so as to reduce his speed. This places him in a new elliptical orbit with apogee equal to the radius of the old orbit, but with a smaller perigee (Figure 9.16). Suppose that the perigee of the new orbit is $7.0 \times 10^3$ km. Compare the periods of the old and new orbits.

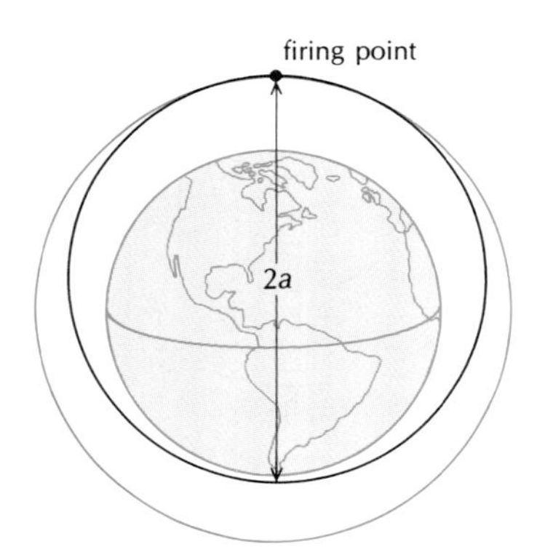


**Fig. 9.16** Old and new orbits of the spacecraft. The old orbit (color) is a circle, the new (black) an ellipse.

SOLUTION: According to Eq. (16), the period of the old, circular orbit is

$$T_{\text{old}} = \sqrt{\frac{4\pi^2}{GM_E} r^3}$$

$$= \sqrt{\frac{4\pi^2 \times (9.6 \times 10^6 \text{ m})^3}{6.67 \times 10^{-11} \text{ N} \cdot \text{m}^2/\text{kg}^2 \times 5.98 \times 10^{24} \text{ kg}}} = 9.4 \times 10^3 \text{ s}$$

According to Kepler's Third Law, the period of the new, elliptical orbit is given by an equation similar to Eq. (16), but with $r$ replaced by the semimajor axis $a$ of the ellipse:

$$T_{\text{new}} = \sqrt{\frac{4\pi^2}{GM_E} a^3}$$

With $a = \frac{1}{2}(9.6 \times 10^3 \text{ km} + 7.0 \times 10^3 \text{ km}) = 8.3 \times 10^3$ km,

$$T_{\text{new}} = \sqrt{\frac{4\pi^2 \times (8.3 \times 10^6 \text{ m})^3}{6.67 \times 10^{-11} \text{ N} \cdot \text{m}^2/\text{kg}^2 \times 5.98 \times 10^{24} \text{ kg}}} = 7.5 \times 10^3 \text{ s}$$

Thus, the period of the new orbit is about 20% shorter than the period of the old orbit — even though the astronaut's maneuver has *reduced* his speed at apogee, he takes a shorter time to complete the orbit! The explanation is, of course, that the maneuver has *increased* his speed at perigee and also has shortened the distance around the orbit.

---

In a slightly generalized form, Kepler's laws also apply to stars orbiting about each other. For example, Figure 9.17 shows the orbits of the two stars in the binary system Krüger 60. The two stars have comparable masses and hence both have comparable accelerations — neither can be regarded as remaining fixed. However, the center of mass of the system can be regarded as remaining fixed. Under the influence of their mutual gravitational attraction, the stars orbit about each other, each star describing an ellipse with the center of mass at the focus.

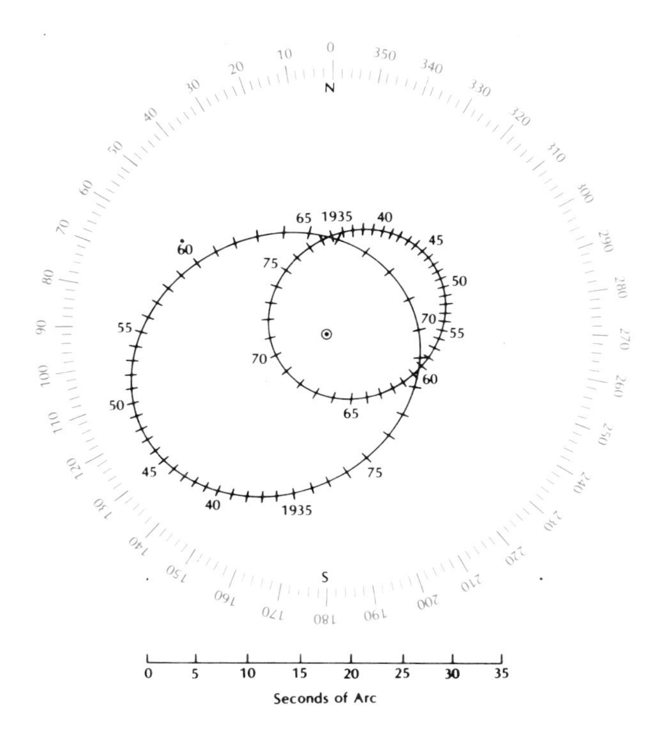


**Fig. 9.17** The orbits of the two stars of the binary system Krüger 60. The center of mass is at the focus of each ellipse.

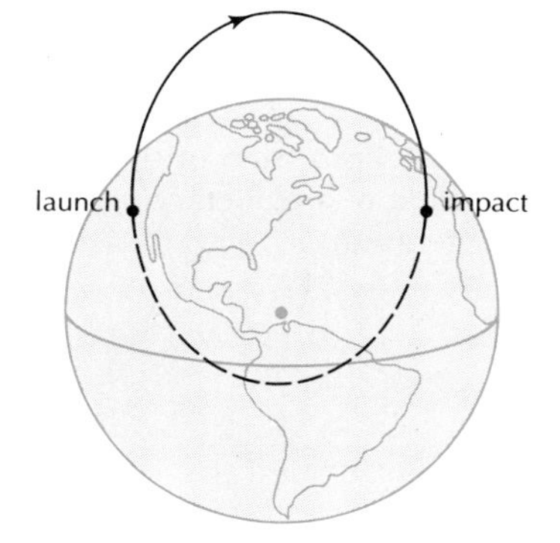


**Fig. 9.18** Orbit of an intercontinental ballistic missile.

Of course, Kepler's laws also apply to the motion of projectiles near the Earth. For instance, Figure 9.18 shows the trajectory of an intercontinental ballistic missile (ICBM). During most of this trajectory, the only force acting on the missile is the gravity of the Earth; the thrust of the engines and the friction of the atmosphere act only during relatively short initial and final segments of the trajectory (on the scale of Figure 9.18 these initial and final segments are too small to show). The trajectory is a portion of an elliptical orbit cut short by impact on the Earth. Likewise, the motion of an ordinary low-altitude projectile, such as a cannon ball, is also a portion of an elliptical orbit (if we ignore atmospheric friction). In Chapter 4 we pretended that gravity was constant in magnitude and in direction; with these approximations, we found that the orbit of a projectile was a parabola. Although the exact orbit of a projectile is an ellipse, the parabola approximates this ellipse quite well over the relatively short distance involved in ordinary projectile motion.

The connection between projectile motion and orbital motion is neatly illustrated by a *Gedankenexperiment*[4] due to Newton: Imagine that we fire a projectile horizontally from a gun emplaced on a high mountain (Figure 9.19). If the muzzle speed is fairly low, the projectile

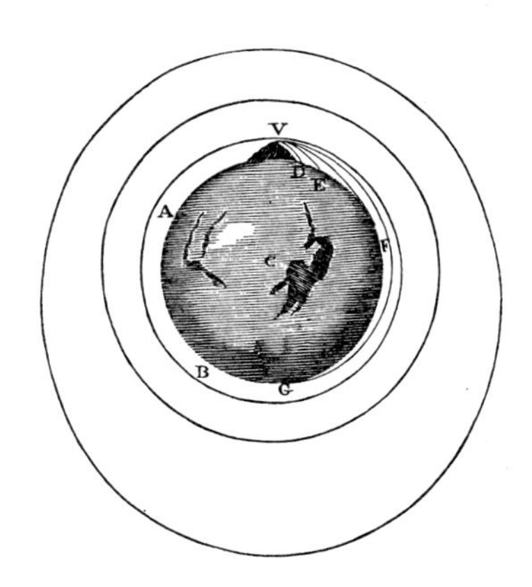


**Fig. 9.19** A *Gedankenexperiment* by Newton. The trajectory of a fast projectile is a circular orbit.

[4] *Gedankenexperiment* is German for "thought experiment," an imaginary experiment that can be done in principle but that has never been done in practice, and whose outcome can be discovered by thought.

will arc toward the Earth and strike near the base of the mountain. The trajectory is a segment of a parabola, or, more precisely, a segment of an ellipse. If we increase the muzzle speed, the projectile will describe larger and larger arcs. Finally, if the muzzle speed is just large enough, the rate at which the trajectory curves downward is precisely matched by the curvature of the surface of the Earth — the projectile never hits the Earth and keeps on falling forever while moving in a circular orbit. This example makes it very clear that orbital motion is free-fall motion.

Note that in our mathematical description of planetary motion we have neglected the gravitational forces that the planets exert on one another. These forces are much smaller than the force exerted by the Sun, but in a precise calculation they must be taken into account. The net force on any one planet is then a function of the position of all the other planets. The solution of the equation of motion involves a **many-body problem:** the motions of all the planets are coupled together, and the calculation of the motion of one planet requires the simultaneous calculation of the motion of all the other planets. No exact solution of this many-body problem exists — it is necessary to have recourse to numerical methods. Thus, Kepler's simple laws describe the planetary motions only in first approximation, and the interplanetary forces generate diverse perturbations in Kepler's laws. The most common such perturbation is a gradual drift of the orientation of the Keplerian elliptical orbit. For example, because of the action of the interplanetary forces, the orientation of the major axis of the elliptical orbit of the Earth drifts 104 seconds of arc in each century (Figure 9.20); most of this perturbation is due to the gravitational force that Jupiter exerts on the Earth.

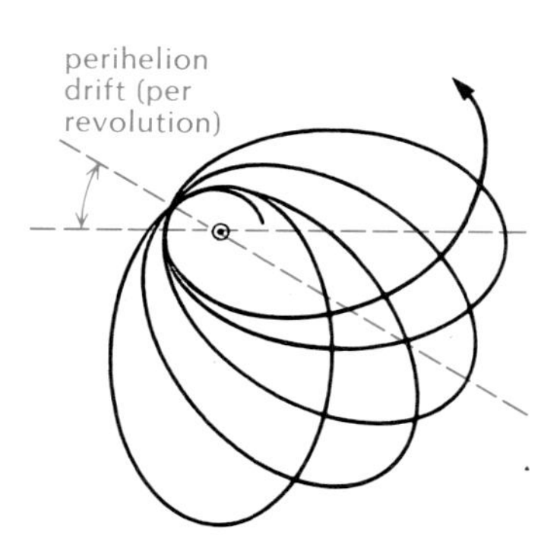


**Fig. 9.20** Precession of the perihelion of a planetary orbit (exaggerated).

## 9.5 Gravitational Potential Energy

The gravitational force is a *conservative* force, that is, the work done by this force on a particle moving from a point $P_1$ to a point $P_2$ depends on the position of these points but not on the shape of the path connecting them. To prove this, consider the gravitational force $GMm/r^2$ that acts between two particles of masses $M$ and $m$. Suppose that $M$ remains stationary at the origin and that $m$ moves. Figure 9.21 shows two alternative paths connecting the initial and final positions of $m$. The work along a path is $\int \mathbf{F} \cdot d\mathbf{r}$. To evaluate this integral, we approximate each path by a series of steplike segments, each step being either a radial line or an arc of a circle (Figure 9.21). The circular segments do not contribute anything to the work; along these segments the work is zero because the force is perpendicular to the displacement. The radial segments do contribute to the work; these segments contribute *equally* for each of the two alternative paths because for every radial segment belonging to the first path there is an equal radial segment belonging to the second path, and the magnitude of the force is exactly the same at equal radial distances from the central mass $M$. Hence the work is completely independent of the shape of the path.

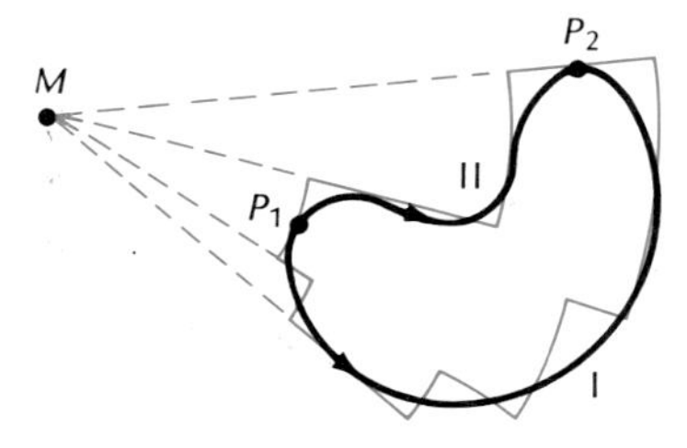


**Fig. 9.21** Two paths (I and II) connecting the points $P_1$ and $P_2$. The paths can be approximated by radial segments and by arcs of circles.

We can then construct the gravitational potential energy by following the recipe of Section 8.2. For the reference point $P_0$ required in this recipe, we take a point at infinite distance[5] from the central mass

[5] The *direction* in which this point $P_0$ lies is irrelevant. All points at infinite distance have the same potential energy.

$M$; for the value of the potential energy at this point, we take $U(P_0) = 0$. Then

$$U(r) = -\int_{\infty}^{r} \mathbf{F} \cdot d\mathbf{r} \tag{20}$$

Any path connecting $\infty$ and $r$ may be used in the evaluation of this integral. The most convenient choice is a straight radial path (Figure 9.22). If we place the $x$ axis along this path, then $\mathbf{F} = -(GMm/x^2)\hat{\mathbf{x}}$ and $d\mathbf{r} = \hat{\mathbf{x}}\, dx$, so that $\mathbf{F} \cdot d\mathbf{r} = -(GMm/x^2)$ and

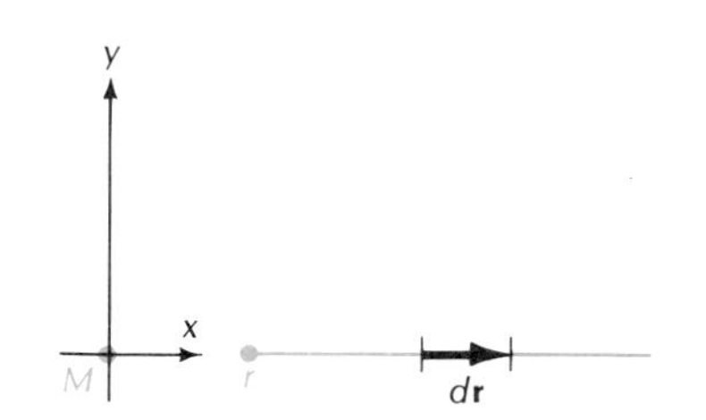


**Fig. 9.22** Path from the point $r$ to $\infty$.

$$U(r) = -\int_{\infty}^{r} -\left(\frac{GMm}{x^2}\right) dx = -\left[\frac{GMm}{x}\right]_{\infty}^{r} \tag{21}$$

or

$$U(r) = -\frac{GMm}{r} \tag{22}$$

Gravitational potential energy

Figure 9.23 is a plot of this potential energy as a function of distance. The potential energy *increases* with distance (it increases from a large negative value toward zero). Such an increase of potential energy with distance is of course characteristic of an attractive force. Note that $U(r)$ is really the *mutual* potential energy of both particles $M$ and $m$. However, if $M$ is a very large mass which does not move (for example, the Sun), then it is convenient to regard $U(r)$ as the potential energy of the mass $m$ which does move.

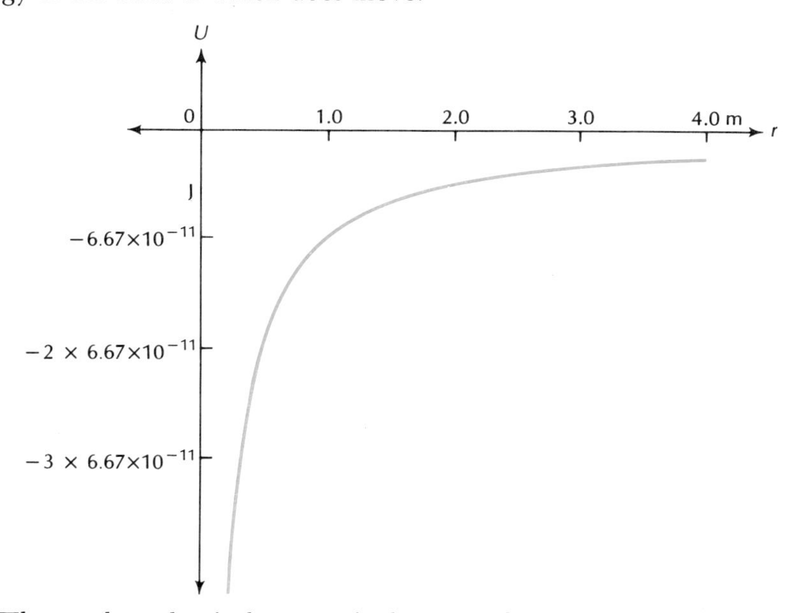


**Fig. 9.23** Potential energy for a particle of 1 kg attracted gravitationally by another particle of 1 kg.

The total mechanical energy is the sum of the potential energy and the kinetic energy. If the mass $M$ is stationary, then the kinetic energy is entirely due to the motion of the mass $m$ and

$$E = K + U = \tfrac{1}{2}mv^2 - \frac{GMm}{r} = [\text{constant}] \tag{23}$$

Conservation of energy

This total energy remains constant during the motion. As in Chapter 8, by examination of the energy, we can draw some general conclusions concerning the motion. Obviously, Eq. (23) implies that if $r$ increases, $v$ must decrease, and conversely.

Let us now investigate the possible orbits around, say, the Sun from the point of view of their energy. For a circular orbit, the orbital speed is [see Eq. (12)]

$$v = \sqrt{GM_S/r} \tag{24}$$

and the kinetic energy is

$$K = \tfrac{1}{2}mv^2 = \frac{GM_S m}{2r} \tag{25}$$

Hence the total energy is

$$E = \tfrac{1}{2}mv^2 - \frac{GM_S m}{r} = \frac{GM_S m}{2r} - \frac{GM_S m}{r} \tag{26}$$

or

*Energy for circular orbit*

$$\boxed{E = -\frac{GM_S m}{2r}} \tag{27}$$

Consequently, the total energy for a circular orbit is negative and is exactly one-half the potential energy.

For an elliptical orbit the total energy is also negative. It can be proven that the energy can still be written in the form of Eq. (27), but the quantity $r$ must be taken equal to the semimajor axis of the ellipse. The total energy does not depend on the shape of the ellipse, but only on its overall size. Figure 9.24 shows several orbits with exactly the same total energy. Incidentally: The angular momenta of these orbits differ; the circular orbit has the highest angular momentum and the very elongated ellipse has the lowest.

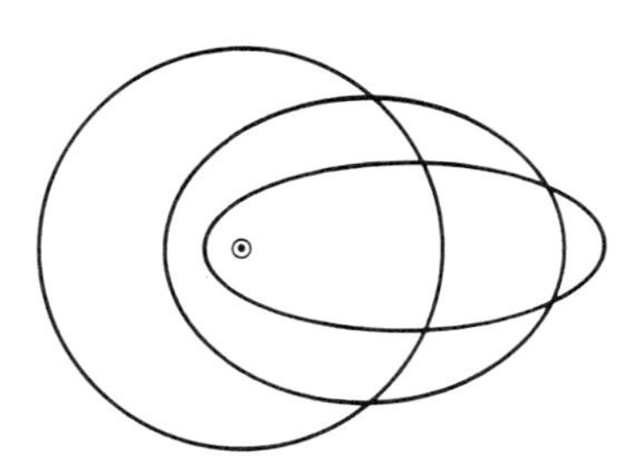

**Fig. 9.24** Orbits of the same total energy. All these orbits have the same semimajor axis.

If the energy is nearly zero, then the size of the orbit is very large. Such orbits are characteristic of comets, many of which have elliptical orbits that extend far beyond the edge of the Solar System (Figure 9.25 and Table 9.4). If the energy is exactly zero, then the "ellipse" extends all the way to infinity and never closes; such an "open ellipse" is actually a parabola (Figure 9.26). Equation (23) indicates that if the energy is zero, the comet will reach infinite distance with zero velocity (if $r = \infty$, then $v = 0$). By considering the reverse of this motion, we recognize that a comet initially at rest at a very large distance from the Sun will fall along this type of parabolic orbit.

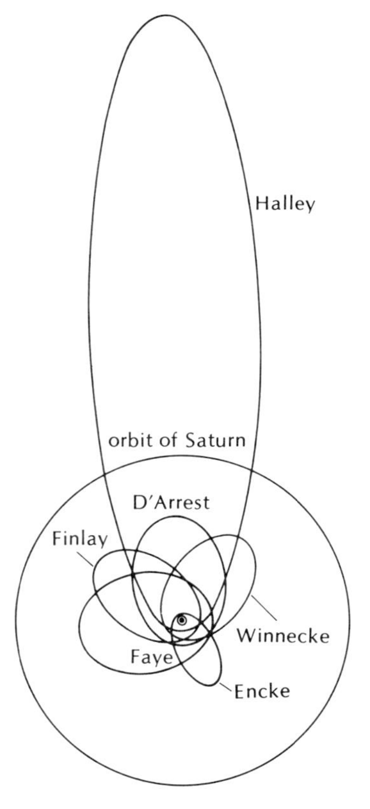


**Fig. 9.25** Orbits of some periodic comets.

**Table 9.4** SOME COMETS

| Comet | Perihelion distance | Aphelion distance | Period |
|---|---|---|---|
| Encke | $50.6 \times 10^6$ km | $612 \times 10^6$ km | 3.3 years |
| Pons–Winnecke | 187 | 838 | 6.34 |
| D'Arrest | 175 | 839 | 6.23 |
| Finlay | 162 | 923 | 6.90 |
| Faye | 242 | 895 | 7.41 |
| Halley | 87.8 | 5280 | 76.1 |

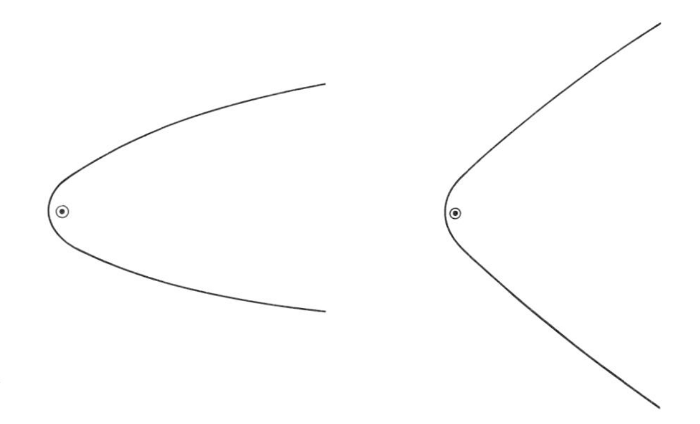

**Fig. 9.26** (left) Orbit of zero energy — a parabola.

**Fig. 9.27** (right) Orbit of positive energy — a hyperbola.

If the energy is positive, then the orbit again extends all the way to infinity and again fails to close; such an open orbit is a hyperbola. The comet will then reach infinite distance with some nonzero velocity and continue moving along a straight line (Figure 9.27).

EXAMPLE 7. A meteoroid (a chunk of rock) is initially at rest in interplanetary space at a large distance from the Sun. Under the influence of gravity, the meteoroid begins to fall toward the Sun along a straight radial line. With what speed does it strike the Sun?

SOLUTION: The energy of the meteoroid is

$$E = \tfrac{1}{2}mv^2 - \frac{GM_S m}{r} = [\text{constant}]$$

Initially, both the kinetic and potential energies are zero ($v = 0$ and $r \cong \infty$). Hence at any later time

$$\tfrac{1}{2}mv^2 - \frac{GM_S m}{r} = 0$$

With $r = R_s$, this leads to the following formula for the speed at the moment of impact:

$$\boxed{v = \sqrt{2GM_S/R_S}} \qquad (28)$$

*Escape velocity*

Since the solar radius is $R_S = 6.96 \times 10^8$ m, we find

$$v = \sqrt{2 \times 6.67 \times 10^{-11}\ \text{N} \cdot \text{m}^2/\text{kg}^2 \times 1.99 \times 10^{30}\ \text{kg}/6.96 \times 10^8\ \text{m}}$$

$$= 6.18 \times 10^5\ \text{m/s} = 618\ \text{km/s} \qquad (29)$$

The quantity given by Eq. (28) is called the **escape velocity,** because it is the minimum initial velocity with which a body must be launched

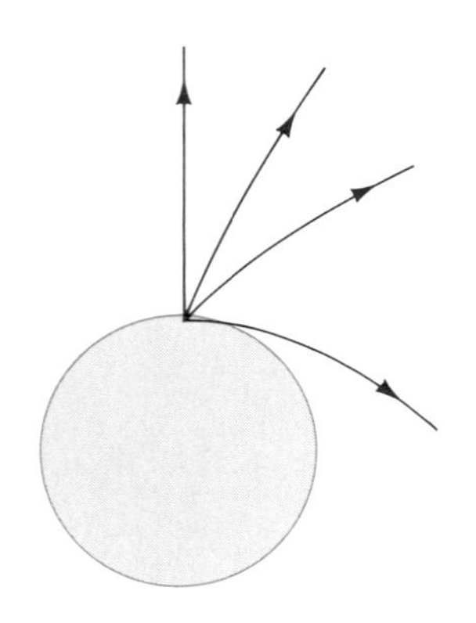

**Fig. 9.28** Different parabolic orbits with the same starting point and initial speed.

upward from the surface of the Sun if it is to escape and never fall back. We can recognize this by looking at the motion of the meteroid in Example 7 in reverse: it starts with a velocity of 618 km/s at the surface of the Sun and gradually slows as it rises, but it never quite stops until it reaches a very large distance ($r \cong \infty$).

Note that the direction in which an escaping body is launched is immaterial — the body will succeed in its escape whenever the direction of launch is above the horizon. Of course, the escape route that the body takes will depend on the direction of launch (Figure 9.28).

The escape velocity for a body launched from the surface of the Earth can be calculated from a formula analogous to Eq. (28), provided that we ignore atmospheric friction and the pull of the Sun on the body. Atmospheric friction will be absent if we launch the body from just above the atmosphere, and the pull of the Sun has only a small effect on the escape from the Earth if we contemplate a body that "escapes" to a distance of, say, $r = 100R_E$ or $200R_E$ rather than $r = \infty$. For such a body the escape velocity is approximately $\sqrt{2GM_E/R_E} = 11.2$ km/s.

---

EXAMPLE 8. From the data given in Table 9.1, calculate the speed of Mercury at perihelion and at aphelion.

SOLUTION: We designate the distances and speeds at perihelion and at aphelion by $r_1$, $v_1$, $r_2$, and $v_2$, respectively. According to Eq. (19),

$$r_1 v_1 = r_2 v_2 \tag{30}$$

Furthermore, by conservation of energy,

$$\tfrac{1}{2}mv_1^2 - \frac{GM_S m}{r_1} = \tfrac{1}{2}mv_2^2 - \frac{GM_S m}{r_2} \tag{31}$$

Obviously, the mass $m$ cancels in this equation. From Eq. (30), $v_2 = v_1 r_1/r_2$, which when substituted into Eq. (31) yields

$$\tfrac{1}{2}v_1^2 - \frac{GM_S}{r_1} = \frac{1}{2}\left(\frac{v_1 r_1}{r_2}\right)^2 - \frac{GM_S}{r_2}$$

or

$$v_1^2 = \frac{2GM_S(1/r_1 - 1/r_2)}{1 - r_1^2/r_2^2} \tag{32}$$

This can be simplified to read

$$v_1^2 = 2GM_S \frac{r_2}{r_1(r_1 + r_2)} \tag{33}$$

so that the perihelion speed is

$$v_1 = \sqrt{2 \times 6.67 \times 10^{-11}\ \text{N}\cdot\text{m}^2/\text{kg}^2 \times 1.99 \times 10^{30}\ \text{kg} \times \frac{69.8}{45.9 \times (69.8 + 45.9) \times 10^9\ \text{m}}}$$

$$= 5.91 \times 10^4\ \text{m/s}$$

The aphelion speed is then

$$v_2 = v_1 \frac{r_1}{r_2} = 5.91 \times 10^4\ \text{m/s} \times \frac{45.9}{69.8}$$

$$= 3.88 \times 10^4\ \text{m/s}$$

COMMENTS AND SUGGESTIONS: This method can, of course, also be applied to the calculation of the speeds at perigee and apogee of a satellite orbiting the Earth, or a satellite orbiting some other planet.

Finally, let us examine the gravitational potential energy of a particle in the vicinity of the Earth. According to Eq. (22),

$$U(r) = -\frac{GM_E m}{r} \tag{34}$$

The *change* in the potential energy between the point $r$ and a point on the surface of the Earth is then

$$\Delta U = U(r) - U(R_E) = -\frac{GM_E m}{r} + \frac{GM_E m}{R_E}$$

$$= GM_E m \frac{r - R_E}{rR_E} \tag{35}$$

If the point $r$ is near the surface of the Earth so that $r \cong R_E$, then we can approximate the product $rR_E$ by $R_E^2$. Furthermore, the difference $r - R_E$ is simply the height $z$ above the surface so

$$\Delta U = \frac{GM_E m}{R_E^2} z \tag{36}$$

But by Eq. (6), $GM_E/R_E^2 = g$ and hence

$$\Delta U = mgz \tag{37}$$

This, of course, is our old expression for the potential energy of gravity near the surface of the Earth (see Section 7.4). Consequently, the old expression is an approximation to the exact potential energy; this approximation is valid if the height $z$ is much smaller than the radius of the Earth ($z \ll R_E$).

## 9.6* Newton's Theorem

In Section 9.1 we claimed that the gravitational attraction exerted by a spherical mass distribution is the same as if all the mass were concentrated at the center. This famous theorem is due to Newton. The theorem is valid as long as the attracted body is *outside* of the spherical mass distribution. In what follows, we will prove this theorem and also investigate what happens if the attracted body is inside the mass distribution.

A spherical mass distribution can be regarded as constructed of a collection of thin spherical shells, each one nested inside the other, like the layers of an onion. We will begin with a calculation of the gravitational attraction that such a shell exerts on a particle outside it. Figure 9.29 shows the shell, of mass $M$ and radius $R$, and a particle of mass $m$ at a distance $r$ from the center of the shell. The net gravitational force exerted by the shell on the particle $m$ is the vector sum of all the forces

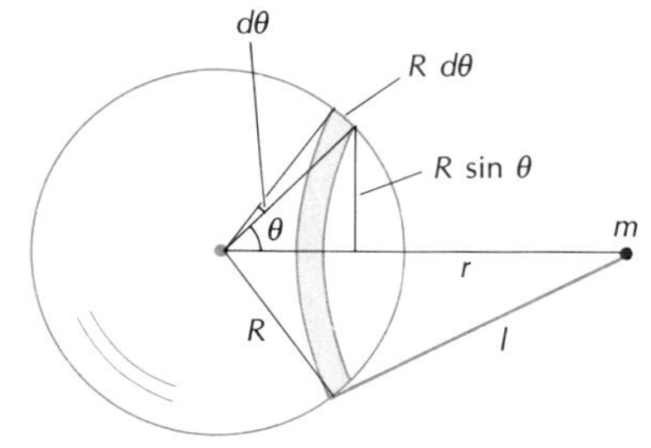


**Fig. 9.29** A thin spherical shell. The particle $m$ is outside the shell.

* This section is optional.

exerted by all the particles of the shell. This vector sum is somewhat messy to evaluate directly. It is easier first to evaluate the net potential energy; the force can be calculated afterward from the potential energy by taking derivatives [see Eqs. (8.20)–(8.22)].

The net potential energy between the shell and the particle $m$ is the sum of all the individual potential energies between the particles in the shell and the particle $m$. If we divide the shell into small mass elements $\Delta M_i$, then this net potential energy is approximately

$$U = \sum_i \left( -\frac{Gm\,\Delta M_i}{r_i} \right) \tag{38}$$

where $r_i$ is the distance between $\Delta M_i$ and $m$. To obtain the potential energy exactly, it is necessary to proceed to the limit $\Delta M_i \to 0$ so that the sum becomes an integral. For the mass element $\Delta M_i$ it is convenient to take a ring cut from the shell as in Figure 9.29; obviously, the entire shell can be constructed out of such rings. The ring shown in Figure 9.29 is at a distance $r_i = l$ from the particle $m$. The ring has a width $R\,d\theta$, a radius $R \sin\theta$, and a circumference $2\pi R \sin\theta$; hence the surface area of the ring is $2\pi R^2 \sin\theta\,d\theta$. The mass of the ring is proportional to this surface area. Since the total mass $M$ is uniformly distributed over the total area $4\pi R^2$ of the shell, the proportionality between mass and area yields

$$\Delta M_i = M \times \frac{2\pi R^2 \sin\theta\,d\theta}{4\pi R^2} = \tfrac{1}{2} M \sin\theta\,d\theta$$

for the mass of the ring. Inserting this in Eq. (38) and proceeding to the limit $\Delta M_i \to 0$, we obtain the integral

$$U = -\int \frac{GmM}{2} \frac{\sin\theta\,d\theta}{l} \tag{39}$$

If we apply the law of cosines to the triangle with sides $R$, $r$, and $l$, we find

$$l^2 = R^2 + r^2 - 2rR\cos\theta \tag{40}$$

Let us take the differential of this, remembering that $r$ and $R$ are constants:

$$2l\,dl = 2rR \sin\theta\,d\theta$$

or

$$\frac{\sin\theta\,d\theta}{l} = \frac{dl}{rR}$$

We can now substitute this expression into Eq. (39) and obtain

$$U = -\int \frac{GmM}{2} \frac{1}{rR}\,dl = -\frac{GmM}{2rR} \int dl$$

The largest value of $l$ is $r + R$ and the smallest is $r - R$. Hence

$$U = -\frac{GmM}{2rR}\int_{r-R}^{r+R} dl = -\frac{GmM}{2rR}\left[l\right]_{r-R}^{r+R} \tag{41}$$

$$= -\frac{GmM}{2rR}(2R)$$

or

$$U = -\frac{GmM}{r} \tag{42}$$

Comparing this with Eq. (22), we see that the potential energy of the shell behaves exactly as though all of the mass of the shell were at its center. Thus, the force between the shell and the particle must also behave exactly as though all the mass were at its center.

A spherical mass distribution is nothing but a collection of shells. Since each shell acts as though its mass were at its center, the complete spherical mass distribution will also act as though its mass were at its center. Note that this conclusion remains unchanged if the density of the mass distribution is some function of radius (as it is in the case of the Earth, where the inner layers are much denser than the outer).

Furthermore, the spherical mass distribution acts as though its mass were at its center, not only in regard to the force it exerts on a particle, but also in regard to the force it feels from this particle. We can deduce that this is so from the equality of action and reaction (Newton's Third Law): since the force felt by the particle acts as though all the mass of the spherical distribution were at its center, the equal and opposite force felt by the spherical mass distribution must act likewise. Thus, it follows that the mutual gravitational forces exerted and experienced by the (nearly) spherical planets are as though the mass of each planet were concentrated in a point at its center.

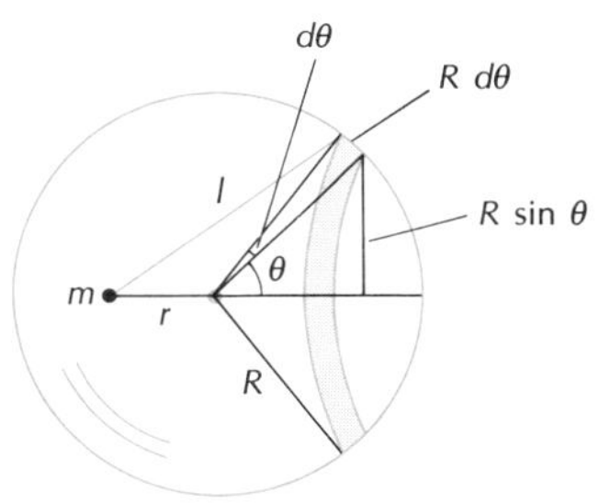

In the above we have assumed that the particle that experiences the force is outside of the spherical mass distribution. To see what happens if the particle is inside, we begin by calculating the gravitational attraction that a shell exerts on a particle inside it (Figure 9.30). The calculation proceeds exactly as above. The only change is in Eq. (41): the largest value of $l$ is now $R + r$ and the smallest is $R - r$. With these new limits of integration, Eq. (41) becomes

**Fig. 9.30** A thin spherical shell. The particle $m$ is inside the shell.

$$U = -\frac{GmM}{2rR}\left[l\right]_{R-r}^{R+r} = -\frac{GmM}{2rR}(2r)$$

or

$$U = -\frac{GMm}{R} \tag{43}$$

This means that inside the shell, the potential energy of the particle $m$ is *independent* of the position $r$. But a constant potential energy implies the absence of any force. We have therefore obtained the surprising result that a spherical shell exerts *no gravitational force* on a particle inside it.

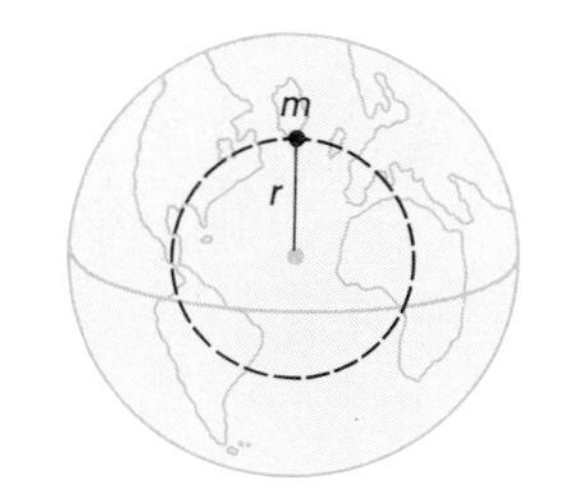


**Fig. 9.31** The particle $m$ is inside the Earth.

For a spherical mass distribution this has the following consequence: consider a particle within the mass distribution; for example, consider a particle inside a mine shaft dug in the Earth (Figure 9.31). All the spherical shells of radius larger than the radius at which the particle is located do not exert any force. Thus the force is entirely due to the mass contained in a radius smaller than the radius at which the particle is located; this mass exerts a force just as though it were concentrated at the center. The force may then be written as

*Gravitational force on a particle inside a spherical mass*

$$F = \frac{GmM(r)}{r^2} \tag{44}$$

where $M(r)$ is the amount of mass contained within the radius $r$.

EXAMPLE 9. A uniform sphere has mass $M$ and radius $R$. Find the gravitational force on a particle of mass $m$ at radius $r < R$.

SOLUTION: The mass contained within the radius $r$ is directly proportional to the volume $4\pi r^3/3$. The total mass $M$ is distributed over a volume $4\pi R^3/3$. Hence the proportionality between mass and volume implies

$$M(r) = M\frac{4\pi r^3/3}{4\pi R^3/3} = \frac{Mr^3}{R^3}$$

and

$$F = \frac{Gm}{r^2}\frac{Mr^3}{R^3} = \frac{GmM}{R^3}r \tag{45}$$

Hence the force *increases* in direct proportion to the radius. Of course, at $r = R$, the force ceases to increase and begins to decrease as $1/r^2$ (Figure 9.32).

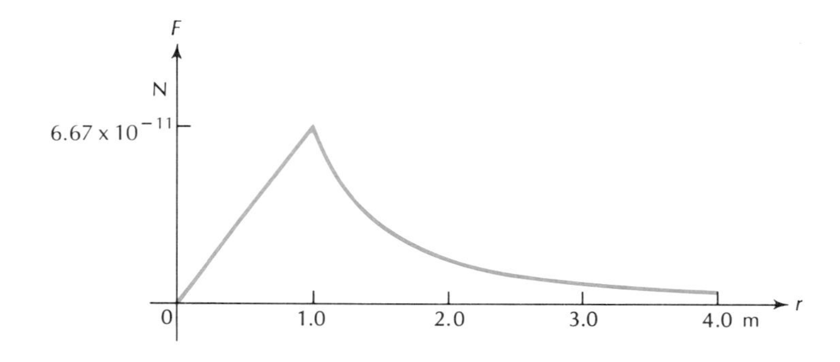


**Fig. 9.32** Magnitude of the gravitational force exerted by a uniform sphere of mass 1 kg and radius 1 m on a particle of mass 1 kg.

## 9.7* Inertial and Gravitational Mass; the Principle of Equivalence

The definition of mass given in Section 5.2 relies on a comparison of the accelerations of two interacting masses, the unknown mass and the standard mass. In brief, the unknown mass and the standard mass are

* This section is optional.

made to exert forces on each other, and the ratio of the consequent accelerations then gives the inverse ratio of the masses [see Eq. (5.2)]. According to this definition, mass is a measure of inertia, i.e., it is a measure of the opposition that a body offers to any attempts at changing its state of motion. Hence the mass measured in this way is often called the **inertial mass.**

*Inertial mass*

However, in everyday practice mass is measured by means of a balance which compares the weights of two masses rather than their inertia. Figure 5.6 shows a beam balance with two equal arms. The unknown mass is placed in one pan of the balance and the standard mass, or a suitable multiple or submultiple of the standard mass, is placed in the other pan. The balance will be in equilibrium if the weights on the two pans are equal. Thus, the balance compares the gravitational force that the Earth exerts on the masses. The mass measured in this way is called the **gravitational mass.**

*Gravitational mass*

It is of fundamental importance to verify that the two methods for the measurement of mass agree, that is, that the inertial mass and the gravitational mass of any body are equal. For the standard mass (standard kilogram) this equality of inertial and gravitational masses holds true by hypothesis, but for any other body the equality must be tested by experiment.

To see what this involves, suppose that two bodies have inertial masses $m$ and $m'$ and weights $w$ and $w'$, respectively. Suppose further that, as tested by a balance, the weights are exactly the same,

$$w = w' \tag{46}$$

With $w = mg$ and $w' = m'g$, this equation becomes

$$mg = m'g \tag{47}$$

Upon cancellation of the factor of $g$ on both sides, we then find that

$$m = m' \tag{48}$$

that is, the inertial masses are also exactly the same. Thus the weights, or gravitational masses, of the two bodies are the same if and only if their inertial masses are the same. If one of the bodies is the standard of mass or a multiple or submultiple of the standard, which has equal gravitational and inertial masses by hypothesis, then it follows that the other body also has equal inertial and gravitational masses.

However, this argument hinges on an implicit assumption: the factors of $g$ on both sides of Eq. (47) can be canceled if and only if these factors are the same, that is, if and only if both masses have exactly the same acceleration of free fall. Hence the equality of inertial and gravitational masses will hold true for a given body if and only if this body has the same acceleration of free fall as the standard mass. The experimental question that must be answered is then this: Are the rates of free fall of different kinds of bodies really exactly the same?

The first experiments that sought to answer this question were performed by Galileo and by Newton. A series of much more precise and much more complete experiments were performed early in this century by Lorand von Eötvös, who compared the rates of free fall of samples of platinum, copper, water, copper sulfate, asbestos, etc., and found that all these bodies fall at the same rate. The most recent and most precise versions of these experiments have verified that the rates of free fall of bodies made of different substances are equal to within better than 1 part in $10^{12}$ (see Section IX.1 for some details on these experiments). Hence the inertial and gravitational masses of a body are equal to within better than 1 part in $10^{12}$ and they are probably exactly equal.

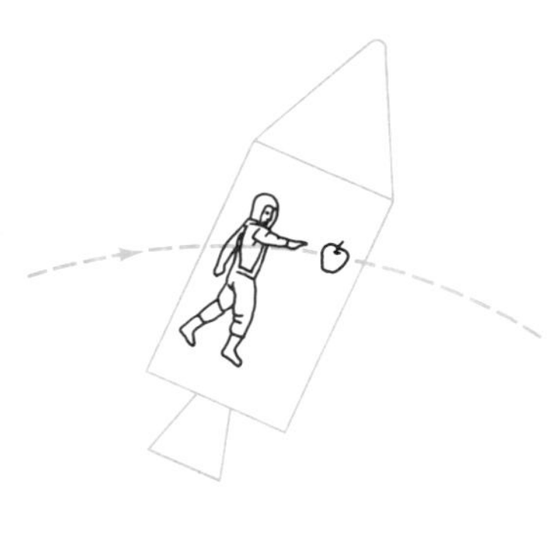

**Fig. 9.33** Astronaut and apple in orbit.

As we already noted in Section 6.2, the universality of the rate of free fall leads to the phenomenon of weightlessness for a freely falling observer. Consider, for example, an astronaut orbiting the Earth in a spacecraft, such as Skylab. The astronaut, and the spacecraft, and any other free body within or near the spacecraft are all in free fall — they accelerate at exactly the same rate and have no acceleration relative to one another. Any body the astronaut releases from his fingers will simply float in midair (Figure 9.33). Hence, in the accelerated reference frame of the astronaut, a condition of apparent weightlessness prevails: bodies behave as though they were at some remote place of the universe, far from the gravity of Earth, Sun, or other stars (Figure 9.34). This shows that a suitably accelerated motion of the reference frame can cancel the effects of gravity.

**Fig. 9.34** Astronauts in Skylab trying to capture floating, weightless trash bags.

More generally, an accelerated motion of the reference frame can either decrease or increase the apparent gravity (Figure 9.35). For example, if a spacecraft far from the Earth, Sun, or other stars accelerates in some direction with a rate of, say, 9.8 m/s$^2$ and an astronaut within this spacecraft releases an apple from his fingers, then the apple will accelerate in the backward direction relative to the astronaut. Hence in the reference frame of the astronaut, the apple behaves exactly as though it were under the influence of ordinary gravity — the accelerated motion of the reference frame simulates the effects of gravity.

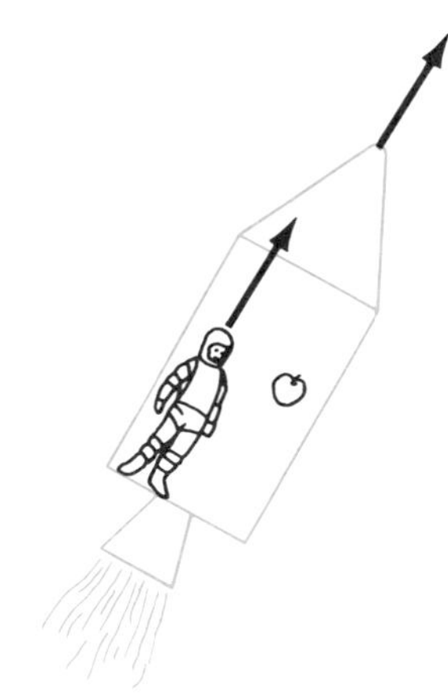

**Fig. 9.35** Astronaut and apple in accelerating spacecraft. Relative to the astronaut, the apple accelerates backward.

*Principle of Equivalence*

The similarity between the effects of gravity and the effects of a suitably accelerated motion of the frame of reference is called the **Principle of Equivalence.** Einstein took this principle as a starting point for his theory of General Relativity (see Interlude IX).

Incidentally: The cancellation of gravity by means of accelerated motion plays an important role in our lives. The Earth, and everything on it, is subjected to the gravitational pull of the Sun, Moon, and other celestial bodies. Yet we do not feel this pull — the Earth is in an accelerated free-fall motion toward these bodies, and bodies on the Earth behave as weightless in regard to the gravitational pull of the celestial bodies.

However, this cancellation of gravitational effects by the free-fall motion is not quite perfect. The **tides** observed on the oceans of the Earth constitute a small residual effect which the free-fall motion fails to cancel (Figure 9.36). For example, the lunar tides arise because the water nearest the Moon experiences a gravitational pull which is a bit too strong, and the water farthest from the Moon a gravitational pull which is a bit too weak, to match the free-fall motion of the Earth's center. Thus the water on the near side bulges toward the Moon and the water on the far side bulges away from the Moon. If we neglect the orbital motion of the Moon, we see that the locations of the tidal bulges remain more or less fixed in space while the Earth, and its oceans, turns relative to them once per day. This gives rise to two high lunar tides and two low lunar tides per day (Figure 9.37). The solar tides involve the same mechanism, but with the Sun as the attracting body.

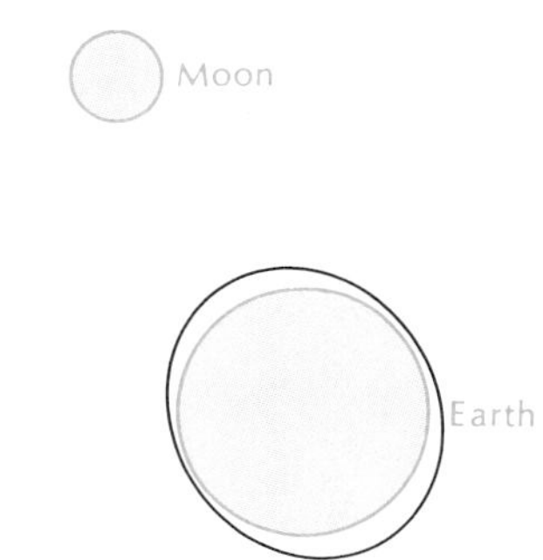


**Fig. 9.36** Tidal bulges on the Earth, generated by the pull of the Moon.

(a)

(b)

**Fig. 9.37** (a) High and (b) low tide in the Bay of Fundy.

The generation of gravity, or pseudo-gravity, by means of acceleration will play an important role in the design of the space stations of the future. For example, Figure 9.38 shows a proposed space station in the shape of a large spinning wheel. Each point on the rim of the wheel has a centripetal acceleration and hence at each such point the apparent gravity is along the radial direction. For the inhabitants of the space station, "up" is toward the center of the wheel and "down" is radially outward. By adjusting the rate of spin of the wheel, the strength of the apparent gravity can be made equal to that of the normal gravity found on Earth.

**Fig. 9.38** A rotating space station.

## SUMMARY

**Law of universal gravitation:** $F = \dfrac{GMm}{r^2}$

$$G = 6.67 \times 10^{-11}\ \mathrm{N \cdot m^2/kg^2}$$

**Acceleration of gravity on Earth:** $g = \dfrac{GM_E}{R_E^2}$

**Circular orbit around Sun:** $v^2 = \dfrac{GM_S}{r}$

**Kepler's First Law:** The orbits of the planets are ellipses with the Sun at one focus.

**Kepler's Second Law:** The radial line segment from the Sun to the planet sweeps out equal areas in equal times.

**Kepler's Third Law:** The square of the period is proportional to the cube of the semimajor axis of the planetary orbit.

**Gravitational potential energy:** $U = -\dfrac{GMm}{r}$

**Energy for circular orbit around the Sun:** $E = -\dfrac{GM_S m}{2r}$

**Escape velocity for Earth:** $v = \sqrt{2GM_E/R_E}$

## QUESTIONS

1. Can you directly feel the gravitational pull of the Earth with your sense organs? (Hint: Would you feel anything if you were in free fall?)

2. According to a tale told by Professor R. Lichtenstein, some apple trees growing in the mountains of Tibet produce apples of negative mass. In what direction would such an apple fall if it fell off its tree? How would such an apple hang on the tree?

3. Eclipses of the Moon can occur only at full Moon. Eclipses of the Sun can occur only at new Moon. Why?

4. Explain why the sidereal day (the time of rotation of the Earth relative to the stars, or 23 h 56 min 4 s) is shorter than the mean solar day (the time between successive passages of the Sun over a given meridian, or 24 h). (Hint: The rotation of the Earth around its axis and the revolution of the Earth around the Sun are in the same direction.)

5. Communications satellites are usually placed in an orbit or radius of $4.2 \times 10^7$ m so that they remain stationary above a point on the Earth's equator (see Example 4). Can we place a satellite in an orbit so that it remains stationary above the North Pole?

6. When an artificial satellite — such as the ill-fated Skylab — experiences friction against the residual atmosphere of the Earth, the radius of the orbit decreases while at the same time the speed of the satellite *increases.* Explain.

7. Suppose that an airplane flies around the Earth along the equator. If this airplane flies *very* fast, it would not need wings to support itself. Why not?

8. The mass of Pluto was not known until 1978 when a moon of Pluto was finally discovered. How did the discovery of this moon help?

9. It is easier to launch an Earth satellite into an eastward orbit than into a westward orbit. Why?

10. Would it be advantageous to launch rockets into space from a pad at very high altitude on a mountain? Why has this not been done?

11. Describe how you could play squash on a small, round asteroid (with no front wall). What rules of the game would you want to lay down?

12. According to an NBC news report of April 5, 1983, a communications satellite launched from the space shuttle went into an orbit as shown in Figure 9.39. Is this believable?

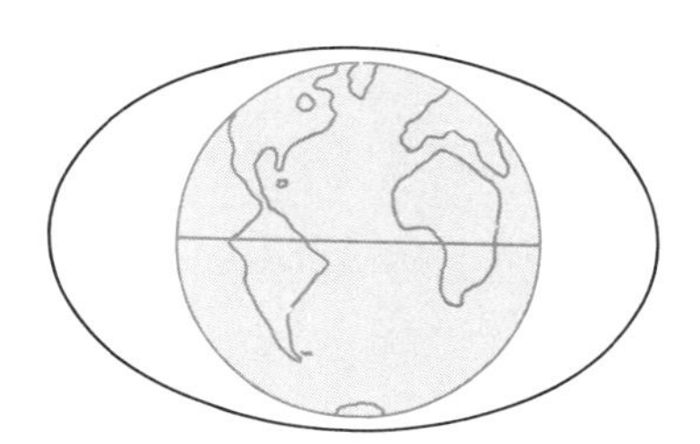

**Fig. 9.39** Communications satellite orbit.

13. Does the radial line from the Sun to Mars sweep out area at the same rate as the radial line from the Sun to the Earth?

14. Why were the Apollo astronauts able to jump much higher on the Moon than on Earth (Figure 9.40)? If they had landed on a small asteroid, could they have launched themselves into a parabolic or hyperbolic orbit by a jump?

**Fig. 9.40** The jump of the astronaut.

15. The Earth reaches perihelion on January 3 and aphelion on July 6. Why is it not warmer in January than in July?

16. When the Apollo astronauts were orbiting around the Moon at low altitude, they detected several mass concentrations ("mascons") below the lunar surface. What is the effect of a mascon on the orbital motion?

17. An astronaut in a circular orbit above the Earth wants to take his spacecraft into a new circular orbit of larger radius. Give him instructions on how to do this.

18. A Japanese and an American astronaut are in two separate spacecrafts in the same circular orbit around the Earth. The Japanese is slightly behind the American and he wants to overtake him. The Japanese fires his thrusters in the *forward* direction, braking for a brief instant. This changes his orbit into an ellipse. One orbital period later, the astronauts return to the vicinity of their initial positions, but the Japanese is now ahead of the American. He then fires his thrusters in the *backward* direction. This restores his orbit to the original circle. Carefully explain the steps of this maneuver, drawing diagrams of the orbits.

19. The gravitational force that a hollow spherical shell of mass exerts on a particle in its interior is zero. Does this mean that such a shell acts as a gravity shield?

20. Consider an astronaut launched in a rocket from the surface of the Earth and then placed in a circular orbit around the Earth. Describe the astronaut's weight (measured in an inertial reference frame) at different times during this trip. Describe the astronaut's *apparent* weight (measured in his own reference frame) at different times.

21. Several of our astronauts suffered severe motion sickness while under conditions of apparent weightlessness aboard Skylab. Since the astronauts were not being tossed about (as in an airplane or a ship in a storm), what caused this motion sickness? What other difficulties does an astronaut face in daily life under conditions of weightlessness?

22. An astronaut on Skylab lights a candle. Will the candle burn like a candle on Earth?

23. Astrology is an ancient superstition according to which the planets influence phenomena on the Earth. The only force that can reach over the large distances between the planets and act on pieces of matter on the Earth is gravitation (planets do not have electric charge and they therefore do not exert electric forces; some planets do have magnetism, but their magnetic forces are too weak to reach the Earth). Given that the Earth is in free fall under the action of the net gravitational force of the planets and the Sun, is there any way

that the gravitational forces of the planets can affect what happens on the Earth?

24. The Sun has a much larger mass than the Moon, and yet the tides generated by the Sun are only about half as large as the tides generated by the Moon. Does this make sense?

25. The two daily lunar tides are usually not of equal height. Why not?

26. The Moon always shows the same face to the Earth, that is, its period of rotation and its period of revolution about the Earth coincide. Can you guess the reason for this coincidence?

27. Extremely high tides ("springs") occur when the Moon is full and when the Moon is new. Why?

## PROBLEMS

### Section 9.1

1. Two supertankers, each with a mass of 700,000 metric tons, are separated by a distance of 2.0 km. What is the gravitational force that each exerts on the other? Treat them as particles.

2. What is the gravitational force between two protons separated by a distance equal to their diameter, $2.0 \times 10^{-15}$ m?

3. Somewhere between the Earth and the Moon there is a point where the gravitational pull of the Earth on a particle exactly balances that of the Moon. At what distance from the Earth is this point?

4. Calculate the value of the acceleration of gravity at the surfaces of Venus, Mercury, and Mars. Use the data on planetary masses and radii given in the table printed on the endpapers.

5. The asteroid Ceres has a diameter of 1100 km and a mass of (approximately) $7 \times 10^{20}$ kg. What is the value of the acceleration of gravity at its surface? What would be your weight (in pounds-force) if you were to stand on this asteroid?

6. What is the magnitude of the gravitational attraction the Sun exerts on the Moon? What is the magnitude of the gravitational attraction the Earth exerts on the Moon? Suppose that the three bodies are aligned, with the Earth between the Sun and the Moon (at full moon). What is the direction of the net force acting on the Moon? Suppose that the three bodies are aligned, with the Moon between the Earth and the Sun (at new moon). What is the direction of the net force acting on the Moon?

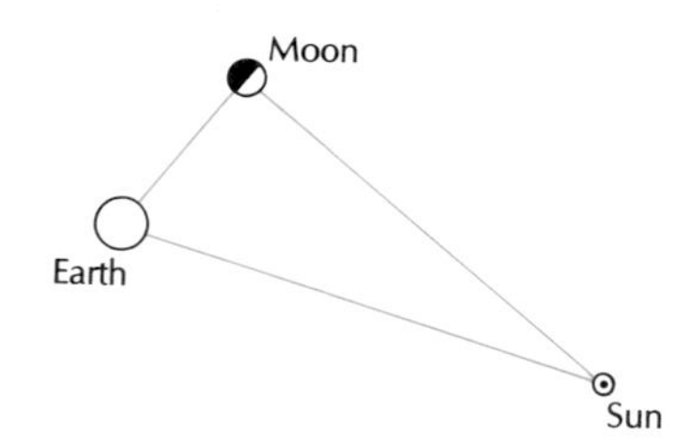


**Fig. 9.41**

7. Suppose that Earth, Sun, and Moon are located at the vertices of a right triangle, with the Moon located at the right angle (at first or last quarter moon; see Figure 9.41). Find the magnitude and direction of the sum of the gravitational forces exerted by the Earth and the Sun on the Moon.

*8. Mimas, a small moon of Saturn, has a mass of $3.8 \times 10^{19}$ kg and a diameter of 500 km. What is the maximum angular velocity with which we can make this moon rotate about its axis if pieces of loose rock sitting on its surface at its equator are not to fly off?

### Section 9.3

9. The Midas II spy satellite was launched into a circular orbit at a height of 500 km above the surface of the Earth. Calculate the orbital period and the orbital speed of this satellite.

10. The Sun is moving in a circular orbit around the center of our Galaxy. The radius of this orbit is $3 \times 10^4$ light-years. Calculate the period of the orbital motion and calculate the orbital speed of the Sun. The mass of our Galaxy is $4 \times 10^{41}$ kg and all of this mass can be regarded as concentrated at the center of the Galaxy.

11. Table 9.5 lists some of the moons of Saturn. Their orbits are circular.
   (a) From the information given, calculate the periods and orbital speeds of all these moons.
   (b) Calculate the mass of Saturn.

**Table 9.5** SOME MOONS OF SATURN

| *Moon* | *Distance from Saturn* | *Period* | *Orbital speed* |
|---|---|---|---|
| Tethys | $2.95 \times 10^5$ km | 1.89 days | — |
| Dione | 3.77 | — | — |
| Rhea | 5.27 | — | — |
| Titan | 12.22 | — | — |
| Iapetus | 35.60 | — | — |

*12. The Discoverer II satellite had an approximately circular orbit passing over both poles of the Earth. The radius of the orbit was about $6.67 \times 10^3$ km. Taking the rotation of the Earth into account, if the satellite passed over New York City at one instant, over what point of the United States would it pass after completing one more orbit?

*13. The binary star system PSR 1913 + 16 consists of two neutron stars orbiting about their common center of mass with a period of 7.75 h. Assume that the stars have equal masses and that their orbits are circular with a radius of $8.67 \times 10^8$ m.
   (a) What are the masses of the stars?
   (b) What are their speeds?

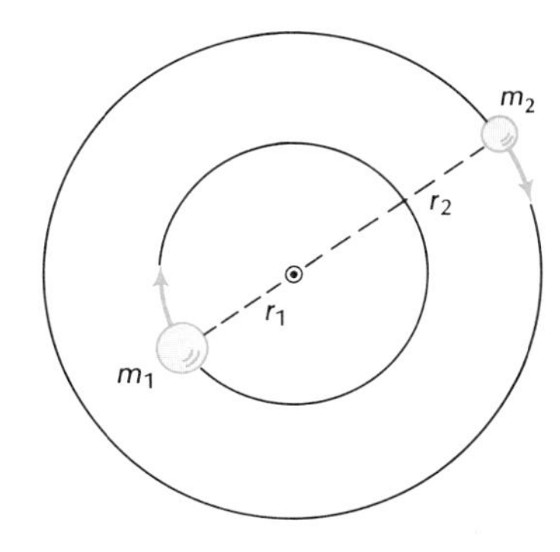

**Fig. 9.42** Binary star system.

*14. Figure 9.17 shows two stars orbiting about their common center of mass in the binary system Krüger 60. From the relative size of their orbits, determine the ratio of their masses.

**15. A binary star system consists of two stars of masses $m_1$ and $m_2$ orbiting about each other. Suppose that the orbits of the stars are circles of radii $r_1$ and $r_2$ centered on the center of mass (Figure 9.42). Show that the period of the orbital motion is given by

$$T^2 = \frac{4\pi^2}{G(m_1 + m_2)}(r_1 + r_2)^3$$

**16. The binary system Cygnus X-1 consists of two stars orbiting about their common center of mass under the influence of their mutual gravitational forces. The orbital period of the motion is 5.6 days. One of the stars is a supergiant with a mass 25 times the mass of the Sun. The other star is believed to be a black hole with a mass of about 10 times the mass of the Sun. From the information given, determine the distance between these stars; assume that the orbits of both stars are circular.

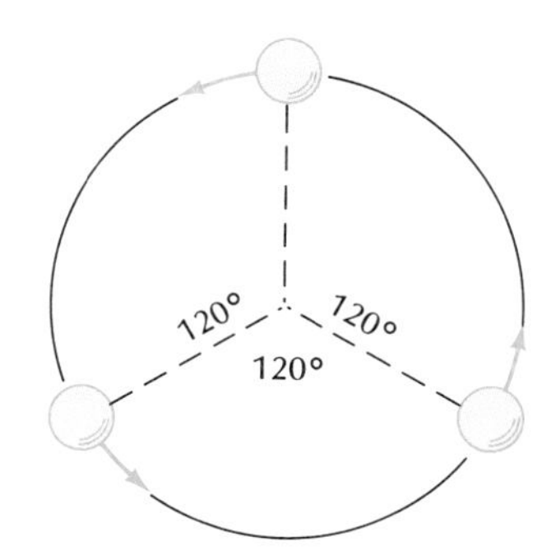

**Fig. 9.43** Three identical stars orbiting around their common center of mass.

**17. A hypothetical triple-star system consists of three stars orbiting about each other. For the sake of simplicity, assume that all three stars have equal masses and that they move along a common circular orbit maintaining an angular separation of 120° (Figure 9.43). In terms of the mass $M$ of each star and the orbital radius $R$, what is the period of the motion?

**18. Take into account the rotation of the Earth in the following problem:
   (a) Cape Canaveral is at a latitude of 28° north. What eastward speed (relative to the ground) must a satellite be given if it is to achieve a low-alti-

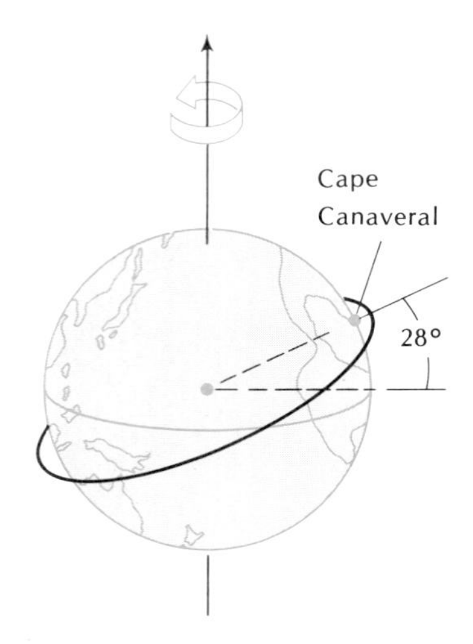


**Fig. 9.44** Orbit of satellite launched from Cape Canaveral.

**Fig. 9.45** Halley's comet, photographed in 1986.

tude circular orbit (Figure 9.44)? What westward speed must the satellite be given if it is to travel along the same orbit in the opposite direction? For the purpose of this problem, pretend that "low altitude" means essentially "zero altitude."

(b) Suppose that the satellite has a mass of 14.0 kg. What kinetic energy must the launch vehicle give to the satellite for an eastward orbit? For a westward orbit?

### Section 9.4

19. Halley's comet (Figure 9.45) orbits the Sun in an elliptical orbit (the comet reached perihelion in 1986). When the comet is at perihelion, its distance from the Sun is $8.78 \times 10^{10}$ m, and its speed is $5.45 \times 10^4$ m/s. When the comet is at aphelion, its distance is $5.28 \times 10^{12}$ m. What is the speed at aphelion?

20. Explorer I, the first American artificial satellite, had an elliptical orbit around the Earth with a perigee distance of $6.74 \times 10^6$ m and an apogee distance of $8.91 \times 10^6$ m. The speed of this satellite was $6.21 \times 10^3$ m/s at apogee. Calculate the speed at perigee.

21. Vanguard I, the second American artificial satellite, moved in an elliptical orbit around the Earth with a perigee distance of $7.02 \times 10^6$ m and an apogee distance of $10.3 \times 10^6$ m. At perigee, the speed of this satellite was $8.22 \times 10^3$ m/s. What was the speed at apogee?

22. The Explorer X satellite had an orbit with perigee 175 km and apogee 181,200 km above the surface of the Earth. What was the period of this satellite?

23. Calculate the orbital periods of Sputnik I and Explorer I from their apogee and perigee distances given in Table 9.3.

### Section 9.5

24. What is the kinetic energy and what is the gravitational potential energy for the orbital motion of the Earth around the Sun? What is the total energy?

25. The Voskhod I satellite, which carried Yuri Gagarin into space in 1961, had a mass of $4.7 \times 10^3$ kg. The radius of the orbit was (approximately) $6.6 \times 10^3$ km. What were the orbital speed, the orbital angular momentum, and the orbital energy of this satellite?

*26. The Andromeda galaxy is at a distance of $2.1 \times 10^{22}$ m from our Galaxy. The mass of Andromeda is $6 \times 10^{41}$ kg and the mass of our Galaxy is $4 \times 10^{41}$ kg.

(a) Gravity accelerates the galaxies toward each other. As reckoned in an inertial reference frame, what is the acceleration of Andromeda? What is the acceleration of our Galaxy? Treat both galaxies as point particles.

(b) The speed of Andromeda *relative to our Galaxy* is 266 km/s.[6] What is the speed of Andromeda and what is the speed of our Galaxy *relative to the center of mass* of the two galaxies?

(c) What is the kinetic energy of each galaxy relative to the center of mass? What is the total energy (kinetic and potential) of the system of the two galaxies? Will the two galaxies eventually escape from each other?

*27. The motor of a Scout rocket uses up all the fuel and stops when the rocket is at an altitude of 200 km above the surface of the Earth and is moving vertically at 8.50 km/s. How high will this rocket rise? Ignore any residual atmospheric friction.

---

[6] This is actually the component of the velocity along the line of sight, i.e., it is the radial component of the velocity. However, for the purpose of this problem, assume that it is the total velocity.

*28. Neglect the gravity of the Moon, neglect atmospheric friction, and neglect the rotational velocity of the Earth in the following problem: A long time ago, Jules Verne, in his book *From Earth to the Moon* (1865) suggested sending an expedition to the Moon by means of a projectile fired from a gigantic gun.

(a) With what muzzle speed must a projectile be fired vertically from a gun on the surface of the Earth if it is to (barely) reach the distance of the Moon?

(b) Suppose that the projectile has a mass of 2000 kg. What energy must the gun deliver to the projectile? The explosion of 1 short ton (2000 lb) of TNT releases $4.2 \times 10^9$ J. How many tons of TNT are required for firing this gun?

(c) If the gun barrel is 500 m long, what must be the average acceleration of the projectile during firing?

29. What is the escape velocity for a projectile launched from the surface of our Moon?

*30. An artificial satellite of 3500 kg made of aluminum is in a circular orbit at a height of 100 km above the surface of the Earth. Atmospheric friction removes energy from the satellite and causes it to spiral downward so that it ultimately crashes into the ground.

(a) What is the initial orbital energy (gravitational plus kinetic) of the satellite? What is the final energy when the satellite comes to rest on the ground? What is the energy change?

(b) Suppose that all of this energy is absorbed in the form of heat by the material of the satellite. Is this enough heat to melt the material of the satellite? To vaporize it? The heats of fusion and of vaporization of aluminum are given in Table 20.4.

*31. According to one theory, glassy meteorites (tektites) found on the surface of the Earth originate in volcanic eruptions on the Moon. With what minimum speed must a volcano on the Moon eject a stone if it is to reach the Earth? With what speed will this stone strike the surface of the Earth? In this problem ignore the orbital motion of the Moon around the Earth; use the data for the Earth–Moon system listed in the tables printed on the endpapers. (Hint: When the rock reaches the intermediate point where the gravitational pulls of the Moon and the Earth cancel, it must have zero velocity.)

*32. A spacecraft is launched with some initial velocity toward the Moon from 300 km above the surface of the Earth.

(a) What is the minimum initial speed required if the spacecraft is to coast all the way to the Moon without using its rocket motors? For this problem pretend that the Moon does not move relative to the Earth. The masses and radii of the Earth and the Moon and their distance are listed in the tables printed on the endpapers. (Hint: When the spacecraft reaches the point in space where the gravitational pulls of the Earth and the Moon cancel, it must have zero velocity).

(b) With what speed will the spacecraft strike the Moon?

33. The Pons–Brooks comet had a speed of 47.30 km/s when it reached its perihelion point, $1.160 \times 10^8$ km from the Sun. Is the orbit of this comet elliptic, parabolic, or hyperbolic?

*34. At a radial distance of $2.00 \times 10^7$ m from the center of the Earth, three artificial satellites (I, II, III) are ejected from a rocket. The three satellites I, II, III are given initial speeds of 5.47 km/s, 4.47 km/s, and 3.47 km/s, respectively; the initial velocities are all in the tangential direction.

(a) Which of the satellites I, II, III will have a circular orbit? Which will have elliptical orbits? Explain your answer.

(b) Draw the circular orbit. Also, on the top of the same diagram draw the elliptical orbits of the other satellites; label the orbits with the names of the satellites. (Note: You need not calculate the exact sizes of the ellipses, but your diagram should show whether the ellipses are larger or smaller than the circle.)

*35. (a) Since the Moon (*our* moon) has no atmosphere, it is possible to place an artificial satellite in a circular orbit that skims along the surface of the Moon (provided that the satellite does not hit any mountains!). Suppose that such a satellite is to be launched from the *surface* of the Moon by means of a gun that shoots the satellite in a horizontal direction. With what velocity must the satellite be shot out from the gun? How long does the satellite take to go once around the Moon?
(b) Suppose that a satellite is shot from the gun with a horizontal velocity of 2.00 km/s. Make a rough sketch showing the Moon and the shape of the satellite's orbit; indicate the position of the gun on your sketch.
(c) Suppose that a satellite is shot from the gun with a horizontal velocity of 3.00 km/s. Make a rough sketch showing the Moon and the shape of the satellite's orbit. Is this a closed orbit?

*36. According to an estimate, a large crater on Wilkes Land, Antarctica, was produced by the impact of a $13 \times 10^9$-ton meteoroid incident on the surface of the Earth at 70,000 km/h. What was the speed of this meteoroid relative to the Earth when it was at a "large" distance from the Earth?

*37. An experienced baseball player can throw a ball with a speed of 140 km/h. Suppose that an astronaut standing on Mimas, a small moon of Saturn of mass $3.76 \times 10^{19}$ kg and radius 195 km, throws a ball with this speed.
(a) If the astronaut throws the ball horizontally, will it orbit around Mimas?
(b) If the astronaut throws the ball vertically, how high will it rise?

*38. An electromagnetic launcher, or rail gun, accelerates a projectile by means of magnetic fields. According to some calculations, it may be possible to attain muzzle speeds as large as 15 km/s with such a device. Suppose that a projectile is launched upward from the surface of the Earth with this speed; ignore air resistance.
(a) Will the projectile escape permanently from the Earth?
(b) Can the projectile escape permanently from the Solar System? (Hint: Take into account the speed of 30 km/s of the Earth around the Sun.)

*39. Sputnik I, the first Russian satellite (1957), had a mass of 83.5 kg; its orbit reached perigee at a height of 225 km and apogee at 959 km. Explorer I, the first American satellite (1958), had a mass of 14.1 kg; its orbit reached perigee at a height of 368 km and apogee at 2540 km. What was the orbital energy of each of these satellites?

*40. The orbits of most meteoroids around the Sun are nearly parabolic.
(a) With what speed will a meteoroid reach a distance from the Sun equal to the distance of the Earth from the Sun? (Hint: In a parabolic orbit the speed at any radius equals the escape velocity at that radius. Why?)
(b) Taking into account the Earth's orbital speed, what will be the speed of the meteoroid *relative to the Earth* in a head-on collision with the Earth? In an overtaking collision? Ignore the effect of the gravitational pull of the Earth on the meteoroid.

41. Calculate the perihelion and the aphelion speeds of Encke's comet. The perihelion and aphelion distances of this comet are $5.06 \times 10^7$ km and $61.25 \times 10^7$ km.

*42. The Explorer XII satellite was given a tangential velocity of 10.39 km/s when at perigee at a height of 457 km above the Earth. Calculate the height of apogee.

*43. Prove that the orbital energy of a planet or a comet in an elliptical orbit around the Sun can be expressed as

$$E = -\frac{GM_S m}{r_1 + r_2}$$

where $r_1$ and $r_2$ are, respectively, the perihelion and aphelion distances. [Hint:

Start with the sum of the kinetic and potential energies at perihelion and use Eq. (33).]

44. Suppose that a comet is originally at rest at a distance $r_1$ from the Sun. Under the influence of the gravitational pull, the comet falls radially toward the Sun. Show that the time it takes to reach a radius $r_2$ is

$$t = \int_{r_2}^{r_1} \frac{dr}{\sqrt{2GM_S/r - 2GM_S/r_1}}$$

*45. Suppose that a projectile is fired horizontally from the surface of the Moon with an initial speed of 2.0 km/s. Roughly sketch the orbit of the projectile. What maximum height will this projectile reach? What will be its speed when it reaches maximum height?

**46. The Earth has an orbit of radius $1.50 \times 10^8$ km around the Sun; Mars has an orbit of radius $2.28 \times 10^8$ km. In order to send a spacecraft from the Earth to Mars, it is convenient to launch the spacecraft into an elliptical orbit whose perihelion coincides with the orbit of the Earth and whose aphelion coincides with the orbit of Mars (Figure 9.46); this orbit requires the least amount of energy for a trip to Mars.

(a) To achieve such an orbit, with what speed (relative to the Earth) must the spacecraft be launched? Ignore the pull of the gravity of the Earth and Mars on the spacecraft.

(b) With what speed (relative to Mars) does the spacecraft approach Mars at the aphelion point? Assume that Mars actually is at the aphelion point when the spacecraft arrives.

(c) How long does the trip from Earth to Mars take?

(d) Where must Mars be (in relation to the Earth) at the instant the spacecraft is launched? Where will the Earth be when the spacecraft arrives at its destination? Draw a diagram showing the relative positions of Earth and Mars at these two times.

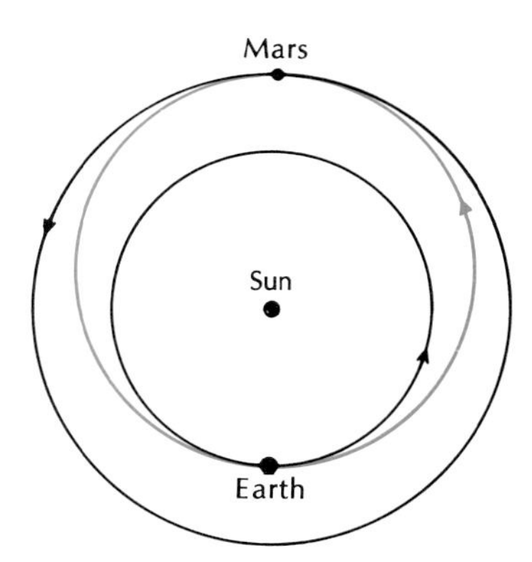


**Fig. 9.46** Orbit for a space probe on trip to Mars.

**47. Repeat the calculations of Problem 46 for the case of a spacecraft launched on a trip to Venus. The orbit of Venus has a radius of $1.08 \times 10^8$ km.

**48. If an artificial satellite, or some other body, approaches a moving planet on a hyperbolic orbit, it can gain some energy from the motion of the planet and emerge with a larger speed than it had initially. This slingshot effect has been used to boost the speeds of the two Voyager spacecraft as they passed near Jupiter. Suppose that the line of approach of the satellite makes an angle $\theta$ with the line of motion of the planet and the line of recession of the satellite is parallel to the line of motion of the planet (Figure 9.47; the planet can be regarded as moving on a straight line during the time interval in question). The speed of the planet is $u$ and the initial speed of the satellite is $v$ (in the reference frame of the Sun).

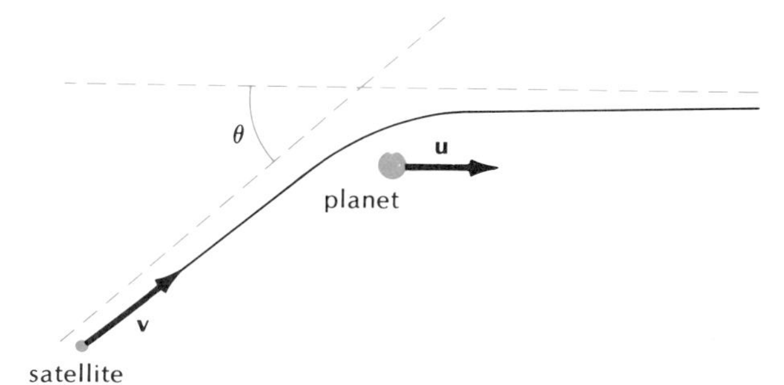


**Fig. 9.47** Trajectory of artificial satellite passing by planet.

(a) Show that the final speed of the satellite is

$$v' = u + \sqrt{v^2 + u^2 - 2uv\cos\theta}$$

(b) Show that the satellite will not gain any speed in this encounter if $\theta = 0$ and show that the satellite will gain maximum speed if $\theta = 180°$.
(c) If a satellite with $v = 3$ km/s approaches Jupiter at an angle of $\theta = 20°$, what will be its final speed?

### Section 9.6

49. The following table lists radii measured from the center of the Earth and the fraction of the Earth's mass that lies within this radius:

| $r$ | Fraction of mass |
|---|---|
| 1400 km | 0.024 |
| 2400 | 0.11 |
| 3400 | 0.31 |
| 4400 | 0.47 |
| 5400 | 0.72 |
| 6400 | 1.00 |

Calculate the value of the acceleration of gravity at each radius. Plot this value as a function of radius.

*50. Calculate the change of the acceleration of gravity between a point on the surface of the sea and a point at a depth of 4000 m below the surface of the sea. For the purposes of this problem, assume that the Earth is a sphere and that it is covered with water to a depth of more than 4000 m. Is the acceleration of gravity at this depth larger or smaller than at the surface?

*51. The gravitational potential energy of a small mass $m$ placed on the surface of a large spherical body of radius $R$ and mass $M$ is $U = -GMm/R$. Find an expression for the gravitational energy if the mass $m$ is placed inside the spherical body, at some radius $r < R$. Assume that the mass of the body is uniformly distributed over the sphere of radius $R$.

### Section 9.7

52. According to one design studied by NASA, a large space colony in orbit around the Earth would consist of a torus of diameter 1.8 km, looking somewhat like a gigantic bicycle wheel (see Figure 9.38). In order to generate artificial gravity of 1 G, how fast must this space colony rotate about its axis?

# THE BIG BANG AND THE EXPANSION OF THE UNIVERSE*

Cosmology studies the universe at large, its size, its shape, and its evolution. Seeking to grasp the universe, the mind of the cosmologist has to wander over distances as great as 10 billion light-years and over times as long as 10 billion years or longer.

Until the early part of this century, astronomers thought the universe to be much smaller. They thought that the farthest stars at the edge of our Galaxy were about 30,000 light-years away and that there was nothing but dark, empty space beyond that distance. But in the 1920s, Edwin Hubble used the 100-inch telescope on Mt. Wilson to establish that the faint, whispy "nebulae" found in all parts of the sky were actually gigantic conglomerations of stars, similar to our own Galaxy but located at a very great distance from us. He developed methods for measuring these enormous cosmic distances and discovered that some of the remote galaxies were more than half a billion light-years away from us.

To begin a theoretical analysis of the universe, cosmologists must make a basic hypothesis: the laws of physics that hold in other parts of the universe are the same as those that hold in our part of the universe. We have some observational evidence in favor of this hypothesis. For instance, the light and the radiowaves reaching us from remote galaxies have the same basic features as light and radiowaves produced in our laboratories on the Earth. But, of course, there is no direct way of verifying the hypothesis in detail. In practice, the hypothesis of the universal applicability of the laws of physics seems to work quite well, and it enables us to make sense out of the observational data that astronomers and radio astronomers collect with their telescopes and radiotelescopes.

Of all the forces of nature, gravity plays the dominant role in shaping the large-scale features and the evolution of the universe. Hence Newton's law of gravitation will be the main tool for the theoretical investigations of this chapter. Strictly speaking, Newton's theory is not entirely adequate for a description of the universe because at the large cosmic distances, relativistic effects become important; thus, we really ought to use Einstein's theory of General Relativity (see Interlude IX) in our investigations. However, it turns out that Newton's theory is surprisingly successful in the study of some of the questions concerning the dynamics of the universe, and in this chapter we will be able to get along quite well without relativity. Of course, there are many questions concerning the geometry of the universe, its size, and its shape that Newton's theory cannot answer. But even if we avoid all the topics that require relativity, there are still many interesting topics we can discuss.

* This chapter is optional.

## II.1 THE EXPANSION OF THE UNIVERSE

In the night sky we can see about 2000 stars with the naked eye. A telescope reveals many more stars and shows that they are part of a large conglomerate or cloud of stars. This cloud is called the **Galaxy,** or the **Milky Way.** It contains about $10^{11}$ stars arranged in an irregular disklike region some $10^5$ light-years in diameter. The disk has a central bulge and it has spiral arms along which stars are concentrated (Figure II.1).

(a)

(b)

**Fig. II.1** These pictures give an impression of what our Galaxy looks like viewed (a) face on and (b) edge on. The pictures are actually photographs of two distant galaxies (NGC 5457 and NGC 4631) similar to our Galaxy.

**Fig. II.2** Spiral galaxy (NGC 3031) in Ursa Major. The plane of this galaxy is inclined to our line of sight; face on, this galaxy would look circular.

**Fig. II.5** Unusual galaxy (NGC 5128) in Centaurus. Note the thick lane of dust surrounding this galaxy.

**Fig. II.3** Spiral galaxy (NGC 4565) in Coma Berenices, seen edge on. Note how thin this galaxy is.

**Fig. II.6** Spiral galaxy (NGC 7217) in Pegasus.

**Fig. II.4** Spiral galaxy (NGC 5194) in Canes Venatici. This beautiful spiral is called the "Whirlpool" galaxy.

**Fig. II.7** Elliptical galaxy (NGC 4486) in Virgo. This is one of the brightest galaxies known.

**Fig. II.8** Peculiar spiral galaxy (NGC 2623) in Cancer.

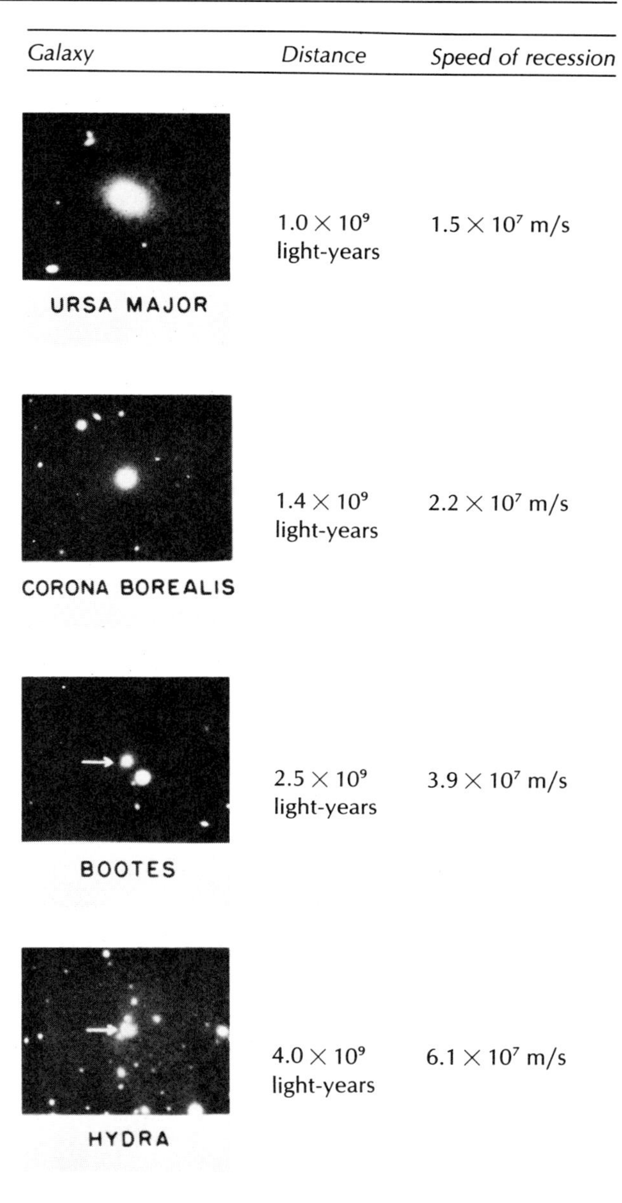

| Galaxy | Distance | Speed of recession |
|---|---|---|
| URSA MAJOR | $1.0 \times 10^9$ light-years | $1.5 \times 10^7$ m/s |
| CORONA BOREALIS | $1.4 \times 10^9$ light-years | $2.2 \times 10^7$ m/s |
| BOOTES | $2.5 \times 10^9$ light-years | $3.9 \times 10^7$ m/s |
| HYDRA | $4.0 \times 10^9$ light-years | $6.1 \times 10^7$ m/s |

**Fig. II.9** Several distant galaxies.

There are many external galaxies beyond our Galaxy. With our large telescopes we can see altogether about $10^{11}$ galaxies. There are supergiant galaxies with $10^{13}$ stars each, and there are dwarf galaxies with "only" $10^6$ stars. There are spherical galaxies and elliptical galaxies, like luminous globes and eggs; there are spiral galaxies and barred-spiral galaxies, like whirling pinwheels; and there are irregular galaxies with the weirdest shapes (Figures II.2–II.8).

All the galaxies are in motion. Many of them congregate in clusters, orbiting about each other, sometimes accidentally colliding. But let us ignore the details of the motion of the galaxies and concentrate on the large-scale features of the motion. We then find that on a large scale, the galaxies have a motion of recession — all the distant galaxies are moving away from us. For example, Figure II.9 shows several distant galaxies, all of which are speeding away from us at more than $10^7$ m/s. There are some objects that are racing away from us even faster than that — the extreme case is that of the quasar Q 1208+1011), which has a speed of about 92% of the speed of light.

The recession velocities of galaxies can be determined by the **red-shift** method. This method hinges on the fact that the light from a receding source will look redder to us than the light from a stationary source (conversely, the light from an approaching source will look bluer). The shift of color, or, more precisely, the shift of wavelength of the light, is directly related to the velocity. We will learn more about this in the discussion of the Doppler effect in Chapter 17. To find the recession velocity, astronomers need only measure how much the color of the light received from atoms in some distant galaxy is shifted relative to the color emitted by similar atoms in our laboratories on Earth.

By measuring the velocities and distances of galaxies, Hubble discovered that the motion of recession obeys a very simple rule: the velocity of each galaxy is directly proportional to its distance, that is, nearby galaxies move slowly and distant galaxies move fast (see Figure II.10). This proportionality is called **Hubble's Law.** Mathematically, it can be expressed as

$$v = H_0 r \tag{1}$$

**Fig. II.10** The motion of recession of distant galaxies (expansion of the universe). The velocities are radial, and they are directly proportional to the distances.

Where $H_0$ is Hubble's constant. If $r$ is expressed in light-years, the numerical value of Hubble's constant is

$$H_0 = 1.7 \times 10^4 \text{ (m/s)/(million light-years)} \qquad (2)$$

For example, the galaxy beyond the constellation Hydra shown in Figure II.9 has a recession velocity of $6.1 \times 10^7$ m/s and it is at a distance of $3.5 \times 10^9$ light-years, in good agreement with Eq. (1).

Astronomers determine the enormous cosmic distance from us to other galaxies by the brightness or "headlight" method. In essence, this method relies on the following: Faraway galaxies look faint to us and nearby galaxies look bright — just as, on a dark road, the headlights of a faraway automobile look faint and the headlights of a nearby automobile look bright. If all galaxies generated precisely the same quantity of light, then differences in their apparent brightness as observed by our telescopes would be entirely due to differences in their distances — there would then be a simple mathematical relationship between apparent brightness and distance. However, in practice, there are some complications; two galaxies can be at the same distance yet differ in apparent brightness because one has more stars and generates more light than the other. Such intrinsic differences between galaxies must be taken into account when using the brightness method. Astronomers have developed clever techniques for selecting galaxies or portions of galaxies of standard brightness, but rather large uncertainties still remain in the distance determinations. Correspondingly, there are uncertainties in the value of the Hubble constant. The uncertainty in Eq. (2) is at least 25% and possibly more.

Although Figure II.10 gives the misleading impression that our Galaxy is at the center of the universe and that all other galaxies are fleeing away from us, our Galaxy does not occupy any special spot in the universe. The other galaxies are not just fleeing away from us; they are fleeing away from each other. All galaxies are receding from all other galaxies — the universe is expanding. An extraterrestrial astronomer sitting on that galaxy beyond Hydra (Figure II.9) would see our Galaxy and all other galaxies fleeing away from her. Hence our spot in the universe is pretty much the same as every other spot. Cosmologists believe that this overall uniformity holds not only in regard to the expansion, but also in regard to all other general features of the universe. For instance, the numbers and types of galaxies that the extraterrestrial astronomer finds in her neighborhood will, on the average, be the same as we find in our neighborhood — the universe is pretty much the same everywhere. This assertion of large-scale uniformity of the universe is called the **Cosmological Principle.**

The motion of recession of the galaxies can be described by a simple analogy. When a grenade explodes in midair, the fragments of shrapnel spurt out in all directions. Different fragments may have different velocities and, in a given amount of time, they will reach different distances. After a time $t$, the position of a fragment having a velocity $v$ will be

$$r = vt \qquad (3)$$

If we rewrite this as

$$v = r/t \qquad (4)$$

we see that at any given time the fragments that are at the greatest distances are those with the highest velocities. This proportionality of velocity and distance has the same form as Hubble's Law [see Eq. (1)]. Thus, Hubble's Law suggests that the galaxies were set in motion by a primordial cosmic explosion many years ago and have been more or less coasting along ever since. The only fault with the analogy between the motions of shrapnel fragments and galaxies is that in the case of the former there is a clearly defined center of explosion, while in the case of the latter the explosion occurred simultaneously everywhere and there is no special center of burst. Incidentally: In the expansion of the universe, only the distances between the galaxies increase; the galaxies themselves do not expand. This is also in agreement with the grenade analogy, where, of course, only the distances between the shrapnel fragments increase while the fragments themselves remain of constant size.

## II.2 THE AGE OF THE UNIVERSE AND THE BIG BANG

The explosion that started the expansion of the universe is called the **Big Bang.** We can reckon how long ago this happened by comparing Eqs. (1) and (4). Clearly, the inverse of the Hubble constant must coincide with the expansion time,

$$t = 1/H_0 \tag{5}$$

or

$$t = \frac{1}{1.7 \times 10^4} \times \frac{\text{million light-years}}{\text{m/s}}$$

$$= \frac{1}{1.7 \times 10^4} \times \frac{9.5 \times 10^{21}\ \text{m}}{\text{m/s}}$$

$$= 5.6 \times 10^{17}\ \text{s} = 1.8 \times 10^{10}\ \text{years} \tag{6}$$

However, in this calculation of the age of the universe, we have ignored the possibility that the velocity of galaxies may change with time. For instance, since gravity pulls the galaxies toward each other, we might expect that gravity tends to inhibit the motion of recession and tends to slow down the expansion of the universe. This would imply that the velocities of all galaxies were somewhat larger in the past and, consequently, the true age of the universe ought to be somewhat smaller than the 18 billion years indicated by our naïve calculation. But in any case, this number gives us a rough estimate of the age.

Since the universe started some finite time ago, only the light from those parts of it that are sufficiently near can have reached us. The speed of light is $c = 3.00 \times 10^8$ m/s $= 1$ light-year/year, and in a time $t$ light travels a distance

$$ct = (1\ \text{light-year/year}) \times (1.8 \times 10^{10}\ \text{year})$$

$$= 1.8 \times 10^{10}\ \text{light-year}$$

This distance is the radius of the **observable universe.**[1] Everything within this radius we can see (given sufficiently powerful telescopes); anything beyond we cannot see because the light has not yet had enough time to reach us. Note that as time increases, the radius $ct$ increases; that is, the observable universe includes more and more of the total universe.

[1] This value of the radius of the observable universe is only an approximation. For an exact calculation we need to use the theory of General Relativity, taking into account that space and time are curved.

**Fig. II.11** The globular cluster (NGC 5272) in Canes Venatici.

The idea of the Big Bang can be put to a direct test: if the universe originated 18 billion or so years ago, then nothing in the universe can be older than that. The Earth and the Sun have an age of only 4.5 billion years — for a serious test of the Big Bang hypothesis, we need to look for much older objects. The oldest objects that we can reliably date are the globular clusters of stars found near our Galaxy. Each of these is an aggregate of $10^5$ or $10^6$ stars, looking rather like a great swarm of bees in the sky (Figure II.11). The theory of stellar evolution permits us to calculate the age of such a cluster from the observed color and luminosity of its stars. As stars reach old age, they enter a red-giant stage, turning a reddish color and swelling to several hundred times normal size. A count of the number of such old red giants in a cluster permits us to deduce the age of the cluster — a young cluster will contain few red giants and an old cluster will contain many. Careful calculations indicate that the oldest globular clusters have ages of about 10 billion to 16 billion years, in good agreement with the expansion time given by Eq. (6).

The age of the oldest globular clusters suggests the date at which the first stars were born in our Galaxy. There is another method by which we can date the birth of stars: by the age of chemical elements. All the elements, with the exception of hydrogen and helium, were synthesized by nuclear reactions in the interior of very massive stars born soon after our Galaxy came into being. These stars survived only a short time and then exploded as supernovas, spurting their chemical elements all over the place, some of these elements eventually winding up in the cloud of gas and dust that was destined to become the Solar System. The ages or the atoms of these elements — for instance, uranium and thorium — can be determined by a method that relies on their radioactivity. These atoms suffer radioactive decay as they grow old and the amount of decay gives an indication of their age. Such radioactive-decay measurements tell us that the atoms are somewhere between 7 billion and 15 billion years old. Thus, the age of the elements agrees with the expansion time to within the experimental uncertainties.

A discordance in the ages deduced from the globular clusters, the chemical elements, and the recessional motion of the galaxies would have proved the Big Bang hypothesis wrong. The concordance of these ages does not prove the hypothesis right, but it encourages us to look for further evidence.

The most decisive item of evidence for the Big Bang is that some of the radiant heat given off by the primordial explosion can still be found in the sky today. At the initial instant of the Big Bang, the universe must have been densely filled with compacted matter, at very high temperature and pressure. The matter must have been expanding at a colossal rate so that even today it continues to coast along because of the momentum it had in the beginning. The matter must initially have been hotter than white hot, emitting intense light. The whole universe must have been filled with a fireball of a brightness much greater than that of the Sun or of a thermonuclear explosion.

Originally, the radiant heat emitted by this primordial fireball was in the form of very penetrating gamma rays and X rays. These rays are essentially light waves of very short wavelength. But as the universe expanded, these light waves expanded with it. Thus, the wavelengths of the fireball light still remaining in the sky at present are much longer than the wavelengths of the original light — they are larger in the same proportion as the present intergalactic distances are larger than the original distances. Theoretical calculations suggest that at present the wavelength should be about 1 mm. This is much longer than the wavelength of visible light — it is the wavelength of radiowaves (microwaves).

This kind of radiation was discovered in 1964 by A. A. Penzias and R. W. Wilson,[2] two scientists working at Bell Laboratories with very sensitive microwave communication equipment (Figure II.12). They found that the entire sky is noisy; there is radiation coming at the Earth from all directions. Physicists at Princeton immediately recognized the cosmological significance of this discovery, and identified the radiation as **cosmic background radiation,** a relic of the Big Bang. Subsequently, many experimenters took careful

**Fig. II.12** Horn antenna at Bell Lab, Holmdel, N.J. This antenna was designed for microwave communication experiments with the Echo and Telstar satellites.

measurements of this radiation and established that it is strongest at a wavelength of about 2 mm; above or below this wavelength, the amount of radiation gradually decreases. The radiation has all the characteristics of radiant heat — it is the same kind of radiant heat as emitted by a "hot" body of a temperature of 3°C above absolute zero. What has happened here is that the extremely hot radiant heat from the primordial fireball has gradually cooled down as the universe expanded and by now its temperature has come pretty close to absolute zero; as the universe continues to expand, the temperature will continue to drop. The cosmic background radiation is residual radiant heat left in the sky by the Big Bang. It is direct material evidence for the Big Bang.

Further evidence for the Big Bang is supplied by the study of the abundance of helium in the universe. The average chemical composition of our universe is about 74% hydrogen (by mass) and 24% helium, with only traces of other elements. All this hydrogen and almost all this helium are primordial, that is, they were formed in the hot primordial fireball, soon after the beginning of the universe.[3] The temperature of the fireball was $10^9$ °C when the universe was 3 minutes old. At this temperature, hydrogen is subject to nuclear fusion leading to the formation of helium. Theoretical calculations show that the fusion reactions lead to an abundance of about 75% hydrogen and 25% helium, in remarkable agreement with the observed abundance. This confirms the picture of a hot Big Bang.

## II.3 THE FUTURE OF THE UNIVERSE

If the galaxies were to continue their present motion with constant velocity, they would gradually recede

[2] **Arno A. Penzias,** 1933–, American astrophysicist and **Robert W. Wilson,** 1936–, American radio astronomer, shared the 1978 Nobel Prize for their discovery of the cosmic background radiation.

[3] Stars obtain their energy from the nuclear fusion of hydrogen, which produces helium. However, the total mass of helium produced by all of the stars since the beginning amounts to only a few percent.

into the distance, looking smaller and smaller, dimmer and dimmer; ultimately they would fade from our sight — and our Galaxy would be left alone.

The motion of recession, however, does not proceed with constant velocity. The force of gravitation attracts every galaxy to every other galaxy. This mutual attraction tends to decelerate the expansion. Of course, the pull of gravity is very weak because the intergalactic distances are so enormously large. But even a very small deceleration can be important in the long run. If the motion ever stops, then the galaxies will begin to fall back toward each other. The universe will then contract. As the galaxies come closer together, the gravitational pull will become stronger and the velocities of contraction will become larger. Finally, the galaxies will collide with one another and the universe will collapse in a terminal cosmic implosion.

The big question in cosmology is this: Will the universe expand forever? Or will it come to a stop and contract? To find the answer to this question, we must calculate the deceleration of the motion of recession.

Since our only concern is the average motion of the galaxies, we will pretend that the galaxies are uniformly distributed throughout the universe, forming a gas of galaxies. The expansion of the universe is then equivalent to the expansion of this gas. To find the law of expansion of the gas, consider a spherical region centered on, say, our Galaxy; the spherical region is supposed to be small compared to the size of the universe, but to contain very many galaxies. Imagine that, by magic, we remove all of the galaxies from this sphere, leaving an empty hole. The rest of the universe has spherical symmetry about this hole. Hence, by the theorem of Section 9.6, gravity in the hole will be exactly zero. It then follows that if we put the galaxies back into the spherical region, their motion will be completely unaffected by the rest of the universe — only the galaxies in the spherical region exert gravitational forces within the spherical region. Consider now one of the galaxies at the surface of this region at a radial distance $r$ from us (Figure II.13). The gravitational force that the mass in the spherical region exerts on that one galaxy is as though all of the mass were concentrated at the center (see Section 9.6). Consequently, the acceleration of that one galaxy is

$$\frac{dv}{dt} = -\frac{GM}{r^2} \tag{7}$$

where $M$ is the mass in the spherical region. This equation determines how the size of the spherical region varies with time.

The equation of motion (7) is of the same form as for a ballistic missile coasting away from the Earth along a radial line. The radial motion continually de-

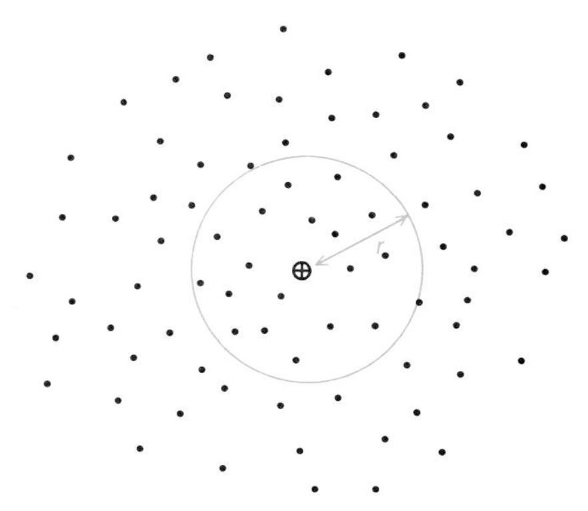

**Fig. II.13** A spherical region of our universe.

celerates, but whether it ever stops and reverses depends on the initial velocity: if the initial velocity is larger than the escape velocity [see Eq. (9.29)], the radial distance continues to increase forever; if the initial velocity is smaller than the escape velocity, the radial motion comes to a halt and then reverses. If we regard the present instant as the initial instant and the present radius as the starting radius, the initial velocity is [see Eq. (1)]

$$v_0 = H_0 r_0 \tag{8}$$

and the escape velocity is

$$v_{crit} = \sqrt{2GM/r_0} \tag{9}$$

In these equations the subscript 0 has been inserted to emphasize that the quantities are evaluated at the present instant. The mass $M$ can be expressed in terms of the average density of mass in the universe:

$$M = \frac{4\pi}{3} r_0^3 \rho_0 \tag{10}$$

and

$$v_{crit} = \sqrt{\frac{8\pi}{3} G r_0^2 \rho_0} \tag{11}$$

The condition for a permanently expanding universe is then

$$v_0 \geq v_{crit} \tag{12}$$

which is equivalent to

$$H_0 r_0 \geq \sqrt{\frac{8\pi}{3} G r_0^2 \rho_0} \tag{13}$$

or

$$\rho_0 \leq \frac{3}{8\pi G} H_0^2 \quad (14)$$

Likewise, the condition for an ultimately contracting universe is

$$\rho_0 > \frac{3}{8\pi G} H_0^2 \quad (15)$$

If we insert the numerical value $H_0 = 1/5.6 \times 10^{17}$ s [see Eq. (6)] and the known numerical value for the gravitational constant on the right sides of Eqs. (14) and (15), we obtain the conditions

$$\rho_0 \leq 6 \times 10^{-27} \text{ kg/m}^3 \quad \text{(permanent expansion)}$$
$$(16)$$
$$\rho_0 > 6 \times 10^{-27} \text{ kg/m}^3 \quad \text{(ultimate contraction)}$$

In principle, this makes it very simple to predict the future evolution of the universe: we only need to measure the average density of mass and check whether it is larger or smaller than the critical value $6 \times 10^{-27}$ kg/m$^3$. In practice, we are handicapped by the uncertainties in the mass density and, to a lesser extent, uncertainties in the value of the right side of Eq. (16).

## II.4 THE SEARCH FOR NUMBERS

The basic cosmological parameters that describe the large-scale behavior of our universe are the velocity of expansion [or the Hubble constant; see Eq. (1)], the rate of change of this velocity, the age of the universe, and the mass of the universe. If we knew the values of these parameters, we could predict the future evolution of the universe. For instance, Eq. (16) permits us to make such a prediction from the observed values of the mass density and the velocity of expansion (or the value of the Hubble constant). Alternative equations involving the age of the universe or the deceleration of the velocity can be used similarly.

The search for numbers has been the great problem of observational cosmology. Although much progress has been made in recent years, the values of the cosmological parameters remain rather uncertain. This makes it hard to decide on a final answer to the big question of the future evolution of the universe.

What do we know about the average mass density? When reckoning the mass of the universe, we must take into account the mass belonging to galaxies and also whatever mass is to be found in the intergalactic space between the galaxies. The mass of the galaxies is rather uncertain — different methods of mass determination give values that differ by as much as a factor of 10. The source of this difficulty is that besides visible stars, a galaxy may contain a large amount of invisible hidden mass in the form of small, very faint stars or black holes. According to the best available estimates, the total visible mass associated with galaxies contributes an average density of about $8 \times 10^{-29}$ kg/m$^3$. This is much below the critical value given in Eq. (16). Hence, unless there is some extra, hidden mass somewhere in the galaxies or in the space between the galaxies, the universe will continue to expand forever.

Recent investigations indicate that galaxies are encased in extragalactic clouds, or halos, of invisible, dark matter. Although astronomers cannot see this dark matter, they can detect its presence by its gravitational effects on the rotational motion of galaxies and on the orbital motion of galaxies around each other. The mass associated with this dark matter contributes an average density of about $1 \times 10^{-27}$ kg/m$^3$, more than 10 times as large as the density of the visible matter contributed by the galaxies. The composition of the dark matter remains a mystery; it might consist of small, almost dark stars, or small black holes, or clouds of neutrinos, or clouds of exotic, not yet discovered particles, or a mixture of all of these. Some circumstantial evidence suggests that not all of this dark matter can consist of the ordinary protons and neutrons that make up the bulk of the mass in the familiar chemical elements. This evidence hinges on studies of element creation in the early universe, in particular, the creation of deuterium. Deuterium is an isotope of hydrogen — the nucleus of hydrogen consists of one proton, whereas the nucleus of deuterium consists of one proton plus one neutron. Deuterium is found wherever hydrogen is found; for example, in water on the Earth, one molecule in 6000 is a molecule of "heavy water" in which one of the hydrogen atoms in $H_2O$ is replaced by a deuterium atom. The Sun contains deuterium mixed with its hydrogen and so do all other stars. Astrophysicists believe that this deuterium was created in nuclear reactions during the Big Bang, long before galaxies and stars were born. From theoretical calculations of deuterium production, we learn that a large density of protons and neutrons in the universe interferes with the formation of deuterium, because it favors some nuclear reactions that destroy deuterium. Since the universe now contains a fair amount of deuterium, the density of protons and neutrons during the Big Bang must have been below a certain limit; consequently, the density of protons and neutrons now must also be below a corresponding limit — at most $6 \times 10^{-28}$ kg/m$^3$. This value is considerably smaller than the total contribution from dark matter, about $1 \times 10^{-27}$ kg/m$^3$. Thus, a good portion of the dark matter must be very unusual stuff.

The deep intergalactic space beyond the galaxies

and their halos may or may not contain extra dark matter in the form of black holes, hydrogen gas at very high temperatures (plasma), neutrinos, or exotic particles. Matter in such forms would be very hard to detect, and we therefore cannot place any tight limits on the amount of such extra mass.

New theories of elementary particles and of the fundamental forces between them have led physicists to speculate about the conditions during the very early stages of the Big Bang, when the age of the universe was less than $10^{-34}$ s. One intriguing theoretical model indicates that at these early moments the universe went through a stage of very rapid expansion, or inflation. According to this inflationary model, the expansion leaves the universe of today with a mass density exactly equal to the critical density, $6 \times 10^{-27}$ kg/m$^3$. This provocative speculation has stimulated much research into the question of dark matter, but it remains unclear where this matter might be hidden.

Will the universe continue to expand or will it ultimately contract? The available data are not yet quite precise enough for a firm prognosis. A forever-expanding universe gives the best fit to all the facts as we know them. But there are enough uncertainties in our measurements and enough loopholes in our arguments that the possibility of a contracting universe cannot be dismissed. It will be a while before we know the ultimate fate of the universe.

## Further Reading

*Galaxies* by T. Ferris (Sierra Book Club, San Francisco, 1980; reprinted in paperback by Stewart, Tabori, and Chang, New York, 1982) is a splendid pictorial survey of the universe with awesome photographs of distant galaxies and with informative, if somewhat fragmented, text.

*Galaxies* by H. Shepley (Harvard University Press, Cambridge, 1972) is a classic introduction to the study of galaxies, including some description of the tools used by astronomers. *The State of the Universe*, edited by G. T. Bath (Clarendon Press, Oxford, 1980), is a collection of lectures on stars, galaxies, and other objects in the sky. *Galaxies: Structure and Evolution* by R. J. Tayler (Wykeham, London, 1978) gives a short course in galactic structure at a slightly more advanced level.

*The Realm of the Nebulae* by E. Hubble (Dover, New York, 1958) is a reprint of Hubble's own account of the first measurements of extragalactic distances and of the discovery of his law. First published in 1936, the details in this account are out of date, but it retains its historical value.

*Cosmology* by E. R. Harrison (Cambridge University Press, Cambridge, 1981) is an excellent introduction to modern cosmological theory at an elementary level; it is very clearly written and each chapter contains an exhaustive list of further references. *The Big Bang* by J. Silk (Freeman, San Francisco, 1980) and *Black Holes, Quasars, and the Universe* by H. L. Shipman (Houghton Mifflin, Boston, 1980) give good, concise introductions to cosmology with a somewhat more observational orientation.

*Modern Cosmology* by D. W. Sciama (Cambridge University Press, Cambridge, 1971) and *Principles of Modern Cosmology* by M. Berry (Cambridge University Press, Cambridge, 1976) are two good books at an intermediate level. Both use some calculus, but avoid the full machinery of Einstein's equations.

*The First Three Minutes* by S. Weinberg (Basic Books, New York, 1977) gives a brilliant description of the early stages of the expansion of the universe. *The Moment of Creation* by J. S. Trefil (Scribner's, New York, 1983), *The Creation of Matter* by H. Fritzsch (Basic Books, New York, 1984), and *The Left Hand of Creation* by J. D. Barrow and J. Silk (Basic Books, New York, 1983) are other good books dealing with the beginning and the evolution of the universe.

*The Red Limit* by T. Ferris (Morrow, New York, 1977), *Violent Universe* by N. Calder (Viking, New York, 1969), and *First Light* by R. Preston (Morgan Entrekin, New York, 1987) were written by journalists; these books deal with the recent discoveries, but they emphasize the astronomers, astrophysicists, and cosmologists who made these discoveries. *Three Degrees Above Zero* by J. Bernstein (Scribner's, New York, 1984) tells the story of the discovery of the cosmic background radiation.

The following are some recent articles dealing with cosmology:

"Will the Universe Expand Forever?" J. R. Gott, J. E. Gunn, D. N. Schramm, and B. M. Tinsley, *Scientific American*, March 1976

"The Curvature of Space in a Finite Universe," J. J. Callahan, *Scientific American*, August 1976

"The Clustering of Galaxies," E. J. Groth, P. J. E. Peebles, M. Seldner, and R. M. Soneira, *Scientific American*, November 1977

"The Cosmic Background Radiation and the New Ether Drift," R. A. Muller, *Scientific American*, May 1978

"The New Inflationary Universe," M. M. Waldrop, *Science*, January 1983

"The Future of the Universe," C. Teplitz and V. L. Teplitz, *Scientific American*, March 1983

"The Structure of the Early Universe," J. D. Barrow and J. Silk, *Scientific American*, April 1980

"The Early Universe and High-Energy Physics," D. N. Schramm, *Physics Today*, April 1983

"Dark Matter in Spiral Galaxies," V. Rubin, *Scientific American*, June 1983

"The Inflationary Universe," A. H. Guth and P. J. Steinhardt, *Scientific American*, May 1984

"Dark Matter in the Universe," L. M. Krauss, *Scientific American*, December 1986

"Particle Physics and Inflationary Cosmology," A. Linde, *Physics Today*, September 1987

The *Resource Letter RC-1* in the *American Journal of Physics*, March 1976, lists some further references.

## Questions

1. Why do astronomers not rely on triangulation to measure the distances of galaxies?

2. If you look at the night sky, you see that the distribution of stars is not uniform. What main deviation from uniformity do you see, and to what is this deviation due?

3. The last figure of Part I of the Prelude shows a plot of the angular distribution of galaxies in a region of the sky. According to this plot, the angular distribution is pretty much uniform (isotropic). However, this plot gives us no direct evidence that the distribution is also uniform in depth (homogeneous). If you assume that the Earth does not occupy a preferred, central spot in the universe, can you conclude that isotropy implies homogeneity?

**Fig. II.14**

4. Figure II.14 shows the positions of four galaxies, with our own Galaxy at the center. Draw a figure showing the positions of these four galaxies at a later time, when the universe is twice as large.

5. Consider the galaxies shown in Figure II.15; the arrows are the velocity vectors of these galaxies in the reference frame of our Galaxy. According to the Galilean addition law for velocities, find the velocity vectors of these galaxies (including our own Galaxy) in the reference frame of the galaxy *P*; do this by graphically subtracting the velocity vectors.

6. If the velocity of recession of the galaxies were not proportional to the distance ($v = H_0 r$) but, rather, proportional to the distance squared ($v = H_0 r^2$) or proportional to some other power of the distance, then our Galaxy would occupy a preferred, central spot in the universe. Explain.

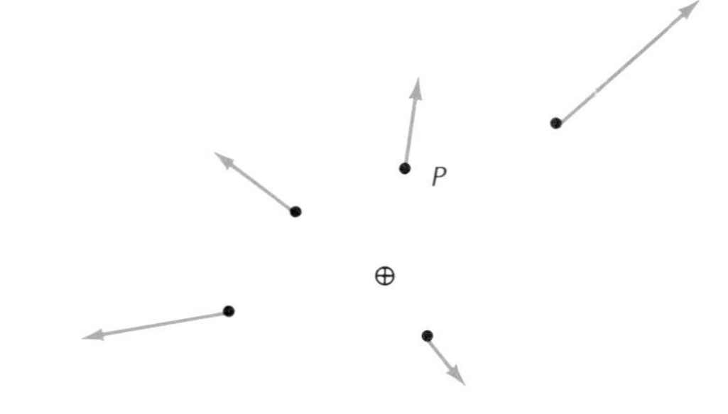


**Fig. II.15**

7. The value of the Hubble constant that Hubble had deduced from the available data in 1935 was $1.6 \times 10^5$ (m/s)/(million light-years). The corresponding expansion time [see Eq. (5)] is $1.9 \times 10^9$ years. How does this compare with the age of the Earth and with the age of globular clusters? With what problem was Hubble faced?

8. If the expansion of the universe is slowing down, can the Hubble constant be truly constant?

9. Imagine a galaxy very close to the edge of the observable universe. This galaxy is moving away from us at a speed nearly equal to the speed of light. Does this mean that this galaxy will reach the edge of the observable universe and disappear from sight?

10. Assume that the universe stops expanding and begins to contract. Would the light from distant galaxies continue to exhibit a red shift?

11. Describe the difference between the final states of the universe for the cases $\rho_0 < 3H_0^2/(8\pi G)$ and $\rho_0 = 3H_0^2/(8\pi G)$.

12. How will life ultimately end if the universe continues to expand forever? If it contracts?

13. Why are black holes, neutrons, or gravitational waves in intergalactic space hard to detect?

14. Where were the atoms in your body made?

CHAPTER 10

# Systems of Particles

So far we have dealt almost exclusively with the motion of a single particle. Now we will begin to study systems of several particles interacting with each other via some forces. Since chunks of ordinary matter are made of particles (electrons, protons, and neutrons), all the macroscopic bodies that we encounter in our everyday environment are in fact many-particle systems containing a very large number of particles. However, for most practical purposes it is not desirable to adopt such an extreme microscopic point of view. For example, in dealing with a collision between two automobiles, we may find it convenient to pretend that each of the automobiles is a particle and not inquire into their internal structure — we then regard the colliding automobiles as a system of two particles which exert forces on each other when in contact. Likewise, in dealing with the Solar System, we may find it convenient to pretend that each planet and each satellite is a particle — we then regard the Solar System as a system of such planet and satellite particles loosely held together by gravitation and orbiting around the Sun and around each other.

The equations of motion of a system of many particles are often hard, and sometimes impossible, to solve. It is therefore necessary to make the most of any information that can be extracted from the general conservation laws. In the following sections we will see how the laws of conservation of momentum, energy, and angular momentum apply to a system of particles.

## 10.1 Momentum of a System of Particles

The momentum of a single particle was defined in Section 5.5; it is the product of the mass and velocity of the particle:

$$\mathbf{p} = m\mathbf{v}$$

Now consider a system of $n$ particles. The total momentum of the system of particles is simply the vector sum of the individual momenta of all the particles. Thus, if $\mathbf{p}_1 = m_1\mathbf{v}_1$, $\mathbf{p}_2 = m_2\mathbf{v}_2$, . . . , $\mathbf{p}_n = m_n\mathbf{v}_n$ are the momenta of the individual particles, then the total momentum is

*Momentum of a system of particles*

$$\boxed{\mathbf{P} = \mathbf{p}_1 + \mathbf{p}_2 + \cdots + \mathbf{p}_n} \tag{1}$$

or

$$\mathbf{P} = \sum_{i=1}^{n} \mathbf{p}_i$$

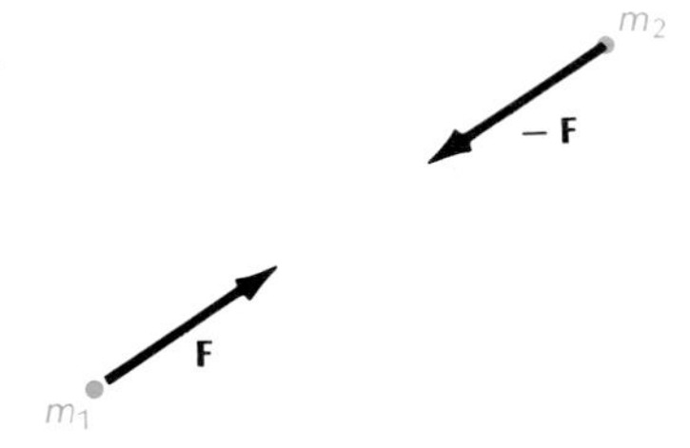


**Fig. 10.1** Two particles exerting forces on each other.

The simplest of all many-particle systems consists of just two particles exerting some mutual forces on one another (Figure 10.1). Let us assume that the two particles are isolated from the rest of the universe so that, besides their mutual forces, they experience no extra forces of any kind. If the force exerted by particle 2 on particle 1 is **F**, then, by Newton's Third Law, the force exerted by particle 1 on particle 2 is $-\mathbf{F}$ and hence the equation of motion of each particle is

$$\frac{d\mathbf{p}_1}{dt} = \mathbf{F} \tag{2}$$

$$\frac{d\mathbf{p}_2}{dt} = -\mathbf{F} \tag{3}$$

If we add these two equations together, we obtain

$$\frac{d\mathbf{p}_1}{dt} + \frac{d\mathbf{p}_2}{dt} = \mathbf{F} + (-\mathbf{F}) \tag{4}$$

that is,

$$\frac{d}{dt}(\mathbf{p}_1 + \mathbf{p}_2) = 0 \tag{5}$$

This shows that the total momentum of the two-particle system is a constant of the motion:

*Law of conservation of momentum*

$$\boxed{\mathbf{P} = \mathbf{p}_1 + \mathbf{p}_2 = [\text{constant}]} \tag{6}$$

This is the **law of conservation of momentum.** Note that this law is a direct consequence of Newton's Third Law: the total momentum is a constant because the equality of action and reaction keeps the momentum changes of the two particles exactly equal in magnitude but opposite in direction — the particles merely exchange some momentum by means of their mutual forces.

Conservation of momentum is a powerful tool which permits us to calculate some general features of the motion even when we are ignorant of the detailed features of the interparticle forces.

EXAMPLE 1. An automobile of mass 1500 kg traveling at 25 m/s (90 km/h) crashes into a similar parked automobile. The two automobiles remain joined after the collision. What is the velocity of the wreck immediately after the collision? Neglect friction against the road.

SOLUTION: Under the assumptions of the problem, the only horizontal forces are the mutual forces of one automobile on the other. Thus, the horizontal component of the momentum is conserved. Before the collision, the (horizontal) velocity of one automobile is $v_1 = 25$ m/s and that of the other is $v_2 = 0$. With the $x$ axis along the direction of motion (Figure 10.2), the total momentum is

$$P_x = m_1 v_1 + m_2 v_2 = m_1 v_1$$

After the collision, both automobiles have the same velocity, $v_1' = v_2' = v'$, and the total momentum is

$$P_x = m_1 v_1' + m_2 v_2' = (m_1 + m_2)v'$$

By momentum conservation, the momentum before the collision must equal the momentum after the collision,

$$m_1 v_1 = (m_1 + m_2)v'$$

or

$$v' = \frac{m_1 v_1}{m_1 + m_2} = \frac{1500 \text{ kg} \times 25 \text{ m/s}}{1500 \text{ kg} + 1500 \text{ kg}} = 12.5 \text{ m/s}$$

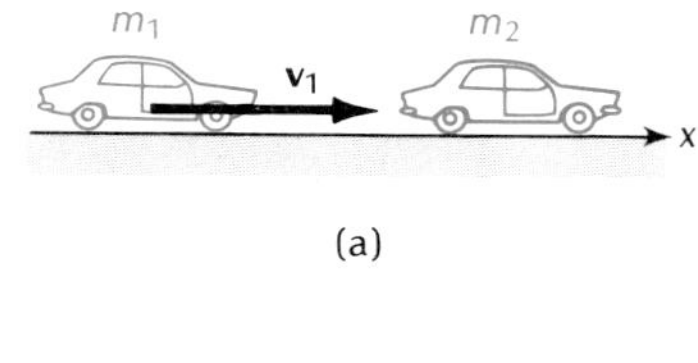


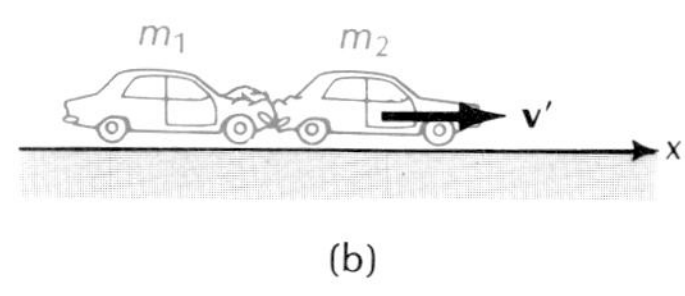


**Fig. 10.2** (a) A speeding automobile crashes into a parked automobile. (b) After the collision, the two automobiles remain joined.

COMMENTS AND SUGGESTIONS: Obviously, the forces acting during this automobile collision are extremely complicated, but momentum conservation permits us to bypass these complications and obtain the answer directly. Note that it is valid to use momentum conservation in this example, but it is *not* valid to use energy conservation. The kinetic energy is *not* conserved in this automobile collision — some of the energy is used up to produce structural changes in the automobiles.

The use of momentum conservation in a problem of motion always involves the familiar three steps: First write an expression for the momentum at one time of the motion, then write an expression for the momentum at another time, and then rely on conservation to equate the two expressions. Since momentum is a vector quantity, it has three components, each of which is conserved separately. In the general case, this leads to three separate conservation equations, which can be solved for three unknown components of one of the velocities. However, in our special example of one-dimensional motion, the only nonzero component of the momentum is the $x$ component, along the line of motion, and we have to deal with only one equation. Whether dealing with one component or with three components of the momentum, always remember to take into account the signs of the components — reversal of signs means reversal of the direction of motion.

EXAMPLE 2. A gun used on board an eighteenth-century warship is mounted on a carriage which allows the gun to roll back each time it is fired (Figure 10.3). The mass of the gun (including the carriage) is 2000 kg. The gun fires a 5.9-kg shot horizontally with a velocity of 490 m/s. What is the recoil velocity of the gun?

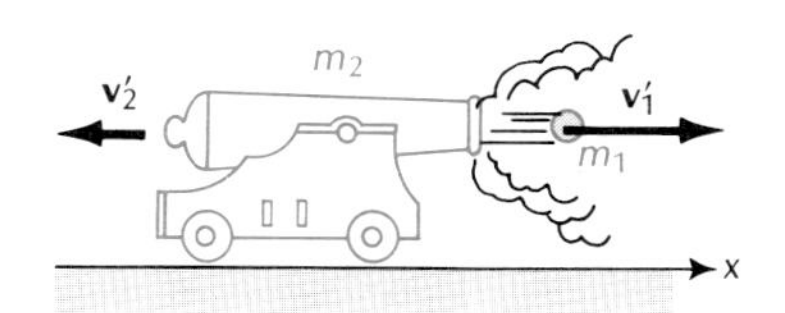


**Fig. 10.3** Recoil of a gun.

SOLUTION: Before the firing, the total momentum is zero. After the firing, the (horizontal) velocity of the shot is $v_1' = 490$ m/s, and that of the gun is $v_2'$; hence the total momentum is

$$P_x = m_1 v_1' + m_2 v_2'$$

where $m_1$ is the mass of the shot and $m_2$ the mass of the gun (including the carriage). By momentum conservation, the total momentum of the shot and gun must be the same before and after the firing. Thus,

$$0 = m_1 v_1' + m_2 v_2'$$

or

$$v_2' = -\frac{m_1}{m_2} v_1'$$

$$= -\frac{5.9 \text{ kg}}{2000 \text{ kg}} \times 490 \text{ m/s} = -1.4 \text{ m/s}$$

The negative sign indicates that the recoil velocity is opposite to the velocity of the shot.

---

EXAMPLE 3. In a demonstration experiment, two pucks of equal masses are made to collide on a nearly frictionless air table (see Figure 10.4a). Before the collision, one of the pucks has a velocity of 0.300 m/s, and the other is at rest. After the (glancing) collision, the first puck has a velocity of 0.260 m/s at an angle of 30° with its initial direction of motion. What is the velocity of the other puck?

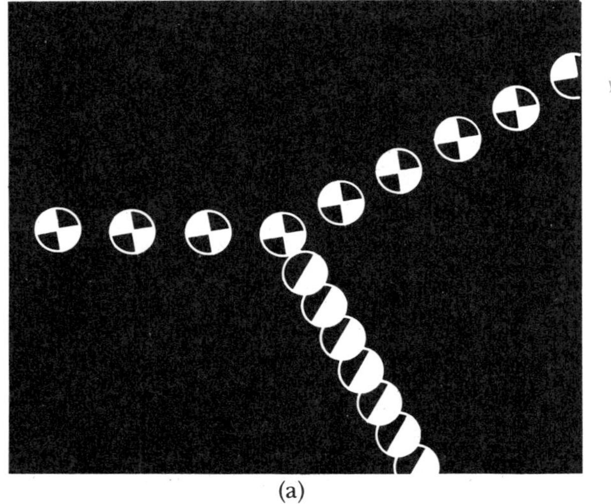

(a)

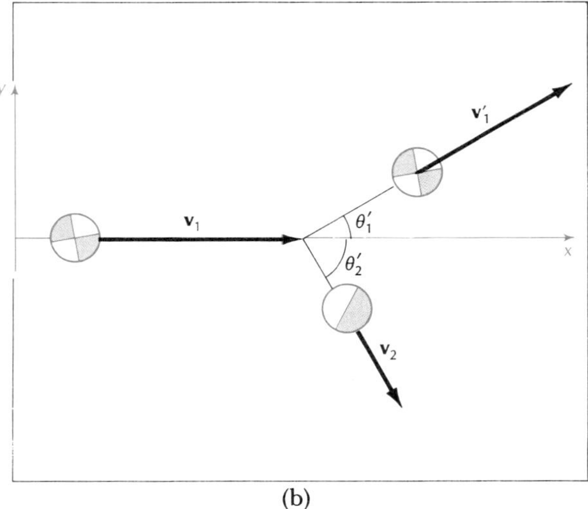


(b)

**Fig. 10.4** (a) Two pucks colliding on an air table (drawing based on multiple-exposure photographs). (b) After the collision, the velocities of the pucks make angles $\theta_1'$ and $\theta_2'$ with the $x$ axis.

SOLUTION: Figure 10.4b shows the velocity vector $\mathbf{v}_1$ before the collision and the velocity vectors $\mathbf{v}_1'$ and $\mathbf{v}_2'$ after the collision. The $x$ axis is arranged along the initial direction of motion. Before the collision, the components of the total momentum are

$$P_x = m_1 v_{1,x}$$

$$P_y = 0$$

After the collision, the components of the total momentum are

$$P_x = m_1 v_{1,x}' + m_2 v_{2,x}'$$

$$P_y = m_1 v_{1,y}' + m_2 v_{2,y}'$$

Momentum conservation tells us that each component of the momentum is unchanged in the collision. We therefore have one equation for the conservation of the $x$ component of the momentum,

$$m_1 v_{1,x} = m_1 v'_{1,x} + m_2 v'_{2,x}$$

and another equation for the conservation of the $y$ component of the momentum,

$$0 = m_1 v'_{1,y} + m_2 v'_{2,y}$$

From these equations, we immediately find the components of the velocity of puck 2 after the collision,

$$v'_{2,x} = \frac{m_1 v_{1,x} - m_1 v'_{1,x}}{m_2} = v_{1,x} - v'_{1,x}$$

$$v'_{2,y} = \frac{-m_1 v'_{1,y}}{m_2} = -v'_{1,y}$$

Since $v_{1,x} = v_1$, $v'_{1,x} = v'_1 \cos\theta'_1$, and $v'_{1,y} = v'_1 \sin\theta'_1$, we obtain

$$\begin{aligned} v'_{2,x} &= v_1 - v'_1 \cos\theta'_1 \\ &= 0.300\ \text{m/s} - 0.260\ \text{m/s} \times \cos 30° = 0.075\ \text{m/s} \end{aligned}$$

and

$$\begin{aligned} v'_{2,y} &= -v'_1 \sin\theta'_1 \\ &= -0.260\ \text{m/s} \times \sin 30° = -0.130\ \text{m/s} \end{aligned}$$

The magnitude of the velocity of puck 2 is then

$$v'_2 = \sqrt{(0.075\ \text{m/s})^2 + (-0.130\ \text{m/s})^2} = 0.150\ \text{m/s}$$

and the angle between the velocity and the $x$ axis is given by

$$\tan\theta'_2 = \frac{-0.130\ \text{m/s}}{0.075\ \text{m/s}} = -1.73$$

or $\theta'_2 = -60°$.

COMMENTS AND SUGGESTIONS: In the case of one-dimensional motion, the line of motion defines a preferred direction, and this direction is the obvious choice for one of the axes, say, the $x$ axis, as in Examples 1 and 2. In the case of two- or three-dimensional motion, there are several lines of motion, and the choice of coordinate axes is not so obvious. Usually, it proves convenient to place one of the axes along the direction of motion of one of the bodies, as in our example of the two pucks.

Instead of dealing with the components of the momentum, we could have solved this example by vector subtraction: the momentum of puck 2 equals the difference between the initial momentum and the final momentum of puck 1, that is, $\mathbf{p}'_2 = \mathbf{p}_1 - \mathbf{p}'_1$. The vector $\mathbf{p}'_2$ can be found by drawing the vector triangle and solving it trigonometrically.

---

The conservation law for the total momentum depends on the absence of "extra" forces. If the particles are not isolated from the rest of the universe, then besides the mutual forces exerted by one particle on the other, there are also forces exerted by other bodies not belong-

*Internal and external forces*

ing to the particle system; the former forces are called **internal forces** and the latter **external forces.** For example, if the two particles are near the Earth, then gravity will act on them and play the role of an external force. To take such external forces into account, we must modify Eqs. (2) and (3). If the external force on particle 1 is $\mathbf{F}_{1,\,\text{ext}}$, then the total force on this particle is $\mathbf{F} + \mathbf{F}_{1,\,\text{ext}}$ and the equation of motion will be

$$\frac{d\mathbf{p}_1}{dt} = \mathbf{F} + \mathbf{F}_{1,\,\text{ext}} \tag{7}$$

Likewise,

$$\frac{d\mathbf{p}_2}{dt} = -\mathbf{F} + \mathbf{F}_{2,\,\text{ext}} \tag{8}$$

These equations lead to

$$\frac{d}{dt}(\mathbf{p}_1 + \mathbf{p}_2) = \mathbf{F}_{1,\,\text{ext}} + \mathbf{F}_{2,\,\text{ext}} \tag{9}$$

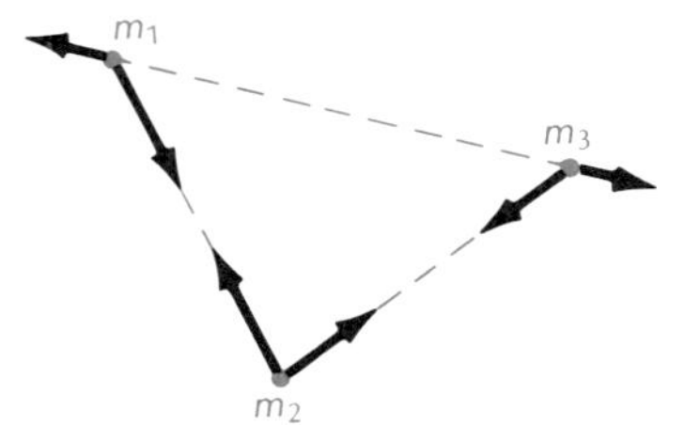

**Fig. 10.5** Three particles exerting forces on each other.

The sum $\mathbf{F}_{1,\,\text{ext}} + \mathbf{F}_{2,\,\text{ext}}$ is simply the total external force on the particle system. Thus, Eq. (9) states that the rate of change of the total momentum of the two-particle system equals the total *external* force on the system.

For a system containing more than two particles, we can obtain similar results. If the system is isolated so that there are no external forces, then the mutual interparticle forces acting between pairs of particles merely transfer momentum from one particle of the pair to the other, just as in the case of two particles. Since all the internal forces necessarily arise from such forces between pairs of particles, these internal forces cannot change the total momentum. For example, Fig. 10.5 shows three particles exerting forces on one another. Consider particle 1; the mutual forces between particles 1 and 2 exchange momentum between these two, while the mutual forces between particles 1 and 3 exchange momentum between those two. But none of these momentum transfers will change the total momentum. The same holds for particles 2 and 3. Consequently, the total momentum of an isolated system obeys the conservation law

$$\boxed{\mathbf{P} = [\text{constant}]} \tag{10}$$

If besides the internal forces, there are external forces, then the latter will change the momentum. The rate of change can be calculated in essentially the same way as for the two-particle system and, again, the rate of change of the total momentum is equal to the total external force. We can write this as

*Rate of change of momentum*

$$\boxed{\frac{d\mathbf{P}}{dt} = \mathbf{F}_{\text{ext}}} \tag{11}$$

where $\mathbf{F}_{\text{ext}}$ is the total external force on the system.

Equations (10) and (11) have exactly the same mathematical form as Eqs. (5.16) and (5.17) and may be regarded as the generalizations for a system of particles of Newton's First and Second Laws. As we will see in Section 10.3, Eq. (11) is an equation of motion for the system of particles — it determines the overall translational motion of the system.

EXAMPLE 4. During a rainstorm the volume of rain falling on 1 $m^2$ of ground in one hour amounts to 0.1 $m^3$. The raindrops hit the ground with a vertical velocity of 10 m/s. What is the average force per unit area (force per square meter) that the impact of the raindrops exerts on the ground?

SOLUTION: We can calculate the force from the rate at which the rain transfers momentum to the ground. The mass of water that falls on 1 $m^2$ in one hour is $m = 0.1\ \text{m}^3 \times 10^3\ \text{kg/m}^3 = 1 \times 10^2\ \text{kg}$, and the momentum contributed by this mass is $\Delta P = mv = 1 \times 10^2\ \text{kg} \times 10\ \text{m/s} = 1 \times 10^3\ \text{kg} \cdot \text{m/s}$. The average rate of momentum transfer to the ground, or the average force, is then

$$F = \frac{\Delta P}{\Delta t} = \frac{1 \times 10^3\ \text{kg} \cdot \text{m/s}}{3600\ \text{s}} = 0.3\ \text{N}$$

## 10.2 Center of Mass

In our study of kinematics and dynamics in the preceding chapters, we have always ignored the sizes of the bodies; even when analyzing the motion of a large body — a railroad car or a ship — we pretended that the motion could be treated as particle motion, position being described by reference to some fiducial point marked on the body. In reality, large bodies are systems of particles and their motion obeys Eq. (11) for a system of particles. This equation can be converted into an equation of motion, containing just one acceleration rather than the rate of change of momentum of the entire system, by making reference to one special point in the body: the **center of mass.** The equation that describes the motion of this special point has the same mathematical form as the equation of motion of a particle, that is, the motion of the center of mass mimics particle motion (see, for example, Fig. 10.6).

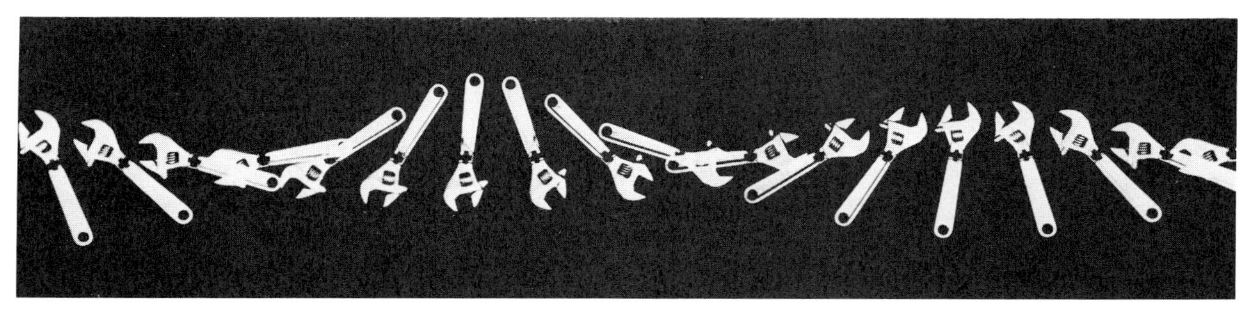

**Fig. 10.6** A wrench sliding on a (nearly) frictionless table. In the absence of external forces, the center of mass, marked with a cross, moves with uniform velocity, just like a free particle.

The position of the center of mass of a system is merely the average position of the mass of the system. For instance, if the system consists of two particles each of a mass of 1 kg, then the center of mass is half-

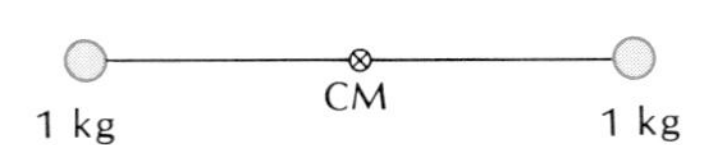


**Fig. 10.7** Two particles of equal mass. Their center of mass is marked by a small circle with a cross.

way between them (Figure 10.7). In any system consisting of $n$ particles of equal mass — such as a piece of pure metal with atoms of only one kind — the position vector of the center of mass is simply the average of the position vectors of all the particles,

$$\mathbf{r}_{\mathrm{CM}} = \frac{\mathbf{r}_1 + \mathbf{r}_2 + \cdots + \mathbf{r}_n}{n} \tag{12}$$

If the system consists of particles of unequal mass, then the position of the center of mass can be calculated by first subdividing the particles into fragments of equal mass. This, of course, means that in the average over the positions of the fragments, the positions of particles of large mass will have to be included more often than the positions of particles of small mass — the number of times the position of a particle must be included in the average is in direct proportion to its mass. This leads to the following general expression for the position of the center of mass of the system of particles:

$$\mathbf{r}_{\mathrm{CM}} = \frac{m_1\mathbf{r}_1 + m_2\mathbf{r}_2 + \cdots + m_n\mathbf{r}_n}{m_1 + m_2 + \cdots + m_n} \tag{13}$$

or

*Center of mass of a system of particles*

$$\boxed{\mathbf{r}_{\mathrm{CM}} = \frac{m_1\mathbf{r}_1 + m_2\mathbf{r}_2 + \cdots + m_n\mathbf{r}_n}{M}} \tag{14}$$

where $M = m_1 + m_2 + \cdots + m_n$ is the total mass.

For numerical calculations it is usually best to treat the $x$, $y$, and $z$ components of Eq. (14) separately:

$$x_{\mathrm{CM}} = \frac{1}{M}(m_1x_1 + m_2x_2 + \cdots + m_nx_n) \tag{15}$$

$$y_{\mathrm{CM}} = \frac{1}{M}(m_1y_1 + m_2y_2 + \cdots + m_ny_n) \tag{16}$$

$$z_{\mathrm{CM}} = \frac{1}{M}(m_1z_1 + m_2z_2 + \ldots + m_nz_n) \tag{17}$$

---

EXAMPLE 5. The distance between the centers of the atoms of potassium and bromine in a potassium bromide (KBr) molecule is 2.82 Å. Since almost the entire mass of each atom is concentrated in its (very small) nucleus, the mass distribution of each atom is that of a pointlike particle located at nucleus, at the center of the atom. Find the center of mass of the molecule.

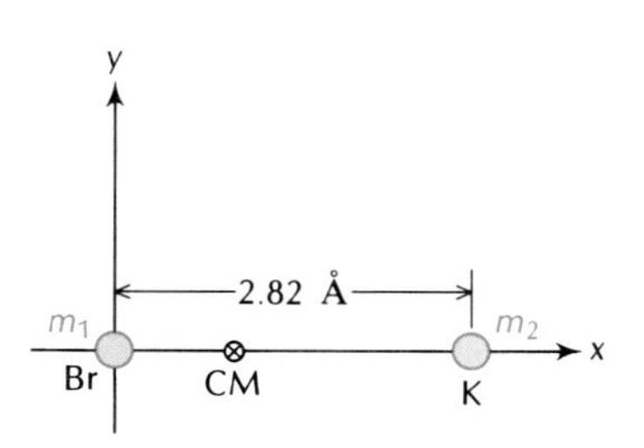


**Fig. 10.8** Atoms of bromine (Br) and potassium (K), regarded as particles.

SOLUTION: In Figure 10.8 the atoms of bromine and potassium are shown as particles. The bromine atom is at $x_1 = 0$ and the potassium atom is at $x_2 = 2.82$ Å. The center of mass is at

$$x_{\mathrm{CM}} = \frac{m_1x_1 + m_2x_2}{m_1 + m_2} \tag{18}$$

or, since $x_1 = 0$,

$$x_{\mathrm{CM}} = \frac{m_2}{m_1 + m_2} x_2 \tag{19}$$

The masses of atoms of bromine and potassium are $m_1 = 79.9$ u and $m_2 = 39.1$ u, respectively (see Appendix 9). This gives

$$x_{\mathrm{CM}} = \frac{39.1\ \mathrm{u}}{79.9\ \mathrm{u} + 39.1\ \mathrm{u}} \times 2.82\ \text{Å} = 0.93\ \text{Å}$$

This center of mass is shown in Figure 10.8.

COMMENTS AND SUGGESTIONS: Note that, according to Eq. (19), the distance of $m_1$ from the center of mass is

$$\frac{m_2 x_2}{(m_1 + m_2)}$$

and the distance of $m_2$ from the center of mass is

$$x_2 - \frac{m_2 x_2}{(m_1 + m_2)} = \frac{m_1 x_2}{(m_1 + m_2)}$$

Thus, the position of the center of mass divides the line segment connecting the two particles in the ratio $m_2 : m_1$. This is a general result for the position of the center of mass of a system of two particles.

The position of the center of mass of a solid body can, in principle, be calculated from Eq. (13) — a solid is a collection of atoms, each of which can be regarded as a particle. However, it would be awkward to deal with the $10^{23}$ or so atoms that make up a chunk of matter the size of a coin. It is more convenient to pretend that matter in bulk has a smooth and continuous distribution of mass over its entire volume. The mass distribution is then described by the **density,** or the mass per unit volume.

To find the center of mass of such a continuous mass distribution, imagine that the volume is divided into small volume elements, each of size $\Delta V$ (Figure 10.9). Then the mass in one of these volume elements is

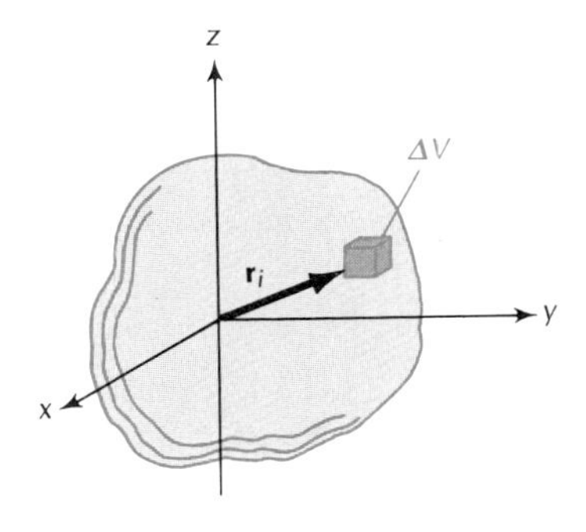


**Fig. 10.9** Volume element $\Delta V$ in a continuous mass distribution.

$$\Delta m_i = \rho\, \Delta V \tag{20}$$

where $\rho$ is the mass density at the position of the volume element (this density may be a function of position). According to Eq. (13), the position of the center of mass is then approximately

$$\mathbf{r}_{\mathrm{CM}} = \frac{1}{M} \sum_{i=1}^{n} \mathbf{r}_i\, \Delta m_i = \frac{1}{M} \sum_{i=1}^{n} \mathbf{r}_i \rho\, \Delta V \tag{21}$$

This approximation becomes exact in the limit $\Delta V \to 0$:

$$\mathbf{r}_{\mathrm{CM}} = \frac{1}{M} \lim_{\Delta V \to 0} \sum_i \mathbf{r}_i \rho\, \Delta V \tag{22}$$

The limit is an integral,

*Center of mass of a continuous distribution of mass*

$$\boxed{\mathbf{r}_{CM} = \frac{1}{M}\int \mathbf{r}\rho \, dV} \tag{23}$$

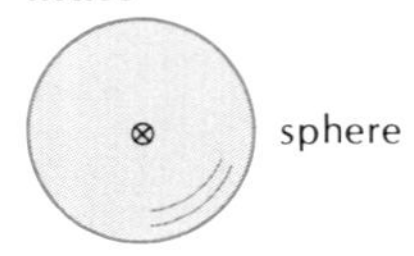


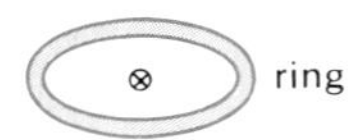

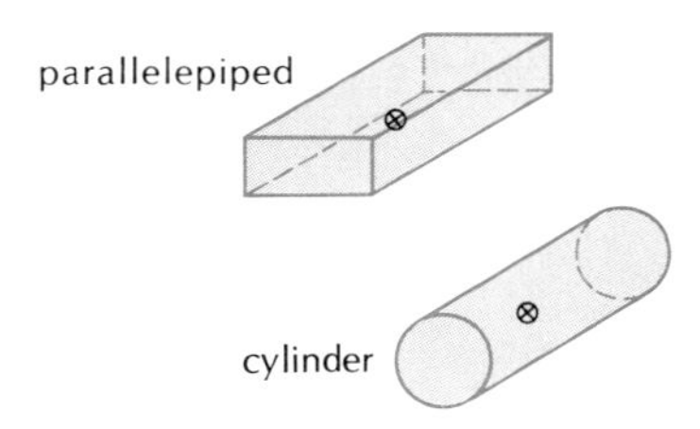


**Fig. 10.10** Several bodies for which the center of mass coincides with the geometric center.

In terms of components, Eq. (23) becomes

$$x_{CM} = \frac{1}{M}\int x\rho \, dV \tag{24}$$

$$y_{CM} = \frac{1}{M}\int y\rho \, dV \tag{25}$$

$$z_{CM} = \frac{1}{M}\int z\rho \, dV \tag{26}$$

If the density of the body is uniform, then the position of the center of mass is simply the average position of all the points making up its volume (this average is usually called the *centroid* in mathematics). If the body has a symmetric shape, this average position will often be obvious by inspection. For instance, a homogeneous sphere, or a ring, or a circular plate, or a parallelepiped, or a cylinder will have its center of mass at the geometrical center (Figure 10.10).

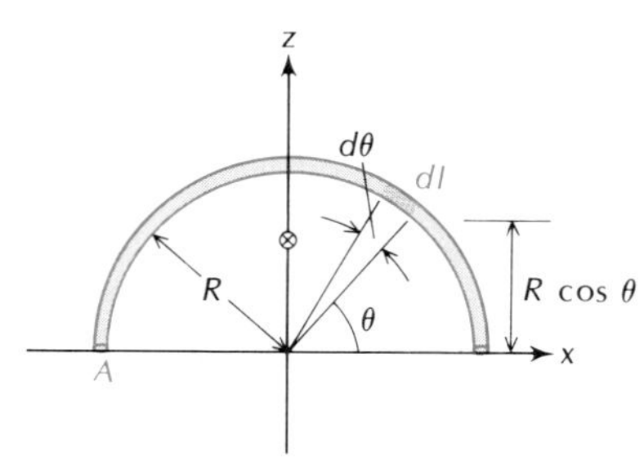


**Fig. 10.11** Thin rod in the shape of a semicircle. The cross-sectional area of the rod is $A$.

EXAMPLE 6. A thin rod of cross-sectional area $A$ is bent in the shape of a semicircle of radius $R$ (Figure 10.11). Where is the center of mass of the rod?

SOLUTION: Assume that the rod is in the $z$–$x$ plane, and that the center of the semicircle is at the origin. In view of the symmetry of the rod, the center of mass will be somewhere on the $z$ axis. To find the $z$ coordinate of the center of mass, we use Eq. (26),

$$z_{CM} = \frac{1}{M}\int z\rho \, dV$$

Consider a small segment $dl$ of the rod (Figure 10.11); this segment subtends a small angle $d\theta$ so

$$dl = R \, d\theta$$

The volume of rod within this small angle is the volume of a small cylinder of base $A$ and height $dl$:

$$dV = A \, dl = AR \, d\theta$$

The $z$ coordinate of this small volume is $z = R \sin \theta$. Hence

$$\begin{aligned} z_{CM} &= \frac{1}{M}\int z\rho \, dV \\ &= \frac{1}{M}\int (R \sin \theta)\rho AR \, d\theta \\ &= \frac{1}{M}\rho AR^2 \int \sin \theta \, d\theta \end{aligned}$$

Taking into account that the limits of integration for the angle $\theta$ are 0° and 180°, we obtain

$$z_{CM} = \frac{1}{M}\rho AR^2 \int_{0°}^{180°} \sin\theta \, d\theta$$

$$= \frac{1}{M}\rho AR^2 \Big[-\cos\theta\Big]_{0°}^{180°}$$

$$= \frac{2}{M}\rho AR^2$$

The total volume of the rod is that of a cylinder of base $A$ and height $\pi R$; therefore the density of the rod is $\rho = M/V = M/(A\pi R)$. Substituting this into our equation for $z_{CM}$, we find the final result

$$z_{CM} = \frac{2}{M}\frac{M}{A\pi R}AR^2 = \frac{2}{\pi}R$$

Thus, the distance of the center of mass from the center of the circle is $(2/\pi)R$, or $0.637R$.

EXAMPLE 7. The Great Pyramid at Giza (Figure 10.12) has a height of 147 m. Assuming that the entire volume is completely filled with stone of uniform density, find the center of mass.

**Fig. 10.12** The Great Pyramid.

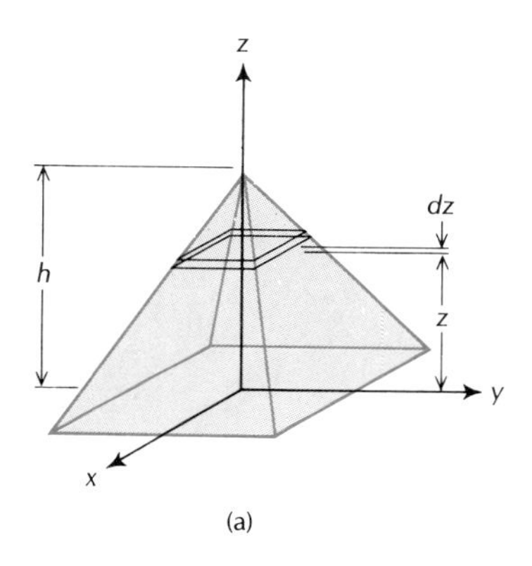


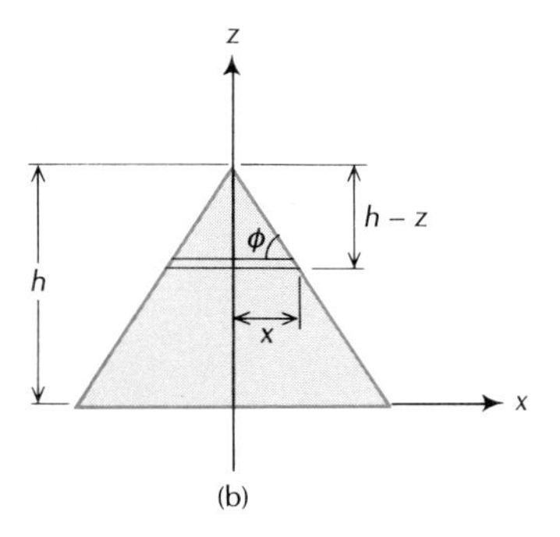


**Fig. 10.13** (a) The pyramid and a thin horizontal slab within the pyramid. (b) Side view ("elevation") of the pyramid.

SOLUTION: Because of symmetry, the center of mass will be on the vertical line through the apex; in Figure 10.13a this line coincides with the $z$ axis. To find the $z$ coordinate of the center of mass, consider a thin horizontal slab of thickness $dz$ at a height $z$ above the ground. The entire pyramid can be regarded as built up of such slabs. The horizontal face of the slab is a square measuring $2x$ by $2x$; thus the volume of the slab is approximately $(2x)^2\, dz$, that is,

$$dV = (2x)^2\, dz$$

From Figure 10.13b we obtain $x = (h - z)/\tan\phi$, where $h$ is the height of the pyramid and $\phi$ the angle between one side and the ground. Hence

$$dV = 4\frac{(h-z)^2}{\tan^2\phi}\, dz$$

Equation (26) then gives

$$z_{CM} = \frac{1}{M}\int z\rho\, dV = \frac{1}{M}\int_0^h 4z\rho\,\frac{(h-z)^2}{\tan^2\phi}\,dz$$

$$= \frac{4\rho}{M\tan^2\phi}\int_0^h z(h-z)^2\,dz$$

$$= \frac{4\rho}{M\tan^2\phi}\int_0^h (zh^2 - 2z^2h + z^3)\,dz$$

$$= \frac{4\rho}{M\tan^2\phi}\left[\tfrac{1}{2}z^2h^2 - \tfrac{2}{3}z^3h + \tfrac{1}{4}z^4\right]_0^h$$

$$= \frac{4\rho}{M\tan^2\phi}\,\frac{h^4}{12} \qquad (27)$$

The density of the pyramid is given by $\rho = M/V$, where $V$ is the total volume,

$$V = \int dV = \int_0^h 4\frac{(h-z)^2}{\tan^2\phi}\,dz = \frac{4}{\tan^2\phi}\,\frac{h^3}{3} \qquad (28)$$

The combination of Eqs. (27) and (28) yields

$$z_{CM} = \tfrac{1}{4}h$$

$$= \tfrac{1}{4}\times 147\text{ m} = 36.8\text{ m} \qquad (29)$$

COMMENTS AND SUGGESTIONS: The final result (29) is independent of the angle $\phi$ — for any pyramid the center of mass is one-quarter of the height above the base. It is easy to demonstrate that the same result also holds for a cone.

The gravitational potential energy of an extended body located near the surface of the Earth can be expressed as a function of the position of the center of mass. According to Eq. (7.45), the potential energy of a single particle at a height $z$ above the ground is $mgz$. For a system of particles, the total potential energy is then

$$U = (m_1z_1 + m_2z_2 + \cdots + m_nz_n)g \qquad (30)$$

Comparison with Eq. (17) shows that the quantity in parentheses is $Mz_{CM}$. Hence

$$\boxed{U = Mgz_{CM}} \qquad (31)$$

This expression for the gravitational potential energy of the system has the same mathematical form as for a single particle — it is as though the entire mass of the system were located at the center of mass.

A simple experimental method for the determination of the center of mass of a body of irregular shape involves suspending the body from a string attached to a point on the surface of the body (see Figure 10.14). The body will then settle into an equilibrium position such that

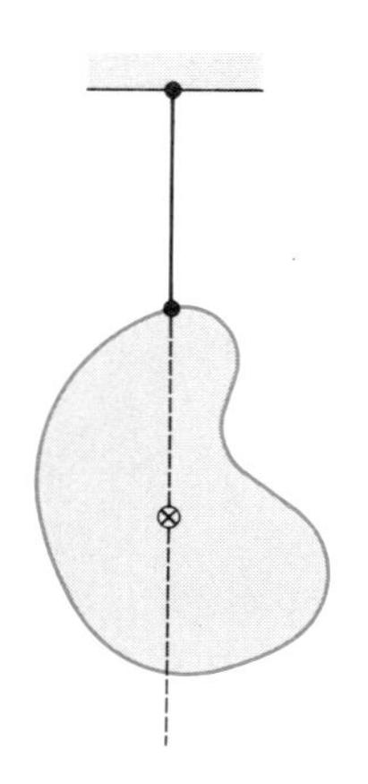

**Fig. 10.14** A body suspended from a string.

its gravitational potential energy is as small as possible. According to Eq. (31), this occurs when the center of mass is located at the smallest possible height. Hence, in equilibrium, the string will be vertical, like a plumb line, and the center of mass of the body will be located directly below the point of attachment of the string (in Chapter 14 we will see that this equilibrium position corresponds to a condition of zero torque). If we draw a line through the body along the prolongation of the line of the string, the center of mass will be somewhere on this line. Repeating this procedure with a different point of attachment, we can draw a second line through the body, and immediately find the center of mass at the intersection of the two drawn lines.

## 10.3 The Motion of the Center of Mass

When a system of particles moves, so does the center of mass. We will now derive the equation of motion of the center of mass. We begin with an expression for the **velocity of the center of mass:**

$$\mathbf{v}_{\mathrm{CM}} = \frac{d\mathbf{r}_{\mathrm{CM}}}{dt} = \frac{1}{M}\left(m_1\frac{d\mathbf{r}_1}{dt} + m_2\frac{d\mathbf{r}_2}{dt} + \cdots + m_n\frac{d\mathbf{r}_n}{dt}\right) \quad (32)$$

*Velocity of the center of mass*

$$= \frac{1}{M}(m_1\mathbf{v}_1 + m_2\mathbf{v}_2 + \cdots + m_n\mathbf{v}_n) \quad (33)$$

Note that this equation has the same mathematical form as Eq. (14), that is, the velocity of the center of mass is an average over the particle velocities, and the number of times each particle velocity is included is in direct proportion to its mass.

From Eq. (33), we obtain

$$m_1\mathbf{v}_1 + m_2\mathbf{v}_2 + \cdots + m_n\mathbf{v}_n = M\mathbf{v}_{\mathrm{CM}} \quad (34)$$

But the left side of this equation is the total momentum of the system. Consequently,

$$\boxed{\mathbf{P} = M\mathbf{v}_{\mathrm{CM}}} \quad (35)$$

*Momentum of a system of particles*

that is, the total momentum of the system equals the total mass times the velocity of the center of mass. This equation is the analog of the familiar equation ($\mathbf{p} = m\mathbf{v}$) for the momentum of a single particle.

The equation of motion now follows from Eq. (11),

$$\mathbf{F}_{\mathrm{ext}} = \frac{d\mathbf{P}}{dt} = \frac{d}{dt}(M\mathbf{v}_{\mathrm{CM}}) = M\frac{d\mathbf{v}_{\mathrm{CM}}}{dt} \quad (36)$$

This can also be written

$$\boxed{M\mathbf{a}_{\mathrm{CM}} = \mathbf{F}_{\mathrm{ext}}} \quad (37)$$

*Acceleration of the center of mass*

where $\mathbf{a}_{\mathrm{CM}} = d\mathbf{v}_{\mathrm{CM}}/dt$ is the **acceleration of the center of mass.** Equation (37) for a system of particles is obviously the analog of Newton's equation of motion for a single particle. The center of mass moves as though it were a particle of mass $M$ under the influence of a force $\mathbf{F}_{ext}$.

This result justifies some of the approximations we made in previous chapters. For instance, in Example 6.5 we treated a ship sliding down a slipway as a particle. Equation (37) shows that this treatment is legitimate: the center of mass of the ship, under the influence of the external forces (gravity and friction), moves parallel to the slipway, just as though it were a particle on an inclined plane with the external forces acting directly on it.

If the net external force vanishes, then the acceleration of the center of mass also vanishes; hence the center of mass remains at rest or moves with uniform velocity.

EXAMPLE 8. During a "space walk" an astronaut floats in space 8.0 m from his Gemini spacecraft orbiting the Earth. He is tethered to the spacecraft by a long umbilical cord; to return, he pulls himself in by this cord. How far does the spacecraft move toward him? The mass of the spacecraft is 3500 kg and the mass of the astronaut, including his space suit, is 140 kg.

SOLUTION: Astronaut and spacecraft exert equal and opposite forces on one another (via the cord); the astronaut is pulled toward the spacecraft and the spacecraft is pulled toward the astronaut. In the absence of external forces, the center of mass of the astronaut–spacecraft system remains at rest. Thus, the spacecraft and the astronaut both move toward the center of mass and there they meet.

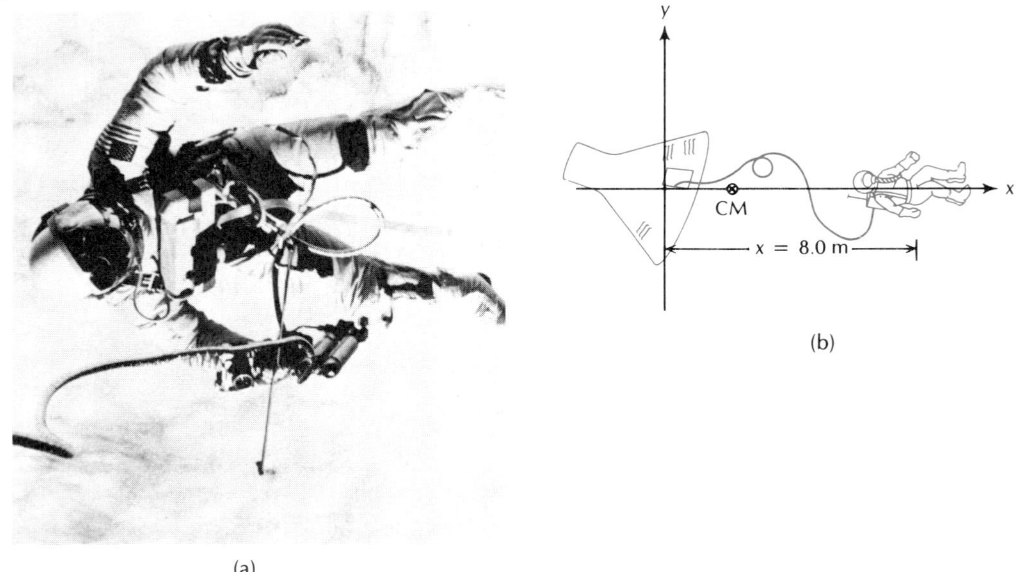


**Fig. 10.15** (a) Astronaut on a "space walk" during the Gemini 4 mission. (b) Astronaut and spacecraft.

With the $x$ axis as in Figure 10.15b, the $x$ coordinate of the center of mass is

$$x_{CM} = \frac{m_1 x_1 + m_2 x_2}{m_1 + m_2}$$

where $m_1 = 3500$ kg is the mass of the spacecraft and $m_2 = 140$ kg is the mass of the astronaut. Strictly, the coordinates $x_1$ and $x_2$ of the spacecraft and the astronaut should correspond to the centers of mass[1] of these bodies, but, for the

[1] The center of mass of a man standing erect is within his body at about the height of the navel.

sake of simplicity, we neglect their size and treat both as particles. The initial values of the coordinates are $x_1 = 0$ and $x_2 = 8.0$ m; hence

$$x_{CM} = \frac{0 + 140 \text{ kg} \times 8.0 \text{ m}}{3500 \text{ kg} + 140 \text{ kg}} = 0.31 \text{ m}$$

During the pulling in, the spacecraft will move from $x_1 = 0$ to $x_1 = 0.31$ m; simultaneously the astronaut will move from $x_2 = 8.0$ m to $x_2 = 0.31$ m.

COMMENTS AND SUGGESTIONS: The crucial ingredient in the solution of this example is to recognize that if two particles meet, the point at which they meet must coincide with their center of mass, and that this center of mass remains at rest or in uniform motion whenever there is no external force. By exploiting these properties of the center of mass, we can predict the meeting point without having to deal with the details of the motion. How the astronaut pulls himself in — fast or slow — does not affect the final result.

The condition of rest or uniform motion for the center of mass is mathematically equivalent to the conservation law for the total momentum, but it often provides a more convenient approach to the solution of a problem than explicit use of momentum conservation.

## 10.4* Energy of a System of Particles

The total kinetic energy of a system of particles is simply the sum of the individual kinetic energies of all the particles,

$$K = \tfrac{1}{2}m_1v_1^2 + \tfrac{1}{2}m_2v_2^2 + \cdots + \tfrac{1}{2}m_nv_n^2 \tag{38}$$

Since Eq. (35) for the momentum of a system of particles resembles the expression for the momentum of a single particle, we might be tempted to guess that the expression for the total kinetic energy of a system of particles also resembles that for a single particle and can be reduced to the form $K = \frac{1}{2}Mv_{CM}^2$. But this is wrong! The total kinetic energy of a system is usually greater than $\frac{1}{2}Mv_{CM}^2$. In what follows, we will see why this is so.

In order to rewrite Eq. (38) in a form involving $\mathbf{v}_{CM}$, we introduce the velocities relative to the center of mass. The particle velocities $\mathbf{v}_1$, $\mathbf{v}_2, \ldots, \mathbf{v}_n$ are reckoned in some given reference frame, say, the reference frame of the ground. We now want to look at these particle velocities from a reference frame moving with the center of mass — the rest frame of the center of mass, or the "CM frame." The velocity of this new reference frame with respect to the old is $\mathbf{v}_{CM}$ and for the particle velocities in the new reference frame the Galilean transformation [Eq. (4.58)] gives the values

$$\mathbf{u}_1 = \mathbf{v}_1 - \mathbf{v}_{CM} \qquad \mathbf{u}_2 = \mathbf{v}_2 - \mathbf{v}_{CM}, \text{ etc.} \tag{39}$$

This yields

$$\mathbf{v}_1 = \mathbf{u}_1 + \mathbf{v}_{CM} \qquad \mathbf{v}_2 = \mathbf{u}_2 + \mathbf{v}_{CM}, \text{ etc.} \tag{40}$$

Inserting Eqs. (40) into Eq. (38), we obtain

* This section is optional.

$$K = \tfrac{1}{2}m_1(\mathbf{u}_1 + \mathbf{v}_{CM})^2 + \tfrac{1}{2}m_2(\mathbf{u}_2 + \mathbf{v}_{CM})^2 + \cdots$$

$$= \tfrac{1}{2}m_1(\mathbf{u}_1^2 + 2\mathbf{u}_1 \cdot \mathbf{v}_{CM} + \mathbf{v}_{CM}^2) + \tfrac{1}{2}m_2(\mathbf{u}_2^2 + 2\mathbf{u}_2 \cdot \mathbf{v}_{CM} + \mathbf{v}_{CM}^2) + \cdots$$

$$= [\tfrac{1}{2}m_1\mathbf{u}_1^2 + \tfrac{1}{2}m_2\mathbf{u}_2^2 + \cdots] + [m_1\mathbf{u}_1 + m_2\mathbf{u}_2 + \cdots] \cdot \mathbf{v}_{CM}$$

$$+ \tfrac{1}{2}[m_1 + m_2 + \cdots]\mathbf{v}_{CM}^2 \qquad (41)$$

The quantity within the first bracket on the right side of Eq. (41) is nothing but the kinetic energy as reckoned in the CM frame; we will call this the **internal kinetic energy** (it is also often called the CM energy):

$$K_{int} = \tfrac{1}{2}m_1u_1^2 + \tfrac{1}{2}m_2u_2^2 + \cdots + \tfrac{1}{2}m_nu_n^2 \qquad (42)$$

The quantity within the second bracket is zero, since it equals

$$m_1(\mathbf{v}_1 - \mathbf{v}_{CM}) + m_2(\mathbf{v}_2 - \mathbf{v}_{CM}) + \cdots = (m_1\mathbf{v}_1 + m_2\mathbf{v}_2 + \cdots)$$

$$- (m_1 + m_2 + \cdots)\mathbf{v}_{CM}$$

$$= (m_1\mathbf{v}_1 + m_2\mathbf{v}_2 + \cdots) - M\mathbf{v}_{CM} \qquad (43)$$

which, indeed, is zero by Eq. (34). The quantity within the last bracket is simply the total mass.

Taking all of this into account, we see that Eq. (41) reduces to

*Kinetic energy of a system of particles*

$$K = K_{int} + \tfrac{1}{2}Mv_{CM}^2 \qquad (44)$$

Thus, the total kinetic energy of a system of particles contains two terms: the translational kinetic energy $\tfrac{1}{2}Mv_{CM}^2$ of the center of mass, calculated just as though the center of mass were a particle of mass $M$ and velocity $\mathbf{v}_{CM}$; and the internal kinetic energy $K_{int}$, which is the energy of motion as reckoned in the CM frame.

If the system of particles is a solid body whose particles are not moving relative to the center of mass, then $K_{int} = 0$. But such absence of "internal" motion requires that the body have no rotational motion about the center of mass (no change of orientation) and also that the particles be rigidly connected to one another (no change of shape, no vibration). The second requirement is very unrealistic. The atoms of a solid body, such as a chunk of metal, are never at rest; they vibrate back and forth about their equilibrium positions at high speed (typically ~400 m/s). Thus, the internal kinetic energy of a solid body is quite large. The kinetic energy associated with the disorganized, random motions of atoms within a body is **heat energy.** If we are interested only in the macroscopic translational and rotational motion of a solid body, we can often ignore the energy of these internal microscopic motions, because it usually remains constant; that is, the heat energy remains constant. However, if friction forces act on the solid body, they will generate heat in the body, and we cannot ignore this energy transfer when we seek to balance the net energy.

EXAMPLE 9. Two automobiles, each of mass 1500 kg, travel in the same direction along a straight road. The speed of one automobile is 25 m/s and the speed of the other automobile is 15 m/s. If we regard these automobiles as a system of two particles, what is the translational kinetic energy of the center of mass? What is the internal kinetic energy?

SOLUTION: With the $x$ axis along the direction of motion, the velocity of the center of mass along this axis is [see Eq. (33)]

$$v_{\mathrm{CM}} = \frac{m_1 v_1 + m_2 v_2}{m_1 + m_2}$$

$$= \frac{1500\ \mathrm{kg} \times 25\ \mathrm{m/s} + 1500\ \mathrm{kg} \times 15\mathrm{m/s}}{3000\ \mathrm{kg}}$$

$$= 20\ \mathrm{m/s}$$

Hence the translational kinetic energy of the center of mass is

$$\tfrac{1}{2}(m_1 + m_2)v_{\mathrm{CM}}^2 = \tfrac{1}{2}(3000\ \mathrm{kg}) \times (20\ \mathrm{m/s})^2$$

$$= 6.0 \times 10^5\ \mathrm{J}$$

The velocities of the automobiles relative to the center of mass are $u_1 = 25\mathrm{m/s} - 20\ \mathrm{m/s} = 5\ \mathrm{m/s}$ and $u_2 = 15\ \mathrm{m/s} - 20\ \mathrm{m/s} = -5\ \mathrm{m/s}$. The internal kinetic energy is then

$$K_{\mathrm{int}} = \tfrac{1}{2}m_1 u_1^2 + \tfrac{1}{2}m_2 u_2^2$$

$$= \tfrac{1}{2} \times 1500\ \mathrm{kg} \times (5\ \mathrm{m/s})^2 + \tfrac{1}{2} \times 1500\ \mathrm{kg} \times (-5\ \mathrm{m/s})^2$$

$$= 3.7 \times 10^4\ \mathrm{J}$$

It is easy to check that the sum of these two kinetic energies has the same value as $\frac{1}{2}m_1 v_1^2 + \frac{1}{2}m_2 v_2^2$.

If the internal and external forces acting on a system of particles are conservative, then the system will have a potential energy. Unless we specify the forces, we cannot write down an explicit formula for the potential energy; but in any case, this potential energy will be some function of the positions of all the particles. The total energy is then the sum of the total kinetic energy [Eq. (44)] and the potential energy (including internal and external contributions). This total energy will be conserved during the motion of the system of particles.

## 10.5* The Motion of a Rocket

The propulsion of any kind of vehicle — automobile, ship, aircraft — depends on reaction forces: the machinery exerts a backward push against its environment — road, water, air — and the reaction of the environment pushes the vehicle forward (see Section 5.4). The propulsion of a spacecraft in empty space is more difficult; since there is nothing in the environment to push against, it is necessary for the ma-

* This section is optional.

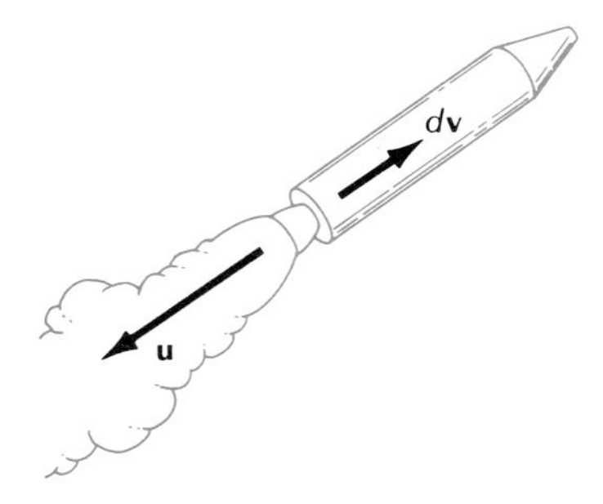


**Fig. 10.16** In the chosen reference frame, the velocity of the rocket is zero at time *t*. A short time *dt* later, the velocity of the rocket is **dv**. The velocity of the exhaust is **u**.

chinery to supply its own medium on which to push. In the operation of a rocket engine, this medium consists of the exhaust gas. The rocket engine produces a large quantity of hot, high-pressure gas from the combustion of liquid or solid fuel in a combustion chamber and then ejects this gas at high speed at the tail end of the rocket. The rocket pushes on the gas, and the reaction force of the gas propels the rocket forward (Figure 10.16). This can be regarded as a recoil mechanism — just as a gun recoils when it ejects a projectile (see Example 2), the rocket recoils when it ejects the particles of gas.

To obtain a simple equation of motion for a rocket, let us assume that the gas particles ejected by the rocket engine all have the same exhaust velocity $u$ (relative to the rocket) and move in an exactly backward direction. This assumption is somewhat unrealistic, since the exhaust actually contains particles with a distribution of speeds and, moreover, the exhaust will tend to spread out laterally, giving the particles a distribution of directions; however, the assumption is not a bad approximation if the numerical value of $u$ is taken to be some average exhaust velocity. Let us further assume that there are no extra forces, that is, the rocket is in deep interstellar space with no nearby gravitating bodies. We can then use the conservation of momentum to obtain the equation of motion.

As gas is ejected from the tail end of the rocket, the mass remaining in the rocket decreases. Suppose that the mass at time $t$ is $M(t)$. To find the acceleration at this time, it is best to examine the conservation of momentum in an inertial reference frame which is at rest relative to the rocket at one chosen instant of time $t$, but which does not participate in the acceleration. In this reference frame, the velocity of the rocket at the time $t$ is zero $[v(t) = 0]$, but the acceleration is not zero $(dv/dt \neq 0)$.

When making use of the conservation of total momentum for a system, we must make sure that the system contains a well-defined number of particles. In the present case, we will take as our system the rocket (with its fuel inside) at the time $t$; the total momentum of this system is zero in the chosen reference frame. A short interval $dt$ later, an amount $-dM$ of fuel has been converted into exhaust gases moving at velocity $-u$ in the chosen reference frame.[2] Our system now consists of the rocket plus these exhaust gases; the total momentum of this system must still be zero. The new velocity of the rocket is $dv$ and hence its momentum is $M\,dv$; the velocity of the exhaust gases is $-u$ and their momentum is $-u(-dM)$, or $u\,dM$. The sum of these momenta must then be zero:

$$M\,dv + u\,dM = 0 \tag{45}$$

Expressing this in terms of time derivatives, we obtain

$$M\frac{dv}{dt} + u\frac{dM}{dt} = 0 \tag{46}$$

[2] Since the mass of the rocket is a decreasing function of time, $dM$ is negative and $-dM$ is positive.

or

$$M \frac{dv}{dt} = -u \frac{dM}{dt} \tag{47}$$

This is the equation of motion of the rocket — it gives the acceleration of the rocket in the absence of extra forces. Although we have derived this equation by means of a rather special reference frame, it is valid in any inertial reference frame because the value of the acceleration is the same in all such reference frames. The quantity $-u\, dM/dt$ appearing on the right side of Eq. (47) plays the role of propulsive force (it equals mass times acceleration); this quantity is called the **thrust** of the rocket.

If $M(t)$ is a known function of time, that is, if the rocket has a known rate of consumption of fuel, then we can solve Eq. (47) to find the motion. However, even without detailed knowledge of the rate of fuel consumption, we can find a general relation between the change of velocity and the amount of fuel that must be consumed to achieve this change of velocity. According to Eq. (45),

$$dv = -u \frac{dM}{M} \tag{48}$$

from which

$$\int dv = -u \int \frac{dM}{M} \tag{49}$$

If the initial velocity, initial mass, final velocity, and final mass are denoted by $v_0$, $M_0$, $v$, and $M$, respectively, then the limits of integration in Eq. (49) are as follows:

$$\int_{v_0}^{v} dv = -u \int_{M_0}^{M} \frac{dM'}{M'} \tag{50}$$

This gives

$$v - v_0 = -u \ln\left(\frac{M}{M_0}\right)$$

or

$$\boxed{v - v_0 = u \ln\left(\frac{M_0}{M}\right)} \tag{51}$$

*Rocket equation*

where ln stands for the natural logarithm. The difference between $M_0$ and $M$ represents the fuel consumed. Thus, Eq. (51) permits the direct calculation of the terminal velocity of a rocket from its terminal mass, amount of fuel, and exhaust velocity. Keep in mind that Eq. (51) is valid only for a rocket moving in empty space, in the absence of extra forces, such as gravity or friction.

**Fig. 10.17** Apollo 11 liftoff.

EXAMPLE 10. The rocket engines of the Saturn V rocket, used for the Apollo and Skylab missions, burn a mixture of kerosene and liquid oxygen (Figure 10.17). Under ideal conditions the exhaust gases from the combustion of this fuel have an exhaust velocity of $3.1 \times 10^3$ m/s. The mass of the rocket at liftoff is $2.45 \times 10^6$ kg, of which $1.70 \times 10^6$ kg are kerosene and liquid oxygen. In the absence of gravity, what would be the terminal velocity of the rocket at burnout?

SOLUTION: At burnout the remaining mass is $2.45 \times 10^6$ kg $- 1.70 \times 10^6$ kg $= 0.75 \times 10^6$ kg. Equation (51) then gives, with $v_0 = 0$,

$$v = u \ln\left(\frac{M_0}{M}\right) = (3.1 \times 10^3 \text{ m/s})\left[\ln\left(\frac{2.45 \times 10^6}{0.75 \times 10^6}\right)\right]$$

$$= 3.7 \times 10^3 \text{ m/s}$$

## SUMMARY

**Momentum of a system of particles:** $\mathbf{P} = \mathbf{p}_1 + \mathbf{p}_2 + \cdots + \mathbf{p}_n$

**Rate of change of momentum:**

$$\frac{d\mathbf{P}}{dt} = \mathbf{F}_{\text{ext}}$$

**Conservation of momentum (in the absence of external forces):**

$$\mathbf{P} = [\text{constant}]$$

**Center of mass:** $\mathbf{r}_{\text{CM}} = \dfrac{m_1\mathbf{r}_1 + m_2\mathbf{r}_2 + \cdots + m_n\mathbf{r}_n}{M}$

$$\mathbf{r}_{\text{CM}} = \frac{1}{M}\int \mathbf{r}\rho \, dV$$

**Momentum of a system of particles:** $\mathbf{P} = M\mathbf{v}_{\text{CM}}$

**Motion of the center of mass:** $M\mathbf{a}_{\text{CM}} = \mathbf{F}_{\text{ext}}$

**Kinetic energy of a system of particles:**

$$K = K_{\text{int}} + \tfrac{1}{2}Mv_{\text{CM}}^2$$

**Rocket equation:** $v - v_0 = u \ln\left(\dfrac{M_0}{M}\right)$

## QUESTIONS

1. When the nozzle of a fire hose discharges a large amount of water at high speed, several strong firemen are needed to hold the nozzle steady. Explain.

2. When firing a shotgun, a hunter always presses it tightly against his shoulder. Why?

3. As described in Example 2, guns on board eighteenth-century warships were often mounted on carriages (see Figure 10.3). What was the advantage of this arrangement?

4. Hollywood movies often show a man being knocked over by the impact of a bullet while the man who shot the bullet remains standing, quite undisturbed. Is this reasonable?

5. Where is the center of mass of this book when it is closed? Mark the center of mass with a cross.

6. Roughly, where is the center of mass of this book when it is open, as it is at this moment?

7. In a high jump (Figure 10.18), is it possible for the body of the jumper to pass over the bar while his center of mass passes under? What would he gain by this?

8. A fountain shoots a stream of water vertically into the air. Roughly, where is the center of mass of the water that is in the air at one instant? Is the center of mass higher or lower than the middle height?

9. Consider the moving wrench shown in Figure 10.6. If the center of mass on this wrench had not been marked, how could you have found it by inspection of this photograph?

10. Is it possible to propel a sailboat by mounting a fan on the deck and blowing air on the sail? Is it better to mount the fan on the stern and blow air toward the rear?

**Fig. 10.18** High jumper passing over the bar.

11. Cyrano de Bergerac's sixth method for propelling himself to the Moon was as follows: "Seated on an iron plate, to hurl a magnet in the air — the iron follows — I catch the magnet — throw again — and so proceed indefinitely." What is wrong with this method (other than the magnet being too weak)?

12. Within the Mexican jumping bean, a small insect larva jumps up and down. How does this lift the bean off the table?

13. Answer the following question, sent by a reader to the *New York Times:*

> A state trooper pulls a truck driver into the weigh station to see if he's overloaded. As the vehicle rolls onto the scales, the driver jumps out and starts beating on the truck box with a club. A bystander asks what he's doing. The trucker says: "I've got five tons of canaries in here. I know I'm overloaded. But if I can keep them flying I'll be OK." If the canaries are flying in that enclosed box, will the truck really weigh any less than if they're on the perch?

14. An elephant jumps off a cliff. Does the Earth move upward while the elephant falls?

15. A juggler stands on a balance, juggling five balls (Figure 10.19). On the average, will the balance register the weight of the juggler plus the weight of the five balls? More than that? Less?

**Fig. 10.19** Juggler on a balance.

16. Suppose you fill a rubber balloon with air and then release it so that the air spurts out of the nozzle. The balloon will fly across the room. Explain.

17. The combustion chamber of a rocket engine is closed at the front and at the sides, but it is open at the rear (Figure 10.20). Explain how the pressure of the gas on the walls of this combustion chamber gives a net forward force that propels the rocket.

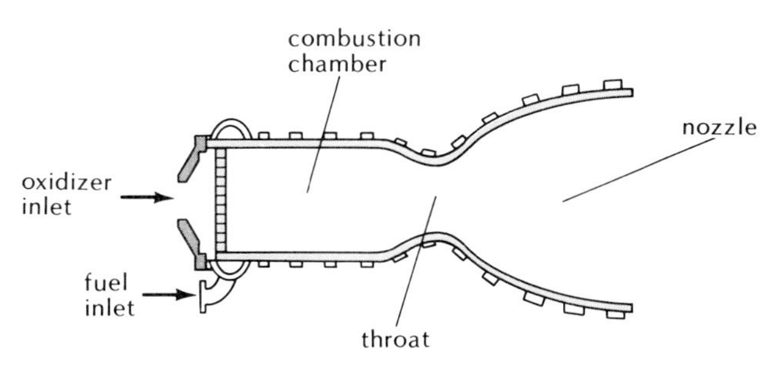


**Fig. 10.20** Combustion chamber of a rocket engine.

18. Can the terminal velocity of a rocket [see Eq. (51)] ever exceed the exhaust velocity $u$ of the gas? What mass ratio $M_0/M$ would this require?

19. What is the advantage of multiple-stage rockets over single-stage rockets?

## PROBLEMS

### Section 10.1

1. Calculate the change of kinetic energy in the collision between the two automobiles described in Example 1.

2. Find the recoil velocity for the gun described in Example 2 if the gun is fired with an elevation angle of 20°.

3. A typical warship built around 1800 (such as the U.S.S. *Constitution*) carried 15 long guns on each side. These guns fired a shot of 11 kg with a muzzle speed of about 490 m/s. The mass of the ship was about 4000 metric tons. Suppose that all of the 15 guns on one side of the ship are fired (almost) simultaneously in a horizontal direction at right angles to the ship. What is the recoil velocity of the ship? Ignore the resistance offered by the water.

4. Two automobiles, moving at 65 km/h in opposite directions, collide head on. One automobile has a mass of 700 kg; the other a mass of 1500 kg. After the collision both remain joined together. What is the velocity of the wreck? What is the change of the velocity of each automobile during the collision?

5. The nucleus of an atom of radium (mass $3.77 \times 10^{-25}$ kg) suddenly ejects an alpha particle (mass $6.68 \times 10^{-27}$ kg) of an energy of $7.26 \times 10^{-16}$ J. What is the velocity of recoil of the nucleus? What is the kinetic energy of the recoil?

6. A lion of mass 120 kg leaps at a hunter with a horizontal velocity of 12 m/s. The hunter has an automatic rifle firing bullets of mass 15 g with a muzzle speed of 630 m/s and he attempts to stop the lion in midair. How many bullets would the hunter have to fire into the lion to stop its horizontal motion? Assume the bullets stick inside the lion.

7. A rifle of 10 kg lying on a smooth table discharges accidentally and fires a bullet of mass 15 g with a muzzle speed of 650 m/s. What is the recoil velocity of the rifle? What is the kinetic energy of the bullet and what is the recoil kinetic energy of the rifle?

*8. Consider the collision between the moving and the initially stationary automobiles described in Example 1. In this example we neglected effects of the friction force exerted by the road during the collision. Suppose that the collision lasts for 0.02 s and suppose that during this time interval the joined automobiles are sliding with locked wheels on the pavement with a coefficient of friction $\mu_s = 0.9$. What change of momentum and what change of speed does the friction produce in the joined automobiles in the interval of 0.02 s? Is this change of speed significant?

*9. A Maxim machine gun fires 450 bullets per minute. Each bullet has a mass of 14 g and a velocity of 630 m/s.

(a) What is the average force that the impact of these bullets exerts on a target? Assume the bullets penetrate the target and remain embedded in it.

(b) What is the average rate at which the bullets deliver their energy to the target?

*10. A vase falls off a table and hits a smooth floor, shattering into three fragments of equal mass which move away horizontally along the floor. Two of the fragments leave the point of impact with velocities of equal magnitude $v$ at right angles. What is the magnitude and direction of the horizontal velocity of the third fragment?

*11. The nucleus of an atom of radioactive copper undergoing beta decay simultaneously emits an electron and a neutrino. The momentum of the electron is $2.64 \times 10^{-22}$ kg · m/s, that of the neutrino is $1.97 \times 10^{-22}$ kg · m/s, and the angle between their directions of motion is 30°. The mass of the residual nucleus is 63.9 u. What is the recoil velocity of the nucleus?

*12. An automobile of mass 1500 kg and a truck of 3500 kg collide at an intersection. Just before the collision the automobile was traveling north at 80 km/h and the truck was traveling east at 50 km/h. After the collision both vehicles remain joined together.

(a) What is the velocity (magnitude and direction) of the vehicles immediately after collision?

(b) How much kinetic energy is lost during the collision?

13. The solar wind sweeping past the Earth consists of a stream of particles, mainly hydrogen ions of mass $1.7 \times 10^{-27}$ kg. There are about $10^7$ ions per cubic meter and their speed is $4 \times 10^5$ m/s. What force does the impact of the solar wind exert on an artificial Earth satellite that has an area of 1.0 m$^2$ facing the wind? Assume that upon impact the ions at first stick to the surface of the satellite.

14. The nozzle of a fire hose ejects 800 l/min of water at a speed of 26 m/s. Estimate the recoil force on the nozzle. By yourself, can you hold this nozzle steady in your hands?

15. The record for the heaviest rainfall is held by Unionville, Maryland, where 3.12 cm of rain (1.23 in.) fell in an interval of 1 min. Assuming that the impact velocity of the raindrops on the ground was 10 m/s, what must have been the average impact force on each square meter of ground during this rainfall?

*16. An automobile is traveling at a speed of 80 km/h through heavy rain. The raindrops are falling vertically at 10 m/s and there are $7.0 \times 10^{-4}$ kg of raindrops in each cubic meter of air. For the following calculation assume that the automobile has the shape of a rectangular box 2 m wide, 1.5 m high, and 4 m long.

(a) At what rate (in kg/s) do raindrops strike the front and top of the automobile?

(b) Assume that when a raindrop hits, it initially sticks to the automobile, although it falls off later. At what rate does the automobile give momentum to the raindrops? What is the horizontal drag force that the impact of the raindrops exerts on the automobile?

*17. A spaceship of frontal area 25 m$^2$ passes through a cloud of interstellar dust at a speed of $1.0 \times 10^6$ m/s. The density of dust is $2.0 \times 10^{-18}$ kg/m$^3$. If all the particles of dust that impact on the spaceship stick to it, find the average decelerating force that the impact of the dust exerts on the spaceship.

**18. A gun mounted on a cart fires bullets of mass $m$ in the backward direction with a horizontal muzzle velocity $u$. The initial mass of the cart, including the mass of the gun and the mass of the ammunition, is $M$ and the initial velocity of the cart is zero. What is the velocity of the cart after firing $n$ bullets? Assume that the cart moves without friction and ignore the mass of the gunpowder.

### Section 10.2

19. A penny coin lies on a table at a distance of 20 cm from a stack of three penny coins. Where is the center of mass of the system of four coins?

20. A 59-kg woman and a 73-kg man sit on a seesaw, 3.6 m long. Where is their center of mass? Neglect the mass of the seesaw.

21. Consider the system Earth–Moon; use the data in the table printed on the endpapers. How far from the center of the Earth is the center of mass of this system?

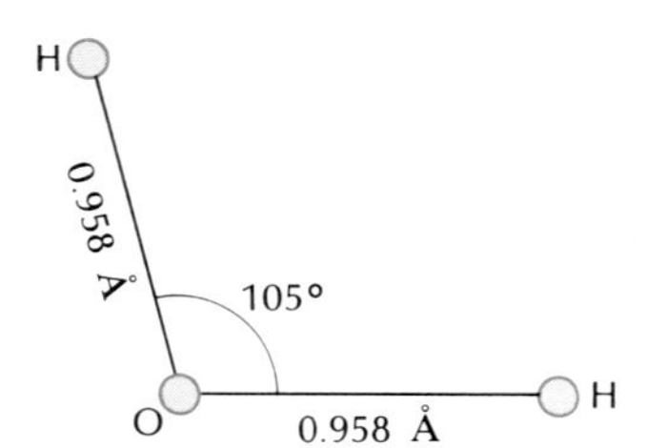


**Fig. 10.21** Water molecule.

*22. In order to balance the wheel of an automobile, a mechanic attaches a piece of lead alloy to the rim of the wheel. The mechanic finds that if he attaches a piece of 40 g at a distance of 20 cm from the center of a wheel of 30 kg, the wheel is perfectly balanced, that is, the center of the wheel coincides with the center of mass. How far from the center of the wheel was the center of mass before the mechanic balanced the wheel?

*23. The distance between the oxygen and each of the hydrogen atoms in a water ($H_2O$) molecule is 0.958 Å; the angle between the two oxygen–hydrogen bonds is 105° (Figure 10.21). Treating the atoms as particles, find the center of mass.

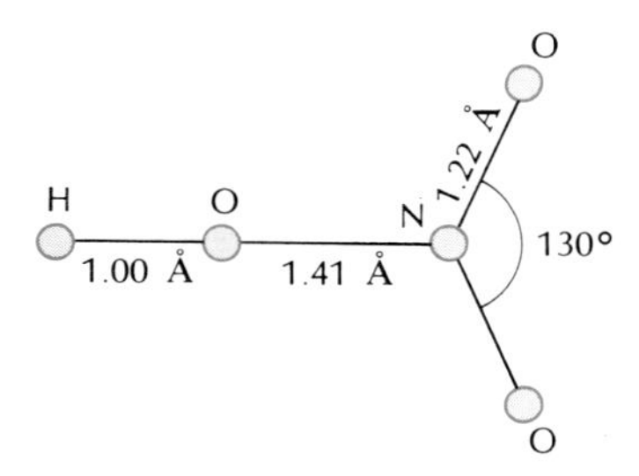


**Fig. 10.22** Nitric acid molecule.

*24. Figure 10.22 shows the shape of a nitric acid ($HNO_3$) molecule and its dimensions. Treating the atoms as particles, find the center of mass of this molecule.

*25. Figure 9.11a shows the positions of the three inner planets (Mercury, Venus, and Earth) on January 1, 1980. Measure angles and distances off this figure and find the center of mass of the system of these planets (ignore the Sun). The masses of the planets are listed in Table 9.1.

*26. The Local Group of galaxies consists of our Galaxy and its nearest neighbors. The masses of the most important members of the Local Group are as follows (in multiples of the mass of the Sun): our Galaxy, $2 \times 10^{11}$; the Andromeda galaxy, $3 \times 10^{11}$; the Large Magellanic cloud, $2.5 \times 10^{10}$; and NGC598, $8 \times 10^9$. The $x$, $y$, $z$ coordinates of these galaxies are, respectively, as follows (in thousands of light-years): (0, 0, 0); (1640, 290, 1440); (8.5, 56.7, −149); and (1830, 766, 1170). Find the coordinates of the center of mass of the Local Group. Treat all the galaxies as point masses.

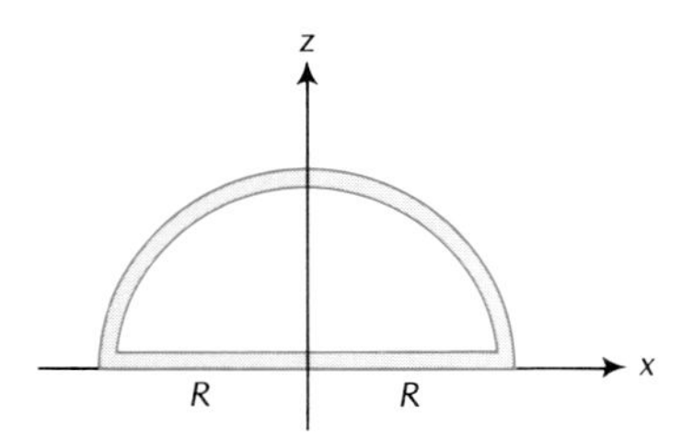


**Fig. 10.23** A semicircular rod and a straight rod joined together.

*27. Suppose we take the semicircular rod described in Example 6 and we add to it a straight rod of length $2R$ fitted between the ends of the semicircular rod (Figure 10.23). Where is the center of mass of this system?

*28. Three uniform square pieces of sheet metal are joined along their edges so as to form three of the sides of a cube (Figure 10.24). The dimensions of the squares are $L \times L$. Where is the center of mass of the joined squares?

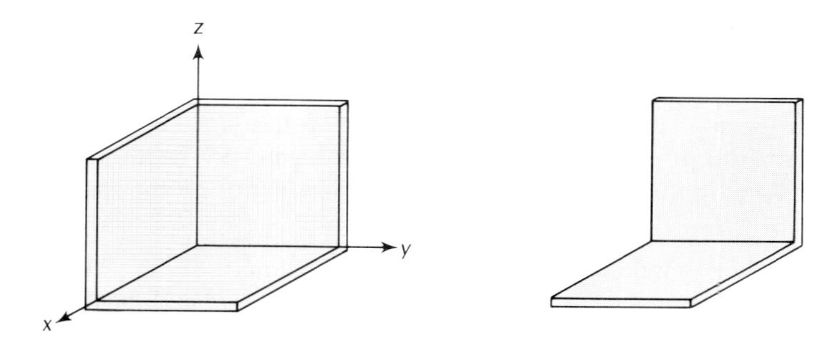


**Fig. 10.24** (left) Three square pieces of sheet metal joined together along their edges.

**Fig. 10.25** (right) Two uniform squares of sheet metal joined along one edge.

*29. Two uniform squares of sheet metal of dimension $L \times L$ are joined at a right angle along one edge (Figure 10.25). One of the squares has twice the mass of the other. Find the center of mass of the combined squares.

*30. Three identical meter sticks are arranged to form a letter **U**. Where is the center of mass of this system?

*31. A box made of plywood has the shape of a cube measuring $L \times L \times L$. The top of the box is missing. Where is the center of mass of the open box?

*32. A cube of iron has dimensions $L \times L \times L$. A hole of radius $\frac{1}{4}L$ has been drilled all the way through the cube, so that one side of the hole is tangent to one face along its entire length (Figure 10.26). Where is the center of mass of the drilled cube?

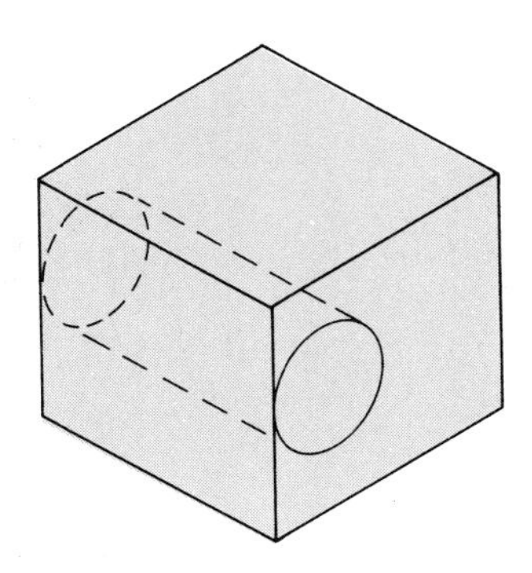

**Fig. 10.26** Iron cube with a hole.

*33. A semicircle of uniform sheet metal has radius $R$ (Figure 10.27). Find the center of mass. (Hint: Regard the semicircle as assembled of many thin, concentric semicircular rods; use the result of Example 6.)

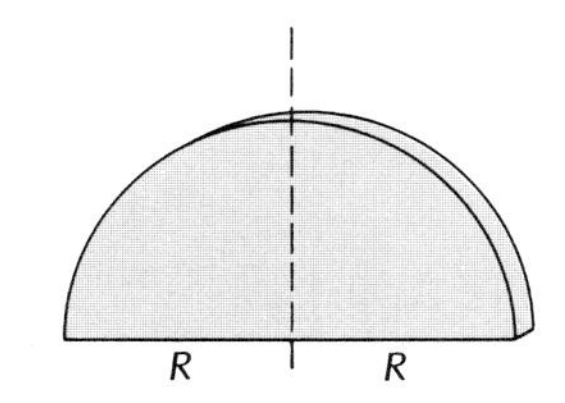


**Fig. 10.27** Semicircle of sheet metal.

**34. Suppose that water drops are released from a point at the edge of a roof with a constant time interval $\Delta t$ between one water drop and the next. The drops fall a distance $l$ to the ground. If $\Delta t$ is very short (so that the number of water drops falling through the air at any given instant is very large), show that the center of mass of the falling drops is at a height of $\frac{2}{3}l$ above the ground. From this, deduce that the time-average height of a projectile launched from the ground and returning to the ground is $\frac{2}{3}$ of its maximum height. (This theorem is useful in the calculation of the average air pressure and air resistance encountered by the projectile.)

35. The great pyramid at Giza has a mass of $6.6 \times 10^6$ metric tons and a height of 147 m (see Example 7). Assume that the mass is uniformly distributed over the volume of the pyramid.
   (a) How much work must the ancient Egyptian laborers have done against gravity to pile up the stones in the pyramid?
   (b) If each laborer delivered work at an average rate of $4 \times 10^5$ J/h, how many man-hours of work have been stored in this pyramid?

*36. Mount Fuji has approximately the shape of a cone. The half-angle at the apex of this cone is 65° and the height of the apex is 3800 m. At what height is the center of mass? Assume that the material in Mount Fuji has uniform density.

*37. Show that the center of mass of a uniform flat triangular plate is at the point of intersection of the lines drawn from the vertices to the midpoints of the opposite sides.

*38. A lock on the Champlain canal is 73 m long and 9.2 m wide; the lock has a lift of 3.7 m, that is, the difference between the water levels of the canal on one side of the lock and on the other side is 3.7 m. How much gravitational potential energy is wasted each time the lock goes through one cycle (involving the filling of the lock with water from the high level and then the spilling of this water to the low level)?

### Section 10.3

39. A proton of energy $1.6 \times 10^{-13}$ J is moving toward a proton at rest. What is the velocity of the center of mass of the system?

40. In a molecule, such as the potassium bromide (KBr) molecule of Example 5, the atoms usually execute a rapid vibrational motion about their equilibrium positions. Suppose that in an isolated KBr molecule the speed of the potassium atom is $5.0 \times 10^3$ m/s at one instant (relative to the center of mass). What is the speed of the bromine atom at the same instant?

41. A tugboat of mass 4000 metric tons and a ship of mass 28,000 metric tons are joined by a long towrope of 400 m. Both vessels are initially at rest in the water. If the tugboat reels in 200 m of towrope, how far does the ship move relative to the water? The tugboat? Ignore the resistance that the water offers to the motion.

42. A fisherman in a boat catches a great white shark with a harpoon. The shark struggles for a while and then becomes limp when at a distance of 300 m

from the boat. The fisherman pulls in the shark by the rope attached to the harpoon. During this operation, the boat (initially at rest) moves 45 m in the direction of the shark. The mass of the boat is 5400 kg. What is the mass of the shark? Pretend that the water exerts no friction.

43. A 75-kg man climbs the stairs from the ground to the fourth floor of a building, a height of 15 m. How far does the Earth recoil in the opposite direction as the man climbs?

*44. A 6000-kg truck stands on the deck of an 80,000-kg ferryboat. Initially the ferry is at rest and the truck is located at its front end. If the truck now drives 15 m along the deck toward the rear of the ferry, how far will the ferry move forward relative to the water? Pretend that the water has no effect on the motion.

*45. While moving horizontally at $5.0 \times 10^3$ m/s at an altitude of $2.5 \times 10^4$ m, a ballistic missile explodes and breaks apart into two fragments of equal mass which fall freely. One of the fragments has zero speed immediately after the explosion and lands on the ground directly below the point of the explosion. Where does the other fragment land? Ignore the friction of air.

**46. Figure 9.11a shows the positions of the three inner planets (Mercury, Venus, Earth) on January 1, 1980. Measuring angles off this figure and using the data on masses, orbital radii, and periods given in Table 9.1, find the velocity of the center of mass of this system of three planets.

### Section 10.4

47. Repeat the calculation of Example 9 if the two automobiles travel in *opposite* directions.

48. A projectile of 45 kg fired from a gun has a speed of 640 m/s. The projectile explodes in flight, breaking apart into a fragment of 32 kg and a fragment of 13 kg (we assume that no mass is dispersed in the explosion). Both fragments move along the original direction of motion. The speed of the first fragment is 450 m/s and that of the second is 1050 m/s.
- (a) Calculate the translational kinetic energy of the center of mass before the explosion. Calculate the internal kinetic energy before the explosion.
- (b) Calculate these quantities after the explosion. Where does the extra internal kinetic energy come from?

49. Consider the automobile collision described in Problem 4. What is the internal kinetic energy before the collision? After the collision?

50. Regard the automobile and the truck described in Problem 12 as a system of two particles.
- (a) What is the translational kinetic energy of the center of mass before the collision? What is the internal kinetic energy?
- (b) What is the translational kinetic energy of the center of mass after the collision? What is the internal kinetic energy?

*51. The typical speed of the vibrational motion of the iron atoms in a piece of iron at room temperature is 360 m/s. What is the total internal kinetic energy of a 1-kg chunk of iron?

### Section 10.5

52. In order to achieve a thrust of $3.3 \times 10^7$ N, at what rate (in metric tons per second) must the engines of the Saturn V rocket consume their fuel? Assume that the exhaust velocity of the hot gas from the engines is 2900 m/s.

53. If a rocket, initially at rest, is to attain a terminal velocity of a magnitude equal to the exhaust velocity, what fraction of the initial mass must be fuel?

54. A rocket burns a kerosene–oxygen mixture. The complete burning of 1.0 kg of kerosene requires 3.4 kg of oxygen; this burning releases about $4.2 \times 10^7$ J of thermal energy. Suppose that all of this thermal energy is converted into kinetic energy of the reaction products (4.4 kg). What will be the exhaust velocity of the reaction products?

*55. Rockets used for space exploration usually consist of several stages sitting on top of one another and fired in sequence. When the first stage has burned out, it is jettisoned and the second stage is ignited, and so on. Show that the terminal velocity of the last stage of such a multiple-stage rocket is never more than $v = u \ln(M_0/M)$, where $M_0$ is the initial mass of the multiple-stage rocket and $M$ is the terminal mass of the last stage at burnout. Assume that all stages have the same exhaust velocity and ignore gravity. (Hint: Compare the multiple-stage rocket with an ideal rocket consisting entirely of fuel except for the terminal mass $M$.)

CHAPTER 11

# Collisions

The collision between two bodies — an automobile and a solid wall (Figure 11.1), a ship and an iceberg, a molecule of oxygen and a molecule of nitrogen, an alpha particle and a nucleus of a gold atom — involves a violent change of the motion, a change brought about by very strong forces that begin to act suddenly when the bodies come into contact, last a short time, and then cease just as suddenly when the bodies separate. The forces that act during a collision are usually rather complicated so that their complete theoretical description is impossible (e.g., in an automobile collision) or at least very difficult (e.g., in a nuclear collision). However, even without exact knowledge of the force law, we can make some predictions about the collision by taking advantage of the general laws of conservation of momentum and energy. In the following sections, we will see what constraints these laws impose on the motion of the colliding bodies.

The study of collisions is an important tool in the experimental investigation of atoms, nuclei, and elementary particles. All subatomic bodies are too small to be made visible with any kind of microscope. Just as a surgeon who cannot see the interior of a wound uses probes to feel the condition of the tissues, a physicist who cannot see the interior of an atom uses probes to "feel" for subatomic structures. The probe used by physicists in the exploration of subatomic structures is simply a stream of fast-moving particles — electrons, protons, alpha particles, or others. These projectiles are aimed at a target containing a sample of the atoms, nuclei, or elementary particles under investigation. From the manner in which the projectiles collide and react with the target, physicists can deduce some of the properties of the subatomic structures in the target.

## 11.1 Impulsive Forces

The force that two colliding bodies exert on one another acts only for a short time, giving a brief but strong push. This force is called an **im-**

**pulsive force.** During the collision, the impulsive force is much stronger than any other forces that may be present; consequently, the impulsive force produces a large change in the motion while the other forces produce only small and insignificant changes. For example, during the automobile collision shown in Figure 11.1, the only important force is the push of the wall on the front end of the automobile; the effects produced by gravity and by the friction force of the road during the collision are insignificant.

*Impulsive force*

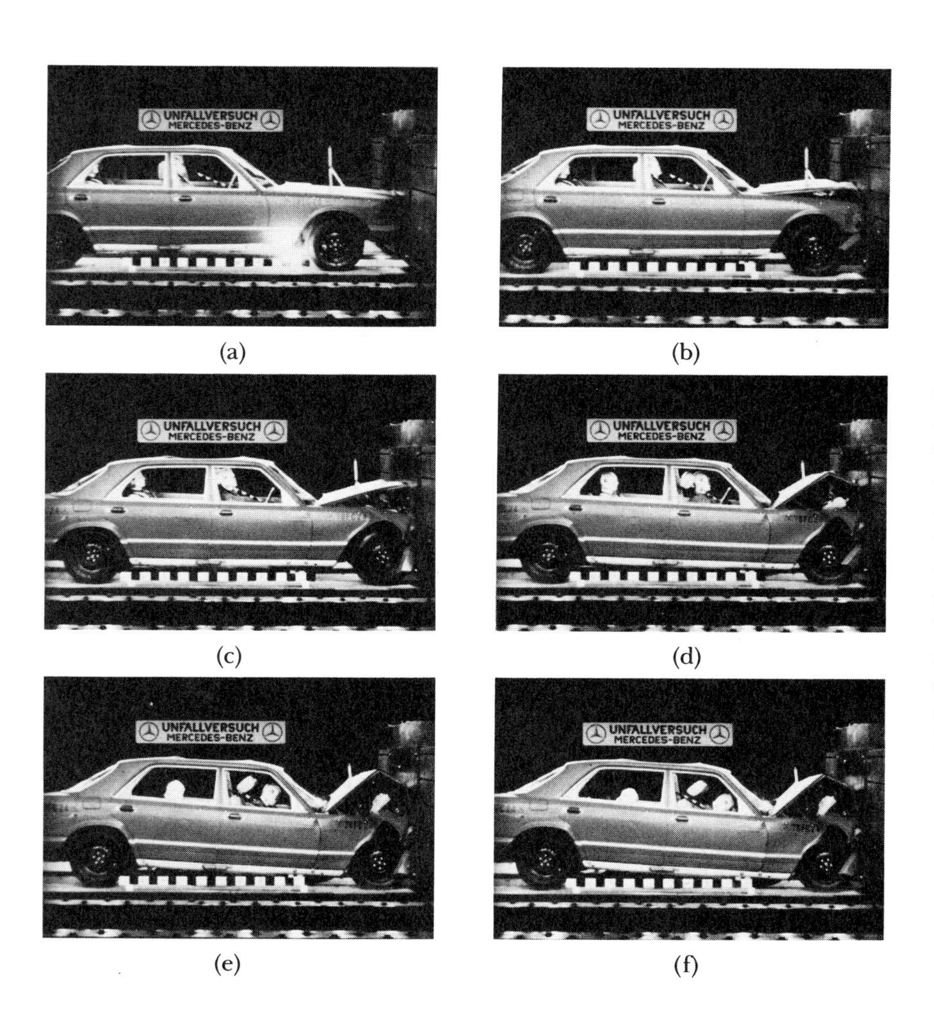


**Fig. 11.1** Crash test of a Mercedes-Benz automobile. The photographs show an impact at 49 km/h on a rigid barrier. The first photograph was taken $5 \times 10^{-3}$ s after the initial contact; the others were taken at intervals of $20 \times 10^{-3}$ s thereafter. The automobile remains in contact with the barrier for 0.120 s; it then recoils from the barrier with a speed of 4.7 km/h. The checkered bar on the ground has a length of 2 m.

Suppose that a collision lasts a time $\Delta t$, say, from $t = 0$ to $t = \Delta t$, and that during this time an impulsive force **F** acts on one of the colliding bodies. The force is zero before $t = 0$ and it is zero after $t = \Delta t$, but is large between these times. For example, Figure 11.2 shows a plot of the force experienced by an automobile in a collision with a solid wall lasting 0.120 s. The force is zero before $t = 0$ and after $t = 0.120$ s, and varies in a complicated way between these times. The **impulse** delivered by such a force **F** to the body is defined as the integral of the force over time,

$$\mathbf{I} = \int_0^{\Delta t} \mathbf{F}\, dt \qquad (1)$$

*Impulse*

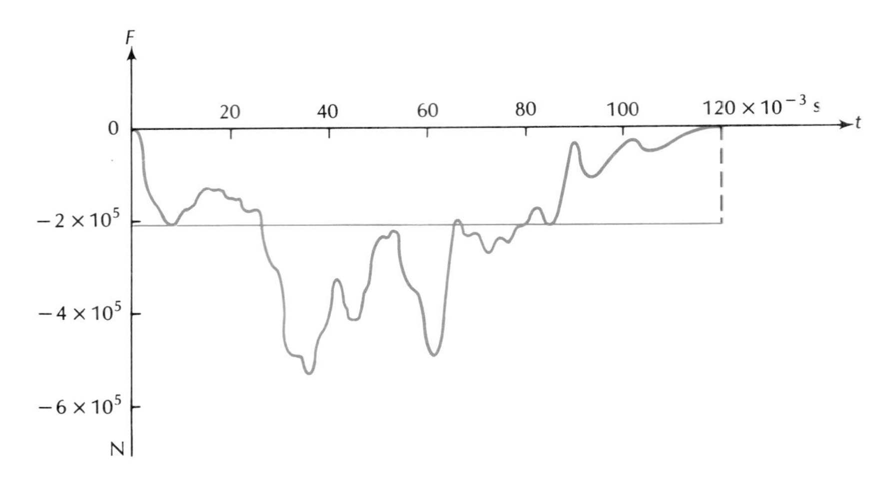


**Fig. 11.2** Force on the automobile as a function of time during the impact shown in Figure 11.1. The colored horizontal line gives the time-average force. (From data supplied by Mercedes-Benz of North America, Inc.)

According to this equation, the $x$ component of the impulse for the force shown in Figure 11.2 is the area between the curve $F_x(t)$ and the $t$ axis, and similarly for the $y$ component and the $z$ component.

The units of impulse are N · s or kg · m/s in the metric system and lbf · s in the British system; these units are the same as those of momentum.

The definition (1) of the impulse is not restricted to forces of short duration — it is equally valid if the duration $\Delta t$ of the impulsive force is long. However, in most of our applications of the concept of impulse in this chapter, the force will be of short duration.

By means of the equation of motion

$$\mathbf{F} = \frac{d\mathbf{p}}{dt} \tag{2}$$

we can transform Eq. (1) into

$$\mathbf{I} = \int_0^{\Delta t} \mathbf{F}\, dt = \int_0^{\Delta t} \frac{d\mathbf{p}}{dt}\, dt = \int d\mathbf{p} = \mathbf{p}' - \mathbf{p} \tag{3}$$

where $\mathbf{p}$ is the momentum before the collision (at time $t = 0$) and $\mathbf{p}'$ is the momentum after the collision (at time $t = \Delta t$). Thus, the impulse of a force is simply equal to the momentum change produced by this force. However, since the force acting during a collision is usually not known in detail, Eq. (3) is not very helpful for calculating momentum changes. It is often best to apply Eq. (3) in reverse for calculating the average force from the known momentum change. The time-average force is defined by

$$\overline{\mathbf{F}} = \frac{1}{\Delta t} \int_0^{\Delta t} \mathbf{F}\, dt \tag{4}$$

In a plot of force as a function of time, such as shown in Figure 11.2, the time-average force simply represents the mean height of the function above the $t$ axis; this mean height is shown by the colored horizontal line in Figure 11.2. By means of Eq. (3) we can write the time-average force as

$$\overline{\mathbf{F}} = \frac{1}{\Delta t} (\mathbf{p}' - \mathbf{p}) \tag{5}$$

This relation gives a quick estimate of the average magnitude of the impulsive force if the duration of the collision and the momentum change are known.

EXAMPLE 1. The collision between the automobile and wall shown in Figure 11.1 lasts 0.120 s. The mass of the automobile is 1700 kg and the initial and final velocities are $v = 13.6$ m/s and $v' = -1.3$ m/s, respectively. Evaluate the impulse and the time-average force from these data.

SOLUTION: With the $x$ axis along the direction of the initial motion, the change in momentum is

$$p'_x - p_x = mv' - mv$$

$$= 1700 \text{ kg} \times (-1.3 \text{ m/s}) - 1700 \text{ kg} \times 13.6 \text{ m/s}$$

$$= -2.53 \times 10^4 \text{ kg} \cdot \text{m/s}$$

Hence the impulse is $I_x = -2.53 \times 10^4$ kg · m/s and the time-average force is

$$\bar{F}_x = \frac{1}{\Delta t}(p'_x - p_x) = \frac{-2.53 \times 10^4 \text{ kg} \cdot \text{m/s}}{0.120 \text{ s}}$$

$$= -2.11 \times 10^5 \text{ N}$$

Note that since the mutual forces on two bodies engaged in a collision are an action–reaction pair of equal magnitudes and of opposite directions, the corresponding impulses are equal and opposite. For instance, in Example 1 the impulse on the automobile is $I_x = -2.53 \times 10^4$ kg · m/s and the impulse on the wall is

$$I_{x,\text{ wall}} = 2.53 \times 10^4 \text{ kg} \cdot \text{m/s}$$

This, of course, expresses momentum conservation: the momentum changes of the two colliding bodies are of equal magnitudes and opposite directions.

*Elastic collision*

A collision in which the kinetic energy is the same before and after the collision is called **elastic.** (The reason for this terminology is that collisions between deformable elastic bodies, which behave like ideal springs, result in the same kinetic energy before and after the collision). Collisions between macroscopic bodies are usually not elastic — during the collision some of the kinetic energy is transformed into heat by the internal friction forces and some is used up in doing work to change the internal configuration of the bodies. For example, the automobile collision shown in Figure 11.1 is highly inelastic; almost the entire kinetic energy is used up in doing work on the automobile parts, changing their shape. On the other hand, the collision of a "superball" and a hard wall or the collision of two billiard balls comes pretty close to being elastic. (The deformation of the ball during the collision involves some internal friction, which absorbs some energy; hence the collision is not *exactly* elastic.)

Collisions between "elementary" particles — such as electrons, protons, and neutrons — are often elastic. These particles have no internal friction forces which could dissipate kinetic energy. A collision between such particles can only be inelastic if it involves the creation of new particles; such new particles may arise either by conversion of some of the available kinetic energy into mass, or else by transmutation of the old particles by means of a change of their internal structure. Collisions of this kind will be discussed in Section 11.4.

Often it is not possible to calculate the motion of the colliding bodies by direct solution of Newton's equation of motion, because the impulsive forces that act during the collision are not known in sufficient detail. Consequently, we must glean whatever information we can from the general laws of conservation of momentum and energy, which do not depend on details of the forces. In some simple instances, these general laws permit the deduction of the motion after the collision from what is known about the motion before the collision.

EXAMPLE 2. A "superball" made of rubberlike plastic is thrown against a hard, smooth wall. The ball strikes the wall from a perpendicular direction with speed $v$. Assuming that the collision is elastic, find the speed of the ball after the collision.

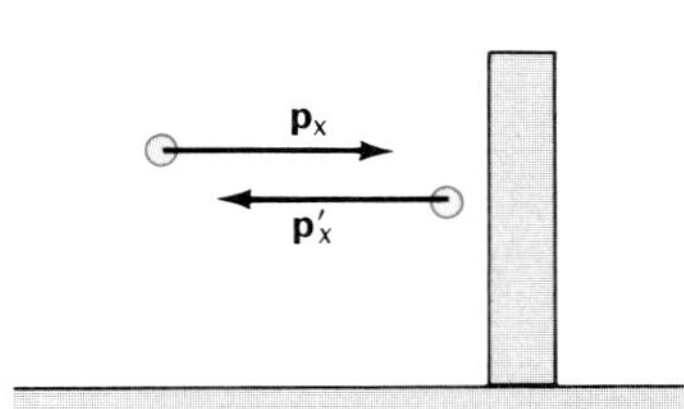


**Fig. 11.3** Recoil of a ball thrown against a wall.

SOLUTION: The only significant force on the ball is the normal force exerted by the wall; this force reverses the motion of the ball (Figure 11.3). Since the wall is very massive, the reaction force of the ball will not give it any appreciable velocity. Hence the energy of the system, both before and after the collision, is merely the kinetic energy of the ball. Conservation of energy then implies that the kinetic energy after the collision is the same as before the collision; this requires that the ball rebound with a speed $v$ equal to the incident speed.

COMMENTS AND SUGGESTIONS: Although the kinetic energy of the ball remains unchanged, the momentum does change (see also Example 5.7). If the $x$ axis is in the direction of the initial motion, then the momentum of the ball before the collision is $p_x = mv$ and after the collision it is $p'_x = -mv$; hence the change of momentum is $p'_x - p_x = -2mv$. The wall suffers an equal and opposite momentum change of $+2mv$ so that the total momentum of the system is conserved. The wall can acquire the momentum $2mv$ without acquiring any appreciable velocity or kinetic energy because its mass is large. For instance, if $m = 0.1$ kg and $v = 10$ m/s, then $2mv = 2 \times 0.1 \text{ kg} \times 10 \text{ m/s} = 2.0 \text{ kg}\cdot\text{m/s}$; and if the mass of the wall is 1000 kg, then its recoil velocity is given by $1000 \text{ kg} \times v_{\text{wall}} = 2.0$ kg m/s, which yields $v_{\text{wall}} = 2 \times 10^{-3}$ m/s. The corresponding kinetic energy is also quite small, $\frac{1}{2} \times 1000 \text{ kg} \times (2 \times 10^{-3} \text{ m/s})^2 = 2 \times 10^{-3}$ J.[1] As a general rule, a body of very large mass can absorb momentum without any appreciable change in its velocity or kinetic energy.

## 11.2 Collisions in One Dimension

The collision of two boxcars on a straight railroad track is an example of a one-dimensional collision. More generally, the collision of any two bodies that approach head-on and recoil along their original line of motion is one-dimensional. Obviously, such collisions will occur only under rather exceptional circumstances; nevertheless, we find it worthwhile to study one-dimensional collisions because they display in a simple way some of the features that reappear in two- and three-dimensional collisions.

*Elastic collision in one dimension*

In an elastic one-dimensional collision between two particles, the laws of conservation of momentum and of energy completely determine the final velocities in terms of the initial velocities. In the follow-

[1] This calculation is not quite accurate. The wall is attached to the Earth, and it therefore cannot move freely — it can at most rock back and forth. Our crude calculation merely provides an upper limit on $v_{\text{wall}}$.

ing calculations we will assume that one particle (the "target") is initially at rest, and the other (the "projectile") is initially in motion.

Figure 11.4a shows the particles before the collision and Figure 11.4b shows them after; the $x$ axis is along the direction of motion. We will designate the $x$ components of the velocity of particle 1 (projectile) and particle 2 (target) by $v_1$ and $v_2$ before the collision, and by $v_1'$ and $v_2'$ after the collision, respectively. Since particle 2 is initially at rest, $v_2 = 0$.

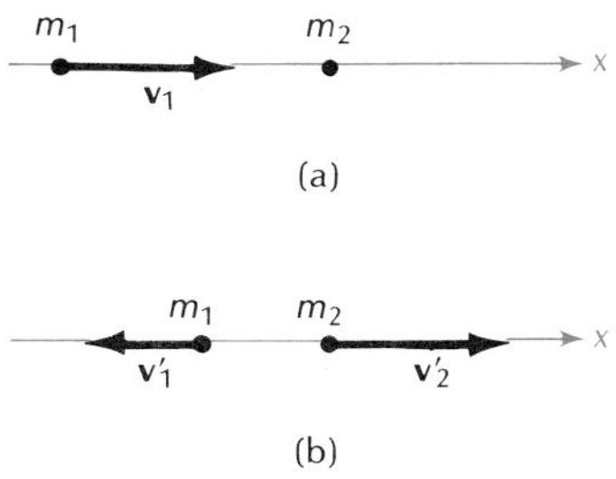

Conservation of momentum states that

$$m_1 v_1 = m_1 v_1' + m_2 v_2' \tag{6}$$

and conservation of energy states that

$$\tfrac{1}{2} m_1 v_1^2 = \tfrac{1}{2} m_1 v_1'^2 + \tfrac{1}{2} m_2 v_2'^2 \tag{7}$$

These two equations can be rearranged as follows:

$$m_1(v_1 - v_1') = m_2 v_2' \tag{8}$$

$$\tfrac{1}{2} m_1 (v_1 - v_1')(v_1 + v_1') = \tfrac{1}{2} m_2 v_2'^2 \tag{9}$$

After dividing Eq. (9) by Eq. (8), we obtain

$$\tfrac{1}{2}(v_1 + v_1') = \tfrac{1}{2} v_2' \tag{10}$$

The advantage of this equation is that it does not contain any squares of the velocities.

We regard the initial velocities $v_1$, $v_2$ as known and the final velocities $v_1'$, $v_2'$ as unknown. Equations (10) and (6) taken together are then a simple (linear) system of two equations for the two unknowns $v_1'$, $v_2'$. The simultaneous solution of these equations leads to

$$\boxed{v_1' = \frac{m_1 - m_2}{m_1 + m_2} v_1} \tag{11}$$

and

$$\boxed{v_2' = \frac{2m_1}{m_1 + m_2} v_1} \tag{12}$$

*Speeds after a one-dimensional elastic collision*

$m_1$ $m_2$ $\mathbf{v}_1$ $x$

(a)

$m_1$ $m_2$ $\mathbf{v}_1'$ $\mathbf{v}_2'$ $x$

(b)

**Fig. 11.4** (a) Before the collision, particle 2 is at rest and particle 1 has velocity $\mathbf{v}_1$. (b) After the collision, particle 1 has velocity $\mathbf{v}_1'$ and particle 2 has velocity $\mathbf{v}_2'$.

---

EXAMPLE 3. An empty boxcar of mass $m_1 = 20$ metric tons rolling on a straight track at 5 m/s collides with a loaded stationary boxcar of mass $m_2 = 65$ metric tons. Assuming that the cars bounce off one another elastically, find the velocities after the collisions.

SOLUTION: With $m_1 = 20$ tons and $m_2 = 65$ tons, Eqs. (11) and (12) yield

$$v_1' = \frac{20 \text{ tons} - 65 \text{ tons}}{20 \text{ tons} + 65 \text{ tons}} \times 5 \text{ m/s} = -2.6 \text{ m/s}$$

$$v_2' = \frac{2 \times 20 \text{ tons}}{20 \text{ tons} + 65 \text{ tons}} \times 5 \text{ m/s} = 2.4 \text{ m/s}$$

---

Note that if the mass of the target is the same as the mass of the projectile ($m_2 = m_1$), then Eqs. (11) and (12) give $v_1' = 0$ and $v_2' = v_1$. Thus, the projectile comes to a halt, and the target moves away with the original speed of the projectile — the target acquires all of the kinetic energy of the projectile.

If the mass of the target is much larger than the mass of the projectile ($m_2 \gg m_1$), then $v_1' \cong -v_1$ and $v_2' \cong 0$, which means that the projectile bounces off with a reversed velocity, and the target remains nearly stationary. Conversely, if the mass of the projectile is much larger than the mass of the target ($m_1 \gg m_2$), then $v_1' = v_1$ and $v_2' = 2v_1$, which means that the projectile plows right on and the target bounces off with *twice* the speed of the projectile. This second case can be understood in terms of the preceding case: in the rest frame of $m_1$, $m_2$ is approaching with velocity $-v_1$ and bounces off with velocity $+v_1$, that is, the velocity of the particle of small mass changes by $2v_1$.

---

EXAMPLE 4. Inside a nuclear reactor containing uranium fuel, the fission reactions produce an abundant flux of fast neutrons of a speed of about $2 \times 10^7$ m/s. These neutrons are used to trigger more fission reactions. But before they can be so used, they must be slowed down to a much lower speed. The slowing down is accomplished by collisions: the uranium fuel in the reactor is surrounded by water or by graphite and the neutrons lose their kinetic energy in collisions with the nuclei of these materials. By what factor is the speed of a neutron reduced in a head-on collision with a stationary carbon nucleus? With a hydrogen nucleus? The masses of a neutron, a carbon nucleus, and a hydrogen nucleus are 1.0087 u, 11.9934 u, and 1.0073 u, respectively.

SOLUTION: The final speed of the neutron is given by Eq. (11):

$$v_1' = \frac{m_1 - m_2}{m_1 + m_2} v_1$$

For a collision with carbon, we substitute the approximate values $m_1 = 1.01$ u for the mass of the neutron and $m_2 = 12.00$ u for the mass of the carbon nucleus:

$$v_1' = \frac{1.01 - 12.00}{1.01 + 12.00} v_1 = -0.84 v_1$$

For a collision with a proton, we need to substitute more precise values of the masses; with $m_1 = 1.0087$ u for the mass of the neutron and $m_2 = 1.0073$ u for the mass of the proton,

$$v_1' = \frac{1.0087 - 1.0073}{1.0087 + 1.0073} v_1 = 6.9 \times 10^{-4} v_1$$

This shows that a single head-on collision with a proton will just about stop a neutron, but a single head-on collision with a carbon nucleus will reduce the speed of the neutron by only 16%. Hence several collisions with carbon nuclei are needed to reduce the speed of a neutron to a small value.

---

Although Eqs. (11) and (12) are based on the assumption of a stationary target ($v_2 = 0$), they can also be useful for solving problems with a moving target ($v_2 \neq 0$). The trick is first to solve the problem in the initial reference frame of the target [where Eqs. (11) and (12) are valid], and then transform the velocities to any other reference frame by means of the Galilean transformation equations.

One interesting feature of an elastic collision is that the *relative velocity* $v_2 - v_1$ reverses during the collision, that is, the relative velocity changes its sign but not its magnitude. This can be readily seen from Eq. (10), which, upon rearrangement, reads

$$v_1' - v_2' = -v_1 \tag{13}$$

The quantity on the left side of this equation is the relative velocity after the collision and the quantity on the right side is the negative of the relative velocity before the collision (remember $v_2 = 0$).

We can gain some insight into this preservation of the relative velocity by looking at the expression for the kinetic energy of the system in terms of the center-of-mass velocity, that is, Eq. (9.44):

$$K = \tfrac{1}{2}Mv_{\text{CM}}^2 + \tfrac{1}{2}m_1u_1^2 + \tfrac{1}{2}m_2u_2^2 \tag{14}$$

where $u_1$ and $u_2$ are the velocities relative to the center of mass:

$$u_1 = v_1 - v_{\text{CM}} = v_1 - \frac{m_1v_1 + m_2v_2}{m_1 + m_2} = \frac{m_2(v_1 - v_2)}{m_1 + m_2} \tag{15}$$

$$u_2 = v_2 - v_{\text{CM}} = v_2 - \frac{m_1v_1 + m_2v_2}{m_1 + m_2} = \frac{m_1(v_2 - v_1)}{m_1 + m_2} \tag{16}$$

In these equations, the velocities may be evaluated either before, during, or after the collision. If we square Eqs. (15) and (16) and substitute them into Eq. (14), we readily find the result

$$K = \tfrac{1}{2}Mv_{\text{CM}}^2 + \frac{1}{2}\frac{m_1m_2}{m_1 + m_2}(v_1 - v_2)^2 \tag{17}$$

The first term on the right side of this equation represents the kinetic energy associated with the translational motion of the center of mass; the second term represents the "internal" kinetic energy associated with motion relative to the center of mass. The first term is necessarily conserved because $v_{\text{CM}}$ is constant in the absence of external forces. The second term must then also be constant, since otherwise the total kinetic energy would *not* be conserved. Thus, the preservation of the magnitude of the relative velocity $v_2 - v_1$ in an elastic collision expresses the conservation of the "internal" kinetic energy.

If the collision is inelastic, then the only conservation law that is applicable is the conservation of momentum. This, by itself, is insufficient to calculate the velocities of both particles after the collision. However, if the collision is **totally inelastic,** so a maximum amount of kinetic energy is lost, then the velocities after the collision can be calculated. If there is a maximum loss of kinetic energy, then the two particles will have zero relative velocity after the collision, that is, they will stick together. This can be understood from Eq. (17): the first term on the right side of this equation remains constant whether the collision is elastic or inelastic; only the second term can change — in a totally inelastic collision this term will become zero, and consequently the relative velocity $v_1' - v_2'$ after the collision becomes zero. If both particles have the same velocity and so remain together, then their velocity must necessarily coincide with the velocity of the center of mass.

*Totally inelastic collision*

EXAMPLE 5. Suppose that the two boxcars of Example 3 couple during the collision and remain locked together. What is the velocity of the coupled cars after the collision? How much kinetic energy is dissipated during the collision?

SOLUTION: With $v_2 = 0$, the velocity of the center of mass, evaluated before the collision, is

$$v_{CM} = \frac{m_1 v_1}{m_1 + m_2}$$

The final velocity of the coupled cars after the collision is equal to this velocity of the center of mass,

$$v_1' = v_2' = \frac{m_1 v_1}{m_1 + m_2}$$

$$= \frac{20 \text{ tons}}{20 \text{ tons} + 65 \text{ tons}} \times 5 \text{ m/s} = 1.2 \text{ m/s}$$

The lost kinetic energy corresponds to the second term on the right side of Eq. (17), evaluated with the initial velocities $v_1 = 15$ m/s and $v_2 = 0$:

$$[\text{loss of kinetic energy}] = \frac{1}{2}\frac{m_1 m_2}{m_1 + m_2}(v_1 - v_2)^2$$

$$= \frac{1}{2}\frac{20 \text{ tons} \times 65 \text{ tons}}{20 \text{ tons} + 65 \text{ tons}}(5 \text{ m/s})^2$$

$$= 1.9 \times 10^2 \text{ tons} \cdot (\text{m/s})^2$$

This lost kinetic energy can, of course, also be calculated by taking the difference between the initial and final total kinetic energies:

$$\tfrac{1}{2}m_1 v_1^2 + \tfrac{1}{2}m_2 v_2^2 - \tfrac{1}{2}(m_1 + m_2)v_{CM}^2$$

$$= \tfrac{1}{2} \times 20 \text{ tons} \times (5 \text{ m/s})^2 + 0 - \tfrac{1}{2}(20 \text{ tons} + 65 \text{ tons}) \times (1.2 \text{ m/s})^2$$

$$= 1.9 \times 10^2 \text{ tons} \cdot (\text{m/s})^2$$

EXAMPLE 6. Figure 11.5a shows a **ballistic pendulum,** a device used in the past century to measure the speeds of bullets. The pendulum consists of a large block of wood of mass $m_2$ suspended from thin wires. Initially, the pendulum is at rest. The bullet of mass $m_1$ strikes the block horizontally and remains stuck in it. The impact of the bullet puts the block in motion, causing it to swing upward to a height $h$ (Figure 11.5b). In a test of a Springfield rifle firing a bullet of 9.7 g, a ballistic pendulum of 4.0 kg swings up to a height of 19 cm. What was the speed of the bullet before impact?

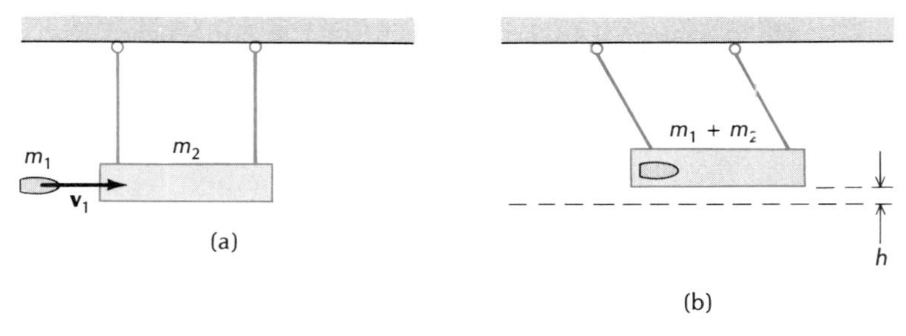


**Fig. 11.5** (a) Impact of a bullet on a ballistic pendulum. (b) After the impact, the pendulum swings to a height $h$.

SOLUTION: The collision of the bullet with the block of wood is totally inelastic. Since the collision takes only a very short time, the external forces (gravity, pull of wires) can be neglected and the momentum is conserved during the collision. Immediately after the collision, bullet and block move horizontally with the velocity of the center of mass (compare Example 5),

$$v_{CM} = \frac{m_1 v_1}{m_1 + m_2}$$

During the subsequent swinging motion of the pendulum, the total mechanical energy (kinetic plus potential) is conserved. At the bottom of the swing, the energy is kinetic, $\frac{1}{2}(m_1 + m_2)v_{CM}^2$, and at the top of the swing it is potential, $(m_1 + m_2)gh$. Hence

$$\tfrac{1}{2}(m_1 + m_2)v_{CM}^2 = (m_1 + m_2)gh$$

from which $v_{CM} = \sqrt{2gh}$ and therefore

$$v_1 = \frac{m_1 + m_2}{m_1}\sqrt{2gh} \qquad (18)$$

$$= \frac{0.0097 \text{ kg} + 4.0 \text{ kg}}{0.0097 \text{ kg}} \sqrt{2 \times 9.8 \text{ m/s}^2 \times 0.19 \text{ m}}$$

$$= 8.0 \times 10^2 \text{ m/s}$$

COMMENTS AND SUGGESTIONS: In this example, the solution hinges on recognizing which conservation law applies during what part of the motion. During the collision, momentum is conserved but not energy (the collision is inelastic); and during the swinging motion, energy is conserved but not momentum (the swinging motion proceeds under the influence of the "external" forces of gravity and the tension of the wires).

## 11.3 Collisions in Two Dimensions

In an elastic two-dimensional collision between two particles, such as the collision of two billiard balls on a billiard table, the laws of conservation of momentum and of energy do not suffice to determine the final velocities in terms of the initial velocities. Conservation gives us three equations: two equations from the conservation of the two separate components of momentum and one equation from the conservation of energy. But there are four unknowns: two components of velocity for each of the two particles. Thus, we cannot calculate the final motion completely. However, the conservation laws provide some useful information by placing severe restrictions on the possible final motions.

We will again assume that one of the particles is initially at rest and the other is initially in motion. Figure 11.6a shows the particles before the collision and Figure 11.6b shows them after. Particle 1 ("projectile") initially moves along a straight line;[2] but when it comes close to particle 2 (target), it feels a force and begins to deflect. Note that the collision shown in Figure 11.6 is not head-on — the initial line of motion of particle 1 passes to one side of particle 2. The particles never quite come into contact, but we will assume that the force reaches from one to the other, bridging the empty space between. The region over which the force acts (shown colored in Figure 11.6) is

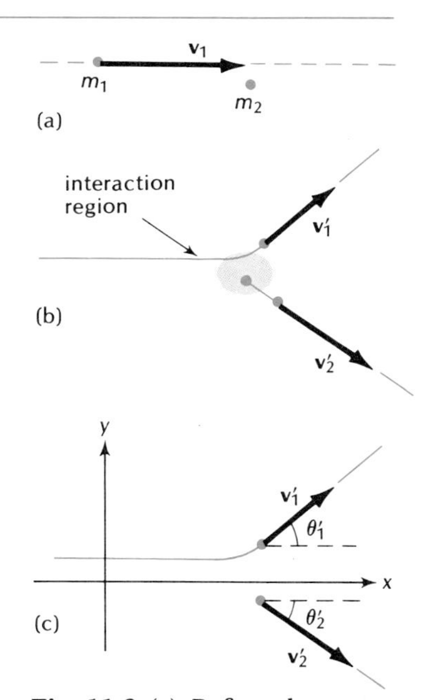


**Fig. 11.6** (a) Before the collision, particle 2 is at rest and particle 1 has velocity $\mathbf{v}_1$. (b) After the collision, particle 1 has velocity $\mathbf{v}'_1$ and particle 2 has velocity $\mathbf{v}'_2$. (c) The velocities $\mathbf{v}'_1$ and $\mathbf{v}'_2$ make angles $\theta'_1$ and $\theta'_2$ with the $x$ axis, respectively.

[2] We ignore external forces.

called the **interaction region.** What processes occur within this region need not concern us — we need to know only that these processes conserve energy and momentum. All the changes of motion occur within the interaction region; outside of this region the particles move with uniform velocity along straight paths. For example, if the projectile is an electron and the target an atom, then the interaction region is about equal to the volume of the atom — nothing happens to the electron until it penetrates the atom and begins to feel the presence of the atomic electrons and the atomic nucleus.

The initial line of motion of particle 1 and the initial position of particle 2 define a plane (the plane of the page in Figure 11.6). If the motion is to be two dimensional, the particles must always remain in this plane. Obviously this will be the case if the force acting between them is in the plane. Among the forces that play an important role in the physical world, both the electric force and the gravitational force between particles (see Chapters 9 and 22) lie along the line joining the particles; thus, these forces are in the plane of the motion, and collisions involving these forces may be treated as two dimensional.

We will designate the velocity of particle 1 before the collision by $\mathbf{v}_1$, and the velocities of particles 1 and 2 after the collision by $\mathbf{v}_1'$ and $\mathbf{v}_2'$. The velocity before the collision is along the $x$ axis and the velocities after the collision make angles $\theta_1'$ and $\theta_2'$ with the $x$ axis (Figure 11.6c).

Conservation of the $x$ component of momentum and of the $y$ component of momentum implies, respectively,

*Conservation of momentum and energy in a two-dimensional elastic collision*

$$m_1v_1 = m_1v_1' \cos \theta_1' + m_2v_2' \cos \theta_2' \qquad (19)$$

$$0 = m_1v_1' \sin \theta_1' - m_2v_2' \sin \theta_2' \qquad (20)$$

Conservation of energy gives

$$\tfrac{1}{2}m_1v_1^2 = \tfrac{1}{2}m_1v_1'^2 + \tfrac{1}{2}m_2v_2'^2 \qquad (21)$$

Note that, according to the present notation, the quantities $v_1$, $v_1'$, and $v_2'$ are speeds, or magnitudes of velocities; that is, they are all positive (this is in contrast to the notation of Section 11.2, where these quantities were positive or negative, depending on direction).

Equations (19), (20), and (21) constitute three restrictions on the four quantities $v_1'$, $v_2'$, $\theta_1'$, and $\theta_2'$; if, besides the initial speed $v_1$, one of these four quantities is known, then the other three can be calculated.

EXAMPLE 7. Figure 11.7 shows an elastic collision of a deuteron (the nucleus of an isotope of hydrogen, with a mass of 2.0 u) and a proton (with a mass of 1.0 u). This collision took place within the emulsion of a photographic film; in this medium the particles produce visible tracks because their passage damages (exposes) the film. The deuteron has an initial speed of $2.7 \times 10^7$ m/s and a final speed of $2.2 \times 10^7$ m/s. The proton is initially at rest. Calculate the final speed of the proton, and the final directions of motion of the deuteron and of the proton.

SOLUTION: The deuteron has a mass $m_1 = 2.0$ u and the proton has a mass $m_2 = 1.0$ u. According to Eq. (21), the final speed $v_2'$ of the proton is then

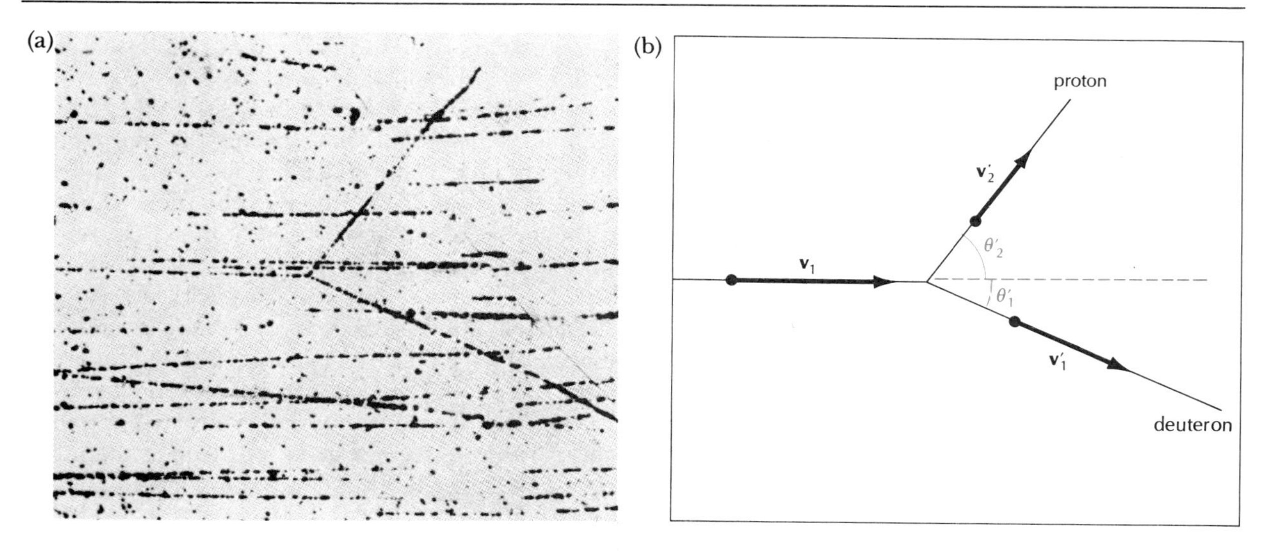


**Fig. 11.7** (a) Tracks of a deuteron and a proton in a photographic emulsion (enlarged about 750×). The deuteron entered from the left, collided with a proton, and was deflected toward the lower right. The proton recoiled toward the upper right. (b) Tracing of the tracks based on the photograph.

$$v_2' = \sqrt{\frac{m_1}{m_2}(v_1^2 - v_1'^2)} = \sqrt{2(v_1^2 - v_1'^2)}$$

$$= \sqrt{2[(2.7 \times 10^7 \text{ m/s})^2 - (2.2 \times 10^7 \text{ m/s})^2]} = 2.2 \times 10^7 \text{ m/s}$$

To find the angle $\theta_1'$ for the deuteron, we must eliminate $v_2'$ and $\theta_2'$ between Eqs. (19), (20), and (21). We begin with Eqs. (19) and (20):

$$m_2 v_2' \cos \theta_2' = m_1 v_1 - m_1 v_1' \cos \theta_1' \tag{22}$$

$$m_2 v_2' \sin \theta_2' = m_1 v_1' \sin \theta_1' \tag{23}$$

We square these equations and then add them together side to side, making use of the identity $\sin^2 \theta + \cos^2 \theta = 1$:

$$m_2^2 v_2'^2 = m_1^2 v_1^2 - 2m_1^2 v_1 v_1' \cos \theta_1' + m_1^2 v_1'^2 \tag{24}$$

Thus, we find

$$\cos \theta_1' = \frac{1}{2v_1 v_1'}\left[v_1^2 + v_1'^2 - \frac{m_2^2}{m_1^2} v_2'^2\right] = \frac{1}{2v_1 v_1'}\left[v_1^2 + v_1'^2 - \frac{1}{4} v_1'^2\right]$$

$$= \frac{1}{2 \times 2.7 \times 10^7 \text{ m/s} \times 2.2 \times 10^7 \text{ m/s}}$$

$$\times \left[(2.7 \times 10^7 \text{ m/s})^2 + \frac{3}{4}(2.2 \times 10^7 \text{ m/s})^2\right]$$

$$= 0.919$$

and $\theta_1' = 23°$.

The angle $\theta_2'$ for the proton is given by Eq. (23),

$$\sin \theta_2' = \frac{m_1}{m_2}\frac{v_1'}{v_2'} \sin \theta_1' = 2 \times \frac{2.2 \times 10^7 \text{ m/s}}{2.2 \times 10^7 \text{ m/s}} \sin 23° = 0.788$$

and $\theta_2' = 52°$.

---

If the collision is inelastic, then the conservation law of Eq. (21) for the energy is not applicable. The conservation laws of Eqs. (19) and

(20) for the $x$ and $y$ components of the momentum remain applicable, and they place restrictions on the possible final motions. Since the number of available equations in the inelastic case is smaller than in the elastic case, we are now even less able to calculate the final motion completely. However, there is an exception: if the collision is totally inelastic, then the laws of conservation of momentum by themselves are sufficient to calculate the velocities after the collision. As we saw in Section 11.2, in a totally inelastic collision, the two particles will stick together and, therefore, their final velocity coincides with the velocity of the center of mass. There are then only two unknowns: the two components of the velocity of the center of mass.

Although in the above discussion we emphasized the calculation of the final motion from the known initial motion, in practical applications of momentum conservation we often need to calculate the initial motion (or some aspect of the initial motion) from the known final motion. As the following example shows, this involves the same set of equations; it is merely a matter of keeping in mind which quantities are known, and which unknown.

EXAMPLE 8. A white automobile of mass 1100 kg and a black automobile of mass 1300 kg collide at an intersection. The investigation of this collision discloses that just before the collision the white automobile was traveling due east, and the black automobile was traveling due north (Figure 11.8). After the collision, the wrecked automobiles remained joined together and their tires made skid marks 18.7 m long in a direction 30° north of east before coming to rest. What was the speed of each automobile before the collision? Was one of them exceeding the legal speed limit of 25 m/s (90 km/h)? Assume that the wheels of both automobiles remained locked after the collision and that the coefficient of kinetic friction between the locked wheels and the pavement is $\mu_k = 0.80$.

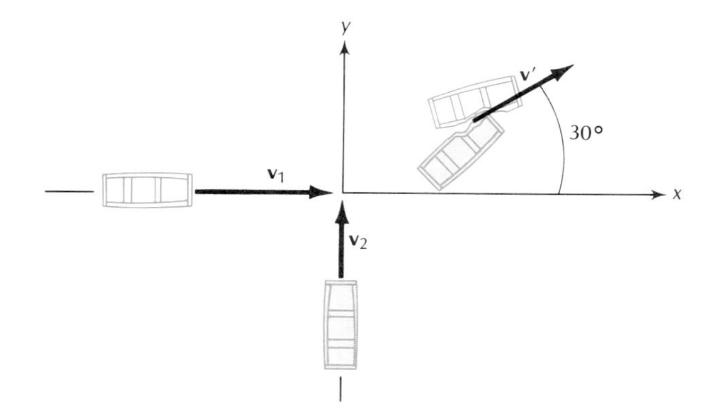


**Fig. 11.8** An automobile collision. Before the collision the velocities of the automobiles were $\mathbf{v}_1$ and $\mathbf{v}_2$; after the collision both velocities were $\mathbf{v}'$.

SOLUTION: We designate the velocities of the white and the black automobile before the collision by $\mathbf{v}_1$ and $\mathbf{v}_2$, respectively. Their velocity after the collision is $\mathbf{v}'$, the same for both. The velocity $\mathbf{v}_1$ is along the $x$ axis; $\mathbf{v}_2$ is along the $y$ axis; and $\mathbf{v}'$ is at an angle $\theta' = 30°$ with the $x$ axis (Figure 11.8). Conservation of the $x$ component and of the $y$ component of momentum gives

$$m_1 v_1 = (m_1 + m_2)v' \cos\theta' \tag{25}$$

$$m_2 v_2 = (m_1 + m_2)v' \sin\theta' \tag{26}$$

In order to calculate the speeds $v_1$ and $v_2$ from these equations, we need to know $v'$, the speed immediately after the collision. This speed is determined by the length of the skid marks. The deceleration of the skidding automobiles is

$$a = \frac{f_k}{m_1 + m_2} = \frac{\mu_k(m_1 + m_2)g}{m_1 + m_2} = \mu_k g$$

According to Eq. (2.25), the speed $v'$ is related to the length $l$ of the skid marks:

$$al = \tfrac{1}{2}v'^2$$

Hence

$$v' = \sqrt{2al} = \sqrt{2\mu_k gl} = \sqrt{2 \times 0.80 \times (9.8 \text{ m/s}^2) \times 18.7 \text{ m}}$$

$$= 17 \text{ m/s}$$

Equations (25) and (26) then yield

$$v_1 = \frac{m_1 + m_2}{m_1} v' \cos\theta' = \frac{1100 \text{ kg} + 1300 \text{ kg}}{1100 \text{ kg}} \times 17 \text{m/s} \times \cos 30°$$

$$= 32 \text{ m/s}$$

and

$$v_2 = \frac{m_1 + m_2}{m_2} v' \sin\theta' = \frac{1100 \text{ kg} + 1300 \text{ kg}}{1300 \text{ kg}} \times 17 \text{m/s} \times \sin 30°$$

$$= 16 \text{ m/s}$$

The white automobile *was* speeding.

## 11.4* Collisions and Reactions of Nuclei and of Elementary Particles

To explore the structure of atoms, nuclei, and elementary particles, physicists use beams of high-energy projectiles as probes. These beams, usually consisting of a stream of electrons or protons produced by a particle accelerator, are made to impact on a target containing the atoms, nuclei, etc., to be explored. The projectiles penetrate deep into the subatomic structures and there they engage in collisions. The manner in which the projectiles bounce off or react gives physicists some clues about the structures responsible for the collisions.

In elastic collisions between nuclei or elementary particles, only the direction of motion changes; the particles themselves do not change. The process of deflection of such particles in a collision is called **scattering.** In inelastic collisions between nuclei or elementary particles, kinetic energy is lost (or gained) by the alteration of the internal structure of the particles and by the conversion of energy into rest mass (or vice versa); this is a process of destruction of old particles and creation of new particles.

*Scattering*

* This section is optional.

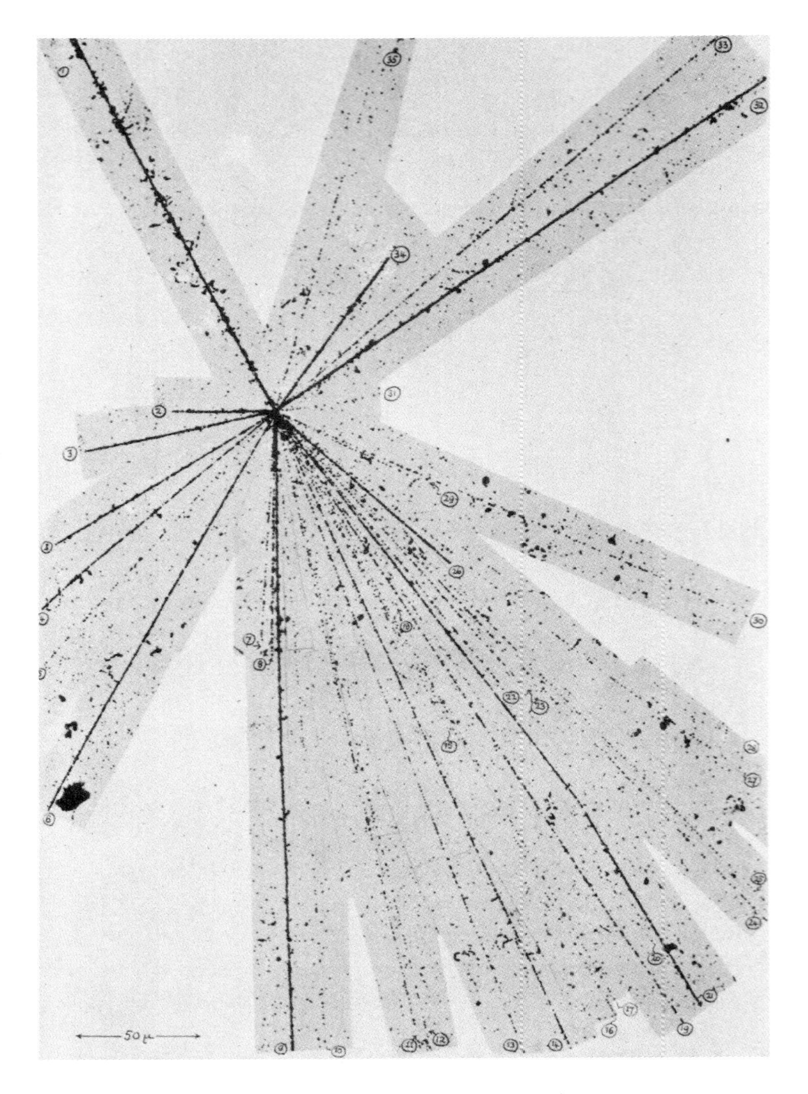


**Fig. 11.9** Tracks of particles in a photographic emulsion (enlarged about 330×). The heavy track on the upper left was made by a cosmic ray. This cosmic ray suffered a collision with the nucleus of an atom. The tracks of many new particles emerge from the scene of the accident.

In some high-energy collisions the number of particles created may be extremely large. Figure 11.9 shows a spectacular inelastic collision between a very energetic cosmic-ray particle and a nucleus. The collision took place within the emulsion of a photographic film, where the particles left visible tracks because their passage exposed the film. The tracks show that a great many new particles were created in the collision and that they sped away from the scene of the accident; obviously a large amount of kinetic energy must have been converted into rest mass.

In order to write down the energy balance for an inelastic collision involving the destruction and creation of particles, we find it useful to introduce a quantity $Q$ which is the net change in the rest-mass energy. If the particles present before the reaction have masses $m_1, m_2, \ldots, m_n$ and those present after the collision have masses $m_1', m_2', \ldots, m_r'$, then

*Rest-mass energy change in an inelastic collision*

$$Q = (m_1 + m_2 + \cdots + m_n)c^2 - (m_1' + m_2' + \cdots + m_r')c^2 \qquad (27)$$

Thus, $Q$ is positive if the net rest-mass energy decreases in the collision and $Q$ is negative if it increases. The conservation of total energy (kinetic and rest-mass energy) then reads

$$\boxed{K + Q = K'} \tag{28}$$

where $K$ and $K'$ represent the kinetic energies before and after the collision.

Obviously, a reaction with a positive $Q$ releases kinetic energy and a reaction with negative $Q$ absorbs kinetic energy. The latter kind of reaction is viable only if the initial kinetic energy available for inelastic processes is larger than $-Q$. According to Eq. (17), the initial kinetic energy for a two-particle collision (in one dimension) is

$$K = \tfrac{1}{2}Mv_{\text{CM}}^2 + \frac{1}{2}\frac{m_1 m_2}{m_1 + m_2}(v_1 - v_2)^2 \tag{29}$$

Only the second term on the right side is available for inelastic processes (compare Example 5). Hence an inelastic reaction involving these two particles is only possible if this second term is at least as large as $-Q$, that is, *if the internal kinetic energy is at least as large as* $-Q$.

*Condition for inelastic reaction*

---

EXAMPLE 9. The bombardment of a $^7$Li nucleus with protons can give rise to the reaction

$$\text{p} + {}^7\text{Li} \rightarrow {}^7\text{Be} + \text{n}$$

What is the minimum kinetic energy that the proton must have if it is to induce this reaction on a stationary $^7$Li nucleus? The relevant masses are $m_\text{p} = 1.0073$ u, $m_\text{Li} = 7.0144$ u, $m_\text{Be} = 7.0147$ u, and $m_\text{n} = 1.0087$ u.

SOLUTION: The $Q$ value for the reaction is

$$\begin{aligned} Q &= (m_\text{p} + m_\text{Li})c^2 - (m_\text{Be} + m_\text{n})c^2 \\ &= (1.0073 + 7.0144)\ \text{u} \times c^2 - (7.0147 + 1.0087)\ \text{u} \times c^2 \\ &= -\,0.0017\ \text{u} \times c^2 \\ &= -\,0.0017\ \text{u} \times 1.66 \times 10^{-27}\ \text{kg/u} \times (3.0 \times 10^8\ \text{m/s})^2 \\ &= -2.6 \times 10^{-13}\ \text{J} \end{aligned}$$

Since the sum of the final masses is larger than the sum of the initial masses, the reaction can proceed only by the conversion of some kinetic energy into rest mass: $2.6 \times 10^{-13}$ J of kinetic energy must be converted into rest mass. With $v_\text{Li} = 0$, the condition on the second term on the right side of Eq. (29) is

$$\frac{1}{2}\frac{m_\text{p} m_\text{Li}}{m_\text{p} + m_\text{Li}} v_\text{p}^2 \geq -Q$$

or

$$\begin{aligned} \tfrac{1}{2}m_\text{p}v_\text{p}^2 &\geq \frac{m_\text{p} + m_\text{Li}}{m_\text{Li}} \times (-Q) \qquad (30) \\ &\geq \frac{1.0073 + 7.0144}{7.0144} \times 2.6 \times 10^{-13}\ \text{J} \\ &\geq 3.0 \times 10^{-13}\ \text{J} \end{aligned}$$

COMMENTS AND SUGGESTIONS: Note that the total initial kinetic energy is simply the kinetic energy of the proton. Equation (30) therefore shows that it is *not sufficient* for the total initial kinetic energy to match $-Q$. The total kinetic energy must be larger than $-Q$ by a factor of $(m_p + m_{Li})/m_{Li}$ because part of this kinetic energy is tied up in the motion of the center of mass and is therefore not available for the inelastic process.

---

The minimum initial kinetic energy that makes a reaction viable is called the **threshold energy.** For instance, in Example 9 the threshold energy is $3.0 \times 10^{-13}$ J. If the reaction happens at exactly the threshold energy, then the reaction products have minimum energy, that is, they have zero relative velocity and remain together at the center of mass.

## SUMMARY

**Impulse:** $\mathbf{I} = \int_0^{\Delta t} \mathbf{F}\, dt$

$$= \mathbf{p}' - \mathbf{p}$$

**Elastic collision:** Kinetic energy is conserved.

**Totally inelastic collision:** Maximum amount of kinetic energy is lost; particles remain joined together.

**Speeds in one-dimensional elastic collision:**

before: $v_1 \neq 0, \quad v_2 = 0$

after: $v_1' = \dfrac{m_1 - m_2}{m_1 + m_2} v_1, \quad v_2' = \dfrac{2m_1}{m_1 + m_2} v_1$

**Change of rest-mass energy in inelastic collision:**

$$Q = (m_1 + m_2 + \cdots + m_n)c^2 - (m_1' + m_2' + \cdots + m_r')c^2$$

**Condition for reaction with negative $Q$:** $K_{\text{int}} > |Q|$

## QUESTIONS

1. According to the data given in Example 1, what percentage of the initial kinetic energy does the automobile retain after the collision?

2. A (foolish) stuntman wants to jump out of an airplane at high altitude without a parachute. He plans to jump while tightly encased in a strong safe which can withstand the impact on the ground. How would you convince the stuntman to abandon this project?

3. In the crash test shown in the photographs of Figure 11.1, anthropomorphic dummies were riding in the automobile. These dummies were (partially) restrained by seat belts, which limited their motion relative to the automobile. How would the motion of the dummies have differed from that shown in these photographs if they had not been restrained by seat belts?

4. For the sake of safety, would it be desirable to design automobiles so that their collisions are elastic or inelastic?

5. Two automobiles have collided at a north–south east–west intersection. The skid marks their tires made after the collision point roughly northwest. One driver claims he was traveling west; the other driver claims he was traveling south. Who is lying?

6. Statistics show that, on the average, the occupants of a heavy ("full-size") automobile are more likely to survive a crash than those of a light ("compact") automobile. Why would you expect this to be true?

7. In Joseph Conrad's tale *Gaspar Ruiz,* the hero ties a cannon to his back and, hugging the ground on all fours, fires several shots at the gate of a fort. How does the momentum absorbed by Ruiz compare with that absorbed by the gate? How does the energy absorbed by Ruiz compare with that absorbed by the gate?

8. Give an example of a collision between two bodies in which *all* of the kinetic energy is lost to inelastic processes.

9. Explain the operation of the five-pendulum toy, called Newton's cradle, shown in Figure 11.10.

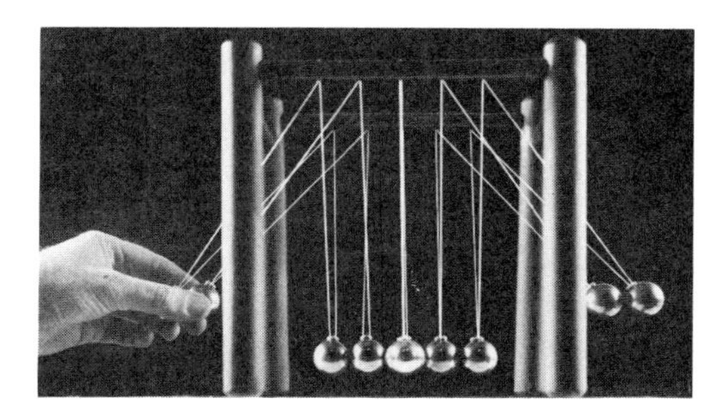

**Fig. 11.10** Newton's cradle.

10. In order to split a log with a small ax, you need a greater impact speed than you would need with a large ax. Why? If the energy required to split the log is the same in both cases, why is it more tiring to use a small ax? (Hint: Think about the kinetic energy of your arms.)

11. If you throw an (elastic) baseball at an approaching train, the ball will bounce back at you with an increased speed. Explain.

12. You are investigating the collision of two automobiles at an intersection. The automobiles remained joined together after this collision and their wheels made measurable skid marks on the pavement before they came to rest. Assume that during skidding all the wheels remained locked so that the deceleration was entirely due to kinetic friction. You know the direction of motion of the automobiles before the collision (drivers are likely to be honest about this), but you do not know the speeds (drivers are likely to be dishonest about this). What do you have to measure at the scene of the accident to calculate the speeds of both automobiles before the collision?

## PROBLEMS

### Section 11.1

1. A stunt man of mass 77 kg "belly-flops" on a shallow pool of water from a height of 11 m. When he hits the pool, he comes to rest in about 0.05 s. What is the average braking force that the water and the bottom of the pool exert on his body during this time interval?

2. A large ship of 700,000 metric tons steaming at 20 km/h runs aground on a reef, which brings it to a halt in 5.0 s. What is the average force on the ship? What is the average deceleration?

3. The photographs of Figure 11.1 show the impact of an automobile on a rigid wall.
   (a) Measure the positions of the automobile on these photographs and calculate the average velocity for each of the $20 \times 10^{-3}$-s intervals between one photograph and the next; calculate the average acceleration for each time interval from the change between one average velocity and the next.
   (b) The mass of this automobile is 1700 kg. Calculate the average force for each time interval.
   (c) Make a plot of this force as a function of time and find the impulse by estimating the area under this curve.

4. The "land divers" of Pentecost Island (New Hebrides) jump from platforms 21 m high. Long liana vines tied to their ankles jerk them to a halt just short of the ground. If the pull of the liana takes 0.02 s to halt the diver, what is the average acceleration of the diver during this time interval? If the mass of the diver is 64 kg, what is the corresponding average force on his ankles?

5. High-speed photography shows that when a golf club hits a golf ball, the club and the ball typically remain in contact for $1.0 \times 10^{-3}$ s and the ball acquires a speed of 70 m/s. The mass of the ball is 45 g. Estimate the magnitude of the force that the club exerts on the ball.

*6. Suppose that in a baseball game, the batter succeeds in hitting the baseball thrown toward him by the pitcher. Suppose that just before the bat hits, the ball is moving toward the batter horizontally with a speed of 35 m/s; and after the bat has hit, the ball is moving away from the batter and upward at an angle of 50° and finally lands on the ground 110 m away. The mass of the ball is 0.15 kg. From this information, calculate the magnitude and direction of the impulse the ball received in the collision with the bat. Neglect air friction and neglect the initial height of the ball above the ground.

*7. Bobsleds racing down a bobsled run often suffer glancing collisions with the vertical walls enclosing the run. Suppose that a bobsled of 600 kg traveling at 120 km/h approaches a wall at an angle of 3° and bounces off at the same angle. Subsequent inspection of the wall shows that the side of the bobsled made a scratch mark of length 2.5 m along the wall. From this data, calculate the time interval the bobsled was in contact with the wall, and calculate the average magnitude of the force that acted on the side of the bobsled during the collision.

### Section 11.2

8. In a lecture demonstration experiment, two masses collide elastically on a frictionless air track. The moving mass (projectile) is 60 g and the initially stationary mass (target) is 120 g. The initial velocity of the projectile is 0.80 m/s.
   (a) What is the velocity of each mass after the collision?
   (b) What is the kinetic energy of each mass before the collision? After the collision?

**Fig. 11.11** Meteor Crater in Arizona.

9. Meteor Crater in Arizona (Figure 11.11), a hole 180 m deep and 1300 m across, was gouged in the surface of the Earth by the impact of a large meteorite. The mass and speed of this meteorite have been estimated at $2 \times 10^9$ kg and 10 km/s, respectively, before impact.
   (a) What recoil velocity did the Earth acquire during this (inelastic) collision?
   (b) How much kinetic energy was released for inelastic processes during the collision? Express this energy in the equivalent of tons of TNT; 1 ton of TNT releases $4.2 \times 10^9$ J upon explosion.

10. It has been reported (fallaciously) that the deer botfly can attain a maximum airspeed of 818 mi/h, that is, 366 m/s. Suppose that such a fly, buzzing along at this speed, strikes a stationary hummingbird and remains stuck in it. What will be the recoil velocity of the hummingbird? The mass of the fly is 2 g; the mass of the hummingbird is 50 g.

*11. A projectile of 45 kg has a muzzle speed of 656.6 m/s when fired horizontally from a gun held in a rigid support (no recoil). What will be the muzzle speed (relative to the ground) of the same projectile when fired from a gun that is free to recoil? The mass of the gun is $6.6 \times 10^3$ kg. (Hint: The kinetic energy of the gun–projectile system is the same in both cases.)

12. An automobile approaching an intersection at 10 km/h bumps into the rear of another automobile standing at the intersection with its brakes off and its gears in neutral. The mass of the moving automobile is 1200 kg and that of the stationary automobile is 700 kg. If the collision is elastic, find the velocities of both automobiles after the collision.

*13. Two automobiles of 540 and 1400 kg collide head-on while moving at 80 km/h in opposite directions. After the collision the automobiles remain locked together.

(a) Find the velocity of the wreck immediately after the collision.
(b) Find the kinetic energy of the two-automobile system before and after the collision.
(c) The front end of each automobile crumples by 0.60 m during the collision. Find the acceleration (relative to the ground) of the passenger compartments of each automobile; make the assumption that these accelerations are constant during the collision.

*14. A speeding automobile strikes the rear of a parked automobile. After the impact the two automobiles remain locked together and they skid along the pavement with all their wheels locked. An investigation of this accident establishes that the length of the skid marks made by the automobiles after the impact was 18 m; the mass of the moving automobile was 2200 kg and that of the parked automobile was 1400 kg, and the coefficient of friction between the wheels and the pavement was 0.95.

(a) What was the speed of the two automobiles immediately after impact?
(b) What was the speed of the moving automobile before impact?

*15. A ship of $3.0 \times 10^4$ metric tons steaming at 40 km/h strikes an iceberg of $8.0 \times 10^5$ metric tons. If the collision is totally inelastic, what fraction of the initial kinetic energy of the ship is converted into inelastic energy? What fraction remains as kinetic energy of the ship–iceberg system? Ignore the effect of the water on the motion of the ship and iceberg.

*16. A cat crouches on the floor, at a distance of 1.2 m from a desk chair of height 0.45 m. The cat jumps onto the chair, landing with zero vertical velocity (this is standard procedure for cat jumps). The desk chair has frictionless coasters, and rolls away when the cat lands. The mass of the cat is 4.5 kg and the mass of the chair is 12 kg. What is the speed of recoil of the chair and cat?

*17. A crude but simple method for measuring the speed of a bullet is to shoot the bullet horizontally into a block of wood resting on a table. The block of wood will then slide until its kinetic energy is expended against the friction of the surface of the table. Suppose that a 3-kg block of wood slides a distance of 6 cm after it is struck by a bullet of 12 g. If the coefficient of sliding friction for the wood on the table is 0.6, what impact speed can you deduce for the bullet?

18. The impact of the head of a golf club on a golf ball can be approximately regarded as an elastic collision. The mass of the head of the golf club is 0.15 kg and that of the ball is 0.045 kg. If the ball is to acquire a speed of 60 m/s in the collision, what must be the speed of the club before impact?

19. In karate, the fighter makes his hand collide at high speed with the target; this collision is inelastic and a large portion of the kinetic energy of the hand becomes available to do damage in the target. According to a crude estimate, the energy required to break a concrete block (28 cm $\times$ 15 cm $\times$ 1.9 cm supported only at its short edges) is of the order of 10 J. Suppose the fighter delivers a downward hammer-fist strike with a speed of 12 m/s to such a concrete block. In principle, is there enough energy to break the block? Assume that the fist has a mass of 0.4 kg.

20. The impact of a hammer on a nail can be regarded as an elastic collision between the head of the hammer and the nail. Suppose that the mass of the head of the hammer is 0.50 kg and it strikes a nail of mass 12 g with an impact speed of 5 m/s. How much energy does the nail acquire in this collision?

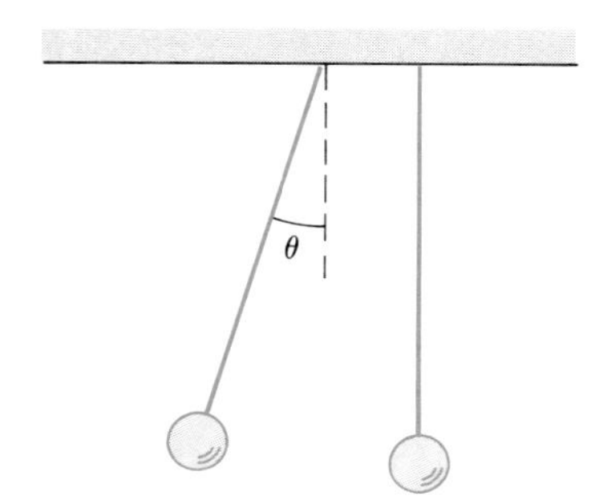

**Fig. 11.12**

21. Suppose that a neutron in a nuclear reactor initially has an energy of $4.8 \times 10^{-13}$ J. How many head-on collisions with carbon nuclei must this neutron make before its energy is reduced to $1.6 \times 10^{-19}$ J?

22. A proton of energy $8.0 \times 10^{-13}$ J collides head-on with a proton at rest. How much energy is available for inelastic reactions between these protons?

*23. (a) Two identical small steel balls are suspended from strings of length $l$ so that they touch when in their equilibrium position (Figure 11.12). If we pull one of the balls back until its string makes an angle $\theta$ with the vertical and then let go, it will collide elastically with the other ball. How high will the other ball rise?

(b) Suppose that instead of steel balls we use putty balls. They will then collide inelastically and remain stuck together. How high will the balls rise?

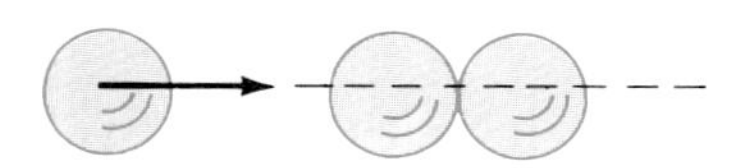
**Fig. 11.13**

*24. Two small balls are suspended side by side from two strings of length $l$ so that they touch when in their equilibrium position. Their masses are $m$ and $2m$, respectively. If the left ball (of mass $m$) is pulled aside and released from a height $h$, it will swing down and collide with the right ball (of mass $2m$) at the lowest point. Assume the collision is elastic.

(a) How high will each ball swing after the collision?

(b) Both balls again swing down and they collide once more at the lowest point. How high will each swing after this second collision?

*25. On a smooth, frictionless table, a billiard ball of velocity $v$ is moving toward two other aligned billiard balls in contact (Figure 11.13). What will be the velocity of each ball after impact? Assume that all balls have the same mass and that the collisions are elastic. (Hint: Treat this as two successive collisions.)

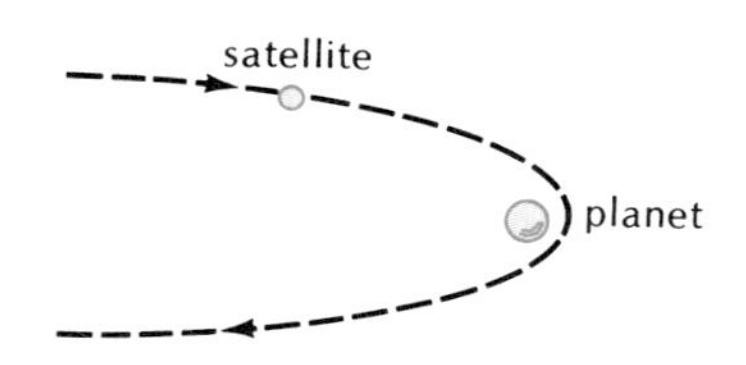

**Fig. 11.14** Satellite "colliding" with a planet.

*26. Repeat Problem 25 but assume that the middle ball has twice the mass of each of the others.

*27. If an artificial satellite, or some other body, approaches a planet at fairly high speed at a suitable angle, it will whip around the planet and recede in a direction almost opposite to the initial direction of motion (Figure 11.14). This can be regarded approximately as a one-dimensional "collision" between the satellite and the planet; the collision is elastic. In such a collision the satellite will gain kinetic energy from the planet, provided that it approaches the planet along a direction opposite to the direction of the planet's motion. This slingshot effect has been used to boost the speed of both Voyager spacecraft as they passed near Jupiter. Consider the head-on "collision" of a satellite of initial speed 10 km/s with the planet Jupiter, which has a speed of 13 km/s. (The speeds are measured in the reference frame of the Sun.) What is the maximum gain of speed that the satellite can achieve?

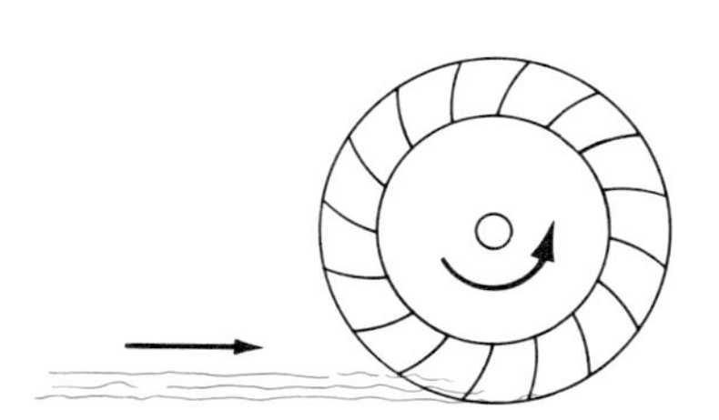
**Fig. 11.15** An undershot turbine wheel.

**28. A turbine wheel with curved blades is driven by a high-velocity stream of water that impinges on the blades and bounces off (Figure 11.15). Under ideal conditions the velocity of the water particles after collision with the blade is exactly zero so that all of the kinetic energy of the water is transferred to the turbine wheel. If the speed of the water particles is 27 m/s, what is the ideal speed of the turbine blade? (Hint: Treat the collision of a water particle and the blade as a one-dimensional elastic collision.)

**29. In Section 11.2 we showed that for an elastic one-dimensional collision the relative velocity reverses during the collision. If $v_2$ is zero, we can express this condition as $v_1' - v_2' = -v_1$ [see Eq. (13)]; if $v_2$ is not zero, we can express this condition as $v_1' - v_2' = -(v_1 - v_2)$. For a partially inelastic collision the relative velocity after the collision will have a smaller magnitude than the relative velocity before the collision. We can express this mathematically as $v_1' - v_2' = -e(v_1 - v_2)$, where $e < 1$ is called the **coefficient of restitution.** For some kinds of bodies, the coefficient $e$ is a constant, independent of $v_1$ and $v_2$.

(a) Show that in this case the final internal energy is less than the initial internal kinetic energy by a factor of $e^2$, that is, $K_{\text{int}} = e^2 K_{\text{int}}$.

(b) Derive formulas analogous to Eqs. (11) and (12) for the velocities $v_1'$ and $v_2'$ in terms of $v_1$ and $v_2$.

### Section 11.3

30. Two hockey players of mass 80 kg collide while skating at 7.0 m/s. The angle between their initial directions of motion is 130°.

(a) Suppose that the players remain entangled and that the collision is totally inelastic. What is their velocity immediately after collision?

(b) Suppose that the collision lasts 0.030 s. What is the magnitude of the average acceleration of each player during the collision?

*31. Two automobiles of equal masses collide at an intersection. One was traveling eastward and the other northward. After the collision, they remain joined together and skid, with locked wheels, before coming to rest. The length of the skid marks is 18 m and the coefficient of friction between the locked wheels and the pavement is 0.80. Each driver claims his speed was less than 14 m/s (50 km/h) before the collision. Prove that at least one driver is lying.

*32. Your automobile of mass $m_1 = 900$ kg collides at a traffic circle with another automobile of mass $m_2 = 1200$ kg. Just before the collision your automobile was moving due east and the other automobile was moving 40° south of east. After the collision the two automobiles remain entangled while they skid, with locked wheels, until coming to rest. Your speed before the collision was 14 m/s. The length of the skid marks is 17.4 m and the coefficient of kinetic friction between the tires and the pavement is 0.85. Calculate the speed of the other automobile before the collision.

*33. On July 27, 1956, the ships *Andrea Doria* (40,000 metric tons) and *Stockholm* (20,000 metric tons) collided in the fog south of Nantucket Island and remained locked together (for a while). Immediately before the collision the velocity of the *Andrea Doria* was 22 knots at 15° east of south and that of the *Stockholm* was 19 knots at 48° east of south (1 knot = 1 nmi/h = 1.85 km/h).

(a) Calculate the velocity (magnitude and direction) of the combined wreck immediately after the collision.

(b) Find the amount of kinetic energy that was converted into other forms of energy by inelastic processes during the collision.

(c) The large amount of energy absorbed by inelastic processes accounts for the heavy damage to both ships. How many kilograms of TNT would have to be exploded to obtain the same amount of energy as was absorbed by inelastic processes in the collision? The explosion of 1 kg of TNT releases $4.6 \times 10^6$ J.

*34. Two billiard balls are placed in contact on a smooth, frictionless table. A third ball moves toward this pair with velocity $v$ in the direction shown in Figure 11.16. What will be the velocity (magnitude and direction) of the three balls after the collision? The balls are identical and the collisions are elastic.

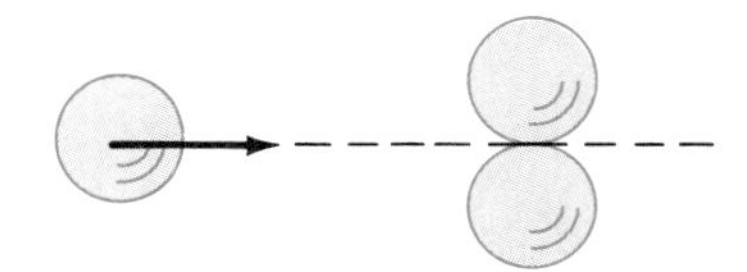

**Fig. 11.16**

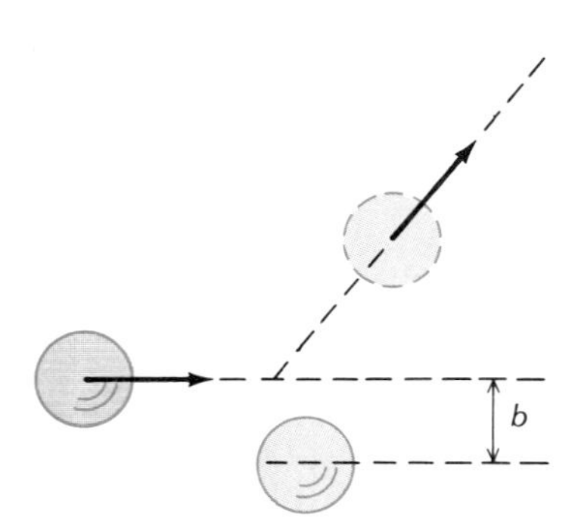


Fig. 11.17

*35. A billiard ball of mass $m$ and radius $R$ moving with speed $v$ on a smooth, frictionless table collides elastically with an identical stationary billiard ball glued firmly to the surface of the table.

(a) Find a formula for the angular deflection suffered by the moving billiard ball as a function of the impact parameter $b$ (defined in Figure 11.17). Assume the billiard balls are very smooth so that the force during contact is entirely along the center-to-center line of the balls.

(b) Find a formula for the magnitude of the momentum change suffered by the billiard ball.

*36. A neutron with an initial speed of $3.0 \times 10^7$ m/s collides elastically with a helium nucleus, which is initially at rest. The neutron is deflected through an angle of 60°. What is its final speed? [Hint: Use Eqs. (21) and (24)].

*37. A nuclear reactor designed and built in Canada (CANDU) contains heavy water ($D_2O$). In this reactor, the fast neutrons are slowed down by collisions with the deuterium nuclei of the heavy-water molecule.

(a) By what factor will the speed of a neutron be reduced in a head-on collision with a deuterium nucleus? The mass of this nucleus is 2.01 u.

(b) After how many head-on collisions with deuterium nuclei will the speed be reduced by the same factor as in a single head-on collision with a proton?

*38. The photograph of Figure 11.18 shows an elastic collision between two protons in a bubble chamber. Initially, one of the protons (track $AP$) has an energy of $8.0 \times 10^{-13}$ J and the other is at rest (at $P$). After the collision both protons are in motion (tracks $PB$ and $PC$). Measure the angles between the initial direction of motion and the final directions of motion and then calculate the final energy of each proton.

Fig. 11.18

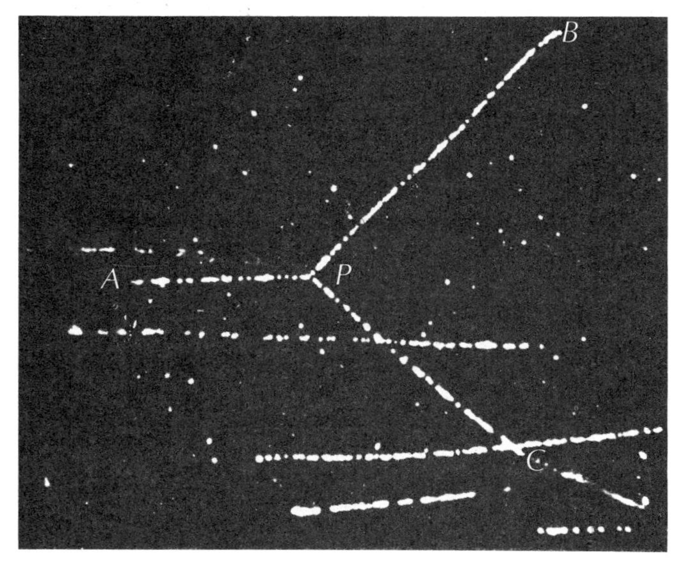


**39. Repeat the calculations of Problem 35 if the second billiard ball is initially stationary but not glued to the table. (Hint: First solve this problem in the center-of-mass frame; there, the motion of the balls is symmetric. Then transform the angle of the deflected motion into the reference frame of the table.)

**40. A proton with kinetic energy $5.0 \times 10^6$ eV collides elastically with a lithium nucleus and is deflected through an angle of 30°. What energy does the lithium nucleus acquire in this collision and what will be its direction of motion? The mass of the proton is 1.0 u and the mass of the lithium nucleus is 7.0 u.

**41. A moving particle collides elastically with another particle of the same mass that is initially at rest. Show that the angle between the tracks of the two particles after the collision is 90°; assume that the collision is not head-on.

Section 11.4

42. What is the $Q$ value for the reaction

$$\nu + \text{n} \rightarrow \text{p} + \text{e}?$$

The neutrino ($\nu$) has zero mass.

43. If struck by a neutron of sufficient energy, the nucleus of helium can split into two nuclei of isotopes of hydrogen according to the reaction

$$\text{n} + {}^4\text{He} \rightarrow {}^3\text{H} + {}^2\text{H}$$

The $Q$ value for this reaction is $-2.80 \times 10^{-12}$ J. What minimum kinetic energy must the neutron have if it is to initiate this reaction by impact on a (initially) stationary helium nucleus?

44. Consider the creation of a pion in a collision between two protons. The reaction is described by the equation

$$\text{p} + \text{p} \rightarrow \text{p} + \text{p} + \pi^0$$

If one of the protons is initially stationary, what minimum kinetic energy must the other proton have to make this reaction possible? The mass of the pion is $2.4 \times 10^{-28}$ kg. Pretend that the Newtonian formulas for kinetic energy remain valid (even though the speed of the proton is near the speed of light).

# AUTOMOBILE COLLISIONS AND AUTOMOBILE STRUCTURE*

Motor-vehicle accidents are the leading cause of death for the youngest third of the population of the industrialized nations. In the United States, 50,000 people are killed each year in such accidents. Besides this high toll in lives, the carnage on the roads exacts a high toll in injuries. In the United States, some 4,000,000 people are injured each year in motor-vehicle accidents. Among these, about 70,000 suffer permanent brain damage or spinal-cord damage, often leading to epilepsy, paraplegia, or quadriplegia. We can bring this stupefying toll of deaths and injuries into sharper focus by asking what these numbers mean for a single individual. An infant born in the United States today faces a chance of 2 in 3 of suffering an injury in a motor-vehicle accident at some time during his life, and he faces a chance of 1 in 60 of ending his life in such an accident.

The death rate from motor-vehicle accidents has been gradually increasing over the years, an increase that can be attributed to the increase in motor-vehicle use (see Figure III.1). The death rate per mile driven actually exhibits a slight decrease, due to improvements in the safety of vehicles and highways. Modern automobiles incorporate a variety of safety features designed to protect the occupants in a crash (see Figure III.2). Further substantial improvements in safety are within the reach of current technology. There is no fundamental reason why the occupant of an automobile should not walk away unscathed from a crash at 80 or 90 km/h (50 or 56 mi/h) into a rigid barrier, and some prototypes of experimental safe vehicles have already attained this level of protection.

Unfortunately, the general public does not perceive the operation of automobiles as a hazardous activity, and safety features rank low among the desiderata sought by buyers of new automobiles. Automobile manufacturers know that "safety does not sell," and they are therefore unwilling to devote much effort to research and development of safer automobiles. The ignorance and apathy of the public are clearly illustrated by the failure of the average American driver to wear a seat belt. Surveys show that only 30% of American drivers bother to wear a seat belt when not required to do so by law, and even when required by

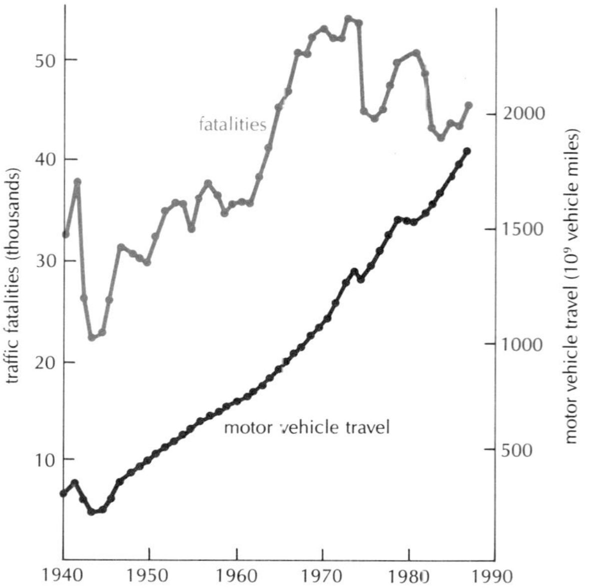

**Fig. III.1** Increasing death rate and increasing motor vehicle use in the United States. However, note the sharp dip in the death rate in 1974, when the federal speed limit of 55 mi/h was imposed. Although some portion of this dip reflects a temporary decrease of vehicle use caused by the oil shortage, most of it reflects lives saved by the lower driving speed.

* This chapter is optional.

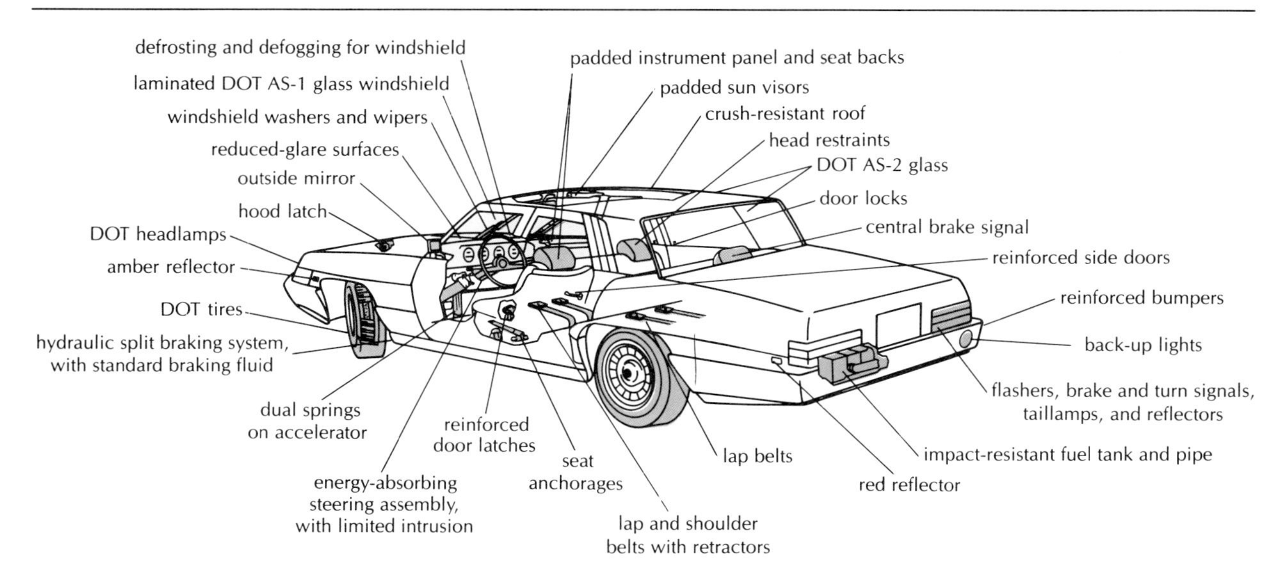


**Fig. III.2** The main safety features required by federal safety standards for automobiles used in the United States.

law, only 42% will comply (although, when caught in an accident, 68% will *claim* they had been wearing a belt).

## III.1 AUTOMOBILE VS. BARRIER

An automobile of 1500 kg travelling at 90 km/h (56 mi/h) has a kinetic energy of $4.7 \times 10^5$ J. We can better appreciate the magnitude of this energy by noting that it is equivalent to the potential energy released in dropping the automobile from a height of 32 m, or eleven floors down from the top of an apartment building. Most automobile collisions are totally inelastic, or nearly so. During a collision, all, or almost all, of this large kinetic energy is dissipated in the body of the automobile (and in the bodies of the occupants), with a consequent large amount of breakage and deformation.

Throughout the following discussion we will concentrate on head-on collisions, since these are the most frequent and usually the most severe, and they are also the easiest to analyze. Head-on collisions between automobiles and stationary, rigid barriers are routinely used for experimental tests of automobile safety. In the United States, such barrier crash tests are required for all new models of automobiles. The sequence of photographs in Figure 11.1 was obtained in one of these crash tests. The automobile was towed into the barrier, a massive concrete wall, at a speed of 49 km/h (30 mi/h). Figure III.3 gives the acceleration, the velocity, and the displacement of the "body" (that is, the passenger compartment) of the automobile during the crash. The instantaneous acceleration was measured by an accelerometer mounted in the body of the automobile; the instantaneous velocity and displacement were obtained from a slow-motion film of the crash (the photographs in Figure 11.1 are frames from this film). The acceleration varies in a rather complicated way; the average value of the acceleration is 13 G, but the instantaneous acceleration reaches a maximum value of 32 G. The impulsive force acting during the collision is quite large. As we saw in Example 11.1, the average impulsive force for this crash is $2.1 \times 10^5$ N, but the maximum instantaneous force is more than twice as large.

During the collision, the front part of the automobile is crushed by the impulsive force. This crushing dissipates the kinetic energy. In a collision at 49 km/h, the typical crush distance for the front part of the automobile is 0.6 or 0.7 m. The crush distance increases with impact speed. Figure III.4 presents data on crush distances obtained in automobile–barrier crash tests. The scatter of the experimental points reflects differences among different automobile models. Heavier, larger automobiles tend to have larger crush distances. In such automobiles, there are larger gaps between the separate parts, and there is more space available for crushing. Thus, in the larger automobile, the crushing of the front provides better cushioning of the collision. As Figure III.4 shows, the crush distance is an approximately linear function of the impact speed. Investigators of vehicle accidents often use the measured crush distance for an approximate determination of the impact speed, if this speed cannot be determined from other data. The approximate speed determined from the crush distance is called the energy-equivalent speed (EES).

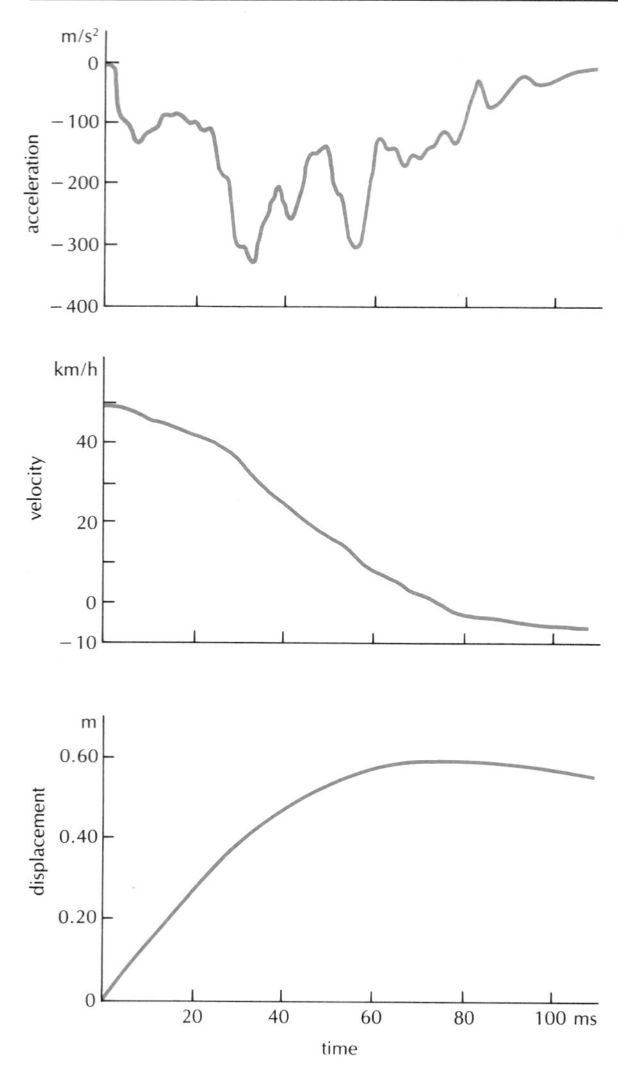


**Fig. III.3** Acceleration, velocity, and displacement of the automobile during the crash shown in Figure 11.1. (From data supplied by Mercedes-Benz of North America, Inc.)

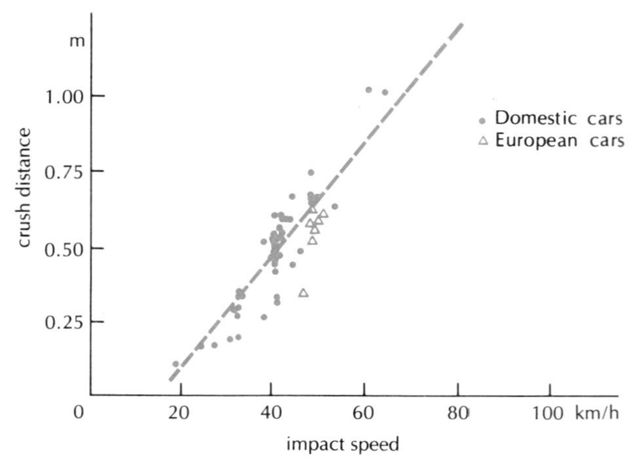


**Fig. III.4** Crush distance vs. impact speed, for frontal impacts on a flat, rigid barrier.

For a detailed analysis of the collision, we must take into account the deformation of the front parts of the automobile and their relative displacements. The crushing proceeds in stages; first the front bumper and the radiator are deformed and pushed back against the engine and the suspension, and then the engine and the attached transmission are pushed back against the body. Engineers have developed schematic models of the collision process. In such a model, the engine (with transmission), the suspension, and the body are regarded as massive, rigid parts within the automobile; these rigid parts are linked by approximately massless, deformable structural components which transmit forces between the massive parts (see Figure III.5). Each deformable structural component acts somewhat like a spring linking the masses; but in the collision, this "spring" is permanently deformed beyond its elastic limit, and the dependence of deformation on force is highly nonlinear. The force-deformation characteristics of the separate structural components can be determined by computer calculations, starting with the geometry of the component and the elastic and plastic properties of its material (usually mild steel). Alternatively, the component can be placed in a mechanical crusher, or a large press, which permits a direct experimental measurement of the deformation produced by a known applied force. The force-deformation charac-

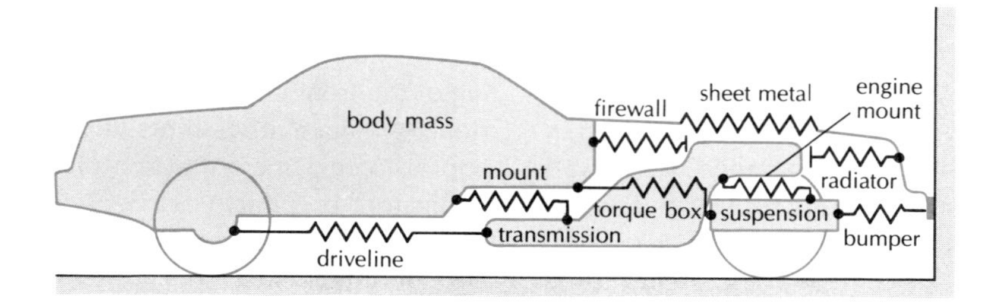


**Fig. III.5** Schematic model of an automobile used for calculations of the behavior in frontal collisions.

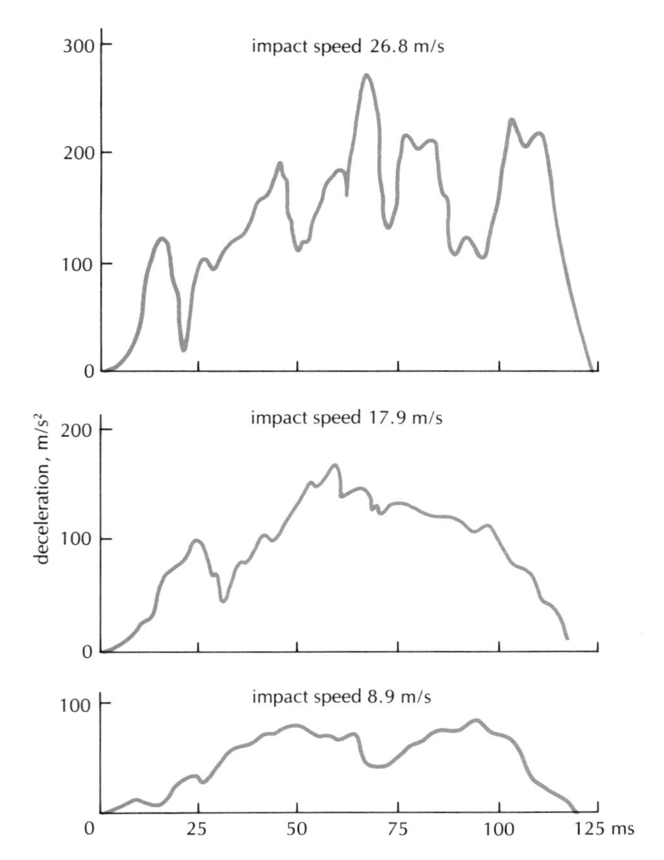


**Fig. III.6** Calculated instantaneous deceleration of the body of an automobile of 1600 kg during a crash into a rigid barrier. The three curves correspond to different impact speeds.

teristics of the individual structural components can then be used to calculate the behavior of the automobile in a collision. The diagrams in Figure III.6 display the results of such calculations for an automobile of medium mass, at several different impact speeds. The deceleration plotted in these diagrams is the deceleration of the body of the automobile, that is, the deceleration of the passenger compartment. Note that in these calculations, the crash duration, from zero acceleration to zero acceleration, increases only slightly with impact speed—in all the cases plotted in Figure III.6, it is near $\Delta t \simeq 0.11$ s. The calculations also indicate that the crash duration increases slightly with the automobile mass; the latter increase reflects the larger crush distance of larger automobiles, already mentioned above. Experimental data from crash tests are in accord with the results of these calculations.

If we ignore the slight variations of crash durations, we can pretend that all collisions have the same duration of about 0.11 s. This means that the severity of the collision, as measured by the average acceleration, can be regarded as directly proportional to the velocity change,

$$a = \frac{\Delta v}{\Delta t} \simeq [\text{constant}] \times \Delta v \tag{1}$$

Automotive engineers and accident investigators often use the velocity change $\Delta v$ as a criterion for the severity of a collision.

## III.2 AUTOMOBILE VS. AUTOMOBILE

Whereas in a totally inelastic collision between an automobile and a barrier all of the initial kinetic energy is dissipated in crushing, in a totally inelastic collision between two automobiles, some of the kinetic energy is locked up in the motion of the center of mass, and is not available for crushing. We know from Section 11.2 [see Eq. (11.17)] that the net kinetic energy of two bodies is the sum of the translational kinetic energy of the center of mass and the "internal" kinetic energy of motion relative to the center of mass:

$$K = \tfrac{1}{2} M v_{CM}^2 + \tfrac{1}{2} \frac{m_1 m_2}{m_1 + m_2} (v_1 - v_2)^2 \tag{2}$$

Only the "internal" kinetic energy is available for dissipation during the crushing; this energy depends on the relative speed, or the closing speed, of the two automobiles. Before the collision, $v_1$ and $v_2$ have opposite signs, and the magnitude of the closing speed $v_1 - v_2$ is therefore the sum of the two individual speeds. For instance, if the velocities are $v_1 = 49$ km/h and $v_2 = -49$ km/h, and the masses are $m_1 = 2000$ kg and $m_2 = 1000$ kg, then the individual kinetic energies of the automobiles are $1.9 \times 10^5$ J and $0.93 \times 10^5$ J, the total kinetic energy is $2.8 \times 10^5$ J, and the "internal," available kinetic energy is

$$\frac{1}{2} \frac{2000 \text{ kg} \times 1000 \text{ kg}}{2000 \text{ kg} + 1000 \text{ kg}} \times (98 \text{ km/h})^2 = 2.5 \times 10^5 \text{ J} \tag{3}$$

Thus, only 89% of the total kinetic energy is available for crushing the fronts of the automobiles. The other 11% of the kinetic energy is locked up in the motion of the center of mass, and this energy is gradually dissipated against the external friction force (and brake friction, if the brakes are still operative) while the wreckage skids along the street after the collision. During the collision, the *average* accelerations of the bodies of the colliding automobiles are in the inverse proportion of the masses,[1] and the net velocity changes are therefore also in the inverse ratio of the masses. Hence, the collision is more severe for the smaller automobile. In the above example, the velocity of the larger automobile decreases from 49 km/h to 16 km/h, but the velocity of the smaller automobile reverses from −49 km/h to +16 km/h, which makes the velocity change, and the severity of the collision, of the smaller automobile twice that of the larger. For the larger automobile, the collision is equivalent to a barrier crash at 33 km/h (20 mi/h); but for the smaller automobile, it is equivalent to a barrier crash at 65 km/h (40 mi/h).

Furthermore, the energy dissipations in the two automobiles will, in general, be unequal. How the available energy is distributed between the two automobiles during the collision depends on the masses and on the crush characteristics of each. Roughly, the automobile with the softer, more yielding front part will tend to acquire energy from the other, and will therefore have to dissipate more energy and suffer more damage. The energy transfer from one automobile to the other poses an additional hazard to the small automobile. The small automobile is less capable of dissipating energy, and an excessive amount of energy transferred to it during a collision is likely to compromise the integrity of the passenger compartment, with consequent additional injuries to the passengers. Engineers concerned with the special hazards of collisions between unequal automobiles have proposed that the front parts of large automobiles be designed for softness, so that more energy dissipation occurs in the large automobile, which can better afford it.

Because of the energy transfer between colliding automobiles, an automobile–barrier crash cannot exactly simulate an automobile–automobile crash. It is of course easy to arrange the automobile–barrier crash so as to obtain the same change of velocity, or change of momentum, as in an automobile–automobile crash. The impact speed of the automobile on the barrier merely must be made equal to the speed of the automobile relative to the center of mass; this guarantees that the automobile striking the barrier will receive the same impulse as the automobile striking another automobile. However, the energy the automobile receives (or delivers) in a collision with a fixed, rigid barrier is necessarily zero, whereas the energy the automobile receives (or delivers) in an automobile–automobile collision is usually not zero. Only in a collision between identical automobiles is the energy transfer exactly zero. But not even such a symmetric collision is exactly equivalent to a barrier collision, since the fronts of the two colliding automobiles will interpenetrate to some extent, and the deformations will therefore be somewhat different from those produced by a barrier crash. In view of this lack of an exact equivalence between barrier collisions and automobile–automobile collisions, the barrier crash tests can give only a rough indication of the damage to be expected in actual collisions. In a few instances, automobile manufacturers have performed automobile–automobile crash tests (see Figure III.7), but these tests are complicated and expensive to set up.

**Fig. III.7** Head-on collision test of two Volvo automobiles.

The general conclusion that large automobiles are safer in a collision is consistent with statistical data on fatalities in automobile collisions. Figure III.8 gives a

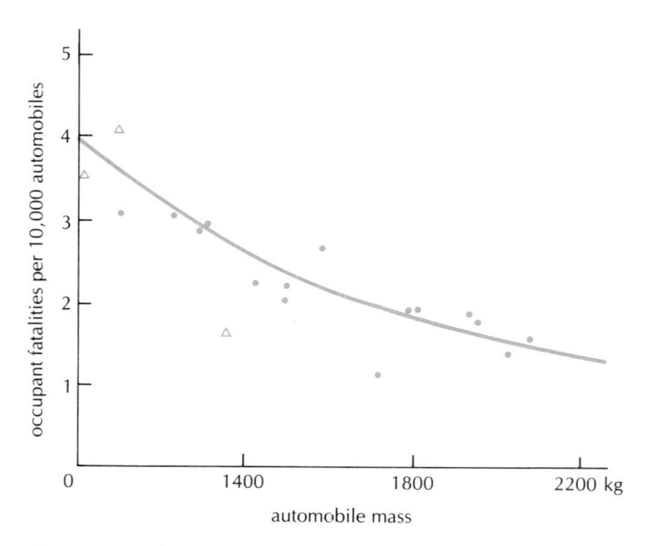


**Fig. III.8** Fatalities vs. automobile mass. This plot is based on data collected by the National Highway Traffic Safety Administration on all fatal crashes in the United States in 1976–1978.

[1] The *instantaneous* accelerations of the bodies of the automobiles are not necessarily in the inverse proportion of the masses, because the crushing leads to a displacement of the bodies relative to the centers of mass of the automobiles (the instantaneous accelerations of the centers of mass are, of course, exactly in inverse proportion to the masses).

plot of number of fatalities per automobile vs. mass of the automobile. Taken at face value, this plot indicates a substantial increase of safety with mass; however, the differences in fatalities among different car models reflect not only the intrinsic structural safety of the cars, but also the differences in driving habits of the drivers and the differences in the prevalent driving environments. Hence the agreement between the statistical data in Figure III.8 and the advantage expected on the basis of the physics of the collision merely means that the other factors do not outweigh the physical factors.

## III.3 THE SECONDARY COLLISION

Injury to the occupant of an automobile usually results from the collision of the occupant with some part of the interior of the automobile, such as the steering wheel, the instrument panel, or the windshield. This collision of the occupant with the interior of the automobile is called the secondary collision, or the human collision. It occurs some time after the beginning of the primary collision of the front of the automobile. Consider, for instance, an automobile crash with an impact speed of 56 km/h (35 mi/h) in which the body of the automobile stops within a time of 0.11 s. If the occupant is not restrained by a seat belt, he will continue to move forward at a constant speed of 56 km/h while the body of the automobile decelerates.[2] The occupant continues to move forward at this speed until he collides with the steering wheel, instrument panel, or windshield. In the typical case, the time the occupant spends flying through the air is comparable to the time the body of the automobile takes to come to a full stop. Hence the impact speed of the occupant on the steering wheel, instrument panel, or windshield is comparable to the primary impact speed of 56 km/h. The crush distance, or yield distance, of a steering wheel struck by a human chest or a windshield struck by a head is considerably shorter that the crush distance of the front of the automobile. Thus, the secondary collision has less cushioning than the primary collision, and it generates larger decelerations. The large forces involved in this secondary collision result in serious or fatal injuries of the occupant.

We can gain some quantitative understanding of the secondary collision by the following simple calculation. Let us pretend that the motion of the occupant is particle motion and that the deceleration of the automobile is uniform. If the velocity of the body of the automobile decreases by 56 km/h in 0.11 s, that is, by 15.6 m/s in 0.11 s, the deceleration of the body is

$$a = -\frac{15.6 \text{ m/s}}{0.11 \text{ s}} = -141 \text{ m/s}^2 = -14 \text{ G} \tag{4}$$

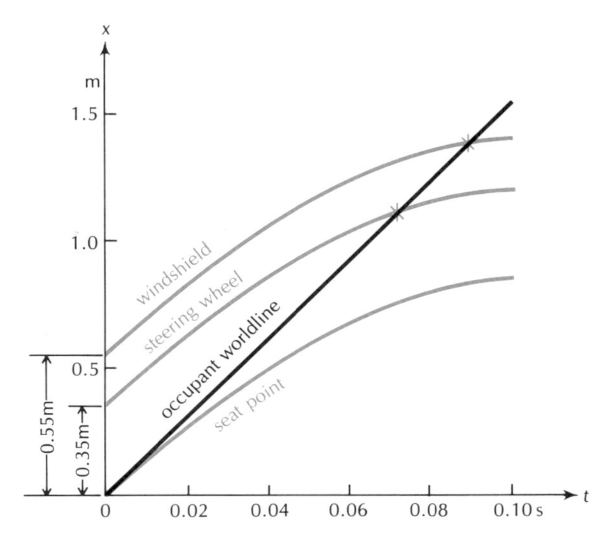


**Fig. III.9** Worldlines of occupant (black) and worldlines of seat point, steering wheel, and windshield (color).

Figure III.9 shows the worldline of the (unrestrained) occupant and the worldline of the seat point in the automobile where the occupant was sitting before the collision began. The worldline of the occupant is a straight line corresponding to a constant velocity of 15.6 m/s, and the worldline of the seat point is a parabola corresponding to an initial velocity 15.6 m/s and a uniform deceleration of 141 m/s². We will assume that the steering wheel is 0.35 m from the seat point and the windshield is 0.55 m from the seat point (these are typical chest-steering wheel and head-windshield distances for a driver of average build in an automobile of medium size). Figure III.9 includes plots of the worldlines of the steering wheel and of the windshield, placed at the appropriate distances from the seat point. The intersections of the occupant worldline with the worldlines of these obstacles tell us when the secondary collisions will occur. The equations for the worldlines of the occupant and the obstacle are, respectively,

$$x = v_0 t \tag{5}$$

and

$$x = v_0 t + \tfrac{1}{2} a t^2 + x_0 \tag{6}$$

[2] Friction between the occupant and the seat is insignificant. Furthermore, the maximum force that the occupant can exert if he braces himself against the impact is considerably smaller than what would be required to hold him in the seat.

where $v_0 = 15.6$ m/s is the initial speed of the occupant and the automobile, and $x_0 = 0.35$ m or 0.55 m is the initial distance between occupant and obstacle. If we set the expressions (5) and (6) equal, the terms $v_0 t$ cancel; this merely indicates that, relative to the automobile, the motion of the occupant is accelerated motion with zero initial velocity. We are then left with

$$0 = \frac{1}{2} at^2 + x_0 \qquad (7)$$

Substituting $a = -141$ m/s² and $x_0 = 0.35$ m or 0.55 m, we find

$$t = 0.070 \text{ s or } t = 0.088 \text{ s} \qquad (8)$$

Thus, the chest of the driver strikes the steering wheel at $t = 0.070$ s after the primary impact. Since the collision with the steering wheel halts or slows the motion of the driver, we cannot rely on our simple kinematic calculation to determine the subsequent motion of the driver's head toward the windshield. However, we can apply our calculation to the motion of a passenger sitting next to the driver in the front seat; according to Eq. (8), the head of this passenger will strike the windshield at $t = 0.088$ s. From these times for the secondary collisions we can immediately evaluate the speeds of impact relative to the steering wheel and relative to the windshield. The results are

9.9 m/s (36 km/h) for impact on steering wheel
12.5 m/s (45 km/h) for impact on windshield (9)

In this example, both of the secondary impacts occur a short time before the automobile comes to a full stop. The impact on the windshield has a higher impact speed (relative to the windshield) because the longer distance and the longer travel time from the seat point to the windshield implies that the automobile is nearly at rest at the time of the secondary impact, and therefore the speed of the passenger relative to the windshield is almost equal to the initial speed of the automobile. This means that, for unrestrained passengers, an automobile with a spacious interior and exceptionally large passenger–obstacle distances is exceptionally hazardous. In the extreme case, the time of free flight of a passenger may exceed the crash time (0.11 s in our example), and then the passenger will suffer a secondary collision at an impact speed equal to the initial speed (15.6 m/s in our example). Such extreme forms of secondary impact have been known to occur when an unrestrained passenger in the *rear* seat (usually the middle of the rear seat) flies all the way forward in a collision and strikes the windshield.

**Table III.1** COMPARISON OF IMPACT SPEEDS AND HEIGHTS OF FALL

| *Speed* | | *Height (number of floors down)*[a] |
|---|---|---|
| 15 km/h | 9.3 mi/h | $\frac{1}{3}$ |
| 30 | 18.6 | 1 |
| 45 | 28.0 | 3 |
| 60 | 37.2 | 5 |
| 75 | 46.6 | 8 |
| 90 | 55.9 | 11 |
| 105 | 65.2 | 15 |

[a] Each floor is 2.9 m.

We can better appreciate the effects of the secondary impact on the human body if we compare the impact speeds given in Eq. (9) with the speed attained by a body in free fall from some height. The impact on the windshield is equivalent to falling three floors down from an apartment building and landing head first on a hard surface. Our intuition tells us that this is likely to be fatal. Since our intuition about the dangers of heights is much better than our intuition about the dangers of speeds, it is often instructive to compare impact speeds with equivalent heights of fall. Table III.1 lists impact speeds and equivalent heights, expressed as the number of floors the body has to fall down to acquire the same speed (keep in mind that according to the American way of labeling floors of buildings, you have to climb to the fourth floor if you want to fall down three floors).

For more detailed calculations of the motion of occupants during an automobile collision, we must take into account the shape and the mass distribution of the human body. The human body can be approximated as several mass segments, or sticks, connected by flexible joints of specified stiffness. Figure III.10 shows the segments and their centers of mass. The mass distribution of each segment is nonuniform, so as to simulate the mass distribution of the human body.[3] The force-deformation characteristics of the joints and segments and the force-deformation characteristics of the obstacles—such as the steering wheel or the windshield—determine the mutual forces and the impact forces on each segment. The motion of the body segments can then be calculated from these forces.

Such calculations permit us to follow the complete motion of the occupant from his initial position in the seat, to his impact on the steering wheel or the windshield. Figure III.11 displays the result of such computer calculations of the motions of the driver and the

[3] Each mass segment is characterized by its length, mass, center of mass, and moment of inertia (as we will see in Chapters 12 and 13, the latter determines the rotational inertia of the mass segment).

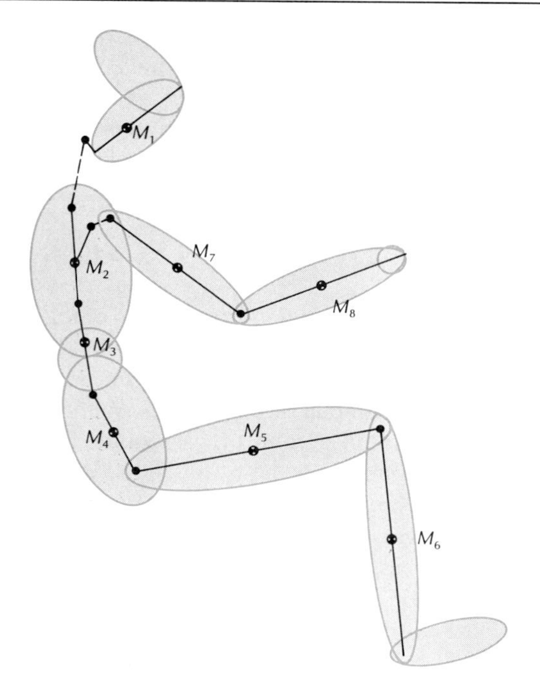


**Fig. III.10** In this mathematical model of the human body, the black lines represent mass segments and the small dots represent joints. The large colored ellipses are zones of influence that determine the external forces acting on the segments when in contact with an external obstacle. The force exerted by an obstacle—such as the steering wheel or the windshield—depends in a specified way on the depth of penetration into the zone.

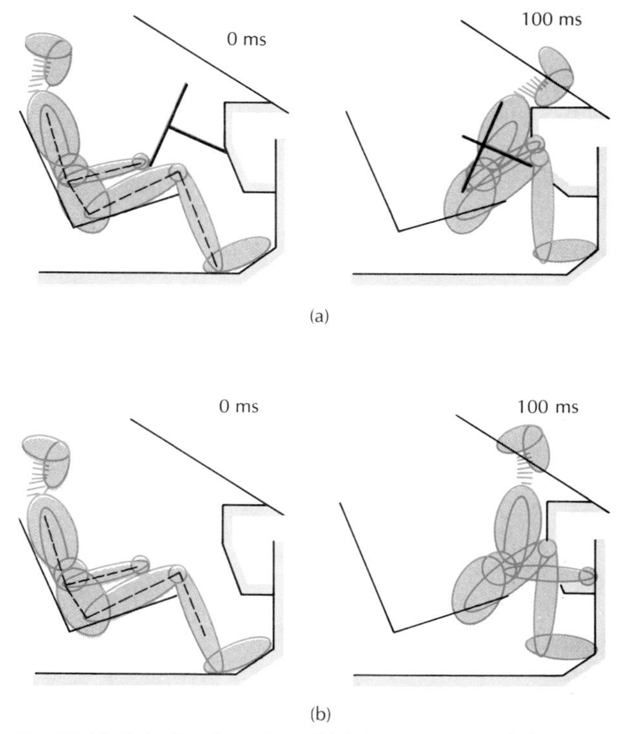


**Fig. III.11** Calculated motion of (a) the unrestrained driver and (b) the unrestrained passenger during a frontal collision at 56 km/h. The pictures display the positions of the driver and the passenger at $t = 0$ and at $t = 100$ m/s.

passenger. In these calculations, and also in the simple calculations above, the penetration, or "intrusion," of the automobile parts into the passenger compartment has been neglected. In collisions at low speeds, this is a fair approximation; but in collisions at high speeds, the steering wheel, instrument panel, and foot controls are often pushed back into the passenger compartment, especially in a small automobile with front-engine and front-wheel drive. Computer calculations can easily take such intrusion effects into account.

Alternatively, the motions of the occupants of an automobile during a collision can be investigated experimentally by means of anthropomorphic dummies. Figure III.12 shows the standard anthropomorphic dummy used in crash tests required for new models of automobiles. The size and mass of this dummy mimics that of a median male human body. The dummy is built with a metallic skeleton, with joints similar to those of the human skeleton. The skeleton is covered with an outer layer of soft plastic. The force-deflection characteristics of the neck and the rib cage match human characteristics, and so do the force-deflection characteristics of the outer layers of the head and of the knees.

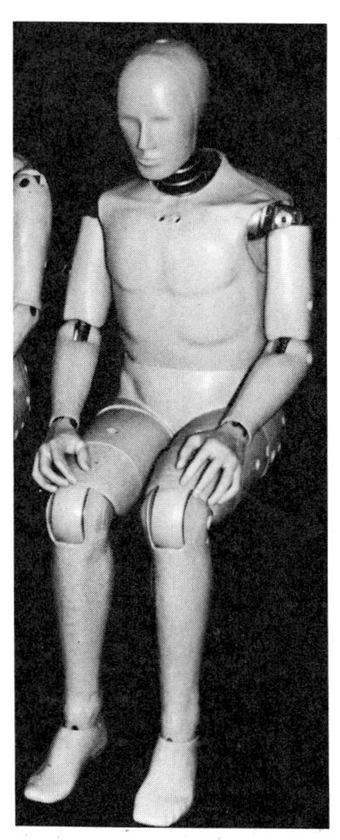

**Fig. III.12** The Hybrid III anthropomorphic dummy. This dummy mimics a male human body of height 170 cm and mass 78 kg. The skeleton is made of metal and the outer, skinlike jacket is made of soft plastic.

In view of the prevalence of chest and head injuries, the dummy is instrumented with accelerometers that monitor head and chest accelerations during the collision, and it also contains sensors that monitor the deformation of the chest. Other sensors monitor the forces on the leg bones (femur), the deformations of the knees, the deformation of the neck, and so on.

Figure III.13 shows a crash test with an anthropomorphic dummy. In this test, the dummy was equipped with seat belts, but they were left loose, to illustrate the poor protection offered by such loose belts. The dummy suffered a violent secondary collision with the steering wheel.

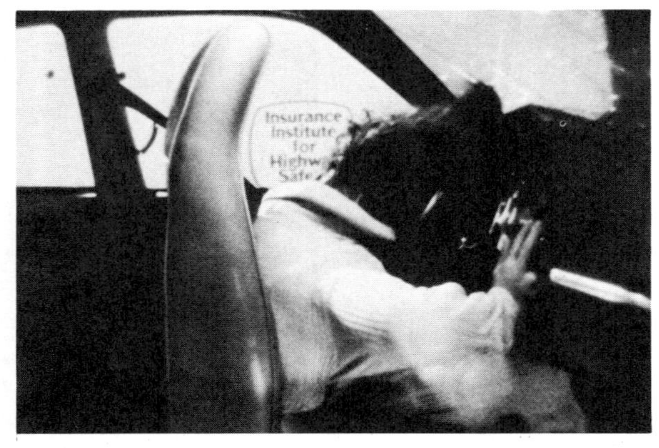

**Fig. III.13** Motion of an anthropomorphic dummy during a frontal collision at 56 km/h. (Insurance Institute for Highway Safety.)

## III.4 SEAT BELTS AND AIR BAGS

If the occupants of a colliding automobile are held snugly by seat belts, they will not suffer a secondary collision. Instead they will decelerate at the same rate as the body of the automobile and come to a stop at the same time as the automobile. Of course, the seat belt must supply the force to decelerate the occupant, and this seat-belt force can itself cause some injuries; but such injuries are usually minor compared with the severe injuries produced by the secondary collision of an unrestrained occupant.

Let us estimate the forces exerted by the lap-and-diagonal-shoulder belt (three-point belt), which has been required equipment in American automobiles since 1968. In a collision at 56 km/h with a constant deceleration of 141 m/s$^2$, or 14 G, the force the seat belt must exert to hold a man of 78 kg is

$$F = ma = 78\ \text{kg} \times 141\ \text{m/s}^2 = 1.1 \times 10^4\ \text{N} \quad (10)$$

For a rough prognosis of traumatic effects of this force on the body of the man, we can assume that the tensions in the lap and the shoulder belt are equal, and that the direction of pull is straight back. With the configuration shown in Figure III.14, each of the four ends then carries a load of $\frac{1}{4} \times 1.1 \times 10^4$ N, or $2.7 \times 10^3$ N; this is the tension in the belt. If the lap belt is correctly placed low over the hips, it will transfer most of its load to the pelvis, which can stand such forces with no more than a bruising of the skin. The torso also can stand forces of this magnitude without injury.

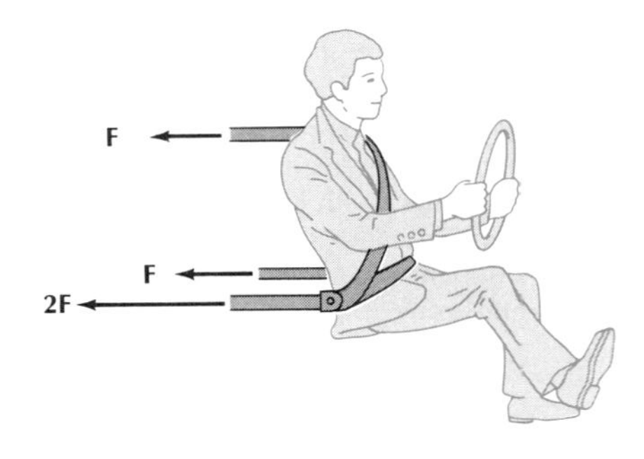


**Fig. III.14** Forces exerted by lap-and-diagonal-shoulder belt (three-point belt). In this simplified arrangement, all the forces are horizontal.

In our simple estimate of the force, we have relied on the optimistic assumption of a constant deceleration (which generates the least force for the specified change of velocity). In an actual collision, the acceleration and the force will fluctuate, reaching instantaneous peak values two to three times as large as their average values. Furthermore, in our simple estimate, we have ignored the initial slack in the belt and the elasticity of the belt. If the belt is initially somewhat slack, the occupant will move forward at constant speed until the belt snubs him, with a sudden jerk. This amounts to a collision of the occupant with the seat belt, and it leads to a momentarily higher force than that required for uniform deceleration. In contrast, the elasticity of the belt provides extra cushioning, and tends to reduce the force on the occupant. From crash tests with anthropomorphic dummies, the actual value of the peak tension in the belt was found to be about $7 \times 10^3$ N; the precise value depends on the exact time history of the acceleration, the configuration and slack in the seat belt, and the characteristics of the material of the belt. But this does not change our conclusion that the 56-km/h crash is readily survivable, in contrast to the same crash without a seat belt, which is likely to be fatal.

Tests of seat belts, with anthropomorphic dummies placed in the seats, are included in the routine crash tests performed on new models of automobiles. At present, safety standards for American automobiles require lap-and-diagonal-shoulder belts for each of the two outboard front seats, but require no more than lap belts for the middle front seat (if any) or the rear seats (see Figure III.2). Crash tests with dummies demonstrate that lap belts alone are much less effective in preventing a secondary collision than lap-and-shoulder belts. During the collision, the lap belt restrains the hips, but the torso jackknifes forward and down, and the head of the occupant suffers a violent secondary collision with the instrument panel. Besides, the sudden snubbing of a slack lap belt is concentrated entirely across the hips and the abdomen, and can lead to pelvic and lumbar fractures and to ruptures of the abdominal organs. The main advantage of the lap belt over no belt is that it prevents ejection of the occupant from the automobile; such ejections often lead to severe or fatal injuries, especially in lateral collisions and in rollovers.

Although crash tests with anthropomorphic dummies accurately simulate the motion of and the forces on the occupants during collisions, these tests do not provide direct information on the injuries that would be suffered by the occupants. Measurements of the forces on the dummy permit some (rough) predictions of likely bone fractures, but internal injuries are unpredictable. Experimental studies of such injuries rely on crash tests with cadavers and with live animals, such as baboons, chimpanzees, or pigs. The anatomy of the trunk and head of a baboon or chimpanzee is fairly similar to that of man, and the pattern of injuries therefore is expected to be about the same as in man. In a series of tests with a deceleration sled, baboons were subjected to impact decelerations ranging from 30 G to 100 G. These tests compared the effectiveness of alternative belt configurations (see Figure III.15): lap belt alone, diagonal shoulder belt alone, lap-and-diagonal-shoulder belt, and the lap-and-double-shoulder harness. The tests clearly demonstrated the superiority of lap-and-shoulder belt or lap-and-shoulder harness over the other configurations. Figure III.16 gives the deceleration of the sled and the measured belt-tension forces for an average baboon during a forward-facing impact. As expected from our simple estimate of forces, the tension for a single-belt configuration (either lap or diagonal) is about twice as large as the tension for a two-belt configuration (lap-and-shoulder belt

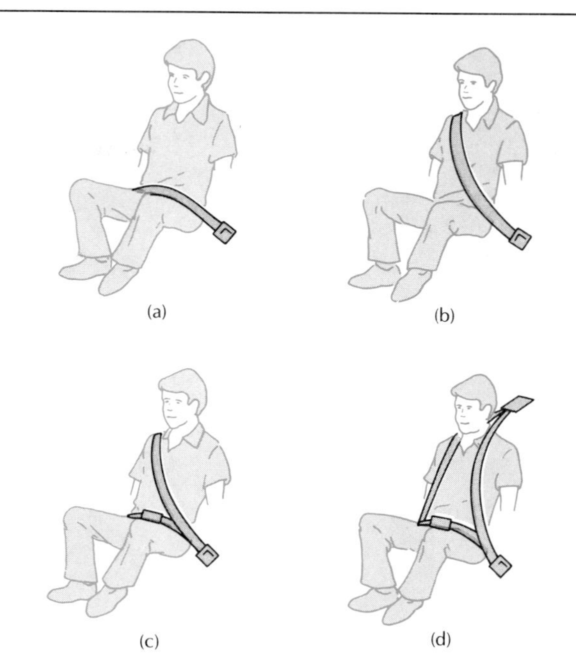


**Fig. III.15** Alternative seat-belt configurations: (a) Lap belt. (b) Diagonal shoulder belt. (c) Lap-and-diagonal-shoulder belt. (d) Lap-and-shoulder harness.

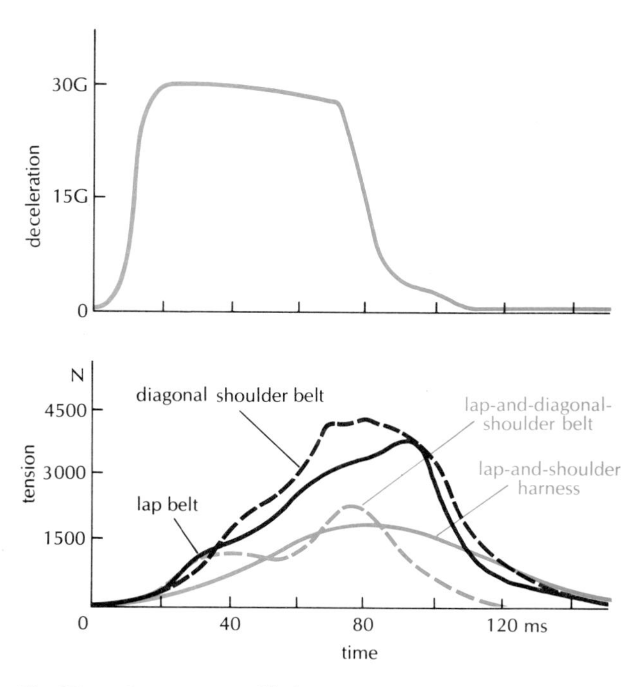


**Fig. III.16** Comparison of belt tensions for baboons subjected to a deceleration of 30 G.

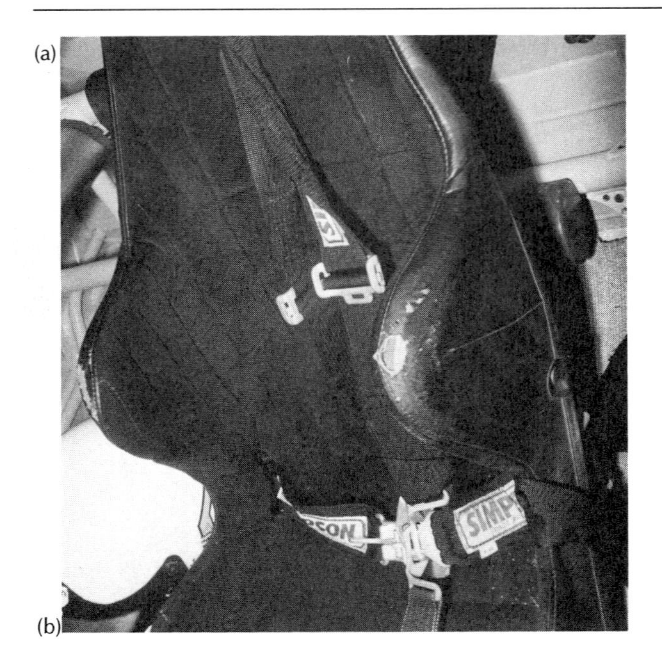

**Fig. III.17** (a) Lap-and-shoulder harness used in a racing car. (b) Richard Petty's crash at the Daytona 500 race provided a dramatic demonstration of the protection offered by the lap-and-shoulder harness. He suffered only minor injuries.

or lap-and-shoulder harness). For a deceleration of 30 G, no significant injuries were found in baboons using either two-belt configuration, but several animals suffered fatal injuries when using the lap belt alone or the diagonal belt alone.

The crash tests with baboons also established the high effectiveness of the lap-and-shoulder harness combination. With this belt configuration, 50% of the animals survived crashes with decelerations of 100 G. Lap-and-shoulder harnesses are routinely used by drivers of racing cars (see Figure III.17) and stunt cars, by pilots and crew of commercial aircraft during takeoff and landing, and by pilots of military aircraft. This configuration gives a lower value of the maximum belt tension than any other belt configuration. It also gives a better limitation of the forward head movement. Furthermore, it is the only belt configuration that provides good restraint in lateral impacts, where the occupant would tend to slip out of a diagonal belt. Unfortunately, the lap-and-shoulder harness has been made commercially available only in some high-performance sports cars.

Some automobile manufacturers now offer air-bag restraint systems. This system consists of a plastic bag packed into a compartment in the hub of the steering wheel or in the glove compartment, a gas generator for inflating the bag, and an electronic crash sensor installed in the body of the automobile. When the sensor experiences an acceleration in excess of a present threshold level, it triggers the explosive combustion of a solid propellant in the gas generator. This combustion produces nitrogen gas, which inflates the air bag in about 0.04 s. The bag balloons out of its compartment and slams into the face and chest of the driver, braking his forward motion and preventing a secondary collision with the steering wheel or the instrument panel. Crash tests with volunteers have shown that such air-bag systems can offer complete protection against injuries in a 48-km/h crash into a barrier. Crash tests with baboons have resulted in nothing but minor injuries during decelerations of up to 120 G.

Figure III.18 shows a test of the air-bag system installed as standard equipment on Mercedes-Benz automobiles. This system is intended to supplement the seat belts, and it consists of a single air bag for the protection of the driver. The passenger is protected by seat belts. Both the passenger's and the driver's seat belts incorporate an automatic seat-belt retractor which tightens the seat belt when activated by the crash sensor. The automatic retractor consists of a cylinder with a piston connected to a cable wound around the seat-belt reel. When the retractor is triggered, the explosion of a small pyrotechnic charge drives the piston up the cylinder and the cable reels in the slack of the seat belt and tightens it to a tension of about 500 N. The tightening of the seat belt significantly reduces the forward displacement of driver or passenger during the collision.

## III.5 TOLERANCE LIMITS OF THE HUMAN BODY; EXPERIMENTAL SAFE VEHICLES

To achieve the best crash protection, the crush characteristics of the automobile and the characteristics of the occupant restraint system should be designed to slow down the motion of the occupant as gradually as possible. In an automobile of ideal design, the survival limits for a crash are set by the maximum acceleration

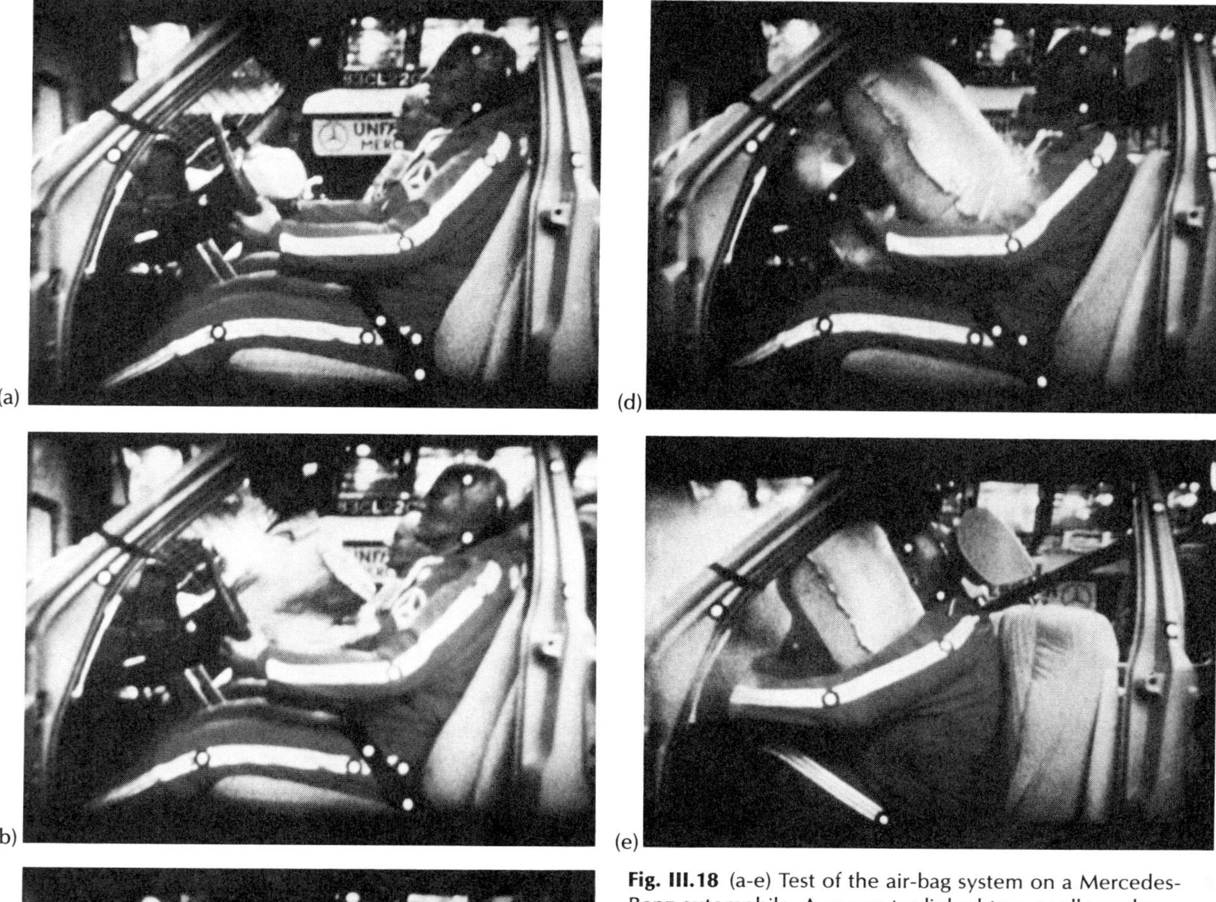


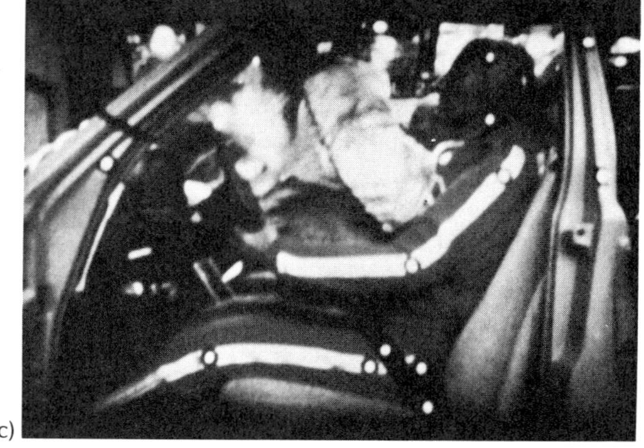


**Fig. III.18** (a-e) Test of the air-bag system on a Mercedes-Benz automobile. A computer linked to a small accelerometer senses the beginning of a crash and triggers the inflation of the bag if the deceleration exceeds a preset level.

that can be tolerated by the human body. Extensive experimental investigations of the effects of high accelerations on the human body were performed in the 1940s and 1950s by Colonel J. Stapp and his associates at the Air Force Medical Corps. These investigations were motivated by the Air Force's concern over survival conditions during aircraft crashes and during explosive-seat ejections from aircraft. In an extensive series of tests, volunteers were placed in a rocket-propelled sled riding on a track, and the sled was brought to sudden stop by friction brakes (see Figure III.19).

**Fig. III.19** In a test in 1950, Colonel Stapp is subjected to a peak acceleration of 20 G while his sled brakes from 143 km/h to 0 within a distance of 5.5 m.

Stapp subjected himself to over 40 G, but the record for acceleration was achieved by Captain E. Beeding, who withstood 83 G for 0.04 s. Extrapolation of results obtained with volunteers and results obtained with animals indicate that accelerations as large as 100 G are survivable.

The tests established that the limits of tolerance depend not only on the peak value of the acceleration and on the duration of this peak value, but also on the rate of onset, that is, the rate at which the acceleration grows from its initial value of zero to its peak value. In automobile collisions, the relevant duration of the acceleration is about 0.1 s. For an individual in good health and in good physical fitness whose body is well supported over a broad area of contact, the maximum forward or backward acceleration that can be tolerated over a time span of 0.1 s without significant injuries is 45 G. The maximum tolerable upward, downward, or lateral accelerations are less than half as large (see Figure III.20). Furthermore, the maximum tolerable rate of onset is about 2000 G per second; this means that the acceleration cannot be permitted to grow to its peak value of, say, 45 G in a time shorter than 45/2000 s, or 0.23 s.

Consider an automobile crashing into a solid barrier at 88 km/h (55 mi/h), and suppose that the crush characteristics of the automobile and the characteristics of the occupant restraint system are such as to ensure an occupant deceleration that grows from zero to 45 G in 0.023 s, and then remains constant at 45 G. A simple calculation shows the occupant will travel 0.51 m while his deceleration grows to 45 G in 0.023 s, and he will then travel a further 0.43 m at a constant deceleration of 45 G before coming to a stop. His net travel distance is 0.94 m. Let us compare this with the distance available in a typical automobile of today. For an impact speed of 88 km/h, the crush distance of the front of the automobile is about 1.4 m (see Figure III.4), to which we can still add the chest–steering wheel distance of 0.4 m. Obviously, this is more than enough to bring the occupant to a nonfatal stop. Thus, we conclude that even one of today's ordinary automobiles will permit the occupant to survive an 88-km/h crash into a rigid barrier, provided the restraint system is finely tuned to impose the correct deceleration of 45 G on the occupant. A restraint system that accomplishes this goal might consist of an air bag with a pressure control programmed to let the occupant gradually move forward, so his deceleration always has the required value.

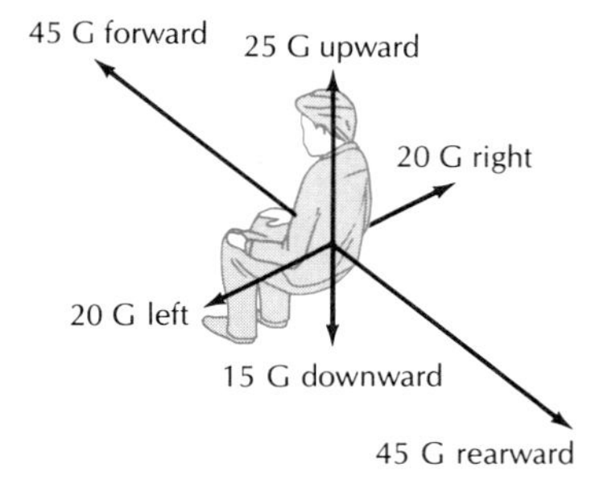


**Fig. III.20** Tolerance limits for the human body.

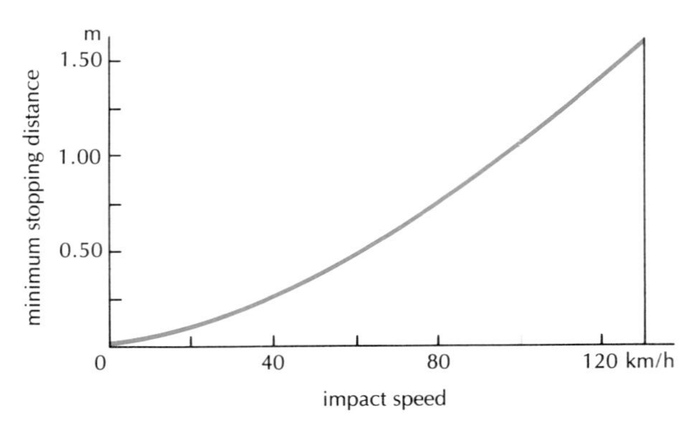


**Fig. III.21** Minimum stopping distance of occupant vs. impact speed.

Figure III.21 shows a plot of minimum stopping distance vs. impact speed, for an occupant subjected to the maximum tolerable deceleration. The plot is cut off at an elapsed time of 0.1 s, the upper limit for tolerance of the deceleration of 45 G. This cutoff does not mean that occupants cannot be safely stopped from speeds in excess of 130 km/h, but that such decelerations must be performed at less than 45 G, and therefore require more than the distance plotted here.

Of course, the values given in Figure III.21 are purely theoretical, and they do not take into account the time required for the air bag, or other restraint system, to come into operation. But such extra factors will not change the qualitative conclusion that survival, and survival without injuries, in an 88-km/h crash into a rigid barrier is possible.

Several prototypes of experimental safe vehicles (ESV) designed for survival in extreme crashes have been built and tested. For instance, Figure III.22 shows a prototype ESV built by Minicars, Inc., under sponsorship of the Department of Transportation. This vehicle is designed to protect the occupant in crashes of up to 80 km/h (50 mi/h). The engine is rear-mounted, and

**Fig. III.22** An experimental safe vehicle, designed to offer protection in high-speed crashes.

thus almost the entire front portion, from bumper to windshield, is available for crushing and for cushioning the impact. The ESV is made of sheet metal, similar to that used in current automobiles, but the hollow spaces are filled with polyurethane foam, which stiffens the sheet metal and permits a low-weight, fuel-efficient construction. The ESV has heavy gull-wing doors, which are less likely to be crushed or opened during a collision than conventional doors, and it has a built-in roll bar and strong roof supports. The occupants of the front seat are restrained by dual-chambered air bags; the occupants of the rear seat are restrained by lap-and-diagonal-shoulder belts. Crash tests of this vehicle against barriers and against other vehicles have shown that the occupants would survive a barrier impact at 80 km/h without serious injuries.

## Further Reading

Brief discussions of automobile crashes and crash injuries are presented in the following articles:

"Four Facets of Automotive Crash Injury Research," R. A. Wolf, *New York State Journal of Medicine,* July 1966

"Forensic Physics of Vehicle Accidents," A. C. Damask, *Physics Today,* March 1987

Statistics regarding automobile collisions, and numbers of injuries and deaths will be found in

*55 MPH Fact Book* (U.S. Department of Transportation, National Highway Traffic Safety Administration, 1981)

"Transportation Safety," E. J. Cantilli, in *Transport and Traffic Engineering Handbook* (Prentice-Hall, Englewood Cliffs, 1982)

*Crash Injury Impairment and Disability* (Society of Automotive Engineers, Warrendale, 1986)

Miscellaneous aspects of automobile crashes and automobile safety are covered in

*Small Car Safety in the 1980s* (U.S. Department of Transportation, National Highway Traffic Safety Administration, 1980)

*Automobile Occupant Crash Protection* (U.S. Department of Transportation, National Highway Traffic Safety Administration, 1980)

*Crash Protection* (Society of Automotive Engineers, Warrendale, 1982)

*The Prevention of Highway Injury* by M. L. Selzer, P. W. Gikas, and D. F. Huelke (Highway Safety Institute, University of Michigan, Ann Arbor, 1967)

*Highway Vehicle Safety, Collected SAE Papers* (Society of Automotive Engineers, Warrendale, 1968)

"Collision Simulation," M. M. Kamal and K. Lin in *Modern Automotive Structural Analysis,* edited by M. M. Kamal and J. A. Wolf (Van Nostrand Reinhold, New York, 1982)

Human tolerance to impacts and to accelerations is summarized in *Human Tolerance to Impact Conditions as Related to Motor Vehicle Design,* SAE Information Report, July 1986. Some specific experiments on human tolerance are described in *Human Impact Response, Measurement and Simulation,* W. F. King and H. J. Mertz (Plenum, New York, 1973).

The yearly *Stapp Car Crash Conferences,* published by the Society of Automotive Engineers, and the *International Technical Conferences on Experimental Safe Vehicles,* published by the National Highway Safety Administration, contain much interesting information on crash tests and crash injuries, but they also contain much useless information, and it is often difficult to sort one from the other.

Behavioral aspects of driving and recommendations and attempts to improve driving habits are discussed in *Human Behavior and Traffic Safety,* edited by L. Evans and R. C. Schwing (Plenum, New York, 1985); *The Causes, Ecology, and Prevention of Traffic Accidents,* H. J. Roberts (C. H. Thomas, Springfield, Ill., 1971); and *Accident Causation* (Society of Automotive Engineers, Warrendale, 1980).

*Unsafe at Any Speed,* R. Nader (Grossman, New York, 1965) is an exposé of the passive and active opposition of automobile manufacturers to the development of safer vehicles.

## Questions

1. The population of the United States is 240 million. If 50,000 people die per year in motor-vehicle accidents, what is the average chance of death per individual per year?

2. If the death rate from accidents continues to climb from year to year as suggested by Figure III.1, what will be the number of fatalities in the year 2000?

3. In 1974, the number of fatalities dropped by about 16% as compared with the preceding year. Examine the data given in Figure III.1 and deduce what fraction of this drop is the result of the concurrent drop in motor vehicle use.

4. From the data in Figure III.3, what is the crush distance of the front of the automobile in this collision? Compare with the crush distances plotted in Figure III.4

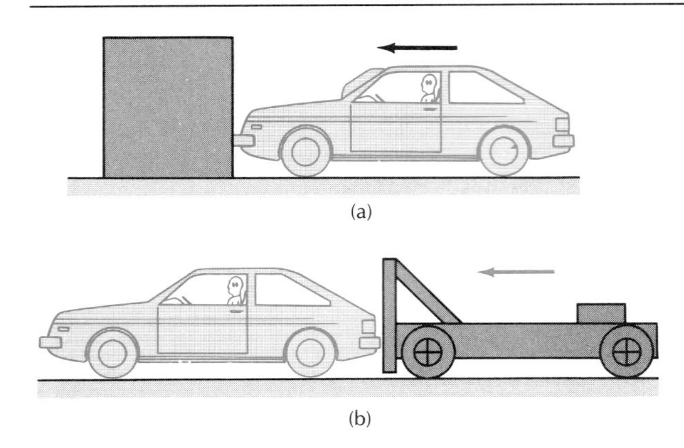


**Fig. III.23** (a) Test procedure for front impact. (b) Test procedure for rear impact.

5. Deduce a rough value of the crash duration from the data given in Figure III.4. Assume that the average speed during the collision is roughly one-half the impact speed.

6. According to the test procedures laid down by the National Highway Traffic Safety Administration, a stationary barrier (of very large mass) and a towed automobile are used for tests of front impacts, but a moving barrier of 1800 kg and a stationary (unbraked) automobile are used for tests of rear impacts (see Figure III.23). Explain how this test with the moving barrier and the stationary automobile could be replaced by an equivalent test with a stationary barrier and an automobile towed *backward* at some appropriate speed. If the automobile has a mass of 1400 kg, and the moving barrier has a speed of 8 km/h, what is the appropriate equivalent speed of the moving automobile incident backward on the stationary barrier?

7. What force would an occupant have to exert on the steering wheel to hold himself away from the wheel during a collision with a deceleration of 12 G? Express your answer as multiple of the body weight.

8. On British military transport aircraft, the passenger seats face *backward*. What are the advantages of such an arrangement in a crash? Why has this arrangement not been adopted in civilian airliners?

9. The laminated safety glass currently used in automobile windshields suffers penetration when struck by a human head at 21 km/h (13 mi/h). What would be the advantages and disadvantages of using a stronger laminate in the windshield?

10. According to federal safety standards, the steering wheel must exert a force of no more than $1.1 \times 10^4$ N (2500 lbf) when the chest of the driver strikes it at 24 km/h (15 mi/h). Estimate how much the steering wheel must yield to meet this standard. Pretend that the chest is rigid.

11. According to federal safety standards, the door lock on an automobile is required to hold the door shut when a forward or backward force of $8.9 \times 10^3$ N (2000 lbf) is applied to it. If the mass of the door is 60 kg, what acceleration will correspond to such a force? Do you think the requirement is adequate?

12. The safety standard for automobile roofs requires that the roof withstand a load of 1.5 times the weight of the automobile. As Figure III.24 shows, the roof of a Volvo automobile is stronger than demanded by this minimum safety standard. Can you imagine an accident in which the minimum safety standard is likely to prove inadequate?

13. Some automobile manufacturers have developed collapsible steering columns which, when struck by the chest of the driver, telescope in a controlled manner while offering a roughly constant resistance. How would you design such a telescoping steering column?

14. If cost were no object and you could have an automobile custom built for yourself, what extra safety features would you incorporate?

15. Some opponents of seat belts have pointed out that in a few cases small children were trapped in burning cars by seat belts. How would you answer this argument against seat belts?

16. In a remarkable incident, a 52-kg woman jumped from the 10th floor of a building, fell 28 m, and landed on her side on soft earth in a freshly dug garden. She fractured her wrist and a rib, but remained conscious and fully alert, and recovered completely after some time in a hospital. The earth was

**Fig. III.24** A truck of 6000 kg placed on top of the roof of a Volvo automobile.

depressed 15 cm by her impact. Calculate the average deceleration during the impact, and calculate the average impact force.

17. When seat belts were first introduced in automobiles in the 1950s, a safety engineer at General Motors declared that he did not believe that the seat belt could afford the driver any significant protection over and above what is available by gripping the wheel firmly and by pushing against the floor with the feet and legs. Discuss this statement. Consider the forces that the arms or legs of a driver would have to support in a collision at 30 G, and consider the reaction time required if the driver is to make a voluntary response to the collision.

18. Suppose that a seat-belted mother riding in an automobile holds a 10-kg baby in her arms. Will she be able to hold on to the baby in a 50-km/h crash?

CHAPTER 12

# Kinematics of a Rigid Body

*Rigid body*

A body is **rigid** if the particles in the body do not move relative to one another. Thus, the body has a fixed shape and all its parts have a fixed position relative to one another. No real body is absolutely rigid. For instance, consider a ball of hardened steel, such as found in a ball bearing. On a microscopic scale this body is certainly not rigid — the atoms in the steel are in a state of continual thermal agitation; they randomly move back and forth about their equilibrium positions. On a macroscopic scale we can ignore these atomic motions because, on the average, they do not change the shape of the body. But even so, the ball is not quite rigid — if subjected to a very heavy load, the ball will suffer a noticeable deformation. In this chapter and in the following chapter, we will ignore such deformations produced by the forces acting on bodies. We will study the motion of bodies under the assumption that rigidity is a good approximation.

## 12.1 Motion of a Rigid Body

*Rotational motion*

A rigid body can simultaneously have two kinds of motion: it can change its position in space and it can change its orientation in space. Change in position is translational motion; as we saw in Chapter 10, this motion can be conveniently described by the motion of the center of mass. Change in orientation is **rotational motion;** it can be proved mathematically that any change of orientation of a rigid body can be regarded as a rotation about some axis.[1]

[1] This is **Euler's theorem.**

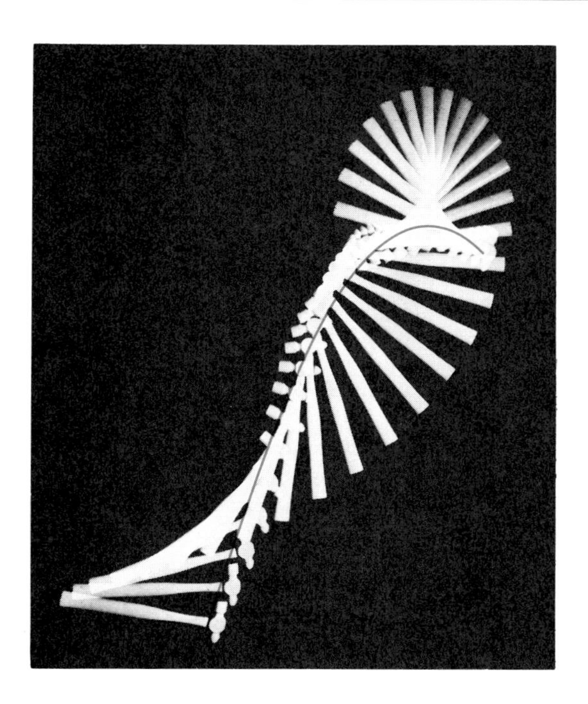

**Fig. 12.1** A hammer in free fall under the influence of gravity. The center of mass of the hammer, marked by the colored dot, moves with constant vertical acceleration $g$, just like a particle in free fall. The colored curve indicates the parabolic path of the center of mass. (Courtesy Harold E. Edgerton, MIT.)

As an example consider the motion of a hammer thrown upward (Figure 12.1). The orientation of the hammer changes relative to fixed coordinates attached to the ground. Instantaneously, the body rotates about a horizontal axis, say, a horizontal axis that passes through the center of mass (in Figure 12.1 this axis sticks out of the plane of the page). The complete motion can be described as a rotation of the hammer about this axis and a simultaneous translation of the axis along a parabolic path. It is worth keeping in mind that the choice of axis is somewhat arbitrary: instead of using a horizontal axis passing through the center of mass of the hammer, we could just as well use a horizontal axis passing through any other point, say, through the end of the handle. The hammer rotates in relation to both of these axes; the only difference lies in their translational motion — the axis through the end of the hammer undergoes a much more complicated translational motion than the axis through the center of mass. This emphasizes that there are two ways in which rotation is relative: first, it is necessary to specify the fixed reference frame with respect to which the motion occurs; and second, it is necessary to specify some point of the body through which the axis of rotation is supposed to pass (by hypothesis, this point then does *not* rotate). The freedom in the choice of axis of rotation will come in handy in the discussion of the motion of a rolling wheel in Section 13.5. However, it is often most convenient to suppose that the axis passes either through the center of mass or through some point which is held fixed by rigid mechanical constraints (e.g., a point on a fixed pivot or an axle).

In the above example of the thrown hammer the axis of rotation always remains horizontal. In the general motion of a rigid body, the axis of rotation can have any direction and can also change its direction. To describe such a complicated motion, it is convenient to separate the rotation into three components along three perpendicular axes. According to nautical and aeronautical terminology, the three components of the rotational motion are called **roll, pitch,** and **yaw** (Figure 12.2). However, in the following sections we will not deal with

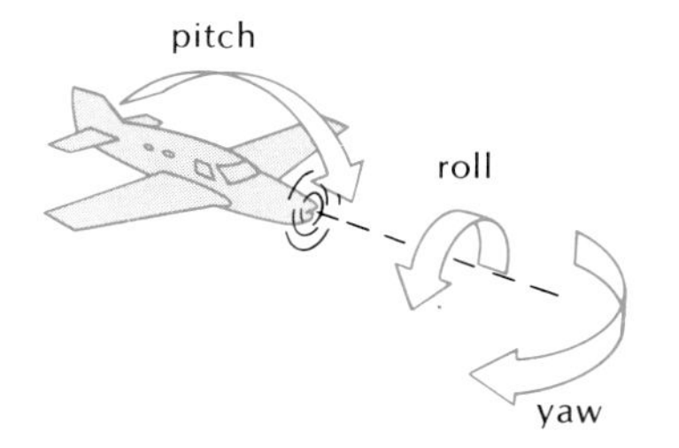


**Fig. 12.2** The three independent rotational motions of an aircraft.

this general complicated three-component rotation; we will deal only with rotation about a single, fixed axis.

EXAMPLE 1. One interesting feature of rotations about different axes is that they *do not commute,* that is, the net result of two successive rotations depends on which rotation is done first, which second. Show that this is so by rotating a book about two different axes.

SOLUTION: Figure 12.3a shows a book lying on a table and facing you. Perform two rotations: first, a clockwise rotation by 90° about the vertical axis; and second, a clockwise rotation by 180° about the horizontal axis pointing away from you (Figures 12.3b and c).

Next, perform the same rotations on a similar book, but in the opposite order: first, a clockwise rotation by 180° about the horizontal axis pointing away from you; and second, a clockwise rotation by 90° about the vertical axis (Figures 12.4b and c). Comparison of Figures 12.3c and 12.4c shows that the final results are different, that is, the results depend on the order in which the rotations were carried out.

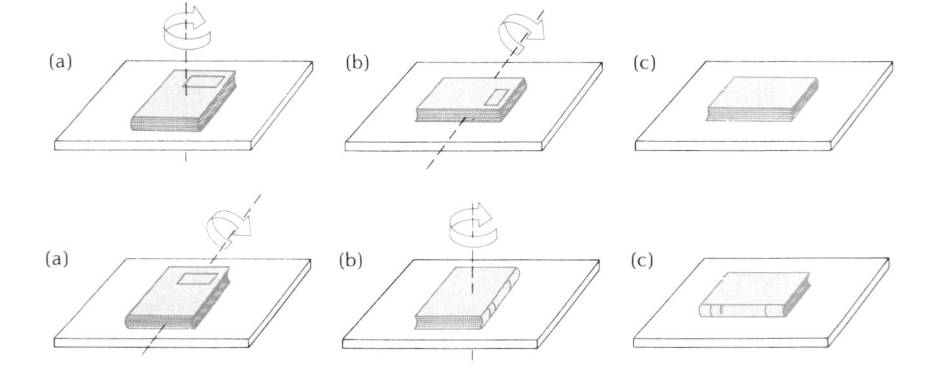


**Fig. 12.3** Rotation through 90° about a vertical axis followed by a rotation through 180° about a horizontal axis.

**Fig. 12.4** Rotation through 180° about a horizontal axis followed by a rotation through 90° about a vertical axis.

## 12.2 Rotation About a Fixed Axis

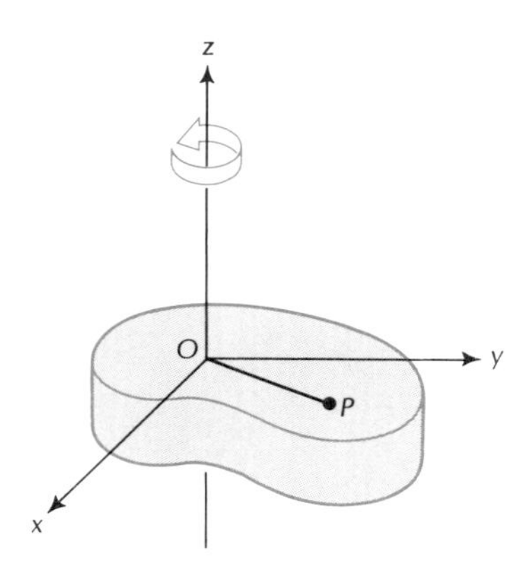


**Fig. 12.5** Rotation of a rigid body about a fixed axis ($z$ axis).

Figure 12.5 shows a rigid body rotating about a fixed axis which coincides with the $z$ axis. During the rotational motion, each point of the body remains at a given distance from this axis and moves along a circle centered on the axis. To describe the orientation of the body at any instant, we select one particle in the body and use it as a reference point; any particle will do as reference point, provided that it is not on the axis of rotation. The circular motion of this fiducial particle (labeled $P$ in Figure 12.5) is then representative of the rotational motion of the entire rigid body, and the angular position of this particle is representative of the angular position of the entire rigid body.

Figure 12.6 shows the rigid body as seen from along the axis of rotation. The coordinates in Figure 12.6 have been chosen so that the $z$ axis coincides with the axis of rotation while the $y$ and $x$ axes are in the plane of the circle traced out by the motion of the fiducial particle. The angular position of the fiducial particle — and hence the angular orientation of the entire rigid body — can be described by the position angle $\phi$ between the radial line $OP$ and the $x$ axis.[2] Conventionally, the angle $\phi$ is reckoned as positive when measured in a counterclockwise direction (as in Figure 12.6). We will always measure the position angle $\phi$ in radians. Consequently, the length of the path between the $x$ axis and the point $P$ is

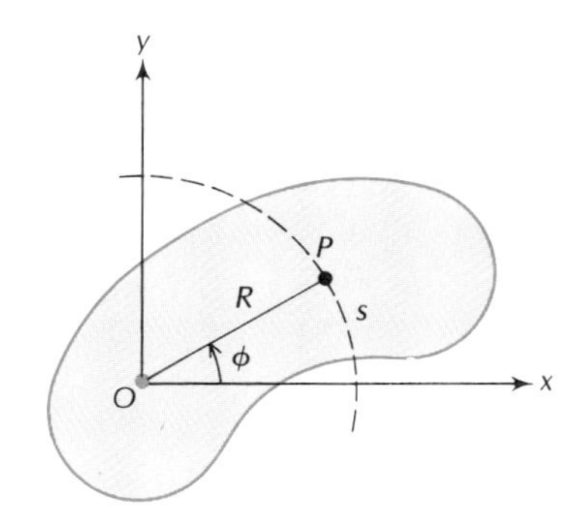


Fig. 12.6 Fiducial particle $P$ on a rigid body. The radius of the circle traced out by the motion is the same as the length of the position vector, $R = \sqrt{x^2 + y^2}$.

$$s = \phi R \tag{1}$$

where $R$ is the radius of the circle traced out by the motion.

For a rotating body, the position angle $\phi$ is some function of time, $\phi = \phi(t)$. The definition of the angular velocity for rotational motion is mathematically analogous to the definition of velocity for one-dimensional motion (see Sections 2.2 and 2.3). The **average angular velocity** is defined as

*Average angular velocity*

$$\overline{\omega} = \frac{\Delta\phi}{\Delta t} \tag{2}$$

where $\Delta\phi$ is the change in the angle and $\Delta t$ the corresponding change in the time. The **instantaneous angular velocity** is defined as

*Instantaneous angular velocity*

$$\omega = \frac{d\phi}{dt} \tag{3}$$

The unit of angular velocity is the radian per second (radian/s). An angle expressed in radians is the ratio of two lengths [compare Eq. (1)] and hence is a pure number, with no dimensions of length, time, or mass; thus, dimensionally, 1 radian/s is the same thing as 1/s. However, to prevent confusion in calculations, it is often useful to retain the dimensionally vacuous label *radian*.

If the body rotates with a constant angular velocity, then we can also measure the rate of rotation in terms of the **frequency,** or number of revolutions per unit time. Since each complete revolution comprises $2\pi$ radians, the frequency of revolution is smaller than the angular velocity by a factor $2\pi$. Designating the frequency by $\nu$, we therefore have

*Frequency*

$$\nu = \frac{\omega}{2\pi} \tag{4}$$

The **period** of the motion, or the time for one revolution, is $1/\nu$. We will designate the period by $T$,

*Period*

$$T = 1/\nu \tag{5}$$

or

$$T = 2\pi/\omega \tag{6}$$

[2] A list of Greek letters and their names appears on the endpapers.

The unit of frequency is revolutions per second (rev/s). Dimensionally, 1 rev/s is the same thing as 1/s; but, as in the case of radians, to prevent confusion it is useful to retain the label *rev.* If the rate of rotation of the body is not uniform, then it is best to avoid the use of the word *frequency;* this word is usually reserved for repetitive motions in which each revolution takes the same time as the preceding one.

---

EXAMPLE 2. A phonograph turntable rotates at $33\frac{1}{3}$ revolutions per minute. Find the frequency of revolution, the angular velocity, and the period of the motion.

SOLUTION: The frequency is

$$\nu = \frac{33.3 \text{ rev}}{1 \text{ min}} = \frac{33.3 \text{ rev}}{60 \text{ s}} = 0.555 \text{ rev/s}$$

The angular velocity is

$$\omega = 2\pi\nu = 2\pi \times 0.555 \text{ rev/s} = 3.49 \text{ radians/s}$$

And the period of the motion is

$$T = \frac{1}{\nu} = \frac{1}{0.555 \text{ rev/s}} = 1.80 \text{ s}$$

COMMENTS AND SUGGESTIONS: Note that in the calculation of the angular velocity, we dropped the label *rev* in the third step and inserted a label *radians.* As remarked above, these labels merely serve to prevent confusion, and they can be inserted and dropped at will once they have served their purpose.

---

The definition of the **average angular acceleration** is

*Average angular acceleration*

$$\bar{\alpha} = \frac{\Delta\omega}{\Delta t} \tag{7}$$

and that of the **instantaneous angular acceleration** is

*Instantaneous angular acceleration*

$$\alpha = \frac{d\omega}{dt} \tag{8}$$

or

$$\alpha = \frac{d^2\phi}{dt^2} \tag{9}$$

The unit of angular acceleration is the radian/s$^2$.

Equations (3) and (8) give the angular velocity and angular acceleration of the rigid body; that is, they give the angular velocity and acceleration of every particle in the body. It is interesting to focus on one of these particles and evaluate its translational speed and acceleration in terms of the corresponding angular quantities. If the particle is at a distance $R$ from the axis (Figure 12.7), then the length along the circular path of the particle is, according to Eq. (1),

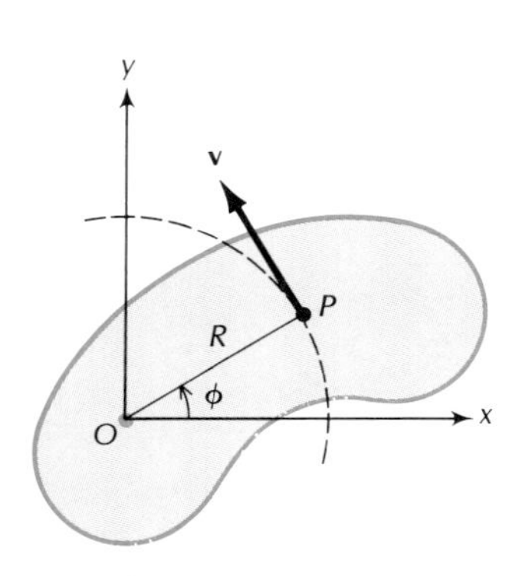

**Fig. 12.7** Instantaneous velocity of a particle in a rotating rigid body.

$$s = \phi R \tag{10}$$

Since $R$ is a constant, differentiation of this equation gives

$$\frac{ds}{dt} = R\frac{d\phi}{dt} \tag{11}$$

Here $ds/dt$ is the speed at which the particle moves along its circular path and $d\phi/dt$ is the angular velocity; hence Eq. (11) is equivalent to

$$\boxed{v = R\omega} \tag{12}$$

This shows that the translational speed of the particle along its circular path around the axis increases in direct proportion to the radius: the farther a particle of the rigid body is from the axis, the faster it moves.

Differentiation of Eq. (12) yields

$$\frac{dv}{dt} = R\frac{d\omega}{dt} \tag{13}$$

Here $dv/dt$ is the rate of change of the speed of the circular motion, i.e., it is the acceleration *along* the circular path. Since this acceleration represents a change in the length of the velocity vector, it is in the same direction as the velocity vector, i.e., it is tangent to the circle (Figure 12.7). If we write this tangential acceleration as $a_{\text{tan}}$ and take into account that $d\omega/dt$ is the angular acceleration, Eq. (13) becomes

$$\boxed{a_{\text{tan}} = R\alpha} \tag{14}$$

Note that the tangential acceleration of Eq. (14) is not the complete acceleration of the particle. As we saw in Section 4.4, a particle in circular motion has a centripetal acceleration of magnitude

$$\boxed{a_{\text{cent}} = \frac{v^2}{R} = R\omega^2} \tag{15}$$

The net acceleration is the vector sum of a tangential vector of magnitude $a_{\text{tan}}$ and a radial vector of magnitude $a_{\text{cent}}$; this is illustrated in Figure 12.8 (and will be further explored in Example 4). The tangential acceleration is due to the change of *magnitude* of the velocity and the centripetal acceleration is due to the change of *direction* of the velocity.

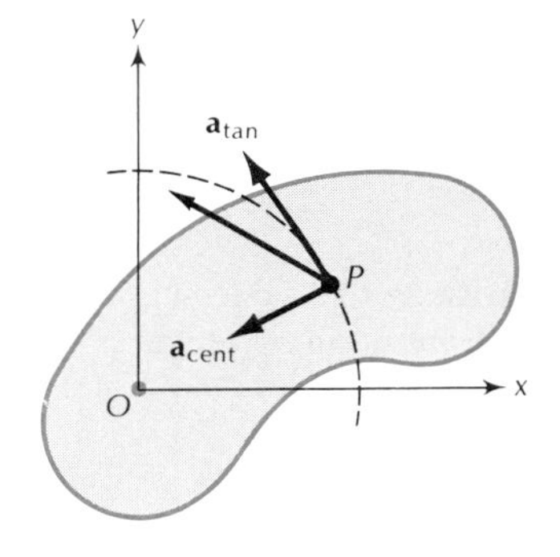


**Fig. 12.8** Instantaneous acceleration of a particle in a rotating rigid body.

## 12.3 Motion with Constant Angular Acceleration

If the rigid body rotates with a constant angular acceleration, then the angular velocity increases at a constant rate and, after a time $t$ has elapsed, the angular velocity will attain the value

*Equations for constant angular acceleration*

$$\boxed{\omega = \omega_0 + \alpha t} \tag{16}$$

where $\omega_0$ is the initial value at $t = 0$.

The angular position $\phi$ can be calculated from this angular velocity by exactly the same methods as were used in Section 2.5 to calculate $x$ from $v$ [see Eqs. (2.17)–(2.21)]. The result is

$$\phi = \phi_0 + \omega_0 t + \tfrac{1}{2}\alpha t^2 \tag{17}$$

Furthermore, the methods of Section 2.5 lead to an identity between acceleration, position, and velocity [see Eqs. (2.23)–(2.25)]:

$$\alpha(\phi - \phi_0) = \tfrac{1}{2}(\omega^2 - \omega_0^2) \tag{18}$$

Equations (16), (17), and (18) provide a complete description of the motion of wheels or other bodies rotating with constant angular acceleration.

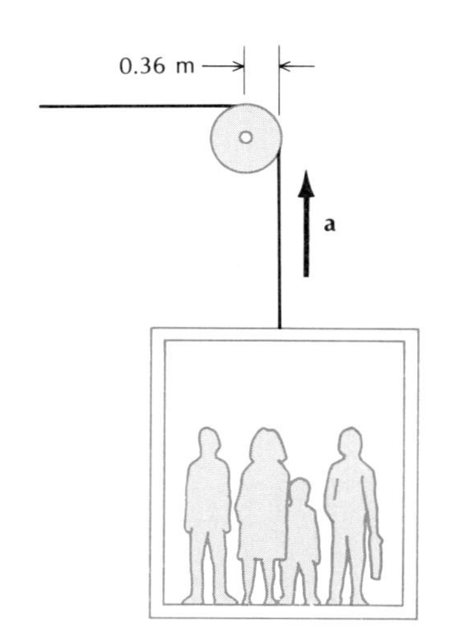


**Fig. 12.9** The cable supporting an elevator runs over a rotating wheel.

EXAMPLE 3. The cable supporting an elevator runs over a wheel of radius 0.36 m (Figure 12.9). If the elevator ascends with an upward acceleration of 0.47 m/s² (as in Example 6.4), what is the angular acceleration of the wheel? How many turns does the wheel make if this accelerated motion lasts 5.0 s starting from rest? Assume that the cable runs over the wheel without slipping.

SOLUTION: If there is no slipping, the speed of the cable must always coincide with the speed of a point on the rim of the wheel. The acceleration $a = 0.47\ \text{m/s}^2$ of the cable must then coincide with the tangential acceleration of a point on the rim of the wheel,

$$a = a_{\tan} = R\alpha$$

where $R = 0.36$ m is the radius of the wheel. Hence

$$\alpha = a/R = (0.47\ \text{m/s}^2)/0.36\ \text{m} = 1.3/\text{s}^2$$
$$= 1.3\ \text{radians/s}^2$$

According to Eq. (17), the angular displacement in 5.0 s is

$$\phi - \phi_0 = \omega_0 t + \tfrac{1}{2}\alpha t^2$$
$$= 0 + \tfrac{1}{2}(1.3\ \text{radians/s}^2) \times (5.0\ \text{s})^2$$
$$= 16\ \text{radians}$$

Each revolution comprises $2\pi$ radians; thus, 16 radians is the same as $16/2\pi$ revolutions, or 2.6 revolutions.

COMMENTS AND SUGGESTIONS: The equation for the angular displacement we used in this example is analogous to the equation for the translational displacement we obtained in Chapter 2. Analogies between rotational and translational quantities can serve as a convenient mnemonic for remembering the equations for rotational motion. The following is a list of analogous equations:

| | | |
|---|---|---|
| $v = dx/dt$ | → | $\omega = d\phi/dt$ |
| $a = d^2x/dt^2$ | → | $\alpha = d^2\phi/dt^2$ |
| $v = v_0 + at$ | → | $\omega = \omega_0 + \alpha t$ |
| $x = x_0 + v_0 t + \frac{1}{2}at^2$ | → | $\phi = \phi_0 + \omega_0 t + \frac{1}{2}\alpha t^2$ |
| $a(x - x_0) = \frac{1}{2}(v^2 - v_0^2)$ | → | $\alpha(\phi - \phi_0) = \frac{1}{2}(\omega^2 - \omega_0^2)$ |

EXAMPLE 4. Find the net instantaneous acceleration (tangential and centripetal) of a point on the rim of the wheel described in the preceding example at the instant $t = 1.0$ s.

SOLUTION: We already know the tangential acceleration,

$$a_{\tan} = 0.47 \text{ m/s}^2$$

To find the centripetal acceleration, we need the angular velocity. According to Eq. (16),

$$\omega = \omega_0 + \alpha t = 0 + (1.3 \text{ radians/s}^2) \times 1.0 \text{ s}$$

$$= 1.3 \text{ radians/s}$$

and, according to Eq. (15), the centripetal acceleration is

$$a_{\text{cent}} = \omega^2 R = (1.3 \text{ radians/s})^2 \times 0.36 \text{ m}$$

$$= 0.61 \text{ m/s}^2$$

The net acceleration of the point is the vector sum of $a_{\tan}$ and $a_{\text{cent}}$ (Figure 12.10); this net acceleration has a magnitude

$$a = \sqrt{a_{\tan}^2 + a_{\text{cent}}^2} = 0.77 \text{ m/s}^2$$

and makes an angle of 38° with the radius.

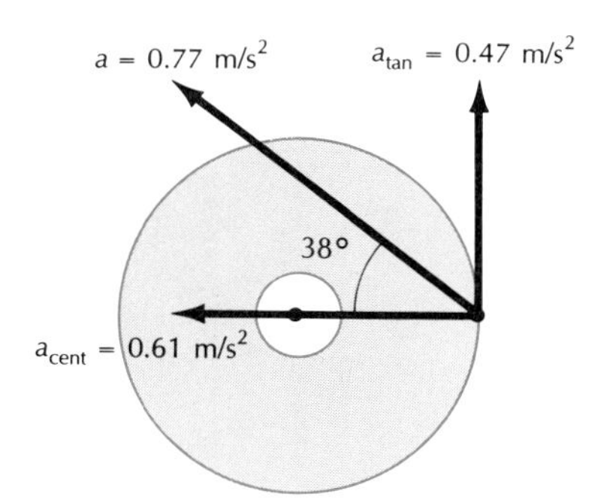


**Fig. 12.10** Acceleration of a point on the rim of the wheel.

## 12.4 Kinetic Energy of Rotation; Moment of Inertia

As for any system of particles, the total kinetic energy $K$ of a rotating rigid body is simply the sum of the individual kinetic energies of all the particles. If the particles have masses $m_i$ and velocities $\mathbf{v}_i$ (where $i = 1, 2, \ldots, n$), then

$$K = \sum_{i=1}^{n} \tfrac{1}{2} m_i v_i^2$$

In a rigid body rotating about a given axis, all the particles move with the same angular velocity $\omega$ along circular paths. The speeds of the particles along their paths are proportional to their radial distances:

$$v_i = R_i \omega \qquad (i = 1, 2, 3, \ldots) \tag{19}$$

and hence the total kinetic energy is

$$K = \sum_{i=1}^{n} \tfrac{1}{2} m_i R_i^2 \omega^2 \tag{20}$$

We will write this as

$$\boxed{K = \tfrac{1}{2} I \omega^2} \tag{21}$$

*Kinetic energy of rotation*

where

*Moment of inertia of a system of particles*

$$I = \sum_{i=1}^{n} m_i R_i^2 \tag{22}$$

is the **moment of inertia** of the rotating body about the given axis. The units of moment of inertia are kg · m² or lb · ft².

Note that Eq. (21) has a mathematical form reminiscent of the familiar expression $\frac{1}{2}mv^2$ for the kinetic energy of a single particle — the moment of inertia replaces the mass and the angular velocity replaces the translational velocity. As we will see in the next chapter, the moment of inertia is a measure of the resistance that a body offers to changes in its rotational motion, just as mass is a measure of the resistance that a body offers to changes in its translational motion. These analogies between rotational and translational quantities make it easy to remember the rotational equations.

Equation (22) shows that the moment of inertia—and consequently the kinetic energy for a given value of $\omega$—is large if most of the mass of the body is at a large distance from the axis of rotation. This is very reasonable: for a given value of $\omega$, particles at a large distance from the axis move with high speeds and therefore have a large kinetic energies.

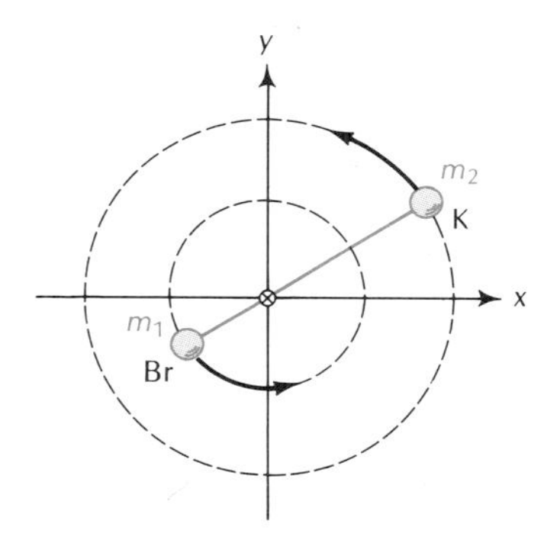


**Fig. 12.11** The bromine and potassium atoms revolve about their common center of mass. The atoms can be regarded as particles joined by a rod.

EXAMPLE 5. Suppose that the molecule of potassium bromide (KBr) described in Example 10.5 rotates rigidly about its center of mass (Figure 12.11). What is the moment of inertia of the molecule about this axis? Suppose that the molecule rotates with an angular velocity of $1.0 \times 10^{12}$ radian/s. What is the rotational kinetic energy?

SOLUTION: We can regard the molecule as a rigid body, consisting of two particles joined by a massless rod. It is best to place the origin of coordinates at the center of mass (see Figure 12.11). According to the results obtained in Example 10.5, the distance of the Br atom is then $R_1 = 0.93$ Å, and that of the K atom is $R_2 = 1.89$ Å. The corresponding masses are $m_1 = 79.9$ u and $m_2 = 39.1$ u. The moment of inertia of the molecule is

$$\begin{aligned} I &= m_1 R_1^2 + m_2 R_2^2 \\ &= 79.9\ \text{u} \times (0.93\ \text{Å})^2 + 39.1\ \text{u} \times (1.89\ \text{Å})^2 \\ &= 2.09 \times 10^2\ \text{u} \cdot \text{Å}^2 \\ &= 2.09 \times 10^2\ \text{u} \cdot \text{Å}^2 \times \frac{1.66 \times 10^{-27}\ \text{kg}}{1\ \text{u}} \times \left(\frac{10^{-10}\ \text{m}}{1\ \text{Å}}\right)^2 \\ &= 3.47 \times 10^{-45}\ \text{kg} \cdot \text{m}^2 \end{aligned}$$

The kinetic energy is

$$\begin{aligned} K &= \tfrac{1}{2} I \omega^2 \\ &= \tfrac{1}{2} \times 3.47 \times 10^{-45}\ \text{kg} \cdot \text{m}^2 \times (1.0 \times 10^{12}\ \text{radians/s})^2 \\ &= 1.73 \times 10^{-21}\ \text{J} \end{aligned}$$

COMMENTS AND SUGGESTIONS: This kinetic energy could equally well have been obtained by first calculating the individual velocities of the atoms ($v_1 = R_1\omega$, $v_2 = R_2\omega$) and then adding the corresponding individual kinetic energies.

If we regard the mass of a solid body as continuously distributed over its volume, then we can calculate the moment of inertia by converting Eq. (22) into an integral. As in the calculation of the position of the center of mass [see Eqs. (10.21)–(10.26)], we imagine that the volume is divided into small volume elements with masses

$$\Delta m_i = \rho \, \Delta V \tag{23}$$

where $\rho$ is the density. Then

$$I = \sum_{i=1}^{n} m_i R_i^2 = \sum_{i=1}^{n} \rho \, R_i^2 \, \Delta V \tag{24}$$

In the limit $\Delta V \to 0$, this becomes the integral

*Moment of inertia of a continuous mass distribution*

$$\boxed{I = \int \rho R^2 \, dV} \tag{25}$$

In some simple cases, it is possible to find the moment of inertia without having recourse to an explicit calculation of the integral in Eq. (25). For example, if the rigid body is a thin hoop or a thin cylindrical shell of radius $R_0$ (Figures 12.12 and 12.13) rotating about its axis of symmetry, then *all* of the mass of the body is at the same distance from the axis of rotation — the moment of inertia is then simply the total mass $M$ of the hoop or shell multiplied by its radius $R_0$ squared,

$$I = MR_0^2 \tag{26}$$

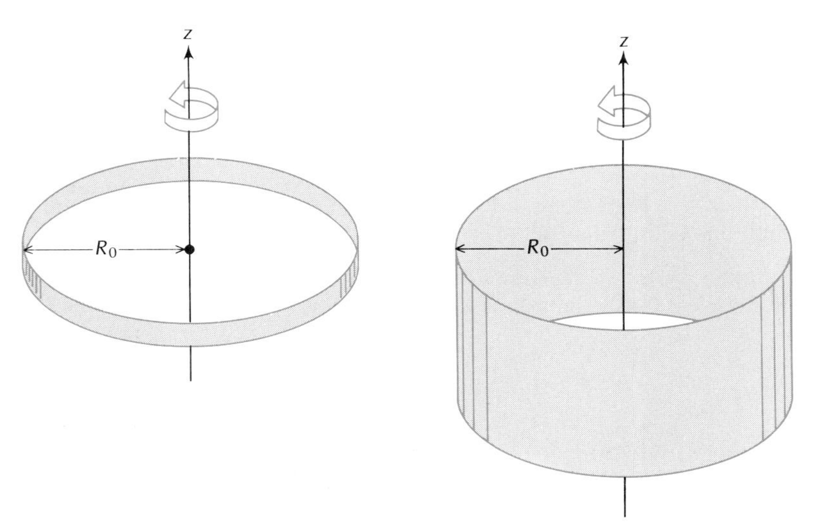


**Fig. 12.12** (left) A thin hoop rotating about its axis of symmetry.

**Fig. 12.13** (right) A thin cylindrical shell rotating about its axis of symmetry.

The following are some examples of explicit evaluations of the integral in Eq. (25).

EXAMPLE 6. Find the moment of inertia of a disk of uniform density of mass $M$ and radius $R_0$ rotating about its axis.

SOLUTION: Figure 12.14 shows the disk with the $z$ axis along the axis of rotation. Suppose the disk has a thickness $l$. The disk can be regarded as made up of a large number of thin concentric hoops fitting one around another. Figure 12.14 shows one such hoop of radius $R$, width $dR$, and thickness $l$. The volume of the hoop is $dV = 2\pi Rl\, dR$; with this, Eq. (25) gives

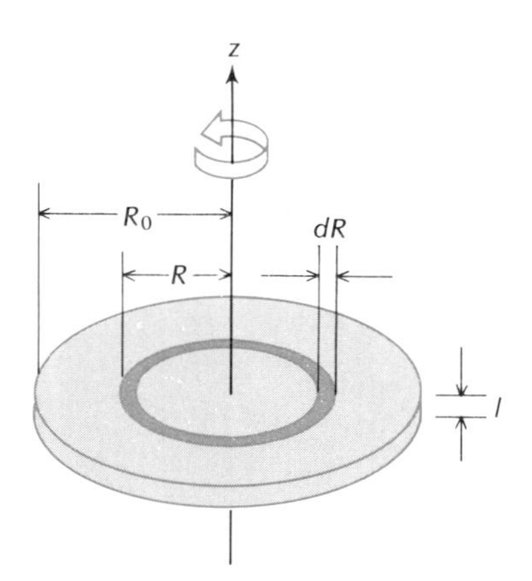


Fig. 12.14 A disk rotating about its axis of symmetry.

$$I = \int_0^{R_0} \rho R^2 2\pi Rl\, dR = 2\pi\rho l \int_0^{R_0} R^3\, dR = 2\pi\rho l \frac{R_0^4}{4} \tag{27}$$

The total mass of the disk is the volume $\pi R_0^2 l$ multiplied by the density $\rho$,

$$M = \pi R_0^2 l \rho$$

Hence $\rho l = M/\pi R_0^2$, which, when substituted into Eq. (27), yields

$$I = \tfrac{1}{2} M R_0^2 \tag{28}$$

for the moment of inertia of the disk.

COMMENTS AND SUGGESTIONS: This formula also gives the moment of inertia of a cylinder, since such a body is simply a very thick disk. The above calculation is equally valid for thin and for thick disks, and the final formula is independent of the thickness $l$ of the disk.

EXAMPLE 7. Find the moment of inertia of a uniform thin rod of length $l$ and mass $M$ rotating about an axis perpendicular to the rod and through its center.

SOLUTION: Figure 12.15 shows the rod lying along the $x$ axis; the axis of rotation is the $z$ axis. Consider a small slice $dx$ of the rod. If the cross-sectional area of the rod is $A$, then $dV = A\, dx$ and Eq. (25) becomes

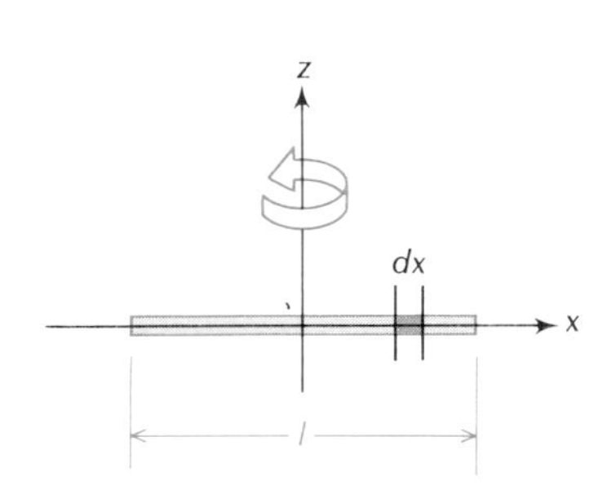


Fig. 12.15 A thin rod rotating about an axis through its midpoint.

$$I = \int_{-l/2}^{l/2} \rho x^2 A\, dx = \rho A \left[\frac{x^3}{3}\right]_{-l/2}^{l/2} = \tfrac{1}{12}\rho A l^3 \tag{29}$$

Since $Al$ is the volume of the rod, $\rho Al$ is its total mass and therefore the moment of inertia can be written

$$I = \tfrac{1}{12} M l^2 \tag{30}$$

EXAMPLE 8. Repeat the calculation of the preceding example for an axis through one end of the rod.

SOLUTION: Figure 12.16 shows the rod and the axis of rotation. The rod extends from $x = 0$ to $x = l$; hence instead of Eq. (29) we obtain

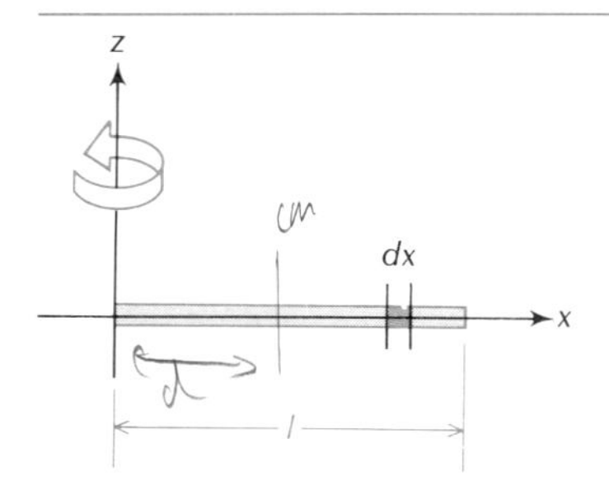


Fig. 12.16 A thin rod rotating about an axis through one end.

$$I = \int_0^l \rho x^2 A\, dx = \rho A \left[\frac{x^3}{3}\right]_0^l = \tfrac{1}{3}\rho A l^3 \tag{31}$$

and instead of Eq. (30), we obtain

$$I = \tfrac{1}{3} M l^2 \tag{32}$$

Comparison of Eqs. (30) and (32) makes it very clear that the value of the moment of inertia depends on the location of the axis of rotation. The moment of inertia is small if the axis passes through the center of mass, and large if it passes through the end of the rod.

We can prove a general theorem, the **parallel-axis theorem,** that relates the moment of inertia $I_{CM}$ about an axis through the center of mass to the moment of inertia $I$ about a parallel axis through some other point. The theorem asserts that

*Parallel-axis theorem*

$$\boxed{I = I_{CM} + Md^2} \tag{33}$$

where $M$ is the total mass of the body and $d$ the distance between the two axes. It is a corollary of Eq. (33) that the moment of inertia about an axis through the center of mass is always less than that about any other parallel axis.

A simple and instructive proof of the theorem is as follows: Figure 12.17 shows the body rotating about a fixed axis and also shows an alternative, parallel axis through the center of mass. The kinetic energy of the body rotating about the fixed axis is

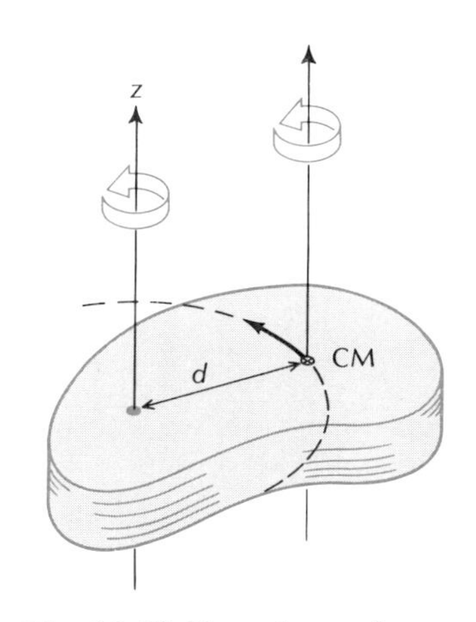


**Fig. 12.17** Two alternative parallel axes of rotation of a rigid body. The $z$ axis is fixed. The center-of-mass axis moves along a circle of radius $d$ around the $z$ axis. The body is in rotational motion relative to each of these axes.

$$K = \tfrac{1}{2}I\omega^2 \tag{34}$$

However, we can express the kinetic energy in an alternative form. According to Eq. (10.44) the kinetic energy of an arbitrary system of particles is the sum of the translational energy of motion of the center of mass and the "internal" energy of motion relative to the center of mass,

$$K = \tfrac{1}{2}Mv_{CM}^2 + K_{int} \tag{35}$$

In Figure 12.17 the center of mass moves along a circle of radius $d$ around the axis. Hence

$$v_{CM} = \omega d \tag{36}$$

and the first term on the right side of Eq. (35) is

$$\tfrac{1}{2}Mv_{CM}^2 = \tfrac{1}{2}M\omega^2 d^2 \tag{37}$$

Let us next look at the second term on the right side of Eq. (35). Although the rotational motion of the body in Figure 12.17 is most simply described as a rotation about the fixed axis, it can also be described as a rotation about the moving axis through the center of mass. Each time the body completes one revolution about the fixed axis, it also completes one revolution about this center-of-mass axis; consequently, the angular velocity of the rotation around the center-of-mass axis is also $\omega$. The kinetic energy of the rotational motion relative to the center of mass is then

$$K_{int} = \tfrac{1}{2}I_{CM}\omega^2 \tag{38}$$

Thus, the total kinetic energy of Eq. (35) is

$$K = \tfrac{1}{2}M\omega^2 d^2 + \tfrac{1}{2}I_{CM}\omega^2 = \tfrac{1}{2}(Md^2 + I_{CM})\omega^2 \tag{39}$$

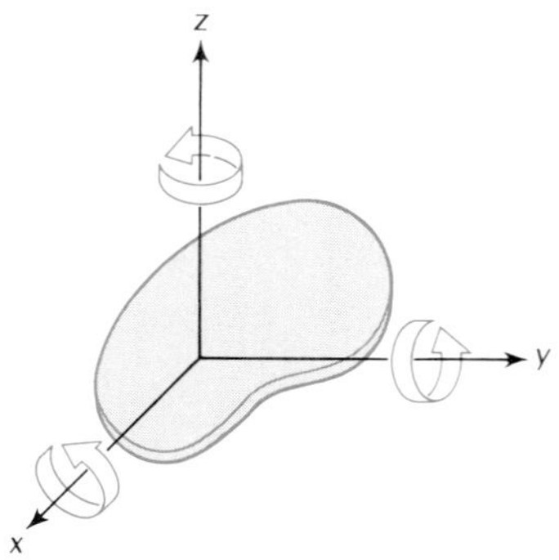

The values given by Eqs. (34) and (39) must be equal; this implies that

$$I = I_{CM} + Md^2$$

which is the result we wanted to obtain.

We can also prove another theorem, the **perpendicular-axis theorem,** which relates the moments of inertia of a thin flat plate (for example, a sheet of metal) about three mutually perpendicular axes. Figure 12.18 shows a flat plate of arbitrary shape in the $x$–$y$ plane. The plate may rotate either about an axis perpendicular to the plate (the $z$ axis), or about an axis in the plane of the plate (the $x$ axis or the $y$ axis). We will call the moments of inertia for rotation about these alternative axes $I_z$, $I_x$, and $I_y$, respectively. Then the perpendicular–axis theorem asserts that

**Fig. 12.18** A thin, flat plate. The plate may rotate about either the $x$ axis, the $y$ axis, or the $z$ axis.

*Perpendicular-axis theorem*

$$I_z = I_x + I_y \tag{40}$$

The proof of the theorem is very simple: For rotation about the $z$ axis, the distance from the axis of rotation to a particle in the plate is

$$R = \sqrt{x^2 + y^2}$$

hence

$$I_z = \int \rho R^2 \, dV = \int \rho(x^2 + y^2) \, dV$$

For rotation about the $x$ axis, $R = y$ and

$$I_x = \int \rho y^2 \, dV$$

Likewise, for rotation about the $y$ axis, $R = x$ and

$$I_y = \int \rho x^2 \, dV$$

Comparison of these three formulas makes it obvious that the sum of $I_x$ and $I_y$ equals $I_z$.

Table 12.1 lists the moments of inertia of a variety of rigid bodies about axes through their centers of mass; all the bodies are assumed to have uniform density. The parallel-axis theorem can be used to find the moment of inertia about any axis parallel to the center-of-mass axis. Furthermore, by means of combinations of the formulas of Table 12.1, it is possible to find the moments of inertia of some other bodies.

---

EXAMPLE 9. Find the moment of inertia of a spherical shell of inner radius $R_1$ and outer radius $R_2$, about an axis through the center (Figure 12.19).

SOLUTION: According to Table 12.1, the moment of inertia of a solid sphere of radius $R_2$ is

$$I = \frac{2}{5} M R_2^2$$

Since $M = \rho V = \rho \times \frac{4}{3}\pi R_2^3$, this is the same as

$$I = \frac{2}{5} \frac{4\pi}{3} \rho R_2^5$$

The shell of Figure 12.19 is a solid sphere with a concentric spherical hole. Correspondingly, the inertia is that of a solid sphere of radius $R_2$ minus that of a concentric sphere of radius $R_1$:

$$I = \frac{2}{5}\frac{4\pi}{3}\rho(R_2^5 - R_1^5)$$

The volume of the shell is

$$\frac{4\pi(R_2^3 - R_1^3)}{3}$$

and hence the density is

$$\rho = M/[4\pi(R_2^3 - R_1^3)/3]$$

where $M$ is now the mass of the shell. This leads to

$$I = \frac{2}{5}\frac{R_2^5 - R_1^5}{R_2^3 - R_1^3}M$$

for the inertia of the spherical shell.

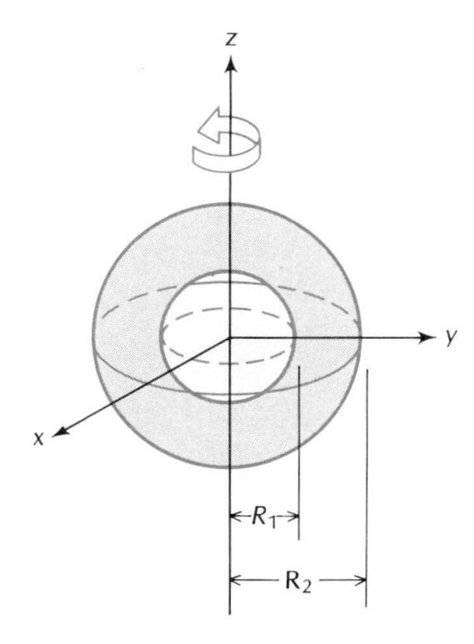


**Fig. 12.19** A thick spherical shell of inner radius $R_1$ and outer radius $R_2$.

**Table 12.1** SOME MOMENTS OF INERTIA

| *Body* | *Moment of inertia* |
|---|---|
| Thin hoop about symmetry axis | $MR^2$ |
| Thin hoop about diameter | $\frac{1}{2}MR^2$ |
| Disk or cylinder about symmetry axis | $\frac{1}{2}MR^2$ |
| Cylinder about diameter through center | $\frac{1}{4}MR^2 + \frac{1}{12}Ml^2$ |
| Thin rod about perpendicular axis through center | $\frac{1}{12}Ml^2$ |
| Thin rod about perpendicular axis through end | $\frac{1}{3}Ml^2$ |
| Sphere about diameter | $\frac{2}{5}MR^2$ |
| Thin spherical shell about diameter | $\frac{2}{3}MR^2$ |

## 12.5 Angular Momentum of a Particle; the Cross Product

In Chapter 10 we saw that in the study of the translational motion of a system of particles, the concept of momentum plays a crucial role — it enables us to formulate the general equation of motion for the system and to formulate the law of conservation of momentum. In the study of rotational motion of a rigid body, the concept that plays an analogous role is the angular momentum — it will enable us to formulate the general equation of rotational motion for the rigid body and to formulate a law of conservation. We will first introduce the concept of angular momentum for the special case of a single particle, and then proceed to the case of a system of particles, such as a rigid body.

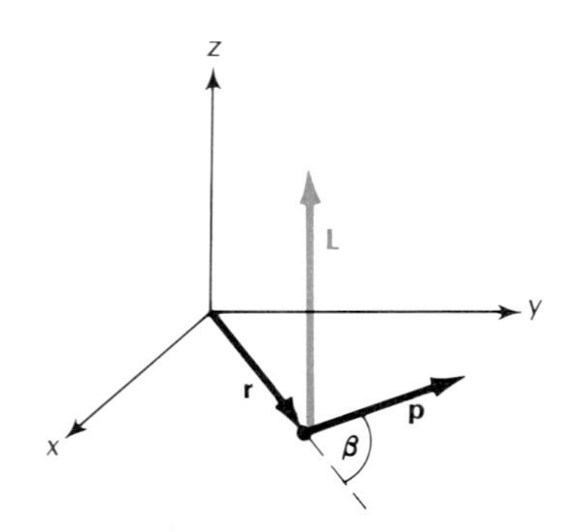


**Fig. 12.20** Position, momentum, and angular momentum vectors of a particle. The vector **L** is perpendicular to both **r** and **p.**

Consider a single particle of mass $m$ which, at one instant of time, has a momentum **p** and is at a distance $r$ from the origin of coordinates. The angular momentum **L** of this particle is defined as a vector of magnitude

$$\boxed{L = rp \sin \beta} \tag{41}$$

where $\beta$ is the angle between the momentum vector **p** and the position vector **r.** The direction of the vector **L** is along the perpendicular to the plane defined by the vectors **p** and **r** (see Figure 12.20). The direction of the vector **L** along this perpendicular is specified by the **right-hand rule:** *place the fingers of your right hand along* **r** *and curl them toward* **p** *in the direction of the smaller angle from* **r** *to* **p** *(see Figures 12.21 a and b); the thumb then points in the direction of* **L.**

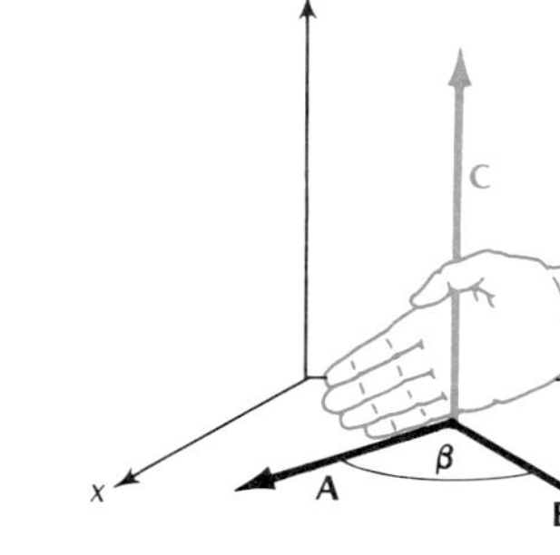


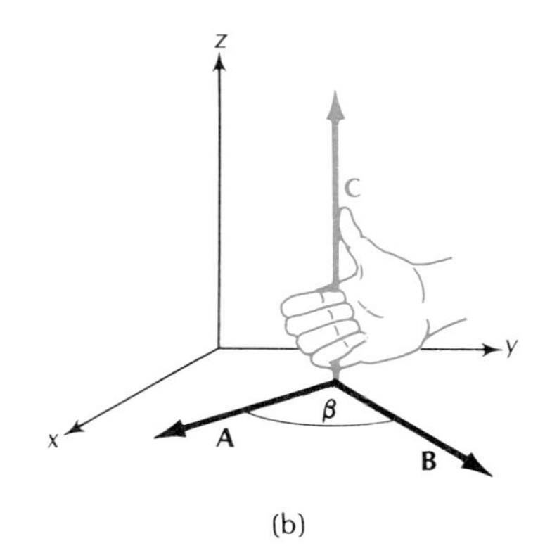


**Fig. 12.21** The right-hand rule. If the fingers curl from **A** toward **B,** the thumb points along **C.**

Given any two arbitrary vectors **A** and **B** making an angle $\beta$ with each other, the vector **C** of magnitude

$$C = AB \sin \beta \tag{42}$$

*Cross product*

and of direction specified by the right-hand rule is called the **cross product** of **A** and **B.** The notation for the cross product consists of the two vector symbols separated by a cross,

$$\mathbf{C} = \mathbf{A} \times \mathbf{B} \tag{43}$$

Note that, according to the right-hand rule, the fingers must be curled from the first vector in the product toward the second; hence the direction of the cross product $\mathbf{A} \times \mathbf{B}$ is opposite to the direction of the cross product $\mathbf{B} \times \mathbf{A}$. This means that the order of the factors affects the value of the cross product (in mathematical language: the cross product is not commutative). Also, note that the cross product of any vector with itself is zero,

$$\mathbf{A} \times \mathbf{A} = 0$$

since in this case the angle $\theta$ between the vector and itself is zero.[3]

In view of the general definition (42), we recognize that the angular-momentum vector of a particle is the cross product of the position vector and the momentum vector:

*Angular momentum of a particle*

$$\boxed{\mathbf{L} = \mathbf{r} \times \mathbf{p}} \tag{44}$$

or

$$\mathbf{L} = m\mathbf{r} \times \mathbf{v} \tag{45}$$

The unit of angular momentum is the product of the units of length and momentum. In the metric system, the unit of angular momentum is therefore m · kg · m/s, or kg · m$^2$/s. In the British system, the unit is lb · ft$^2$/s.

The angular momentum of a particle moving with constant velocity in the absence of forces is constant. Figure 12.22 makes this clear. It shows the line of motion of the particle and the origin of coordinates. Obviously, the direction of **L** is constant (out of the plane of Figure 12.22). The magnitude of **L** is $rp \sin \beta$. In this expression the quantity $p$ is constant because the motion proceeds at uniform velocity; and the quantity $r \sin \beta$ is also constant because it merely represents the shortest distance between the origin and the line of motion, and is therefore not affected by the motion along the line. Hence both the magnitude and the direction of **L** remain constant.

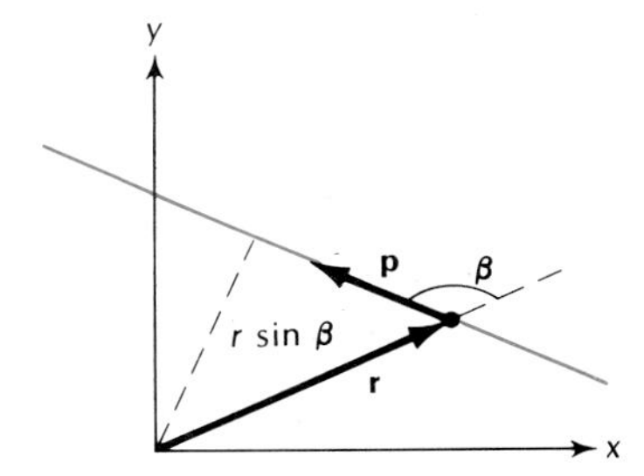


**Fig. 12.22** The distance between the origin and the line of motion is $r \sin \beta$.

This conservation of angular momentum of a free particle is of no great interest — it tells us nothing new about the motion of the particle. For a somewhat more interesting case of conservation of angular momentum, consider a particle in uniform circular motion, such as a stone being whirled along a circle at the end of a string, or the Earth moving along its circular orbit around the Sun. Figure 12.23 shows the origin at the center of the circle; since the position vector is always perpendicular to the velocity vector, the magnitude of the angular-momentum vector is

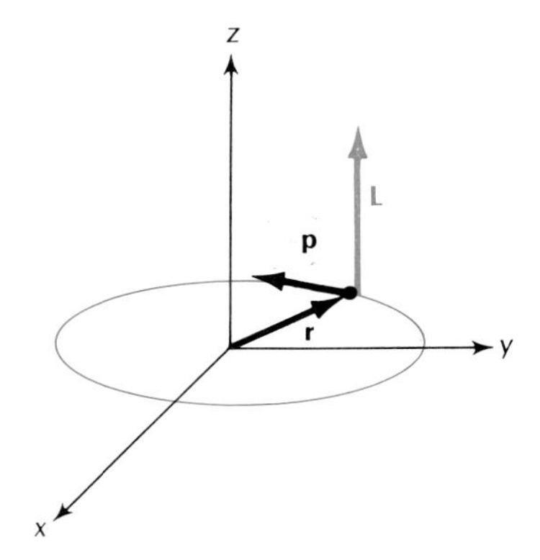


**Fig. 12.23** A particle in uniform circular motion. The vector **L** is perpendicular to the plane of the circle.

$$\boxed{L = rp = mrv} \tag{46}$$

The direction of the angular-momentum vector is perpendicular to the plane of the circle (Figure 12.23). As the particle moves around the circle, **L** remains constant in magnitude and direction.

[3] For more details on the cross product, see Section 3.4.

EXAMPLE 10. The Earth moves around the Sun with (approximately) uniform circular motion. If the origin of coordinates is placed on the Sun, what is the angular momentum of the Earth?

SOLUTION: The radius of the circle is $1.5 \times 10^{11}$ m and the speed of the Earth is $3.0 \times 10^4$ m/s (see Example 2.2); the mass of the Earth is $6.0 \times 10^{24}$ kg. Hence, the magnitude of the angular momentum is

$$L = mrv = 6.0 \times 10^{24}\ \text{kg} \times 1.5 \times 10^{11}\ \text{m} \times 3.0 \times 10^4\ \text{m/s}$$

$$= 2.7 \times 10^{40}\ \text{kg} \cdot \text{m}^2/\text{s}$$

The direction of the angular momentum is perpendicular to the plane of the orbit, upward (as in Figure 12.23).

We must remember that the angular momentum in Eqs. (44) and (46) is reckoned relative to a specific origin of coordinates. Obviously, any shift of the origin of coordinates alters the position vector **r** and hence the angular momentum. It is therefore meaningless to speak of the angular momentum without first specifying the origin of the coordinates. Thus, the angular momentum of a particle in uniform circular motion is constant if the origin of coordinates is placed at the center of the circle, as in Figure 12.23, but not if the origin is placed elsewhere.

*Angular momentum and central forces*

We recall from the discussion of Kepler's law of areas in Section 9.4, that the quantity $rv \sin \beta$ is constant for the orbital motion of a particle under the influence of gravity or any other central force (here it is assumed that the origin used for the evaluation of the radial distance $r$ coincides with the center of force). Thus, the angular momentum $L = mrv \sin \beta$ is constant, and Kepler's law of areas is equivalent to a law of conservation of angular momentum for motion with a central force.

## 12.6 Angular Momentum of a Rigid Body

In the previous section we defined the angular momentum of a single particle as

$$\mathbf{L} = \mathbf{r} \times \mathbf{p} = m\mathbf{r} \times \mathbf{v}$$

where **r** is the position vector of the particle (relative to a given origin) and **p** is the momentum. The total **angular momentum of a rigid body** is simply the sum of the angular momenta of all the particles in the body. If these particles have masses $m_i$, velocities $\mathbf{v}_i$, and position vectors $\mathbf{r}_i$ (relative to a given origin of coordinates), then the total angular momentum is

*Angular momentum of a rigid body*

$$\mathbf{L} = \sum_{i=1}^{n} m_i \mathbf{r}_i \times \mathbf{v}_i \tag{47}$$

As in the case of a single particle, the value of the angular momentum obtained from this formula depends on the choice of origin of coordinates. For the calculation of the angular momentum of a rigid body rotating about a fixed axis, it is usually convenient to choose an origin on the axis of rotation or at the center of mass. Since the angular momentum is a vector, it has three separate components; in general, these components are complicated functions of the geometry, orientation, and angular velocity of the body.

Incidentally: Making use of the metric unit of energy, we see that the unit of angular momentum can be conveniently written as joule-second. Since $1\ \text{J} = 1\ \text{kg}\cdot\text{m}^2/\text{s}^2$, this unit agrees with the unit $\text{kg}\cdot\text{m}^2/\text{s}$ of angular momentum that we introduced in Section 12.5. Table 12.2 gives the values of some typical angular momenta.

**Table 12.2** SOME ANGULAR MOMENTA

| | |
|---|---|
| Orbital motion of all planets in Solar System | $3.2 \times 10^{43}$ J · s |
| Orbital motion of Earth | $2.7 \times 10^{40}$ J · s |
| Rotation of Earth | $5.8 \times 10^{33}$ J · s |
| Helicopter rotor (320 rev/min) | $5 \times 10^{4}$ J · s |
| Automobile wheel (90 km/h) | $1 \times 10^{2}$ J · s |
| Electric fan | 1 J · s |
| Frisbee | $1 \times 10^{-1}$ J · s |
| Toy gyroscope | $1 \times 10^{-1}$ J · s |
| Phonograph record ($33\frac{1}{3}$ rev/min) | $6 \times 10^{-3}$ J · s |
| Bullet fired from rifle | $2 \times 10^{-3}$ J · s |
| Orbital motion of electron in atom | $1.05 \times 10^{-34}$ J · s |
| Spin of electron | $0.53 \times 10^{-34}$ J · s |

EXAMPLE 11. Figure 12.24 shows a dumbbell, a rigid body consisting of two particles of mass $m$ attached to the ends of a (nearly) massless rigid rod of length $2r$. The body rotates with angular velocity $\omega$ about a perpendicular axis through the center of the rod. Find the angular momentum.

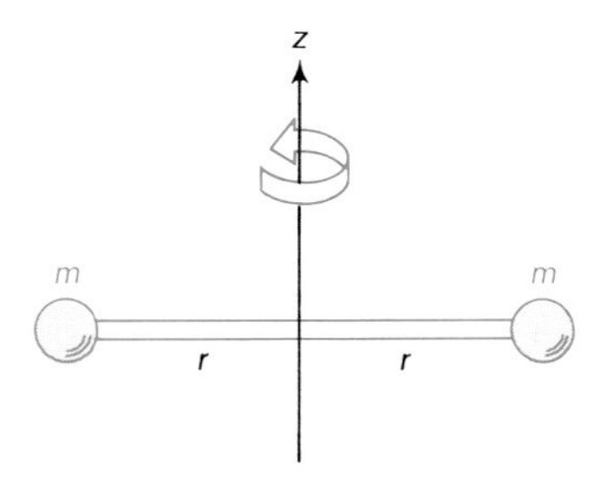


**Fig. 12.24** Two particles of mass $m$ joined by a massless rod.

SOLUTION: Since the instantaneous position vector of each particle is perpendicular to the instantaneous velocity vector, the magnitude of the angular momentum vector $m\mathbf{r} \times \mathbf{v}$ of each particle is simply $mrv$. The direction of the angular momentum vector of each particle is parallel to the axis of rotation. The total angular momentum is

$$L = mrv + mrv = 2mrv = 2mr^2\omega$$

where we have taken into account that $v = r\omega$.

EXAMPLE 12. Suppose that the rod of the dumbbell described in the preceding example is welded to an axle inclined at an angle $\theta$ with respect to the rod. The dumbbell rotates with angular velocity $\omega$ on this axle, which rests in fixed bearings (see Figure 12.25). Find the angular momentum.

SOLUTION: It is intuitively obvious that, without bearings to hold to axle in a fixed orientation, the dumbbell would not continue to rotate about such an inclined axis. At the instant shown in Figure 12.25, the centripetal force required to keep the upper mass moving in a circle is toward the left and, consequently, the axle exerts a push on the upper bearing toward the right. At the same instant, the axle exerts a push on the lower bearing toward the left. These pushes of the axle against its bearings represent a **dynamic imbalance** of the rotating body. The pushes are associated with the rotation, and they cease when the rotation ceases (under static conditions, the rigid body in Figure 12.25 is in equilibrium and exerts no lateral push on the bearings). Dynamic imbalance often occurs when a rotating body is not symmetric about the

*Dynamic imbalance*

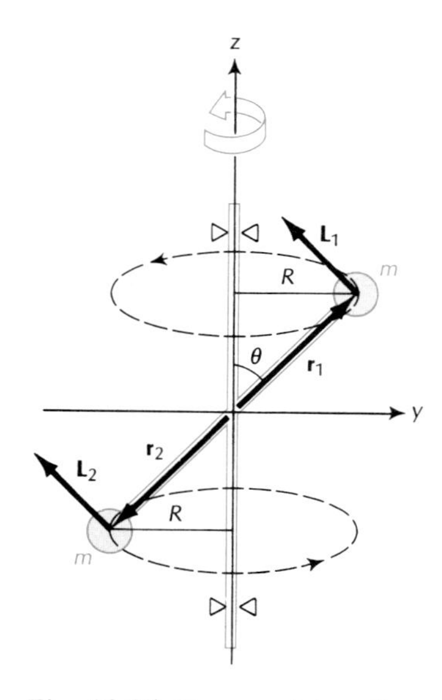


**Fig. 12.25** Two particles of mass $m$ at the ends of a rod that makes an angle $\theta$ with the axis of rotation. At the instant shown here the angular momentum vectors $\mathbf{L}_1$ and $\mathbf{L}_2$ of the two particles are both in the $y$–$z$ plane.

axis of rotation; at high speeds of rotation, it can lead to severe vibration in the bearings.

For the purposes of the present example, we will suppose that the bearings are strong and frictionless, so the dumbbell can continue to rotate on the inclined axle indefinitely. The distance between each mass and the axis of rotation is

$$R = r \sin \theta \tag{48}$$

and the velocity of each mass has a magnitude

$$v = \omega R = \omega r \sin \theta \tag{49}$$

The direction of the velocity is perpendicular to the position vector. Hence the angular momentum of each mass has a magnitude

$$m\,|\mathbf{r} \times \mathbf{v}| = mrv = m\omega r^2 \sin \theta \tag{50}$$

The direction of the angular momentum of each mass is perpendicular to both the velocity and the position vectors, as specified by the right-hand rule. The angular momentum vector of each mass is shown in Figure 12.25; these vectors are parallel to each other, they are in the plane of the axis and rod, and they make an angle of $90° - \theta$ with the axis. The total angular momentum is then the vector sum of these individual angular momenta. This vector is in the same direction as the individual angular momentum vectors and has a magnitude twice as large as Eq. (50):

$$L = 2m\omega r^2 \sin \theta \tag{51}$$

At the instant shown in Figure 12.25, the rod is in the $z$–$y$ plane. Hence the instantaneous angular momentum is also in this plane and has components

$$L_z = L \cos(90° - \theta) = 2m\omega r^2 \sin^2 \theta \tag{52}$$

$$L_y = -L \cos \theta = -2m\omega r^2 \sin \theta \cos \theta \tag{53}$$

As the body rotates, so does the angular-momentum vector; that is, the components of this vector change but its magnitude remains fixed at the value given by Eq. (51).

COMMENTS AND SUGGESTIONS: Note that Eq. (52) can also be written

$$L_z = 2m\omega R^2 \tag{54}$$

where $R = r \sin \theta$ is the distance between each mass and the axis of rotation. Since $2mR^2$ is simply the moment of inertia of our bodies about the $z$ axis, Eq. (54) is the same as

$$L_z = I\omega \tag{55}$$

As we will see below, this formula is of general validity.

The preceding example shows that the angular momentum vector of a rotating body need not always lie along the axis of rotation. However, if the axis of rotation is also an axis of symmetry of the body, then the angular momentum vector will lie along this axis. For such a symmetric body, each particle on one side of the axis has a counterpart on the other side of the axis. The angular momenta of these two particles have the same components along the axis, but opposite components transverse to the axis. For instance, Figure 12.26 shows a pair of particles in a body rotating about its axis of symmetry, which coincides with the $z$ axis. The $z$ components of the angular momenta of these particles are equal, and the $y$ components are opposite. Thus, the net angular momentum contributed by this pair of particles — and any other pair of particles — has a component only along the axis of symmetry.

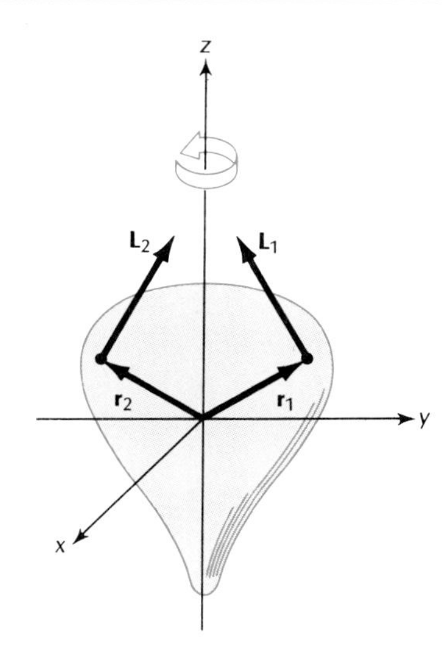


**Fig. 12.26** A pair of particles in a rigid body rotating about its axis of symmetry.

In any case, for the remainder of this section, we will focus our attention exclusively on the component of the angular momentum *along the axis of rotation.* Figure 12.27 shows an arbitrary rigid body rotating about a fixed axis, which coincides with the $z$ axis. The relevant component of the angular momentum is then the $z$ component. To evaluate this component, we consider one of the particles of the body (see Figure 12.27). The position vector $\mathbf{r}$ of the particle makes an angle $\theta$ with the $z$ axis. By a calculation similar to that which led to Eqs. (51) and (54), we find that the magnitude of the angular momentum of this particle is $m\omega r^2 \sin\theta$, and the $z$ component of the angular momentum is $m\omega R^2$, where, as in Eq. (54), $R$ is the distance between the particle

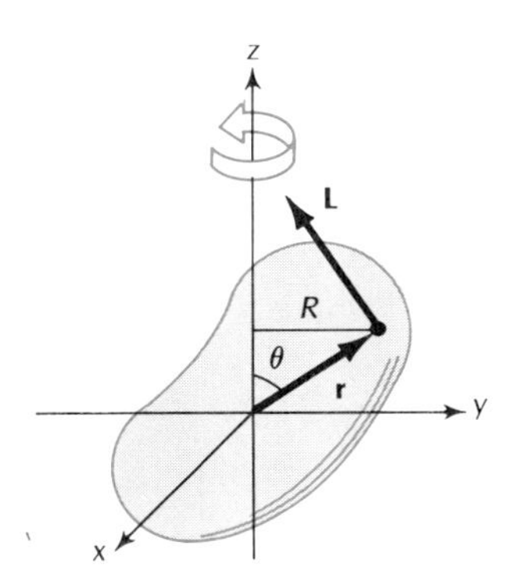


**Fig. 12.27** An arbitrary rigid body rotating about a fixed axis.

and the axis of rotation. Hence the $z$ component of the total angular momentum of all the particles in the rotating body is

$$L_z = \sum_{i=1}^{n} m_i R_i^2 \omega \tag{56}$$

or

$$\boxed{L_z = I\omega} \tag{57}$$

*Component of angular momentum along the axis of rotation*

where $I = \Sigma m_i R_i^2$ is the usual expression [Eq. (22)] for the total moment of inertia of the body (about the $z$ axis). This establishes the general validity of the simple formula [Eq. (55)] that we found in Example 11: the component of the angular momentum along the axis of rotation is the product of moment of inertia and angular velocity. This result will be very useful in the next chapter when we study the dynamics of rotational motion. Equation (57) is reminiscent of the familiar expression $p = mv$ for the translational momentum of a particle — the moment of inertia again replaces the mass and the angular velocity replaces the translational velocity.

If we solve Eq. (57) for $\omega$,

$$\omega = L_z/I \tag{58}$$

and substitute this into the formula for the rotational kinetic energy, we obtain

$$K = \tfrac{1}{2}I\omega^2 = \tfrac{1}{2}I(L_z/I)^2 \tag{59}$$

or

$$K = \tfrac{1}{2}L_z^2/I \tag{60}$$

This is the analog of Eq. (7.43).

*Spin*

The angular momentum associated with the rotational motion of a rigid body is often called the **spin** to distinguish it from the angular momentum associated with the translational motion. For example, the Earth has spin angular momentum because of its rotation about its axis and it has orbital angular momentum because of its motion around the Sun.

---

EXAMPLE 13. The moment of inertia of the Earth about an axis through the poles is $8.1 \times 10^{37}$ kg · m². Calculate the spin angular momentum. Calculate the rotational kinetic energy.

SOLUTION: The frequency of rotation is one revolution per day.[4]

$$\nu = 1/\text{day} = 1/8.6 \times 10^4 \text{ s} = 1.2 \times 10^{-5}/\text{s} \tag{61}$$

and

$$\omega = 2\pi\nu = 7.3 \times 10^{-5} \text{ radian/s} \tag{62}$$

The angular momentum is then

$$\begin{aligned} L_z &= I\omega \\ &= 8.1 \times 10^{37} \text{ kg} \cdot \text{m}^2 \times 7.3 \times 10^{-5} \text{ radian/s} \\ &= 5.9 \times 10^{33} \text{ kg} \cdot \text{m}^2/\text{s} \end{aligned}$$

and the kinetic energy

$$\begin{aligned} K &= \tfrac{1}{2}L_z^2/I \\ &= \tfrac{1}{2}(5.9 \times 10^{33} \text{ kg} \cdot \text{m}^2/\text{s})^2/(8.1 \times 10^{37} \text{ kg} \cdot \text{m}^2) \\ &= 2.1 \times 10^{29} \text{ J} \end{aligned}$$

---

The subatomic particles — electrons, protons, and neutrons — all have intrinsic spin angular momenta. The magnitudes of the spin angular momenta of electrons, protons, and neutrons are equal: they all are $0.53 \times 10^{-34}$ J · s. This magnitude of the spin angular momentum is usually expressed as $\frac{1}{2}\hbar$, where $\hbar = 1.05 \times 10^{-34}$ J · s is a fundamental unit of angular momentum which plays a pervasive role in modern quantum physics.[5] (The fundamental constant $\hbar$, pronounced *aitch bar,*

[4] Strictly, the frequency is one revolution per sidereal day (23 h 56 min). But within the accuracy of our present calculation, we can ignore this small correction.

[5] See Appendix 8 for a more precise value of $\hbar$.

equals Planck's constant $h$ divided by $2\pi$, that is, $\hbar = h/2\pi$). The spin angular momentum of a subatomic particle is analogous to the spin angular momentum that the Earth possesses because of its rotation about its axis. However, this analogy between the spin of a subatomic particle and the classical angular momentum of rotation of a rigid body must not be taken too seriously — the spin of subatomic particles hinges on quantum physics in a very essential way. In contrast to a classical angular momentum, the spin of a subatomic particle has an absolutely fixed and immutable magnitude. According to modern quantum physics, external forces cannot change the magnitude of the spin of such a particle; they can only change the direction of this spin.

## SUMMARY

**Average angular velocity:** $\overline{\omega} = \dfrac{\Delta\phi}{\Delta t}$

**Instantaneous angular velocity:** $\omega = \dfrac{d\phi}{dt}$

**Average angular acceleration:** $\overline{\alpha} = \dfrac{\Delta\omega}{\Delta t}$

**Instantaneous angular acceleration:** $\alpha = \dfrac{d\omega}{dt}$

**Speed of particle on rotating body:** $v = R\omega$

**Acceleration of particle on rotating body:**

$a_{\text{tan}} = R\alpha$

$a_{\text{cent}} = R\omega^2$

**Motion with constant angular acceleration:**

$$\omega = \omega_0 + \alpha t$$

$$\phi - \phi_0 = \omega_0 t + \tfrac{1}{2}\alpha t^2$$

$$\alpha(\phi - \phi_0) = \tfrac{1}{2}(\omega^2 - \omega_0^2)$$

**Moment of inertia:** $I = \sum_{i=1}^{n} m_i R_i^2$

$$I = \int \rho R^2 \, dV$$

**Kinetic energy of rotation:** $K = \tfrac{1}{2}I\omega^2$

**Parallel-axis theorem:** $I = I_{\text{CM}} + Md^2$

**Perpendicular-axis theorem (for a flat plate in the *x–y* plane):**

$I_z = I_x + I_y$

**Cross product:** magnitude of $\mathbf{A} \times \mathbf{B}$ is $AB \sin \beta$
direction is given by right-hand rule

**Angular momentum of a particle:** $\mathbf{L} = \mathbf{r} \times \mathbf{p} = m\mathbf{r} \times \mathbf{v}$

**Angular momentum for uniform circular motion:** $L = mrv$

**Angular momentum of rigid body rotating about z axis:** $L_z = I\omega$

## QUESTIONS

1. A spinning flywheel in the shape of a disk suddenly shatters into many small fragments. Draw the trajectories of a few of these small fragments; assume that the fragments do not interfere with each other.

2. You may have noticed that in some old movies the wheels of moving carriages or stagecoaches seem to rotate backward. How does this come about?

**Fig. 12.28** Gears.

3. Relative to an inertial reference frame, what is your angular velocity right now about an axis passing through your center of mass?

4. Consider the wheel of an accelerating automobile. Draw the instantaneous acceleration vectors for a few points on the rim of the wheel.

5. When engineers design the teeth for gears, they usually make the sides of the teeth round (Figure 12.28). Can you guess why?

6. The hands of a watch are small rectangles with a common axis passing through one end. The minute hand is long and thin; the hour hand is short and thicker. Both hands have the same mass. Which has the greater moment of inertia? Which has the greater kinetic energy and angular momentum?

7. What configuration and what axis would you choose to give your body the smallest possible moment of inertia? The greatest?

8. About what axis through the center of mass is the moment of inertia of this book largest? Smallest? (Assume the book is closed.)

9. A circular hoop made of thin wire has a radius $R$ and mass $M$. About what axis must you rotate this hoop to obtain the minimum moment of inertia? What is the value of this minimum?

10. Automobile engines and other internal combustion engines have flywheels attached to their crankshafts. What is the purpose of these flywheels? (Hint: Each explosive combustion in one of the cylinders of such an engine gives a sudden push to the crankshaft. How would the crankshaft respond to this push if it had no flywheel?)

11. Suppose you pump a mass $M$ of sea water into a pond on land at the equator. How does this change the moment of inertia of the Earth?

12. An automobile travels at constant speed along a road consisting of two straight segments connected by a curve in the form of an arc or circle. Taking the center of the circle as origin, what is the direction of the angular momentum of the automobile? Is the angular momentum constant as the automobile travels along this road?

13. Is the angular momentum of the orbital motion of a planet constant if we choose an origin of coordinates *not* centered on the Sun?

14. A pendulum is swinging back and forth. Is the angular momentum of the pendulum bob constant?

15. What is the direction of the angular momentum of rotation of the Earth?

16. A bicycle is traveling east along a level road. What is the direction of the angular momentum of its wheels?

## PROBLEMS

### Section 12.2

1. The minute hand of a wall clock has a length of 20 cm. What is the speed of the tip of this hand?

2. Quito is on the Earth's equator; New York is at latitude 41° north. What is

the angular velocity of each city about the Earth's axis of rotation? What is the linear speed of each?

3. An automobile has wheels with a radius of 30 cm. What is the angular velocity (in radians per second) and the frequency (in revolutions per second) of the wheels when the automobile is traveling at 88 km/h?

*4. An automobile has wheels of diameter 0.63 m. If the automobile is traveling at 80 km/h, what is the instantaneous velocity vector (relative to the ground) of a point at the top of a wheel? At the bottom? At the front?

*5. An aircraft passes directly over you with a speed of 900 km/h at an altitude of 10,000 m. What is the angular velocity of the aircraft (relative to you) when directly overhead? Three minutes later?

*6. The outer edge of the grooved area of a long-playing record is at a radial distance of 14.6 cm from the center; the inner edge is at a radial distance of 6.35 cm. The record rotates at $33\frac{1}{3}$ rev/min. The needle of the pick-up arm takes 25 min to play the record and in that time interval it moves uniformly and radially from the outer edge to the inner edge. What is the radial speed of the needle? What is the speed of the outer edge relative to the needle? What is the speed of the inner edge relative to the needle?

*7. Consider the phonograph record described in Problem 6. What is the total length of the groove in which the needle travels?

*8. Because of the rotation of the Earth about its axis, a stone resting on a tower at, say, the equator has an eastward velocity in an inertial reference frame centered on the Earth but not rotating with it. In this reference frame, the foot of the tower also has an eastward velocity, but slightly smaller than that of the stone (Figure 12.29).

(a) Show that if the height of the tower is $h$, the difference in the magnitude of these velocities is $\omega h$, where $\omega$ is the angular velocity of rotation of the Earth.

(b) If the stone drops from the tower, its excess eastward velocity will cause it to land at some distance from the foot of the tower. Show that it will land at a distance of about $\omega h\sqrt{2h/g}$ eastward of the foot of the tower. Evaluate for $h = 30$ m.

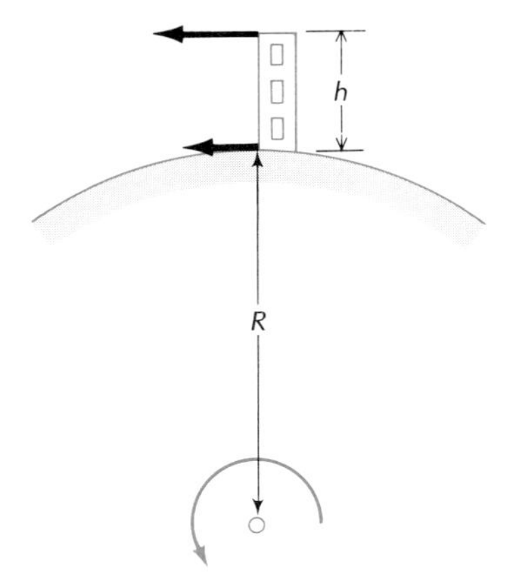


**Fig. 12.29**

Section 12.3

9. The blade of a circular saw of diameter 20 cm accelerates uniformly from rest to 7000 rev/min in 1.2 s. What is the angular acceleration? How many revolutions will the blade have made by the time it reaches full speed?

10. An automobile accelerates uniformly from 0 to 80 km/h in 6.0 s. The automobile has wheels of radius 30 cm. What is the angular acceleration of the wheels?

11. When you turn off the motor, a phonograph turntable initially rotating at $33\frac{1}{3}$ rev/min makes 25 revolutions before it stops. Calculate the angular deceleration of this turntable; assume it is constant.

*12. The rotation of the Earth is slowing down. In 1977, the Earth took 1.01 s longer to complete 365 rotations than in 1900. What was the average angular deceleration of the Earth in the time interval from 1900 to 1977?

*13. An automobile engine accelerates at a constant rate from 200 rev/min to 3000 rev/min in 7.0 s and then runs at constant speed.

(a) Find the angular velocity and the angular acceleration at $t = 0$ (just after acceleration begins) and at $t = 7.0$ s (just before acceleration ends).

(b) A flywheel with a radius of 18 cm is attached to the shaft of the engine. Calculate the tangential and the centripetal acceleration of a point on the rim of the flywheel at the times given above.

(c) What angle does the net acceleration vector make with the radius at $t = 0$ and at $t = 7.0$ s? Draw diagrams showing the wheel and the acceleration vector at these times.

Section 12.4

14. According to spectroscopic measurements, the moment of inertia of an oxygen molecule about an axis through the center of mass and perpendicular to the line joining the atoms is $1.95 \times 10^{-46}$ kg · m². The mass of an oxygen atom is $2.66 \times 10^{-26}$ kg. What is the distance between the atoms? Treat the atoms as pointlike particles.

15. The moment of inertia of the Earth about its polar axis is $0.331 M_E R_E^2$, where $M_E$ is the mass and $R_E$ the equatorial radius. Why is this moment of inertia smaller than that of a sphere of uniform density? What would the radius of a sphere of uniform density have to be if its mass and moment of inertia are to coincide with those of the Earth?

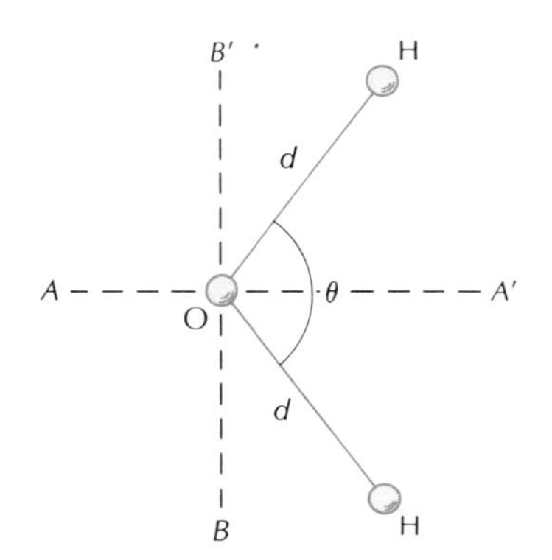


**Fig. 12.30** Water molecule.

16. Problem 10.24 gives the dimensions of a molecule of nitric acid ($HNO_3$). What is the moment of inertia of this molecule when rotating about the symmetry axis passing through the H, O, and N atoms? Treat the atoms as pointlike particles.

17. The water molecule has the shape shown in Figure 12.30. The distance between the oxygen and the hydrogen atoms is $d$ and the angle between the hydrogen atoms is $\theta$. From spectroscopic investigations it is known that the moment of inertia of the molecule is $1.93 \times 10^{-47}$ kg · m² for rotation about the axis $AA'$ and $1.14 \times 10^{-47}$ kg · m² for rotation about the axis $BB'$. From this information and the known values of the masses of the atoms, determine the values of $d$ and $\theta$. Treat the atoms as pointlike.

18. What is the moment of inertia (about the axis of symmetry) of a bicycle wheel of mass 4.0 kg, radius 0.33 m? Neglect the mass of the spokes.

*19. An empty beer can has a mass of 50 g, a length of 12 cm, and a radius of 3.3 cm. Find the moment of inertia of the can about its axis of symmetry. Assume that the can is a perfect cylinder of sheet metal with no ridges, indentations, or holes.

*20. Suppose that a supertanker transports $4.4 \times 10^8$ kg of oil from a storage tank in Venezuela (latitude 10° north) to a storage tank in Holland (latitude 53° north). What is the change of the moment of inertia of the Earth–oil system?

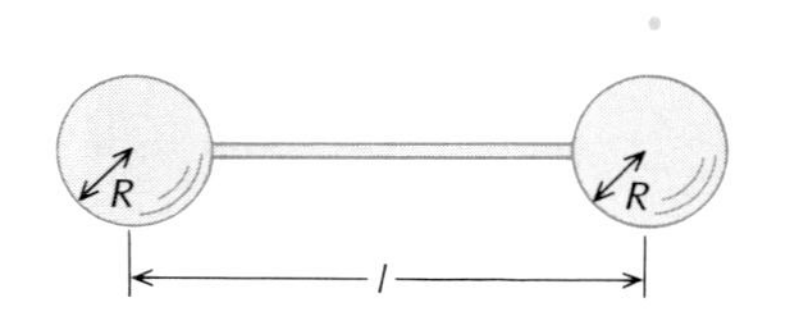


**Fig. 12.31** A dumbbell.

*21. A dumbbell consists of two uniform spheres of mass $M$ and radius $R$ joined by a thin rod of mass $m$ and length $l$ (Figure 12.31). What is the moment of inertia of this device about an axis through the center of the rod perpendicular to the rod? About an axis along the rod?

*22. Suppose that the Earth consists of a spherical core of mass $0.22M_E$ and radius $0.54R_E$ and a surrounding mantle (a spherical shell) of mass $0.78M_E$ and outer radius $R_E$. Suppose that the core is of uniform density and the mantle is also of uniform density. According to this simple model, what is the moment of inertia of the Earth? Express your answer as a multiple of $M_E R_E^2$.

*23. In order to increase her moment of inertia about a vertical axis, a spinning figure skater stretches out her arms horizontally; in order to reduce her moment of inertia, she brings her arms down vertically along her sides. Calculate the change of moment of inertia between these two configurations of the arms. Assume that each arm is a thin, uniform rod of length 0.60 m and mass 2.8 kg hinged at the shoulder at a distance of 0.20 m from the axis of rotation.

*24. Find the moment of inertia of a thin rod of mass $M$ and length $L$ about an axis through the center inclined at an angle $\theta$ with respect to the rod.

25. Use the parallel-axis theorem to derive Eq. (32) from Eq. (30).

26. Given that the moment of inertia of a sphere about a diameter is $\frac{2}{5}MR^2$, show that the moment of inertia about an axis tangent to the surface is $\frac{7}{5}MR^2$.

27. Find a formula for the moment of inertia of a uniform, thin, square plate (mass $m$, dimension $l \times l$) rotating about an axis that coincides with one of its edges.

*28. Find the moment of inertia of the flywheel shown in Figure 12.32 rotating about its axis. The flywheel is made of material of uniform thickness; its mass is $M$.

*29. A solid cylinder capped with two solid hemispheres rotates about its axis of symmetry (Figure 12.33). The radius of the cylinder is $R$, its height is $h$, and the total mass (hemispheres included) is $M$. What is the moment of inertia?

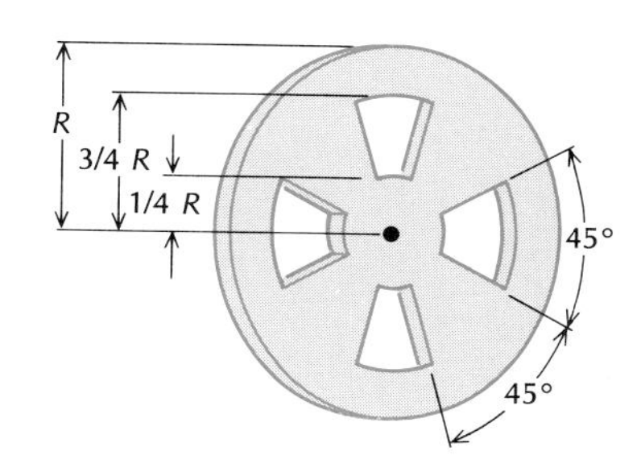


**Fig. 12.32**

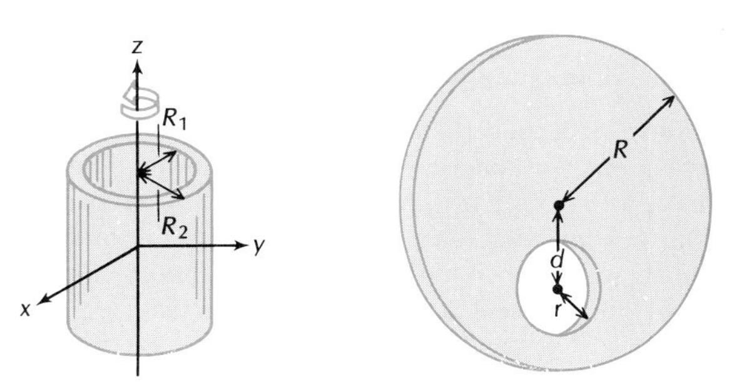


**Fig. 12.33** **Fig. 12.34**

*30. A hole of radius $r$ has been drilled in a circular, flat plate of radius $R$ (Figure 12.34). The center of the hole is at a distance $d$ from the center of the circle. The mass of this body is $M$. Find the moment of inertia for rotation about an axis through the center of the circle, perpendicular to the plate.

*31. A piece of steel pipe (Figure 12.35) has an inner radius $R_1$, an outer radius $R_2$, and a mass $M$. Find the moment of inertia about the axis of the pipe.

*32. Find the moment of inertia of a flywheel of mass $M$ made by cutting four large holes of radius $r$ out of a uniform disk of radius $R$ (Figure 12.36). The holes are centered at a distance $R/2$ from the center of the flywheel.

*33. Show that the moment of inertia of a long, very thin cone (Figure 12.37) about an axis through the apex and perpendicular to the centerline is $\frac{3}{5}Ml^2$, where $M$ is the mass and $l$ the height of the cone.

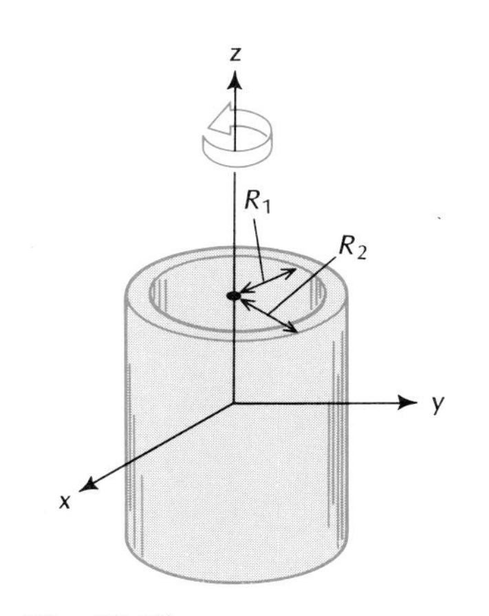


**Fig. 12.35**

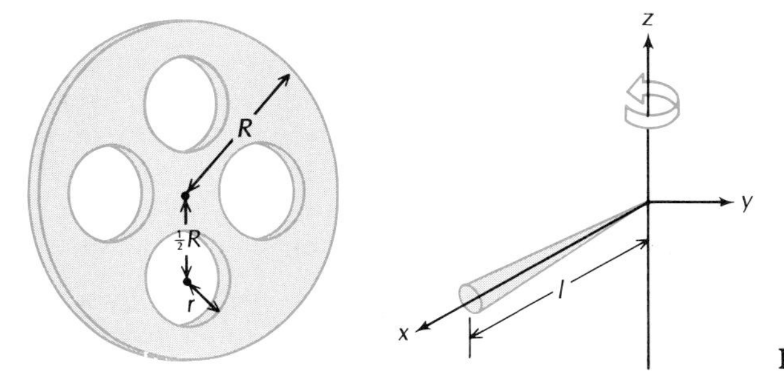


**Fig. 12.36** **Fig. 12.37**

*34. The mass distribution within the Earth can be roughly approximated by several concentric spherical shells, each of constant density. The following table gives the outer and the inner radius of each shell and its mass (expressed as a fraction of the Earth's mass):

| *Shell* | *Outer radius* | *Inner radius* | *Fraction of mass* |
|---|---|---|---|
| 1 | 6400 km | 5400 km | 0.28 |
| 2 | 5400 | 4400 | 0.25 |
| 3 | 4400 | 3400 | 0.16 |
| 4 | 3400 | 2400 | 0.20 |
| 5 | 2400 | 0 | 0.11 |

Use these data to calculate the moment of inertia of the Earth about its axis.

35. An airplane propeller consists of three radial blades, each of length 1.8 m and mass 20 kg. What is the kinetic energy of this propeller when rotating at 2500 rev/min? Assume that each blade is (approximately) a uniform rod.

36. The drilling pipe of an oil rig is 2 km long, 15 cm in diameter, and has a mass of 20 kg per meter of length. Assume that the wall of the pipe is very thin.

(a) What is the moment of inertia of this pipe rotating about its longitudinal axis?

(b) What is the kinetic energy when rotating at 1 rev/s?

37. Engineers have proposed that large flywheels be used for the temporary storage of surplus energy generated by electric power plants. A suitable flywheel would have a diameter of 3.6 m, a mass of 300 metric tons, and spin at 3000 revolutions per minute. What is the kinetic energy of rotation of this flywheel? Give the answer in both joules and kilowatt-hours. Assume that the moment of inertia of the flywheel is that of a uniform disk.

38. An automobile of mass 1360 kg has wheels 76.2 cm in diameter of mass 27.2 kg each. Taking into account the rotational kinetic energy of the wheels about their axles, what is the total kinetic energy of the automobile when traveling at 80.0 km/h? What percentage of the kinetic energy belongs to the rotational motion of the wheels about their axles? Pretend that each wheel has a mass distribution equivalent to that of a uniform disk.

39. What is the kinetic energy of rotation of a phonograph record of mass 170 g and radius 15.2 cm rotating at $33\frac{1}{3}$ revolutions per minute?

*40. The Oerlikon Electrogyro bus uses a flywheel to store energy for propelling the bus. At each bus stop, the bus is briefly connected to an electric power line, so that an electric motor on the bus can spin up the flywheel to 3000 revolutions per minute. If the flywheel is a disk of radius 0.6 m and mass 1500 kg, and if the bus requires an average of 40 hp for propulsion at an average speed of 20 km/h, how far can it move with the energy stored in the rotating flywheel?

41. Pulsars are rotating stars made almost entirely of neutrons closely packed together. The rate of rotation of most pulsars gradually decreases because rotational kinetic energy is gradually converted into other forms of energy by a variety of complicated "frictional" processes. Suppose that a pulsar of mass $1.5 \times 10^{30}$ kg and radius 20 km is spinning at the rate of 2.1 rev/s and is slowing down at the rate of $1.0 \times 10^{-15}$ rev/s$^2$. What is the rate (in joules per second or watts) at which the rotational energy is decreasing? If this rate of decrease of the energy remains constant, how long will it take the pulsar to come to a stop? Treat the pulsar as a sphere of uniform density.

42. For the sake of directional stability, the bullet fired by a rifle is given a spin angular velocity about its axis by means of spiral grooves ("rifling") cut into the barrel. The bullet fired by a Lee-Enfield rifle is (approximately) a uniform cylinder of length 3.18 cm, diameter 0.790 cm, and mass 13.9 g. The bullet emerges from the muzzle with a translational velocity of 628 m/s and a spin angular velocity of $2.47 \times 10^3$ rev/s. What is the translational kinetic energy of the bullet? What is the rotational kinetic energy? What fraction of the total kinetic energy is rotational?

43. Derive the formula for the moment of inertia of a thin disk of mass $M$ and radius $R$ rotating about a diameter. (Hint: Use the perpendicular-axis theorem.)

44. Derive the formula for the moment of inertia of a thin hoop of mass $M$ and radius $R$ rotating about a diameter.

*45. Find a formula for the moment of inertia of a uniform, thin, square plate (mass $M$, dimension $l \times l$) rotating about an axis through the center and perpendicular to the plate.

*46. Find the moment of inertia of a uniform cube of mass $M$ and edge $l$. Assume the axis of rotation passes through the center of the cube and is perpendicular to two of the faces.

*47. What is the moment of inertia of a thin, flat plate in the shape of a semicircle rotating about the straight side (Figure 12.38)? The mass of the plate is $M$ and the radius is $R$.

**48. Find the moment of inertia of the *thin* disk with two semicircular cutouts shown in Figure 12.39 rotating about its axis. The disk is made of material of uniform thickness; its mass is $M$.

**49. A cone of mass $M$ has a height $h$ and a base diameter $R$. Find its moment of inertia about its axis of symmetry.

**50. Derive the formula given in Table 12.1 for the moment of inertia of a sphere.

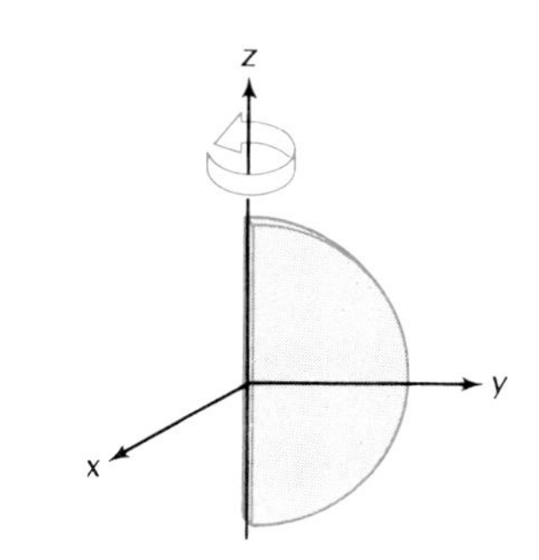


**Fig. 12.38**

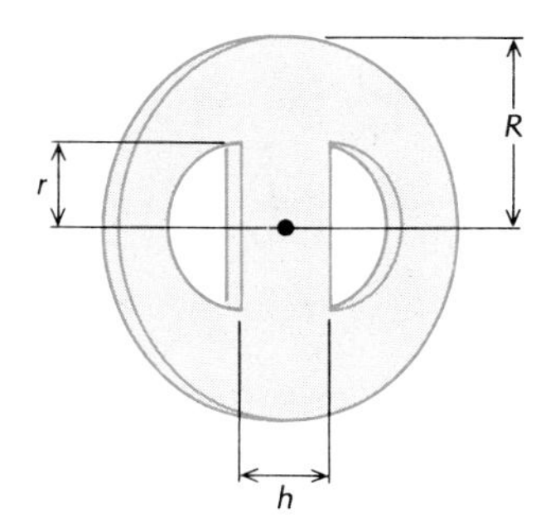


**Fig. 12.39**

### Section 12.5

51. The Moon moves around the Earth in an (approximately) circular orbit of radius $3.8 \times 10^8$ m in a time of 27.3 days. Calculate the magnitude of the orbital angular momentum of the Moon. Assume that the origin of coordinates is centered on the Earth.

52. At the Fermilab accelerator, protons of momentum $5.3 \times 10^{-16}$ kg · m/s travel around a circular path of diameter 2.0 km. What is the orbital angular momentum of one of these protons? Assume that the origin is at the center of the circle. Draw a diagram showing the direction of the angular momentum in relation to the path of the proton.

53. Prior to launching a stone from a sling, a Bolivian native whirls the stone at 3.0 revolutions per second around a circle of radius 0.75 m. The mass of the stone is 0.15 kg. What is the angular momentum of the stone relative to the center of the circle? Draw a diagram showing the direction of motion and the direction of the angular momentum.

54. A communications satellite of mass 100 kg is in a circular orbit of radius $4.22 \times 10^7$ m around the Earth. The orbit is in the equatorial plane of the Earth and the satellite moves along it from west to east with a speed of $4.90 \times 10^2$ m/s. What are the magnitude and the direction of the angular momentum of this satellite?

55. According to Bohr's (oversimplified) theory, the electron in the hydrogen atom moves in one or another of several possible circular orbits around the nucleus. The radii and the orbital velocities of the three smallest orbits are, respectively, $0.529 \times 10^{-10}$ m, $2.18 \times 10^6$ m/s; $2.12 \times 10^{-10}$ m, $1.09 \times 10^6$ m/s; and $4.76 \times 10^{-10}$ m, $7.27 \times 10^5$ m/s. For each of these orbits calculate the orbital angular momentum of the electron, with the origin at the center. How do these angular momenta compare?

56. A high-speed meteoroid moves past the Earth along an (almost) straight line. The mass of the meteoroid is 150 kg, its speed relative to the Earth is 60 km/s, and its distance of closest approach to the center of the Earth is $1.2 \times 10^4$ km.

(a) What is the angular momentum of the meteoroid in the reference frame of the Earth (origin at the center of the Earth)?

(b) What is the angular momentum of the Earth in the reference frame of the meteoroid (origin at the center of the meteoroid)?

57. A train of mass 1500 metric tons runs along a straight track at 85 km/h. What is the angular momentum (magnitude and direction) of the train about a point 50 m to the side of the track, left of the train? About a point on the track?

*58. Consider the motion of the Earth around the Sun as described in Example 9. Take as origin the point at which the Earth is today and treat the Earth as a particle.

(a) What is the angular momentum of the Earth about this origin today?
(b) What will be the angular momentum of the Earth about the same origin three months later? Six months later? Nine months later? Is the angular momentum conserved?

*59. Consider a projectile of mass $m$ launched with speed $v_0$ at an elevation angle of 45°. If the launch point is the origin of coordinates, what is the angular momentum of the projectile at the instant of launch? At the instant it reaches maximum height? At the instant it strikes the ground? Is the angular momentum conserved in this motion with this choice of origin?

*60. The electron in a hydrogen atom moves around the nucleus under the influence of the electric force of attraction, a central force pulling the electron toward the nucleus. According to the Bohr theory, one of the possible orbits of the electron is an ellipse of angular momentum $2\hbar$ with a distance of closest approach $(1 - 2\sqrt{2}/3)a_0$ and a distance of farthest recession $(1 + 2\sqrt{2}/3)a_0$, where $\hbar$ and $a_0$ are two atomic constants with the numerical values $1.05 \times 10^{-34}$ kg · m²/s ("Planck's constant," see Section 12.6) and $5.3 \times 10^{-11}$ m ("Bohr radius"), respectively. In terms of $\hbar$ and $a_0$, find the speed of the electron at the points of closest approach and farthest recession; then evaluate numerically.

### Section 12.6

61. According to a simple (but erroneous) model, the proton is a uniform rigid sphere of mass $1.67 \times 10^{-27}$ kg and radius $1.0 \times 10^{-15}$ m. The spin angular momentum of the proton is $5.3 \times 10^{-35}$ J · s. According to this model, what is the angular velocity of rotation of the proton? What is the linear velocity of a point on its equator? What is the rotational kinetic energy? How does this rotational energy compare with the rest-mass energy $mc^2$?

62. A phonograph turntable is a uniform disk of radius 15 cm and mass 1.4 kg. If this turntable accelerates from 0 rev/min to 78 rev/min in 2.5 s, what is the average rate of change of the angular momentum in this time interval?

63. The propeller shaft of a cargo ship has a diameter of 8.8 cm, a length of 27 m, and a mass of 1200 kg. What is the rotational kinetic energy of this propeller shaft when it is rotating at 200 rev/min? What is the angular momentum?

64. The Sun rotates about its axis with a period of about 25 days. Its moment of inertia is $0.20M_SR_S^2$, where $M_S$ is its mass and $R_S$ its radius. Calculate the angular momentum of rotation of the Sun. Calculate the total orbital angular momentum of all the planets; make the assumption that each planet moves in a circular orbit of radius equal to its mean distance from the Sun listed in Table 9.1. What percentage of the angular momentum of the Solar System is in the rotational motion of the Sun?

*65. The friction of the tides on the coastal shallows and the ocean floors gradually slows down the rotation of the Earth. The period of rotation (length of a sidereal day) is gradually increasing by 0.0016 s per century. What is the angular deceleration (in radians/s²) of the Earth? What is the rate of decrease of the rotational angular momentum? What is the rate of decrease of the rotational kinetic energy? The moment of inertia of the Earth about its axis is $0.331M_ER_E^2$, where $M_E$ is the mass of the Earth and $R_E$ its equatorial radius.

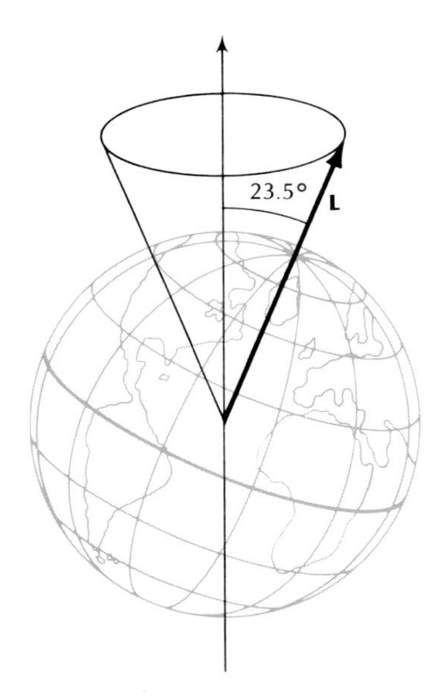


Fig. 12.40

**66. The spin angular momentum of the Earth has a magnitude of $5.9 \times 10^{33}$ kg · m²/s. Because of forces exerted by the Sun and the Moon, the spin angular momentum gradually changes direction, describing a cone of half angle 23.5° (Figure 12.40). The angular momentum vector takes 26,000 years to swing once around this cone. What is the magnitude of the rate of change of the angular momentum vector, i.e., what is the value of $|d\mathbf{L}/dt|$?

CHAPTER 13

# Dynamics of a Rigid Body

As we saw in Chapter 6, Newton's Second Law is the equation that determines the translational motion of a body. In this chapter we will derive an equation that determines the rotational motion of a rigid body. Just as Newton's equation of motion gives us the translational acceleration and permits us to calculate the change in velocity and position, the analogous equation for rotational motion gives us the angular acceleration and permits us to calculate the change in angular velocity and angular position. The equation for rotational motion is not a new law of physics, distinct from Newton's three laws. Rather, it is a consequence of these laws.

## 13.1 Torque

Before we lay down the laws for the rotational motion of a rigid body, we must discuss the concept of **torque** or **moment of a force.** We will see in the next section that this quantity plays a role in rotational motion analogous to that played by the force in translational motion — it produces angular acceleration, just as force produces linear acceleration. For instance, the push of your hand against a crank on a wheel (Figure 13.1) exerts a torque or "twist" that produces angular acceleration of the wheel. For a rigorous definition of the torque, suppose that a force **F** acts on a particle having a position vector **r** with respect to a given origin of coordinates. Then the torque $\tau$ of this force with respect to the origin is defined as the cross product of **r** and **F**,

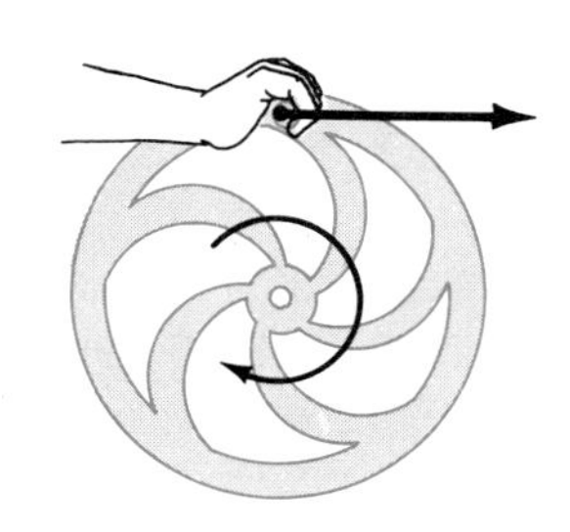

**Fig. 13.1** Push of hand against crank of a wheel.

$$\boxed{\tau = \mathbf{r} \times \mathbf{F}} \qquad (1)$$

*Torque of a force*

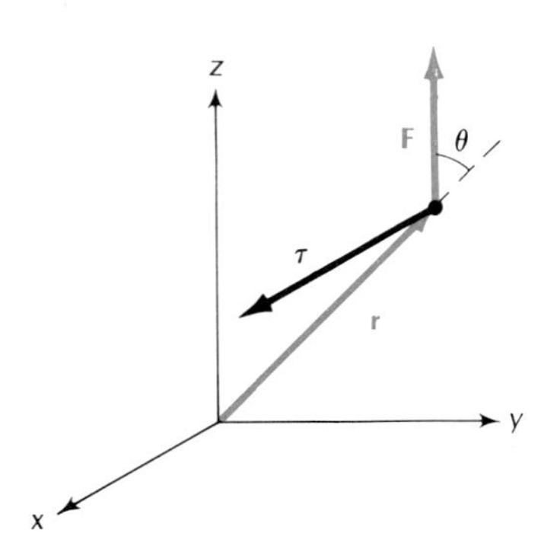


**Fig. 13.2** Position vector, force vector, and torque vector. Here **r** and **F** are in the *y-z* plane and $\tau$ is parallel to the *x* axis.

Thus, the torque is a vector. In terms of the angle $\theta$ between the position vector and the force (see Figure 13.2), the magnitude of the torque is

$$\tau = rF \sin \theta \tag{2}$$

and the direction of the torque is perpendicular to both the position vector and the force, as indicated by the right-hand rule (Figure 13.2).

The quantity $r \sin \theta$ appearing in Eq. (2) has a simple geometric interpretation: it is the perpendicular distance between the line of action of the force and the origin of coordinates (Figure 13.3a); this perpendicular distance is called the **moment arm** of the force. Hence Eq. (2) states that the magnitude of the torque equals the magnitude of the force multiplied by the moment arm.

Alternatively, the quantity $F \sin \theta$ can be given an interpretation: it is the component of the force along a direction perpendicular to the position vector, i.e., it is the transverse component of the force (Figure 13.3b). Hence Eq. (3) states the magnitude of the torque equals the radial distance multiplied by the transverse component of the force.

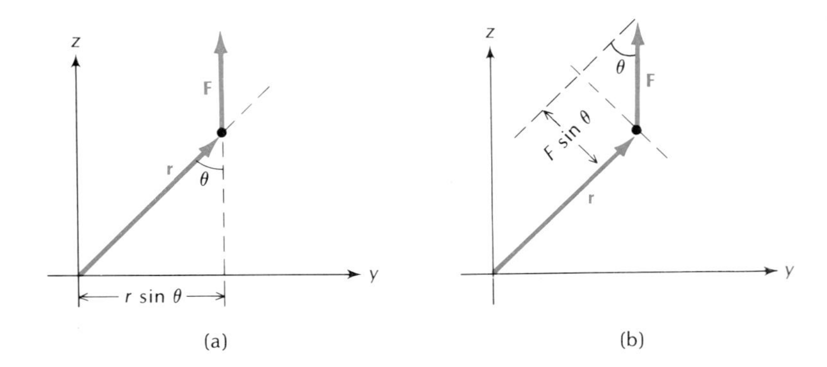


**Fig. 13.3** (a) Here $r \sin \theta$ is the perpendicular distance between the line of action of the force and the origin of coordinates. (b) Here $F \sin \theta$ is the component of **F** perpendicular to **r**.

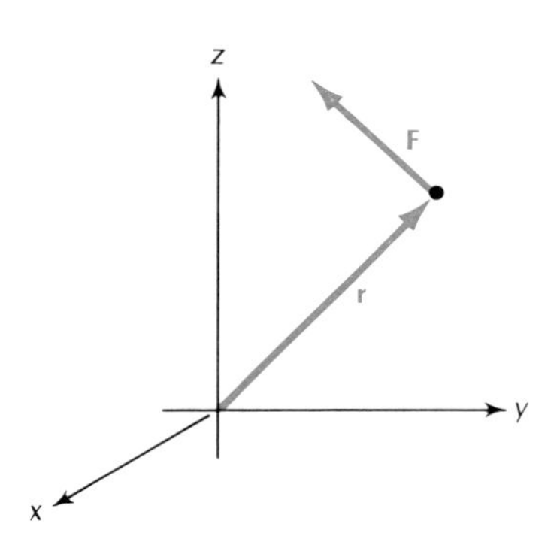


**Fig. 13.4** An example of a force perpendicular to the position vector.

If the force is parallel to the position vector, then $\tau = 0$; if the force is perpendicular to the position vector, then $\tau = rF$ (Figure 13.4). Thus, as a function of the angle $\theta$, a force of given magnitude generates a maximum torque if it is perpendicular to the position vector. If the particle on which the force acts belongs to a rigid body, then we can readily recognize that this dependence of the torque on the angle $\theta$ agrees with our intuition about the rotational effect of a force. For instance, if we want to make a wheel rotate about an origin at the center of the wheel, we must push on some point of the wheel transversely to the radial line; if we push along the radial line, we will merely displace the wheel in the direction of this push, without rotation.

It is important to keep in mind that the torque of a given force depends on the choice of origin; for instance, if we were to change the origin of coordinates of Figure 13.2 so that it sits on the line of action of the force, then the torque would be zero. This is why we call the expression (1) the torque *with respect to the origin* or the torque *about the origin.*

In the metric system the unit of torque is the newton-meter (N · m). This means that the unit of torque is the same as the unit of work. Since 1 N · m = 1 J, we could in principle measure torque in joules; however, in order to maintain at least some distinction between the units of torque and work, we will insist on the exclusive use of newton-meters for the former and joules for the latter.

In the British system the unit of torque is the pound-force-foot (lbf–ft).

EXAMPLE 1. During a sudden stop, the horizontal braking force exerted by the road on the front wheel of a bicycle with a front-wheel brake is 600 N. What is the torque of this force about the center of mass of the bicycle and its rider? The center of mass is 95 cm above the road and 70 cm behind the point of contact of the front wheel with the ground.

SOLUTION: In Figure 13.5 the origin of coordinates is at the center of mass. The distance between the origin and the point of application of the force is

$$r = \sqrt{(0.70\text{ m})^2 + (0.95\text{ m})^2} = 1.18\text{ m}$$

The sine of the angle between the position vector and the force is $\sin\theta = 0.95\text{ m}/1.18\text{ m} = 0.81$. Hence the magnitude of the torque is

$$\tau = rF\sin\theta = 1.18\text{ m} \times 600\text{ N} \times 0.81 = 570\text{ N}\cdot\text{m}$$

To obtain the direction of $\boldsymbol{\tau}$ from the right-hand rule, we must place our fingers along the angle from $\mathbf{r}$ to $\mathbf{F}$ (see Figure 13.5); the direction of $\boldsymbol{\tau}$ is along the thumb, into the plane of the page. It is intuitively obvious that the torque exerted by the braking force will tend to flip the bicycle over its front wheel, that is, this torque tends to produce clockwise rotation. Sudden braking with a front-wheel brake is hazardous!

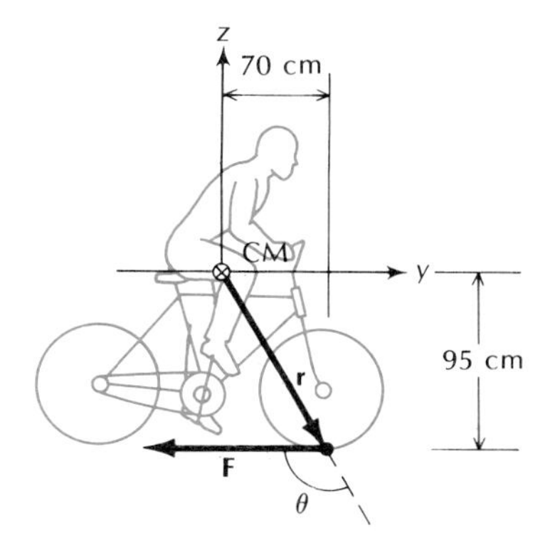

**Fig. 13.5** Bicycle with front-wheel brake. The braking force **F** produces a torque about the center of mass (CM).

## 13.2 The Equation of Rotational Motion

Suppose that a system of particles, such as a rigid body, is subjected to some external forces. The forces acting on the particles change their momenta, and thus change their angular momenta. We want to find the rate of change of the total angular momentum. Let us begin by evaluating the rate of change of the angular momentum of one particle in the body. The angular momentum of one particle is

$$\mathbf{L} = \mathbf{r} \times \mathbf{p} \tag{3}$$

and the derivative of this angular momentum is

$$\frac{d\mathbf{L}}{dt} = \frac{d}{dt}(\mathbf{r} \times \mathbf{p}) \tag{4}$$

We can evaluate the derivative of the vector product on the right side of this equation by the usual rules for differentiating a product, but we must take care not to change the order of the factors, since this would affect the value of the cross product. Hence

$$\frac{d\mathbf{L}}{dt} = \frac{d\mathbf{r}}{dt} \times \mathbf{p} + \mathbf{r} \times \frac{d\mathbf{p}}{dt} \tag{5}$$

The first term on the right side is

$$\frac{d\mathbf{r}}{dt} \times \mathbf{p} = \mathbf{v} \times (m\mathbf{v}) = m(\mathbf{v} \times \mathbf{v})$$

This is zero, since the cross product of a vector with itself is always zero. According to Newton's Second Law, the second term on the right side of Eq. (5) is

$$\mathbf{r} \times \frac{d\mathbf{p}}{dt} = \mathbf{r} \times \mathbf{F}$$

where **F** is the force acting on the particle. Therefore Eq. (5) becomes

$$\frac{d\mathbf{L}}{dt} = \mathbf{r} \times \mathbf{F} \tag{6}$$

For a system of particles, the total angular momentum is the sum of the angular momenta of the individual particles, and the rate of change of the total angular momentum is the sum of the rates of change of the individual angular momenta. Hence, the rate of change of the total angular momentum is a sum of terms such as given in Eq. (6),

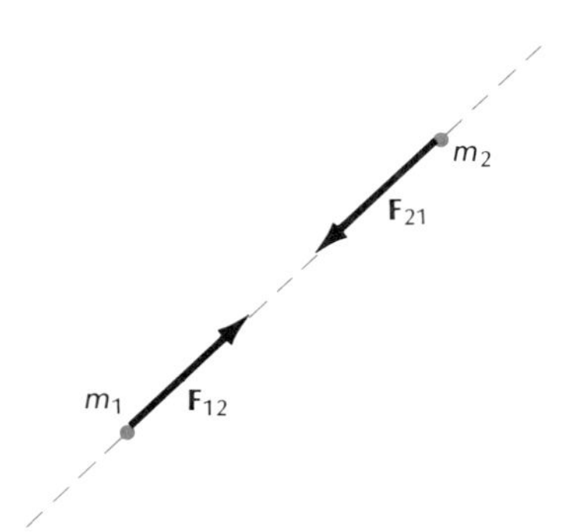


**Fig. 13.6** The forces that two particles exert on one another are of equal magnitudes and of opposite directions. Furthermore, we assume that the forces act along the line joining the particles.

$$\frac{d\mathbf{L}}{dt} = \sum_{i=1}^{n} \mathbf{r}_i \times \mathbf{F}_i \tag{7}$$

where $\mathbf{F}_i$ is the force on particle $i$. According to Newton's Third Law, the mutual forces of the particles of the system occur in action–reaction pairs, so the mutual forces of a pair of particles are equal in magnitude and opposite in direction. Let us now introduce the extra assumption that the two forces in an action–reaction pair are not only equal in magnitude and opposite in direction, but also lie along the line joining the particles (Figure 13.6). Then the corresponding values of $\mathbf{r} \times \mathbf{F}$ are also equal in magnitude and opposite in direction. This can be seen from Figure 13.7, which shows one pair of particles: the magnitudes of the vectors $\mathbf{r}_1 \times \mathbf{F}_{12}$ and $\mathbf{r}_2 \times \mathbf{F}_{21}$ are $F_{12}r_1 \sin\theta_1$ and $F_{21}r_2 \sin\theta_2$, respectively; these magnitudes are equal because the magnitudes of the forces $F_{12}$ and $F_{21}$ are equal, and the factors $r_1 \sin\theta_1$ and $r_2 \sin\theta_2$ are also equal (each of these factors represents the perpendicular distance from the origin to the line joining the particles). The directions of the vectors $\mathbf{r}_1 \times \mathbf{F}_{12}$ and $\mathbf{r}_2 \times \mathbf{F}_{21}$ are opposite because the forces are opposite. Hence the mutual forces between pairs of particles contribute equal and opposite terms to the right side of Eq. (7). In the net sum, all these terms cancel, and only the terms involving external forces remain:

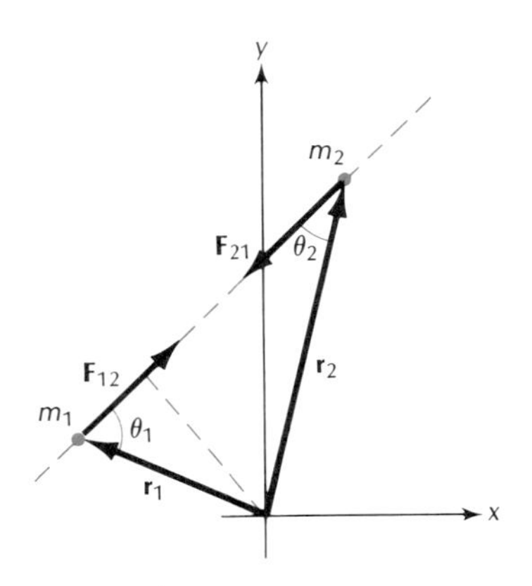


**Fig. 13.7** The perpendicular distance from the origin to the line joining the particles is shown in color.

$$\frac{d\mathbf{L}}{dt} = \sum_{i=1}^{n} \mathbf{r}_i \times \mathbf{F}_{i,\text{ext}} \tag{8}$$

According to the definition given in Section 13.1, the quantity $\mathbf{r}_i \times \mathbf{F}_i$ is the **torque** of the force $\mathbf{F}_i$ on particle $i$. Hence the terms appearing on the right side of Eq. (8) are the torques caused by the external forces, and the net sum of all these terms is the total external torque. If we write this total external torque as $\tau_{\text{ext}}$, then Eq. (8) takes the form

$$\frac{d\mathbf{L}}{dt} = \tau_{\text{ext}} \tag{9}$$

*Rate of change of angular momentum*

This equation for the rate of change of the angular momentum of a system of particles is analogous to the equation $d\mathbf{P}/dt = \mathbf{F}_{\text{ext}}$ for the rate of change of the (translational) momentum of a system.

If the external forces are such that the total external torque is zero, then the angular momentum of the system is conserved,

$$\mathbf{L} = [\text{constant}] \tag{10}$$

*Law of conservation of angular momentum*

This is the law of **conservation of angular momentum.** We will see some applications of this law in Section 13.4.

If the system of particles we are dealing with is a rigid body, then Eq. (9) is the dynamical equation that determines the rotational motion of this rigid body. Since this equation is a vector equation, it has three components:

$$\frac{dL_x}{dt} = \tau_{\text{ext},x} \tag{11}$$

$$\frac{dL_y}{dt} = \tau_{\text{ext},y} \tag{12}$$

$$\frac{dL_z}{dt} = \tau_{\text{ext},z} \tag{13}$$

These three equations are exactly what is necessary and sufficient to determine all three components of the rotational motion of a rigid body; in the terminology introduced in Section 12.1, the three equations given above determine the angular accelerations of the roll, pitch, and yaw motions of the rigid body.

As already mentioned, Eq. (9) for rotational motion is the analog of the equation $d\mathbf{P}/dt = \mathbf{F}_{\text{ext}}$ for translational motion. Together, these equations entirely determine the translational and rotational motion of a rigid body. It is worth emphasizing that our equations for the rotational motion of a rigid body are not new laws, distinct from Newton's laws; our equations are nothing but consequences of Newton's laws as applied to a system of many particles. Incidentally: Our derivation of Eqs. (9)–(13) implies that the validity of these equations is restricted to inertial reference frames, since Newton's laws are restricted to such reference frames.

In most of the following examples, we will be dealing with a rigid body that rotates about a fixed axis. We then need only one of Eqs. (11), (12), and (13). If the axis of rotation is along the $z$ axis, the relevant equation is

$$\frac{dL_z}{dt} = \tau_z$$

Here we have omitted the label *ext* for the sake of brevity; this will cause no confusion, because in the following examples we will never concern ourselves with anything but external torques. According to Eq. (12.57),

$$L_z = I\omega \tag{14}$$

and hence

$$I\frac{d\omega}{dt} = \tau_z \tag{15}$$

or

*Equation of rotational motion*

$$\boxed{I\alpha = \tau_z} \tag{16}$$

This equation relates the angular acceleration about the axis of rotation to the torque about this axis. Obviously, this is the analog of Newton's equation $ma = F$.

---

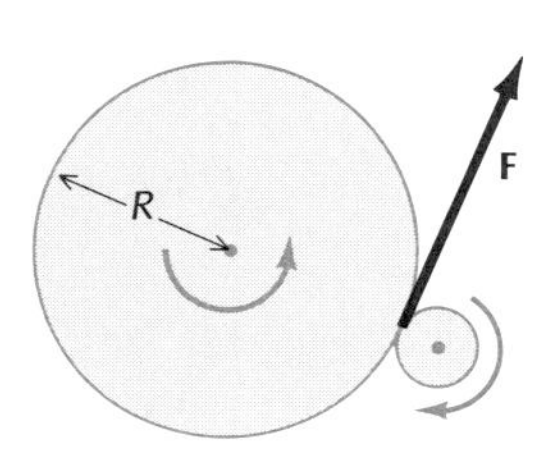


**Fig. 13.8** The driving wheel exerts a force on the turntable.

EXAMPLE 2. A phonograph turntable driven by an electric motor accelerates at a constant rate from 0 to $33\frac{1}{3}$ rev/min in a time of 2.0 s. The turntable is a uniform disk of metal, of a mass 1.5 kg and a radius 13 cm. What torque about the axis is required to drive this turntable? If the driving wheel makes contact with the turntable at its outer rim (Figure 13.8), what is the force that it must exert?

SOLUTION: The final angular velocity is $\omega = 2\pi \times 33.3$ radians/60 s = 3.49 radians/s and the angular acceleration is

$$\alpha = \Delta\omega/\Delta t = (3.49 \text{ radians/s})/2.0 \text{ s} = 1.75 \text{ radians/s}^2$$

The moment of inertia of the turntable is that of a disk,

$$I = \tfrac{1}{2}MR^2 = \tfrac{1}{2} \times 1.5 \text{ kg} \times (0.13 \text{ m})^2$$

$$= 1.27 \times 10^{-2} \text{ kg} \cdot \text{m}^2$$

Hence the required torque is

$$\tau_z = I\alpha = 1.27 \times 10^{-2} \text{ kg} \cdot \text{m}^2 \times 1.75 \text{ radians/s}^2$$

$$= 2.22 \times 10^{-2} \text{ N} \cdot \text{m}$$

The driving force is perpendicular to the radius vector so that $\tau_z = RF$ and

$$F = \tau_z/R = 2.22 \times 10^{-2} \text{ N} \cdot \text{m}/0.15 \text{ m} = 0.147 \text{ N}$$

COMMENTS AND SUGGESTIONS: This calculation of the torque from the equation $I\alpha = \tau_z$ is analogous to the calculation of the force from $ma = F$ in a problem of translational motion. The analogy between rotational and translational equations mentioned in Section 12.3 can be extended as follows:

$$\begin{array}{lcl} ma = F & \rightarrow & I\alpha = \tau_z \\ p = mv & \rightarrow & L_z = I\omega \\ K = \tfrac{1}{2}mv^2 & \rightarrow & K = \tfrac{1}{2}I\omega^2 \end{array}$$

As we will see in the next section, this analogy also encompasses the equations for work and power:

$$W = \int F\,dx \quad \rightarrow \quad W = \int \tau_z\,d\phi$$
$$P = Fv \quad \rightarrow \quad P = \tau_z\omega$$

EXAMPLE 3. Two masses $m_1$ and $m_2$ are suspended from a string which runs, without slipping, over a pulley (Figure 13.9a). The pulley has a radius $R$ and a moment of inertia $I$ about its axle, and rotates without friction. Find the acceleration of the masses.

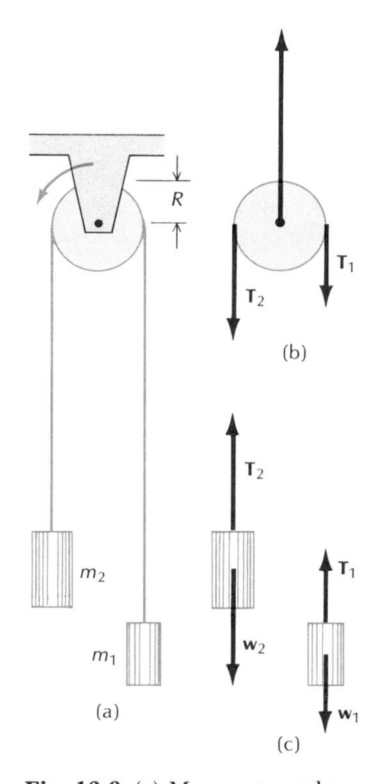


**Fig. 13.9** (a) Masses $m_1$ and $m_2$ hanging from a string passing over a pulley of radius $R$ and moment of inertia $I$. (b) "Free-body" diagram for the pulley. (c) "Free-body" diagram for the masses $m_1$ and $m_2$.

SOLUTION: We have already found the motion of this system in Section 6.3, where we neglected the inertia of the pulley. Now we will take this inertia into account.

Suppose that the tensions in the two parts of the string attached to the masses are $T_1$ and $T_2$. These tensions are shown in Figure 13.9b. (Note that $T_1$ and $T_2$ are not equal. If the moment of inertia of the pulley were zero, then these tensions would be equal. But if the moment of inertia is not zero, then a difference between $T_1$ and $T_2$ is required for the acceleration of the pulley.) The equations of motion of the two masses $m_1$ and $m_2$ are

$$T_1 - m_1 g = m_1 a$$

and

$$T_2 - m_2 g = -m_2 a$$

where the acceleration $a$ is reckoned as positive if the mass $m_1$ moves upward.

Figure 13.9c shows the pulley and the forces acting on it. (Since the string does not slip, it behaves as though it were attached to the pulley at the points of first contact; this is why the tension forces are shown acting at the points of first contact.) The upward supporting force at the axle generates no torque about the axle. The tension forces $T_1$ and $T_2$ generate torques $-RT_1$ and $RT_2$ about the axle, the torque being reckoned as positive if it tends to produce counterclockwise acceleration. The equation of rotational motion of the pulley is then

$$I\alpha = \tau = -RT_1 + RT_2$$

The angular and linear accelerations are related by $\alpha = a/R$. Hence

$$Ia/R = -R(m_1 g + m_1 a) + R(m_2 g - m_2 a)$$

from which

$$a = \frac{m_2 - m_1}{m_1 + m_2 + I/R^2}\,g \qquad (17)$$

COMMENTS AND SUGGESTIONS: If the mass of the pulley is small, then $I/R^2$ can be neglected; with this approximation Eq. (17) does reduce to Eq. (6.15), which was obtained without taking into account the inertia of the pulley.

A device of this kind, called **Atwood's machine,** can be used to determine the value of $g$ (see Figure 6.35). For this purpose it is best to use masses $m_1$ and $m_2$ that are nearly equal. Then $a$ is much smaller than $g$ and easier to measure; the value of $g$ can be calculated from the measured value of $a$ according to Eq. (17).

## 13.3 Work, Energy, and Power in Rotational Motion

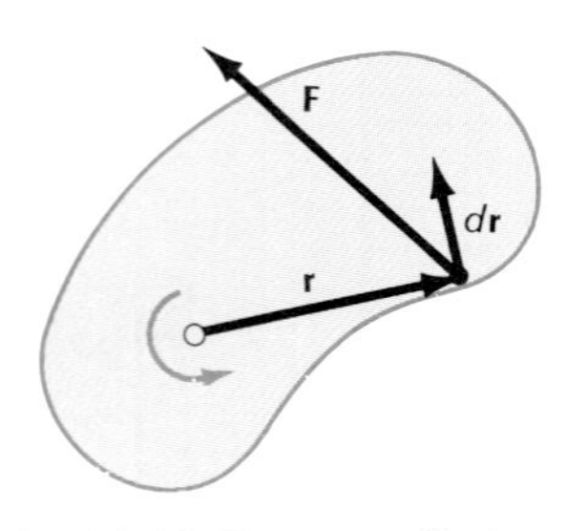


**Fig. 13.10** Force applied to a rigid body rotating about a fixed axis.

We will now calculate the work done by an external force on a rigid body rotating about a fixed axis. Figure 13.10 shows the body as seen from along the axis; the force is applied at some point of the body at a distance $R$ from the axis of rotation. For a start we will assume that the force has no component parallel to the axis; in Figure 13.10 the force is therefore entirely in the plane of the page. The work done by the force is

$$dW = \mathbf{F} \cdot d\mathbf{r} \tag{18}$$

where $d\mathbf{r}$ is the displacement of the particle on which the force acts, i.e., it is the displacement of the point on the rigid body at which the force is applied. The dot product $\mathbf{F} \cdot d\mathbf{r}$ can also be expressed as the magnitude of the displacement multiplied by the component of the force along the direction of the displacement. When the body rotates through an angle $d\phi$, the displacement has a magnitude $|d\mathbf{r}| = R\,d\phi$ and a tangential direction. Hence we can express Eq. (18) in terms of the tangential component of the force,

$$dW = F_{\tan} R\,d\phi \tag{19}$$

According to the definition of torque [see the discussion following Eq. (2)], the quantity $F_{\tan}R$ is the torque of the force about the axis of rotation. If, as always, we take the $z$ axis along the axis of rotation, then $F_{\tan}R = \tau_z$ and

$$dW = \tau_z\,d\phi \tag{20}$$

This is the rotational analog of our familiar equation $dW = F_x\,dx$ for translational motion. The work done during a finite angular displacement is obtained by integrating Eq. (20):

*Work done by torque*

$$\boxed{W = \int \tau_z\,d\phi} \tag{21}$$

Although Eqs. (20) and (21) were obtained under the assumption that the force has no component parallel to the axis of rotation, this restriction is actually superfluous: any component of the force along the direction of the axis neither does work (the displacement in this direction is zero) nor generates any torque about the axis; consequently, such a component can be ignored in the above calculation, and Eqs. (20) and (21) are of general validity.

The work done by the torque changes the rotational kinetic energy of the body. If the initial and final angular velocities are $\omega_1$ and $\omega_2$, respectively, then the change in the kinetic energy of the body is

$$W = \tfrac{1}{2}I\omega_2^2 - \tfrac{1}{2}I\omega_1^2 \tag{22}$$

This is the work–energy theorem for rotational motion. The proof of Eq. (22) is entirely analogous to the proof of the corresponding equation for translational motion [Eq. (7.42)].

If the force acting on the body is conservative — such as the force of gravity or the force of a spring — then the work equals the negative of the change of the potential energy, and Eq. (22) becomes

$$-U_2 + U_1 = \tfrac{1}{2}I\omega_2^2 - \tfrac{1}{2}I\omega_1^2$$

or

$$\tfrac{1}{2}I\omega_1^2 + U_1 = \tfrac{1}{2}I\omega_2^2 + U_2$$

This equation expresses the **conservation of mechanical energy in rotational motion,**

$$\boxed{E = \tfrac{1}{2}I\omega^2 + U = [\text{constant}]} \qquad (23)$$

*Conservation of mechanical energy*

EXAMPLE 4. A meter stick is initially standing vertically on the floor. If the meter stick falls over, with what angular velocity will it hit the floor? Assume that the end in contact with the floor does not slip.

SOLUTION: The motion of the meter stick is rotation about a fixed axis passing through the point of contact with the floor (Figure 13.11). The stick is a uniform rod of mass $M$ and length $l = 1\text{m}$. Its moment of inertia about the point of contact is $Ml^2/3$ (see Table 12.1) and the rotational kinetic energy is therefore $Ml^2\omega^2/6$. The gravitational potential energy is $Mgz_{CM}$, where $z_{CM}$ is the height of the center of mass. When the meter stick is standing vertically, $z_{CM} = \frac{1}{2}l$, and the energy is

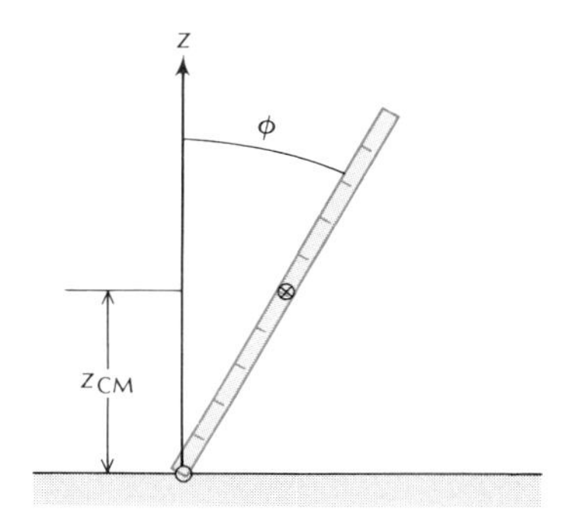


**Fig. 13.11** Meter stick rotating about its lower end.

$$E = \tfrac{1}{6}Ml^2\omega_1^2 + Mgz_{CM} = 0 + Mgl/2$$

When the meter stick hits the floor, $z_{CM} = 0$, and

$$E = \tfrac{1}{6}Ml^2\omega_2^2 + Mgz_{CM} = \tfrac{1}{6}Ml^2\omega_2^2 + 0$$

Conservation of energy therefore implies

$$\tfrac{1}{6}Ml^2\omega_2^2 = Mgl/2$$

from which

$$\omega_2^2 = 3g/l$$

or

$$\omega_2 = \sqrt{3g/l} = \sqrt{(3 \times 9.8\ \text{m/s}^2)/1.0\ \text{m}} = 5.4\ \text{radians/s}$$

From Eq. (20) we find that the **power** delivered by a torque is

$$P = \frac{dW}{dt} = \tau_z \frac{d\phi}{dt}$$

or

$$\boxed{P = \tau_z \omega} \qquad (24)$$

*Power delivered by torque*

Thus, in rotational motion the power is torque times angular velocity, just as in translational motion it is force times velocity.

EXAMPLE 5. Consider the motion of the turntable of Example 2. Calculate the work done by the torque during the acceleration. Calculate the average power.

SOLUTION: The total angular displacement during the 2.0-s interval is

$$\Delta\phi = \tfrac{1}{2}\alpha t^2 = \tfrac{1}{2} \times (1.75 \text{ radians/s}^2) \times (2.0 \text{ s})^2$$

$$= 3.49 \text{ radians}$$

Since the torque is constant, the work is

$$W = \tau_z \, \Delta\phi = 2.22 \times 10^{-2} \text{ N} \cdot \text{m} \times 3.49 \text{ radians}$$

$$= 7.72 \times 10^{-2} \text{ J}$$

The average power is

$$\overline{P} = W/t = 7.72 \times 10^{-2} \text{ J}/2.0 \text{ s} = 3.86 \times 10^{-2} \text{ W}$$

COMMENTS AND SUGGESTIONS: Alternatively, we can find the work done from the change in the rotational kinetic energy, via the work–energy theorem. The initial kinetic energy is zero and the final kinetic energy is $\frac{1}{2}I\omega^2$ so

$$W = \tfrac{1}{2}I\omega^2 - 0$$

$$= \tfrac{1}{2} \times 1.27 \times 10^{-2} \text{ kg} \cdot \text{m}^2 \times (3.49 \text{ radians/s})^2$$

$$= 7.72 \times 10^{-2} \text{ J}$$

EXAMPLE 6. A Prony **dynamometer,** used for the measurement of the power output of an engine, consists of two equal beams clamped loosely around a flywheel mounted to the shaft of the engine (Figure 13.12). The frictional drag between the spinning flywheel and the clamp tends to turn the beams counterclockwise; a weight attached to the right beam tends to turn them clockwise. The weight must be adjusted so that the opposing torques of weight and friction are in balance and the beams remain in a static, horizontal position. In a test of an automobile engine driving a flywheel of 0.305-m radius at 2500 rev/min, a weight of 33.1 kg balances the beams when hung at a distance of 0.910 m from the axis of the flywheel. What is the power output of the engine?

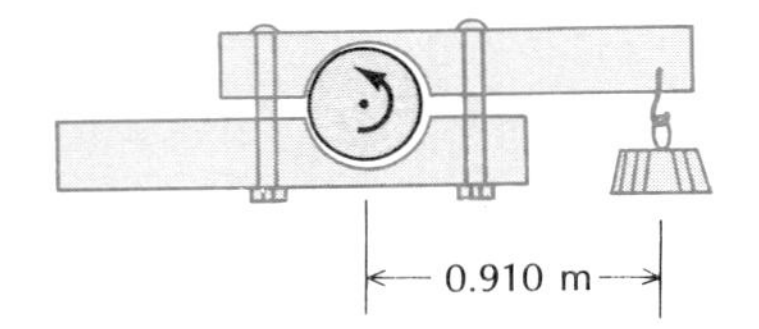


**Fig. 13.12** Prony dynamometer.

SOLUTION: The torque of the weight $mg$ acting at a distance $l$ from the axis of the rotation is $mgl$. In equilibrium this torque on the beams must have the same magnitude as the frictional torque of the flywheel on the beams. By Newton's Third Law, the frictional torque of the beams *on the flywheel* must then also have the same magnitude. The power of this latter frictional torque is then

$$P = \tau_z\omega = mgl\omega$$

where $\omega$ is the angular velocity of the flywheel. The mechanical power removed by friction from the flywheel is equal to the mechanical power delivered by the engine to the flywheel. Hence the above equation gives the power output of the engine:

$$P = mgl\omega$$

$$= 33.1 \text{ kg} \times 9.81 \text{ m/s}^2 \times 0.910 \text{ m} \times \left(2500 \frac{\text{rev}}{\text{min}} \times \frac{2\pi \text{ radians}}{1 \text{ rev}} \times \frac{1 \text{ min}}{60 \text{ s}}\right)$$

$$= 7.74 \times 10^4 \text{ W} = 104 \text{ hp}$$

This quantity is usually called the **brake horsepower** of the engine.

## 13.4 The Conservation of Angular Momentum

If the external torque on a system of particles is zero, then Eq. (10) asserts that the total angular momentum of the system is conserved,

$$\mathbf{L} = [\text{constant}] \tag{25}$$

This conservation law can be used in much the same way as the conservation of total (translational) momentum to calculate those features of the motion of the system that do not depend on the details of the internal forces.

A gyroscope, consisting of a flywheel mounted in gimbals (Figure 13.13), provides us with a nice illustration of the conservation law. The angular momentum of the flywheel lies along its axis of rotation. (As we saw in Section 12.6, the angular momentum of a symmetric body rotating about its axis of symmetry always lies along this axis.) Since there are no torques on the flywheel, except for the very small and negligible frictional torques in the pivots of the gimbals, the angular momentum remains constant both in magnitude and direction. Hence the angular velocity remains constant and the orientation of the axis remains fixed in space — the gyroscope can be carried about, its base twisted and turned in any way, and yet the axis always continues to point in its original direction. High-precision gyroscopes are used in inertial guidance systems for ships, aircraft, missiles, and spacecraft (Figure 13.14); they provide an absolute reference direction relative to which the direction of travel of the vehicle can be reckoned. In such applications, three gyroscopes aimed along mutually perpendicular axes establish the orientation of an absolute $x$, $y$, $z$ coordinate grid.

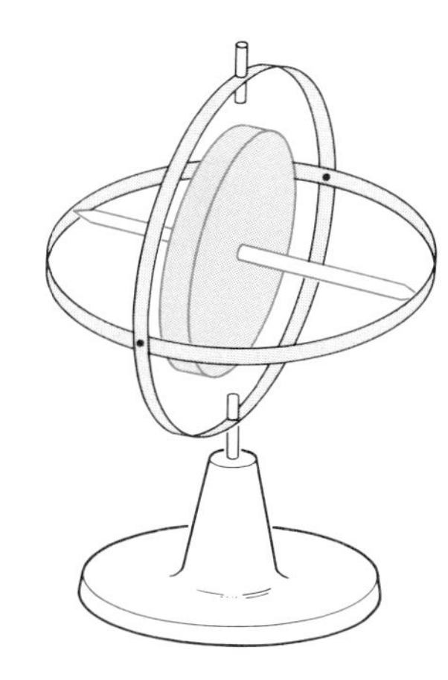

**Fig. 13.13** Gyroscope in gimbals.

**Fig. 13.14** A Sperry C-12 directional gyroscope, such as currently used aboard many commercial aircraft.

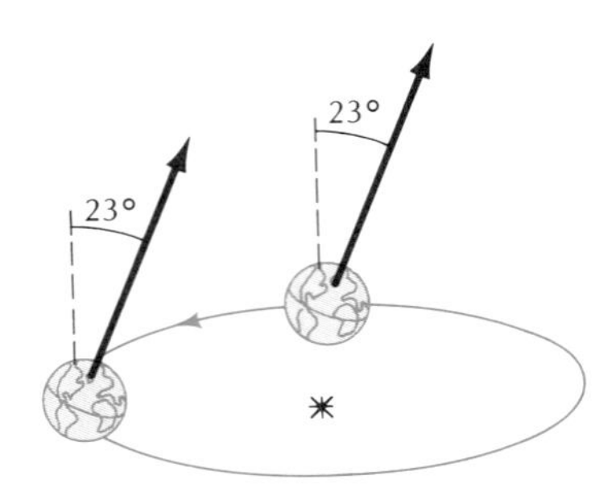


**Fig. 13.15** The axis of rotation of the Earth maintains a fixed direction as the Earth moves around the Sun.

The rotation of the Earth about its axis also illustrates the conservation of angular momentum. The axis of rotation makes an angle of 23.5° with the plane of the Earth's orbit around the Sun (Figure 13.15). There is no external torque on the Earth, and its angular momentum remains constant.[1] Hence, as the Earth moves along its orbit, the orientation of the axis of rotation remains fixed; the axis always remains parallel to itself.

For a rigid body rotating about a fixed axis (the $z$ axis) with $\tau_z = 0$, the law of **conservation of angular momentum** reduces to $L_z =$ [constant], or

$$\boxed{I\omega = [\text{constant}]} \tag{26}$$

The following example involves the application of this conservation law.

---

EXAMPLE 7. Suppose that the phonograph turntable described in Example 2 is coasting (with the motor disengaged) at $33\frac{1}{3}$ rev/min when a stack of 10 phonograph records suddenly drops down on it. What is the angular velocity after the drop? What is the kinetic energy before and after the drop? Each record has a mass of 0.17 kg and a radius of 15.2 cm.

SOLUTION: Since the records drop suddenly, the external (frictional) torque can be ignored and the angular momentum of the complete system (turntable plus records) is conserved. The angular momentum before the drop is $I\omega$ (where $\omega$ is the initial angular velocity and $I$ moment of inertia of the turntable), and the angular momentum after the drop is $I'\omega'$ (where $\omega'$ is the final angular velocity and $I'$ the moment of inertia of turntable and records combined). Hence,

$$I\omega = I'\omega'$$

and

$$\omega' = \frac{I}{I'}\omega \tag{27}$$

The moment of inertia of the turntable is (see Example 2)

$$I = 1.27 \times 10^{-2}\ \text{kg}\cdot\text{m}^2$$

and the moment of inertia of the 10 records is

$$I_r = 10 \times \tfrac{1}{2}M_r R_r^2 = 10 \times \tfrac{1}{2} \times 0.17\ \text{kg} \times (0.152\ \text{m})^2$$

$$= 1.96 \times 10^{-2}\ \text{kg}\cdot\text{m}^2$$

Accordingly,

$$\omega' = \frac{I}{I'}\omega = \frac{I}{I + I_r}\omega$$

[1] Actually, both the Moon and the Sun exert some small torque on the Earth by their gravitational attraction on the equatorial bulge of the Earth. This torque gradually changes the direction of the angular momentum of the Earth and the orientation of the Earth. This is what astronomers call the **precession of the equinoxes.**

$$= \frac{1.27 \times 10^{-2}}{1.27 \times 10^{-2} + 1.96 \times 10^{-2}} \times 3.49 \text{ radians/s}$$

$$= 1.37 \text{ radians/s}$$

The initial kinetic energy is simply that of the turntable,

$$K = \tfrac{1}{2}I\omega^2$$

$$= \tfrac{1}{2} \times 1.27 \times 10^{-2} \text{ kg} \cdot \text{m}^2 \times (3.49 \text{ radians/s})^2$$

$$= 7.73 \times 10^{-2} \text{ J}$$

The final kinetic energy is that of turntable and records together,

$$K' = \tfrac{1}{2}I'\omega'^2$$

$$= \tfrac{1}{2} \times 3.23 \times 10^{-2} \text{ kg} \cdot \text{m}^2 \times (1.37 \text{ radians/s})^2$$

$$= 3.03 \times 10^{-2} \text{ J}$$

COMMENTS AND SUGGESTIONS: Note that some kinetic energy is lost. When the records drop on the turntable, they slide for a few instants until friction within the system brings their speed up to that of the turntable; during this process friction dissipates some of the kinetic energy of the turntable. The impact of the records on the turntable is the rotational analog of a totally inelastic collision.

EXAMPLE 8. Water flowing along an open channel drives an undershot waterwheel of radius 2.2 m (Figure 13.16). The water approaches the wheel with a speed of 5.0 m/s and leaves with a speed of 2.5 m/s; the amount of water passing by is 300 kg per second. At what rate does the water deliver angular momentum to the wheel? What is the torque that the water exerts on the wheel? If the speed of the rim of the wheel is 3 m/s, what is the power delivered to the wheel?

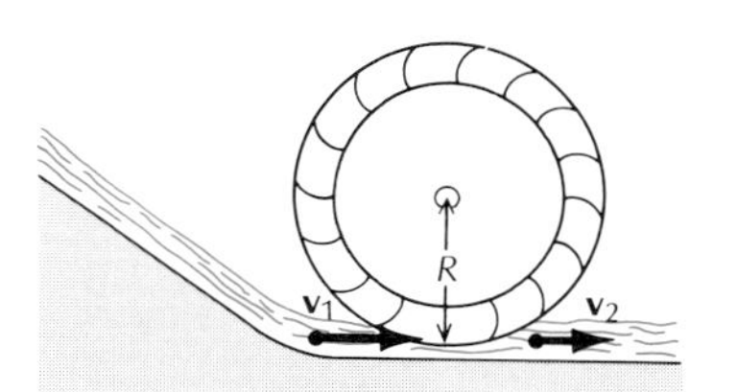


**Fig. 13.16** Undershot waterwheel.

SOLUTION: Consider a small mass $dm$ of water which approaches the wheel with a speed $v_1$ and leaves with a speed $v_2$. The angular momentum of this mass, reckoned about an origin at the center of the wheel at a radial distance $R$, decreases from $(dm)v_1R$ to $(dm)v_2R$ as it passes through the wheel. Hence the angular momentum lost by the water is $(dm)(v_1 - v_2)R$ and, by conservation, this must be the angular momentum gained by the wheel (we ignore friction of the water against the channel). The rate at which the wheel acquires angular momentum from the water is then

$$\frac{dL}{dt} = \frac{dm}{dt}(v_1 - v_2)R \tag{28}$$

where $dm/dt$ is the rate at which water flows through the channel (in kilograms of water per second). Inserting the relevant numbers, we obtain

$$\frac{dL}{dt} = 300 \text{ kg/s} \times (5.0 \text{ m/s} - 2.5 \text{ m/s}) \times 2.2 \text{ m}$$

$$= 1.7 \times 10^3 \text{ kg} \cdot \text{m}^2/\text{s}^2 = 1.7 \times 10^3 \text{ J}$$

The torque exerted by the water on the wheel equals this rate of transfer of angular momentum,

$$\tau_z = dL/dt = 1.7 \times 10^3 \text{ N} \cdot \text{m}$$

(This assumes that friction against the channel is negligible.)

The angular velocity of the wheel is $\omega = (3 \text{ m/s})/2.2 \text{ m} = 1.36$ radians/s and the power is therefore

$$P = \tau_z \omega = 1.7 \times 10^3 \text{ N} \cdot \text{m} \times 1.36 \text{ radians/s}$$

$$= 2.3 \times 10^3 \text{ J/s} = 2.3 \text{ kW}$$

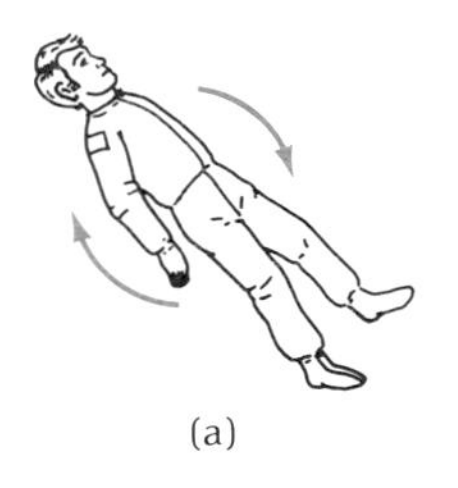

(a)

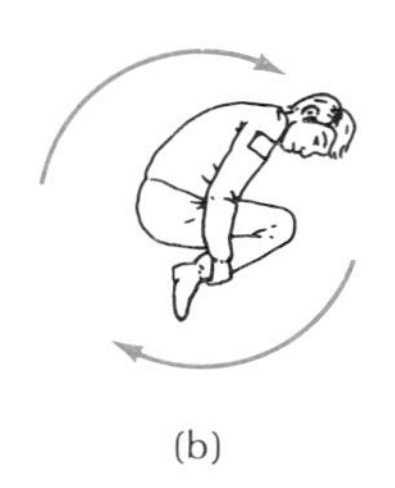

(b)

**Fig. 13.17** Astronaut in Skylab.

EXAMPLE 9. In an experiment first performed during the Skylab mission, an astronaut floating freely in weightless conditions in his orbiting spacecraft demonstrates how he can change his rotational velocity by changing his moment of inertia. Initially the astronaut holds his body erect while rotating about his center of mass at a rate of 0.17 rev/s (Figure 13.17a). He then contracts his body to a fetal position (Figure 13.17b). What is his new rate of rotation? What is the change in rotational kinetic energy? Assume that the moment of inertia is 18 kg · m² in the erect position and 5.5 kg · m² in the fetal position.

SOLUTION: The initial angular momentum is $I\omega$ and the final angular momentum is $I'\omega'$. Since there are no external torques, the angular momentum is conserved:

$$I\omega = I'\omega'$$

With $\nu = \omega/2\pi$, this yields

$$\nu' = \frac{I}{I'}\nu$$

$$= \frac{18 \text{ kg} \cdot \text{m}^2}{5.5 \text{ kg} \cdot \text{m}^2} \times 0.17 \text{ rev/s} = 0.56 \text{ rev/s}$$

The change in kinetic energy is

$$\tfrac{1}{2}I'\omega'^2 - \tfrac{1}{2}I\omega^2 = \tfrac{1}{2}I'(2\pi\nu')^2 - \tfrac{1}{2}I(2\pi\nu)^2$$

$$= \tfrac{1}{2} \times 5.5 \text{ kg} \cdot \text{m}^2 \times (2\pi \times 0.56/\text{s})^2$$

$$-\tfrac{1}{2} \times 18 \text{ kg} \cdot \text{m}^2 \times (2\pi \times 0.17/\text{s})^2$$

$$= 33.6 \text{ J} - 10.3 \text{ J} = 23.3 \text{ J}$$

The increase of kinetic energy comes from the work the astronaut does while contracting his body.

COMMENTS AND SUGGESTIONS: Divers in free fall after jumping off a diving board and figure skaters pirouetting on (nearly) frictionless ice also use this method of changing their rotational velocity by changing their moment of inertia.

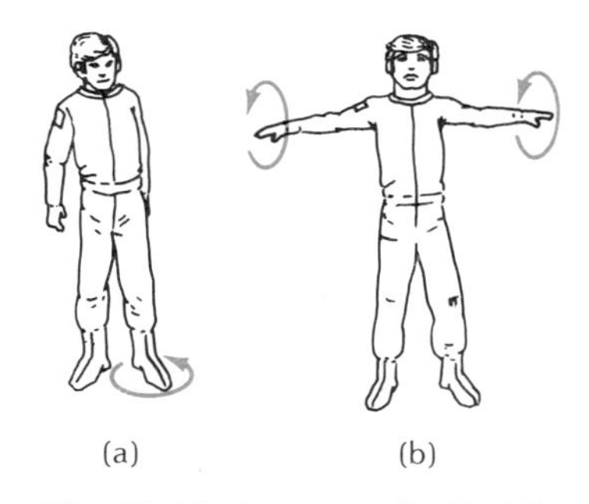


**Fig. 13.18** Astronaut in Skylab.

EXAMPLE 10. In another experiment the astronaut demonstrates how he can change the orientation of his body witout having recourse to external forces. How can the astronaut turn his body about-face? How can the astronaut turn his body upside down?

SOLUTION: To set his body rotating about a longitudinal axis, the astronaut swings one of his feet (or both) in a circle (Figure 13.18a). His foot then has an angular momentum and, by conservation, the rest of his body must have an opposite angular momentum, that is, it must turn in the opposite direction. This turning motion continues as long as, and only as long as, the astronaut keeps his foot swinging.

To set his body rotating about a transverse axis, the astronaut swings his arms in a lateral circle (Figure 13.18b).

COMMENTS AND SUGGESTIONS: These experiments explain how a cat always manages to land on its feet. When dropped toward the floor from an upside-down position, the cat turns its body by means of very quick swinging motions of his hind legs and tail.

## 13.5 Rolling Motion

In our study of the rotational motion of a rigid body in the preceding sections, we dealt only with rotation about a fixed axis. This restriction made it easy to convert the equation for the rate of change of angular momentum ($d\mathbf{L}/dt = \tau$) into an equation for the rate of change of the angular velocity ($I\ d\omega/dt = \tau_z$), from which the motion can be calculated. Of course, we can relax this restriction somewhat and deal with rotation about an axis that has a uniform translational motion; such a motion of the axis makes no essential difference, since it can be eliminated by a simple change of reference frame.

Now we will deal with a somewhat more complicated case of combined translational and rotational motion which is of considerable practical importance: the motion of a body rolling on some surface without slipping. A typical example is the motion of an automobile wheel rolling along the street. The motion can be described as a rotation of the wheel about an axis through its center plus a translation of the axis along the street (Figure 13.19a). If there is no slipping, then these two motions are coupled: the translational speed $v$ and the rotational velocity $\omega$ are related by

$$v = \omega R \tag{29}$$

where $R$ is the radius of the wheel.

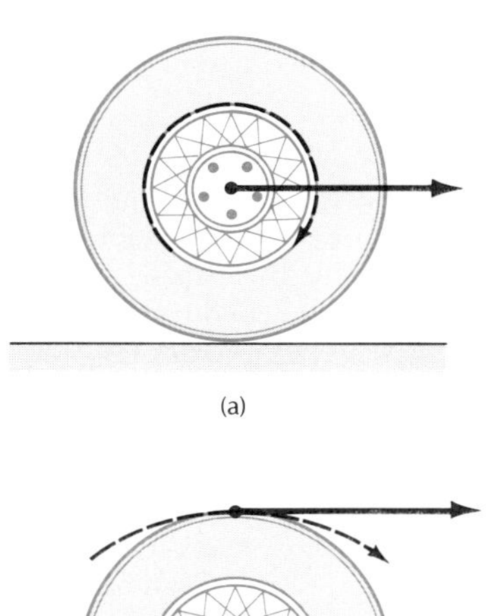


**Fig. 13.19** The motion of a rolling wheel can be regarded either as (a) rotation about an axis through the center plus a translation of this center or else as (b) rotation about an axis through the instantaneous point of contact.

Alternatively, the motion of the wheel can be described as pure rotation about an instantaneous axis that passes through the instantaneous point of contact (Figure 13.19b). As we have seen in Section 12.1, there is always some freedom in the choice of the axis of rotation — if the wheel rotates about an axis through its center, it also rotates about any parallel axis through any other point; furthermore, the angular velocities about all these alternative axes are the same. Of all these alternative axes, the one passing through the point of contact is the most interesting: if there is no slipping, then the point of contact of the wheel with the street is instantaneously at rest on the street; hence the axis passing through this point is an **instantaneously fixed axis.**

*Instantaneously fixed axis for rolling motion*

The instantaneous velocities of the points shown in Figure 13.19b, as well as the instantaneous velocities of any other points on the wheel, can be obtained by means of the usual addition procedure for velocities. For example, the uppermost point of the wheel has a forward velocity $\omega R$ relative to the center, which itself has a velocity $v$ relative to the street; hence the net forward velocity of the uppermost point is

$$v + \omega R = v + v = 2v \tag{30}$$

Likewise, the net forward velocity of the lowermost point of the wheel is

$$v - \omega R = v - v = 0 \tag{31}$$

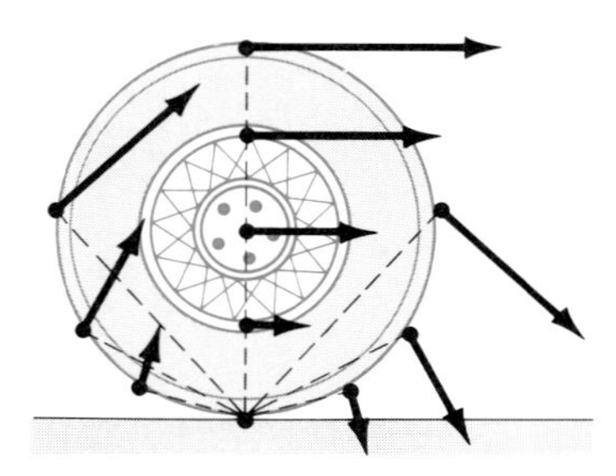

**Fig. 13.20** Instantaneous velocity vectors of points on a rolling wheel.

This, of course, means that the lowermost point is at rest, just as required by the condition of no slipping. Figure 13.20 shows the instantaneous velocities of selected points on the wheel. The magnitudes of the velocities increase in direct proportion to their distance from the point of contact; such an increase is a characteristic feature of rotation about a point.

Figure 13.21 is a time-exposure photograph showing the path traced out by a small light bulb attached to the rim of a rolling wheel. Note that whenever the light bulb touches the ground, the horizontal component of the motion vanishes; furthermore, the vertical component of the motion reverses direction, that is, this component also vanishes instantaneously. The curve traced out in Figure 13.21 is called a **cycloid.**

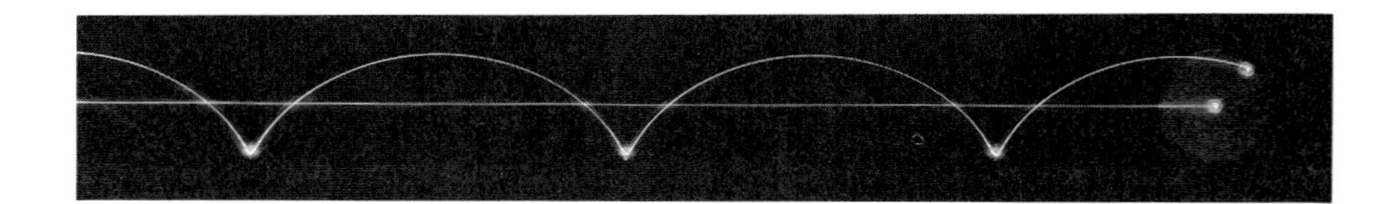

**Fig. 13.21** Path of a point on the rim of a rolling wheel, made visible by a small light-bulb attached to the rim.

The existence of an instantaneous fixed axis in rolling motion (without slipping) enables us to deal with this motion by the methods developed in the preceding sections. The rotation about the instantaneous fixed axis obeys our old equation $I\, d\omega/dt = \tau_z$. From this, we can calculate the motion of a rolling body on which external forces and torques act.

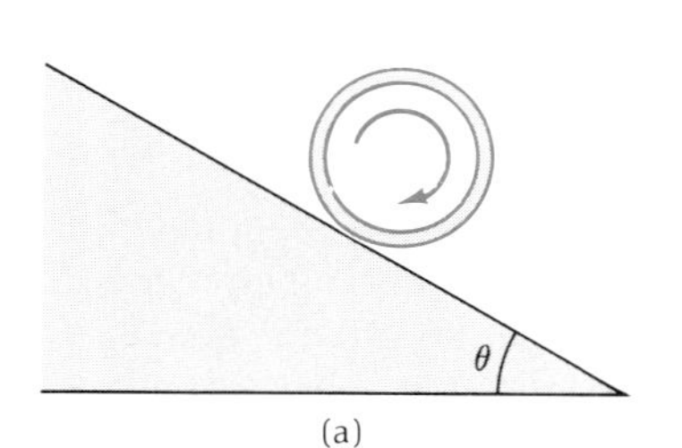


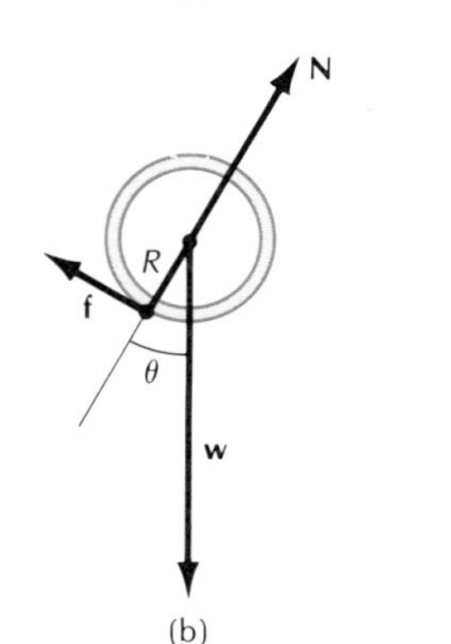


**Fig. 13.22** (a) A pipe rolling down a ramp. (b) "Free-body" diagram for the pipe.

EXAMPLE 11. A piece of steel pipe of mass 360 kg rolls down a ramp inclined at 30° to the horizontal. What is the acceleration if the pipe rolls without slipping? What is the magnitude of the friction force that acts at the point of contact between the pipe and the ramp?

SOLUTION: Figure 13.22a shows the pipe on the ramp; Figure 13.22b shows the forces in a "free-body" diagram. The only force that generates a torque about the instantaneous axis through the point of contact is the weight. The weight $Mg$ effectively acts at the center of mass of the pipe,[2] at a distance $R$ from the point of contact; since the angle between the weight and the radial line is $\theta$, the torque of the weight about the point of contact is

$$\tau_z = MgR \sin\theta$$

The moment of inertia of the pipe about the center of mass is $I_{CM} = MR^2$ and, by the parallel-axis theorem, the moment of inertia about the point of contact is $I = I_{CM} + MR^2 = MR^2 + MR^2 = 2MR^2$. Consequently the equation of motion

$$I\alpha = \tau_z$$

becomes

$$2MR^2\alpha = MgR \sin\theta$$

or

$$\alpha = \frac{1}{2R} g \sin\theta$$

[2] We will prove this in the next section.

The linear acceleration down the ramp is

$$a = \alpha R = \tfrac{1}{2} g \sin \theta \tag{32}$$

$$= \tfrac{1}{2} \times 9.81 \text{ m/s}^2 \times \sin 30° = 2.45 \text{ m/s}^2$$

Note that Eq. (32) shows that the acceleration of the rolling pipe is exactly half as large as that of a particle sliding down a frictionless ramp of the same angle.

The friction force $f$ can be obtained from the equation for the translational motion along the ramp:

$$Ma = Mg \sin \theta - f \tag{33}$$

from which we have

$$f = M(g \sin \theta - a)$$

$$= M(g \sin \theta - \tfrac{1}{2} g \sin \theta) = \tfrac{1}{2} Mg \sin \theta \tag{34}$$

$$= \tfrac{1}{2} \times 360 \text{ kg} \times 9.81 \text{ m/s}^2 \times \sin 30° = 883 \text{ N}$$

---

EXAMPLE 12. Suppose that the pipe of the preceding example starts from rest and rolls a distance of 3.00 m along the ramp. What is the total kinetic energy at this instant? What is the translational kinetic energy? What is the kinetic energy of rotation about the center of mass?

SOLUTION: The total kinetic energy is simply the rotational kinetic energy about the instantaneous axis through the point of contact,

$$K = \tfrac{1}{2} I \omega^2 = \tfrac{1}{2}(2MR^2)(v/R)^2 = Mv^2$$

In terms of the distance and acceleration, the change of velocity is given by

$$v^2 - v_0^2 = 2ax$$

or

$$v = \sqrt{2ax} = \sqrt{2 \times 2.45 \text{ m/s}^2 \times 3.00 \text{ m}} = 3.84 \text{ m/s}$$

so that

$$K = Mv^2 = 360 \text{ kg} \times (3.84 \text{ m/s})^2$$

$$= 5.30 \times 10^3 \text{ J}$$

The translational kinetic energy is

$$\tfrac{1}{2} M v_{\text{CM}}^2 = \tfrac{1}{2} M v^2 = 2.65 \times 10^3 \text{ J}$$

The rotational kinetic energy about the center of mass is

$$K_{\text{int}} = \tfrac{1}{2} I_{\text{CM}} \omega^2 = \tfrac{1}{2}(MR^2)(v/R)^2 = \tfrac{1}{2} M v^2$$

$$= 2.65 \times 10^3 \text{ J}$$

COMMENTS AND SUGGESTIONS: Note that the translational and rotational kinetic energies add up to the total kinetic energy, as they should according to Eq. (12.35). Note, furthermore, that the total kinetic energy of $5.3 \times 10^3$ J is exactly equal to the change in gravitational potential energy,

$$Mgh = Mg \times 3.00 \text{ m} \times \sin 30°$$

$$= 360 \text{ kg} \times 9.81 \text{ m/s}^2 \times 3.00 \text{ m} \times \sin 30°$$

$$= 5.30 \times 10^3 \text{ J}$$

This indicates that although the force of friction has a magnitude of 883 N, it *does no work.* The reason why friction does no work in the rolling motion is that the point of the wheel at which the friction acts is the point of contact — and that point is always instantaneously at rest, that is, it has no displacement in the direction of the friction force.

---

*Rate of change of angular momentum relative to center of mass*

The solution of the problem of the rolling pipe in Example 11 relied on a very special trick: we reckoned the angular momentum and the torque about an instantaneous fixed axis. But this trick only applies to rolling without slipping. The general translational and rotational motion of a rigid body subjected to arbitrary forces has neither an axis that is fixed nor an axis that is in uniform translational motion. For example, the axis of rotation of the hammer shown in Figure 12.1 has an *accelerated* translational motion. Since our derivation of the equation $d\mathbf{L}/dt = \boldsymbol{\tau}$ in Section 13.2 assumed an inertial reference frame, the equation is not necessarily valid for rotational motion about an axis that has an accelerated translational motion. To deal with this general motion, we can begin by calculating the acceleration of the center of mass according to $M\mathbf{a}_{CM} = \mathbf{F}_{ext}$. Then we can calculate the rotational motion about an axis through the center of mass according to the following theorem: *If both the angular momentum* $\mathbf{L}$ *and the torque* $\boldsymbol{\tau}$ *are reckoned with respect to the center of mass, then the equation* $d\mathbf{L}/dt = \boldsymbol{\tau}$ *remains valid even when the center of mass has some arbitrary acceleration.* We will not attempt the proof of this theorem but only remark that the motion of the thrown hammer shown in Figure 12.1 agrees with the consequences of the theorem: since gravity exerts no torque about the center of mass of the hammer,[3] the angular momentum about the center of mass ought to remain constant; inspection of Figure 12.1 indicates that the angular velocity of the hammer, and consequently also its angular momentum about the center of mass, does indeed remain constant.

If we apply the theorem quoted in the preceding paragraph to the special case of rolling motion, the final results will be in agreement with those obtained by our method of the instantaneous fixed axis. The following example illustrates this agreement.

---

EXAMPLE 13. Recalculate the acceleration of the pipe described in Example 11 by starting with the torque about the center of mass.

SOLUTION: The torque about the center of mass of the pipe is entirely due to the friction force. From Figure 13.22b we see that this torque is

$$\tau_{z,CM} = Rf$$

The moment of inertia of the pipe about the center of mass is $I_{CM} = MR^2$. By the theorem quoted above, the torque about the center of mass equals the rate of change of the angular momentum about the center of mass, that is,

$$I_{CM}\alpha = \tau_{z,CM}$$

[3] We will prove this in the next section.

or

$$MR^2\alpha = Rf$$

With $\alpha = a/R$, this becomes

$$Ma = f \tag{35}$$

The equation for the translational motion is of course still Eq. (33),

$$Ma = Mg \sin \theta - f \tag{36}$$

Equations (35) and (36) are a simultaneous system of equations for the unknowns $a$ and $f$. The solution of these equations gives

$$f = \tfrac{1}{2}Mg \sin \theta$$

and

$$a = \tfrac{1}{2} \sin \theta$$

These results agree with the results in Eqs. (32) and (34) that we obtained in our previous calculation.

COMMENTS AND SUGGESTIONS: Whether we formulate the equation of motion for a rolling body by taking the torque about the point of contact (Example 11) or about the center of mass (Example 13) is a question of convenience. The advantage of using the torque about the point of contact is that this immediately yields the acceleration of the rotational motion [see Eq. (32)], without any need to examine the equation of translational motion. The disadvantage is that this method can be used only for rolling without slipping, whereas the other method can be used even when there is slipping.

## 13.6* Precession of a Gyroscope

In Section 13.2 we asserted that Eq. (9) is sufficient to determine the rotational motion of a rigid body. Let us apply this equation to a gyroscope or symmetric top which consists of a symmetric flywheel mounted on an axle. Figure 13.23 shows such a flywheel within a rigid framework which helps in the manipulation of the flywheel; the framework is nearly massless and plays no role in the following calculations. The flywheel spins at high speed in nearly frictionless bearings and is supported at its left end by a pivot. For a symmetric rigid body, such as the flywheel, the angular momentum is along the axis of rotation and has the direction shown in the figure.

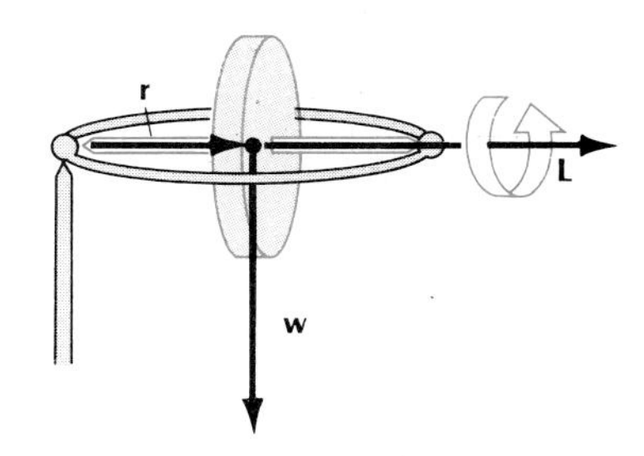


**Fig. 13.23** Spinning gyroscope supported at one end.

Suppose now that after placing the left end of the axis on the pivot, we release the flywheel. Intuitively we expect that the force of gravity pulling on the center of mass will cause the right end of the axis to swing downward. But in this our intuition misleads us — what actually happens is quite different. To find out, we must use Eq. (9). We can write this equation as

$$d\mathbf{L} = \tau \, dt \tag{37}$$

* This section is optional.

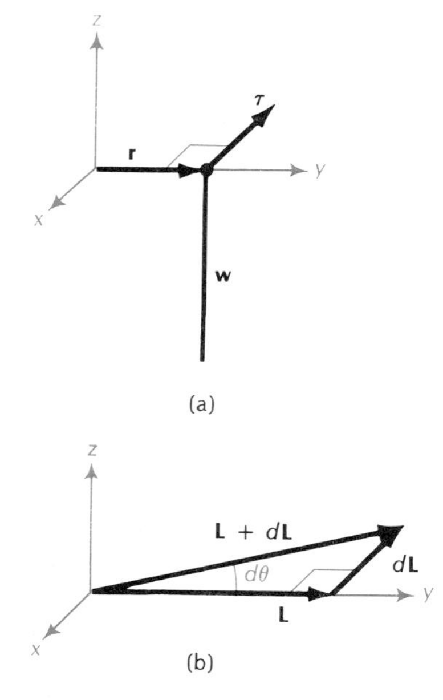


**Fig. 13.24** (a) The torque $\tau$ is perpendicular to **r** and **w.** (b) Hence $d\mathbf{L}$ is perpendicular to **L.**

The torque is caused by the weight **w** acting at the center of mass. For the calculation of this torque, we will take the pivot as origin. According to the right-hand rule, the direction of the torque $\mathbf{r} \times \mathbf{w}$ is then into the plane of Figure 13.24a. Hence the change $d\mathbf{L}$ of the angular momentum is also into the plane of the figure. This means that the weight does not dip the axis downward; rather it deflects the axis horizontally in a direction perpendicular to the applied force. More generally, whenever we exert a lateral push or pull against the axis of a gyroscope, the resulting motion of the axis will be perpendicular to the push or pull — a vertical push will deflect the axis horizontally and a horizontal push will deflect the axis vertically.

From Eq. (37) we can calculate the rate at which gravity causes the gyroscope to move horizontally. If the mass of the flywheel is $M$ and its distance from the pivot is $r$, then

$$\tau = rMg \tag{38}$$

and

$$dL = rMg\ dt \tag{39}$$

From Figure 13.24b we see that the angle $d\theta$ through which the axis swings in a time $dt$ is

$$d\theta = \frac{dL}{L} = \frac{rMg\ dt}{L}$$

or

$$\frac{d\theta}{dt} = \frac{rMg}{L} \tag{40}$$

This is the angular velocity with which the axis swings around in the horizontal plane. This motion of the axis is called **precession,** and the angular velocity of Eq. (40) is called the **precession frequency:**

*Precession frequency*

$$\boxed{\omega_p = \frac{rMg}{L}} \tag{41}$$

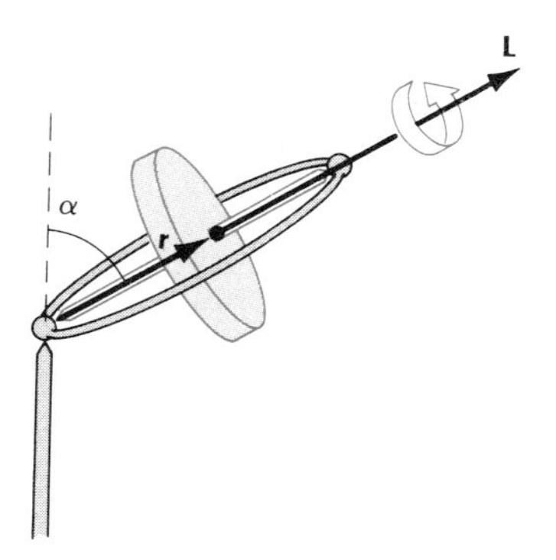


**Fig. 13.25** Spinning gyroscope with axis inclined at an angle $\alpha$.

Although we have obtained this result under the assumption that the axis is initially horizontal, it is easy to see that the precession proceeds with the same frequency if the axis is inclined at some arbitrary angle with the vertical (Figure 13.25). The result can therefore be applied to an ordinary top, whose angle of inclination can of course never reach 90° (Figure 13.26).

One crucial ingredient in the above calculations is the assumption that the gyroscope spins at high speed. If the gyroscope does not spin at all or spins only slowly, then it will flop down when released from the horizontal position shown in Figure 13.23. Even when the gyroscope spins at high speed, it will at first swing downward a little bit and then swing back up; and it will continue to repeat this down-and-up motion as it precesses in a horizontal direction. For a fast-spinning gyroscope this oscillatory vertical motion, or nutation, is so small that it is hardly noticeable. Our calculation gave no indication of this vertical motion because we assumed that the angular momentum is entirely associated with the spin of the gyroscope on its axis. But this is not quite true — the precession involves a gradual motion of the mass around a horizontal circle, and there is some small amount of extra angular momentum associated with this motion. Because of this extra angular momentum, the total angular momentum vector does not lie exactly along the axis of the gyroscope. A precise calculation which takes this small deviation into account can explain the small up-and-down motions of the gyroscope, but this is a rather complicated calculation and we will not attempt it here.

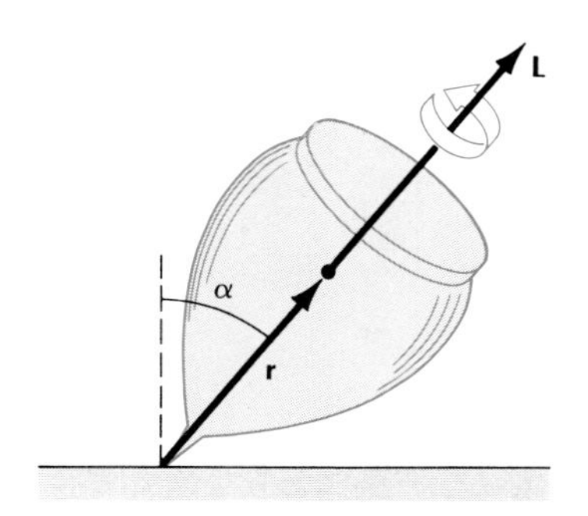


**Fig. 13.26** Spinning top.

## SUMMARY

**Torque:** $\boldsymbol{\tau} = \mathbf{r} \times \mathbf{F}$

**Equation of rotational motion (fixed axis or axis through the center of mass):**

$$I\alpha = \tau_z$$

**Work done by torque:** $W = \int \tau_z \, d\phi$

**Conservation of energy:** $E = \frac{1}{2}I\omega^2 + U = [\text{constant}]$

**Power delivered by torque:** $P = \tau_z \omega$

**Conservation of angular momentum:** $I\omega = [\text{constant}]$

**Rolling motion (without slipping):** The axis through the point of contact is instantaneously a fixed axis.

**Precession of gyroscope:** $\omega_p = rMg/L$

## QUESTIONS

1. Suppose you push down on the rim of a stationary phonograph turntable. What is the direction of the torque you exert about the center of the turntable?

2. Many farmers have been injured when their tractors suddenly flipped over backward while pulling a heavy piece of farm equipment. Can you explain how this happens?

3. According to popular belief, a falling piece of bread always lands on its buttered side. Why or why not?

4. Rifle bullets are given a spin about their axis by spiral grooves ("rifling") in the barrel of the gun. What is the advantage of this?

5. You are standing on a frictionless turntable (like a phonograph turntable, but sturdier). How can you turn 180° without leaving the turntable or pushing against any exterior body?

6. If you give a hard-boiled egg resting on a table a twist with your fingers, it will continue to spin. If you try doing the same with a raw egg, it will not. Why?

7. A tightrope walker uses a balancing pole to keep steady (Figure 13.27). How does this help?

**Fig. 13.27** A tightrope walker.

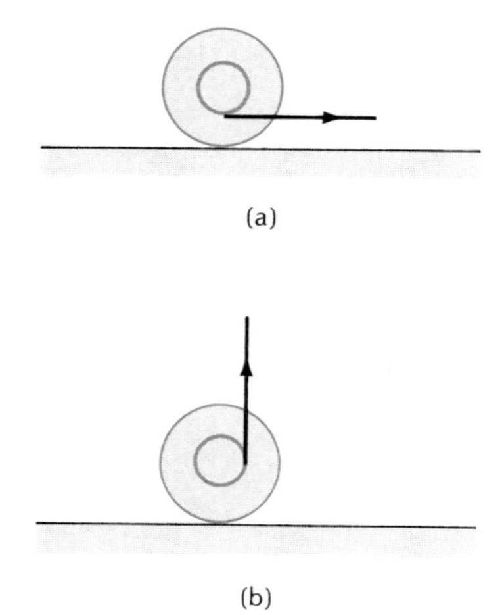


**Fig. 13.28** Yo-yo resting on a table: (a) string pulls horizontally; (b) string pulls vertically.

8. Why do helicopters need a small vertical propeller on their tail?

9. The rotation of the Earth is subject to small seasonal variations (see Section 1.4). Does this mean that angular momentum is not conserved?

10. Why does the front end of an automobile dip down when the automobile is braking sharply?

11. The friction of the tides against the ocean coasts and the ocean shallows is gradually slowing down the rotation of the Earth. What happens to the lost angular momentum?

12. An automobile is traveling on a straight road at 90 km/h. What is the speed, relative to the ground, of the lowermost point on one of its wheels? The topmost point? The midpoint?

13. A sphere and a hoop of equal masses roll down an inclined plane without slipping. Which will get to the bottom first? Will they have equal kinetic energies when they reach the bottom?

14. A yo-yo rests on a table (Figure 13.28). If you pull the string horizontally, which way will it move? If you pull vertically?

15. Stand a pencil vertically on its point on a table and let go. The pencil will topple over.

(a) If the table is very smooth, the point of the pencil will slip in the direction opposite to that of the toppling. Why?

(b) If the table is somewhat rough, or covered with a piece of paper, the point of the pencil will jump in the direction of the toppling. Why? (Hint: During the early stages of the toppling, friction holds the point of the pencil fixed; thus the pencil acquires a horizontal momentum.)

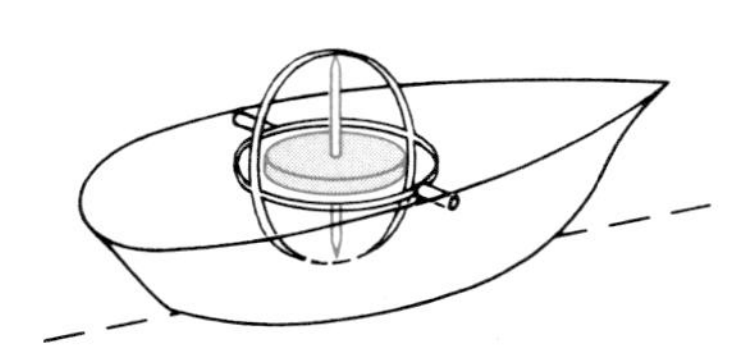

**Fig. 13.29** Schick stabilizer.

16. The Schick stabilizer, which was installed on a luxury yacht, consists of a large flywheel mounted in a frame within the ship (Figure 13.29). The frame permits the gyroscope to rotate freely about the transverse horizontal axis and about the vertical axis, but not about the longitudinal axis. Describe what happens to the flywheel and to the ship if a wave tries to heel the ship to the left or to the right.

## PROBLEMS

### Section 13.1

1. The operating instructions for a small crane specify that when the boom is at an angle of 20° above the horizontal (Figure 13.30), the maximum safe load for the crane is 500 kg. Assuming that this maximum load is determined by the maximum torque that the pivot can withstand, what is the maximum torque for 20°? What is the maximum safe load for 40°? For 60°?

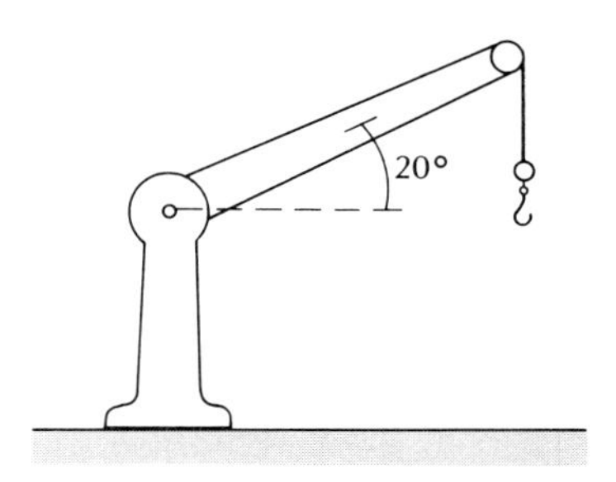


**Fig. 13.30** Small crane.

2. A simple manual winch consists of a drum of radius 4.0 cm to which is attached a handle of radius 25 cm (Figure 13.31). When you turn the handle, the rope winds up on the drum and pulls the load. Suppose that the load carried by the rope is 2500 N. What force must you exert on the handle to hold this load?

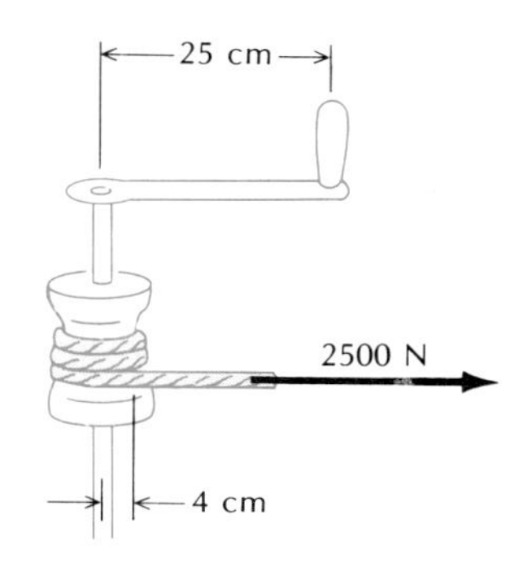


**Fig. 13.31** Manual winch.

3. The repair manual for an automobile specifies that the cylinder-head bolts are to be tightened to a torque of 62 N · m. If a mechanic uses a wrench of length 20 cm on such a bolt, what perpendicular force must he exert on the end of this wrench to achieve the correct torque?

*4. While braking, a 1500-kg automobile decelerates at the rate of 8.0 m/s$^2$. What is the magnitude of the braking force that the road exerts on the automobile? What torque does this force generate about the center of mass of the automobile? Will this torque tend to lift the front end of the automobile or tend to depress it? Assume that the center of mass of the automobile is 60 cm above the surface of the road.

### Section 13.2

5. Because of friction, the turntable of a phonograph gradually slows down when the motor is disengaged and finally stops. The initial angular speed of the turntable is $33\frac{1}{3}$ rev/min and the time it takes to stop is 1.2 min. The turntable is a disk of mass 2.0 kg and radius 15 cm. What is the average frictional torque on the turntable?

6. The center span of a revolving drawbridge consists of a uniform steel girder of mass 300 metric tons and length 25 m. This girder can be regarded as a uniform thin rod. The bridge opens by rotating about a vertical axis through its center. What torque is required to open this bridge in 60 s? Assume that the bridge first accelerates uniformly through an angular interval of 45° and then the torque is reversed, so the bridge decelerates uniformly through an angular interval of 45° and comes to rest after rotating by 90°.

7. The original Ferris wheel, built by George Ferris, had a radius of 38 m and a mass of $1.9 \times 10^6$ kg (see Figure 13.32). Assume that all of its mass was uniformly distributed along the rim of the wheel. If the wheel was initially rotating at 0.05 rev/min, what constant torque had to be applied to bring it to a full stop in 30 s? What force exerted on the rim of the wheel would have given such a torque?

**Fig. 13.32** The Ferris wheel.

8. The pulley of an Atwood machine is a brass disk of mass 120 g. When using masses $m_1 = 450.0$ g and $m_2 = 455.0$ g, an experimenter finds that the larger mass descends 1.6 m in 8.0 s, starting from rest. What is the value of $g$?

*9. An automobile of mass 1200 kg has four brake drums of diameter 25 cm. The brake drums are rigidly attached to the wheels of diameter 60 cm. The braking mechanism presses brake pads against the rim of each drum and the friction between the pad and the rim generates a torque that slows the rotation of the wheel. Assume that all four wheels contribute equally to the braking. What torque must the brake pads exert on each drum in order to decelerate the automobile at 0.8 G? If the coefficient of friction between the pad and the drum is $\mu_k = 0.6$, what normal force must the brake pad exert on the rim of the drum? Ignore the masses of the wheels.

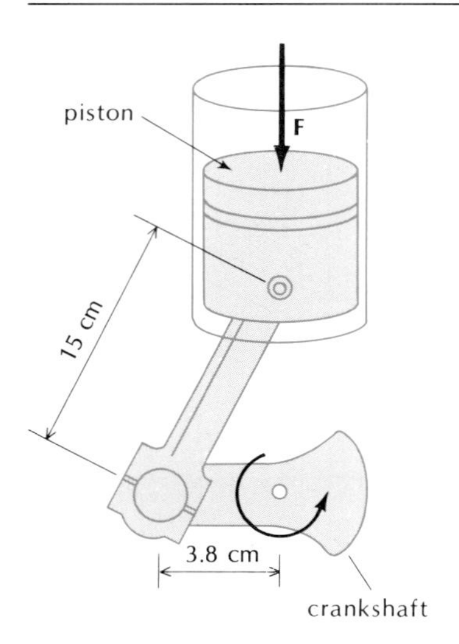


**Fig. 13.33**

*10. In one of the cylinders of an automobile engine, the gas released by internal combustion pushes on the piston, which, in turn, pushes on the crankshaft by means of a piston rod (Figure 13.33). If the crankshaft experiences a torque of 31 N · m and if the dimensions of the crankshaft and piston rod are as in Figure 13.33, what must be the force of the gas on the piston when the crankshaft is in the horizontal position as in Figure 13.33? Ignore friction, and ignore the masses of the piston and rod.

*11. A disk of mass $M$ is free to rotate about a fixed horizontal axis. A string is wrapped around the rim of this disk and a mass $m$ is attached to this string (Figure 13.34). What is the downward acceleration of the mass?

*12. The wheel of an automobile has a mass of 25 kg and a diameter of 70 cm. Assume that the wheel can be regarded as a uniform disk.

(a) What is the angular momentum of the wheel when the automobile is traveling at 25 m/s (90 km/h) on a straight road?

(b) What is the rate of change of the angular momentum of the wheel when the automobile is traveling at the same speed along a curve of radius 80 m?

(c) For this rate of change of angular momentum, what must be the torque on the wheel? Draw a diagram showing the path of the automobile, the angular momentum vector of the wheel, and the torque vector.

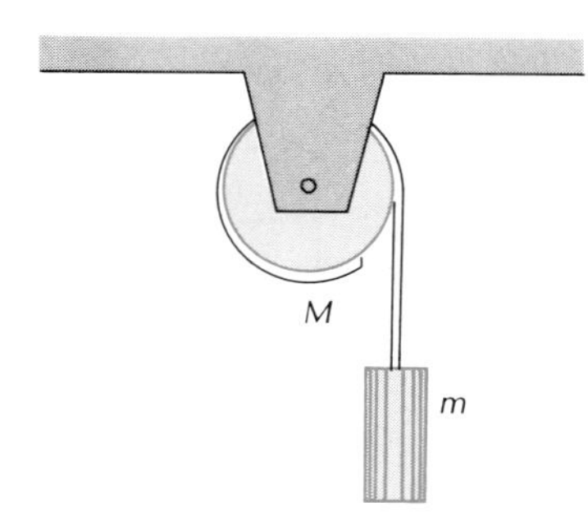


**Fig. 13.34**

*13. Consider the airplane propeller described in Problem 12.35. If the airplane is flying around a curve of radius 500 m at a speed of 360 km/h, what is the rate of change of the angular momentum of the propeller? What torque is required to change the angular momentum at this rate? Draw a diagram showing **L**, $d\mathbf{L}/dt$, and $\tau$.

*14. A heavy hatch on a ship is made of a uniform plate of steel that measures 1.2 m × 1.2 m and has a mass of 400 kg. The hatch is hinged along one side; it is horizontal when closed and opens upward. A torsional spring assists in the opening of the hatch. The spring exerts a torque of $2.0 \times 10^3$ N · m when the hatch is horizontal and a torque of $0.3 \times 10^3$ N · m when the hatch is vertical; in the range of angles between horizontal and vertical, the torque decreases linearly with the angle (e.g., the torque is $1.15 \times 10^3$ N · m when the hatch is at 45°).

(a) At what angle will the hatch be in equilibrium so that the spring exactly compensates the torque due to the weight?

(b) What minimum push must a sailor exert on the hatch to open it from the closed position? To close it from the open position? Assume the sailor pushes perpendicularly on the hatch at the edge that is farthest from the hinge.

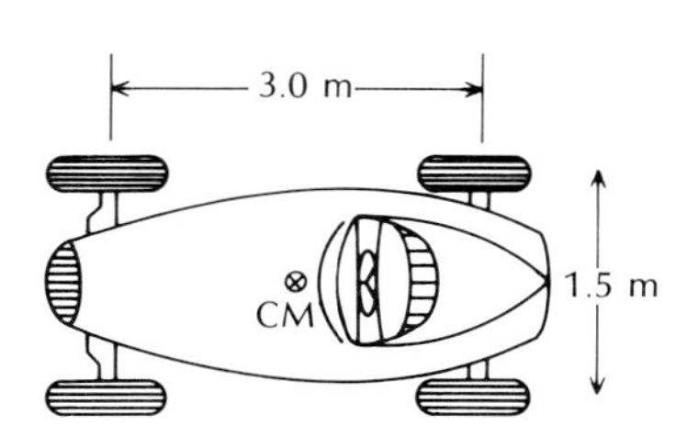


**Fig. 13.35**

*15. An automobile has the arrangement of wheels shown in Figure 13.35. The mass of this automobile is 1800 kg, the center of mass is at the midpoint of the rectangle formed by the wheels, and the moment of inertia about a vertical axis through the center of mass is 2200 kg · m². Suppose that during braking in an emergency, the left front and rear wheels lock and begin to skid while the right wheels continue to rotate just short of skidding. The coefficient of static friction between the wheels and the road is $\mu_s = 0.90$ and the coefficient of kinetic friction is $\mu_k = 0.50$. Calculate the instantaneous angular acceleration of the automobile about the vertical axis.

*16. A large flywheel designed for energy storage at a power plant has a moment of inertia of $5 \times 10^5$ kg · m² and spins at 3000 rev/min. Suppose that this flywheel is mounted on a horizontal axle oriented in the east–west direction. What is the magnitude and direction of its angular momentum? What is the rate of change of this angular momentum due to the rotational motion of the Earth and the consequent motion of the axle of the flywheel? What is the torque that the axle of the flywheel exerts against the bearings supporting it? If the bearings are at a distance of 0.6 m from the center of the flywheel on each side, what are the forces associated with this torque?

Section 13.3

17. The engine of an automobile delivers a maximum torque of 203 N · m when running at 4600 rev/min and it delivers a maximum power of 142 hp when running at 5750 rev/min. What power does the engine deliver when running at maximum torque? What torque does it deliver when running at maximum power?

18. The flywheel of a motor is connected to the flywheel of a pump by a drive belt (Figure 13.36). The first flywheel has a radius $R_1$, and the second a radius $R_2$. While the motor wheel is rotating at a constant angular velocity $\omega_1$, the tensions in the upper and the lower portions of the drive belt are $T$ and $T'$, respectively. Assume that the drive belt is massless.

(a) What is the angular velocity of the pump wheel?
(b) What is the torque of the drive belt on each wheel?
(c) By taking the product of torque and angular velocity, calculate the power delivered by the motor to the drive belt, and the power removed by the pump from the drive belt. Are these powers equal?

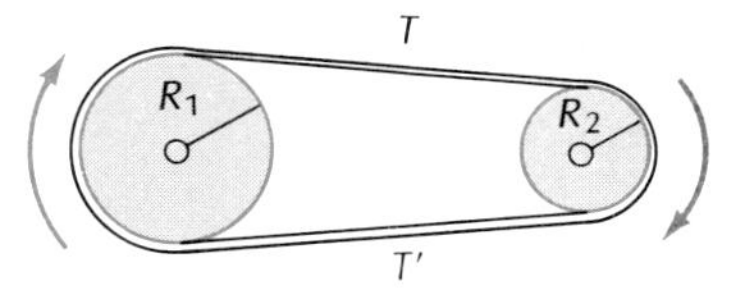


**Fig. 13.36**

19. The Wright Cyclone engine on a DC-3 airplane delivers a power of 850 hp with the propeller revolving steadily at 2100 rev/min. What is the torque exerted by air resistance on the propeller?

20. A woman on an exercise bicycle has to exert an (average) tangential push of 35 N on each pedal to keep the wheel turning at constant speed. Each pedal has a radial length of 0.18 m. If she pedals at the rate of 60 revolutions per minute, what is the power she expends against the exercise bicycle? Express your answer in kilocalories per minute.

*21. A tractor of mass 4500 kg has rear wheels of radius 0.80 m. What torque and what power must the engine supply to the rear axle to move the tractor up a road of slope 1:3 at a constant speed of 4.0 m/s?

*22. A bicycle and its rider have a mass of 90 kg. While accelerating from rest to 12 km/h, the rider turns the pedals through three full revolutions. What torque must the rider exert on the pedals? Assume that the torque is constant during the acceleration and ignore friction within the bicycle mechanism.

*23. A meter stick is initially standing vertically on the floor. If the meter stick falls over, with what angular velocity will it hit the floor? Assume that the end in contact with the floor experiences no friction and slips freely.

*24. A meter stick is held to a wall by a nail passing through the 60-cm mark (Figure 13.37). The meter stick is free to swing about this nail, without friction. If the meter stick is released from an initial horizontal position, what angular velocity will it attain when it swings through the vertical position?

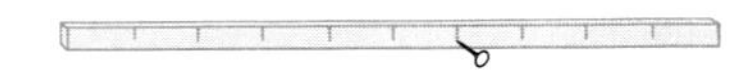

**Fig. 13.37**

*25. A uniform solid sphere of mass $M$ and radius $R$ hangs from a string of length $R/2$. Suppose the sphere is released from an initial position making an angle of 45° with the vertical (Figure 13.38).

(a) Calculate the angular velocity of the sphere when it swings through the vertical position.
(b) Calculate the tension in the string at this instant.

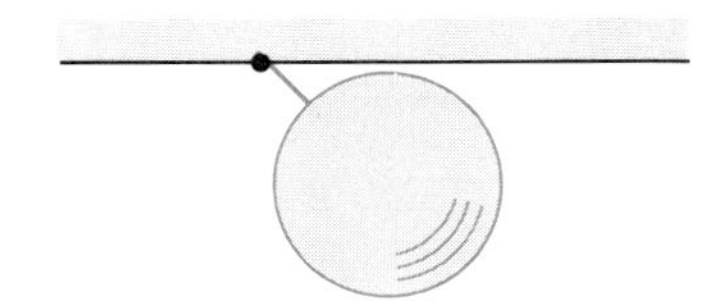

**Fig. 13.38**

*26. The maximum (positive) acceleration an automobile can achieve on a level road depends on the maximum torque the engine can deliver to the wheels.

(a) The engine of a Maserati sports car delivers a maximum torque of 441 N · m to the gearbox. The gearbox steps down the rate of revolution by a factor of 2.58; that is, whenever the engine makes 2.58 revolutions, the wheels make 1 revolution. What is the torque delivered to the wheels? Ignore frictional losses in the gearbox.
(b) The mass of the car (including fuel, driver, etc.) is 1770 kg and the radius of its wheels is 0.30 m. What is the maximum acceleration? Ignore the moment of inertia of the wheels and frictional losses.

Section 13.4

27. A very heavy freight train made up of 250 cars has a total mass of 7700 metric tons. Suppose that such a train accelerates from 0 to 65 km/h on a track running exactly east from Quito, Ecuador (on the equator). The force that the engine exerts on the Earth will slow down the rotational motion of the Earth. By how much will the angular velocity of the Earth have decreased when the train reaches its final speed? Express your answer in revolutions per day. The moment of inertia of the Earth is $0.33M_ER_E^2$.

28. There are $1.1 \times 10^8$ automobiles in the United States, each of an average mass of 2000 kg. Suppose that one morning all these automobiles simultaneously start to move in an eastward direction and accelerate to a speed of 80 km/h.

(a) What total angular momentum about the axis of the Earth do all these automobiles contribute together? Assume that the automobiles travel at an average latitude of 40°.

(b) How much will the rate of rotation of the Earth change because of the action of these automobiles? Assume that the axis of rotation of the Earth remains fixed. The moment of inertia of the Earth is $8.1 \times 10^{37}$ kg · m$^2$.

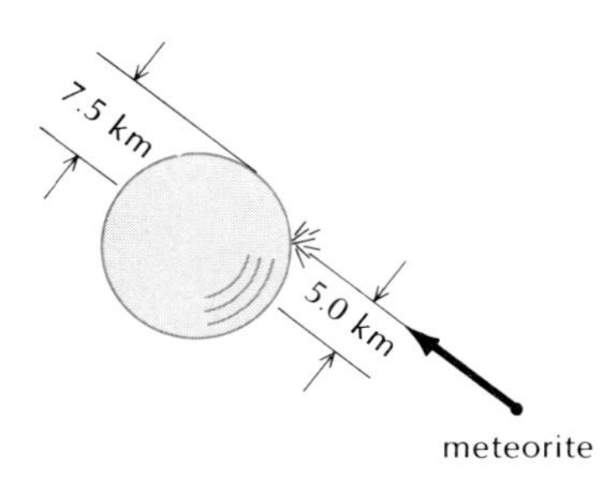


**Fig. 13.39**

*29. Phobos is a small moon of Mars. For the purposes of the following problem, assume that Phobos has a mass of $5.8 \times 10^{15}$ kg and that it has the shape of a uniform sphere of radius $7.5 \times 10^3$ m. Suppose that a meteorite strikes Phobos $5.0 \times 10^3$ m off center (Figure 13.39) and remains stuck. If the momentum of the meteorite was $3 \times 10^{13}$ kg · m/s before impact and the mass of the meteorite is negligible compared to the mass of Phobos, what is the change in the rotational angular velocity of Phobos?

*30. A woman stands in the middle of a small rowboat. The rowboat is floating freely and experiences no friction against the water. The woman is initially facing east. If she turns around 180° so that she faces west, through what angle will the rowboat turn? Assume that the woman performs her turning movement at constant angular velocity and that her moment of inertia remains constant during this movement. The moment of inertia of the rowboat about the vertical axis is 20 kg · m$^2$ and that of the woman is 0.80 kg · m$^2$.

*31. Two automobiles both of 1200 kg and both traveling at 30 km/h collide on a frictionless icy road. They were initially moving on parallel paths in opposite directions, with a center-to-center distance of 1.0 m (Figure 13.40). In the collision, the automobiles lock together, forming a single body of wreckage; the moment of inertia of this body about its center of mass is $2.5 \times 10^3$ kg · m$^2$.

(a) Calculate the angular velocity of the wreck.

(b) Calculate the kinetic energy before the collision and after the collision. What is the change of kinetic energy?

**Fig. 13.40**

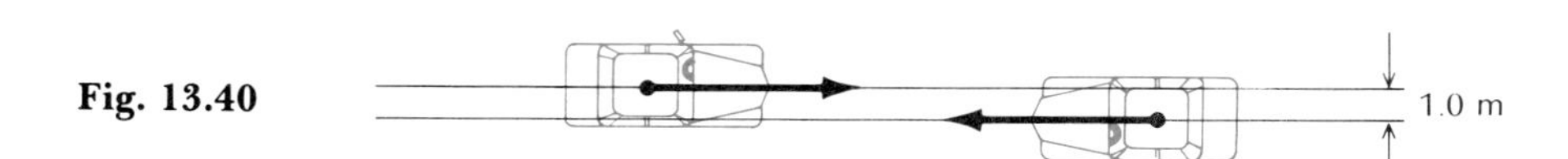


*32. In one experiment performed under weightless conditions in Skylab, the three astronauts ran around a path on the inside wall of the spacecraft so as to generate artificial gravity for their bodies (Figure 13.41). Assume that the center of mass of each astronaut moves around a circle of radius 2.5 m; treat the astronauts as particles.

(a) With what speed must each astronaut run if the average normal force on his feet is to equal his normal weight ($mg$)?

(b) Suppose that before the astronauts begin to run, Skylab is floating in its orbit without rotating. When the astronauts begin to run clockwise,

**Fig. 13.41** Three astronauts about to start running around the inside of Skylab.

Skylab will begin to rotate counterclockwise. What will be the angular velocity of Skylab when the astronauts are running steadily with the speed calculated above? Assume that the mass of each astronaut is 70 kg and that the moment of inertia of Skylab about its longitudinal axis is $3 \times 10^5$ kg · m$^2$.

(c) How often must the astronauts run around the inside if they want Skylab to rotate through an angle of 30°?

*33. A flywheel rotating freely on a shaft is suddenly coupled by means of a drive belt to a second flywheel sitting on a parallel shaft (Figure 13.42). The initial angular velocity of the first flywheel is $\omega$; that of the second is zero. The flywheels are uniform disks of masses $M_1$, $M_2$, and of radii $R_1$, $R_2$, respectively. The drive belt is massless and the shafts are frictionless.

(a) Calculate the final angular velocity of each flywheel.

(b) Calculate the kinetic energy lost during the coupling process. What happens to this energy?

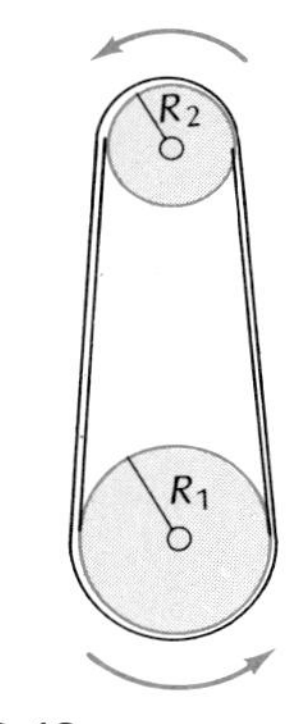


**Fig. 13.42**

*34. A thin rod of mass $M$ and length $l$ hangs from a pivot at its upper end. A ball of clay of mass $m$ and of horizontal velocity $v$ strikes the lower end at right angles and remains stuck (a totally inelastic collision). How high will the rod swing after this collision?

**35. If the melting of the polar ice caps were to raise the water level on the Earth by 10 m, by how much would the day be lengthened? Assume that the moment of inertia of the ice in the polar ice caps is negligible (they are very near the axis), and assume that the extra water spreads out uniformly over the entire surface of the Earth (that is, neglect the area of the continents compared with the area of the oceans). The moment of inertia of the Earth (now) is $8.1 \times 10^{37}$ kg · m$^2$.

**36. A rod of mass $M$ and length $l$ is lying on a flat, frictionless surface. A ball of putty of mass $m$ and initial velocity $v$ at right angles to the rod strikes the rod at a distance $l/4$ from the center (Figure 13.43). The collision is inelastic and the putty adheres to the rod. Find the translational and rotational motions of the rod after this collision.

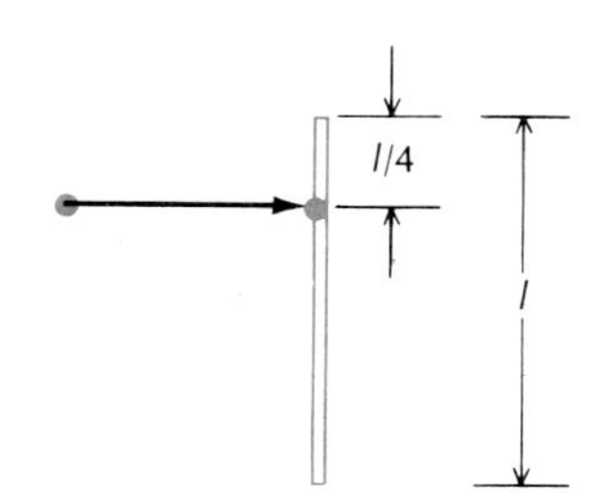


**Fig. 13.43** A ball of putty strikes a rod.

### Section 13.5

37. A hula-hoop rolls down a slope of 1:10 without slipping. What is the (linear) acceleration of the hoop?

38. A uniform cylinder rolls down a plane inclined at an angle $\theta$ with the horizontal. Show that if the cylinder rolls without slipping, the acceleration is $a = \frac{2}{3}g \sin \theta$.

ing down a steep hill, rolls down the hill without slipping. The mass of the wheel is 60 kg and its radius is 0.40 m; the mass distribution of the wheel is approximately that of a uniform disk. At the bottom of the hill, at a vertical distance of 120 m below the point of release, the wheel slams into a telephone booth. What is the total kinetic energy of the wheel just before impact? How much of this kinetic energy is translational energy of the center of mass of the wheel? How much is rotational kinetic energy about the center of mass? What is the speed of the wheel?

40. Galileo measured the acceleration of gravity by rolling a sphere down an inclined plane. Suppose that, starting from rest, a sphere takes 1.6 s to roll a distance of 3.00 m down a 20° inclined plane. What value of $g$ can you deduce from this?

41. A yo-yo consists of a uniform disk with a string wound around the rim. The upper end of the string is held fixed. The yo-yo unwinds as it drops. What is its downward acceleration?

*42. An automobile with rear-wheel drive is accelerating at 8.0 m/s$^2$. The wheels of the automobile are uniform disks of mass 25 kg and radius 0.38 m. What *horizontal* force does the front wheel exert on the road? What *horizontal* force does the axle exert on the wheel?

*43. A barrel of mass 200 kg and radius 0.5 m rolls down a 40° ramp without slipping. What is the value of the friction force acting at the point of contact between barrel and ramp? Treat the barrel as a cylinder with uniform mass density.

*44. A bowling ball sits on the smooth floor of a subway car. If the car has a horizontal acceleration $a$, what is the acceleration of the ball? Assume that the ball rolls without slipping.

*45. A hoop rolls down an inclined ramp. The coefficient of static friction between the hoop and the ramp is $\mu_s$. If the ramp is very steep, the hoop will slip while rolling. Show that the critical angle of inclination at which the hoop begins to slip is given by $\tan\theta = 2\mu_s$.

*46. A solid cylinder rolls down an inclined plane. The angle of inclination $\theta$ of the plane is large so that the cylinder slips while rolling. The coefficient of kinetic friction between the cylinder and the plane is $\mu_k$. Find the rotational and the translational acceleration of the cylinder. Show that the translational acceleration is the same as that of a block sliding down the plane.

**Fig. 13.44** A landing airliner

**47. Suppose that a tow truck applies a horizontal force of 4000 N to the front end of an automobile similar to that described in Problem 12.38. Taking into account the rotational inertia of the wheels and ignoring frictional losses, what is the acceleration of the automobile? What is the percentage difference between this value of the acceleration and the value calculated by neglecting the rotational inertia of the wheels?

**48. A cart consists of a body and four wheels on frictionless axles. The body has a mass $m$. The wheels are uniform disks of mass $M$ and radius $R$. Taking into account the moment of inertia of the wheels, find the acceleration of this cart if it rolls without slipping down an inclined plane making an angle $\theta$ with the horizontal.

**49. When the wheels of a landing airliner touch the runway, they are not rotating initially. The wheels first slide on the runway (and produce clouds of smoke and burn marks on the runway, which you may have noticed; see Figure 13.44), until the sliding friction force has accelerated the wheels to the rotational speed required for rolling without slipping. From the following data, calculate how far the wheel of an airliner slips before it begins to roll without slipping: the wheel has a radius of 0.60 m and a mass of 160 kg, the normal force acting on the wheel is $2 \times 10^5$ N, the speed of the airliner is 200 km/h; the coefficient of sliding friction for the wheel on the runway is 0.8 Treat the wheel as a uniform disk.

Section 13.6

50. A child's toy top consists of a uniform thin disk of radius 5.0 cm and mass 0.15 kg with a thin spike passing through its center. The lower part of the spike protrudes 6.0 cm from the disk. If you stand this top on a table and start it spinning at 200 rev/s, what will be its precession frequency?

51. In order to stabilize a ship against rolling, an inventor proposes that a large flywheel spinning at high speed should be mounted within the ship (Figure 13.45). The axis of the flywheel lies across the ship and the bearings are rigidly attached to the side of the ship.

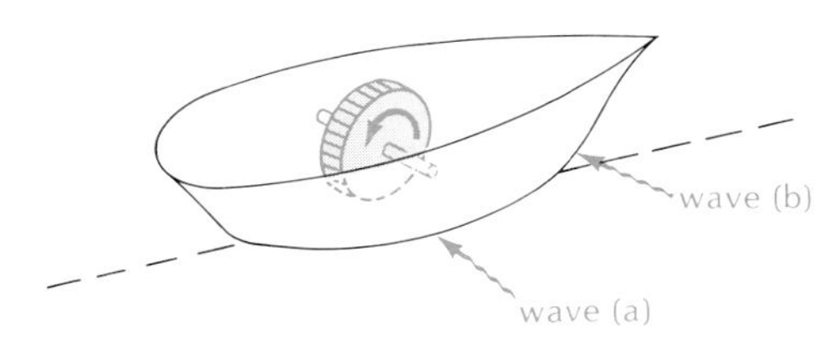


**Fig. 13.45** Ship with a flywheel.

(a) If a wave hits the ship broadside and attempts to roll (capsize) the ship to the left, what will be the response of the ship? In answering this question, assume that the response of the ship is dominated by the effect of the flywheel. With words or with a diagram, carefully describe how the orientation of the ship will change.
(b) If a wave hits the bow of the ship and attempts to push the bow to the left, what will be the response of the ship?

52. Suppose that the flywheel of the gyroscope shown in Figure 13.25 is a uniform disk of mass 250 g and radius 3.5 cm. The distance of this flywheel from the point of support is 4.0 cm. What is the precession frequency if the flywheel is spinning at 120 rev/s?

*53. The angular momentum of rotation (spin) of a neutron is

$$\tfrac{1}{2}\hbar = 5.3 \times 10^{-35}\ \mathrm{J \cdot s}$$

Suppose that, when placed in a certain magnetic field, the spin of the neutron precesses at the rate of $2.4 \times 10^7$ radian/s about a line perpendicular to the spin, just like the gyroscope of Figure 13.23. What is the magnitude of the torque on the neutron?

CHAPTER 14

# Statics and Elasticity

The basic problem of statics is the calculation of the forces necessary to hold a body in equilibrium. This is a problem of practical importance for engineers and architects who need to know what forces act on the structural members of bridges, buildings, the hulls of ships, the fuselages or wings of aircraft, and all kinds of machinery. As a first approximation, we will assume that the "rigid" structural members—such as beams and columns—indeed remain rigid, that is, they do not deform. We will rely on this assumption in the first three sections. However, in the last section, we will take a brief look at the deformations produced in solid bodies when subjected to the action of large forces.

## 14.1 Statics of Rigid Bodies

*Condition for static equilibrium*

If a rigid body is to remain at rest, its translational and rotational accelerations must be zero. Hence, the condition for the static equilibrium of a rigid body is that *the sum of external forces and the sum of external torques on the body must be zero.* For the purposes of statics, any line through the body or any line passing near the body can be thought of as a (conceivable) axis of rotation; and since the angular acceleration about every such line is zero, the torque about every such line must be zero. This means we have complete freedom in the choice of the axis of rotation, and we can make whatever choice seems convenient. With some practice, one learns to recognize which choice of axis will most quickly lead to the solution of a problem in statics.

The force of gravity plays an important role in many problems of statics. The force of gravity on a body is distributed over all parts of the body, each part being subjected to a force proportional to its mass. However, for the calculation of the torque exerted by gravity on a rigid body, *the entire gravitational force may be regarded as acting on the center of mass.* The proof of this rule is easy: Suppose that we release the body and permit it to fall freely from an initial condition of rest. Since all the particles in the body fall at the same rate, the body will not change its orientation as it falls. This absence of angular acceleration implies that gravity does not generate any torque about the center of mass. Hence, if we want to simulate gravity by a single force acting at one point of the rigid body, that point will have to be the center of mass, so that this single force does not generate any torque either. Both gravity and the single force that replaces it then produce exactly the same rotational motion about the center of mass (namely, no motion), and they are therefore equivalent in regard to the equations of rotational motion of the body. Note, however, that this equivalence holds only for a rigid body—for an elastic body or a fluid, the distribution of gravity over all parts of the body plays an essential role in the equations of motion of all these parts.

*Effect of gravity on rigid body*

## 14.2 Examples of Static Equilibrium

The following are some examples of solutions of problems in statics. The first step in the solution is always a careful enumeration of all the forces acting on the body. To keep track of these forces, it helps to draw them on a "free-body" diagram. The next step is the application of the equilibrium conditions for forces and torques. The equilibrium condition for forces is most conveniently expressed in terms of components; thus, there are three separate equilibrium equations for the $x$, $y$, and $z$ components of the forces. The equilibrium condition for the torques also involves three separate components; however, in all the examples of this section, the relevant axis of rotation is perpendicular to the plane of the page, and hence the only component of the torque that needs to be considered is the single component along this axis.

EXAMPLE 1. A locomotive of 80 metric tons is one-third of the way across a bridge 90 m long. The bridge consists of a uniform iron girder of 800 metric tons, which rests on two piers (see Figure 14.1a). What is the load on each pier?

SOLUTION: Fig. 14.1b is a "free-body" diagram for the bridge, showing all the forces acting on it. The weight of the bridge is shown acting at the center of mass. The bridge is static, and hence the net torque on the bridge reckoned about any point must be zero. Let us first consider the torques about the point $P_2$. These torques are generated by the weight of the bridge acting at a distance of 45 m, the weight of the locomotive acting at a distance of 30 m, and the upward thrust $F_1$ of the pier at $P_1$ acting at a distance of 90 m. Setting the sum of the three torques equal to zero, we have

$$45 \text{ m} \times 800 \text{ tons} \times g + 30 \text{ m} \times 80 \text{ tons} \times g - 90 \text{ m} \times F_1 = 0 \qquad (1)$$

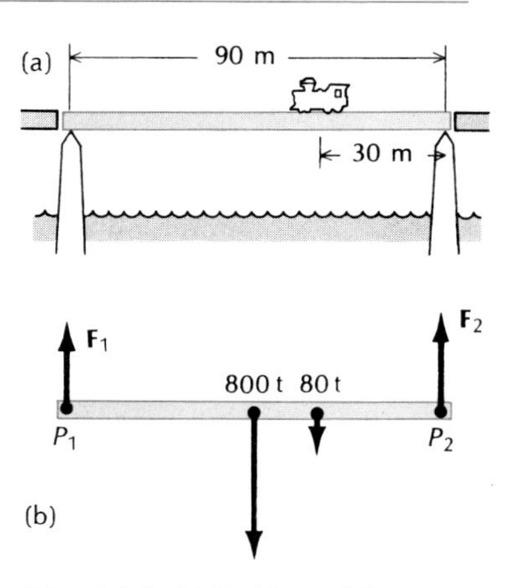


**Fig. 14.1** (a) Bridge with a locomotive. (b) "Free-body" diagram for the bridge.

Here, the torque has been reckoned as positive if it tends to produce counterclockwise rotation and negative if it tends to produce clockwise rotation. Solving this equation for $F_1$, we find

$$F_1 = 427 \text{ tons} \times g$$

$$= 427 \times 1000 \text{ kg} \times 9.81 \text{ m/s}^2 = 4.2 \times 10^6 \text{ N}$$

Next, consider the torques about the point $P_1$. These torques are generated by the weight of the bridge, the weight of the locomotive, and the upward thrust $F_2$ at point $P_2$. Setting the sum of these three torques about the point $P_1$ equal to zero, we have

$$-45 \text{ m} \times 800 \text{ tons} \times g - 60 \text{ m} \times 80 \text{ tons} \times g + 90 \text{ m} \times F_2 = 0 \quad (2)$$

from which

$$F_2 = 4.4 \times 10^6 \text{ N}$$

COMMENTS AND SUGGESTIONS: Note that the net vertical upward force exerted by the piers is $F_1 + F_2 = 8.6 \times 10^6$ N. It is easy to check that this matches the sum of the weights of the bridge and the locomotive; thus, the condition for zero net vertical force, as required for translational equilibrium, is automatically satisfied. This automatic result for the equilibrium of vertical forces came about because we used the condition for rotational equilibrium twice. Instead, we could have used the condition for rotational equilibrium once [Eq. (1)], and then evaluated $F_2$ by means of the condition for translational equilibrium. The result for zero net torque about the point $P_1$ would then have emerged automatically.

However, this automatic equivalence between the conditions for rotational and translational equilibrium is not of general validity—it occurs in this example because of the special configuration of the forces (it can be shown to occur whenever the points of application of all the forces lie on a single straight line). In the other examples of this section, we will need to invoke the condition for translational equilibrium explicitly.

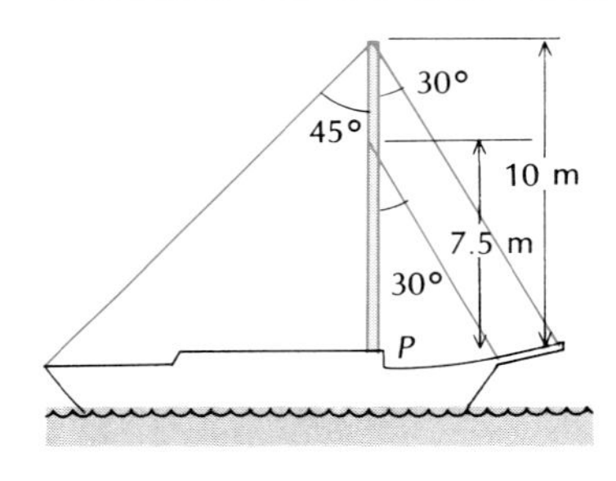


(a)

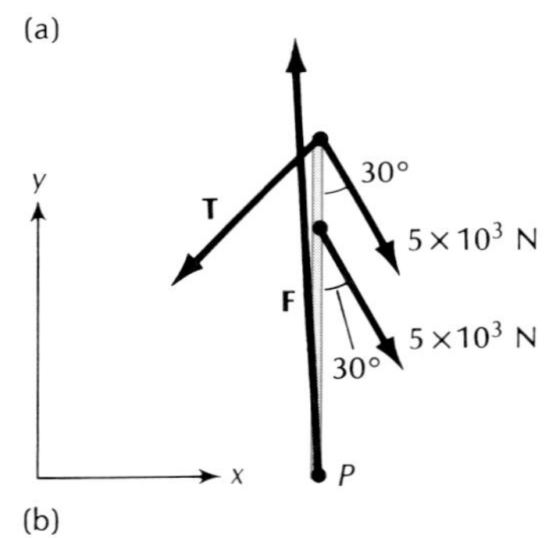


(b)

**Fig. 14.2** (a) Steel cables staying a mast. (b) "Free-body" diagram for the mast.

EXAMPLE 2. The mast of a sailboat is held fore and aft by three steel cables attached as shown in Figure 14.2a. Each of the fore cables has a tension of $5.0 \times 10^3$ N. What is the tension in the aft cable? What force does the foot of the mast exert on the sailboat? Neglect the weight of the mast and assume that the foot of the mast is hinged.

SOLUTION: Figure 14.2b is a "free-body" diagram displaying the forces on the mast. To find the tension $T$ in the aft cable it is convenient to reckon the net torque on the mast about the point $P$, the foot of the mast. This net torque must be zero:

$$10 \text{ m} \times T \times \sin 45° - 7.5 \text{ m} \times 5.0 \times 10^3 \text{ N} \times \sin 30°$$
$$- 10 \text{ m} \times 5.0 \times 10^3 \text{ N} \times \sin 30° = 0$$

from which

$$T = 6.2 \times 10^3 \text{ N}$$

The net force on the mast must also be zero. Figure 14.2b shows the forces on the mast; the force **F** exerted by the boat against the mast has horizontal and vertical components $F_x$ and $F_y$. The net horizontal force is

$$F_x + 5.0 \times 10^3 \cos 60° + 5.0 \times 10^3 \cos 60° - 6.2 \times 10^3 \cos 45° = 0$$

and the net vertical force is

$$F_y - 5.0 \times 10^3 \cos 30° - 5.0 \times 10^3 \cos 30° - 6.2 \times 10^3 \cos 45° = 0$$

which gives

$$F_x = -0.62 \times 10^3 \text{ N}$$

$$F_y = 13.0 \times 10^3 \text{ N}$$

The force exerted by the mast on the boat is opposite to this.

EXAMPLE 3. The bottom end of a meter stick rests on the floor and the top end rests against a wall (Figure 14.3a). If the coefficient of static friction between the stick and the floor and wall is $\mu_s = 0.4$, what is the maximum angle that the stick can make with the wall without slipping?

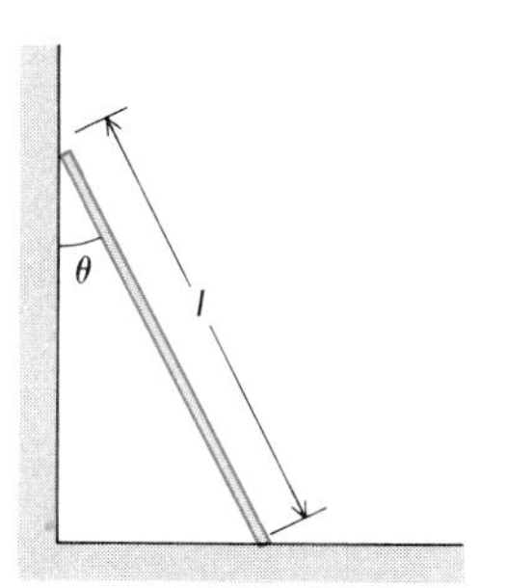


(a)

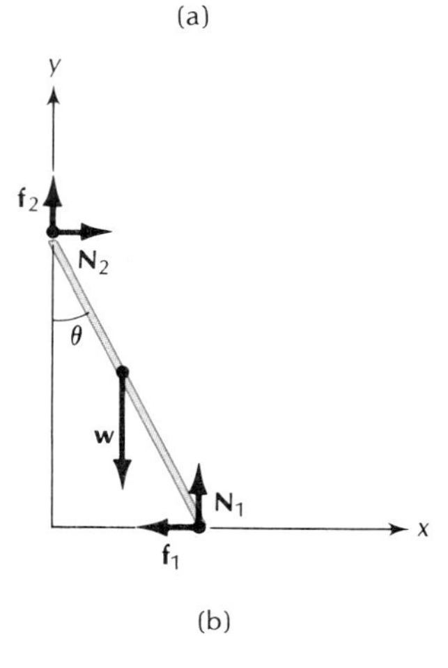


(b)

**Fig. 14.3** (a) Meter stick leaning against a wall. (b) "Free-body" diagram for the meter stick.

SOLUTION: Figure 14.3b shows a "free-body" diagram. The forces acting on the ends of the meter stick are shown resolved into horizontal ($x$) and vertical ($y$) components. If the stick is about to slip, the friction forces have their maximum values

$$f_1 = \mu_s N_1 \qquad f_2 = \mu_s N_2$$

The weight of the stick acts vertically downward at the center of mass.

The sum of the torques of these forces must be zero. If we reckon the torques about the point of contact with the floor, this condition becomes

$$lN_2 \cos\theta + l\mu_s N_2 \sin\theta - \frac{l}{2}Mg \sin\theta = 0 \tag{3}$$

Furthermore, the sums of the horizontal and of the vertical components of these forces must be zero,

$$-\mu_s N_1 + N_2 = 0 \tag{4}$$

$$\mu_s N_2 + N_1 - Mg = 0 \tag{5}$$

Solving the last two equations for $N_1$ and $N_2$, we obtain

$$N_1 = Mg/(\mu_s^2 + 1)$$

$$N_2 = \mu_s Mg/(\mu_s^2 + 1)$$

and substituting these results into Eq. (3), we obtain

$$\frac{\mu_s Mgl}{\mu_s^2 + 1}\cos\theta + \frac{\mu_s^2 Mgl}{\mu_s^2 + 1}\sin\theta - \frac{Mgl}{2}\sin\theta = 0$$

Here, we can cancel $Mgl$ and then solve for $\sin\theta/\cos\theta$:

$$\frac{\sin\theta}{\cos\theta} = \tan\theta = \frac{2\mu_s}{1 - \mu_s^2} \tag{6}$$

With $\mu_s = 0.4$ this gives

$$\tan\theta = \frac{2 \times 0.4}{1 - (0.4)^2} = 0.95$$

or $\theta = 44°$.

EXAMPLE 4. A rectangular box 2 m high, 1 m wide, and 1 m deep stands on the platform of a truck (Figure 14.4a). What is the maximum forward acceleration of the truck that the box can withstand without toppling over? Assume that the box is loaded with a material of uniform density and that it does not slide.

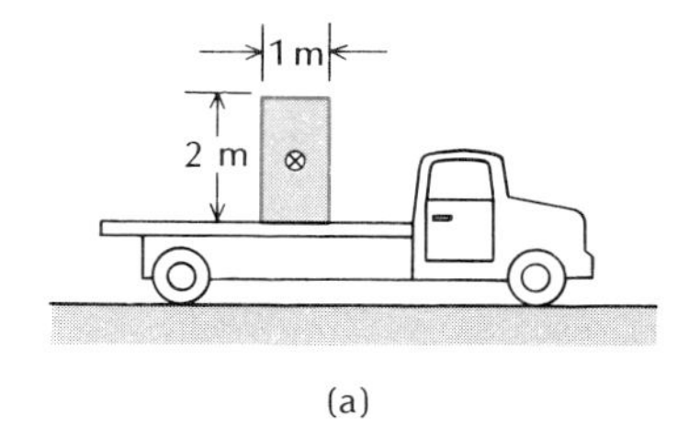


(a)

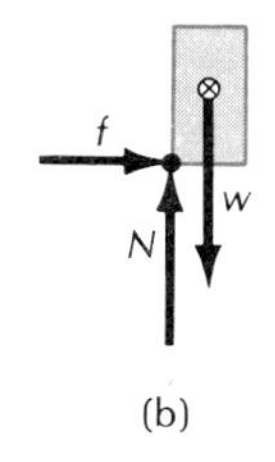


(b)

**Fig. 14.4** (a) Box on a truck. (b) "Free-body" diagram for the box.

SOLUTION: Strictly, this is not a problem of statics, since the translational motion is accelerated; however, the rotational motion involves a question of equilibrium and can be treated by the methods of this section. Under the condition of the problem, the forces on the box are as shown in Figure 14.4b. Both the normal force **N** and the friction force **f** act at the corner. (When the box is about to topple, it only makes contact with the platform at the rear corner.)

Gravity acts at the center of mass, which is 1 m above and 0.5 m in front of the corner. If the box is to remain upright, the net torque about the center of mass must be zero:

$$1.0 \text{ m} \times f - 0.5 \text{ m} \times N = 0 \tag{7}$$

Expressions for $f$ and $N$ can be obtained from the equations for the horizontal and vertical translational motions; the horizontal acceleration is $a$ and the vertical acceleration is zero:

$$f = Ma \tag{8}$$

$$N - Mg = 0 \tag{9}$$

Thus, Eq. (7) becomes

$$1.0 \text{ m} \times Ma - 0.5 \text{ m} \times Mg = 0$$

or

$$a = 0.5 \text{ g} = 4.91 \text{ m/s}^2$$

If the acceleration exceeds this value, rotational equilibrium fails and the box topples.

*Truss bridge*

EXAMPLE 5. The truss supporting the roadway of a **truss bridge** consists of straight beams joined together so as to form two triangular and one square panel (see Figure 14.5a). Assume that at the joints the beams are held together by pins, so the rigidity of the structure is entirely due to the compressional or tensional loading of the beams (in practice, the beams of bridges are riveted or welded together, but the joints exert next to no bending forces on the beams, and are therefore equivalent to pins). The roadway of the bridge has a mass $M$, whose weight is uniformly distributed along the bridge. Assume that the weight of the roadway is entirely supported by the truss, and assume that the beams are massless. Find the compressional or tensional forces in each beam.

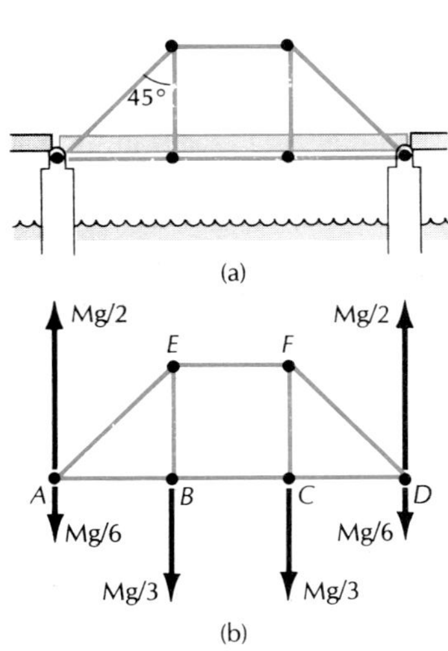


**Fig. 14.5** (a) A truss bridge. (b) "Free-body" diagram for the truss.

SOLUTION: The roadway may be regarded as consisting of three segments of equal length, each of which is a rigid body, since it is supported by one of the rigid horizontal beams. The weight $Mg/3$ of each segment may therefore be regarded as concentrated at each center of mass or, equivalently, one-half of this weight may be regarded as concentrated at each end. The weights acting on the truss are therefore $Mg/6$ at the left end ($A$), $Mg/6 + Mg/6$ at the first pin ($B$), $Mg/6 + Mg/6$ at the second pin ($C$), and $Mg/6$ at the right end ($D$). Figure 14.5b displays these weights on a "free-body" diagram of the truss. However, to determine the forces within a beam, we need to examine the "free-body" diagrams of parts of the truss, instead of the "free-body" diagram of the truss as a whole. For instance, Figure 14.6 shows the forces acting on pin $A$. The forces $T_{AB}$ and $T_{AE}$ are the tension or compression in beams $AB$ and $AE$. The condition for the equilibrium of the pin $A$ (a point mass) is that the sum of all the forces is zero. For the vertical and the horizontal components of these forces this implies

$$Mg/2 - Mg/6 + T_{AE} \cos 45° = 0 \tag{10}$$

$$T_{AE} \cos 45° + T_{AB} = 0 \tag{11}$$

Here, we have adopted the convention that all forces in the beams are to be regarded as tension; if the force is actually a compression, it is to be regarded as a negative tension. The first of these equations yields

$$T_{AE} = -\frac{Mg}{3\cos 45^\circ} = -\frac{\sqrt{2}}{3}Mg$$

and then the second equation yields

$$T_{AB} = -T_{AE}\cos 45^\circ = \frac{1}{3}Mg$$

The negative value of $T_{AE}$ indicates that this force is a compression, and its direction at pin $A$ is opposite to the direction drawn in Figure 14.6.

Likewise, at pin $E$,

$$-T_{AE}\cos 45^\circ - T_{BE} = 0 \tag{12}$$

$$-T_{AE}\cos 45^\circ + T_{EF} = 0 \tag{13}$$

which leads to

$$T_{BE} = \frac{1}{3}Mg$$

$$T_{EF} = -\frac{1}{3}Mg$$

The signs indicate that $T_{EF}$ is a compression and $T_{BE}$ is a tension.

The forces in the remaining beams can be evaluated in the same way (obviously, because of the symmetry of the truss, it suffices to compute the forces in the left half of the truss). The forces in all the beams are summarized in Figure 14.7. Bridges usually have a truss on each side of the roadway, and the diagram schematically represents the combined forces in the paired trusses. The forces in the beams in each separate truss are then one-half of the forces indicated in the diagram.

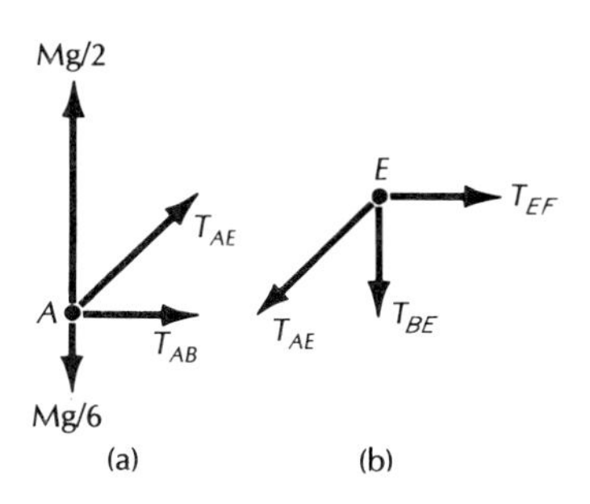


**Fig. 14.6** (a) Forces acting on pin $A$. (b) Forces acting on pin $E$.

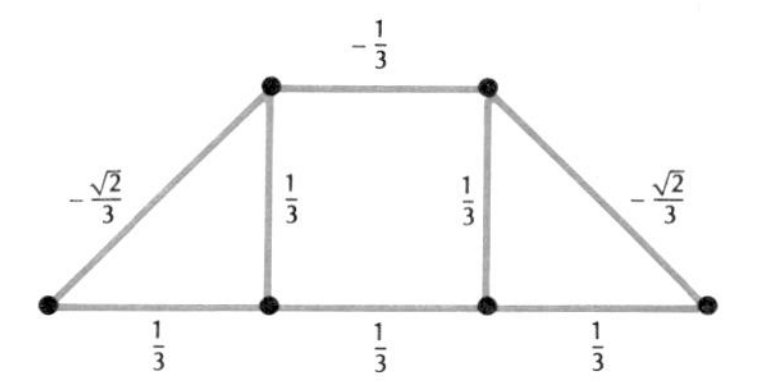


**Fig. 14.7** Forces in the beams of the truss. The forces are expressed as multiples of $Mg$. Positive values indicate tension, negative values compression.

Comments and Suggestions: The above joint-by-joint calculation of the forces in the beams of a truss is called the **method of joints.** It can be used to calculate the forces in a truss of any number of panels of any shape, and also works for three-dimensional trusses. The advantage of this method over other methods for dealing with trusses is that we do not have to bother explicitly with the equilibrium condition for torques; the disadvantage is that to find the forces acting at some pin at the middle of the truss, we have to start at one end and work our way joint-by-joint toward the middle, which is very tedious. Besides, for more complicated trusses, the equilibrium conditions at the pins lead to complicated systems of equations in several unknowns.

*Method of joints for forces in a truss*

EXAMPLE 6. A **suspension bridge** consists of a pair of cables hung between two towers with the roadway suspended from these cables by means of closely spaced vertical wires (see Figures 14.8a). Assume that the weight of the roadway is uniformly distributed along its length and neglect the weight of the cables. What is the shape of the cable?

*Suspension bridge*

(a)

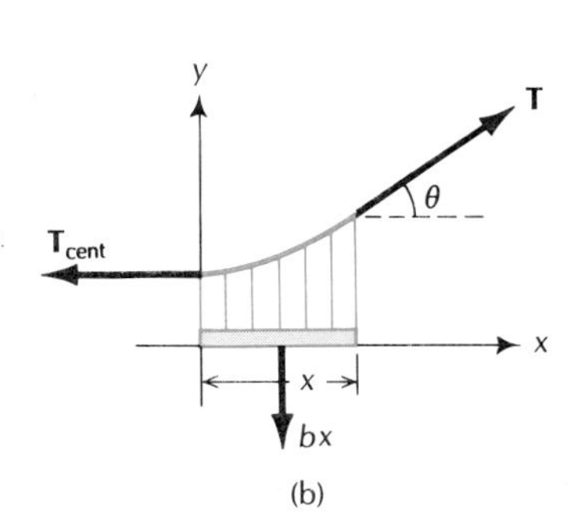


**Fig. 14.8** (a) A suspension bridge. (b) "Free-body" diagram for a segment of the bridge. The weight of the segment is $bx$.

SOLUTION: Consider a segment of the bridge, from the midpoint of the bridge to a distance $x$. Figure 14.8b shows the "free-body" diagram for this segment. The external forces are the weight $bx$ of the segment of roadway (where $b$ is the weight per unit length) and the tension forces at the right and the left of the segment of cable. These tension forces, labeled $\mathbf{T}_{\text{cent}}$ and $\mathbf{T}$ in Figure 14.8b, are *not* of equal magnitudes, since the pull of the vertical wires has a component along the cable, which changes the tension. Assuming that in the "free-body" diagram the tensions represent the combined tensions of the two paired cables of the bridge, the condition for the horizontal equilibrium of the forces is then

$$T \cos \theta = T_{\text{cent}} \tag{14}$$

and the condition for the vertical equilibrium of the forces is

$$T \sin \theta = bx \tag{15}$$

If we divide the second of these equations by the first, we obtain

$$\tan \theta = \frac{bx}{T_{\text{cent}}}$$

But $\tan \theta$ is the slope dy/dx of the cable; thus

$$\frac{dy}{dx} = \frac{bx}{T_{\text{cent}}} \tag{16}$$

or

$$dy = \frac{b}{T_{\text{cent}}} x dx$$

Integrating both sides of this equation, we immediately obtain

$$y = \frac{b}{2T_{\text{cent}}} x^2 + \text{constant} \tag{17}$$

which is the equation of a parabola with vertex at $x = 0$.

Real suspension bridges are only approximately parabolic, since their supporting cables are rather heavy, and the weight affects the shape assumed by the cable. Figure 14.9 shows a photograph of the Golden Gate Bridge, with a parabolic curve superimposed; the fit of the parabolic curve is quite good.

**Fig. 14.9** The Golden Gate Bridge, at the entrance to San Francisco Bay. The colored curve is a parabola; it has been displaced slightly upward, for the sake of clarity

EXAMPLE 7. In order to hold a mooring rope against the strong pull of a ship, a sailor wraps the rope several times around a cylindrical bollard (Figure 14.10). He then finds that by pulling on the tail end with a fairly small tension $\mathbf{T}_1$, he can hold the rope steady against the much larger tension $\mathbf{T}_2$ on the other end. Assume that the coefficient of friction of the rope against the bollard is $\mu_s = 0.5$. Assume that $T_2 = 3.0 \times 10^4$ N and assume that the sailor can exert at most a pull $T_1 = 500$ N. How many turns must the sailor wrap around the bollard?

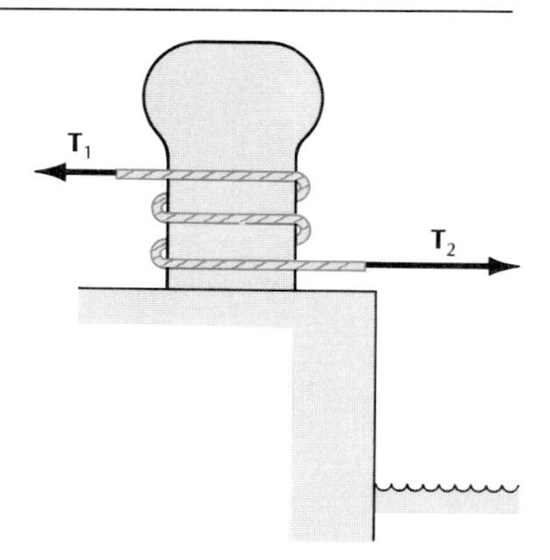


**Fig. 14.10** Rope wrapped around a bollard.

SOLUTION: Figure 14.11a shows a cross section through the bollard, with the rope wrapped around. Consider a small segment of rope subtending a small angle $d\theta$. Figure 14.11b shows a "free-body" diagram of this small segment of rope. The forces on this small segment are the tensions **T** and **T′** at the ends (which are of approximately equal magnitude), the normal force **N**, and the friction force $\mathbf{f}_s$. Under the conditions of the problem, the rope is about to slip, so that the friction force has its maximum value, $f_s = \mu_s N$. The vertical component of each of the tension forces in Figure 14.11b is $T \sin(d\theta/2)$; since $d\theta$ is small, this equals $T(d\theta/2)$.[1] The equilibrium of the radial forces (vertical forces in Figure 14.11b) requires

$$N - T(d\theta/2) - T(d\theta/2) = 0 \tag{18}$$

from which

$$N = T\, d\theta$$

The friction force is then

$$f_s = \mu_s N = \mu_s T\, d\theta \tag{19}$$

The equilibrium of the tangential forces (horizontal forces in Figure 14.11b) then requires that the tension on the right of the segment of rope be slightly larger than the tension on the left so that the increment $dT$ of tension matches the friction force,

$$dT = \mu_s T\, d\theta$$

or

$$\frac{dT}{T} = \mu_s\, d\theta \tag{20}$$

To find the total change of tension along the entire wrapped portion of the rope, we must integrate Eq. (20):

$$\int_{T_1}^{T_2} \frac{dT}{T} = \mu_s \int_{\theta_1}^{\theta_2} d\theta$$

This yields

$$\Big[\ln T\Big]_{T_1}^{T_2} = \mu_s \Big[\theta\Big]_{\theta_1}^{\theta_2}$$

or

$$\ln T_2 - \ln T_1 = \mu_s(\theta_2 - \theta_1) \tag{21}$$

We can also write this as

$$\theta_2 - \theta_1 = \frac{1}{\mu_s} \ln\left(\frac{T_2}{T_1}\right) \tag{22}$$

and, with the numbers specified in our problem,

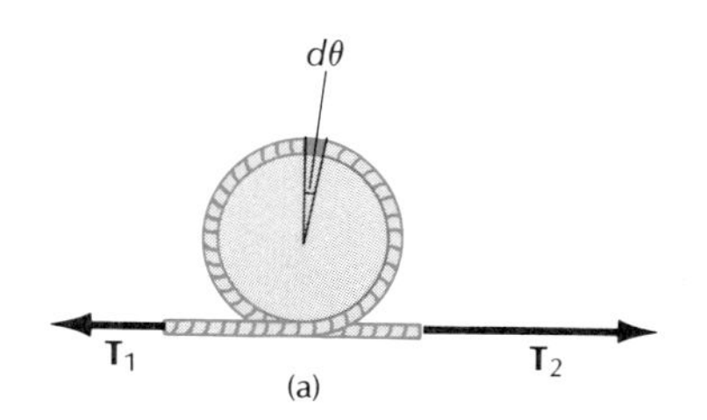


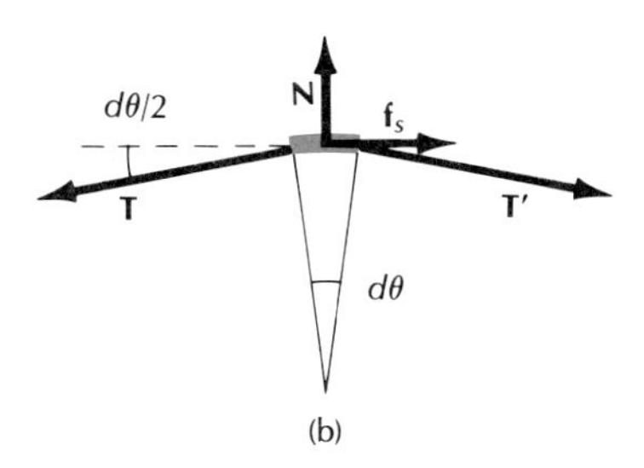


**Fig. 14.11** (a) Cross section through the bollard, with the rope wrapped around. (b) "Free-body" diagram for the rope segment.

[1] $\sin \alpha \cong \alpha$ if $\alpha$ is small.

$$\theta_2 - \theta_1 = \frac{1}{0.5} \ln\left(\frac{3.0 \times 10^4}{500}\right) = 8.2 \text{ radians}$$

or

$$\theta_2 - \theta_1 = 8.2/2\pi = 1.3 \text{ turns}$$

COMMENTS AND SUGGESTIONS: Note that we can rewrite Eq. (21) as

$$\ln \frac{T_2}{T_1} = \mu_s(\theta_2 - \theta_1)$$

or, if we take the exponential function of each side of this equation,

$$\frac{T_2}{T_1} = e^{\mu_s(\theta_2 - \theta_1)} \tag{23}$$

This shows that the tension varies exponentially along the portion of the rope in contact with the bollard.

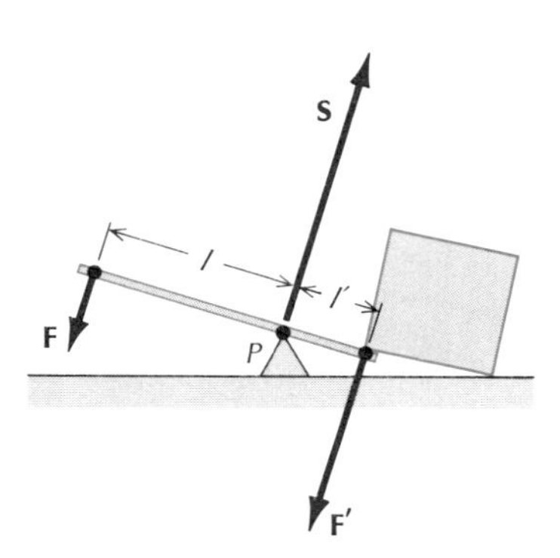


**Fig. 14.12** A lever. The vectors show the forces acting on the lever; **F′** is our push, **F** is the push of the load, and **S** is the supporting force of the pivot. The force exerted by the lever on the load is of the same magnitude as **F′**, but of opposite direction.

## 14.3 Levers and Pulleys

A **lever** consists of a rigid bar swinging on a pivot (see Figure 14.12). If we apply a force $F$ at the long end, the short end of the bar pushes against a load with a larger force $F'$. The relation between $F$ and $F'$ follows from the usual condition of static equilibrium: the net torque about the pivot point $P$ must be zero. If the forces are at right angles to the lever, as in Figure 14.12, then the condition on the net torque is $Fl - F'l' = 0$, and

$$\boxed{\frac{F'}{F} = \frac{l}{l'}} \tag{24}$$

Thus, the forces are in the inverse ratio of the distances from the pivot point. For a powerful lever, we must make the lever arm $l$ as long as possible and the lever arm $l'$ as short as possible. The ratio $F'/F$ of the force delivered by the lever to the force we must supply is called the **mechanical advantage.**

*Mechanical advantage*

The principle of the lever finds application in many hand tools, such as pliers and bolt cutters. The handles of these tools are long, and the working ends are short, yielding an enhancement of the force exerted by the hand. A simple manual winch also relies on the principle of the lever. The crank of the winch is long and the drum of the winch, which acts as the short lever arm, is small (see Figure 14.13). The force the winch delivers to the rope attached to the drum is then larger than the force exerted by the hand pushing on the crank. Compound winches, used for trimming sails on sailboats, have internal sets of gears which provide a larger mechanical advantage; in essence, such compound winches stagger one winch within another, so the force ratio generated by one winch is further multiplied by the force ratio of the other.

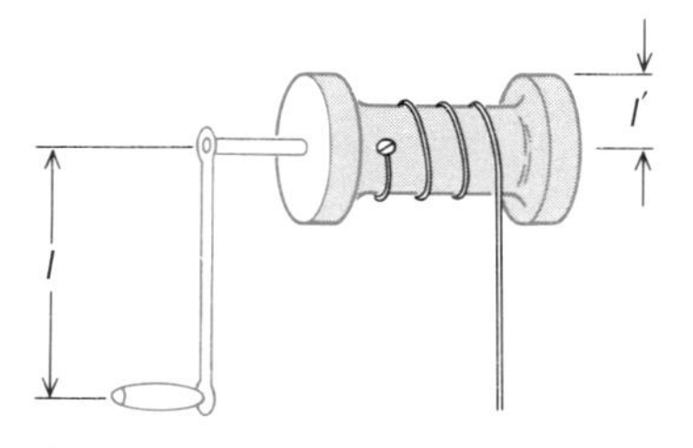


**Fig. 14.13** A manual winch.

In the human skeleton, many bones play the roles of levers which permit muscles or groups of muscles to support or to move the body. For example, Figure 14.14 shows the bones of the human foot; these act as a lever, hinged at the ankle. The rear end of this lever, at the heel, is tied to the muscles of the calf by the Achilles tendon, and the front end of the lever is in contact with the ground, at the ball of the foot. When the muscle contracts, it rotates the heel about the ankle, and presses the ball of the foot against the ground, thereby lifting the entire body on tiptoe. Note that muscle is attached to the short end of this lever—the muscle must provide a larger force than the force generated at the ball of the foot. From an engineering point of view, it would be advantageous to install a longer projecting spur at the heel of the foot and attach the Achilles tendon to the end of this spur, but this would make for a very clumsy foot. In most of the levers found in the human skeleton, the muscle is attached to the short end of the lever.

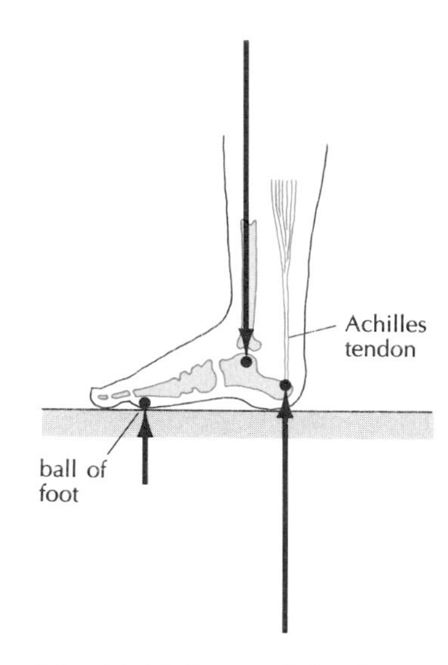


**Fig. 14.14** Bones of the foot acting as a lever.

Equation (15) is valid only if the forces are applied at right angles to the lever. A similar equation is valid if the forces are applied at some other angle, but instead of the lengths $l$ and $l'$ of the lever, we must substitute the lengths of the moment arms of the forces, that is, the perpendicular distances between the pivot point and the lines of action of the forces. These moment arms play the role of effective lengths of the lever.

EXAMPLE 8. To help his horses drag a heavy wagon up a hill, a teamster pushes forward at the top of one of the wheels (see Figure 14.15a). If he pushes with a force of 800 N, what force does he generate on the wagon?

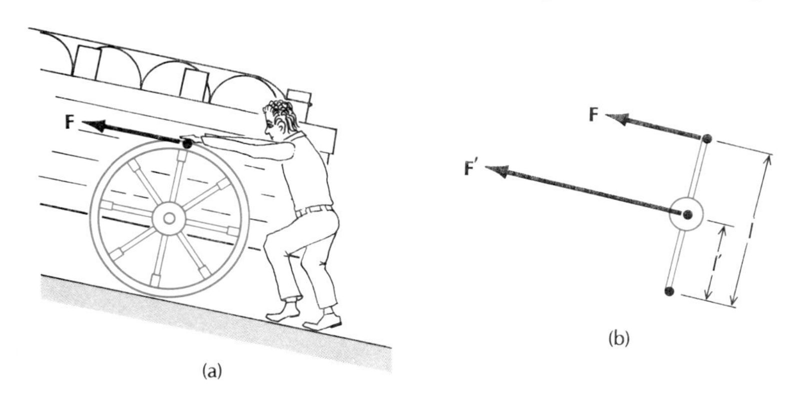


**Fig. 14.15** (a) Teamster pushing top of wheel forward. (b) The diameter of the wheel is a lever.

SOLUTION: From section 13.5 we know that the point of contact of the wheel with the ground is instantaneously at rest. Consequently, the vertical diameter of the wheel acts as a lever, pivoted at the ground (see Figure 14.15b). The ratio of the lengths $l$ and $l'$ of this lever is $l/l' = 2/1$, giving a mechanical advantage of 2, and thus

$$F' = 2 \times F = 2 \times 800 \text{ N} = 1600 \text{ N}$$

This is the force with which the wheel pushes the wagon forward at the axle. By pushing on the top of the wheel, instead of pushing directly on the wagon, the teamster effectively doubles his force.

When forces are applied to a rigid body by means of flexible strings or ropes, a pulley is often used to change the direction of the string or rope, and the direction of the force exerted on the body. If the pulley is frictionless, the tension at each point of a flexible rope passing over the pulley is the same. Thus, there is no gain of mechanical advantage to be had from a single pulley; the only benefit is that it permits us to pull more comfortably than if we attempted to lift the load directly.

*Block and tackle*

However, an arrangement of several pulleys linked together, called a **block and tackle,** can provide a large gain of mechanical advantage. For example, consider the arrangement of three pulleys shown in Figure 14.16a; the axles of the two upper pulleys are bolted together, and they are linked to each other and to the third pulley by a single rope. If the rope segments linking the pulleys are parallel and there is no friction, then the mechanical advantage of this arrangement is 3, that is, the magnitudes of the forces $F$ and $F'$ are in the ratio of 1 to 3. This can be most easily understood by drawing the "free-body" diagram for the lower portion of the pulley system, including the load (see Figure 14.16b). In this diagram, the three ropes leading upward have been cut off and replaced by the forces exerted on them by the external (upper) portions of ropes. Since the tension is the same everywhere along the rope, the forces pulling upward on each of the three rope ends shown in the "free-body" diagram all have the same magnitude $F$, and thus the net upward force is $3F$.

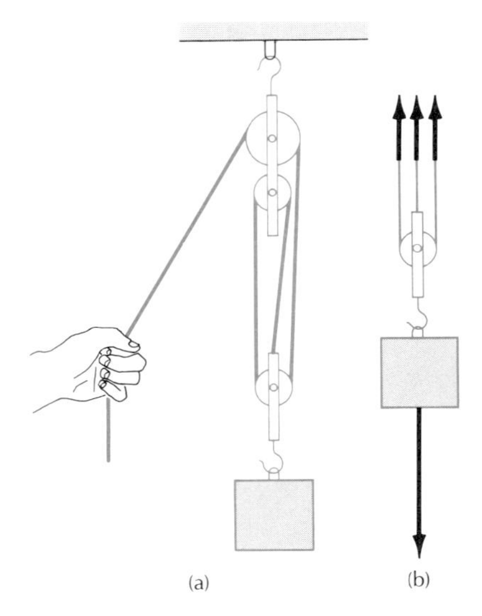


**Fig. 14.16** (a) Block and tackle. (b) "Free-body" diagram for lower portion of system.

Block and tackle arrangements have many practical applications. They are used to provide the proper tension in overhead power cables for electric trains and trams (see Figure 14.17); without such an arrangement, the cables would sag on warm days when thermal expansion increases their length, and they would be stretched excessively tight and perhaps snap on cold days, when they contract. One common cause of power failures on cold winter nights is the snapping of power lines lacking such compensating pulleys.

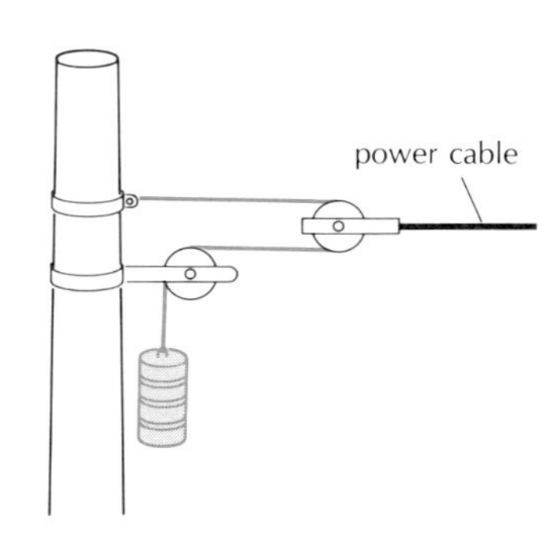


**Fig. 14.17** Block and tackle used for tensioning power line.

The mechanical advantage provided by levers, arrangements of pulleys, or other devices can be calculated in a general and elegant way by appealing to the law of conservation of energy. A lever merely transmits the work we supply at one end to the load at the other end. We can express this equality of work input and work output by

$$F'\,\delta x' = F\,\delta x \tag{25}$$

where $\delta x$ is a small displacement of our hand in the direction of the force and $\delta x'$ the displacement of the load. According to this equation, the forces $F'$ and $F$ are in the inverse ratio of the displacements,

$$\frac{F'}{F} = \frac{\delta x}{\delta x'} \tag{26}$$

Consider, now, the rotation of the lever by a small angle (see Figure 14.18). The distances $\delta x$ and $\delta x'$ are in the same ratio as the lever arms $l$ and $l'$; thus, we immediately recognize from Eq. (26) that the mechanical advantage of the lever is $l/l'$.

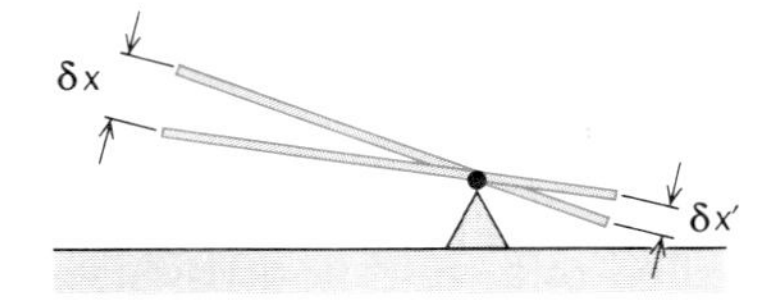


**Fig. 14.18** Rotation of lever by a small angle produces displacements $\delta x$ and $\delta x'$ of the ends.

Likewise, we immediately recognize from Eq. (26) that the mechanical advantage of the arrangement of pulleys shown in Figure 14.16 is

3, since whenever our hand pulls a length $\delta x$ of rope out of the upper pulley, the load moves upward by a distance of only $\delta x/3$.

## 14.4 Elasticity of Materials

Although solid bodies, such as bars or blocks of steel, are nearly rigid, they are not exactly rigid, and they will deform if a large enough force is applied to them. A solid bar may be thought of as a very stiff spring. If the force is fairly small, this "spring" will suffer only an insignificant deformation, but if the force is large, it will suffer a noticeable deformation. Provided that the force and the deformation remain within some limits, the deformation of a solid body is **elastic,** which means that the body returns to its original shape once the force ceases to act. Such elastic deformations of a solid body obey Hooke's Law: the deformation is proportional to the force. But the constant of proportionality is small, giving a small deformation unless the force is large.

A solid block of material can suffer several kinds of deformation, depending on how the force is applied. If one end of the body is held fixed, and the force pulls on the other end, the deformation is a simple **elongation** of the body (see Figure 14.19). If one side of the body is held fixed, and the force pushes tangentially along the other side, then the deformation is a **shear,** which changes the shape of the body from a rectangular parallelepiped to a rhomboidal parallelepiped (see Figure 14.20a). During this deformation, the parallel layers of the body slide past one another just as the pages of a book slide past one another when we push along its cover (see Figure 14.20b). If the force is applied from all sides simultaneously, by immersing the body in a fluid and subjecting it to some pressure, then the deformation is a **compression** of the volume of the body, without any change of the geometrical shape (see Figure 14.21).

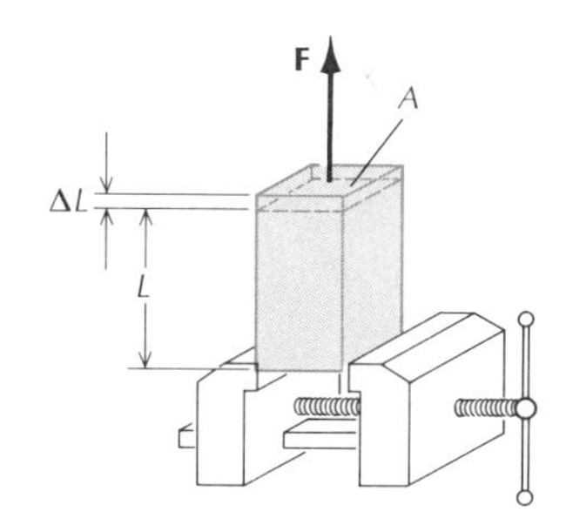


**Fig. 14.19** Tension applied to the end of a block of material causes elongation.

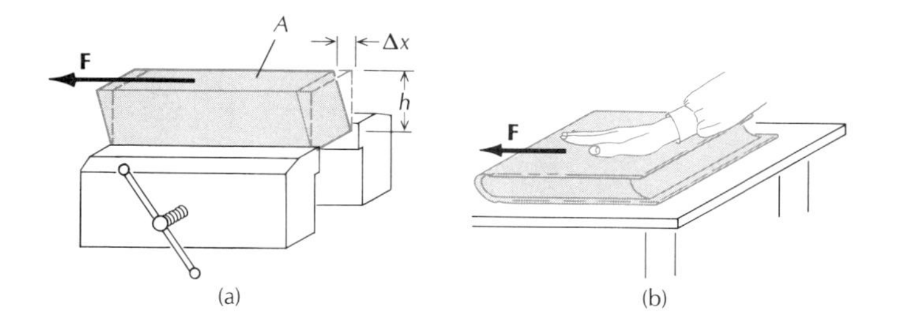


**Fig. 14.20** (a) Tangential force applied to the side of a block of material causes shear. (b) When such a tangential force is applied to the cover of a book, the pages slide past one another.

In all of these cases, the fractional deformation is directly proportional to the applied force and inversely proportional to the area over which the force is distributed. For instance, if a given force produces an elongation of 1% when pulling on the end of a block, then the same force pulling on the end of a block of, say, twice the cross-sectional area will produce an elongation of $\frac{1}{2}$%. This can be readily understood if we think of the block as consisting of parallel rows of atoms linked by springs, which represent the interatomic forces that hold the atoms in their places. When we pull on the end of the block with a given force, we stretch the interatomic springs by some amount; and when we pull on a block of twice the cross-sectional area, we have to stretch twice as many springs, and therefore the force acting on each spring is only half as large and produces only half the elongation in each spring.

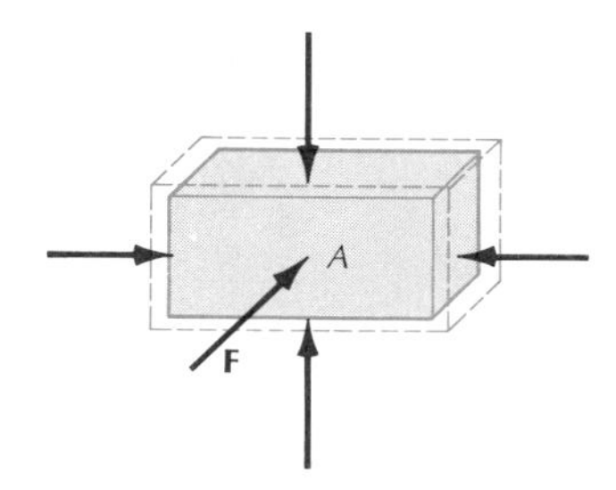


**Fig. 14.21** Pressure applied to all sides of a block of material causes compression.

Furthermore, since the force applied to the end of a row of atoms is communicated to all the interatomic springs in that row, a given force produces a given elongation in each spring in a row. The net elongation of the block is therefore proportional to the number of atoms in the row, which means it is proportional to the (initial) length of the block. Thus, if a block elongates by 0.1 mm when subjected to a given force, then a block, say, twice as long will elongate by 0.2 mm when subjected to the same force, which means the fractional elongation, or the percent elongation, remains the same.

To express the relationships among elongation, force, and area mathematically, let us consider a block of initial length $L$ and cross-sectional area $A$. If a force $F$ pulls on the end of this block, the elongation is $\Delta L$, and the fractional elongation is $\Delta L/L$. This fractional elongation is directly proportional to the force and inversely proportional to the area $A$:

$$\boxed{\frac{\Delta L}{L} = \frac{1}{Y}\frac{F}{A}} \tag{27}$$

*Young's modulus*

Here the quantity $Y$ is the constant of proportionality (this constant is traditionally written as $1/Y$, so it divides the right side, instead of multiplying it; thus a stiff material has a large value of $Y$). The constant $Y$ is called **Young's modulus.** Table 14.1 lists values of Young's moduli for some solid materials. Note that if, instead of exerting a pull on the end of the block, we exert a push, then $F$ in Eq. (27) must be reckoned as negative, and the change $\Delta L$ of length will then likewise be negative—the block becomes shorter.

**Table 14.1** ELASTIC MODULI OF SOME MATERIALS[a]

| *Material* | *Bulk modulus* | *Shear modulus* | *Young's modulus* | *Ultimate tensile strength* |
|---|---|---|---|---|
| Aluminum | $7.5 \times 10^{10} N/m^2$ | $2.5 \times 10^{10} N/m^2$ | $7.0 \times 10^{10} N/m^2$ | $0.78 \times 10^{8} N/m^2$ |
| Copper | 13.5 | 4.4 | 11.0 | 2.4 |
| Gold | 16.5 | 2.8 | 8.0 | 2.5 |
| Iron, | | | | |
| cast | 9.5 | 5.0 | 11.0 | 1.1 |
| wrought | 16.0 | 7.7 | 21.0 | 3.3 |
| Lead | 4.1 | 0.6 | 1.6 | 2.1 |
| Nickel | 17.0 | 7.8 | 21.0 | 4.1 |
| Steel, | | | | |
| cast | 17.0 | 7.5 | 20.0 | 5.2 |
| mild | 16.0 | 8.0 | 22.0 | 3.8 |
| Brass | 6.0 | 3.5 | 9.0 | 3.8 |
| Bronze | 9.0 | 3.7 | 10.5 | 2.9 |
| Nylon | 0.59 | 0.12 | 0.36 | 3.2 |
| Bone (long) | 3.1 | 1.2 | 3.2 | — |
| Water | 0.22 | — | — | — |
| Glycol | 0.27 | — | — | — |
| Carbon tetrachloride | | | | |
| Methanol | 0.10 | — | — | — |
| | 0.083 | — | — | — |

[a] The values of the elastic constants of solids depend markedly on the preparation and treatment of the sample; values in this table are representative, but not entirely reliable.

In engineering language, the force per unit area is usually called the **stress** and the fractional deformation is called the **strain.** In this terminology, Eq. (27) simply states that the strain is proportional to the stress.

*Stress and strain*

This proportionality of strain and stress is also valid for shearing deformations and compressional deformations, provided we adopt a suitable definition of strain, or fractional deformation, for these cases. For shear, the fractional deformation is defined as the ratio of the sideway displacement $\Delta x$ of the edge of the block to the height $h$ of the block (see Figure 14.20a). This fractional deformation is directly proportional to the force $F$ and inversely proportional to the area $A$ (note that the relevant area $A$ is now the top area of the block, where the force is applied):

$$\boxed{\frac{\Delta x}{h} = \frac{1}{S}\frac{F}{A}} \tag{28}$$

Here, the constant of proportionality $S$ is called the **shear modulus.** Table 14.1 includes values of shear moduli of solids.

*Shear modulus*

For compression, the fractional deformation is defined as the ratio of the decrease $\Delta V$ of the volume to the initial volume, and this fractional deformation is, again, proportional to the force $F$ pressing on each face of the block and inversely proportional to the area $A$ of that face:

$$\boxed{\frac{\Delta V}{V} = \frac{1}{B}\frac{F}{A}} \tag{29}$$

The constant of proportionality $B$ in this equation is called the **bulk modulus.** Table 14.1 includes values of bulk moduli for solids. This table also includes values of bulk moduli for some liquids. The formula (20) is equally valid for solids and for liquids—when we squeeze a liquid from all sides, it will suffer a compression. Note that Table 14.1 does *not* include values of Young's moduli and of shear moduli for liquids; elongation and shear stress are not supported by a liquid—we can elongate or shear a block of liquid as much as we please without having to exert any significant force.

*Bulk modulus*

EXAMPLE 9. A piano wire of steel of length 1.8 m and radius 0.3 mm is subjected to a tension of 70 N by means of a weight attached to its lower end (see Figure 14.22). By how much does this wire stretch in excess of its initial length? Young's modulus for the piano wire is $1.9 \times 10^{11}\ \text{N/m}^2$.

SOLUTION: The cross-sectional area of the wire is

$$A = \pi r^2 = \pi \times (0.0003\ \text{m})^2 = 2.8 \times 10^{-7}\text{m}^2$$

and the force per unit area, or the stress, is

$$\frac{F}{A} = \frac{70\ \text{N}}{2.8 \times 10^{-7}\ \text{m}^2} = 2.5 \times 10^{8}\ \text{N/m}^2$$

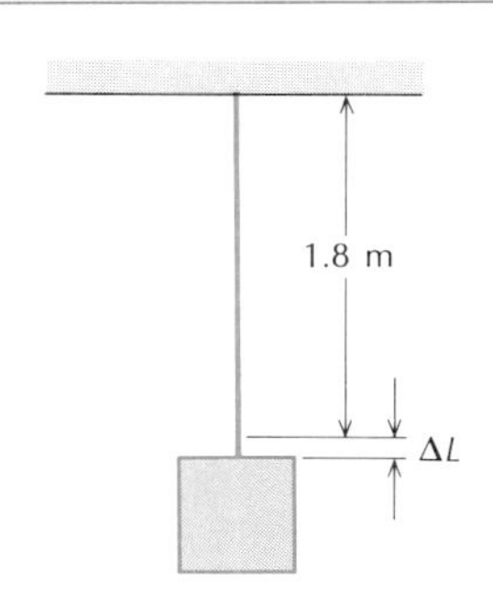


**Fig. 14.22** Elongation of a piano wire.

With $Y = 1.9 \times 10^{11}$ N/m², Eq. (27) gives a strain

$$\frac{\Delta L}{L} = \frac{1}{Y}\frac{F}{A} = \frac{1}{1.9 \times 10^{11}\ \text{N/m}^2} \times 2.5 \times 10^8\ \text{N/m}^2$$

$$= 1.3 \times 10^{-3}$$

from which

$$\Delta L = 1.3 \times 10^{-3} \times L = 1.3 \times 10^{-3} \times 1.8\ \text{m}$$

$$= 2.3 \times 10^{-3}\ \text{m} = 2.3\ \text{mm}$$

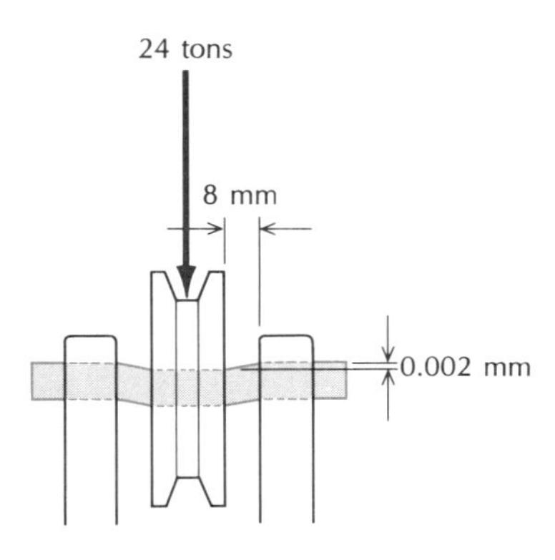


**Fig. 14.23** Deformation of axle of pulley.

EXAMPLE 10. The axle of the pulley at the end of the boom of a crane rests on two bearings. The axle is made of cast steel. The distance between each bearing and the adjacent side of the pulley is 8 mm. If the axle is to sag downward by no more than 0.002 mm when subjected to a load of 24 metric tons (see Figure 14.23), what must be the least diameter of the axle?

SOLUTION: The deformation of each side of the axle is a shear. Each side of the axle has to support half of the load, that is, $F/2$. If the radius of the axle is $R$, its cross-sectional area is $\pi R^2$, and the shearing stress is $F/2A = F/2\pi R^2$. The strain is therefore

$$\frac{\Delta x}{h} = \frac{1}{S}\frac{F}{2\pi R^2} \tag{30}$$

The shear modulus of the steel is $7.5 \times 10^{10}$ N/m². With $x = 0.002$ mm and $h = 8$ mm, we then find a radius

$$R = \sqrt{\frac{Fh}{2\pi S \Delta x}}$$

$$= \sqrt{\frac{24 \times 10^3\ \text{kg} \times 9.8\ \text{m/s}^2 \times 8 \times 10^{-3}\ \text{m}}{2\pi \times 7.5 \times 10^{10}\ \text{N/m}^2 \times 2 \times 10^{-6}\ \text{m}}} = 4.5 \times 10^{-2}\ \text{m} = 4.5\ \text{cm}$$

The simple uniform deformations of elongation, shear, and compression described above require a rather special arrangement of forces. In general, the forces applied to a solid body will produce non-uniform elongation, shear, and compression. For instance, when a beam supported at its ends sags in the middle because of its own weight or the weight of a load, it is elongated along its lower edge, and compressed along its upper edge.

Finally, note that the formulas (27)–(29) are valid only as long as the deformation is reasonably small—a fraction of a percent or so. If the deformation becomes excessive, the material will deviate from the simple proportionality of strain and stress. At some larger deformation, the material will exceed its **elastic limit,** that is, the material will suffer permanent damage (by plastic flow), and it will *not* return to its original size and shape when the force ceases. If the deformation is even larger, the material will finally break apart or crumble. The maximum stress that the material can tolerate is called the **ultimate strength** of the material. Figure 14.24 is a schematic plot of stress vs. strain for a typical sample of metal; the different critical points are identified on this plot. The values of the elastic limit and of the ultimate strength depend on the material and on the kind of deformation to which it is

*Elastic limit*

*Ultimate strength*

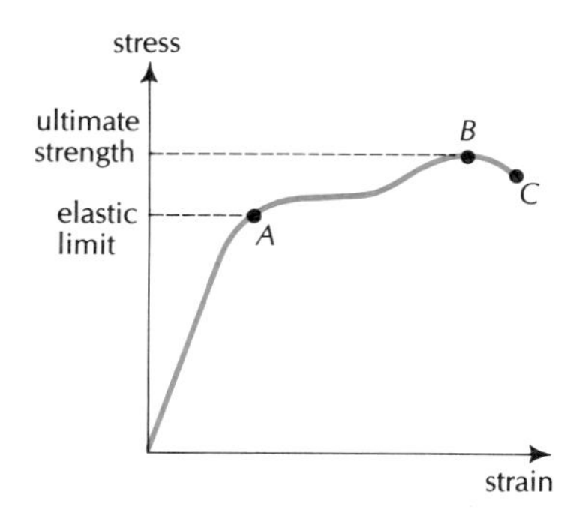


**Fig. 14.24** Stress vs. strain for a typical sample of metal. *A* is the elastic limit, or yield point. *B* is the ultimate strength. If a stress larger than this is applied, the sample cannot attain equilibrium (that is, it cannot provide a large enough restoring force to balance the stress); thus, the sample continues to stretch catastrophically, and finally breaks apart at point *C*.

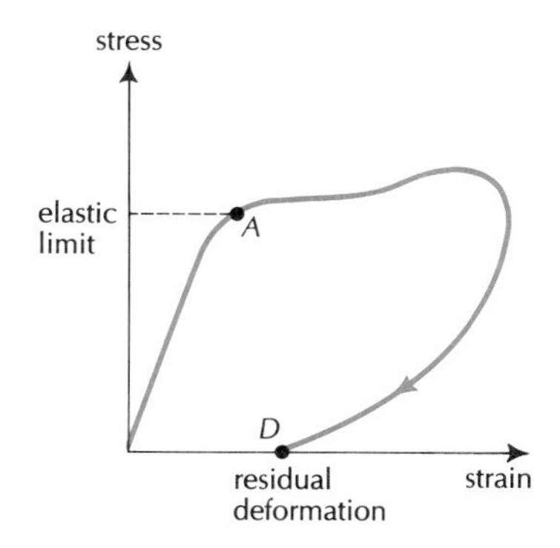


**Fig. 14.25** Stress vs. strain for a sample of metal that was first subjected to a gradually increasing stress, which deformed the sample beyond its elastic limit. The stress was then gradually reduced to zero. The point *D* indicates the residual, permanent deformation of the sample.

subjected. For instance, cast steel has an ultimate tensile strength of about $5.2 \times 10^8$ N/m$^2$, and an ultimate shearing strength of $2.5 \times 10^8$ N/m$^2$. Table 14.1 lists the ultimate tensile strengths of some other solids.

If the stress on the sample of metal is reduced before the sample breaks, then the deformation of the sample will decrease. However, for a sample that has been deformed beyond its elastic limit, the decrease of deformation fails to match the original increase in deformation. For the plot of stress vs. strain, this means that during the reduction of stress the plot does not retrace itself toward the origin; instead, the plot curves downward, below the original plot, and reaches zero stress at some finite, residual value of the strain (see Figure 14.25). This residual value of the strain represents the permanent deformation suffered by the sample. The failure of the plot of stress vs. strain to retrace itself during the reduction of stress implies that, for a sample deformed beyond its elastic limit, the relationship between stress and strain is not unique—the strain depends not only on the current value of the stress, but also on the past history of stress to which the sample has been subjected. Such a dependence on past history is called **hysteresis.**

*Hysteresis*

## SUMMARY

**Static equilibrium:** The sums of the external forces and of the external torques on a rigid body are zero. Gravity effectively acts at the center of mass.

**Mechanical advantage of lever:** $\dfrac{F'}{F} = \dfrac{l}{l'}$

**Elongation of elastic material:** $\dfrac{\Delta L}{L} = \dfrac{1}{Y}\dfrac{F}{A}$

**Shear:** $\dfrac{\Delta x}{h} = \dfrac{1}{S}\dfrac{F}{A}$

**Compression of volume:** $\dfrac{\Delta V}{V} = \dfrac{1}{B}\dfrac{F}{A}$

## QUESTIONS

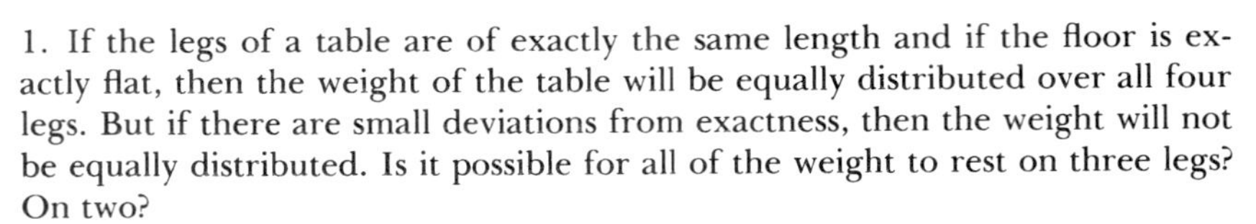

1. If the legs of a table are of exactly the same length and if the floor is exactly flat, then the weight of the table will be equally distributed over all four legs. But if there are small deviations from exactness, then the weight will not be equally distributed. Is it possible for all of the weight to rest on three legs? On two?

**Fig. 14.26** The Bayonne Bridge, a steel arch bridge connecting Staten Island and New Jersey.

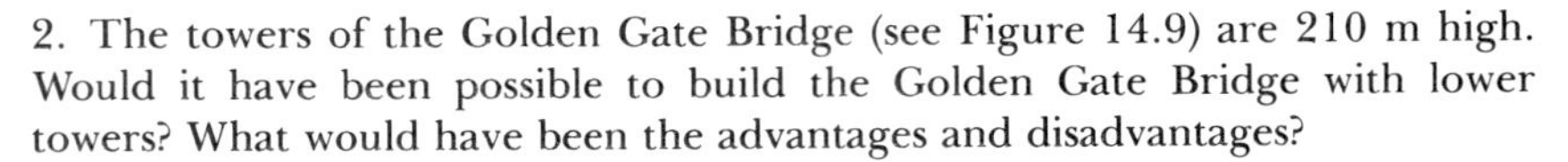

2. The towers of the Golden Gate Bridge (see Figure 14.9) are 210 m high. Would it have been possible to build the Golden Gate Bridge with lower towers? What would have been the advantages and disadvantages?

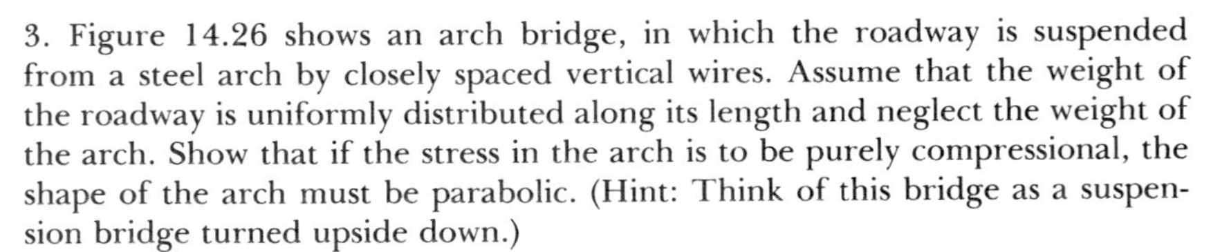

3. Figure 14.26 shows an arch bridge, in which the roadway is suspended from a steel arch by closely spaced vertical wires. Assume that the weight of the roadway is uniformly distributed along its length and neglect the weight of the arch. Show that if the stress in the arch is to be purely compressional, the shape of the arch must be parabolic. (Hint: Think of this bridge as a suspension bridge turned upside down.)

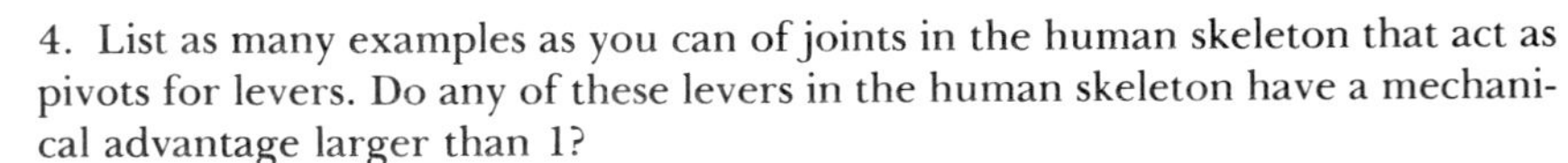

4. List as many examples as you can of joints in the human skeleton that act as pivots for levers. Do any of these levers in the human skeleton have a mechanical advantage larger than 1?

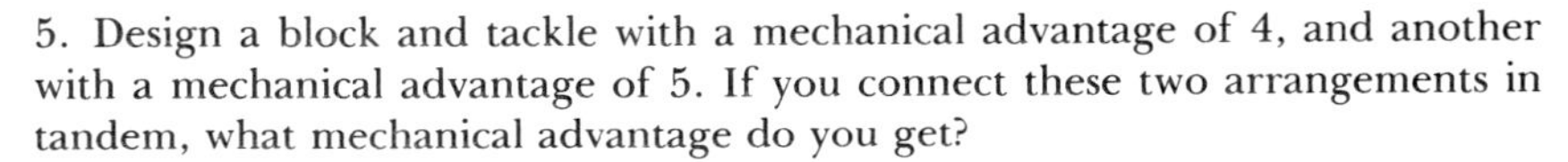

5. Design a block and tackle with a mechanical advantage of 4, and another with a mechanical advantage of 5. If you connect these two arrangements in tandem, what mechanical advantage do you get?

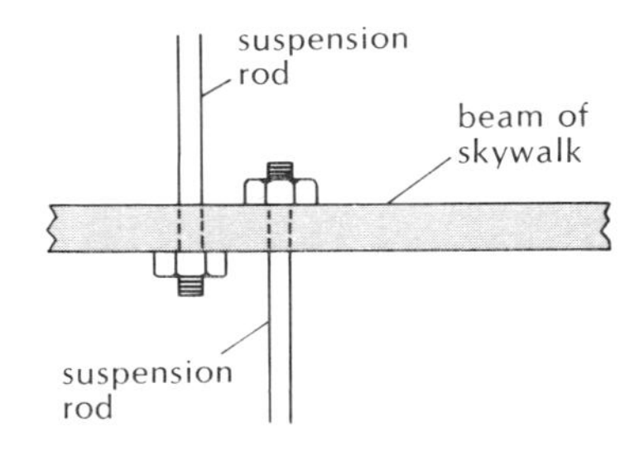


**Fig. 14.27** Beams of skywalk.

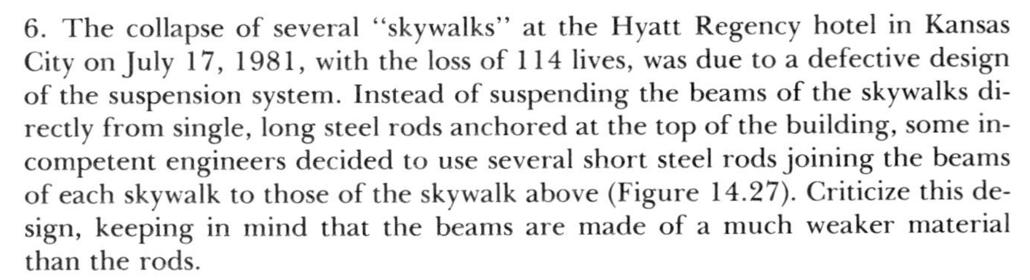

6. The collapse of several "skywalks" at the Hyatt Regency hotel in Kansas City on July 17, 1981, with the loss of 114 lives, was due to a defective design of the suspension system. Instead of suspending the beams of the skywalks directly from single, long steel rods anchored at the top of the building, some incompetent engineers decided to use several short steel rods joining the beams of each skywalk to those of the skywalk above (Figure 14.27). Criticize this design, keeping in mind that the beams are made of a much weaker material than the rods.

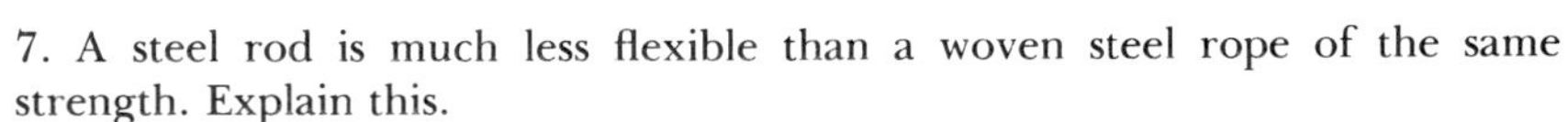

7. A steel rod is much less flexible than a woven steel rope of the same strength. Explain this.

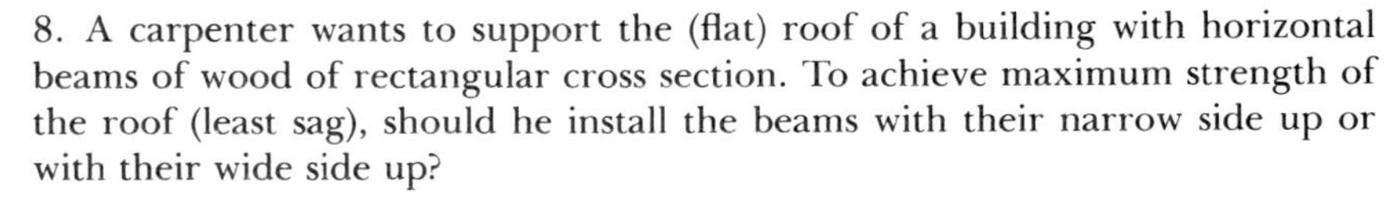

8. A carpenter wants to support the (flat) roof of a building with horizontal beams of wood of rectangular cross section. To achieve maximum strength of the roof (least sag), should he install the beams with their narrow side up or with their wide side up?

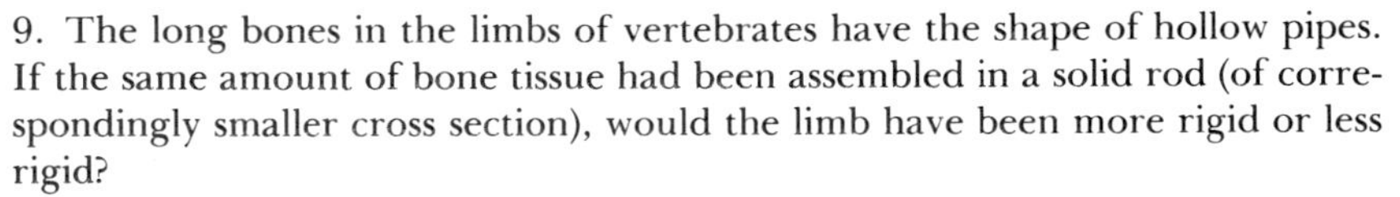

9. The long bones in the limbs of vertebrates have the shape of hollow pipes. If the same amount of bone tissue had been assembled in a solid rod (of correspondingly smaller cross section), would the limb have been more rigid or less rigid?

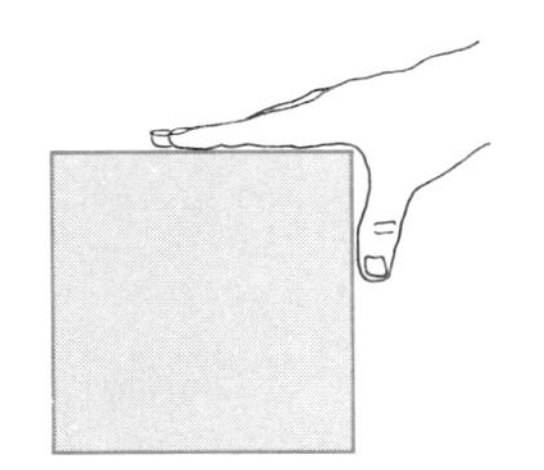

**Fig. 14.28**

## PROBLEMS

### Section 14.2

1. You want to pick up a nearly massless rectangular cardboard box by grabbing its top and side between your forefinger and thumb (see Figure 14.28).

Show that this is impossible unless the coefficient of friction between your fingers and the box is at least 1.

2. Figure 14.29 shows cargo hanging from the loading boom of a ship. If the boom is inclined at the angle shown in this figure and the cargo has a mass of 2.5 metric tons, what is tension in the upper cable? What is compressional force in the boom? Neglect the mass of the boom.

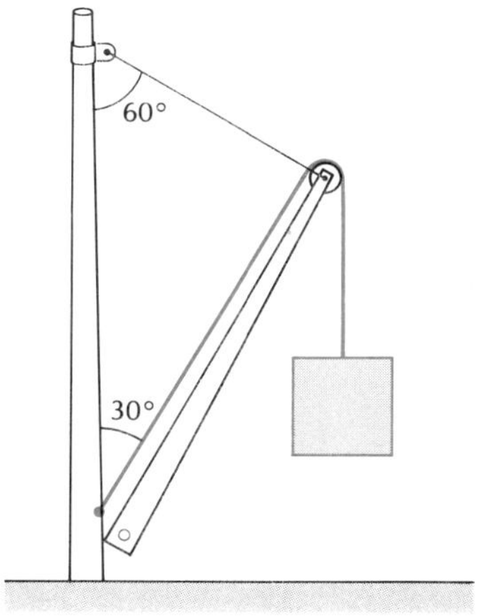

**Fig. 14.29**

3. Figure 14.30 shows the arrangement of wheels on a passenger engine of the Caledonian Railway. The numbers give the distances between the wheels in feet and the downward forces that each wheel exerts on the track in short tons (1 short ton = 2000 lbf; the numbers for the forces include both the right and the left wheels). From the information given, find how far the center of mass of the engine is behind the front wheel.

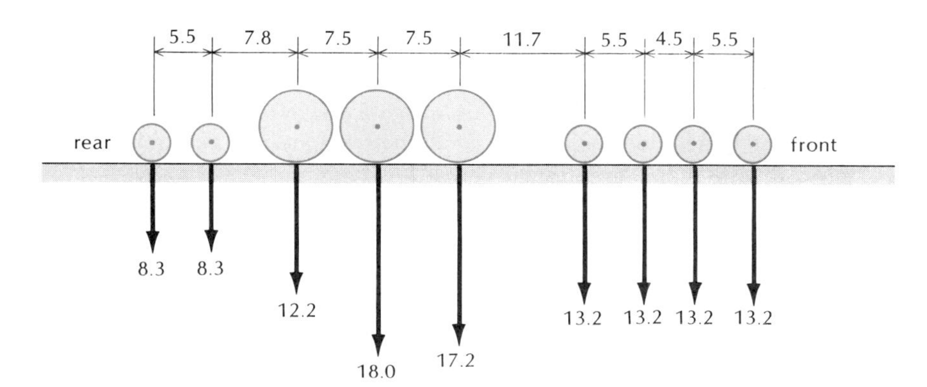


**Fig. 14.30**

4. A door made of a uniform piece of wood measures 1 m by 2 m and has a mass of 18 kg. The door is entirely supported by two hinges, one at the bottom corner and one at the top corner. Find the force (magnitude and direction) that the door exerts on each hinge. Assume that the *vertical* force on each hinge is the same.

5. Repeat the calculations of Example 1 assuming that the bridge has a slope of 1:7, with the left end higher than the right.

6. A meter stick of wood of 0.40 kg is nailed to the wall at the 75-cm mark. If the stick is free to rotate about the nail, what horizontal force must you exert at the upper (short) end to deflect the stick 30° to one side?

*7. Consider a heavy cable of diameter $d$ and density $\rho$ from which hangs a load of mass $M$. What is the tension in the cable as a function of the distance from the lower end?

*8. Figure 14.31 shows two methods for supporting the mast of a sailboat against the lateral force exerted by the pull of the sail. In Figure 14.31a, the shrouds (wire ropes) are led directly to the top of the mast; in Figure 14.31b, the shrouds are led around a rigid pair of spreaders. Suppose that the dimensions of the mast and the boat are as indicated in this figure, and that the pull of the sail is equivalent to a horizontal force of 2400 N acting from the left at half the height of the mast. The foot of the mast permits the mast to tilt, so the only lateral support of the mast is that provided by the shrouds. What is the excess tension in the left shroud supporting the mast in case (a)? In case (b)? Which arrangement is preferable?

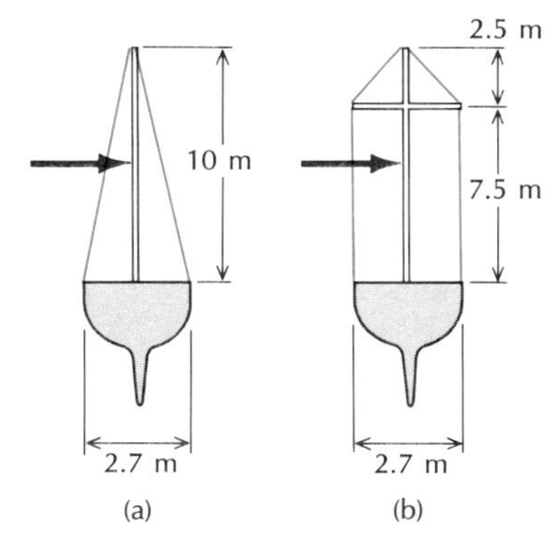


**Fig. 14.31** Two methods for supporting the mast of a sailboat.

*9. The plant of the foot of an average male is 26 cm and the height of the center of mass above the floor is 1.03 m. When standing upright, the center of mass is vertically aligned with the center of the foot. Without losing his equilibrium, at what angle can the man lean forward or backward while keeping his body straight and his feet stiff and immobile?

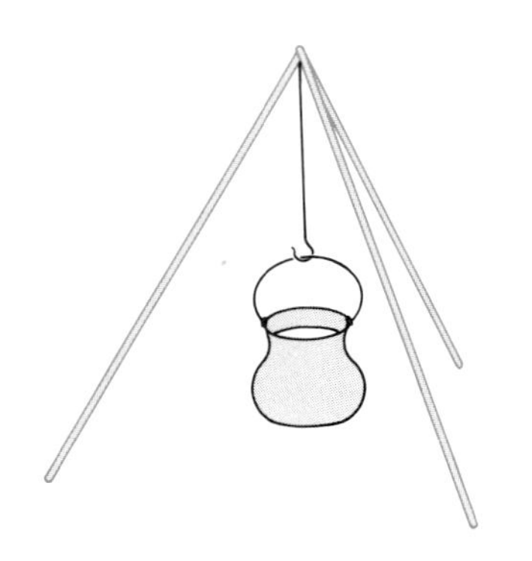

Fig. 14.32

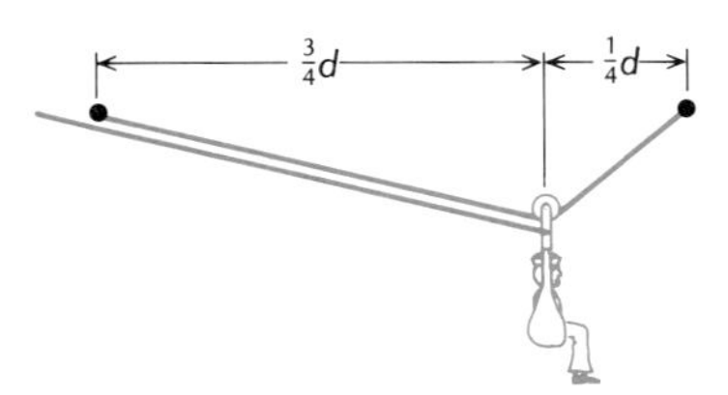


Fig. 14.33

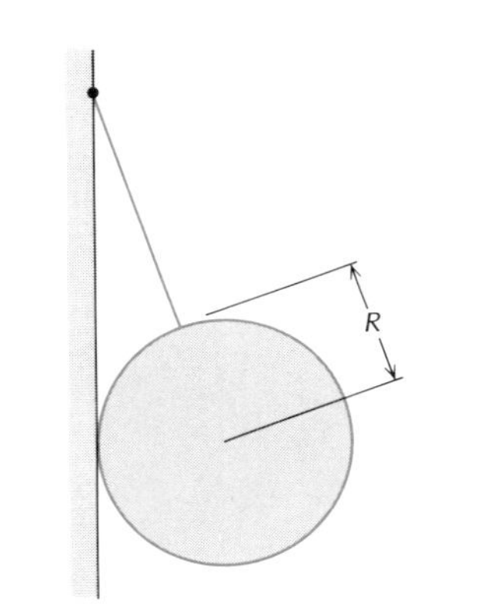


Fig. 14.34

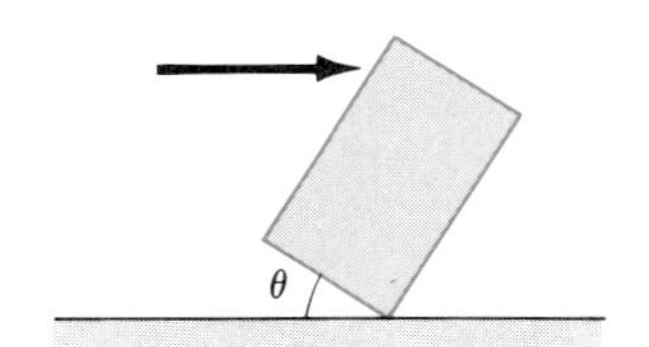


Fig. 14.35

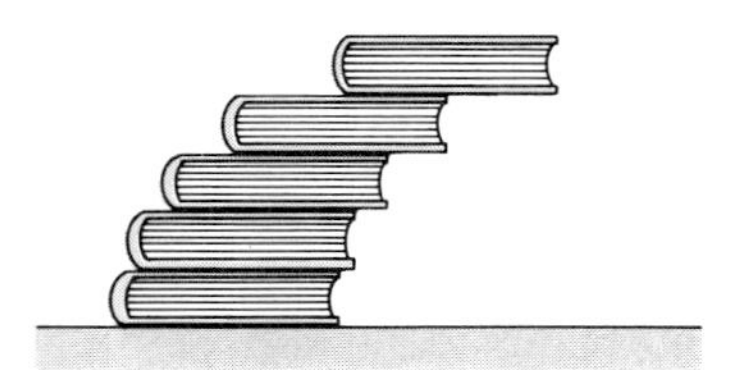

Fig. 14.36

*10. A tetrahedral tripod consist of three massless legs (see Figure 14.32). A mass $M$ hangs from the apex of the (regular) tetrahedron. What are the compressional forces in the three legs?

*11. A sailor is being transferred from one ship to another by means of a bosun's chair (see Figure 14.33). The chair hangs from a roller riding on a rope strung between the two ships. The distance between the ships is $d$, and the rope has a length $1.2d$. The mass of the sailor plus chair is $m$. If the sailor is at a (horizontal) distance $0.25d$ from one ship, find the force that must be exerted on the pull rope to keep the sailor in equilibrium. Ignore the masses of the ropes.

*12. A uniform solid disk of mass $M$ and radius $R$ hangs from a string of length $l$ attached to a smooth vertical wall (see Figure 14.34). Calculate the tension in the string and the normal force acting at the point of contact of disk and wall.

*13. Three traffic lamps of equal masses of 20 kg hang from a wire stretched between two telephone poles, 15 m apart. The horizontal spacing of the traffic lamps is uniform. At each pole, the wire makes a downward angle of 10° with the horizontal line. Find the tensions in all the segments of wire, and find the distance of each lamp below the horizontal line.

*14. Consider the meter stick leaning against a wall described in Example 3. If the stick makes an angle of 40° with the wall, how hard can you press down vertically on the top of the stick with your hand before slipping begins?

*15. An automobile with a wheelbase (distance from the front wheels to the rear wheels) of 3.0 m has its center of mass at a point midway between the wheels at a height of 0.65 m above the road. When the automobile is on a level road, the force with which each wheel presses on the road is 3100 N. What is the normal force with which each wheel presses on the road when the automobile is standing on a steep road of slope 3:10 with all the wheels locked?

*16. A wooden box is filled with material of uniform density. The box (with its contents) has a mass of 80 kg; it is 0.6 m wide, 0.6 m deep, and 1.2 m high. The box stands on a level floor. By pushing against the box, you can tilt it over (Figure 14.35). Assume that when you do this, one edge of the box remains in contact with the floor without sliding.

(a) Plot the gravitational potential energy of the box as a function of the angle $\theta$ between the bottom of the box and the floor.
(b) What is the critical angle beyond which the box will topple over when released?
(c) How much work must you do to push the box to this critical angle?

*17. A meter stick of mass $M$ hangs from a 1.5-m string tied to the meter stick at the 80-cm mark. If you push the bottom end of the meter stick to one side with a horizontal push of magnitude $Mg/2$, what will be the equilibrium angles of the meter stick and the string?

*18. Five identical books are to be stacked one on top of the other. Each book is to be shifted sideways by some variable amount, so as to form a curved leaning tower with maximum protrusion (see Figure 14.36). How much must each book be shifted? What is the maximum protrusion? If you had an infinite number of books, what would be the limiting maximum protrusion? (Hint: Try this experimentally; start with the top book, and insert the others underneath, one by one.)

**19. A wooden box, filled with a material of uniform density, stands on a concrete floor. The box has a mass of 75 kg and is 0.5 m wide, 0.5 m long, and 1.5 m high. The coefficient of friction between the box and the floor is $\mu_s = 0.80$. If you exert a (sufficiently strong) horizontal push against the side of the box, it will either topple over or start sliding without toppling over, depending on how high above the level of the floor you push. What is the maximum height at which you can push if you want the box to slide? What is the magnitude of the force you must exert to start the sliding?

*20. The left and right wheels of an automobile are separated by a transverse distance of $l = 1.5$ m. The center of mass of this automobile is $h = 0.60$ m above the ground. If the automobile is driven around a flat (no banking) curve of radius $R = 25$ m with an excessive speed, it will topple over sideways. What is the speed at which it will begin to topple? Express your answer in terms of $l$, $h$, and $R$; then evaluate numerically. Assume that the wheels do not skid.

*21. An automobile has a wheelbase (distance from front wheels to rear wheels) of 3.0 m. The center of mass of this automobile is at a height of 0.60 m above the ground. Suppose that this automobile has rear-wheel drive and that it is accelerating along a level road at 6 $m/s^2$. When the automobile is parked, 50% of its weight rests on the front wheels and 50% on the rear wheels. What is the weight distribution when it is accelerating? Pretend that the body of the automobile remains parallel to the road at all times.

Fig. 14.37

*22. Consider a bicycle and rider with the dimensions described in Example 13.1. The bicycle has (only) a front-wheel brake. During braking, what is the maximum deceleration that this bicycle can withstand without flipping over its front wheel?

*23. A bicycle and its rider are traveling around a curve of radius 6.0 m at a constant speed of 20 km/h. What is the angle at which the rider must lean the bicycle toward the center of the curve (see Figure 14.37)?

**24. An automobile is braking on a flat, dry road with a coefficient of static friction of 0.90 between its wheels and the road. The wheelbase (the distance between the front and the rear wheels) is 3.0 m, and the center of mass is midway between the wheels, at a height of 0.60 m above the road.

(a) What is the deceleration if all four wheels are braked with the maximum force that avoids skidding?

(b) What is the deceleration if the rear-wheel brakes are disabled? Take into account that during braking, the normal force on the front wheels is larger than that on the rear wheels.

(c) What is the deceleration if the front-wheel brakes are disabled?

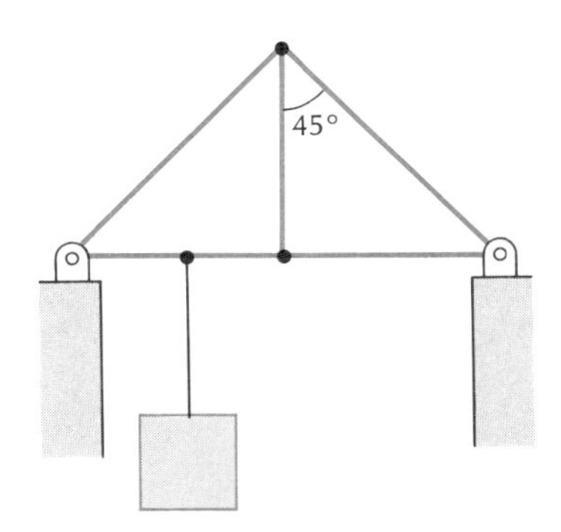


Fig. 14.38

*25. A mass $M$ hangs from a truss consisting of massless beams arranged as shown in Figure 14.38. Find the compressional or tensional force in each beam of the truss. Which of the beams is subjected to the largest force?

*26. A truss bridge with two triangular and two square panels (see Figure 14.39) supports a roadway of mass $M$ whose weight is uniformly distributed along the length of the bridge. The beams of the truss are massless. Find the compressional or tensional force in each beam. Which beams are subjected to the largest forces?

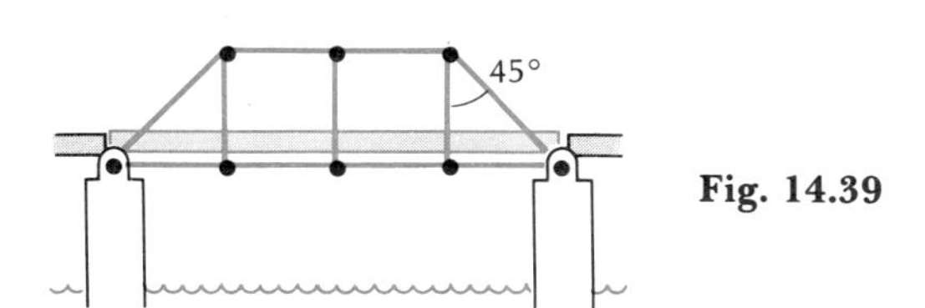


Fig. 14.39

*27. Suppose that a heavy locomotive of mass $M/4$ is passing over the truss bridge discussed in Example 5. If the locomotive is at a distance $2/9\ l$ from the center of the bridge (where $l$ is the total length of the bridge), what are the forces in the beams?

*28. A square framework of steel hangs from a crane by means of cables attached to the upper corners making an angle of 60° with each other (see Fig-

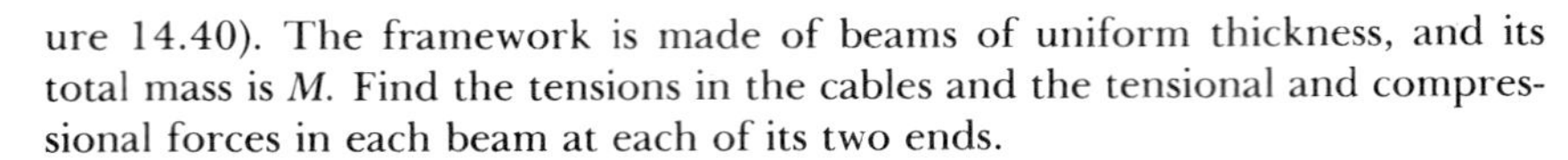
ure 14.40). The framework is made of beams of uniform thickness, and its total mass is $M$. Find the tensions in the cables and the tensional and compressional forces in each beam at each of its two ends.

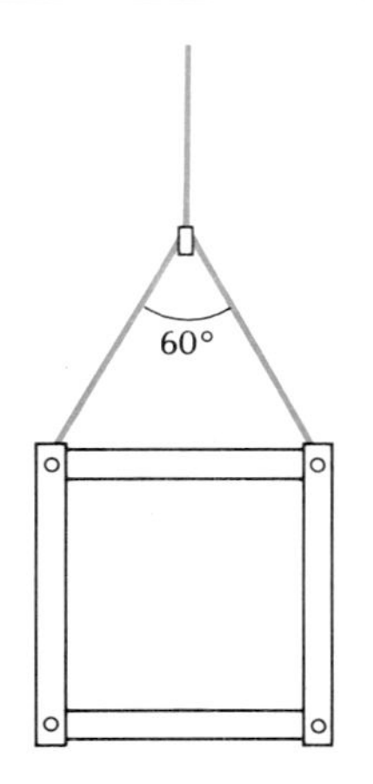

**Fig. 14.40**

**29. Suppose that the suspension bridge discussed in Example 6 has a length $l$ and the towers have a height $h$ (above the lowest point of the suspension cable). In terms of $b$, $l$, and $h$, what is the tension in the cables at the towers? What is the tension at the center? [Hint: Show that $T^2 = (bx)^2 + (T_{\text{cent}})^2$.]

**30. Two smooth balls of steel of mass $m$ and radius $R$ are sitting inside a tube of radius $1.5R$. The balls are in contact with the bottom of the tube and with the wall (at two points; see Figure 14.41). Find the contact force at the bottom and at the two points on the wall.

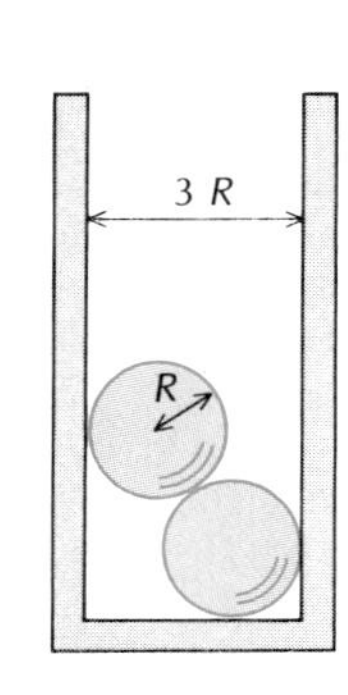

**Fig. 14.41**

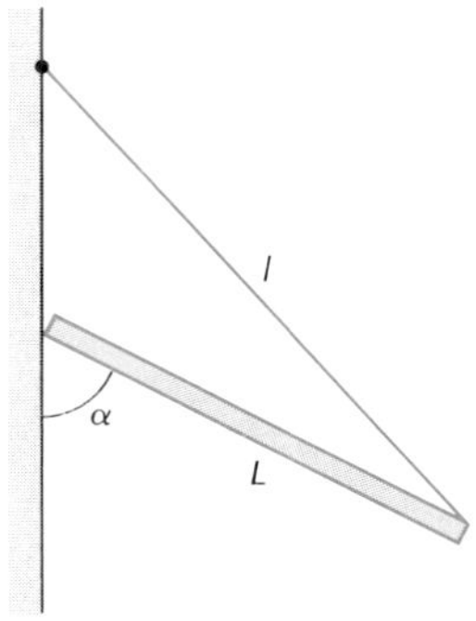

**Fig. 14.42**

**31. One end of a uniform beam of length $L$ rests against a smooth, frictionless vertical wall, and the other end is held by a string of length $l = \frac{3}{2} L$ attached to the wall (see Figure 14.42). What must be the angle of the beam with the wall if it is to remain at rest without slipping?

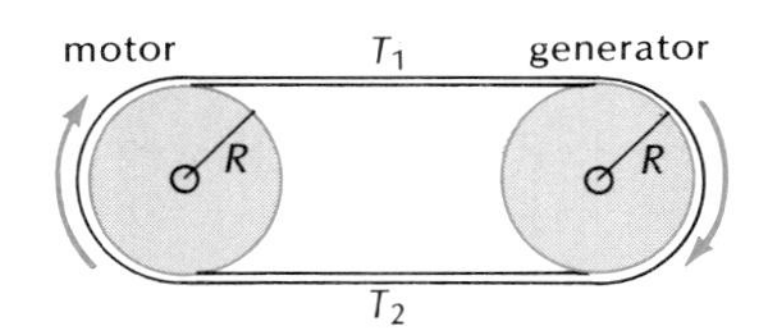

**Fig. 14.43**

**32. Two playing cards stand on a table leaning against each other so as to form an "A-frame" roof. The frictional coefficient between the bottoms of the cards and the table is $\mu_s$. What is the maximum angle that the cards can make with the vertical without slipping?

**33. A rope is draped over the round branch of a tree, and unequal masses $m_1$ and $m_2$ are attached to its ends. The coefficient of sliding friction for the rope on the branch is $\mu_k$. What is the acceleration of the masses?

**34. The flywheel of a motor is connected to the flywheel of an electric generator by a drive belt (Figure 14.43). The flywheels are of equal size, each of radius $R$. While the flywheels are rotating, the tensions in the upper and the lower portions of the drive belt are $T_1$ and $T_2$, respectively, so the drive belt exerts a torque $\tau = (T_2 - T_1)R$ on the generator. The coefficient of static friction between each flywheel and the drive belt is $\mu_s$. Assume that the tension in the drive belt is as low as possible with no slipping, and that the drive belt is massless. Show that under these conditions

$$T_1 = \frac{\tau}{R} \frac{1}{e^{\mu_s \pi} - 1}$$

$$T_2 = \frac{\tau}{R} \frac{1}{1 - e^{-\mu_s \pi}}$$

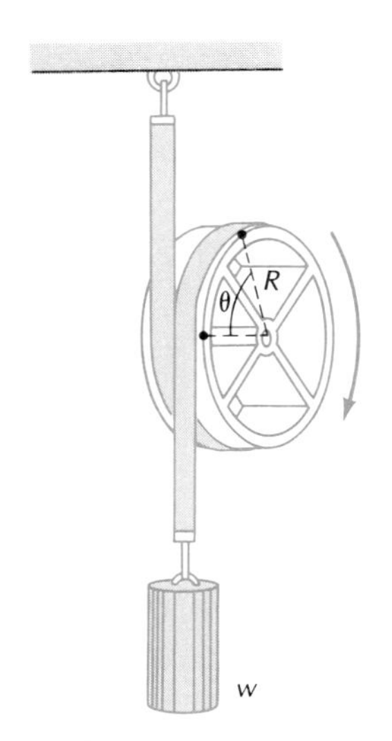

**Fig. 14.44**

**35. A power brake invented by Lord Kelvin consists of a strong flexible belt wrapped once around a spinning flywheel (Figure 14.44). One end of the belt is fixed to an overhead support; the other end carries a weight $w$. The coefficient of kinetic friction between the belt and the wheel is $\mu_k$. The radius of the wheel is $R$ and its angular velocity is $\omega$.

(a) Show that the tension in the belt is

$$T = we^{-\mu_k \theta}$$

as a function of the angle of contact (Figure 14.44).

(b) Show that the net frictional torque the belt exerts on the flywheel is

$$\tau = wR(1 - e^{-2\pi\mu_k})$$

(c) Show that the power dissipated by friction is

$$P = wR\omega(1 - e^{-2\pi\mu_k})$$

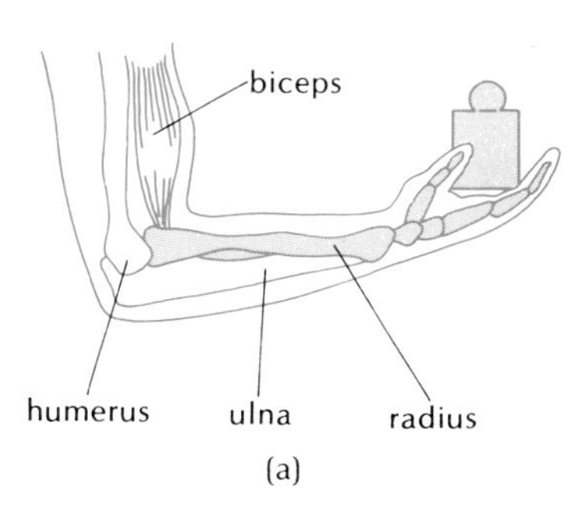


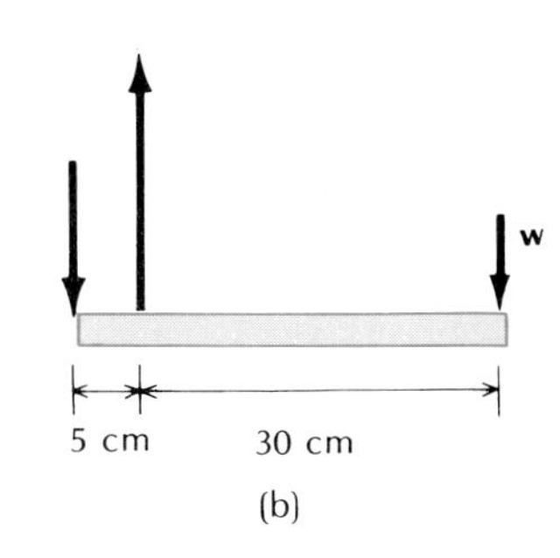


**Fig. 14.45**

## Section 14.3

36. The human forearm (including the hand) can be regarded as a lever pivoted at the joint of the elbow and pulled upward by the tendon of the biceps (Figure 14.45a). The dimensions of this lever are given in Figure 14.45b. Suppose that a load of 25 kg rests in the hand. What upward force must the biceps exert to keep the forearm horizontal? What is the downward force at the elbow joint? Neglect the weight of the forearm.

37. Repeat the preceding problem if, instead of being vertical, the upper arm is tilted, so as to make an angle of 135° with the (horizontal) forearm.

38. When you bend over to pick up something from the floor, your backbone acts as a lever pivoted at the sacrum (see Figure 14.46). The weight of the trunk pulls downward on this lever, and the muscles attached along the upper part of the backbone pull upward. The actual arrangement of the muscles is rather complicated, but for a simple mechanical model we can pretend that the muscles are equivalent to a string attached to the backbone at an angle of about 12° at a point beyond the center of mass (the other end of the "string" is attached to the pelvis). Assume that the mass of the trunk, including head and arms, is 48 kg, and that the dimensions are as shown in the diagram. What force must the muscles exert to balance the weight of the trunk when bent over horizontally?

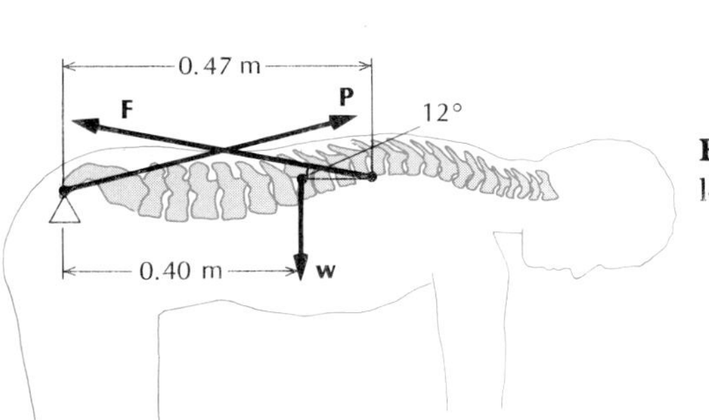


**Fig. 14.46** Backbone acting as lever.

39. A man of 73 kg stands on one foot, resting all of his weight on the ball of the foot. As described in Section 14.3, the bones of the foot play the role of a lever. The short end of the lever (to the heel) measures 5.0 cm and the long end (to the ball of the foot) 14 cm. Calculate the force exerted by the Achilles tendon and the force at the ankle.

40. A rope hoist consists of four pulleys assembled in two pairs with rigid straps, with a rope wrapped around as shown in Figure 14.47. A load of 300 kg hangs from the lower pair of pulleys. What tension must you apply to the rope to hold the load steady? Treat the pulleys and the rope as massless, and ignore any friction in the pulleys.

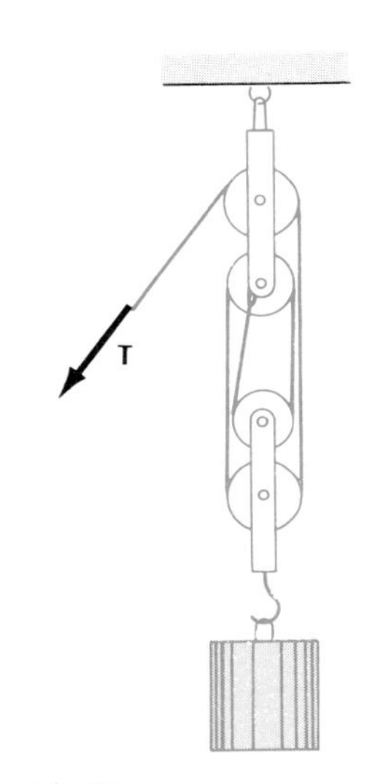


**Fig. 14.47**

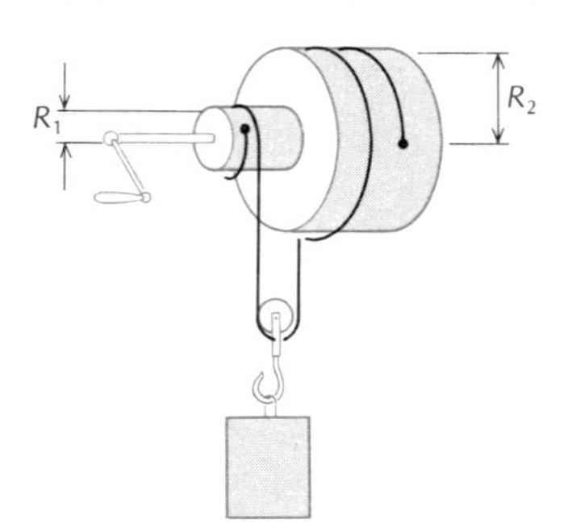

**Fig. 14.48**

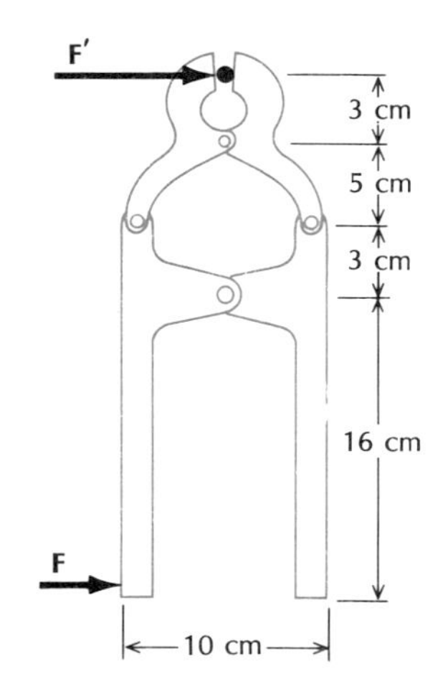

**Fig. 14.49**

*41. Figure 14.48 shows a differential windlass consisting of two rigidly joined drums of radii $R_1$ and $R_2$ around which is wound the cable or rope that holds the load. What clockwise torque must be applied to the joined drums to lift a load of mass $m$?

*42. Figure 14.49 shows a compound bolt cutter. If the dimensions are as indicated in this figure, what is the mechanical advantage?

*43. The drum of a winch is rigidly attached to a concentric large gear, which is driven by a small gear attached to a crank. The dimensions of the drum, the gears, and the crank are given in Figure 14.50. What is the mechanical advantage of this geared winch?

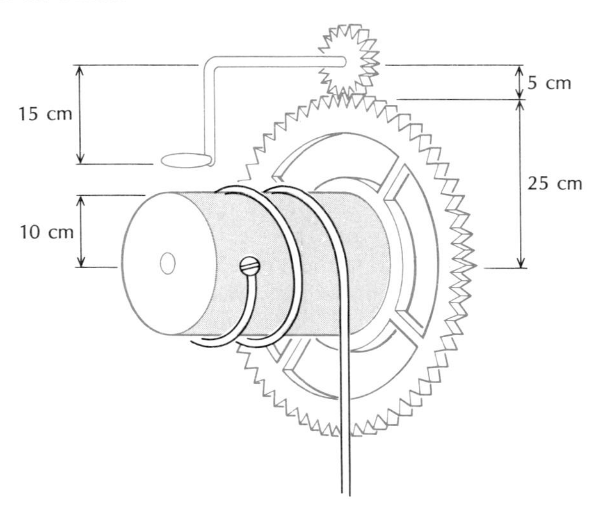

**Fig. 14.50**

*44. The screw of a vise has a "pitch" of 4 mm, that is, it advances 4 mm when given one full turn. The handle of the vise is 25 cm long, measured from the screw to the end of the handle. What is the mechanical advantage when you push perpendicularily on the end of the handle?

*45. A scissors jack has the dimensions shown in Figure 14.51. The screw of the jack has a "pitch" of 5 mm (as stated in the previous problem, this is the distance the screw advances when given one full turn). Suppose the scissors jack is partially extended, with an angle of 55° between its upper sides. What is the mechanical advantage provided by the jack?

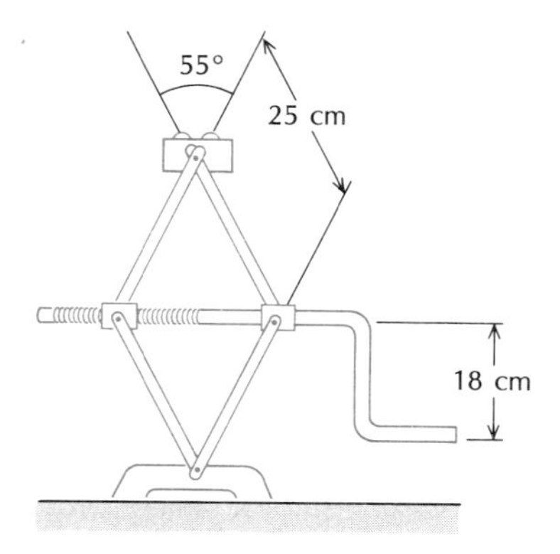

**Fig. 14.51**

**46. Figure 14.52 shows a tensioning device used to tighten the rear stay of the mast of a sailboat. The block and tackle pulls down a rigid bar with two rollers that squeeze together the two branches of the split rear stay. If the angles are as given in the figure, what is the mechanical advantage?

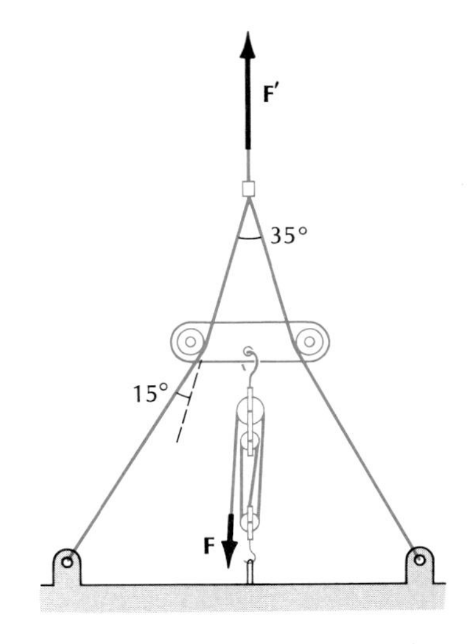

Fig. 14.52

## Section 14.4

47. The anchor rode of a sailboat is a nylon rope of length 60 m and diameter 1.3 cm. While anchored during a storm, the sailboat momentarily pulls on this rope with a force of $1.8 \times 10^4$ N. How much does the rope stretch?

48. The piano wire described in Example 9 can be regarded as a spring. What is the effective spring constant of this spring?

49. The recommended diameter for the nylon anchor rope of a 3500-kg sailboat is 1.3 cm (1/2 in.). What is the maximum force this rope can withstand? Compare with the weight of the sailboat.

50. The length of the femur of a woman is 38 cm and the average cross section is 10 $cm^2$. How much will the femur be compressed in length if the woman lifts another woman of 68 kg and carries her piggyback? Assume that the extra weight is evenly distributed over both legs.

51. Suppose you drop an aluminum sphere of radius 10 cm into the ocean and it sinks to a depth of 5000 m, where the pressure is $5.7 \times 10^7$ N/$m^2$. Calculate by how much the diameter of this sphere will shrink.

52. At the bottom of the Marianas Trench in the Pacific Ocean, at a depth of 10,900 m, the pressure is $1.24 \times 10^8$ N/$m^2$. What is the percent increase of the density of water at this depth as compared with the density at the surface?

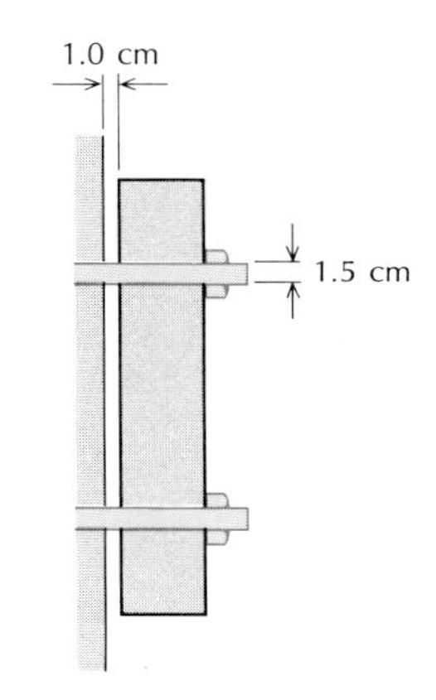

Fig. 14.53

53. A slab of stone of mass 1200 kg is attached to the wall of a building by two bolts of iron of diameter 1.5 cm (see Figure 14.53). The distance between the wall and the slab of stone is 1.0 cm. Calculate by how much the bolts will sag downward because of the shear stress they are subjected to.

54. According to (somewhat oversimplified) theoretical considerations, Young's modulus, the shear modulus, and the bulk modulus are related by

$$Y = \frac{9BS}{3B + S}$$

Check this for the first four materials listed in Table 14.1.

55. A nylon rope of diameter 1.3 cm is to be spliced to a steel rope. If the steel rope is to have the same ultimate breaking strength as the nylon, what diameter should it have? The ultimate tensile strength for the wire in the steel rope is $2.0 \times 10^9$ N/$m^2$.

*56. A rod of aluminum has a diameter of 1.000002 cm. A ring of cast steel has an inner diameter of 1.000000 cm. If the rod and the ring are placed in a liquid under high pressure, at what value of the pressure will the aluminum rod fit inside the steel ring?

*57. A rope of length 12 m consists of an upper half of nylon of diameter 1.9 cm spliced to a lower half of steel of diameter 0.95 cm. How much will this rope stretch if a mass of 4000 kg is suspended from it? Young's modulus for steel wire is $19 \times 10^{10}$ N/$m^2$.

*58. A long rod of wrought iron hangs straight down into a very deep mine shaft. For what length will the rod break off at the top because of its own weight? The density of iron is $7.8 \times 10^3$ kg/$m^3$.

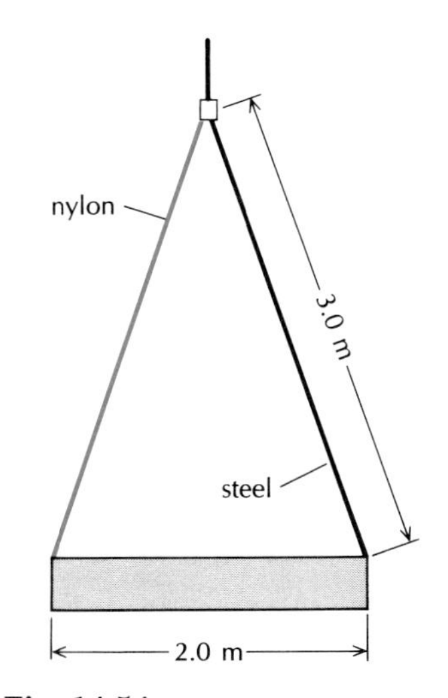

Fig. 14.54

*59. A heavy uniform beam of mass 8000 kg and length 2.0 m is suspended at one end by a nylon rope of diameter 2.5 cm and at the other end by a steel rope of diameter 0.64 cm. The ropes are tied together above the beam (see Figure 14.54). The unstretched lengths of the ropes are 3.0 m each. What angle will the beam make with the horizontal?

*60. A rod of cast iron is welded to the upper edge of a plate of copper whose

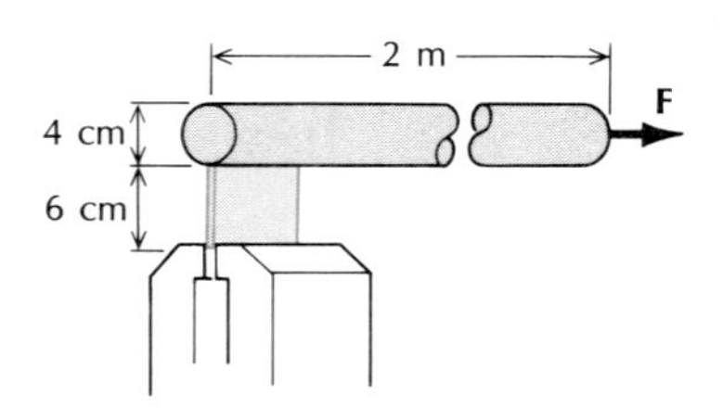


**Fig. 14.55**

lower edge is held in a vise (see Figure 14.55). The rod has a diameter of 4.0 cm and a length of 2 m. The copper plate measures 6 cm × 6 cm × 1 cm. If we pull the free end of the iron rod forward by 3.0 mm, what is the shear strain ($\Delta x/h$) of the copper plate?

*61. When a bar of steel is heated, it expands in length by 0.0012% for each degree centigrade of temperature increase. If the length of the heated bar is to be reduced to its original value, a compressive stress must be applied to it. The compressive stress required to cancel the thermal expansion is called **thermal stress.** What is its value for a cylindrical bar of cast steel of cross section 4 $cm^2$ heated by 150°C?

*62. A power cable of copper is stretched straight between two fixed towers. If the temperature decreases, the cable tends to contract (compare the preceding problem). The amount of contraction for a free copper cable or rod is 0.0017% per degree centigrade. Estimate what temperature decrease will cause the cable to snap. Pretend that the cable obeys Eq. (27) until it reaches its breaking point. Ignore the weight of the cable and the sag and stress produced by the weight.

**63. A meter stick of mild steel, of density $7.8 \times 10^3$ kg/m$^3$, is made to rotate about a perpendicular axis passing through its middle. What is the maximum angular velocity with which the stick can rotate if its center is to hold?

**64. The wall of a pipe of diameter 60 cm is constructed of a sheet of mild steel of thickness 0.3 cm. The pipe is filled with water under high pressure. What is the maximum pressure, that is, force per unit area, that the pipe can withstand?

**65. A hoop of aluminum of radius 40 cm is made to spin about its axis of symmetry at high speed. The density of aluminum is $2.7 \times 10^3$ kg/m$^3$ and the ultimate tensile breaking strength is $7.8 \times 10^7$ N/m$^2$. At what angular velocity will the hoop break apart?

**66. A pipe of mild steel with a wall 0.4 cm thick and a diameter of 50 cm contains a liquid at a pressure of $2 \times 10^4$ N/m$^2$. How much will the diameter of the pipe expand due to this pressure?

CHAPTER 15

# Oscillations

The motion of a particle or of a system of particles is **periodic,** or cyclic, if it repeats again and again at regular intervals of time. The orbital motion of a planet around the Sun, the uniform rotational motion of a phonograph turntable, the back-and-forth motion of a piston in an automobile engine, the swinging motion of a pendulum in a grandfather clock, and the vibration of a guitar string are examples of periodic motions. A periodic back-and-forth or swinging motion of a body is called an **oscillation.** Thus, the motion of the piston is an oscillation, and so is the motion of the pendulum, and the motion of the individual particles of the guitar string.

*Oscillation*

In this chapter we will examine in great detail the motion of a mass oscillating back and forth under the push and pull exerted by an ideal, massless spring. The mathematical equations that we will develop for the description of this mass–spring system are of great importance because analogous equations recur in the description of all other oscillating systems.

## 15.1 Simple Harmonic Motion

Let us consider the special case of one-dimensional periodic motion involving a particle that oscillates back and forth along a straight path. For example, Figure 15.1 shows the worldline of a particle oscillating back and forth along the $x$ axis between $x = -0.3$ m and $x = +0.3$ m. The worldline in this special example has a simple mathematical form:

$$(0.3\text{ m}) \times \cos\left(\frac{\pi}{4}\frac{\text{radian}}{\text{s}} \times t\right) \qquad (1)$$

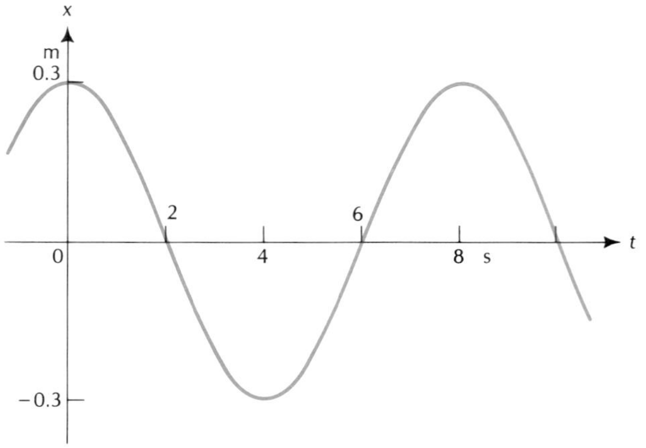

**Fig. 15.1** Worldline of a particle oscillating along the $x$ axis with simple harmonic motion.

Here, the units radian/s have been included with the constant $\pi/4$ in the argument of the cosine, so that, when a value of the time $t$ is inserted into the equation, the units of time cancel, and the argument of the cosine is left in radians. For instance, if $t = 2$ s, then $\cos(\pi/4 \text{ radian/s} \times 2 \text{ s}) = \cos(\pi/2 \text{ radian}) = 0$.

Cosines and sines are called **harmonic functions;** accordingly, the position of the particle specified by Eq. (1) is a harmonic function of time. At $t = 0$ the particle is at maximum distance from the origin ($x = 0.3$ m) and is just starting to move; at $t = 2.0$ s, it passes through the origin ($x = 0$); at $t = 4.0$ s, it reaches maximum distance, but on the negative $x$ axis ($x = -0.3$ m); at $t = 6.0$ s, it again passes through the origin. Finally, at $t = 8.0$ s, it returns to maximum distance on the positive $x$ axis, exactly as at $t = 0$ — it has completed one cycle of the motion and is ready to begin the next cycle. Thus, the period of the motion is

$$8.0 \text{ s} \tag{2}$$

and the frequency of the motion, or the rate of repetition of the motion, is

$$\frac{1}{8.0 \text{ s}} = 0.125/\text{s} \tag{3}$$

The points $x = 0.3$ m and $x = -0.3$ m, at which the $x$ coordinate attains its maximum and minimum, are the turning points of the motion; and the point $x = 0$ is the midpoint, or the equilibrium point.

Equation (1) is an example of **simple harmonic motion.** More generally, a motion is simple harmonic if the position as a function of time has the form

*Simple harmonic motion*

$$x = A \cos(\omega t + \delta) \tag{4}$$

*Amplitude*

*Angular frequency*

The quantities $A$, $\omega$, and $\delta$ are constants. The quantity $A$ is called the **amplitude** of the motion; it is simply the distance between the midpoint ($x = 0$) and either turning point ($x = +A$ or $x = -A$). The quantity $\omega$ is called the **angular frequency;** it is directly related to the period and to the frequency of the motion. To establish this relationship, note that an increase of $t$ by $2\pi/\omega$ changes the function in Eq. (4) into

$$x = A\cos[\omega(t + 2\pi/\omega) + \delta]$$
$$= A\cos(\omega t + \delta + 2\pi) = A\cos(\omega t + \delta) \tag{5}$$

that is, such an increase of $t$ does not change the function at all because the function merely repeats itself. The **period** of the motion, designated by the symbol $T$, is the time interval for one such repetition:

*Period*

$$\boxed{T = \frac{2\pi}{\omega}} \tag{6}$$

and **frequency** of the motion, designated by the symbol $\nu$, is

*Frequency*

$$\boxed{\nu = \frac{1}{T} = \frac{\omega}{2\pi}} \tag{7}$$

The units of angular frequency are radians per second (radian/s). The units of frequency are cycles per second (cycle/s). Dimensionally, both of these units are equivalent to 1/s; but, to prevent confusion between them, it is useful to retain the labels *radian* and *cycle*. The unit of frequency is often called a **hertz** (Hz):

*Hertz,* Hz

$$1 \text{ hertz} = 1 \text{ Hz} = 1 \text{ cycle per second}$$

Note that the above equations connecting *angular frequency,* period, and frequency are formally the same as the equations connecting *angular velocity,* period, and frequency of uniform rotational motion [see Eqs. (12.4) and (12.6)]. As we will see later in this section, this coincidence arises from a special geometrical relationship between simple harmonic motion and uniform circular motion. Aside from this special relationship, angular frequency has nothing to do with angular velocity, even though both quantities are labeled by the same letter $\omega$ and are measured in the same units.

In terms of the period or the frequency, Eq. (4) can be written as

$$x = A\cos\left(\frac{2\pi}{T}t + \delta\right) \tag{8}$$

or as

$$x = A\cos(2\pi\nu t + \delta) \tag{9}$$

The argument of the cosine function is called the **phase** of the oscillation and the quantity $\delta$ is called the **phase constant.** This constant determines at what time the particle reaches the point of maximum displacement. At this instant

*Phase and phase constant*

$$\omega t_{\max} + \delta = 0 \tag{10}$$

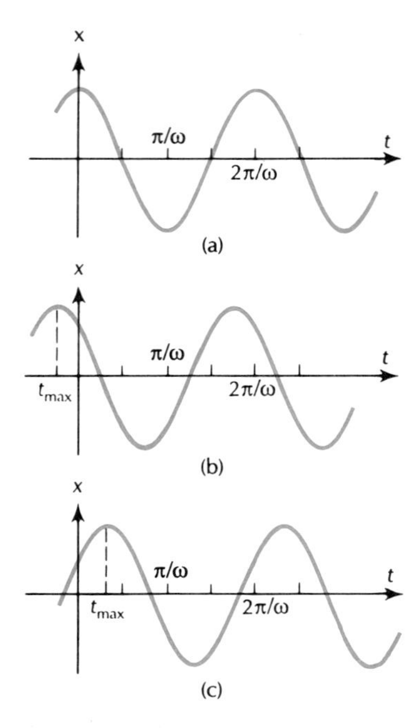


**Fig. 15.2** Examples of worldlines of particles with simple harmonic motion with different phase constants. (a) $\delta = 0$. The particle has maximum displacement at $t = 0$. (b) $\delta = \pi/4$ (or 45°). The particle has maximum displacement before $t = 0$. (c) $\delta = -\pi/3$ (or −60°). The particle has maximum displacement after $t = 0$.

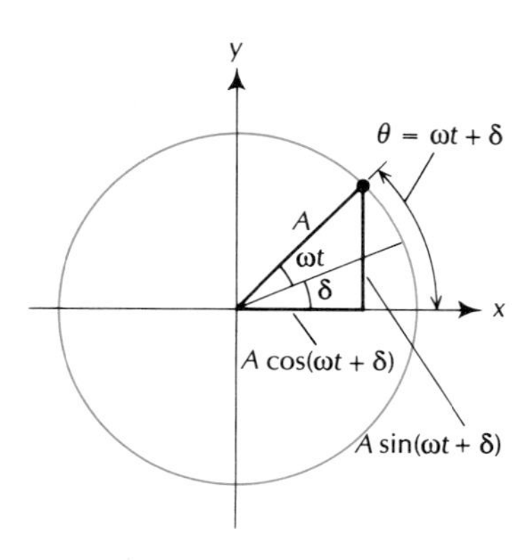


**Fig. 15.3** Reference particle with uniform circular motion on a reference circle of radius $A$.

that is,

$$t_{\text{max}} = -\delta/\omega \tag{11}$$

Hence the particle reaches the point of maximum displacement at a time $\delta/\omega$ *before* $t = 0$ (Figure 15.2). Of course, the particle also passes through this point at periodic intervals before and after this time, when

$$t_{max} = -\delta/\omega \pm 2\pi/\omega, -\delta/\omega \pm 4\pi/\omega, -\delta/\omega \pm 6\pi/\omega, \text{etc.} \tag{12}$$

The values of phase constants are sometimes expressed in degrees by converting the value in radians into an equivalent value in degrees.

Note that since

$$A \cos(\omega t + \delta) = A \sin(\omega t + \delta + \pi/2)$$
$$= A \sin[\omega t + (\delta + \pi/2)]$$

the cosine function in Eq. (4) can be replaced by a sine function by merely changing the phase constant by $\pi/2$ radians, or 90°. Thus, simple harmonic motion can be equally well expressed in terms of the sine function. Alternatively, the trigonometric identity for the cosine of the sum of two angles gives

$$A \cos(\omega t + \delta) = A \cos \delta \cos \omega t - A \sin \delta \sin \omega t$$
$$= (A \cos \delta)\cos \omega t - (A \sin \delta)\sin \omega t \tag{13}$$

This expresses simple harmonic motion as a superposition of sine and cosine functions, each with zero phase constant and with amplitudes $A \cos \delta$ and $-A \sin \delta$.

There is a simple geometric relationship between simple harmonic motion and uniform circular motion. Consider a particle moving with angular velocity $\omega$ along a circle of radius $A$, equal in magnitude to the amplitude of the simple harmonic motion (Figure 15.3). This particle is called the reference particle, and the circle is called the **reference circle.** If at $t = 0$ the angular position of the reference particle is $\theta = \delta$, then the position angle at any later time is

$$\theta = \omega t + \delta \tag{14}$$

and the $x$ and $y$ coordinates of the reference particle are (see Figure 15.3)

$$x = A \cos \theta = A \cos(\omega t + \delta)$$
$$y = A \sin \theta = A \sin(\omega t + \delta) = A \cos(\omega t + \delta - \pi/2) \tag{15}$$

Hence, while the reference particle performs uniform circular motion, its $x$ coordinate performs simple harmonic motion with a phase constant $\delta$, and the angular velocity of the circular motion coincides with the angular frequency of the harmonic motion. Likewise, the $y$ coordinate of this particle performs simple harmonic motion, but the phase differs by $\pi/2$ (or 90°) from the motion of the $x$ coordinate.

This geometric relationship is often exploited mechanically to generate simple harmonic motion from circular motion. It is only necessary to place a slotted arm over a peg which is attached to a wheel in uniform circular motion. In Figure 15.4, the slot is vertical and the arm is constrained to move horizontally. The peg then drags the arm back and forth, and makes it move with simple harmonic motion.

Finally, let us compute the velocity and acceleration corresponding to simple harmonic motion. If the displacement is

$$x = A\cos(\omega t + \delta) \tag{16}$$

then differentiation of this displacement gives the velocity:

$$\frac{dx}{dt} = -A\omega\sin(\omega t + \delta) \tag{17}$$

and differentiation of this velocity gives the acceleration:

$$\frac{d^2x}{dt^2} = -A\omega^2\cos(\omega t + \delta) \tag{18}$$

Here we have used the standard formulas for the derivatives of the sine function and the cosine function (see Appendix 5). Bear in mind that the arguments of the sine and cosine functions in this chapter (and also the next) are always expressed in radians, as is required for the validity of the standard formulas for the derivatives.

Comparison of Eqs. (16) and (18) shows that

$$\boxed{\frac{d^2x}{dt^2} = -\omega^2 x} \tag{19}$$

that is, the acceleration is always proportional to the displacement, but oppositely directed. This is a characteristic feature of simple harmonic motion, a fact that will be useful in the next section.

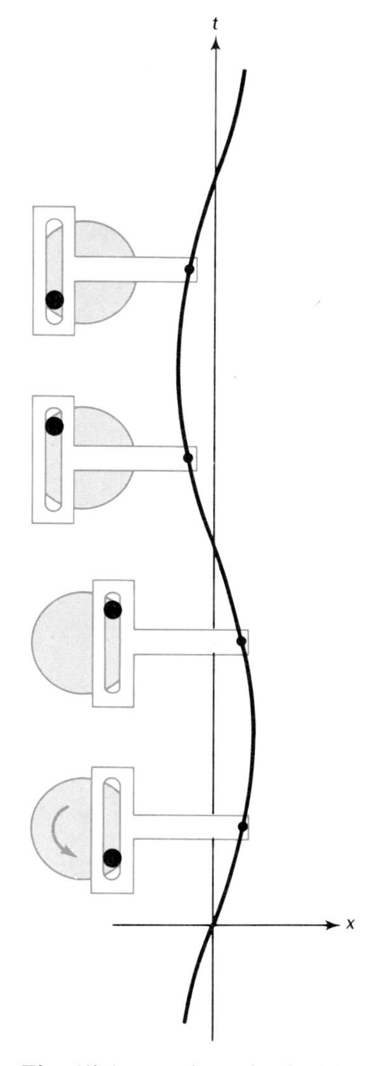


**Fig. 15.4** Rotating wheel with a peg driving a slotted arm back and forth. The drawing shows several "snapshots" of the wheel at different times.

## 15.2 The Simple Harmonic Oscillator

The **simple harmonic oscillator** consists of a mass coupled to an ideal, massless spring which obeys Hooke's Law. One end of the spring is attached to the mass and the other is held fixed (Figure 15.5). We will ignore gravity and friction, so the spring force is the only force acting on the mass. This system has an equilibrium position corresponding to the relaxed length of the spring. If the system is initially not in this equilibrium position, then the spring supplies a restoring force which impels the mass toward the equilibrium position. But inertia causes the mass to overshoot this equilibrium position, and the mass then oscillates back and forth — forever if there is no friction.

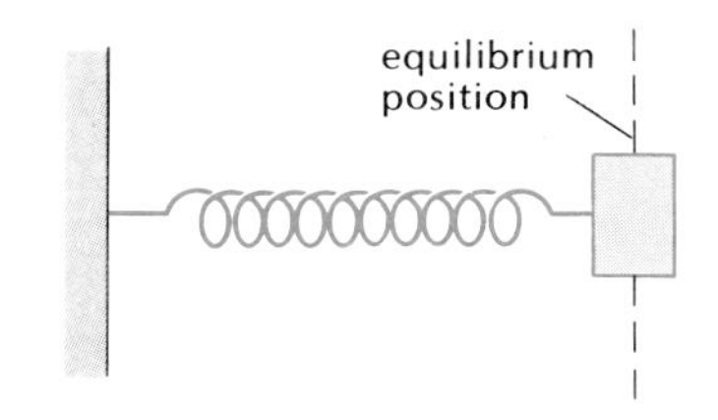


**Fig. 15.5** Mass attached to a spring.

A practical example of such a mass–spring system is the device used aboard Skylab for the daily measurement of the masses of the astronauts (Figure 15.6); in this device, the astronaut and the chair into which he is strapped play the role of the mass of Figure 15.5, and a torsional spring plays the role of the coil spring of Figure 15.5. Aside from such direct practical applications, the great importance of the simple harmonic oscillator is that many physical systems are mathema-

**Fig. 15.6** Body-mass measurement device used on Skylab. The torsional spring is attached to the corners of the triangular frame on the left.

tically equivalent to simple harmonic oscillators, that is, these systems have an equation of motion of the same mathematical form as the simple harmonic oscillator. A pendulum, the balance wheel of a watch, a tuning fork, the air in an organ pipe, and the atoms in a diatomic molecule are systems of this kind; the restoring force and the inertia are of the same mathematical form in these systems as in the simple harmonic oscillator, and we can transcribe the general mathematical results from the latter to the former.

To derive the equation of motion of the simple harmonic oscillator, we begin with Hooke's Law for the force exerted by the spring on the mass (compare Section 6.5):[1]

$$F = -kx \tag{20}$$

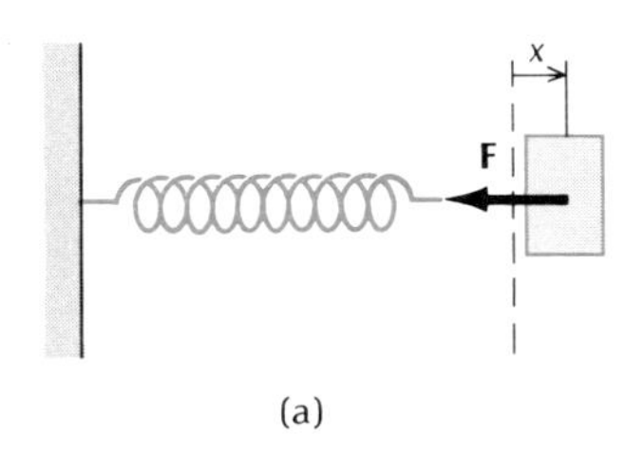


(a)

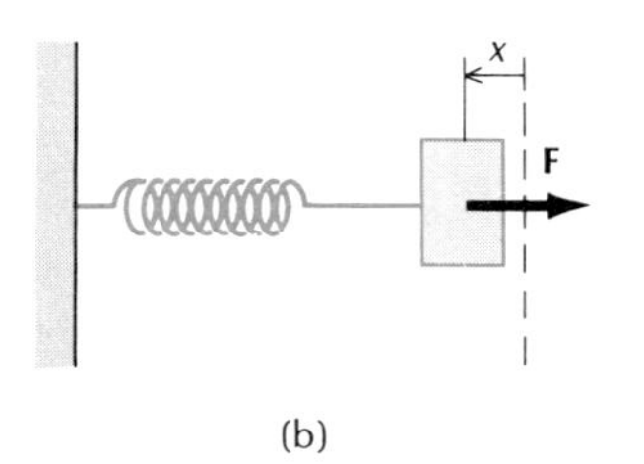


(b)

**Fig. 15.7** (a) Positive displacement of the mass; the force is negative. (b) Negative displacement of the mass; the force is positive.

Here the displacement is measured from the equilibrium position. The constant $k$ is the spring constant. The force is negative if $x$ is positive (stretched spring; see Figure 15.7a), and the force is positive if $x$ is negative (compressed spring; see Figure 15.7b). Of course, Eq. (20) is valid only as long as the spring is not stretched or compressed too much. Excessive stretch would ultimately break the spring, and excessive compression would bring the coils of the spring into contact — in either case Eq. (20) would fail. We will assume that the displacement remains within limits acceptable to the spring.

With the force as given by Eq. (20), the equation of motion of the mass is

*Equation of motion of a simple harmonic oscillator*

$$m\frac{d^2x}{dt^2} = -kx \tag{21}$$

Rather than attempt to solve this equation by the standard mathematical techniques for the solution of differential equations, let us make use of our knowledge of simple harmonic motion. First we rewrite Eq. (21) as

[1] Note that in Section 6.5 we were dealing with static conditions ($x$ was constant or varied only very slowly), whereas now we are dealing with dynamic conditions ($x$ varies with time). We will make the assumption that even when $x$ varies with time, the force law of Eq. (20) remains applicable. This is equivalent to the assumption that the spring is massless, because, if so, no force is required to accelerate the spring and all of the restoring force remains available to accelerate the mass attached to the spring.

$$\frac{d^2x}{dt^2} = -\frac{k}{m}x \tag{22}$$

This says that the acceleration is directly proportional to the displacement. As we saw in the preceding section, such a proportionality is a characteristic feature of simple harmonic motion. In fact, comparison of Eqs. (19) and (22) shows that they are identical, provided that

$$\omega^2 = \frac{k}{m} \tag{23}$$

We know from Chapter 6 that, for given initial conditions, the equation of motion completely determines the motion. Since Eqs. (19) and (22) are identical, we can conclude that the motion of a mass on a spring is simple harmonic motion with an angular frequency

$$\boxed{\omega = \sqrt{k/m}} \tag{24}$$

*Angular frequency of a simple harmonic oscillator*

and with a period

$$T = \frac{2\pi}{\omega} = 2\pi\sqrt{\frac{m}{k}}$$

The position of the mass as a function of time is then

$$x = A\cos(\sqrt{k/m}t + \delta) \tag{25}$$

As in Eq. (13), we can also write this in the form

$$x = (A\cos\delta)\cos(\sqrt{k/m}t) - (A\sin\delta)\sin(\sqrt{k/m}t) \tag{26}$$

The constants $A$ and $\delta$ remain to be determined. These constants can be expressed in terms of the initial conditions of the motion, that is, the initial position $x_0$ and velocity $v_0$ at $t = 0$. According to Eqs. (17) and (25),

$$x_0 = A\cos(0 + \delta) = A\cos\delta \tag{27}$$

$$v_0 = -\sqrt{k/m}\,A\sin(0 + \delta) = -\sqrt{k/m}\,A\sin\delta \tag{28}$$

If we take the values of $A\cos\delta$ and $A\sin\delta$ from Eqs. (27) and (28) and insert them into Eq. (26), we obtain

$$x = x_0\cos(\sqrt{k/m}t) + v_0\sqrt{m/k}\sin(\sqrt{k/m}t) \tag{29}$$

which expresses the motion in terms of the initial conditions.

Note that the frequency of the motion of the simple harmonic oscillator depends *only* on the spring constant and the mass. The frequency of the oscillator will always be the same, regardless of the amplitude with which the oscillator has been set swinging; this property of the oscillator is called **isochronism.**

*Isochronism*

Figure 15.8 is a tracing based on a multiple-exposure photograph of the oscillations of a mass on a spring; the photograph demonstrates the variation of the speed of the mass. The motion of the mass is simple harmonic.

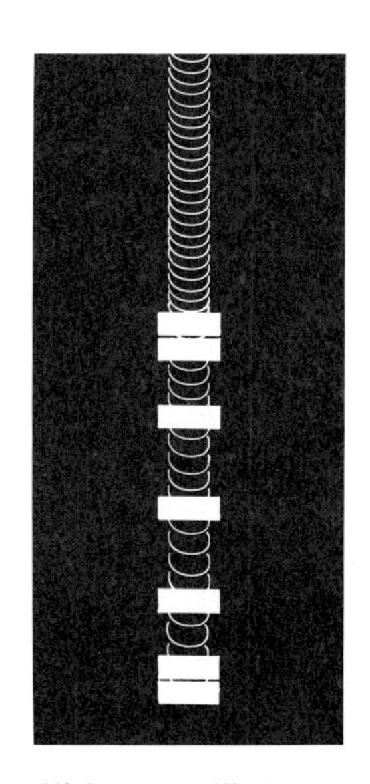

**Fig. 15.8** An oscillating mass on a spring (tracing based on a stroboscopic photograph). Note that the mass moves slowly at the extremes of its motion.

EXAMPLE 1. A mass of 400 kg is moving along the $x$ axis under the influence of the force of a spring with $k = 3.5 \times 10^4$ N/m. There are no other forces acting on the mass. The equilibrium point is at $x = 0$. Suppose that at $t = 0$, the mass is at $x = 0$ and has a velocity of 2.4 m/s in the positive direction. What is the frequency of oscillation? What is the amplitude? Where will the mass be at $t = 0.60$ s?

SOLUTION: The angular frequency of oscillation is

$$\omega = \sqrt{\frac{k}{m}} = \sqrt{\frac{3.5 \times 10^4 \text{ N/m}}{4.0 \times 10^2 \text{ kg}}} = 9.4 \text{ radians/s}$$

and the frequency is

$$\nu = \frac{\omega}{2\pi} = 1.5/\text{s} = 1.5 \text{ Hz}$$

Since $x_0 = 0$ and $v_0 = 2.4$ m/s, Eq. (29) gives

$$x = 2.4 \text{ m/s} \times \sqrt{\frac{m}{k}} \sin \sqrt{\frac{k}{m}}\, t$$

$$= 0.26 \text{ m} \times \sin (9.4t \text{ radians/s})$$

The amplitude of oscillation is 0.26 m and, at $t = 0.60$ s, the position of the mass will be

$$x = 0.26 \text{ m} \times \sin(9.4 \times 0.60 \text{ radians}) = -0.16 \text{ m}$$

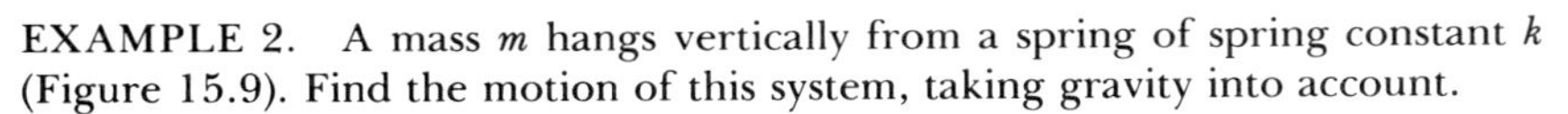

EXAMPLE 2. A mass $m$ hangs vertically from a spring of spring constant $k$ (Figure 15.9). Find the motion of this system, taking gravity into account.

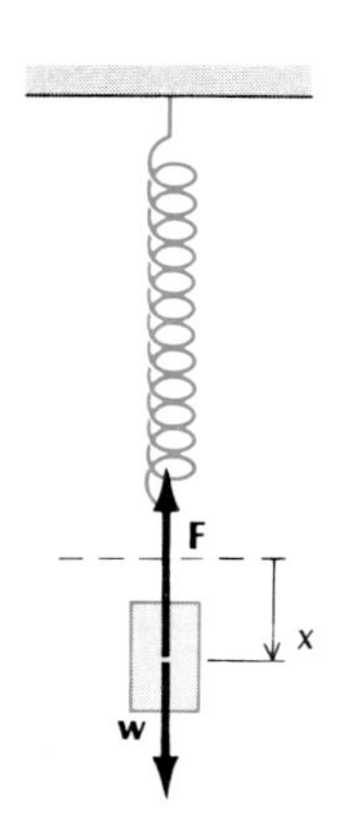


**Fig. 15.9** Mass hanging from a spring. At the equilibrium point, the spring force must match the weight.

SOLUTION: We arrange the $x$ axis downward with origin at the point corresponding to the relaxed length of the spring. The spring force is then $-kx$. Besides this force, there is also the force $mg$ due to gravity. Hence the equation of motion is

$$m \frac{d^2x}{dt^2} = -kx + mg \tag{30}$$

It is convenient to express this equation in terms of a new variable $x'$ which represents the displacement from the new equilibrium position. When the mass is in equilibrium, the spring must be stretched just enough so that the spring force matches the force of gravity, that is,

$$k\, \Delta x = mg \tag{31}$$

Hence at equilibrium the spring is stretched an amount

$$\Delta x = mg/k \tag{32}$$

If $x'$ represents the displacement measured from this equilibrium position, then

$$x' = x - \Delta x = x - mg/k \tag{33}$$

or

$$x = x' + mg/k \tag{34}$$

Inserting this expression for $x$ in Eq. (30), we obtain

$$m\frac{d^2}{dt^2}\left(x' + \frac{mg}{k}\right) = -k\left(x' + \frac{mg}{k}\right) + mg \tag{35}$$

which simplifies to

$$m\frac{d^2}{dt^2}x' = -kx' \tag{36}$$

This is the usual equation of motion for the simple harmonic oscillator [see Eq. (21)]. Hence the solution for the coordinate $x'$ is the usual harmonic function characteristic of simple harmonic motion,

$$x' = A\cos(\omega t + \delta) \tag{37}$$

with

$$\omega = \sqrt{k/m} \tag{38}$$

The solution for the coordinate $x$ is then

$$x = x' + mg/k = A\cos(\omega t + \delta) + mg/k \tag{39}$$

COMMENTS AND SUGGESTIONS: The mass executes simple harmonic motion around the new equilibrium point. Note that gravity does not affect the frequency of oscillation of this motion; it only affects the location of the equilibrium point. With gravity, the equilibrium point is a distance $mg/k$ lower than what it would be without gravity [see Eq. (34)].

## 15.3 Kinetic Energy and Potential Energy

The kinetic energy of a mass with simple harmonic motion is

$$K = \tfrac{1}{2}mv^2 = \tfrac{1}{2}m[-A\omega\sin(\omega t + \delta)]^2 \tag{40}$$

$$= \tfrac{1}{2}m\omega^2A^2\sin^2(\omega t + \delta) \tag{41}$$

This can also be written

$$K = \tfrac{1}{2}kA^2\sin^2(\omega t + \delta) \tag{42}$$

The potential energy associated with the force $F = -kx$ is [see Eq. (8.10)]

$$U = \tfrac{1}{2}kx^2 \tag{43}$$

For simple harmonic motion this becomes

$$U = \tfrac{1}{2}k[A\cos(\omega t + \delta)]^2$$

$$= \tfrac{1}{2}kA^2\cos^2(\omega t + \delta) \tag{44}$$

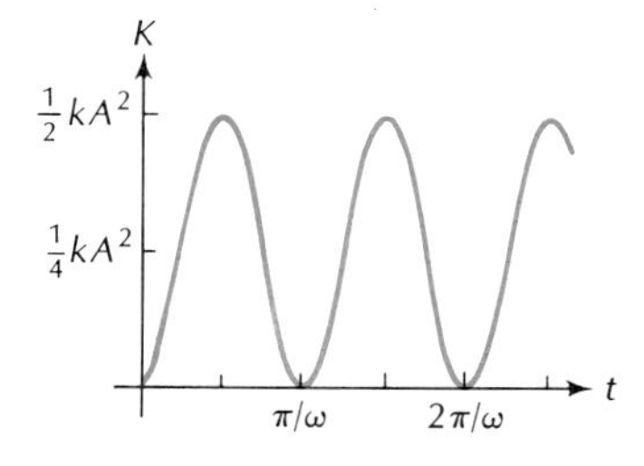


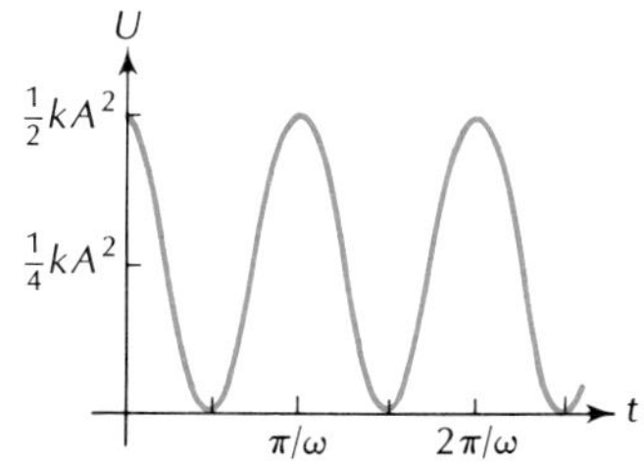


**Fig. 15.10** Kinetic energy and potential energy of a simple harmonic oscillator as a function of time (for $\delta = 0$).

The kinetic energy and the potential energy are both functions of time. Each oscillates between a minimum value of 0 and a maximum value of $\frac{1}{2}kA^2$ (Figure 15.10). When the mass passes the equilibrium position ($x = 0$), the kinetic energy is maximum (maximum speed) and the potential energy is zero; when the mass reaches the turning point, the kinetic energy is zero and the potential energy is maximum (maximum displacement).

Since the force $F = -kx$ is conservative, the total energy $E = K + U$ is a constant of the motion. This conservation law for the energy can be verified explicitly by taking the sum of Eqs. (42) and (44),

$$E = K + U$$

$$= \tfrac{1}{2}kA^2 \sin^2(\omega t + \delta) + \tfrac{1}{2}kA^2 \cos^2(\omega t + \delta)$$

$$= \tfrac{1}{2}kA^2[\sin^2(\omega t + \delta) + \cos^2(\omega t + \delta)]$$

or

*Energy of a simple harmonic oscillator*

$$\boxed{E = \tfrac{1}{2}kA^2} \qquad (45)$$

This establishes that the energy is constant and is proportional to the square of the amplitude of oscillation.

Note that by means of Eq. (45), the maximum displacement can be expressed in terms of the energy:

$$x_{\text{max}} = A = \sqrt{2E/k} \qquad (46)$$

Likewise, the maximum speed can be expressed in terms of the energy:

$$\boxed{E = \tfrac{1}{2}mv_{\text{max}}^2}$$

or

$$v_{\text{max}} = \sqrt{2E/m} \qquad (47)$$

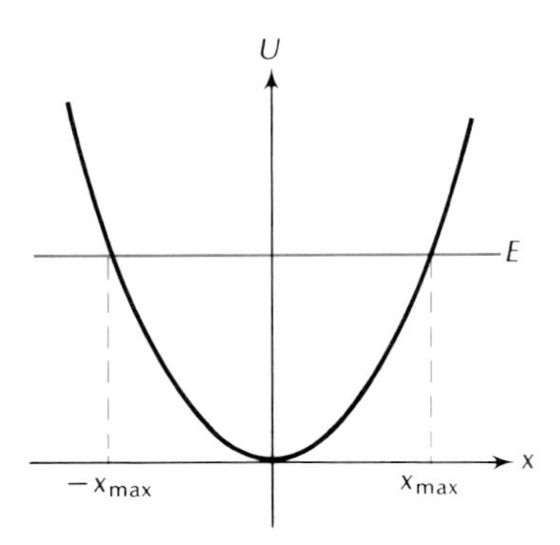


**Fig. 15.11** Potential energy of simple harmonic oscillator as a function of $x$. The horizontal line shows the energy level $E$.

According to Eq. (46), the amplitude of oscillation is large whenever the energy is large. This increase of the amplitude of oscillation with energy can be readily understood by examining the curve of potential energy. For the simple harmonic oscillator, the potential energy is $U = \tfrac{1}{2}kx^2$; Figure 15.11 shows a plot of this potential energy as a function of $x$. The horizontal line shows an energy level $E$. The turning points of the motion are at $+x_{\text{max}}$ and $-x_{\text{max}}$. If we were to increase the height of the energy level, we would spread these turning points farther apart, and we would increase the amplitude of oscillation.

---

EXAMPLE 3. The hydrogen molecule ($H_2$) may be regarded as two masses joined by a spring (Figure 15.12). The center of the spring (center of mass) can be regarded as fixed, so the molecule consists of two identical simple harmonic oscillators vibrating in opposite directions. The spring constant of each oscillator is $1.1 \times 10^3$ N/m and the mass of each oscillator is $1.67 \times 10^{-27}$ kg. Suppose that the vibrational energy of the molecule is $1.3 \times 10^{-19}$ J. Find the corresponding amplitude of oscillation and the maximum velocity.

SOLUTION: Each atom has half the energy of the molecule; thus the energy per atom is

$$E = 6.5 \times 10^{-20}\ \text{J}$$

and the amplitude of oscillation and the maximum speed for each atom are

$$x_{\max} = \left(\frac{2E}{k}\right)^{1/2} = \left(\frac{2 \times 6.5 \times 10^{-20}\ \text{J}}{1.1 \times 10^{3}\ \text{N/m}}\right)^{1/2}$$

$$= 1.1 \times 10^{-11}\ \text{m}$$

$$v_{\max} = \left(\frac{2E}{m}\right)^{1/2} = \left(\frac{2 \times 6.5 \times 10^{-20}\ \text{J}}{1.67 \times 10^{-27}\ \text{kg}}\right)^{1/2}$$

$$= 8.8 \times 10^{3}\ \text{m/s}$$

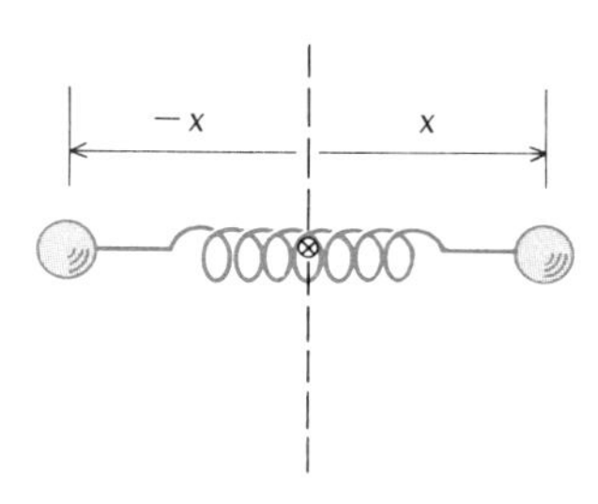


**Fig. 15.12** A hydrogen molecule, represented as two masses joined by a spring. The masses move symmetrically relative to the center of mass.

## 15.4 The Simple Pendulum

A **simple pendulum** consists of a small bob suspended from a fixed point by a string or a rod (Figure 15.13). The bob is assumed to behave like a pointlike particle of mass $m$, and the string is assumed massless. Gravity acting on the bob provides the restoring force. When in equilibrium, the pendulum hangs vertically, just like a plumb line. When released at some angle with the vertical, the pendulum will swing back and forth along an arc of circle in a fixed vertical plane containing the equilibrium position and the initial position of the string (Figure 15.14).[2] The motion is two dimensional; however, the position of the pendulum can be completely described by a single variable: the angle between the string and the vertical (Figure 15.13). We will reckon this angle as positive on the right side of the vertical, and negative on the left side.

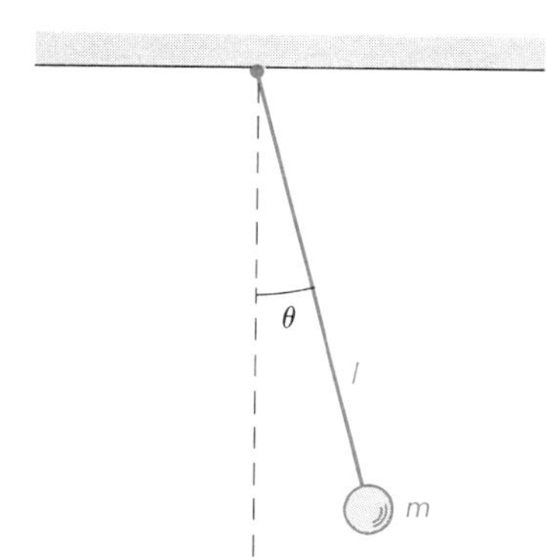


**Fig. 15.13** A pendulum. The angle $\theta$ is reckoned as positive if the deflection of the pendulum is toward the right, as in this figure.

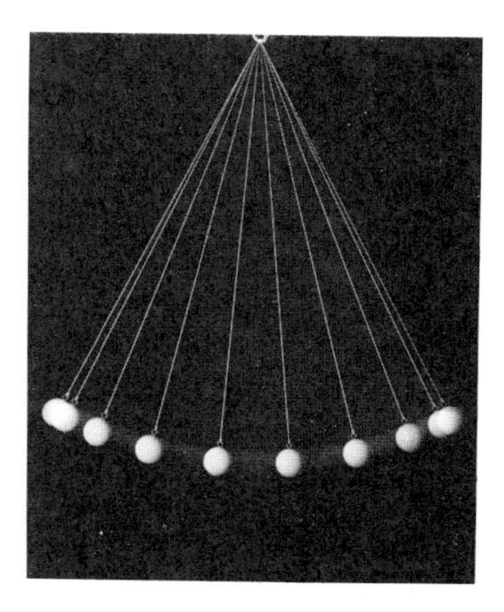

**Fig. 15.14** Stroboscopic photograph of swinging pendulum. The pendulum moves slowly at the extremes of its motion.

Since the bob and the string swing as a rigid unit, the motion can be

[2] We ignore the Foucault effect, that is, the slow rotation of the plane of oscillation of the pendulum caused by the rotation of the Earth (see Section 5.1).

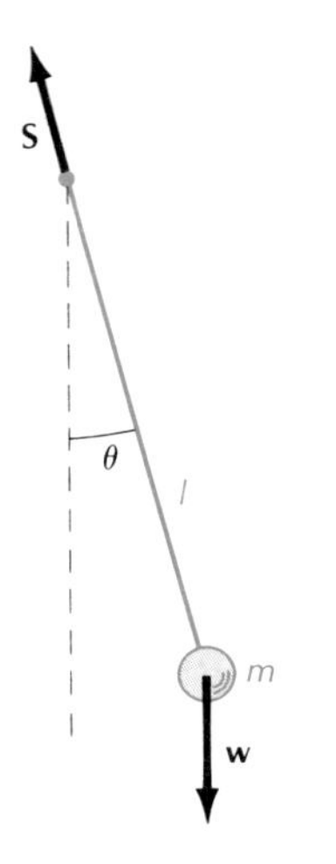


**Fig. 15.15** "Free-body" diagram for the string–particle system.

regarded as rotation about a horizontal axis through the fixed point of suspension, and the equation of motion is that of a rigid body [see Eq. (13.16)]:

$$I\alpha = \tau \tag{48}$$

Here the moment of inertia and the torque are reckoned about the horizontal axis through the point of suspension.

Figure 15.15 shows the external forces acting on the string–bob system. The external forces are the weight **w** acting on the mass $m$ and the suspension force **S** acting on the string at the point of suspension. The suspension force exerts no torque, since its point of application is on the axis of rotation. The weight exerts a torque

$$\tau = -mgl \sin\theta \tag{49}$$

where $l$ is the length of the pendulum, measured from the fixed point of suspension to the center of the bob. The minus sign in Eq. (49) indicates that this is a restoring torque which tends to pull the pendulum to its equilibrium position.

The moment of inertia is simply that of the particle of mass $m$ at a distance $l$ from the axis of rotation,

$$I = ml^2 \tag{50}$$

Hence the equation of motion is

$$ml^2 \frac{d^2\theta}{dt^2} = -mgl \sin\theta \tag{51}$$

or

$$l \frac{d^2\theta}{dt^2} = -g \sin\theta \tag{52}$$

We will solve this equation of motion only in the special case of small oscillations of the pendulum. If $\theta$ is small, then we can make the approximation[3]

$$\sin\theta \cong \theta \tag{53}$$

where the angle is measured in radians. The equation of motion then becomes

$$l \frac{d^2\theta}{dt^2} = -g\theta \tag{54}$$

This differential equation has the same form as Eq. (21); $\theta$ replaces $x$, $l$ replaces $m$, and $g$ replaces $k$. Hence the angular motion is simple harmonic,

$$\theta = A\cos(\omega t + \delta) \tag{55}$$

with an angular frequency

[3] See Appendix 5.

$$\boxed{\omega = \sqrt{g/l}} \tag{56}$$

*Angular frequency of a simple pendulum*

The period of the pendulum is then

$$\boxed{T = 2\pi/\omega = 2\pi\sqrt{l/g}} \tag{57}$$

*Period of a simple pendulum*

Note that this period depends only on the length of the pendulum and on the acceleration of gravity; it does not depend on the mass of the pendulum bob or on the amplitude of oscillation.

---

EXAMPLE 4. What is the length of the "seconds" pendulum at a place where $g = 9.81\ \text{m/s}^2$? The "seconds" pendulum has a period of exactly 2.0 s so that each one-way swing takes exactly 1.0 s.

SOLUTION: From Eq. (57)

$$l = \left(\frac{T}{2\pi}\right)^2 g = \left(\frac{2.0\ \text{s}}{2\pi}\right)^2 \times 9.81\ \text{m/s}^2 = 0.994\ \text{m}$$

---

The kinetic energy of the pendulum can be calculated from Eq. (12.21):[4]

$$K = \tfrac{1}{2}I\left(\frac{d\theta}{dt}\right)^2 = \tfrac{1}{2}ml^2[-\omega A\ \sin(\omega t + \delta)]^2$$

$$= \tfrac{1}{2}mglA^2\ \sin^2(\omega t + \delta) \tag{58}$$

The potential energy is simply the gravitational potential energy $mgh$, where $h$ represents the height of the mass above the equilibrium point. From Figure 15.16 we see that

$$h = l - l\cos\theta \tag{59}$$

so that

$$U = mgh = mgl(1 - \cos\theta) \tag{60}$$

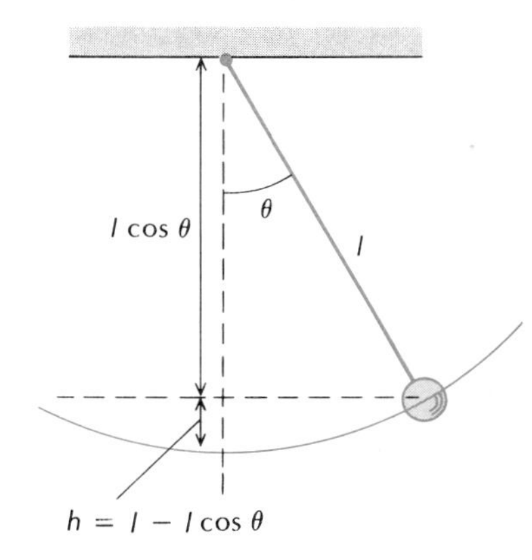


**Fig. 15.16** The height $h$ of the mass above its equilibrium point is the difference between $l$ and $l\cos\theta$.

If the angle $\theta$ is small, then we can take advantage of the approximation[5]

$$\cos\theta \cong 1 - \tfrac{1}{2}\theta^2$$

This leads to the expression

$$U = mgl(1 - 1 + \tfrac{1}{2}\theta^2) = \tfrac{1}{2}mgl\theta^2 \tag{61}$$

$$= \tfrac{1}{2}mglA^2\ \cos^2(\omega t + \delta) \tag{62}$$

---

[4] Keep in mind that the angular velocity $\omega = d\theta/dt$ of Chapter 12 is not the same thing as the angular frequency $\omega$ of the present chapter. Only for uniform circular motion are these two $\omega$'s the same.

[5] See Appendix 5.

The total energy $E = K + U$ is of course constant, as the following formulas show explicitly:

$$E = K + U$$

$$= \tfrac{1}{2}mglA^2 \sin^2(\omega t + \delta) + \tfrac{1}{2}mglA^2 \cos^2(\omega t + \delta)$$

$$= \tfrac{1}{2}mglA^2 \qquad (63)$$

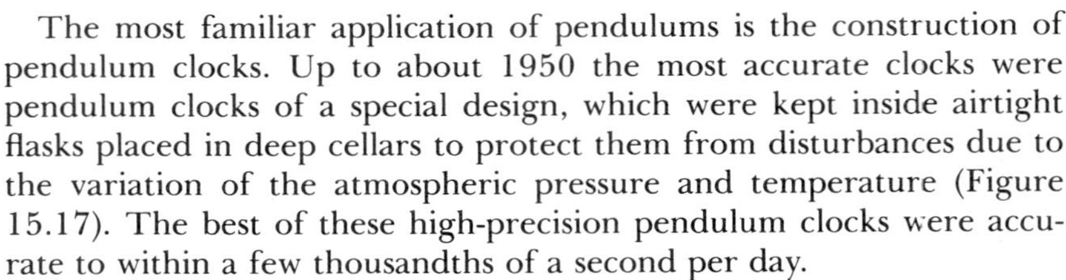

The most familiar application of pendulums is the construction of pendulum clocks. Up to about 1950 the most accurate clocks were pendulum clocks of a special design, which were kept inside airtight flasks placed in deep cellars to protect them from disturbances due to the variation of the atmospheric pressure and temperature (Figure 15.17). The best of these high-precision pendulum clocks were accurate to within a few thousandths of a second per day.

**Fig. 15.17** This electromechanical clock designed by W. H. Shortt served as the U.S. frequency standard from 1924 to 1929. The clock has two pendulums: a master pendulum (in the evacuated cannister at left) and a slave pendulum (at right) that drives the clock mechanism. The master controls the slave by means of periodic electric signals.

Another important application of pendulums is the measurement of the acceleration of gravity $g$. For this purpose it is only necessary to time the swings of a pendulum of known length; the value of $g$ can then be calculated from Eq. (57). Instead of a string and bob, precise determinations of $g$ rely on a rigid pendulum consisting of a solid bar swinging about one end (we will discuss this kind of pendulum in the next section). This pendulum method is very convenient and very precise, but modern electronic instrumentation permits the attainment of even higher precision by direct timing of the free-fall motion of a small projectile in an evacuated chamber (see Section 2.6).

Pendulums can also be used to test whether the acceleration of gravity is the same for masses made of different materials. If two pendulum bobs made of different materials experience slightly different values of $g$, then their periods of oscillation would be slightly different, even if the lengths of the strings suspending them were exactly the same; if the two pendulums are released simultaneously, they would gradually get out of step. This method was used by Galileo and by Newton in the early experiments testing the universality of the rate of free fall.

Finally, we emphasize that the approximation contained in Eq. (53) is valid only for small angles. If the amplitude of oscillation is more than a few degrees, then Eq. (53) fails, and the motion of the pendulum deviates from simple harmonic motion. For instance, at large amplitudes the period of the pendulum depends on the amplitude — the larger the amplitude, the larger the period. This lack of isochronism indicates that the pendulum does not behave any more as a simple harmonic oscillator. Table 15.1 shows how the period of a pendulum increases with amplitude.

**Table 15.1** INCREASE OF PERIOD OF SIMPLE PENDULUM

| *Amplitude* | *Increase of period* |
|---|---|
| 0° | 0.0% |
| 15° | 0.4 |
| 30° | 1.7 |
| 45° | 4.0 |
| 60° | 7.3 |
| 90° | 18.0 |
| 135° | 52.8 |
| 175° | 187.7 |
| 180° | $\infty$ |

## 15.5 Other Oscillating Systems

As we saw in the preceding section, the simple pendulum behaves approximately as a simple harmonic oscillator. The crucial requirement for this approximation is that the amplitude of the oscillations around the equilibrium position be small; if so, then the effective restoring force is directly proportional to the (angular) displacement, and the equation of motion has the same mathematical form as that of the simple harmonic oscillator [see Eqs. (52)–(54)]. Many other physical systems behave approximately as simple harmonic oscillators when oscillating with small amplitude around an equilibrium position. The reason is that near the equilibrium position, the effective force is usually directly proportional to the displacement. This can easily be seen by writing a Taylor-series[6] expansion for the force as a function of displacement. Suppose that the position of the system (a particle, rigid body, atom, or whatever) is described by the variable $x$. If we place the origin at the equilibrium point, then the Taylor series for the force as a function of $x$ is of the form

$$F(x) = F(0) + \left(\frac{dF}{dx}\bigg|_{x=0}\right)x + \frac{1}{2}\left(\frac{d^2F}{dx^2}\bigg|_{x=0}\right)x^2 + \cdots \qquad (64)$$

We are assuming here that the motion is in one dimension. It is also possible to deal with motion in three dimensions — then *each* component of the force has a series expansion and, furthermore, each of the series must include $x$, $y$, and $z$ terms; but we will not deal with these complications. Incidentally: If the motion is described by an angular variable, then it is best to write a Taylor series involving this angular variable rather than the $x$ variable [see Eq. (53)].

By hypothesis, the point $x = 0$ is an equilibrium point, that is, a point at which the force vanishes,

$$F(0) = 0 \qquad (65)$$

Furthermore, if the displacement is small, then the second-order and higher-order terms in Eq. (64) can be neglected compared to the first-order term. Consequently, Eq. (64) leads to the approximation

$$F(x) \cong \left(\frac{dF}{dx}\bigg|_{x=0}\right)x \qquad (66)$$

which says that the force is directly proportional to the displacement. Let us write this equation as

$$F(x) \cong -kx \qquad (67)$$

with

$$k = -\frac{dF}{dx}\bigg|_{x=0} \qquad (68)$$

Equation (67) has the familiar form of Hooke's Law. We therefore see that this law is a general approximation formula which describes forces near equilibrium points.

[6] See Appendix 5 for some examples of Taylor series.

*Stable and unstable equilibria*

If the derivative of the force is negative, then $k > 0$ and Eq. (67) shows that the force is a *restoring* force analogous to the familiar spring force. Under these conditions the equilibrium is **stable:** when the system is displaced slightly from $x = 0$, the force tends to pull it back. If the derivative of the force is positive, then $k < 0$ and then Eq. (67) shows that the force is *repulsive*. This means that the equilibrium is **unstable:** when the system is displaced slightly from $x = 0$, the force tends to push it away even farther.

Figure 15.18 gives some examples of equilibria. The automobile shown in Figure 15.18a rests at the bottom of a valley; this is stable equilibrium: if we push the automobile forward or backward, it tends to return to its original position.[7] The automobile shown in Figure 15.18b rests on the top of a hill; this is unstable equilibrium: if we push the automobile ever so slightly, it will roll down the hill. Finally, the automobile shown in Figure 15.18b rests on a level road; this is neutral equilibrium.

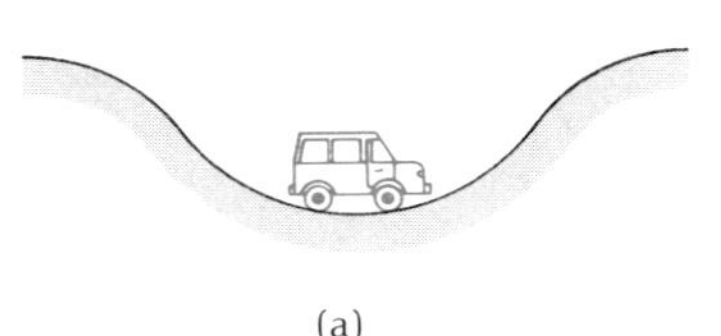

(a)

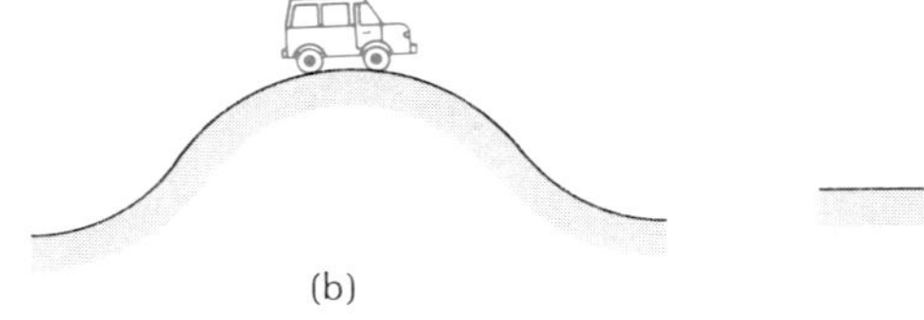

(b)

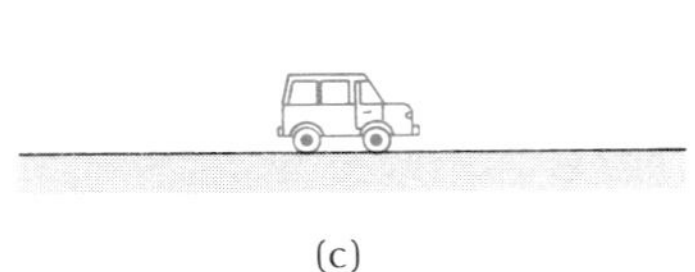

(c)

**Fig. 15.18** Stationary automobile in (a) stable, (b) unstable, and (c) neutral equilibrium.

We may then conclude that systems in motion near a stable equilibrium point will usually behave as simple harmonic oscillators — such systems will oscillate back and forth through the equilibrium point.[8] In the rest of this section we will look at several examples of such systems.

If the motion is a rotation of a rigid body on some axis, then we can formulate equations analogous to Eqs. (65)–(68) using the angular variable $\theta$ in place of $x$, and the torque $\tau$ in place of $F$. Near an (angular) equilibrium point, the torque will then be approximately proportional to the angular displacement. The approximation $\tau = -mgl \sin \theta \simeq -mgl\theta$ we made in connection with the equations (52)–(54) for the simple pendulum is an instance of such a proportionality of torque and angular displacement near an equilibrium point.

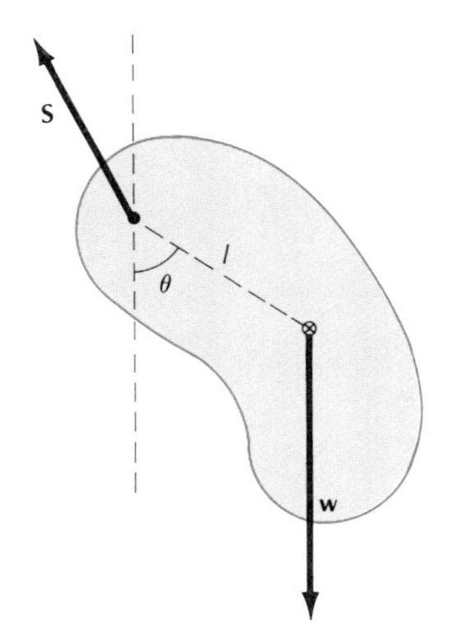


**Fig. 15.19** "Free-body" diagram for a physical pendulum.

THE PHYSICAL PENDULUM The **physical pendulum** consists of a solid body which is suspended from a horizontal axis (Figure 15.19). Under the influence of gravity, the body will swing back and forth. The position of the pendulum can be described by the angle $\theta$. The theory of the physical pendulum is much the same as that of the simple pendulum. The equation of motion is that of a rigid body,

$$I \frac{d^2\theta}{dt^2} = \tau \tag{69}$$

---

[7] The motion is two dimensional, instead of one dimensional as in the discussion of Eqs. (64)–(68). However, the motion can be described by a single variable — for instance, by the angle measured along the circular arc of the bottom of the valley.

[8] However, this conclusion fails if $dF/dx = 0$ at the equilibrium point. Then the higher order terms in Eq. (64) must be taken into account.

where the torque and the moment of inertia are reckoned about the horizontal axis. The only external torque on the system is that exerted by the force of gravity. This torque can be evaluated by pretending that the point of action of the force is the center of mass. If the distance between the axis and the center of mass is $l$, then (see Figure 15.19)

$$\tau = -Mgl \sin \theta \tag{70}$$

where $M$ is the mass of the body. Hence

$$I \frac{d^2\theta}{dt^2} = -Mgl \sin \theta$$

For small oscillations about the vertical, we can make the same approximation as in the case of the simple pendulum. We then obtain the approximate equation

$$I \frac{d^2\theta}{dt^2} = -Mgl\theta \tag{71}$$

This equation is, again, analogous to Eq. (21). The moment of inertia $I$ plays the role of mass and $Mgl$ plays the role of force constant. Consequently, the motion is simple harmonic with a frequency

$$\boxed{\omega = \sqrt{Mgl/I}} \tag{72}$$

*Angular frequency of a physical pendulum*

Comparing this with Eq. (56), we see that the simple pendulum is a special case of a physical pendulum with $I = Ml^2$, the moment of inertia of a pointlike particle. Figure 15.20 shows the physical pendulum used in an accurate pendulum clock.

As we mentioned in the preceding section, measurements of the acceleration of gravity are commonly done with a physical pendulum. Figure 15.21 shows a pendulum designed for such measurements.

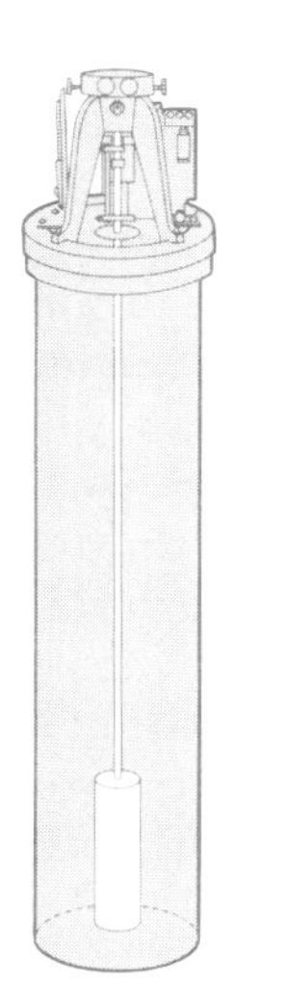

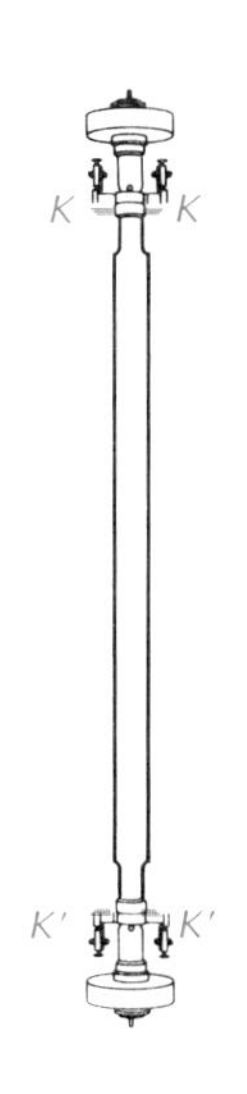


**Fig. 15.20** (left) The pendulum of the Shortt clock (see Figure 15.17) in its cannister. As in most high-precision pendulum clocks, the bob is a cylinder attached to a rod.

**Fig. 15.21** (right) Reversible pendulum used for the very accurate determination of the acceleration of gravity at Potsdam, Germany, in 1905, which for many years served as the basis of comparison for determinations of the acceleration of gravity at other stations. When in use, the pendulum swings on the knife edges $KK$ resting on a support (not shown). The acceleration of gravity is calculated from the measured period of swing. The pendulum can be reversed, so it swings on the knife edges $K'K'$. This reversal permits elimination of air friction from the calculation.

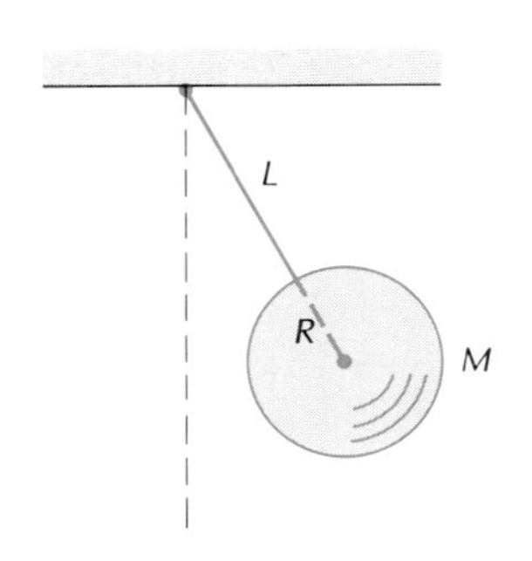


**Fig. 15.22** Physical pendulum consisting of a sphere suspended by a massless string.

EXAMPLE 5. A physical pendulum consists of a uniform spherical bob of mass $M$ and radius $R$ suspended from a massless string of length $L$. Taking into account the size of the bob, what is the period of small oscillations of this pendulum?

SOLUTION: The moment of inertia of a sphere about its center of mass is $\frac{2}{5}MR^2$. In Figure 15.22, the sphere is rotating about the point of suspension, at a distance $R + L$ from the center of mass. According to the parallel-axis theorem [see Eq. (12.33)], the moment of inertia of the sphere about this point is

$$I = I_{CM} + M(R + L)^2 = \tfrac{2}{5}MR^2 + M(R + L)^2$$

$$= M[\tfrac{2}{5}R^2 + (R + L)^2]$$

The pull of gravity acts at the center of mass of the sphere. Hence the length $l$ in Eq. (72) is $l = R + L$ and the frequency of oscillation is

$$\omega = \sqrt{\frac{Mgl}{I}} = \sqrt{\frac{Mg(R + L)}{M[\frac{2}{5}R^2 + (R + L)^2]}}$$

$$= \sqrt{\frac{g(R + L)}{\frac{2}{5}R^2 + (R + L)^2}} \tag{73}$$

COMMENTS AND SUGGESTIONS: This formula shows how the frequency of such a physical pendulum depends on the size of the bob. If we substitute $R = 0$ into Eq. (73), we recover the familiar result (56) for the frequency of a simple pendulum, that is, a pendulum with a bob consisting of a pointlike particle. From Eq. (73) we can easily verify that the frequency of the physical pendulum is lower than the frequency of a simple pendulum of length $L$, and is also lower than the frequency of a simple pendulum of length $L + R$, the distance from the point of suspension to the center of the bob. But if $R \ll L$, then the difference between the frequencies of the physical pendulum and either of these simple pendulums is small.

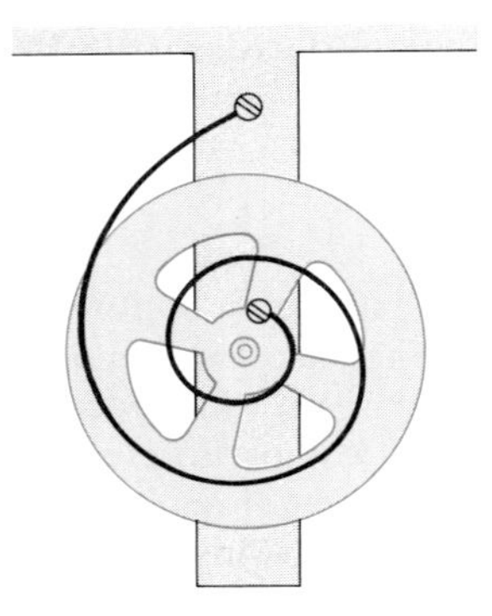

**Fig. 15.23** Torsional pendulum consisting of a wheel with a spiral spring.

THE TORSIONAL PENDULUM The **torsional pendulum** consists of a rigid body swinging back and forth about a fixed axis — in this regard the torsional pendulum is quite similar to the physical pendulum. However, the restoring force is supplied by a spring rather than by gravity. Figure 15.23 shows a torsional pendulum with a spiral spring; one end of the spring is attached to the body; the other end is held fixed by a support. Such a spiral spring has an equilibrium configuration and exerts a restoring torque if disturbed from this equilibrium configuration. Figure 15.24 shows an example of a torsional pendulum with a torsional suspension fiber. The pendulum has the shape of a symmetric dumbbell, and the fiber on which it hangs serves both as axis of rotation and as "spring." If the pendulum is turned through some arc about the vertical axis, the fiber is twisted and exerts a restoring torque as it tries to untwist.

Under the assumption that the angular displacement of the torsional pendulum from its equilibrium position is small, the restoring torque provided by a spiral spring or a torsional fiber is proportional to this angular displacement,

$$\tau = -\kappa\theta \tag{74}$$

In this context, what is meant by a "small" displacement from the equilibrium position depends on the design of the spring. Some torsional pendulums (with long spiral springs or long, thin suspension fibers) can be twisted through several turns without introducing appreciable deviations into the proportionality in Eq. (74). For such a pendulum, several turns can be regarded as "small."

The quantity $\kappa$ in Eq. (74) is the **torsional constant** of the spring or fiber; its units are N · m/radian.

Equation (74) is the angular analog of Eq. (20). The equation of motion of the rigid body on which the torque acts is then

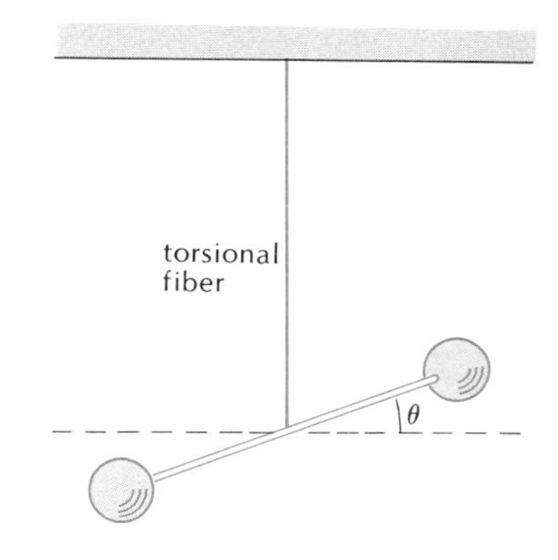


**Fig. 15.24** Torsional pendulum consisting of a dumbbell suspended from a fiber.

$$I\frac{d^2\theta}{dt^2} = -\kappa\theta$$

This, again, has the form of the equation for a simple harmonic oscillator. The angular motion is then simple harmonic with a frequency

$$\boxed{\omega = \sqrt{\kappa/I}} \qquad (75)$$

*Angular frequency of a torsional pendulum*

Torsional pendulums with spiral springs ("hairsprings") find an important application in mechanical watches and chronometers. The rate of such a watch is controlled by the oscillations of the balance wheel, a torsional pendulum. Figure 15.25 shows the balance wheel of a wristwatch and its hairspring; the screws on the rim of the wheel are used for fine adjustments of the moment of inertia.

**Fig. 15.25** Balance wheel of a watch.

The Cavendish balance of Section 9.2 is essentially a torsional pendulum with a suspension fiber. In modern versions of this balance, the suspension fiber is usually made of a very thin strand of quartz which has a very low restoring torque (small value of $\kappa$); correspondingly, the period of the rotational oscillations is very long.

---

EXAMPLE 6. The dumbbell of a modern Cavendish balance consists of two equal masses of 0.025 kg connected by a nearly massless rod 0.40 m long. When set in motion, the balance rotates back and forth with a period of 3.8 min. Find the value of the torsional constant from these data.

SOLUTION: The angular frequency of the oscillations is

$$\omega = \frac{2\pi}{T} = \frac{2\pi}{3.8 \times 60 \text{ s}} = 2.76 \times 10^{-2} \text{ radian/s}$$

and the moment of inertia of the dumbbell is

$$I = mR^2 + mR^2 = 2mR^2 = 2 \times 0.025 \text{ kg} \times (0.20 \text{ m})^2 = 2.0 \times 10^{-3} \text{ kg} \cdot \text{m}^2$$

Hence, from Eq. (75),

$$\kappa = I\omega^2 = 2.00 \times 10^{-3} \text{ kg} \cdot \text{m}^2 \times (2.76 \times 10^{-2}/\text{s})^2 = 1.52 \times 10^{-6} \text{ N} \cdot \text{m/radian}$$

COMMENTS AND SUGGESTIONS: This illustrates how the torsional constant can be determined from a simple measurement of the period of oscillation. Since $\kappa$ is very small, it would be extremely difficult to determine $\kappa$ by a direct measurement of torque and angular displacement.

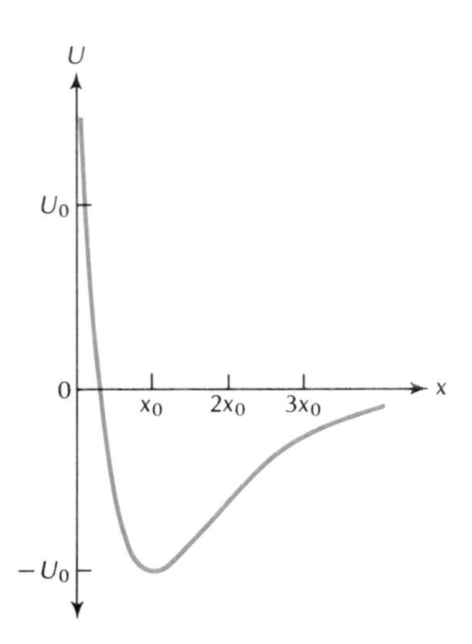


**Fig. 15.26** Potential energy of one H atom in an $H_2$ molecule. Here $x$ is the distance of the atom from the center of mass.

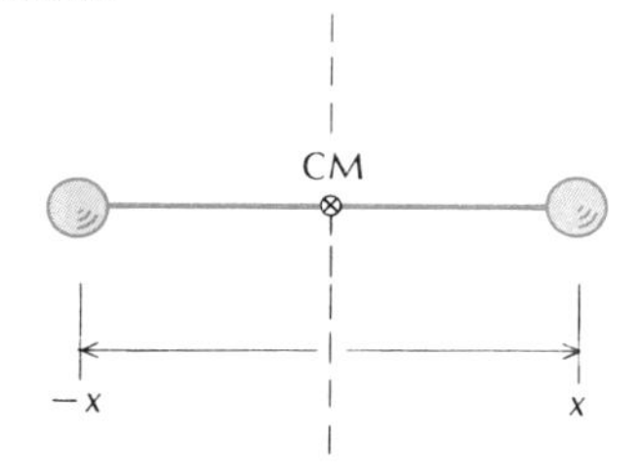


**Fig. 15.27** The atoms of a hydrogen molecule oscillate symmetrically relative to the center of mass.

THE DIATOMIC MOLECULE The atoms of a **diatomic molecule,** such as the hydrogen atoms in $H_2$, are bound to each other; but they are not rigidly bound. The interatomic force that holds the atoms in the molecule permits some freedom of movement (this force arises from the attractions and repulsions between the positive and negative electric charges within one atom and those within the other). If the atoms are widely separated, the net force between them is attractive; but if they are very close together, the force is repulsive and opposes their interpenetration. At one particular distance the force is zero — the atoms are then in equilibrium.

Figure 15.26 shows a plot of the potential energy of a hydrogen atom in an $H_2$ molecule. The $x$ coordinate represents the position of the atom relative to the center of mass. Since the two atoms have equal masses, the center of mass is always midway between the two atoms (Figure 15.27). If the coordinate of one atom is $x$, that of the other must be $-x$; the distance between the atoms is therefore $2x$. The atoms overlap completely when $x = 0$; as Figure 15.26 shows, the potential energy is then very large and positive, that is, it takes a very large amount of work to squeeze the atoms together. The potential energy has a minimum $U = -U_0$ at $x = x_0$, and the potential energy tends to zero as $x$ tends to infinity. The quantity $U_0$ represents the work that must be done to separate two atoms that are initially at the minimum of potential energy.

The mathematical expression for the potential energy is of the form

$$U = U_0\,(e^{-2(x-x_0)/b} - 2e^{-(x-x_0)/b}) \tag{76}$$

(This formula, which has already been mentioned in Chapter 8, is based on an approximate theory of interatomic forces, a theory that goes beyond the scope of our discussion.) If we take Eq. (76) for granted, we can calculate the force according to Eq. (8.20),

$$F_x = -\frac{\partial U}{\partial x} \tag{77}$$

This gives (compare Example 8.4)

$$F_x = \frac{2U_0}{b}\,(e^{-2(x-x_0)/b} - e^{-(x-x_0)/b}) \tag{78}$$

Figure 15.28 is a plot of this force as a function of $x$. As expected, the force is attractive (negative) if $x$ is large and it is strongly repulsive (positive) if $x$ is small. The force is zero at $x = x_0$; this is the equilibrium point.

In the vicinity of the equilibrium point, the motion of the atom will be simple harmonic. According to the general discussion at the beginning of this section, the "spring" constant for motion is

$$k = -\left.\frac{dF_x}{dx}\right|_{x=x_0} = -\frac{2U_0}{b}\left(-\frac{2}{b}-\frac{1}{b}\right) = \frac{6U_0}{b^2} \tag{79}$$

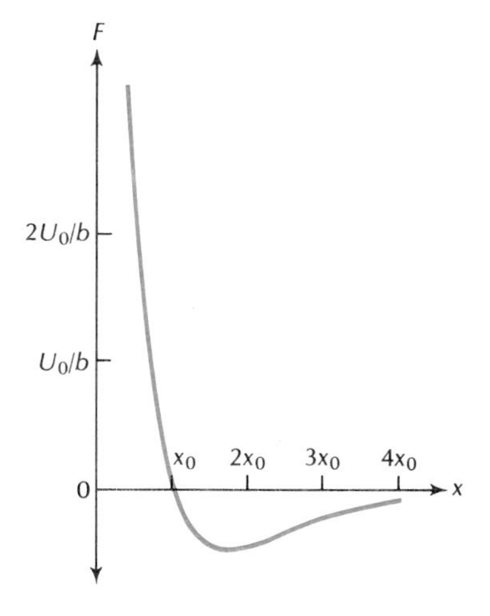


**Fig. 15.28** Force on one H atom in an $H_2$ molecule.

Thus, for small vibrations about the equilibrium point, the interatomic force behaves like a spring force and the hydrogen molecule behaves like two masses joined by a spring. If the molecule is left undisturbed, the atoms sit at their equilibrium positions. But if the molecule suffers a collision or some other disturbance that shakes the atoms, they will vibrate back and forth about their equilibrium positions.

---

EXAMPLE 7. The spring constant for the interatomic "spring" in an $H_2$ molecule cannot be measured directly. However, the frequency of vibration and the mass of each atom can be measured — the frequency is $1.3 \times 10^{14}$ Hz. The mass of each atom is $1.67 \times 10^{-27}$ kg. What value of the spring constant does this imply?

SOLUTION: According to Eq. (24),

$$k = m\omega^2 = m(2\pi\nu)^2$$

$$= 1.67 \times 10^{-27}\ \text{kg} \times (2\pi \times 1.3 \times 10^{14}/\text{s})^2$$

$$= 1.1 \times 10^3\ \text{N/m}$$

This is the spring constant for the motion of one atom relative to the center of mass of the molecule, i.e., it is the spring constant for one-half of the total length of the spring. The spring constant for the entire spring is smaller by a factor of 2; it is $0.55 \times 10^3$ N/m.

---

## 15.6 Damped Oscillations and Forced Oscillations

So far we have proceeded on the assumption that the only force acting on an oscillator is the restoring force. However, in a real oscillator, say, a pendulum, there is always an extra force of friction. If the pendulum starts its swinging motion with some initial amplitude, then the friction against the air and against the point of suspension will gradually brake the pendulum, reducing its amplitude of oscillation.

Figure 15.29 shows the displacement as a function of time for a harmonic oscillator with friction. Such a gradually decreasing oscillation is called **damped harmonic motion.** Mathematically, the motion can be represented by the function

$$x = A_0 e^{-\gamma t/2} \cos(\omega t + \delta) \tag{80}$$

***Damped harmonic motion***

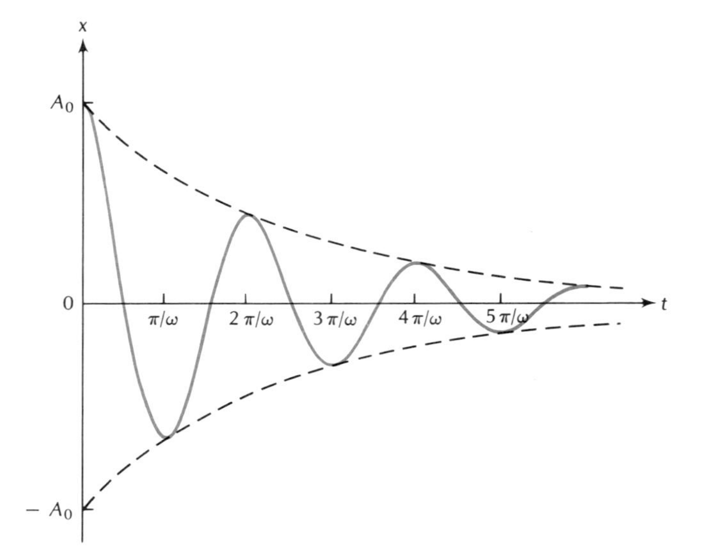


**Fig. 15.29** Worldline of a particle with damped harmonic motion.

where $\gamma$ is a constant which represents the frictional effects. The exponential factor

$$A = A_0 e^{-\gamma t/2} \tag{81}$$

may be regarded as the (time-dependent) amplitude of the motion; the dashed line in Figure 15.29 is a plot of this amplitude factor. The motion given by Eq. (80) may then be regarded as simple harmonic motion with an exponentially decreasing amplitude. Incidentally: The frequency $\omega$ of Eq. (80) is not quite the same as the frequency $\sqrt{k/m}$ of a simple harmonic oscillator without friction. As one might expect intuitively, the friction slows the oscillations and the frequency $\omega$ is somewhat less than $\sqrt{k/m}$, but we will not worry about this correction.

Since the oscillator must do work against the friction, the mechanical energy gradually decreases. This can also be seen from the relation in Eq. (45) between energy and amplitude: the energy is proportional to the square of the amplitude and as the latter decreases, so does the former. It is instructive to calculate the fractional energy loss in a time equal to one period. According to Eq. (45),

$$E = [\text{constant}] \times A^2 = [\text{constant}] \times e^{-\gamma t} \tag{82}$$

Hence

$$\ln E = -\gamma t + \ln[\text{constant}] \tag{83}$$

The time derivative of this equation gives us

$$\frac{1}{E}\frac{dE}{dt} = -\gamma \tag{84}$$

For a motion such as shown in Figure 15.29, the damping is not excessively strong and the energy loss per period is small; then the energy loss $\Delta E$ per period divided by the period is approximately equal to the derivative $dE/dt$,

$$\frac{\Delta E}{T} \cong \frac{dE}{dt} \qquad (85)$$

or

$$\frac{\Delta E}{T} \cong -\gamma E \qquad (86)$$

Accordingly, the fractional energy loss per period is

$$\frac{\Delta E}{E} = -\gamma T \qquad (87)$$

This relation is usually written in terms of the angular frequency:

$$\boxed{\frac{\Delta E}{E} = -2\pi \frac{\gamma}{\omega}} \qquad (88)$$

*Energy loss of a damped oscillator*

The quantity $2\pi|E/\Delta E|$ is called the "$Q$" (or the quality factor) of the damped oscillator:

$$Q = 2\pi|E/\Delta E|$$

or

$$\boxed{Q = \omega/\gamma} \qquad (89)$$

$Q$

In an oscillator of high $Q$, the oscillations continue for a long time with only a small loss of amplitude. In an oscillator of low $Q$, the amplitude is quickly damped by friction. Essentially, $Q$ is the **ringing time,** or the time it takes for the oscillations to dampen appreciably, expressed in units of one period. To see this, note that the amplitude given by Eq. (81) only approaches zero asymptotically — the oscillations never stop completely. However, after a characteristic time

$$t^* = 2/\gamma \qquad (90)$$

the amplitude will have decreased to

$$A = A_0 e^{-\gamma t^*/2} = A_0 e^{-\gamma(2/\gamma)/2} = A_0 e^{-1} \qquad (91)$$

that is, the amplitude will have decreased by a factor of e. We can therefore regard the characteristic time given by Eq. (90) as the ringing time of the oscillator. If we express this time in multiples of the period, we obtain

$$\frac{t^*}{T} = \frac{2}{\gamma T} = \frac{2}{\gamma(2\pi/\omega)}$$

According to Eq. (89), this is

$$\frac{t^*}{T} = \frac{2}{2\pi(1/Q)} = \frac{Q}{\pi}$$

Thus, except for a factor of $\pi$, $Q$ is the number of periods contained in $t^*$.

Mechanical oscillators of low friction, such as tuning forks or piano strings, have $Q$'s of a few thousand, that is, they "ring" for a few thousand periods before their oscillations fade so much that they become hardly noticeable.

If the oscillations of a damped oscillator are to be maintained at a constant level, it is necessary to exert some extra force on the oscillator, so the energy fed into the oscillator by this new force compensates for the energy lost to friction. An extra force is also needed to start the oscillations of any oscillator, damped or not, by supplying the initial energy for the motion. Any such extra force exerted on an oscillator is called a **driving force.** A familiar example is the "pumping" force that must be exerted on a swing (a pendulum) to start it moving and to keep it going at a constant amplitude. This is an example of a periodic driving force.

*Driving force*

If the period of the driving force coincides with the period of the natural oscillations of the oscillator, then even a quite small driving force can gradually build up large amplitudes. Essentially, what happens is that under these conditions the driving force steadily feeds energy into the oscillations, and the amplitude of these grows until the friction becomes so large that it inhibits further growth. Thus, the ultimate amplitude that is reached depends on friction; in an oscillator of low friction, or high $Q$, this ultimate amplitude can be extremely large. The buildup of a large amplitude by the action of a driving force in tune with the natural frequency of an oscillator is called **resonance,** or **sympathetic oscillation.**

*Resonance*

The phenomenon of resonance plays a crucial role in many pieces of mechanical machinery — if one vibrating part of a machine is driven at resonance by a perturbing force originating from some other part, then the amplitude of oscillation can build up to a violent level and shake the machine apart. Such dangerous resonant effects can occur not only in moving pieces of machinery, but also in structures that are normally regarded as static. In a famous accident that took place in 1850 in Angers (France), the stomping of 487 soldiers marching over a suspension bridge excited a resonant swinging motion of the bridge; the motion quickly rose to a disastrous level and broke the bridge apart, causing the death of 226 of the soldiers.

## SUMMARY

**Simple harmonic motion:** $x = A\cos(\omega t + \delta)$

**Period:** $T = 2\pi/\omega$

**Frequency:** $\nu = 1/T = \omega/2\pi$

**Equation of motion of simple harmonic oscillator:**

$$m\frac{d^2x}{dt^2} = -kx$$

**Angular frequency of simple harmonic oscillator:** $\omega = \sqrt{k/m}$

**Energy of simple harmonic oscillator:** $E = \frac{1}{2}kA^2$

$$= \frac{1}{2}mv_{max}^2$$

**Angular frequency and period of simple pendulum:**

$$\omega = \sqrt{g/l} \qquad T = 2\pi\sqrt{l/g}$$

**Angular frequency of physical pendulum:** $\omega = \sqrt{Mgl/I}$

**Angular frequency of torsional pendulum:** $\omega = \sqrt{\kappa/I}$

**Fractional energy loss per period of damped oscillator:**

$$\frac{\Delta E}{E} = -2\pi\frac{\gamma}{\omega}$$

**$Q$ of damped oscillator:** $Q = \omega/\gamma$

## QUESTIONS

1. Is the motion of the piston of an automobile engine simple harmonic motion? How does it differ from simple harmonic motion?

2. According to Section 15.1, the superposition of two simple harmonic motions along the $x$ and $y$ axes gives uniform circular motion if their amplitudes are equal and their phases differ by $\pi/2$.
   (a) Draw a picture of the orbit if the amplitudes of the $x$ and $y$ motions are not equal.
   (b) Draw a picture of the orbit if the amplitudes are equal but the frequency of the $x$ motion is twice as large as the frequency of the $y$ motion (this orbit is a Lissajous figure).

3. In our calculation of the frequency of the simple harmonic oscillator, we ignored the mass of the spring. Qualitatively, how does the mass of the spring affect the frequency?

4. A simple harmonic oscillator has a frequency of 1.5 Hz. What will happen to the frequency if we cut the spring in half and attach both halves to the mass so that both springs push jointly?

5. A grandfather clock is regulated by a pendulum. If the clock is running late, how must we readjust the length of the pendulum?

6. Figure 15.30 shows the escapement of a pendulum clock, i.e., the linkage that permits the pendulum to control the rotation of the wheels of the clock. Explain how the wheel turns as the pendulum swings.

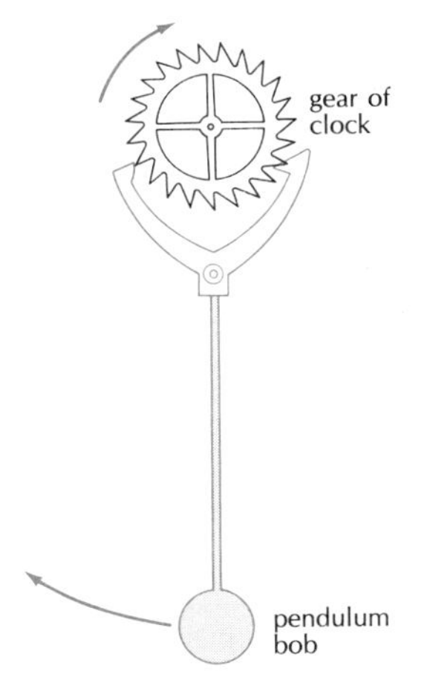


**Fig. 15.30** Escapement of a pendulum clock.

7. Would a pendulum clock keep good time on a ship?

8. Galileo claimed that the oscillations of a pendulum are isochronous, even for an amplitude of oscillation as large as 30°. What is your opinion of this claim?

9. Why would you expect a pendulum oscillating with an amplitude of nearly (but not quite) 180° to have a very long period?

10. Can a pendulum oscillate with an amplitude of more than 180°?

11. Figure 15.31 shows a "tilted pendulum" designed by Christiaan Huygens in the seventeenth century. When the pendulum is tilted, its period is longer than when the pendulum is vertical. Explain.

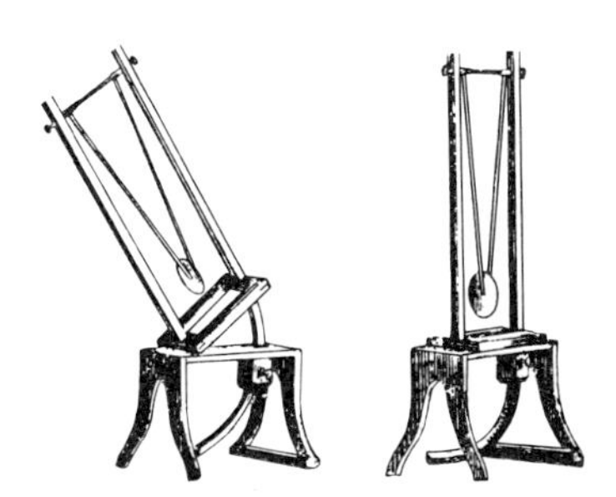

**Fig. 15.31** Huygens' tilted pendulum.

12. An "interrupted" pendulum consists of a simple pendulum of length $l$ with a nail placed at a distance $\frac{3}{4}l$ below the point of support. If this pendulum is released from one side, it will begin to wrap around the nail as soon as it

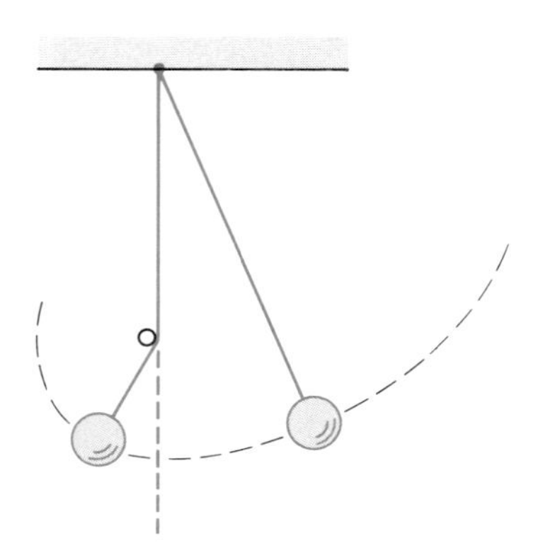

**Fig. 15.32** An "interrupted" pendulum.

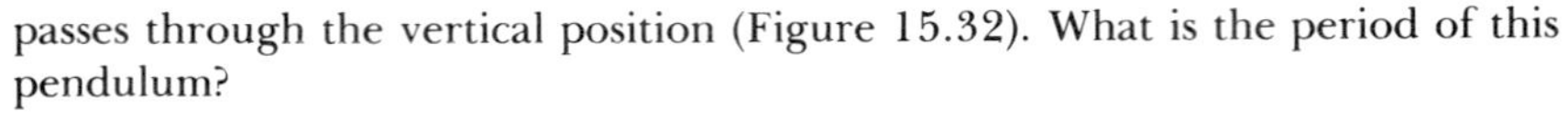

passes through the vertical position (Figure 15.32). What is the period of this pendulum?

13. Most grandfather clocks have a lenticular pendulum bob which supposedly minimizes friction by "slicing" through the air. However, experience has shown that a cylindrical pendulum bob experiences less air friction. Can you suggest an explanation?

14. Galileo described an experiment to compare the acceleration of gravity of lead and of cork:

> I took two balls, one of lead and one of cork, the former being more than a hundred times as heavy as the latter, and suspended them from two equal thin strings, each four or five bracchia long. Pulling each ball aside from the vertical, I released them at the same instant, and they, falling along the circumferences of the circles having the strings as radii, passed thru the vertical and returned along the same path. This free oscillation, repeated more than a hundred times, showed clearly that the heavy body kept time with the light body so well that neither in a hundred oscillations, nor in a thousand, will the former anticipate the latter by even an instant, so perfectly do they keep step.

Since air friction affects the cork ball much more than the lead ball, do you think Galileo's results are credible?

Newton reported a more careful experiment that avoided the inequality of friction:

> I tried the thing in gold, silver, lead, glass, sand, common salt, wood, water, and wheat. I provided two equal wooden boxes. I filled the one with wood, and suspended an equal weight of gold (as exactly as I could) in the centre of oscillation of the other. The boxes, hung by equal threads of 11 feet, made a couple of pendulums perfectly equal in weight and figure . . . and, placing the one by the other, I observed them to play together forwards and backwards for a long while, with equal vibrations. . . . And by these experiments, in bodies of the same weight, one could have discovered a difference of matter less than the thousandth part of the whole.

Explain how Newton's experiment was better than Galileo's.

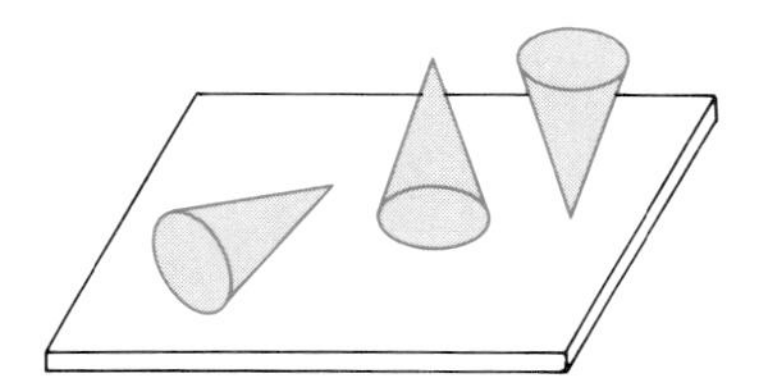

**Fig. 15.33** A cone lying on its side, standing on its base, and standing on its apex.

15. A simple pendulum hangs below a table, with its string passing through a small hole in the tabletop. Suppose you gradually pull the string while the pendulum is swinging. What happens to the frequency of oscillation? To the (angular) amplitude?

16. Consider a cone (a) lying on its side, (b) standing on its base, and (c) standing on its apex on a flat table (Figure 15.33). For which of these positions is the equilibrium stable, unstable, or neutral?

17. Shorter people have a shorter length of stride, but a higher rate of step when walking "naturally." Explain.

18. A girl sits on a swing whose ropes are 1.5 m long. Is this a simple pendulum or a physical pendulum?

19. A simple pendulum consists of a particle of mass $m$ attached to a string of length $l$. A physical pendulum consists of a body of mass $m$ attached to a string in such a way that the center of mass is at a distance $l$ from the point of support. Which pendulum has the shorter period?

20. Is Eq. (73) valid for a negative value of $L$? What is the meaning of negative $L$?

21. All metals expand slightly with increasing temperature. The balance wheel of a watch is made of metal. What will happen to the rate of the watch when the temperature increases? Can you think of some way to compensate for such temperature errors?

22. Suppose that the spring in the front-wheel suspension of an automobile has a natural frequency of oscillation equal to the frequency of rotation of the wheel at, say, 80 km/h. Why is this bad?

23. When marching soldiers are about to cross a bridge, they break step. Why?

## PROBLEMS

### Section 15.1

1. A particle moves back and forth along the $x$ axis between the points $x = 0.20$ m and $x = -0.20$ m. The period of the motion is 1.2 s, and it is simple harmonic. At the time $t = 0$, the particle is at $x = 0$ and its velocity is positive.
   (a) What is the frequency of the motion? The angular frequency?
   (b) What is the amplitude of the motion?
   (c) What is the phase constant?
   (d) At what time will the particle reach the point $x = 0.20$ m? At what time will it reach the point $x = -0.10$ m?
   (e) What is the speed of the particle when it is at $x = 0$? What is the speed of the particle when it reaches the point $x = -0.10$ m?

2. The motion of the piston in an automobile engine is approximately simple harmonic. Suppose that the piston travels back and forth over a distance of 8.50 cm and that the piston has a mass of 1.2 kg. What are its maximum acceleration and maximum speed if the engine is turning over at its highest safe rate of 6000 rev/min? What is the maximum force on the piston?

3. A given point on a guitar string (say, the midpoint of the string) executes simple harmonic motion with a frequency of 440 Hz and an amplitude of 1.2 mm. What is the maximum speed of this motion? The maximum acceleration?

4. A particle moves as follows as a function of time:

$$x = 3.0 \text{ m} \times \cos\left(2.0 \frac{\text{radian}}{\text{s}} t + \frac{\pi}{3}\right)$$

   where distance is measured in meters and time in seconds.
   (a) What is the amplitude of this simple harmonic motion? The frequency? The angular frequency? The period?
   (b) At what earliest positive time does the particle reach the equilibrium point? The turning point?
   (c) At what later times does the particle reach the equilibrium point and the turning point again?

*5. Experience shows that from one-third to one-half of the passengers in an airliner can be expected to suffer motion sickness if the airliner bounces up and down with a peak acceleration of 0.4 G and a frequency of about 0.3 Hz. Assume that this up-and-down motion is simple harmonic. What is the amplitude of the motion?

*6. The frequency of oscillation of a mass attached to a spring is 3.0 Hz. At time $t = 0$, the mass has an initial displacement of 0.20 m and an initial velocity of 4.0 m/s.
   (a) What is the position of the mass as a function of time?
   (b) When will the mass first reach a turning point? What will be its acceleration at that time?

## Section 15.2

7. The body-mass measurement device used aboard Skylab consisted of a chair supported by a spring. The device was calibrated before the space flight by placing a standard mass of 66.91 kg in the chair; with this mass the period of oscillation of the chair was 2.088 s. During the space flight astronaut Lousma sat in the chair; the period of oscillation was then 2.299 s. What was the mass of the astronaut? Ignore the mass of the chair.

8. The body of an automobile of mass 1100 kg is supported by four vertical springs attached to the axles of the wheels. In order to test the suspension, a man pushes down on the body of the automobile and then suddenly releases it. The body rocks up and down with a period of 0.75 s. What is the spring constant of each of the springs? Assume that all the springs are identical and that the compressional force on each spring is the same; also assume that the shock absorbers of the automobile are completely worn out so that they do not affect the oscillation frequency.

9. Deuterium is an isotope of hydrogen. The mass of the deuterium atom is 1.998 times larger than the mass of the hydrogen atom. Given that the frequency of vibration of the $H_2$ molecule is $1.31 \times 10^{14}$ Hz (see Example 7), calculate the frequency of vibration of the $D_2$ molecule. Assume the "spring" connecting the atoms is the same in $H_2$ and $D_2$.

10. Calculate the frequency of vibration of the HD molecule consisting of one atom of hydrogen and one of deuterium. See Problem 9 for necessary data.

*11. A mass $m = 2.5$ kg hangs from the ceiling by a spring with $k = 90$ N/m. Initially, the spring is in its unstretched configuration and the mass is held at rest by your hand. If, at time $t = 0$, you release the mass, what will be its position as a function of time?

*12. The wheel of a sports car is suspended below the body of the car by a vertical spring with a spring constant $1.1 \times 10^4$ N/m. The mass of the wheel is 14 kg and the diameter of the wheel is 61 cm.

(a) What is the frequency of up-and-down oscillations of the wheel? Regard the wheel as a mass on one end of a spring and regard the body of the car as a fixed support for the other end of the spring.

(b) Suppose that the wheel is slightly out of round, having a bump on one side. As the wheel rolls on the street it receives a periodic push each time the bump comes in contact with the street. At what speed of the translational motion of the car will the frequency of this push coincide with the natural frequency of the up-and-down oscillations of the wheel? What will happen to the car at this speed? (Note: This problem is not quite realistic because the elasticity of the tire also contributes a restoring force to the up-and-down motion of the wheel.)

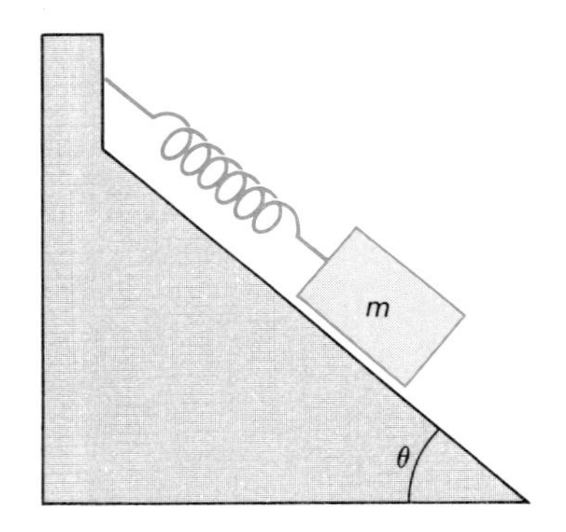


**Fig. 15.34** Mass sliding on an inclined frictionless plane.

*13. A mass $m$ slides on a frictionless plane inclined at an angle $\theta$ with the horizontal. The mass is attached to a spring, parallel to the plane (Figure 15.34); the spring constant is $k$. How much is the spring stretched at equilibrium? What is the frequency of the oscillations of this mass up and down the plane?

**14. Two identical masses slide with one-dimensional motion on a frictionless plane under the influence of three identical springs attached as shown in Figure 15.35. The magnitude of each mass is $m$ and the spring constant of each spring is $k$.

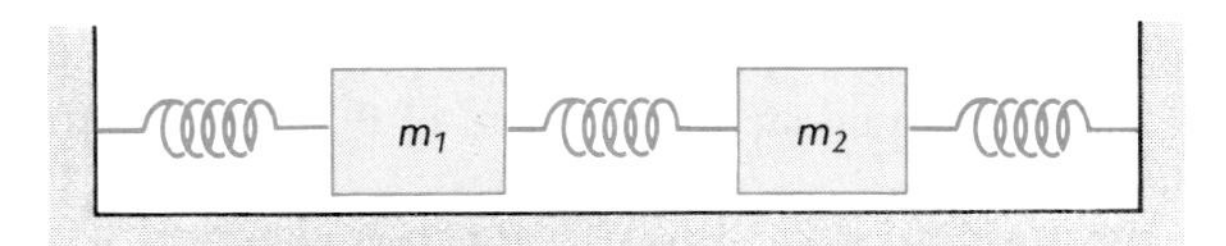


**Fig. 15.35** Two masses sliding on a frictionless plane.

(a) Suppose that at time $t = 0$, the masses are at their equilibrium positions and their instantaneous velocities are $v_1 = -v_2$. Find the position of each mass as a function of time. What is the frequency of the motion?

(b) Suppose that at time $t = 0$, the masses are at their equilibrium positions and their instantaneous velocities are $v_1 = v_2$. Find the position of each mass as a function of time. What is the frequency of the motion?

***15. A cart consists of a body and four wheels on frictionless axles. The body has a mass $m$. The wheels are uniform disks of mass $M$ and radius $R$. The cart rolls, without slipping, back and forth on a horizontal plane under the influence of a spring attached to one end of the cart (Figure 15.36). The spring constant is $k$. Taking into account the moment of inertia of the wheels, find a formula for the frequency of the back-and-forth motion of the cart.

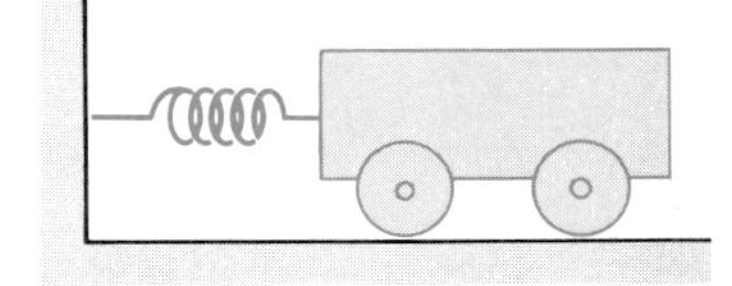

**Fig. 15.36** A cart attached to a spring.

## Section 15.3

16. Suppose that a particle of mass 0.24 kg acted upon by a spring undergoes simple harmonic motion with the parameters given in Problem 1.

(a) What is the total energy of this motion?

(b) At what time is the kinetic energy zero? At what time is the potential energy zero?

(c) At what time is the kinetic energy equal to the potential energy?

17. A mass of 8.0 kg is attached to a spring and oscillates with an amplitude of 0.25 m and a frequency of 0.60 Hz. What is the energy of the motion?

18. The separation between the equilibrium positions of the two atoms of a hydrogen molecule is 1.0 Å. Using the data given in Example 7, calculate the value of the vibrational energy that corresponds to an amplitude of vibration of 0.5 Å for each atom. Is it valid to treat the motion as a small oscillation if the energy has this value?

*19. A mass of 3.0 kg sliding along a frictionless floor at 2.0 m/s strikes and compresses a spring of constant $k = 300$ N/m. The spring stops the mass. How far does the mass travel while being slowed by the spring? How long does the mass take to stop?

*20. Two masses $m_1$ and $m_2$ are joined by a spring of spring constant $k$. Show that the frequency of vibration of these masses along the line connecting them is

$$\omega = \sqrt{\frac{k(m_1 + m_2)}{m_1 m_2}}$$

(Hint: The center of mass remains at rest.)

*21. Although it is usually a good approximation to neglect the mass of a spring, sometimes this mass must be taken into account. Suppose that a uniform spring has a relaxed length $l$ and a mass $m'$; a mass $m$ is attached to the end of the spring. The mass $m'$ is uniformly distributed along the spring. Suppose that if the moving end of the spring has a speed $v$, all other points of the spring have speeds directly proportional to their distance from the fixed end, for instance, a point midway between the moving and the fixed end has a speed $\frac{1}{2}v$.

(a) Show that the kinetic energy in the spring is $\frac{1}{6}m'v^2$ and that the kinetic energy of the mass $m$ and the spring is

$$K = \tfrac{1}{2}mv^2 + \tfrac{1}{6}m'v^2 = \tfrac{1}{2}(m + \tfrac{1}{3}m')v^2$$

Consequently, the effective mass of the combination is $m + \frac{1}{3}m'$.

(b) Show that the frequency of oscillation is $\omega = \sqrt{k/(m + \frac{1}{3}m')}$.

(c) Suppose that the spring described in Example 1 has a mass of 5 kg. The frequency of oscillation of the 400-kg mass attached to this spring will then be somewhat smaller than calculated in Example 1. How much smaller? Express your answer as a percentage of the answer of Example 1.

### Section 15.4

22. A mass suspended from a parachute descending at constant velocity can be regarded as a pendulum. What is the frequency of pendulum oscillations of a human body suspended 7 m below a parachute?

23. A "seconds" pendulum is a pendulum that has a period of exactly 2.0 s: — each one-way swing of the pendulum therefore takes exactly 1.0 s. What is the length of the seconds pendulum in Paris ($g = 9.809$ m/s$^2$), Buenos Aires ($g = 9.797$ m/s$^2$), and Washington, D.C. ($g = 9.801$ m/s$^2$)?

24. A grandfather clock controlled by a pendulum of length 0.9932 m keeps good time in New York ($g = 9.803$ m/s$^2$).
   (a) If we take this clock to Austin ($g = 9.793$ m/s$^2$), how many minutes per day will it fall behind?
   (b) In order to adjust the clock, by how many millimeters must we shorten the pendulum?

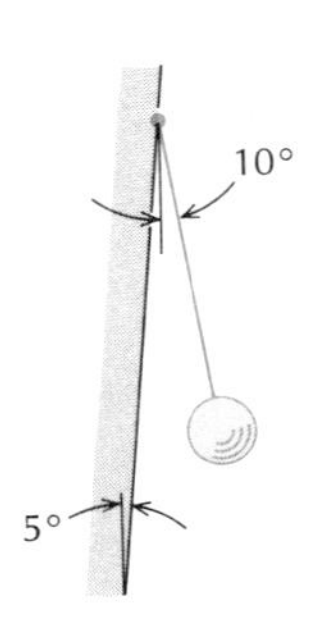

**Fig. 15.37** Pendulum hanging from an inclined wall.

25. The pendulum of a grandfather clock has a length of 0.994 m. If the clock runs late by 1 minute per day, how much must you shorten the pendulum to make it run on time?

*26. A pendulum hangs from a wall inclined at an angle of 5° with the vertical (see Figure 15.37). Suppose that this pendulum is released at an initial angle of 10° and it bounces off the wall elastically whenever it hits. What is the period of this pendulum?

*27. The pendulum of a pendulum clock consists of a rod of length 0.99 m with a bob of mass 0.40 kg. The pendulum bob swings back and forth along an arc of length 20 cm.
   (a) What are the maximum velocity and the maximum acceleration of the pendulum bob along the arc?
   (b) What is the force that the pendulum exerts on its support when it is at the midpoint of its swing? At the endpoint? Neglect the mass of the rod in your calculations.

*28. The pendulum of a regulator clock consists of a mass of 120 g at the end of a (massless) wooden stick of length 44 cm.
   (a) What is the total energy (kinetic plus potential) of this pendulum when oscillating with an amplitude of 4°?
   (b) What is the speed of the mass when at its lowest point?

**29. Galileo claimed to have verified experimentally that a pendulum oscillating with an amplitude as large as 30° has the same period as a pendulum of identical length oscillating with a much smaller amplitude. Suppose that you let two pendulums of length 1.5 m oscillate for 10 min. Initially, the pendulums oscillate in step. If the amplitude of one of them is 30° and the amplitude of the other is 5°, by what fraction of a (one-way) swing will the pendulums be out of step at the end of the 10-min interval? What can you conclude about Galileo's claim?

### Section 15.5

30. In windup clocks a strong torsional spring is used to store mechanical energy. Suppose that each week a clock requires four full turns of the winding key to keep running. The initial turn requires a torque of 0.30 N · m and the final turn a torque of 0.45 N · m.
   (a) What amount of mechanical energy do you store in the spring when winding the clock?
   (b) What is the consumption of mechanical power by the clock?
   (c) What is the torsional spring constant?

*31. The balance wheel of a watch, such as that shown in Figure 15.25, can be approximately described as a hoop of diameter 1.0 cm and mass 0.60 g. Each of the screws, whose masses are included in the mass given for the hoop, has a mass of 0.020 g. Suppose the watch runs fast by 1.2 minutes per day. To adjust the watch so that it keeps perfect time, by how much must we increase the moment of inertia of the balance wheel? If we want to achieve this increment by moving one of the screws outward in a radial direction, how far must we move the screw?

*32. Show that the potential energy of a torsional pendulum is $U = \frac{1}{2}\kappa\theta^2$. [Hint: Begin with Eq. (13.21) for the work done by the torque.]

33. At the National Bureau of Standards in Washington, D.C., the value of the acceleration of gravity is 9.80095 m/s$^2$. Suppose that at this location a very precise physical pendulum, designed for measurements of the acceleration of gravity, has a period of 2.10356 s. If we take this pendulum to a new location at the U.S. Coast and Geodetic Survey, also in Washington, D.C., it has a period of 2.10354 s. What is the value of the acceleration of gravity at this new location? What is the percentage change of the acceleration between the two locations?

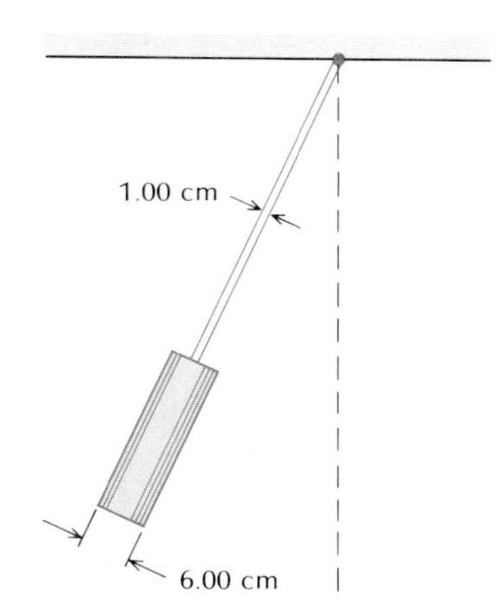


**Fig. 15.38** A physical pendulum.

*34. A pendulum consists of a brass rod with a brass cylinder attached to the end (Figure 15.38). The diameter of the rod is 1.00 cm and its length is 90.00 cm; the diameter of the cylinder is 6.00 cm and its length is 20.00 cm. What is the period of this pendulum?

*35. To test that the acceleration of gravity is the same for a piece of iron and a piece of brass, an experimenter takes a pendulum of length 1.800 m with an iron bob and another pendulum of the same length with a brass bob and starts them swinging in unison. After swinging for 12 min, the two pendulums are no more than one-quarter of a (one-way) swing out of step. What is the largest difference between the values of $g$ for iron and for brass consistent with these data? Express your answer as a fractional difference.

*36. Calculate the natural period of the swinging motion of a human leg. Treat the leg as a rigid physical pendulum with axis at the hip joint. Pretend that the mass distribution of the leg can be approximated as two rods joined rigidly end to end. The upper rod (thigh) has a mass of 6.8 kg and a length of 43 cm; the lower rod (shin plus foot) has a mass of 4.1 kg and a length of 46 cm. Using a watch, measure the period of the natural swinging motion of *your* leg when you are standing on one leg and letting the other dangle freely. Alternatively, measure the period of the swinging motion of your leg when you walk at a normal rate (this approximates the natural swinging motion). Compare with the calculated number.

*37. A hole has been drilled through a meter stick at the 30-cm mark and the meter stick has been hung on a wall by a nail passing through this hole. If the meter stick is given a push so that it swings about the nail, what is the period of the motion?

*38. A physical pendulum has the shape of a disk of radius $R$. The pendulum swings about an axis perpendicular to the plane of the disk and at distance $l$ from the center of the disk.

(a) Show that the frequency of the oscillations of this pendulum is

$$\omega = \sqrt{\frac{gl}{\frac{1}{2}R^2 + l^2}}$$

(b) For what value of $l$ is this frequency at a maximum?

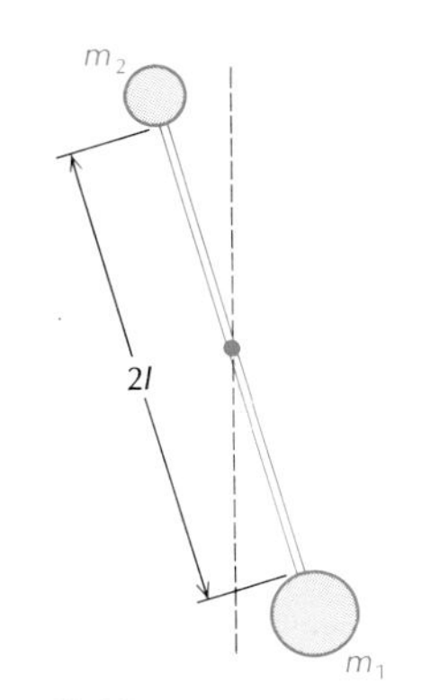


**Fig. 15.39**

*39. A physical pendulum consists of a massless rod of length $2l$ rotating about an axis through its center. A mass $m_1$ is attached at the lower end of the rod and a smaller mass $m_2$ at the upper end (see Figure 15.39). What is the period of this pendulum?

**40. A thin vertical rod of steel is clamped at its lower end. When you push the upper end to one side, bending the rod, the upper end moves (approximately) along an arc of circle[9] of radius $R$ and the rod opposes your push with

[9] The radius $R$ of the approximating (osculating) circle is somewhat shorter than the length of the rod.

a restoring force $F = -\kappa\theta$, where $\theta$ is the angular displacement and $\kappa$ is a constant. If you attach a mass $m$ to the upper end, what will be the frequency of small oscillations? For what value of $m$ does the rod become unstable, that is, for what value of $m$ is $\omega = 0$? Treat the rod as massless in your calculations. (Hint: Think of the rod as an inverted pendulum of length $R$, with an extra restoring force $-\kappa\theta$.)

*41. Suppose that the physical pendulum in Figure 15.19 is a thin rigid rod of mass $m$ suspended at one end. Suppose that this rod has an initial position $\theta = 20°$ and an initial angular velocity $\omega = 0$. Calculate the force **F** that the support exerts on the pendulum at this initial instant (give horizontal and vertical components).

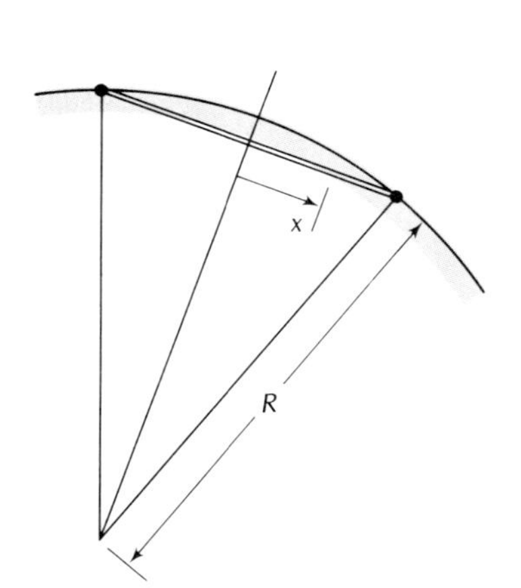


**Fig. 15.40**

*42. The door of a house is made of wood of uniform thickness. The door has a mass of 27 kg and measures 1.90 m × 0.91 m. The door is held shut by a torsional spring with $\kappa = 30$ N · m/radian arranged so that it exerts a torque of 54 N · m when the door is fully open (at right angles to the wall of the house). What angular speed does the door attain if it slams shut from the fully open position? What linear speed does the edge of the door attain?

**43. According to a proposal described in Example 1.5, very fast trains could travel from one city to another in straight subterranean tunnels (see Figure 15.40). For the following calculations, assume that the density of the Earth is constant so that, according to Eq. (9.45), the acceleration of gravity as a function of the radial distance $r$ from the center of the Earth is $g = (GM/R^3)r$.

(a) Show that the component of the acceleration of gravity along the track of the train is

$$g_x = -(GM/R^3)x$$

where $x$ is measured from the midpoint of the track (see Figure 15.40).

(b) Ignoring friction, show that the motion of the train along the track is simple harmonic motion with a period independent of the length of the track,

$$T = 2\pi\sqrt{\frac{R^3}{GM}}$$

(c) Starting from rest, how long would a train take to roll freely along its track from San Francisco to Washington, D.C.? What would be its maximum speed (at the midpoint)? Use the numbers calculated in Example 1.5 for the length and depth of the track.

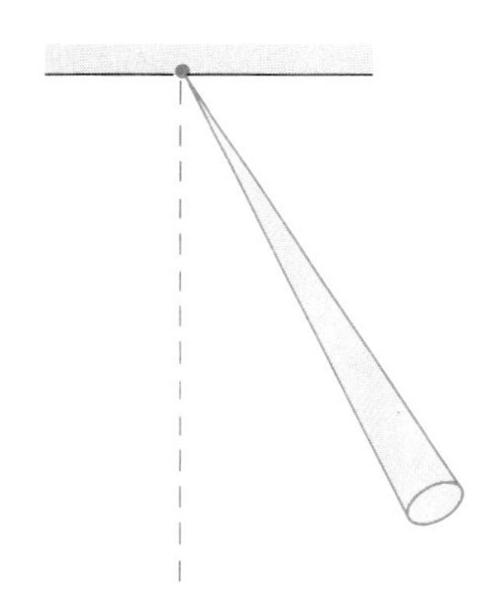

**Fig. 15.41**

**44. A physical pendulum consists of a long, thin cone suspended at its apex (Figure 15.41). The height of the cone is $l$. What is the period of this pendulum?

**45. The net gravitational force on a particle placed midway between two equal spherical bodies is zero. However, if the particle is placed some distance away from this equilibrium point, then the gravitational force is not zero.

(a) Show that if the particle is at a distance $x$ from the equilibrium point in a direction toward one of the bodies, then the force is approximately $4GMmx/r^3$, where $M$ is the mass of each spherical body, $m$ is the mass of the particle, and $2r$ is the distance between the spherical bodies. Assume $x \ll r$.

(b) Show that if the particle is at a distance $x$ from the equilibrium point in a direction perpendicular to the line connecting the bodies, then the force is approximately $-2GMmx/r^3$, where the negative sign indicates that the direction of the force is toward the equilibrium point.

(c) What is the frequency of small oscillations of the mass $m$ about the equilibrium point when moving in a direction perpendicular to the line connecting the bodies? Assume that the bodies remain stationary.

**46. The motion of a simple pendulum is given by

$$\theta = A \cos(\sqrt{g/l}\,t)$$

(a) Find the tension in the string of this pendulum; assume that $\theta \ll 1$.
(b) The tension is a function of time. At what time is the tension maximum? What is the value of this maximum tension?

***47. The total energy of a body of mass $m$ orbiting the Sun is

$$E = \left[\tfrac{1}{2}m\left(\frac{dr}{dt}\right)^2 + \tfrac{1}{2}mr^2\left(\frac{d\theta}{dt}\right)^2\right] - \frac{GmM_S}{r}$$

where the quantity in brackets represents the kinetic energy written in terms of the radial and the tangential component of the velocity.

(a) Show that in terms of the (constant) angular moment $L$, the energy can be written

$$E = \tfrac{1}{2}m\left(\frac{dr}{dt}\right)^2 + \frac{1}{2}\frac{L^2}{mr^2} - \frac{GmM_S}{r}$$

(b) For a circular orbit the radius is constant: $r = r_0$. For a nearly circular orbit, the radius differs from $r_0$ only by a small quantity: $r = r_0 + x$. Show that in terms of the small quantity $x$, the energy for a nearly circular orbit is approximately

$$E = \tfrac{1}{2}m\left(\frac{dx}{dt}\right)^2 + \frac{3}{2}\frac{L^2}{mr_0^4}x^2 - \frac{GmM_S}{r_0^3}x^2 - \frac{GmM_S}{2r_0}$$

The last term in this expression is a constant (independent of $x$).

(c) Show that, except for an additive constant, this expression for the energy coincides with the equation for the energy of a harmonic oscillator, provided that we identify

$$k = 3L^2/mr_0^4 - 2GmM_S/r_0^3$$

Show that this equals $k = GmM_S/r_0^3$.

(d) What is the frequency of small radial oscillations about the circular orbit? Show that this frequency equals the frequency of the circular orbit. Make a sketch of the shape of the orbit that results from the combination of revolution around the circle and small oscillations along the radius.

### Section 15.6

48. A pendulum of length 1.50 m is set swinging with an initial amplitude of 10°. After 12 min, friction has reduced the amplitude to 4°. What is the value of $\gamma$ for this pendulum?

49. The pendulum of a grandfather clock has a length of 0.994 m and a mass of 1.2 kg.

(a) If the pendulum is set swinging, the friction of the air reduces its amplitude of oscillation by a factor of 2 in 13.0 min. What is the value of $\gamma$ for this pendulum?
(b) If we want to keep this pendulum swinging at a constant amplitude of 8°, we must supply mechanical energy to it at a rate sufficient to make up for the frictional loss. What is the required mechanical power?

*50. If you stand on one leg and let the other dangle freely back and forth starting at an initial amplitude of, say, 20° or 30°, the amplitude will decay to one-half of the initial amplitude after about four swings. Regarding the dangling leg as a damped oscillator, what value of "$Q$" can you deduce from this?

CHAPTER 16

# Waves

In Chapters 12 and 13 we studied the rotational motion of a rigid body. This is a collective motion of all the particles in the body. Although the number of particles in the body may be very large, the equations for the rotational motion are fairly simple because all the particles are rigidly connected, so they do not move relative to one another. We will now study the **wave motion** of a deformable body — water, air, strings, elastic solids, and so on. This is a collective motion of the particles in the body, but here the particles do move relative to one another, and they exert time-dependent forces on one another. Nevertheless, the equations of motion remain fairly simple because each particle only interacts with its nearest neighbors. Wave motion is a disturbance propagating from one particle to the next in a stepwise manner.

*Wave motion*

For the sake of simplicity, in this chapter we will concentrate on the mathematical analysis of wave motion in a string. However, most of our results also apply to wave motion in other elastic bodies. In the next chapter we will examine some details of wave motion in water (sea waves and tidal waves), in air (sound waves), and in the crust of the Earth (seismic waves).

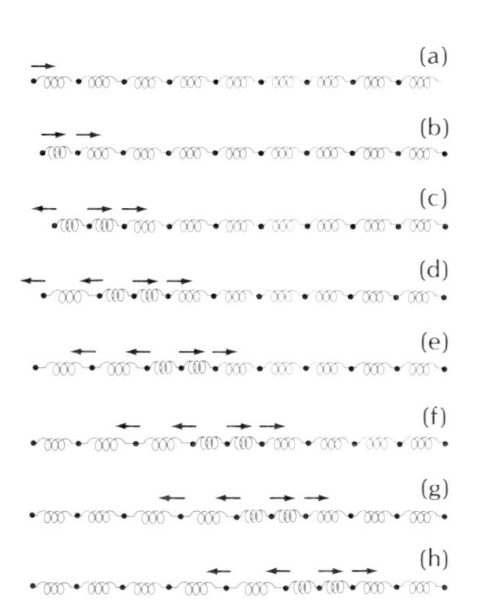


**Fig. 16.1** Particles joined by springs. A transverse disturbance propagates from left to right. The particles move up and down. The vectors indicate the instantaneous velocities of the particles.

## 16.1 Wave Pulses

Consider a tightly stretched elastic string, such as a long rubber cord. If we snap one end of the string up and down with a flick of the wrist, a disturbance travels along the string. Figure 16.1 shows in detail how such a traveling disturbance comes about. The string may be regarded as a row of particles joined by small, massless springs. When we jerk the first particle to one side, it will pull the second particle to the same side, and this will pull the third, and so on. If we then jerk the first particle back to its original position, it will pull the second particle back, and this will likewise pull the third, and so on. As the motion is transmitted from one particle to the next particle, the disturbance propagates along the row of particles. Such a disturbance is called a **transverse** wave pulse.

Alternatively, we can generate a disturbance by suddenly pushing the first particle toward the second. Figure 16.2 shows such a compressional disturbance propagating along the row of particles. This kind of disturbance is called a **longitudinal** wave pulse.

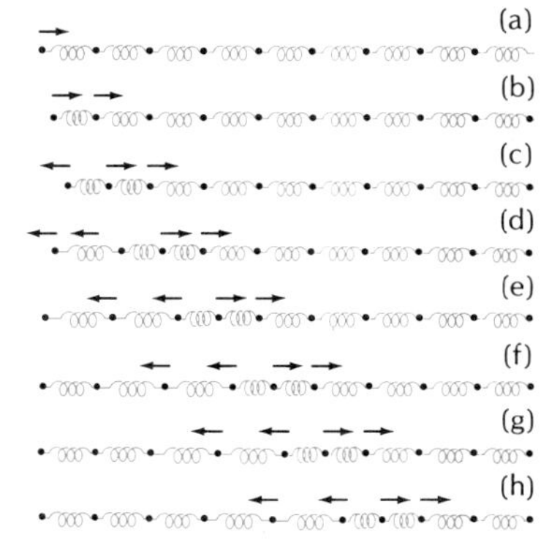


**Fig. 16.2** Particles joined by springs. A longitudinal disturbance propagates from left to right. The particles move back and forth.

Note that although the wave pulse travels along the string, the particles do not — they merely move back and forth around their equilibrium positions. Also note that in the region of the wave pulse, the string has kinetic energy (due to the back-and-forth motion of particles) and potential energy (due to the stretching of the springs between the particles). Hence a wave pulse traveling along the string carries energy with it — the wave transports energy from one end of the string to the other.

Wave motion in water, air, or any other elastic body displays the same general features. The wave is a propagating deformation or compression in the body communicated by pushes and pulls from one particle to the next. The wave transports energy without transporting particles.

For the mathematical analysis of the propagation of a wave pulse, we need to examine the behavior of the function representing the shape of the wave, or the **wavefunction.** Consider a transverse wave pulse traveling along a string with some speed $v$. Let us make the assumption that the shape of the wave pulse remains constant as it travels (we will justify this assumption in Section 16.3). The wavefunction that represents the wave pulse at any time is then directly related to the wavefunction at the initial time, $t = 0$. If the string is stretched in the $x$ direction and its deformation is in the $y$ direction, then its initial shape can be represented by some function of $x$:

*General wavefunction*

$$y = f(x) \quad \text{at initial time } t = 0 \qquad (1)$$

For example, Figure 16.3a shows a wavefunction with a peak at $x = 0$. The deformation of the string at the peak is $y = f(0)$. At some later time $t$, the entire wave pulse will have traveled a distance $vt$ to the right, and the peak will have shifted from $x = 0$ to $x = vt$. The initial wavefunction $f(x)$ must then be replaced by a new wavefunction in which the value of the argument is shifted by a distance $vt$. Hence, the new wavefunction representing the wave is of the form

$$y = f(x - vt) \quad \text{at later time } t \qquad (2)$$

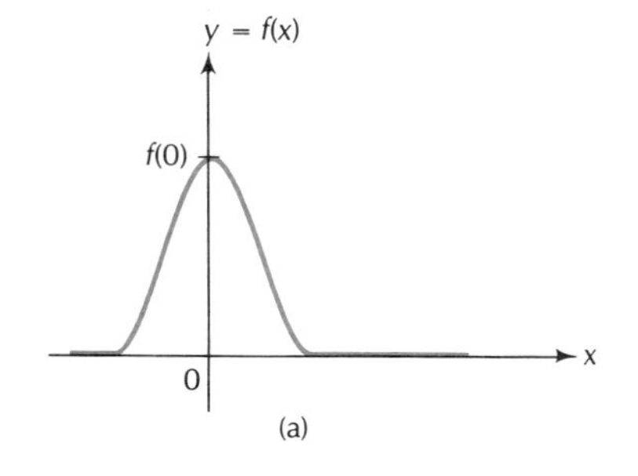


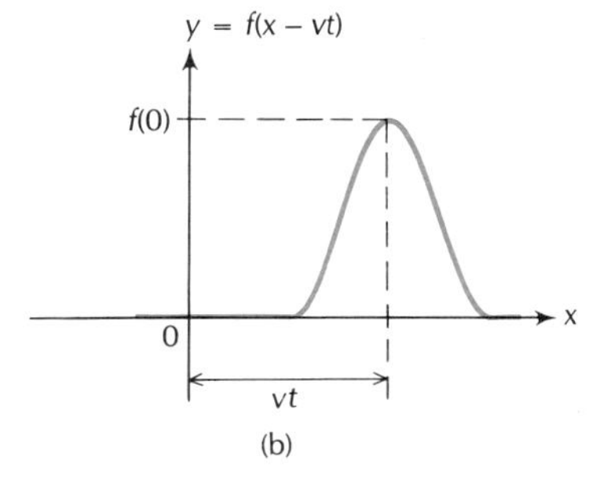


**Fig. 16.3** (a) Wave pulse at $t = 0$. The peak of the wave pulse is at $x = 0$. (b) Wave pulse at $t > 0$. The peak has shifted to $x = vt$.

Note that according to Eq. (1), the peak $f(0)$ is at $x = 0$, whereas according to Eq. (2), the peak $f(0)$ is at $x = vt$; this is exactly as it should be (see Figure 16.3b). Equation (2) is a general wavefunction for a wave traveling in the positive $x$ direction.

Similar reasoning shows that a wave traveling in the negative $x$ direction is represented by a wavefunction of the form

$$y = f(x + vt) \quad \text{at later time } t \tag{3}$$

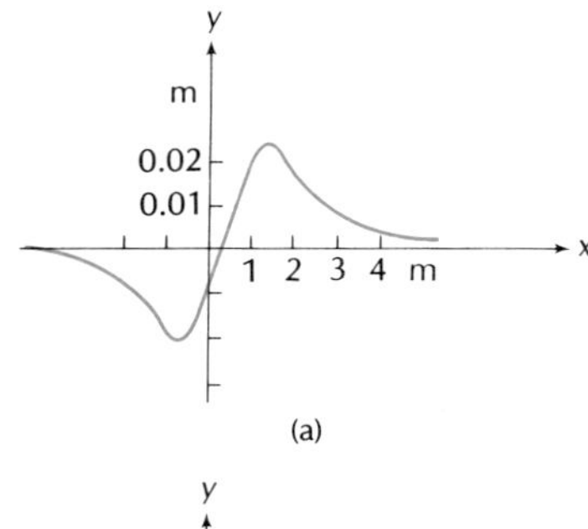


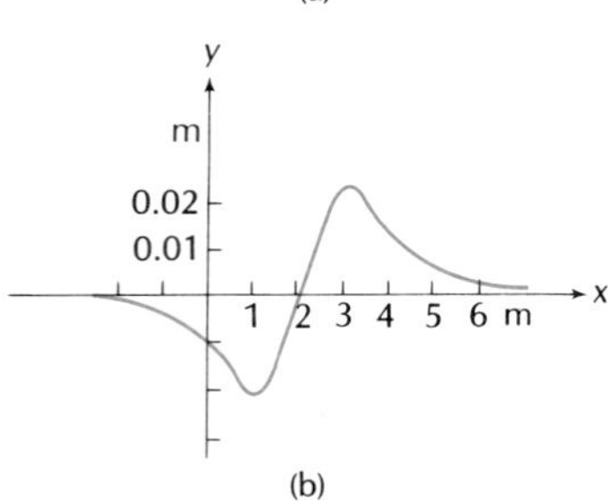


**Fig. 16.4** (a) Plot of wave pulse at $t = 0$. (b) Plot of wave pulse at $t = 1$s. The pulse has traveled 2m toward the right.

EXAMPLE 1. Suppose that at an initial time $t = 0$, the shape of a wave pulse on a string is represented by the wavefunction

$$y = f(x) = \frac{0.03x}{1 + x^4} \quad \text{at initial time } t = 0$$

where $y$ and $x$ are in meters. Suppose that this wave pulse has a velocity $v =$ 2 m/s toward the positive $x$ direction. What function represents the wave pulse at time $t$? Plot this function when $t = 1$ s.

SOLUTION: Figure 16.4(a) shows the plot of the wavefunction specified above. This function has a crest at $x = 1$ and a trough at $x = -1$. The plot gives the shape of the wave pulse at $t = 0$. To find the shape at $t > 0$, we must shift the wave pulse a distance $vt$ toward the right. According to our general discussion, we accomplish this by replacing $x$ in $f(x)$ by $x - vt$:

$$y = f(x - vt) = \frac{0.03(x - vt)}{1 + (x - vt)^4} \quad \text{at later time } t$$

If $v = 2$ m/s and $t = 1$ s, this becomes

$$y = \frac{0.03(x - 2)}{1 + (x - 2)^4}$$

Figure 16.4(b) shows the plot of this new wavefunction. As expected, this new plot has the same shape as the old plot of Figure 16.4(a), but it is shifted toward the right by 2 meters.

## 16.2 Periodic Waves

If we shake the end of a long elastic string up and down, and we continue shaking it steadily, we will generate a **periodic wave** on the string. Such a wave can be regarded as consisting of a regular succession of positive and negative wave pulses. Figure 16.5a shows such a periodic wave at one instant of time. The maxima of the wave are called **wave crests,** and the minima are called **wave troughs.** The distance from one crest to the next or from one trough to the next is called the **wavelength,** designated by the symbol $\lambda$. The wavelength is the repeat distance of the wave pattern—a shift of the wave pattern by one wavelength to the right (or the left) reproduces the original wave pattern.

With the passing of time, the wave crests and wave troughs travel toward the right at a speed $v$. As the wave travels, the entire wave pattern shifts toward the right, that is, the wave pattern performs a rigid translational motion. Figures 16.5b–16.5h display the wave at successive instants of time. These pictures span one **period** of the wave, that is, they span the interval of time required for the wave pattern to travel exactly one wavelength to the right. The period is the repeat time of the wave pattern—after one period, each wave crest or trough will have traveled to the position previously occupied by the adjacent wave crest or trough, and the wave will have attained exactly the same configuration as it had at the initial time.

*Period and frequency*

Since in one period, the wave travels a distance equal to one wavelength, the ratio of wavelength to period must equal the wave speed,

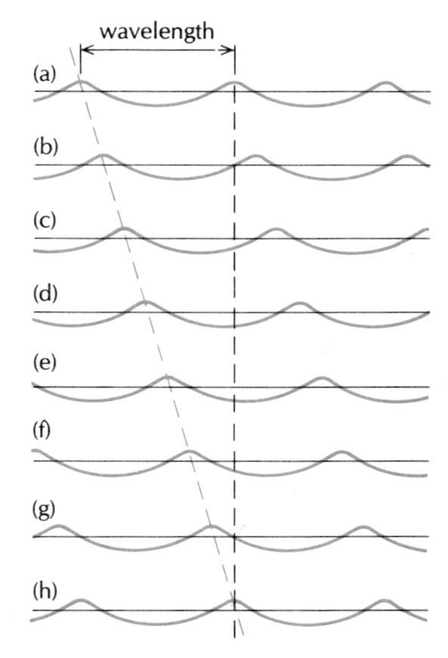


**Fig. 16.5** A periodic wave traveling to the right. The diagrams show "snapshots" of the wave at successive instants of time. The wave pattern (h) coincides with the wave pattern (a) because the wave has moved exactly one wavelength to the right.

$$\boxed{\frac{\lambda}{T} = v} \tag{4}$$

As in the case of simple harmonic motion, we define the **frequency** of the wave as the inverse of the period,

$$\boxed{\nu = \frac{1}{T} = \frac{v}{\lambda}} \tag{5}$$

The frequency of the wave is simply the number of wave crests arriving at some point on the string per second. In terms of the frequency, Eq. (4) becomes

$$\boxed{\lambda\nu = v} \tag{6}$$

An important special case of a periodic wave is a **harmonic wave.** This kind of a wave has the shape of a sine or cosine curve. At the initial time $t = 0$, the wavefunction is

$$y = A \cos kx \tag{7}$$

The constant $A$, which represents the height of the wave crests (or the depth of the wave troughs), is called the **amplitude** of the wave, and the constant $k$ is called the **wave number.**[1] Note that here, as in the preceding chapter, the argument of the cosine function is supposed to be expressed in radians.

*Amplitude*
*Wave number*

Figure 16.6 is a plot of the wavefunction (7). The wave crests (maxima) occur at

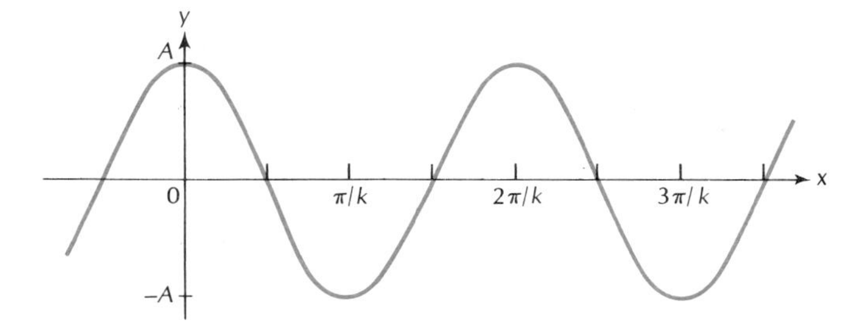


**Fig. 16.6** A harmonic wave at $t = 0$.

[1] Do not confuse the wave number $k$ with the spring constant of the preceding chapter. The former has nothing to do with the latter!

$$kx = 0,\ 2\pi,\ 4\pi,\ 6\pi,\ \text{etc.} \tag{8}$$

and the wave troughs (minima) occur at

$$kx = \pi,\ 3\pi,\ 5\pi,\ \text{etc.} \tag{9}$$

From these equations we see that the distance from one crest to the next, or from one trough to the next, is $2\pi/k$. This is the wavelength of the harmonic wave,

*Wavelength of harmonic wave*

$$\boxed{\lambda = \frac{2\pi}{k}} \tag{10}$$

At any later time, the harmonic wave will have traveled some distance to the right or the left. According to the general discussion in Section 16.1, the initial wavefunction (7) must then be replaced by a new, shifted wavefunction,

*Harmonic wavefunctions*

$$\boxed{y = A \cos k(x - vt)} \quad \text{for a wave traveling in positive } x \text{ direction}$$

or else (11)

$$\boxed{y = A \cos k(x + vt)} \quad \text{for a wave traveling in negative } x \text{ direction}$$

For harmonic waves, it is customary to introduce the **angular frequency,**

*Angular frequency*

$$\boxed{\omega = 2\pi\nu = kv} \tag{12}$$

In terms of wavelength, frequency, and angular frequency, we can express the first of the wavefunctions in Eq. (11) in the alternative forms

$$y = A \cos\left(\frac{2\pi}{\lambda}x - 2\pi\nu t\right) \tag{13}$$

and

$$y = A \cos(kx - \omega t) \tag{14}$$

Note that a given particle on the string, at some given position $x_0$, has a transverse displacement

$$y = A \cos(kx_0 - \omega t) = A \cos(\omega t - kx_0) \tag{15}$$

If we compare this with Eq. (15.4), we recognize that this given particle executes simple harmonic motion with amplitude $A$ and frequency $\omega$. The quantity $kx_0$ has a fixed value for this particle and it acts as phase constant in the simple harmonic motion. Different particles, at different positions, with different values of $x_0$, oscillate with the same amplitudes and frequencies, but with different phase constants; for instance, two particles separated by half a wavelength along the $x$ axis oscillate with phase constants differing by $\pi$—when one of these particles reaches maximum transverse displacement, the other reaches minimum displacement, and conversely. Figure 16.7 shows a plot of the transverse displacement of a typical particle as a function of time.

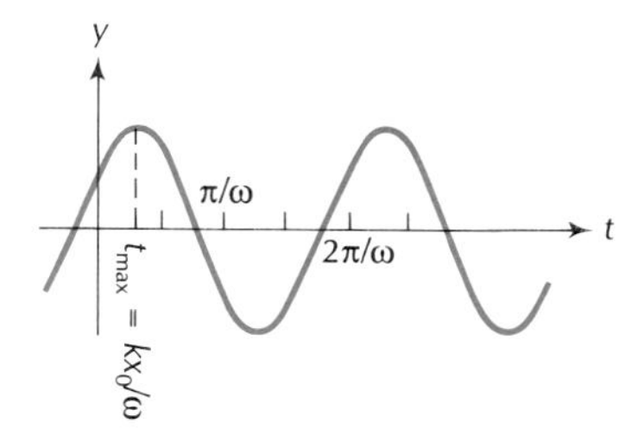


**Fig. 16.7** Transverse displacement $y$ vs. time for a particle at the position $x_0$ on a string.

Harmonic waves play a central role in the study of wave motion because, as we will see in a later section, any periodic wave of arbitrary shape can be regarded as a superposition, or sum, of several harmonic waves of suitably chosen amplitudes and wavelengths. Thus, if we understand the motion of harmonic waves, we understand the motion of any kind of periodic wave. Hereafter, we will concentrate on harmonic waves.

Although we have derived the formulas in this section in the context of waves on a string, we can also apply these formulas to other kinds of waves, provided that we give the "deformation" $y$ a suitable interpretation. For instance, to describe waves on the surface of the sea, we must regard $y$ as the height of the water above the mean sea level; to describe seismic waves, we must regard $y$ as the displacement of a lump of earth from its original position, and so on.

---

EXAMPLE 2. Ocean waves of a period of 10 s have a speed of 16 m/s. What is the wavelength of these waves? What is the (horizontal) distance between a wave crest and a wave trough? If the wave height is 1.2 m measured from trough to crest, what is the maximum vertical speed of a point on the water surface at a given horizontal position? Assume the waves are harmonic.

SOLUTION: From Eq. (9), the wavelength is

$$\lambda = Tv = 10\text{ s} \times 16\text{ m/s} = 160\text{ m}$$

The distance between a wave crest and a wave trough is one-half a wavelength, that is, 80 m.

According to Eq. (15), for a point at a given horizontal position, the vertical displacement is

$$y = A \cos(\omega t - kx_0)$$

and therefore the vertical speed is

$$v_y = \frac{dy}{dt} = -A\omega \sin(\omega t - kx_0)$$

This has a maximum magnitude $A\omega$. Since the amplitude is one-half of the height from trough to crest, we obtain a maximum speed

$$A\omega = 0.60\text{ m} \times \frac{2\pi}{10\text{ s}} = 0.38\text{ m/s}$$

COMMENTS AND SUGGESTIONS: As we will see in the next chapter, in a water wave the motion of the particles is not purely transverse, but simultaneously transverse and longitudinal (the particles move simultaneously up and down and back and forth, tracing out a circle). However, this does not affect the above calculation of the vertical speed of the water surface.

---

The wavefunction of a wave on a string, or of some other kind of wave, satisfies an interesting equation involving the second derivatives with respect to time and space. To obtain this equation, we begin by evaluating the second derivative of the wavefunction (14) with respect to time. For a wave on a string, this second derivative is the transverse acceleration of a particle at a given position,

$$\frac{d^2y}{dt^2} = -\omega^2 A \cos(kx - \omega t)$$

Now, look at the second derivative with respect to $x$,

$$\frac{d^2y}{dx^2} = -k^2 A \cos(kx - \omega t)$$

The comparison of these equations shows that $d^2y/dt^2$ and $d^2y/dx^2$ are proportional:

$$\frac{d^2y}{dt^2} = \frac{\omega^2}{k^2}\frac{d^2y}{dx^2}$$

But $\omega/k$ equals $v$, the speed of the wave; hence,

*Wave equation*

$$\boxed{\frac{\partial^2 y}{\partial t^2} = v^2 \frac{\partial^2 y}{\partial x^2}} \tag{16}$$

Here we have introduced the standard notation $\partial$ for partial derivatives to indicate that on the left side of Eq. (16), the wavefunction must be differentiated with respect to $t$ only, and on the right side, with respect to $x$ only. Note that although we have obtained this result with a wave traveling in the positive $x$ direction, it is equally valid for a wave traveling in the negative $x$ direction [a change of the sign in front of $\omega$ in the argument of the cosine in Eq. (14) makes no difference to the final result].

Equation (16) is called the **wave equation.** It is satisfied not only by harmonic waves, but also by any other kind of wave that propagates without change of shape. For instance, it is easy to check that the general expressions (2) and (3) for traveling waves obey the wave equation, regardless of the details of the wavefunction.

In essence, the wave equation is the equation of motion of the wave, just as Newton's Second Law is the equation of motion of a particle. For the simple case of wave propagation in one dimension, such as we are considering in this chapter, the wave equation provides us with no new information. But for the more complicated case of wave propagation in three dimensions—and especially for propagation through and around obstacles—the generalized three-dimensional wave equation forms the basis for the mathematical investigation of the behavior of waves.

## 16.3 Speed of Waves on a String

The speed of waves depends on the characteristics of the medium. In some cases the speed also depends on the wavelength; for instance, ocean waves of long wavelength — such as tidal waves — have a larger speed than waves of short wavelength. In this section we will derive an expression for the speed of waves on a string. These waves are simple to treat because their speed does not depend on their wavelength.

Figure 16.8a shows an elastic string tightly stretched between two end points. The tension in the string is $F$ and the mass per unit length of the string is $\mu$ kilograms per meter of length. Of course, the string is a system of a large number of particles, but for our present purposes we will regard the mass as continuously distributed along the string. Figure 16.8b shows a wave pulse propagating along the string. We will assume that the amplitude of the wave pulse is very small (that is, very small compared to the length of the string); then the wave pulse only produces a small perturbation in the tension and, to a good approximation, the tension everywhere along the string is constant, as in the static case.

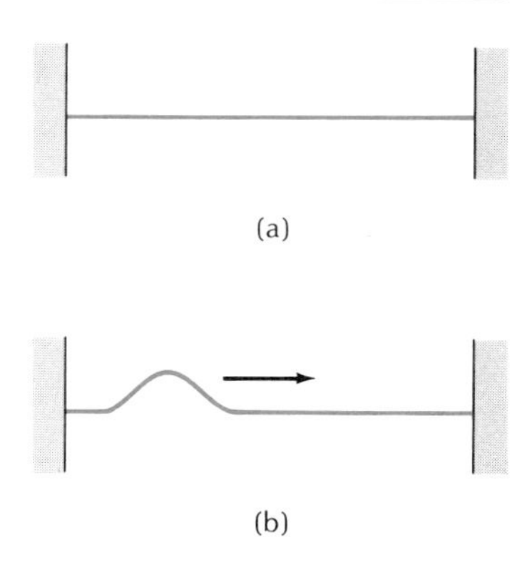


**Fig. 16.8** (a) A tightly stretched string; the tension is $F$. (b) A wave pulse on the string; the speed of this wave pulse is $v$.

It is convenient to analyze the motion of the string in a reference frame moving to the right with a velocity equal to that of the wave pulse. In this reference frame the wave pulse is at rest, and the entire string travels to the left. Each segment of string travels along a curved path having the shape of the wave pulse. Figure 16.9a shows one short segment $\Delta l$ rounding this curved path at speed $v$. Over the short length $\Delta l$, the path can be approximated as an arc of a circle of radius $R$; the segment $\Delta l$ subtends a small angle $\Delta\theta$ of this circle. Instantaneously, the segment $\Delta l$ is in uniform circular motion. The mass of the segment is $\mu\,\Delta l$ and its centripetal acceleration is $v^2/R$. The forces on the segment are the tensions acting at its ends (Figure 16.9a); the vector sum of these forces is a centripetal force of magnitude $F\,\Delta\theta$ (Figure 16.9b). Hence the equation of motion is

$$\mu\,\Delta l\,\frac{v^2}{R} = F\,\Delta\theta$$

or

$$\mu\,\Delta l v^2 = FR\,\Delta\theta \tag{17}$$

Since $R\,\Delta\theta$ equals the length $\Delta l$ of the segment, Eq. (17) can be simplified to

$$\mu v^2 = F \tag{18}$$

which yields a speed

$$\boxed{v = \sqrt{F/\mu}} \tag{19}$$

*Speed of a wave on a string*

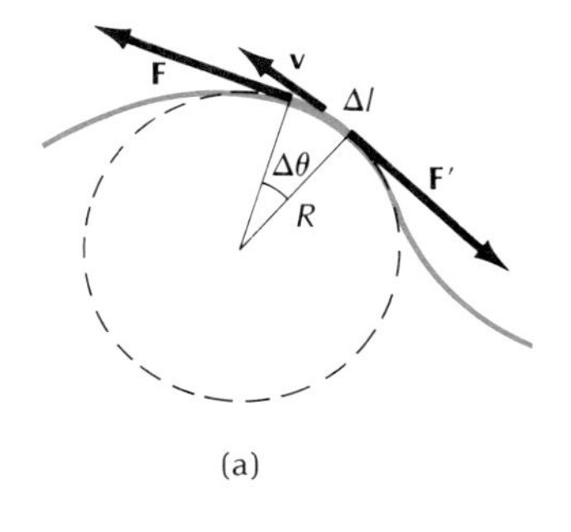


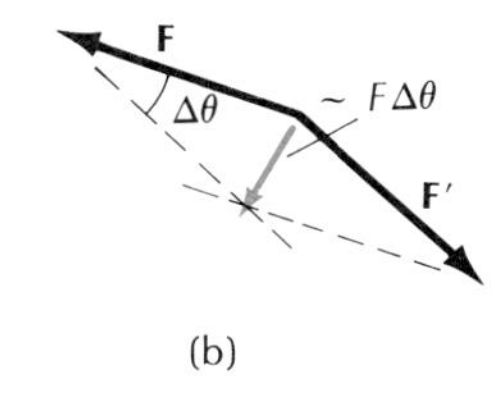


**Fig. 16.9** (a) Forces on a short segment $\Delta l$ of the string. (b) Resultant force.

This equation expresses the speed in terms of the tension and the density of the string. Note that the speed is large if the tension is large and the mass density small. This is intuitively reasonable — a large tension can move a small mass very quickly. Of course, Eq. (19) applies not only to strings but also to tightly stretched wires and cables.

---

EXAMPLE 3. A long piece of piano wire of radius 0.4 mm is made of steel of density $7.8 \times 10^3$ kg/m³. The wire is under a tension of $1.0 \times 10^3$ N. What is the speed of transverse waves on this wire? What is the wavelength of a wave on this wire if its frequency is 262 Hz?

SOLUTION: Consider a 1-m-long piece of this wire. The volume of this piece is $\pi \times (0.4 \times 10^{-3}\ \text{m})^2 \times 1\ \text{m} = 5.0 \times 10^{-7}\ \text{m}^3$, and the mass is $5.0 \times 10^{-7}\ \text{m}^3$

$\times 7.8 \times 10^3$ kg/m$^3 = 3.9 \times 10^{-3}$ kg. Hence, the mass per unit length of the wire is $3.9 \times 10^{-3}$ kg/m. From Eq. (19), the wave speed is then

$$v = \sqrt{\frac{F}{\mu}} = \sqrt{\frac{1.0 \times 10^3 \text{ N}}{3.9 \times 10^{-3} \text{ kg/m}}} = 5.1 \times 10^2 \text{ m/s}$$

Consequently, the wavelength is

$$\lambda = v/\nu = (5.1 \times 10^2 \text{ m/s})/(262/\text{s}) = 1.9 \text{ m}$$

---

Since in our derivation of Eq. (19) we made no special assumption about the shape of the wave, it is clear that the wave speed is independent of the shape. Consequently, all the different portions of a wave pulse (different shapes) propagate at the same speed, that is, the wave pulse propagates as though it were in rigid translation along the string.

A harmonic wave can be regarded as a succession of alternating positive and negative wave pulses. Since all such pulses have the same speed, all harmonic waves on a string have the same speed, independent of wavelength.

Although a wave on a string is a rather special and simple case of wave motion, Eq. (19) exhibits a general feature of wave motion. In broad terms, this equation states that the speed of the wave depends on the restoring force and on the inertia of the medium. This is true for any kind of wave propagating in any kind of medium. In all cases some force within the medium opposes its deformation — tension tends to keep the string straight, gravity tends to keep the surface of the sea smooth, elastic forces tend to keep the crust of the Earth as it is. But if something provides an initial disturbance, then the restoring force will cause it to propagate (compare Figure 16.1) with a speed depending on the magnitude of the restoring force and on the magnitude of the inertia or, equivalently, the density of mass. In general, the speed will be large if the restoring force is large and the density of mass is small.

The lack of dependence of speed on shape is a rather special feature of waves on a string. For many other kinds of waves, the speed does depend on the shape, so different parts of a wave pulse travel at different speeds. Consequently, the wave pulse changes shape as some of its parts get ahead, or fall behind, other parts. Because of this, the wave pulse usually tends to spread out, becoming more and more shallow. A medium that gives rise to such behavior is called a **dispersive medium.**

*Dispersive medium*

In contrast to the case of a harmonic wave on a string, a harmonic wave in a dispersive medium cannot be regarded as simply a succession of wave pulses, because the pulses change their shape, whereas a harmonic wave does not. There is then no simple connection between the speed of a wave pulse and the speed of a harmonic wave. To distinguish between these speeds, we call the speed of the peak of a wave pulse the **group velocity,** and the speed of a harmonic wave the **phase velocity.**

*Group velocity and phase velocity*

The group velocity is also sometimes called the signal velocity because it describes the propagation of signals in the medium. If we want to send a signal by means of a wave, we must use a wave pulse rather than a harmonic wave; the latter has no beginning and no end — it lasts forever and is therefore useless as a signal.

The phase velocity of harmonic waves in a dispersive medium depends on wavelength. For instance, ocean waves of long wavelength have a higher velocity than those of short wavelength. We will discuss this characteristic of ocean waves in the next chapter.

## 16.4* Energy in a Wave; Power

A transverse wave on a string has kinetic energy because the particles are in motion, and it has potential energy because work is required to stretch the string. Consider a small interval $dx$ along the string. The mass of string within this interval is $\mu\, dx$ and the velocity of the piece of string is $dy/dt$; hence the kinetic energy associated with this piece of string is

$$dK = \tfrac{1}{2}\mu\, dx\left(\frac{dy}{dt}\right)^2 \tag{20}$$

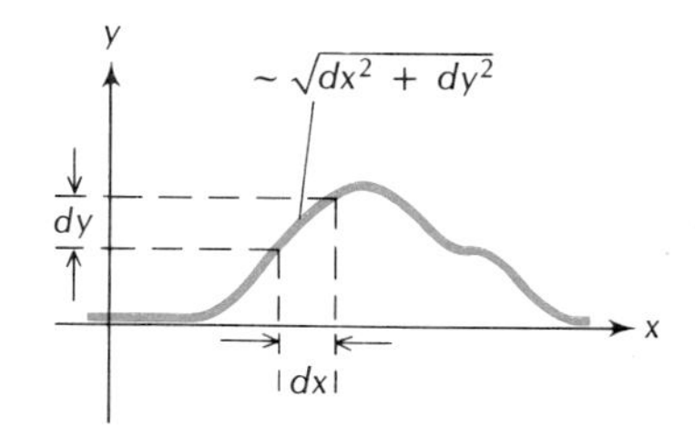


**Fig. 16.10** A short segment of string included between $x$ and $x + dx$.

To find the potential energy, we note that the piece of string is slightly elongated while the wave passes. The wave stretches the string from its original length $dx$ to a new length $\sqrt{dx^2 + dy^2}$ (Figure 16.10). Hence the change of length is

$$\delta l = \sqrt{dx^2 + dy^2} - dx \tag{21}$$

or

$$\delta l = dx\left[\sqrt{1 + \left(\frac{dy}{dx}\right)^2} - 1\right] \tag{22}$$

We will assume that $dy/dx$ is small (so the string only makes a small angle with its original, horizontal direction). Then we can take advantage of approximation

$$\sqrt{1 + \left(\frac{dy}{dx}\right)^2} \cong 1 + \frac{1}{2}\left(\frac{dy}{dx}\right)^2 \tag{23}$$

to obtain

$$\delta l = \frac{1}{2}\left(\frac{dy}{dx}\right)^2 dx \tag{24}$$

The potential energy associated with the interval $dx$ is simply the work that must be done against the tension $F$ to stretch the string by the amount $\delta l$, that is,

$$dU = F\,\delta l = \tfrac{1}{2}F\left(\frac{dy}{dx}\right)^2 dx \tag{25}$$

* This section is optional.

The total energy associated with the interval $dx$ is the sum of Eqs. (20) and (25),

$$dE = dK + dU = \tfrac{1}{2}\mu\left(\frac{\partial y}{\partial t}\right)^2 dx + \tfrac{1}{2}F\left(\frac{\partial y}{\partial x}\right)^2 dx \tag{26}$$

Here we have again introduced the standard notation for partial derivatives to indicate that in the first term on the right side of Eq. (26) the function $y$ must be differentiated with respect to $t$ only, and in the second term with respect to $x$ only.

If we divide Eq. (26) by $dx$, we obtain the amount of energy per unit length, or the **energy density,** of the wave:

*Energy density of a wave*

$$\frac{dE}{dx} = \tfrac{1}{2}\mu\left(\frac{\partial y}{\partial t}\right)^2 + \tfrac{1}{2}F\left(\frac{\partial y}{\partial x}\right)^2 \tag{27}$$

If we are dealing with a harmonic wave, then the required derivatives are

$$\frac{\partial y}{\partial t} = \frac{\partial}{\partial t}[A\cos(kx - \omega t)] = \omega A\sin(kx - \omega t) \tag{28}$$

$$\frac{\partial y}{\partial x} = \frac{\partial}{\partial x}[A\cos(kx - \omega t)] = -kA\sin(kx - \omega t) \tag{29}$$

so that

$$\frac{dE}{dx} = \tfrac{1}{2}(\mu\omega^2 + Fk^2)A^2\sin^2(kx - \omega t) \tag{30}$$

In view of Eqs. (12) and (19),

$$Fk^2 = F\omega^2/v^2 = \mu\omega^2 \tag{31}$$

and therefore

$$\frac{dE}{dx} = \mu\omega^2 A^2\sin^2(kx - \omega t) \tag{32}$$

Figure 16.11 is a plot of this energy density at one instant of time. Note that there is no energy at the wave crests — the velocity of the

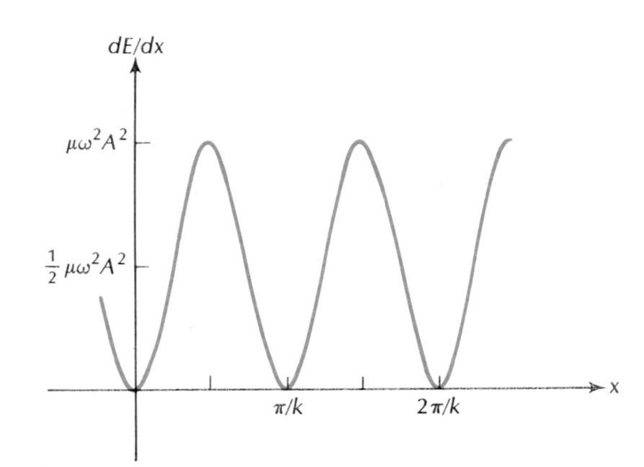


**Fig. 16.11** Energy density in a harmonic wave at one instant of time. At the wave crests ($x = 0, \pi/k, 2\pi/k, \ldots$), the energy density is zero.

particles is zero at the crests (no kinetic energy), and the string is unstretched at the crests (no potential energy).

The power transported by the wave is simply the amount of energy that moves past a given point on the string in one unit of time. The energy must travel with the wave at the wave speed $v$. Hence the energy in the interval $dx$ takes a time

$$dt = \frac{dx}{v} \tag{33}$$

to move out of this interval (of course, as fast as the energy moves out of one end of the segment, it is replaced with energy moving in at the other end of the segment). Combining Eqs. (32) and (33), we obtain an expression for the (instantaneous) **power** transported by the harmonic wave:

*Power of a wave*

$$P = \frac{dE}{dt} = v\frac{dE}{dx} = v\mu\omega^2 A^2 \sin^2(kx - \omega t) \tag{34}$$

The power arriving at any given position oscillates as a function of time between a maximum value $v\mu\omega^2 A^2$ and a minimum value zero.

Note that the power is proportional to the wave velocity, the square of the frequency, and the square of the amplitude. This proportionality turns out to be true not only for waves on a string, but also for other kinds of waves.

## 16.5 The Superposition of Waves

*Superposition principle*

Waves on a string and waves in other elastic bodies usually obey a **superposition principle**: when two or more waves arrive at any given point simultaneously, *the resultant instantaneous deformation is the sum of the individual instantaneous deformations.* Such a superposition means that the waves do not interact; they have no effect on one another. Each wave propagates as though the other were not present, and the contribution that each makes to the displacement of a particle in the elastic body is as though the other were not present. For instance, if the sound waves from a violin and a flute reach us simultaneously, then each of these waves produces a displacement of the air molecules just as though it were acting alone, and the net displacement of the air molecules is simply the (vector) sum of these individual displacements.

For waves of low amplitude on a string and for sound waves of ordinary intensity in air, the superposition principle is very well satisfied. However, for waves of very large amplitude or intensity, the superposition principle fails. When a wave of very large amplitude is propagating on a string, it alters the tension of the string, and therefore affects the behavior of a second wave propagating on the same string. Likewise, a very intense sound wave (a shock wave) produces significant alterations of the temperature and the pressure of the air, and therefore affects the behavior of a second wave propagating through this same region. In this section, we will not worry about such extreme conditions, and we will assume that the superposition principle is a good approximation.

As a first example of superposition, let us consider two waves propagating in the same direction with the same frequency and amplitude, but different phases. These waves might be waves on a string, in air, on the surface of water, or whatever. The wavefunctions describing the individual waves are

$$y_1 = A\cos(kx - \omega t) \tag{35}$$

$$y_2 = A\cos(kx - \omega t + \delta) \tag{36}$$

The second of these waves has a phase constant $\delta$, whereas the first of these waves has a phase constant zero (as did all the waves in the preceding sections). The phase constant plays the same role for a wave as it does for the simple harmonic oscillator—it determines the position of the wave at the initial time $t = 0$. According to the superposition principle, the wavefunction for the resultant wave is

$$y = y_1 + y_2 = A\cos(kx - \omega t) + A\cos(kx - \omega t + \delta) \tag{37}$$

With the trigonometric identity

$$\cos\alpha + \cos\beta = 2\cos\tfrac{1}{2}(\alpha+\beta)\cos\tfrac{1}{2}(\alpha-\beta) \tag{38}$$

we can transform Eq. (37) into

$$y = 2A\cos(kx - \omega t + \tfrac{1}{2}\delta)\cos\tfrac{1}{2}\delta \tag{39}$$

*Constructive and destructive interference*

The resultant wave has the same frequency as the original waves, and its amplitude [the factor multiplying $\cos(kx - \omega t + \frac{1}{2}\delta)$] is $2A\cos\frac{1}{2}\delta$. If the phase difference $\delta$ between the two waves is zero, the two waves are said to be in phase; they meet crest to crest and trough to trough, reinforcing each other (see Figure 16.12). This is **constructive interference.** If the phase difference is $\delta = \pi$ radians, or 180°, the two waves are said to be out of phase; they meet crest to trough, canceling each other completely (see Figure 16.13). This is **destructive interference.**

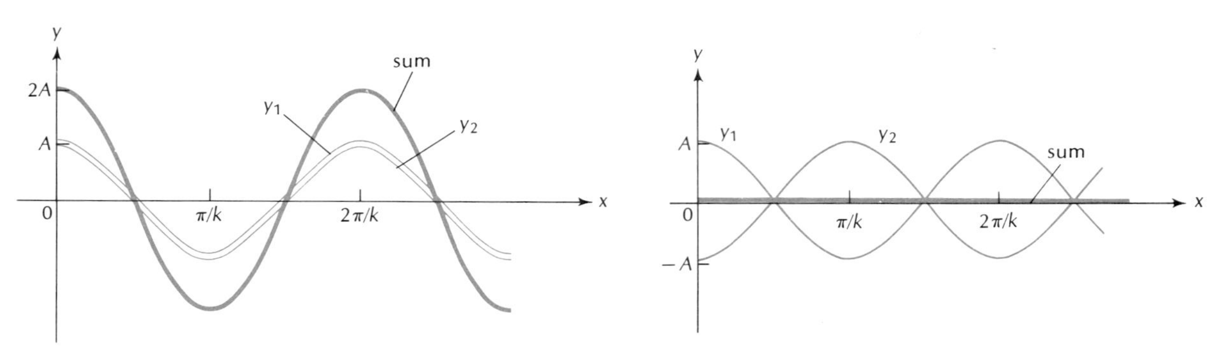


**Fig. 16.12** Constructive interference of two waves.

**Fig. 16.13** Destructive interference of two waves; the waves cancel everywhere.

The complete mutual cancellation of two waves raises a question: If the waves cancel, what happens to the energy they carry? To answer this question, we must examine in detail how the two waves were brought together. For instance, if the two waves were initially propagating on two separate strings which merge into a single string at a junction or knot, then the cancellation of the waves beyond the junction is necessarily associated with a strong backward reflection of the two incident waves at the junction, and the waves reflected backward from the junction account for the missing energy.

If the two waves are out of phase, but their amplitudes are not equal, then their cancellation will not be complete; some portion of the wave that has the larger amplitude will be left over (see Figure 16.14).

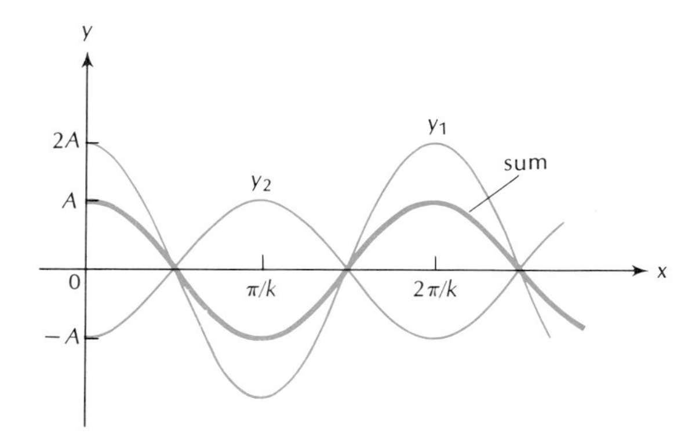


**Fig. 16.14** Destructive interference of two waves of different amplitude. The sum is not zero.

As another example of superposition, let us consider two waves of the same amplitude, but slightly different frequencies and, therefore, slightly different wavelengths. Figure 16.15 shows the two waves at one instant of time and their superposition. At $x = 0$, the waves are in phase, and they interfere constructively, giving a large resultant amplitude. But farther along the $x$ axis, the difference in wavelengths gradually causes the waves to acquire a phase difference. At some point, the waves will be out of phase by half a cycle, and they will interfere destructively, giving a resultant amplitude of zero. Beyond this point, the phase difference will exceed one half cycle. Farther along, the phase difference will grow to one cycle; but since a phase difference of one cycle means that the crests of the two waves coincide, they interfere

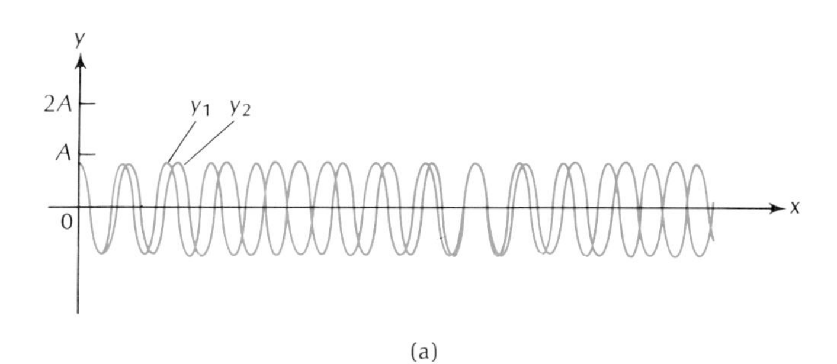


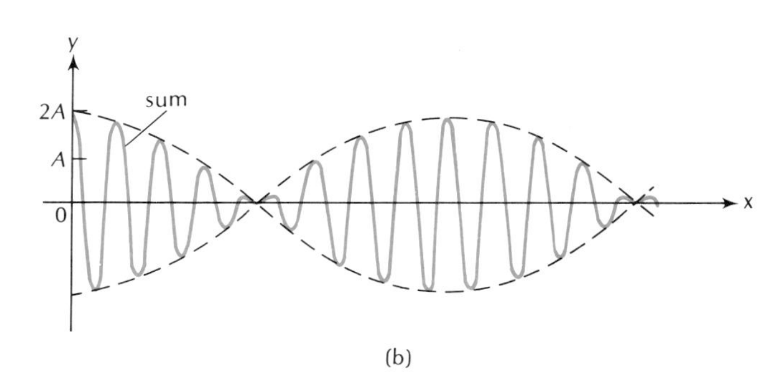


**Fig. 16.15** Superposition of two waves of slightly different wavelengths and frequencies. The dashed line shows the wave envelope, or the average amplitude.

constructively, again giving a large amplitude, and so on. The superposition of the two waves displays regularly alternating regions of constructive and destructive interference, that is, alternating regions of large amplitude and small amplitude. A wave with such a gradually varying amplitude is said to be **modulated.**

*Modulation of wave*

For the mathematical analysis of this kind of superposition, suppose that the two individual wavefunctions are

$$y_1 = A\cos(k_1x - \omega_1t) \tag{40}$$

$$y_2 = A\cos(k_2x - \omega_2t) \tag{41}$$

Using the same trigonometric identity as before, we obtain the wavefunction for the resultant wave:

$$\begin{aligned} y &= y_1 + y_2 \\ &= 2A\cos[\tfrac{1}{2}(k_1 + k_2)x - \tfrac{1}{2}(\omega_1 + \omega_2)t]\cos[\tfrac{1}{2}(k_1 - k_2)x - \tfrac{1}{2}(\omega_1 - \omega_2)t] \end{aligned} \tag{42}$$

Let us examine this wavefunction at an initial time $t = 0$. We can then write Eq. (42) as

$$y = 2A\cos[\tfrac{1}{2}(\Delta k)x]\cos(\bar{k}x) \tag{43}$$

where $\Delta k$ is the difference in wave numbers,

$$\Delta k = k_1 - k_2 \tag{44}$$

and $\bar{k}$ is the average wave number,

$$\bar{k} = \tfrac{1}{2}(k_1 + k_2) \tag{45}$$

If $\Delta k$ is small compared to $\bar{k}$, the expression on the right side of Eq. (43) can be interpreted as a cosine wave of wave number $\bar{k}$ with an amplitude $2A\cos[\frac{1}{2}(\Delta k)x]$, that is, the amplitude of the wave is a slowly varying function of position. From Eq. (43), we see that the amplitude is large at $x = 0$ where the two superposed waves interfere constructively; it then gradually decreases and becomes zero at $x = \pi/\Delta k$, where the two waves interfere destructively. Then again it increases and becomes large at $x = 2\pi/\Delta k$, and so on.

With the passing of time, the entire pattern of Figure 16.15 moves to the right with the wave velocity. This gives rise to the phenomenon of **beats.** At any given position the amplitude of the wave pulsates — first the amplitude of the oscillations is large, then it becomes small, then again large, and so on. The frequency with which the amplitude pulsates is called the **beat frequency.** Since the time interval between one amplitude maximum and the next in Figure 16.15 is $\Delta t = \Delta x/v = 2\pi/(\Delta k v)$, the beat frequency is

$$\nu_{\text{beat}} = \frac{1}{\Delta t} = \frac{v\,\Delta k}{2\pi} = \frac{vk_1}{2\pi} - \frac{vk_2}{2\pi}$$

or

*Beat frequency*

$$\boxed{\nu_{\text{beat}} = \nu_1 - \nu_2} \tag{46}$$

Thus the beat frequency is simply the difference between the two wave frequencies.

EXAMPLE 4. Suppose that two flutes generate sound waves of frequencies 264 Hz and 262 Hz, respectively. What is the beat frequency?

SOLUTION: According to Eq. (46),

$$\nu_{\text{beat}} = 264\ \text{Hz} - 262\ \text{Hz} = 2\ \text{Hz}$$

Hence a listener will hear a tone of average frequency 263 Hz, but with an amplitude pulsating 2 times per second.

COMMENTS AND SUGGESTIONS: Beats are a sensitive indication of small frequency differences, and they are very useful in the tuning of musical instruments. For example, to bring the two flutes in tune, the musicians listen to the beats and, by trial, adjust one of the flutes so as to reduce the beat frequency; when the beat disappears entirely (zero beat frequency), the two flutes will be generating waves of exactly equal frequencies.

*Fourier's theorem and Fourier series*

By the superposition of harmonic waves of different amplitude and frequencies, we can construct some rather complicated wave shapes. In fact, it can be shown that any arbitrary periodic wave can be constructed by the superposition of a sufficiently large number of sinusoidal harmonic waves. This is **Fourier's theorem.** For example, Figure 16.16a shows a square wave (plotted at $t = 0$) that alternates periodically between positive and negative values. The wavelength of this wave is $L$. To construct this wave by the superposition of harmonic waves we must take the following combination, which is called a **Fourier series:**

$$y = \frac{4A}{\pi}\sin\frac{2\pi x}{L} + \frac{4A}{3\pi}\sin\frac{6\pi x}{L} + \frac{4A}{5\pi}\sin\frac{10\pi x}{L} + \cdots \qquad (47)$$

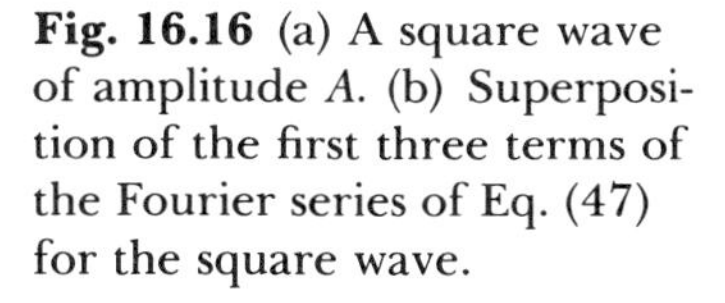

**Fig. 16.16** (a) A square wave of amplitude $A$. (b) Superposition of the first three terms of the Fourier series of Eq. (47) for the square wave.

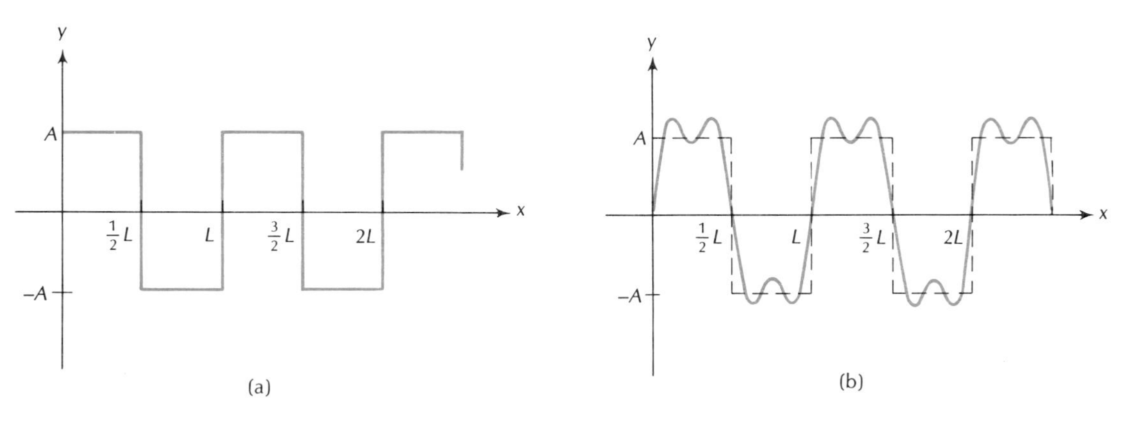


These are harmonic waves of wavelength $L$, $L/3$, $L/5$, and so on. The three dots on the right side stand for extra harmonic waves of even shorter wavelength; for an accurate representation of the square wave we need an infinite series of extra harmonic waves of smaller and smaller wavelengths, but, as Figure 16.16b shows, the three waves on the right side of Eq. (47) already give a rough approximation to the square wave.

## 16.6 Standing Waves

Let us now consider the superposition of two waves of the same amplitudes and frequencies but of opposite directions of propagation. If the individual wavefunctions are

$$y_1 = A\cos(kx - \omega t) \tag{48}$$

$$y_2 = A\cos(kx + \omega t) \tag{49}$$

then the resultant wavefunction is

*Standing wave*

$$\boxed{y = y_1 + y_2 = 2A\cos kx \cos \omega t} \tag{50}$$

where again we have used the trigonometric identity of Eq. (38).

The expression on the right side of Eq. (50) describes a **standing wave.** This wave travels neither right nor left; its wave crests remain at fixed positions while the entire wave increases and decreases in unison. The entire wave pulsates with a frequency $\omega$, as indicated by the overall factor $\cos \omega t$. Figure 16.17 shows the standing wave at successive instants of time.

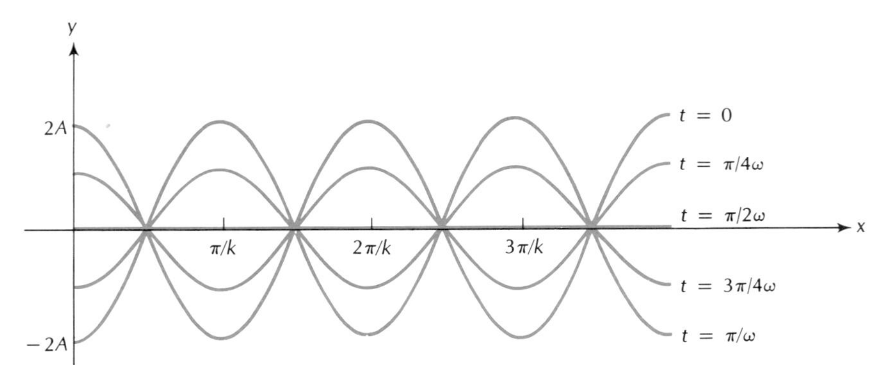


**Fig. 16.17** A standing wave at successive instants of time.

If the wavefunction of Eq. (50) represents the motion of a string, then each particle of this string executes simple harmonic motion. However, in contrast to the case of a traveling wave, where the amplitudes of the harmonic oscillations of all the particles are the same and the phases depend on position, the amplitudes of the harmonic oscillations in the standing wave depend on position whereas all the phases are the same. From Eq. (50), we see that the amplitude of oscillation of a particle at the position $x$ is $A \cos kx$.

The positions at which the amplitude of oscillation is maximum are given by

$$kx = 0,\ \pi,\ 2\pi,\ \text{etc.} \tag{51}$$

With $k = 2\pi/\lambda$ this becomes

$$x = 0,\ \frac{\lambda}{2},\ \lambda,\ \frac{3\lambda}{2},\ \text{etc.} \tag{52}$$

The maxima are due to constructive interference between the two waves of Eqs. (48) and (49).

Likewise, the positions at which the amplitude is zero are given by

$$kx = \frac{\pi}{2}, \frac{3\pi}{2}, \frac{5\pi}{2}, \text{ etc.} \tag{53}$$

or

$$x = \tfrac{1}{4}\lambda, \tfrac{3}{4}\lambda, \tfrac{5}{4}\lambda, \text{ etc.} \tag{54}$$

The minima are due to destructive interference between the two waves. The minima of the standing waves are called **nodes** and the maxima are called **antinodes.** Note that distance between each node and the next node is $\frac{1}{2}\lambda$, the distance between each antinode and the next antinode is also $\frac{1}{2}\lambda$, and the distance between each node and the adjacent antinode is $\frac{1}{4}\lambda$.

*Nodes and antinodes*

So far, in our discussion of the waves on a string, we have assumed that the string is very long, and we have ignored the endpoints of the string. When a traveling wave arrives at an endpoint, something drastic will have to happen to it: the wave will either have to be absorbed at the endpoint or it will have to be reflected, with a reversal of its direction of motion. If the endpoint is a fixed point (the string is attached to a rigid support), then the endpoint cannot absorb the energy of the wave, and the wave will be completely reflected. This results in the simultaneous presence of two waves of equal amplitudes and opposite directions of travel—that is, it results in a standing wave.

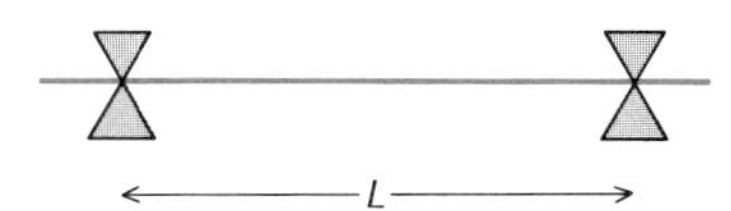


**Fig. 16.18** A tightly stretched string with fixed ends. The next figures shows possible standing waves on this string.

If the other endpoint of the string is also fixed (see Figure 16.18), then the possible standing waves on this string are subject to the restriction that the wave must be zero at each endpoint at all times. Such a restriction on what happens at the endpoints of a wave is called a **boundary condition.** Obviously, the boundary condition for our standing wave requires that the endpoints be nodes. Figures 16.19–16.21 show possible standing waves on a string with fixed endpoints (the waves are plotted at the initial time $t = 0$). All these standing waves have nodes at the endpoints.

*Boundary condition*

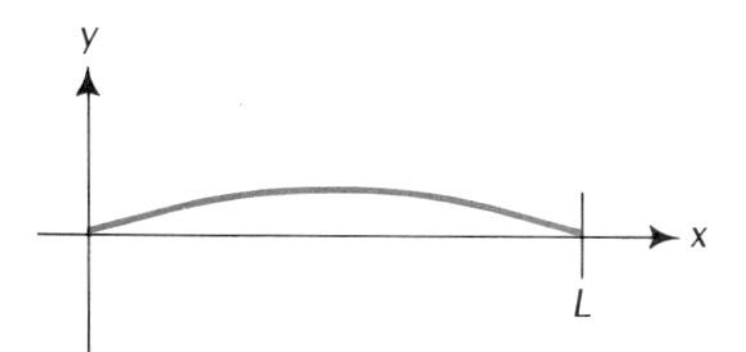


**Fig. 16.19** The fundamental mode.

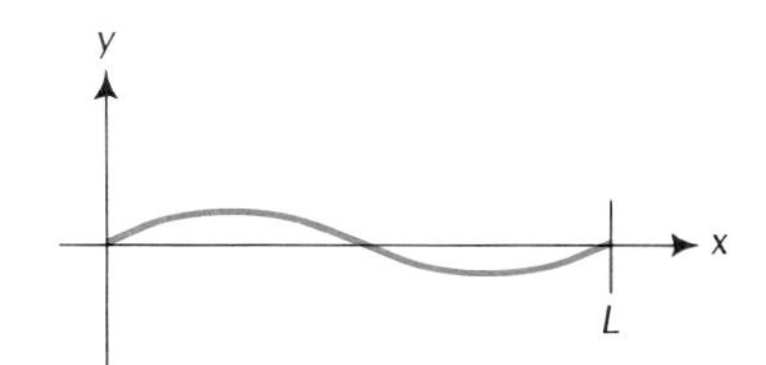


**Fig. 16.20** The first overtone.

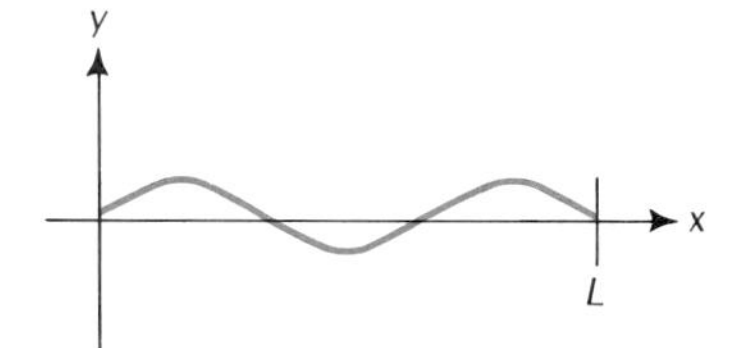


**Fig. 16.21** The second overtone.

With one end of the string is at $x = 0$ and the other at $x = L$, the wavefunctions corresponding to Figures 16.19–16.21 are, respectively,

$$y_1 = A \sin\left(\frac{\pi}{L}x\right)\cos\left(\frac{\pi}{L}vt\right) \tag{55}$$

$$y_2 = A \sin\left(\frac{2\pi}{L}x\right)\cos\left(\frac{2\pi}{L}vt\right) \tag{56}$$

$$y_3 = A \sin\left(\frac{3\pi}{L}x\right)\cos\left(\frac{3\pi}{L}vt\right) \tag{57}$$

*Normal modes*

These possible standing-wave motions of the string are called the **normal modes** of the string. Figure 16.19 shows the **fundamental** mode; Figure 16.20, the **first overtone;** Figure 16.21, the **second overtone,** and so on. In all these modes, some number of half wavelengths exactly fits the length of the string. In the fundamental mode, one half wavelength fits the string; in the first overtone, two half wavelengths fit the string; in the second overtone, three half wavelengths fit the string, and so on. Thus, the wavelengths for the modes are

*Wavelengths of normal modes*

$$\lambda_1 = 2L, \qquad \lambda_2 = L, \qquad \lambda_3 = \tfrac{2}{3}L, \text{ etc.} \tag{58}$$

and the frequencies are

$$\nu_1 = \frac{v}{2L}, \qquad \nu_2 = \frac{v}{L}, \qquad \nu_3 = \frac{3v}{2L}, \text{ etc.}$$

*Eigenfrequencies*

The frequencies of these modes are called the normal frequencies, proper frequencies, or **eigenfrequencies** of the string. We can write the following general formula for these frequencies:

$$\nu_n = \frac{nv}{2L} \qquad n = 1, 2, 3, \ldots \tag{59}$$

This formula clearly shows that all the eigenfrequencies are multiples of the fundamental frequency $v/2L$.

In general, any arbitrary motion of a string with fixed endpoints will be some superposition of several of the above normal modes. Which modes will be present in the superposition depends on how the motion is started. For instance, when a guitar player plucks a string on her guitar near the middle, she will excite the fundamental mode and also the second overtone and (to a lesser extent) some of the higher, even-numbered overtones.

---

EXAMPLE 5. The middle C string of a piano vibrates with a frequency of 261.6 Hz when excited in its fundamental mode. What are the frequencies of the first, second, and third overtones of this string?

SOLUTION: According to Eq. (59), $\nu_2$ equals $2\nu_1$, $\nu_3$ equals $3\nu_1$, and $\nu_4$ equals $4\nu_1$. Hence the frequencies of the first, second, and third overtones are, respectively, $2 \times 261.6$ Hz, $3 \times 261.6$ Hz, and $4 \times 261.6$ Hz.

---

The normal modes of vibration of a long, thin elastic rod or beam fixed at both ends are mathematically similar to the normal modes of a string. However, such an elastic body can experience transverse deformations (like those of a string), longitudinal deformations (compression), and rotational deformations (torsion). Figure 16.22 shows a spectacular example of a torsional standing wave in the span of a bridge at Tacoma, Washington. This standing wave was excited by a wind blowing across the bridge, which generated vortices in resonance with one of the normal modes of vibration of the span. The bridge oscillated for several hours, with increasing amplitude, and then broke apart.

**Fig. 16.22** Standing wave on the Tacoma Narrows bridge, July 1, 1940. The bridge broke a short time after this picture was taken.

## SUMMARY

**General wave function:** $y = f(x \mp vt)$

**Wavelength and frequency:** $\lambda \nu = v$

**Harmonic wave:** $y = A \cos k(x \mp vt)$
$= A \cos(kx \mp \omega t)$

**Wave number:** $k = 2\pi/\lambda$

**Angular frequency:** $\omega = 2\pi\nu$

**Wave equation:** $\dfrac{\partial^2 y}{\partial t^2} = v^2 \dfrac{\partial^2 y}{\partial x^2}$

**Speed of wave on a string:** $v = \sqrt{F/\mu}$

**Power of wave on a string:** $P \propto v\omega^2 A^2$

**Superposition principle for two or more waves:** The net instantaneous deformation is the sum of the individual instantaneous deformations.

**Constructive interference:** Waves meet crest to crest

**Destructive interference:** Waves meet crest to trough

**Beat frequency:** $\nu_{beat} = \nu_1 - \nu_2$

**Standing harmonic wave:** $y = A \cos kx \cos \omega t$

**Node:** Point of zero oscillation

**Antinode:** Point of maximum oscillation

**Wavelengths of normal modes of string:** $\lambda = 2L, L, \frac{2}{3}L, \ldots$

**Eigenfrequencies:** $\nu = \dfrac{v}{2L}, \dfrac{v}{L}, \dfrac{3v}{2L}, \ldots$

## QUESTIONS

1. You have a long, thin steel rod and a hammer. How must you hit the end of the rod to generate a longitudinal wave? A transverse wave?

2. A wave pulse on a string transports energy. Does it also transport momentum? To answer this question, imagine a washer loosely encircling the string at some place; what happens to the washer when the wave pulse strikes it?

3. The strings of a guitar are made of wires of different thicknesses (the thickest wires are manufactured by wrapping copper or brass wire around a strand of steel). Why is it impractical to use wire of the same thickness for all the strings?

4. According to Eq. (19), the speed of a wave on a string increases by a factor of 2 if we increase the tension by a factor of 4. However, in the case of a rubber string, the speed increases by more than a factor of 2 if we increase the tension by a factor of 4. Why are rubber strings different?

5. A harmonic wave is traveling along a string. Where in this wave is the kinetic energy at maximum? The potential energy? The total energy?

6. Suppose that two strings of different densities are knotted together to make a single long string. If a wave pulse travels along the first string, what will happen to the wave pulse when it reaches the junction? (Hint: If the second string had the same density as the first string, the wave pulse would proceed without interruption; if the second string were much denser than the first, the wave pulse would be totally reflected.)

7. Figure 16.17 shows a standing wave on a string. At time $t = \pi/2\omega$, the amplitude of the wave is everywhere zero. Does this mean the wave has zero energy at this instant?

8. After an arrow has been shot from a bow, the bowstring will oscillate back and forth, forming a standing wave. Which of the overtones shown in Figures 16.19–16.21 do you expect to be present?

9. In tuning a guitar or violin, by what means do you change the frequency of a string?

10. A mechanic can make a rough test of the tension in the spokes of a wire wheel by striking the spokes with a wrench or a small hammer. A spoke under tension will ring, but a loose spoke will not. Explain.

11. What is the purpose of the frets on the neck of a guitar or a mandolin?

## PROBLEMS

### Section 16.1

1. Suppose that at time $t = 0$ a wave pulse on a string has a shape described by the wavefunction

$$y = \frac{9 \times 10^{-2}}{9 + x^2}$$

where $y$ and $x$ are measured in meters.

(a) If this wave travels in the positive $x$ direction at a speed of 2 m/s, what is the wavefunction that describes the wave at a time $t$?

(b) Make a rough plot of the shape of the wave at $t = 0$ and at $t = 3$ s.

### Section 16.2

2. An ocean wave has a wavelength of 120 m and a period of 8.77 s. Calculate the frequency, angular frequency, wave number, and speed of this wave.

3. The speed of tidal waves in the Pacific is about 740 km/h.
   (a) How long does a tidal wave take to travel from Japan to California, a distance of 8000 km?
   (b) If the wavelength of the wave is 300 km, what is its frequency?

4. Suppose that the function $y = 6.0 \times 10^{-3} \cos(20x + 4.0t + \pi/3)$ describes a wave on a long string (distance is measured in meters and time in seconds).
   (a) What are the amplitude, wavelength, wave number, frequency, angular frequency, direction of propagation, and speed of this wave?
   (b) At what time does this wave have a maximum at $x = 0$?

5. A harmonic wave on a string has an amplitude of 2.0 cm, a wavelength of 1.2 m, and a velocity of 6.0 m/s in the positive $x$ direction. At time $t = 0$, this wave has a crest at $x = 0$.
   (a) What are the period, frequency, angular frequency, and wave number of this wave?
   (b) What is the mathematical equation describing this wave as a function of $x$ and $t$?

6. Ocean waves smash into a breakwater at the rate of 12 per minute. The wavelength of these waves is 39 m. What is their speed?

7. The velocity of sound in fresh water at 15°C is 1440 m/s and at 30°C it is 1530 m/s. Suppose that a sound wave of frequency 440 Hz penetrates from a layer of water at 30°C into a layer of water at 15°C. What will be the change in the wavelength? Assume that the frequency remains unchanged.

8. A light wave of frequency $5.5 \times 10^{14}$ Hz penetrates from air into water. What is its wavelength in air? In water? The speed of light is $3.0 \times 10^{8}$ m/s in air and $2.3 \times 10^{8}$ m/s in water; assume that the frequency remains the same.

*9. Ocean waves of wavelength 100 m have a speed of 6.2 m/s; ocean waves of wavelength 20 m have a speed of 2.8 m/s.[2] Suppose that a sudden storm at sea generates waves of all wavelengths. The long-wavelength waves travel fastest and reach the coast first. A fisherman standing on the coast first notices the arrival of 100-m waves; 10 hours later he notices the arrival of 20-m waves. How far is the storm from the coast?

*10. Prove that $f(x \pm vt)$, where $f$ is any arbitrary differentiable function, is a solution of the wave equation (16).

### Section 16.3

11. A string has a length of 3.0 m and a mass of 12 g. If this string is subjected to a tension of 250 N, what is the speed of transverse waves?

12. A clothesline of length 10 m is stretched between a house and a tree. The clothesline is under a tension of 50 N and it has a mass per unit length of $6.0 \times 10^{-2}$ kg/m. How long does a wave pulse take to travel from the house to the tree and back?

13. A wire rope used to support a radio mast has a length of 20 m and a mass per unit length of 0.8 kg/m. When you give the wire rope a sharp blow at the lower end and generate a wave pulse, it takes 1 s for this wave pulse to travel to the upper end and to return. What is the tension in the wire rope?

14. A nylon rope of length 24 m is under a tension of $1.3 \times 10^{4}$ N. The total mass of this rope is 2.7 kg. If a wave pulse starts at one end of this rope, how long does it take to reach the other end?

*15. A string of mass per unit length $\mu$ is tied to a second string of mass per unit length $\mu'$. A harmonic wave of speed $v$ traveling along the first string reaches the junction and enters the second string. What will be the speed $v'$ of this wave in the second string? Your answer should be a formula involving $\mu$, $\mu'$, and $v$.

---

[2] These values are group velocities, or signal velocities.

*16. A steel wire of length 5 m and radius 0.3 mm is knotted to another steel wire of length 5 m and radius 0.1 mm. The wires are stretched with a tension of 150 N. How long does a transverse wave pulse take to travel the distance of 10 m from the beginning of the first wire to the end of the other? The density of steel is $7.8 \times 10^3$ kg/m$^3$.

*17. A long, uniform rope of length $l$ hangs vertically. The only tension in the rope is that produced by its own weight. Show that, as a function of the distance $z$ from the lower end of the rope, the speed of a transverse wave pulse on the rope is $\sqrt{gz}$. What is the time the wave pulse takes to travel from one end of the rope to the other?

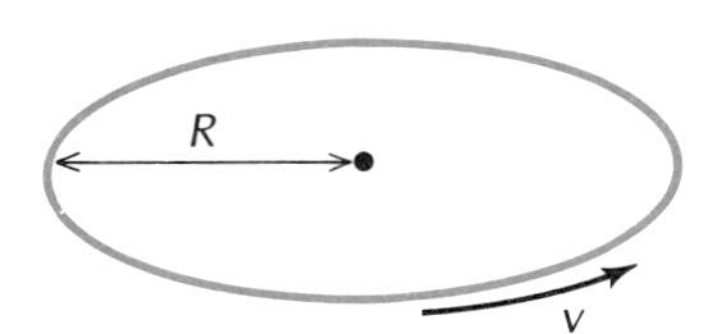


**Fig. 16.23**

**18. Suppose you take a loop of rope and make it rotate about its center at speed $V$. The centrifugal tendency of the segments of rope will then stretch it out along a circle of some radius $R$ (see Figure 16.23). What is the tension in the rope under these conditions? Show that the speed of transverse waves on the rope (relative to the rope) coincides with the speed of rotation $V$.

**19. A flexible rope of length $l$ and mass $m$ hangs between two walls. The length of the rope is more than the distance between the walls (see Figure 16.24), and the rope sags downward. At the ends, the rope makes an angle of $\alpha$ with the walls. At the middle, the rope approximately has the shape of an arc of circle; the radius of the approximating (osculating) circle is $R$. What is the tension in the rope at its ends? What is the tension in the rope at its middle? What is the speed of transverse waves at the ends? At the middle?

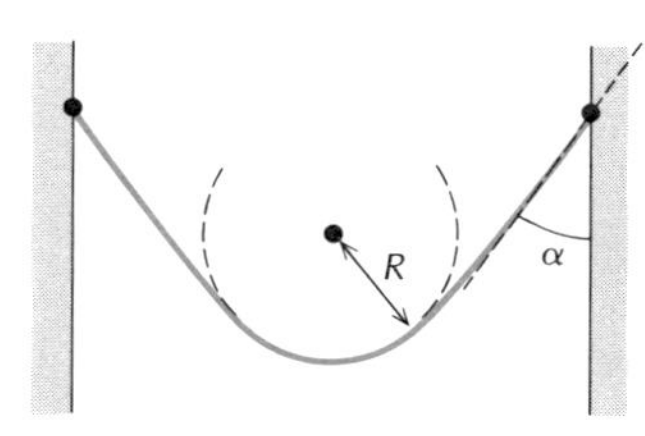


**Fig. 16.24**

**20. The end of a long string of mass per unit length $\mu$ is knotted to the beginning of another long string of mass per unit length $\mu'$ (the tensions in these strings are equal). A harmonic wave travels along the first string toward the knot. This incident wave will be partially transmitted into the second string, and partially reflected. The frequencies of all these waves are the same. With the knot at $x = 0$, we can write the following expressions for the incident, reflected, and transmitted waves:

$$y_1 = A_{\text{in}} \cos(kx - \omega t)$$

$$y_2 = A_{\text{ref}} \cos(kx + \omega t)$$

$$y_3 = A_{\text{trans}} \cos(k'x - \omega t)$$

Show that

$$A_{\text{ref}} = \frac{k - k'}{k + k'} A_{\text{in}} = \frac{\sqrt{\mu} - \sqrt{\mu'}}{\sqrt{\mu} + \sqrt{\mu'}} A_{\text{in}}$$

$$A_{\text{trans}} = \frac{2k}{k + k'} A_{\text{in}} = \frac{2\sqrt{\mu}}{\sqrt{\mu} + \sqrt{\mu'}} A_{\text{in}}$$

[Hint: At $x = 0$, the displacement of the string must be continuous, $y_1 + y_2 = y_3$; if not, the string would break at the knot. Furthermore, the slope of the string must be continuous, $dy_1/dx + dy_2/dx = dy_3/dx$; if not, the string would have a kink and the (massless) knot would receive an infinite acceleration.]

## Section 16.4

21. A harmonic wave travels along a string of mass per unit length $5.0 \times 10^{-3}$ kg/m. The amplitude of the wave is $2.0 \times 10^{-2}$ m, its frequency is 60 Hz, and the tension in the string is 20 N. What is the average power transported by this wave?

22. The average waves generated by a wind of 30 knots blowing on the open sea for a long time (steady-state conditions) have a height of 3.4 m, a period of 11 s, and a speed of 17 m/s. The average waves generated by a wind of 40

knots have a height of 6.1 m, a period of 15 s, and a speed of 24 m/s. What is the ratio of the powers carried by these two kinds of waves?

*23. Consider the wave on a string described in Problem 4. What is the energy density as a function of time at $x = 0.5$ m? The mass per unit length of the string is $6.0 \times 10^{-3}$ kg/m.

### Section 16.5

24. Two transverse waves on a string have equal amplitudes of 1.5 cm, equal wavelengths, and they travel in the same direction. Their phases differ by $\pi/4$ radians. What is the amplitude of the resultant wave?

25. The wavefunctions for two transverse waves on a string are

$$y_1 = 0.03 \cos (6.0x - 18t + 1.5)$$

$$y_2 = 0.03 \cos (6.0x - 18t - 2.3)$$

where $y$ and $x$ are measured in meters and $t$ in seconds.
(a) What is the phase difference between these waves?
(b) What is the amplitude and the phase constant of the resultant wave?
(c) What is the transverse displacement of the string at $x = 0$ at $t = 0$?

*26. Consider the wavefunction $y = 3.0 \cos(5.0x - 8.0t) + 4.0 \sin(5.0x - 8.0t)$ which is a superposition of two wavefunctions expressed in some suitable units. Show that this wavefunction can be written in the form $y = A \cos(5.0x - 8.0t + \delta)$. What are the values of $A$ and $\delta$?

*27. A thin wire of length 1.0 m vibrates in a superposition of the fundamental mode and the second harmonic. The wavefunction is

$$y = 0.006 \sin \pi x \cos 400\pi t - 0.004 \sin 3\pi x \cos 1200\pi t$$

where $y$ and $x$ are measured in meters and $t$ in seconds.
(a) What is the deformation at $x = 0.5$ m as a function of time?
(b) Plot this deformation as a function of time in the interval $0 \text{ s} \leq t \leq 0.005$ s.

28. Two ocean waves with $\lambda = 100$ m, $\nu = 0.125$ Hz and $\lambda = 90$ m, $\nu = 0.132$ Hz arrive at a seawall simultaneously. What is the beat frequency of these waves?

29. A guitar player attempts to tune her instrument perfectly with the help of a tuning fork. If the guitar player sounds the tuning fork and a string on her guitar simultaneously, she perceives beats at a frequency of 4 per second. The tuning fork is known to have a frequency of 294 Hz. What fractional increase (or decrease) of the tension of the guitar string is required to bring the guitar in tune with the tuning fork? From the available information, can you tell whether an increase or decrease of tension is required?

*30. Figure 16.25 shows the height of tide at Pakhoi. These tides can be regarded as a wave. The shape of the curve in Figure 16.25 indicates that the

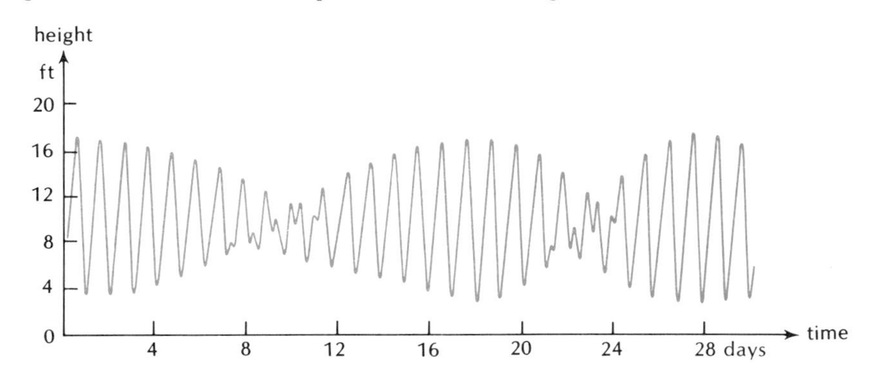


**Fig. 16.25** Height of the tide at Pakhoi as a function of time.

wave consists of two periodic waves of slightly different frequencies beating against each other. What are the frequencies of the two periodic waves? What are the periods? Which is caused by the Moon, which by the Sun?

*31. A piano wire of length 1.5 m fixed at its end vibrates in its second overtone. The frequency of vibration is 440 Hz and the amplitude at the midpoint of the wire is 0.4 mm. Express this standing wave as a superposition of traveling waves. What are the amplitudes and speeds of the traveling waves?

### Section 16.6

32. The fundamental mode of the G string of a violin has a frequency of 196 Hz. What are the frequencies of the first, second, third, and fourth overtones?

33. Suppose that a vibrating mandolin string of length 0.34 m vibrates in a mode with five nodes (including the nodes at the ends) and four antinodes. What overtone is this? What is its wavelength?

34. A telegraph wire made of copper is stretched tightly between two telephone poles 50 m apart. The tension in the wire is 500 N and the mass per unit length is $2 \times 10^{-2}$ kg/m. What is the frequency of the fundamental mode? The first overtone?

35. A violin has four strings; all the strings have (approximately) equal tensions and lengths but they have different masses per unit length (kg/m), so that when excited in their fundamental modes they vibrate at different frequencies. The fundamental frequencies of the four strings are 196, 294, 440, and 659 Hz. What must be the ratios of the densities of the strings?

36. The fundamental mode of the G string in a mandolin has a frequency of 196 Hz. The length of this string is 0.34 m and its mass per unit length is $4.0 \times 10^{-3}$ kg/m. What is the tension on this string?

37. Some automobiles are equipped with wire wheels. The spokes of these wheels are made of short segments of thick wire installed under large tension. Suppose that one of these wires is 9.0 cm long, 0.40 cm in diameter, and under a tension of 2200 N. The wire is made of steel; the density of steel is 7.8 g/cm$^3$. To check the tension, a mechanic gives the spoke a light blow with a wrench near its middle. With what frequency will the spoke ring? Assume that the frequency is that of the fundamental mode.

38. A light wave of wavelength $5.0 \times 10^{-7}$ m strikes a mirror perpendicularly. The reflection of the wave by the mirror makes a standing wave with a node at the mirror. At what distance from the mirror is the nearest antinode? The nearest node?

39. A wave on the surface of the sea with a wavelength of 3.0 m and a period of 4.4 s strikes a seawall oriented perpendicularly to its path. The reflection of the wave by the seawall sets up a standing wave. For such a wave, there is an antinode at the seawall. How far from the seawall will there be nodes?

*40. The D string of a violin vibrates in its fundamental mode with a frequency of 294 Hz and an amplitude of 2.0 mm. What are the maximum velocity and the maximum acceleration of the midpoint of the string?

*41. The middle C string of a piano is supposed to vibrate at 261.6 Hz when excited in its fundamental mode. A piano tuner finds that in a piano that has tension of 900 N on this string, the frequency of vibration is too low (flat) by 15 Hz. How much must he increase the tension of the string to achieve the correct frequency?

*42. The wire rope supporting the mast of a sailboat from the rear is under a large tension. The rope has a length of 9.0 m and a mass per unit length of 0.22 kg/m.

(a) If a sailor pushes on the rope sideways at its midpoint with a force of 150 N, he can deflect it by 7.0 cm. What is the tension in the rope?

(b) If the sailor now plucks the rope near its midpoint, the rope will vibrate back and forth like a guitar string. What is the frequency of the fundamental mode?

*43. Many men enjoy singing in shower stalls because their voice resonates in the cavity of the shower stall. Consider a shower stall measuring 1 m $\times$ 1 m $\times$ 2.5 m. What are the four lowest resonant frequencies of standing sound waves in such a shower stall?

*44. A mandolin has a string of length 0.34 m. Suppose that the string vibrates in its second overtone at a frequency of 588 Hz and an amplitude of 1.2 mm. The mass per unit length of the string is $1.8 \times 10^{-3}$ kg/m. What is the energy density along the string? Where is the energy density maximum? Where minimum? What is the value of the energy density at its maximum?

**45. A piano wire of length 0.18 m vibrates in its fundamental mode. The frequency of vibration is 494 Hz; the amplitude is $3.0 \times 10^{-3}$ m. The mass per unit length of the wire is $2.2 \times 10^{-3}$ kg/m. What is the energy of vibration of the entire wire?

CHAPTER 17

# Sound and Other Wave Phenomena

In this chapter we will study some important examples of wave propagation in two and in three dimensions: water waves on the surface of water, sound waves in the volume of air, and seismic waves in the volume of the Earth. In a later chapter we will study the propagation of light waves in a transparent medium and in vacuum. All these waves can be described graphically by their **wave fronts,** that is, the locations of the wave crests at a given instant of time. For example, Figures 17.1 and 17.2 show the circular wave fronts of a water wave and the spherical wave fronts of a sound wave.

*Wave front*

As time passes, the wave fronts spread outward, expanding as they move away from their source. This spreading of the waves is a characteristic feature of wave propagation in two or three dimensions. It implies that the intensity at a given wave front decreases as the wave front increases in size. Consider the case of circular water waves

**Fig. 17.1** (left) Water wave spreading out on the surface of a pond. The wave fronts are concentric circles.

**Fig. 17.2** (right) Sound wave spreading out in air. The wave fronts are concentric spheres. The sound wave has been made visible by means of a small electric light bulb attached to a microphone that controlled the brightness of the bulb. The bulb and microphone were swept through the space in front of the telephone earpiece along arcs, as indicated by the fine pattern of ridges.

spreading out from a pointlike disturbance caused by, say, the impact of a pebble on the water's surface. At some time, a wave front has a radius $r_1$ and its energy is distributed along its circumference $2\pi r_1$. At a later time, the wave front has a larger radius $r_2$ and the same amount of energy is distributed along a larger circumference $2\pi r_2$. The intensity of the wave is proportional to the energy delivered by the wave per unit length along the circumference; hence the intensity decreases in inverse proportion to the length of the circumference, that is, in inverse proportion to the radial distance from the source. This decrease of intensity is clearly noticeable in Figure 17.1. By a similar argument we will find that the intensity of a sound wave, or some other spherical three-dimensional wave, decreases in inverse proportion to the square of the radius.

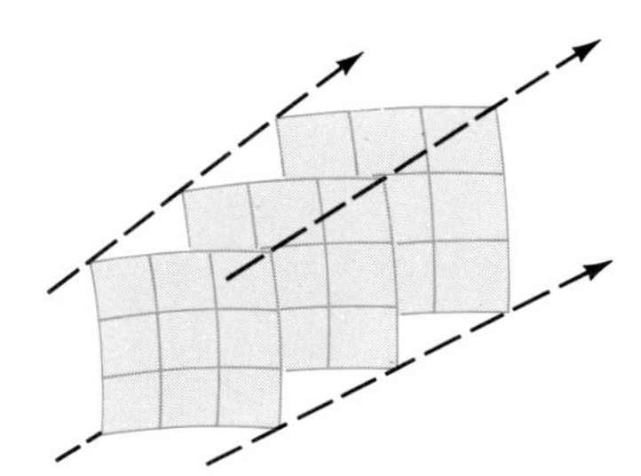

**Fig. 17.3** Parallel, nearly flat wave fronts.

At a large distance from a pointlike source, portions of the spherical wave fronts of a three-dimensional wave are nearly flat, provided the portions of interest are of a size small compared to the radius (see Figure 17.3). Waves with such flat, parallel wave fronts are called **plane waves.**

*Plane wave*

## 17.1 Sound Waves in Air

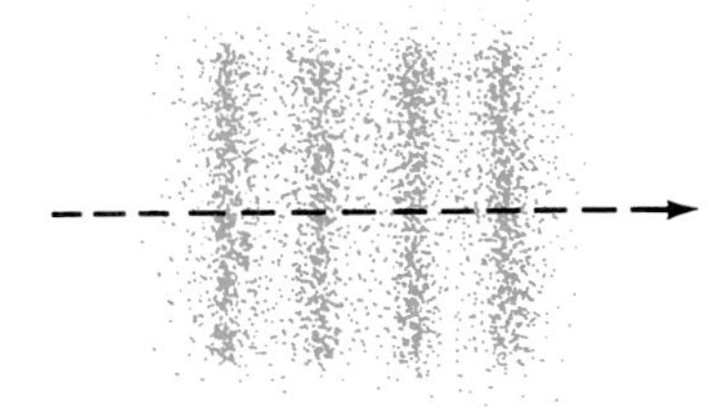

**Fig. 17.4** Density changes in air in a sound wave.

A sound wave in air is a longitudinal wave in which pushes are communicated from molecule to molecule. The restoring force for such a wave is supplied by the pressure of air — wherever the density of molecules is higher than normal, the pressure also is higher than normal and pushes the molecules apart. Figure 17.4 shows a sound wave in air. The wave consists of alternating zones of low density and high density. Such zones of alternating density are generated by the vibrating diaphragm of a loudspeaker or the vibrating prong of a tuning fork which exerts successive pushes on the surrounding air. At the zones of low density, the molecules have suffered maximum displacement — they have been displaced away from the zone. At the zones of high density, the molecules have suffered minimum displacement—they have remained stationary while molecules from the adjacent zones of maximum displacement have converged upon them. In Figure 17.4, the displacements and the density enhancements are shown much exaggerated. Even in an extremely intense sound wave, such as that produced by a jet airliner at takeoff, the displacements are only about $10^{-4}$ m and the density changes only about 1%.

The frequency of the sound determines the pitch we hear; that is, it determines whether the tone is perceived as high or low by our ears. The range of frequencies audible to the human ear extends from 20 to 20,000 Hz. These limits are somewhat variable; for instance, the ears of older people are less sensitive to high frequencies. Sound waves above 20,000 Hz are called **ultrasound;** some animals — dogs, cats, and bats — can hear these frequencies.

*Ultrasound*

*Intensity of sound*

The **intensity** of a sound wave is the power transported by this wave per square meter of wave front; the units of intensity are W/m$^2$. At a frequency of 1000 Hz, the minimum intensity audible to the human ear is $2.5 \times 10^{-12}$ W/m$^2$. This intensity is called the **threshold of hearing.** There is no upper limit for the audible intensity of sound; however, an intensity above 1 W/m$^2$ produces a painful sensation in the ear.

*Intensity level of sound*
*Decibel*

The intensity of sound is often expressed on a logarithmic scale called the **intensity level.** The unit of intensity level is the **decibel** (dB); this is a dimensionless unit. The definition of intensity level is as follows: We take an intensity of $0.937 \times 10^{-12}$ W/m$^2$ as our standard of intensity which corresponds to 0 dB. An intensity 10 times as large corresponds to 10 dB; an intensity 100 times as large corresponds to 20 dB; an intensity 1000 times as large corresponds to 30 dB, and so on. (This logarithmic scale is intended to agree with our subjective perception of the loudness of sounds — our ears tend to underestimate sounds of great intensity.) Mathematically, the relationship between intensities in watts per square meter and intensity levels in decibels is

$$[\text{intensity level in dB}] = (10 \text{ dB}) \times \log\left(\frac{[\text{intensity in W/m}^2]}{0.937 \times 10^{-12} \text{ W/m}^2}\right) \quad (1)$$

Thus, the threshold of hearing ($2.5 \times 10^{-12}$ W/m$^2$) corresponds to 4 dB and the threshold of pain (1 W/m$^2$) corresponds to 120 dB. Note that in consequence of the mathematical properties of logarithms, a multiplicative increase of intensity by a factor of 2 implies an additive increase in intensity level of 10 log 2 dB, or 3.01 dB. Table 17.1 lists some examples of sounds of different intensities.

**Table 17.1** SOME SOUND INTENSITIES

| *Sound* | *Intensity* | *Intensity level* |
|---|---|---|
| Loudest noise achieved in laboratory | $10^9$ | 210 |
| Rupture of eardrum | $10^4$ W/m$^2$ | 160 dB |
| Jet engine (at 30 m) | 10 | 130 |
| Threshold of pain | 1 | 120 |
| Thunder (loud) | $10^{-1}$ | 110 |
| Subway train (New York City) | $10^{-2}$ | 100 |
| Heavy street traffic | $10^{-5}$ | 70 |
| Normal conversation | $10^{-6}$ | 60 |
| Whisper | $10^{-10}$ | 20 |
| Normal breathing | $10^{-11}$ | 10 |
| Threshold of hearing | $2.5 \times 10^{-12}$ | 4 |

According to Fourier's theorem (see Section 16.5), a periodic sound wave of arbitrary shape can be regarded as a superposition of harmonic waves. The relative intensity of the harmonic waves in this superposition determines the perceived timbre (or quality) of the sound. Pure noise, or **white noise,** consists of a mixture of harmonic waves of all frequencies with equal intensities. The musical notes emitted by a musical instrument consist of a mixture of just a few harmonic waves: the fundamental and its first few overtones. Figure 17.5 shows the wave forms emitted by a violin, a trumpet, and a clarinet when the musical note A is played on these instruments. In both cases the wave is periodic, repeating at the rate of 440 cycles per second; but the shape of the wave and the intensities of the overtones are quite different in each case. This difference in the shapes of the waves permits the ear to distinguish between diverse musical instruments.

(a)
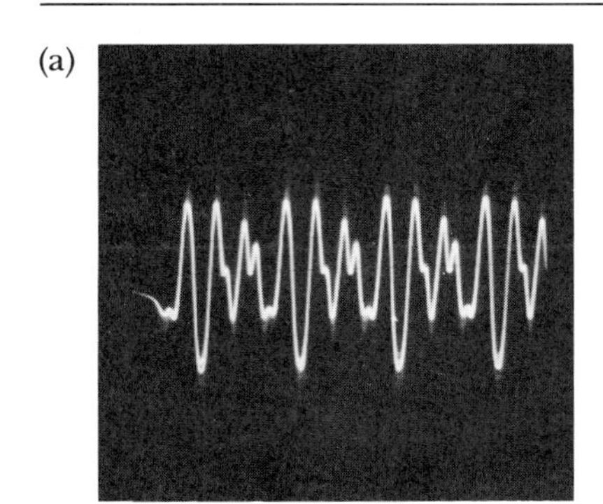
(b)
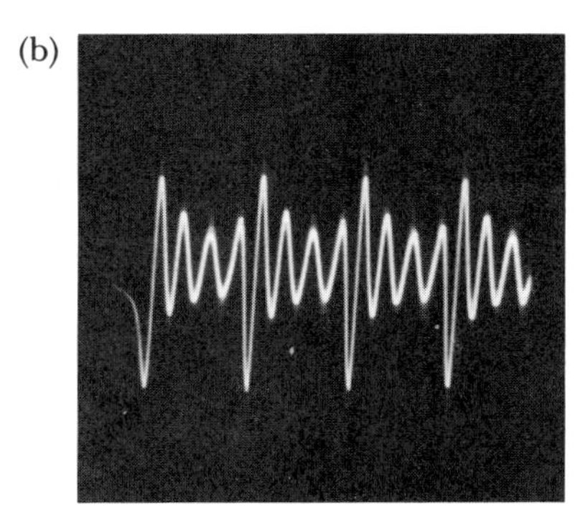
(c)
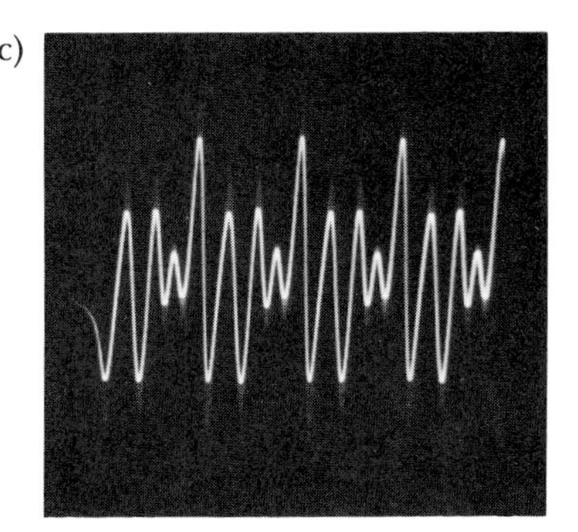

**Fig. 17.5** Wave forms emitted by (a) a violin, (b) a trumpet, and (c) a clarinet playing the note C. The wave pattern repeats 262 times per second.

**Table 17.2** THE CHROMATIC MUSICAL SCALE

| *Note* | *Frequency*[a] |
|---|---|
| C | 261.7 Hz |
| C# | 277.2 |
| D | 293.7 |
| D# | 311.2 |
| E | 329.7 |
| F | 349.2 |
| F# | 370.0 |
| G | 392.0 |
| G# | 415.3 |
| A | 440.0 |
| A# | 466.2 |
| B | 493.9 |

[a] Based on a frequency of 440 Hz for A.

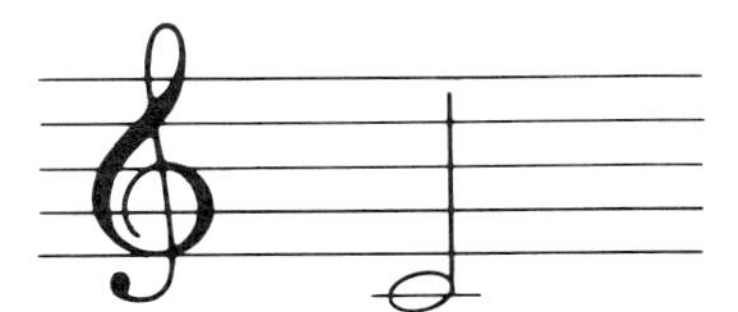

**Fig. 17.6** Middle C.

Table 17.2 gives the frequencies of the 12 notes of the chromatic musical scale. These frequencies are based on the system of equal temperament; successive frequencies in this scale differ by a factor of $(2)^{1/12}$. The first entry in this table is middle C, with a frequency of 261.7 Hz (Figure 17.6). Any musical note not listed in Table 17.2 can be obtained by multiplying (or dividing) the listed frequencies by a factor of 2, or 4, or 8, etc. Musical notes that differ by a factor of 2 in frequency are said to be separated by an **octave.** For example, C one octave above middle C has a frequency of 523.3 Hz, C two octaves above middle C has a frequency of 1046.6 Hz, etc.

*Octave*

Incidentally: For a musician the absolute values of these frequencies are not as important as the ratios of the frequencies. If an orchestra tunes its instruments so that their middle C has a frequency of, say, 255 Hz, this will not do any noticeable harm to the music, provided that the frequencies of all the other notes are also decreased in proportion.

As a sound wave spreads out from its source, its intensity falls off because the area of the wave front grows larger and therefore the wave energy per unit area grows smaller. In a homogeneous medium the intensity of the wave decreases with the inverse square of the distance from the source. Figure 17.7 helps to make this clear; it shows a spherical wave front at successive instants of time. The wave front grows from an old radius $r_1$ to a new radius $r_2$; correspondingly, its area grows from $4\pi r_1^2$ to $4\pi r_2^2$. The total power carried by this wave front remains the same; hence the intensity, or power per unit area, varies as the inverse square of the distance:

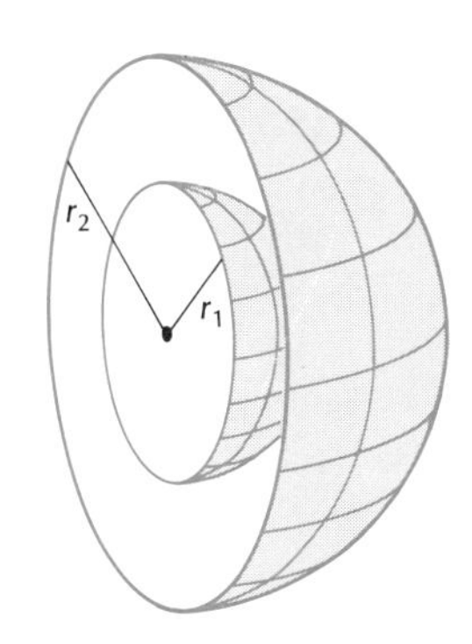


**Fig. 17.7** Concentric spherical wave fronts of a sound wave in air.

*Decrease of intensity of a sound wave*

$$I_2 = \frac{r_1^2}{r_2^2} I_1 \tag{2}$$

In the derivation of this simple result, we have taken for granted that the medium is homogeneous, so the wave speed does not depend on position. We know from Section 16.4 that the power carried by the wave is proportional to the speed; hence, if the speed were to vary, we would have to take this extra complication into account. Furthermore, we have taken for granted that the medium is perfectly elastic, so it does not absorb any of the wave energy by friction.

Although Figure 17.7 shows the wave fronts spreading out uniformly in all directions, Eq. (2) remains valid if the wave is a beam, such as the beam emitted by a loudspeaker with a horn, aimed in some preferential direction. In this case, the intensity falls off as the inverse square of the distance along the direction of the beam.

---

EXAMPLE 1. At a distance of 60 m from a jet airliner engaged in takeoff, the intensity level of sound is 120 dB. What is the intensity (in $W/m^2$) and the intensity level (in decibels) at a distance of 180 m?

SOLUTION: According to Eq. (1), an intensity level of 120 dB corresponds to

$$\begin{aligned} I_1 &= 9.37 \times 10^{-12}\ \text{W/m}^2 \times 10^{(\text{intensity level}/10\ \text{dB})} \\ &= 0.937 \times 10^{-12}\ \text{W/m}^2 \times 10^{120/10} \\ &= 0.937\ \text{W/m}^2 \end{aligned}$$

In Eq. (2) the intensities are measured in $W/m^2$. At a distance of 180 m the intensity is

$$\begin{aligned} I_2 &= (r_1^2/r_2^2)I_1 = (60\ \text{m}/180\ \text{m})^2 \times 0.937\ \text{W/m}^2 = \tfrac{1}{9} \times 0.937\ \text{W/m}^2 \\ &= 0.104\ \text{W/m}^2 \end{aligned}$$

When converted to decibels, this gives an intensity level of

$$10 \log\left(\frac{0.104\ \text{W/m}^2}{0.937 \times 10^{-12}\ \text{W/m}^2}\right) = 110.5\ \text{dB}$$

COMMENTS AND SUGGESTIONS: Note that we could have evaluated the change in intensity level directly, without first evaluating the intensity. A decrease in intensity by a factor of 9 implies a decrease in intensity level of (10 dB) × log 9, or 9.5 dB.

---

Ultrasonic waves of very high frequency do not propagate well through air—they are rapidly absorbed and dissipated by air molecules. However, these waves propagate readily through liquids and solids; in recent years, this has led to the development of some interesting practical applications of ultrasonic waves. For instance, such waves are now used in place of X rays to take pictures of the interior of the human body; this permits the examination of the fetus in the body of a pregnant woman, and avoids the damage that X rays might do to the very sensitive tissues of the fetus. The ultrasonic "cameras" that take such pictures employ sound waves of a frequency of about $10^6$ Hz. Further development of this technique has led to the construction of acoustic microscopes. The most powerful of these devices employ ultrasound waves of a frequency in excess of $10^9$ Hz to make highly magnified pictures of small samples of materials. The wavelength of sound waves of such extremely high frequency is about $10^{-6}$ m, roughly the same as the wavelength of ordinary light waves. The micrographs made by experimental acoustic microscopes compare favorably with micrographs made by ordinary optical microscopes (Figures 17.8 and 17.9).

**Fig. 17.8** Acoustic micrograph of reverse surface of a penny coin. This picture was made with sound waves of 50 megahertz, which were sent through the coin (the obverse surface of the coin was machined smooth). Note the mottling; this is due to crystal grains in the copper. (Courtesy R. S. Gilmore, General Electric Research and Development Center.)

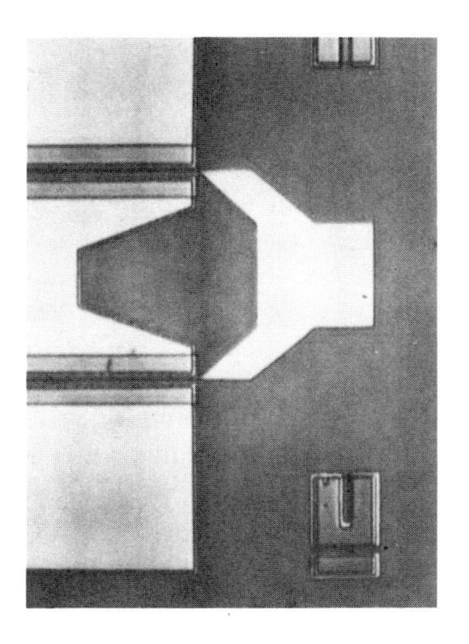

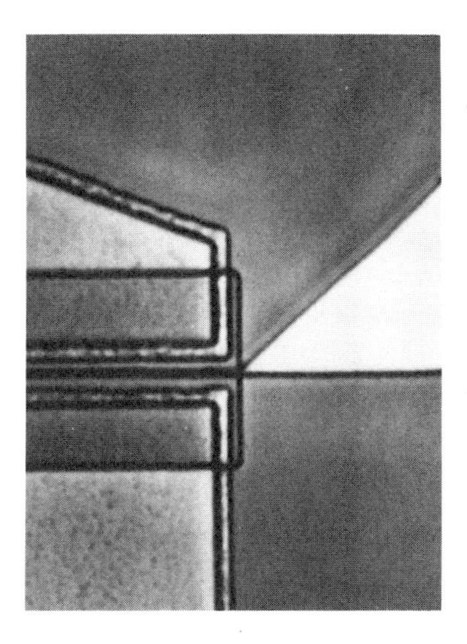

**Fig. 17.9** (left) Acoustic micrograph of a portion of a transistor (250×) and (right) detail (1000×). This picture was made with sound waves of 2.7 gigahertz. Note that these sound waves give a clear view through several layers of material. (Courtesy C. F. Quate and L. Lam, Stanford University.)

## 17.2 The Speed of Sound

As in the case of a wave on a string, the speed of sound in air depends on the restoring force and on the density. Since the restoring force is provided by the pressure of air, we expect the speed of sound to be a function of the air pressure and the air density. A somewhat involved calculation shows that the theoretical formula for the speed of sound is

$$v = \sqrt{1.40\frac{p_0}{\rho_0}} \qquad (3)$$

where $p_0$ and $\rho_0$ designate the unperturbed pressure and density, respectively. Note that this formula does display the expected increase of speed with restoring force and decrease with density. Under standard conditions (temperature 0°C, $p_0 = 1$ atm $= 1.01 \times 10^5$ N/m$^2$, $\rho_0 =$ 1.29 kg/m$^3$), the value of the **speed of sound** is 331.29 m/s.

*Speed of sound in air*

The speed of sound in liquids and in solids is often higher than that in air because the restoring force is much larger — liquids and solids offer much more opposition to compression than do gases. Table 17.3 gives the values of the speed of sound in some materials.

**Table 17.3** THE SPEED OF SOUND IN SOME MATERIALS

| *Material* | $v$ |
|---|---|
| Air | |
| 0°C, 1 atm | 331 m/s |
| 20°C, 1 atm | 344 |
| 100°C, 1 atm | 386 |
| Helium, 0°C, 1 atm | 965 |
| Water (distilled) | 1497 |
| Water (sea) | 1531 |
| Aluminum | 5104 |
| Iron | 5130 |
| Glass | 5000–6000 |
| Granite | 6000 |

A simple method for the measurement of the speed of sound in air takes advantage of standing waves in a tube open at one end and closed at the other (Figure 17.10). The standing sound wave must then obviously have a displacement node at the closed end, since the motion of the air is restricted by the wall at this end. The wave must have a displacement antinode at the open end. This is not so obvious, but can be understood by first considering the pressure. The pressure at the open end must remain constant because the open end is accessible to the atmosphere, and hence any incipient decrease or increase of pressure would immediately lead to an inflow or outflow of air from the surrounding atmosphere, canceling the pressure change. Thus, the atmosphere behaves as a reservoir of constant pressure. This is not quite exact, because some sound waves are radiated from the open end and



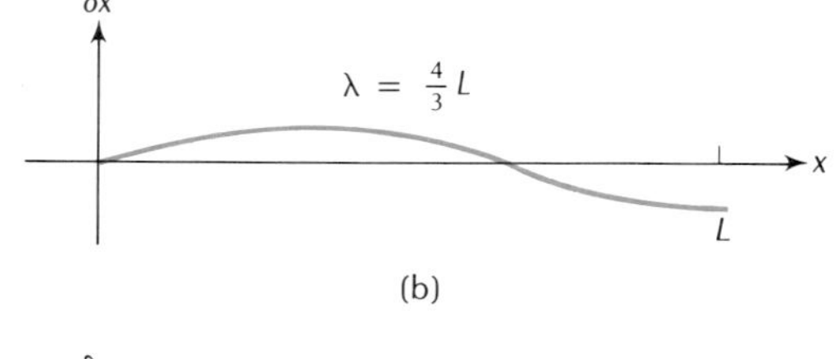


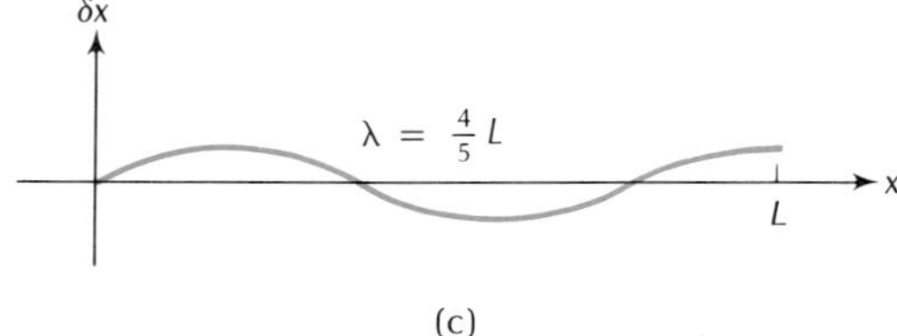


**Fig. 17.10** Possible standing waves in a tube open at one end: (a) the fundamental mode, (b) the first overtone, and (c) the second overtone.

produce a periodic oscillation of the pressure of the surrounding atmosphere, but it is a good approximation because the pressure oscillations of the surrounding atmosphere are much smaller than the oscillations within the tube. The open end is therefore a pressure node; as we saw at the beginning of the preceding section, minima of pressure are maxima of displacement, and the pressure node is therefore a displacement antinode.

*Wavelengths and eigenfrequencies for standing waves in a tube open at one end*

With these boundary conditions at the ends of the tube, the possible standing waves are those shown in Figure 17.10. If the length of the tube is $L$, then the wavelengths of these standing waves, or normal modes, are

$$\lambda_1 = 4L, \qquad \lambda_2 = \tfrac{4}{3}L, \qquad \lambda_3 = \tfrac{4}{5}L, \ldots \tag{4}$$

and the corresponding eigenfrequencies are

$$\nu_1 = \frac{v}{4L}, \qquad \nu_2 = \frac{3v}{4L}, \qquad \nu_3 = \frac{5v}{4L}, \ldots \tag{5}$$

or, in general,

$$\nu_n = n\frac{v}{4L} \qquad n = 1, 3, 5, \ldots \tag{6}$$

Note that the expressions (4) for the wavelengths of the normal modes of a tube differ from the expressions (16.58) for the wavelengths of the normal modes of a string. This is due to the difference in boundary conditions: the tube has a node at one end and an antinode at the other end, whereas the string has nodes at both ends.

The speed of sound can be determined by measuring the resonant frequency of a tube of known length. This method is convenient but not entirely accurate, because the walls of the pipe exert a certain amount of friction on the air, which slightly affects the speed of sound waves.

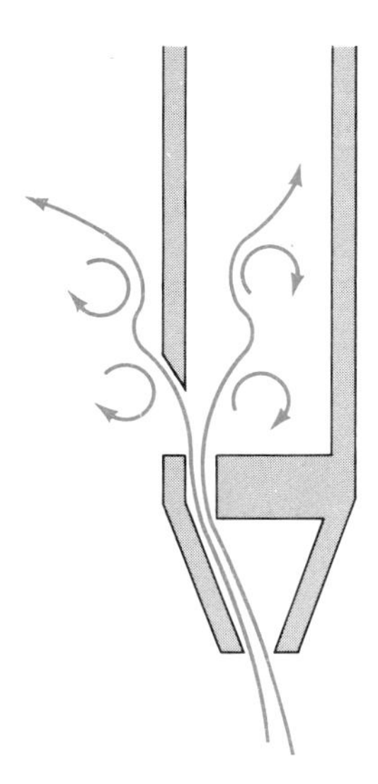

**Fig. 17.11** Vortices at the blowhole of an organ pipe.

Most musical instruments involve standing waves in cavities of diverse shapes. Organs, flutes, trumpets, and other wind instruments are essentially tubes within which standing waves are excited by a stream of air blown across a blowhole. That a steady stream of air can excite vibrations depends on a subtle phenomenon occurring at the edges of the blowhole. As the air streams past the edge, it forms a vortex (Figure 17.11); this vortex soon breaks away from the edge and is replaced by another vortex, and another, and so on. The regular succession of vortices constitutes a vibration of the stream of air, and this can excite the vibration of a standing wave by resonance.

The resonant frequency of the tube is determined by its length. In many wind instruments — flutes, trumpets, bassoons — the effective length of the tube can be varied by opening or closing valves.

Stringed instruments — violins, guitars, mandolins — use a resonant cavity to amplify and modify the sound of the string. The cavity is mechanically coupled to the string, and the vibrations of the latter excite resonant vibrations of the former. The resonant vibrations involve not only standing waves in the air in the cavity, but also standing waves in the solid material (wood) of the walls. Because the area of the body of, say, a violin is much larger than the area of its strings, the body pushes against much more air and radiates sound more efficiently than the strings. Hence most of the sound from a violin emerges from its body.

**Christian Doppler,** *1803–1853, Austrian physicist, professor at Vienna. After Doppler discovered his formula for the frequency shift of sound, he recognized that light from a moving source should also be subject to a frequency shift, and he speculated that this might explain the visible differences in the colors of the nearby stars. This explanation proved untenable—most of the color differences among nearby stars are intrinsic, and the Doppler shifts are too slight to be perceived by eye. However, the light from all distant galaxies displays substantial Doppler shifts toward low frequencies (red shifts), which arise from the general motion of expansion of the universe (see Interlude II).*

## 17.3 The Doppler Effect

Under standard conditions, the speed of a sound wave is 331 m/s when measured in the rest frame of the air. But when measured in a reference frame moving through the air, the speed of the sound wave will be larger or smaller, depending on the direction of motion of the reference frame. For example, if a train moving at 30 m/s approaches a stationary siren (Figure 17.12a), the speed of the sound waves relative to the train will be 361 m/s; and if the train moves away from the siren (Figure 17.12b), the speed of the sound waves will be 301 m/s.

The motion of the train affects not only the speed of the sound waves, but also their frequency. For instance, if the train approaches the siren, it runs head on into the sound waves (Figure 17.12a), and hence encounters more wave fronts per unit time than if it were stationary; and if the train recedes from the siren, it runs with the sound waves (Figure 17.12b), and hence encounters fewer wave fronts. Consequently, a receiver on the train will detect a higher frequency when approaching the siren, and a lower frequency when receding. This frequency change due to motion of the receiver (or of the emitter; see below) is called the **Doppler shift.**

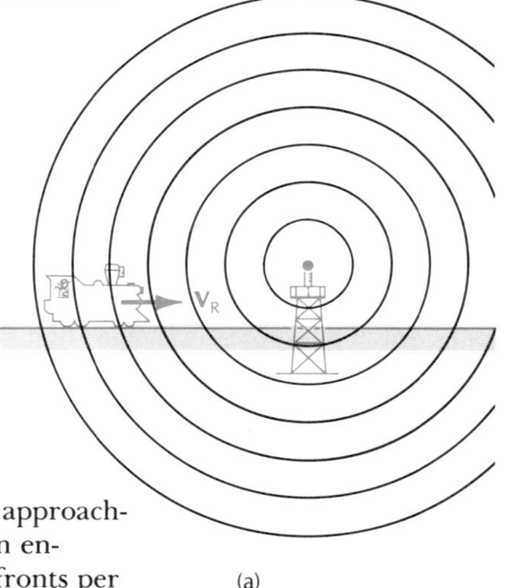


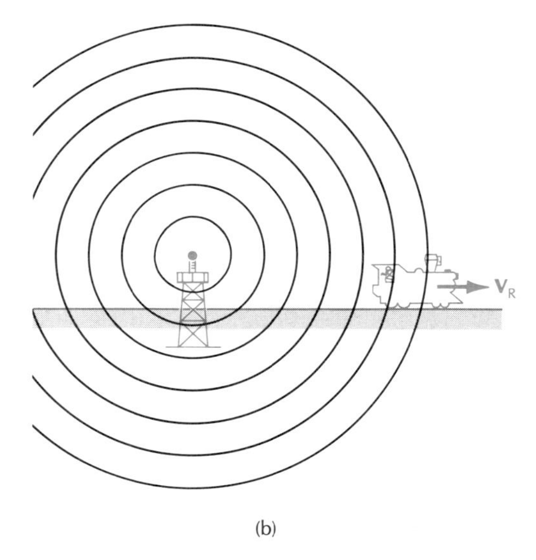


**Fig. 17.12** (a) Train approaching a siren. The train encounters *more* wave fronts per unit time than when stationary. (b) Train receding from a siren. The train encounters *fewer* wave fronts per unit time than when stationary.

To calculate the frequency shift, we note that in the reference frame of the ground we have the usual relation between the frequency, speed, and wavelength of sound:

$$\nu = v/\lambda \qquad (7)$$

and in the reference frame of the train we have a corresponding relation:

$$\nu' = v'/\lambda \tag{8}$$

The wavelengths in Eqs. (7) and (8) are exactly the same because the distance between wave crests does not depend on the reference frame. Dividing Eq. (8) by Eq. (7), we obtain

$$\frac{\nu'}{\nu} = \frac{v'}{v} \tag{9}$$

We will designate by $V_R$ the speed of the train acting as receiver of sound waves. In the reference frame of the train, the speed of sound is $v' = v \pm V_R$, where the positive sign corresponds to motion of the train toward the source of sound and the negative sign to motion away from the source of sound. Consequently, Eq. (9) yields

$$\nu' = \nu\left(\frac{v \pm V_R}{v}\right)$$

or

$$\boxed{\nu' = \nu\left(1 \pm \frac{V_R}{v}\right)} \qquad \begin{array}{l} + \text{ for approaching receiver} \\ - \text{ for receding receiver} \end{array} \tag{10}$$

*Frequency at moving receiver*

The fractional difference between the frequencies in the two reference frames is then

$$\boxed{\frac{\Delta\nu}{\nu} = \frac{\nu' - \nu}{\nu} = \pm\frac{V_R}{v}} \tag{11}$$

---

EXAMPLE 2. Suppose that the stationary siren emits a tone of a frequency 440 Hz as the train approaches it at 30 m/s. What is the frequency received on the train?

SOLUTION: From Eq. (10)

$$\nu' = \nu\left(1 + \frac{V_R}{v}\right) = 440\ \text{Hz}\left(1 + \frac{30\ \text{m/s}}{331\ \text{m/s}}\right) = 480\ \text{Hz}$$

---

Note that if the receiver is moving away from the source at a speed $V_R = v$, then the frequency $\nu'$ is zero; this simply means that the receiver is moving exactly with the waves so no wave fronts catch up with it. If the receiver is moving away at a speed greater than $v$, then Eq. (10) gives a negative frequency; this means that the receiver overruns the wave fronts from behind. Of course, Eq. (10) applies not only to sound waves, but also to water waves and other waves.

---

EXAMPLE 3. A motorboat speeding at 15 m/s is moving in the same direction as a group of water waves of frequency 0.17/s and speed 9.3 m/s (relative to the water). What is the frequency with which wave crests pound on the motorboat?

SOLUTION: We again use Eq. (10):

$$\nu' = \nu\left(1 - \frac{V_R}{v}\right) = \left(0.17/\text{s}\right)\left(1 - \frac{15 \text{ m/s}}{9.3 \text{ m/s}}\right)$$

$$= -0.10/\text{s}$$

The negative sign indicates that the motorboat overtakes the waves at the rate of 0.10/s, that is, one wave every 10 seconds.

A shift between the frequency emitted by a source of sound waves (or other waves) and the frequency detected by a receiver will also occur if the emitter is in motion and the receiver is at rest. For example, if a train approaching a station blows its whistle, the successive wave fronts emitted by the whistle are centered at regular intervals along the path of the whistle, and they will be crowded together in the forward direction and spaced apart in the rearward direction (Figure 17.13). Consequently, a stationary receiver will detect a higher frequency when in front of the train, and a lower frequency when in rear. As the train rushes by the stationary receiver, the detected frequency suddenly changes from high to low. This explains the sudden drop in pitch that you hear when standing next to the railroad track as a whistling train passes by you.

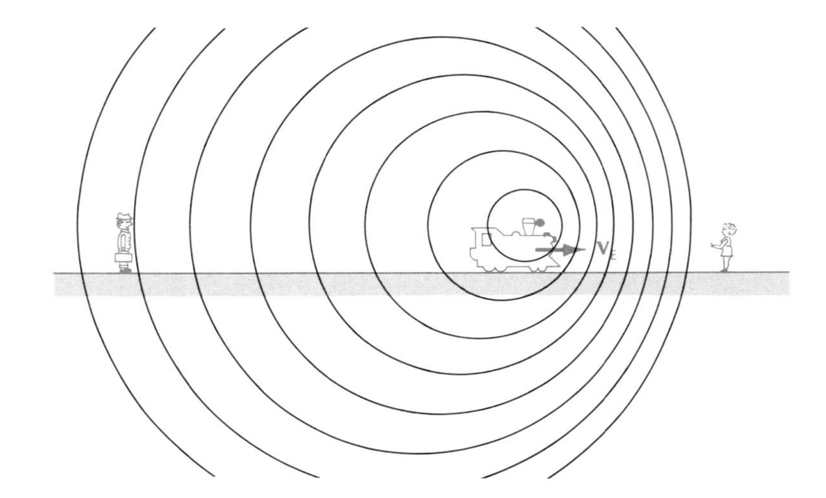


**Fig. 17.13** Train emitting sound waves while in motion. The wavelength ahead of the train is shorter, and that behind the train is longer than when the train is stationary.

We will designate by $V_E$ the speed of the train when it acts as an emitter of sound waves. To calculate the frequency change produced by the motion of the emitter, we begin by noting that in the time $1/\nu$ corresponding to one period, the train travels a distance $(1/\nu)V_E$ and hence the wavelength is shortened or lengthened from its normal value $\lambda$ to a new value $\lambda' = \lambda \mp V_E/\nu$, where the negative sign corresponds to the wavelength ahead of the train and the positive sign to the wavelength behind. The new frequency is therefore

$$\nu' = \frac{v}{\lambda'} = \frac{v}{\lambda \mp V_E/\nu} = \frac{v}{v/\nu \mp V_E/\nu} \tag{12}$$

or

$$\nu' = \nu\left(\frac{1}{1 \mp V_E/v}\right) \qquad \begin{array}{l} - \text{ for approaching emitter} \\ + \text{ for receding emitter} \end{array} \qquad (13)$$

***Frequency at receiver for moving emitter***

This gives the Doppler shift for a moving emitter.

EXAMPLE 4. Suppose that the whistle of a train emits a tone of a frequency 440 Hz as the train approaches a stationary observer at 30 m/s. What frequency does the observer hear?

SOLUTION: According to Eq. (13),

$$\nu' = \nu \frac{1}{1 - V_E/v} = \frac{440 \text{ Hz}}{1 - (30 \text{ m/s})/(331 \text{ m/s})}$$

$$= 484 \text{ Hz}$$

If we compare the results of Examples 2 and 4, we see that motion of the emitter and motion of the receiver have nearly the same effect on the frequency — in both examples the frequency is increased by about 10%. This symmetry of the Doppler shifts has to do with the low speed of the motion. Whenever $V_E$ is small compared to $v$, we can use the approximation

$$\frac{1}{1 - x} = 1 + x + \cdots \qquad (14)$$

in Eq. (13) to obtain the approximation

$$\frac{1}{1 \mp V_E/v} \cong 1 \pm V_E/v + \cdots \qquad (15)$$

As a consequence

$$\nu' \cong \nu(1 \pm V_E/v) \qquad (16)$$

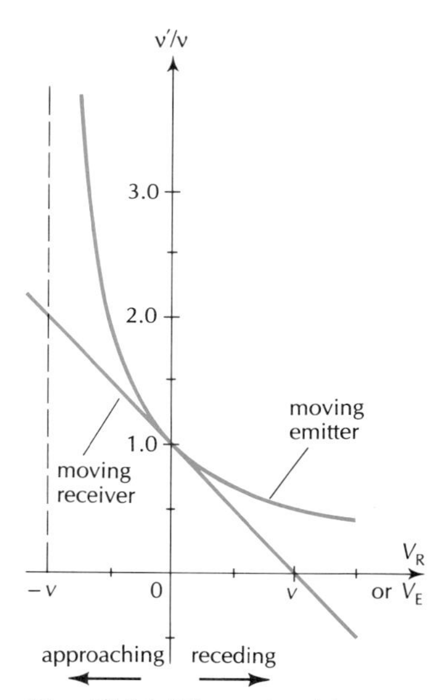


**Fig. 17.14** The ratio $\nu'/\nu$ as a function of velocity.

This equation has the same form as Eq. (10) and shows that, at low speeds, the frequency shift is the same whether the emitter or the receiver is moving. Figure 17.14 is a plot of $\nu'/\nu$ as a function of $V_R$ or $V_E$.

Finally, let us consider the case of an emitter, such as a fast aircraft, moving with a speed nearly equal to the speed of sound. If the aircraft emits sound of some frequency $\nu$, then Eq. (13) indicates that the frequency received at points just ahead of the airplane is very large — in the limiting case $V_E \to v$, the frequency becomes infinite. This is so because all the wave fronts are infinitely bunched together, and they all arrive at almost the same instant as the aircraft (Figure 17.15). If the speed of the aircraft exceeds the speed of sound, then the aircraft will overtake the wave fronts (Figure 17.16). In this case the sound is always confined to a conical region that has the aircraft as its apex and moves with the aircraft at the speed $V_E$; ahead of this region, the air has not yet been disturbed, although it will be disturbed when the air-

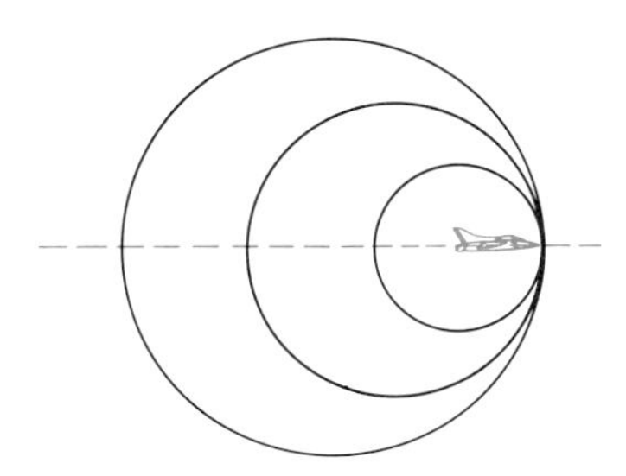

**Fig. 17.15** A subsonic aircraft of a speed very close to the speed of sound emitting sound waves.

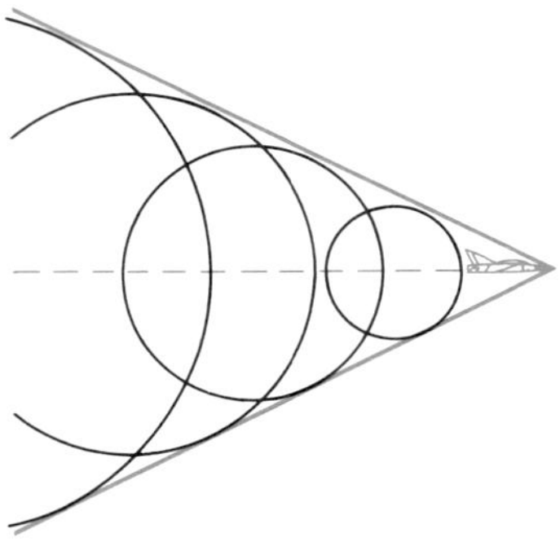

**Fig. 17.16** (left) A supersonic aircraft emitting sound waves.

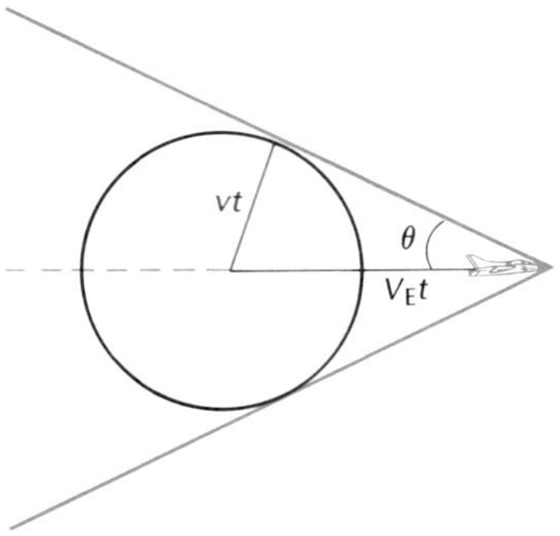

**Fig. 17.17** (right) Mach cone.

*Mach cone*

craft moves sufficiently far to the right. The cone is called the **Mach cone.**

The half angle of the apex of the Mach cone is given by

$$\sin\theta = v/V_E \tag{17}$$

This can readily be seen from Figure 17.17, which shows the aircraft at a time $t$ and the wave front that was emitted by the aircraft at time zero. In the time $t$, the sound wave has traveled a distance $vt$, while the aircraft has traveled a distance $V_E t$. Thus the radius of the wave front is $vt$. This radius is the opposite side of a right triangle of angle $\theta$ and of hypotenuse $V_E t$. Consequently,

$$\sin\theta = vt/V_E t \tag{18}$$

which is equivalent to Eq. (17).

Any supersonic aircraft or other body moving through air will generate a Mach cone, whether or not it carries an artificial source of sound aboard (Figure 17.18). The motion of the aircraft through the air creates a pressure disturbance which spreads outward with the speed of sound and forms the cone. The cone trails behind the aircraft much as the wake trails behind a ship. The sharp pressure disturbance at the surface of the cone is heard as a loud bang whenever the cone sweeps over the ear. This is the **sonic boom.** For a large aircraft, such as the Concorde SST, the noise level of the sonic boom reaches the pain threshold even if the aircraft is 20 km away.

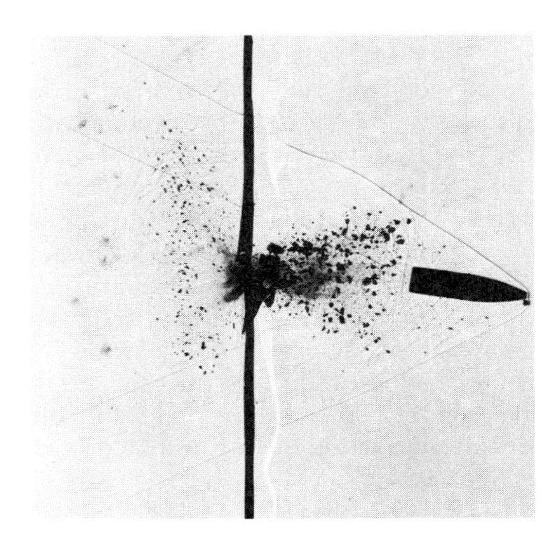

**Fig. 17.18** A .30-caliber bullet and its Mach cone. The bullet has just smashed through a plate of Plexiglas. Note the faint Mach cones of the many Plexiglas fragments. (Courtesy H. E. Edgerton, MIT.)

## 17.4 Water Waves

*Capillary waves and gravity waves*

Waves on the surface of water are of two kinds: **capillary waves** and **gravity waves.** The former are ripples of fairly short wavelength — no more than a few centimeters — and the restoring force that produces them is the surface tension of water. This is nothing but the attraction between neighboring water molecules. Surface tension tends to keep the surface of, say, a pond flat just as though a (weak) rubber sheet were stretched along the surface.

Gravity waves have wavelengths above half a meter; typically, their wavelengths range from several meters to several hundred meters. The restoring force that produces these waves is the pull of gravity, which tends to keep the water surface at its lowest level. In the following discussion we will deal exclusively with gravity waves.

The speed of these waves depends on their wavelength and also on the depth $h$ of the water. We will consider two extreme cases: deep water ($h \gg \lambda$) and shallow water ($h \ll \lambda$). If the water is *deep,* the speed is independent of the depth, but is dependent on the wavelength. It can be shown that the formula for the speed is[1]

*Speed of wave in deep water*

$$\boxed{v = \sqrt{g\lambda/2\pi}} \tag{19}$$

This formula is exact if the depth is infinite, and it is a good approximation whenever the depth is more than three wavelengths. The frequency of the wave is

$$\nu = v/\lambda = \sqrt{g/2\pi\lambda} \tag{20}$$

and the period is

$$T = \lambda/v = \sqrt{2\pi\lambda/g} \tag{21}$$

Note that this expression for the period resembles the expression for the period of a simple pendulum [$T = 2\pi\sqrt{l/g}$; see Eq. (15.57)]; the resemblance is no coincidence, since the motion of the particles in a water wave is essentially a swinging motion under the influence of gravity. Figure 17.19 is a plot of the speed of waves in deep water as a function of their wavelength.

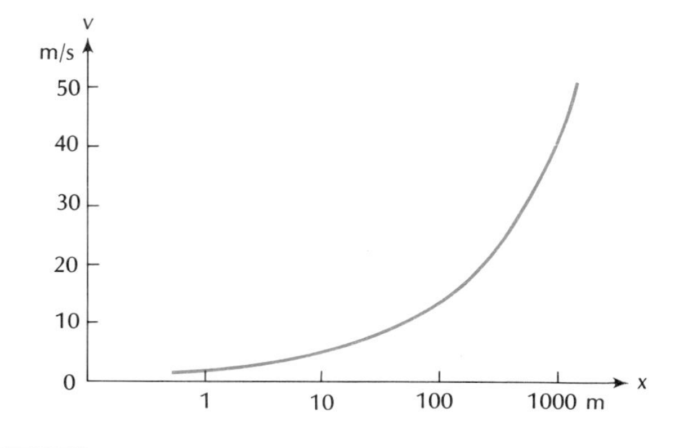


**Fig. 17.19** Speed of ocean waves in deep water.

[1] Actually, this is the phase velocity (see the discussion in Section 16.3). The value of the group velocity of these waves is half as large.

If the water is *shallow,* the speed is dependent on the depth, but is independent of the wavelength. The formula for the speed is[2]

*Speed of wave in shallow water*

$$v = \sqrt{gh} \tag{22}$$

This formula is a good approximation whenever the depth is less than $\frac{1}{10}$ of the wavelength.

In a water wave, whether in deep or in shallow water, the motion of the water particles is not only up and down but also back and forth. Figure 17.20 shows the orbits of a few water particles in a wave in deep water. The particles on the surface move in circles of a radius equal to the amplitude of the wave. The particles below the surface also move in circles, but of smaller radius. You can readily detect the longitudinal back-and-forth motion at the surface by watching small pieces of flotsam, such as chips of wood or styrofoam; as a wave crest passes by, the flotsam will first surge forward, and then back.

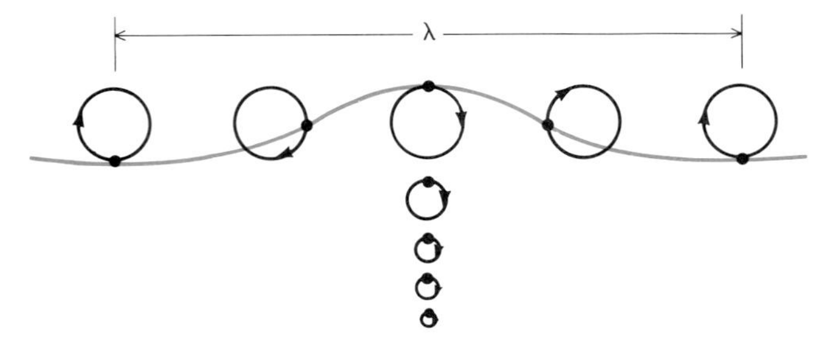


**Fig. 17.20** Motion of water particles in an ocean wave in deep water.

The shape of the water surface is approximately sinusoidal. This is a good approximation for waves of small amplitude (compared with their wavelength). Waves of larger amplitude tend to form sharper peaks and flatter troughs (Figure 17.21). If the amplitude is excessively large, then the crest tends to break (Figure 17.22); empirically, one finds that a wave breaks when its height (measured from trough to crest) is more than one-seventh of its length.

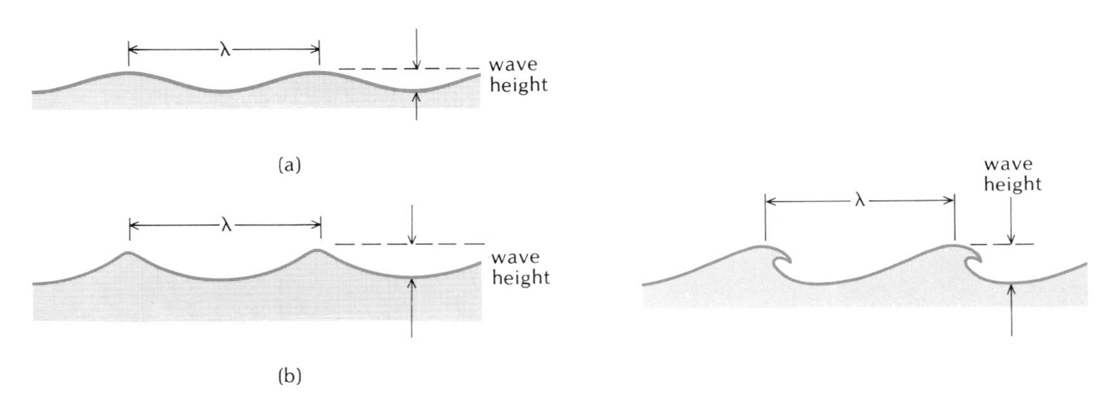


**Fig. 17.21** (a) Ocean wave of small amplitude. (b) Ocean wave of large amplitude.

**Fig. 17.22** Breaking wave crests.

[2] This is the phase velocity and it is also the group velocity. For these waves, the two velocities coincide.

The waves on the ocean are generated by the wind. Turbulence and eddies in the wind play a large role in this, giving the water repeated pushes that build up waves. The typical amplitudes and wavelengths that will be generated depend on the speed of the wind, how long it blows, and over how large an area. Oceanographers can forecast the sea conditions from the wind conditions. Strong winds blowing for a long time over a large area will generate gigantic waves. In one recorded incident, the U.S.S. *Ramapo* encountered a 34-m wave (from trough to crest) during a 7-day storm in the North Pacific.

Waves of extremely long wavelength are produced by earthquakes of the ocean floor. The sudden motion of the ocean floor disturbs the water and generates waves on the surface, called **tidal waves** or **tsunamis.** The wavelengths of these waves are typically 100–400 km. This means that the wavelength is much larger than the depth of the water (the mean depth of the Pacific Ocean is 4.3 km). Under these conditions the water of the ocean must be regarded as shallow; the formula of Eq. (22) for the speed is then applicable. For tsunamis in the Pacific Ocean, this gives a speed of

$$v = \sqrt{9.8 \text{ m/s}^2 \times 4.3 \times 10^3 \text{ m}} = 205 \text{ m/s}$$

$$= 740 \text{ km/h!}$$

In the open sea, tsunamis have a height of only a meter or less; they pass under ships without being noticed. But when they run into the shallow water of the continental shelf, the leading waves slow down [see Eq. (22)] — this causes the waves to pile up, and their height then reaches tens of meters. After the great Alaskan earthquake of 1964, a tidal wave 70 m high appeared at Valdez on the Gulf of Alaska. Figure 17.23 shows a tsunami, about 5 m high, reaching the shore at Hilo, Hawaii. Tsunamis easily travel from one end of the Pacific basin to the other. For example, in 1960 a tsunami originating in southern Chile damaged harbors as far away as Mexico and California (Figure 17.24).

**Fig. 17.23** Tsunami striking the beach near the Putumaile Hospital at Hilo, Hawaii, on April 1, 1946.

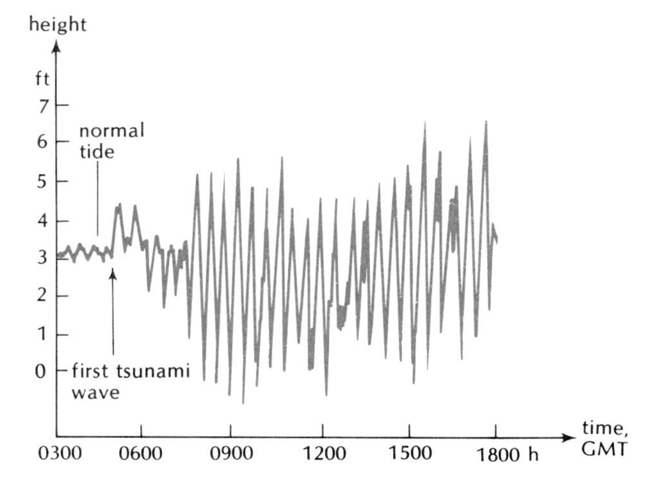


**Fig. 17.24** Height of water as a function of time at Acapulco, Mexico, May 22, 1960. The arrow indicates the arrival of a tsunami from southern Chile.

## 17.5 Seismic Waves

Seismic waves are propagating vibrations in the body of the Earth. Where these vibrations reach the surface of the Earth, they are felt as a trembling of the ground, or an earthquake (Figure 17.25). The

**Fig. 17.25** Cracks in the pavement on Union Street in San Francisco after the great earthquake of 1906.

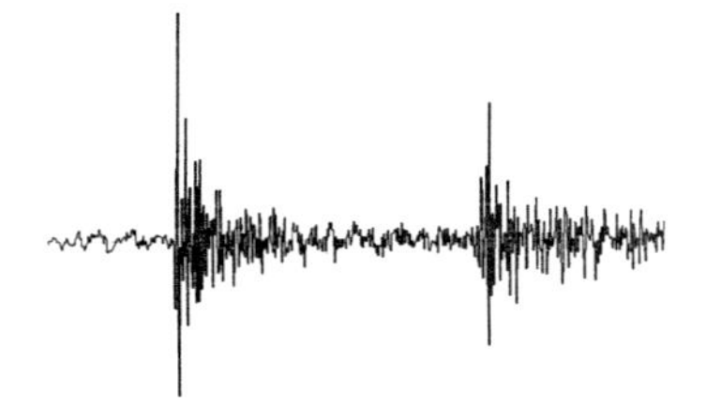

**Fig. 17.26** Seismic wave recorded by a seismometer. The sudden increase in the amplitude of the wriggles indicates the arrival of the wave

propagation of seismic waves in earth involves a mechanism similar to the propagation of sound waves in the air — the restoring force is provided by the elasticity of earth. However, the periods of seismic waves are usually longer than $\frac{1}{10}$ s, that is, the periods are longer than those of audible sound. The amplitudes of seismic waves range from as small as $10^{-6}$ m to as large as several meters (Figure 17.26).

The initial disturbance that generates the seismic wave is a sudden rupture within the solid crust of the Earth. The place of rupture may be several kilometers or several hundred kilometers below the surface. The point on the surface directly above the rupture is called the **epicenter** of the earthquake. The waves that spread outward from the initial disturbance are of two kinds: transverse waves and longitudinal waves. Seismologists designate these as **S waves** and **P waves.** The former (*S*hear waves) can only propagate through the solid crust and mantle of the Earth, whereas the latter (*P*ressure waves) can also penetrate through the liquid core. When these waves reach the surface of the Earth or the core of the Earth, they are reflected, that is, they bounce off at an angle (Figure 17.27). During this reflection the S and P waves can generate yet another kind of wave: a **surface wave.** This wave propagates along the surface of the Earth or along the surface of the core of the Earth; it is similar to a water wave propagating along the surface of the ocean.

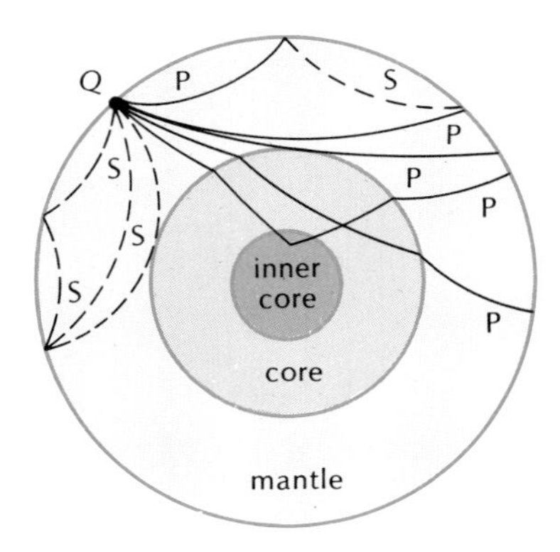


**Fig. 17.27** Propagation of S and P waves within the Earth. The waves originate at point *Q* near the surface of the Earth. The S waves cannot propagate in the (liquid) core.

The total energy of the seismic waves generated during a large earthquake may be as much as $10^{17}$ or $10^{18}$ J. The energy is usually expressed by the **Richter magnitude scale,** according to which the magnitude $M$ of the earthquake is

$$M = 0.67 \log E - 2.9 \tag{23}$$

where $E$ is the energy in joules. For example, an earthquake with an energy of $10^{17}$ J has a Richter magnitude $M = 0.67 \times 17 - 2.9 = 8.5$, which was roughly the magnitude of the great San Francisco earthquake of 1906.

The speed of seismic waves depends on the characteristics of the material within the Earth. P waves have speeds of about 5 km/s in the crust and speeds as great as 14 km/s in the deepest part of the mantle; S waves have speeds of 3–8 km/s. The difference in speed between P and S waves implies that the P waves reach the surface ahead of the S waves. The time lag between these waves can be used to calculate the distance they have traveled.

Careful analysis of the propagation of seismic waves reveals the characteristics of the material through which they have traveled. What we know about the interior of the Earth is largely due to the analysis of seismic waves. For instance, we know that the Earth has a liquid core, because S waves (transverse waves) do not propagate through this core (see Figure 17.27). In recent years the study of seismic waves has acquired military significance. Underground explosions of nuclear bombs generate seismic waves that can be detected by sensitive seismometers over large distances. This permits us to monitor underground nuclear explosions all around the globe.

## 17.6 Diffraction

**Fig. 17.28** Ocean waves incident on breakwater.

It is a characteristic feature of waves that they will deflect around the edges of obstacles placed in their path and penetrate into the "shadow" zone behind the obstacle. For example, Figure 17.28 shows water waves incident on a breakwater at the entrance to a harbor. The region directly behind the breakwater is outside of the direct path of the waves, but nevertheless waves reach this region because each wave front spreads sideways once it has passed the entrance to the harbor. This lateral spreading of the wave fronts can be easily understood: The breakwater cuts a segment out of each wave front, and the segments of wave front cannot just keep moving straight on as though nothing had happened — the end of the segment is a vertical wall of water where the breakwater has chopped off the wave front. The water at the ends will immediately begin to spill out sideways, producing a disturbance at the edge of the segment. This disturbance continues to spread and gradually forms the curved wave fronts to the left and to the right of the main beam of the wave.

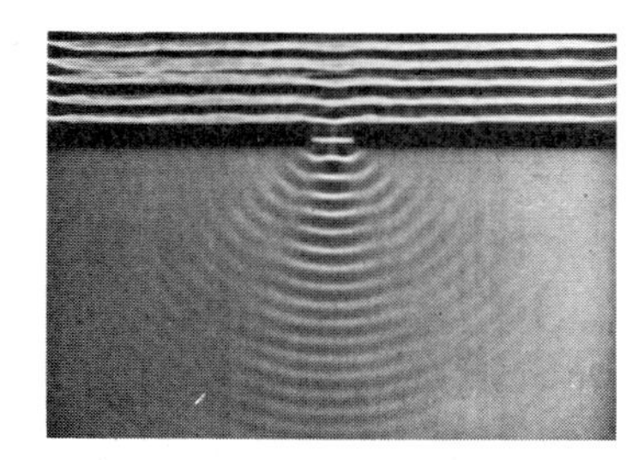

**Fig. 17.29** Water waves of fairly long wavelength in a ripple tank exhibit strong diffraction when passing through a gap.

Such a deflection of waves by the edge of an obstacle is called **diffraction.** It is a general rule that the amount of diffraction suffered by a wave passing through a gap depends on the ratio of wavelength to the size of the gap. An increase of the wavelength (or a decrease of the gap size) makes the diffraction effects more pronounced; a decrease of the wavelength (or an increase of the gap size) makes the diffraction effects less pronounced. For instance, Figure 17.29 shows diffraction of waves of relatively long wavelength by a small gap; the waves spread out very strongly, forming divergent, fanlike beams of concentric wave fronts. Figure 17.30 shows diffraction of waves of shorter wavelength; here the wave spreads out only slightly; most of the wave remains within a straight beam of nearly parallel wave fronts. Note that in Figure 17.30 the beam has a fairly well-defined edge — the region in the "shadow" of the breakwater remains nearly undisturbed, while the region facing the gap receives the full impact of the waves.

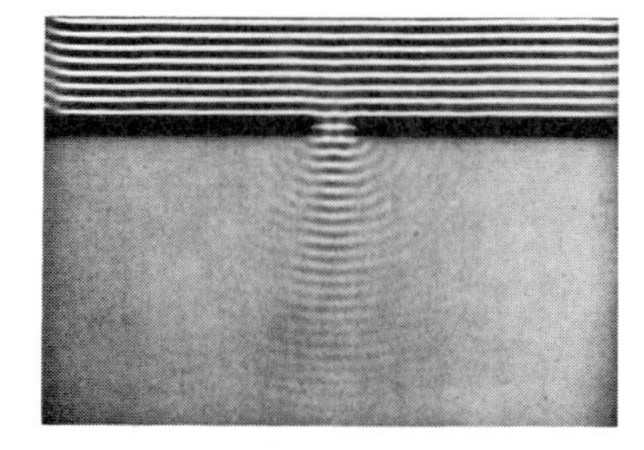

**Fig. 17.30** Water waves of shorter wavelength exhibit less diffraction.

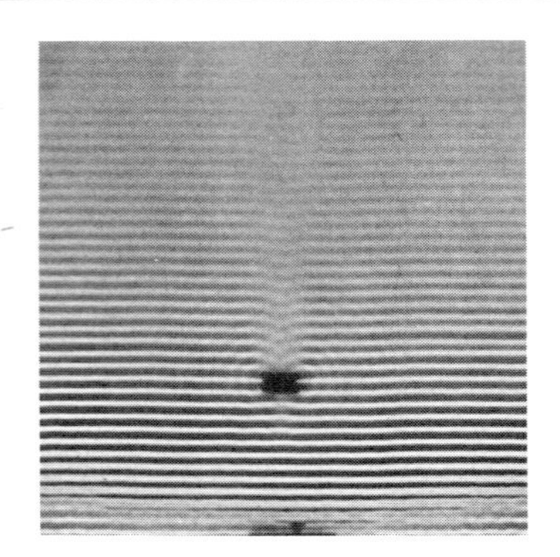

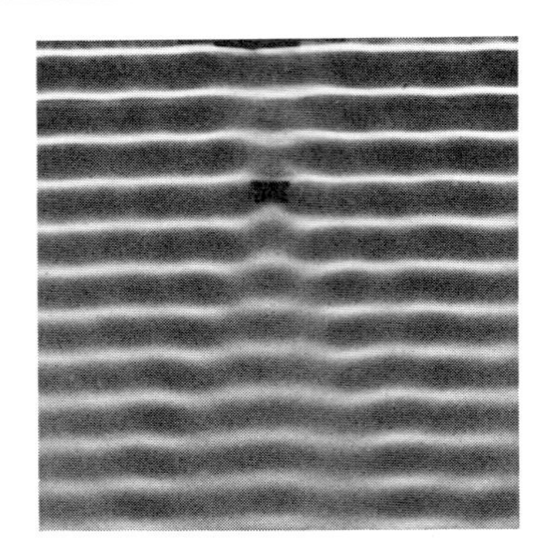

**Fig. 17.31** (left) Diffraction pattern generated by a small island; short wavelength.

**Fig. 17.32** (right) Diffraction pattern generated by a small island; long wavelength.

The fanlike beams of waves spreading out from the gap constitute a **diffraction pattern.** In Figure 17.29, the diffraction pattern consists of a central beam and two clearly recognizable secondary beams on each side. The beams are separated by nodal lines along which the wave amplitude is zero. Figures 17.31 and 17.32 show the diffraction patterns generated by a small island. Note that if the wavelength is large compared to the size of the island, then there exists no shadow zone; instead, the island merely produces some distortion of the waves.

The diffraction pattern generated by some obstacle depends in a complicated way on the shape of the obstacle. In Chapter 40 we will develop some mathematical formulas for the intensity of the beams of diffracted waves. Although the main concern of Chapter 40 is the behavior of light waves, the same mathematical treatment applies to other kinds of waves. Incidentally: One of the crucial experiments that convinced physicists early in the nineteenth century that light is a wave was a diffraction experiment. Augustin Fresnel, following up the earlier investigations of Thomas Young, showed that light exhibits diffraction effects quite similar to those of any other kind of wave.

Diffraction also plays an important role in the propagation of sound. We can listen to, and talk to, a person standing out of our line of sight beyond the corner of, say, a house because the sound waves diffract around the corner.

## SUMMARY

**Decrease of intensity of sound wave with distance:**

$$I_2 = I_1 \frac{r_1^2}{r_2^2}$$

**Speed of sound in air:** 331 m/s

**Standing waves in tube open at one end:**

$$\lambda_1 = 4L, \qquad \tfrac{4}{3}L, \qquad \tfrac{4}{5}L, \ldots$$

**Doppler shift:**

stationary emitter: $\nu' = \nu(1 \pm V_R/v)$

stationary receiver: $\nu' = \dfrac{\nu}{1 \mp V_E/v}$

**Doppler shift for small speed:** $\dfrac{\Delta\nu}{\nu} = \pm\dfrac{V_R}{v}$ or $\pm\dfrac{V_E}{v}$

**Mach cone:** $\sin\theta = v/V_E$

**Speed of water waves:**

$$\text{deep water:} \quad v = \sqrt{g\lambda/2\pi}$$

$$\text{shallow water:} \quad v = \sqrt{gh}$$

## QUESTIONS

1. Could an astronaut be heard playing the violin while standing on the surface of the Moon?

2. You can estimate your distance from a bolt of lightning by counting the seconds between seeing the flash and hearing the thunder, and then dividing by three to obtain the distance in kilometers. Explain this rule.

3. A hobo can hear a very distant train by placing an ear against the rail. How does this help? (Hint: Ignoring frictional losses, how does the intensity of sound decrease with distance in air? In the rail?)

4. If you speak while standing in a corner with your face toward the wall, you will sometimes notice that your voice sounds unusually loud. Explain.

5. What happens to the frequency of musical notes if you play a $33\frac{1}{3}$ r.p.m. record at 45 r.p.m.?

6. Why does the wind whistle in the rigging of a ship or in the branches of a tree?

7. How does a flutist play different musical notes on a flute?

8. When inside a boat, you can often hear the engine noises of another boat much more loudly than when on deck. Can you guess why?

9. Many men like singing in shower stalls because the stall somehow enhances their voice. How does this happen? Would the effect be different for men and women?

10. According to a novel proposal for the reduction of engine noise inside aircraft cabins, loudspeakers installed along each side of the cabin are to cancel the noise by "antinoise," that is, sound waves of an amplitude exactly opposite to that of the noise (see Figure 17.33). The loudspeakers would be controlled by sensors and electronic circuits that detect the arriving engine noise and continuously adjust the amplitude and the phase of the required antinoise. Can such a noise cancellation system eliminate the noise throughout the cabin? In what part of the cabin would it be most effective? What happens to the energy in the arriving sound waves?

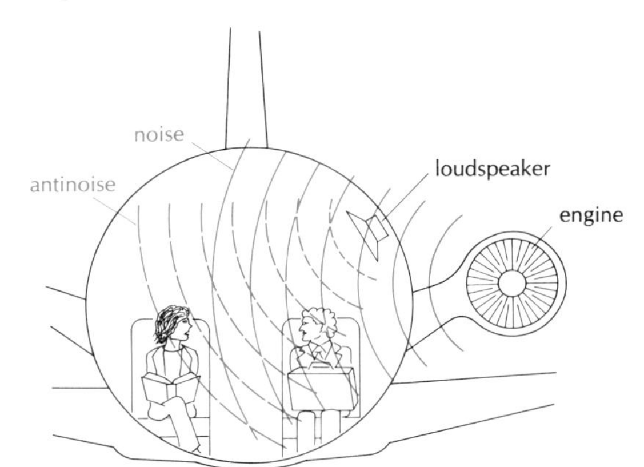


**Fig. 17.33** A system for the reduction of engine noise in an aircraft cabin.

11. The pitch of the vowels produced by the human voice is determined by the frequency of standing waves in several resonant cavities (larynx, pharynx, mouth, and nose). In an amusing demonstration experiment, a volunteer inhales helium gas and then speaks a few words. As long as his resonant cavities are filled with helium gas, the pitch of his voice will be much higher than normal. Given that the speed of sound in helium is about three times as large as in air, calculate the factor by which the eigenfrequencies of his resonant cavities will be higher than normal.

12. The pipes that produce the lowest frequencies in a great organ are very long, usually 16 ft. Why must they be long? These pipes are also very thick. Why would a thin pipe give a poor performance?

13. Some of the old European opera houses and concert halls renowned for their acoustic excellence have very irregular walls, heavily encrusted with an abundance of stucco ornamentation which reflects sound waves in almost all directions. How does the sound reaching a listener in such a hall differ from the sound reaching a listener in a modern concert hall with four flat, plain walls?

14. Electric guitars amplify the sound of the strings electronically. Do such guitars need a body?

15. Does the temperature of the air affect the pitch of a flute? A guitar?

16. The human auditory system is very sensitive to small differences between the arrival times of a sound signal at the right and left ears. Explain how this permits us to perceive the direction from which a sound signal arrives.

17. The depth finder (or "fish finder") on a boat sends a pulse of sound toward the bottom and measures the time an echo takes to return. The dial of the depth finder displays this echo time on a scale directly calibrated in distance units. Experienced operators can tell whether the bottom is clean rock, or rock covered by a layer of mud, or whether a school of fish is swimming somewhere above the bottom. What echo times would you expect to see displayed on the dial of the depth finder in each of these instances?

18. The helmsman of a fast motorboat heading toward a cliff sounds his horn. A woman stands on the top of the cliff and listens. Compare the frequency of the horn, the frequency heard by the woman, and the frequency heard by the helmsman in the echo from the cliff. Which of these three is the highest frequency? Which the lowest?

19. Two automobiles are speeding in opposite directions while sounding their horns. Describe the changes of pitch that each driver hears as they pass by one another.

20. A man is standing north of a woman while a strong wind is blowing from the south. If the man and the woman yell at each other, how does the wind affect the pitch of the voice of each as heard by the other?

21. A Concorde SST passing overhead at an altitude of 20 km produces a sonic boom with an intensity level of 120 dB lasting about half a second. How does this compare with some other loud noises? Would it be acceptable to let this aircraft make regular flights over populated areas?

22. Many people have reported seeing UFOs traveling through air noiselessly at speeds much greater than the speed of sound. If the UFO consisted of a solid impenetrable body, would you expect its motion to produce a sonic boom? What can you conclude from the absence of sonic booms?

23. When an ocean wave approaches a beach, its height increases. Why?

24. Occasionally ocean waves passing by a harbor entrance will excite very high standing waves ("seiche") within the harbor. Under what conditions will this happen?

25. Seismic waves of the S and P types have different speeds. Explain how a scientist at a seismometer station can take advantage of this difference in speed to determine the distance between his station and the point of origin of the waves.

26. The amplitude of an ocean wave initially decreases as the wave travels outward from its point of origin; but when the wave has traveled a quarter of the distance around the Earth, its amplitude *increases.* Explain how this comes about. (Hint: If the wave were to travel half the distance around the Earth, it would converge on a point, if no continents block its progress.)

27. Underground nuclear explosions generate seismic waves. How could you discriminate between the seismic waves received from such an explosion and the seismic waves from an earthquake? (Hint: Would you expect an explosion to produce mainly S waves or mainly P waves?)

28. You are standing on the surface of the Earth; a seismic wave approaches from below. If the wave is an S wave, how will it shake you? If the wave is a P wave?

29. Why did the Japanese use to build their houses out of wood slats and paper?

30. Figure 17.34 shows a seismometer, an instrument used to detect and measure seismic waves. A vertical post is firmly set in the ground and a large mass is suspended from it by a rigid horizontal beam and a diagonal wire. The beam ends in a sharp point that rests against the post; the beam is therefore free to swing in the horizontal plane. Describe how the beam will swing if the ground moves and tilts the post. For what direction of motion of the post is this seismometer most sensitive?

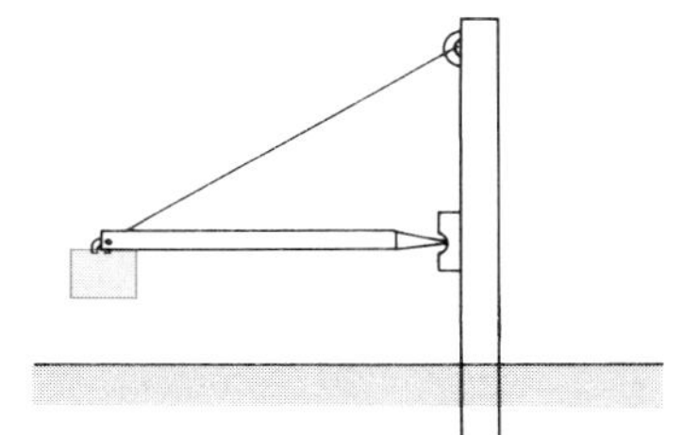

**Fig. 17.34** Seismometer.

31. If you are standing on the south side of a house, you can speak to a friend standing on the east side, out of sight around the corner. How do your sound waves reach into the shadow zone?

## PROBLEMS[3]

### Section 17.1

1. The range of frequencies audible to the human ear extends from 20 to 20,000 Hz. What is the corresponding range of wavelengths?

2. The lowest musical note available on a piano is A, three octaves below that listed in Table 17.2; and the highest note available is C, four octaves above that listed in Table 17.2. What are the frequencies of these notes?

*3. A violin has four strings, each of them 0.326 m long. When vibrating in their fundamental modes, the four strings have frequencies of 196, 294, 440, and 659 Hz, respectively.

(a) What is the wavelength of the standing wave on each string? What is the wavelength of the sound wave generated by the string?
(b) What is the frequency and the wavelength of the first overtone on each string? What is the corresponding wavelength of the sound wave generated by each string?
(c) According to Table 17.2, to what musical tones do the frequencies calculated above correspond?

*4. A mandolin has strings 34.0 cm long fixed at their ends. When the mandolin player plucks one of these strings, exciting its fundamental mode, this

[3] In all the problems assume that the speed of sound in air is 331 m/s, unless otherwise stated.

string produces the musical note D (293.7 Hz; see Table 17.2). In order to produce other notes of the musical scale, the player shortens the string by holding a portion of the string against one or another of several frets (small transverse metal bars) placed underneath the string. The player shortens the string by one fret to produce the note D#, by two frets to produce the note E, by three frets to produce the note F, etc. Calculate the correct spacing between the successive frets of the mandolin for one complete octave. Assume that the string always vibrates in its fundamental mode and assume that the tension in the string is always the same.

5. The noise level in a quiet automobile is 50 dB. Find the sound intensity in $W/m^2$.

6. The intensity level of sound near a loud rock band is 120 dB. What is the intensity level of sound near two such rock bands playing together?

7. In a screaming contest, a Japanese woman achieved 115 dB. How many such women would have to scream at you to bring you to the threshold of pain?

8. Suppose that a whisper has an intensity level of 20 dB at a distance of 0.5 m from the speaker's mouth. At what distance will this whisper be below your threshold of hearing?

*9. A loudspeaker receives 8 W of electric power from an audio amplifier and converts 3% of this power into sound waves. Assuming that the loudspeaker radiates the sound uniformly over a hemisphere (a vertical and horizontal angular spread of 180°), what will be the intensity and the intensity level at a distance of 10 m in front of the loudspeaker?

*10. An old-fashioned hearing trumpet has the shape of a flared funnel, with a diameter of 8 cm at its wide end and a diameter of 0.7 cm at its narrow end. Suppose that all of the sound energy that reaches the wide end is funneled into the narrow end. By what factor does this hearing trumpet increase the intensity of sound (measured in $W/m^2$)? By how many decibels does it increase the intensity level of sound?

### Section 17.2

11. Spectators at soccer matches often notice that they hear the sound of the impact of the ball on the player's foot (or head) sometime after seeing this impact. If a spectator notices that the delay time is about 0.5 s, how far is he from the player?

12. In the past century a signal gun was fired at noon at most harbors so that the navigators of the ships at anchor could set their chronometers. This method is somewhat inaccurate, because the sound signal takes some time to travel the distance from gun to ship. If this distance is 3.0 km, how long does the signal take to reach the ship? Can you suggest a better method for signaling noon?

13. In fresh water, sound travels at a speed of 1460 m/s. In air, sound travels at a speed of 331 m/s. Suppose that an explosive charge explodes on the surface of a lake. A woman with her head in the water hears the bang of the explosion and, lifting her head out of the water, she hears the bang again 5 s later. How far is she from the site of the explosion?

14. In order to measure the depth of a ravine, a physicist standing on a bridge drops a stone and counts the seconds between the instant he releases the stone and the instant he hears it strike some rocks at the bottom. If this time interval is 6.0 s, how deep is the ravine? Take into account the travel time of the sound signal, but ignore air friction.

15. A bat can sense its distance from the wall of a cave (or whatever) by emitting a sharp ultrasonic pulse that is reflected by the wall. The bat can tell the distance from the time the echo takes to return.

(a) If a bat is to determine the distance of a wall 10 m away with an error of less than $\pm 0.5$ m, how accurately must it sense the time interval between emission and return of the pulse?

(b) Suppose that a bat flies into a cave filled with methane (swamp gas). By what factor will this gas distort the bat's perception of distances? The speed of sound in methane is 432 m/s.

16. The ultrasonic range finder on a new automatic camera sends a pulse of sound to the target and determines the distance by the time an echo takes to return.

(a) If the range finder is to determine a distance of 50 cm with an error no larger than $\pm 2$ cm, how accurately (in seconds) must it measure the travel time?

(b) If you aim this camera at an object placed beyond a sheet of glass (a window or a glass door), on what will the camera focus?

17. The commonly accepted value for the speed of sound in dry air under standard conditions is 331.45 m/s. However, a scientist at the National Research Council of Canada recently discovered an error in the earlier determinations of the speed of sound, and he concluded that the correct value for the speed of sound is 331.29 m/s. What is the percent difference between the old and the new values? According to Eq. (3), what percent change of the pressure or of the density of the air will produce an equal change in the speed of sound?

18. Estimate the wavelength and frequency of the sound waves made visible in Figure 17.2.

*19. The mass per unit length of a steel wire of diameter 1.3 mm is 0.010 kg/m and the yield strength, or maximum tension that the wire can withstand, is $3.6 \times 10^3$ N. Is it possible to apply enough tension to the wire so that the speed of a transverse wave on this wire exceeds the speed of sound in the steel of the wire, 5000 m/s?

*20. In the eighteenth century, members of the French Academy organized the first careful measurement of the speed of sound in air. To compensate for wind speed, they adopted a reciprocal method. Cannons were fired alternately at Montmartre and Montlhéry, 29.0 km apart. Observers at each station measured the time delay between the muzzle flash seen at the *other* station and the arrival of the sound. Show that from the measurements of these travel times $t_1$ and $t_2$ of sound in both directions, the speed of sound in still air can be calculated as follows, independently of the speed of the wind blowing from one station to the other:

$$v = \frac{d}{2}\left(\frac{1}{t_1} + \frac{1}{t_2}\right)$$

where $d$ is the distance between the stations. Evaluate numerically for the measured travel times of 87.4 s and 84.8 s.

*21. Because the human auditory system is very sensitive to small differences between the arrival times of a sound signal at the right and the left ear, we can perceive the direction from which a sound signal arrives to within about 5°. Suppose that a source of sound (a ringing bell) is 10 m in front of and 5° to the left of a listener. What is the difference in the arrival times of sound signals at the left and right ears? The separation between the ears is about 15 cm.

*22. Consider a tube of length $L$ open at both ends. Show that the eigenfrequencies of standing sound waves in this tube are

$$\nu_n = n\frac{v}{2L} \qquad n = 1, 2, 3, \ldots$$

Draw diagrams similar to those of Figure 17.10 showing the displacement amplitude for each of the first four standing waves.

*23. The largest pipes in a great organ usually have a length of about 16 ft (4.8 m). These pipes are open at both ends so that a standing sound wave will have a displacement antinode at each end. What is the frequency of the fundamental mode of such a pipe?

*24. A flute can be regarded as a tube open at both ends. It will emit a musical note if the flutist excites a standing wave in the air column in the tube.

(a) The lowest musical note that can be played on a flute is C (261.7 Hz; see Table 17.2). What must be the length of the tube? Assume that the air column is vibrating in its fundamental mode (see Problem 22).

(b) In order to produce higher musical notes, the flutist opens valves arranged along the side of the tube. Since the holes in these valves are large, an open valve has the same effect as shortening the tube. The flutist opens one valve to play C#, two valves to play D, etc. Calculate the successive spacings between the valves of a flute for one complete octave. (The actual spacings used on flutes differ slightly from the results of this simple theoretical evaluation because the mouth cavity of the flutist also resonates and affects the frequency.)

25. The human ear canal is approximately 2.7 cm long. The canal can be regarded as a tube open at one end and closed at the other. What are the eigenfrequencies of standing waves in this tube? The ear is most sensitive at a frequency of about 3000 Hz. Would you expect that resonance plays a role in this?

*26. Consider a tube of length $L$ closed at both ends. Show that the eigenfrequencies of standing sound waves in this tube are

$$\nu_n = n\frac{v}{2L} \qquad n = 1, 2, 3, \ldots$$

Draw diagrams similar to those of Figures 17.10 showing the displacement amplitude for each of the first four standing waves.

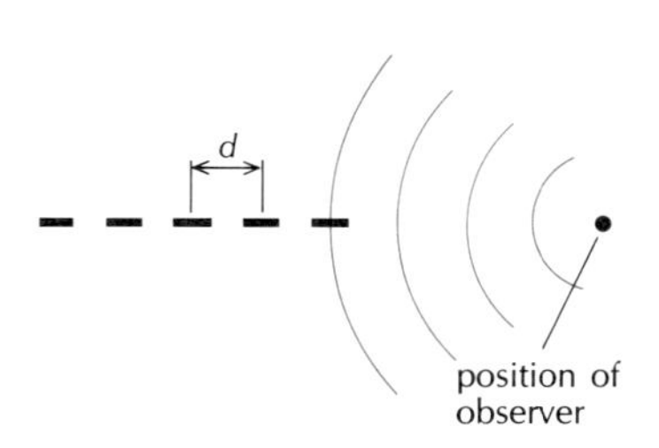


**Fig. 17.35**

*27. If you stand in the vicinity of a picket fence and clap your hands, you will notice that the sound waves reflected by the fence and reaching your ear are strongly reinforced at a selected wavelength, that is, the picket fence seems to ring with a musical tone. This selective reinforcement of sound waves occurs whenever the wavelength is such that waves reflected by different boards in the fence arrive at your ear in phase, giving constructive interference. Suppose that the picket fence consists of boards separated by a distance $d$ and you stand in line with this fence (see Figure 17.35). Show that in this case the condition for constructive interference for waves traveling from you to the boards and back to you is that the wavelength be equal to $2d$.

**Section 17.3**

28. The horn of a stationary automobile emits a sound wave of 580 Hz. What frequency will you hear if you are driving toward this automobile at 80 km/h?

29. In an experiment performed shortly after Doppler proposed his theoretical formula for the frequency shift, several trumpeters were placed on a train and told to play a steady musical tone. As the train sped by, listeners standing on the side of the track judged the pitch of the tone received from the trumpets. Suppose that the train had a speed of 60 km/h and that the trumpets on the train sounded the note of E (329.7 Hz; see Table 17.2). What was the frequency of the note perceived by a listener on the ground when the train was approaching? When the train was receding? Approximately to what musical notes do these Doppler-shifted frequencies correspond?

30. Ocean waves with a wavelength of 100 m have a period of 8 s. A motorboat, with a speed of 9 m/s, heads directly into the waves. What is the speed of the waves relative to the motorboat? With what frequency do wave crests hit the front of the motorboat?

31. The horn of an automobile emits a tone of frequency 520 Hz. What frequency will a pedestrian hear when the automobile is approaching at a speed of 85 km/h? Receding at the same speed?

*32. Two automobiles are driving on the same road in opposite directions. The speed of the first automobile is 90.0 km/h and that of the second is 60.0 km/h. The horns of both automobiles emit tones of frequency 524 Hz. Calculate the frequency that the driver of each automobile hears coming from the other automobile. Assume that there is no wind blowing along the road.

*33. Repeat Problem 32 under the assumption that a wind of 40 km/h blows along the road in the same direction as that of the faster automobile.

*34. A train approaches a mountain at a speed of 75 km/h. The train's engineer sounds a whistle that emits a frequency of 420 Hz. What will be the frequency of the echo that the engineer hears reflected off the mountain?

*35. Suppose that a moving train carries a source of sound and also a receiver of sound so that both have the same velocity relative to the air. Show that in this case the Doppler shift due to motion of the source cancels the Doppler shift due to motion of the receiver — the frequency detected by the receiver is the same as the frequency generated by the source.

*36. The whistle on a train generates a tone of 440 Hz as the train approaches a station at 30 m/s. A wind blows at 20 m/s in the same direction as the motion of the train. What is the frequency that an observer standing at the station will hear?

37. Figure 17.18 shows the shock wave of a bullet speeding through air. Measure the angle of the Mach cone and calculate the speed of the bullet.

*38. The Concorde SST has a cruising speed of 2160 km/h.

(a) What is the half angle of the Mach cone generated by this aircraft?

(b) If the aircraft passes directly over your head at an altitude of 12,000 m, how long after this instant will the shock wave strike you?

*39. You may have noticed that at the instant a fast (but subsonic) jet aircraft passes directly over your head, the sound it makes seems to come from a point behind the aircraft.

(a) Show that the direction from which the sound seems to come makes an angle $\theta$ with the vertical such that $\sin\theta = V_E/v$, where $V_E$ is the speed of the aircraft and $v$ the speed of sound.

(b) If you hear the sound from an angle of 30° behind the aircraft, what is the speed of the aircraft?

### Section 17.4

40. What is the speed and the frequency of waves in deep water if the wavelength is 10 m? If the wavelength is 100 m?

41. Show that the speed of water waves in deep water can be expressed as $v = gT/2\pi$, where $T$ is the period of the wave.

42. A giant, freak wave encountered by a weather ship in the North Atlantic was 23.5 m high from trough to crest; its wavelength was 350 m and its period 15.0 s. Calculate the maximum vertical acceleration of the ship as the wave passed underneath; calculate the maximum vertical velocity. Assume that the motion of the ship was purely vertical.

43. A rule of thumb known to naval architects is that the speed of a ship through the water cannot exceed the speed of an ocean wave of wavelength equal to the length of the hull of the ship.[4] This limiting speed is called the **hull speed** of the ship. The reason for this rule is that a ship passing through

[4] This rule does not apply to boats that skim along the surface of the water, relying on hydrodynamic lift to keep most of their hull out of the water.

the water makes waves; when the speed of the ship reaches the hull speed, the crest of one of these waves will be at the bow of the ship, and the crest of the next at the stern. The ship is then permanently trapped in a wave trough — any increase of engine power (or sail power) will merely push the ship against the wave crest at its bow; this will feed more energy into the wave, making it higher, but produce only an insignificant increase in the speed of the ship.

(a) Show that the hull speed (in meters per second) is given by the formula $v = 1.2\sqrt{L}$, where $L$ is the length of the ship (in meters). Naval architects prefer to measure speed in knots (1 knot = 1 nautical mi/h = 1.85 km/h) and length in feet. Show that in these units the formula becomes $v = 1.3\sqrt{L}$.

(b) What is the hull speed for a sailboat of length 10 m? A clipper ship of length 60 m?

*44. The passenger of an airplane flying over a ship notices that ocean waves are smashing into the ship regularly at the rate of 10 per minute. He also notices that the length of the ship is about the same as three wavelengths. Deduce the length of the ship from this information; assume the ship is at rest.

*45. The National Ocean Survey has deployed buoys off the Atlantic Coast to measure ocean waves. Such a buoy detects waves by the vertical acceleration that it experiences as it is lifted and lowered by the waves. In order to calibrate the device that measures the acceleration, scientists placed the buoy on a Ferris wheel at an amusement park. If the *vertical* acceleration (as a function of time) of a buoy riding on a Ferris wheel of a radius 6.1 m rotating at 6 rev/min is to simulate the vertical acceleration of a buoy riding a wave, what would be the amplitude and wavelength of this wave? Assume that the waves are in deep water and that the buoy always rides on the surface of the water.

*46. Figure 17.24 is a record of a tsunami striking the coast of Mexico.

(a) Approximately, what was the frequency of this wave?

(b) What was its wavelength and its speed while still in the open sea? Assume that the depth of the open sea (Pacific) is 4.3 km.

*47. In the open sea, a tsunami usually has an amplitude less than 30 cm and a wavelength longer than 80 km. Assume that the speed of the tsunami is 740 km/h. What is the maximum vertical velocity and acceleration that such a tsunami will give to a ship floating on the water? Will the crew of the ship notice the passing of the tsunami?

*48. The Bay of Fundy (Nova Scotia) is about 250 km long and about 100 m deep.

(a) What is the speed of waves of long wavelength in the bay?

(b) What are the frequency and the period of the fundamental mode of oscillation of the bay? Use the speed calculated in part (a) and treat the bay as a long, narrow tube open at one end and closed at the other.

(c) The period of the tidal pull exerted by the Moon is about 12 h. Would you expect that the very large tidal oscillations (with a height of up to 50 ft) observed in the Bay of Fundy are due to resonance?

**49. Suppose that a tsunami spreads out along the surface of the Pacific Ocean in concentric circles from its source. If the distance from the source is less than 1000 or 2000 km, the surface of the Earth can be (approximately) regarded as flat and the intensity of the wave therefore decreases as the inverse of the distance (see the argument given in the introduction to this chapter). However, if the distance is several thousand kilometers, then the curvature of the surface of the Earth must be taken into account in the calculation of the intensity.

(a) Show that the intensity of the wave varies as $1/\sin(r/R)$, where $r$ is the distance measured along the surface of the Earth and $R$ is the radius of the Earth.

(b) Show that if $r \ll R$, this variation of intensity is proportional to $1/r$, as expected.

(c) Show that for $r > 10{,}000$ km, the intensity *increases* with increasing distance.

(d) Suppose that the tsunami has an amplitude of 30 cm when at a distance of 4000 km from its source. What will be the amplitude at 8000 km? At 10,000 km? At 12,000 km? (Hint: The intensity is proportional to the square of the amplitude.)

**50. In the open sea, where the depth $h$ of the water is large, the amplitude of a tsunami is small; but when the tsunami enters the coastal shallows, where $h$ is small, the amplitude becomes large.

(a) Show that the amplitude varies as $A \propto 1/h^{1/4}$. (Hint: As in the case of a wave on a string, the power carried by a tsunami is proportional to $v\omega^2A^2$. Assume that the power in the wave remains constant as it enters the coastal shallows; also note that the frequency remains constant.)

(b) Suppose that a tsunami has an amplitude of 35 cm when in the open sea, where $h = 4.3$ km. If this wave reaches a coastal region where $h = 10$ m, what will be its amplitude? Its height from trough to crest?

**51. Experienced sailors know that when waves traveling in one direction encounter a tidal stream flowing in the opposite direction, the waves become dangerously steep and high. Show that if a wave of speed $v$ in still water runs into a stream of water flowing in the opposite direction at speed $V$, then the wavelength will decrease by a factor $1 - V/v$. Qualitatively, explain why the amplitude of the wave will increase.

### Section 17.5

52. In the crust of the Earth, seismic waves of the P type have a speed of about 5 km/s; waves of the S type have a speed of about 3 km/s. Suppose that after an earthquake, a seismometer placed at some distance first registers the arrival of P waves and 9 min later the arrival of S waves. What is the distance between the seismometer and the source of the waves?

53. By what factor is the energy in the seismic waves of an earthquake of magnitude 8.0 on the Richter scale larger than in those of an earthquake of magnitude 4.0?

54. The shock of an underground nuclear explosion generates seismic waves which can be detected at a large distance with sensitive seismometers. What is the magnitude, on the Richter scale, of the earthquake generated by an underground explosion equivalent to 1 megaton of TNT? The explosion of 1 ton of TNT releases an energy of $4.2 \times 10^9$ J. Assume that 30% of the energy of the explosion goes into seismic waves.

55. The great earthquake that struck Lisbon, Portugal, in 1755 had an (estimated) magnitude of 9 on the Richter scale. What was the energy of this earthquake? Express your answer in the equivalent of megatons of TNT; the explosion of 1 ton of TNT releases $4.2 \times 10^9$ J.

*56. Many inhabitants of Tangshan, China, reported that during the catastrophic earthquake of July 28, 1976, they were thrown 2 m into the air as if by a "huge jolt from below."

(a) With what speed must a body be thrown upward to reach a height of 2 m?

(b) Assume that the vertical motion of the ground was simple harmonic with a frequency of 1 Hz. What amplitude of the vertical motion is required to generate a speed equal to that calculated in part (a)?

# RADIATION AND LIFE*

Ordinary light is the most familiar of all forms of radiation. It consists of waves of electric and magnetic energy streaming out from a source — the flame of a candle, the hot filament of a light bulb, the Sun, or whatever. A pulse of light, or light ray, propagates outward from its source in a straight line. Light rays can propagate through thousands of kilometers of air, and hundreds of meters of water, or glass, or other transparent materials. Yet light rays are easily stopped by a thin layer of opaque material such as a sheet of paper, aluminum foil, or human skin. Light has only a small penetrating power.

Several forms of invisible radiation — X rays, alpha rays, beta rays, gamma rays, cosmic rays — have a large penetrating power. These rays can easily pass through thick layers of material opaque to light. Some of these penetrating radiations are closely related to ordinary light — they consist of extremely energetic waves of electric and magnetic energy; others are corpuscular — they consist of a stream of fast-moving particles. When these radiations hit a human body, their impact is painless yet their effect can be lethal.

## IV.1 PENETRATING RADIATIONS

**X rays**[1] are generated in collisions between electrons and atoms. In an X-ray tube (Figure IV.1), a stream of energetic electrons, moving at high speed in a vacuum, strikes a block or target of metal. X rays are then emitted by the incident electrons when these are suddenly decelerated and stopped by collisions with the target atoms, and they are also emitted by the atomic electrons within the target atoms when these are disturbed (excited) by the impact of the incident electrons. X rays are waves of electric and magnetic energy. These waves are qualitatively similar to light waves, but the wavelength of X-ray waves is typically 10,000 times shorter than the wavelength of light waves. X rays readily penetrate through the soft tissues of the human body, but they are partially absorbed by the denser tissues in bone. Hence the bones throw shadows when illuminated by X rays, and these shadows can be exploited to form images on a photographic plate (Figure IV.2).

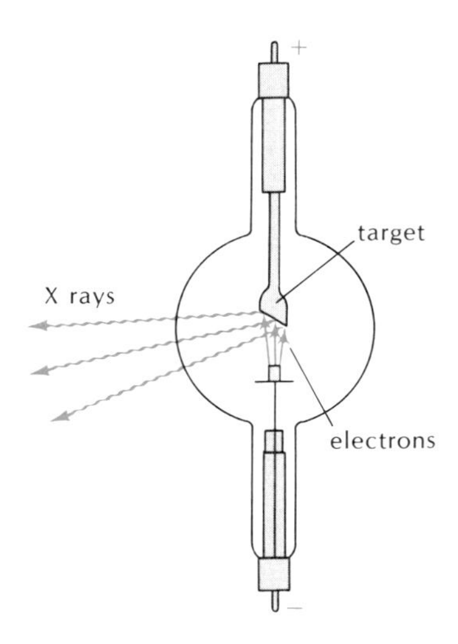


**Fig. IV.1** In an X-ray tube, energetic electrons emit X rays when they strike a target.

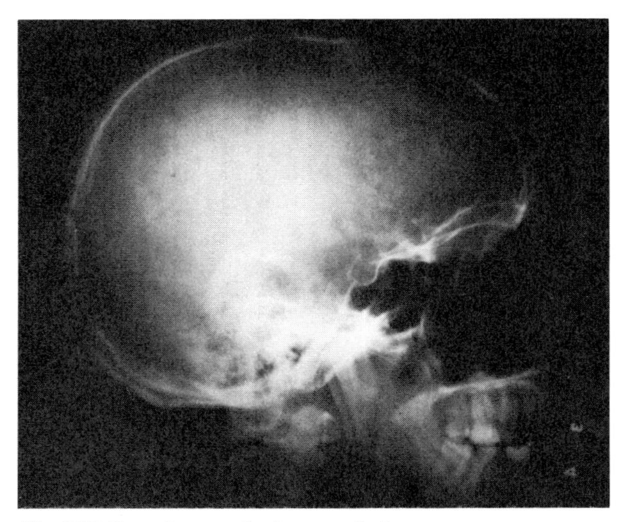

**Fig. IV.2** X-ray image of a human skull.

* This chapter is optional.

[1] Discovered by the German physicist **Wilhelm Konrad Röntgen,** 1845–1923. For this discovery, Röntgen was awarded the first Nobel Prize in 1901.

**Fig. IV.3** Tracks of alpha rays emitted by a radioactive sample in a cloud chamber containing air with an excess of water vapor. The alpha rays trigger the formation of fine trails of water droplets, which make their tracks visible.

Alpha rays ($\alpha$), beta rays ($\beta$), and gamma rays ($\gamma$) are nuclear radiations — these rays are emitted by spontaneous processes that take place inside the nuclei of atoms. For example, the nuclei of atoms of radiocarbon and of radiostrontium emit beta rays, nuclei of radiocobalt emit beta rays and gamma rays, nuclei of radium emit alpha rays and gamma rays, etc. (Figure IV.3). Atoms that emit alpha, beta, or gamma rays are said to be **radioactive.** Among these rays, the most penetrating are the gamma rays — some of them will penetrate one meter of concrete. The least penetrating are the alpha rays — most of them will be stopped by a sheet of heavy paper.

**Gamma rays** are essentially high-energy X rays (megavolt X rays). The only difference between the former and the latter is their place of origin: X rays are produced by the motion of electrons in the volume of the atom, whereas gamma rays are produced by the motion of protons and neutrons inside the nucleus.

**Beta rays** are particles — they are particles of either negative or positive electric charge that are created by a spontaneous reaction inside nuclei and then ejected at high speed. Experiments with beta rays show that the negative beta particles ($\beta^-$) are nothing but high-speed electrons, and the positive beta particles ($\beta^+$) are nothing but high-speed antielectrons, or positrons. When a beta-ray antielectron is stopped by some absorbing material, it sooner or later collides with an electron, and both annihilate; this mutual annihilation of matter and antimatter generates a flash of gamma rays.

**Alpha rays** also are particles — they are particles of a positive electric charge and of a relatively large mass (about 7300 times the mass of a beta particle). Experiments reveal that the alpha particle has exactly the same structure as the nucleus of a helium atom. To produce an alpha ray, the radioactive nucleus somehow assembles part of its material into a heliumlike structure and ejects this fragment at high speed.

**Cosmic rays** are another kind of penetrating radiation. These rays come from outer space and they continually bombard the Earth from all directions. Cosmic rays are particles, mainly high-energy protons. Some of these particles originate on the Sun; others originate somewhere in the depths of interstellar space. The solar contribution is highly variable — during solar flares it can become extremely large and cause disturbances in the upper atmosphere ("magnetic storms" and aurora borealis). The interstellar contribution is constant in time and it reaches the Earth from all directions with equal strength; it includes particles of extremely high energies — some of these cosmic rays have an energy much in excess of any energy ever achieved by our most potent accelerating machines. We do not know exactly where the interstellar cosmic rays come from; probably they are generated by supernova explosions of stars. When cosmic-ray particles strike the top of our atmosphere, large numbers of a variety of new particles are produced in the violent collisions between the incident cosmic rays and the nuclei of air atoms (Figure IV.4). The particles produced in such collisions are called **secondary cosmic rays** to distinguish them from the incident particles, or

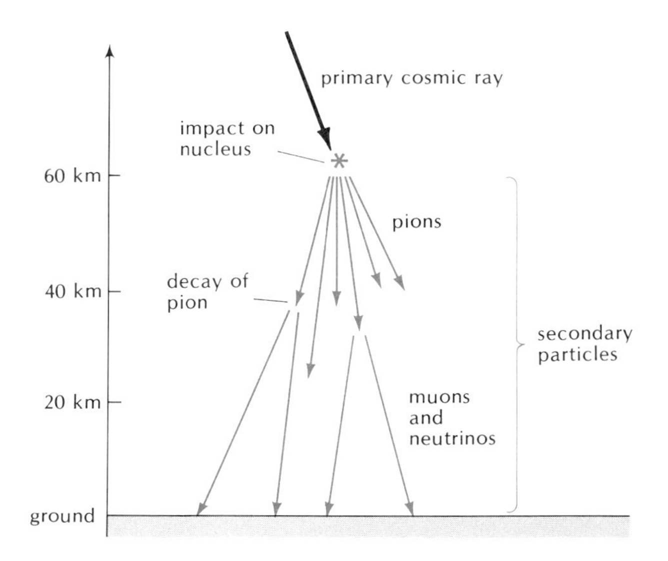


**Fig. IV.4** Primary cosmic rays (black) striking the upper atmosphere generate showers of secondary cosmic rays (color), which reach the ground.

**primary cosmic rays.** Most of the cosmic rays that transverse the atmosphere and reach sea level are secondary particles — about 75% of these are muons, particles similar to electrons, but heavier.

Because of their exceedingly high energy, cosmic rays are by far the most penetrating kind of radiation. They transverse not only the thickness of the atmosphere, but sometimes penetrate thousands of meters into the ground — cosmic-ray muons have been known to penetrate more than a kilometer of rock and to reach the deepest mine pits.

Table IV.1 summarizes some of the properties of penetrating radiations. Neutrons have been included in this table even though they are not emitted spontaneously by radioactive materials. However, neutrons are emitted in great abundance during the fission reactions that take place in a nuclear reactor or in a nuclear bomb; hence these artificial devices are intense sources of neutron "rays." Neutrons are also emitted by some nuclei when suitably stimulated with other kinds of radiation; for example, beryllium releases neutrons when exposed to bombardment by alpha rays. Small neutron sources ("neutron howitzers") can be constructed by simply mixing beryllium with radium or some other alpha emitter; within such a mixture the beryllium is then under continuous alpha bombardment, which provokes it to emit neutrons.

## IV.2 RADIOACTIVE DECAY

Although all atoms of a given chemical element have exactly the same chemical properties, some atoms have more mass than others, and they therefore do not all have exactly the same physical properties. The differences in mass among the atoms of a given chemical element hinge on the numbers of neutrons in the nuclei — all the atoms have the same number of protons in their nuclei, but they have different numbers of neutrons. As already mentioned in the Prelude, atoms with the same number of protons but with different numbers of neutrons are called **isotopes.** For instance, the isotope $^{14}C$ of carbon has six protons and eight neutrons in its nucleus, $^{13}C$ has six protons and seven neutrons, and $^{12}C$ (which is the most abundant isotope in naturally occurring samples of carbon) has six protons and six neutrons. The superscript on the chemical symbol is the sum of the number of protons and neutrons, called the **mass number.** Carbon has several other isotopes; Table P.2 lists all the isotopes of carbon, as well as the isotopes of a few other elements.

Most isotopes are unstable, that is, they decay by a spontaneous nuclear reaction and transmute themselves into another, more stable element. The unstable isotopes are radioactive — their spontaneous nuclear reactions involve the emission of alpha rays, beta rays,

**Table IV.1** PENETRATING RADIATIONS

| | | | *Typical penetration depth*[a] | | | |
|---|---|---|---|---|---|---|
| *Radiation* | *Mass* | *Electric charge*[b] | *Air* | *Water* | *Lead* | *Description* |
| X rays | 0 | 0 | 100 m | 10 cm | 1 mm | Very energetic form of light |
| Alpha rays | 4.0 u | +2e | 10 cm | 0.1 mm | 0.01 mm | Same structure as helium nucleus |
| Beta rays | 0.00055 u | ±e | 1 m | 1 mm | 0.1 mm | High-energy electrons or positrons |
| Gamma rays | 0 | 0 | 1 km | 1 m | 10 cm | Extremely energetic form of light |
| Cosmic rays, primary | Mixed | Mixed | | Very large | | Mostly high-energy protons |
| Cosmic rays, secondary | Mixed | Mixed | | Very large | | Mostly high-energy muons |
| Neutrons | 1.0 u | 0 | 1 km | 1 m | 10 cm | — |

[a] X rays, gamma rays, and neutrons do not have a sharply defined penetration depth — they are attenuated gradually; the numbers in the table give the thickness of material that attenuates beams of these rays by a factor of 1000. All these numbers are rough approximations; exact values depend on the energy of the radiation.

[b] The quantity e appearing in this column represents the electric charge of the proton. Thus, an alpha particle has twice the electric charge of the proton, etc.

or gamma rays. For example, the isotope $^{14}C$ is unstable, and it decays into $^{14}N$ with the emission of a beta ray,

$$^{14}C \rightarrow {}^{14}N + \beta^- \tag{1}$$

The decays of the following isotopes of cobalt, strontium, and radium are further examples of radioactive decays with emission of beta rays and alpha rays:

$$^{60}Co \rightarrow {}^{60}Ni + \beta^- \tag{2}$$

$$^{90}Sr \rightarrow {}^{90}Y + \beta^- \tag{3}$$

$$^{226}Ra \rightarrow {}^{222}Rn + \alpha \tag{4}$$

Note that all of the reactions (1)–(4) involve **transmutation of elements:** carbon into nitrogen, cobalt into nickel, strontium into yttrium, and radium into radon. Such transmutations are a characteristic feature of alpha and beta decays.

In many cases of alpha and beta decays, the nucleus emerges from the reaction with an excess of internal energy; the nucleus will then eliminate this energy by emitting a gamma ray. For instance, the $^{60}Ni$ nucleus formed by the beta decay of $^{60}Co$ contains extra internal energy and emits such a gamma ray. Note that although a transmutation of elements [Eq. (2)] precedes the emission of the gamma ray, the emission itself does not involve transmutation. Gamma-ray emission from a nucleus is analogous to light emission from an atom; in both cases energy is released in consequence of a sudden change of motion of the constituents of a system (protons and neutrons of a nucleus, or electrons of an atom), but the constituents themselves do not change.

If we initially have a given amount of some radioactive element, then this amount will gradually decrease in time as more and more of the atoms decay. Measurements show that the quantitative law that describes this decay process is very simple: if a certain fraction of the initial amount of radioactive material decays in a certain time interval, then the same fraction of the remainder decays in the next (equal) time interval, and the same fraction of the new remainder decays in the next (equal) time interval, etc. For example, suppose we initially have 1 g of radioactive strontium. This decays by beta decay into yttrium, according to the reaction (3). In this reaction, strontium is called the **parent** material and yttrium is the **daughter** material. Measurements show that it takes 29 years for one-half of the initial amount of parent to decay. The law of radioactive decay then asserts that during the next 29

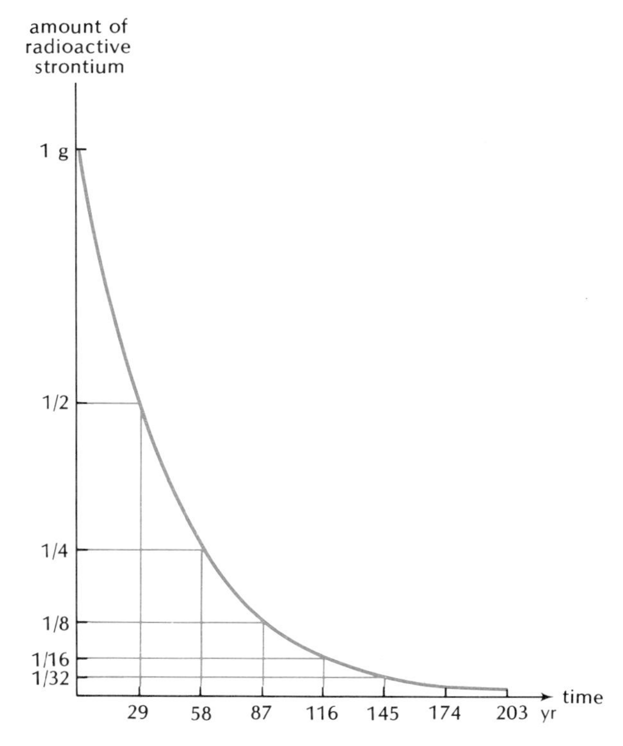


**Fig. IV.5** Decay of radioactive strontium. At $t = 0$ the initial amount of strontium is 1 g; the curve shows the amount at later times.

years, one-half of the remainder will decay, and so on. Hence the amounts of parent material left at times $t = 0$, 29, 58, and 87 years will be 1, $\frac{1}{2}$, $\frac{1}{4}$, and $\frac{1}{8}$ g, respectively (Figure IV.5). This means that the amounts left after the lapse of equal time intervals form a simple geometric progression. The time required for the decay of one-half of the amount of parent material is called the **half-life.** For radioactive strontium, the half-life is 29 years.

In a sample of radioactive material, a certain number of decays will occur during each second. This decay rate is called the **activity** of the sample. Usually, the activity is expressed in **curies:**[2]

$$\begin{aligned} 1 \text{ curie} &= 1 \text{ Ci} \\ &= 3.7 \times 10^{10} \text{ disintegrations per second} \end{aligned} \tag{5}$$

or in **becquerels,**

$$1 \text{ becquerel} = 1 \text{Bq} = 1 \text{ disintegration per second} \tag{6}$$

[2] After the French scientists **Pierre Curie,** 1859–1906, and **Marie Sklodowska Curie,** 1867–1934. Both are famous for their work on radioactivity, and their discovery of radium and polonium. They shared a Nobel Prize with **A. H. Becquerel** in 1903. Marie Curie was also awarded a Nobel Prize in chemistry in 1911.

**Fig. IV.6** This warning sign is usually magenta on a yellow background.

For example, the activity of 1 g of pure strontium is about 140 Ci, or $5.2 \times 10^{12}$ Bq. Commercially available sources of penetrating radiation carry labels indicating their strength in curies and a label with a warning symbol (Figure IV.6). Any source of a strength more than about $10^{-6}$ Ci is regarded as potentially dangerous. Health authorities require that such sources be handled only by qualified persons. Table IV.2 lists the strengths of some typical radioactive sources used in diverse applications.

**Table IV.2** STRENGTHS OF RADIOACTIVE SOURCES

| *Source* | *Radiation* | *Strength* |
|---|---|---|
| Fission products released by 20-kiloton atomic bomb | $\alpha, \beta, \gamma$ | $6 \times 10^{11}$ Ci |
| Nuclear power reactor | $\alpha, \beta, \gamma$ | $10^{10}$ |
| Radioactivity released by Chernobyl reactor | $\alpha, \beta, \gamma$ | $10^{8}$ |
| $^{144}$Ce as heat source for thermoelectric generator | $\beta, \gamma$ | $10^{6}$ |
| $^{60}$Co in high-level industrial irradiation cell | $\beta, \gamma$ | $10^{6}$ |
| $^{60}$Co in teletherapy unit | $\beta, \gamma$ | $\sim 10^{3}$ |
| $^{131}$I for eradication of thyroid | $\beta, \gamma$ | 0.1 |
| $^{131}$I for thyroid scan | $\beta, \gamma$ | $5\text{–}50 \times 10^{-6}$ |
| $^{226}$Ra in fluorescent paint in wristwatch | $\beta, \gamma$ | $\sim 10^{-6}$ |
| Natural $^{40}$K in human body | $\beta, \gamma$ | $10^{-7}$ |

## IV.3 RADIATION DAMAGE IN ATOMS

When radiation penetrates a material, it disrupts the atoms, that is, it disturbs the motion of the electrons in the atoms and sometimes it even ejects electrons entirely from the atoms. The latter kind of disruption is called **ionization** — the ejection of electrons from their atoms leaves behind charged atoms, or ions. Because of this, alpha, beta, and gamma rays are often called **ionizing radiation.**

Alpha, beta, or gamma rays penetrating through a material gradually spend their energy in the disruption of the atoms along their tracks. The larger the initial energy of the ray, the farther it will penetrate before all its energy is exhausted. The details of the mechanism of energy loss depend on the type of radiation.

**Alpha rays** or other electrically charged particles passing through matter exert electric forces on the atomic electrons, and the latter exert electric forces on the former. Since the alpha ray passes by the electron very quickly, the mutual interaction can be regarded as a collision. An alpha ray is so massive that such collisions with atomic electrons, will not deflect it very much — the alpha ray violently pushes the electrons aside and continues on its straight path (Figure IV.7).[3]

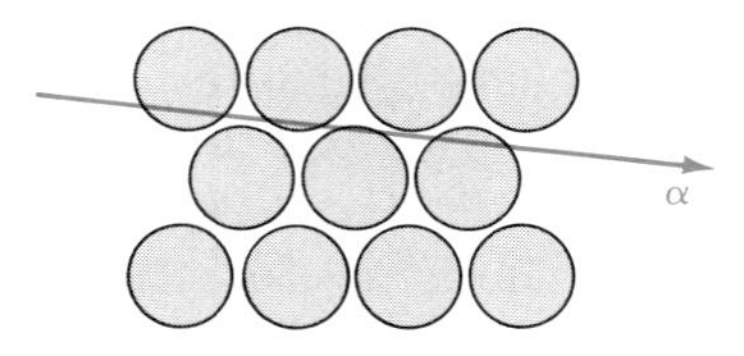


**Fig. IV.7** An alpha ray passing through atoms.

Although each collision with an electron slows the alpha ray down only very slightly and produces only a small loss of kinetic energy, the number of electrons that the alpha ray encounters along its path is very large, and hence the cumulative loss of kinetic energy is rapid. Alpha rays produce intense ionization along their path (Figure IV.8); thus, they soon lose their energy, and they are stopped by a thin layer of material.

**Beta rays** penetrating matter also suffer collisions with atomic electrons. However, since their mass is equal to the mass of atomic electrons, they cannot shoulder these electrons aside without suffering large deflections themselves (Figure IV.9). Thus the path of a beta ray is not straight, but tortuous, full of erratic twists and turns (see Figure IV.8). Besides, since the electric charge of a beta ray is of lesser magnitude than that of an alpha ray, the electric forces and disturbances that it exerts on nearby atomic electrons are of a correspondingly lesser magnitude. The beta ray generates less intense ionization than an alpha ray, dissipates its energy more slowly, and penetrates a deeper layer of material.

[3] Collisions with the atomic nuclei of the matter are so rare as to be insignificant.

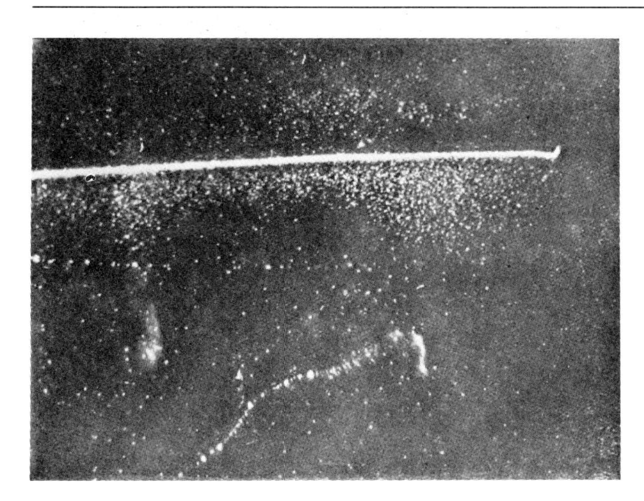

**Fig. IV.8** Tracks of an alpha and beta ray in a cloud chamber. The track of the alpha ray is heavy and straight. The track of the beta ray is faint and tortuous.

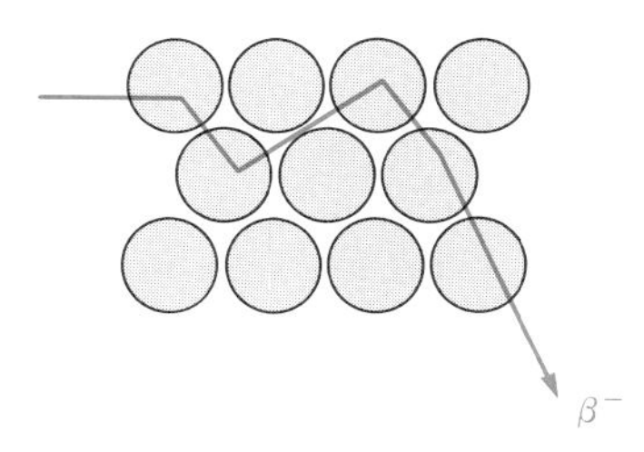


**Fig. IV.9** A beta ray passing through atoms.

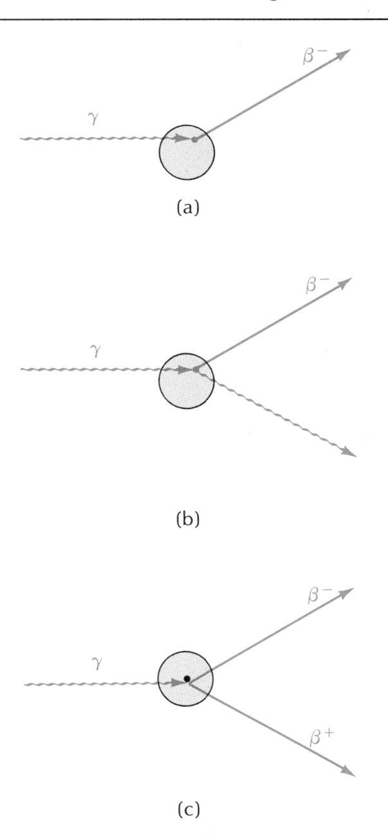


**Fig. IV.10** A gamma ray passing through an atom may (a, b) lose all or some of its energy to an electron or (c) create an electron–positron pair.

**Gamma rays,** in passing through matter, behave quite differently from alpha and beta rays. Whereas the latter engage in a great many collisions with atomic electrons and thereby dissipate their energy more or less continuously along their path, a gamma ray will sometimes penetrate a thick layer of material without any energy loss. What happens to an individual gamma ray is a matter of chance: as a gamma ray attempts to pass through a layer of material, there is some chance that the gamma ray will collide with an electron and lose its energy, and there is some chance that it will pass through without collision and without energy loss. When a gamma ray finally does interact with an atomic electron, it loses all, or a large part, of its energy; and the electron is then ejected at high speed, leaving the atom (Figure IV.10a and b). If the gamma ray has a sufficiently large energy, an energy of at least twice the electron rest-mass energy (twice $m_ec^2$), then it can also lose energy by an alternative mechanism: when passing near a nucleus, such a gamma ray can spend its energy in the creation of an electron–positron pair (Figure IV.10c). Note that neither of these primary energy transfer mechanisms causes much direct damage in the material — only one atom is involved in each case. However, the high-energy electrons (and positrons) that inherit the original gamma-ray energy behave exactly like beta rays, and they will damage a large number of atoms along their path.

**Neutrons** passing through matter do not interact with the atomic electrons — the neutrons have no electric charge and electrons scarcely feel them.[4] The attenuation of a neutron beam by a layer of material is due to direct contact collisions between neutrons and atomic nuclei. Since the nuclei are small, such collisions are not very frequent, and therefore neutrons, just like gamma rays, can penetrate very thick layers of material without stopping. The impact of a neutron on a nucleus is a matter of chance; when such an impact finally occurs, the neutron may either bounce off the nucleus, losing some energy, or else be absorbed by the nucleus. Under the impact, the nucleus recoils and, if the recoil energy is large, forms a "nuclear ray," which behaves similarly to an alpha ray, but causes

[4] Even in a head-on collision, an electron and a neutron will usually just pass through each other with no effect on either.

much heavier ionization. If the nucleus absorbs the neutron, it becomes radioactive — the capture of a neutron gives the nucleus extra internal energy, which it will eliminate by undergoing one or more radioactive transformations. The secondary alpha, beta, or gamma rays emitted in this process will then damage the material in the usual way.

When a material is exposed to alpha, beta, or gamma radiation for some length of time, it will absorb a certain amount of radiation energy. What alterations this absorbed energy will produce in the properties of the material depends on how much energy is absorbed per unit mass of the material. The amount of absorbed radiation energy per kilogram of material is called the **absorbed dose;** it is measured in **rads,**

$$1 \text{ rad} = 0.01 \text{ J/kg} \qquad (7)$$

For example, if a human body of 75 kg receives a dose of 600 rads over the entire volume of the body (a "whole-body exposure"), then the total amount of absorbed energy is

$$\begin{aligned}[\text{absorbed energy}] &= [\text{absorbed dose}] \times [\text{mass of body}] \\ &= 600 \text{ rads} \times 75 \text{ kg} \\ &= 600 \times 0.01 \text{ J/kg} \times 75 \text{ kg} \\ &= 450 \text{ J} \qquad (8)\end{aligned}$$

Absorption of such a dose of radiation is very likely to be lethal.

## IV.4 RADIATION DAMAGE IN MOLECULES AND IN LIVING CELLS

Radiation damages atoms by the forcible ejection of electrons from the atom (ionization). In biological tissues, the damaged atoms or molecules will then engage in chemical reactions that disrupt the normal chemical processes. Such an interference with the normal processes of living tissues can have grave consequences.

Since biological tissues are composed of 70% to 90% water, most of the primary impacts of radiation on molecules will be impacts on water molecules. These impacts lead to the decomposition of water (radiolysis of water) and the subsequent formation of a variety of extremely corrosive molecules.

The first step in the decomposition of water is the ionization of a water molecule by the impact of radiation; this results in a positive water ion and a free electron,

$$[\text{radiation}] + H_2O \rightarrow H_2O^+ + e^- \qquad (9)$$

The electron travels some distance and is then captured by another water molecule;[5] this results in a negative water ion,

$$e^- + H_2O \rightarrow H_2O^- \qquad (10)$$

Both the positive and the negative water ions ($H_2O^+$ and $H_2O^-$) are unstable; they dissociate into radicals and ions of hydroxyl and hydrogen,

$$H_2O^+ \rightarrow OH^{\cdot} + H^+ \qquad (11)$$

$$H_2O^- \rightarrow OH^- + H^{\cdot} \qquad (12)$$

These radicals $OH^{\cdot}$ and $H^{\cdot}$ are electrically neutral but they have "loose ends," that is, they have unsaturated chemical bonds; these radicals are therefore very eager to attach themselves to something — they are very reactive.

The reactive hydroxyl and hydrogen radicals attack the complicated organic molecules — proteins, enzymes, nucleic acids — that are the basic building blocks of living cells. Furthermore, by reacting with one another and with molecules of their environment, the hydroxyl and hydrogen radicals produce some other corrosive compounds that inflict additional damage on the organic molecules. The damage done to the organic molecules can take several forms: breakage of bonds, leading to fragmentation of the long molecular chains that serve as backbones for the molecules; rearrangement of bonds, leading to changes in the chemical behavior of molecules; and formation of extra bonds, leading to cross-linking either within one molecule or between neighboring molecules. Such damage to the organic molecules in a living cell disrupts the normal chemical processes that sustain life. If the disruption is severe, the life of the cell ends.

Massive doses of radiation (many thousands of rads) can cause an almost immediate cessation of cellular metabolism followed by cellular disintegration. However, even a much smaller dose of radiation (a few hundred rads or less) can cause "reproductive death" of cells by destroying their ability to undergo cell division and to reproduce. Since the life of any complex organism depends on the continual replacement of old cells by new ones, the "reproductive death" of cells will ultimately bring about the physiological death of the organism, even though the damaged cells continue to metabolize and, except for their inability to reproduce, may seem chemically and morphologicaly quite normal.

[5] If the electron has sufficiently high energy, it can, of course, produce some ionization before it is captured.

There exists considerable evidence that reproductive cell death is due to damage to the chromosomes of the cell — specifically, damage to the DNA molecules within the chromosomes. DNA is the most crucial molecule in the cell; it is the molecule that contains all the information regarding the construction and operation of the cell, and it therefore directs all the processes within the cell. The DNA molecules and the chromosomes that contain them are both quite radiosensitive — both DNA and chromosomes are subject to breakage when exposed to radiation. Furthermore, experiments show that even very low levels of radiation, which produce no overt morphological changes in the cell, interfere with DNA synthesis and consequently inhibit cell division.

The sensitivity of a cell to radiation depends on the phase of the life of the cell. The cell is most susceptible to suffer lethal damage if the radiation strikes it during the period of cell division (mitosis) just when the DNA and the chromosomes are attempting duplication. Obviously, cells that divide frequently are more likely to get caught in this delicate stage than cells that divide rarely or not at all. This leads to a general rule: the cells with the highest radiosensitivity are those that have the highest rate of cell division (law of Bergonié and Tribondeau). For example, in mammals, the cells in the bone marrow, lymphatic tissues, lining of the intestine, ovaries and testes, and the cells in the embryo undergo very frequent cell divisions — all these cells are very easily damaged by radiation. In contrast, the cells in the brain, muscles, bones, liver, and kidneys undergo little or no cell division — these cells are relatively resistant to damage by radiation.

The damage that a given absorbed dose of radiation inflicts on a tissue also depends on the kind of radiation. Some kinds of radiation, for instance, alpha rays, are more efficient than others in damaging and killing cells. The kill efficiency of alpha rays is high because they produce very intense ionization along their path. When an alpha ray passes through tissues, it will damage many molecules in each cell; this concentrated damage is likely to have a lethal effect. The kill efficiency of gamma rays is much lower. When a gamma ray passes through tissues, it will damage only a few molecules in each cell; hence the damage will be spread out over a larger number of cells, and the injuries to individual cells are not as likely to be lethal. Injured cells have a chance to repair themselves gradually and to recover.

Because of such differences in kill efficiency, the absorbed dose (in rads) is not a good indicator of the permanent biological damage. To obtain a measure of the biological damage, the absorbed dose must be multiplied by a correction factor that accounts for the kill efficiency of the radiation. The correction factor is called the RBE (*R*elative *B*iological *E*ffectiveness). Table IV.3 gives the RBE for diverse kinds of radiation; the values in this table are approximate because the exact RBE depends on the energy of the rays. The RBE of X rays is 1 by definition; according to Table IV.3, alpha rays are 10 to 20 times more damaging than X rays, fast neutrons are 10 times more damaging than X rays, and so on.

**Table IV.3** RBE VALUES

| *Radiation* | *RBE* |
|---|---|
| Alpha rays | 10–20 |
| Beta rays | 1 |
| Gamma rays | 1 |
| X rays | 1 |
| Neutrons (fast) | 10 |
| Neutrons (slow) | 5 |

A reasonably good indicator of the biological damage in human tissues is the **"equivalent" absorbed dose,** which is the absorbed dose multiplied by the RBE:

$$[\text{"equivalent" absorbed dose}] = \text{RBE} \times [\text{absorbed dose}]$$

Since RBE is a pure number, the unit of "equivalent" absorbed dose is the same as that of absorbed dose (rads); however, in order to keep track of these two kinds of doses, the unit of "equivalent" absorbed dose is not written as rad, but as (*r*ad *e*quivalent *m*an). For example, if a human body receives a 100-rad absorbed dose of fast neutrons, the "equivalent" absorbed dose is 10 × 100 rems = 1000 rems (incidentally: this is an absolutely lethal dose). Note that doses of 100 rads of fast neutrons, 1000 rads of X rays, 1000 rads of gamma rays, and so on, all yield the same "equivalent" dose.

## IV.5 THE PHYSIOLOGICAL EFFECTS OF RADIATION

The effect of radiation on man depends not only on the total "equivalent" dose, but also on how it is delivered. A dose that is delivered all at once (acute) is more dangerous than a dose spread out over a long period of time (chronic); a dose delivered to the entire volume of the body (whole-body exposure) is more dangerous than a dose delivered to only some part of the body; a dose delivered to a radiosensitive part of the body (bone marrow, lymphatic tissue, gastrointestinal tract) is more dangerous than a dose delivered to a radioresistant part (brain, muscles, bones, etc.).

Table IV.4 summarizes the effects of absorption of a whole-body dose delivered all at once. The informa-

**Table IV.4** EFFECTS OF ACUTE RADIATION DOSES

| *Dose (whole body)* | *Critical organ* | *Effect* | *Incidence of death* |
|---|---|---|---|
| 0–100 rems | Blood | Some blood-cell destruction | None |
| 100–200 | Blood-forming tissue (bone marrow, lymphatic tissue) | Decrease in white blood-cell count | None |
| 200–600 | Blood-forming tissue | Severe decrease of white blood-cell count, internal hemorrhage, infection, loss of hair | 0–80% within 2 months |
| 600–1000 | Blood-forming tissue | Same | 80–100% within 2 months |
| 1000–5000 | Gastrointestinal tract | Diarrhea, fever, electrolyte imbalance | Nearly 100% within 2 weeks |
| 5000 and above | Central nervous system | Convulsions, tremor, lack of coordination, lethargy | 100% within 2 days |

tion on radiation sickness and radiation death contained in this table is based on observations of the victims of the bombings of Hiroshima, Nagasaki, and the Bikini atoll; observations of the victims of nuclear-reactor accidents; and extensive laboratory studies of animals subjected to irradiation.

Doses below 100 rems do not produce radiation sickness, although there may be some temporary changes in the blood-cell count.

Doses between 100 and 200 rems produce radiation sickness, with nausea and vomiting. There is a reduction of the number of white blood cells due to damage to the bone marrow which generates these cells. If no secondary complications ensue, the victim recovers in a few weeks.

Doses from 200 to 600 rems lead to severe radiation sickness, with a possibly lethal outcome. A dose of 300 to 350 rems is sufficient to kill 50% of the exposed population if no medical treatment is available. With an exposure in this range there is severe damage to the bone marrow and a concomitant drastic reduction of the white blood-cell count; this renders the body very susceptible to infections.

Doses between 600 and 1000 rems involve qualitatively similar, but more massive, damage. Survival is unlikely, but not impossible.

Doses above 1000 rems are almost invariably fatal. Exposure at this level destroys the lining of the intestine; this eliminates the control of the body over its fluid balance, and also lays the interior of the body open to massive bacterial and viral invasion.

Doses of more than 5000 rems result in cerebral death by direct damage to the brain; at extreme doses, death can be nearly instantaneous.

Although low doses of radiation do not cause radiation sickness, they can cause other damage within the body. Irradiation of a pregnant woman, even at a low level, is likely to interfere with the normal development of the embryo. The embryonic tissues are extremely radiosensitive, and a dose of 15 rems or even less during the first 2 months of pregnancy may induce monstrous deformations of the embryonic organs.

Furthermore, radiation exposure produces delayed effects such as cancers, cataracts, and genetic defects, which only make their appearance years, or generations, later. For example, among the survivors of the bombings in Japan, leukemia was two or three times as frequent as normally expected.

Even very small doses of radiation carry with them some risk of cancer. It has been estimated that a yearly dose of 200 millirems delivered to every inhabitant of the United States would lead to an average of 7100 extra cancer deaths each year. Besides, such doses of radiation increase the incidence of mutations in cells, and therefore carry with them an increased risk of genetic defects. A yearly dose of 200 millirems delivered to every inhabitant of the United States would lead to an average of 6000 extra genetic defects in newborn children each year.

These numbers are directly relevant because they correspond to the average amount of radiation that inhabitants of the United States actually receive from diverse sources (see Table IV.5).

Note that one-third of the yearly dose is due to medical practice; this contribution could be substantially reduced (by a factor of about two) by eliminating unnecessary diagnostic X-ray procedures and by restricting the size of the X-ray beam to the size of the film that is being exposed. Table IV.6 gives the doses received from various medical X-ray procedures.

**Table IV.5** AVERAGE ANNUAL RADIATION DOSES PER INHABITANT OF THE UNITED STATES[a]

| *Source* | *Dose (per year)* |
|---|---|
| Medical and dental X rays | 70 millirems |
| Cosmic rays | 50 |
| Radioactivity of ground and buildings | 50 |
| Natural $^{40}K$ in human body | 20 |
| Other natural radioactivity in human body | 4 |
| Radioactivity of air | 5 |
| Total | ~200 millirems |

[a] From B. L. Cohen, *Nuclear Science and Society*.

**Table IV.6** RADIATION DOSES FROM DIAGNOSTIC X RAYS[a]

| *Diagnostic procedure* | *Doses*[b] |
|---|---|
| Mammography (breast) | 250–300 millirem |
| Upper gastrointestinal series | 150–400 |
| Middle spine | 150–400 |
| Lower gastrointestinal series | 90–250 |
| Lower back and spine | 70–250 |
| Upper spine | 40–80 |
| Gallbladder | 25–60 |
| Skull | 20–50 |
| Chest | 5–35 |
| Dental (whole mouth) | 10–30 |

[a] From P. W. Laws and The Public Citizen Health Research Group, *The X-Ray Information Book*.
[b] This is the effective dose of whole-body exposure that produces the same average damage as the actual localized exposure of the target organ and the surrounding organs.

The maximum permissible dose for persons occupationally exposed to radiation — radiologists, nuclear-reactor operators, experimental physicists — has been arbitrarily set at 5 rems/year.

## IV.6 MEDICAL APPLICATIONS OF RADIOISOTOPES

In medical practice, radioactive isotopes, or **radioisotopes,** are used both for diagnosis and for therapy — they are used both as tracers to investigate bodily organs that are suspected of disease and as sources of intense radiation to destroy the diseased tissues of, say, a malignant tumor.

As tracers, radioisotopes are injected into the body and their subsequent flow through the bloodstream and their accumulation in different organs, or in different parts of one organ, gives an indication of some of the physical and chemical processes that take place deep within the body.

Tracers are routinely used to measure blood flow, total blood volume, metabolic rates of diverse organs, and so on. One of the most interesting and informative tracer techniques is **scanning**; this technique produces a picture of the distribution of radioisotope within the body. The radioisotopes suited to this purpose are gamma emitters; since gamma rays have great penetrating power, the gamma emissions of a radioisotope in an organ deep within the body can easily reach a detector placed outside of the body — the detector can "see" the organ by the gamma-ray "light" that it emits. Unfortunately, gamma rays cannot be focused on a screen or photographic plate in the same way that light rays can be focused. When a gamma ray encounters a lens made of glass (or any other material), it is likely to pass straight through; and if it is deflected, then the deflection is random. However, the gamma-ray detector can be given directional sensitivity by means of a **collimator,** a lead block with many finely drilled holes, all pointed at one spot (Figure IV.11). Rays from the aiming spot pass through the collimator and reach the detector; rays from elsewhere strike the lead and are lost. To build up a picture of the radioisotope distribution, the detector and its collimator must

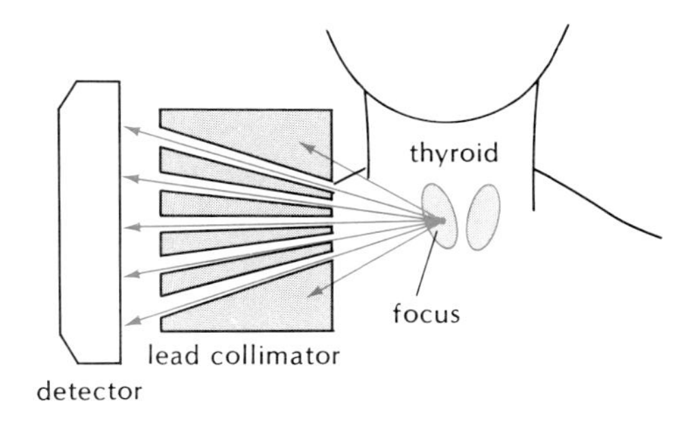


**Fig. IV.11** A collimator.

be moved from spot to spot and the intensity at each place must be recorded; this procedure is called scanning. Alternatively, a picture can be made by means of a **pinhole camera** consisting of a plate of lead with a fine hole, behind which is placed a sheet of photographic film (Figure IV.12). The gamma rays that pass

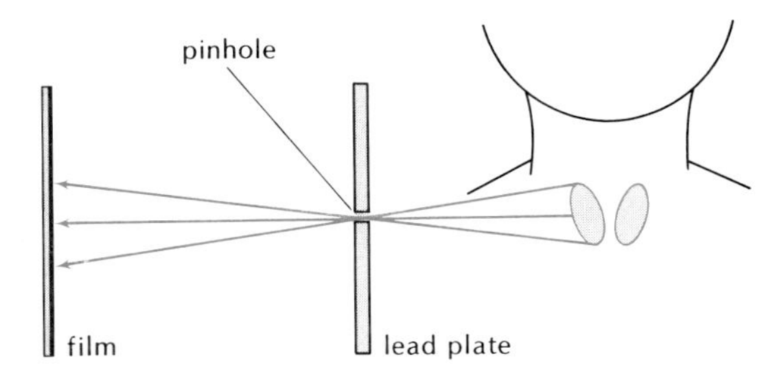


**Fig. IV.12** A pinhole camera.

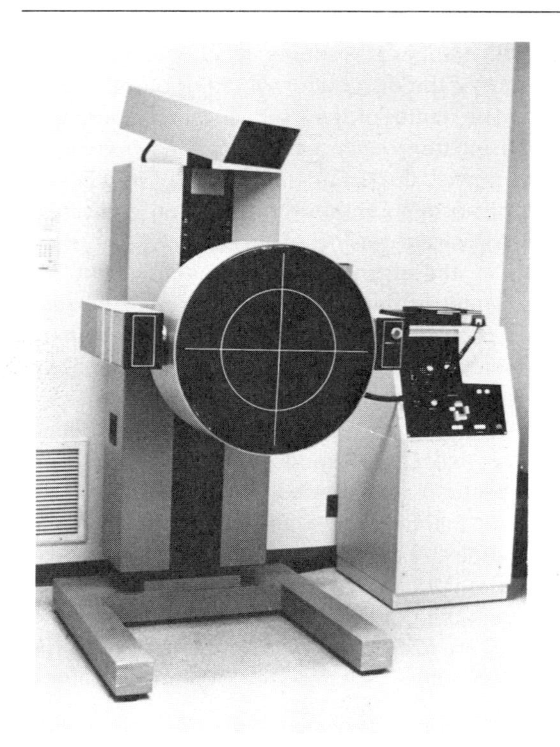

**Fig. IV.13** Siemens gamma camera. (Courtesy Nuclear Medicine Section, Hospital of the University of Pennsylvania.)

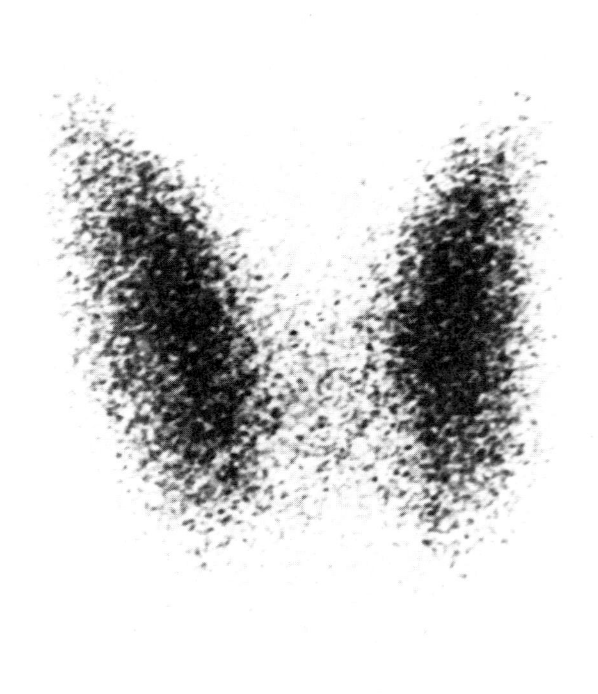

**Fig. IV.14** Pinhole-camera scan of normal thyroid showing uniform concentration of $^{123}$I. (Courtesy Michael G. Velchik, M.D., Nuclear Medicine Section, Hospital of the University of Pennsylvania.)

through the hole form an image on the film. Figure IV.13 shows such a camera used in a hospital.

Different radioisotopes in conjunction with different chemical carriers are used to take pictures of different parts of the body. The choice of radioisotope and the choice of carrier substance depends on the organ that one wishes to "see." For example, Figure IV.14 shows a scan of the thyroid gland of a healthy person; this scan was made with a $^{123}$I compound injected into the bloodstream. The thyroid metabolizes iodine, and hence this element tends to concentrate there — the thyroid then glows in its gamma ray "light." Figure IV.15 shows a scan of an abnormal thyroid of a patient suffering from cancer of the thyroid.

Brain scans can be made either with $^{123}$I or with $^{99}$Tc.[6] These radioisotopes are incorporated in certain protein molecules (albumin) that are not absorbed by normal brain tissue; however, the tissue of a brain tumor is abnormal, and it absorbs these protein molecules with their radioactive $^{123}$I or $^{99}$Tc. Hence the tumor begins to emit gamma "light" and becomes visible on a scan. Figure IV.16 is a scan of an abnormal brain with malignant tumors.

[6] The $^{99}$Tc used for this purpose is the daughter of $^{99}$Mo. This daughter is a gamma-ray emitter.

**Fig. IV.15** Pinhole-camera scan of abnormal thyroid. The left lobe displays a localized area of low concentration of $^{123}$I, indicating the presence of a tumor. (Courtesy Michael G. Velchik, M.D., Nuclear Medicine Section, Hospital of the University of Pennsylvania.)

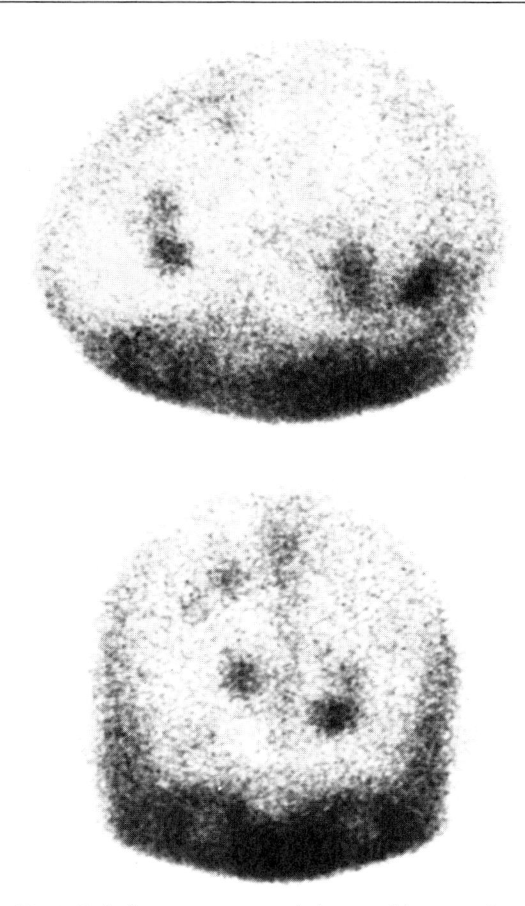

**Fig. IV.16** Pinhole-camera scan of abnormal brain with several metastatic tumors caused by breast cancer: (a) lateral view; (b) anterior view. (Courtesy Michael G. Velchik, M.D., Nuclear Medicine Section, Hospital of the University of Pennsylvania.)

Scans of other organs — lungs, heart, liver, kidneys, bones — can be made with appropriate radioisotopes. For example, cancer of the bone can be detected with $^{90}$Sr; this radioisotope is chemically similar to Ca and therefore tends to concentrate in the bones. Cancerous tissues are regions of high metabolic activity and they absorb the $^{90}$Sr more greedily than adjacent normal tissues; the consequent high concentration of radioisotope shows up in the scan picture.

When radioisotopes are used as sources of intense radiation for the destruction of diseased tissue, they may be placed either outside or inside the body. *External* sources of gamma rays are used for radiation therapy in much the same way as intense X rays. The advantage of gamma rays over megavolt X rays lies in the energy distribution of the rays — all gamma rays emitted from a particular radioisotope source have the same energy, whereas the X rays generated in X-ray tubes always have mixed energies. This uniformity of the gamma-ray energy permits a more precise delivery into the deep layers of the body. Figure IV.17 shows a $^{60}$Co teletherapy unit; this machine contains a pellet of $^{60}$Co (several hundred or a thousand curies) in a heavy lead container. The gamma rays emerge through a small channel in the lead, forming a beam that can be aimed at the diseased tissue. This machine is designed so that during operation it can be continuously rotated around the patient; the gamma-ray beam then swings around the target spot but always remains centered on it. With this arrangement, the gamma rays are highly concentrated at the "focal point" and they achieve a maximum amount of damage at this point; at the same time the gamma rays are spread out over the adjacent layers of the body and they do less damage to healthy tissues.

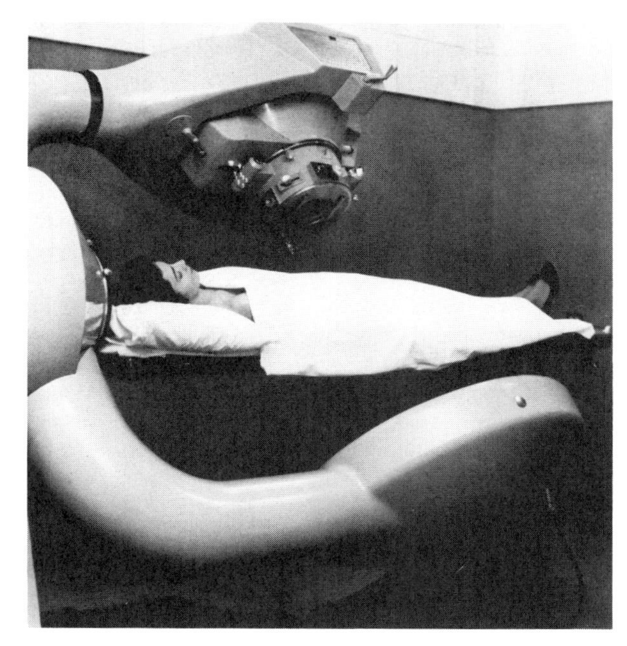

**Fig. IV.17** A $^{60}$Co radiation therapy unit.

*Internal* sources of beta and gamma rays can be used to deliver radiation to specific locations at short range. This can be accomplished by taking advantage of body chemistry to concentrate a radioisotope compound in a specific organ. For example, the partial destruction of an overactive thyroid can be accomplished with a radioiodine compound similar to that used in thyroid scanning, but administered in a much larger dose.

## Further Reading

*Radiation and Life* by E. J. Hall (Pergamon Press, Oxford, 1976) and, at a somewhat more advanced level, *Ionizing Ra-*

*diation and Life* by V. Arena (Mosby, Saint Louis, 1971) are introductions to radiation, its effects on living tissues, and its application in medicine. *Biological Effects of Radiation* by J. Coggle (Wykeham, London, 1973) provides a very clear and concise survey of the action of radiation at the cellular level, with emphasis on mammals. *The X-Ray Information Book* by P. W. Laws and The Public Citizen Health Research Group (Farrar, Straus, and Giroux, New York, 1983) gives very sensible advice on how to keep your exposure to medical X rays to a minimum.

*The Effects of Nuclear Weapons,* edited by S. Glasstone (U.S. Government Printing Office, Washington, D.C., 1964) includes a description of severe radiation injury (radiation sickness). *Energy, Ecology, and the Environment* by R. Wilson and W. J. Jones (Academic Press, New York, 1974) contains an informative chapter on low-level radiation injuries and radiation hazards to man.

The following are some recent articles dealing with radiation and radioactivity:

*Sources and Effects of Ionizing Radiation,* Scientific Committee on the Effects of Atomic Radiation, United Nations, 1977

"The Biological Effects of Low-Level Ionizing Radiation," A. C. Upton, *Scientific American,* February 1982

"Neutrons in Science and Technology," D. A. Bromley, *Physics Today,* December 1983

"Radiocarbon Dating by Accelerator Mass Spectrometry," R.E.M. Hedges and J.A.J. Gowlett, *Scientific American,* January 1986

## Questions

1. The penetration depth of alpha, beta, and gamma rays, and neutrons in water is about 10 times as large as in lead. To what is this difference due?

2. In 1968 a team of physicists under the direction of Luis Alvarez used cosmic rays to "X-ray" the pyramid of Kephren. The aim of this project was to detect secret chambers hidden in the volume of stone. Explain how a cosmic-ray detector placed under the pyramid in a subterranean vault, discovered long ago, can be used to detect empty cavities in the stone above. (No cavities were found.)

3. Use Table P.2 to make a list of all the isotopes of oxygen. What are the number of protons and the number of neutrons in the nucleus of each of these isotopes?

4. According to Table P.2, what ionizing radiations are emitted by the isotopes of carbon?

5. Why do alpha and beta emissions involve transmutation of elements, but gamma emission does not?

6. The isotope $^{24}$Na decays by emission of a negative beta ray. What is the daughter isotope produced in this decay?

7. Nuclear fission bombs produce a long-lived radioactive isotope of strontium ($^{90}$Sr), an element that is chemically similar to calcium. Explain why radioactive strontium in the environment poses a severe hazard to man.

8. Suppose that a Geiger counter placed next to a sample of radioactive isotope initially registers 3620 counts per second, and 30 minutes later it registers 450 counts per second. What is the half-life of the radioactive sample?

9. The isotope $^{14}$C is used in the radioactive dating of samples of material of biological origin, such as wood. The half-life of this isotope is 5570 years. As long as a tree is alive and breathes, the concentration of $^{14}$C relative to $^{12}$C in the wood will be the same as in the atmosphere, about 1 part in $10^{12}$. But when the tree dies, the $^{14}$C is not replenished anymore, and its decay leads to a gradual decrease of its concentration relative to that of the stable isotope $^{12}$C. Suppose that in a sample of wood found in an ancient tomb, 75% of the original amount of $^{14}$C has decayed. How old is the tomb?

10. Tritium ($^{3}$H) is a radioactive isotope of hydrogen which occurs naturally in small concentrations in ordinary water in the environment. The half-life of this isotope is 12 years. Describe how you could take advantage of this isotope to determine the age of a bottle full of wine that your wine merchant claims is 25 years old.

11. If you irradiate a sample of material with alpha, beta, or gamma rays, is the sample likely to become radioactive? What if you irradiate it with neutrons?

12. Some artificial satellites carry small nuclear reactors as power sources. What danger does this pose for people on Earth?

13. Can you guess why fast neutrons have a higher RBE than slow neutrons?

14. Insects are extremely resistant to radiation — most insect species can tolerate a radiation dose of up to 2000 or even 100,000 rads. Explain this radiation resistance. (Hint: The cells of mature insects divide only rarely.)

15. How many dental X rays were you subjected to during the course of the year? According to the numbers given in Table IV.6, how many millirems did you receive from these X rays?

16. The stone used in the construction of Grand Central Station, New York, is slightly radioactive. The radiation dose to people in the station is between 90 and 550 millirems per year. How does this compare with the average dose for inhabitants of the United States (see Table IV.5)?

17. According to Table IV.5, the natural $^{40}$K in your body gives you a radiation dose of 20 millirems per year. Estimate what extra dose you receive if your body is in close contact with other human bodies (in a crowded room) for 24 hours.

18. Astronauts in a spacecraft above the atmosphere receive a radiation dose of 100 millirems per day from cosmic rays. How many days can an astronaut stay in such a spacecraft before he attains the maximum permissible occupational dose of 5 rems (within any given year)?

19. Some new "antipersonnel" bombs discharge a large number of plastic pellets upon explosion. The material in

these pellets has been chosen so that an X ray of a wound will not reveal the pellets. Discuss this technological achievement from the point of view of physics, medicine, and morals.

CHAPTER 18

# Fluid Mechanics

A fluid is a system of particles loosely held together by their own cohesive forces or by the restraining forces exerted by the walls of a container. In contrast to the particles in a rigid body, which are permanently locked into fixed positions, the particles in a fluid body are more or less free to wander about within the volume of the body. A fluid will flow, that is, it will change its shape in response to external forces. Both liquids and gases are fluids — a body of water or a body of air will change its shape in response to the forces exerted by gravity and the container. The difference between these two kinds of fluids is that liquids are incompressible, or nearly so, whereas gases are compressible. A body of water has a constant volume independent of the shape and size of the container, whereas a body of air has a variable volume — the air always spreads so as to entirely fill the container, and it can be made to expand or contract by increasing or decreasing the size of the container.

**Fig. 18.1** The Barnard Glacier on the Alaska–Canada frontier.

Solids can sometimes behave like fluids. For example, in the manufacture or "drawing" of wire, a metal is forced to flow plastically through a small nozzle; such plastic deformation of a metal requires a very strong pull on the wire as it emerges through the nozzle. The photograph in Figure 18.1 shows another spectacular example of flow of a solid: here the ice in a glacier flows down a valley like a thick, viscous liquid. This flow of the glacier is brought about by melting and refreezing of grains of ice when subjected to large pressures.

What distinguishes the flow of a solid from that of a liquid or gas is the large pressure and large stress required to initiate the flow. The solid will not flow at all unless the applied forces are in excess of some threshold value, whereas the liquid or gas will flow even if the forces are very small. This is also true of liquids of high viscosity, such as molasses or pitch; even if the force on them is small, they will flow — albeit slowly.

In this chapter we will concentrate on the motion of **perfect fluids,** that is, those that do not offer any resistance to flow except through their inertia.

## 18.1 Density and Flow Velocity

Although a fluid is a system of particles, the number of particles in, say, a cubic centimeter of water is so large that it is not feasible to describe the state of the fluid microscopically in terms of the masses, positions, and velocities of all the individual particles. Instead, we will describe the state of the fluid in terms of its **density, velocity of flow,** and **pressure.** These quantities give us a macroscopic description of the fluid — they tell us the *average* behavior of the particles within regions of the fluid. For example, if the velocity of flow of water in a firehose is 4 m/s, this does not mean that all the water molecules have this velocity. The water molecules have a high-speed thermal motion of about 900 m/s; they move in short zigzags because they frequently collide with one another. This thermal motion of a water molecule is random; the motion is as likely to be in a direction opposite to the flow as along the flow (Figure 18.2). The flow of the water molecules along the fire hose at 4 m/s represents a slow drift superimposed on the much faster random zigzag motion. However, on a macroscopic scale we notice only the drift and not the random small-scale motion — we notice only the *average* motion of the water molecules.

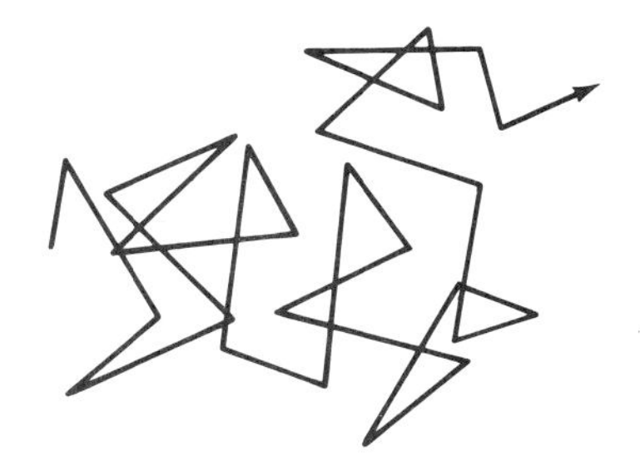

**Fig. 18.2** Motion of a water molecule in water. The straight segments are typically about $10^{-10}$ m long.

Table 18.1 lists the density, or amount of mass per unit volume, for a few liquids and gases. The densities of gases depend on the temperature and pressure (see Chapter 19); unless otherwise noted, the values listed in Table 18.1 are for a standard temperature of 0°C and a stan-

**Table 18.1** DENSITIES OF SOME FLUIDS[a]

| *Fluid* | $\rho$ |
|---|---|
| Water | |
| 0°C | 999.8 kg/m³ |
| 4°C | 1,000.0 |
| 20°C | 998.2 |
| 100°C | 958.4 |
| Sea water, 15°C | 1,025 |
| Mercury | 13,600 |
| Sodium, liquid at 98°C | 929 |
| Texas crude oil, 15°C | 875 |
| Gasoline, 15°C | 739 |
| Olive oil, 15°C | 920 |
| Human blood, 25°C | 1,060 |
| Air | |
| 0°C | 1.29 |
| 20°C | 1.20 |
| Water vapor, 100°C | 0.598 |
| Hydrogen | 0.0899 |
| Helium | 0.178 |
| Nitrogen | 1.25 |
| Oxygen | 1.43 |
| Carbon dioxide | 1.98 |
| Propane | 2.01 |

[a] At 0°C and 1 atm, unless otherwise noted.

dard pressure of 1 atm. The densities of liquids depend only slightly on pressure, but they do depend appreciably on temperature. For instance, water has a maximum density at about 4°C.

In general, both the density $\rho$ and the flow velocity $\mathbf{v}$ in a fluid will be functions of position and of time,

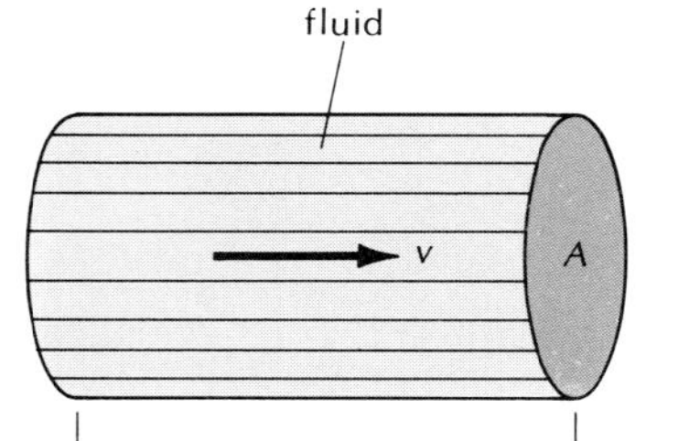


**Fig. 18.3** Flow of fluid across an area $A$.

$$\rho = \rho(x,y,z,t) \tag{1}$$

$$\mathbf{v} = \mathbf{v}(x,y,z,t) \tag{2}$$

The magnitude of $\mathbf{v}$ equals the volume of fluid that flows across a unit area perpendicular to $\mathbf{v}$ in unit time. Figure 18.3 helps to make this clear; it shows a stationary area $A$ and a volume of fluid about to cross this area. The fluid that crosses the area in a time $\Delta t$ is initially in a cylinder of base $A$ and length $v\ \Delta t$. The amount of fluid volume that crosses the area $A$ is therefore

$$\Delta V = Av\ \Delta t \tag{3}$$

and the amount that crosses per unit area and per unit time is

$$\frac{1}{A}\frac{\Delta V}{\Delta t} = v \tag{4}$$

EXAMPLE 1. The water in a fire hose of diameter 6.4 cm ($2\frac{1}{2}$ in.) has a speed of flow of 4.0 m/s. At what rate does this hose deliver water? Give the answer both in $m^3/s$ and in kg/s.

SOLUTION: The cross-sectional area of the hose is $A = \pi \times (3.2 \times 10^{-2}\ m)^2 = 3.2 \times 10^{-3}\ m^2$. Hence the rate of delivery is

$$\frac{\Delta V}{\Delta t} = Av = 3.2 \times 10^{-3}\ m^2 \times 4.0\ m/s \tag{5}$$

$$= 1.3 \times 10^{-2}\ m^3/s$$

To find the rate of delivery in terms of the mass of water, we must multiply $\Delta V/\Delta t$ by $\rho$:

$$\frac{\Delta m}{\Delta t} = \rho\frac{\Delta V}{\Delta t} = 1.0 \times 10^3\ kg/m^3 \times 1.3 \times 10^{-2}\ m^3/s \tag{6}$$

$$= 13\ kg/s$$

## 18.2 Steady Incompressible Flow; Streamlines

*Velocity field*

Graphically, we can represent the velocity of flow in a fluid by drawing the velocity vector at each point of space; for example, Figure 18.4 shows the velocity vectors for the tidal flow of the water in a bay. Such a picture of all these velocity vectors is often called the **velocity field** of the fluid. If the velocity field is a function of time, then we must draw separate diagrams for different times.

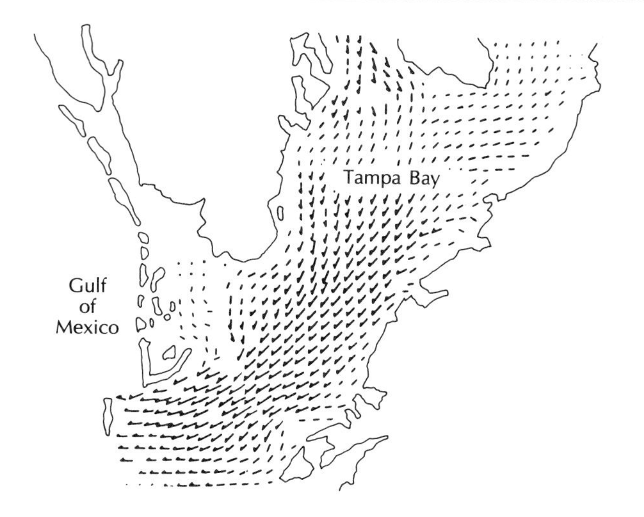


**Fig. 18.4** The velocity vectors show the tidal flow of water in Tampa Bay at one stage of the tidal cycle.

*Steady flow*

Most of the examples in this chapter will deal with **steady flow,** for which the velocity at each point of space is independent of time. Obviously, steady flow requires a source that can continuously supply fluid and a sink that can continuously withdraw an equal amount of fluid. The tidal flow shown in Figure 18.4 is not steady—the amount of water that can enter or leave the bay is limited, and the flow depends on the stage of the tide. Figure 18.5 shows an example of steady flow of water around a cylindrical obstacle. The water enters the picture in a broad stream from a source on the left, and disappears in a similar broad stream toward a sink on the right.[1]

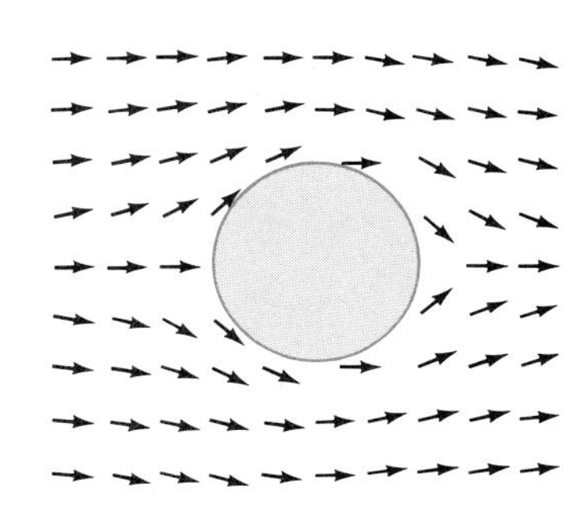

**Fig. 18.5** Velocity vectors for water flowing around a cylinder.

For the steady flow of an incompressible fluid, such as water, the picture of velocity vectors can be replaced by an alternative graphical representation. Suppose we focus our attention on a small volume element of water, say, 1 mm³ of water, and we observe the path of this 1 mm³ from the source to the sink. The path traced out by the small volume element is called a **streamline**. Neighboring volume elements will trace out neighboring streamlines. Figure 18.6 shows the pattern of streamlines for the same steady flow of water that we already represented in Figure 18.5 by means of velocity vectors. The streamlines on the far left (and far right) of Figure 18.6 are evenly spaced to indicate the uniform and parallel flow in this region.

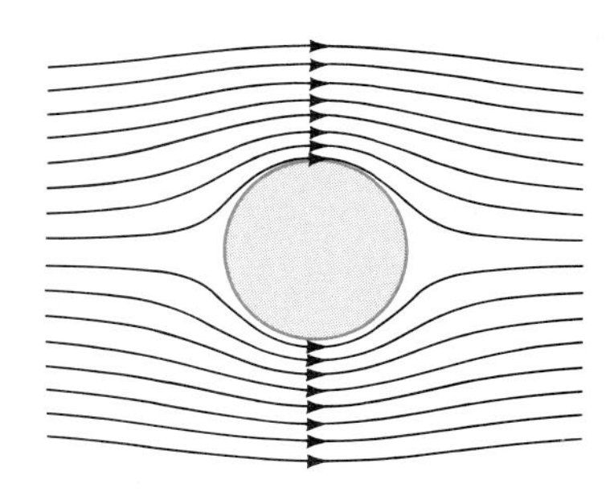

**Fig. 18.6** Streamlines for water flowing around a cylinder.

The information contained in the pattern of streamlines is equivalent to that in the velocity field. If we know the velocity field, we can trace out the motion of a small volume element and therefore construct the streamlines. But the converse is also true — if we know the streamlines, we can reconstruct the velocity field. We can do this by means of the following rule: The direction of the velocity at any one point is tangent to the streamline, and the magnitude of the velocity is proportional to the density of the streamlines. The first part of the rule is self-evident. To establish the second part, consider a bundle of streamlines forming a pipelike region, called a **stream tube.** Any fluid inside the stream tube will have to move along the tube; it cannot cross the surface of the tube because streamlines never cross. The tube

[1] Figure 18.5 is based on the pretense that water is free of friction (free of viscosity). This is a reasonable approximation, except at the surface of the obstacle. There, the friction between the water and the surface slows the water to a standstill and produces a thin layer in which the flow of the water is very complicated (turbulent). Figure 18.5 does not show this layer.

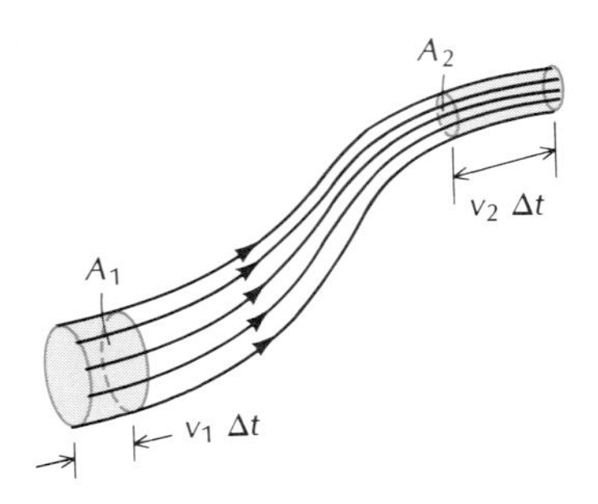


**Fig. 18.7** A stream tube. A volume $A_1v_1\Delta t$ enters at the left, and a volume $A_2v_2\Delta t$ leaves at the right in a time $\Delta t$.

therefore plays the same role as a pipe made of some impermeable material — it serves as a conduit for the fluid. If we assume that the tube is very narrow, so that its cross-sectional area is infinitesimal, the velocity of flow will vary along the length of the tube but it will be nearly the same at all points on a given cross-sectional area. For instance, on the area $A_1$ (Figure 18.7) the velocity is $v_1$, and on the area $A_2$ the velocity is $v_2$. In a time $\Delta t$, the fluid volume that enters across the area $A_1$ is $v_1A_1\ \Delta t$, and the fluid volume that leaves across the area $A_2$ is $v_2A_2\ \Delta t$. The amount of fluid that enters must match the amount that leaves, since, under steady conditions, fluid cannot accumulate in the segment of tube between $A_1$ and $A_2$. Hence

$$v_1A_1\ \Delta t = v_2A_2\ \Delta t \tag{7}$$

or

*Continuity equation*

$$\boxed{v_1A_1 = v_2A_2} \tag{8}$$

This relation is called the **continuity equation.** It shows that along any thin stream tube, the speed of flow is inversely proportional to the cross-sectional area of the stream tube. The density of streamlines inside the tube is the number of such lines divided by the cross-sectional area; since the number of stream lines entering $A_1$ is necessarily the same as that leaving $A_2$, the density of streamlines is inversely proportional to the cross-sectional area. This implies that the speed at any point in the fluid is directly proportional to the density of streamlines at that point. For example, in Figure 18.6 the speed of the water is large at the top and bottom of the obstacle (large density of streamlines) and smaller to the left and right (smaller density of streamlines).

In experiments on fluid flow, the streamlines of a fluid can be made directly visible by several clever techniques. If the fluid is water, we can place grains of dye at diverse points within the volume of water; the dye will then be carried along by the flow and it will mark the streamlines. The photograph of Figure 18.8 shows a pattern of streamlines made visible with this technique. The water emerges from a pointlike source on the left and disappears into a pointlike sink on the right. Small grains of potassium permanganate create the streamers of dye while dissolving in the water.

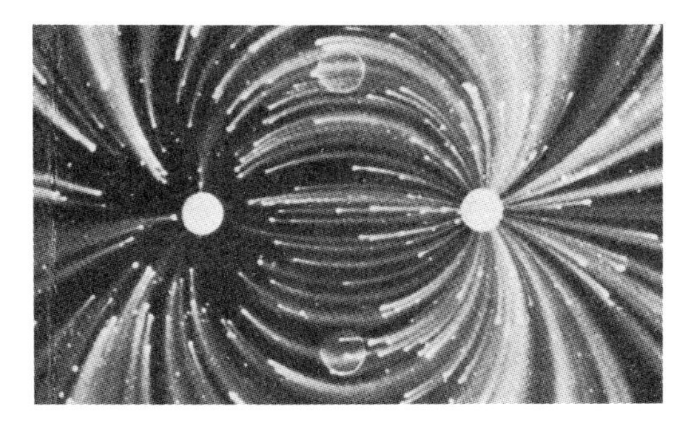

**Fig. 18.8** Streamers of dye indicate the streamlines in water flowing from a source (left) to a sink (right).

If the fluid is air, we can make the streamlines visible by releasing smoke from small jets at diverse points within the flow of air. The photograph of Figure 18.9 shows fine trails of smoke marking the stream-

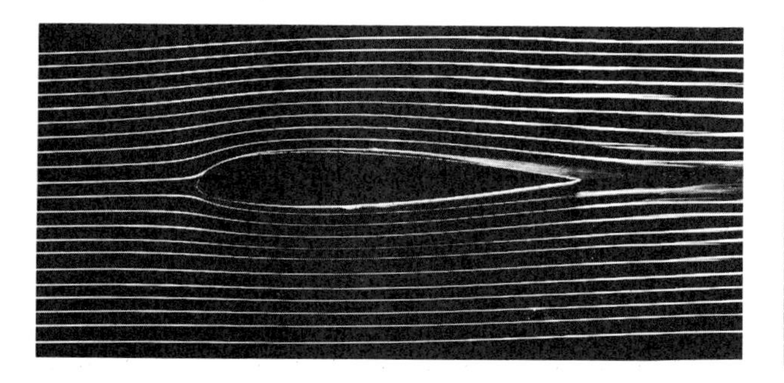

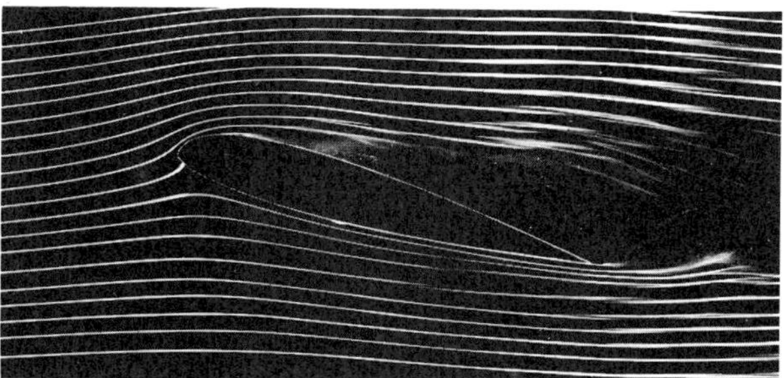

**Fig. 18.9** (left) Fine trails of smoke indicate the steamlines in air flowing around the wing of an airplane.

**Fig. 18.10** (right) Here the wing is in a partial stall, and the flow behind the wing has become turbulent.

lines in air flowing past a scale model of the wing of an airplane in a wind tunnel. The experimental investigation of such streamline patterns plays an important role in airplane design. Incidentally: The flow of air can be regarded as nearly incompressible provided that the speed of flow is well below the speed of sound (331 m/s). Although the air will suffer some changes of density in its flow around obstacles, the changes are usually small enough to be neglected in comparison with the standard density.

Finally, Figure 18.10 shows an example of turbulent flow. In the region behind the wing, the streamers of smoke become twisted and chaotic. This is due to the generation of vortices in this region. As the vortices form, grow, break away, and disappear in quick succession, the velocity of flow fluctuates violently. The flow of the fluid becomes very unsteady and very irregular. The formation of vortices and the onset of turbulence have to do with the viscosity of the fluid. It is a general rule that vortices and turbulence will develop in a fluid of given viscosity whenever the velocity of flow, the size of the obstacle, or both exceed a certain limit.

---

EXAMPLE 2. The world's tallest fountain (at Fountain Hills, Arizona; Figure 18.11) shoots water to a height of 170 m at the rate of 26,000 liters/min. What is the flow velocity at the base of the fountain? At a height of 100 m? What is the diameter of the water column at the base? At a height of 100 m? Assume that the flow is steady and incompressible, and that the falling water does not interfere with the rising water. Ignore friction.

**Fig. 18.11** The world's tallest fountain.

SOLUTION: The initial speed required to attain a height of 170 m is determined by Eq. (2.29),

$$\tfrac{1}{2}v_0^2 = g(x - x_0)$$

with $x - x_0 = 170$ m, this gives

$$v_0 = \sqrt{2g(x - x_0)} = \sqrt{2 \times 9.8 \text{ m/s}^2 \times 170 \text{ m}}$$

$$= 58 \text{ m/s}$$

The speed at a height of 100 m is also determined by Eq. (2.29),

$$\tfrac{1}{2}v^2 = \tfrac{1}{2}v_0^2 - g(x - x_0)$$

with $x - x_0 = 100$ m and $v_0 = 58$ m/s this gives

$$v = \sqrt{v_0^2 - 2g(x - x_0)} = \sqrt{(58 \text{ m/s})^2 - 2 \times 9.8 \text{ m/s}^2 \times 100 \text{ m}}$$

$$= 37 \text{ m/s}$$

To calculate the cross-sectional area at the base, we use Eq. (4),

$$A_0 = \frac{1}{v_0}\frac{\Delta V}{\Delta t}$$

The rate of delivery of the fountain is

$$\frac{\Delta V}{\Delta t} = \frac{2.6 \times 10^4 \text{ liters}}{\text{min}} = \frac{26 \text{ m}^3}{60 \text{ s}}$$

and therefore

$$A_0 = \frac{1}{58 \text{ m/s}} \times \frac{26 \text{ m}^3}{60 \text{ s}} = 7.5 \times 10^{-3} \text{ m}^2$$

The diameter of a circle of this area is $2\sqrt{A_0/\pi} = 9.8 \times 10^{-2}$ m.

To calculate the cross-sectional area at a height of 100 m, we use the continuity equation,

$$vA = v_0 A_0$$

or

$$A = \frac{v_0}{v}A_0 = \frac{58 \text{ m/s}}{37 \text{ m/s}} \times 7.5 \times 10^{-3} \text{ m}^2 = 1.2 \times 10^{-2} \text{ m}^2$$

The diameter of a circle of this area is $2\sqrt{A_0/\pi} = 1.2 \times 10^{-1}$ m.

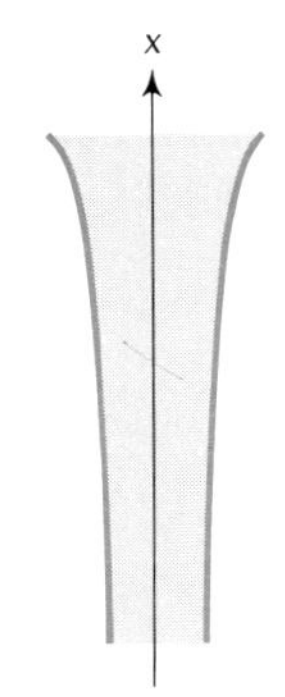


**Fig. 18.12** Increase of the width of a cylindrical water column with height. The vertical scale in this picture has been compressed.

COMMENTS AND SUGGESTIONS: Note that the water column is narrow at the base and widens at the top. The photograph in Figure 18.11 does not show this widening because the column of rising water is hidden in a curtain of falling water; besides, the emerging water column does not have a uniform velocity across the width of the nozzle, and it is affected by friction and by entrainment of air. Figure 18.12 shows the calculated width of the water column as a function of height, under ideal conditions. Such a change of width as a function of height is more easily observed in a falling water column than in a rising column. For instance, you can clearly see the narrowing of a falling water column when you let water stream out of a faucet.

## 18.3 Pressure

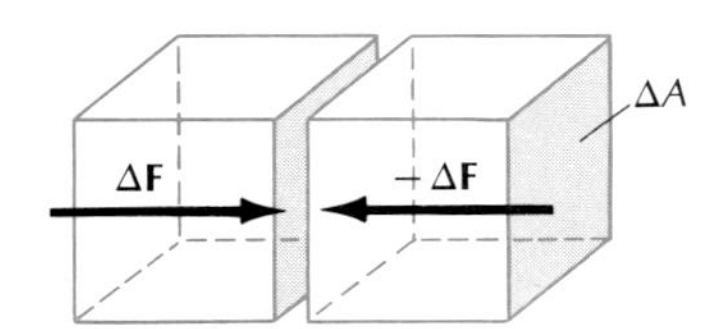


**Fig. 18.13** Adjacent small cubes of fluid exerting forces on each other.

The **pressure** within a fluid is defined in terms of the force that a small volume of fluid exerts on an adjacent volume or on the adjacent wall of a container. Figure 18.13 shows two small adjacent volumes of fluid of cubical shape. The cube of fluid on the left presses against the cube on the right, and vice versa. Suppose that the magnitude of the perpendicular force between the two cubes is $\Delta F$ and that the area of one face of one of the cubes is $\Delta A$; then the pressure is defined by

*Pressure*

$$\boxed{p = \frac{\Delta F}{\Delta A}} \qquad (9)$$

where the limit $\Delta A \to 0$ is to be taken. According to this definition, pressure is simply the force per unit area. Note that the pressure $p$ is a scalar quantity, without direction. We cannot associate a direction with the pressure, because each small volume exerts pressure forces in all directions and the pressure forces in all directions are equal (in the limit $\Delta A \to 0$).

In the metric system, the unit of pressure is the N/m², which has been baptized with the name **pascal** (Pa):

$$1 \text{ Pa} = 1 \text{ N/m}^2 \tag{10}$$

The British unit of pressure is the lbf/in.². Other units in common use are the **atmosphere** (atm),

$$1 \text{ atm} = 1.013 \times 10^5 \text{ N/m}^2 = 14.7 \text{ lbf/in.}^2 \tag{11}$$

the millimeter of mercury (mmHg), or **torr,**

$$1 \text{ mmHg} = 1 \text{ torr} = \tfrac{1}{760} \text{ atm} \tag{12}$$

and the **millibar** (mbar),

$$1 \text{ mbar} = 10^2 \text{ N/m}^2 = 0.750 \text{ mmHg} \tag{13}$$

Table 18.2 gives diverse examples of values of pressure.

**Blaise Pascal,** *1623–1662, French scientist. He made important contributions to mathematics and is regarded as the founder of modern probability theory. In physics, he performed experiments on atmospheric pressure and on the equilibrium of fluids.*

**Table 18.2** SOME PRESSURES

| | |
|---|---|
| Core of neutron star | $1 \times 10^{38}$ N/m² |
| Center of Sun | $2 \times 10^{16}$ N/m² |
| Highest shock pressure achieved with nuclear explosion | $7 \times 10^{12}$ N/m² |
| Highest sustained pressure achieved in laboratory | $5 \times 10^{11}$ N/m² |
| Center of Earth | $4 \times 10^{11}$ N/m² |
| Bottom of Pacific Ocean (5.5-km depth) | $6 \times 10^{7}$ N/m² |
| Water in core of nuclear reactor | $1.6 \times 10^{7}$ N/m² |
| Overpressure[a] in automobile tire | $2 \times 10^{5}$ N/m² |
| Air at sea level (1 atm) | $1.0 \times 10^{5}$ N/m² |
| Overpressure at 7 km from 1-megaton explosion | $3 \times 10^{4}$ N/m² |
| Air in funnel of tornado | $2 \times 10^{4}$ N/m² |
| Overpressure in human heart | |
| systolic | $1.6 \times 10^{4}$ N/m² |
| diastolic | $1.1 \times 10^{4}$ N/m² |
| Lowest vacuum achieved in laboratory | $10^{-12}$ N/m² |

[a] The *overpressure* is the amount of pressure in excess of normal atmospheric pressure.

EXAMPLE 3. In 1934 C. W. Beebe and O. Barton descended to a depth of 923 m below the surface of the ocean in a steel bathysphere of diameter 1.45 m (Figure 18.14). The pressure at this depth is $9.3 \times 10^6$ N/m². Under these conditions, what is the force pressing one-half of the steel sphere toward the opposite half?

SOLUTION: Figure 18.14(b) is a diagram of the right half of the steel sphere. The arrows show the pressure force exerted by the water. On a small area element $\Delta A$ of the sphere, the pressure force is $p\, \Delta A$, and the horizontal compo-

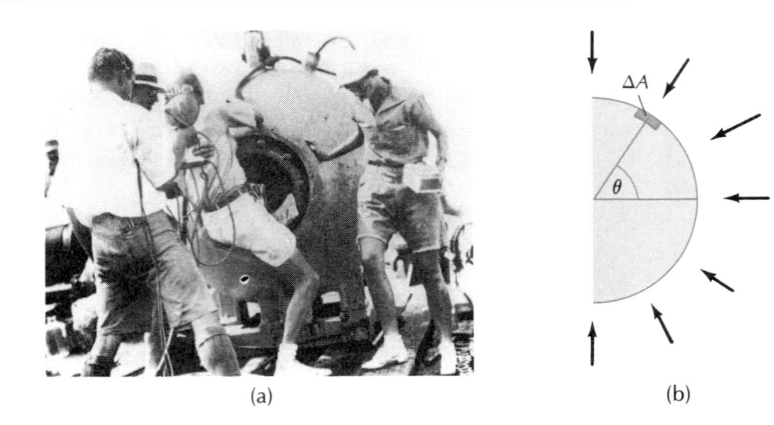


**Fig. 18.14** (a) Explorer C. W. Beebe entering the bathysphere. (b) Pressure forces on the sphere.

nent of this force is $p\ \Delta A \cos\theta$. But $\Delta A \cos\theta$ is the projection of the area $\Delta A$ on a vertical plane, i.e., it is the fraction of the area facing to the right, in silhouette. For the complete half sphere, the net area facing to the right is $\pi R^2$ and the net force toward the left is therefore

$$p\pi R^2 = 9.3 \times 10^6\ \text{N/m}^2 \times \pi \times (0.72\ \text{m})^2$$

$$= 1.5 \times 10^7\ \text{N}$$

## 18.4 Pressure in a Static Fluid

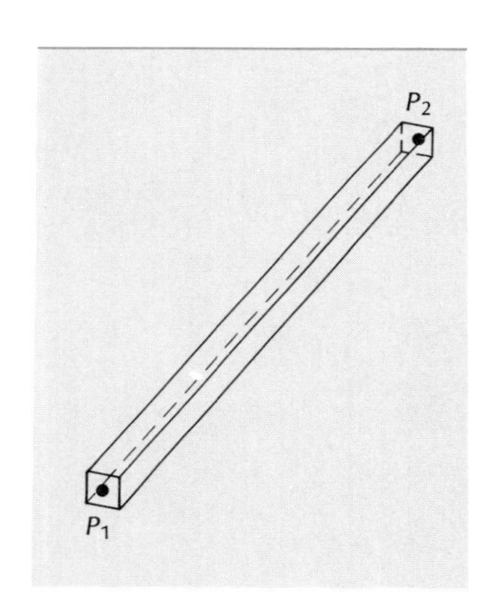


**Fig. 18.15** Parallelepiped in a static fluid.

We will now consider a fluid in static equilibrium so that the velocity of flow is everywhere zero. An example of static fluid is the air in a closed room with no air currents, or the water in the pipes of a house when there is no flow. At first we will neglect gravity and pretend that the only forces acting on the fluid are those exerted by the walls of the container. Under these conditions, the pressure at all points within the fluid must be the same. To see that this is so, consider two points $P_1$ and $P_2$ in the fluid and imagine a long, thin parallelepiped with bases at these two points (Figure 18.15). The fluid outside the parallelepiped exerts pressure forces on the fluid inside the parallelepiped. The component of the net force along the long direction of the parallelepiped is entirely due to the forces on the bases at $P_1$ and $P_2$. Since the parallelepiped of fluid is to remain static, these forces on the opposite bases must be equal in magnitude. Thus, the pressures at $P_1$ and $P_2$ must be equal. If the shape of the fluid does not permit the points $P_1$ and $P_2$ to be connected by a single straight parallelepiped lying entirely within the fluid, then we have to use several such parallelepipeds placed end to end at different angles, and apply the above argument to each in succession. For example, the pressure of the air is the same at all points of a room — if the pressure is 1 atm in one corner of the room, it will be the same at any other point of the room.

The uniformity of pressure throughout a fluid implies that if we apply a pressure to some part of the surface of a confined fluid by means of a piston or a weight pushing against the surface, then this pressure will be transmitted without change to all parts of the fluid.

*Pascal's Principle*

This rule for the transmission of pressure is **Pascal's Principle,** which finds widespread application in the design of hydraulic presses, jacks,

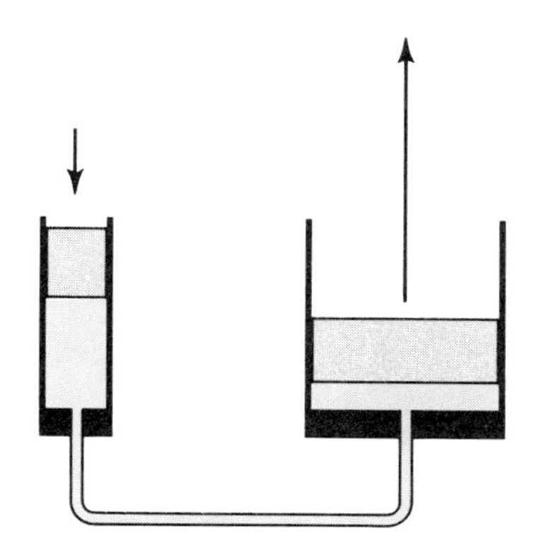

**Fig. 18.16** Hydraulic press (schematic).

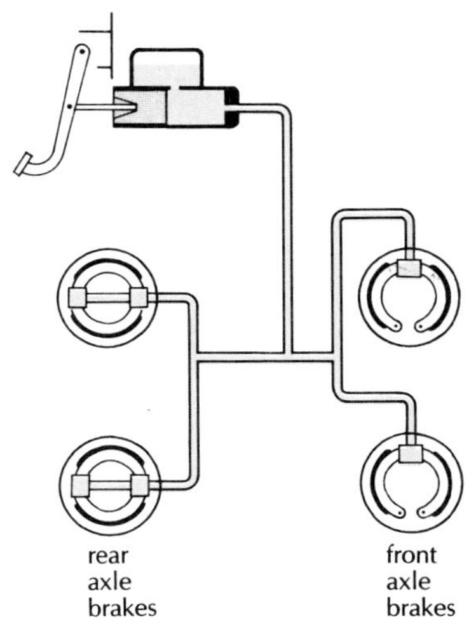


**Fig. 18.17** Hydraulic brake system of an automobile.

and remote controls. Figure 18.16 is a schematic diagram of a hydraulic press consisting of two cylinders with pistons, one small and one large. The cylinders are filled with an incompressible fluid, and they are connected by a pipe. By pushing down on the small piston, we increase the pressure in the fluid and transmit this increase of pressure to the large piston. Since the pressure is the same on both pistons, the forces on the pistons are in the ratio of the areas of their faces; thus, a small force on the small piston will generate a large force on the large piston. The brake systems and other control systems on automobiles, trucks, and aircraft employ such hydraulic mechanisms. Figure 18.17 shows the hydraulic brake system of an automobile. The brake pedal pushes on the small master piston, and the resultant pressure of the brake fluid pushes the large slave pistons and activates the brakes.

EXAMPLE 4. Suppose that the diameter of the small piston in Figure 18.16 is 1.9 cm and that of the large piston is 8.9 cm. If you exert a force of 130 N on the master piston, what force will this generate on the slave piston?

SOLUTION: The forces are in the ratio of the areas of the pistons:

$$F_2 = F_1 \frac{A_2}{A_1} = 130 \text{ N} \times \frac{(8.9)^2}{(1.9)^2} = 2.9 \times 10^3 \text{ N}$$

Next, we want to take into account the effect of gravity on the pressure in a fluid. For a static fluid in the gravitational field of the Earth, such as the water of a calm lake, the pressure force at any given depth must support the weight of all the overlaying mass of fluid; consequently, the pressure must increase as a function of depth. To derive the dependence of pressure on depth, we must consider the condition for the equilibrium of a small volume of fluid. Figure 18.18 shows a small cube at some depth below the surface of the fluid. As always, the $z$ coordinate is reckoned as positive in the upward direction; hence the depth of the cube corresponds to some negative value of $z$. The dimension of the cube is $dz \times dz \times dz$ and hence its weight is

$$g\, dm = g\rho\, dz \times dz \times dz \tag{14}$$

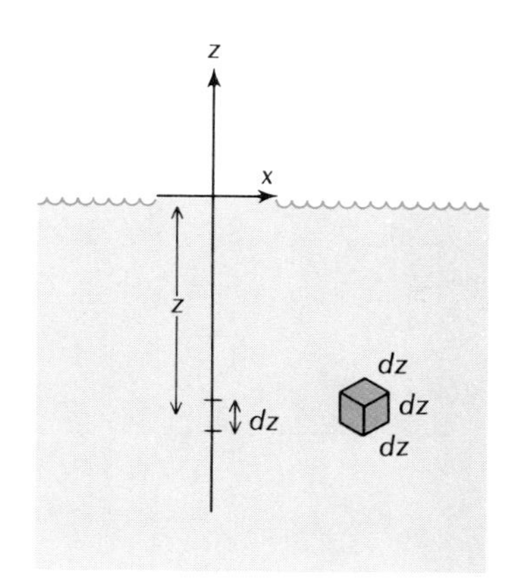


**Fig. 18.18** A small cube of fluid at a depth $z$ below the surface.

where $\rho$ is the density of the fluid. The weight of the cube must be balanced by the vertical forces contributed by the pressure. Suppose that the pressure at the midpoint of the cube is $p$; the pressure at the top of the cube is then $p - \frac{1}{2}dp$ and the pressure at the bottom is $p + \frac{1}{2}dp$, where $dp$ represents the increase of pressure in the interval $dz$. Since the area of the faces is $dz \times dz$, the vertical pressure force on the cube will be

$$dz \times dz(p + \tfrac{1}{2}dp) - dz \times dz(p - \tfrac{1}{2}dp) = dz \times dz \times dp \qquad (15)$$

Setting Eqs. (14) and (15) equal to each other, we find

$$dz \times dz \times dp = -g\rho \, dz \times dz \times dz \qquad (16)$$

or

$$dp = -\, g\rho \, dz \qquad (17)$$

This formula gives the small increase in pressure for a small increase in depth. For a compressible fluid the density $\rho$ will be a function of $p$ and, consequently, a function of $z$, and we cannot proceed further without knowledge of the explicit form of this function. However, for an incompressible fluid the density is constant and therefore Eq. (17) shows that the small increment in pressure is proportional to the small increment in depth; the total change in pressure is therefore directly proportional to the total depth,

*Hydrostatic pressure for incompressible fluid*

$$\boxed{p - p_0 = -g\rho z} \qquad (18)$$

Here $p_0$ is the pressure at the surface of the fluid.

According to Eq. (18), if we increase the pressure $p_0$ applied at the surface of the fluid, the pressure $p$ at a given depth will increase by the same amount. This transmission of pressure from the surface to any other point within the fluid is, of course, in agreement with Pascal's Principle.

---

EXAMPLE 5. What is the pressure at a depth of 10 m below the surface of a lake? Assume that the pressure of air at the surface of the lake is 1.0 atm.

SOLUTION: With $p_0 = 1$ atm $= 1.01 \times 10^5$ N/m², $\rho = 1.00 \times 10^3$ kg/m³, and $z = -10$ m, Eq. (18) gives

$$\begin{aligned} p &= p_0 - g\rho z \\ &= 1.01 \times 10^5 \text{ N/m}^2 + 9.81 \text{ m/s}^2 \times 1.00 \times 10^3 \text{ kg/m}^3 \times 10 \text{ m} \\ &= 1.99 \times 10^5 \text{ N/m}^2 \end{aligned}$$

Thus, the pressure is 1 atm at the surface and about 2 atm at a depth of 10 m, that is, the pressure increases by about 1 atm per 10 m of water.

---

EXAMPLE 6. What is the change in atmospheric pressure between the basement of a house and the attic, at a height of 10 m above the basement? The density of air is 1.29 kg/m³.

SOLUTION: Although air is compressible fluid, the change in its density is small if the change in altitude is small, as it is in the present example. Therefore Eq. (18) is a good approximation:

$$\begin{aligned} p - p_0 &= -g\rho z = -9.8 \text{ m/s}^2 \times 1.29 \text{ kg/m}^3 \times 10 \text{ m} \\ &= -\,1.3 \times 10^2 \text{ N/m}^2 = -0.95 \text{ mmHg} \end{aligned}$$

Hence the pressure decreases by about 1 mmHg per 10 m of air. This decrease of pressure can be readily detected by carrying an ordinary barometer from the basement to the attic of the house.

Several simple instruments for the measurement of pressure make use of a column of liquid. Figure 18.19 shows a **mercury barometer** consisting of a tube of glass, about 1 m long, closed at the upper end and open at the lower end. The tube is filled with mercury, except for a small empty space at the top. The bottom of the tube is immersed in an open bowl filled with mercury. The atmospheric pressure acting on the exposed surface of mercury in the bowl prevents the mercury from flowing out of the tube. At the level of the exposed surface, the pressure exerted by the column of mercury is $g\rho h$, where $\rho = 1.3595 \times 10^4$ kg/m$^3$ is the density of mercury and $h$ the height of the mercury column. For equilibrium, this pressure must match the atmospheric pressure:

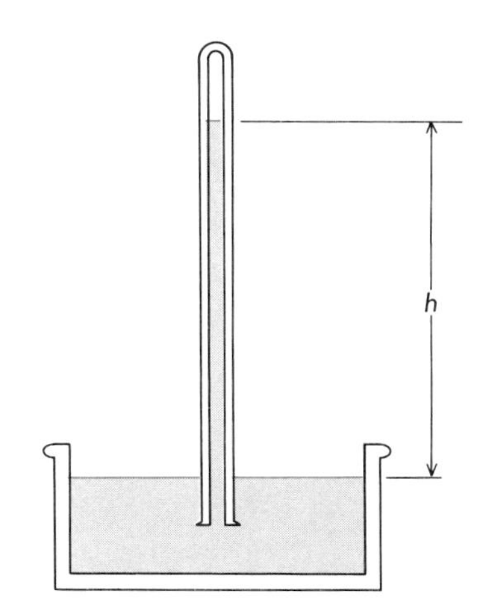


**Fig. 18.19** Mercury barometer.

$$p_{\rm at} = g\rho h \tag{19}$$

This equation permits a simple determination of the atmospheric pressure from a measurement of the height of the mercury column.

In view of the direct correspondence between the atmospheric pressure and the height of the mercury column, the pressure is often quoted in terms of this height, usually expressed in millimeters [see Eq. (12)]. However, for accurate measurements we must take into account that the acceleration of gravity depends on location and that the density of mercury depends on temperature. The convention that we will follow is that the height of the mercury column to be quoted as an indicator of pressure is not the actual height of the column but, rather, the height that the column would have if $g$ had the standard value 9.8066 m/s$^2$ and $\rho$ had the standard value $1.3595 \times 10^4$ kg/m$^3$. The average value of the **atmospheric pressure** at sea level is 760 mmHg; by definition, this is 1 atm. Hence

*Atmospheric pressure*

$$\begin{aligned} 1 \text{ atm} &= 760 \text{ mmHg} = 0.760 \text{ m} \times \rho \times g \\ &= 0.760 \text{ m} \times 1.3595 \times 10^5 \text{ kg/m}^3 \times 9.8066 \text{ m/s}^2 \\ &= 1.013 \times 10^5 \text{ N/m}^2 \end{aligned} \tag{20}$$

This value for 1 atm has already been quoted in Eq. (11).

Figure 18.20 shows an open-tube **manometer,** a device for the measurement of the pressure of a fluid contained in a tank. The tube contains mercury, or water, or oil. One side of the tube is in contact with the fluid in the tank; the other is in contact with the air. The fluid in the tank therefore presses down on one end of the mercury column and the air presses down on the other end. The difference $h$ in the heights of the levels of mercury at the two ends gives the difference in the pressure at the two ends,

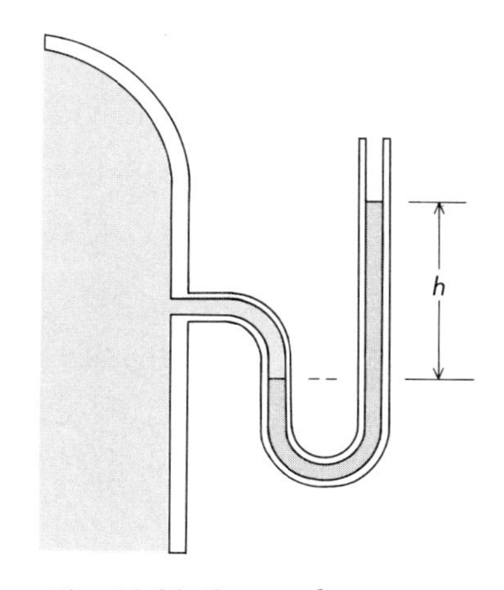


**Fig. 18.20** Open-tube manometer.

$$p - p_{\rm at} = g\rho h \tag{21}$$

Hence this kind of manometer indicates the amount of pressure in the tank in excess of the atmospheric pressure. This excess is called the **overpressure,** or **gauge pressure.** It is well to keep in mind that many pressure gauges used in engineering practice are calibrated in terms of overpressure rather than absolute pressure. For instance, the pressure gauges used for automobile tires read overpressure.

## 18.5 Archimedes' Principle

*Buoyant force*

If a body is partially or totally immersed in a static fluid in gravitational equilibrium, the pressure of the fluid exerts a vertically upward force on the body. This force is called the **buoyant force.** The magnitude of this force is given by **Archimedes' Principle:**

*Archimedes' Principle*

*The buoyant force on an immersed body has the same magnitude as the weight of the fluid displaced by the body.*

The proof of this famous principle is very simple. Imagine that we replace the immersed volume of a body by an equal volume of fluid (Figure 18.21). The volume of fluid will then be in static equilibrium. Obviously, this requires a balance between the weight of the fluid and the resultant of all the pressure forces acting on the surface enclosing this volume of fluid. But the pressure forces on the surface of the original immersed body are exactly the same as the pressure forces on the surface of the volume of fluid by which we have replaced it. Hence the magnitude of the resultant of the pressure forces acting on the original body must equal the weight of the displaced fluid.

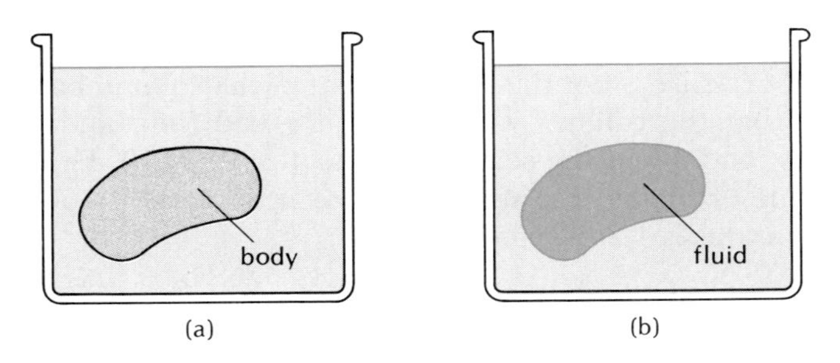


**Fig. 18.21** (a) Submerged body. (b) Volume of fluid of the same shape as the body.

EXAMPLE 7. A chunk of ice floats in water (Figure 18.22). What percentage of the volume of ice will be above the level of the water? The density of ice is 917 kg/m$^3$.

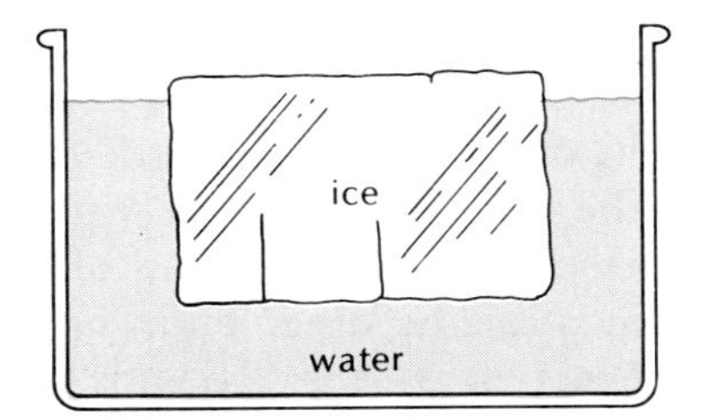


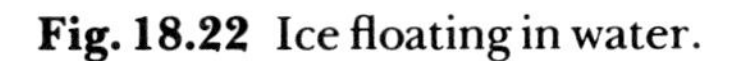

**Fig. 18.22** Ice floating in water.

SOLUTION: If the mass of the chunk of ice is, say, 1000 kg, it must displace 1 m$^3$. The volume of ice below the water level is then 1 m$^3$, whereas the total volume of ice is 1000 kg/(917 kg/m$^3$) = 1.091 m$^3$. The fraction of ice above the water level is therefore 0.091 m$^3$/1.091 m$^3$ = 0.083, or 8.3%.

EXAMPLE 8. A hot-air balloon (Figure 18.23) has a volume of $2.20 \times 10^3$ m$^3$; it is filled with hot air of density 0.96 kg/m$^3$. What maximum load can this balloon lift when surrounded by cold air of density 1.29 kg/m$^3$?

SOLUTION: The mass of cold air displaced by the balloon is 1.29 kg/m$^3$ $\times$ 2.20 $\times$ 10$^3$ m$^3$ = 2.84 $\times$ 10$^3$ kg. The weight of this cold air is $g \times 2.84 \times 10^3$ kg, and this is therefore the buoyant force on the balloon. The buoyant force must support both the weight of the hot air and the load. The weight of the hot air is $g \times 0.96$ kg/m$^3$ $\times$ 2.20 $\times$ 10$^3$ m$^3$ = $g \times 2.11 \times 10^3$ kg. Hence the weight of the load can be at most

$$g \times 2.84 \times 10^3 \text{ kg} - g \times 2.11 \times 10^3 \text{ kg} = g \times 730 \text{ kg}$$

that is, the mass of the load can be at most 730 kg. Note that in this context, the "load" includes the hull of the balloon, the gondola, and everything attached to it.

**Fig. 18.23** A hot-air balloon.

## 18.6 Fluid Dynamics; Bernoulli's Equation

Although it is not difficult to derive an equation of motion for a fluid corresponding to Newton's Second Law for a particle, the solution of this equation is usually quite laborious. One of the troubles is that the pressure affects the motion of the fluid, and in turn the motion affects the pressure. Hence we must deal with simultaneous equations for motion and for pressure. When faced with such an awkward set of equations, we find it very profitable to make use of conservation theorems. In the following discussion we will formulate the conservation theorem for energy in the special case of steady flow of an incompressible fluid with no friction.

We know from Section 18.2 that steady incompressible flow can be described by streamlines. As in the derivation of the equation of continuity, we consider a bundle of streamlines forming a thin stream tube. The fluid flows inside this tube as though the surface of the tube were an impermeable pipe. Figure 18.24a shows a segment of this "pipe"; this segment contains some mass of fluid. Figure 18.24b shows the same mass of fluid at a slightly later time — the fluid has moved toward the right. During this movement the pressure at the left and at the right end of the segment does some work on the mass of fluid. By energy conservation, this work must equal the change of kinetic and potential energy. To express this mathematically, we begin by calculating the work done by the pressure. As the left end of the mass of fluid moves through a distance $\Delta l_1$, the work done by the pressure is

$$\Delta W_1 = p_1 A_1 \, \Delta l_1 \tag{22}$$

Since the product $A_1 \, \Delta l_1$ is the volume $\Delta V$ vacated by the retreat of the fluid on the left end, we can also write this as

$$\Delta W_1 = p_1 \, \Delta V \tag{23}$$

Likewise, the work done by the pressure at the right end is

$$\Delta W_2 = -p_2 \, \Delta V \tag{24}$$

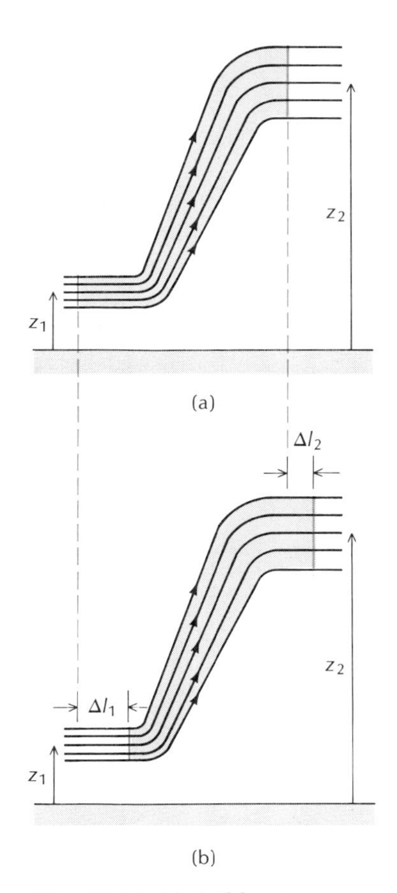


**Fig. 18.24** (a) A thin stream tube. (b) Motion of fluid along the stream tube.

Note that the *same* volume $\Delta V$ appears in Eqs. (23) and (24) — since the fluid is incompressible, the volume vacated by the retreat of the fluid at one end must equal the volume occupied by the advance of the fluid at the other end. The net work is then

$$\Delta W = \Delta W_1 + \Delta W_2 = p_1\, \Delta V - p_2\, \Delta V \tag{25}$$

**Daniel Bernoulli** (bernooyee), *1700–1782, Swiss physician, physicist, and mathematician. He was professor of anatomy and natural philosophy at Basel, where he wrote his great treatise* Hydrodynamica, *which included the equation named after him. Several other members of the Bernoulli family made memorable contributions to mathematics and physics.*

The change of kinetic and potential energy is entirely due to the changes at the ends of the mass of fluid; everywhere else the shift of the fluid merely replaces fluid of some kinetic and potential energy with fluid of exactly the same kinetic and potential energy. The change at the ends involves replacing a mass $\Delta m$ of fluid, of speed $v_1$ at height $z_1$, by an equal mass $\Delta m$, of speed $v_2$ at height $z_2$. The corresponding change of kinetic and potential energy is

$$\Delta K + \Delta U = \tfrac{1}{2}\Delta m v_2^2 - \tfrac{1}{2}\Delta m v_1^2 + \Delta m g z_2 - \Delta m g z_1 \tag{26}$$

Setting the expressions on the right sides of Eqs. (25) and (26) equal to each other, we find

$$p_1\, \Delta V - p_2\, \Delta V = \tfrac{1}{2}\Delta m v_2^2 - \tfrac{1}{2}\Delta m v_1^2 + \Delta m g z_2 - \Delta m g z_1 \tag{27}$$

Upon rearrangement this becomes

$$\frac{1}{2}\frac{\Delta m}{\Delta V}v_2^2 + \frac{\Delta m}{\Delta V}g z_2 + p_2 = \frac{1}{2}\frac{\Delta m}{\Delta V}v_1^2 + \frac{\Delta m}{\Delta V}g z_1 + p_1 \tag{28}$$

Since $\Delta m/\Delta V$ is the density $\rho$ of the fluid, we can also express this as

$$\tfrac{1}{2}\rho v_2^2 + \rho g z_2 + p_2 = \tfrac{1}{2}\rho v_1^2 + \rho g z_1 + p_1 \tag{29}$$

or as

*Bernoulli's equation*

$$\boxed{\tfrac{1}{2}\rho v^2 + \rho g z + p = [\text{constant}]} \tag{30}$$

This is **Bernoulli's equation.** It states that along any streamline, the sum of the density of kinetic energy, density of potential energy, and pressure is a constant. The mathematical form of Bernoulli's equation is similar to that of the equation for the energy of a particle moving under the influence of gravity [see Eq. (7.51)]. But besides the kinetic term and the potential term, it also contains the pressure term. Bernoulli's equation tells us how an element of fluid moving along a streamline trades speed for height or for pressure.

If the fluid is static, so $v = 0$, Bernoulli's equation reduces to

$$p - [\text{constant}] = -\rho g z \tag{31}$$

This coincides with Eq. (18) for the pressure in a static fluid.

According to Bernoulli's equation, the velocity along any given streamline must *decrease* wherever the pressure *increases.* Intuitively, we might have expected that where the pressure is high, the velocity is large; but energy conservation demands exactly the opposite. The relation between pressure and velocity plays an important role in the design of wings for airplanes. Figure 18.25 shows an airfoil and the

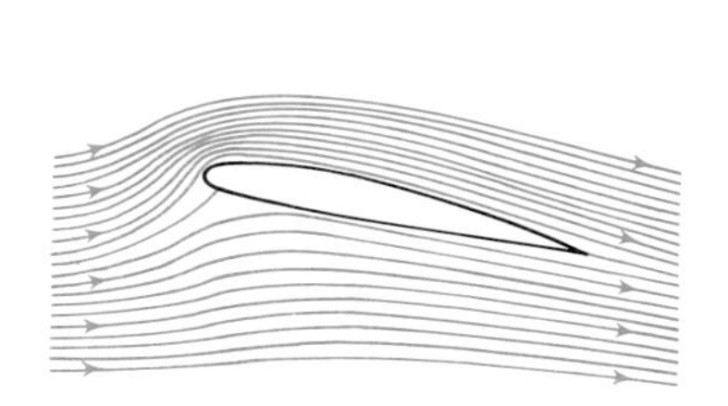

**Fig. 18.25** Flow of air around an airfoil.

streamlines of air flowing around it. The shape of the airfoil has been designed so that along its upper part the velocity is large (high density of streamlines), and along its lower part the velocity is small (low density of streamlines). Consider now one streamline passing just over the airfoil and one just under the airfoil. At a large distance to the left, the fluid on all streamlines has the same pressures and the same velocities. Bernoulli's equation applied to each of the two streamlines therefore tells us that in the region just above the airfoil, the pressure is low; and in the region just below the airfoil, the pressure is high. This leads to a net upward force, or **lift,** on the airfoil — this lift force supports the airplane in flight. As Figure 18.25 shows, at a large distance to the right, the streamlines have a slight downward trend; what has happened is that the air gave some upward momentum to the airfoil and, in return, acquired an equal amount of downward momentum. Ultimately, the downward flow of air presses against the ground and transmits the weight of the airplane to the ground.

Bear in mind that Bernoulli's equation is *not* valid for the flow of a fluid through a pump or a turbine wheel with a moving piston or moving blades, which do work on the fluid and add energy or remove energy. Since Bernoulli's equation expresses energy conservation for the fluid, it cannot be valid when such an external device adds energy to or removes energy from the fluid. Mathematically, the failure of Bernoulli's equation under these conditions arises from the failure of the assumption of steady flow, which entered the above derivation. The motion of the piston of the pump or the motion of the blades of the turbine requires a time-dependence of the pattern of flow of the fluid; this means the flow is *not* steady.

To conclude this chapter we will work out some examples involving Bernoulli's equation.

---

EXAMPLE 9. A water tank has a (small) hole near its bottom at a depth $h$ from the top surface (Figure 18.26). What is the speed of the stream of water emerging from the hole?

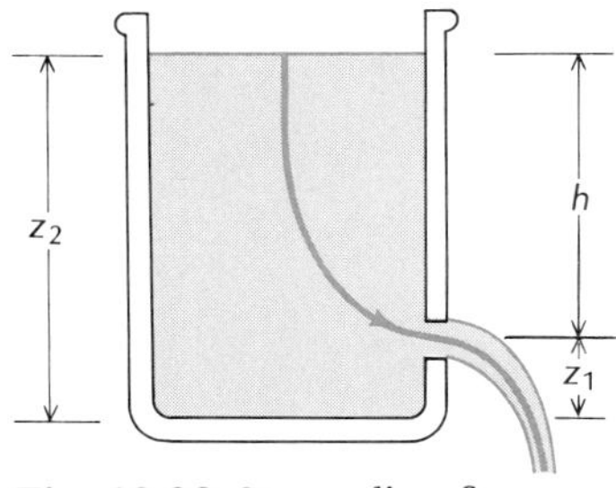

**Fig. 18.26** Streamline for water flowing out of a tank.

SOLUTION: Qualitatively, one of the streamlines for the water flowing out of this tank will look as shown in Figure 18.26. Since the hole is small, the water level at the top of the tank drops only very slowly; we can therefore take $v_1 = 0$ at the top of the tank. Furthermore, the pressures at the top and in the emerging stream of water are the same: $p_1 = p_2 = p_{atm}$. Thus, the left side of Eq. (30) is $\rho g z_1 + p_{atm}$ at the top of the tank, and $\frac{1}{2}\rho v_2^2 + \rho g z_2 + p_{atm}$ at the hole. By Bernoulli's equation, these two expressions must be equal,

$$\tfrac{1}{2}\rho v_2^2 + \rho g z_2 + p_{at} = \rho g z_1 + p_{at} \tag{32}$$

which yields

$$v_2 = \sqrt{2g(z_1 - z_2)} = \sqrt{2gh} \tag{33}$$

COMMENTS AND SUGGESTIONS: The speed of the emerging water is exactly what it would be if the water were to fall freely from a height $h$, a result known as **Torricelli's theorem.** This result, of course, expresses conservation of energy: when a drop of water flows out at the bottom, the loss of potential energy of the water in the tank is equivalent to the removal of a drop of water from the top; the conversion of this potential energy into kinetic energy will give the drop the speed of free fall.

Note that the use of Bernoulli's equation in this example and in the next examples involves the familiar three steps we learned in Chapter 7: first write an expression for the left side of Eq. (30) at one point on a streamline, then write an expression at another point of the (same) streamline, and then equate the two expressions.

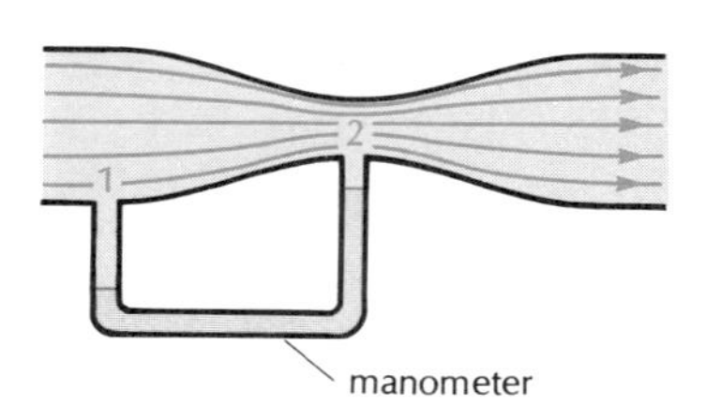


**Fig. 18.27** Venturi flowmeter. The manometer measures the pressure difference between points 1 and 2.

EXAMPLE 10. The **Venturi flowmeter** is a simple device that measures the speed of a fluid flowing in a pipe. It consists of a constriction in the pipe with a cross-sectional area $A_2$ that is smaller than the cross-sectional area $A_1$ of the pipe itself (Figure 18.27). Small holes in the constriction and in the pipe permit the measurement of the pressures at these points by means of a manometer. Express the speed of flow in terms of the pressure difference registered by the manometer.

SOLUTION: The speeds at points 1 and 2 are related to the cross-sectional areas by the continuity equation [Eq. (8)],

$$v_2 = v_1 A_1/A_2$$

With $z_1 = z_2 = 0$, Bernoulli's equation then gives us the following expression for the pressure difference:

$$p_1 - p_2 = \tfrac{1}{2}\rho v_2^2 - \tfrac{1}{2}\rho v_1^2 = \tfrac{1}{2}\rho v_1^2[(A_1/A_2)^2 - 1]$$

so that

$$v_1 = \sqrt{\frac{2(p_1 - p_2)}{\rho[(A_1/A_2)^2 - 1]}}$$

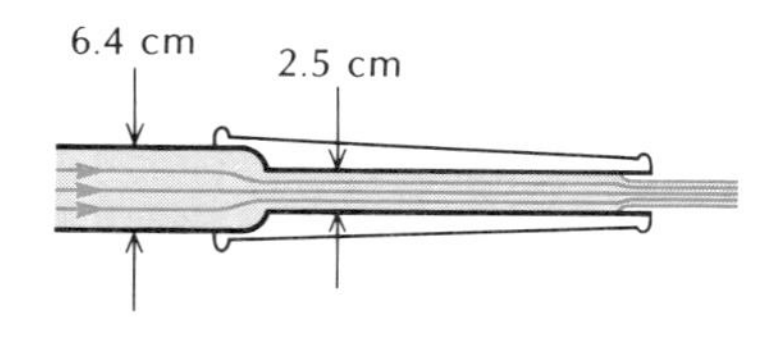


**Fig. 18.28** A fire-hose tip.

EXAMPLE 11. The overpressure in a fire hose of diameter 6.4 cm is $3.5 \times 10^5$ N/m² and the speed of flow is 4.0 m/s. The fire hose ends in a metal tip of diameter 2.5 cm (Figure 18.28). What are the pressure and speed of the water in the tip?

SOLUTION: As in the preceding example, the speed of the water in the tip is determined by the equation of continuity:

$$v_2 = v_1 \frac{A_1}{A_2} = 4.0 \text{ m/s} \times \frac{(6.4 \text{ cm})^2}{(2.5 \text{ cm})^2} = 26.2 \text{ m/s}$$

The pressure of water in the tip can then be calculated from Bernoulli's equation with $z_1 = z_2 = 0$:

$$p_2 = p_1 + \tfrac{1}{2}\rho v_1^2 - \tfrac{1}{2}\rho v_2^2$$

To convert this into an equation involving overpressures, we subtract $p_{at}$ from both sides:

$$p_2 - p_{at} = p_1 - p_{at} + \tfrac{1}{2}\rho v_1^2 - \tfrac{1}{2}\rho v_2^2$$

With $p_1 - p_{at} = 3.5 \times 10^5$ N/m², this yields

$$\begin{aligned} p_2 - p_{at} &= 3.5 \times 10^5 \text{ N/m}^2 + \tfrac{1}{2} \times 1.0 \times 10^3 \text{ kg/m}^3 \times (4.0 \text{ m/s})^2 \\ &\quad - \tfrac{1}{2} \times 1.0 \times 10^3 \text{ kg/m}^3 \times (26.2 \text{ m/s})^2 \\ &= 1.4 \times 10^4 \text{ N/m}^2 \end{aligned}$$

EXAMPLE 12. What is the speed of the water just outside of the tip of the fire hose of Example 11?

SOLUTION: The speed of the water just outside the tip can be directly calculated from Bernoulli's equation. We again set $z_1 = z_3 = 0$ and we set the pressure $p_3$ equal to the atmospheric pressure $p_{at}$:

$$v_3^2 = v_1^2 + \frac{2p_1}{\rho} - \frac{2p_{at}}{\rho} = v_1^2 + \frac{2}{\rho}(p_1 - p_{at})$$

$$= (4.0 \text{ m/s})^2 + \frac{2 \times 3.5 \times 10^5 \text{ N/m}^2}{1.0 \times 10^3 \text{ kg/m}^3}$$

$$= 716 \text{ (m/s)}^2$$

and

$$v_3 = 26.8 \text{ m/s}$$

COMMENTS AND SUGGESTIONS: The diameter of the stream of water just outside of the tip is slightly smaller than the diameter in the tip. This follows from the equation of continuity,

$$A_3 = A_2 \frac{v_2}{v_3} = A_2 \frac{26.2 \text{ m/s}}{26.8 \text{ m/s}} = A_2 \times 0.98$$

Since the cross-sectional area $A_2$ has a diameter of 2.5 cm, the cross-sectional area $A_3$ will have a diameter $2.5 \text{ cm} \times \sqrt{0.98} = 2.5 \text{ cm} \times 0.99$. Thus, the stream of water contracts by about 1% as it leaves the nozzle (Figure 18.28). Intuitively, we might have expected that the stream expands as it leaves the nozzle, but the opposite is the case.

## SUMMARY

**Continuity equation (for incompressible fluid):** $vA = [\text{constant}]$

**Definition of pressure:** $p = \dfrac{\Delta F}{\Delta A}$

**Atmospheric pressure (average):** 1 atm = 760 mmHg
= $1.0 \times 10^5$ N/m² or Pa

**Hydrostatic pressure (for incompressible fluid):** $p - p_0 = -g\rho z$

**Archimedes' Principle:** The buoyant force has the same magnitude as the weight of displaced fluid.

**Bernoulli's equation:** $\frac{1}{2}\rho v^2 + \rho g z + p = [\text{constant}]$

## QUESTIONS

1. According to popular belief, blood is denser than water. Is this true?

2. The sheet of water of a waterfall is thick at the top and thin at the bottom. Explain.

3. During construction, three lanes of a four-lane highway are closed. Suppose that bumper-to-bumper traffic on the four lanes funnels into the single open lane. If the cars on the four approaching lanes proceed at a crawl, say, 15 km/h, what will be their speed when they proceed along the single lane? (Hint: Think of the continuity equation.)

4. If you place a block of wood on the bottom of a swimming pool, why does the pressure of the water not keep it there?

5. Explain what holds a suction cup on a smooth surface.

6. Newton gave the following description of his classic experiment with a rotating bucket:

> If a vessel, hung by a long cord, is so often turned about that the cord is strongly twisted, then filled with water, and held at rest together with the water; thereupon, by the sudden action of another force, it is whirled about the contrary way, and while the cord is untwisting itself, the vessel continues for some time in this motion; the surface of the water will at first be plain, as before the vessel began to move; but after that, the vessel, by gradually communicating its motion to the water, will make it begin sensibly to revolve, and recede by little and little from the middle, and ascend to the sides of the vessel, forming itself into a concave figure (as I have experienced), and the swifter the motion becomes, the higher will the water rise, till at last, performing its revolutions in the same times with the vessel, it becomes relatively at rest in it.

Explain why hydrostatic equilibrium requires that the rotating water be higher at the rim of the bucket than at the center.

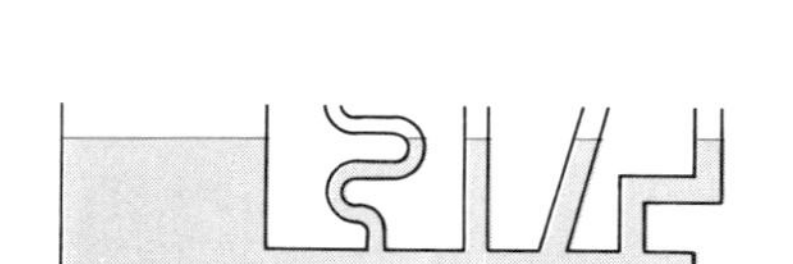

**Fig. 18.29** Glass vessel with vertical tubes.

7. When a doctor measures the blood pressure of a patient he places the cuff on the arm, at the same vertical level as the heart. What would happen if he were to place the cuff around the leg?

8. Scuba divers have survived short intervals of free swimming at depths of 430 m. Why does the pressure of the water at this depth not crush them?

9. Figure 18.29 shows a glass vessel with vertical tubes of different shapes. Explain why the water level in all the tubes is the same.

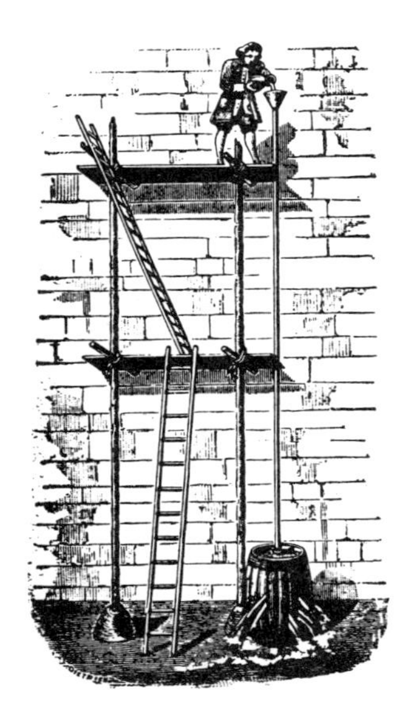

**Fig. 18.30** Blaise Pascal's experiment.

10. In a celebrated experiment, Blaise Pascal attached a long metal funnel to a tight cask (Figure 18.30). When he filled the cask by pouring water into this funnel, the cask burst. Explain.

11. Face masks for scuba divers have two indentations at the bottom into which the diver can stick thumb and forefinger to pinch his nose shut. What is the purpose of this arrangement?

12. Figure 18.31 shows grain silos held together by circumferential steel bands. Why has the farmer placed more bands near the bottom than near the top?

**Fig. 18.31** Grain silos.

13. The level of a large oil slick on the sea is slightly higher than the level of the surrounding water. Explain.

14. Flooding on the low coasts of England and the Netherlands is most severe when the following three conditions are in coincidence: onshore wind, full or new moon, and low barometric pressure. Can you explain this?

15. Why do some men or women float better than others? Why do they all float better in salt water than in plain water?

16. Would you float if your lungs were full of water?

17. Will a stone float in a tub full of mercury?

18. An ice cube floats in a glass full of water. Will the water level rise or fall when the ice melts?

19. The density of a solid body can be determined by first weighing the body in air and then weighing it again when it is immersed in water (Figure 18.32). How can you deduce the density from these two measurements?

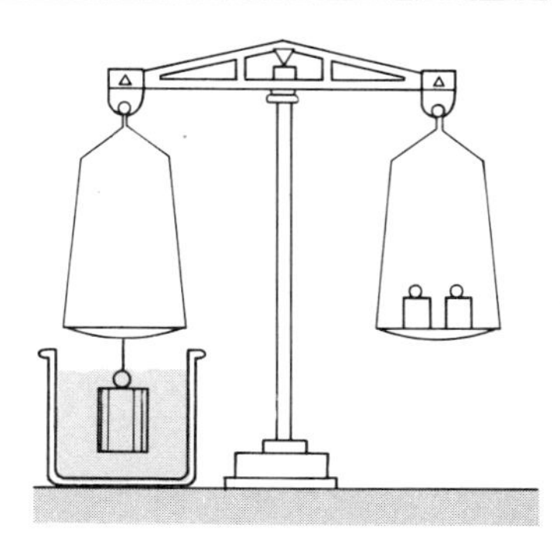

**Fig. 18.32** Weighing a body underwater.

20. A buoy floats on the water. Will the flotation level of the buoy change when the atmospheric pressure changes?

21. Will the water level in a canal lock rise or fall if a ship made of steel sinks in the lock? What if the ship is made of wood?

22. While training for the conditions of weightlessness they would encounter in an orbiting spacecraft, NASA astronauts were made to float submerged in a large water tank (Figure 18.33). Small weights attached to their spacesuits gave the astronauts neutral buoyancy. To what extent does such simulated weightlessness imitate true weightlessness?

**Fig. 18.33** Astronaut floating underwater.

23. A girl standing in a subway car holds a helium balloon on a string. Which way will the balloon move when the car accelerates?

24. A slurry of wood chips is used in some tanning operations. What would happen to a man were he to fall into a vat filled with such slurry?

25. Figure 18.34 shows a wide pipe with a constriction, and a thin tube that connects the constricted part of the pipe with the wide part. Suppose that the fluid in the thin tube is initially at rest, and the fluid in the wide pipe is flowing from left to right. What will happen to the fluid in the tube? What if the fluid in the pipe were flowing from right to left?

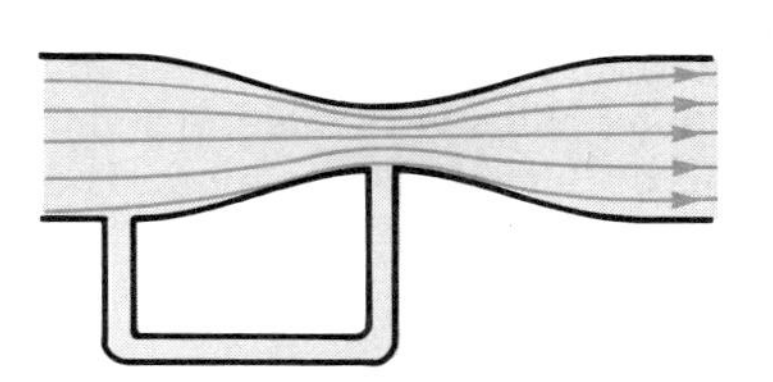

**Fig. 18.34**

26. The bathyscaphe *Trieste,* which set a record of 10,917 m in a deep dive in the Marianas Trench in 1960, consists of a large tank filled with gasoline below which hangs a steel sphere for carrying the crew (Figure 18.35). Can you guess the purpose of the tank of gasoline?

**Fig. 18.35** The bathyscaphe *Trieste.*

**Fig. 18.36** A Cartesian diver.

27. Gasoline vapor from, say, a small leak in the fuel system poses a very serious hazard in a motorboat, but only a minor hazard in an automobile. Explain. (Hint: Gasoline vapor is denser than air.)

28. How does a submarine dive? Ascend?

29. How does a balloonist control the ascent and descent of a hot-air balloon?

30. When you release a bubble of air while underwater, the bubble grows in size as it ascends. Explain.

31. A "Cartesian diver" consists of a small inverted bottle floating inside a larger bottle whose mouth is covered by a rubber membrane (Figure 18.36). By depressing the membrane, you can increase the water pressure in the large bottle. How does this affect the buoyancy of the small bottle?

32. Why are the continuity equation and Bernoulli's equation only *approximately* valid for the flow of air?

33. To throw a curve ball, the baseball pitcher gives the ball a spinning motion about a vertical axis. The air on the left and right sides of the ball will then be dragged along by the rotation and acquire slightly different speeds. Using Bernoulli's equation, explain how this creates a lateral deflecting force on the ball.

34. Hurricanes with the lowest central barometric pressures have the highest wind speeds. Observations show that the square of the wind speed in a hurricane is roughly proportional to the difference between the barometric pressure outside the hurricane and the barometric pressure in the hurricane. Show that this proportionality is expected from Bernoulli's equation. (Hint: Consider a streamline of air that starts outside the hurricane and gradually spirals in toward the "eye.")

35. If you place a ping-pong ball in the jet of air from a vacuum cleaner hose aimed vertically upward, the ping-pong ball will be held in stable equilibrium within this jet. Explain this by means of Bernoulli's equation. (Hint: The speed of air is maximum at the center of the jet.)

## PROBLEMS

### Section 18.1

1. The following table, taken from a firefighter's manual, lists the rate of flow of water (in liters per minute) through a fire hose of diameter 3.81 cm (1.5 in.) connected to a nozzle of given diameter; the listed rate of flow will maintain a pressure of 3.4 atm (50 lbf/in.$^2$) in the nozzle:

RATE OF FLOW FOR A 3.81-CM HOSE

| *Rate of flow* | *Nozzle diameter* | *Nozzle pressure* |
|---|---|---|
| 95 liters/min | 0.95 cm | 3.4 atm |
| 190 | 1.27 | 3.4 |
| 284 | 1.59 | 3.4 |

For each case calculate the speed of flow (in meters per second) in the hose and in the nozzle.

*2. A fountain shoots a stream of water vertically upward. Assume that the stream is inclined very slightly to one side so that the descending water does not interfere with the ascending water. The upward velocity at the base of the column of water is 15 m/s.
(a) How high will the water rise?
(b) The diameter of the column of water is 7.0 cm at the base. What is the diameter at the height of 5 m? At the height of 10 m?

### Section 18.3

3. Within the funnel of a tornado, the air pressure is much lower than normal — about 200 mbar as compared to the normal value of 1010 mbar. Suppose that such a tornado suddenly envelops a house; the air pressure inside the house is 1010 mbar and the pressure outside suddenly drops to 200 mbar. This will cause the house to burst explosively. What is the net outward pressure force on a 12 m × 3 m wall of this house? Is the house likely to suffer less damage if all the windows and doors are open?

4. What is the downward force that air pressure (1 atm) exerts on the upper surface of a sheet of paper ($8\frac{1}{2}$ in. × 11 in.) lying on a table? Why does this force not squash the paper against the table?

5. At a distance of 7 km from a 1-megaton nuclear explosion, the blast wave has an overpressure of $3 \times 10^4$ N/m$^2$. Calculate the force that this blast wave exerts on the front of a standing man; the frontal area of the man is 0.7 m$^2$. (The actual force on a man exposed to the blast wave is larger than the result of this simple calculation because the blast wave will be reflected by the man and this leads to a substantial increase of pressure.)

6. The overpressure in the tires of a 1300-kg automobile is 2.4 atm. If each tire supports one-fourth the weight of the automobile, what must be the area of each tire in contact with the ground? Pretend the tires are completely flexible.

7. The shape of the wing of an airplane is carefully designed so that, when the wing moves through the air, a pressure difference develops between the bottom surface of the wing and the top surface; this supports the weight of the airplane. A fully loaded DC-3 airplane has a mass of 10,900 kg. The (bottom) surface area of its wings is 92 m$^2$. What is the average pressure difference between the top and bottom surfaces when the airplane is in flight?

8. Pressure gauges used on automobile tires read the overpressure, that is, the amount of pressure in excess of atmospheric pressure. If a tire has an overpressure of $2.4 \times 10^5$ N/m$^2$ on a day when the barometric pressure is 72.4 cm Hg, what will be the overpressure when the barometric pressure increases to 77.0 cm Hg? Assume that the volume and the temperature of the tire remain constant.

9. Commercial jetliners have pressurized cabins enabling them to carry passengers at a cruising altitude of 10,000 m. The air pressure at this altitude is 210 mmHg. If the air pressure inside the jetliner is 760 mmHg, what is the net outward force on a 1 m $\times$ 2 m door in the wall of the cabin?

10. In 1654 Otto von Guericke, the inventor of the air pump, gave a public demonstration of air pressure (see Figure 5.22). He took two hollow hemispheres of copper, whose rims fitted tightly together, and evacuated them with his air pump. Two teams of 15 horses each, pulling in opposite directions, were unable to separate these hemispheres. If the evacuated sphere had a radius of 40 cm and the pressure inside it was nearly zero, what force would each team of horses have to exert to pull the hemispheres apart?

11. The baggage compartment of the DC-10 airliner is under the floor of the passenger compartment. Both compartments are pressurized at a normal pressure of 1 atm. In a disastrous accident near Orly, France, in 1974, a faulty lock permitted the baggage compartment door to pop open in flight, depressurizing this compartment. The normal pressure in the passenger compartment then caused the floor to collapse, jamming the control cables. At the time the airliner was flying at an altitude of 3800 m, where the air pressure is 0.64 atm. What was the net pressure force on a 1 m $\times$ 1 m square of the floor?

### Section 18.4

12. Porpoises dive to a depth of 520 m. What is the water pressure at this depth?

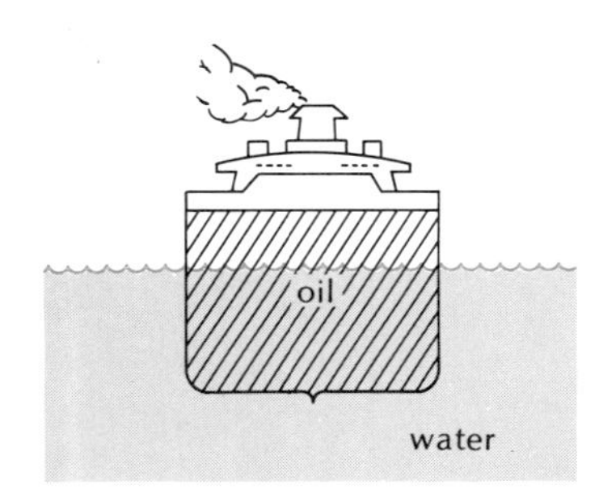


**Fig. 18.37** Cross section of a tanker.

13. A tanker is full of oil of density 880 kg/m$^3$. The flat bottom of the hull is at a depth of 26 m below the surface of the surrounding water. Inside the hull, oil is stored with a depth of 30 m (Figure 18.37). What is the pressure of the water on the bottom of the hull? The pressure of the oil? What is the net vertical pressure force on 1 m$^2$ of bottom?

14. A pencil sharpener is held to the surface of a desk by means of a rubber "suction" cup measuring 6 cm $\times$ 6 cm. The air pressure under the suction cup is zero and the air pressure above the suction cup is 1 atm.
   (a) What is the magnitude of the pressure force pushing the cup against the table?
   (b) If the coefficient of static friction between the rubber and the table is 0.9, what is the maximum transverse force the suction cup can withstand?

15. (a) Calculate the mass of air in a column of base 1 m$^2$ extending from sea level to the top of the atmosphere. Assume that the pressure at sea level is 760 mmHg and that the value of the acceleration of gravity is 9.81 m/s$^2$, independent of height.
   (b) Multiply your result by the surface area of the Earth to find the total mass of the entire atmosphere.

16. Suppose that a zone of low atmospheric pressure (a "low") is at some place on the surface of the sea. The pressure at the center of the "low" is 64 mm Hg less than the pressure at a large distance from the center. By how much will this cause the water level to rise at the center?

17. (a) Under normal conditions the human heart exerts a pressure of 110 mmHg on the arterial blood. What is the arterial blood pressure in the feet of a man standing upright? What is the blood pressure in the brain? The feet are 140 cm below the level of the heart; the brain is 40 cm above the level of the heart; the density of human blood is 1055

kg/m³. Neglect the speed of flow of the blood, i.e., pretend it is at rest.

(b) Under conditions of stress the human heart can exert a presure of up to 190 mmHg. Suppose that an astronaut were to land on the surface of a large planet where the acceleration of gravity is 61 m/s². Could the astronaut's heart maintain a positive blood pressure in his brain while he is standing upright? Could the astronaut survive?

18. A **"suction" pump** consists of a piston in a cylinder with a long pipe leading down into a well (Figure 18.38). What is the maximum height to which such a pump can "suck" water?

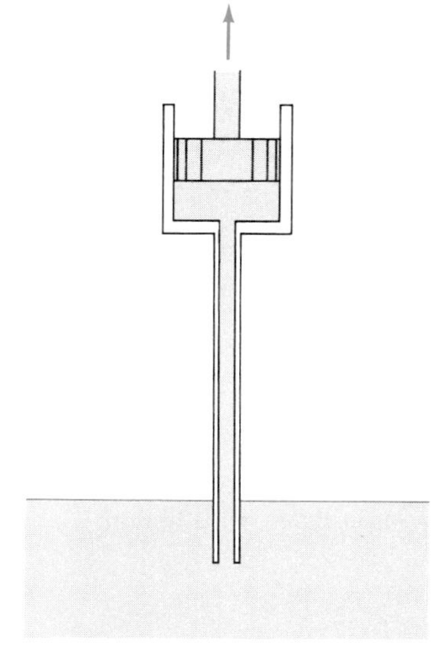

**Fig. 18.38** Suction pump.

*19. A man whirls a bucket full of water around a vertical circle at the rate of 0.70 rev/s. The surface of the water is at a radial distance of 1.0 m from the center.

(a) What is the pressure difference between the surface of the water and a point 1.0 cm below the surface when the bucket is at the lowest point of the circle? You may assume that a point on the surface and a point 1.0 cm below have practically the same centripetal acceleration.

(b) What if the bucket is at the highest point of the circle?

*20. A test tube filled with water is being spun around in an ultracentrifuge with angular velocity $\omega$. The test tube is lying along a radius and the free surface of the water is at radius $r_0$ (Figure 18.39).

(a) Show that the pressure at radius $r$ within the test tube is

$$p = \tfrac{1}{2}\rho\omega^2(r^2 - r_0^2)$$

where $\rho$ is the density of the water. Ignore gravity and ignore atmospheric pressure.

(b) Suppose that $\omega = 3.8 \times 10^5$ radian/s and $r_0 = 10$ cm. What is the pressure at $r = 13$ cm?

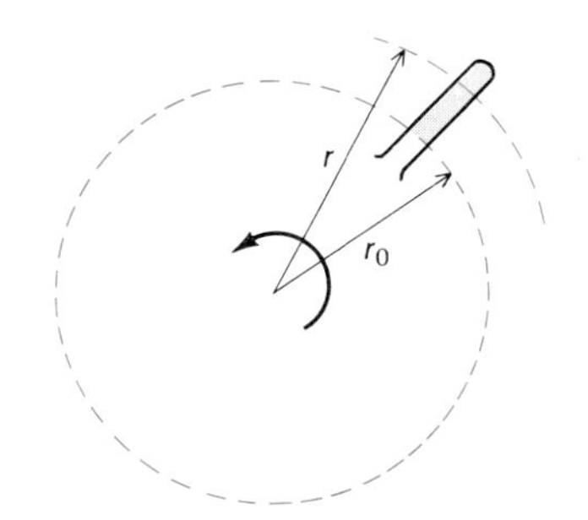


**Fig. 18.39** Test tube in an ultracentrifuge.

*21. In a test in a centrifuge, a NASA scientist tolerated (suffered?) a sustained centripetal acceleration of 25 G. Estimate the pressure difference between the front and the back of his brain during the ordeal. Measure the relevant distance on your own head, and assume this distance was radial during the test.

*22. (a) Figure 18.40a shows a round conical flask filled with water of a depth $h$. The radius of the upper water surface is $R_1$ and that of the lower surface is $R_2$. What is the net force that the water exerts on the sides of the flask? On the bottom of the flask? What is the sum of these forces? Ignore atmospheric pressure.

(b) Figure 18.40b shows another round conical flask, with the same radii. Answer the same questions for this flask.

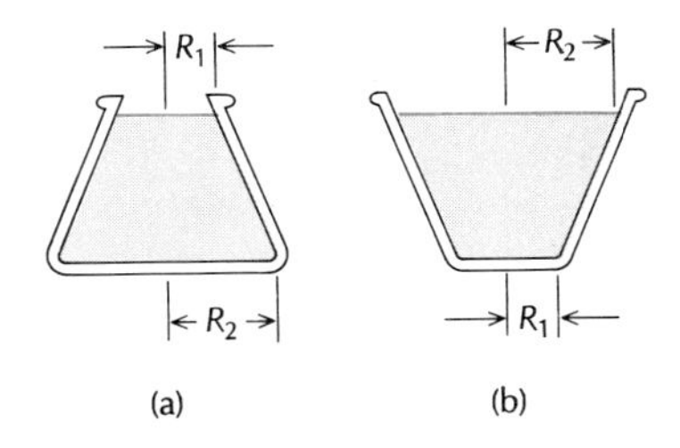


**Fig. 18.40** Conical flasks.

Section 18.5

23. What is the buoyant force on a human body of volume $7.4 \times 10^{-2}$ m³ when totally immersed in air? In water?

24. Icebergs commonly found floating in the North Atlantic (Figure 18.41) are 30 m high (above the water) and 400 m × 400 m across. The density of ice is 920 kg/m³ and the density of sea water is 1025 kg/m³.

**Fig. 18.41** An iceberg.

**18.42** A research balloon being inflated with helium.

**Fig. 18.43** Fully loaded and empty supertankers.

(a) What is the total volume of such an iceberg (including the volume below the water)?
(b) What is the total mass?

25. You can walk on water if you wear very large shoes shaped like boats. Calculate the length of the shoes that will support you; assume that each shoe is 30 cm $\times$ 30 cm in cross section.

26. A gasoline barrel, made of steel, has a mass of 20 kg when empty. The barrel is filled with 0.12 $m^3$ of gasoline with a density of 730 $kg/m^3$. Will the full barrel float in water? Neglect the volume of the steel.

27. A typical medium-size balloon used for scientific research (Figure 18.42) is designed to attain an altitude of 40 km, at which altitude the helium in the balloon will have expanded to 570,000 $m^3$. What is the buoyant force on the balloon under these conditions? The density of air at this altitude is $4.3 \times 10^{-3}$ $kg/m^3$.

*28. The supertanker *Globtik London* has a mass of 220,000 metric tons when empty and it can carry up to 440,000 metric tons of oil when fully loaded. Assume that the shape of its hull is approximately that of a rectangular parallelepiped 380 m long, 60 m wide, and 40 m high (Figure 18.43).
(a) What is the draft of the empty tanker, that is, how deep is the hull submerged in the water? Assume that the density of (sea) water is $1.02 \times 10^3$ $kg/m^3$.
(b) What is the draft of the fully loaded tanker?

*29. A supertanker has a draft (submerged depth) of 30 m when in sea water (density $\rho = 1.02 \times 10^3$ $kg/m^3$). What will be the draft of this tanker when it enters a river estuary with fresh water (density $\rho = 1.00 \times 10^3$ $kg/m^3$)? Assume that the sides of the ship are vertical.

*30. The Raven S-66A hot-air balloon has a volume of 4000 $m^3$ and a height of 27 m. Fully loaded, its mass is 1400 kg. If the density of the air outside the balloon is 1.29 $kg/m^3$, what must be the density of the air inside the balloon to achieve lift-off? What is the inside overpressure at the top of the balloon? (Hint: The bottom of the balloon is open, and therefore at the same pressure as the exterior air. Treat the air outside and the air inside the balloon as fluids of uniform densities.)

*31. A round log of wood of density 600 $kg/m^3$ floats in water. The diameter of the log is 30 cm. How high does the upper surface of the log protrude from the water?

*32. A child's rubber balloon of mass 2.5 grams is filled with helium gas of density 0.33 $kg/m^3$. The balloon is spherical, with a radius of 12 cm. A long cotton string with a mass of 2.0 grams per meter hangs from the bottom of the balloon. Initially, this string lies loosely on the floor, but when the balloon ascends, it pulls the string upward and straightens it out. At what height will the balloon stop ascending, having reached equilibrium with the hanging portion of the string? Assume that the surrounding air has the standard density 1.29 $kg/m^3$.

*33. When a force accelerates a body immersed in a fluid, some of the fluid must also be accelerated, since it must be pushed out of the way of the body and flow around it. Thus, the force must overcome not only the inertia of the body, but also the inertia of the fluid pushed out of the way. It can be shown that for a spherical body completely immersed in a nonviscous fluid, the extra inertia is that of a mass of fluid half as large as the fluid displaced by the body.
(a) From this, deduce that the downward acceleration of a spherical body of density $\rho$ failing through a fluid of density $\rho'$ is

$$a = \frac{\rho - \rho'}{\rho + \frac{1}{2}\rho'} g$$

(b) Find the upward acceleration of an empty bubble in the fluid.
(c) What value would you have found for the acceleration of an empty bubble if you had not taken into account the extra inertia of the displaced fluid?

*34. When a body is weighed in air on an analytical balance, a correction must be made for the buoyancy contributed by the air. Suppose that the weights used in the balance are made of brass, of density $8.7 \times 10^3$ kg/m$^3$, and that the density of air is 1.3 kg/m$^3$. By how many percent must you increase (or decrease) the mass indicated by the balance in order to obtain the true mass of a body of density $1.0 \times 10^3$ kg/m$^3$? $5.0 \times 10^3$ kg/m$^3$? $10.0 \times 10^3$ kg/m$^3$?

*35. The bottom half of a tank is filled with water ($\rho = 1.0 \times 10^3$ kg/m$^3$) and the top half is filled with oil ($\rho = 8.5 \times 10^2$ kg/m$^3$). Suppose that a rectangular block of wood of mass 5.5 kg, 30 cm long, 20 cm wide, and 10 cm high, is placed in this tank. How deep will the bottom of the block be submerged in the water?

**36. A **hydrometer,** the device used to determine the density of battery acid, wine, or other fluids, consists of a bulb with a long vertical stem (Figure 18.44). The device floats in the liquid with the bulb submerged and the stem protruding above the surface. The density of the liquid is directly related to the length $h$ of the protruding portion of the stem. Suppose that the bulb is a sphere of radius $R$ and that the stem is a cylinder of radius $R'$ and length $l$; the mass of both together is $M$. Derive a formula for the density of the liquid in terms of $h$, $R$, $R'$, $l$, and $M$.

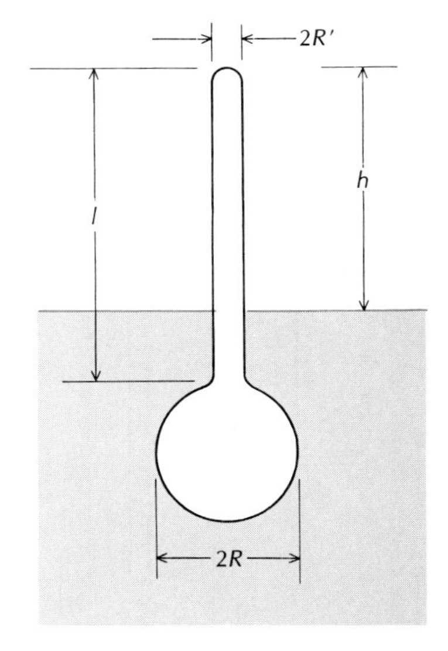


**Fig. 18.44** Hydrometer.

**37. A ship that displaces a weight of water equal to its own weight is in equilibrium with regard to vertical motion. If the ship were placed lower in the water, the buoyant force would exceed gravity and the ship would surge upward. If the ship were placed higher in the water, the buoyant force would be less than gravity and the ship would sink downward. Suppose that the sides of a ship are vertical above and below its normal waterline.
(a) Show that the frequency of small up-and-down oscillations of the ship is roughly $\omega = \sqrt{A\rho g/M}$, where $A$ is the horizontal area bounded by the waterline, $M$ is the mass of the ship, and $\rho$ is the density of water. (This formula is only a rough aproximation because, as the ship moves up and down, the water also has to move; hence the effective inertia is greater than $M$.)
(b) What is the frequency of such up-and-down oscillations for the fully loaded supertanker described in Problem 28?

**38. You drop a pencil, point down, into the water. If you release the pencil from a height of 4.0 cm (measured from the point to the water level), how far will it dive? Treat the pencil as a cylinder of radius 0.40 cm, length 19 cm, and mass 4.0 g. Ignore the frictional and inertial resistance of the water, but take into account the buoyant force.

Section 18.6

39. A thin stream of water emerges vertically from a small hole on the side of a water pipe and ascends to a height of 1.2 m. What is the pressure inside the pipe? Assume the water inside the pipe is nearly static.

40. If you blow a thin stream of air with a speed of 7 m/s out of your mouth, what must be the overpressure in your mouth? Assume that the speed of the air in your mouth is (nearly) zero.

41. A pump has a horizontal intake pipe at a depth $h$ below the surface of a lake. What is the maximum speed for steady flow of water into this pipe? (Hint: Inside the pipe the maximum speed of flow corresponds to zero pressure.)

*42. To fight a fire on the fourth floor of a building, firemen want to use a hose of diameter 6.35 cm ($2\frac{1}{2}$ in.) to shoot 950 liters/min of water to a height of 12 m.

(a) With what minimum speed must the water leave the nozzle of the fire hose if it is to ascend 12 m?

(b) What pressure must the water have inside the fire hose? Ignore friction.

*43. Streams of water from fire hoses are sometimes used to disperse crowds. Suppose that the stream of water emerging from the fire hose described in Example 11 impinges horizontally on a man. The collision of the water with the man is totally inelastic.

(a) What is the force that the stream of water exerts on the man, i.e., what is the rate at which the water delivers momentum to the man?

(b) What is the rate at which the water delivers energy?

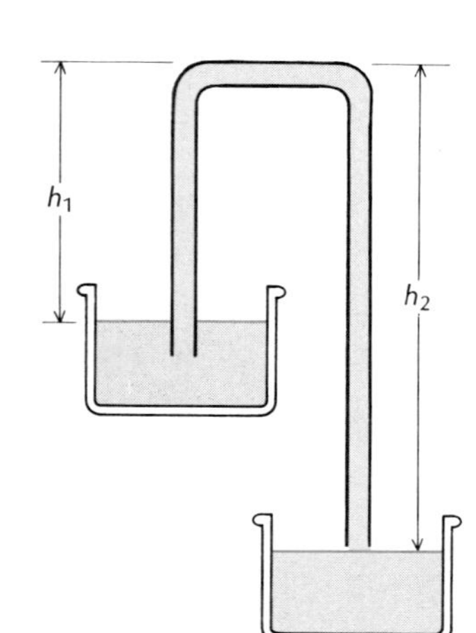


**Fig. 18.45** Siphon.

*44. A **siphon** is an inverted U-shaped tube that is used to transfer liquid from a container at a high level to a container at a low level (Figure 18.45).

(a) Using the lengths shown in Figure 18.45 and the density $\rho$ of the liquid, find a formula for the speed with which the liquid emerges from the lower end of the siphon. Assume that the containers are large, and that the lower end of the tube is *not* immersed in the container.

(b) Find the pressure at the highest point of the siphon.

(c) By setting this pressure equal to zero, find the maximum height $h_1$ with which the siphon can operate.

*45. A pump of a fire engine draws 1100 liters of water per minute from a well with a water level 4.5 m below the pump and discharges this water into a fire hose of diameter 6.35 cm ($2\frac{1}{2}$ in.) at a pressure of 5.4 atm. In the absence of friction, what power (in hp) does this pump require?

*46. In the calculation of pressures in fire hoses, it is often necessary to take into account the frictional losses suffered by the water as it flows along the hose. Consider a horizontal hose of diameter 3.81 cm ($1\frac{1}{2}$ in.) and length 30 m carrying 380 liters/min of water. According to tables used by firemen, a pressure of 6.1 atm at the upstream end of this hose will result in a pressure of only 3.4 atm at the downstream end; the loss of 2.7 atm is attributed to friction. At what rate (in hp) does friction remove energy from the water?

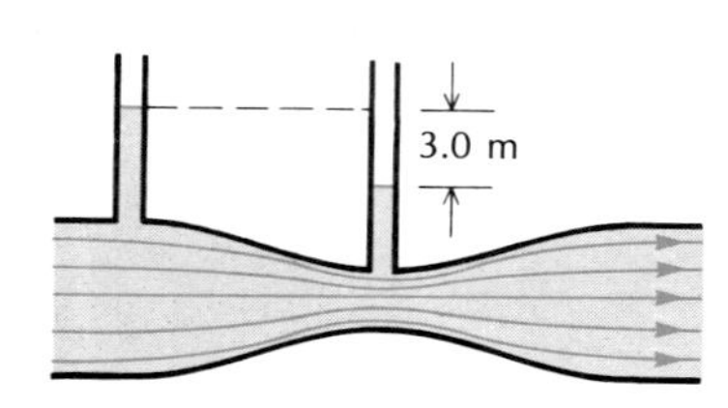


**Fig. 18.46** Venturi flowmeter.

*47. A Venturi flowmeter in a water main of diameter 30 cm has a constriction of diameter 10 cm. Vertical pipes are connected to the water main and to the constriction (Figure 18.46); these pipes are open at their upper ends and the water level within them indicates the pressure at their lower ends. Suppose that the difference in the water levels in these two pipes is 3.0 m. What is the velocity of flow in the water main? What is the rate (in liters per second) at which water is delivered?

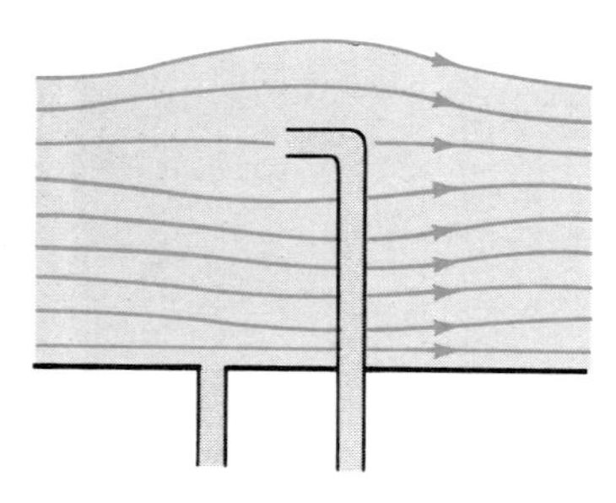

**Fig. 18.47** Pitot tube.

*48. The **Pitot tube** is used for the measurement of the flow speeds of fluids, such as the flow speed of air past the fuselage of an airplane. It consists of a bent tube protruding into the airstream (Figure 18.47) and another tube opening flush with the fuselage. The pressure difference between the air in the two tubes can be measured with a manometer. Show that in terms of the pressures $p_1$ and $p_2$ in the two tubes, the speed of airflow is

$$v = \sqrt{\frac{2(p_2 - p_1)}{\rho}}$$

where $\rho$ is the density of air. (Hint: $p_1$ is simply the static air pressure. To find $p_2$, consider a streamline reaching the opening of the bent tube; at this point, the velocity of the air is zero.)

*49. The central front window of the cockpit of an airliner measures 30 cm $\times$ 30 cm. Estimate the force with which the air presses against this window when the airliner is flying at 900 km/h. (Hint: At the window, the air (almost) stops relative to the airliner.)

*50. A water tank filled to a height $h$ has a small hole at a height $z$ (see Figure 18.48). Show that the stream of water emerging from this hole strikes the ground at a horizontal distance $2\sqrt{(h-z)z}$ from the base of the tank. What choice of $z$ gives the largest horizontal distance? What is the largest horizontal distance?

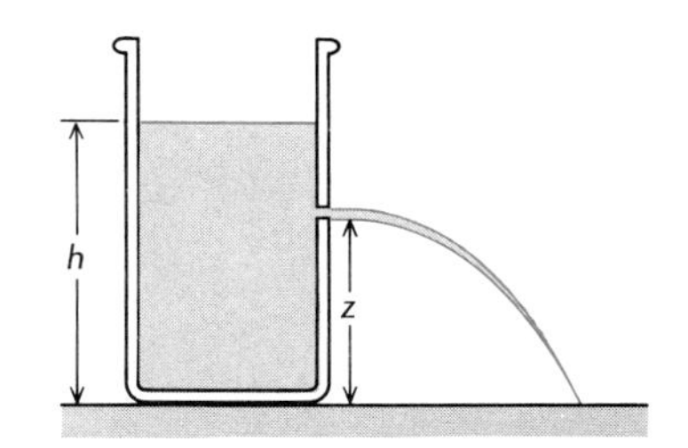


**Fig. 18.48** Tank with small hole.

*51. Very high pressures can be generated (for a brief instant) by launching a projectile at high speed against a rigid target. According to one proposal, an electromagnetic launcher, or rail gun, could be used to give the projectile a speed of 15 km/s. Estimate the maximum pressure generated by the impact of such a projectile on a rigid target. Assume that the density of the projectile is $10^4$ kg/m$^3$. (Hint: At extreme pressures, the solid projectile will flow like a liquid of approximately constant density; hence Bernoulli's equation is approximately valid.)

**52. A cylindrical tank has a base area $A$ and a height $l$; it is initially full of water. The tank has a small hole of area $A'$ at its bottom. Calculate how long it will take all the water to flow out of this hole.

**53. A rectangular opening on the side of a tank has a width $l$. The top of the opening is at a depth $h_1$ below the surface of the water and the bottom is at a depth $h_2$ (Figure 18.49). Show that the volume of water that emerges from the opening in a unit time is

$$\tfrac{2}{3}l\sqrt{2g}(h_2^{3/2} - h_1^{3/2})$$

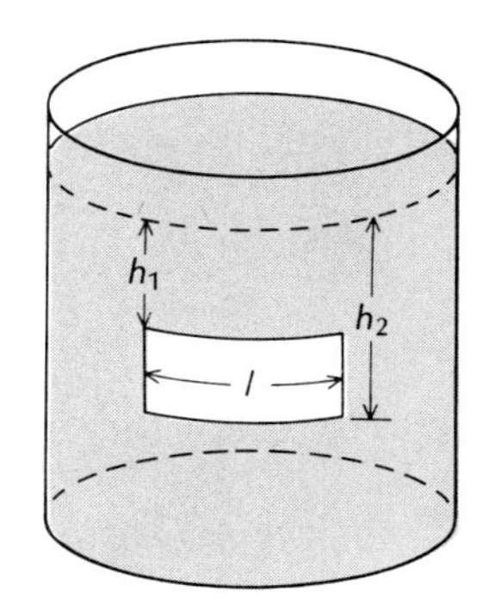


**Fig. 18.49** Tank with a rectangular opening.

CHAPTER 19

# The Ideal Gas and Kinetic Theory

One cubic centimeter of air contains about $2.7 \times 10^{19}$ molecules. As already mentioned in the preceding chapter, such an enormous number precludes a microscopic description of the individual motions of the molecules. We have no way of ascertaining the initial positions and velocities of all the molecules; and even if we had, the calculation of the simultaneous motions of $2.7 \times 10^{19}$ molecules is far beyond the capabilities of even the fastest conceivable computer.

In lack of a microscopic description involving the individual positions and velocities of the molecules of the gas, we must be content with a macroscopic description involving just a few variables that characterize the *average* conditions in the volume of gas. For example, we can regard the mass as such a macroscopic variable. The assertion that the mass of a cubic centimeter of air is $1.29 \times 10^{-6}$ kg does not mean that every cubic centimeter of air has exactly the same mass as every other cubic centimeter. The air molecules wander about and sometimes a few more, or a few less, molecules will be within any given cubic centimeter of space. We can only say what the mass will be on the average. In practice the fluctuations of the mass in a cubic centimeter of air are quite small — typically about $10^{-8}\%$.

Other macroscopic parameters characterizing the average conditions of a gas that can be measured with large-scale laboratory instruments are the number of moles, the volume, the density, the pressure, and the temperature.

Kinetic theory attempts to derive relations between the macroscopic parameters by investigating the average behavior of the microscopic parameters that characterize the individual molecules. Thermodynamics also attempts to derive relations between macroscopic parameters, but without making any use of microscopic properties. Instead, thermodynamics relies only on a few general principles — such as conservation of energy — that do not depend on the details of the structure of matter, and do not even depend on the existence of atoms. In the present chapter we will study some aspects of the kinetic theory of gases, and in Chapter 21 we will study the laws of thermodynamics and their consequences.

## 19.1 The Ideal-Gas Law

**William Thomson, Lord Kelvin,** *1824–1907, British physicist and engineer, professor at the University of Glasgow. Besides inventing the absolute temperature scale, he made many other contributions to the theory of heat. He was first to state the principle of dissipation of energy incorporated in the Second Law of Thermodynamics. In engineering, Kelvin made crucial improvements in the design of cables for submarine telegraphy and in the design of the mariner's compass.*

The pressure, volume, and temperature of a gas obey some simple laws. Before we state these laws, let us recall the definition of pressure (see Section 18.3). Imagine that the gas is divided into small adjacent cubical volumes. The pressure is the force that one of these cubes exerts on an adjacent cube, or on an adjacent wall, divided by the area of one face of the cube; that is, the pressure is the force per unit area. We know from Section 18.4 that the pressure is the same throughout the entire volume of a container of gas (within a container of gas, gravity causes a small decrease of pressure from the bottom to the top, but this decrease of pressure can usually be ignored).

Consider now a given amount of gas, say, $n$ moles of gas. We saw in Chapter 1 that a mole of any chemical element (or chemical compound) is that amount of matter that contains exactly as many atoms (or molecules) as there are atoms in 12.0 g of carbon. The "atomic mass" of a chemical element (or the "molecular mass" of a compound) is the mass of 1 mole. Thus, according to the table of atomic masses (see Appendix 9), 1 mole of carbon has a mass of 12.0 g, 1 mole of oxygen molecules ($O_2$) has a mass of 32.0 g, 1 mole of nitrogen molecules ($N_2$) has a mass of 28.0 g, and so on.

Suppose we place this amount of gas in a container of volume $V$ at a temperature $T$. The gas will then exert a pressure $p$. Experiments show that — to a good approximation — the pressure $p$, the volume $V$, and the temperature $T$ of the $n$ moles of gas are related by the **ideal-gas law:**

*Ideal-gas law*

$$pV = nRT \tag{1}$$

Here $R$ is the **universal gas constant** with the value

*Universal gas constant, R*

$$R = 8.31 \text{ J/K} \tag{2}$$

The temperature in Eq. (1) is measured on the **Kelvin temperature scale,** and the unit of temperature is the **kelvin,** or **degree absolute,** abbreviated K. We have not previously given the definition of temperature because Eq. (1) plays a dual role: it is a law of physics and also serves for the definition of temperature. This is by now a familiar story — in Chapter 5 we already came across laws that play such a dual role.

We will give the details concerning the definition of temperature in the next section. For now, it will suffice to note that the freezing point of water corresponds to a temperature of 273.15 K, and the boiling point of water corresponds to 373.15 K; hence, there is an interval of exactly 100 K between the freezing and the boiling points. The zero of temperature on the Kelvin scale is the absolute zero, $T = 0$ K. According to Eq. (1), the pressure of the gas vanishes at this point. Actually, the gas may liquefy or even solidify before the absolute zero of temperature can be reached; when this happens, Eq. (1) becomes inapplicable.

The ideal-gas law is a simple relationship among the macroscopic parameters that characterize a gas. At normal densities and pressures, real gases obey this law quite well; but if a real gas is compressed to an excessively high density, then its behavior will deviate from this law. For instance, at 1-atm pressure the volume of 1 mole of oxygen gas is only about 0.13% smaller than predicted by the ideal-gas law, but at 20-atm pressure the volume is about 2.3% smaller than predicted. In the following, we will always take it for granted that the pressures and densities are low enough so that deviations from Eq. (1) can be neglected. An **ideal gas** is a gas that obeys Eq. (1) *exactly.* The ideal gas is a limiting case of a real gas when the density and the pressure of the latter tend to zero. The ideal gas may be thought of as consisting of atoms of infinitesimal size which only exert forces over an infinitesimal range; the atoms then exert no forces at all, except for the instantaneous impulsive forces during collisions with the walls of the container holding the gas.

*Ideal gas*

*Boyle's Law and Gay-Lussac's Law*

The ideal-gas law incorporates two laws: **Boyle's Law** and **Gay-Lussac's Law** (or **Charles' Law**). Boyle's Law asserts that if the temperature is constant, then the product of pressure and volume must remain constant as a given amount of gas is compressed or expanded,

$$pV = [\text{constant}] \qquad \text{for } T = [\text{constant}] \tag{3}$$

Gay-Lussac's Law asserts that if the pressure is constant, the ratio of volume to temperature remains constant as a given amount of gas is heated or cooled,

$$V/T = [\text{constant}] \qquad \text{for } p = [\text{constant}] \tag{4}$$

The relation given by Eq. (3) can be tested experimentally by placing a sample of gas in a cylinder with a movable piston, surrounded by some substance at a constant temperature (say, a water bath; see Figure 19.1). The relation of Eq. (4) can be tested with a similar cylinder–piston arrangement with a weight mounted on the piston and a heat source below the cylinder (see Figure 19.2). Together, the experimental tests of Eqs. (3) and (4) amount to a test of the ideal-gas law.

**Robert Boyle,** *1627–1691, English experimental physicist. He was one of the younger sons of the first earl of Cork, and he used the financial means at his disposal to set up laboratories of his own at Oxford and at London. Boyle invented a new air pump, with which he performed the experiments on gases that led to discovery of the law named after him.*

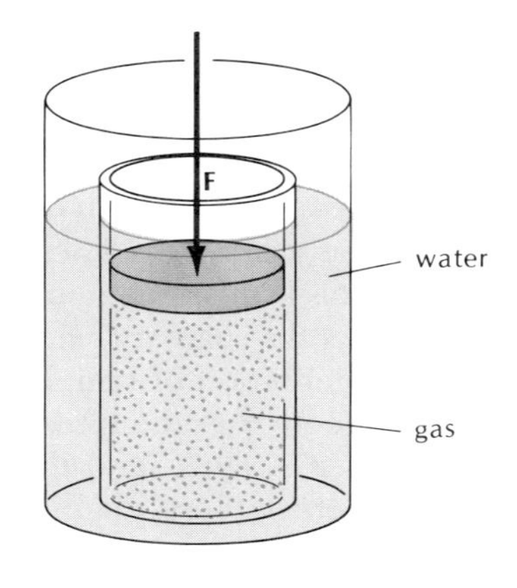


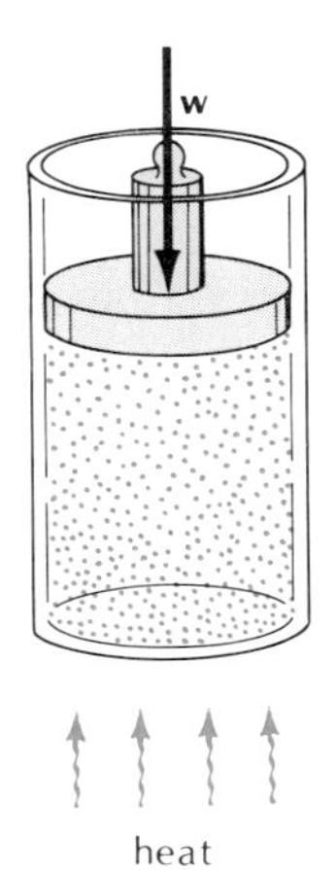


**Fig. 19.1** (left) Compression of a gas by a force applied to the piston. The container with the gas is surrounded by a water bath that keeps the temperature constant.

**Fig. 19.2** (right) Expansion of a gas kept at constant pressure by a weight on the piston. The gas is being heated.

EXAMPLE 1. Suppose you heat 1 kg of water and convert it into steam at the boiling temperature of water, 373 K, and at normal atmospheric pressure, 1 atm, or $1.01 \times 10^5$ N/m$^2$. What is the volume of the steam?

SOLUTION: The molecular mass of water ($H_2O$) is the sum of the atomic masses of two hydrogen atoms and one oxygen atom, that is, 18.0 g. Thus, the number of moles of steam in 1 kg is

$$n = \frac{1 \text{ kg}}{18.0 \text{ g}} = \frac{1000 \text{ g}}{18.0 \text{ g}} = 55.6 \text{ moles}$$

According to Eq. (1), the volume of the steam is then

$$V = \frac{nRT}{p} = \frac{55.6 \times 8.31 \text{ J/K} \times 373 \text{ K}}{1.01 \times 10^5 \text{ N/m}^2}$$

$$= 1.71 \text{ m}^3$$

COMMENTS AND SUGGESTIONS: Keep in mind that the "atomic masses" listed in tables of atoms, such as the table in Appendix 9, are given in *grams*. Since most examples and problems in this book are stated in kilograms, calculation of the number of moles requires conversion from grams to kilograms, or vice versa.

EXAMPLE 2. Early in the morning at the beginning of a trip, the tires of an automobile are cold (280 K) and their air is at a pressure of 3.0 atm. Later in the day, after a long trip on hot pavements, the tires are hot (330 K). What is the pressure? Assume that the volume of the tires remains constant.

SOLUTION: At constant volume the pressure is proportional to the temperature [see Eq. (1)]. Hence

$$p_2 = p_1 \times \frac{T_2}{T_1} = 3.0 \text{ atm} \times \frac{330 \text{ K}}{280 \text{ K}} = 3.5 \text{ atm}$$

COMMENTS AND SUGGESTIONS: The pressure gauges for automobile tires are commonly calibrated to read **overpressure,** that is, the excess above atmospheric pressure. Thus, the pressure gauge would read 2.0 atm in the morning; and 2.5 atm later in the day (if the atmospheric pressure remains constant at 1 atm).

The ideal-gas law can also be written in terms of the number of molecules, instead of the number of moles. The number of molecules per mole is **Avogadro's number,**[1]

$$\boxed{N_A = 6.0221 \times 10^{23} \text{ molecules/mole}} \qquad (5)$$

*Avogadro's number, $N_A$*

Thus, if the number of moles is $n$, the number of molecules is

$$N = N_A n \qquad (6)$$

With this, Eq. (1) becomes

$$pV = \frac{N}{N_A} RT$$

[1] See Appendix 8 for a more precise value of this constant.

which we can write as

*Ideal-gas law*

$$pV = NkT \tag{7}$$

with

$$k = \frac{R}{N_A} = \frac{8.314 \text{ J/K}}{6.022 \times 10^{23}} = 1.381 \times 10^{-23} \text{ J/K} \tag{8}$$

*Boltzmann's constant, k*

The constant $k$ is called **Boltzmann's constant.**[2] As we will see, this constant tends make an appearance in equations relating macroscopic quantities (such as $p$ or $V$) to microscopic quantities (such as the number $N$ of molecules).

*Standard temperature and pressure (STP)*

EXAMPLE 3. (a) What is the number of molecules in 1 $cm^3$ of air at a temperature of 273 K and a pressure of 1 atm, called **standard temperature and pressure,** or **STP**? (b) What is the mass of air in 1 $cm^3$ under these conditions? The mean molecular mass of air is 29.0 g.[3]

SOLUTION: (a) In metric units, 1 atm is $1.01 \times 10^5$ N/$m^2$ and 1 $cm^3$ is $10^{-6}$ $m^3$. Hence, the number of molecules in 1 $cm^3$ is

$$N = \frac{pV}{kT} = \frac{1.01 \times 10^5 \text{ N/m}^2 \times 10^{-6} \text{ m}^3}{1.38 \times 10^{-23} \text{ J/K} \times 273 \text{ K}}$$

$$= 2.68 \times 10^{19} \text{ molecules}$$

Note that this result is valid for any kind of gas — the number of molecules in 1 $cm^3$ of any kind of gas under STP conditions is $2.68 \times 10^{19}$.

(b) The number of moles in 1 $cm^3$ is

$$\frac{2.68 \times 10^{19}}{N_A} = \frac{2.68 \times 10^{19}}{6.02 \times 10^{23}} = 4.45 \times 10^{-5} \text{ moles}$$

Since the mass per mole is 29.0 g, the amount of mass is

$$4.45 \times 10^{-5} \times 29.0 \text{ g} = 1.29 \times 10^{-3} \text{ g} = 1.29 \times 10^{-6} \text{ kg}$$

This implies that the mass in 1 $m^3$ of air is 1.29 kg, in agreement with the density of air listed in Table 18.1.

## 19.2 The Ideal-Gas Temperature Scale

To use Eq. (1) for the definition of temperature, we take a fixed amount of some gas, say, helium, and place it in an airtight, nonexpanding container, say, a Pyrex glass bulb. The amount of gas should

[2] See Appendix 8 for a more precise value of this constant.

[3] Air consists, on the average, of 75.54% nitrogen, 23.1% oxygen, and 1.3% argon by mass.

be small, so the density and pressure are low, and the gas behaves like an ideal gas. According to Eq. (1), the pressure of an ideal gas kept in such a constant volume is directly proportional to the temperature. Thus, a simple measurement of pressure gives us the temperature.

To calibrate the scale of this ideal-gas thermometer, we must choose a standard reference temperature. The standard adopted in the SI system of units is the temperature of the **triple point of water,** that is, the temperature at which water, ice, and water vapor coexist when placed in a closed vessel. Figure 19.3 shows a triple-point cell used to achieve the standard temperature. This standard temperature has been assigned the value 273.16 kelvin, or 273.16 K. If the bulb of the gas thermometer is placed in thermal contact with this cell so that it attains a temperature of 273.16 K, it will read some pressure $p_{\text{tri}}$. If the bulb is then placed in thermal contact with some body at an unknown temperature $T$, it will read a pressure $p$ which is greater or smaller than $p_{\text{tri}}$ by some factor. The unknown temperature $T$ is then greater or smaller than 273.16 K by this same factor; for instance, if the pressure $p$ is half as large as $p_{\text{tri}}$, then $T = \frac{1}{2} \times 273.16$ K.

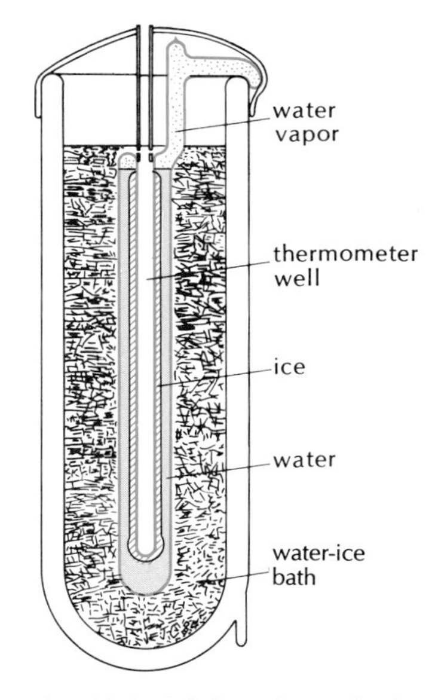


**Fig. 19.3** Triple-point cell of the National Bureau of Standards. The inner tube (colored) contains water, water vapor, and ice in equilibrium.

In general, the temperature $T$ may be expressed as

$$T = 273.16 \text{ K} \times \frac{p}{p_{\text{tri}}} \tag{9}$$

This equation calibrates our thermometer. The temperature scale defined in this way is called the **ideal-gas temperature scale.**

*Ideal-gas temperature scale*

When connecting a pressure gauge to the bulb of gas, we must take special precautions to ensure that the operation of the pressure gauge does not alter the volume available to the gas. Figure 19.4 shows a device that will serve our purposes; this device is called a **constant-volume gas thermometer.** The pressure gauge used in this thermometer consists of a closed-tube manometer; one branch of the manometer is connected to the bulb of gas, and the other branch consists of a closed, evacuated tube. The difference $h$ in the heights of the levels of mercury in these two branches is proportional to the pressure of the gas. The manometer is also connected to a mercury reservoir. During the operation of the thermometer, this reservoir must be raised or lowered so that the level of mercury in the left branch of the manometer tube always remains at a constant height; this keeps the gas in the bulb at a constant volume. The bulb of this thermometer may be put in thermal contact with any body whose temperature we wish to measure, and the pressure registered by the manometer then gives us the ideal-gas temperature.

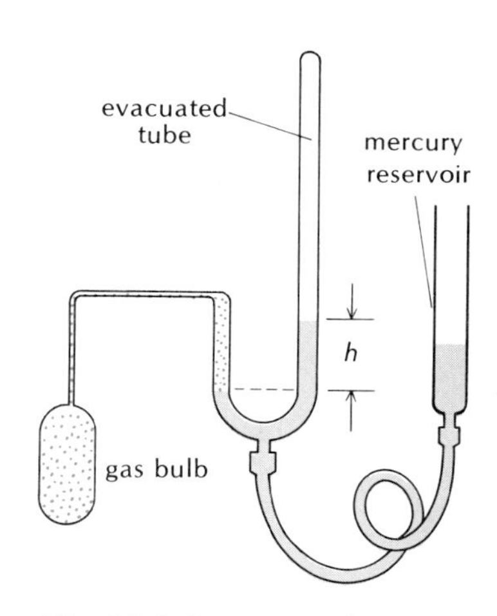


**Fig. 19.4** Constant-volume gas thermometer.

Table 19.1 lists some examples of temperatures of diverse bodies.

The ideal-gas thermometer plays a primary role in the measurement of temperature because, as we will see in Chapter 21, the ideal-gas temperature scale coincides with the Kelvin temperature scale, also called the absolute thermodynamic temperature scale, which is the fundamental temperature scale for the study of thermodynamic processes. The name *kelvin,* which we introduced for the unit of temperature in Section 19.1, anticipates this coincidence of the ideal-gas temperature scale and the Kelvin scale. To simplify the terminology, we will hereafter use the single name *Kelvin scale* for both of these scales.

For practical applications, the ideal-gas thermometer is somewhat in-

**Table 19.1** SOME TEMPERATURES

| | *Kelvin temperature* | *Celsius temperature* |
|---|---|---|
| Interior of hottest stars | $10^9$ K | $10^9$ °C |
| Center of H-bomb explosion | $10^8$ K | $10^8$ °C |
| Highest temperature attained in laboratory (plasma) | $6 \times 10^7$ K | $6 \times 10^7$ °C |
| Center of Sun | $1.5 \times 10^7$ K | $1.2 \times 10^7$ °C |
| Surface of Sun | $4.5 \times 10^3$ K | $4.2 \times 10^3$ °C |
| Center of Earth | $4 \times 10^3$ K | $3.7 \times 10^3$ °C |
| Acetylene flame | $2.9 \times 10^3$ K | $2.6 \times 10^3$ °C |
| Melting of iron | $1.8 \times 10^3$ K | $1.5 \times 10^3$ °C |
| Melting of lead | $6.0 \times 10^2$ K | $3.3 \times 10^2$ °C |
| Boiling of water | 373 K | 100 °C |
| Human body | 310 K | 37 °C |
| Surface of Earth (average) | 287 K | 14 °C |
| Freezing of water | 273 K | 0 °C |
| Liquefaction of nitrogen | 77 K | −196 °C |
| Liquefaction of hydrogen | 20 K | −253 °C |
| Liquefaction of helium | 4.2 K | −269 °C |
| Interstellar space | 3 K | −270 °C |
| Lowest temperature attained in laboratory | $3 \times 10^{-8}$ K | $\cong -273.15$ °C |

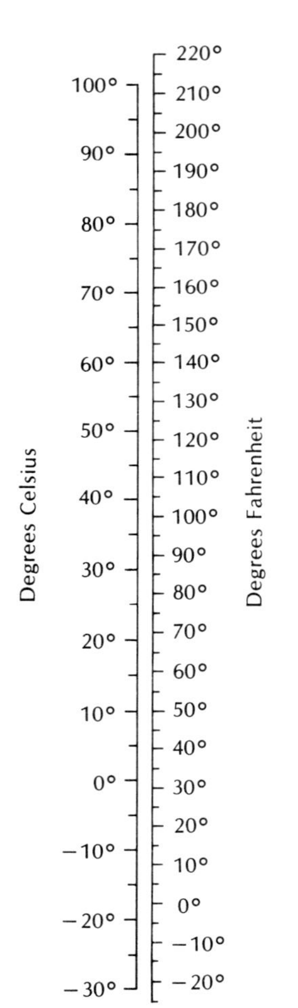


**Fig. 19.5** The correspondence between the Fahrenheit scale and the Celsius scale.

convenient and is often replaced by mercury-bulb thermometers, bimetallic strips, electrical-resistance thermometers, or thermocouples. These must be calibrated in terms of the ideal-gas thermometer so they, too, will read Kelvin temperature. We will deal with some details of these secondary thermometers in the next chapter.

Although the Kelvin temperature scale is the only scale of fundamental significance, several other temperature scales are in practical use. The **Celsius scale** (formerly known as the *centigrade scale*) is shifted by 273.15 degrees relative to the Kelvin scale,

$$T_C = T - 273.15\,°\text{C} \tag{10}$$

Note that on the Celsius scale, absolute zero is at −273.15°C. The triple point of water is then at 0.01°C, and the boiling point at 100°C. The freezing point of water, at atmospheric pressure, is at 0°C. (The slight difference between the temperatures of the freezing point and the triple point is due to the difference in the pressure of the water. Normal freezing occurs at normal atmospheric pressure, whereas the upper portion of the triple-point cell is evacuated and contains only a small amount of water vapor of very low pressure. In water, a lowering of the pressure causes a rising of the freezing point.)

The **Fahrenheit scale** is shifted relative to the Celsius scale and, furthermore, uses degrees of smaller size, each degree Fahrenheit corresponding to $\frac{5}{9}$ degree Celsius:

$$T_F = \tfrac{9}{5}T_C + 32\,°\text{F} \tag{11}$$

On this scale, the freezing point of water is at 32°F and the boiling point at 212°F. Figure 19.5 can be used for rough conversions between the Fahrenheit and Celsius scales.

## 19.3 Kinetic Pressure: The Maxwell Distribution

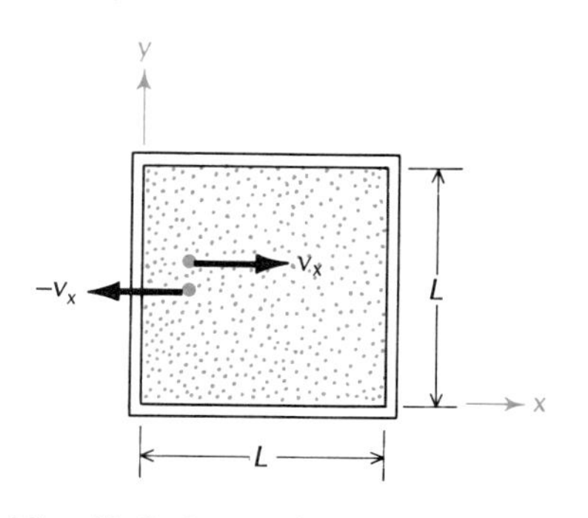


**Fig. 19.6** Gas molecules in a container. In a collision with the wall at $x = 0$, the $x$ velocity of a molecule reverses.

The pressure of a gas against the walls of its container is due to the impacts of the molecules on the walls. We will now calculate this pressure by means of kinetic theory by considering the average motion of the molecules of gas. We will assume that the container is a cube of side $L$, that the gas molecules collide only with the walls but not with each other, and that the collisions are elastic. These assumptions are not necessary, but they simplify the calculations.

Figure 19.6 shows the container filled with gas molecules. The motion of each molecule can be resolved into $x$, $y$, and $z$ components. Consider one molecule, and consider the component of its motion in the $x$ direction. The component of velocity in this direction is $v_x$, and the magnitude of this velocity remains constant, since the collisions with the wall are elastic. The time that the molecule takes to move from the wall of the cube at $x = 0$ to $x = L$ and back to $x = 0$ is

$$\Delta t = 2L/v_x$$

This is therefore the time between one collision with the wall at $x = 0$ and the next collision with the same wall. When the molecule strikes the wall, its $x$ velocity is reversed from $-v_x$ to $+v_x$ (see Figure 19.6). Hence during each collision at $x = 0$, the $x$ momentum changes by

$$mv_x - (-mv_x) = 2mv_x \tag{12}$$

where $m$ is the mass of the molecule. The average rate at which the molecule transfers momentum to the wall at $x = 0$ is then

$$\frac{2mv_x}{\Delta t} = \frac{2mv_x}{2L/v_x} = \frac{mv_x^2}{L} \tag{13}$$

This is therefore the average force that the impacts of this one molecule exert on the wall. To find the total force exerted by the impacts of all the $N$ molecules, we must multiply the force given in Eq. (13) by $N$; and to find the pressure, we must divide by the area $L^2$ of the wall. This leads to a pressure

$$p = \frac{N}{L^2}\frac{mv_x^2}{L} \tag{14}$$

or, in terms of the volume $V = L^3$,

$$p = \frac{Nmv_x^2}{V} \tag{15}$$

In this calculation, we made the implicit assumption that all the molecules have the same speed. This is, of course, not true; the molecules of the gas have a distribution of speeds — some molecules have high speeds, some low. To account for this, we must replace the force in Eq. (13) due to one given molecule by the average over all the molecules. Consequently, we must replace $v_x^2$ by an average over all the molecules in the container. We will designate this average by $\overline{v_x^2}$. Equation (15) then becomes

$$p = \frac{Nm\overline{v_x^2}}{V} \tag{16}$$

To proceed further, we note that on the average, molecules are just as likely to move in the $x$, $y$, or $z$ direction. Hence the average values of $v_x^2$, $v_y^2$, and $v_z^2$ are equal,

$$\overline{v_x^2} = \overline{v_y^2} = \overline{v_z^2} \tag{17}$$

The sum of the squares of the components of the velocity is the square of the magnitude of the velocity,

$$\overline{v_x^2} + \overline{v_y^2} + \overline{v_z^2} = \overline{v^2} \tag{18}$$

and therefore each of the terms on the left side of Eq. (18) must equal $\frac{1}{3}\overline{v^2}$. We can then write Eq. (16) as

$$pV = \frac{Nm\overline{v^2}}{3} \tag{19}$$

This expresses the pressure in terms of the average square of the speed of the molecules.

Let us now compare this result with the ideal-gas law [Eq. (7)],

$$pV = NkT \tag{20}$$

Obviously, the agreement between the right sides of Eqs. (19) and (20) demands

$$\frac{m\overline{v^2}}{3} = kT \tag{21}$$

This shows that the average of the square of the speed is proportional to the temperature. The square root of $\overline{v^2}$ is called the **root-mean-square speed,** or rms speed, and it is usually designated by $v_{rms}$. If we take the square root of both sides of Eq. (21), we find

*Root-mean-square speed*

$$v_{rms} = \sqrt{\overline{v^2}} = \sqrt{\frac{3kT}{m}} \tag{22}$$

This rms speed may be regarded as the typical speed of the molecules of the gas. There are several other ways of calculating a typical speed; for example, we may want to know the average of all the molecular speeds or the most probable of all the molecular speeds. These other typical speeds turn out to be approximately the same as $v_{rms}$, but their calculation requires some further knowledge of the distribution of molecular speeds.

---

EXAMPLE 4. What is the rms speed of nitrogen molecules in air at 0°C? Of oxygen molecules?

SOLUTION: The molecular mass of $N_2$ molecules is 28.0 g. Hence the mass of one molecule is

$$m = \frac{28.0\ \text{g}}{N_A} = \frac{28.0\ \text{g}}{6.02 \times 10^{23}} = 4.65 \times 10^{-26}\ \text{kg}$$

and Eq. (22) yields

$$v_{\rm rms} = \sqrt{\frac{3kT}{m}} = \sqrt{\frac{3 \times 1.38 \times 10^{-23}\ \text{J/K} \times 273\ \text{K}}{4.65 \times 10^{-26}\ \text{kg}}}$$

$$= 493\ \text{m/s}$$

Likewise, for $O_2$ molecules the molecular mass is 32.0 g, the mass of one molecule is $5.32 \times 10^{-26}$ kg, and $v_{\rm rms}$ is 461 m/s.

COMMENTS AND SUGGESTIONS: Note that the rms speed of nitrogen molecules is slightly larger than that of oxygen molecules. In general, Eq. (22) shows that the rms speed is inversely proportional to the square root of the mass of the molecule — at a given temperature, the molecules of lowest mass have the highest speeds.

---

*Maxwell distribution*

The distribution of molecular speeds in a sample of gas at some given temperature can be deduced by means of kinetic theory; it is called the **Maxwell distribution** of molecular speeds. We will not attempt to deduce this distribution here, but only examine some of its qualitative features. The distribution of speeds in a sample of gas is described mathematically by the distribution function $N_v$, which specifies the number of molecules per unit speed interval. Thus, if $dv$ is a small interval of speeds centered on a given speed $v$, then the number of molecules that have speeds in the interval $dv$ is

$$dN = N_v dv \tag{23}$$

Keep in mind that when dealing with the distribution of speeds, or the distribution of any other kind of physical quantity with a continuous range of variation, it is not reasonable to ask how many molecules have a speed $v$, since it is unlikely that any molecule has a speed *exactly* equal to $v$. The only reasonable question is how many molecules have speeds in some specified interval of speeds.

Figure 19.7 shows plots of the function $N_v$ for the Maxwell distributions of speeds at two different temperatures. As we can see from these plots, the molecular speeds are spread over broad ranges. The distribution functions fade away as $v \rightarrow 0$ and as $v \rightarrow \infty$; thus, there are few

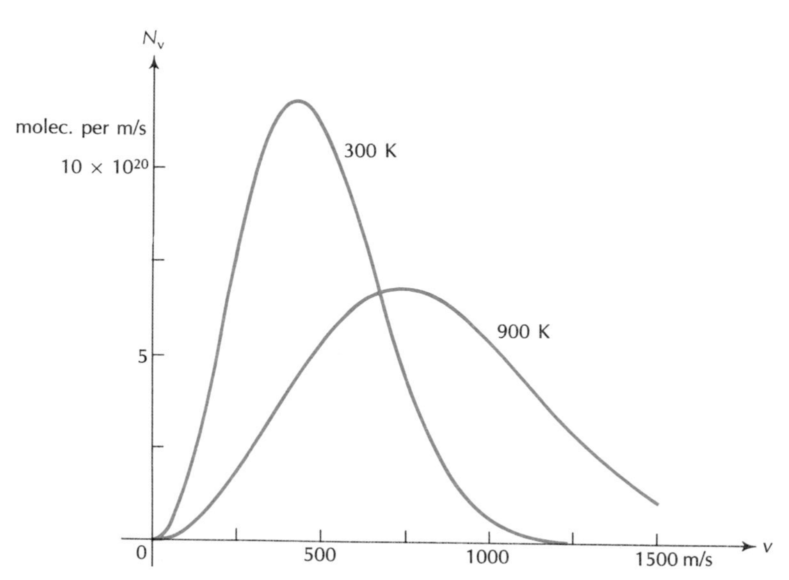


**Fig. 19.7** Maxwell distribution of speeds for one mole of $N_2$ molecules. The higher curve corresponds to a temperature of 300 K; the lower curve corresponds to 900 K. The most probable speed for the Maxwell distribution is $0.82v_{\rm rms}$ and the average speed is $0.92v_{\rm rms}$. The most probable speed is indicated by the peak of the curve.

molecules near zero speed, and few molecules of very large speed. The peaks of the distribution functions indicate the **most probable speed.** Comparing the peaks at the two different temperatures, we see that the most probable speed increases with temperature. The average speed and the rms speed for the Maxwell distribution are somewhat larger than the most probable speed; the values of these speeds are marked in Figure 19.7.

*Most probable speed*

From Eq. (23) we recognize that the integral of $N_v$ from $v = 0$ to $v = \infty$ is the total number of molecules in the sample. Graphically, in a plot of the distribution function, this integral is the area under the curve. The two distributions plotted in Figure 19.7 contain the same numbers of molecules, and the areas under these two curves are equal (the reduction of the overall height of the second distribution in Figure 19.7 is a consequence of this requirement of equal areas).

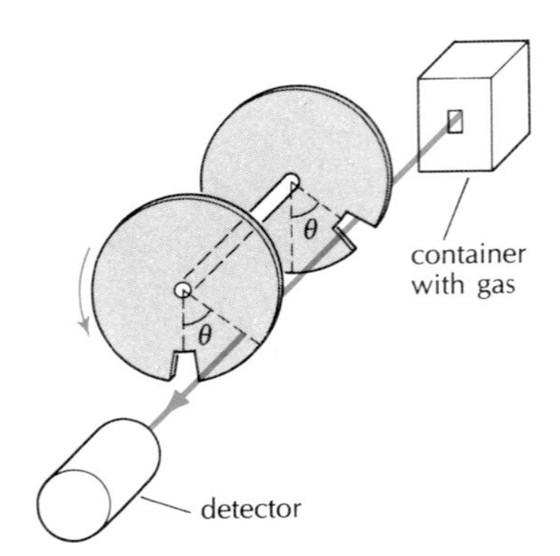


**Fig. 19.8** Velocity filter for the measurement of molecular speeds.

For the experimental determination of the distribution of molecular speeds, a thin beam of molecules is allowed to stream out of a small hole in the wall of the container holding the sample of gas. The speeds in this thin beam are then measured with a velocity filter, or velocity selector, consisting of a pair of slotted disks rotating on an axis (see Figure 19.8). The slots in the disks are offset by a fixed angle $\theta$. A molecule that passes through the slot in the first disk will be able to pass through the slot in the second disk if and only if its travel time matches the time taken by the disks to rotate through the angle $\theta$. Thus, the disks act as a filter: they permit the passage of molecules of one speed, but block molecules of all other speeds. The speed accepted by the filter can be increased or decreased by adjusting the angular velocity of the disks. The molecules that succeed in passing through the filter are registered by a detector. Measurement of the intensity registered by the detector vs. speed accepted by the filter yields the distribution function of molecular speeds in the beam. The distribution functions measured by such experiments are in perfect agreement with the Maxwell distribution.[4]

## 19.4 The Energy of an Ideal Gas

According to Eq. (21), the average kinetic energy of a molecule of an ideal gas at temperature $T$ is

$$\tfrac{1}{2}m\overline{v^2} = \tfrac{3}{2}kT \tag{24}$$

If we multiply this by the total number of molecules, we will obtain the total translational kinetic energy of all the molecules jointly. In an ideal gas, the molecules exert no forces on one another (they do not collide), and hence there is no intermolecular potential energy. The kinetic energy is then the total energy,

[4] Actually, the distribution of speeds in the beam cannot be compared directly with the Maxwell distribution. A correction is required because, when molecules escape from a hole in a container, the molecules most likely to escape are those of high speeds. Thus, the distribution of speeds in the beam is skewed toward higher speeds, and differs in a predictable way from the distribution of speeds found inside the container. This difference must be taken into account when testing the measured distribution of speeds in the beam against the theoretical Maxwell distribution.

$$\boxed{E = K = \tfrac{3}{2}NkT} \tag{25}$$

*Kinetic energy of an ideal monoatomic gas*

This formula tells us how much energy is stored in the microscopic thermal motions of the gas. We will call this energy the **internal energy** of the gas to distinguish it from any extra energy the gas might have because of an overall translational motion, caused by, say, a translational motion of the container. Since $Nk = nR$ [see Eqs. (6) and (8)], we can also write this energy as

$$E = \tfrac{3}{2}\, nRT \tag{26}$$

Note that here we have assumed that the molecules behave like pointlike particles — each molecule has translational kinetic energy, but no internal energy within the molecule. This assumption is valid for monoatomic gases, such as helium, argon, and krypton.

---

EXAMPLE 5. What is the thermal kinetic energy in 1 kg of He gas at 0°C? How much extra energy must be supplied to this gas to increase its temperature to 60°C (at a constant volume)?

SOLUTION: The atomic mass of He is 4.0 g. The number of moles in 1 kg of He is

$$n = \frac{1\text{ kg}}{4.0\text{ g}} = 250\text{ moles}$$

Hence Eq. (25) yields

$$E = \tfrac{3}{2}nRT = \tfrac{3}{2} \times 250 \times 8.3\text{ J/K} \times 273\text{ K} = 8.5 \times 10^5\text{ J}$$

The extra energy needed to increase the temperature to 60°C is

$$\Delta E = \tfrac{3}{2}nR\,\Delta T = \tfrac{3}{2} \times 250 \times 8.3\text{ J/K} \times 60\text{ K} = 1.9 \times 10^5\text{ J}$$

---

Diatomic gases, such as $N_2$ and $O_2$, store additional amounts of energy in the internal motions of the atoms within each molecule. The molecules of these gases may be regarded as two pointlike particles rigidly connected together (a dumbbell; see Figure 19.9). If such a molecule collides with another molecule or with the wall of the container, it will usually start rotating about its center of mass. We therefore expect that, on the average, an appreciable fraction of the energy of the gas will be in the form of this kind of rotational kinetic energy.

The molecule may rotate about either of the two axes through the center of mass perpendicular to the line joining the atoms (Figure 19.9). If the moments of inertia about these axes are $I_1$ and $I_2$, and if the corresponding angular velocities are $\omega_1$ and $\omega_2$, then the kinetic energy for these two rotations is

$$\tfrac{1}{2}I_1\omega_1^2 + \tfrac{1}{2}I_2\omega_2^2$$

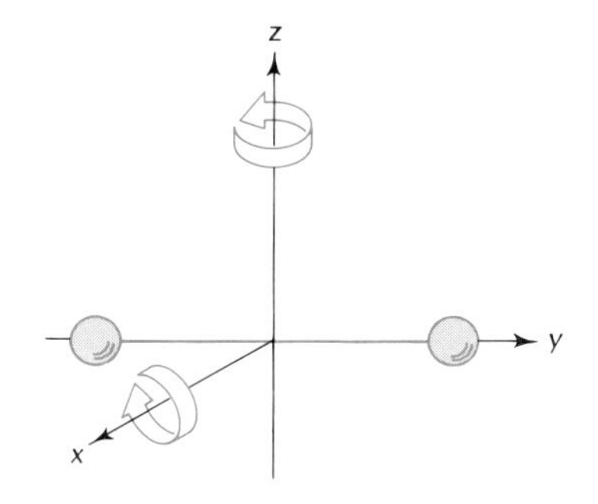

**Fig. 19.9** A diatomic molecule represented as two pointlike particles joined by a rod. The possible axes of rotation of the molecule are marked with curved arrows.

The average kinetic energy is

$$\tfrac{1}{2}I_1\overline{\omega_1^2} + \tfrac{1}{2}I_2\overline{\omega_2^2} \tag{27}$$

where, as in the preceding section, the overbars denote the average over all the molecules of the gas.

To find out the value of these average rotational energies, let us return to Eq. (24) and write it in terms of the *x*, *y*, and *z* components of velocity:

$$\tfrac{1}{2}m\overline{v_x^2} + \tfrac{1}{2}m\overline{v_y^2} + \tfrac{1}{2}m\overline{v_z^2} = \tfrac{3}{2}kT \tag{28}$$

We know that the average *x*, *y*, and *z* speeds are equal. Hence Eq. (28) asserts that the kinetic energy for each component of the motion has the value $\frac{1}{2}kT$,

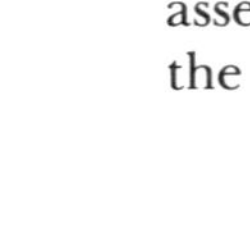

$$\tfrac{1}{2}m\overline{v_x^2} = \tfrac{1}{2}kT, \qquad \tfrac{1}{2}m\overline{v_y^2} = \tfrac{1}{2}kT, \qquad \tfrac{1}{2}m\overline{v_z^2} = \tfrac{1}{2}kT \tag{29}$$

It turns out that this is true not only for the components of translational motion, but also for rotational motion. The general result is known as the **equipartition theorem:**

*Equipartition theorem*

> *Each translational or rotational component of the random thermal motion of a molecule has an average kinetic energy of* $\frac{1}{2}kT$.

We will not prove this theorem, but we will make use of it.[5]

According to this theorem, each of the terms in Eq. (27) has a value $\frac{1}{2}kT$. The total average rotational kinetic energy is then $kT$, and when we add this to the translational kinetic energy $\frac{3}{2}kT$, we obtain a total kinetic energy

$$kT + \tfrac{3}{2}kT = \tfrac{5}{2}kT \tag{30}$$

for one molecule. The energy of all the molecules taken together is then

$$E = \tfrac{5}{2}NkT \tag{31}$$

**Ludwig Boltzmann,** *1844–1906, Austrian theoretical physicist, professor at Munich, Vienna, and Leipzig. He made crucial contributions in the kinetic theory of gases and in statistical mechanics.*

Note that in this calculation we have ignored the possibility of rotation about the longitudinal axis of the molecule. This means we have ignored the rotation of the atoms about an axis through them, just as we have ignored this kind of rotation of the atoms of a monoatomic gas. Furthermore, we have ignored the vibrational motion of the atoms of the diatomic molecule. The interatomic forces do not really hold these atoms rigidly; rather, the forces act somewhat like springs (see Example 15.3) and they permit a restricted back-and-forth vibration of the atoms about their equilibrium positions. The reason why we have ignored these motions in our calculation lies beyond the realm of classical physics; it lies in the realm of quantum physics. There it is established that rotations of atoms about their own axes and vibrations of atoms in a molecule do not occur unless the temperature is rather high, 400°C or more. As we will see in Section 20.6, the energies calculated from Eqs. (25) and (31) actually agree pretty well with experiments, provided that we do not exceed this temperature limit.

---

[5] Each translational or rotational component of the motion is called a **degree of freedom.** With this terminology, the equipartition theorem can be stated as: Each degree of freedom has an average energy of $\frac{1}{2}kT$.

## 19.5 The Mean Free Path

In our discussion of kinetic pressure we have pictured the molecules as pointlike particles that suffer collisions with the walls of the container, but not with each other. This is a somewhat unrealistic picture — the molecules have a finite size and, under typical conditions prevailing in a gas, they collide quite frequently. Although these collisions are of little consequence for the kinetic pressure the gas exerts on the walls, the collisions play a crucial role in transport phenomena, such as heat transport in a gas with nonuniform temperatures, and momentum transport across layers of a gas with nonuniform streaming velocities (the momentum transport across the gas generates the viscous drag force between adjacent layers), and transport of one kind of gas through another kind by diffusion.

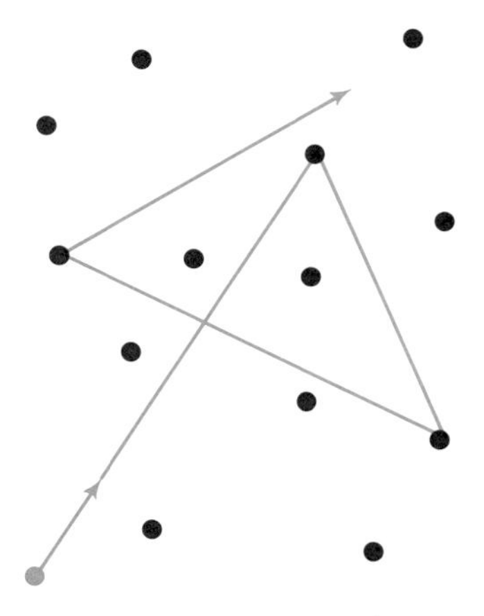

Fig. 19.10 Path of a molecule through a gas. Each collision changes the direction of motion of the molecule. In this figure, the lengths of the straight segments have been scaled down to fit the available space (in air under standard conditions, the average length of the straight segments is about a hundred times the size of a molecule).

To investigate the collision rate of the molecules of a gas, we must make some assumption about the intermolecular forces and the sizes and shapes of the molecules. We will adopt the crude model that the molecules are hard elastic spheres of some given radius $R$. This means that two molecules exert no mutual forces unless their centers come to within a distance $2R$ of each other. Let us now focus on one molecule, and inquire how many collisions this molecule suffers as it moves a distance $L$ through the gas. Between collisions, the molecule moves with constant velocity; but at each collision, the molecule suddenly changes its direction of motion. Thus, the path of the molecule consists of a series of zigzags (see Figure 19.10). Since the molecule must come to within a distance $2R$ of another molecule to suffer a collision, we can imagine that the molecule sweeps out a tube of radius $2R$ and engages in a collision whenever the center of another molecule lies within this tube (see Figure 19.11). Thus, the number of collisions equals the number of molecules found in a tube of radius $2R$ and total length $L$. The volume of this tube is $\pi(2R)^2L$, and the product of this volume and the average number $N/V$ of molecules per unit volume gives us the average number of collisions:

$$[\text{number of collisions}] = 4\pi R^2 L(N/V) \qquad (32)$$

The average distance the molecule must travel to suffer *one* collision is called the **mean free path,** usually designated by $\lambda$. According to Eq. (32),

*Mean free path*

$$\boxed{\lambda = \frac{1}{4\pi R^2(N/V)}} \qquad (33)$$

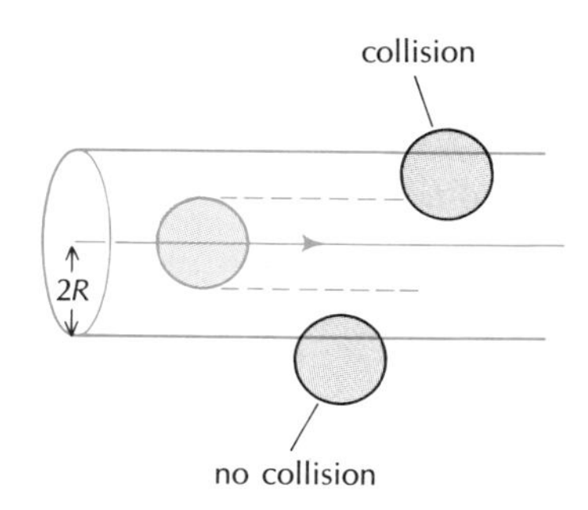


Fig. 19.11 Imaginary tube of radius $2R$ surrounding the path of the molecule. A collision will occur whenever the center of another molecule lies within this tube.

In this simple calculation we have not taken into account that the target molecules are themselves moving. The motion of the target molecules increases the rate of collisions. For instance, a target molecule can enter the swept volume after the sweeping molecule has passed, and hit the sweeping molecule at an angle from behind. From this we see that the motion of the target molecules makes the mean free path somewhat shorter than stated in Eq. (33). Other alterations of the equation for the mean free path arise from alterations in the definition of this quantity. A commonly used definition states that the mean free path is the average distance the molecule travels between one collision and the next. In our calculation, we took a *random* starting point for the molecule and found the distance to the first collision; whereas if we wanted to find the distance from one collision to the next, we would have to take a collision as the starting point. This makes the mean free path somewhat longer. Since we do not wish to worry about all these fine details, we will accept Eq. (33) as a rough estimate of the mean free path, correct to within a factor of 2 or so, regardless of the fine details in the definition. Also, we must keep in mind that the actual free path of a molecule may differ widely from the mean free path. The collisions are a matter of chance — sometimes a molecule will travel much farther than $\lambda$ without any collision, sometimes it will suffer a collision at a distance shorter than $\lambda$.

---

EXAMPLE 6. What is the mean free path of a nitrogen molecule in air at STP? What is the average collision frequency, or the number of collisions per unit time? The radii of the molecules are about $1.5 \times 10^{-10}$ m.

SOLUTION: The number of molecules per unit volume is $2.7 \times 10^{19}/\text{cm}^3$ (see Example 3), or $2.7 \times 10^{25}/\text{m}^3$. Hence, the mean free path is

$$\lambda = \frac{1}{4\pi R^2 (N/V)} = \frac{1}{4\pi \times (1.5 \times 10^{-10}\ \text{m})^2 \times 2.7 \times 10^{25}/\text{m}^3}$$

$$= 1.3 \times 10^{-7}\ \text{m}$$

The average speed of the molecules is about 490 m/s (see Example 4). The average time per collision is the mean free path divided by the average speed $\bar{v}$,

$$\frac{\lambda}{\bar{v}} = \frac{1.3 \times 10^{-7}\ \text{m}}{490\ \text{m/s}} = 2.7 \times 10^{-10}\ \text{s}$$

and the average collision frequency is

$$\frac{\bar{v}}{\lambda} = \frac{1}{2.7 \times 10^{-10}\ \text{s}} = 3.7 \times 10^{9}/\text{s}$$

---

As the preceding example shows, the mean free path in air at sea level is quite short. However, at high altitudes, where the density of air is small, the mean free path is tens of meters, and more.

If we know the mean free path, we can calculate how far a molecule will wander from some given initial position in some given time. The molecule moves in straight steps of average length $\lambda$, with random angles between one step and the next. This kind of motion is called a **random walk.** Suppose the displacement vectors for the successive steps are $\boldsymbol{\lambda}_1$, $\boldsymbol{\lambda}_2$, $\boldsymbol{\lambda}_3$, etc. Then the net displacement vector after $n$ steps is

*Random walk*

$$\mathbf{r} = \boldsymbol{\lambda}_1 + \boldsymbol{\lambda}_2 + \boldsymbol{\lambda}_3 + \cdots + \boldsymbol{\lambda}_n \tag{34}$$

and the square of the magnitude of this net displacement is

$$\mathbf{r} \cdot \mathbf{r} = (\boldsymbol{\lambda}_1 + \boldsymbol{\lambda}_2 + \boldsymbol{\lambda}_3 + \cdots + \boldsymbol{\lambda}_n) \cdot (\boldsymbol{\lambda}_1 + \boldsymbol{\lambda}_2 + \boldsymbol{\lambda}_3 + \cdots + \boldsymbol{\lambda}_n) \tag{35}$$

When we multiply out the two parentheses on the right side, we obtain terms of the kind $\lambda_1^2$, $\lambda_2^2$, $\cdots$ and terms of the kind $\boldsymbol{\lambda}_1 \cdot \boldsymbol{\lambda}_2$, $\boldsymbol{\lambda}_1 \cdot \boldsymbol{\lambda}_3$, $\cdots$

On the average, the terms of the second kind are zero, because the angles between the displacements vectors are random, and the dot product is equally likely to be positive or negative. The terms of the first kind are all of the same average magnitude; they all are equal to the square of the mean free path. Since there are $n$ such terms, we obtain an average result

$$r^2 = n\lambda^2 \tag{36}$$

From this we see that the average distance traveled by the molecule in a random walk of $n$ steps is

$$\boxed{r = \sqrt{n}\,\lambda} \tag{37}$$

This distance is not proportional to the number of steps, but to the square root of the number of steps — to reach twice a given distance, the molecule must take four times as many steps!

The random walk determines the rate at which a gas diffuses through another gas. For example, consider the diffusion of molecules of a smelly gas, such as perfume, through the volume of (calm) air of a room. If an open bottle of perfume is placed in the middle of the room, molecules will emerge from the bottle and execute a random walk, gradually wandering away from the bottle. For an estimate of the time required for the molecules to diffuse a distance $r$ across the room, we use Eq. (36) for the number of steps,

$$n = r^2/\lambda^2 \tag{38}$$

If the average speed of the molecules is $\bar{v}$, each step of length $\lambda$ takes a time $\lambda/\bar{v}$, and $n$ steps take a time

$$n\frac{\lambda}{\bar{v}} = \frac{r^2}{\bar{v}\lambda} \tag{39}$$

For instance, with $\lambda = 1.3 \times 10^{-7}$ m, $\bar{v} = 490$ m/s, and $r = 1$ m, the time calculated from this equation is about $1.6 \times 10^4$ s, or 4.4 h. Even though the molecules move at high speeds, they take a long time to wander any significant distance. However, there will always be a few molecules that manage to travel the given distance in a much shorter time, because, by chance, they suffer fewer than the expected number of collisions. Hence, you will be able to get a faint whiff of the perfume within a much shorter time than that calculated from Eq. (39).

## SUMMARY

**Ideal-gas law:** $pV = nRT$ $\quad R = 8.31$ J/K
$pV = NkT$ $\quad k = 1.38 \times 10^{-23}$ J/K

**Avogadro's number:** $N_A = 6.02 \times 10^{23}$ molecules/mole

**Temperature scales:** Absolute: $T$
Celsius: $T_C = T - 273.15$
Fahrenheit: $T_F = \frac{9}{5}T_C + 32$

**Root-mean-square speed:** $v_{rms} = \sqrt{3kT/m}$

**Kinetic energy of ideal monoatomic gas:** $K = \frac{3}{2}NkT$

**Equipartition theorem:** Each translational or rotational component of the random thermal motion of a molecule has an average kinetic energy of $\frac{1}{2}kT$.

**Mean free path:** $\lambda = \dfrac{1}{4\pi R^2(N/V)}$

**Random walk:** $r = \sqrt{n}\ \lambda$

## QUESTIONS

1. Why do meteorologists usually measure the temperature in the shade rather than in the sun?

2. Why are there no negative temperatures on the absolute temperature scale?

3. The temperature of the ionized gas in the ionosphere of the Earth is about 2000 K, but the density of this gas is extremely low, only about $10^5$ gas particles per cubic centimeter. If you were to place an ordinary mercury thermometer in the ionosphere, would it register 2000 K? Would it melt?

4. The temperature of the intergalactic space is 3 K. How can empty space have a temperature?

5. At the airport of La Paz, Bolivia, one of the highest in the world, pilots of aircraft find it preferable to take off early in the morning or late at night, when the air is very cold. Why?

6. If you release a rubber balloon filled with helium, it will rise to a height of a few thousand meters and then remain stationary. What determines the height reached? Is there an optimum pressure to which you should inflate the balloon to reach greatest height?

7. How can you use a barometer as an altimeter?

8. Explain why a real gas behaves like an ideal gas at low densities but not at high densities.

9. Helium and neon approach the behavior of an ideal gas more closely than do any other gases. Why would you expect this?

10. Ultrasound waves of extremely short wavelength cannot propagate in air. Why not?

11. In our calculation of the pressure on the walls of a box (see Section 19.3) we have ignored gravity. If we take gravity into account, the pressure on the bottom of the box will be greater than that at the top. Show that the pressure difference is $p - p_0 = (N/V)\ mgL$. (Hint: When a molecule falls from the top to the bottom, its speed increases according to $v_y^2 - v_{0y}^2 = 2gL$.)

12. Prove that it is impossible for all of the molecules in a gas to have the same speeds and to keep these speeds forever. (Hint: Consider an elastic collision between two molecules with the same speed. Will the speeds remain constant if the initial lines of motion are not parallel?)

13. If you increase the absolute temperature by a factor of 2, by what factor will you increase the average speed of the molecules of gas?

14. Air consists of a mixture of nitrogen ($N_2$), oxygen ($O_2$), and argon (A). Which of these molecules has the highest average speed? The lowest?

15. The **median speed** for a distribution of speeds, such as the Maxwell distribution, is the speed such that one-half of the molecules have larger speeds

than the median, and one-half have smaller speeds. Determine the median speed for the 900-K Maxwell distribution plotted in Figure 19.7. (Hint: Make a paper cutout with the shape of the plot, and balance it on the edge of a ruler.)

16. Equipartition of energy applies not only to atoms and molecules, but also to macroscopic "particles" such as, golf balls. If so, why do golf balls remain at rest on the ground instead of flying through the air like molecules?

17. If you open a bottle of perfume in one corner of a room, it takes a rather long time for the smell to reach the opposite corner (assuming that there are no air currents in the room). Explain why the smell spreads slowly, even though the typical speeds of perfume molecules are 300–400 m/s.

## PROBLEMS

### Sections 19.1 and 19.2

1. Express the last six temperatures listed in Table 19.1 in terms of degrees Fahrenheit.

2. The hottest place on Earth is Al' Aziziyah, Libya, where the temperature has soared to 136.4°F. The coldest place is Vostok, Antarctica, where the temperature has plunged to −126.9°F. Express these temperatures in degrees Celsius and in kelvin.

3. In summer when the temperature is 30°C, the overpressure within an automobile tire is 2.2 atm. What will be the overpressure within this tire in winter when the temperature is 0°C? Assume that no air is added to the tire and that no air leaks from the tire; assume that the volume of the tire remains constant and that the atmospheric pressure remains at 1 atm.

4. What is the number of molecules in 1 liter of air at 273 K and 1 atm?

5. Show that the volume of 1 mole of ideal gas at 273 K and 1-atm pressure is 22.4 liters.

6. Repeat the calculation of Example 2 assuming that, because of the increase of pressure, the volume of the tire increases by 5%.

7. The storage tank of a small air compressor holds 0.3 $m^3$ of air at a pressure of 5.0 atm and a temperature of 20°C. How many moles of air is this?

8. On a warm day, the outdoor temperature is 32°C and the indoor temperature in an air-conditioned house is 17°C. What is the difference between the densities of the air outdoors and indoors? Assume the pressure is 1 atm.

9. The lowest pressure attained in a "vacuum" in a laboratory on the Earth is $1 \times 10^{-16}$ atm. Assuming a temperature of 20°C, what is the number of molecules per cubic centimeter in this vacuum?

10. Clouds of interstellar hydrogen gas have densities of up to $10^{10}$ atoms/$m^3$ and temperatures of up to $10^4$ K. What is the pressure in such a cloud?

11. The following table gives the pressure and density of the Earth's upper atmosphere as a function of altitude:

| *Altitude* | *Pressure* | *Density* |
|---|---|---|
| 20,000 m | 56 mbar | $9.2 \times 10^{-2}$ kg/$m^3$ |
| 40,000 | 3.2 | $4.3 \times 10^{-3}$ |
| 60,000 | 0.28 | $3.8 \times 10^{-4}$ |
| 80,000 | 0.013 | $2.5 \times 10^{-5}$ |

Calculate the temperature at each altitude. The mean molecular mass for air is 29.0 g.

12. What is the density (in kilograms per cubic meter) of helium gas at 1 atm at the temperature of boiling helium liquid (see Table 19.1)?

13. The volume of an automobile tire is $2.5 \times 10^{-2}$ m$^3$. The pressure of the air in this tire is 3.0 atm and the temperature is 17°C. What is the mass of air? The mean molecular mass of air is 29.0 g.

*14. Suppose you pour 10 g of water into a 1-liter jar and seal it tightly. You then place the jar into an oven and heat it to 500°C (a dangerous thing to do!). What will be the pressure of the vaporized water?

*15. How much does the frequency of middle C (see Table 17.2) played on a flute change when the air temperature drops from 20°C to −10°C? [Hint: The speed of sound in air is given by Eq. (17.3).]

*16. A scuba diver releases an air bubble of diameter 1.0 cm at a depth of 15 m below the surface of a lake. What will be the diameter of this bubble when it reaches the surface? Assume that the temperature of the bubble remains constant.

*17. The helium atom has a volume of about $3 \times 10^{-30}$ m$^3$. What fraction of a volume of helium gas at STP is actually occupied by atoms?

18. A carbon dioxide ($CO_2$) fire extinguisher has an interior volume of $2.8 \times 10^{-3}$ m$^3$. The extinguisher has a mass of 5.9 kg when empty and a mass of 8.2 kg when fully loaded with $CO_2$. At a temperature of 20°C, what is the pressure of $CO_2$ in the extinguisher?

*19. (a) When you heat the air in a house, some air escapes because the pressure inside the house must remain the same as the pressure outside. Suppose you heat the air from 10°C to 30°C. What fraction of the mass of air originally inside will escape?
(b) If the house were completely airtight, the pressure would have to increase as you heat the house. Suppose that the initial pressure inside the house is 1.0 atm. What is the final pressure? What force does the excess inside pressure exert on a window 1.0 m high and 1.0 m wide? Do you think the window can withstand this force?

*20. During the volcanic eruption of Mt. Pelée on the island of Martinique in 1902, a *nuée ardente* (burning cloud) of very hot gas rolled down the side of the volcano and killed the 30,000 inhabitants of Saint-Pierre. The temperature in the cloud has been estimated at 700°C. Assume that this cloud consisted of a gas of high molecular mass. What must have been this molecular mass to make the cloud as dense as, or denser than, the surrounding air (at 20°C)?

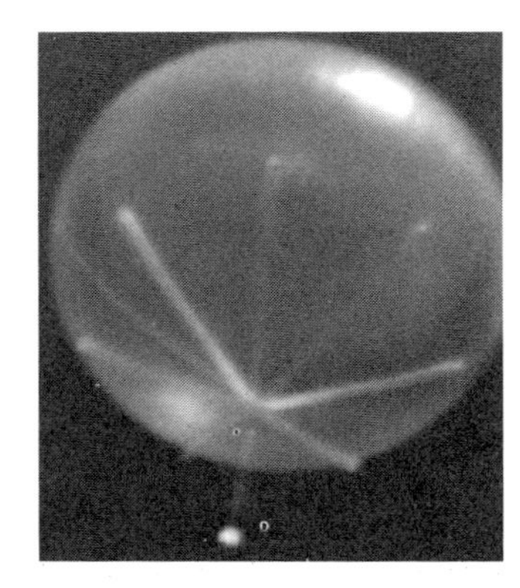

**Fig. 19.12** A research balloon floating at 40 km in the upper atmosphere.

*21. A typical hot-air balloon has a volume of 2200 m$^3$ and a mass of 730 kg (including balloon, gondola, four passengers, and a propane tank). Since the balloon is open at the bottom, the pressures of the internal and the external air are (approximately) equal. If the temperature of the external air is 20°C, what must be the minimum temperature of the internal air in the balloon to achieve lift-off? The density of the external air is 1.20 kg/m$^3$.

*22. A research balloon ascends to an altitude of 40 km and floats in equilibrium (Figure 19.12). The pressure (outside and also inside the balloon) is $3.2 \times 10^2$N/m$^2$ and the temperature is −13°C. The volume of the balloon is $8.5 \times 10^5$ m$^3$ and it is filled with helium. What payload (including the mass of the fabric but excluding the helium) can this balloon carry? What was the volume of the balloon on the ground (at STP), before it was released?

*23. A sunken ship of steel is to be raised by making the upper part of the hull airtight and then pumping compressed air into it while letting the water escape through holes in the bottom. The mass of the ship is 50,000 metric tons and it is at a depth of 60 m. How much compressed air (in kilograms) must be pumped into the ship? The temperature of the air and the water is 15°C.

*24. A **diving bell** is a cylinder closed at the top and open at the bottom; when it is immersed in the water, any air initially in the cylinder remains trapped in the cylinder. Suppose that such a diving bell, 2 m high and 1.5 m across, is immersed to a depth of 15 m measured from water level to water level (see Figure 19.13).

(a) How high will the water have risen within the diving bell?

(b) If compressed air is pumped into the bell, water will be expelled from the bell. How much air (in kilograms) must be pumped into the bell, and at what pressure, to get rid of all the water? Assume that the temperature of the air is 15°C.

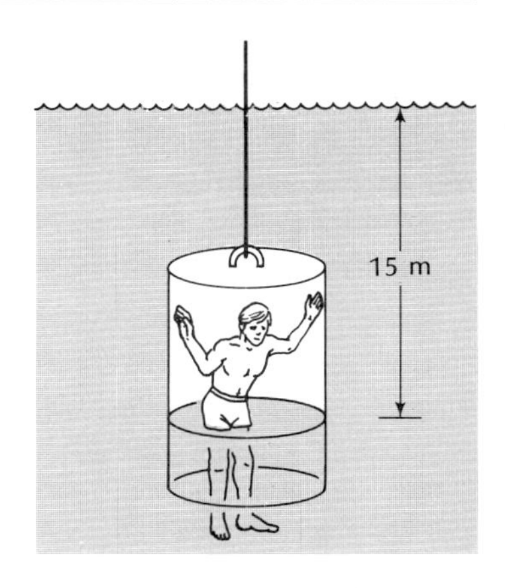


**Fig. 19.13** Submerged diving bell.

*25. At high altitudes, pilots and mountain climbers must breathe an enriched mixture containing more oxygen than the standard concentration of 21% found in ordinary air at sea level. At an altitude of 11,000 m, the atmospheric pressure is 0.22 atm. What oxygen concentration is required at this altitude if with each breath the same number of oxygen molecules is to enter the lungs as for ordinary air at sea level?

*26. Air is 75.54% nitrogen ($N_2$), 23.1% oxygen ($O_2$), and 1.3% argon (A) by mass. From this information and from the molecular masses of $N_2$, $O_2$, and A, deduce the mean molecular mass of air.

**27. (a) The gas at the center of the Sun is 38% hydrogen and 62% helium at a temperature of $15.0 \times 10^6$ K and a density of $1.48 \times 10^5$ kg/m$^3$. What is the pressure?

(b) The gas at a distance of 20% of the solar radius from the center of the Sun is 71% hydrogen and 29% helium at a temperature of $9.0 \times 10^6$ K and a density of $3.6 \times 10^4$ kg/m$^3$. What is the pressure?

**28. Show that if the temperature in the atmosphere is independent of altitude, then the pressure as a function of altitude is

$$p = p_0 e^{-mgh/kT}$$

where $m$ is the average mass per molecule of air. (This formula is applicable only for altitudes less than about 2 km; higher up, the temperature depends on the altitude.)

### Section 19.3

29. According to Eq. (17.3) the speed of sound in air is $\sqrt{1.4p/\rho}$.

(a) Show by means of the ideal-gas law that this expression equals $\sqrt{1.4kT/m}$, where $m$ is the average mass per molecule of air.

(b) Show that, in terms of the rms speed, the latter expression equals $\sqrt{1.4/3}\,v_{rms}$ or $0.68v_{rms}$.

(c) Calculate the speed of sound in air at 0°C, 10°C, 20°C, and 30°C.

30. What is the average kinetic energy of an oxygen molecule in air at STP? A nitrogen molecule?

31. What is the rms speed of a helium atom at 0°C? At −269°C?

32. At the top of the stratosphere, at an altitude of 30 km, the temperature is −38°C. What is the rms speed of an oxygen molecule at −38°C? Of an ozone ($O_3$) molecule? What are the average kinetic energies for these molecules?

33. The rms speed of nitrogen molecules in air at some temperature is 493 m/s. What is the rms speed of hydrogen molecules in air at the same temperature?

34. One method for the separation of the rare isotope $^{235}U$ (used in nuclear bombs and reactors) from the abundant isotope $^{238}U$ relies on diffusion through porous membranes. Both isotopes are first made into a gas of uranium hexafluoride ($UF_6$). The molecules of $^{235}UF_6$ have a higher rms speed and they will diffuse faster through a porous membrane than the molecules of

$^{238}UF_6$. Their molecular masses are 349 g and 352 g, respectively. What is the percent difference between their rms speeds at a given temperature?

*35. Estimate the number of impacts of air molecules on the palm of your hand per second. Assume that the air is at 20°C and 1 atm, and assume that it consists entirely of nitrogen molecules.

### Section 19.4

36. What is the thermal kinetic energy in 1 kg of oxygen gas at 20°C? What fraction of this energy is translational? What fraction is rotational?

37. Assume that air consists of the diatomic gases $O_2$ and $N_2$. How much must we increase the thermal energy of 1 kg of air in order to increase its temperature by 1°C?

*38. A container is divided into two equal compartments by a partition. One compartment is initially filled with helium at a temperature of 250 K; the other is filled with oxygen at a temperature of 310 K. Both gases are at the same pressure. If we remove the partition and allow the gases to mix, what will be their final temperature?

### Section 19.5

39. At an altitude of 160 km, the density of air is $1.5 \times 10^{-9}$ kg/m$^3$ and the temperature is approximately 500 K. What is the mean free path of a molecule? What is the average collision frequency? Assume the molecular radius is $1.5 \times 10^{-10}$ m.

40. What is the mean free path of water molecules in (liquid) water? Pretend the water molecules are spheres of a radius of $1 \times 10^{-10}$ m. Does the result make sense?

41. For each of the altitudes listed in the table in Problem 11, calculate the mean free path for a molecule in air. The molecular radius is $1.5 \times 10^{-10}$ m.

42. You place 50 marbles in a closed can and shake them violently. The marbles have a radius of 0.6 cm, and the volume of the can is 1000 cm$^3$. What is the mean free path of a marble?

*43. A wind is blowing at 60 km/h. Consider the motion of a molecule of nitrogen in this wind. The random thermal motion has a speed of about 500 m/s; this motion consists of zigzags, each leg of which typically takes $1 \times 10^{-10}$s. How many zigzags does the molecule make as it moves 1 cm downwind? What total distance does the molecule travel as it moves 1 cm downwind?

*44. A hunter standing in a pine forest fires a bullet horizontally. The trees in the forest have trunks of average diameter 30 cm and the average density of trees is 1 tree per 5 m$^2$. What is the mean free path of the bullet?

*45. A container of volume $V$ holds $N$ molecules of gas. The mean free path of each molecule is $\lambda$ and the average speed is $\bar{v}$.

(a) In terms of the given quantities, find an expression for the total number of collisions of *all* the $N$ molecules per unit time. (Hint: Beware of double counting.)

(b) Evaluate numerically for 1 cm$^3$ of air at STP. Use the values for the mean free path and the average speed given in Example 6.

46. A room measures 3 m $\times$ 3 m $\times$ 3 m. Estimate how long a nitrogen molecule takes to wander from the center of the room to the wall.

47. A nitrogen molecule is initally in a large volume of calm air at STP.

(a) How many collisions does the nitrogen molecule suffer in 10 minutes?

(b) What is the length of the path traveled by the molecule?

(c) On the average, how far will the molecule be from its starting point at the end of the 10 minutes?

# CHAPTER 20

# Heat

Heat is a form of energy: it is the kinetic and potential energy of the random microscopic motion of molecules, atoms, ions, electrons, and other particles. When we say that a body absorbs some amount of heat, we mean that the mechanical energy of the random microscopic motions of the atoms and molecules in the body increases. As we found in the preceding chapter, the increase of the kinetic energy of the random microscopic motions of the molecules in a gas is directly proportional to the increase of temperature [see Eq. (19.23)]. In a solid or a liquid, the kinetic energy of the random microscopic motions increases with temperature in the same way as in a gas. Furthermore, the atoms and molecules in a solid or liquid have potential energies associated with the forces they exert on one another; these potential energies also increase with temperature. Thus, the absorption of heat is associated with an increase of the temperature of the body.

**Benjamin Thompson, Count Rumford,** *1753–1814, American-British scientist, minister of war and of police in Bavaria. On the basis of experimental observations that he collected while supervising the boring of cannon, Rumford argued against the prevailing view that heat is a substance, and he proposed that heat is nothing but the random microscopic motion of the particles within a body. Robert von Mayer, 1814–1878, German physician and physicist, calculated the mechanical equivalent of heat by comparing the work done on a gas during adiabatic compression with the consequent increase of temperature. Finally, J. P. Joule measured this quantity directly by means of his famous experiment.*

Today, heat is often called **thermal energy.** But until well into the nineteenth century, scientists did not have a clear understanding of the concept of energy and they thought that heat was an invisible, imponderable fluid, called "caloric." The first experiments to give conclusive evidence on the nature of heat were performed by Count Rumford, who showed that the mechanical energy lost in friction is converted into heat. The practical development of steam engines for the industrial generation of (macroscopic) mechanical energy from heat motivated a careful examination of the theoretical principles underlying the operation of such engines, which led to the discovery of the law of conservation of energy and to the recognition that heat is a form of energy. Steam engines and other heat engines do not create energy; they merely convert thermal energy into mechanical energy.

## 20.1 Heat as a Form of Energy

Long before physicists recognized that heat is the kinetic and potential energy of the random microscopic motion of atoms, they had defined

*Calorie*, cal

heat in terms of the temperature changes it produces in a body. The traditional unit of heat is the **calorie** (cal), which is the amount of heat needed to raise the temperature of 1 g of water by 1°C. The kilocalorie is 1000 cal:

$$1 \text{ kcal} = 10^3 \text{ cal}$$

*British thermal unit*

In the British system the unit of heat is the **British thermal unit** (Btu), which is the heat needed to raise the temperature of 1 lb of water by 1°F. The relationship between these units is

$$1 \text{ Btu} = 0.252 \text{ kcal}$$

Incidentally: The "calories" marked on some packages of food in grocery stores are actually kilocalories, sometimes also called large calories.

*Specific heat capacity*

The heat necessary to raise the temperature of 1 kg of a material by 1°C is called the **specific heat capacity**, or the **specific heat**, usually designated by the symbol $c$. Thus, by definition, water has a specific heat capacity

$$\boxed{c = 1 \text{ kcal/kg} \cdot {}^\circ\text{C}}$$

The specific heat varies from substance to substance; Table 20.1 lists the specific heats of some common substances. The specific heat also varies with the temperature. For example, the specific heat of water varies by about 1% between 0°C and 100°C, reaching a minimum at 35°C. (This variation must be taken into account for a precise definition of the calorie: a calorie is the heat needed to raise the temperature of 1 g of water from 14.5°C to 15.5°C.) Finally, the specific heat depends on the pressure to which the material is subjected during the heating. All the values in Table 20.1 were obtained at room temperature (20°C) and at a constant pressure of 1.0 atm.

**Table 20.1** SOME SPECIFIC HEATS[a]

| *Substance* | *c* |
|---|---|
| Aluminum | 0.214 kcal/kg · °C |
| Brass | 0.092 |
| Copper | 0.092 |
| Iron, steel | 0.11 |
| Lead | 0.031 |
| Tin | 0.054 |
| Silver | 0.056 |
| Invar | 0.120 |
| Mercury | 0.033 |
| Water | 1.00 |
| Ice (− 10°C) | 0.530 |
| Ethyl alcohol | 0.581 |
| Glycol | 0.571 |
| Mineral oil | 0.5 |
| Glass, thermometer | 0.20 |
| Marble | 0.21 |
| Granite | 0.19 |
| Sea water | 0.93 |

[a]At room temperature and 1 atm, unless otherwise noted.

The values in Table 20.1 give the amount of heat required to increase the temperature of 1 kg of a given substance by 1 °C. For a mass $m$ of this substance, the absorbed heat $\Delta Q$ and the increase of temperature $\Delta T$ are related by

$$\Delta Q = mc\,\Delta T \tag{1}$$

*Relation between heat and temperature changes*

Since heat is a form of energy, it can be transformed into macroscopic mechanical energy, and vice versa. The transformation of heat into work is accomplished by a steam engine, a steam turbine, or a similar device; we will examine the theory of such heat engines in the next chapter. The transformation of work into heat requires no complicated machinery — any kind of friction will convert work into heat. Because heat is a form of energy, the calorie is a unit of energy and it must be possible to express it in joules. The conversion factor between these units is called the **mechanical equivalent of heat.**

The traditional method for the measurement of the mechanical equivalent of heat is **Joule's experiment.** A set of falling weights drives a paddle wheel which churns the water in a thermally insulated bucket (Figure 20.1). The churning raises the temperature of the water by a measurable amount, converting a known amount of gravitational potential energy into a known amount of heat. The best available experimental results give

$$1 \text{ cal} = 4.186 \text{ J}$$

*Mechanical equivalent of heat*

for the mechanical equivalent of heat.[1]

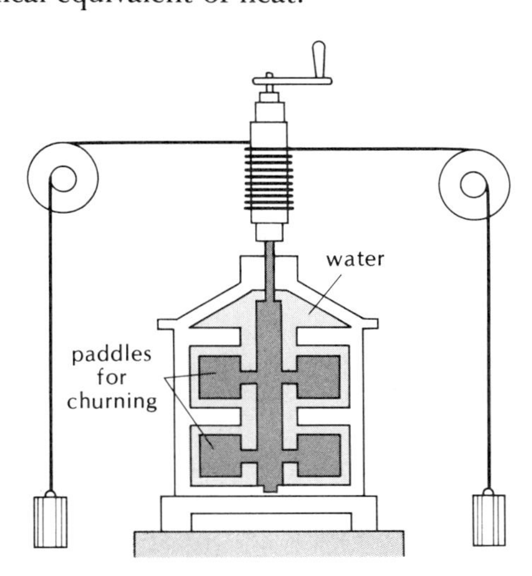


**Fig. 20.1** Joule's apparatus. The falling weights turn the drum and the attached paddles.

EXAMPLE 1. When an automobile is braking, the friction between the brake drums and the brake shoes converts translational kinetic energy into heat. If a 2000-kg automobile brakes from 25 m/s (55 mi/h) to 0 m/s, how

[1] This is the conventional calorie. Several other calories, defined in slightly different ways, are in use, for instance the **"International Table" calorie** (4.1868 J) and the **thermochemical calorie** (4.1840 J), often used by chemists. When reading chemistry textbooks, always check what calorie is being used.

much heat is generated by the brakes? If each of the four brake drums has a mass of 9.0 kg of iron of specific heat 0.11 kcal/kg · °C, how much does the temperature of the brake drums rise? Assume that all the heat accumulates in the brake drums (there is not enough time for the heat to leak away into the air) and that the heat in all brake drums is the same.

SOLUTION: The initial kinetic energy of the automobile is

$$K = \tfrac{1}{2}mv^2 = \tfrac{1}{2} \times 2000 \text{ kg} \times (25 \text{ m/s})^2 = 6.3 \times 10^5 \text{ J}$$

Expressed in kilocalories, this gives the amount of heat

$$\Delta Q = 6.3 \times 10^5 \text{ J} \times \frac{1 \text{ kcal}}{4.19 \times 10^3 \text{ J}} = 1.5 \times 10^2 \text{ kcal}$$

The total mass of iron to be heated is $4 \times 9.0$ kg. Hence the corresponding temperature increase is [see Eq. (1)]

$$\Delta T = \frac{\Delta Q}{mc} = \frac{1.5 \times 10^2 \text{ kcal}}{4 \times 9.0 \text{ kg} \times 0.11 \text{ kcal/kg} \cdot {}^\circ\text{C}} = 38^\circ\text{C}$$

## 20.2 Thermal Expansion of Solids and Liquids

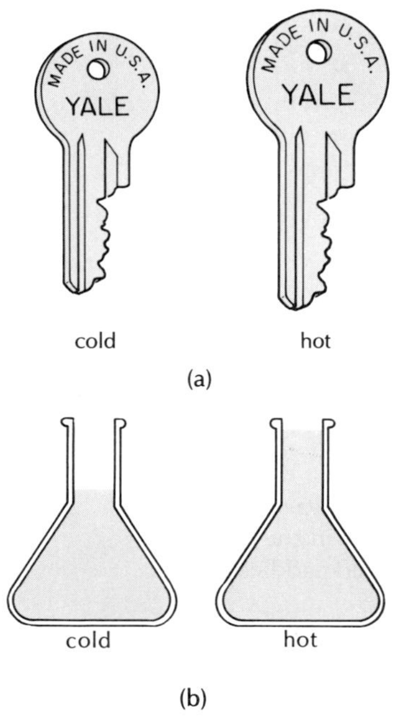


**Fig. 20.2** (a) Thermal expansion of a solid. (b) Thermal expansion of a liquid. The expansion of the flask has been neglected.

As we saw in the preceding chapter, if the pressure is held constant, the volume of a given amount of gas will increase with the temperature. Such an increase in volume with temperature also occurs for solids and liquids; this phenomenon is called **thermal expansion.** However, the thermal expansion of solids and liquids is much less than that of gases. For example, if we raise the temperature of a piece of iron by 100°C, we will increase its volume by only 0.36%. During the thermal expansion the solid retains its shape, but all its dimensions increase in proportion. Figure 20.2a shows the thermal expansion of a piece of metal; for the sake of clarity, the expansion has been exaggerated. An expanding liquid does not, of course, retain its shape; the liquid will merely fill more of the container that holds it. Figure 20.2b shows the thermal expansion of a liquid.

The thermal expansion of a solid and a liquid arises from an increase of the random microscopic motions. The atoms or molecules of the solid are held together by interatomic or intermolecular forces which act rather like springs (see Figure 20.3). The atoms oscillate about

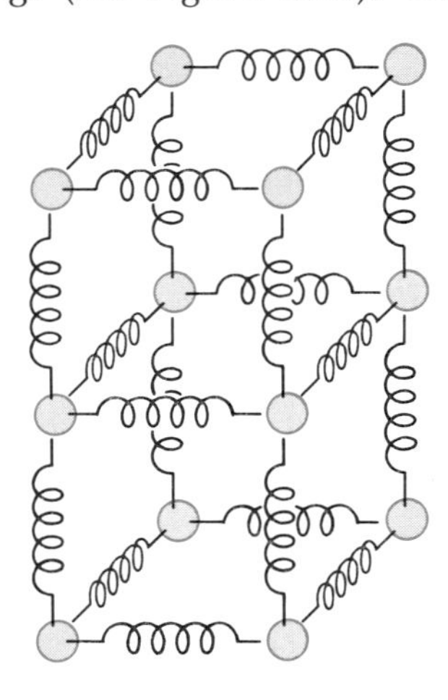

**Fig. 20.3** The atoms or molecules in a solid may be regarded as held together by springs.

their equilibrium positions with an average amplitude that depends on the thermal energy. An increase of temperature leads to vibrational motions of larger amplitude. This, in itself, does not necessarily result in an expansion of the solid, since the size of the solid is determined by the *average* separation between the atoms, not by their amplitude of motion. However, the interatomic "springs" deviate somewhat from Hooke's Law — they are somewhat easier to stretch than to compress. (This asymmetry of the interatomic force can be recognized from the plot of the interatomic potential energy given in Figure 8.5. The curve of potential energy is asymmetric; it is steeper on the near side of the equilibrium point than on the far side). Consequently, when an increase of temperature brings about an increase of the amplitude of motion, the maximum displacement attained during stretching of the spring exceeds the maximum displacement attained during compression, and therefore the average separation between the atoms increases, resulting in an expansion of the solid.

The thermal expansion of a solid can be described mathematically by the increase in the linear dimensions of the solid. The increment in the length is directly proportional to the increment of temperature and to the original length,

$$\Delta L = \alpha L \, \Delta T \tag{2}$$

The constant of proportionality $\alpha$ is called the **coefficient of linear expansion.** Table 20.2 lists the values of this coefficient for a few materials.

*Coefficient of linear expansion*

**Table 20.2** COEFFICIENTS OF EXPANSION[a]

| *Solids* | $\alpha$ |
|---|---|
| Aluminum | $24 \times 10^{-6}/°C$ |
| Brass | 19 |
| Concrete | ~12 |
| Copper | 17 |
| Glass | |
| commercial | 11 |
| Pyrex | 3.3 |
| Invar | 0.9 |
| Iron, steel | ~12 |
| Lead | 29 |
| Quartz, fused | 0.50 |

| *Liquids* | $\beta$ |
|---|---|
| Alcohol, | |
| ethyl (99%) | $1.01 \times 10^{-3}/°C$ |
| Carbon | |
| tetrachloride | 1.18 |
| Ether | 1.51 |
| Gasoline | 0.95 |
| Glycerine | 0.49 |
| Olive oil | 0.68 |
| Mercury | 0.18 |

[a] At room temperature.

The increment in the volume of the solid is directly proportional to the increment in the temperature and to the original volume,

$$\Delta V = \beta V \, \Delta T \tag{3}$$

Here the constant of proportionality $\beta$ is called the **coefficient of cubical expansion.** This coefficient is three times the coefficient of linear expansion,

*Coefficient of cubical expansion*

$$\beta = 3\alpha$$

To see how this relationship comes about, consider a solid in the shape of a cube of edge $L$ and volume $V = L^3$. A small increment $\Delta L$ in the length can be treated as a differential and consequently $\Delta V = 3L^2\,\Delta L$, which gives

$$\Delta V = 3L^3(\Delta L/L) = 3V(\alpha\,\Delta T)$$

Comparing this with Eq. (3), we see that, indeed, $\beta = 3\alpha$.

The increment in the volume of a liquid can be described by the same equation [Eq. (3)] as the increment in the volume of a solid. Table 20.2 also lists values of coefficients of cubical expansion for some liquids.

Water has not been included in this table because its behavior is rather peculiar: from 0°C to 3.98°C, the volume *decreases* with temperature, but not uniformly; above 3.98°C, the volume increases with temperature. Figure 20.4 plots the volume of 1 kg of water as a function of the temperature. The strange behavior of the density of water at low temperatures can be traced to the crystal structure of ice. Water molecules have a rather angular shape that prevents a tight fit of these molecules; when they assemble in a solid, they adopt a very complicated crystal structure with large gaps. As a result, ice has a lower density than water — the density of ice is 917 kg/m$^3$, and the volume of 1 kg of ice is 1091 cm$^3$. At a temperature slightly above the freezing point, water is liquid, but some of the water molecules already have assembled themselves into microscopic (and ephemeral) ice crystals; these microscopic crystals give the cold water an excess volume.

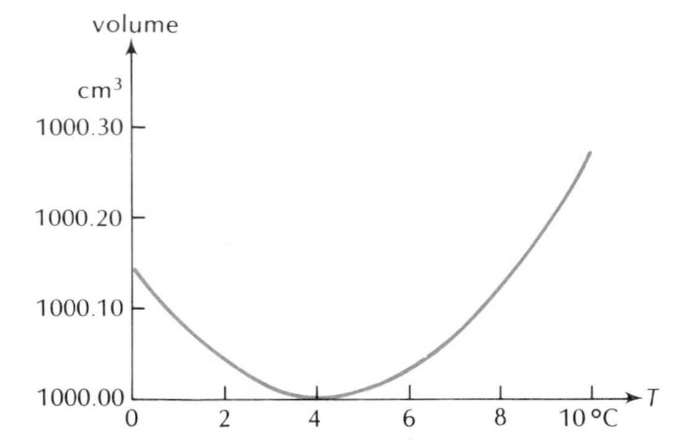


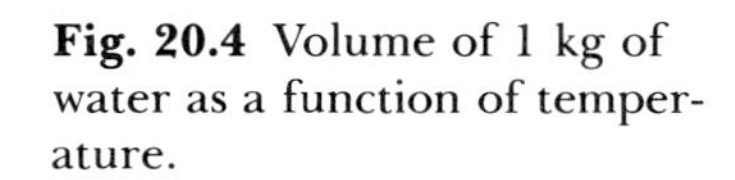

**Fig. 20.4** Volume of 1 kg of water as a function of temperature.

The maximum in the density of water at about 4°C has an important consequence for the ecology of lakes. In winter the layer of water on the surface of the lake cools, becomes denser than the lower layer, and sinks to the bottom. This process continues until the temperature of the entire body of the lake reaches 4°C. Beyond this point, the cooling of the surface layer will make it *less dense* than the lower layers; thus the surface layer stays in place, floating on top of the lake. Ultimately, this surface layer freezes, becoming a solid sheet of ice while the body of the lake remains at 4°C. The sheet of ice inhibits the heat loss from the lake, especially if covered with a blanket of snow. Besides, any further heat loss merely causes some thickening of the sheet of ice, without disturbing the deeper layers of water, which remain at a stable temperature of 4°C — fish and other aquatic life can survive the winter in reasonable comfort.

EXAMPLE 2. A glass vessel of volume 200 cm³ is filled to the rim with mercury. How much of the mercury will overflow the vessel if we raise the temperature by 30°C?

SOLUTION: The volume of mercury will increase by

$$\Delta V_{\text{Hg}} = \beta_{\text{Hg}} V \, \Delta T = 0.18 \times 10^{-3}/^{\circ}\text{C} \times 200 \text{ cm}^3 \times 30^{\circ}\text{C} = 1.08 \text{ cm}^3$$

The volume of the glass vessel will increase just as though all of the vessel were filled with glass (solid glass):

$$\Delta V_{\text{glass}} = \beta_{\text{glass}} V \, \Delta T = 3\alpha_{\text{glass}} V \, \Delta T$$

$$= 3 \times 11 \times 10^{-6}/^{\circ}\text{C} \times 200 \text{ cm}^3 \times 30^{\circ}\text{C} = 0.20 \text{ cm}^3$$

The difference $1.08 \text{ cm}^3 - 0.20 \text{ cm}^3 = 0.88 \text{ cm}^3$ is the volume of mercury that will overflow.

Thermal expansion must be taken into account in the design of long structures, such as bridges or railroad tracks. The decks of bridges usually have several expansion joints with gaps (see Figure 20.5) that permit changes of length and prevent the bridge from buckling. Likewise, gaps are left between the segments of rail in a railroad track; but if the temperature changes exceed the expectations of the designers, the results can be disastrous (see Figure 20.6).

**Fig. 20.5** Expansion joints in deck of a bridge.

**Fig. 20.6** These rails buckled on an exceptionally hot day.

Incidentally: Our ability to erect large buildings and other structures out of reinforced concrete hinges on the fortuitous coincidence of the coefficients of expansion of iron and concrete (see Table 20.2). Reinforced concrete consists of iron rods in a concrete matrix. If the coefficients of expansion for these two materials were appreciably different, then the daily and seasonal temperature changes would cause the iron rods to move relative to the concrete — ultimately, the iron rods would work loose, and the reinforcement would come to an end.

## 20.3 Thermometers and Thermal Equilibrium

Although the ideal-gas thermometer is the primary thermometer for the definition of the temperature scale, many other kinds of thermometers are employed in practice. Most ordinary thermometers and thermostats make use of thermal expansion to sense changes of temperature. The mercury-bulb thermometer (Figure 20.7) consists of a glass bulb filled with mercury connected to a capillary tube. Thermal expansion makes the mercury overflow into the capillary tube and increase the length of the mercury column; this length indicates the temperature. The bimetallic-strip thermometer (Figure 20.8) consists of two adjoining strips of different metals, such as aluminum and iron, bonded together and curled into a spiral or a helix. The differential thermal expansion increases the length of one side of the bonded strip more than that of the other side; this causes the strip to curl up more tightly and rotates the upper end of the helix relative to the lower end; a pointer attached to the upper end indicates the temperature.

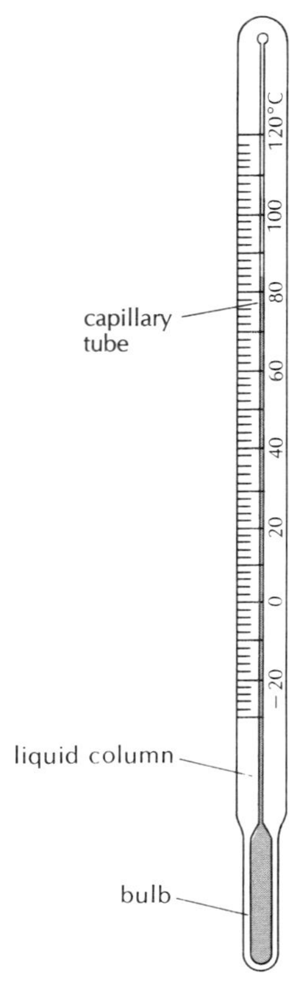


**Fig. 20.7** Mercury-bulb thermometer. The thermal expansion of the mercury in the bulb causes it to rise in the capillary tube, indicating the temperature.

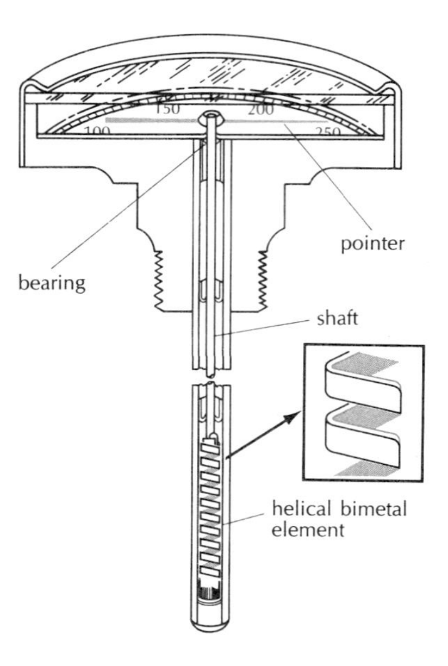


**Fig. 20.8** Bimetallic-strip thermometer. The helix consists of two strips of different metals bonded together. Thermal expansion curls or uncurls the helix and rotates the pointer.

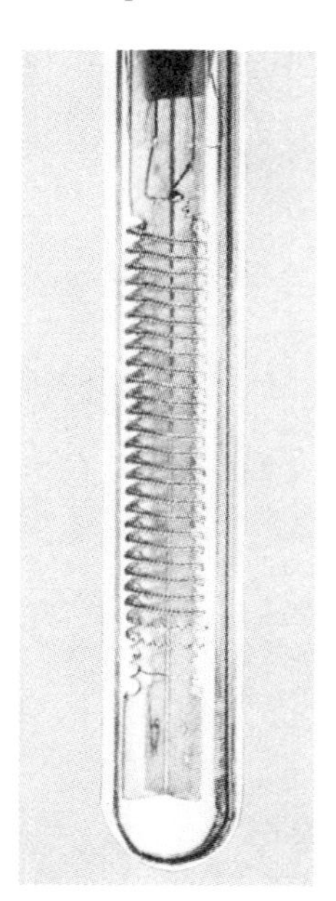

**Fig. 20.9** Platinum-resistance thermometer. The resistance that the fine coil of platinum wire offers to an electric current serves as indicator of temperature.

Other kinds of thermometers sense changes of temperature by a concomitant change of some electrical property of a material. For instance, an increase of temperature of a wire leads to an increase of the resistance that this wire offers to the passage of an electric current — a warm wire exerts more "friction" on the electric current than a cold wire. This effect is exploited by resistance thermometers, such as the platinum resistance thermometer shown in Figure 20.9. Modern resistance thermometers are often constructed out of a piece of semiconductor material; in a suitably prepared semiconductor, the electrical resistance is exceptionally sensitive to temperature changes, and therefore the semiconductor serves as a good indicator of temperature. Such semiconductors, called thermistors, are widely used in electronic thermometers with digital readouts.

Another electrical property exploited in the construction of thermometers is the generation of an electric current by junctions of two dissimilar metals. If two wires of dissimilar metals are connected in a loop with two junctions, a (weak) electric current will flow around the loop whenever there is a temperature difference between the junctions (see Figure 20.10). The magnitude of the current, which can be measured with a suitable "current" meter inserted in the loop, is proportional to the temperature difference between the junctions.

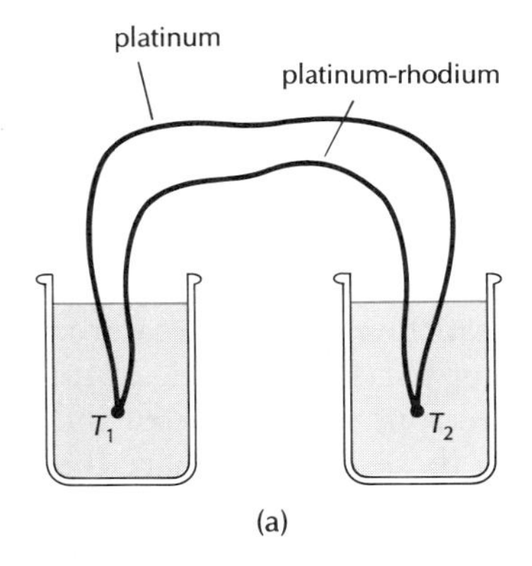


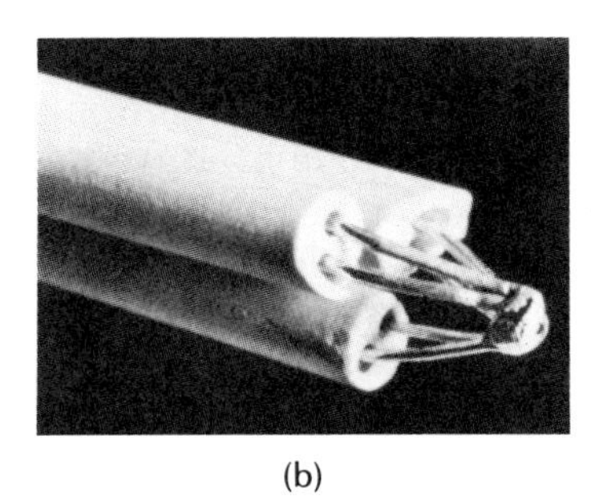


**Fig. 20.10** (a) A thermocouple consisting of two wires of dissimilar metals with two junctions. (b) Each of the three thermocouple junctions shown here consists of one wire of platinum and one of platinum-rhodium alloy joined at their ends. By using three pairs of wire loops connected in a suitable way, we obtain a net current three times as large as for a single loop.

To achieve agreement with the absolute temperature scale, thermometers based on thermal expansion or on electric properties of materials must be calibrated in terms of the ideal-gas thermometer. For high accuracy, this calibration must be performed at a large number of points over the full range of temperatures measured by the thermometer. Ordinary thermometers for everyday use, such as mercury thermometers, are often calibrated at only two points of their scale, say 0°C and 100°C, and the intermediate points are extrapolated by dividing the scale into 100 equal intervals. This crude procedure is adequate for everyday use, but does not accurately reproduce the ideal-gas temperature scale.

So far, the physical significance we have attached to temperature is that of a parameter that determines the average kinetic and potential energies of the atoms and molecules in a body [see Eq. (19.23)]. However, temperature has another important physical interpretation: it determines whether or not two bodies will be in thermal equilibrium. Consider two bodies in thermal contact, that is, two bodies that touch directly and intimately, without any intervening layer of insulator. Two such bodies are said to be in **thermal equilibrium** if no heat flows from one to the other. It is a fact of experience that *two bodies will be in thermal equilibrium if and only if their temperatures are equal.* We can regard the thermometer used for the temperature measurements as a third body, which, during the measurements, is brought into thermal equilibrium successively with each measured body; and we can then rephrase the equilibrium condition as follows: two bodies are in thermal equilibrium with each other if and only if they are in thermal equilibrium with a third body.[2]

*Thermal equilibrium*

[2] This equilibrium condition is sometimes called the "Zeroth" Law of Thermodynamics. However, as we will see in the next chapter, this law is implicitly contained in the Second Law of Thermodynamics, and therefore hardly deserves a title of its own.

The temperatures of two bodies determine their thermal equilibrium, and also determine the direction of the heat flow between the bodies if they are not in thermal equilibrium. When two bodies at different temperatures are put into thermal contact, heat always flows from the hotter body to the colder body. This heat transfer gradually lowers the temperature of the former and raises the temperature of the latter; ultimately, the two bodies will attain thermal equilibrium at some intermediate temperature. In the next section we will examine the quantitative relationship between the heat flow and the temperature difference.

## 20.4 Conduction of Heat

*Conduction*

If you put one end of an iron poker or rod into the fire and hold the other end in your hand, you will feel the end in your hand gradually become warmer. This is an example of heat transport by **conduction.** The atoms and electrons in the hot end of the rod have greater kinetic and potential energies than those in other parts of the rod. In random collisions, these energetic atoms and electrons share some of their energy with their less energetic neighbors; these, in turn, share their energy with their neighbors, and so on. The result is a gradual diffusion of thermal energy from the hot end of the rod to the cold end.

Metals are excellent conductors of heat and also excellent conductors of electricity. The high thermal and electric conductivities of a metal are due to an abundance of "free" electrons within the volume of the metal; these are electrons that have become detached from their atoms — they move at high speeds and they wander all over the volume of the metal with little hindrance, but they are held back by the surface of the metal. The free electrons behave like the particles of a gas, and the metal acts like a bottle holding this gas. Typically, a free electron will move past a few hundred atoms before it suffers a collision. Because the electrons move such fairly large distances between collisions, they can travel very quickly from one end of a metallic rod to the other. Thus the motion of the free electrons transports the thermal energy much more efficiently than does the back-and-forth vibrational motion of the atoms.

*Heat flow*

We can describe the transport of heat quantitatively by the **heat flow,** or the **heat current**; this is the amount of heat that passes by some given place on the rod per unit time. We will use the symbol $\Delta Q/\Delta t$ for heat flow. The metric unit of heat flow is the joule per second (J/s); however, in practice the preferred unit is the calorie per second (cal/s) or, in the British system, the British thermal unit per second (Btu/s).

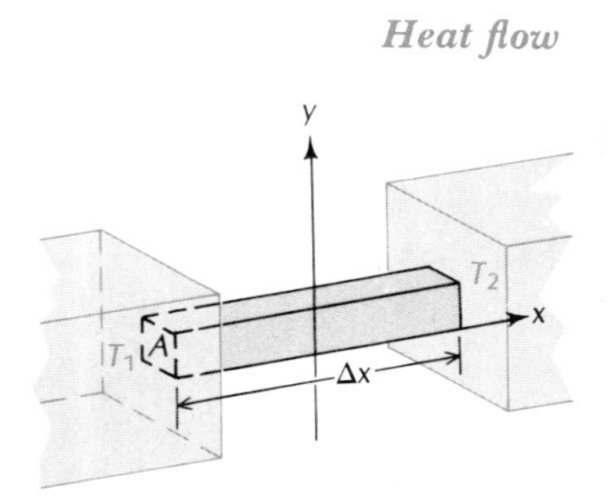


**Fig. 20.11** A rod of cross-sectional area $A$ conducting heat from a high-temperature reservoir ($T_2$) to a low-temperature reservoir ($T_1$).

Consider a rod of cross-sectional area $A$ and length $\Delta x$ (Figure 20.11). Assume that the cold end of the rod is kept at a constant temperature $T_1$ and the hot end at a constant temperature $T_2$, so that the difference of temperature between the ends is $\Delta T = T_2 - T_1$. If the ends are kept at these constant temperatures for a while, the temperatures at all other points of the rod will settle to final steady values. Under these steady-state conditions, the heat flow along the rod is found to be directly proportional to the temperature difference and to the cross-sectional area, and inversely proportional to the length,

$$\frac{\Delta Q}{\Delta t} = -kA\frac{\Delta T}{\Delta x} \qquad (4)$$

The direction of the heat flow is, of course, from the hot end of the rod toward the cold end. The negative sign in our equation indicates this direction of flow: if the rod lies along the $x$ axis and $T$ increases in the positive $x$ direction ($\Delta T/\Delta x > 0$), then the heat flows in the negative $x$ direction ($\Delta Q/\Delta t < 0$). The constant of proportionality $k$ is called the **thermal conductivity.** Table 20.3 lists values of $k$ for some materials.[3]

*Thermal conductivity*

**Table 20.3** SOME THERMAL CONDUCTIVITIES[a]

| *Substance* | *k* |
|---|---|
| Aluminum | 49 cal/(s · m · °C) |
| Copper | 92 |
| Iron, cast | 11 |
| Steel | 11 |
| Lead | 8.3 |
| Silver | 97 |
| Ice, 0°C | 0.3 |
| Snow, 0°C, compact | 0.05 |
| Glass, crown | 0.25 |
| Porcelain | 0.25 |
| Concrete | 0.2 |
| Brick | 0.15 |
| Fiber glass insulation | 0.01 |
| Styrofoam | 0.002 |
| Wood, pine | 0.03 |
| Down | 0.0046 |

[a] At room temperature, unless otherwise noted.

Equation (4) can be regarded as an empirical law that has been verified by many experiments. Alternatively, this equation can be derived by a detailed study of the process of diffusion of thermal energy along the rod. For the sake of brevity, we will accept the equation as an empirical law and not attempt any derivation. Equation (4) also applies to heat conduction through a wide slab or a plate; such a piece of material can be regarded as a rod of very short length and very large cross-sectional area.

For a rod or a slab, the temperature decreases linearly from the hot end to the cold end; hence the ratio $\Delta T/\Delta x$ equals the temperature gradient $dT/dx$ and we can write our equation for heat flow as

$$\frac{\Delta Q}{\Delta t} = -kA\frac{dT}{dx} \qquad (5)$$

*Equation of heat conduction*

In this form the equation is valid for a rod or a slab with a cross section or a conductivity that depends on position — the equation gives us the

[3] The thermal conductivity constant must not be confused with the Boltzmann constant; both are designated with the same letter $k$, but they are not related.

heat flow at a given position $x$ along the rod, in terms of the values of $k$, $A$, and $dT/dx$ at that position.

EXAMPLE 3. The wall of a room a house measures 3 m × 5 m. It consists of a layer of wood of thickness 2.5 cm and an adjacent layer of fiber glass insulation of thickness 10 cm. What is the heat flow through this wall if the inside temperature is 21°C and the outside temperature −18°C?

SOLUTION: The heat flows in the wood and the fiber glass must be the same, that is

$$k_{wd}A\left.\frac{dT}{dx}\right|_{wd} = k_{fi}\,A\left.\frac{dT}{dx}\right|_{fi}$$

Furthermore, the sum of the temperature changes across the wood and the fiber glass must equal the net temperature change of $\Delta T = 39°\text{C}$,

$$\left.\frac{dT}{dx}\right|_{wd}\Delta x_{wd} + \left.\frac{dT}{dx}\right|_{fi}\Delta x_{fi} = \Delta T$$

By combining these equations, we can solve for $dT/dx|_{wd}$:

$$\left.\frac{dT}{dx}\right|_{wd} = \frac{\Delta T}{\Delta x_{wd} + \Delta x_{fi}k_{wd}/k_{fi}}$$

This gives a heat flow

$$\frac{\Delta Q}{\Delta t} = -k_{wd}A\left.\frac{dT}{dx}\right|_{wd}$$

$$= \frac{-k_{wd}A\,\Delta T}{\Delta x_{wd} + \Delta x_{fi}k_{wd}/k_{fi}} = \frac{-A\,\Delta T}{\Delta x_{wd}/k_{wd} + \Delta x_{fi}/k_{fi}}$$

$$= \frac{-3\text{ m} \times 5\text{ m} \times 39°\text{C}}{0.025\text{ m}/0.03\text{ cal}/(\text{s}\cdot\text{m}\cdot°\text{C}) + 0.10\text{ m}\ /\ 0.01\text{ cal}/(\text{s}\cdot\text{m}\cdot°\text{C})}$$

$$= -54\text{ cal/s}$$

COMMENTS AND SUGGESTIONS: This example illustrates the solution of a problem of heat flow through adjacent layers of material. It is easy to generalize the calculation to three or more adjacent layers (the result for an arbitrary number of layers is stated in Problem 42).

*Convection and radiation*

Besides conduction, there are two other mechanisms of heat transport: convection and radiation. In **convection,** the heat is stored in a moving fluid and it is carried from one place to another by the motion of this fluid. In **radiation,** the heat is carried from one place to another by electromagnetic waves (light waves, infrared waves, radio waves, etc.). All three mechanisms of heat transport are neatly illustrated by the operation of a hot-water heating system in a house. In this system, the heat is carried from the boiler to the radiators in the rooms by means of water flowing in pipes (convection); the heat then diffuses through the metallic walls of the radiators (conduction); and finally it spreads from the surface of the radiator into the volume of the room (radiation, supplemented by some convection of air heated by direct contact with the radiator).

Radiation is the only mechanism of heat transport that can carry heat through a vacuum; for instance, the heat of the Sun reaches the Earth by radiation. We will study thermal radiation in Chapter 42.

## 20.5 Changes of State

Heat absorbed by a body will not only increase the temperature, but it will also bring about a change of state from solid to liquid or from liquid to gas when the body reaches its melting point or its boiling point. At the melting temperature or the boiling temperature, the thermal motion of the atoms and molecules becomes so violent that the bonds holding them in the solid or liquid loosen or break. The loosening of the bonds in a solid transforms it into a liquid, and the breaking of bonds in a liquid transforms it into a gas.

While the body is melting or boiling, it absorbs some amount of heat without any increase of temperature. This heat represents the energy required to loosen or break the bonds. The heat absorbed during the change of state is called the **heat of transformation,** or, more specifically, the **heat of fusion** or the **heat of vaporization,** as the case may be. Table 20.4 lists the heats of fusion and of vaporization for a few substances (at a pressure of 1 atm).

*Heat of fusion and of vaporization*

The values listed in Table 20.4 depend on the pressure. The decrease of the boiling point of water with a decrease of pressure is a phenomenon familiar to people living at high altitude; for instance, in Denver, Colorado, at an altitude of 1600 m, the mean pressure is 0.96 atm, and the boiling point of water is 90°C.

**Table 20.4** HEATS OF FUSION AND VAPORIZATION[a]

| *Substance* | *Melting point* | *Heat of fusion* | *Boiling point* | *Heat of vaporization* |
|---|---|---|---|---|
| Water | 0°C | 79.7 kcal/kg | 100°C | 539 kcal/kg |
| Nitrogen | −210 | 6.2 | −196 | 47.8 |
| Oxygen | −218 | 3.3 | −183 | 51 |
| Helium | — | — | −269 | 5.97 |
| Hydrogen | −259 | 15.0 | −253 | 107 |
| Aluminum | 660 | 95.3 | 2467 | 2520 |
| Copper | 1083 | 48.9 | 2567 | 1240 |
| Iron | 1535 | 65 | 2750 | 1620 |
| Lead | 328 | 6.8 | 1740 | 203 |
| Tin | 232 | 14.2 | 2270 | 463 |
| Silver | 962 | 23.7 | 2212 | 563 |
| Tungsten | 3410 | 44 | 5660 | 1180 |
| Mercury | −39 | 2.7 | 357 | 69.7 |
| Carbon dioxide[b] | −79 | — | — | 138 |

[a]At a pressure of 1 atm.
[b]Undergoes direct vaporization (sublimation) from solid to gas.

EXAMPLE 4. How many ice cubes must be added to a bowl containing 1 liter of boiling water at 100°C so that the resulting mixture reaches a temperature of 40°C? Assume that each ice cube has a mass of 20 g and that the bowl and the environment do not exchange any heat with the water.

SOLUTION: The heat released by the hot water during cooling from 100°C to 40°C is

$$\Delta Q = 1\ \text{kg} \times 1\ \text{kcal}/(\text{kg}\cdot{}^\circ\text{C}) \times 60^\circ\text{C} = 60\ \text{kcal}$$

If the total mass of ice is $m$, then the heat absorbed by this mass during fusion and subsequent heating from 0°C to 40°C is

$$\Delta Q = m \times 79.7\ \text{kcal/kg} + m \times 1\ \text{kcal}/(\text{kg}\cdot{}^\circ\text{C}) \times 40^\circ\text{C}$$

These amounts of heat must be equal,

$$m \times 79.7\ \text{kcal/kg} + m \times 40\ \text{kcal/kg} = 60\ \text{kcal}$$

which yields

$$m = 0.50\ \text{kg}$$

This is 500/20 or 25 ice cubes.

*Superheating*

In exceptional circumstances, a liquid may sometimes fail to boil at its normal boiling temperature. It may briefly remain quiescent at a temperature several degrees higher than its normal boiling temperature. Such a liquid is said to be in a **superheated** state. The formation of bubbles during normal boiling is triggered by the presence of impurities and disturbances in the liquid. To superheat a liquid, we must make sure there are no impurities and no disturbances. Since heating of a liquid itself tends to disturb the liquid, the preferred method for reaching the superheated state is by the sudden decompression of a liquid initially just below its boiling point. The reduction of pressure lowers the boiling point, and leaves the liquid in a superheated state. This is an unstable state — any small disturbance within the liquid will trigger violent boiling. At accelerator laboratories, physicists investigating the collisions of high-energy elementary particles take advantage of superheated liquids to make the tracks of these particles visible. The particles are aimed into a "bubble chamber" filled with superheated liquid. While a particle passes through the chamber, it ionizes some of the molecules that it encounters along its path, and this provides enough of a disturbance to trigger the formation of a fine trail of bubbles in the superheated liquid. The trail of bubbles makes the path of the particle visible. Figure 20.12a shows a large bubble chamber of

(a)

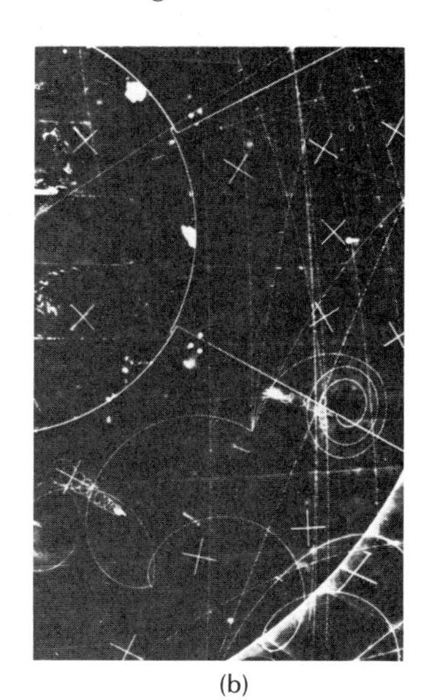

(b)

**Fig. 20.12** (a) The big bubble chamber at the Organization Européenne pour la Recherche Nucléaire (CERN), near Geneva. Here we see the chamber during construction, before it became hidden by surrounding equipment.
(b) Trails of bubbles reveal the paths of several high-energy particles that have passed through the bubble chamber.

this kind, holding about 38 $m^3$ of superheated liquid hydrogen. Figure 20.12b is a photograph of trails of bubbles marking the paths of several particles that have passed through the bubble chamber.

Similarly, a gas may sometimes fail to condense at a temperature below the normal boiling point, and a liquid may sometimes fail to freeze below the normal freezing point. Such a gas or liquid is said to be in a **supercooled** state. A disturbance then triggers sudden condensation or freezing. Chambers filled with a supercooled gas can also be used to make the paths of high-energy particles visible. When a particle passes through such a "cloud chamber," it triggers the formation of a fine trail of condensation droplets along its path.

*Supercooling*

## 20.6 The Specific Heat of a Gas

If we heat a gas, the increase of temperature causes an increase of the pressure, and this tends to bring about an expansion of the gas. The value of the specific heat of the gas depends on whether the container permits this expansion. If the container is perfectly rigid, the heating proceeds at constant volume (Figure 20.13). For gases it is customary to reckon the specific heat per mole, rather than per kilogram. The **specific heat at constant volume** is designated by $C_V$; it is the heat needed to raise the temperature of 1 mole of gas by 1°C. If we are dealing with $n$ moles of gas, the heat absorbed and the temperature increase at constant volume are related by

*Specific heat at constant volume*

$$\Delta Q = nC_V\,\Delta T \tag{6}$$

If the container is fitted with a vertical piston, the heating proceeds at a constant pressure determined by the weight of the piston (Figure 20.14). The **specific heat at constant pressure** is designated by $C_p$. For $n$ moles of gas, the heat absorbed and the temperature increase at constant pressure are related by

*Specific heat at constant pressure*

$$\Delta Q = nC_p\,\Delta T \tag{7}$$

We expect $C_p$ to be larger than $C_V$ because, if we supply some amount of heat to the container of Figure 20.14, only part of this heat will go into a temperature increase of the gas; the rest will be converted into work as the expanding gas lifts the piston. Let us calculate the difference between the two heat capacities.

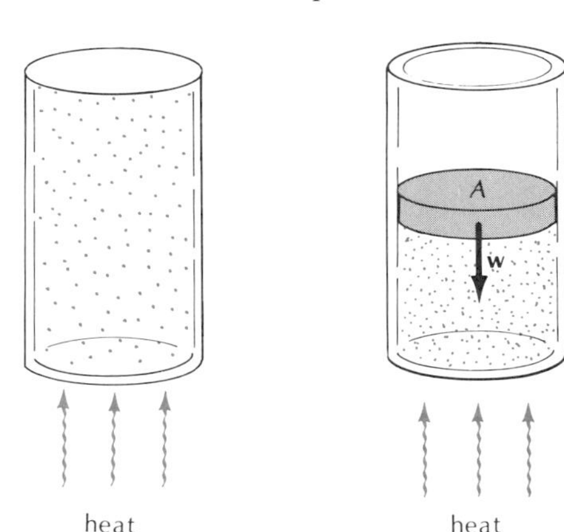


**Fig. 20.13** (left) A gas kept at constant volume while being heated.

**Fig. 20.14** (right) A gas kept at constant pressure by the weight of a piston while being heated.

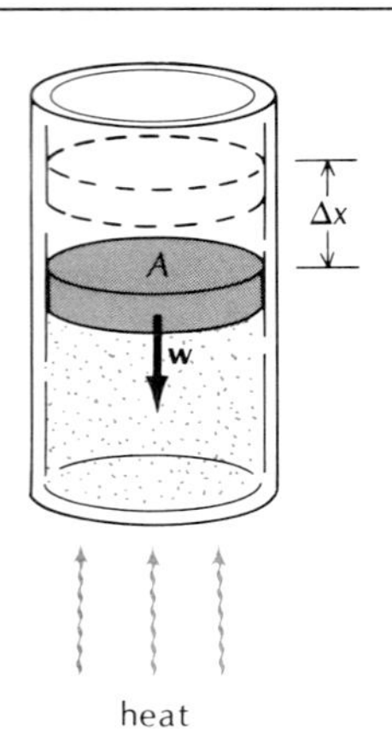


**Fig. 20.15** Displacement of the piston by the expanding gas.

At constant volume, the gas does no work. Hence all the heat absorbed will go into the energy of the gas,

$$\Delta Q = \Delta E$$

or, according to Eq. (6),

$$nC_V\,\Delta T = \Delta E \tag{8}$$

At constant pressure, the gas does work against the moving piston. Suppose that the piston has an area $A$ and is displaced by a small distance $\Delta x$ (Figure 20.15). The force of the gas on the piston is $pA$ and the work done by the gas is $pA\,\Delta x$,

$$\Delta W = pA\,\Delta x \tag{9}$$

The product $A\Delta x$ is simply the change of volume $\Delta V$. Hence

$$\Delta W = p\,\Delta V \tag{10}$$

The heat absorbed must provide both the energy increase of the gas and the work done by the gas:

$$\Delta Q = \Delta E + \Delta W = \Delta E + p\,\Delta V \tag{11}$$

so that, according to Eq. (7),

$$nC_p\,\Delta T = \Delta E + p\,\Delta V \tag{12}$$

In an ideal gas, the energy $E$ is a function of the temperature only [see Eqs. (19.25) and (19.31)]; consequently, if the temperature increment at constant pressure has the same value as the temperature increment at constant volume, the increases $\Delta E$ in the energies must be the same. We can therefore insert Eq. (8) into Eq. (12) and obtain

$$nC_p\,\Delta T = nC_V\,\Delta T + p\,\Delta V \tag{13}$$

Hence

$$C_p - C_V = \frac{p\,\Delta V}{n\,\Delta T} \tag{14}$$

The quantity on the right side can be evaluated by appealing to the ideal-gas law. At constant pressure, Eq. (19.1) gives

$$p\,\Delta V = nR\,\Delta T \tag{15}$$

so that

*Relation between specific heats*

$$\boxed{C_p - C_V = R} \tag{16}$$

The numerical value of $R$ is 8.31 J/K · mole, or 1.99 cal/K · mole. Hence Eq. (16) shows that $C_p$ is larger than $C_V$ by about 2 cal/K · mole.

Note that although the above general argument does permit us to evaluate the difference $C_p - C_V$, it does not permit us to find the individual values of $C_p$ and $C_V$. For this we must turn to kinetic theory. According to Eq. (19.26), we then find, for a monoatomic gas,

$$\Delta E = \tfrac{3}{2}nR\ \Delta T \tag{17}$$

By the definition of $C_V$ [Eq. (8)], this yields

$$C_V = \tfrac{3}{2}R \tag{18}$$

and by Eq. (16)

$$C_p = C_V + R = \tfrac{5}{2}R \tag{19}$$

Likewise, for a diatomic gas

$$C_V = \tfrac{5}{2}R \tag{20}$$

and

$$C_p = \tfrac{7}{2}R \tag{21}$$

Table 20.5 lists the values of the specific heats of some gases. In the cases of monoatomic and diatomic gases, these values are in reasonable agreement with Eqs. (18)–(21), but there are some minor deviations due to our oversimplifications in kinetic theory. Note that in all cases the difference $C_p - C_V$ agrees quite precisely with Eq. (16).

**Table 20.5** SPECIFIC HEATS OF SOME GASES[a]

| *Gas* | $C_V$ | $C_p$ | $C_p - C_V$ |
|---|---|---|---|
| Helium (He) | 3.00 cal/K · mole | 4.98 cal/K · mole | 1.98 cal/K · mole |
| Argon (Ar) | 3.00 | 5.00 | 2.00 |
| Nitrogen ($N_2$) | 4.96 | 6.95 | 1.99 |
| Oxygen ($O_2$) | 4.96 | 6.95 | 1.99 |
| Carbon monoxide (CO) | 4.93 | 6.95 | 2.02 |
| Carbon dioxide ($CO_2$) | 6.74 | 8.75 | 2.01 |
| Methane ($CH_4$) | 6.48 | 8.49 | 2.01 |

[a] At STP.

## 20.7 The Adiabatic Equation

If an amount of gas at high pressure and temperature is placed in a container fitted with a piston, the gas will push the piston outward and do work on it. Such a process of expansion converts thermal energy into useful mechanical energy — the temperature of the gas decreases as it delivers work to the piston. This process is at the core of the operation of steam engines, automobile engines, and other heat engines.

In this section we will investigate the equations for the expansion of a gas. We will assume that the gas is thermally insulated, so it neither receives heat from its environment nor loses any. The temperature decrease of the gas is then entirely due to the work that the gas does on its environment. Such a process occurring without exchange of heat with the environment is called **adiabatic.**

*Adiabatic process*

If the volume of the gas increases by a small amount $dV$, the work done by the gas on this piston is [see Eq. (10)]

$$dW = p\,dV \tag{22}$$

The heat absorbed is zero; hence the change of energy of the gas is

$$dE = -dW = -p\,dV \tag{23}$$

The change of energy can also be expressed in terms of the change of temperature [see Eq. (8)]:

$$dE = nC_V\,dT \tag{24}$$

Combining Eqs. (23) and (24), we have

$$nC_V\,dT = -p\,dV \tag{25}$$

By differentiating the ideal-gas law $nRT = pV$, we find

$$nR\,dT = p\,dV + V\,dp \tag{26}$$

If we eliminate $dT$ between Eqs. (25) and (26), we obtain

$$-Rp\,dV = C_V p\,dV + C_V V\,dp \tag{27}$$

or

$$-\frac{R + C_V}{C_V}\frac{dV}{V} = \frac{dp}{p} \tag{28}$$

Let us write this as

$$-\gamma\frac{dV}{V} = \frac{dp}{p} \tag{29}$$

where

$\gamma$

$$\boxed{\gamma = \frac{R + C_V}{C_V} = \frac{C_p}{C_V}} \tag{30}$$

The quantity $\gamma$ is the ratio of the two kinds of heat capacities; for instance, for an ideal monoatomic gas, $\gamma = C_p/C_V = \frac{5}{3}$. To obtain an equation linking $p$ and $V$, we must integrate Eq. (29). The integrals of each side are natural logarithms with an additive constant of integration:

$$-\gamma \ln V = \ln p + [\text{constant}] \tag{31}$$

Using the properties of logarithms, we can rewrite this as

$$-\ln(pV^{\gamma}) = [\text{constant}] \tag{32}$$

that is,

$$pV^{\gamma} = [\text{constant}] \tag{33}$$

*Adiabatic equation*

This is the equation for the adiabatic expansion of a gas. From this we can readily calculate the drop of temperature that the gas suffers as it expands. According to the ideal-gas law, $pV \propto T$, and thus Eq. (33) becomes

$$TV^{\gamma-1} = [\text{constant}] \tag{34}$$

which shows that the temperature is inversely proportional to $V^{\gamma-1}$. Figure 20.16 shows the adiabatic curves $pV^{\gamma} = [\text{constant}]$ in a $p$–$V$ diagram for an ideal monoatomic gas, with $\gamma = \frac{5}{3}$. For comparison the figure also shows the isothermal curves $T = [\text{constant}]$, or $pV = [\text{constant}]$. Note that the adiabatic curves are steeper than the isothermal curves. When a gas expands adiabatically, it evolves downward along one of the adiabatic curves, and it crosses into regions of lower temperature.

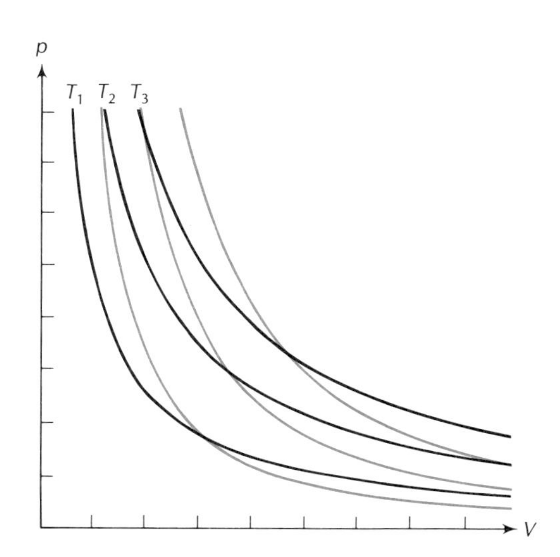


**Fig. 20.16** Adiabatic curves (colored) and isothermal curves (black) for an ideal monoatomic gas. Each curve is characterized by a different value of the constant $pV^{\gamma}$ or $T$ (the numerical values of these constants have not been listed; they depend on the number of moles of the gas).

The decrease of temperature of an expanding gas can be perceived quite readily when air is allowed to rush out of the valve of an automobile tire; this expanding air feels quite cool. The expansion process is approximately adiabatic because the rushing air, although not insulated from its surroundings, expands so quickly that it does not have time to exchange heat with the surrounding atmospheric air. Conversely, the increase of temperature during the adiabatic compression of air can be perceived when operating a manual air pump. The compression of air in the barrel of the pump produces a noticeable warming of the pump.

EXAMPLE 5. In one of the cylinders of an automobile engine, after the explosive combustion of the fuel, the gas has an initial pressure of $3.4 \times 10^6$ N/m$^2$ and an initial volume of 50 cm$^3$. As the piston moves outward, the gas expands nearly adiabatically to a final volume of 250 cm$^3$. What is the final pressure? Assume that $\gamma = 1.40$.

SOLUTION: By Eq. (33),

$$p_1V_1^\gamma = p_2V_2^\gamma$$

so that

$$p_2 = p_1\left(\frac{V_1}{V_2}\right)^\gamma = 3.4\times10^6\ \text{N/m}^2\times\left(\frac{50\ \text{cm}^3}{250\ \text{cm}^3}\right)^{1.4}$$

$$= 3.6\times10^5\ \text{N/m}^2$$

---

EXAMPLE 6. How much work does the gas described in the preceding example do during its adiabatic expansion?

SOLUTION: The work done during a small increase of volume is $p\,dV$ and the work done during the entire expansion is

$$W = \int_{V_1}^{V_2} p\,dV \qquad (35)$$

According to Eq. (33),

$$p = p_1V_1^\gamma/V^\gamma \qquad (36)$$

so that

$$W = \int_{V_1}^{V_2}\frac{p_1V_1^\gamma}{V^\gamma}dV = p_1V_1^\gamma\left(-\frac{1}{\gamma-1}\frac{1}{V_2^{\gamma-1}}+\frac{1}{\gamma-1}\frac{1}{V_1^{\gamma-1}}\right)$$

$$= \frac{1}{\gamma-1}p_1V_1\left[1-\left(\frac{V_1}{V_2}\right)^{\gamma-1}\right] \qquad (37)$$

With the numbers of Example 5, this gives

$$W = \frac{1}{1.40-1}\times3.4\times10^6\ \text{N/m}^2\times5.0\times10^{-5}\ \text{m}^3\left[1-\left(\frac{50\ \text{cm}^3}{250\ \text{cm}^3}\right)^{1.40-1}\right]$$

$$= 2.0\times10^2\ \text{J}$$

---

## SUMMARY

**Specific heat of water:** $c = 1$ kcal/kg · °C

**Mechanical equivalent of heat:** 1 cal = 4.186 J

**Thermal expansion:** $\Delta L = \alpha L\,\Delta T$

$$\Delta V = \beta V\,\Delta T$$

**Conduction of heat:** $\dfrac{\Delta Q}{\Delta t} = -kA\dfrac{dT}{dx}$

**Relation between specific heats of a mole of ideal gas:** $C_p - C_V = R$

**Adiabatic equation for gas:** $pV^\gamma = [\text{constant}]$

$$\gamma = C_p/C_V$$

## QUESTIONS

1. Can the body heat from a crowd of people produce a significant temperature increase in a room?

2. The expression "cold enough to freeze the balls off a brass monkey" originated aboard ships of the British Navy where cannonballs of lead were kept in brass racks ("monkeys"). Can you guess how the balls might fall off a "monkey" on a very cold day?

3. If the metal lid of a glass jar is stuck, it can usually be loosened by running hot water over the lid. Explain.

4. On hot days, bridges expand. How do bridge designers prevent this expansion from buckling the road?

5. At regular intervals, oil pipelines have lateral loops (shaped like a **U**; see Figure 20.17). What is the purpose of these loops?

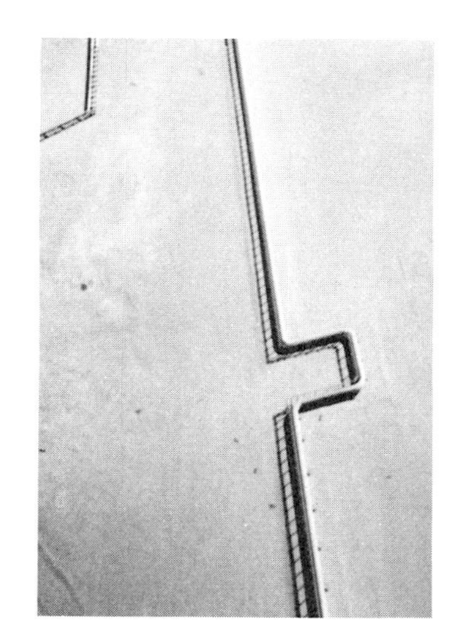

**Fig. 20.17** Oil pipeline.

6. When you heat soup in a metal pot, sometimes the soup rises at the rim of the pot and falls at the center. Explain.

7. A sheet of glass will crack if heated in one spot. Why?

8. When aluminum wiring is used in electrical circuits, special terminal connectors are required to hold the ends of the wires securely. If an ordinary brass screw were used to hold the end of an aluminum wire against a brass plate, what is likely to happen during repeated heating and cooling of the circuit?

9. The frequency of the ticking of a mechanical wristwatch depends on the moment of inertia of its balance wheel. If the watch is subjected to an increase of temperature, the balance wheel expands and its moment of inertia increases, which makes the watch run slow. Good wristwatches have a built-in temperature compensation in their balance wheels, so that the moment of inertia stays constant. Design such a compensated balance wheel using two metals of different coefficients of thermal expansion.

10. Suppose that a piece of metal and a piece of wood are at the same temperature. Why does the metal feel colder to the touch than the wood?

11. In lack of better, some nineteenth-century explorers in Africa measured altitude by sticking a thermometer into a pot of boiling water. Explain.

12. Can you guess why an alloy of two metals usually has a lower melting point than either pure metal?

13. In the cooling system of an automobile, how is the heat transferred from the combustion cylinder to the cooling water (conduction, convection, or radiation)? How is the heat transferred from the water in the engine to the water in the radiator? How is the heat transferred from the radiator to the air?

14. A fan installed near the ceiling of a room blows air down toward the floor. How does such a fan help to keep you cool in summer and warm in winter?

15. It is often said that an open fireplace sends more heat up the chimney than it delivers to the room. What is the mechanism for heat transport to the room? For heat transport up the chimney?

16. A large fraction of the heat lost from a house escapes through the windows (Figure 20.18). This heat is carried to the windowpane by convection — hot air at the top of the room descends along the windowpane, giving up its heat. Suppose that the windows are equipped with venetian blinds. In order to minimize the heat loss, should you close the blinds so that the slats are oriented down and away from the window or up and away?

**Fig. 20.18** Thermal photograph of a house.

17. Fiber glass insulation used in the walls of houses has a shiny layer of aluminum foil on one side. What is the purpose of this layer?

18. Is it possible to add heat to a system without changing its temperature? Give an example.

19. Why is boiling oil much more likely to cause severe burns on skin than boiling water?

20. Would you expect the melting point of ice to increase or decrease with an increase of pressure?

21. A very cold ice cube, fresh out of the freezer, tends to stick to the skin of your fingers. Why?

22. What is likely to happen to the engine of an automobile if there is no antifreeze in the cooling system and the water freezes?

23. If an evacuated glass vessel, such as a TV tube, fractures and implodes, the fragments fly about with great violence. From where does the kinetic energy of these fragments come?

24. When you boil water and convert it into water vapor, is the heat you supply equal to the change of the internal energy of the water?

25. A gas is in a cylinder fitted with a piston. Does it take more work to compress the gas isothermally or adiabatically?

26. If you let some air out of the valve of an automobile tire, it feels cold. Why?

27. According to the result of Section 11.2, when a particle of small mass collides elastically with a body of very large mass, the particle gains kinetic energy if the body of large mass was approaching the particle before the collision. Using this result, explain how the collisions between the particles of gas and the moving piston lead to an increase of temperature during an adiabatic compression.

28. The air near the top of a mountain is usually cooler than that near the bottom. Explain this by considering the adiabatic expansion of a parcel of air carried from the bottom to the top by an air current.

29. When a gas expands adiabatically, its temperature decreases. How could you take advantage of this effect to design a refrigerator?

30. A sample of ideal gas is initially confined in a bottle at some given temperature. If we break the bottle and let the gas expand freely into an evacuated chamber of larger volume, will the temperature of the gas change?

## PROBLEMS

### Section 20.1

1. The immersible electric heating element in a coffee maker converts 620 W of electric power into heat. How long does this coffee maker take to heat 1.0 liter of water from 20°C to 100°C? Assume that no heat is lost to the environment.

2. The body heat released by children in a school makes a contribution toward heating the building. How many kilowatts of heat do 1000 children release? Assume that the daily food intake of each child has a chemical energy of 2000 kcal and that this food is burned at a steady rate throughout the day.

3. In 1847 Joule attempted to measure the frictional heating of water in a waterfall near Chamonix in the French Alps. If the water falls 120 m and all of its gravitational energy is converted into thermal energy, how much does the temperature of the water increase? Actually, Joule found no increase of temperature because the falling water cools by evaporation.

4. A nuclear power plant takes in $5 \times 10^6$ m$^3$ of cooling water per day from a river and exhausts 1200 megawatts of waste heat into this water. If the temperature of the inflowing water is 20°C, what is the temperature of the outflowing water?

5. Your metabolism extracts about 100 kcal of chemical energy from one apple. If you want to get rid of all this energy by jogging, how far must you jog? At a speed of 12 km/h, jogging requires about 750 kcal/h.

6. For basic subsistence a human body requires a diet with about 2000 kcal/day. Express this power in watts.

7. By turning a crank, you can do mechanical work at the steady rate of 0.15 hp. If the crank is connected to paddles churning 4 liters of water, how long must you churn the water to raise its temperature by 5°C?

*8. You can warm the surfaces of your hands by rubbing one against the other. If the coefficient of friction between your hands is 0.6 and if you press your hands together with a force of 60 N while rubbing them back and forth at an average speed of 0.5 m/s, at what rate (in calories per second) do you generate heat on the surfaces of your hands?

*9. Problem 8.64 gives the relevant numbers for frictional losses in the Tennessee River. If all the frictional heat were absorbed by the water and if there were no heat loss by evaporation, how much would the water temperature rise per mile?

*10. The first quantitative determination of the mechanical equivalent of heat was made by Robert von Mayer, who compared available data on the amount of mechanical work needed to compress a gas and the amount of heat generated during the compression. From this comparison, Mayer deduced that the energy required for warming 1 kilogram of water by 1°C is equivalent to the potential energy released when a mass of 1 kilogram falls from a height of 365 m. By how many percent does Mayer's result differ from the modern result given in Section 1?

*11. On a hot summer day, the use of air conditioners raises the consumption of electric power in New York City to 22,400 megawatts. All of this electric power ultimately produces heat. Compare the heat produced in this way with the solar heat incident on the city. Assume that the incident flux of solar energy is 1 kilowatt per m$^2$ and the area of the city is 850 km$^2$. Would you expect that the consumption of electric power significantly increases the ambient temperature?

*12. A simple gadget for heating water for showers consists of a black plastic bag holding 10 liters of water. When hung in the sun, the bag absorbs heat. On a clear, sunny day, the power delivered by sunlight per unit area facing the Sun is $1.0 \times 10^3$ W/m$^2$. The bag has an area of 0.10 m$^2$ facing the Sun. How long does it take for the water to warm from 20°C to 50°C? Assume that the bag loses no heat.

*13. The beam dump at the Stanford Linear Accelerator consists of a large tank with 12 m$^3$ of water into which the accelerated electrons can be aimed when they are not wanted elsewhere. The beam carries $3.0 \times 10^{14}$ electrons/s; the kinetic energy per electron is $3.2 \times 10^{-9}$ J. In the beam dump this energy is converted into heat.

(a) What is the rate of production of heat?

(b) If the water in the tank does not lose any heat to the environment, what is the rate of increase of temperature of the water?

*14. A solar collector consists of a flat plate that absorbs the heat of sunlight. A water pipe attached to the back of the plate carries away the absorbed heat (Figure 20.19). Assume that the solar collector has an area of 4.0 m$^2$ facing the Sun and that the power per unit area delivered by sunlight is $1.0 \times 10^3$ W/m$^2$. What is the rate at which water must circulate through the pipe if the temperature of the water is to increase by 40°C as it passes through the collector?

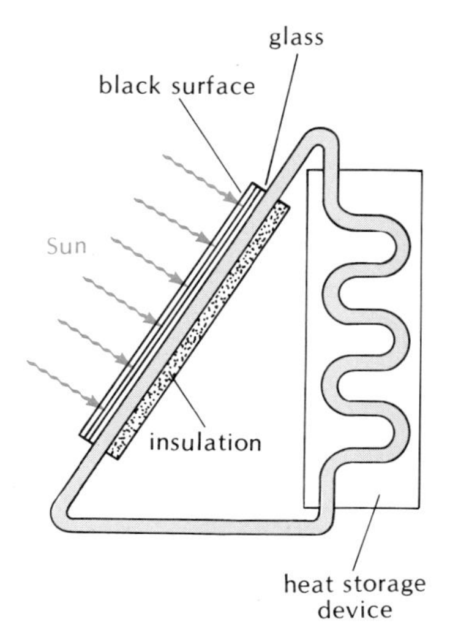


**Fig. 20.19** Collector of solar heat.

*15. Consider the waterwheel described in Example 13.8. How much does the temperature of the water increase as it passes through the wheel?

### Section 20.2

16. The tallest building in the world is the Sears Tower in Chicago, which is 443 m high. It is made of concrete and steel. How much does its height change between a day when the temperature is 35°C and a day when the temperature is −29°C?

17. The height of the Eiffel Tower is 321 m. What increase of temperature will lead to an increase of height by 10 cm?

18. Machinists use gauge blocks of steel as standards of length. A one-inch gauge block is supposed to have a length of 1 in., to within $\pm 10^{-6}$ in. In order to keep the length of the block within this tolerance, how precisely must the machinist control the temperature of the block?

19. A mechanic wants to place a sleeve (pipe) of copper around a rod of steel. At a temperature of 18°C, the sleeve of copper has an inner diameter of 0.998 cm and the rod of steel has a diameter of 1.000 cm. To what temperature must the mechanic heat the copper to make it fit round the steel?

20. (a) Segments of steel railroad rails are laid end to end. In an old railroad, each segment is 18 m long. If they are originally laid at a temperature of −7°C, how much of a gap must be left between adjacent segments if they are to just barely touch at a temperature of 43°C?
(b) In a modern railroad, each segment is 790 m long, with a special expansion joint at each end. How much of a gap must be left between adjacent segments in this case?

*21. Suppose you heat a 1-kg cube of iron from 20°C to 80°C while it is surrounded by air at a pressure of 1 atm. How much work does the iron do against the atmospheric pressure while expanding? Compare this work with the heat absorbed by the iron. (The density of iron is $7.9 \times 10^3$ kg/m$^3$.)

*22. When a solid expands, the increment of the area of one of its faces is directly proportional to the increment of temperature and to the original area. Show that the coefficient of proportionality for this expansion of area is two times the coefficient of linear expansion.

*23. A spring made of steel has a relaxed length of 0.316 m at a temperature of 20°C. By how much will the length of this spring increase if we heat it to 150°C? What compressional force must we apply to the hot spring to bring it back to its original length? The spring constant is $3.5 \times 10^4$ N/m.

*24. A wheel of metal has a moment of inertia $I$ at some given temperature. Show that if the temperature increases by $\Delta T$, the moment of inertia will increase by approximately $\Delta I = 2\alpha I\, \Delta T$.

*25. The pendulum (rod and bob) of a pendulum clock is made of brass.
(a) What will be the fractional increment in the length of this pendulum if the temperature increases by 20°C? What will be the fractional increase in the period of the pendulum?
(b) The pendulum clock keeps good time when its temperature is 15°C. How much time (in seconds per day) will the clock lose when its temperature is 35°C?

*26. (a) The density of gasoline is 730 kg/m$^3$ when the temperature is 0°C. What will be the density of gasoline when the temperature is 30°C?
(b) The price of gasoline is 30 cents per liter. What is the price per kilogram at 0°C? What is the price per kilogram at 30°C? (Note that 1 gal. = $3.80 \times 10^{-3}$ m$^3$.) Is it better to buy cold gasoline or warm gasoline?

**27. In order to compensate for deviations caused by temperature changes, a pendulum clock built during the last century for an astronomical observatory uses a large cylindrical glass tube filled with mercury as a pendulum bob. This tube is held by a brass rod and bracket (Figure 20.20); the combined length of the rod and bracket is $l$ (measured from the point of suspension of the pendulum). Neglecting the mass of the brass and the glass and neglecting the expansion of the glass, show that the height of the mercury in the glass tube must be

$$h = (2\alpha_{\text{brass}}/\beta_{\text{mercury}})l$$

if the center of mass of the mercury is to remain at a fixed distance from the point of suspension, regardless of temperature.

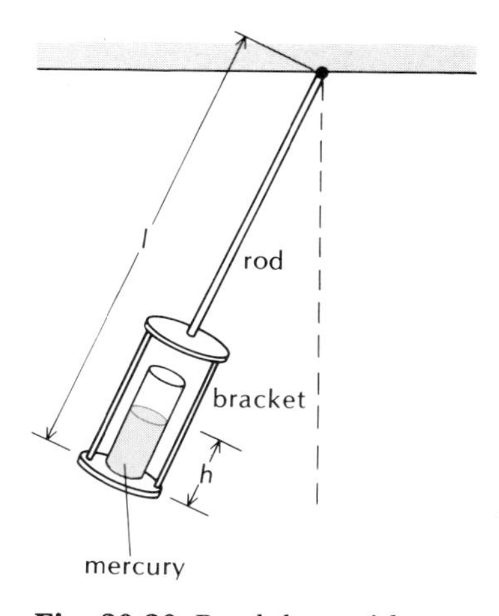


**Fig. 20.20** Pendulum with a temperature compensator.

### Section 20.3

28. An ordinary mercury thermometer consists of a glass bulb to which is attached a fine capillary tube. As the mercury expands, it rises up the capillary tube. Given that the bulb has a volume of 0.20 cm$^3$ and that the capillary tube has a diameter of $7.0 \times 10^{-3}$ cm, how far will the mercury column rise up the capillary tube for a temperature increase of 10°C? Ignore the expansion of the glass and ignore the expansion of the mercury in the capillary tube.

### Section 20.4

29. The walls of an igloo are made of compacted snow, 30 cm thick. What thickness of styrofoam would provide the same insulation as the snow?

30. A pan of aluminum, filled with boiling water, sits on a hot plate. The bottom area of the pan is 300 cm$^2$ and the thickness of the aluminum is 0.10 cm. If the hot plate supplies 2000 W of heat to the bottom of the pan, what must be the temperature of the upper surface of the hot plate?

31. A rod of steel 0.70 cm in diameter is surrounded by a tight copper sleeve of inner diameter 0.70 cm and outer diameter 1.00 cm. What will be the heat flow along this compound rod if the temperature gradient along the rod is 50°C/cm? What fraction of the heat flows in the copper? What fraction in the steel?

32. A window in a room measures 1 m $\times$ 1.5 m. It consists of a single sheet of glass of thickness 2.5 mm. What is the heat flow through this window if the temperature difference between the inside surface of the glass and the outside is 39°C? Compare the heat loss through the window with the heat loss through the wall calculated in Example 3.

33. The bottom of a tea kettle consists of a layer of stainless steel 0.050 cm thick welded to a layer of copper 0.030 cm thick. The area of the bottom of the kettle is 300 cm$^2$. The copper sits in contact with a hot plate at a temperature of 101.2°C and the steel is covered with boiling water at 100.0°C. What is the rate of heat transfer through the bottom of the kettle from the hot plate to the water?

34. According to Eq. (4), the heat flow through a rod or a slab of cross-sectional area $A$ can be expressed as

$$\frac{\Delta Q}{\Delta t} = -\frac{A\,\Delta T}{R}$$

where $R = \Delta x/k$ is called the thermal resistance, or the **R-value.** Since the heat flow is inversely proportional to the R-value, a good insulator has a high R-value.

(a) What is the R-value of a slab of figer glass insulation, 10 cm thick?
(b) In the United States, R-values of commercially available insulation are commonly expressed in units of ft$^2\cdot$°F$\cdot$h/Btu. What is the R-value of 10-cm slab of fiber glass in these units?

*35. The wall of a house is to be built of a layer of wood, an adjacent layer of fiber glass insulation, and an adjacent layer of brick. The thickness of the wood is 1.2 cm and that of the brick is 10 cm. The heat loss per square meter of wall is to be no more than 10 kcal/hr when the temperatures inside and outside the house differ by 30°C. How thick must the fiber glass insulation be?

*36. A styrofoam box, used for the transportation of medical supplies, is filled with dry ice (carbon dioxide) at a temperature of −79°C. The box measures 30 cm × 30 cm × 40 cm and its walls are 4 cm thick. If the outside surface of the box is at a temperature of 20°C, what is the rate of loss of dry ice by vaporization?

*37. The icebox on a sailboat measures 60 cm × 60 cm × 60 cm. The contents of this icebox are to be kept at a temperature of 0°C for 4 days by the gradual melting of a block of ice of 20 kg, while the temperature of the outside of the box is 30°C. What minimum thickness of styrofoam insulation is required for the walls of the icebox?

*38. A man has a skin area of 1.8 $m^2$; his skin temperature is 34°C. On a cold winter day, the man wears a whole-body suit insulated with down. The temperature of the outside surface of his suit is −25°C. If the man can stand a heat loss of no more than 100 kcal/h, what is the minimum thickness of down required for his suit?

*39. The end of a rod of copper 0.50 cm in diameter is welded to a rod of silver of half the diameter. Each rod is 6.0 cm long. What is the heat flow along these rods if the free end of the copper rod is in contact with boiling water and the free end of the silver rod is in contact with ice? What is the temperature of the junction? Assume that there is no heat loss through the lateral surfaces of the rods.

*40. On a cold winter day, the water of a shallow pond is covered with a layer of ice 6.0 cm thick. The temperature of the air is −20°C and the temperature of the water is 0°C. What is the (instantaneous) rate of growth of the thickness of the ice (in centimeters per hour)? Assume that wind chill keeps the top surface of the ice at exactly the temperature of the air, and assume that there is no heat transfer through the bottom of the pond.

*41. Suppose that the pond described in the preceding problem has a layer of compacted snow 3.0 cm thick on top of the ice. What is the rate of growth of the thickness of the ice?

*42. Several slabs of different materials are piled one on top of another. All slabs have the same face area $A$, but their thickness and conductivities are $\Delta x_i$ and $k_i$, respectively. Show that the heat flow through the pile of slabs is

$$\frac{\Delta Q}{\Delta t} = \frac{-A\,\Delta T}{\sum_{i=1}^{n} \Delta x_i / k_i}$$

where $\Delta T$ is the temperature difference between the bottom of the first slab and the top of the last slab.

**43. A cable used to carry electric power consists of a metallic conductor of radius $r_1$ encased in an insulator of inner radius $r_1$ and outer radius $r_2$. The metallic conductor is at a temperature $T_1$ and the outer surface of the insulator is at a temperature $T_2$. Show that the radial heat flow across a length $l$ of the insulator is given by

$$\frac{\Delta Q}{\Delta t} = \frac{2\pi k (T_1 - T_2) l}{\ln(r_2/r_1)}$$

where $k$ is the conductivity of the insulator.

### Section 20.5

44. Thunderstorms obtain their energy by condensing the water vapor contained in humid air. Suppose that a thunderstorm succeeds in condensing *all*

the water vapor in 10 $km^3$ of air.

(a) How much heat does this release? Assume the air is initially at 100% humidity and that each cubic meter of air at 100% humidity (at 20°C and 1 atm) contains $1.74 \times 10^{-2}$ kg of water vapor. The heat of vaporization of water is 585 kcal/kg at 20°C.

(b) The explosion of an A bomb releases an energy of $2 \times 10^{10}$ kcal. How many A bombs does it take to make up the energy of one thunderstorm?

45. You place 1 kg of ice (at 0°C) in a pot and heat it until the ice melts and the water boils off, making steam. How much heat must you supply to achieve this?

*46. The heat of vaporization of water at 100°C and 1 atm is 539 kcal/kg. How much of this energy is due to the work the water vapor does against atmospheric pressure? What would be this work at a pressure 0.1 atm? At (nearly) zero pressure?

*47. During a rainstorm lasting 2 days, 7.6 cm of rain fell over an area of $2.6 \times 10^3$ $km^2$.

(a) What is the total mass of the rain (in kilograms)?

(b) Suppose that the heat of vaporization of water in the rain clouds is 580 kcal/kg. How many calories of heat are released during formation of the total mass of rain by condensation of the water vapor in those clouds?

(c) Suppose that the rain clouds are at a height of 1500 m above the ground. What is the gravitational potential energy of the total mass of rain before it falls? Express your answer in calories.

(d) Suppose that the raindrops hit the ground with a speed of 10 m/s. What is the total kinetic energy of all the raindrops taken together? Express your answer in calories. Why does your answer to part (c) not agree with this?

*48. If you pour 0.50 kg of molten lead at 328°C into 2.5 liters of water at 20°C, what will be the final temperatures of the water and the lead? The specific heat of (solid) lead has an average value of $3.4 \times 10^{-2}$ kcal/kg · °C over the relevant temperature range.

*49. Suppose you drop a cube of titanium of mass 0.25 kg into a Dewar flask (a thermos bottle) full of liquid nitrogen at −196°C. The initial temperature of the titanium is 20°C. How many kilograms of nitrogen will boil off as the titanium cools from 20°C to −196°C? The specific heat of titanium is $8.2 \times 10^{-2}$ kcal/kg · °C.

*50. While jogging on a level road, your body generates heat at the rate of 750 kcal/h. Assume that evaporation of sweat removes 50% of this heat and convection and radiation the remainder. The evaporation of 1 kg (or 1 liter) of sweat requires 580 kcal. How many kilograms of sweat do you evaporate per hour?

*51. The Mediterranean loses a large volume of water by evaporation. This loss is made good, in part, by currents flowing into the Mediterranean through the straits joining it to the Atlantic and the Black Sea. Calculate the rate of evaporation (in $km^3$/h) of the Mediterranean on a clear summer day from the following data: the area of the Mediterranean is $2.9 \times 10^6$ $km^2$, the power per unit area supplied by sunlight is $1 \times 10^3$ W/$m^2$, and the heat of vaporization of water is 580 kcal/kg (at a temperature of 21°C). Assume that all the heat of sunlight is used for evaporation.

### Section 20.6

52. A TV tube of glass with zero pressure inside and atmospheric pressure outside suddenly cracks and implodes. The volume of the tube is $2.5 \times 10^{-2}$ $m^3$. During the implosion, the atmosphere does work on the fragments of the tube and on the layer of air immediately adjacent to the tube. This amount of work represents the energy released in the implosion. Calculate this

energy. If all of this energy is acquired by the fragments of glass, what will be the mean speed of the fragments? The total mass of the glass is 2.0 kg.

53. The rear end of an air conditioner dumps 3000 kcal/h of waste heat into the air outside a building. A fan assists in the removal of this heat. The fan draws in 15 $m^3$/min of air at a temperature of 30°C and ejects this air after it has absorbed the waste heat. With what temperature does the air emerge?

54. What are the specific heats $C_V$ and $C_p$ for air consisting of 75% nitrogen, 24% oxygen, and 1% argon at STP? Use the values for $C_V$ and $C_p$ of nitrogen, oxygen, and argon listed in Table 20.5.

55. Table 20.5 gives the specific heat per mole for some gases at STP. For each of these gases, calculate the specific heat per kilogram.

56. If we heat 1.00 kg of hydrogen gas from 0°C to 50.0°C in a cylinder with a piston keeping the gas at a constant pressure of 1.0 atm, we must supply $1.69 \times 10^2$ kcal of heat. How much work does the gas deliver to the cylinder during this process? How many kilocalories of heat must we supply to heat the same amount of gas from 0°C to 50.0°C in a container of constant volume?

57. The theoretical expression for the speed of sound in a gas is $\sqrt{\gamma p/\rho}$ [compare Eq. (17.3)]. Calculate the speed of sound in carbon dioxide at STP.

58. A helium balloon consists of a large bag loosely filled with 600 kg of helium at an initial temperature of 10°C. While exposed to the heat of the Sun, the helium gradually warms to a temperature of 30°C. The heating proceeds at a constant pressure of 1.0 atm. How much heat does the helium absorb during this temperature change?

*59. On a winter day you inhale cold air at a temperature of −30°C and at 0% humidity. The amount of air you inhale is 0.45 kg per hour. Inside your body you warm and humidify the air; you then exhale the air at a temperature of 37°C and 100% relative humidity. At a temperature of 37°C, each kilogram of air at 100% relative humidity contains 0.041 kg of water vapor. How many calories are carried out of your body by the air that passes through your lungs in one hour? Take into account both the heat needed to warm the air at constant pressure and the heat needed to vaporize the moisture that the exhaled air carries out of your body. The specific heat of air at constant pressure is 0.25 kcal/°C · kg; the heat of vaporization of water at 37°C is 576 kcal/kg.

*60. An air conditioner removes heat from the air of a room at the rate of 2000 kcal/h. The room measures 5.0 m × 5.0 m × 2.5 m and the pressure is constant at 1.0 atm.

(a) If the initial temperature of the air in the room is 30.0°C, how long does it take the air conditioner to reduce the temperature of the air by 5.0°C? Pretend that the mass of air in the room is constant.

(b) As the air in the room cools, it contracts slightly and draws in some extra air from the outside; hence the mass of air is not exactly constant. Repeat your calculation taking into account this increase of the mass of air. Assume that the extra air enters with an initial temperature of 30°C. Does the result of your second calculation differ appreciably from that of your first calculation?

### Section 20.7

61. A fire extinguisher is filled with 1.0 kg of compressed nitrogen gas at a pressure of $1.2 \times 10^6$ N/$m^2$ and at a temperature of 20°C.

(a) What is the volume of this gas?

(b) If you open the nozzle of the fire extinguisher, the gas will escape expanding adiabatically to atmospheric pressure ($1.01 \times 10^5$ N/$m^2$). What will be the volume and the temperature of the expanded gas? The value of $\gamma$ for nitrogen is 1.4.

*62. Suppose that a submarine suddenly breaks up at a depth of 300 m below sea level. The air in the submarine will then be suddenly (adiabatically) compressed. If the initial pressure of the air is 1.0 atm and the initial temperature is 20°C, what will be the final pressure and temperature? For air, $\gamma = 1.4$.

*63. In one of the combustion chambers of a diesel engine, the piston compresses the air–fuel mixture from an initial volume of 630 $cm^3$ to a final volume of 30 $cm^3$. The initial temperature of the mixture is 40°C. Assuming that the compression is adiabatic with $\gamma = 1.4$, what is the final temperature?

*64. Suppose that you have a sample of oxygen gas at a pressure of 300 atm and a temperature of −29°C. Assume that oxygen behaves as an ideal gas with $\gamma = 1.4$. If you suddenly (adiabatically) let this gas expand to a final pressure of 1 atm, what will be the final temperature? This method of cooling by adiabatic expansion was used by Cailletet in 1877 to liquify oxygen (the liquefaction point of oxygen at 1 atm is −183°C).

*65. The air in an automobile tire is at an overpressure of 2.4 atm and at a temperature of 20°C. The atmospheric pressure outside the tire is 1.0 atm. If you let some air escape through the valve, what will be the temperature of the emerging air? Assume that the escaping air expands adiabatically with $\gamma = 1.4$.

*66. Under normal conditions, the temperature of air in the atmosphere decreases with altitude, that is, the air temperature at the top of a mountain is lower than that at the foot of the mountain. This temperature difference is maintained by winds that move the air from one place to another. When a wind carries a parcel of air up the side of a mountain, the parcel expands adiabatically and cools; when the wind carries the parcel of air down, it is compressed adiabatically and warms. Calculate the temperature difference between the bottom and the top of a mountain 100 m high. The temperature at the bottom is 20°C. The pressure at the bottom is 1 atm and the pressure at the top is 0.988 atm. For air, $\gamma = 1.4$.

**67. One safety device for stopping the accidental fall of an elevator cage relies on an "air cushion," which consists of air trapped in the bottom portion of the elevator shaft and compressed by the falling cage. To achieve this cushioning, the bottom portion of the elevator shaft must be tapered to fit snugly around the sides of the cage, and any doors in this portion of the elevator shaft must be airtight. (Some openings or valves are needed to permit the controlled escape of some of the air, to prevent a rebound of the falling cage; but we will ignore this complication for our purposes). In a test of such a safety device, an elevator cage was allowed to fall freely from the 20th floor of a building, at a height of 87 m from the bottom of the shaft, and brought to rest by the air cushion, without damage to a basket of eggs placed in the cage. The airtight portion of the shaft extended from the bottom of the shaft to a height of 15 m. Calculate the height at which the cage stops in the elevator shaft, given that the shaft measures 1.5 m $\times$ 1.5 m across, and the mass of the cage is 400 kg. Calculate the pressure and the temperature of the compressed air when the cage stops. Assume that the compression of the air proceeds adiabatically with $\gamma = 1.4$ and that the initial temperature is 25°C and the initial pressure 1 atm.

**68. By means of a hand pump, you inflate an automobile tire from 0 atm to 2.4 atm (overpressure). The volume of the tire remains constant at 0.10 $m^3$ during this operation. How much work must you do on the air with the pump? Assume that each stroke of the pump is an adiabatic process and that the air is initially at STP.

**69. In an air pressure gun (BB gun), a projectile is driven along a barrel by the pressure of air suddenly released into the barrel. Initially, the air is held at high pressure in a small reservoir; when this air is suddenly released into the barrel, it expands adiabatically and does work on the projectile, giving it kinetic energy. You are told to design a gun that shoots a projectile of 1.0 g with a muzzle speed of 100 m/s from a barrel of length 25 cm and diameter 0.5 cm. The volume of the reservoir is $1.0 \times 10^{-5}$ $m^3$. What must be the minimum initial pressure in the reservoir? How much air (in kilograms, at 20°C) must this reservoir contain initially? For air, $\gamma = 1.4$. Ignore friction in the barrel.

CHAPTER 21

# Thermodynamics

Thermodynamics is the study of the relationships among the purely macroscopic parameters describing the behavior of physical systems. Such a macroscopic, large-scale description is necessarily somewhat crude, since it overlooks all of the small-scale, microscopic details. However, in practical applications, these microscopic details are often irrelevant. For instance, an engineer investigating the behavior of the combustion gases in the cylinder of an automobile engine can get by reasonably well with such macroscopic quantities as temperature, pressure, density, and heat capacity.

The most important application of thermodynamics concerns the conversion of one form of energy into another, especially the conversion of heat into other forms of energy. These conversions are governed by the two fundamental laws of thermodynamics. The first of these is essentially a general statement of the law of conservation of energy and the second is a statement about the maximum efficiency attainable in the conversion of heat into work.

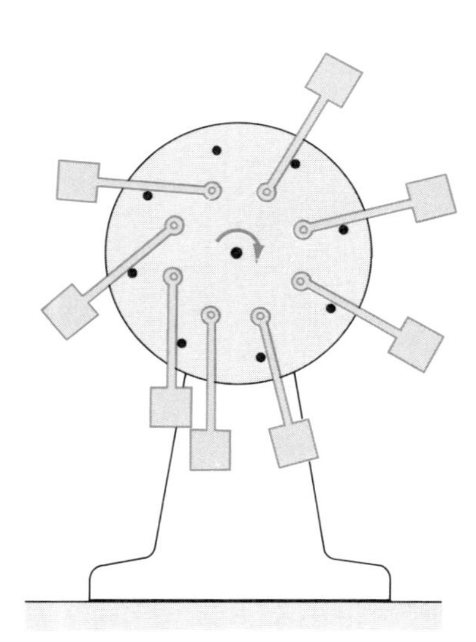

**Fig. 21.1** A hypothetical perpetual motion machine.

The study of thermodynamics was inaugurated by nineteenth-century engineers who wanted to know what ultimate limitations the laws of physics impose on the operation of steam engines and other machines that generate mechanical energy. They soon recognized that perpetual motion machines are impossible. A **perpetual motion machine of the first kind** is a (hypothetical) device that supplies an endless output of work without any input of fuel or any other input of energy. Figure 21.1 shows a proposed design for such a machine. Weights are attached to the rim of a wheel by short pivoted rods resting against pegs. With the rods in the position shown, there is an imbalance in the weight distribution causing a clockwise torque on the wheel; as the wheel turns, the rod coming to the top presumably flips over, maintaining the imbalance. This perpetual torque would not only keep the wheel turning, but would also continually deliver energy to the axle of the wheel. Of course, a detailed analysis demonstrates that the machine will not perform as intended — the wheel actually settles in a static equilibrium configuration such that the top rod just barely fails to flip over. The First Law of Thermodynamics, or the law of conservation of energy, directly tells us of the failure of this machine: after one revolution of the wheel, the masses all return to their initial positions, their potential energy returns to its initial value, and they will not have delivered any net energy to the motion of the wheel.

A **perpetual motion machine of the second kind** is a device that extracts thermal energy from air or from the water of the oceans and converts it into mechanical energy. Such a device is not forbidden by conservation laws. The oceans are an enormous reservoir of thermal energy; if we could extract this thermal energy, a temperature drop of just 1°C of the oceans would supply the energy needs of the United States for the next 50 years. But, as we will see, the Second Law of Thermodynamics tells us that conversion of heat into work requires not only a heat source, but also a heat sink. Heat flows out of a warm body only if there is a colder body that can absorb it. If we want heat to flow from the ocean into our machine, we must provide a low-temperature heat sink toward which the heat will tend to flow spontaneously. Without a low-temperature sink, the extraction of heat from the oceans is impossible. We cannot build a perpetual motion engine of the second kind.

## 21.1 The First Law of Thermodynamics

Consider some amount of gas with a given initial volume $V_1$, pressure $p_1$, and temperature $T_1$. The gas is in a container fitted with a piston (Figure 21.2). Suppose we compress the gas to some smaller volume $V_2$ and also cool the gas and lower its temperature to some smaller value $T_2$; the pressure will then reach some new value $p_2$. (Such a compression and cooling process is of practical importance in the liquefaction of gases, say, oxygen or nitrogen — before the gas can be liquefied it must be compressed and cooled.) Obviously we can reach the new state $V_2$, $p_2$, $T_2$ from the old state $V_1$, $p_1$, $T_1$ in a variety of ways. For instance, we may first compress the gas and then cool it. Or else we may first cool it and then compress it. Or we may go through small alternating steps of compressing and cooling. In order to compress the gas, we must do work on it [see Eq. (20.10)]; and in order to cool the gas, we must remove heat from it.

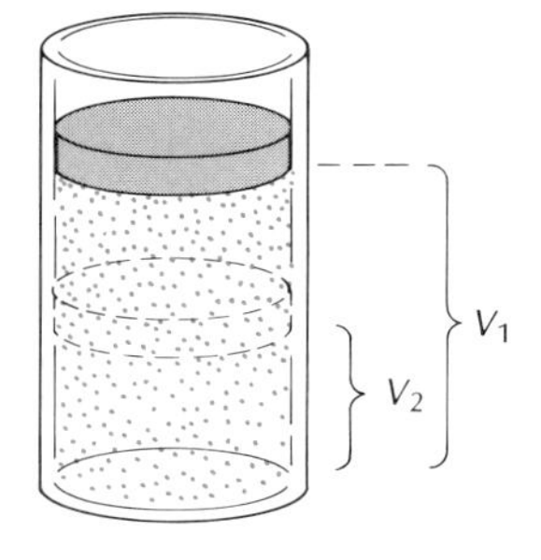


**Fig. 21.2** Compression of a gas.

The work done on or by the gas, and the heat transferred from or to the gas result in a change of the internal energy of the gas. We can express this change of energy as

$$\Delta E = \Delta Q - \Delta W \tag{1}$$

where $\Delta Q$ is the amount of heat transferred to the gas, and $\Delta W$ is the amount of work performed by the gas. Note the sign conventions in this equation: $\Delta Q$ is positive if we add heat to the gas and negative if we remove heat; $\Delta W$ is positive if the gas does work on us and negative if we do work on the gas.

The values of $\Delta Q$ and $\Delta W$ depend on the process. If we first compress the gas and subsequently cool it, then during the first step $\Delta W$ is negative and $\Delta Q$ zero; and during the second step $\Delta W$ is zero and $\Delta Q$ is negative. If we first cool the gas and then compress it, the values of $\Delta W$ and $\Delta Q$ will be quite different. Yet, it turns out that regardless of what sequence of operations we use to transform the gas from its initial state $V_1$, $p_1$, $T_1$ to its final state $V_2$, $p_2$, $T_2$, the net change $\Delta E$ in the internal energy is always the same: $\Delta Q$ and $\Delta W$ vary, but the sum of $\Delta Q$ and $-\Delta W$ remains fixed. This is the **First Law of Thermodynamics:**

> *Whenever we employ some process involving heat and work to change a system from an initial state characterized by certain values of the macroscopic parameters to a final state characterized by new values of the macroscopic parameters, the change in the internal energy of the system has a fixed value*

*First Law of Thermodynamics*

$$\Delta E = \Delta Q - \Delta W \tag{2}$$

> *which does not depend on the details of the process.*

Note that the First Law tells us that energy is conserved — the change of the internal energy equals the input of heat and work. But the First Law tells us more than that. If we describe a system in terms of the detailed microscopic positions and velocities of all its constituent particles, then energy conservation is a theorem of mechanics. But if we describe a system in terms of nothing but macroscopic parameters, then it is not at all obvious that we have available enough information to determine the energy and to formulate a conservation law. The First Law of Thermodynamics tells us that a knowledge of the macroscopic parameters is indeed sufficient to determine the internal energy of the system.

The First Law sometimes permits us to calculate the unknown amount of heat or work required for a process, if we know the amount of heat and work for a different process that takes the system from the same initial state to the same final state. The following example illustrates how the First Law can be exploited in such a calculation.

EXAMPLE 1. At a pressure of 1 atm, the heat of vaporization of water is 539 kcal/kg; this is the heat required to convert 1 kg of water at 100°C into water vapor at the same temperature. Given that the specific heat of water is approximately 1 kcal/kg · °C and the specific heat of water vapor is $c_p = 0.48$ kcal/kg · °C , calculate the heat of vaporization of water at 20°C.

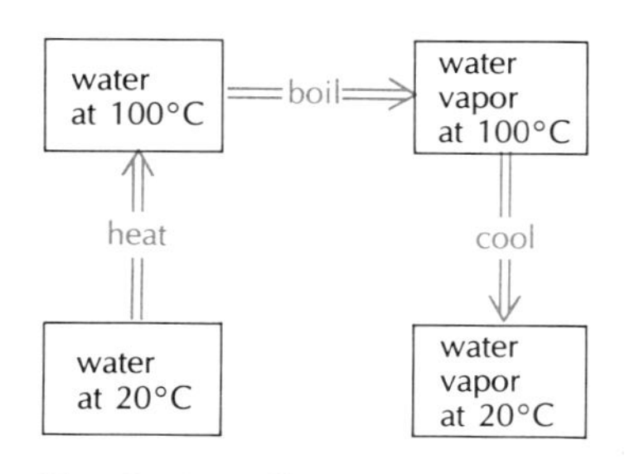


**Fig. 21.3** Indirect process for the evaporation of water.

SOLUTION: Instead of directly converting water at 20°C into water vapor, we can first heat the water to 100°C, then vaporize it, and then cool the vapor to 20°C (without condensation). This sequence of steps is summarized in Figure 21.3. The net work done against atmospheric pressure is the same during this indirect process as during the direct process, since the changes of volume are the same. The net changes of the internal energies are also the same for both processes. Hence, according to the First Law, the heats absorbed must be the same. For the indirect process indicated in Figure 21.3, the value of $\Delta Q$ is

$$\Delta Q = (1\ \text{kcal/kg} \cdot {}^\circ\text{C}) \times 80^\circ\text{C} + 539\ \text{kcal/kg} - (0.48\ \text{kcal/kg} \cdot {}^\circ\text{C}) \times 80^\circ\text{C}$$

$$= 581\ \text{kcal/kg}$$

and this must then be the heat of vaporization of water at 20°C.

COMMENTS AND SUGGESTIONS: In this calculation, we have used the sign convention required by our statement of the First Law: heat added to the system is positive, and heat removed from the system is negative.

The result of the calculation agrees quite well with the experimental value, 585 kcal/kg. The slight discrepancy is not to be blamed on a failure of the First Law, but rather on slight variations of the specific heats of water and of water vapor with temperature. These variations ought to be taken into account for a more accurate calculation.

Our result indicates that the heat of vaporization of water at a temperature below the boiling point is larger than the heat of vaporization at the boiling point. The microscopic explanation of this is that the bonds among the water molecules are tighter at low temperature.

The First Law sometimes permits us to draw some general qualitative conclusions about the behavior of a physical system. For instance, consider the following experiment: Take a thermally insulated bottle with ideal gas at some temperature $T_1$ and, by means of a pipe with a stopcock, connect this to another insulated bottle which is evacuated (Figure 21.4). If we suddenly open the stopcock, the gas will rush from the first bottle into the second until the pressures are equalized. Experimentally, we find that this process of free expansion does not change the temperature of the gas — when the gas attains equilibrium and stops flowing, the final temperature of both bottles are equal to the initial temperature $T_1$. What can we deduce from this experimental observation? Since the bottles are thermally insulated from their environment, the expansion process neither adds nor removes heat from the gas, that is, $\Delta Q = 0$. Furthermore, the expansion process involves no work (except for an insignificant amount required for turning the stopcock), that is $\Delta W = 0$. Consequently, Eq. (2) tells us that the energy of the gas does not change,

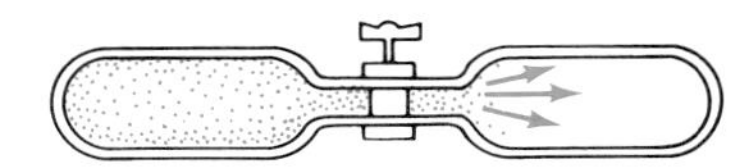

**Fig. 21.4** Free expansion of a gas.

$$\Delta E = \Delta Q - \Delta W = 0 \tag{3}$$

This shows that a change of volume does not affect the energy; that is, the internal energy of the ideal gas is *not* a function of the volume. According to the First Law, the energy of the gas is supposed to be some function of the macroscopic parameters $V$, $p$, and $T$. Since the ideal-gas law allows us to express $p$ in terms of $V$ and $T$, the energy may be regarded as a function of the two variables $V$ and $T$. But Eq. (3) shows that a change of volume does not affect the energy; consequently, the internal energy of the ideal gas is a function of the *temperature* alone.

Note that this conclusion is consistent with Eqs. (19.24) and (19.31), obtained from kinetic theory, which tell us explicitly what function of the temperature is involved in the energy. However, the calculations of kinetic theory depend on some questionable assumptions concerning the behavior of an ideal gas, whereas the arguments of thermodynamics are free of these. The conclusions of thermodynamics are therefore of very general validity, but the price we pay for this is that the information we obtain from thermodynamics is not as detailed as that which we obtain from kinetic theory.

The conclusions of thermodynamics apply only to the equilibrium states of a system, that is, those static states into which the system settles when mass transfer, heat transfer, and all chemical and other reactions have come to an end. For instance, for gas in the two bottles shown in Figure 21.4, the initial state (gas confined to one bottle, with stopcock closed) is an equilibrium state, and the final state (gas evenly distributed over both bottles) is also an equilibrium state. However, the intermediate state, when gas is rushing from the full bottle into the empty bottle immediately after we open the stopcock, is not an equilibrium state. Thermodynamics does not tell us anything about this intermediate, time-dependent state of the gas. In the following discussions we will always suppose that the system under consideration is in an equilibrium state.

## 21.2 The Carnot Engine

If an engine is continually to convert thermal energy into mechanical energy, it must operate cyclically. At the end of each cycle it must return to its initial configuration, so it can repeat the process of conversion of heat into work over and over again. Steam engines and automobile engines are obviously cyclic — after one (or sometimes two) revolutions, they return to their initial configuration. These engines are not 100% efficient. The condenser of a steam engine and the radiator and exhaust of an automobile engine eject a substantial amount of heat into the environment; this waste heat represents lost energy. Besides, there are frictional losses.

*Heat engine*

Any device that converts heat into work by means of a cyclic process is called a **heat engine.** The engine absorbs heat from a heat reservoir at high temperature, converts this heat partially into work, and ejects the remainder as waste heat into a reservoir at low temperature. In this context, a **heat reservoir** is simply a body that remains at constant temperature, even when heat is removed from or added to it. In practice, the high-temperature heat reservoir is often a boiler whose temperature is kept constant by the controlled combustion of some fuel, and the low-temperature reservoir is usually a condenser in contact with a body of water or in contact with the atmosphere of the Earth, whose large volume permits it to absorb the waste heat without appreciable change of temperature.

*Heat reservoir*

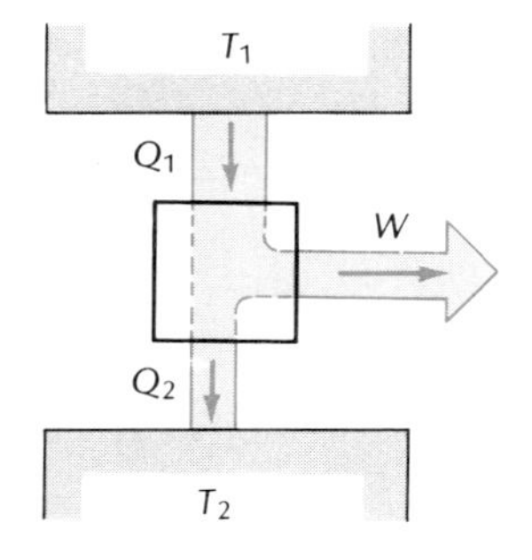


**Fig. 21.5** Flow chart for a heat engine. The square represents the engine.

Figure 21.5 is a flow chart for the energy, showing the heat $Q_1$ flowing into the engine from the high-temperature reservoir, the heat $Q_2$ (waste heat) flowing out of the engine into the low-temperature reservoir, and the work generated. The work generated is the difference between $Q_1$ and $Q_2$,

$$W = Q_1 - Q_2 \tag{4}$$

The **efficiency** of the engine is defined as the ratio of this work to the heat absorbed from the high-temperature reservoir,

*Efficiency*

$$e = \frac{W}{Q_1} = \frac{Q_1 - Q_2}{Q_1} = 1 - \frac{Q_2}{Q_1} \tag{5}$$

This says that if $Q_2 = 0$ (no waste heat), then the efficiency would be $e = 1$, or 100%. If so, the engine would convert the high-temperature heat *totally* into work. As we will see later, this extreme efficiency is unattainable. Even under ideal conditions, the engine will produce some waste heat. It turns out that the efficiency of an ideal engine depends only on the temperatures of the heat reservoirs.

EXAMPLE 2. The steam engine of a locomotive delivers $5.4 \times 10^8$ J of work per minute and receives $3.6 \times 10^9$ J of heat per minute from its boiler. What is the efficiency of this engine? How much heat is wasted per minute?

SOLUTION: From Eq. (5),

$$e = \frac{W}{Q_1} = \frac{5.4 \times 10^8 \text{ J}}{3.6 \times 10^9 \text{ J}} = 0.15$$

Expressed in percent, this is 15%.

The wasted heat is the difference between the heat received and the work:

$$Q_2 = Q_1 - W = 3.6 \times 10^9 \text{ J} - 5.4 \times 10^8 \text{ J} = 3.1 \times 10^9 \text{ J}$$

We will now calculate the efficiency of an ideal heat engine that converts heat into work with maximum efficiency. As we will see in the next section, for maximum efficiency, the theromodynamic process within the engine should be **reversible.** This means that the engine can, in principle, be operated in reverse, and it then converts work into heat at the same rate as it converts heat into work when operating in the forward direction. The simplest kind of reversible engine is the **Carnot engine,** consisting of some amount of ideal gas enclosed in a cylinder with a piston (see Figure 21.6). The cylinder can be heated by being placed in contact with the high-temperature reservoir, and it then delivers work as the hot gas expands and pushes the piston outward. To achieve reversibility with this engine, the motion of the piston must be sufficiently slow, so that the gas is always in an equilibrium configuration. If the piston were to have a sudden, fast motion, a pressure wave would travel through the gas, and the motion of this pressure wave could not be reversed by giving the piston a sudden motion in the opposite direction — this would merely create a second pressure wave. Furthermore, the temperature of the gas must coincide with the temperature of the heat reservoir during contact. If the gas were to have, say, a lower temperature than that of the heat reservoir with which it is in contact, heat would rush from the reservoir into the gas, and this flow of heat could not be reversed by any manipulation of the piston. For reversibility, the gas must remain in mechanical equilibrium with the piston and in thermal equilibrium with the heat reservoir.

*Carnot engine*

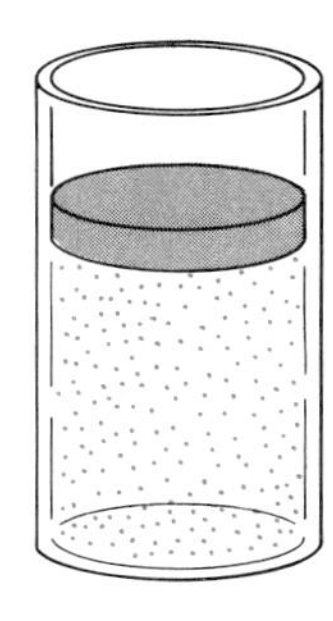

**Fig. 21.6** Carnot engine: a gas-filled cylinder with a piston.

The Carnot engine shown in Figure 21.6 can be put in thermal contact with either of the heat reservoirs of temperatures $T_1$ or $T_2$, so that heat can flow from the reservoir into the cylinder, or vice versa. The operation of the Carnot engine takes the gas through a sequence of four steps with varying volume and pressure, but at the end of the last step the gas returns to its initial volume and pressure. The four steps are illustrated in Figure 21.7. This sequence of four steps is called the **Carnot cycle:**

*Carnot cycle*

a. We begin the cycle by placing the cylinder in contact with the high-temperature heat reservoir. This maintains the temperature of the gas at the constant value $T_1$. The gas is now allowed to expand from the initial volume $V_1$ to a new volume $V_2$. During this isothermal expansion the gas does work on the piston, that is, the engine absorbs heat from the reservoir and converts it into work.
b. When the gas has reached volume $V_2$ and pressure $p_2$, we remove it

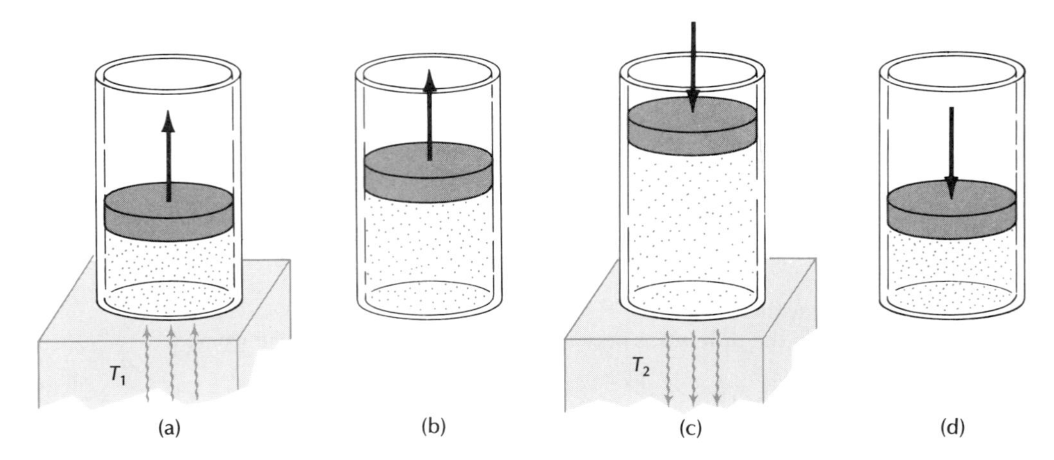


**Fig. 21.7** The Carnot cycle. The arrows indicate the displacements of the piston. (a) Expansion at constant temperaure $T_1$ while in contact with high-temperature heat reservoir. (b) Adiabatic expansion. (c) Compression at constant temperature $T_2$ while in contact with low-temperature reservoir. (d) Adiabatic compression to initial volume and pressure.

from the heat reservoir and allow it to continue the expansion adiabatically; during this expansion the temperature falls.

c. When the temperature has fallen to $T_2$, we stop the piston and place the gas in contact with the low-temperature heat reservoir. The volume at this instant is $V_3$ and the pressure is $p_3$. We now begin to push the piston back toward its starting position, that is, we compress the gas. During this isothermal compression, the engine converts work into heat and ejects this heat into the low-temperature reservoir.

d. When the gas has reached volume $V_4$ and pressure $p_4$, we remove it from the reservoir and continue to compress it adiabatically until the volume and the pressure return to their initial values $V_1$ and $p_1$.

For the mathematical description of the Carnot cycle, it is best to use a $p$–$V$ diagram (see Figure 21.8). Each point in this diagram represents an equilibrium configuration of the gas; $p$ and $V$ can be read directly from the diagram, and $T$ can then be calculated from the idea-gas equation. The initial volume and pressure of the gas are $V_1$ and $p_1$. In the first step of the Carnot cycle, the gas expands to $V_2$, $p_2$, in the second step, it expands to $V_3$, $p_3$. In the third step, the gas is compressed to $V_4$, $p_4$; and in the final step, it is compressed to the initial values $V_1$, $p_1$.

To find the efficiency, we need to calculate the heat $Q_1$ that the en-

**Sadi Carnot** (karno), *1796–1832, French engineer and physicist. In his book* On the Motive Power of Heat *he formulated the theory of the conversion of heat into work.*

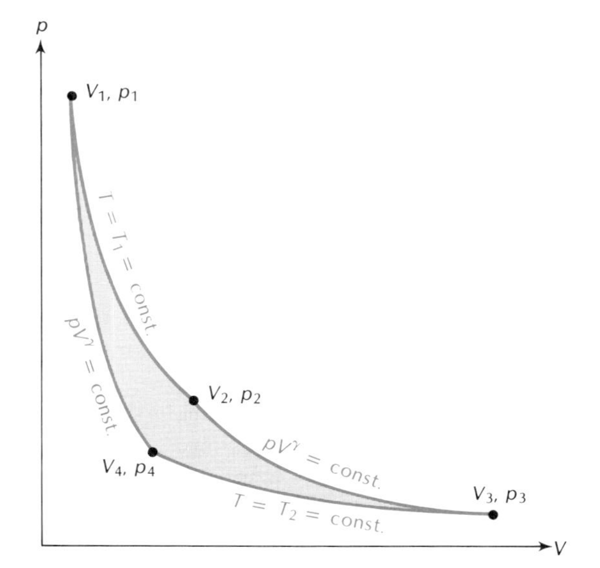


**Fig. 21.8** The Carnot cycle shown on a $p$–$V$ diagram.

gine absorbs from the high-temperature reservoir in step (a) and the heat $Q_2$ that it ejects into the low-temperature reservoir in step (c). During the heat-absorption process, the temperature is constant; since the energy $E$ of the ideal gas is a function of only the temperature, we see that $\Delta E = 0$ and, by Eq. (2), the absorbed heat $Q_1$ equals the work done by the gas,

$$Q_1 = \Delta W \tag{6}$$

During the expansion from $V_1$ to $V_2$, the pressure is

$$p = nRT_1/V \tag{7}$$

and the work done by the gas during a small displacement is

$$p\,dV = nRT_1\,dV/V \tag{8}$$

By integrating this, we find

$$\Delta W = \int p\,dV = \int_{V_1}^{V_2} nRT_1 \frac{dV}{V} = nRT_1 \ln\left(\frac{V_2}{V_1}\right) \tag{9}$$

Thus, the absorbed heat is

$$Q_1 = nRT_1 \ln\left(\frac{V_2}{V_1}\right) \tag{10}$$

Likewise, we readily find that during the compression, from $V_3$ to $V_4$, the ejected heat is

$$Q_2 = nRT_2 \ln\left(\frac{V_3}{V_4}\right) \tag{11}$$

The ratio of these heats is

$$\frac{Q_2}{Q_1} = \frac{T_2}{T_1}\frac{\ln(V_3/V_4)}{\ln(V_2/V_1)} \tag{12}$$

During the adiabatic expansion and the adiabatic compression of the gas, the pressure and volume obey the relation given in Eq. (20.33):

$$pV^\gamma = [\text{constant}] \tag{13}$$

By the ideal-gas law this is equivalent to

$$TV^{\gamma-1} = [\text{constant}] \tag{14}$$

Applying this to the adiabatic expansion and compression, we obtain, respectively,

$$T_1V_2^{\gamma-1} = T_2V_3^{\gamma-1} \tag{15}$$

and

$$T_1V_1^{\gamma-1} = T_2V_4^{\gamma-1} \tag{16}$$

Dividing these equations into each other, we find

$$\left(\frac{V_2}{V_1}\right)^{\gamma-1} = \left(\frac{V_3}{V_4}\right)^{\gamma-1} \tag{17}$$

or

$$\frac{V_2}{V_1} = \frac{V_3}{V_4} \tag{18}$$

It follows from this that the two logarithms in Eq. (12) are equal and they cancel. This leaves us with

$$\frac{Q_2}{Q_1} = \frac{T_2}{T_1} \tag{19}$$

The efficiency of the Carnot engine is therefore

*Efficiency of a Carnot engine*

$$\boxed{e = 1 - \frac{Q_2}{Q_1} = 1 - \frac{T_2}{T_1}} \tag{20}$$

This expresses the efficiency in terms of the temperatures of the heat reservoirs. Note that an efficiency of $e = 1$ (or 100%) can be achieved only if $T_2 = 0$, that is, if the low-temperature reservoir is at the absolute zero of temperature. Unfortunately, we have no such absolutely cold reservoir available; even if we could devise some clever method for using interstellar space as a heat dump, the temperature would still be $T_2 = 3$ K and not zero.

---

EXAMPLE 3. The boiler of a steam engine produces steam at a temperature of 500°C. The engine exhausts its waste heat into the atmosphere where the temperature is 20°C. The actual efficiency of this steam engine is 0.15. Compare this with the efficiency of a Carnot engine operating between the same temperatures.

SOLUTION: According to Eq. (20), the efficiency of a Carnot engine with $T_1 = (500 + 273)$ K and $T_2 = (20 + 273)$ K is

$$e = 1 - \frac{293\text{ K}}{773\text{ K}} = 0.62 \tag{21}$$

Thus, the ideal Carnot engine is more efficient than the actual steam engine by 0.47, or 47%.

---

In contrast to the Carnot engine, which uses gas as working fluid, practical steam engines use gas and liquid (steam and water) as working fluid, and the cycle of a steam engine differs from the Carnot cycle. Figure 21.9 is a schematic diagram of the main parts of a simple steam engine with boiler, cylinder, and condenser. The boiler produces hot, high-pressure steam which enters the cylinder and pushes against the piston and does work. The low-pressure, spent steam is then exhausted from the cylinder and sent to a condenser where an external coolant (air or flowing water) condenses the steam into liquid water. This liquid water is pumped back to the boiler. Each completed

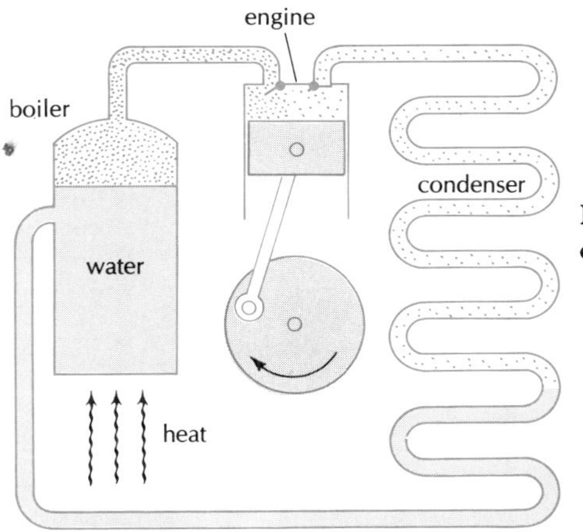


**Fig. 21.9** Schematic diagram of a steam engine.

circulation of the fluid through the circuit can be regarded as one cycle of operation of the steam engine. Such simple steam engines have actual efficiencies of only 5–18%. Most modern steam engines employ a turbine wheel instead of the cylinder and piston; large engines of this kind achieve efficiencies of up to 40%.

If the Carnot engine is operated in reverse, it uses up work to transfer heat from the low-temperature reservoir to the high-temperature reservoir. This is the principle involved in the operation of refrigerators, air conditioners, and "heat pumps." The amount of work required to operate a Carnot engine in reverse can be calculated from Eq. (19), which is valid whether the engine is operated in the forward direction or the reverse direction.

---

EXAMPLE 4. Suppose a homeowner uses a Carnot engine operating in reverse as a heat pump to extract heat from the outside air and inject it into his home. If the outside temperature is $-10°C$, and the inside temperature is $20°C$, what is the amount of work that must be supplied to pump 1 kcal of heat from the outside to the inside?

SOLUTION: According to Eq. (19), the ratio of the heats exchanged at the low-temperature reservoir and at the high-temperature reservoir is

$$\frac{Q_1}{Q_2} = \frac{T_2}{T_2} = \frac{263 \text{ K}}{293 \text{ K}} = 0.90 \tag{22}$$

If $Q_2 = 1$ kcal, then $Q_1 = Q_2/0.90 = 1.11$. The difference $Q_1 - Q_2$ represents the work that must be supplied; hence the work is 0.11 kcal.

COMMENTS AND SUGGESTIONS: Note that by the expenditure of 0.11 kcal of work, the heat pump delivers a total of 1.11 kcal of heat into the house. This is obviously a much more economical heating method than the expenditure of 1.11 kcal of fuel or electric energy in a conventional furnace or electric heater.

---

Practical refrigerators use gas and liquid freon (dichlorodifluoromethane) as working fluid, and their cycle differs from the Carnot cycle. Freon, and similar substances employed as refrigerants, have a boiling point near room temperature when at high pressure, but a boiling point below 0°C when at low pressure. Figure 21.10 is a schematic diagram of the parts of a practical refrigerator. Liquid freon at low pressure enters the cooling coils in the refrigerator box and absorbs heat while evaporating into freon gas. This gas flows to the com-

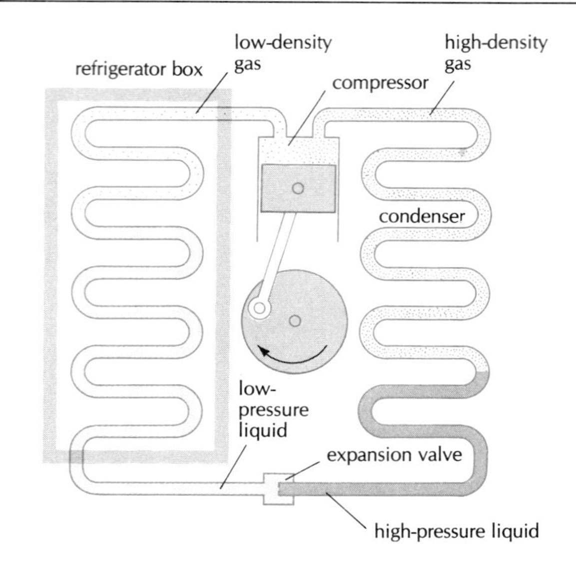


**Fig. 21.10** Schematic diagram of a refrigerator.

pressor, where its pressure and density are increased by the push of a piston; this heats the gas. The high-pressure, warm gas then circulates through the condensor coils, which are exposed to the atmospheric air. The freon gas loses its heat and condenses into a liquid. This high-pressure liquid then passes through an expansion valve (a small orifice) where its pressure is reduced to match the low pressure in the cooling coils. This return of the fluid to the cooling coils completes the cycle.

An air conditioner employs a similar refrigeration cycle; and a "heat pump" is, in essence, an air conditioner turned around, so its cold end is outdoors and its warm end indoors.

We can use the relation given by Eq. (20) between the efficiency of a Carnot engine and the temperatures of heat reservoirs for a determination of temperature. We need only to operate the Carnot engine between a heat reservoir at the unknown temperature and a heat reservoir at a known temperature (say, the temperature of the triple point of water); a direct measurement of the efficiency then determines the unknown temperature. The temperature scale defined by this procedure is the **Kelvin temperature scale,** also called the **absolute thermodynamic temperature scale.** The fundamental significance of this temperature scale lies in that it is independent of any special properties of any material. It depends only on the general laws of thermodynamics, and can be realized with any kind of reversible heat engine, because, as we will see in Section 21.4, any such heat engine has an efficiency equal to that of the Carnot engine.

*Absolute thermodynamic temperature scale*

The Kelvin scale coincides with the ideal-gas temperature scale, since, as is clear from the derivation leading to Eq. (20), the variables $T_1$ and $T_2$ are ideal-gas temperatures.

## 21.3 Entropy

The Carnot cycle described in the preceding section involves several steps of expansion and compression of an ideal gas. During each step the gas is kept either at constant temperature (isothermal process) or at constant energy (adiabatic process). It is easy to see that any reversible cycle involving any processes whatsoever can be regarded as a collec-

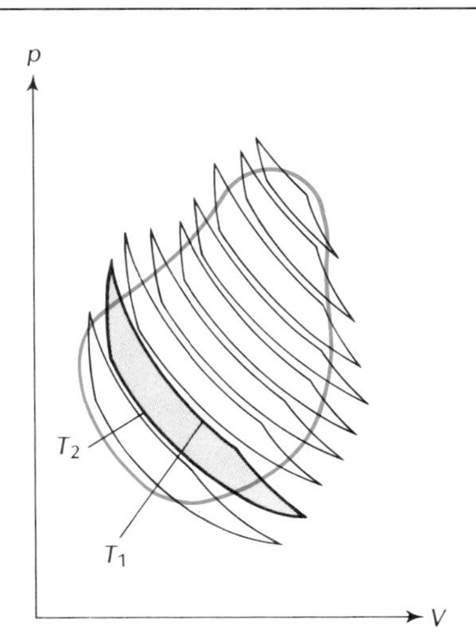


**Fig. 21.11** A reversible cycle shown on a $p$–$V$ diagram. This cycle can be approximated as a collection of small Carnot cycles. One of these small cycles operates between temperatures $T_1$ and $T_2$.

tion of Carnot cycles. In a reversible cycle, the gas goes through a sequence of equilibrium configurations, and the state of the gas at each instant can be represented by a point in a $p$–$V$ diagram. The complete cycle is then represented by a closed curve in the $p$–$V$ diagram. Figure 21.11 shows how this cycle can be approximated by a collection of Carnot cycles. Obviously, the approximation becomes exact in the limit of very many, very small Carnot cycles. The sum of the amounts of heat and work associated with the small cycles equals the amount of heat and work associated with the original cycle.

**Rudolph Clausius,** *1822–1888, German mathematical physicist, professor at Zurich and at Bonn. He was one of the creators of the science of thermodynamics. He contributed the concept of entropy, as well as the restatement of the Second Law of Thermodynamics.*

This equivalence between an arbitrary reversible cycle and a collection of small Carnot cycles leads to the following **theorem of Clausius:**

*The integral of $dQ/T$ around any reversible cycle is zero,*

*Theorem of Clausius (reversible process)*

$$\int \frac{dQ}{T} = 0 \tag{23}$$

In this equation, the change of heat is reckoned as positive if heat flows into the system and negative if it flows out. The proof of the theorem is simple: the cycle can be regarded as a collection of small Carnot cycles. Hence Eq. (23) will be true, provided that it is true for each Carnot cycle. Consider one of the Carnot cycles; it involves the absorption of heat $Q_1$ at temperature $T_1$ and the rejection of heat $Q_2$ at temperature $T_2$. Thus, the integral $\int dQ/T$ for this cycle may be expressed as

$$\int \frac{dQ}{T} = \frac{Q_1}{T_1} - \frac{Q_2}{T_2} \tag{24}$$

where the negative sign has been inserted in front of $Q_2$ because this is heat that *leaves* the system. According to Eq. (19), the right side of Eq. (24) is zero — this confirms Eq. (23) for each of the small Carnot cycles and hence establishes the Clausius theorem.

In the $p$–$V$ diagram, a reversible cycle is represented by a closed curve. The Clausius theorem therefore asserts that the integral of $dQ/T$ around any closed curve is zero. From this we can immediately

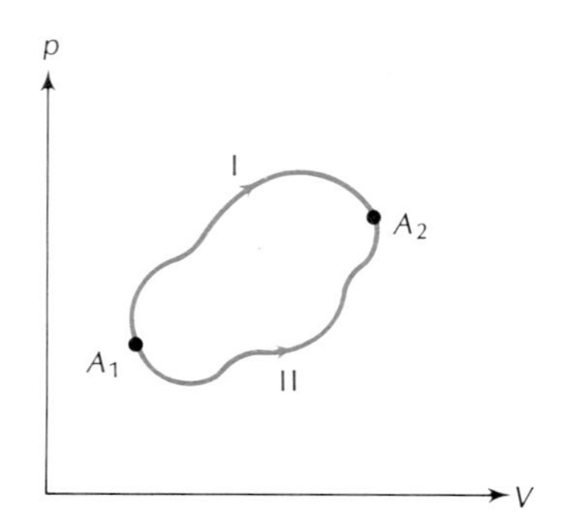


**Fig. 21.12** Two paths connecting the points $A_1$ and $A_2$ in a $p$–$V$ diagram.

deduce the following: if two points $A_1$ and $A_2$ in the $p$–$V$ diagram are connected by a curve (representing a reversible process), then the integral

$$\int_{A_1}^{A_2} \frac{dQ}{T} \tag{25}$$

depends only on the points $A_1$ and $A_2$, but not on the shape of the curve connecting these points. For a proof, consider two curves I and II connecting $A_1$ and $A_2$ (Figure 21.12). By the Clausius theorem the integral around the closed curve that starts at $A_1$, goes to $A_2$, and returns to $A_1$ must be zero:

$$0 = \underset{\text{(curve I)}}{\int_{A_1}^{A_2} \frac{dQ}{T}} + \underset{\text{(curve II)}}{\int_{A_2}^{A_1} \frac{dQ}{T}} \tag{26}$$

that is,

$$\underset{\text{(curve I)}}{\int_{A_1}^{A_2} \frac{dQ}{T}} = -\underset{\text{(curve II)}}{\int_{A_2}^{A_1} \frac{dQ}{T}} = \underset{\text{(curve II)}}{\int_{A_1}^{A_2} \frac{dQ}{T}} \tag{27}$$

Hence the integrals along curves I and II are indeed equal.

Obviously, this mathematical argument is quite analogous to that given in Section 8.1 in the study of conservative forces. There, the path independence of the integral of $\mathbf{F} \cdot d\mathbf{r}$ permitted us to define a potential-energy function [see Eq. (8.4)]. Likewise, the curve independence of the integral of $dQ/T$ permits us to define a new function whose value depends on position in the $p$-$V$ plane:

$$\boxed{S(A) = \int_{A_0}^{A} \frac{dQ}{T} + S(A_0)} \tag{28}$$

Here $A$ is an arbitrary point in the $p$–$V$ plane at which the function $S$ is being evaluated and $A_0$ is a fixed reference point at which the function $S$ has some prescribed standard value. The new function defined by Eq. (28) is called the **entropy.** The units of entropy are calories per kelvin (cal/K) or joules per kelvin (J/K). The entropy in thermodynamics plays a role somewhat analogous to that of the potential energy in mechanics. Just as the potential energy allows us to make some predictions about the possible motions of a mechanical system (see Section 8.4), the entropy allows us to make some predictions about the possible behavior of a thermodynamic system. And just as we can calculate the force in a mechanical system by differentiating the potential (see Section 8.3), we can calculate certain forcelike quantities in a thermodynamic system by differentiating the entropy.

*Entropy (reversible process)*

For an ideal gas, we can express the entropy as an explicit function of $p$ and $V$ [see Eq. (33) below]. Before we give a derivation of this function, we will look at two examples of how to use Eq. (28) for the calculation of changes of entropy.

EXAMPLE 5. One mole of ideal gas in a cylinder fitted with a piston is made to expand slowly (reversibly) from an initial volume $V_1 = 10^3$ cm$^3$ to a final volume $V_2 = 2 \times 10^3$ cm$^3$ (Figure 21.13). The cylinder is in contact with a heat reservoir so that, throughout the expansion process, the gas is held at constant temperature. What is the change of entropy of the gas?

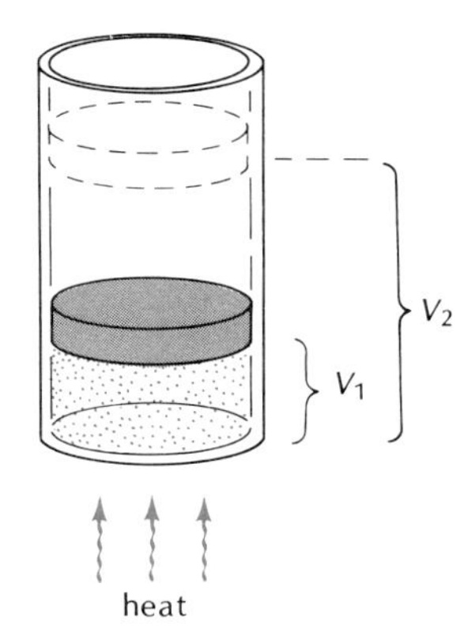


**Fig. 21.13** Expansion of a gas at constant temperature. The gas absorbs heat from the heat reservoir.

SOLUTION: The internal energy of an ideal gas is a function of the temperature only. Consequently, during an isothermal expansion the internal energy remains constant, that is, $dE = 0$. By Eq. (2), the heat absorbed from the reservoir will be

$$dQ = dW = p\, dV$$

and

$$S_2 - S_1 = \int \frac{dQ}{T} = \int_{V_1}^{V_2} \frac{p}{T} dV$$

From the ideal-gas law, $p/T = nR/V = R/V$ and hence

$$S_2 - S_1 = \int_{V_1}^{V_2} R \frac{dV}{V} = R \ln\left(\frac{V_2}{V_1}\right)$$

Numerically, this gives

$$S_2 - S_1 = 1.99 \text{ cal/K} \times \ln\left(\frac{2 \times 10^3 \text{cm}^3}{10^3 \text{cm}^3}\right)$$

$$= 1.38 \text{ cal/K}$$

EXAMPLE 6. One mole of ideal gas is initially contained in a thermally insulated bottle of volume $V_1 = 10^3$ cm$^3$. A pipe connects this bottle to a second evacuated bottle; the combined volume of both bottles is $V_2 = 2 \times 10^3$ cm$^3$ (Figure 21.14). If we suddenly open the stopcock and let the gas rush (irreversibly) into the second bottle so that it finally fills the volume of both bottles, what is the change of entropy of the gas?

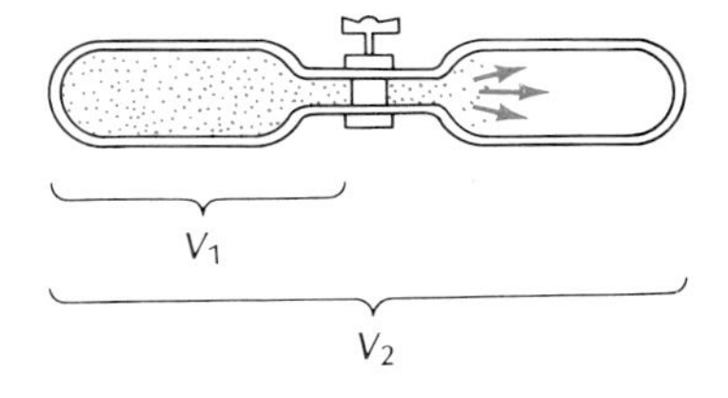


**Fig. 21.14** Free expansion of a gas.

SOLUTION: During the free expansion of an ideal gas, the internal energy remains constant and hence the temperature remains constant. Therefore the initial state and the final state of the freely expanding gas are exactly the same as for the isothermally expanding gas of Example 5. Since the entropy is a function of only the initial and final states, the entropy change of the freely expanding gas must be the same as in Example 5,

$$S_2 - S_1 = 1.38 \text{ cal/K}$$

COMMENTS AND SUGGESTIONS: Note that during a free expansion, the gas is not in equilibrium and the pressure and volume are not well defined. Hence this process cannot be represented by a curve in the $p$–$V$ plane, and we cannot directly apply Eq. (28) to this process. To calculate the entropy change, we must first find some reversible process that transforms the gas from the same initial state to the same final state as the free-expansion process. In the present example, isothermal expansion provides us with a suitable reversible process so that, by making a comparison, we can calculate the change of the entropy.

Let us now evaluate the entropy change of an ideal gas that expands from an arbitrary pressure and volume $p_1$, $V_1$ to $p_2$, $V_2$. For this evaluation, we need to consider a reversible process that transforms the gas from the initial state to the final state. Such a reversible process can be represented by a curve in the $p$–$V$ diagram, starting at the initial point $p_1$, $V_1$ and ending at the final point $p_2$, $V_2$. The entropy change is the integral of $dQ/T$ along this curve:

$$S_1 - S_2 = \int_{p_1,V_1}^{p_2,V_2} \frac{dQ}{T} \tag{29}$$

By Eqs. (2) and (20.8),

$$dQ = dE + p\,dV = nC_V\,dT + p\,dV \tag{30}$$

where $n$ is the number of moles of gas. Hence

$$S_2 - S_1 = \int nC_V \frac{dT}{T} + \int \frac{p}{T}\,dV \tag{31}$$

Substituting $p/T = nR/V$ into this, we obtain

$$S_2 - S_1 = \int nC_V \frac{dT}{T} + \int nR \frac{dV}{V} \tag{32}$$

Provided that $C_V$ is constant, both of these integrals are easy to do:

*Entropy of ideal gas*

$$S_2 - S_1 = nC_V \ln\left(\frac{T_2}{T_1}\right) + nR \ln\left(\frac{V_2}{V_1}\right) \tag{33}$$

The temperatures $T_2$ and $T_1$ can be expressed in terms of the pressures and volumes $(T = pV/nR)$, but it is often convenient to leave Eq. (33) as it is.

Note that Eq. (33) shows quite explicitly that the change of entropy does not depend on the detailed steps by which the gas changes from its initial state to its final state. It only depends on the initial and final values of the macroscopic parameters $T$ and $V$ (or $p$ and $V$).

We have so far discussed the entropy of only an ideal gas. But the Clausius theorem and the definition Eq. (28) of the entropy are valid in general, for any thermodynamic system consisting of any combination of solids, liquids, and gases. The state of any such system can be described by some macroscopic parameters analogous to $p$ and $V$, and reversible processes in the system can be described by curves in the "space" of these macroscopic parameters. The integral in Eq. (28) can then be evaluated along one of the curves connecting the initial and final states. Any curve or any process will give us the same value for the integral; we must only make sure that the process is reversible.

---

EXAMPLE 7. A heat reservoir at a temperature of $T_1 = 400$ K is briefly put in thermal contact with a reservoir at $T_2 = 300$ K. If 1 cal of heat flows from the hot reservoir to the cold reservoir, what is the change of the entropy of the system consisting of both reservoirs?

SOLUTION: Obviously, the heat flow from the hot to the cold reservoir is irreversible, and hence Eq. (28), which is restricted to reversible processes, is not directly applicable. To evaluate the entropy change by means of Eq. (28), we must invent some process that reversibly takes the system from the initial state to the final state. We can imagine that 1 cal flows from the hot reservoir into an auxiliary reservoir of a temperature just barely below 400 K, and that simultaneously 1 cal flows from another auxiliary reservoir of a temperature just barely above 300 K into the cold reservoir. Then all processes are reversible, and Eq. (28) gives us the change of entropy

$$S(A) - S(A_0) = -\frac{\Delta Q}{T_1} + \frac{\Delta Q}{T_2} = \Delta Q\left(\frac{1}{T_2} - \frac{1}{T_1}\right) \tag{34}$$

$$= 1 \text{ cal}\left(\frac{1}{300 \text{ K}} - \frac{1}{400 \text{ K}}\right) = 8.3 \times 10^{-4} \text{ cal/K}$$

COMMENTS AND SUGGESTIONS: Here we found that the entropy *increases* when heat flows from a hot reservoir to a cold reservoir. As we will see in the next section, all natural processes have a tendency to increase the entropy.

## 21.4 The Second Law of Thermodynamics

As we saw in Section 21.2, a heat engine operating with an ideal gas as working fluid has a limited efficiency — it fails to convert all the heat into work and instead produces some waste heat. The **Second Law of Thermodynamics** asserts that this is a limitation from which all heat engines suffer. As formulated by Kelvin and Max Planck,[1] this law simply states:

*Second Law of Thermodynamics (Kelvin–Planck)*

> *An engine operating in a cycle cannot transform heat into work without some other effect on its environment.*

It is an immediate corollary of this law that the efficiency of any heat engine operating between two heat reservoirs of high and low temperature is never greater than the efficiency of a Carnot engine. Furthermore, the efficiency of any reversible engine equals the efficiency of a Carnot engine. The proof of these statements, known as **Carnot's theorem,** is by contradiction. Imagine that a heat engine of very high efficiency converts heat from a reservoir at a high temperature into work and ejects only a small amount of waste heat into a reservoir at a low temperature. We can then use the work output of this engine to drive a Carnot engine in reverse, pumping the waste heat from the low-temperature reservoir back into the high-temperature reservoir. By hypothesis, the given engine is more efficient than the Carnot engine; hence only part of its work output will be needed to drive the reversed Carnot engine and return all of the waste heat to the high-temperature reservoir. The remainder of the output constitutes available work (see Figure 21.15 for a flow chart). The net effect of the joint operation of both engines is then the complete conversion of heat into work, without waste heat, in contradiction with the Second Law. We can avoid this contradiction only if the efficiency of any engine is never greater than that of a Carnot engine.

*Carnot's theorem*

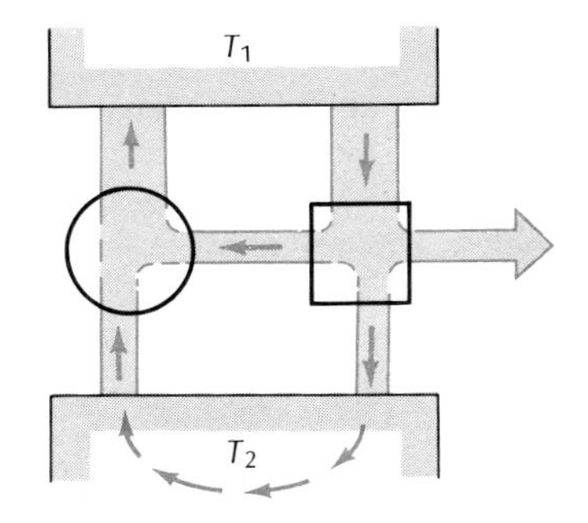


**Fig. 21.15** Flow chart for an arbitrary reversible engine (box) and a Carnot engine (circle) connected together. The arbitrary engine drives the Carnot engine in reverse.

To prove that the efficiency of any reversible engine equals that of a Carnot engine, we again consider the net effect of the joint operation

[1] See Section 42.2 for a short biography of Max Planck.

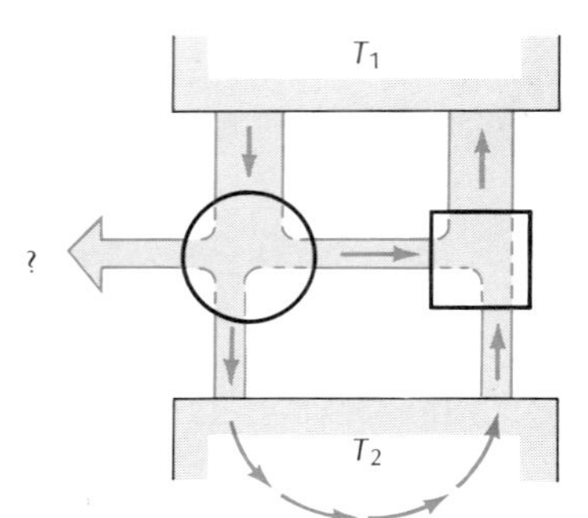


**Fig. 21.16** Flow chart for an arbitrary reversible engine (box) and a Carnot engine (circle) connected together. The Carnot engine drives the arbitrary engine in reverse.

of the two engines with the Carnot engine running in the forward direction and its work output driving the other reversible engine in the backward direction (Figure 21.16). By an argument similar to that given above, it now follows that the efficiency of the Carnot engine cannot be greater than that of the other engine. Thus, the efficiency of each engine can be no greater than that of the other, that is, they both must have exactly the same efficiency.

We recall that a perpetual motion machine of the second kind is an engine that takes heat energy from a reservoir and completely converts it into work. Thus, the Second Law asserts that no perpetual motion machines of the second kind exist. Essentially, the operation of any heat engine hinges on the temperature difference between two heat reservoirs. The heat in the high-temperature reservoir has a high "potential" for work, and that in the low-temperature reservoir has a low "potential." If we were to place the two reservoirs in thermal contact, heat would rush from the high-temperature to the low-temperature reservoir. By interposing a heat engine in the path of this rush of heat, we can force the heat to do useful work. Thus, heat engines depend on the tendency of heat to flow from a hot reservoir to a colder reservoir. The Second Law can be formulated in an alternative form that is based on this characteristic of the flow of heat. This formulation, due to Rudolf Clausius, states:

*Second Law of Thermodynamics (Clausius)*

> *An engine operating in a cycle cannot transfer heat from a cold reservoir to a hot reservoir without some other effect on its environment.*

This means that a refrigerator cannot be operated without some work input.

The Clausius and Kelvin–Planck formulations of the Second Law are equivalent — each implies the other. To see this, suppose that the Clausius statement is false and that there exists a refrigerator that can make heat flow from a cold to a hot reservoir without any work input. If such a device is operated jointly with an ordinary heat engine, it can be used to remove the waste heat from the cold reservoir and return it to the hot reservoir. The net effect would be the complete conversion of heat from the hot reservoir into work (Figure 21.17), in contradiction to the Kelvin–Planck statement. Conversely, suppose the Kelvin–Planck statement is false and a heat engine exists that converts heat from a hot reservoir completely into work. Then the work output of this engine can be used to operate an ordinary refrigerator and the net effect would be the transfer of heat from a cold reservoir to the hot reservoir (Figure 21.18), in contradiction to the Clausius statement.

As we saw in Example 7, a flow of heat from a hot reservoir to a cold reservoir leads to an increase of entropy. This suggests that we express

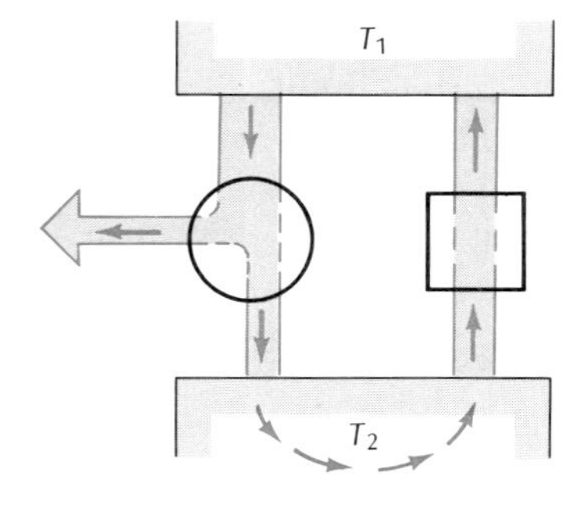


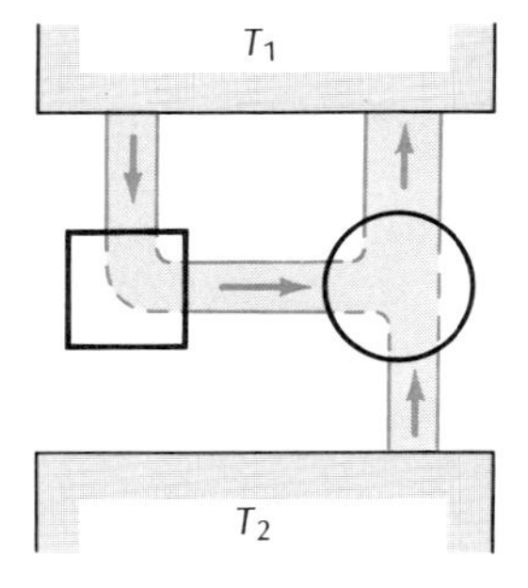


**Fig. 21.17** (left) Flow chart for a hypothetical refrigerator (box) and a Carnot engine (circle). This refrigerator pumps heat from the cold to the hot reservoir without work input.

**Fig. 21.18** (right) Flow chart for a hypothetical engine (box) driving a Carnot engine (circle) in reverse. This engine converts heat completely into work, and the Carnot engine acts as a refrigerator.

the Second Law in terms of changes of the entropy. To do that, we need the following theorem, a generalization of the theorem of Clausius:

*The integral of $dQ/T$ for any cyclic irreversible process is less than or equal to zero,*

$$\oint \frac{dQ}{T} \leq 0 \tag{35}$$

The proof is quite similar to the proof for reversible cycles given in Section 21.3. Figure 21.19 shows an arbitrary irreversible cycle; the axes in this figure have been labeled with $p$ and $V$, but they could equally well represent some other macroscopic parameters describing an arbitrary thermodynamic system. The cycle consists of a reversible portion indicated by a solid line and an irreversible portion indicated symbolically by the dotted line; this latter portion involves nonequilibrium processes and cannot be represented as a curve in the $p$–$V$ plane. The cycle can be approximated by a collection of small irreversible cycles consisting of adiabatic portions, isothermal portions, and irreversible (nonequilibrium) portions (Figure 21.20). Consider one of these cycles involving the absorption of heat $Q_1$ at temperature $T_1$ and the rejection of heat $Q_2$ at temperature $T_2$. As in Eq. (24), we can express the integral $\oint dQ/T$ for this cycle as

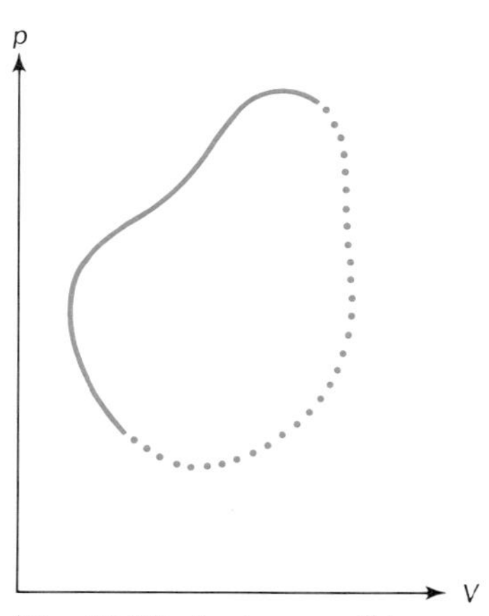


**Fig. 21.19** An irreversible cycle. The dotted portion of the path cannot be represented as a curve in the $p$–$V$ diagram.

$$\oint \frac{dQ}{T} = \frac{Q_1}{T_1} - \frac{Q_2}{T_2} \tag{36}$$

By the Second Law of Thermodynamics, the efficiency of this cycle can at most equal that of a Carnot cycle operating between the same temperatures. For the Carnot cycle, $Q_1/T_1 - Q_2/T_2 = 0$ [see Eq. (19)]. For any other cycle, such as our cycle, the ratio $Q_2/Q_1$ or the ratio $T_1/T_2$ which determines the efficiency [see Eq. (20)] is no smaller than that for the Carnot cycle, and therefore $Q_1/T_1 - Q_2/T_2 \leq 0$. Thus, the right side of Eq. (36) is less than or equal to zero. This establishes the inequality in Eq. (35) for each small cycle, and hence it also establishes this inequality for the entire irreversible cycle.

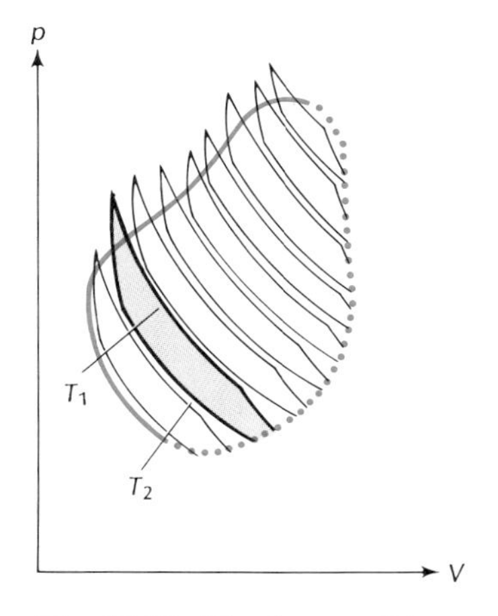


**Fig. 21.20** The irreversible cycle of Figure 21.19 can be approximated by a collection of small irreversible cycles.

Let us now see what Eq. (35) says about the entropy change associated with an irreversible process. Suppose that a system in a state $A$ suffers some irreversible process that brings it to a state $B$. We can then imagine some reversible process that takes the system back to state $A$; in Figure 21.21 this reversible process is represented by a solid line. For the complete cycle $A \to B \to A$, we have

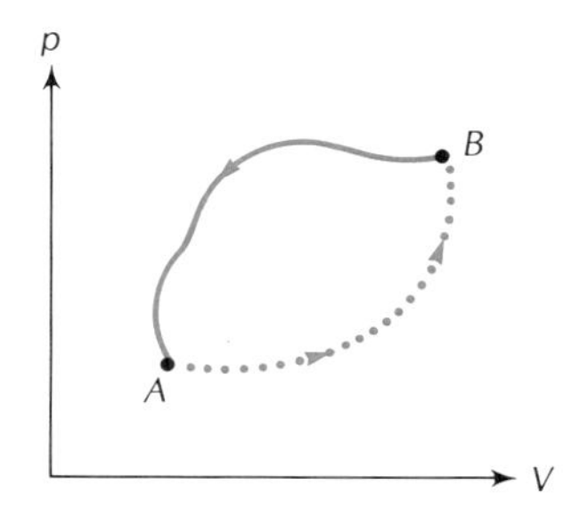


**Fig. 21.21** An irreversible process takes the system from the point $A$ to the point $B$. A reversible process returns the system from $B$ to $A$.

$$\int \frac{dQ}{T} \leq 0 \tag{37}$$

or

$$\underset{\text{(irreversible)}}{\int_A^B \frac{dQ}{T}} + \underset{\text{(reversible)}}{\int_B^A \frac{dQ}{T}} \leq 0 \tag{38}$$

But, by the definition of entropy,

$$\underset{\text{(reversible)}}{\int_B^A \frac{dQ}{T}} = S(A) - S(B) \tag{39}$$

so the inequality in Eq. (38) becomes, with some rearrangement,

*Entropy (irreversible process)*

$$\boxed{S(B) - S(A) \geq \underset{\text{(irreversible)}}{\int_A^B \frac{dQ}{T}}} \tag{40}$$

This inequality may be regarded as a reformulation of the Second Law of Thermodynamics. In the particular case of an isolated system which exchanges no heat with its surroundings ($dQ = 0$), the inequality becomes

$$\boxed{S(B) - S(A) \geq 0} \tag{41}$$

This says that *the entropy of an isolated system never decreases* — it either increases or stays constant.

We have already come across two examples of such an increase of entropy in irreversible processes in isolated systems. The first was the spontaneous expansion of a gas released from a bottle (see Example 6) and the second was the spontaneous flow of heat from a hot reservoir to a cold reservoir (see Example 7). In both of these cases, we found by explicit calculation that the entropy of the final configuration was larger than that of the initial configuration.

It is easy to think of other examples. For instance, consider the effect of friction forces on a mechanical system, such as a stone sliding down a hill (in this case the system consists of the stone and the environment). Friction converts mechanical energy into heat; this is obviously an irreversible process and it makes a positive contribution to the right side of Eq. (40). Thus, it brings about an increase of the entropy of the system.

---

EXAMPLE 8. A large stone, of mass 80 kg, slides down a hill of a vertical height of 100 m and stops at the bottom. What is the increase of entropy of the stone plus the environment? Assume that the temperature of the environment (hill and air) is 270 K.

SOLUTION: All of the initial mechanical energy of the stone is converted into heat:

$$\Delta Q = mgz = 80 \text{ kg} \times 9.8 \text{ m/s}^2 \times 100 \text{ m}$$

$$= 7.8 \times 10^4 \text{ J} = 1.87 \times 10^4 \text{ cal} \qquad (42)$$

This heat is delivered to the environment at a temperature of 270 K.

Obviously, the process described in this example is irreversible. To calculate the entropy from Eq. (28), we must imagine a reversible process that brings the stone down the hill and delivers the heat into the environment. In principle, we can use an elevator that lets the stone down slowly without friction and extracts work while removing the potential energy; and afterward we can use a heat reservoir of a temperature barely above 270 K to reversibly supply the correct amount of heat [Eq. (42)] to the stone's environment. The first of these processes makes no contribution to the entropy and the second makes a contribution

$$S(B) - S(A) = \frac{\Delta Q}{T} = \frac{1.87 \times 10^4 \text{ cal}}{270 \text{ K}} = 69 \text{ cal/K}$$

---

From a microscopic point of view, the increase of the entropy of a system is an increase of disorder. This is very obvious in the preceding example of the conversion of mechanical energy into heat by friction. The translational kinetic energy of a macroscopic body is ordered energy — all the particles in the body move in the same direction with the same speed. Heat is disordered energy — the particles move in random directions with a mixture of speeds.

The increase of disorder also holds true in our other examples of spontaneous expansion of a gas and spontaneous flow of heat. When a gas is confined to a small volume, it has more order than when confined to a larger volume; we can best recognize this if we imagine that the small volume is *very* small — if all the particles of gas are at just about one point, the system clearly has a very high degree of order. When a given amount of heat is added to a reservoir of low temperature, it causes more disorder than when the same amount of heat is added to a reservoir of high temperature; in both reservoirs the added heat generates extra random motions and extra disorder, but in the cold reservoir the percent increment of the random motions is larger than in the hot reservoir, and consequently the extra disorder is larger.

The connection between entropy and disorder can be given a precise mathematical meaning, but the details would require a discussion of statistical mechanics and of information theory — and here we cannot go into this. The Second Law can be reformulated to say that the processes in a closed, isolated system always tend to increase the disorder.

The human activities on the Earth, like any other processes in nature, cause an increase of disorder. Of course, some of our activities result in an increase of order in some portion of the system. For instance, when we extract dispersed bits of metal from ores and assemble them into a watch, we are obviously increasing the order of the bits of metal. However, we can do this only by simultaneously generating disorder somewhere else — the smelting of ores and the machining of metals demands an input of energy which is converted into waste heat, and increases the disorder of our environment. The net result is always an increase of disorder. All our activities depend on a supply of highly ordered energy in the form of chemical or nuclear fuels or light from the Sun that can "soak up" disorder while becoming degraded into waste heat. Our activities continually convert useful, ordered energy into useless, disordered energy.

We end this chapter with a brief statement of the **Third Law of Thermodynamics.** This law, as formulated by Walther Nernst, asserts that

*Third Law of Thermodynamics*

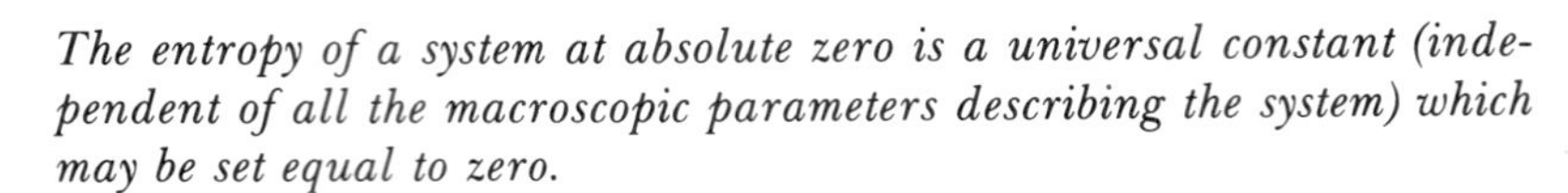

> *The entropy of a system at absolute zero is a universal constant (independent of all the macroscopic parameters describing the system) which may be set equal to zero.*

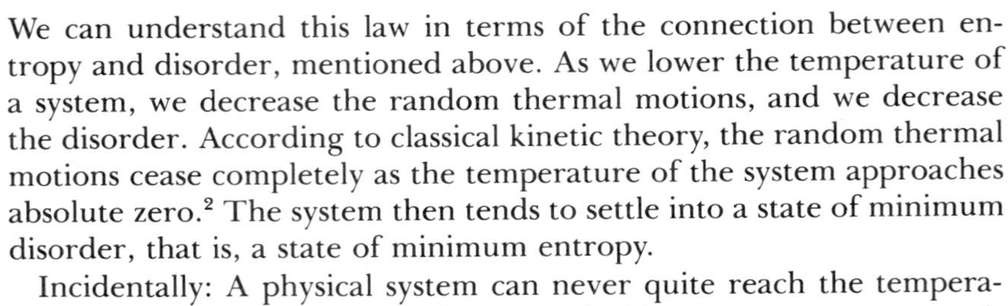

We can understand this law in terms of the connection between entropy and disorder, mentioned above. As we lower the temperature of a system, we decrease the random thermal motions, and we decrease the disorder. According to classical kinetic theory, the random thermal motions cease completely as the temperature of the system approaches absolute zero.[2] The system then tends to settle into a state of minimum disorder, that is, a state of minimum entropy.

**Walther Hermann Nernst,** *1864–1941, German physicist and chemist. He was a pioneer in physical chemistry and received the Nobel Prize in chemistry in 1920 for his discovery of the Third Law of Thermodynamics.*

Incidentally: A physical system can never quite reach the temperature of absolute zero; it can only approach this temperature asymptotically. This unattainability of the absolute zero can be shown to be mathematically equivalent to the vanishing of the entropy at absolute zero. Hence, in an alternative formulation, the Third Law of Thermodynamics asserts that the absolute zero of temperature is unattainable. The lowest temperature achieved in laboratory experiments to date is $3 \times 10^{-8}$ K.

## SUMMARY

**First Law of Thermodynamics:** $\Delta E = \Delta Q - \Delta W$

**Efficiency of heat engine:** $e = 1 - \dfrac{Q_2}{Q_1}$

**Efficiency of Carnot engine:** $e = 1 - \dfrac{T_2}{T_1}$

**Carnot's theorem:** The efficiency of any engine cannot exceed that of a Carnot engine; the efficiency of any reversible engine equals that of a Carnot engine.

**Clausius theorem (reversible cycle):** $\displaystyle\int \frac{dQ}{T} = 0$

**Definition of entropy:** $S(A) = \underset{\text{(reversible)}}{\displaystyle\int_{A_0}^{A} \frac{dQ}{T}} + S(A_0)$

**Second Law of Thermodynamics:** An engine operating in a cycle cannot transform heat into work without some other effect on its environment *or* an engine operating in a cycle cannot transfer heat from a

[2] Actually, owing to quantum effects, the motion never ceases completely — at absolute zero there remains a zero-point motion. But this zero-point motion is not a disordered motion, and it does not contribute to the entropy.

cold reservoir to a hot reservoir without some other effect on its environment.

**Clausius theorem (irreversible cycle):** $\int \frac{dQ}{T} \leq 0$

**Change of entropy in irreversible process:** $S(B) - S(A) \geq \underset{\text{(irreversible)}}{\int_A^B \frac{dQ}{T}}$

**Third Law of Thermodynamics:** At absolute zero the entropy is zero.

## QUESTIONS

1. Figure 21.22 shows a perpetual motion machine designed by M. C. Escher. Exactly at what point is there a defect in this design?

**Fig. 21.22** Waterfall. Lithograph by M. C. Escher.

2. An inventor proposes the following scheme for the propulsion of ships on the ocean without an input of energy: cover the hull of the ship with copper sheets and suspend an electrode of zinc in the water at some distance from the hull; since sea water is an electrolyte, the hull of the ship and the electrode will then act as the terminals of a battery which can deliver energy to an electric motor propelling the ship. Will this scheme work? Does it violate the First Law of Thermodynamics?

3. The **"Zeroth" Law of Thermodynamics,** found in some textbooks, states that *whenever two bodies are individually in thermal equilibrium with a third body, then they are also in thermal equilibrium with each other.* If we take as the third body the ideal-gas thermometer described in Section 19.2, then the "Zeroth" Law states that two bodies at the same ideal-gas temperature are in thermal equilibrium with each other. Show that if this assertion were not true, then the Second Law could not be true; thus the "Zeroth" Law is implicitly contained in the Second Law. (Hint: Consider two bodies of the same initial temperatures in thermal contact; if heat flows from one to the other, one will cool down and the other will heat up.)

4. An electric power plant consists of a coal-fired boiler that makes steam, a turbine, and an electric generator. The boiler delivers 90% of the heat of combustion of the coal to the steam; the turbine converts 50% of the heat of the steam into mechanical energy, and the electric generator converts 99% of this mechanical energy into electric energy. What is the overall efficiency of generation of electric power?

5. The lowest-temperature heat sink available in nature is interstellar and intergalactic space with a temperature of 3 K. Why don't we use this heat sink in the operation of a Carnot engine?

6. In some showrooms, salesmen demonstrate air conditioners by simply plugging them into an outlet without bothering to install them in a window or a wall. Does such an air conditioner cool the showroom or heat it?

7. Can mechanical energy be converted completely into heat? Give some examples.

8. If you leave the door of a refrigerator open, will it cool the kitchen?

9. Which of the following processes are irreversible, which are reversible? (a) Burning of a piece of paper. (b) Slow descent of an elevator attached to a perfectly balanced counterweight. (c) Breaking of a windowpane. (d) Explosion of a stick of dynamite. (e) Lifting of water by an electric pump from a low-level reservoir to a high-level reservoir. (f) Slow pushing of a block up a frictionless inclined plane.

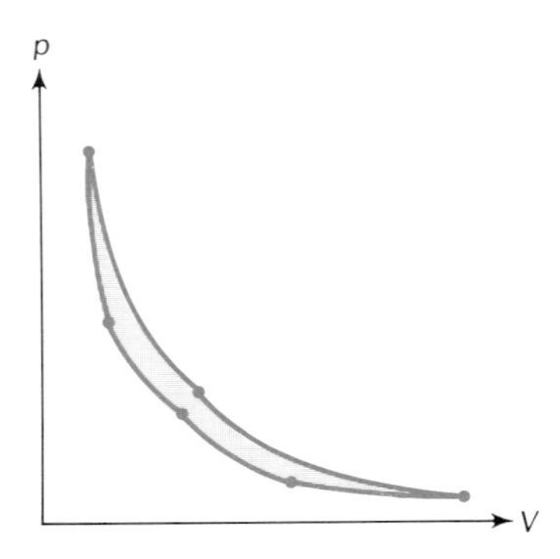


**Fig. 21.23** A modified Carnot cycle.

10. Consider a modified Carnot cycle consisting of an isothermal expansion, an adiabatic expansion, an isothermal compression, an adiabatic compression, another isothermal compression, and another adiabatic compression (see Figure 21.23). Does the efficiency of this cycle differ from that of the standard Carnot cycle?

11. If we measure the efficiency of a Carnot engine directly, we can use Eq. (20) to calculate the temperature of one of the heat reservoirs. What are the advantages and the disadvantages of such a thermodynamic determination of temperature?

12. In lack of a heat sink, energy cannot be extracted from an ocean at uniform temperature. However, within the oceans there are small temperature differences between the warm water on the surface and the cooler water in the depths. Design a Carnot engine that exploits this temperature difference.

13. Why are heat pumps used for heating houses in mild climates but not in very cold climates?

14. Does the Second Law of Thermodynamics forbid the spontaneous flow of heat between two bodies of equal temperature?

15. The inside of an automobile parked in the sun becomes much hotter than the surrounding air. Does this contradict the Second Law of Thermodynamics?

16. According to a story by George Gamow,[3] on one occasion Mr. Tompkins was drinking a highball when all of a sudden one small part at the surface of the liquid became hot and boiled with violence, releasing a cloud of steam, while the remainder of the liquid became cooler. Is this consistent with the First Law of Thermodynamics? With the Second Law?

17. A vessel is divided into two equal volumes by a partition. One of these volumes contains helium gas, and the other contains argon at the same temperature and pressure. If we remove the partition and allow the gases to mix, does the entropy of the system increase?

18. Does the motion of the planets around the Sun generate entropy?

19. Does static friction generate entropy?

20. Consider the process of emission of light by the surface of the Sun followed by absorption of this light by the surface of the Earth. Does this entail an increase of entropy?

21. Suppose that a box contains gas of extremely low density, say, only 50 molecules of gas altogether. The molecules move at random and it is possible that once in a while all of the 50 molecules are simultaneously in the left half of the box, leaving the right half empty. At this instant the entropy of the gas is less than when the gas is more or less uniformly distributed throughout the box. Is this a violation of the Second Law of Thermodynamics? What are the implications for the range of validity of this law?

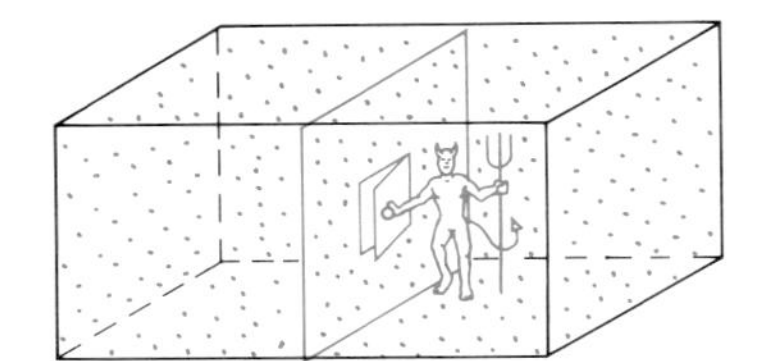

**Fig. 21.24** Maxwell's demon.

22. **Maxwell's demon** is a tiny hypothetical creature that can see individual molecules. The demon can make heat flow from a cold body to a hot body as follows: Suppose that a box initially filled with gas at uniform temperature and pressure is divided into two equal volumes by a partition, equipped with a small door which is closed but can be opened by the demon (Figure 21.24). Whenever a molecule of above-average speed approaches the door from the left, the demon quickly opens the door and lets it through. Whenever a molecule of below-average speed approaches from the right, the demon also lets it through. This selective action of the demon accumulates hot gas in the right volume, and cool gas in the left volume. Does this violate the Second Law of Thermodynamics? Do any of the activities of the demon involve an *increase* of entropy?

[3] G. Gamow, *Mr. Tompkins in Paperback.*

23. According to one cosmological model based on Einstein's theory of General Relativity, the universe oscillates — it expands, then contracts, then expands, and then contracts, etc., etc. If the Second Law of Thermodynamics is valid, can each cycle of oscillation be the same as the preceding cycle?

24. **Negentropy** is defined as the negative of the entropy ([negentropy] $= -S$). Explain the following statement: "In our everyday activities on the Earth, we do not consume energy, but we consume negentropy."

25. The amount of energy dissipated in the United States per year is $8 \times 10^{19}$ J. Roughly, what is the entropy increase that results from this dissipation?

26. If you tidy up a messy room you are producing a decrease of disorder. Does this violate the Second Law?

## PROBLEMS

### Section 21.1

1. A large closed bag of plastic contains 0.10 m$^3$ of an unknown gas at an initial temperature of 10°C and at the same pressure as the surrounding atmosphere, 1.0 atm. You place this bag in the sun and let the gas warm up to 38°C and expand to 0.11 m$^3$. During this process, the gas absorbs 840 cal of heat. Assume the bag is large enough so the gas never strains against it, and therefore remains at a constant pressure of 1.0 atm.
   (a) How many moles of gas are in the bag?
   (b) What is the work done by the gas in the bag against the atmosphere during the expansion?
   (c) What is the change in the internal energy of the gas in the bag?
   (d) Is the gas a monoatomic gas? A diatomic gas?

2. A ball of lead of mass 0.25 kg drops from a height of 0.80 m, hits the floor, and remains there at rest. Assume that all the heat generated during the impact remains within the lead. What are the values of $\Delta Q$, $\Delta W$, and $\Delta E$ for the lead during this process? What is the increase of temperature of the lead?

*3. Consider 1 kg of water at 100°C surrounded by air at a constant pressure of 1 atm. If we vaporize all of this water and convert it into steam at 100°C and 1 atm pressure, what is the heat absorbed by the water? The work done by the water vapor against atmospheric pressure? The change of the internal energy of the water? Assume that steam obeys the ideal-gas law.

*4. Calculate the heat of vaporization of water at a temperature of 140°C, at a pressure of 1 atm. Use the data given in Example 1.

*5. Calculate the heat of vaporization of water at a temperature of 100°C, at a pressure of 2 atm. Assume that water vapor behaves as an ideal gas.

### Section 21.2

6. A coal-burning power plant uses thermal energy at a rate of 850 megawatts and produces 300 megawatts of mechanical power for the generation of electricity. What is the efficiency of this power plant?

7. In an automobile proceeding at medium speed, the engine delivers 20 hp of mechanical power. The engine burns gasoline which provides thermal energy at the rate of 15 kilocalories per second. What is the efficiency of the engine under these conditions? What is the rate at which the engine ejects waste heat?

8. While running up stairs at a (vertical) rate of 0.30 m/s, a man of 70 kg generates waste heat at a rate of 300 cal/s. What efficiency for the human body can you deduce from this?

9. A steam turbine in a power plant converts the thermal energy of hot steam into mechanical energy. The turbine consists of a high-pressure stage and a low-pressure stage operating in tandem. The steam first passes through the high-pressure stage and gives up some of its thermal energy. The waste heat from this stage is used to reheat the steam, which then enters the low-pressure stage where it gives up some of its remaining thermal energy.

(a) Draw a flow chart showing the two stages of the turbine and the heat and work inputs and outputs.

(b) If the efficiency of the high-pressure stage is 40% and the efficiency of the low-pressure stage is 20%, what is the overall efficiency of the turbine?

10. Each of the two engines of a DC-3 airplane produces 1100 hp. The engines consume gasoline; the combustion of 1 kg of gasoline yields $44 \times 10^6$ J. If the efficiency of the engines is 20%, at what rate do the two engines consume gasoline?

*11. A nuclear power plant generates 1000 megawatts of electric (or mechanical) power. If the efficiency of this plant is 33%, at what rate does the plant generate waste heat? If this waste heat is to be removed by passing water from a river through the plant, and if the water is to suffer a temperature increase of at most 8°C, how many cubic meters of water per second are required?

*12. Equation 20.37 gives the work done during the adiabatic expansion (or contraction) of a gas. Use this equation to calculate the amount of work done by the gas during the adiabatic expansion and the contraction in the Carnot cycle described in Figure 21.8. Verify explicitly that the amount of the work for the complete Carnot cycle coincides with $Q_1 - Q_2$.

*13. A gun may be regarded as a thermodynamic engine. The barrel of the gun is a cylinder; the shot moving along the barrel is a piston; and the hot gases released by the combustion of the explosive charge do work on this piston, giving it a large kinetic energy. During the first few instants after the detonation of the charge, there is simultaneous combustion and expansion; but after the combustion is complete, the expansion proceeds adiabatically. Experiments with a gun with a barrel of length 3.7 m and diameter 15 cm firing a shot of mass 45 kg have shown that when the combustion ends, the shot is 1.0 m along the barrel and has a speed of 311 m/s; the pressure of the gas at this instant is $2.2 \times 10^8$ N/m$^2$.

(a) How much work does the adiabatically expanding gas do on the shot as it travels the remaining 2.7 m along the barrel? Assume that $\gamma = 1.2$.

(b) What will be the muzzle speed of the shot? Ignore friction.

*14. A Carnot engine has 40 g of helium gas as its working substance. It operates between two reservoirs of temperatures $T_1 = 400$°C and $T_2 = 60$°C. The initial volume of the gas is $V_1 = 2.0 \times 10^{-2}$ m$^3$, the volume after isothermal expansion is $V_2 = 4.0 \times 10^{-2}$ m$^3$, and the volume after the subsequent adiabatic expansion is $V_3 = 5.0 \times 10^{-2}$ m$^3$.

(a) Calculate the work generated by this engine during one cycle.

(b) Calculate the heat absorbed and the waste heat ejected during one cycle.

15. A Carnot engine operates between two heat reservoirs of temperature 500°C and 30°C, respectively.

(a) What is the efficiency of this engine?

(b) If the engine generates $1.5 \times 10^3$ J of work, how many calories of heat does it absorb from the hot reservoir? Eject into the cold reservoir?

16. In principle, nuclear reactions can achieve temperatures of the order of $10^{11}$ K. What is the efficiency of a Carnot engine taking in heat from such a nuclear reaction and exhausting waste heat at 300 K?

17. An automobile engine takes heat from the combustion of gasoline, converts part of this heat into mechanical work, and ejects the remainder into the atmosphere. The temperature attainable by the combustion of gasoline is

about 2100°C and the temperature of the atmosphere is 20°C. What is the efficiency of a Carnot engine operating between these temperatures? (The actual efficiency attained by an automobile engine is typically 0.2, much lower than the theoretical maximum.)

18. On a hot day a house is kept cool by an air conditioner. The outside temperature is 32°C and the inside temperature is 21°C. Heat leaks into the house at the rate of 9000 kcal/h. If the air conditioner has the efficiency of a Carnot engine, what is the mechanical power that it requires to hold the inside temperature constant?

19. A scheme for the extraction of energy from the oceans attempts to take advantage of the temperature difference between the upper and lower layers of ocean water. The temperature at the surface in tropical regions is about 25°C; the temperature at a depth of 300 m is about 5°C.

(a) What is the efficiency of a Carnot engine operating between these temperatures?

(b) If a power plant operating at the maximum theoretical efficiency generates 1 megawatt of mechanical power, at what rate does this power plant release waste heat?

(c) The power plant obtains the mechanical power and the waste heat from the surface water by cooling this water from 25°C to 5°C. At what rate must the power plant take in surface water?

20. In a nuclear power plant, the reactor produces steam at 520°C and the cooling tower eliminates waste heat into the atmosphere at 30°C. The power plant generates 500 megawatts of electric (or mechanical) power.

(a) If the efficiency is that of a Carnot engine, what is the rate of release of waste heat (in megawatts)?

(b) Actual efficiencies of nuclear power plants are about 33%. For this efficiency, what is the rate of release of heat?

21. An air conditioner removes 2000 kcal/h of heat from a room at a temperature of 21°C and ejects this heat into the ambient air at a temperature of 27°C. This air conditioner requires 950 W of electric power.

(a) How much mechanical power would a Carnot engine, operating in reverse, require to remove this heat at the same rate?

(b) By what factor is the power required by the air conditioner larger than that required by the Carnot engine?

*22. An ice-making plant consists of a reversed Carnot engine extracting heat from a well-insulated ice box. The temperature in the icebox is −5°C and the temperature of the ambient air is 30°C. Water, of an initial temperature of 30°C, is placed in the icebox and allowed to freeze and to cool to −5°C. If the ice-making plant is to produce 10,000 kg of ice per day, what mechanical power is required by the Carnot engine?

*23. The boiler of a power plant supplies steam at 540°C to a turbine which generates mechanical power. The steam emerges from the turbine at 260°C and enters a steam engine that generates extra mechanical power. The steam is finally released into the atmosphere at a temperature of 38°C. Assume that the conversion of heat into work proceeds with the efficiency of a Carnot engine.

(a) What is the efficiency of the turbine? Of the steam engine?

(b) What is the net efficiency of both engines acting together? How does it compare with the efficiency of a single engine operating between 540°C and 38°C?

**24. One liter of water is initially at the same temperature as the surrounding air, 30°C. You wish to cool this water to 5°C by transferring heat from the water into the air. What is the minimum amount of work that you must supply in order to accomplish this? (Hint: Take a sequence of Carnot engines operating in reverse; use each engine to reduce the temperature by an infinitesimal amount.)

### Sections 21.3 and 21.4

25. Using Eq. (33), prove that in the adiabatic expansion of an ideal gas, the final entropy equals the initial entropy.

26. On a winter day heat leaks out of a house at the rate of $2.5 \times 10^4$ kcal/h. The temperature inside the house is 21°C and the temperature outside is −5°C. At what rate does this process produce entropy?

27. Consider the air conditioner described in Problem 21. Calculate the rate of increase of entropy contributed by the operation of this air conditioner.

28. Your body generates about 2000 kcal of heat per day. Estimate how much entropy you generate per day. Neglect the (small) amount of entropy that enters your body in the food you consume.

29. A steam engine operating between reservoirs at temperatures of 480°C and 27°C has an efficiency of 40%. The engine delivers 2000 hp of mechanical power. At what rate does this engine generate entropy?

30. Suppose that 1.0 kg of water freezes while at 0°C. What is the change of entropy of the water during this freezing process?

31. What is the increase of entropy of 1.0 kg of water when it vaporizes at 100°C and 1 atm?

32. A parachutist of 80 kg descends at a constant speed of 5 m/s. What is the rate of increase of entropy of the parachute and the environment? The air temperature is 20°C.

33. For an automobile moving at a constant speed of 65 km/h on a level road, rolling friction, air friction, and friction in the drive train absorb a mechanical power of 12 kW. At what rate do these processes generate entropy? The temperature of the environment is 20°C.

34. An automobile of 2100 kg moving at 80 km/h brakes to a stop. In this process the kinetic energy of the automobile is first converted into thermal energy of the brake drums; this thermal energy later leaks away into the ambient air. Suppose that the temperature of the brake drums is 60°C when the automobile stops, and that the temperature of the air, and the final temperature of the brake drums, is 20°C.
   (a) How much entropy is generated by the conversion of mechanical energy into thermal energy of the brake drums?
   (b) How much extra entropy is generated as the heat leaks away into the air?

35. At Niagara Falls, 5700 $m^3$/s of water fall through a vertical distance of 50 m, dissipating all of their gravitational energy. Calculate the rate of increase of entropy contributed by this falling water. The temperature of the environment is 20°C.

*36. (a) Consider a material with a constant specific heat capacity $c$. Show that if we gradually supply heat to a mass $m$ of this material, increasing its temperature from $T_1$ to $T_2$, the increase of entropy of the mass $m$ is

$$S_2 - S_1 = mc \ln(T_2/T_1)$$

(b) What is the increase of entropy in 1.0 kg of water that is heated from 20°C to 80°C?

*37. Consider the process of dissolving ice cubes in water as described in Example 20.4. What is the change of entropy of the system? (Hint: Use the formula derived in Problem 36.)

*38. You mix 1 liter of water at 20°C with 1 liter of water at 100°. What is the increase of entropy?

# ENERGY, ENTROPY, AND ENVIRONMENT*

Ours is an age of mechanization. We use electric motors and internal combustion engines to run gadgetry and machinery ranging from electric toothbrushes to supertankers. The intensive use of power from purely electric and mechanical sources is a fairly recent development in the history of our civilization. Up to the middle of the nineteenth century, human and animal muscles were the main sources of power. The laborers of ancient Egypt, ancient Rome, and medieval Europe tilled and harvested their fields and erected their pyramids, aqueducts, and cathedrals by hand; oxen, horses, and mules pulled plows or carts and carried loads. The only purely mechanical sources of power were windmills, waterwheels, and sailing ships. The power developed by windmills and waterwheels was modest — about 2 kW for a waterwheel and about 10 kW for a windmill, although the large water machine constructed at Marly in 1684 to pump water for the fountains of Versailles developed a power of about 100 kW. Sailing ships were the most powerful machines prior to the mechanical age — in a gale of wind, the sails of a clipper ship running at its maximum speed of 20 knots developed a propulsive power of about 15,000 kW (or about 20,000 hp).

Around the middle of the nineteenth century, steam engines became the principal sources of power in the industrial world. Later, turbines and internal combustion engines came into use, and these remain at present our principal sources of power. All these engines consume prodigious amounts of fossil fuel in the form of oil, coal, and natural gas — in the next 30 years the United States will consume about as much fuel as it has in the 200 years since its birth. This voracious consumption raises the question of how much longer our reserves of fuel will last and what will replace them when they finally do run out.

## V.1 THE CONVERSION OF ENERGY

Energy is always conserved; it cannot be consumed, but only converted from one form to another. Our civilization does not face a problem of energy consumption, but one of the energy dissipation. We are converting useful forms of energy into useless forms of energy.

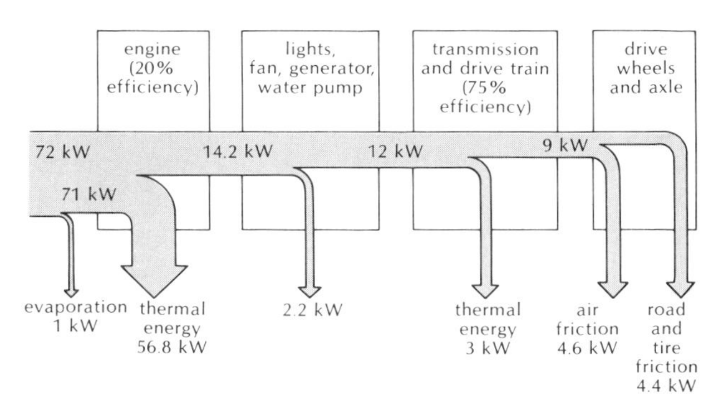


**Fig. V.1** Flow chart summarizing energy conversions in automobile. (Based on R. H. Romer, *Energy: An Introduction to Physics.*)

Consider for instance, the sequence of energy conversions that take place in an automobile propelled by a gasoline engine. The gasoline contains chemical energy (gasoline releases $3.4 \times 10^7$ J/liter when burned with oxygen at atmospheric pressure), and the engine partially converts this chemical energy into mechanical energy. Figure V.1 is a flow chart showing the successive energy conversions for a typical automobile traveling at 65 km/h on a level road. The engine delivers the mechanical energy to the transmission and axle of the automobile. From there, one part of the energy goes into the body of the automobile, compensating for the losses that the body suffers because of friction against the air, and another part goes into the wheels, compensating for the losses due to rolling friction against the ground. The engine consumes fuel at the rate of 7.6 liter/h, corresponding to an input of 72 kW of chemical energy, but only 9 kW of this emerges as propulsive power doing useful mechanical work against the external friction forces of air and road. Most of the energy is dissipated in the conversion process. Note that the largest waste occurs in the engine itself — in a gasoline engine most of the chemical energy of the fuel is ejected into the exhaust gases. Diesel engines achieve a somewhat higher efficiency because they burn their fuel more completely.

* This chapter is optional.

However, in part, the inefficiency of the engine is a consequence of the basic laws of thermodynamics. The gasoline engine is a heat engine — it converts the heat released by the burning fuel into mechanical energy. The efficiency of an ideal heat engine is [see Eq. (21.20)]

$$e = 1 - \frac{T_2}{T_1} \tag{1}$$

where $T_1$ is the temperature of the heat source and $T_2$ that of the heat sink. In the gasoline engine, the heat source is the hot gas produced by combustion and the heat sink is the air surrounding the engine. The complete combustion of hydrocarbon fuel gives the residual gas a temperature of about 2400 K. The air surrounding the engine typically has a temperature of 300 K. Hence the maximum attainable efficiency is

$$e = 1 - \frac{300\text{ K}}{2400\text{ K}} = 0.88 \tag{2}$$

Gasoline engines, and other engines that burn hydrocarbon fuels, do not come anywhere near this maximum efficiency of an ideal engine, because they do not burn the fuel completely, eject exhaust gases at temperatures much higher than 300 K, and suffer from internal friction.

**Table V.1** ENERGY AND ENTROPY

| *Kind of energy* | *Entropy per joule* |
|---|---|
| Kinetic (macroscopic) | 0 J/K |
| Gravitational potential | 0 |
| Electric potential | 0 |
| Thermal, from nuclear reactions | $10^{-11}$ |
| Thermal, within Sun | $10^{-7}$ |
| Sunlight | $10^{-5}$ |
| Thermal, from chemical reactions | $\sim 10^{-4}$ |
| Thermal, in Earth's environment | $3 \times 10^{-3}$ |
| Thermal, cosmic background radiation | $3 \times 10^{-1}$ |

In any case, Eq. (2) tells us there is a limit to the conversion of chemical energy of fuel into mechanical energy. An appreciable part of the chemical energy of fuel is disordered energy. If we want to convert disordered energy into ordered energy, we have to pay a price. In general, the disorder inherent in a given kind of energy can be measured by the corresponding entropy. Table V.1 lists some kinds of energy and the corresponding entropy per joule. The kinetic and potential energies of macroscopic bodies are the most ordered kinds of energy (least entropy), and the thermal cosmic background radiation is the most disordered kind (most entropy). Energy of low entropy can be totally converted into energy of higher entropy. For instance, the gravitational potential energy of the

**Table V.2** PRACTICAL ENERGY CONVERSION EFFICIENCIES

| *Conversion from* | *Conversion to* | | | | | |
|---|---|---|---|---|---|---|
| | *Mechanical* | *Gravitational potential* | *Electric* | *Radiant (light)* | *Chemical* | *Thermal* |
| *Mechanical* | — | — | 99% (electric generator) | — | — | 100% (brake drum) |
| *Gravitational potential* | 86% (water turbine) | — | 85% (hydroelectric plant) | — | — | — |
| *Electric* | 93% (electric motor) | 80% (pumped storage) | — | 40% (gas laser) | 72% (storage battery) | 100% (heating coil) |
| *Radiant (light)* | — | — | 27% (solar cell) | — | 0.6% (photosynthesis) | 100% (solar furnace) |
| *Chemical* | 45% (animal muscle) | — | 91% (dry cell battery) | 15% (chemical laser) | — | 88% (furnace of steam boiler) |
| *Thermal* | 47% (steam turbine, rocket engine) | — | 7% (thermocouple) | 3% (light bulb) | — | — |

water stored behind a dam can, in principle, be converted into an equal amount of mechanical or electric energy (in practice, hydroelectric power stations attain an efficiency of 85%). Energy of high entropy can be only partially converted into energy of low entropy; some of the energy must be converted into energy of even higher entropy so as to guarantee the net increase of entropy required by the Second Law of Thermodynamics.

In practice, the efficiency of energy conversion is often much below the theoretical limit set by thermodynamics. Table V.2 lists the maximum efficiencies attainable with present-day devices.

## V.2 THE DISSIPATION OF ENERGY

Most of the mechanical and electric energy supplied to our machinery as well as the heat used in our homes and factories is generated from fossil fuels — oil, coal, and natural gas. A small part is contributed by hydroelectric dams and nuclear reactors. Figure V.2 summarizes the rate of energy dissipation in the United States since 1850; the figure includes an extrapolation up to the year 2000.

Almost all of this energy is dissipated into our environment, ultimately becoming degraded heat at a temperature of about 300 K. The example of the automobile mentioned in the preceding section illustrates this dissipation process. The engine of the automobile releases thermal energy with the exhaust gases; as these gases mix with the air, their temperature gradually drops to 300 K, and the thermal energy becomes degraded. Simultaneously, the engine supplies mechanical energy to the body and wheels of the car; air and road friction convert this energy into heat, and this also becomes degraded as it spreads into the environment. The net result is the conversion of chemical energy into degraded thermal energy. The unburned hydrocarbons in the exhaust gases represent some small amount of chemical energy that is not degraded directly. But natural reactions in the atmosphere gradually decompose these hydrocarbons, so that ultimately their energy will also be degraded. In Figure V.2 no distinction has been made between such delayed dissipation and immediate dissipation, that is, Figure V.2 is based on the yearly rate of consumption of fuel.

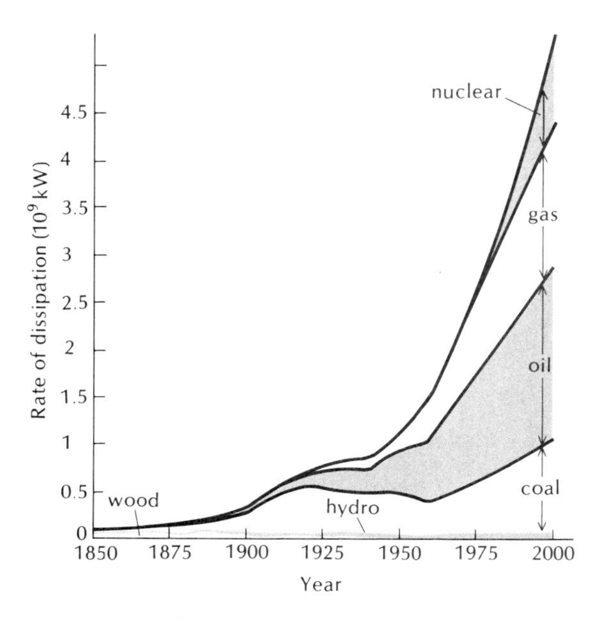


**Fig. V.2** Rate of energy dissipation in the United States. (Based on *Scientific American*, September 1971.)

At present (1980) the rate of dissipation of energy in the United States is $2.2 \times 10^6$ MW.[1] Worldwide, the rate of dissipation is about four times as large. The United States has by far the highest per capita rate of energy dissipation — about 14 kW per capita. Although this is in part due to the very high standard of living enjoyed by Americans and to the prodigious expenditure of energy in heavy industry, it is also in part due to the careless use of energy — inefficient trucks and automobiles, wasteful industrial installations, poor thermal insulation in homes. The standard of living enjoyed by Swedes and by Germans is also very high, but their per capita rate of dissipation of energy is much lower — about 6 kW per capita.

Figure V.3 shows how the dissipation of energy is distributed among transportation, industrial, commercial, and residential users. At present, the overall efficiency of energy use in the United States is only 0.41; this means that 41% of the total energy is exploited as useful heat and work before it is dissipated, and 59% is wasted or dissipated directly.

In almost all the uses of energy, substantial savings are possible. According to a study by the National Bureau of Standards, savings as high as 30% could be attained in industry. We could also significantly increase the efficiency of transportation by heavier use of trains. Trains are more energy efficient than airplanes, trucks, or automobiles — a train can transport a given amount of freight over a given distance with only $\frac{1}{4}$ of the energy expenditure required by a truck, or $\frac{1}{60}$ the energy expenditure required by an airplane. With improved insulation in residential homes, we could cut the consumption of heating fuel — with insulation a

[1] Large amounts of power are usually measured in megawatts,

$$1 \text{ MW} = 10^6 \text{ W}$$

Large amounts of energy are often measured in Quads,

$$1 \text{ Quad} = 10^{15} \text{ Btu} = 1.05 \times 10^{18} \text{ J}$$

or in Q,

$$1 \text{ Q} = 10^{18} \text{ Btu} = 1.05 \times 10^{21} \text{ J}$$

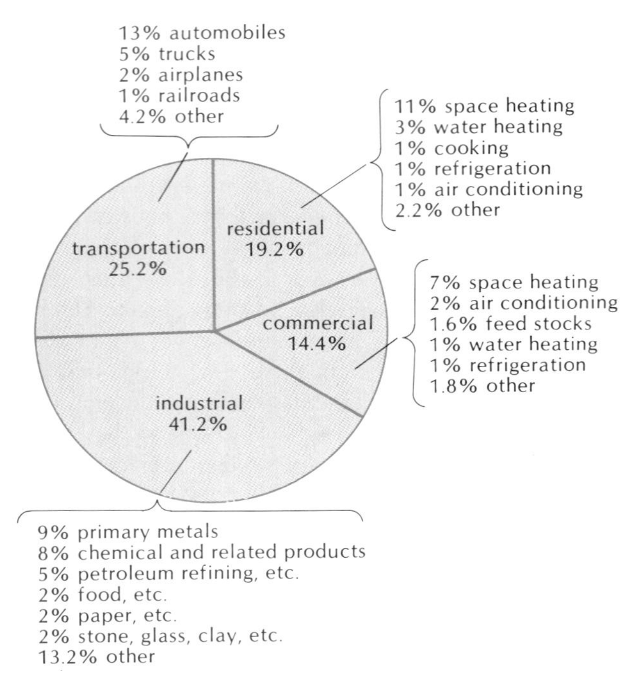


**Fig. V.3** Distribution of energy dissipation in the United States. (Based on R. L. Loftness, *Energy Handbook*.)

typical home requires 45% less heating fuel than a home without any insulation. The added insulation can pay for itself out of savings on the fuel bill in just a few years. Furthermore, the simple expedient of lowering the thermostat in winter can save 15 to 20% of the heating fuel for a 3°C (5°F) reduction of indoor temperature.[2]

## V.3 THE LIMITS OF OUR RESOURCES

Worldwide, the rate of dissipation of energy is increasing by a factor of 2 every 10 years. Most of this energy is being generated from fossil fuels. The total amount of such fuels stored in natural deposits under the surface of the Earth is limited. Hence, within the foreseeable future the rising demand will exhaust our supplies of fossil fuels. How much longer will these fuels last?

The supply of oil and natural gas will be exhausted fairly soon; we have already consumed almost a fourth of the original supply of these fuels (see Table V.3). Only a small fraction of the oil will be left 50 years from now; next to none will be left 100 years from now (Figure V.4). The exploitation of oil shale and of tar sand could provide extra oil — up to 10 times more oil than the total obtained from liquid petroleum deposits. However, it is not yet clear to what extent we can economically extract this oil from the stone in which it is absorbed.

[2] Incidentally: Such a reduction of indoor temperature is good for the health of most people.

**Table V.3** THE WORLDWIDE SUPPLY OF FOSSIL FUELS

| Fuel | *Original supply*[a] | *Remaining supply (1980)* |
|---|---|---|
| Oil | $12 \times 10^{21}$ J | $9.4 \times 10^{21}$ J |
| Natural gas | $11 \times 10^{21}$ | $9.8 \times 10^{21}$ |
| Coal | $188 \times 10^{21}$ | $184 \times 10^{21}$ |
| Oil shales (rich deposits) | $90 \times 10^{21}$ (?) | $90 \times 10^{21}$ |

[a] From M. K. Hubbert, *American Journal of Physics*, November 1981; data on shales from *The Global 2000 Report to the President* by the Council on Environmental Quality and the Department of State.

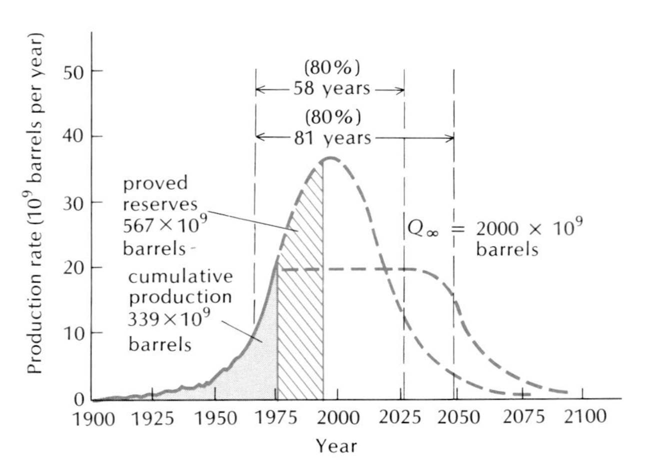


**Fig. V.4** Estimated world production of oil. Beyond the year 2000, production is expected to decline because oil wells will run dry and few new wells will be discovered. The total supply is $Q_\infty = 2000 \times 10^9$ barrels. The upper curve assumes that 80% of this total is consumed in 58 years; the lower curve assumes 81 years. One barrel of crude oil is equivalent to $5.9 \times 10^9$ J. (Based on M. K. Hubbert, *American Journal of Physics*, November 1981.)

Our supply of coal will last much longer; we have consumed only about 2% of the original supply. The rate of production of coal will probably continue to increase well into the next century; at current rates of consumption, it will take 300 or 400 years before coal becomes scarce (Figure V.5).

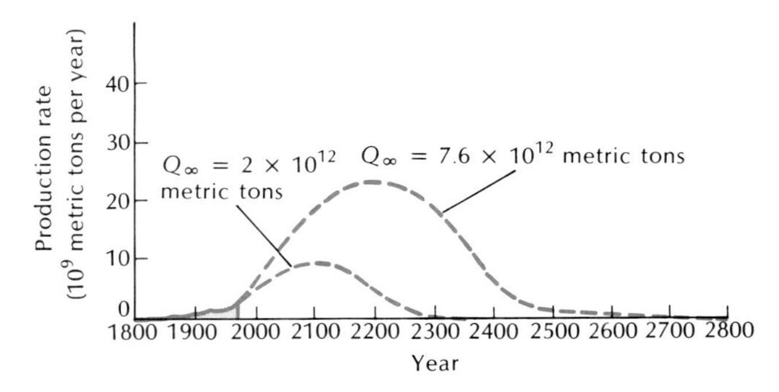


**Fig. V.5** Estimated world production of coal. The upper curve assumes a total supply $Q_\infty = 7.6 \times 10^{12}$ metric tons; the lower curve assumes $Q_\infty = 2 \times 10^{12}$ metric tons. One metric ton of coal is equivalent to $2.9 \times 10^{10}$ J. (Based on M. K. Hubbert, *American Journal of Physics*, November 1981.)

## V.4 ALTERNATIVE ENERGY SOURCES

Upon the exhaustion of fossil fuels, we will have to rely on alternative sources of energy to satisfy the increasing demand. Some of the most interesting alternatives are nuclear energy, solar energy, geothermal energy, and tidal energy.

### Nuclear Energy from Fission

At present, nuclear reactors (Figure V.6) supply about 12% of the worldwide electric power. In these nuclear reactors, energy is generated by the **fission** of uranium — the nuclei of uranium are made to split into two or more pieces, a reaction that releases a large amount of energy in the form of heat. The same kind of reaction also generates the explosive energy of an atomic bomb; but whereas in an atomic bomb the energy is released in a sudden burst, in a nuclear reactor the energy is released gradually, under controlled conditions (see Interlude XI for details on nuclear fission.)

The total fission of 1 kg of uranium releases as much heat as the combustion of $3 \times 10^6$ kg of coal. However, commercial reactors "burn" only a few percent of the fresh uranium fuel placed in them; the rest must be reprocessed chemically before it can be used again. Furthermore, the efficiency of nuclear power plants is lower than that of conventional power plants — in a nuclear power plant only about 30% of the thermal energy released by the uranium is converted into electric energy, whereas in a modern conventional power plant about 40% of the thermal energy of coal or oil is converted into electric energy.

Naturally occurring uranium is a mixture of two isotopes: 99.3% of $^{238}$U and 0.7% of $^{235}$U. Only the rare isotope $^{235}$U will burn in an ordinary nuclear reactor. Since it is rather difficult to separate the isotopes completely, most nuclear reactors have been designed to operate with "enriched" uranium (several percent $^{235}$U) rather than pure uranium $^{235}$U; in this mixture, only the $^{235}$U burns, while the $^{238}$U remains behind. The worldwide supply of high-grade ores suitable for the (partial) extraction of $^{235}$U is rather limited (Table V.4); and this supply is likely to become exhausted before the end of the century.

**Fig. V.6** Salem Generating Station at Alloways Creek, New Jersey.

Fortunately, the abundant isotope $^{238}$U can be converted, by nuclear alchemy, into $^{239}$Pu, an artificial isotope of plutonium, which will burn in a nuclear reactor. The conversion is accomplished in a breeder reactor. In such a reactor, an initial supply of $^{239}$Pu is surrounded by a blanket of $^{238}$U. As the plutonium fuel burns, it releases not only heat, but it also releases an intense stream of fast neutrons which impinge on the uranium and convert it into plutonium. Thus, the reactor produces not only energy, but also new fuel — in an efficient breeder, the amount of new fuel generated by conversion of uranium exceeds the amount of initial fuel consumed. By taking advantage of this breeding process, we gain access to the large amount of energy locked in the worldwide supply of $^{238}$U (see Table V.4). Our supply of fuel for nuclear reactors would then last more than a hundred years. Because of the high efficiency of breeders, we can even exploit low-grade uranium ores that are good enough for $^{238}$U extraction but not good enough for $^{235}$U extraction. Our supply of fuel for nuclear reactors would then last several hundred years. Furthermore, breeding is also possible with $^{232}$Th, an isotope of thorium. There are extensive deposits of low-grade thorium ores in granite, such as the Conway granite of New Hampshire. In principle there is enough fuel for breeder reactors to last us several thousand years.

For the immediate future, nuclear fission appears to be the most practical source of energy to replace oil and coal. According to estimates based on recent trends, nuclear fission is expected to contribute about 50% of electric power requirements of the United

**Table V.4** THE WORLDWIDE SUPPLY OF NUCLEAR FUELS

| Fuel | *Original supply*[a] | *Remaining supply (1980)* |
|---|---|---|
| Uranium-235 (high-grade ores) | $20.7 \times 10^{21}$ J | $20.6 \times 10^{21}$ J |
| Uranium-238 (high-grade ores) | $1 \times 10^{24}$ | $1 \times 10^{24}$ |
| Uranium-238 and thorium-232 (low-grade ores) | $3 \times 10^{27}$ | $3 \times 10^{27}$ |
| Deuterium | $7 \times 10^{30}$ | $7 \times 10^{30}$ |

[a] Uranium data from *The Global 2000 Report to the President;* other data from article by J. L. Tuck in *Cosmology, Fusion & Other Matters.*

States in the year 2000. Unfortunately, nuclear fission yields rather dirty energy — the fission reactions generate dangerous radioactive residues. Nuclear power plants must be carefully designed to hold these radioactive residues in confinement. The cumbersome safety features that must be incorporated in the design make the construction and maintenance of nuclear power plants extremely expensive. Furthermore, when the load of fuel of a nuclear reactor has been spent, the residual radioactive wastes must be removed to a safe place to be held in storage for hundreds of years until their radioactivity has died away.

**Nuclear Energy from Fusion**

Nuclear fusion is an attractive source of energy because it bypasses some of the safety problems associated with nuclear fission. In fusion reactions, energy is generated by merging nuclei of hydrogen, deuterium, or tritium;[3] these nuclei combine to form helium. For instance, the heat emitted by the Sun is generated by a fusion reaction that burns hydrogen into helium; and the explosive energy of a hydrogen bomb is generated by a fusion reaction that burns deuterium or tritium into helium. Note that the process of fusion is the reverse of fission: light nuclei (such as hydrogen) release energy when they merge; heavy nuclei (such as uranium) release energy when they split.

In order to initiate a fusion reaction, the nuclei must be subjected to extremely high temperatures and pressures. For instance, the fusion reactions at the center of the Sun involve a temperature of $15 \times 10^6$ K and a pressure of $10^9$ atm. Nuclear reactions that require such extreme temperatures are called **thermonuclear.** Schemes for the development of fusion power on Earth intend to use deuterium or tritium because their reaction rates are much faster than that of hydrogen; however, the temperature required for the fusion of these nuclei is about $10^8$ K — even higher than in the Sun! Under these conditions, the deuterium or tritium will be in the form of plasma and it cannot be contained by a conventional reactor vessel of steel or ceramic material. One scheme for fusion attempts to suspend the plasma in the middle of an evacuated vessel by means of magnetic fields; the development of this scheme is the objective of most of our modern plasma research (discussed in Interlude VII). Another scheme for fusion attempts to extract the energy by exploding small pellets of a deuterium–tritium mixture in a combustion chamber by hitting them with an intense laser beam or an intense electron beam (Figure V.7); the beam suddenly heats the pellet to such a high temperature that fusion begins — the pellet explodes like a miniature hydrogen bomb. After the thermal energy from the exploded pellet has been extracted from the combustion chamber, the next pellet is dropped in, and so on (see Interlude X).

(a)

(b)

**Fig. V.7** (a) Small pellets consisting of glass microballoons filled with deuterium–tritium mixture and (b) the target chamber within which the pellets are exploded at the Lawrence Livermore Laboratory.

One advantage of fusion over fission is that we have available an enormous supply of deuterium. In the oceans of the Earth, 1 water molecule in 6000 is heavy water, with one deuterium atom replacing one of the hydrogen atoms in the molecule. The total supply of deuterium is about $2 \times 10^{16}$ kg — enough fusion fuel for millions of years. Furthermore, the fusion reaction does not produce any appreciable amounts of radioactive residues; in contrast to fission, fusion yields very clean energy.

**Solar Energy, Indirect**

Solar energy is an inexhaustible source of energy; it will remain available as long as the Sun lasts — 5 billion years or longer. The total solar energy intercepted by the Earth is $2.8 \times 10^{10}$ MW, but so far we are using only a small fraction of this total. Aside from agriculture and forestry, our only large-scale exploitation of

[3] Deuterium and tritium are isotopes of hydrogen: $^2$H and $^3$H.

solar power is in hydroelectric power stations. This is an indirect exploitation of solar energy. The oceans may be regarded as the boilers of hydroelectric plants and the Sun is the source of heat. The water evaporated by the Sun falls as rain on mountains, collects behind a dam, and drives the turbines of the power stations (Figure V.8). At present, hydroelectric power contributes about 25% to the worldwide generation of electric power. The maximum conceivable contribution from hydroelectric plants is limited by suitable sites for dams and could reach at most $3 \times 10^6$ MW sometime in the future. Since this represents only a small fraction of the solar energy reaching the Earth, it will pay to develop other means of capturing this energy.

Windmills are another indirect exploitation of solar energy. Winds obtain their energy from the solar heat absorbed by the surface of the Earth — the atmosphere acts as a giant heat engine converting this heat into kinetic energy of motion of the air. A crude estimate suggests that the total available wind power is about $10^5$ MW. Unfortunately, the power of wind is rather dilute. In a wind of 35 km/h, a vertical area of 1 $m^2$ intercepts kinetic energy at the rate of 1.3 kW.

**Fig. V.8** Hydroelectric power station at the Hoover dam on the Colorado River.

**Fig. V.9** Large windmill generator on the island of Culebra. This generator supplies about 20% of the electric power used on the island.

Even if it were possible to extract all of this energy,[4] the generation of large amounts of power would require a windmill, or an array of windmills, presenting a very large frontal area to the wind. Figure V.9 shows a windmill on the island of Culebra, Puerto Rico, which generates up to 200 kW. Small windmills for pumping water and grinding grain were in common use in the United States during the last century and the early part of this century; around 1900 these windmills supplied about 25% of the total power generated for purposes other than transportation. The availability of cheap electric power led to the neglect of these very useful machines.

Other schemes for the indirect exploitation of solar energy try to take advantage of the energy in ocean waves or the energy in warm ocean water. The energy in ocean waves can be captured by allowing the waves to push against paddles or wheels linked to electric generators. An experimental wave power plant in Norway has produced 850 kW. Proposals for the extraction of thermal energy from ocean water hinge on the temperature difference between the warm water on the surface and the colder water at some depth below the surface. We know from Chapter 21 that the operation of a heat engine requires both a heat source and a heat sink, of different temperatures. In the tropical regions, the temperature difference between the surface water and the water at a depth of 1000 m is 20° to 25°C. A steam engine operating with a working fluid of low boiling point — such as ammonia or freon — can use the warm water as heat source to evaporate the fluid, and it can use the cold water, pumped to the surface, as heat sink to condense the fluid. Besides generating electricity, this kind of power plant could

[4] Careful analysis shows that at best 59% can be extracted.

also provide some additional benefits: after the water pumped up by the power plant emerges from the condenser of the steam engine, it could be diverted to other profitable purposes, such as refrigeration, air conditioning, or fish farming (the water in the deep layers of the ocean is rich in nutrients).

**Solar Energy, Direct**

When concentrated by mirrors or lenses, sunlight can be used directly to heat water or other materials. Figure V.10 is a view of the large parabolic mirror of the solar "furnace" at Odeillo, Font-Romeu, France. Sixty-three flat mirrors throw the sunlight at the parabolic mirror which concentrates about 600 kW of solar energy into a focal spot 30 cm across. The intense solar heat — up to 4000°C — has been used for high-temperature research, but this solar furnace has not been designed to generate power.

**Fig. V.10** Solar "furnace" at Odeillo.

A similar arrangement of mirrors can be used to collect solar heat for a power plant; the heat can be focused on a boiler that generates steam for a steam turbine. Figure V.11 shows an experimental solar power station in California's Mojave Desert.

Instead of concentrating the solar energy with large mirrors, it is possible to collect the energy by circulating water through heat pipes in a large array of absorbers. Small solar collectors circulating water through heat pipes are now available for installation on the roofs of private homes; these solar collectors can supply hot water for kitchens and bathrooms and also for radiators to warm the house (Figure V.12).

In principle, the simplest method for obtaining electric power from sunlight is the direct conversion of the energy of light into electric energy. This can be done with a **solar cell** (see Section 29.2), a thin wafer of two types of silicon that generates an electric current when exposed to light. Large numbers of such solar cells as-

**Fig. V.11** The solar power station "Solar One" in California's Mojave Desert consists of collecting mirrors spread over an area of about 0.5 km$^2$. The boiler on the central tower generates enough steam for a turbine delivering $10^4$ kW of electric power.

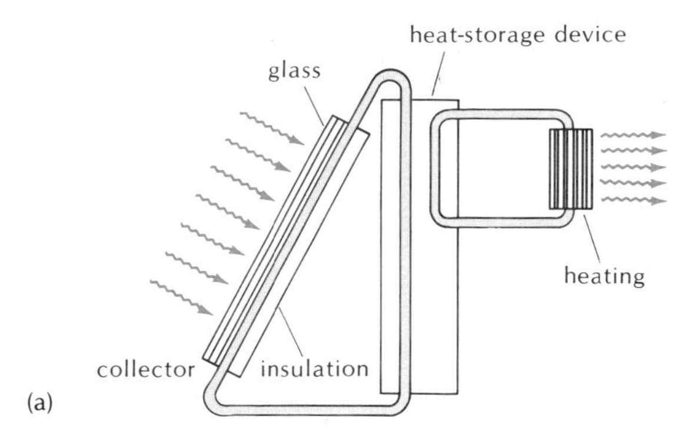

**Fig. V.12** Solar heating system for a home: (a) schematic diagram and (b) collectors on roof. (Courtesy Advance Energy Technologies, Inc.)

**Fig. V.13** Solar panels on Skylab.

**Fig. V.14** The "Sunraycer" that won the race for solar-powered automobiles across Australia in 1987. Its body is covered with 7200 solar cells, generating a net power of 1000 W.

sembled in panels have been used to supply electricity for artificial satellites. Figure V.13 shows the solar panels of the Skylab spacecraft; these supplied most of the electric power required by the astronauts living aboard the spacecraft. Solar cells have been used to supply power for experimental solar automobiles. Figure V.14 shows such a solar automobile used for a race across Australia.

Solar cells are rather inefficient; they convert at best about 27% of the energy of sunlight into electric energy. They are also very expensive and, until we develop some cheap technique of mass production, we will not be able to build large-scale power plants with solar cells. However, some experimental power plants of modest size are already in operation. Figure V.15 shows the array of solar cells of a power plant on the Carrisa Plains in California. This array delivers 7200 kW of electric power.

The suggestion has been made that a large power plant with solar panels 12 km long might be placed in orbit high above the Earth. It would then receive direct sunlight 24 hours a day and it would not be affected by clouds. The electric power could be converted into microwaves and beamed to a large receiving antenna on the surface of the Earth. The output of such a power station would be about 4000 MW at the Earth terminal.

## Geothermal Energy

The interior of the Earth is hot. The temperature in the Earth increases gradually with depth and reaches 4500°C at the center of the Earth. At some exceptional locations, hot rocks at a temperature of several-hundred degrees are found at a depth of just a few kilometers. Hot springs, steam vents, and geysers bring some of the heat from the interior of the Earth to the surface. These natural sources of geothermal energy

**Fig. V.15** Arrays of solar cells mounted on panels at the Carrisa Plains power plant, California. These solar cells now deliver 7200 kW of electric power, but have a capacity for 160,000 kW.

are being exploited in a few places such as Larderello (near Pisa, Italy), the Geysers (California), Cerro Prieto (Mexico), and Wairakei (New Zealand). Figure V.16 shows one of the geothermal power plants at the Geysers. These power plants operate with steam blowing out from wells drilled down to depths of as much as 2100 m. The power plant shown in Figure V.16 delivers 110 MW. The total power generated by the Geysers is 700 MW, which makes it the largest of all the geothermal sources in operation.

Geothermal sources near populated areas can be used directly to provide heat and steam for homes and for industrial processes. In Iceland, nearly half the population lives in houses heated by geothermal sources; almost all the houses in Reykjavík are heated in this manner.

It may be possible to supplement the natural supply of geothermal steam by the artificial exploitation of the thermal energy of hot rocks. In some sites, rocks at a

**Fig. V.16** Unit 18 of Pacific Gas and Electric Company's geothermal power station at the Geysers.

temperature of several hundred degrees are near enough to the surface of the Earth to be reached by drilling. If the rock can be fractured with high-pressure water or with explosives, then water can be circulated through it. The water would absorb the heat of the rock and make steam. Intensive development of such projects could contribute perhaps 10% of the electric power that will be required in the United States 20 or so years from now.

### Tidal Energy

Along the coasts of the oceans, the tides caused by the Moon (and, to a lesser extent, by the Sun) regularly raise and lower the level of the water. In some exceptional locations, such as the Bay of Fundy (Nova Scotia) and the estuary of the La Rance river (France), the lift of the water is more than 10 m. We can generate hydroelectric power by capturing the water in a reservoir behind a dam at high tide and letting it run out through turbines at low tide. This method has been employed with great success at La Rance (Figure V.17), where special low-speed turbines generate an average power of 250 MW; these turbines are reversible — they generate power both when the water rushes in to fill the reservoir and when it rushes out.

Like windmills or solar cells, tidal power is completely free of pollution. Of course, the dam has some impact on aquatic life, but much less than the very high dam used for a conventional hydroelectric power station. Unfortunately, there are only a few potential sites for the development of tidal power — we need high tides, and we need a bay or estuary that can be closed off by a dam. One suitable site that has been discussed sporadically for many years is Passamaquody Bay on the border between Maine and Canada. This site could generate as much as 1800 MW. Another suitable site is the Minas Basin in the Bay of Fundy; this could generate about three times as much. But altogether, tidal power could only supply a few percent of our total energy requirements.

**Fig. V.17** Tidal power plant at La Rance.

Incidentally: The tidal energy derives from the rotational energy of the Earth. The friction of the tides against the shores and the bottoms of the oceans gradually slows the rate of rotation of the Earth. A tidal power station increases the effective "friction" of the tide and therefore slows the rotation even more. But the effect of a tidal power station on the rate of rotation of the Earth is insignificant compared to the effect of natural friction.

## V.5 POLLUTION AND THE ENVIRONMENT

Pollution is a by-product of the operation of our power plants. Coal-fired power plants spew out ash and stack gases, internal combustion engines emit exhaust gases, and nuclear reactors leak small amounts of radioactive materials. Besides, all power plants release large amounts of waste heat, which constitutes thermal pollution. What is most noticeable about all this pollution are the immediate, local effects. Sometimes these effects reach catastrophic levels, as in the great London "killer smog" of December 1952, when nearly 4000 people died from heart trouble, pneumonia, and bronchial diseases triggered by the accumulation of pollutants, from coal-burning furnaces, in the air held stagnant in the city by exceptional atmospheric conditions. However, although the dispersed effects of pollution are less noticeable than such concentrated effects, they may be even more serious in the long run.

**Table V.5** ATMOSPHERIC POLLUTANTS FROM FOSSIL FUELS IN THE UNITED STATES (1980)[a]

| *Pollutant* | *Amount per year* | *Major source* |
|---|---|---|
| Carbon monoxide | $85 \times 10^6$ metric tons | Automobiles |
| Hydrocarbons | $22 \times 10^6$ | Automobiles |
| Fly ash | $8 \times 10^6$ | Industrial plants |
| Sulfur oxides | $24 \times 10^6$ | Power plants |
| Nitrogen oxides | $21 \times 10^6$ | Automobiles, power plants |

[a] From *Environmental Quality 1981,* 12th Annual Report of the Council on Environmental Quality.

The combustion of fossil fuels releases millions of tons of pollutants into the air each year (see Table V.5). In the United States, several tens of thousands of deaths each year can be attributed to air pollution. Furthermore, the sulfur oxides and nitrogen oxides react with oxygen and water vapor in air to form a fine mist of sulfuric acid and of nitric acid, which is ultimately washed out of the atmosphere in the form of acid rain. The effects of acid rain on the environment are especially severe in the northeastern United States and in eastern Canada, where acid rain has damaged forests and eradicated aquatic life in many small lakes. The main sources of the sulfur oxides that cause this acid rain are large power plants in the Midwest and in the Ohio Valley, which burn coal containing sulfur. Much of the American coal has a high sulfur content, and power companies find it too expensive to remove this sulfur from the fuel before combustion or from the stack gases after combustion. Instead, they prefer to build very tall smokestacks (of up to 360 m) to send the pollutants high into the atmosphere and disperse them. But the prevailing winds then carry the pollutants eastward, and they ultimately settle on the ground hundreds and even thousands of kilometers downwind from their source.

Both nitrogen oxides and hydrocarbons are released by internal combustion engines. Reactions between these pollutants and the oxygen in air produce the brownish photochemical smog for which Los Angeles is famous. Fortunately, although this smog is very harmful for plant life, it is merely an irritant for animal life; the Los Angeles smog does not kill directly, at least not in the short run.

An important worldwide effect of the combustion of fossil fuels is the increase of the carbon dioxide content of the atmosphere. The combustion of hydrocarbons releases mainly carbon monoxide, but this combines with oxygen in the atmosphere to form the dioxide. The concentration of carbon dioxide in the atmosphere has increased by about 10% in the last 50 years. The increase would have been even larger than this were it not for the absorption of carbon dioxide by the water of the oceans. The oceans act as a reservoir of carbon dioxide, removing a large fraction of the carbon dioxide injected into the air.

Some climatologists worry that the continuing combustion of hydrocarbon fuels might lead to a drastic increase of the temperature of the Earth's surface. This is the "greenhouse effect." Like the glass on a greenhouse, carbon dioxide is transparent to sunlight, but absorbs heat radiation (infrared); thus, it permits sunlight to enter the atmosphere, but prevents the escape of heat. The accumulation of carbon dioxide in the atmosphere therefore causes an increase of temperature. Even a small increase of the Earth's temperature could have a drastic impact on the climate. It could possibly lead to a catastrophic "runaway greenhouse effect" by triggering the release of extra carbon dioxide from the oceans, which would lead to extra heating, and further release of carbon dioxide, and so on. However, the physics of the atmosphere is rather complicated, and a compensation mechanism might come into play: a slight increase of temperature would encourage the evaporation of water and the formation of clouds; since clouds reflect sunlight, they tend to keep the temperature down. According to a recent (1983) report of the Environmental Protection Agency, the average global temperature is likely to increase by about 5°C by the year 2100. This will lead to major disruptions in the climate patterns, inflicting severe droughts on some regions and severe floods on others. The temperature increase over the polar regions will be even larger, with substantial melting of polar ice and snow. This will raise the sea level by about 2 meters and cause the flooding of many low-lying coastal areas.

The thermal pollution from our power plants is of little consequence on a global scale — the total input of waste heat into our environment is small compared to the input of solar heat. A simple estimate indicates that the total heat from all our dissipated energy has only increased the temperature of the Earth by about 0.01°C. However, on a local scale, the disposal of the waste heat poses severe problems. A nuclear or a coal-burning power plant generating 500 MW of electricity will also generate 1200 MW of waste heat. This heat must be dumped into the environment. The most convenient heat sink is flowing water. The amount of cooling water required by a large power plant is enormous, about $10^7$ m$^3$/day; for this reason, many power plants are located on river banks, lake shores, or ocean coasts. Figure V.18 is an infrared photograph displaying the heat released into a river. The temperature of the plume of warm water emerging from the outlet is

**Fig. V.18** Infrared photograph showing plume of hot water released into a river by a power plant.

**Fig. V.19** (above) Mechanical-draft cooling towers with fans and (right) natural-draft cooling towers.

several degrees higher than the normal temperature; this makes the water unsuitable for some species of fish. Besides, the water intake often sucks in and kills huge quantities of fish and fish larvae.

In an effort to avoid the thermal pollution of rivers, lakes, and estuaries, some new power plants make use of cooling towers in which the air is the heat sink. These cooling towers act much like the radiator in an automobile. The heat is brought out of the power plant by water circulating in pipes; in the cooling tower a draft of air passes over these pipes, taking away their heat. Sometimes the cooling is assisted by water sprayed over the pipes. Figure V.19 shows cooling towers of different types.

## V.6 ACCIDENTS AND SAFETY

The generation of electric power — whether by means of coal-burning, oil-burning, hydroelectric, or nuclear power plants — always involves the manipulation of potentially dangerous concentrations of energy: chemical energy in coal or oil, gravitational energy in water held behind a high dam, or nuclear energy in fissionable and radioactive materials. If this energy overwhelms our controls, the result is disaster: fires and explosions in coal mines, oil tankers, and oil storage depots; dam failures and flash floods; and reactor fires, meltdowns, and explosions.

The worst nuclear-reactor accidents to date are listed in Table V.6. The greatest hazard in a nuclear-reactor accident does not lie in the direct effects of the explosion or fire, but in the consequent release of the large amount of radioactive material contained in the reactor. Radioactivity is extremely efficient at destroying life (see Interlude IV). If radioactive material, in the form of smoke and dust, ascends into the atmosphere, it will be carried away by winds, and it will spread over a large area and ultimately descend to the ground as lethal fallout. The most disastrous nuclear-reactor accident happened at Chernobyl, near Kiev in the U.S.S.R., in 1986, where the explosion and fire in a reactor released a plume of fallout which spread north and west into Europe and produced hazardous contamination thousands of kilometers away. Crops of vegetables (contaminated with radioactive iodine) and milk (contaminated with radioactive cesium) had to be destroyed as far away as Italy. An area of 30-km radius

**Table V.6** REACTOR ACCIDENTS

| *Location and date* | *Reactor type* | *Accident* |
|---|---|---|
| Chalk River, Canada, December 12, 1952 | Heavy water, experimental | Control rods jammed. Reactor overheated, with fuel meltdown, hydrogen explosion, and bursting of reactor vessel. Some radioactive contamination escaped into atmosphere; no casualties. |
| Windscale, England, October 7, 1957 | Gas-cooled graphite, converter | Intentional overheating (for maintenance) ignited a fire in the uranium–graphite core. Radioactive iodine escaped and contaminated neighboring cattle pastures. Thirty-three delayed casualties from cancer have been attributed to this accident. |
| Idaho Falls, Idaho, January 3, 1961 | Light water, experimental | A control rod was mishandled by operators, and reactor went momentarily supercritical. All three operators were killed by massive radiation overexposure. |
| Lagoona Beach, Michigan, October 5, 1966 | Breeder, experimental | A loose piece of metal blocked flow of coolant (liquid sodium), leading to overheating and partial meltdown of fuel. Core was damaged beyond repair. No radioactive contamination; no casualties. |
| Brown's Ferry, Alabama, March 22, 1975 | Boiling water, power | A fire destroyed the electrical cables connecting reactor to control room; emergency cooling system went out of control and shut itself off; only a makeshift arrangement of pumps succeeded in maintaining sufficient cooling. No radioactive contamination; no casualties. |
| Three Mile Island, Pennsylvania, March 28, 1979 | Pressurized water, power | A failure in pump system for cooling water triggered automatic shutdown of reactor, but heat released by radioactivity overheated core. Pressure relief valve opened and stuck, permitting water in core to boil and to spill out. Contaminated water flooded reactor housing; hydrogen bubble formed in reactor vessel, creating an explosion hazard and interfering with cooling for several days. Large parts of fuel rods melted, and molten fuel burned through inner layer of wall of reactor vessel and cascaded to bottom of vessel. Some radioactive steam vented into atmosphere; no casualties. |
| Chernobyl, USSR, April 26, 1986 | Water-cooled graphite, power | An explosion in reactor and fire in graphite released large amount of radioactive fallout (for details, see text). Thirty-one immediate casualties among operators and fire fighters. About 3000 delayed casualties expected from cancer within 20 years. |

around the reactor was so heavily contaminated it had to be permanently evacuated. The reactor was completely demolished by the explosion and fire (Figure V.20), and its remains had to be entombed in concrete. Besides the 31 reactor operators and firefighters who lost their lives (most through radiation overexposure), some 3000 delayed casualties are expected from cancers induced by the radiation.

The immediate cause of the Chernobyl disaster was operator error. But the design of this particular reactor set the stage for the accident. The Chernobyl reactor was a graphite-moderated reactor. The core of such a reactor consists of rods of uranium fuel inserted in a large block of graphite. The graphite plays the role of catalyst — it enhances the rate of the fission reactions (see Interlude XI). Water circulating through channels in the graphite carries away the heat of the reaction. In some reactor designs, water will also act as a catalyst for the fission reaction, but in the Chernobyl design, the dominant effect of the water is to inhibit the reaction. The trouble with the Chernobyl design is that the presence of water leads to an instability: if the core ac-

**Fig. V.20** The Chernobyl reactor after the explosion and fire.

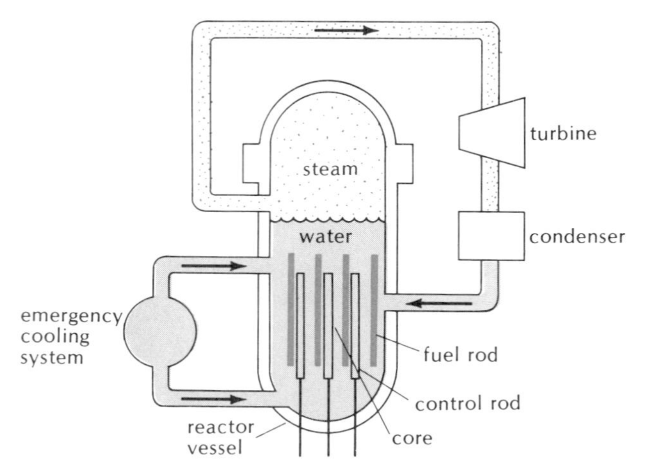


**Fig. V.21** Schematic diagram of a water-moderated nuclear reactor.

cidentally overheats, bubbles form in the water, and the inhibiting effect of the water is reduced; this tends to speed up the reaction, leading to further overheating, and so on. Under normal operating conditions, the reactor is prevented from runaway overheating by control rods that block the reaction and adjust its rate.

The accident was triggered by a series of almost incredible blunders by the operators. During a test run at low power, they disconnected the automatic shutoff system and the emergency cooling system, and accidentally allowed the power to drop to low, unsafe values, where reactor control becomes sluggish. Attempting to restore power, the operators then withdrew an excessive number of control rods. This left the reactor in an uncontrolled and unstable configuration, "free to do as it wished." The power spontaneously shot up to extreme values (100 times normal), shattering the fuel rods into small fragments. The sudden release of heat boiled the water in the core, generating a large volume of steam which blew the lid off the reactor vault. The explosion also started an intense fire in the graphite. The fire threatened to spread to an adjacent reactor, but was contained by firefighters, at the expense of heavy casualties. Gas, dust, and smoke released by the explosion and the fire carried a large amount of radioactive material into the atmosphere.

In the water-moderated reactors commonly employed in the United States, the core consists of rods of uranium fuel immersed in a volume of water (Figure V.21). The water carries away the heat and also acts as catalyst. If the core accidentally overheats, the fission reactions will tend to slow down and come to a halt. The reason is twofold. First, the reaction is intrinsically less efficient at high temperature. And second, the overheating will drive water out of the core; without a catalyst, the nuclear reaction will slow down. In contrast to a conventional fire, the overheated nuclear "fire" simply dies out. Thus, a water-moderated reactor is not afflicted with the instability that initiated the explosion of the Chernobyl reactor.

If the overheating of the core of a water-moderated reactor is extreme, so the fuel rods melt and form a puddle in the bottom of the reactor vessel, then the concentration of uranium could cause an explosion. Estimates of the magnitude of such an explosion are controversial; the explosion would not be anything like the explosion of a nuclear bomb, but, according to pessimistic estimates, it might be sufficiently large to rupture the reactor vessel and perhaps even the containment shell enclosing the reactor. A reactor with plutonium fuel (breeder reactor) is much more dangerous, because plutonium is much more explosive than uranium. In the United States, plans for the construction of plutonium reactors have been abandoned, in part because of the explosion hazard.

The greatest hazard in a water-moderated reactor, as in a graphite-moderated reactor, lies in the large amount of radioactive material contained in the reactor. To prevent the accidental release of this radioactive material, most reactors are encased in a strong, gas-tight containment shell, made of steel or concrete (see Figure V.22).

The amount of radioactive material within a reactor is so large that even when the fission reactions stop, the reactor continues to generate heat because of the energy released in the delayed radioactive decay of

**Fig. V.22** Construction of the containment shell for the nuclear reactor at Wolf Creek, Kansas.

the fission residues. While the reactor is in operation the energy released by this delayed radioactivity is about 7% of the total fission energy. Once the reactor is shut down, this radioactivity gradually declines. If the cooling system fails, the heat released by radioactivity will melt down the core of the reactor — and it might even melt a hole in the reactor vessel and in the containment shell.

Scenarios for conceivable reactor accidents have been analyzed in great detail. For an ordinary nuclear reactor, the worst that might happen is a loss-of-coolant accident, that is, the loss of the water that normally circulates through the core. Such an accident could occur if there were a sudden break in the pipe feeding the water into the core; high-pressure water mixed with steam would then blow out of the break, leaving the core with insufficient cooling. As we have seen above, both the increase of temperature and the loss of water tend to cut off the fission chain. However, the heat flow from the decay of radioactive fission residues does not cut off; and as this heat accumulates in the core, the temperature escalates catastrophically.

To prevent this, reactors are equipped with emergency cooling systems. When the primary coolant escapes, the emergency cooling system pumps water into the core. This must be done very fast, within seconds — otherwise the temperature of the fuel rods might reach 1500°C, at which point a violent reaction begins between steam and the zirconium metal used as a lining on the fuel rods; this not only releases extra heat, but also hydrogen gas, which can cause a (chemical) hydrogen explosion. If the emergency cooling system succeeds in bringing the temperature under control, there still remains the problem of the radioactivity that escapes with the water and steam from the reactor vessel. If all goes well, this radioactive mixture will be held in check by the containment shell.

However, if by mischance the emergency cooling system were to fail, the fuel rods would melt and their radioactive material would drip down into the bottom of the reactor vessel. The hot, molten mass of radioactive material endowed with its own internal source of heat would then gradually melt through the bottom of the reactor and, in a few hours, spill on the floor below. And then it would melt its way through the floor, through the containment shell, and into the earth. The motion of the molten mass of radioactive material digging its way through the earth as though it were heading to the antipode has been called the "China syndrome." Some recent calculations indicate that it will run out of heat and solidify after penetrating no more than 15 m into the ground.

The breach of containment would release some radioactive contamination into the atmosphere, mainly in the form of radioactive iodine, krypton, and xenon. In the worst conceivable case — involving substantial radioactive contamination caused by a meltdown of a reactor located near a densely populated area — several thousand people might die from acute radiation sickness, and those that survive a heavy exposure of radiation are susceptible to develop fatal cancer at a later time (see Interlude IV). According to a reactor safety study of the Atomic Energy Commission [the WASH-740 Report (updated), completed in 1965 but kept secret for many years], a major reactor accident near a city could conceivably kill up to 45,000 people and injure 100,000. According to a more recent study (the WASH-1400 Report, or the Rasmussen Report, released in 1975), the casualty figures are reduced to 3300 killed and 45,000 severely injured. The difference between these casualty estimates reflects the difference in the assumed fallout: the earlier study assumed that secondary explosions (for example, hydrogen explosions) would blow a substantial part of the radioactive material out of the containment; whereas the later, more optimistic, study supposes it much more likely that most radioactive materials will remain trapped in the molten mass at the reactor site and only some iodine, krypton, and xenon gas will escape. In any case, regardless of the precise numbers of casualties, the possible consequences of a major reactor accident are appalling.

Promoters of the nuclear energy program argue that a major disaster could only happen through an extremely improbable concatenation of mischances. They estimate that the probability for a major disaster is extremely small. For example, according to one calculation, the chance of death for any one individual in a given year through a nuclear reactor accident is only 1 in $5 \times 10^9$ — much smaller than the chance of death through automobile accident (1 in 4000) or airplane accident (1 in 100,000). Opponents of the nuclear energy program have disputed this calculation by pointing out that nuclear power plants have only been in operation for a few years and we have not yet accumulated enough statistical data for any accurate evaluation of chances of accident. Although the exact numbers are in dispute, it is undoubtedly true that the risk of death from a nuclear-reactor accident is less than that from many other kinds of accidents. What conclusion are we to draw from such a comparison of risks of accidents? The answer depends on the point of view that we adopt. From the point of view of an administrator in charge of national energy policies, the nuclear-energy program makes sense; the cost–benefit ratio for a nuclear power plant is favorable, or at least more favorable than for airplanes or automobiles. But from the point of view of a citizen behind whose backyard the power company is erecting a nuclear reactor, a cost–benefit analysis is of little consolation. The citizen knows he can avoid airplane and automobile accidents by avoiding airplanes and automobiles, but there is nothing he can do to avoid a nuclear-reactor accident except to insist that the greatest and most conscientious safety precautions be taken in the design, construction, and operation of the reactor.

## Further Reading

*Energy, Ecology, and the Environment* by R. Wilson and W. J. Jones (Academic, New York, 1974) deals with the energy problem from the point of view of physics; *Energy: An Introduction to Physics* by R. H. Romer (Freeman, San Francisco, 1976) deals with physics from the point of view of energy. Both books contain a wealth of information on energy resources, conversion, dissipation, and costs. The valuable statistical and numerical information in the appendices of Romer's book has been reprinted in an expanded and updated form in *Energy Facts and Figures* by R. H. Romer (Spring Street Press, Amherst, 1983).

The article "The World's Evolving Energy System" by M. K. Hubbert in the *American Journal of Physics,* November 1981, gives an authoritative, up-to-date review of our fossil fuel resources. *Ecoscience* by P. R. Ehrlich, A. H. Ehrlich, and J. H. Holdren (Freeman, San Francisco, 1977) is an encyclopedic textbook on population, resources, and environment; it includes excellent annotated lists of further books and articles. *Energy Resources* by J. T. McMullan, R. Morgan, and R. B. Murray (Wiley, New York, 1977) is a short and neat textbook which emphasizes energy technology. *Man, Energy, Society* by E. Cook (Freeman, San Francisco, 1976) gives a broad survey of global energy resources, as well as a brilliant outline of the history of man's use of energy and the consequent effects on social fabric and lifestyle; it includes extensive lists of further books and articles.

*Energy from Heaven and Earth* by E. Teller (Freeman, San Francisco, 1979) is an informal discussion of the energy problem and its possible solutions, enlivened by many a shrewd remark; it is noteworthy that although Teller believes in the safety of existing reactors, he would prefer to bury reactors deep in the ground. *The Global 2000 Report to the President* by the Council of Environmental Quality and the Department of State (U.S. Government Printing Office, Washington, D.C., 1981) is a study of the probable changes in the world's population, resources, and environment in the next 20 years.

The entire issue of *Scientific American,* September 1971, is devoted to diverse aspects of the energy problem. The reprint volume *Energy* by S. F. Singer is a selection of more recent articles from *Scientific American,* among them an incisive article by H. Bethe with forceful arguments for the necessity of fission power. Another useful reprint volume is *Energy and Man: Technical and Social Aspects of Energy,* edited by M. G. Morgan (IEEE Press, New York, 1975), with a wide selection of articles on energy technology and on social issues.

*Nuclear Science and Society* by B. L. Cohen (Doubleday, Garden City, N.Y., 1974) is a nicely written introduction to nuclear energy and its benefits and risks. *Cosmology, Fusion, and Other Matters,* edited by F. R. Reines (University of Colorado Press, Boulder, 1972), contains an excellent chapter by J. L. Tuck on fusion. *The Accident Hazards of Nuclear Power Plants* by R. E. Webb (University of Massachusetts Press, Amherst, 1976) and *The Menace of Atomic Energy* by R. Nader and J. Abbots (Norton, New York, 1977) give warning of the dangers of nuclear reactors. *Nuclear Power: Both Sides,* edited by M. Kaku and J. Trainer (Norton, New York, 1982), is a collection of informative essays by supporters and opponents of nuclear power, dealing with safety, waste disposal, economics, and alternative energy sources. *Chernobyl: The End of the Nuclear Dream* by N. Hawkes et al. (Vintage Books, New York, 1986) gives a vivid account of the series of operator errors that led to the explosion of the Chernobyl reactor, the courageous struggle to bring the fire and the radiation under control, and the aftermath.

The following are some recent articles not included in the above reprint volumes:

"The Reprocessing of Nuclear Fuels," W. P. Bebbington, *Scientific American,* December 1976
"World Coal Production," E. D. Griffith and A. W. Clarke, *Scientific American,* January 1979
"Progress Towards a Tokamak Fusion Reactor," H. P. Furth, *Scientific American,* August 1979
"Acid Rain," G. E. Likens, R. F. Wright, J. N. Galloway, and T. J. Butler, *Scientific American,* October 1979
"Energy-Storage Systems," F. R. Kalhammer, *Scientific American,* December 1979
World Uranium Resources," K. S. Deffeyes and I. D. Mac-

Gregor, *Scientific American,* January 1980
"The Safety of Fission Reactors," H. W. Lewis, *Scientific American,* March 1980
"Energy," W. Sassin, *Scientific American,* September 1980
"Oil Mining," R. A. Dick and S. P. Wimpfen, *Scientific American,* October 1980
"Toward a Rational Strategy for Oil Exploration," H. W. Menard, *Scientific American,* January 1981
"Catastrophic Releases of Radioactivity," S. A. Fetter and K. Tsipis, *Scientific American,* April 1981
"The Fuel Economy of Light Vehicles," C. L. Gray and F. von Hippel, *Scientific American,* May 1981
"Gas-Cooled Nuclear Reactors," H. M. Agnew, *Scientific American,* June 1981
"Carbon Dioxide and World Climate," R. Revelle, *Scientific American,* August 1982
"A Debate on Radioactive-Waste Disposal," F. A. Donath and R. O. Pohl, *Physics Today,* December 1982
"The Engineering of Magnetic Fusion Reactors," R. W. Conn, *Scientific American,* October 1983
"Modern Windmills," P. M. Moretti and L. V. Divone, *Scientific American,* June 1986
"Power from the Sea," T. R. Penney and D. Bharathan, *Scientific American,* January 1987
"Photovoltaic Power," Y. Hamakawa, *Scientific American,* April 1987
"Coal-fired Power Plants for the Future," R. E. Balzshiser and K. E. Yeager, *Scientific American,* September 1987
"Modeling Tidal Power," D. A. Greenberg, *Scientific American,* November 1987

## Questions

1. According to the flow chart for the energy conversions in an automobile (see Figure V.1), where is the greatest potential for possible increases in efficiency?

2. Figure V.1 shows the flow chart for the energy conversions in an automobile traveling on a level road. Qualitatively, what changes would you expect in the flow chart if the automobile were traveling uphill? Downhill?

3. Give an example of conversion of mechanical kinetic energy into gravitational potential energy. How efficient is this conversion in your example?

4. Can you think of any device that converts chemical energy directly into mechanical energy without an intermediate stage of thermal energy?

5. If you convert solar energy into electric energy (by means of a solar cell) and then convert this into mechanical energy (by means of an electric motor), what is the overall conversion efficiency? Use the numbers given in Table V.1.

6. Throughout the nineteenth century most of the power required for agriculture was supplied by horses. What are the advantages and disadvantages of operating a farm on horsepower?

7. Sailing ships obtain their power from the wind, which is free. Nevertheless, early in the twentieth century, shipowners found that sailing ships could not compete economically with steamships. What are the economic disadvantages of sailing ships?

8. According to Figure V.2, roughly what percentage of our energy supply will be due to nuclear power by the year 2000?

9. When Dmitrii Mendeleev made his first acquaintance with petroleum, he declared, "This material is too valuable to burn," a sentiment echoed in more recent times by the late Shah of Iran. Explain.

10. Why is it difficult to separate the isotopes $^{238}U$ and $^{235}U$?

11. A fusion reactor would produce a large amount of tritium ($^{3}H$), a radioactive isotope of hydrogen. If the tritium were accidentally released into the environment, it is likely to contaminate the water. Explain.

12. A temperature of 60°F indoors feels quite chilly. But the same temperature outdoors feels comfortable. Why?

13. In terms of the law of heat conduction [Eq. (20.5)] explain why less heat is lost through the walls of a house if the internal temperature is lowered.

14. According to a simple rule, the amount of heat required to keep a house at a comfortable temperature over a number of days is proportional to the number of days multiplied by the deviation of the average outside temperature from 65°F. This product is called the number of **degree-days,**

$$[\text{number of degree-days}] = [\text{number of days}] \times (65\,^\circ\text{F} - \overline{T})$$

For instance, if the average temperature is 45°F over a period of 3 days, the number of degree-days is $3 \times 20$, or 60. What would the average temperature have to be to give 60 degree-days in a single day?

15. Some people believe that they can save energy by lowering the thermostat in their home at night, while others argue that the extra heating required to rewarm the home in the morning more than balances what is saved at night. Who is right?

16. According to recommendations by experts, houses in the colder parts of the United States would derive maximum benefits from triple-glazed windows on the north side, but only double-glazed windows on the south side. Explain this, taking into account that a sheet of glass partially reflects sunlight.

17. The dissipation of the power obtained from burning fossil fuels or nuclear fuels delivers *extra* heat to the surface of the Earth, but the dissipation of the power obtained from hydroelectric generators or from windmills does not. Explain.

18. In what kind of orbit would we have to place the satellite power station mentioned in Section V.4?

19. Acid rain contains nitric and sulfuric acids. From what do these acids come?

20. The fly ash spewed out by smokestacks in the United States was 35 million metric tons in 1968, but only 8 million tons in 1980. What brought about this decrease?

21. Compare the consequences of the Chernobyl disaster with the consequences of the major reactor accidents envisioned in the WASH reports.

22. Edward Teller has recommended that all nuclear reactors be buried 60 m under the ground. What would be the advantages?

# APPENDIX 1: INDEX TO TABLES

*Note:* In the two-volume edition, all pages after 570 are in Vol. 2. Interlude V is the last in Vol. 1.

# APPENDIX 2: MATHEMATICAL SYMBOLS AND FORMULAS

## A2.1 Symbols

$a = b$ means $a$ equals $b$
$a \neq b$ means $a$ is not equal to $b$
$a > b$ means $a$ is greater than $b$
$a < b$ means $a$ is less than $b$
$a \geq b$ means $a$ is not less than $b$
$a \leq b$ means $a$ is not greater than $b$
$a \propto b$ means $a$ is proportional to $b$
$a \cong b$ means $a$ is approximately equal to $b$
$a \sim b$ means $a$ is of the order of magnitude of $b$, i.e., $a$ is within a factor of 10 or so of $b$
$a \gg b$ means $a$ is much larger than $b$
$a \ll b$ means $a$ is much less than $b$
$\Sigma_i a_i$ stands for the sum $a_1 + a_2 + a_3 + a_4 + \cdots$
$n!$ (or "$n$ factorial") stands for the product $1 \cdot 2 \cdot 3 \cdots n$
$\pi = 3.14159\ldots$
$\mathrm{e} = 2.71828\ldots$

## A2.2 The Quadratic Equation

The quadratic equation $ax^2 + bx + c = 0$ has two solutions:

$$x = \frac{-b \pm \sqrt{b^2 - 4ac}}{2a} \tag{1}$$

## A2.3 Some Approximations

The following approximations are valid for small values of $x$, that is, $x \ll 1$:

$$(1 + x)^n \cong 1 + nx \tag{2}$$

$$(1 + x)^{\frac{1}{2}} \cong 1 + \frac{x}{2} \tag{3}$$

$$\frac{1}{(1 + x)} \cong 1 - x \tag{4}$$

$$\frac{1}{(1 + x)^{\frac{1}{2}}} \cong 1 - \frac{x}{2} \tag{5}$$

These approximations can be derived from the binomial expansion

$$(1+x)^n = 1 + nx + \frac{n(n-1)}{2!}x^2 + \frac{n(n-1)(n-2)}{3!}x^3 + \cdots \quad (6)$$

by neglecting all powers of $x$, except the first power.

## A2.4 The Exponential and the Logarithmic Functions

The **exponential function** $\exp(x)$ is defined by the following infinite series:

$$\exp(x) = 1 + x + \frac{x^2}{2!} + \frac{x^3}{3!} + \frac{x^4}{4!} + \cdots \quad (7)$$

This function is equivalent to raising the constant $e = 2.71828 \ldots$ to the power $x$,

$$\exp(x) = e^x \quad (8)$$

The **natural logarithm** Ln $x$ is the inverse of the exponential function, so

$$x = e^{\ln x} \quad (9)$$

and

$$x = \ln(e^x) \quad (10)$$

Natural logarithms obey the usual rules for logarithms,

$$\ln(x \cdot y) = \ln x + \ln y \quad (11)$$

$$\ln\left(\frac{x}{y}\right) = \ln x - \ln y \quad (12)$$

$$\ln(x^a) = a \ln x \quad (13)$$

Note that

$$\ln e = 1 \quad (14)$$

and

$$\ln 10 = 2.3026 \ldots \quad (15)$$

If we designate the base-10, or **common logarithm,** by log $x$, then the relationship between the two kinds of logarithm is as follows:

$$\ln x = \ln(10^{\log x}) = (\log x)(\ln 10) = 2.3026 \log x \quad (16)$$

# APPENDIX 3: PERIMETERS, AREAS, AND VOLUMES

[perimeter of a circle of radius $r$] $= 2\pi r$
[area of a circle of radius $r$] $= \pi r^2$
[area of a triangle of base $b$, altitude $h$] $= hb/2$
[surface of a sphere of radius $r$] $= 4\pi r^2$
[volume of a sphere of radius $r$] $= 4\pi r^3/3$
[curved surface of a cylinder of radius $r$, height $h$] $= 2\pi rh$
[volume of a cylinder of radius $r$, height $h$] $= \pi r^2 h$

# APPENDIX 4: BRIEF REVIEW OF TRIGONOMETRY

## A4.1 Angles

The angle between two intersecting straight lines is defined as the fraction of a complete circle included between these lines (Figure A4.1). To express the angle in **degrees,** we assign an angular magnitude of 360° to the complete circle; any arbitrary angle is then an appropriate fraction of 360°. To express the angle in **radians,** we assign an angular magnitude of $2\pi$ radian to the complete circle; any arbitrary angle is then an appropriate fraction of $2\pi$. For example, the angle shown in Figure A4.1 is $\frac{1}{8}$ of a complete circle, that is, 45°, or $\pi/4$ radian. In view of the definition of angle, the length of arc included between the two intersecting straight lines is proportional to the angle $\theta$ between these lines; if the angle is expressed in radians, then the constant of proportionality is simply the radius:

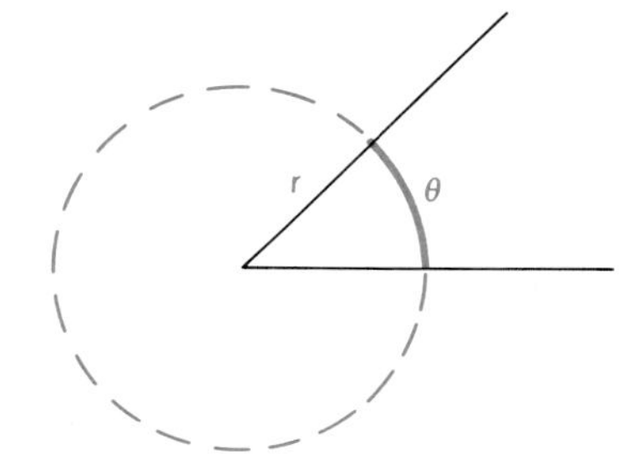

**Fig. A4.1** The angle $\theta$ in this diagram is $\theta = 45°$.

$$s = r\theta \tag{1}$$

Since $2\pi$ radian $= 360°$, it follows that

$$1 \text{ radian} = \frac{360°}{2\pi} = \frac{360°}{2 \times 3.14159} = 57.2958° \tag{2}$$

Each degree is divided into 60 minutes of arc (arcminutes), and each of these into 60 seconds of arc (arcseconds). In degrees, minutes of arc, and seconds of arc, the radian is

$$1 \text{ radian} = 57°\ 17'\ 44.8'' \tag{3}$$

## A4.2 The Trigonometric Functions

The trigonometric functions of an angle are defined as ratios of the lengths of the sides of a right triangle erected on this angle. Figure A4.2 shows an acute angle $\theta$ and a right triangle, one of whose angles coincides with $\theta$. The adjacent side $OQ$ has a length $x$, the opposite side

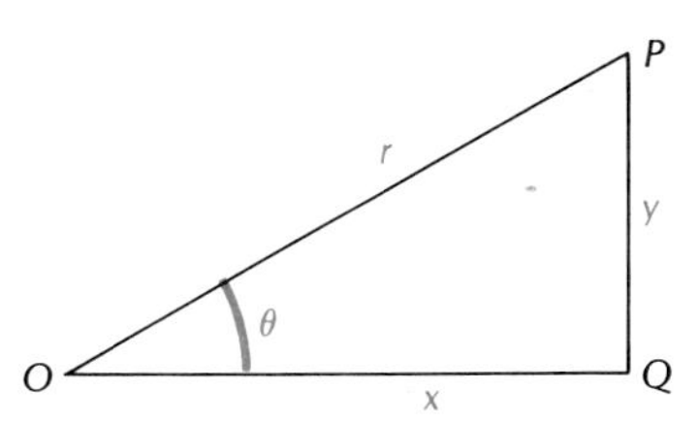

**Fig. A4.2** A right triangle.

*QP* a length *y*, and the hypothenuse *OP* a length *r*. The **sine, cosine, tangent, cotangent, secant,** and **cosecant** of the angle $\theta$ are then defined as follows:

$$\sin\theta = y/r \tag{4}$$

$$\cos\theta = x/r \tag{5}$$

$$\tan\theta = y/x \tag{6}$$

$$\cot\theta = x/y \tag{7}$$

$$\sec\theta = r/x \tag{8}$$

$$\csc\theta = r/y \tag{9}$$

---

EXAMPLE 1. Find the sine, cosine, and tangent for angles of 0°, 90°, and 45°.

SOLUTION: For an angle of 0°, the opposite side is zero ($y = 0$), and the adjacent side coincides with the hypothenuse ($x = r$). Hence

$$\sin 0° = 0 \qquad \cos 0° = 1 \qquad \tan 0° = 0 \tag{10}$$

For an angle of 90°, the adjacent side is zero ($x = 0$), and the opposite side coincides with the hypothenuse ($y = r$). Hence

$$\sin 90° = 1 \qquad \cos 90° = 0 \qquad \tan 90° = \infty \tag{11}$$

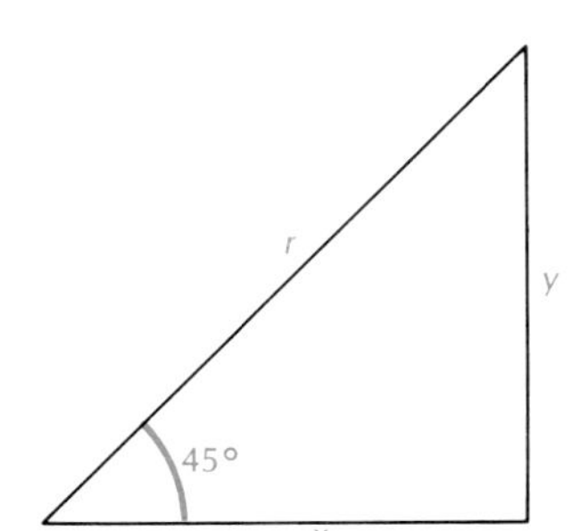


**Fig. A4.3** A right triangle with an angle of 45°.

Finally, for an angle of 45° (Figure A4.3), the adjacent and the opposite sides have the same length ($x = y$) and the hypothenuse has a length of $\sqrt{2}$ times the length of either side ($r = \sqrt{2}x = \sqrt{2}y$). Hence

$$\sin 45° = \frac{1}{\sqrt{2}} \qquad \cos 45° = \frac{1}{\sqrt{2}} \qquad \tan 45° = 1 \tag{12}$$

---

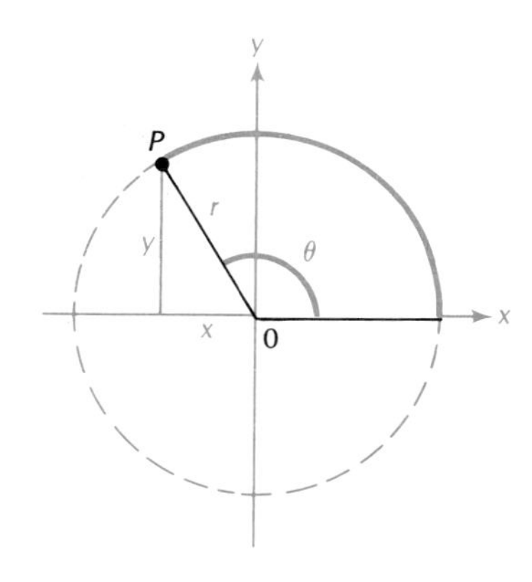


**Fig. A4.4** The angle $\theta$ in this diagram is larger than 90°.

The definitions (4)–(9) are also valid for angles greater than 90°, such as the angle shown in Figure A4.4. In the general case, the quantities *x* and *y* must be interpreted as the rectangular coordinates of the point *P*. For any angle larger than 90°, one or both of the coordinates *x* and *y* are negative. Hence some of the trigonometric functions will also be negative. For instance,

$$\sin 135° = \frac{1}{\sqrt{2}} \qquad \cos 135° = -\frac{1}{\sqrt{2}} \qquad \tan 135° = -1 \tag{13}$$

Figure A4.5 shows plots of the sine, cosine, and tangent vs. $\theta$ for the range 0°–360°.

If the angle $\theta$ is small, say, $\theta < 0.2$ radian, or $\theta < 10°$, then the length of the opposite side is approximately equal to the length of the circular arc (Figure A4.6). If we express the angle in radians, this length is [see Eq. (1)]

$$y \cong s = r\theta \tag{14}$$

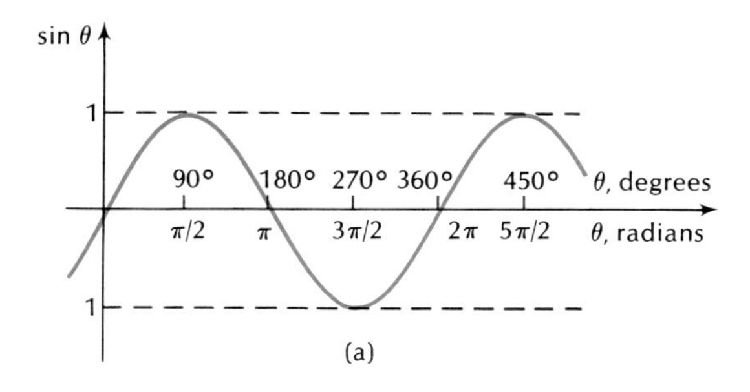


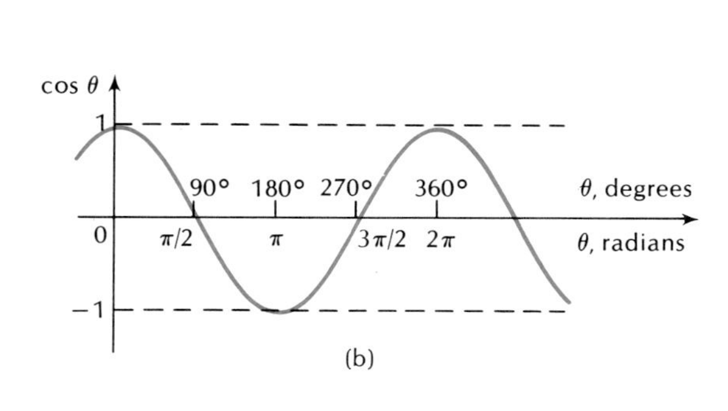


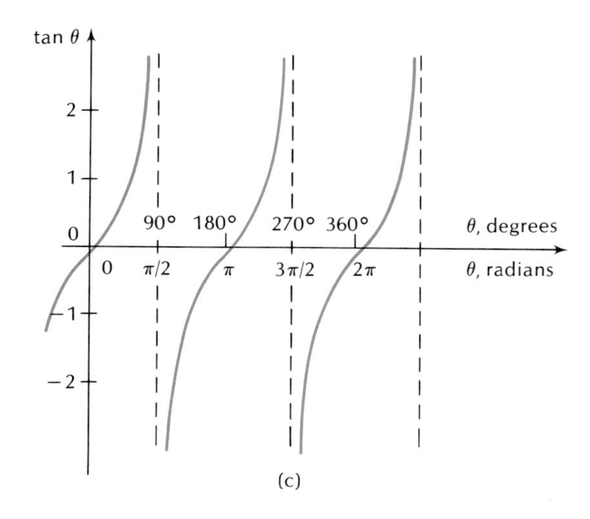


**Fig. A4.5**

and therefore

$$\sin \theta \cong \theta \tag{15}$$

To obtain a corresponding approximation for $\cos \theta$, we use the Pythagorean theorem to express the adjacent side in terms of the hypothenuse $r$ and the opposite side $y \cong r\theta$,

$$x = \sqrt{r^2 - y^2} \cong \sqrt{r^2 - r^2\theta^2} \tag{16}$$

$$\cong r\sqrt{1 - \theta^2} \tag{17}$$

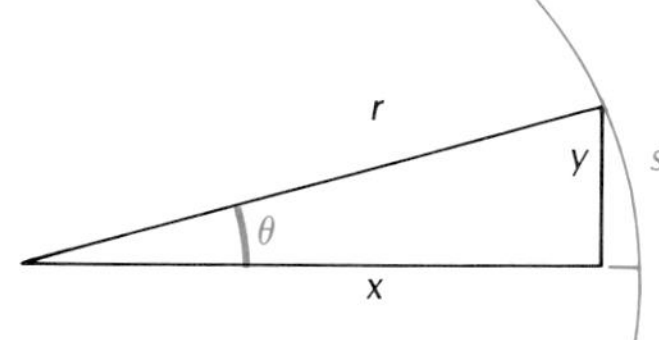


**Fig. A4.6** A small angle.

For small $\theta$, Eq. (A2.3) gives us the approximation

$$\sqrt{1 - \theta^2} \cong 1 - \theta^2/2 \tag{18}$$

so

$$\cos \theta = x/r \cong 1 - \theta^2/2 \tag{19}$$

When using the approximate formulas (15) and (19), we must always remember that the angle $\theta$ is expressed in radians!

The inverse trigonometric functions $\mathbf{sin^{-1}}$, $\mathbf{cos^{-1}}$, and $\mathbf{tan^{-1}}$ give the angle that corresponds to a specified value of the sine, cosine, or tangent, respectively. Thus, if

$$\sin \theta = u$$

then

$$\theta = \sin^{-1} u$$

## A4.3 Trigonometric Identities

From the definitions (4)–(9) we immediately find the following identities:

$$\tan\theta = \sin\theta/\cos\theta \tag{20}$$

$$\cot\theta = 1/\tan\theta \tag{21}$$

$$\sec\theta = 1/\cos\theta \tag{22}$$

$$\csc\theta = 1/\sin\theta \tag{23}$$

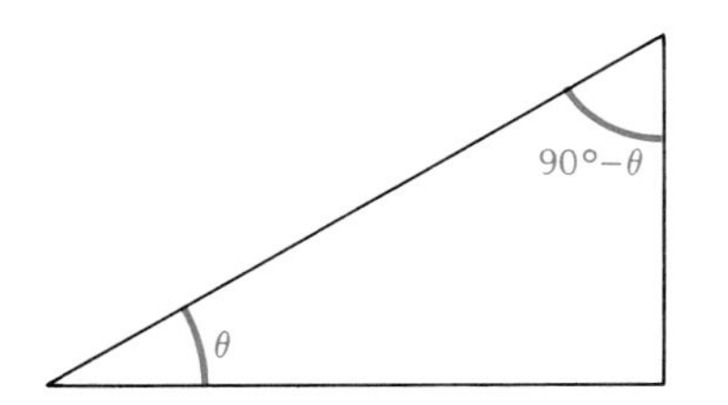


**Fig. A4.7** A right triangle with angles $\theta$ and $90° - \theta$.

Figure A4.7 shows a right triangle with angles $\theta$ and $90° - \theta$. Since the adjacent side for the angle $\theta$ is the opposite side for the angle $90° - \theta$ and vice versa, we see that the trigonometric functions also obey the following identities:

$$\sin(90° - \theta) = \cos\theta \tag{24}$$

$$\cos(90° - \theta) = \sin\theta \tag{25}$$

$$\tan(90° - \theta) = \cot\theta = 1/\tan\theta \tag{26}$$

According to the Pythagorean theorem, $x^2 + y^2 = r^2$. With $x = r\cos\theta$ and $y = r\sin\theta$ this becomes $r^2\cos^2\theta + r^2\sin^2\theta = r^2$, or

$$\cos^2\theta + \sin^2\theta = 1 \tag{27}$$

The following are a few other trigonometric identities, which we state without proof:

$$\sec^2\theta = 1 + \tan^2\theta$$

$$\csc^2\theta = 1 + \cot^2\theta$$

$$\sin 2\theta = 2\sin\theta\cos\theta$$

$$\cos 2\theta = 2\cos^2\theta - 1$$

$$\sin\tfrac{1}{2}\theta = \sqrt{(1 - \cos\theta)/2}$$

$$\cos\tfrac{1}{2}\theta = \sqrt{(1 + \cos\theta)/2}$$

$$\sin(\alpha + \beta) = \sin\alpha\cos\beta + \cos\alpha\sin\beta$$

$$\cos(\alpha + \beta) = \cos\alpha\cos\beta - \sin\alpha\sin\beta$$

$$\tan(\alpha + \beta) = \frac{\tan \alpha + \tan \beta}{1 - \tan \alpha \tan \beta}$$

$$\sin \alpha + \sin \beta = 2 \sin \tfrac{1}{2}(\alpha + \beta) \cos \tfrac{1}{2}(\alpha - \beta)$$

$$\cos \alpha + \cos \beta = 2 \cos \tfrac{1}{2}(\alpha + \beta) \cos \tfrac{1}{2}(\alpha - \beta)$$

## A4.4 The Laws of Cosines and of Sines

In an arbitrary triangle the lengths of the sides and the angles obey the laws of cosines and of sines. The **law of cosines** states that if the lengths of two sides are $A$ and $B$ and the angle between them is $\gamma$ (Figure A4.8), then the length of the third side is given by

$$C^2 = A^2 + B^2 - 2AB \cos \gamma \tag{28}$$

The **law of sines** states that the sines of the angles of the triangle are in the same ratio as the lengths of the opposite sides (Figure A4.8):

$$\frac{\sin \alpha}{A} = \frac{\sin \beta}{B} = \frac{\sin \gamma}{C} \tag{29}$$

Both of these laws are very useful in the calculation of unknown lengths or angles of a triangle.

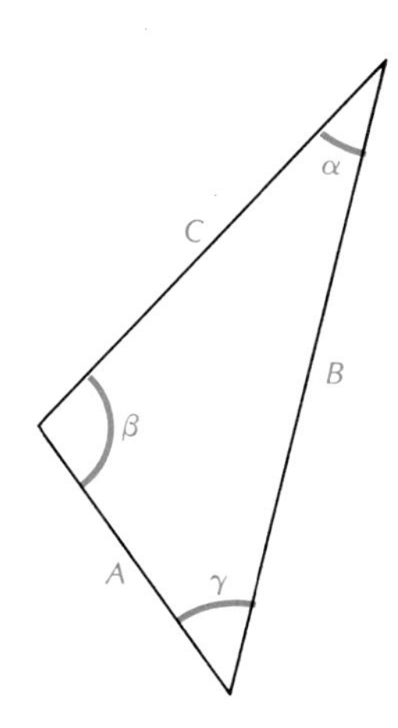


**Fig. A4.8** An arbitrary triangle.

# APPENDIX 5: BRIEF REVIEW OF CALCULUS

## A5.1 Derivatives

We saw in Section 2.3 that if the position of a particle is some function of time, say $x = x(t)$, then the instantaneous velocity of the particle is the derivative of $x$ with respect to $t$,

$$v = \frac{dx}{dt} \tag{1}$$

This derivative is defined by first looking at a small increment $\Delta x$ that results from a small increment $\Delta t$, and then evaluating the ratio $\Delta x/\Delta t$, in the limit when both $\Delta x$ and $\Delta t$ tend toward zero. Thus

$$\frac{dx}{dt} = \lim_{\Delta t \to 0} \frac{\Delta x}{\Delta t} \tag{2}$$

Graphically, in a plot of the worldline, the derivative $dx/dt$ is the slope of the straight line tangent to the (curved) worldline at the time $t$ (see Figure 2.6).

In general, if $f = f(u)$ is some given function of a variable $u$, the **derivative** of $f$ with respect to $u$ is defined by

$$\frac{df}{du} = \lim_{\Delta u \to 0} \frac{\Delta f}{\Delta u} \tag{3}$$

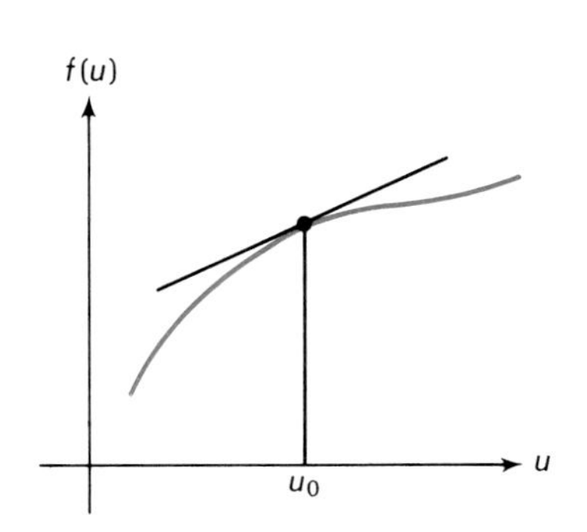


**Fig. A5.1** The derivative of $f(u)$ at $u_0$ is the slope of the straight line tangent to the curve at $u_0$.

In a plot of $f$ vs. $u$, this derivative is the slope of the straight line tangent to the curve representing $f(u)$ (see Figure A5.1).

Starting with the definition (3) we can find the derivative of any function (provided the function is sufficiently smooth so the derivative exists!). For example, consider the function $f(u) = u^2$. If we increase $u$ to $u + \Delta u$, the function $f(u)$ increases to

$$f + \Delta f = (u + \Delta u)^2 \tag{4}$$

and therefore

$$\Delta f = (u + \Delta u)^2 - f = (u + \Delta u)^2 - u^2$$

$$= 2u\,\Delta u + (\Delta u)^2 \tag{5}$$

The derivative $df/du$ is then

$$\frac{df}{du} = \lim_{\Delta u \to 0} \frac{\Delta f}{\Delta u} = \lim_{\Delta u \to 0} \frac{2u\,\Delta u + (\Delta u)^2}{\Delta u} \tag{6}$$

$$= \lim_{\Delta u \to 0} (2u) + \lim_{\Delta u \to 0} (\Delta u) \tag{7}$$

The second term on the right side vanishes in the limit $\Delta u \to 0$; the first term is simply $2u$. Hence

$$\frac{df}{du} = 2u \tag{8}$$

or

$$\frac{d}{du}(u^2) = 2u \tag{9}$$

This is one instance of the general rule for the differentiation of $u^n$:

$$\frac{d}{du}(u^n) = nu^{n-1} \tag{10}$$

This general rule is valid for any positive or negative number $n$, including zero. The proof of this rule can be constructed by an argument similar to that above. Table A5.1 lists the derivatives of the most common functions.

**Table A5.1** SOME DERIVATIVES

$\frac{d}{du} u^n = nu^{n-1}$

$\frac{d}{du} \ln u = \frac{1}{u}$

$\frac{d}{du} e^u = e^u$

$\frac{d}{du} \sin u = \cos u$ (where $u$ is in *radians*)

$\frac{d}{du} \cos u = -\sin u$ (where $u$ is in *radians*)

$\frac{d}{du} \tan u = \sec^2 u$ (where $u$ is in *radians*)

$\frac{d}{du} \cot u = -\csc^2 u$ (where $u$ is in *radians*)

$\frac{d}{du} \sec u = \tan u \sec u$ (where $u$ is in *radians*)

$\frac{d}{du} \csc u = -\cot u \csc u$ (where $u$ is in *radians*)

$\frac{d}{du} \sin^{-1} u = 1/\sqrt{1-u^2}$ (where $u$ is in *radians*)

$\frac{d}{du} \cos^{-1} u = -1/\sqrt{1-u^2}$ (where $u$ is in *radians*)

$\frac{d}{du} \tan^{-1} u = \frac{1}{1+u^2}$ (where $u$ is in *radians*)

## A5.2 Other Important Rules for Differentiation

**i. Derivative of a constant times a function:**

$$\frac{d}{du}(cf) = c\frac{df}{du} \tag{11}$$

For instance,

$$\frac{d}{du}(6u^2) = 6\frac{d}{du}(u^2) = 6 \times 2u = 12u$$

**ii. Derivative of the sum of two functions:**

$$\frac{d}{du}(f+g) = \frac{df}{du} + \frac{dg}{du} \tag{12}$$

For instance,

$$\frac{d}{du}(6u^2 + u) = \frac{d}{du}(6u^2) + \frac{d}{du}(u) = 12u + 1$$

**iii. Derivative of the product of two functions:**

$$\frac{d}{du}(f \times g) = g\frac{df}{du} + f\frac{dg}{du} \tag{13}$$

For instance,

$$\frac{d}{du}(u^2 \sin u) = \sin u \frac{d}{du}u^2 + u^2\frac{d}{du}\sin u$$

$$= \sin u \times 2u + u^2 \times \cos u$$

**iv. Chain rule for derivatives:** If $f$ is a function of $g$ and $g$ is a function of $u$, then

$$\frac{d}{du}f(g) = \frac{df}{dg}\frac{dg}{du} \tag{14}$$

For instance, if $g = 2u$ and $f(g) = \sin g$, then

$$\frac{d}{du}\sin(2u) = \frac{d\sin(2u)}{d(2u)}\frac{d(2u)}{du}$$

$$= \cos(2u) \times 2$$

## A5.3 Integrals

We have learned that if the position of a particle is known as a function of time, then we can find the instantaneous velocity by differentiation. What about the converse problem: if the instantaneous velocity is known as a function of time, how can we find the position? In Section 2.5 we learned how to deal with this problem in the special case of motion with constant acceleration. The velocity is then a fairly simple function of time [see Eq. (2.17)]

$$v = v_0 + at \tag{15}$$

and the position deduced from this velocity is [see Eq. (2.22)]

$$x = x_0 + v_0 t + \tfrac{1}{2}at^2 \tag{16}$$

where $x_0$ and $v_0$ are the initial position and velocity at the initial time $t_0 = 0$. Now we want to deal with the general case of a velocity that is an arbitrary function of time,

$$v = v(t) \tag{17}$$

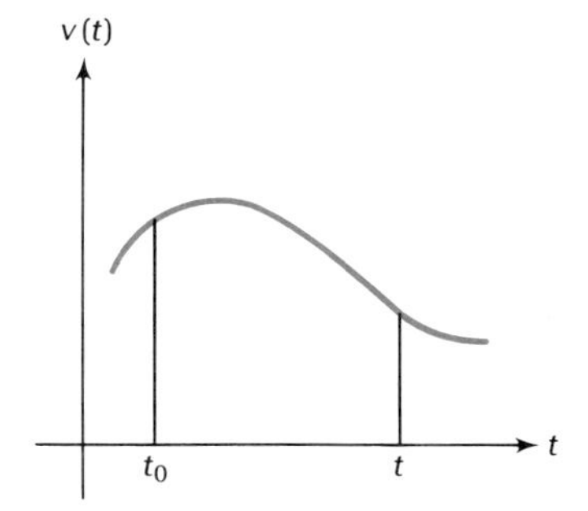


**Fig. A5.2** Plot of a function $v(t)$.

Figure A5.2 shows what a plot of $v$ vs. $t$ might look like. At the initial time $t_0$, the particle has an initial position $x_0$ (for the sake of generality we now assume that $t_0 \neq 0$). We want to find the position at some later time $t$. For this purpose, let us divide the time interval $t - t_0$ into a large number of small time intervals, each of the duration $\Delta t$. The total number of intervals is $N$, so $t - t_0 = N\,\Delta t$. The first of these intervals lasts from $t_0$ to $t_0 + \Delta t$; the second from $t_0 + \Delta t$ to $t_0 + 2\,\Delta t$; etc.

In Figure A5.3 the beginnings and the ends of these intervals have been marked $t_0$, $t_1$, $t_2$, etc., with $t_1 = t_0 + \Delta t$, $t_2 = t_0 + 2\Delta t$, etc. If $\Delta t$ is sufficiently small, then during the first time interval the velocity is approximately $v(t_0)$; during the second, $v(t_1)$; etc. This amounts to replacing the smooth function $v(t)$ by a series of steps (see Figure A5.3). Thus, during the first time interval, the displacement of the particle is approximately $v(t_0)\,\Delta t$; during the second interval, $v(t_1)\,\Delta t$; etc. The net

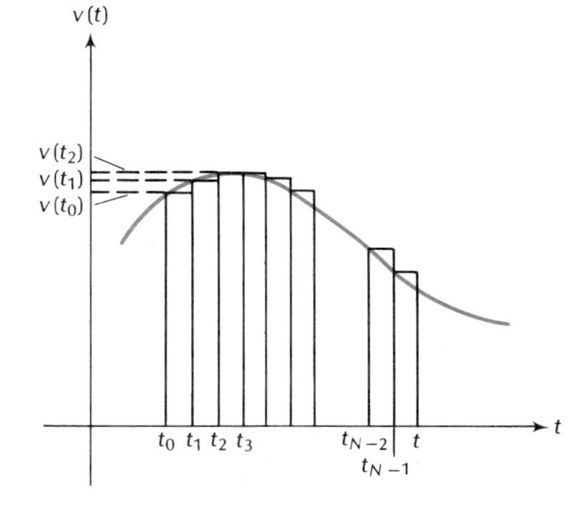


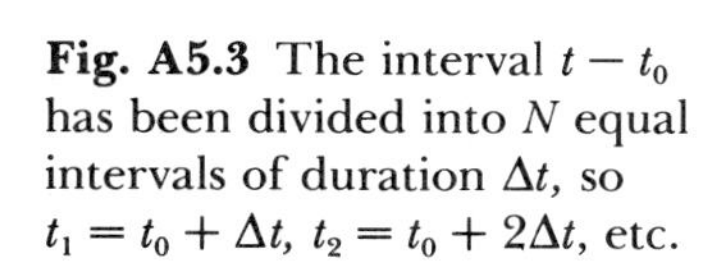
**Fig. A5.3** The interval $t - t_0$ has been divided into $N$ equal intervals of duration $\Delta t$, so $t_1 = t_0 + \Delta t$, $t_2 = t_0 + 2\Delta t$, etc.

displacement of the particle during the entire interval $t - t_0$ is the sum of all these small displacements,

$$x(t) - x_0 \cong v(t_0)\,\Delta t + v(t_1)\,\Delta t + v(t_2)\,\Delta t + \cdots \tag{18}$$

Using the standard mathematical notation for summation, we can write this as

$$x(t) - x_0 \cong \sum_{i=0}^{N-1} v(t_i)\,\Delta t \tag{19}$$

We can give this sum the following graphical interpretation: since $v(t_i)\,\Delta t$ is the area of the rectangle of height $v(x_i)$ and width $\Delta t$, the sum is the net area of all the rectangles shown in Figure A5.3, i.e., it is approximately the area under the velocity curve. Note that if the velocity is negative, the area must be reckoned as negative!

Of course, Eq. (19) is only an approximation. To find the exact displacement of the particle we must let the step size $\Delta t$ tend to zero (while the number of steps $N$ tends to infinity). In this limit, the step-like horizontal and vertical line segments in Figure A5.3 approach the smooth curve. Thus,

$$x(t) - x_0 = \lim_{\substack{\Delta t \to 0 \\ N \to \infty}} \sum_{i=0}^{N-1} v(t_i)\,\Delta t \tag{20}$$

In the notation of calculus, the right side of Eq. (20) is usually written in the following fashion:

$$x(t) - x_0 = \int_{t_0}^{t} v(t')\,dt' \tag{21}$$

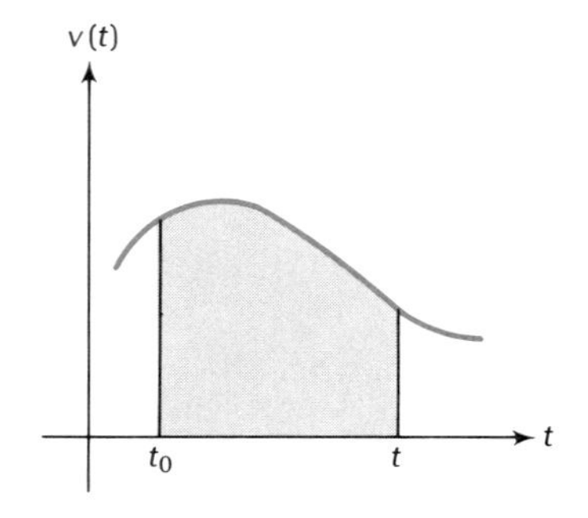


**Fig. A5.4** The area under the velocity curve.

The right side is called the **integral** of the function $v(t)$. The subscript and the superscript on the integration symbol $\int$ are called, respectively, the lower and the upper limit of integration; and $t'$ is called the variable of integration (the prime on the variable of integration $t'$ merely serves to distinguish that variable from the limit of integration $t$). Graphically, the integral is the exact area under the velocity curve between the limits $t_0$ and $t$ in a plot of $v$ vs. $t$ (see Figure A5.4). Areas below the $t$ axis must be reckoned as negative.

In general, if $f(u)$ is a function of $u$, then the integral of this function is defined by a limiting procedure similar to that described above for special case of the function $v(t)$. The integral over an interval from $u = a$ to $u = b$ is

$$\int_a^b f(u)\,du = \lim_{\substack{\Delta u \to 0 \\ N \to \infty}} \sum_{i=0}^{N-1} f(u_i)\,\Delta u \tag{22}$$

where $u_i = a + N\,\Delta u$. As in the case of the integral of $v(t)$, this integral can again be interpreted as an area: it is the area under the curve between the limits $a$ and $b$ in a plot of $f$ vs. $u$.

For the explicit evaluation of integrals we can take advantage of the connection between integrals and antiderivatives. An **antiderivative** of a function $f(u)$ is simply a function $F(u)$ such that $dF/du = f$. For example, if $f(u) = u^n$ and $n \neq -1$, then an antiderivative of $f(u)$ is $F(u) = u^{n+1}/(n+1)$. The fundamental theorem of calculus states that the integral of any function $f(u)$ can be expressed in terms of antiderivatives:

$$\int_a^b f(u)\,du = F(b) - F(a) \tag{23}$$

In essence, this means that integration is the inverse of differentiation. We will not prove this theorem here, but we remark that such an inverse relationship between integration and differentiation should not come as a surprise. We have already run across an obvious instance of such a relationship: we know that velocity is the derivative of the position, and we have seen above that the position is the integral of the velocity.

We will sometimes write Eq. (23) as

$$\int_a^b f(u)\, du = [F(u)]_a^b \tag{24}$$

where the notation $[\ \ ]_a^b$ means that the function enclosed in the brackets is to be evaluated at $a$ and at $b$, and these values are to be subtracted. For example, if $n \neq -1$,

$$\int_a^b u^n\, du = \left[\frac{u^{n+1}}{n+1}\right]_a^b = \frac{b^{n+1}}{n+1} - \frac{a^{n+1}}{n+1} \tag{25}$$

Table A5.2 lists some frequently used integrals. In this table, the limits of integration belonging with Eq. (24) have been omitted for the sake of brevity.

**Table A5.2** SOME INTEGRALS

$$\int u^n\, du = \frac{u^{n+1}}{n+1} \qquad \text{for } n \neq -1$$

$$\int \frac{1}{u}\, du = \ln u \qquad \text{for } u > 0$$

$$\int e^{ku}\, du = \frac{e^{ku}}{k}$$

$$\int \ln u\, du = u \ln u - u$$

$$\int \sin(ku)\, du = -\frac{1}{k}\cos(ku) \qquad \text{(where } ku \text{ is in } \textit{radians}\text{)}$$

$$\int \cos(ku)\, du = \frac{1}{k}\sin(ku) \qquad \text{(where } ku \text{ is in } \textit{radians}\text{)}$$

$$\int \frac{du}{1+ku} = \frac{1}{k}\ln(1+ku)$$

$$\int \frac{du}{\sqrt{k^2-u^2}} = \sin^{-1}\left(\frac{u}{k}\right)$$

$$\int \frac{du}{\sqrt{u^2 \pm k^2}} = \ln(u + \sqrt{u^2 \pm k^2})$$

$$\int \sqrt{k^2-u^2}\, du = \frac{1}{2}\left[u\sqrt{k^2-u^2} + k^2 \sin^{-1}\left(\frac{u}{k}\right)\right]$$

$$\int \frac{du}{k^2+u^2} = \frac{1}{k}\tan^{-1}\left(\frac{u}{k}\right)$$

$$\int \frac{du}{u\sqrt{k^2 \pm u^2}} = -\frac{1}{k}\ln\left(\frac{k+\sqrt{k^2 \pm u^2}}{u}\right)$$

## A5.4 The Most Important Rules for Integration

**i. Integral of a constant times a function:**

$$\int_a^b cf(u)\,du = c\int_a^b f(u)\,du$$

For instance,

$$\int_a^b 5u^2\,du = 5\int_a^b u^2\,du = 5\left(\frac{b^3}{3} - \frac{a^3}{3}\right)$$

**ii. Integral of a sum of two functions:**

$$\int_a^b [f(u) + g(u)]\ du = \int_a^b f(u)\,du + \int_a^b g(u)\,du$$

For instance,

$$\int_a^b (5u^2 + u)\,du = \int_a^b 5u^2\,du + \int_a^b u\,du = 5\left(\frac{b^3}{3} - \frac{a^3}{3}\right) + \left(\frac{b^2}{2} - \frac{a^2}{2}\right)$$

**iii. Change of limits of integration:**

$$\int_a^b f(u)\,du = \int_a^c f(u)\,du + \int_c^b f(u)\,du$$

$$\int_a^b f(u)\,du = -\int_b^a f(u)\,du$$

**iv. Change of variable of integration:**
If $u$ is a function of $v$, then

$$\int_a^b f(u)\,du = \int_{v(a)}^{v(b)} f(u)\,\frac{du}{dv}\,dv$$

For instance, with $u = v^2$,

$$\int_a^b u^3\,du = \int_a^b v^6\,du = \int_{\sqrt{a}}^{\sqrt{b}} v^6(2v)\,dv$$

Finally, let us apply these general results to some specific examples of integration of the velocity.

EXAMPLE 1. A particle with constant acceleration has the following velocity as a function of time [compare Eq. (15)]:

$$v(t) = v_0 + at$$

where $v_0$ is the velocity at $t = 0$.
By integration, find the position as a function of time.

SOLUTION: According to Eq. (21), with $t_0 = 0$,

$$x(t) - x_0 = \int_0^t v(t')\,dt' = \int_0^t (v_0 + at')\,dt'$$

Using rule ii and rule i, we find that this equals

$$x(t) - x_0 = \int_0^t v_0 dt' + \int_0^t at'\,dt' = v_0 \int_0^t dt' + a \int_0^t t'\,dt'$$

The first entry listed in Table A5.2 gives $\int dt' = t'$ (for $n = 0$) and $\int t'\,dt' = t'^2/2$ (for $n = 1$). Thus,

$$x(t) - x_0 = v_0[t']_0^t + a[t'^2/2]_0^t$$

$$= v_0 t + at^2/2 \qquad (27)$$

This, of course, agrees with Eq. (16).

---

EXAMPLE 2. The instantaneous velocity of a projectile traveling through air is the following function of time:

$$v(t) = 655.9 - 61.14t + 3.26t^2$$

where $v(t)$ is measured in meters per second and $t$ is measured in seconds. Assuming that $x = 0$ at $t = 0$, what is the position as a function of time? What is the position at $t = 3.0$ s?

SOLUTION: With $x_0 = 0$ and $t_0 = 0$, Eq. (21) becomes

$$x(t) = \int_0^t (655.9 - 61.14t' + 3.26t'^2)\,dt'$$

$$= 655.9 \int_0^t dt' - 61.14 \int_0^t t'\,dt' + 3.26 \int_0^t t'^2\,dt'$$

$$= 655.9\,[t']_0^t - 61.14[t'^2/2]_0^t + 3.26\,[t'^3/3]_0^t$$

$$= 655.9t - 61.14t^2/2 + 3.26t^3/3$$

When evaluated at $t = 3.0$ s, this yields

$$x(1.0) = 655.9 \times 3.0 - 61.14 \times (3.0)^2/2 + 3.26 \times (3.0)^3/3$$

$$= 1722 \text{ m}$$

---

EXAMPLE 3. The acceleration of a mass pushed back and forth by an elastic spring is

$$a(t) = B \cos \omega t \tag{28}$$

where $B$ and $\omega$ are constants. Find the position as a function of time. Assume $v = 0$ and $x = 0$ at $t = 0$.

SOLUTION: The calculation involves two steps: first we must integrate the acceleration to find the velocity, then we must integrate the velocity to find the position. For the first step we use an equation analogous to Eq. (21),

$$v(t) - v_0 = \int_{t_0}^{t} a(t')\, dt' \tag{29}$$

This equation becomes obvious if we remember that the relationship between acceleration and velocity is analogous to that between velocity and position. With $v_0 = 0$ and $t_0 = 0$, we obtain from Eq. (29)

$$\begin{aligned} v(t) &= \int_0^t B \cos \omega t' dt' = B\left[\frac{1}{\omega} \sin \omega t'\right]_0^t \\ &= \frac{B}{\omega} \sin \omega t \end{aligned} \tag{30}$$

Next,

$$\begin{aligned} x(t) &= \int_0^t v(t')\, dt' = \int_0^t \frac{B}{\omega} \sin \omega t'\, dt' = \frac{B}{\omega}\left[-\frac{1}{\omega} \cos \omega t'\right]_0^t \\ &= -\frac{B}{\omega^2} \cos \omega t + \frac{B}{\omega^2} \end{aligned} \tag{31}$$

## A5.5 The Taylor Series

Suppose that $f(u)$ is a smooth function of $u$ in some neighborhood of a given point $u = a$, so the function has continuous derivatives of all orders. Then the value of the function at an arbitrary point near $a$ can be expressed in terms of the following infinite series, where all the derivatives are evaluated at the point $a$:

$$f(u) = f(a) + \frac{df}{du}(x - a) + \frac{1}{2!}\frac{d^2f}{du^2}(x - a)^2 + \frac{1}{3!}\frac{d^3f}{du^3}(x - a)^3 + \cdots \tag{32}$$

This is called the **Taylor series** for the function $f(u)$ about the point $a$. The series converges, and is valid, provided $x$ is sufficiently close to $a$. How close is "sufficiently close" depends on the function $f$ and on the point $a$. Some functions, such as $\sin u$, $\cos u$, and $e^u$, are extremely well behaved, and their Taylor series converge for any choice of $x$ and of $a$. The Taylor series gives us a convenient method for the approximate evaluation of a function.

EXAMPLE 4. Find the Taylor series for $\sin u$ about the point $u = 0$.

SOLUTION: The derivatives of $\sin u$ at $u = 0$ are

$$\frac{d}{du}\sin u = \cos u = 1$$

$$\frac{d^2}{du^2}\sin u = \frac{d}{du}\cos u = -\sin u = 0$$

$$\frac{d^3}{du^3}\sin u = \frac{d}{du}(-\sin u) = -\cos u = -1$$

$$\frac{d^4}{du^4}\sin u = \frac{d}{du}(-\cos u) = \sin u = 0, \qquad \text{etc.}$$

Hence Eq. (32) gives

$$\sin u = 0 + 1 \times (u - 0) + \frac{1}{2!} \times 0 \times (u - 0)^2 + \frac{1}{3!} \times (-1) \times (u - 0)^3$$

$$+ \frac{1}{4!} \times 0 \times (u - 0)^4 + \cdots$$

$$= u - \tfrac{1}{6}u^3 + \cdots$$

Note that for very small values of $u$, we can neglect all higher powers of $u$, so $\sin u \cong u$, which agrees with the approximation given in Appendix 4.

# APPENDIX 6: THE INTERNATIONAL SYSTEM OF UNITS (SI)

## A6.1 Base Units

The SI system of units is the modern version of the metric system. The SI system recognizes seven fundamental, or base, units for length, mass, time, electric current, thermodynamic temperature, amount of substance, and luminous intensity.[a] The following definitions of the base units were adopted by the Conférence Générale des Poids et Mesures in the years indicated:

**Meter** (m) "The metre is the length of the path travelled by light in vacuum during a time interval of 1/299 792 458 of a second." (Adopted in 1983.)

**Kilogram** (kg) "The kilogram is . . . the mass of the international prototype of the kilogram." (Adopted in 1889 and in 1901.)

**Second** (s) "The second is the duration of 9 192 631 770 periods of the radiation corresponding to the transition between the two hyperfine levels of the ground state of the cesium-133 atom." (Adopted in 1967.)

**Ampere** (A) "The ampere is that constant current which, if maintained in two straight parallel conductors of infinite length, of negligible circular cross section, and placed one meter apart in vacuum, would produce between these conductors a force equal to $2 \times 10^{-7}$ newton per meter of length." (Adopted in 1948.)

**Kelvin** (K) "The kelvin . . . is the fraction 1/273.16 of the thermodynamic temperature of the triple point of water." (Adopted in 1967.)

**Mole** "The mole is the amount of substance of a system which contains as many elementary entities as there are atoms in 0.012 kilogram of carbon-12." (Adopted in 1967.)

**Candela** (cd) "The candela is the luminous intensity, in a given direction, of a source that emits monochromatic radiation of frequency $540 \times 10^{12}$ Hz and that has a radiant intensity in that direction of $\frac{1}{683}$ watt per steradian." (Adopted in 1979.)

Besides these seven base units, the SI system also recognizes two supplementary units of angle and solid angle:

**Radian** (rad) "The radian is the plane angle between two radii of a circle which cut off on the circumference an arc equal in length to the radius."

**Steradian** (sr) "The steradian is the solid angle which, having its vertex in the center of a sphere, cuts off an area equal to that of a [flat] square with sides of length equal to the radius of the sphere."

---

[a] At least two of the seven base units of the SI system are redundant. The mole is merely a certain number of atoms or molecules, in the same sense that a dozen is a number; there is no need to designate this number as a unit. The candela is equivalent to $\frac{1}{683}$ watt per steradian; it serves no purpose that is not served equally well by watt per steradian. Two other base units could be made redundant by adopting new definitions of the unit of temperature and of the unit of electric charge. Temperature could be measured in energy units because, according to the equipartition theorem, temperature is proportional to the energy per degree of freedom. Hence the kelvin could be defined as a derived unit, with 1 K $= \frac{1}{2} \times 1.38 \times 10^{-23}$ joule per degree of freedom. Electric charge could also be defined as a derived unit, to be measured with a suitable combination of the units of force and distance, as is done in the cgs system.

Furthermore, the definitions of the supplementary units — radian and steradian — are gratuitous. These definitions properly belong in the province of mathematics and there is no need to include them in a system of physical units.

## A6.2 Derived Units

The derived units are formed out of products and ratios of the base units. Table A6.1 lists those derived units that have been glorified with special names. (Other derived units are listed in the tables of conversion factors in Appendix A.7.)

**Table A6.1** NAMES OF DERIVED UNITS

| *Quantity* | *Derived Unit* | *Name* | *Symbol* |
|---|---|---|---|
| frequency | $1/s$ | hertz | Hz |
| force | $kg \cdot m/s^2$ | newton | N |
| pressure | $N/m^2$ | pascal | Pa |
| energy | $N \cdot m$ | joule | J |
| power | $J/s$ | watt | W |
| electric charge | $A \cdot s$ | coulomb | C |
| electric potential | $J/C$ | volt | V |
| electric capacitance | $C/V$ | farad | F |
| electric resistance | $V/A$ | ohm | Ω |
| conductance | $A/V$ | siemens | S |
| magnetic flux | $V \cdot s$ | weber | Wb |
| magnetic field | $V \cdot s/m^2$ | tesla | T |
| inductance | $V \cdot s/A$ | henry | H |
| temperature | K | degree Celsius | °C |
| luminous flux | $cd \cdot sr$ | lumen | lm |
| illuminance | $cd \cdot sr/m^2$ | lux | lx |
| radioactivity | $1/s$ | becquerel | Bq |
| absorbed dose | $J/kg$ | gray | Gy |
| dose equivalent | $J/kg$ | sievert | Sv |

## A6.3 Prefixes

Multiples and submultiples of SI units are indicated by prefixes, such as the familiar *kilo, centi,* and *milli* used in *kilometer, centimeter,* and *millimeter,* etc. Table A6.2 lists all the accepted prefixes. Some enjoy more popularity than others; it is best to avoid the use of uncommon prefixes, such as *atto* and *exa,* since hardly anybody will recognize those.

**Table A6.2** PREFIXES FOR UNITS

| *Multiplication factor* | *Prefix* | *Symbol* |
|---|---|---|
| $10^{18}$ | exa | E |
| $10^{15}$ | peta | P |
| $10^{12}$ | tera | T |
| $10^{9}$ | giga | G |
| $10^{6}$ | mega | M |
| $10^{3}$ | kilo | k |
| $10^{2}$ | hecto | h |
| 10 | deka | da |
| $10^{-1}$ | deci | d |
| $10^{-2}$ | centi | c |
| $10^{-3}$ | milli | m |
| $10^{-6}$ | micro | $\mu$ |
| $10^{-9}$ | nano | n |
| $10^{-12}$ | pico | p |
| $10^{-15}$ | femto | f |
| $10^{-18}$ | atto | a |

# APPENDIX 7: CONVERSION FACTORS

The units for each quantity are listed alphabetically, except that the SI unit is always listed first. The numbers are based on "American National Standard; Metric Practice" published by the Institute of Electrical and Electronics Engineers, 1982.

## ANGLE

**1 radian** $= 57.30° = 3.438 \times 10^3\,' = \frac{1}{2\pi}$ rev $= 2.063 \times 10^5\,''$

**1 degree** (°) $= 1.745 \times 10^{-2}$ **radian** $= 60' = 3600'' = \frac{1}{360}$ rev

**1 minute of arc** (′) $= 2.909 \times 10^{-4}$ **radian**$= \frac{1}{60}° = 4.630 \times 10^{-5}$ rev $= 60''$

**1 revolution** (rev) $= 2\pi$ **radian** $= 360° = 2.160 \times 10^4\,' = 1.296 \times 10^6\,''$

**1 second of arc** (″) $= 4.848 \times 10^{-6}$ **radian** $= \frac{1}{3600}° = \frac{1}{60}' = 7.716 \times 10^{-7}$ rev

## LENGTH

**1 meter** (m) $= 1 \times 10^{10}$ Å $= 6.685 \times 10^{-12}$ AU $= 100$ cm $= 1 \times 10^{15}$ Fm $= 3.281$ ft $= 39.37$ in. $= 1 \times 10^{-3}$ km $= 1.057 \times 10^{-16}$ light-year $= 1 \times 10^6\ \mu$m $= 5.400 \times 10^{-4}$ nmi $= 6.214 \times 10^{-4}$ mi $= 3.241 \times 10^{-17}$ pc $= 1.094$ yd

**1 angstrom** (Å) $= 1 \times 10^{-10}$ m $= 1 \times 10^{-8}$ cm $= 1 \times 10^{-5}$ fm $= 3.281 \times 10^{-10}$ ft $= 1 \times 10^{-4}\ \mu$m

**1 astronomical unit** (AU) $= 1.496 \times 10^{11}$ m $= 1.496 \times 10^{13}$ cm $= 1.496 \times 10^8$ km $= 1.581 \times 10^{-5}$ light-year $= 4.848 \times 10^{-6}$ pc

**1 centimeter** (cm) $= 0.01$ m $= 1 \times 10^8$ Å $= 1 \times 10^{13}$ fm $= 3.281 \times 10^{-2}$ ft $= 0.3937$ in. $= 1 \times 10^{-5}$ km $= 1.057 \times 10^{-18}$ light-year $= 1 \times 10^4\ \mu$m

**1 fermi** (fm) $= 1 \times 10^{-15}$ m $= 1 \times 10^{-13}$ cm $= 1 \times 10^5$ Å

**1 foot** (ft) $= 0.3048$ m $= 30.48$ cm $= 12$ in. $= 3.048 \times 10^5\ \mu$m $= 1.894 \times 10^{-4}$ mi $= \frac{1}{3}$ yd

**1 inch** (in.) $= 2.540 \times 10^{-2}$ m $= 2.54$ cm $= \frac{1}{12}$ ft $= 2.54 \times 10^4\ \mu$m $= \frac{1}{36}$ yd

**1 kilometer** (km) $= 1 \times 10^3$ m $= 1 \times 10^5$ cm $= 3.281 \times 10^3$ ft $= 0.5400$ nmi $= 0.6214$ mi $= 1.094 \times 10^3$ yd

**1 light-year** $= 9.461 \times 10^{15}$ m $= 6.324 \times 10^4$ AU $= 9.461 \times 10^{17}$ cm $= 9.461 \times 10^{12}$ km $= 5.879 \times 10^{12}$ mi $= 0.3066$ pc

**1 micron, or micrometer** ($\mu$m) $= 1 \times 10^{-6}$ m $= 1 \times 10^4$ Å $= 1 \times 10^{-4}$ cm $= 3.281 \times 10^{-6}$ ft $= 3.937 \times 10^{-5}$ in.

**1 nautical mile** (nmi) $= 1.852 \times 10^3$ m $= 1.852 \times 10^5$ cm $= 6.076 \times 10^3$ ft $= 1.852$ km $= 1.151$ mi

**1 statute mile** (mi) $= 1.609 \times 10^3$ m $= 1.609 \times 10^5$ cm $= 5280$ ft $= 1.609$ km $= 0.8690$ nmi $= 1760$ yd

**1 parsec** (pc) $= 3.086 \times 10^{16}$ m $= 2.063 \times 10^5$ AU $= 3.086 \times 10^{18}$ cm $= 3.086 \times 10^{13}$ km $= 3.262$ light-years

**1 yard** (yd) $= 0.9144$ m $= 91.44$ cm $= 3$ ft $= 36$ in. $= \frac{1}{1760}$ mi

TIME

**1 second** (s) = $1.157 \times 10^{-5}$ day = $\frac{1}{3600}$ h = $\frac{1}{60}$ min = $1.161 \times 10^{-5}$ sidereal day = $3.169 \times 10^{-8}$ yr
**1 day** = $8.640 \times 10^{4}$ s = 24 h = 1440 min = 1.003 sidereal days = $2.738 \times 10^{-3}$ yr
**1 hour** (h) = 3600 s = $\frac{1}{24}$ day = 60 min = $1.141 \times 10^{-4}$ yr
**1 minute** (min) = 60 s = $6.944 \times 10^{-4}$ day = $\frac{1}{60}$ h = $1.901 \times 10^{-6}$ year
**1 sidereal day** = $8.616 \times 10^{4}$ s = 0.9973 day = 23.93 h = $1.436 \times 10^{3}$ min = $2.730 \times 10^{-3}$ yr
**1 year** (yr) = $3.156 \times 10^{7}$ s = 365.24 days = $8.766 \times 10^{3}$ h = $5.259 \times 10^{5}$ min = 366.24 sidereal days

MASS

**1 kilogram** (kg) = $6.024 \times 10^{26}$ u = 5000 carats = $1.543 \times 10^{4}$ grains = 1000 g = $1 \times 10^{-3}$ t = 35.27 oz. = 2.205 lb = $1.102 \times 10^{-3}$ short ton = $6.852 \times 10^{-2}$ slug
**1 atomic mass unit** (u) = $1.6605 \times 10^{-27}$ kg = $1.6605 \times 10^{-24}$ g
**1 carat** = $2 \times 10^{-4}$ kg = 0.2 g = $7.055 \times 10^{-3}$ oz. = $4.409 \times 10^{-4}$ lb
**1 grain** = $6.480 \times 10^{-5}$ kg = $6.480 \times 10^{-2}$ g = $2.286 \times 10^{-3}$ oz. = $\frac{1}{7000}$ lb
**1 gram** (g) = $1 \times 10^{-3}$ kg = $6.024 \times 10^{23}$ u = 5 carats = 15.43 grains = $1 \times 10^{-6}$ t = $3.527 \times 10^{-2}$ oz. = $2.205 \times 10^{-3}$ lb = $1.102 \times 10^{-6}$ short ton = $6.852 \times 10^{-5}$ slug
**1 metric ton, or tonne** (t) = $1 \times 10^{3}$ kg = $1 \times 10^{6}$ g = $2.205 \times 10^{3}$ lb = 1.102 short ton = 68.52 slugs
**1 ounce** (oz.) = $2.835 \times 10^{-2}$ kg = 141.7 carats = 437.5 grains = 28.35 g = $\frac{1}{16}$ lb
**1 pound** (lb)[b] = 0.4536 kg = 453.6 g = $4.536 \times 10^{-4}$ t = 16 oz. = $\frac{1}{2000}$ short ton = $3.108 \times 10^{-2}$ slug
**1 short ton** = 907.2 kg = $9.07 \times 10^{5}$ g = 0.9072 t = 2000 lb
**1 slug** = 14.59 kg = $1.459 \times 10^{4}$ g = 32.17 lb

AREA

**1 square meter** ($m^2$) = $1 \times 10^{4}$ $cm^2$ = 10.76 $ft^2$ = $1.550 \times 10^{3}$ $in.^2$ = $1 \times 10^{-6}$ $km^2$ = $3.861 \times 10^{-7}$ $mi^2$ = 1.196 $yd^2$
**1 barn** = $1 \times 10^{-28}$ $m^2$ = $1 \times 10^{-24}$ $cm^2$
**1 square centimeter** ($cm^2$) = $1 \times 10^{-4}$ $m^2$ = $1.076 \times 10^{-3}$ $ft^2$ = 0.1550 $in.^2$ = $1 \times 10^{-10}$ $km^2$ = $3.861 \times 10^{-11}$ $mi^2$
**1 square foot** ($ft^2$) = $9.290 \times 10^{-2}$ $m^2$ = 929.0 $cm^2$ = 144 $in.^2$ = $3.587 \times 10^{-8}$ $mi^2$ = $\frac{1}{9}$ $yd^2$
**1 square inch** ($in.^2$) = $6.452 \times 10^{-4}$ $m^2$ = 6.452 $cm^2$ = $\frac{1}{144}$ $ft^2$
**1 square kilometer** ($km^2$) = $1 \times 10^{6}$ $m^2$ = $1 \times 10^{10}$ $cm^2$ = $1.076 \times 10^{7}$ $ft^2$ = 0.3861 $mi^2$
**1 square statute mile** ($mi^2$) = $2.590 \times 10^{6}$ $m^2$ = $2.590 \times 10^{10}$ $cm^2$ = $2.788 \times 10^{7}$ $ft^2$ = 2.590 $km^2$
**1 square yard** ($yd^2$) = 0.8361 $m^2$ = $8.361 \times 10^{3}$ $cm^2$ = 9 $ft^2$ = 1296 $in.^2$

VOLUME

**1 cubic meter** ($m^3$) = $1 \times 10^{6}$ $cm^3$ = 35.31 $ft^3$ = 264.2 gal. = $6.102 \times 10^{4}$ $in.^3$ = $1 \times 10^{3}$ liters = 1.308 $yd^3$
**1 cubic centimeter** ($cm^3$) = $1 \times 10^{-6}$ $m^3$ = $3.531 \times 10^{-5}$ $ft^3$ = $2.642 \times 10^{-4}$ gal. = $6.102 \times 10^{-2}$ $in.^3$ = $1 \times 10^{-3}$ liter
**1 cubic foot** ($ft^3$) = $2.832 \times 10^{-2}$ $m^3$ = $2.832 \times 10^{4}$ $cm^3$ = 7.481 gal. = 1728 $in.^3$ = 28.32 liters = $\frac{1}{27}$ $yd^3$
**1 gallon** (gal.)[c] = $3.785 \times 10^{-3}$ $m^3$ = 0.1337 $ft^3$

---

[b] This is the "avoirdupois" pound. The "troy" or "apothecary" pound is 0.3732 kg, or 0.8229 lb avoirdupois.

[c] This is the U.S. gallon; the U.K. and the Canadian gallon are $4.546 \times 10^{-3}$ $m^3$, or 1.201 U.S. gallons.

**1 cubic inch** (in.$^3$) = $1.639 \times 10^{-5}$ m$^3$ = 16.39 cm$^3$ = $5.787 \times 10^{-4}$ ft$^3$
**1 liter** (l) = $1 \times 10^{-3}$ m$^3$ = 1000 cm$^3$ = $3.531 \times 10^{-2}$ ft$^3$
**1 cubic yard** (yd$^3$) = 0.7646 m$^3$ = $7.646 \times 10^5$ cm$^3$ = 27 ft$^3$ = 202.0 gal.

### DENSITY

**1 kilogram per cubic meter** (kg/m$^3$) = $1 \times 10^{-3}$ g/cm$^3$ = $6.243 \times 10^{-2}$ lb/ft$^3$ = $8.345 \times 10^{-3}$ lb/gal. = $3.613 \times 10^{-5}$ lb/in.$^3$ = $8.428 \times 10^{-4}$ short ton/yd$^3$ = $1.940 \times 10^{-3}$ slug/ft$^3$
**1 gram per cubic centimeter** (g/cm$^3$) = $1 \times 10^3$ kg/m$^3$ = 62.43 lb/ft$^3$ = 8.345 lb/gal. = $3.613 \times 10^{-2}$ lb/in.$^3$ = 0.8428 short ton/yd$^3$ = 1.940 slug/ft$^3$
**1 lb per cubic foot** (lb/ft$^3$) = 16.02 kg/m$^3$ = $1.602 \times 10^{-2}$ g/cm$^3$ = 0.1337 lb/gal. = $1.350 \times 10^{-2}$ short ton/yd$^3$ = $3.108 \times 10^{-2}$ slug/ft$^3$
**1 pound-mass per gallon** (1 lb/gal.) = 119.8 kg/m$^3$ = 7.481 lb/ft$^3$ = 0.2325 slug/ft$^3$
**1 short ton per cubic yard** (short ton/yd$^3$) = $1.187 \times 10^3$ kg/m$^3$ = 74.07 lb/ft$^3$
**1 slug per cubic foot** (slug/ft$^3$) = 515.4 kg/m$^3$ = 0.5154 g/cm$^3$ = 32.17 lb/ft$^3$ = 4.301 lb/gal.

### SPEED

**1 meter per second** (m/s) = 100 cm/s = 3.281 ft/s = 3.600 km/h = 1.944 knot = 2.237 mi/h
**1 centimeter per second** (cm/s) = 0.01 m/s = $3.281 \times 10^{-2}$ ft/s = $3.600 \times 10^{-2}$ km/h = $1.944 \times 10^{-2}$ knot = $2.237 \times 10^{-2}$ mi/h
**1 foot per second** (ft/s) = 0.3048 m/s = 30.48 cm/s = 1.097 km/h = 0.5925 knot = 0.6818 mi/h
**1 kilometer per hour** (km/h) = 0.2778 m/s = 27.78 cm/s = 0.9113 ft/s = 0.5400 knot = 0.6214 mi/h
**1 knot, or nautical mile per hour** = 0.5144 m/s = 51.44 cm/s = 1.688 ft/s = 1.852 km/h = 1.151 mi/h
**1 mile per hour** (mi/h) = 0.4470 m/s = 44.70 cm/s = 1.467 ft/s = 1.609 km/h = 0.8690 knot

### ACCELERATION

**1 meter per second squared** (m/s$^2$) = 100 cm/s$^2$ = 3.281 ft/s$^2$ = 0.1020 G
**1 centimeter per second squared, or Gal** (cm/s$^2$) = 0.01 m/s$^2$ = $3.281 \times 10^{-2}$ ft/s$^2$ = $1.020 \times 10^{-3}$ G
**1 foot per second squared** (ft/s$^2$) = 0.3048 m/s$^2$ = 30.48 cm/s$^2$ = $3.108 \times 10^{-2}$ G
**1 G** = 9.807 m/s$^2$ = 980.7 cm/s$^2$ = 32.17 ft/s$^2$

### FORCE

**1 newton** (N) = $1 \times 10^5$ dynes = 0.1020 kp = 0.2248 lbf = $1.124 \times 10^{-4}$ short ton-force
**1 dyne** = $1 \times 10^{-5}$ N = $1.020 \times 10^{-6}$ kp = $2.248 \times 10^{-6}$ lbf = $1.124 \times 10^{-9}$ short ton-force
**1 kilopond, or kilogram force** (kp) = 9.807 N = $9.807 \times 10^5$ dynes = 2.205 lbf = $1.102 \times 10^{-3}$ short ton-force
**1 pound-force** (lbf) = 4.448 N = $4.448 \times 10^5$ dynes = 0.4536 kp = $\frac{1}{2000}$ short ton-force
**1 short ton-force** = $8.896 \times 10^3$ N = $8.896 \times 10^8$ dynes = 907.2 kp = 2000 lbf

### ENERGY

**1 joule** (J) = $9.478 \times 10^{-4}$ Btu = 0.2388 cal = $1 \times 10^7$ ergs = $6.242 \times 10^{18}$ eV = 0.7376 ft · lbf = $2.778 \times 10^{-7}$ kW · h

**1 British thermal unit** (Btu)[d] = $1.055 \times 10^3$ J = 252.0 cal = $1.055 \times 10^{10}$ ergs = 778.2 ft · lbf = $2.931 \times 10^{-4}$ kW · h
**1 calorie** (cal)[e] = 4.187 J = $3.968 \times 10^{-3}$ Btu = $4.187 \times 10^7$ ergs = 3.088 ft · lbf = $1 \times 10^{-3}$ kcal = $1.163 \times 10^{-6}$ kW · h
**1 erg** = $1 \times 10^{-7}$ J = $9.478 \times 10^{-7}$ Btu = $2.388 \times 10^{-8}$ cal = $6.242 \times 10^{11}$ eV = $7.376 \times 10^{-8}$ ftf · lb = $2.778 \times 10^{-14}$ kW · h
**1 electron-volt** (eV) = $1.602 \times 10^{-19}$ J = $1.602 \times 10^{-12}$ erg = $1.182 \times 10^{-19}$ ft · lbf
**1 foot-pound-force** (ft · lbf) = 1.356 J = $1.285 \times 10^{-3}$ Btu = 0.3239 cal = $1.356 \times 10^7$ ergs = $8.464 \times 10^{18}$ eV = $3.766 \times 10^{-7}$ kW · h
**1 kilocalorie** (kcal), or **large calorie** (Cal) = $4.187 \times 10^3$ J = $1 \times 10^3$ cal
**1 kilowatt-hour** (kW · h) = $3.600 \times 10^6$ J = 3412 Btu = $8.598 \times 10^5$ cal = $3.6 \times 10^{13}$ ergs = $2.655 \times 10^6$ ft · lbf

POWER

**1 watt** (W) = 3.412 Btu/h = 0.2388 cal/s = $1 \times 10^7$ ergs/s = 0.7376 ft · lbf/s = $1.341 \times 10^{-3}$ hp
**1 British thermal unit per hour** (Btu/h) = 0.2931 W = $7.000 \times 10^{-2}$ cal/s = 0.2162 ft · lbf/s = $3.930 \times 10^{-4}$ hp
**1 calorie per second** (cal/s) = 4.187 W = 14.29 Btu/h = $4.187 \times 10^7$ erg/s = 3.088 ft · lbf/s = $5.615 \times 10^{-3}$ hp
**1 erg per second** (erg/s) = $1 \times 10^{-7}$ W = $2.388 \times 10^{-8}$ cal/s = $7.376 \times 10^{-8}$ ft · lbf/s = $1.341 \times 10^{-10}$ hp
**1 foot-pound per second** (ft · lbf/s) = 1.356 W = 0.3238 cal/s = 4.626 Btu/h = $1.356 \times 10^7$ ergs/s = $1.818 \times 10^{-3}$ hp
**1 horsepower** (hp)[f] = 745.7 W = $2.544 \times 10^3$ Btu/h = 178.1 cal/s = 550 ft · lbf/s
**1 kilowatt** (kW) = $1 \times 10^3$ W = $3.412 \times 10^3$ Btu/h = 238.8 cal/s = 737.6 ft · lbf/s = 1.341 hp

PRESSURE

**1 newton per square meter** ($N/m^2$), or **pascal** (Pa) = $9.869 \times 10^{-6}$ atm = $1 \times 10^{-5}$ bar = $7.501 \times 10^{-4}$ cmHg = 10 dynes/$cm^2$ = $2.953 \times 10^{-4}$ in. Hg = $2.089 \times 10^{-2}$ lbf/$ft^2$ = $1.450 \times 10^{-4}$ lbf/in.$^2$ = $7.501 \times 10^{-3}$ torr
**1 atmosphere** (atm) = $1.013 \times 10^5$ $N/m^2$ = 76.00 cmHg = $1.013 \times 10^6$ dynes/$cm^2$ = 29.92 in. Hg = $2.116 \times 10^3$ lbf/$ft^2$ = 14.70 lbf/in.$^2$
**1 bar** = $1 \times 10^5$ $N/m^2$ = 0.9869 atm = 75.01 cmHg
**1 centimeter of mercury** (cmHg) = $1.333 \times 10^3$ $N/m^2$ = $1.316 \times 10^{-2}$ atm = $1.333 \times 10^{-2}$ bar = $1.333 \times 10^4$ dynes/$cm^2$ = 0.3937 in. Hg = 27.85 lbf/$ft^2$ = 0.1934 lbf/in.$^2$ = 10 torr
**1 dyne per square centimeter** (dyne/$cm^2$) = 0.1 $N/m^2$ = $9.869 \times 10^{-7}$ atm = $7.501 \times 10^{-5}$ cmHg = $2.089 \times 10^{-3}$ lbf/$ft^2$ = $1.450 \times 10^{-5}$ lbf/in.$^2$
**1 inch of mercury** (in. Hg) = $3.386 \times 10^3$ $N/m^2$ = $3.342 \times 10^{-2}$ atm = 2.540 cmHg = 0.4912 lbf/in.$^2$
**1 kilopond per square centimeter** (kp/$cm^2$) = $9.807 \times 10^4$ $N/m^2$ = 0.9678 atm = $9.807 \times 10^5$ dynes/$cm^2$ = 14.22 lbf/in.$^2$
**1 pound per square inch** (lbf/in.$^2$, or psi) = $6.895 \times 10^3$ $N/m^2$ = $6.805 \times 10^{-2}$ atm = $6.895 \times 10^4$ dynes/$cm^2$ = 2.036 in. Hg = $7.031 \times 10^{-2}$ kp/$cm^2$
**1 torr,** or **millimeter of mercury** (mmHg) = $1.333 \times 10^2$ $N/m^2$ = 0.1 cmHg

---

[d] This is the "International Table" Btu; there are several other Btus.

[e] This is the "International Table" calorie, which equals exactly 4.1868 J. There are several other calories; for instance, the thermochemical calorie, which equals 4.184 J.

[f] There are several other horsepowers; for instance, the metric horsepower, which equals 735.5 J.

ELECTRIC CHARGE[g]

**1 coulomb** (C) ⇔ $2.998 \times 10^9$ statcoulombs, or esu of charge ⇔ 0.1 abcoulomb, or emu of charge

ELECTRIC CURRENT

**1 ampere** (A) ⇔ $2.998 \times 10^9$ statamperes, or esu of current ⇔ 0.1 abampere, or emu of current

ELECTRIC POTENTIAL

**1 volt** (V) ⇔ $3.336 \times 10^{-3}$ statvolt, or esu of potential ⇔ $1 \times 10^8$ abvolts, or emu of potential

ELECTRIC FIELD

**1 volt per meter** (V/m) ⇔ $3.336 \times 10^{-5}$ statvolt/cm ⇔ $1 \times 10^6$ abvolts/cm

MAGNETIC FIELD

**1 tesla** (T), or **weber per square meter** ($Wb/m^2$) ⇔ $1 \times 10^4$ gauss

ELECTRIC RESISTANCE

**1 ohm** ($\Omega$) ⇔ $1.113 \times 10^{-12}$ statohm, or esu of resistance ⇔ $1 \times 10^9$ abohms, or emu of resistance

ELECTRIC RESISTIVITY

**1 ohm-meter** ($\Omega \cdot m$) ⇔ $1.113 \times 10^{-10}$ statohm-cm ⇔ $1 \times 10^{11}$ abohm-cm

CAPACITANCE

**1 farad** (F) ⇔ $8.988 \times 10^{11}$ statfarads, or esu of capacitance ⇔ $1 \times 10^{-9}$ abfarad, or emu of capacitance

INDUCTANCE

**1 henry** (H) ⇔ $1.113 \times 10^{-12}$ stathenry, or esu of inductance ⇔ $1 \times 10^9$ abhenrys, or emu of inductance

---

[g] The dimensions of the electric quantities in SI units, electrostatic units (esu), and electromagnetic units (emu) are different; hence the relationships among these units are correspondences (⇔) rather than equalities (=).

# APPENDIX 8: BEST VALUES OF FUNDAMENTAL CONSTANTS

The values in the following table were taken from the report of the CODATA Task Group on fundamental constants by E. R. Cohen and B. N. Taylor, *The 1986 Adjustment of the Fundamental Physical Constants,* CODATA Bulletin No. 63, November 1986. The digits in parentheses are the one-standard deviation uncertainty in the last digits of the given value.

| *Quantity* | *Symbol* | *Value* | *Units* | *Relative uncertainty (parts per million)* |
|---|---|---|---|---|
| **UNIVERSAL CONSTANTS** | | | | |
| speed of light in vacuum | $c$ | 299792458 | $ms^{-1}$ | (exact) |
| permeability of vacuum | $\mu_o$ | $4\pi \times 10^{-7}$ | $NA^{-2}$ | |
| | | $=12.566370614\ldots$ | $10^{-7}\ NA^{-2}$ | (exact) |
| permittivity of vacuum | $\varepsilon_o$ | $1/\mu_o c^2$ | | |
| | | $=8.854187817\ldots$ | $10^{-12}\ Fm^{-1}$ | (exact) |
| Newtonian constant of gravitation of gravitation | $G$ | 6.67259(85) | $10^{-11}\ m^3\ kg^{-1}s^{-2}$ | 128 |
| Planck constant | $h$ | 6.6260755(40) | $10^{-34}$ Js | 0.60 |
| in electron volts | | 4.1356692(12) | $10^{-15}$ eVs | 0.30 |
| | $\hbar = h/2\pi$ | 1.05457266(63) | $10^{-34}$ Js | 0.60 |
| in electron volts | | 6.5821220(20) | $10^{-16}$ eVs | 0.30 |
| **ELECTROMAGNETIC CONSTANTS** | | | | |
| elementary charge | $e$ | 1.60217733(49) | $10^{-19}$ C | 0.30 |
| magnetic flux quantum, $h/2e$ | $\Phi_o$ | 2.06783461(61) | $10^{-15}$ Wb | 0.30 |
| Josephson frequency–voltage ratio | $2e/h$ | 4.8359767(14) | $10^{14}\ HzV^{-1}$ | 0.30 |
| quantized Hall conductance | $e^2/h$ | 3.87404614(17) | $10^{-5}\ \Omega^{-1}$ | 0.045 |
| Bohr magneton, $e\hbar/2m_e$ | $\mu_B$ | 9.2740154(31) | $10^{-24}\ JT^{-1}$ | 0.34 |
| in electron volts | | 5.78838263(52) | $10^{-5}\ eVT^{-1}$ | 0.089 |
| nuclear magneton, $e\hbar/2m_p$ | $\mu_N$ | 5.0507866(17) | $10^{-27}\ JT^{-1}$ | 0.34 |
| in electron volts | | 3.15245166(28) | $10^{-8}\ eVT^{-1}$ | 0.089 |
| **ATOMIC CONSTANTS** | | | | |
| fine-structure constant, $e^2/4\pi\varepsilon_o\hbar c$ | $\alpha$ | 7.29735308(33) | $10^{-3}$ | 0.045 |
| inverse fine-structure constant | $\alpha^{-1}$ | 137.0359895(61) | | 0.045 |
| Rydberg constant, $\frac{1}{2}m_e c\alpha^2/h$ | $R\infty$ | 10973731.534(13) | $m^{-1}$ | 0.0012 |
| Bohr radius, $4\pi\varepsilon_o\hbar^2/m_e e^2$ | $a_o$ | 0.529177249(24) | $10^{-10}$ m | 0.045 |
| quantum of circulation | $h/2m_e$ | 3.63694807(33) | $10^{-4}\ m^2s^{-1}$ | 0.089 |
| | $h/m_e$ | 7.27389614(65) | $10^{-4}m^2s^{-1}$ | 0.089 |
| **Electron** | | | | |
| electron mass | $m_e$ | 9.1093897(54) | $10^{-31}$ kg | 0.59 |
| | | 5.48579903(13) | $10^{-4}$ u | 0.023 |
| $m_e c^2$ in electron volts | | 0.51099906(15) | MeV | 0.30 |
| electron–proton mass ratio | $m_e/m_p$ | 5.44617013(11) | $10^{-4}$ | 0.020 |
| electron specific charge | $-e/m_e$ | $-1.75881962(53)$ | $10^{11}\ Ckg^{-1}$ | 0.30 |
| Compton wavelength, $h/m_e c$ | $\lambda_C$ | 2.42631058(22) | $10^{-12}$ m | 0.089 |
| $\lambda_C/2\pi = \alpha a_o$ | $\bar{\lambda}_C$ | 3.86159323(35) | $10^{-13}$ m | 0.089 |
| classical electron radius, $e^2/4\pi\varepsilon_o m_e c^2$ | $r_e$ | 2.81794092(38) | $10^{-15}$ m | 0.13 |
| Thomson cross section, $(8\pi/3)r_e^2$ | $\sigma_e$ | 0.66524616(18) | $10^{-28}\ m^2$ | 0.27 |
| electron magnetic moment | $\mu_e$ | 928.47701(31) | $10^{-26}\ JT^{-1}$ | 0.34 |
| in Bohr magnetons | $\mu_e/\mu_B$ | 1.001159652193(10) | | $1\times10^{-5}$ |
| in nuclear magnetons | $\mu_e/\mu_N$ | 1838.282000(37) | | 0.020 |
| electron magnetic moment anomaly, $\mu_e/\mu_B - 1$ | $a_e$ | 1.159652193(10) | $10^{-3}$ | 0.0086 |
| electron g-factor, $2(1 + a_e)$ | $g_e$ | 2.002319304386(20) | | $1 \times 10^{-5}$ |

| Quantity | Symbol | Value | Units | Relative uncertainty (parts per million) |
|---|---|---|---|---|
| **Muon** | | | | |
| muon mass | $m_\mu$ | 1.8835327(11) | $10^{-28}$ kg | 0.61 |
| | | 0.113428913(17) | u | 0.15 |
| $m_\mu c^2$ in electron volts | | 105.658389(34) | MeV | 0.32 |
| muon–electron mass ratio | $m_\mu/m_e$ | 206.768262(30) | | 0.15 |
| muon magnetic moment | $\mu_\mu$ | 4.4904514(15) | $10^{-26}$ $JT^{-1}$ | 0.33 |
| in Bohr magnetons, | $\mu_\mu/\mu_B$ | 4.84197097(71) | $10^{-3}$ | 0.15 |
| muon magnetic moment anomaly | | | | |
| $[\mu_\mu/(e\hbar/2m_\mu)]-1$ | $a_\mu$ | 1.1659230(84) | $10^{-3}$ | 7.2 |
| muon g-factor, $2(1+a_\mu)$ | $g_\mu$ | 2.002331846(17) | | 0.0084 |
| **Proton** | | | | |
| proton mass | $m_p$ | 1.6726231(10) | $10^{-27}$ kg | 0.59 |
| | | 1.007276470(12) | u | 0.012 |
| $m_p c^2$ in electron volts | | 938.27231(28) | MeV | 0.30 |
| proton–electron mass ratio | $m_p/m_e$ | 1836.152701(37) | | 0.020 |
| proton specific charge | $e/m_p$ | 9.5788309(29) | $10^7$ $Ckg^{-1}$ | 0.30 |
| proton Compton wavelength, $h/m_p c$ | $\lambda_{C,p}$ | 1.32141002(12) | $10^{-15}$ m | 0.089 |
| proton magnetic moment | $\mu_p$ | 1.41060761(47) | $10^{-26}$ $JT^{-1}$ | 0.34 |
| in Bohr magnetons | $\mu_p/\mu_B$ | 1.521032202(15) | $10^{-3}$ | 0.010 |
| in nuclear magnetons | $\mu_p/\mu_N$ | 2.792847386(63) | | 0.023 |
| **Neutron** | | | | |
| neutron mass | $m_n$ | 1.6749286(10) | $10^{-27}$ kg | 0.59 |
| | | 1.008664904(14) | u | 0.014 |
| $m_n c^2$ in electron volts | | 939.56563(28) | MeV | 0.30 |
| neutron–electron mass ratio | $m_n/m_e$ | 1838.683662(40) | | 0.022 |
| neutron–proton mass ratio | $m_n/m_p$ | 1.001378404(9) | | 0.009 |
| neutron Compton wavelength, $h/m_n c$ | $\lambda_{C,n}$ | 1.31959110(12) | $10^{-15}$ m | 0.089 |
| neutron magnetic moment | $\mu_n$ | 0.96623707(40) | $10^{-26}$ $JT^{-1}$ | 0.41 |
| in Bohr magnetons | $\mu_n/\mu_B$ | 1.04187563(25) | $10^{-3}$ | 0.24 |
| in nuclear magnetons | $\mu_n/\mu_N$ | 1.91304275(45) | | 0.24 |
| **Deuteron** | | | | |
| deuteron mass | $m_d$ | 3.3435860(20) | $10^{-27}$ kg | 0.59 |
| | | 2.013553214(24) | u | 0.012 |
| $m_d c^2$ in electron volts | | 1875.61339(57) | MeV | 0.30 |
| deuteron–electron mass ratio | $m_d/m_e$ | 3670.483014(75) | | 0.020 |
| deuteron–proton mass ratio | $m_d/m_p$ | 1.999007496(6) | | 0.003 |
| deuteron magnetic moment | $\mu_d$ | 0.43307375(15) | $10^{-26}$ $JT^{-1}$ | 0.34 |
| in Bohr magnetons, | $\mu_d/\mu_B$ | 0.4669754479(91) | $10^{-3}$ | 0.019 |
| in nuclear magnetons, | $\mu_d/\mu_N$ | 0.857438230(24) | | 0.028 |
| **PHYSICO-CHEMICAL CONSTANTS** | | | | |
| Avogadro constant | $N_A$ | 6.0221367(36) | $10^{23}$ $mol^{-1}$ | 0.59 |
| atomic mass constant, $m_u = \frac{1}{12}m(^{12}C)$ | $m_u$ | 1.6605402(10) | $10^{-27}$ kg | 0.59 |
| $m_u c^2$ in electron volts | | 931.49432(28) | MeV | 0.30 |
| Faraday constant | $F$ | 96485.309(29) | C $mol^{-1}$ | 0.30 |
| molar gas constant | $R$ | 8.314510(70) | J $mol^{-1}K^{-1}$ | 8.4 |
| Boltzmann constant, $R/N_A$ | $k$ | 1.380658(12) | $10^{-23}$ $JK^{-1}$ | 8.5 |
| in electron volts | | 8.617385(73) | $10^{-3}$ $eVK^{-1}$ | 8.4 |
| molar volume (ideal gas), $RT/p$ | | | | |
| $T = 273.15K$, $p = 101325Pa$ | $V_m$ | 22.41410(19) | liter/mol | 8.4 |
| Loschmidt constant, $N_A/V_m$ | $n_o$ | 2.686763(23) | $10^{25}$ $m^{-3}$ | 8.5 |
| Stefan–Boltzmann constant, | | | | |
| $(\pi^2/60)k^4/\hbar^3c^2$ | $\sigma$ | 5.67051(19) | $10^{-8}$ $Wm^{-2}K^{-4}$ | 34 |
| Wien displacement law constant | $b$ | 2.897756(24) | $10^{-3}$ mK | 8.4 |

# APPENDIX 9: THE CHEMICAL ELEMENTS AND THE PERIODIC TABLE

**Table A9.1** THE PERIODIC TABLE OF THE CHEMICAL ELEMENTS[h]

| IA | IIA | IIIB | IVB | VB | VIB | VIIB | | VIII | | IB | IIB | IIIA | IVA | VA | VIA | VIIA | 0 |
|---|---|---|---|---|---|---|---|---|---|---|---|---|---|---|---|---|---|
| 1<br>H<br>1.00794 | | | | | | | | | | | | | | | | | 2<br>He<br>4.00260 |
| 3<br>Li<br>6.941 | 4<br>Be<br>9.01218 | | | | | | | | | | | 5<br>B<br>10.81 | 6<br>C<br>12.011 | 7<br>N<br>14.0067 | 8<br>O<br>15.9994 | 9<br>F<br>18.998403 | 10<br>Ne<br>20.179 |
| 11<br>Na<br>22.98977 | 12<br>Mg<br>24.305 | | | | | | | | | | | 13<br>Al<br>26.98154 | 14<br>Si<br>28.0855 | 15<br>P<br>30.97376 | 16<br>S<br>32.06 | 17<br>Cl<br>35.453 | 18<br>Ar<br>39.948 |
| 19<br>K<br>39.0983 | 20<br>Ca<br>40.08 | 21<br>Sc<br>44.9559 | 22<br>Ti<br>47.88 | 23<br>V<br>50.9415 | 24<br>Cr<br>51.996 | 25<br>Mn<br>54.9380 | 26<br>Fe<br>55.847 | 27<br>Co<br>58.9332 | 28<br>Ni<br>58.69 | 29<br>Cu<br>63.546 | 30<br>Zn<br>65.38 | 31<br>Ga<br>69.72 | 32<br>Ge<br>72.59 | 33<br>As<br>74.9216 | 34<br>Se<br>78.96 | 35<br>Br<br>79.904 | 36<br>Kr<br>83.80 |
| 37<br>Rb<br>85.4678 | 38<br>Sr<br>87.62 | 39<br>Y<br>88.9059 | 40<br>Zr<br>91.22 | 41<br>Nb<br>92.9064 | 42<br>Mo<br>95.94 | 43<br>Tc<br>(98) | 44<br>Ru<br>101.07 | 45<br>Rh<br>102.9055 | 46<br>Pd<br>106.42 | 47<br>Ag<br>107.8682 | 48<br>Cd<br>112.41 | 49<br>In<br>114.82 | 50<br>Sn<br>118.69 | 51<br>Sb<br>121.75 | 52<br>Te<br>127.60 | 53<br>I<br>126.9045 | 54<br>Xe<br>131.29 |
| 55<br>Cs<br>132.9054 | 56<br>Ba<br>137.33 | 57–71<br>Rare<br>Earths | 72<br>Hf<br>178.49 | 73<br>Ta<br>180.9479 | 74<br>W<br>183.85 | 75<br>Re<br>186.207 | 76<br>Os<br>190.2 | 77<br>Ir<br>192.22 | 78<br>Pt<br>195.08 | 79<br>Au<br>196.9665 | 80<br>Hg<br>200.59 | 81<br>Tl<br>204.383 | 82<br>Pb<br>207.2 | 83<br>Bi<br>208.9804 | 84<br>Po<br>(209) | 85<br>At<br>(210) | 86<br>Rn<br>(222) |
| 87<br>Fr<br>(223) | 88<br>Ra<br>226.0254 | 89-103<br>Acti-<br>nides | 104<br>Rf<br>(261) | 105<br>Ha<br>(260) | 106<br>(263) | 107<br>(262) | 108<br>(265) | 109<br>(266) | | | | | | | | | |

| | | | | | | | | | | | | | | | |
|---|---|---|---|---|---|---|---|---|---|---|---|---|---|---|---|
| Rare Earths (Lanthanides) | 57<br>La<br>138.9055 | 58<br>Ce<br>140.12 | 59<br>Pr<br>140.9077 | 60<br>Nd<br>144.24 | 61<br>Pm<br>(145) | 62<br>Sm<br>150.36 | 63<br>Eu<br>151.96 | 64<br>Gd<br>157.25 | 65<br>Tb<br>158.9254 | 66<br>Dy<br>162.50 | 67<br>Ho<br>164.9304 | 68<br>Er<br>167.26 | 69<br>Tm<br>168.9342 | 70<br>Yb<br>173.04 | 71<br>Lu<br>174.967 |
| Actinides | 89<br>Ac<br>227.0278 | 90<br>Th<br>232.0381 | 91<br>Pa<br>231.0359 | 92<br>U<br>238.0289 | 93<br>Np<br>237.0482 | 94<br>Pu<br>(244) | 95<br>Am<br>(243) | 96<br>Cm<br>(247) | 97<br>Bk<br>(247) | 98<br>Cf<br>(251) | 99<br>Es<br>(252) | 100<br>Fm<br>(257) | 101<br>Md<br>(258) | 102<br>No<br>(259) | 103<br>Lr<br>(260) |

[h] In each box, the upper number is the *atomic number.* The lower number is the *atomic mass,* i.e., the mass (in grams) of one mole or, alternatively, the mass (in atomic mass units) of one atom. Numbers in parentheses denote the atomic masses of the most stable or best-known isotope of the element; all other numbers represent the average masses of a mixture of several isotopes as found in naturally occurring samples of the element.

**Table A9.2** THE CHEMICAL ELEMENTS

| *Element* | *Chemical symbol* | *Atomic number* | *Atomic mass*[i] |
|---|---|---|---|
| Hydrogen | H | 1 | 1.00794 u |
| Helium | He | 2 | 4.00260 |
| Lithium | Li | 3 | 6.941 |
| Beryllium | Be | 4 | 9.01218 |
| Boron | B | 5 | 10.81 |
| Carbon | C | 6 | 12.011 |
| Nitrogen | N | 7 | 14.0067 |
| Oxygen | O | 8 | 15.9994 |
| Fluorine | F | 9 | 18.998403 |
| Neon | Ne | 10 | 20.179 |
| Sodium | Na | 11 | 22.98977 |
| Magnesium | Mg | 12 | 24.305 |
| Aluminum | Al | 13 | 26.98154 |
| Silicon | Si | 14 | 28.0855 |
| Phosphorus | P | 15 | 30.97376 |
| Sulfur | S | 16 | 32.06 |
| Chlorine | Cl | 17 | 35.453 |
| Argon | Ar | 18 | 39.948 |
| Potassium | K | 19 | 39.0983 |
| Calcium | Ca | 20 | 40.08 |
| Scandium | Sc | 21 | 44.9559 |
| Titanium | Ti | 22 | 47.88 |
| Vanadium | V | 23 | 50.9415 |
| Chromium | Cr | 24 | 51.996 |
| Manganese | Mn | 25 | 54.9380 |
| Iron | Fe | 26 | 55.847 |
| Cobalt | Co | 27 | 58.9332 |
| Nickel | Ni | 28 | 58.69 |
| Copper | Cu | 29 | 63.546 |
| Zinc | Zn | 30 | 65.38 |
| Gallium | Ga | 31 | 69.72 |
| Germanium | Ge | 32 | 72.59 |
| Arsenic | As | 33 | 74.9216 |
| Selenium | Se | 34 | 78.96 |
| Bromine | Br | 35 | 79.904 |
| Krypton | Kr | 36 | 83.80 |
| Rubidium | Rb | 37 | 85.4678 |
| Strontium | Sr | 38 | 87.62 |
| Yttrium | Y | 39 | 88.9059 |
| Zirconium | Zr | 40 | 91.22 |
| Niobium | Nb | 41 | 92.9064 |
| Molybdenum | Mo | 42 | 95.94 |
| Technetium | Tc | 43 | (98) |
| Ruthenium | Ru | 44 | 101.07 |
| Rhodium | Rh | 45 | 102.9055 |
| Palladium | Pd | 46 | 106.42 |
| Silver | Ag | 47 | 107.8682 |
| Cadmium | Cd | 48 | 112.41 |
| Indium | In | 49 | 114.82 |
| Tin | Sn | 50 | 118.69 |
| Antimony | Sb | 51 | 121.75 |
| Tellurium | Te | 52 | 127.60 |
| Iodine | I | 53 | 126.9045 |

[i] Numbers in parentheses denote the atomic masses of the most stable or best-known isotope of the element; all other numbers represent the average masses of a mixture of several isotopes as found in naturally occurring samples of the element.

**Table A9.2** THE CHEMICAL ELEMENTS

| *Element* | *Chemical symbol* | *Atomic number* | *Atomic mass* |
|---|---|---|---|
| Xenon | Xe | 54 | 131.29 |
| Cesium | Cs | 55 | 132.9054 |
| Barium | Ba | 56 | 137.33 |
| Lanthanum | La | 57 | 138.9055 |
| Cerium | Ce | 58 | 140.12 |
| Praseodymium | Pr | 59 | 140.9077 |
| Neodymium | Nd | 60 | 144.24 |
| Promethium | Pm | 61 | (145) |
| Samarium | Sm | 62 | 150.36 |
| Europium | Eu | 63 | 151.96 |
| Gadolinium | Gd | 64 | 157.25 |
| Terbium | Tb | 65 | 158.9254 |
| Dysprosium | Dy | 66 | 162.50 |
| Holmium | Ho | 67 | 164.9304 |
| Erbium | Er | 68 | 167.26 |
| Thulium | Tm | 69 | 168.9342 |
| Ytterbium | Yb | 70 | 173.04 |
| Lutetium | Lu | 71 | 174.967 |
| Hafnium | Hf | 72 | 178.49 |
| Tantalum | Ta | 73 | 180.9479 |
| Tungsten | W | 74 | 183.85 |
| Rhenium | Re | 75 | 186.207 |
| Osmium | Os | 76 | 190.2 |
| Iridium | Ir | 77 | 192.22 |
| Platinum | Pt | 78 | 195.08 |
| Gold | Au | 79 | 196.9665 |
| Mercury | Hg | 80 | 200.59 |
| Thallium | Tl | 81 | 204.383 |
| Lead | Pb | 82 | 207.2 |
| Bismuth | Bi | 83 | 208.9804 |
| Polonium | Po | 84 | (209) |
| Astatine | At | 85 | (210) |
| Radon | Rn | 86 | (222) |
| Francium | Fr | 87 | (223) |
| Radium | Ra | 88 | 226.0254 |
| Actinium | Ac | 89 | 227.0278 |
| Thorium | Th | 90 | 232.0381 |
| Protactinium | Pa | 91 | 231.0359 |
| Uranium | U | 92 | 238.0289 |
| Neptunium | Np | 93 | 237.0482 |
| Plutonium | Pu | 94 | (244) |
| Americium | Am | 95 | (243) |
| Curium | Cm | 96 | (247) |
| Berkelium | Bk | 97 | (247) |
| Californium | Cf | 98 | (251) |
| Einsteinium | Es | 99 | (252) |
| Fermium | Fm | 100 | (257) |
| Mendelevium | Md | 101 | (258) |
| Nobelium | No | 102 | (259) |
| Lawrencium | Lr | 103 | (260) |
| Rutherfordium | Rf | 104 | (261) |
| Hahnium | Ha | 105 | (260) |
| ? | | 106 | (263) |
| ? | | 107 | (262) |
| ? | | 108 | (265) |
| ? | | 109 | (266) |

# APPENDIX 10: FORMULA SHEETS

## Chapters 1–21:

$v = dx/dt$

$a = dv/dt = d^2x/dt^2$

$x = x_0 + v_0 t + \frac{1}{2}at^2$

$a(x - x_0) = \frac{1}{2}(v^2 - v_0^2)$

$A_x = A \cos \theta_x$

$A = \sqrt{A_x^2 + A_y^2 + A_z^2}$

$\mathbf{A} \cdot \mathbf{B} = AB \cos \phi = A_x B_x + A_y B_y + A_z B_z$

$|\mathbf{A} \times \mathbf{B}| = AB \sin \phi$

$\mathrm{a} = \mathrm{v}^2/r$

$\mathbf{v}' = \mathbf{v} - \mathbf{V_O}$

$m\mathbf{a} = \mathbf{F}$

$\mathbf{p} = m\mathbf{v}$

$w = mg$

$f_k = \mu_k N$

$f_s \leq \mu_s N$

$F = -kx$

$W = \mathbf{F} \cdot \Delta\mathbf{r}$

$W = \int \mathbf{F} \cdot d\mathbf{r}$

$K = \frac{1}{2}mv^2$

$U = mgz$

$E = K + U = [\text{constant}]$

$U(P) = -\int_{P_0}^{P} \mathbf{F} \cdot d\mathbf{r} + U(P_0)$

$U = \frac{1}{2}kx^2$

$E = mc^2$

$P = dW/dt$

$\mathrm{P} = \mathbf{F} \cdot \mathbf{v}$

$F = GMm/r^2$

$v^2 = GM/r$

$g = GM_E/R_E^2$

$U = -GMm/r$

$r_{CM} = \frac{1}{M}\int \mathbf{r}\rho \, dV$

$\mathbf{I} = \int_0^{\Delta t} \mathbf{F} \, dt$

$v_1' = \frac{m_1 - m_2}{m_1 + m_2} v_1; \quad v_2' = \frac{2m_1}{m_1 + m_2} v_1$

$\omega = d\phi/dt$

$\alpha = d\omega/dt = d^2\phi/dt^2$

$v = R\omega$

$K = \frac{1}{2}I\omega^2$

$I = \int \rho R^2 \, dV$

$I_{CM} = MR^2$ (hoop); $\frac{1}{2}MR^2$ (disk); $\frac{2}{5}MR^2$ (sphere); $\frac{1}{12}ML^2$ (rod)

$I = I_{CM} + Md^2$

$\mathbf{L} = \mathbf{r} \times \mathbf{p}$

$\frac{d\mathbf{L}}{dt} = \mathbf{r} \times \mathbf{F}$

$L_z = I\omega$

$I\alpha = \tau_z$

$P = \tau_z \omega$

$x = A \cos(\omega t + \delta)$

$T = 2\pi/\omega; \quad \nu = 1/T = \omega/2\pi$

$m \, d^2x/dt^2 = -kx$

$\omega = \sqrt{k/m}$

$\omega = \sqrt{g/l}; \; T = 2\pi\sqrt{l/g}$

$\omega = \sqrt{mgl/I}$

$\omega = \sqrt{\kappa/I}$

$y = A \cos k(x - vt) = A \cos(kx - \omega t)$

$\lambda = 2\pi/k; \quad \nu = v/\lambda; \quad \omega = 2\pi\nu$

$v = \sqrt{F/\mu}$

$P \propto v\omega^2 A^2$

$\nu_{beat} = \nu_1 - \nu_2$

$\nu' = \nu(1 \pm V_R/v)$

$\nu' = \nu/(1 \mp V_E/v)$

$\sin \theta = v/V_E$

$p - p_0 = -\rho g z$

$\frac{1}{2}\rho v^2 + \rho g z + p = [\text{constant}]$

$pV = NkT$

$T_C = T - 273.15$

$v_{rms} = \sqrt{3kT/m}$

$\lambda = \frac{1}{4\pi R^2 (N/V)}$

$r = \sqrt{n}\,\lambda$

$pV^\gamma = [\text{constant}]; \quad \gamma = C_p/C_V$

$\Delta E = \Delta Q - \Delta W$

$e = 1 - T_2/T_1$

$S(A) = \int_{A_0 \text{ rev.}}^{A} \frac{dQ}{T} + S(A_0)$

$S(B) - S(A) \geq \int_{A \text{ irrev.}}^{B} \frac{dQ}{T}$

---

$g = 9.81 \text{ m/s}^2$
$G = 6.67 \times 10^{-11} \text{ N} \cdot \text{m}^2/\text{kg}^2$
$M_E = 5.98 \times 10^{24} \text{ kg}$
$R_E = 6.37 \times 10^6 \text{ m}$

$m_e = 9.11 \times 10^{-31} \text{ kg}$
$m_p = 1.67 \times 10^{-27} \text{ kg}$
$c = 3.00 \times 10^8 \text{ m/s}$

$N_A = 6.02 \times 10^{23}/\text{mole}$
$k = 1.38 \times 10^{-23} \text{ J/K}$
$1 \text{ cal} = 4.19 \text{ J}$

## Chapters 22–46:

$F = \frac{1}{4\pi\varepsilon_0}\frac{qq'}{r^2}$

$E = \frac{1}{4\pi\varepsilon_0}\frac{q'}{r^2}$

$E = \sigma/2\varepsilon_0$

$p = lQ$

$\tau = \mathbf{p} \times \mathbf{E}$

$U = -\mathbf{p} \cdot \mathbf{E}$

$\oint E_n\, dS = \frac{Q}{\varepsilon_0}$

$V(P_2) - V(P_1) = -\int_{P_1}^{P_2} \mathbf{E} \cdot d\mathbf{l}$

$V = \frac{1}{4\pi\varepsilon_0}\frac{q'}{r}$

$\frac{\partial V}{\partial x} = -E_x, \quad \frac{\partial V}{\partial y} = -E_y, \quad \frac{\partial V}{\partial z} = -E_z$

$U = \frac{1}{2}Q_1V_1 + \frac{1}{2}Q_2V_2 + \frac{1}{2}Q_3V_3 + \cdots$

$u = \frac{1}{2}\varepsilon_0 E^2$

$C = Q/\Delta V$

$C = \varepsilon_0 A/d$

$\mathrm{E} = \mathrm{E}_{\text{free}}/\kappa$

$\oint \kappa E_n\, dS = \frac{Q_{\text{free}}}{\varepsilon_0}$

$u = \frac{1}{2}\kappa\varepsilon_0 E^2$

$I = \Delta V/R$

$R = \rho l/A$

$P = I\mathscr{E}$

$P = I\,\Delta V$

$\mathbf{F} = \frac{\mu_0}{4\pi}\frac{qq'}{r^2}\mathbf{v} \times (\mathbf{v}' \times \hat{\mathbf{r}})$

$\mathbf{B} = \frac{\mu_0}{4\pi}\frac{q'}{r^2}(\mathbf{v}' \times \hat{\mathbf{r}})$

$\mathbf{F} = q\mathbf{v} \times \mathbf{B}$

$d\mathbf{B} = \frac{\mu_0}{4\pi} I \frac{d\mathbf{l} \times \hat{\mathbf{r}}}{r^2}$

$\oint \mathbf{B} \cdot d\mathbf{l} = \mu_0 I$

$B = \mu_0 I_0 n$

$r = \frac{p}{qB}$

$d\mathbf{F} = I\, d\mathbf{l} \times \mathbf{B}$

$\mu = I \cdot [\text{area of loop}]$

$\tau = \mu \times \mathbf{B}$

$U = -\mu \cdot \mathbf{B}$

$\mathscr{E} = vBl$

$\mathscr{E} = -\frac{d\Phi_B}{dt}$

$\Phi_B = \int \mathbf{B} \cdot d\mathbf{S}$

$\Phi_B = LI$

$\mathscr{E} = -L\frac{dI}{dt}$

$U = \frac{1}{2}LI^2$

$u = \frac{1}{2\mu_0}B^2$

$\mu = \frac{e}{2m_e}L$

$B = \kappa_m B_{\text{free}}$

$\omega_0 = 1/\sqrt{LC}$

$Z = \sqrt{R^2 + \left(\omega L - \frac{1}{\omega C}\right)^2}$

$Z = 1\Big/\sqrt{\frac{1}{R^2} + \left(\frac{1}{\omega L} - \omega C\right)^2}$

$\mathscr{E}_2 = \mathscr{E}_1\frac{N_2}{N_1}$

$\oint \mathbf{B} \cdot d\mathbf{l} = \mu_0 I + \mu_0\varepsilon_0\frac{d\Phi}{dt}$

$E_\theta = \frac{1}{4\pi\varepsilon_0}\frac{qa\sin\theta}{c^2 r}$

$B = E_\theta/c$

$\mathbf{S} = \frac{1}{\mu_0}\mathbf{E} \times \mathbf{B}$

$P_x = U/c$

$v = \sqrt{\frac{1 - v/c}{1 + v/c}}\,v_0$

$v = c/n$

$\sin\theta = n\sin\theta'$

$f = \pm\frac{1}{2}R$

$\frac{1}{s} + \frac{1}{s'} = \frac{1}{f}$

Interference minima:

$d\sin\theta = \frac{1}{2}\lambda, \frac{3}{2}\lambda, \frac{5}{2}\lambda, \ldots$

Interference maxima:

$d\sin\theta = 0, \lambda, 2\lambda, \ldots$

Diffraction minima:

$a\sin\theta = \lambda, 2\lambda, 3\lambda, \ldots$

$a\sin\theta = 1.22\lambda$

$x' = \frac{x - V_O t}{\sqrt{1 - V_O^2}}; \quad t' = \frac{t - V_O x}{\sqrt{1 - V_O^2}}$

$\Delta t = \frac{\Delta t'}{\sqrt{1 - V_O^2}}$

$\Delta x = \sqrt{1 - V_O^2}\,\Delta x'$

$v'_x = \frac{v_x - V_O}{1 - v_x V_O}$

$\mathbf{p} = \frac{m\mathbf{v}}{\sqrt{1 - v^2/c^2}}; \quad E = \frac{mc^2}{\sqrt{1 - v^2/c^2}}$

$E = h\nu$

$p = h\nu/c$

$\Delta y\,\Delta p_y \geq \hbar/2$

$L = n\hbar$

$E_n = -\frac{m_e e^4}{2(4\pi\varepsilon_0)^2\hbar^2}\frac{1}{n^2} = -\frac{13.6\text{ eV}}{n^2}$

$\lambda = h/p$

$\mu_{\text{spin}} = \frac{e\hbar}{2m_e}$

$E = \frac{n^2\hbar^2}{2I}$

$R = (1.2 \times 10^{-15}\text{ m}) \times A^{1/3}$

$n = n_0\, e^{-t/\tau}; \tau = t_{1/2}/0.693$

---

$e = 1.60 \times 10^{-19}$ C
$\varepsilon_0 = 8.85 \times 10^{-12}$ F/m

$\mu_0 = 1.26 \times 10^{-6}$ H/m
$c = 1/\sqrt{\mu_0\varepsilon_0} = 3.00 \times 10^8$ m/s

$m_e = 9.11 \times 10^{-31}$ kg
$m_p = 1.67 \times 10^{-27}$ kg
$h = 2\pi\hbar = 6.63 \times 10^{-34}$ J·s

# APPENDIX 11: ANSWERS TO PROBLEMS

### Chapter 1

2. $6 \times 10^{-5}$ m
4. $8 \times 10^{-7}$, $4 \times 10^{-9}$, $3 \times 10^{-13}$, and $8 \times 10^{-14}$ in.
6. $6.9 \times 10^{8}$ m
8. $7.5 \times 10^{5}$ m
10. $1.00 \times 10^{7}$ m; $9.01 \times 10^{6}$ m
12. $6.3 \times 10^{6}$ m
14. $1.4 \times 10^{17}$ s
16. 0.25 minute of arc, 0.46 km
18. 0.134%, 99.866%
20. 0.021%, 99.979%
22. (a) $8.4 \times 10^{24}$; (b) $4.3 \times 10^{46}$; (c) $1.6 \times 10^{3}$
24. $9.22 \times 10^{56}$
26. $6.7 \times 10^{27}$
28. 8.3 light-minutes; 1.3 light-seconds
30. $8.9 \times 10^{3}$ kg/m$^3$, $5.6 \times 10^{2}$ lb/ft$^3$, 0.32 lb/in.$^3$
32. $73 \times 10^{-3}$ m$^3$
34. $5.0 \times 10^{-3}$ m$^3$/s; 5.0 kg/s
36. $2.3 \times 10^{8}$ tons/cm$^3$
38. $6.0 \times 10^{7}$ tons/cm$^3$
40. $5.4 \times 10^{3}$, $5.2 \times 10^{3}$, $5.5 \times 10^{3}$, $3.9 \times 10^{3}$, $1.2 \times 10^{3}$, $0.63 \times 10^{3}$, $1.3 \times 10^{3}$, $0.17 \times 10^{3}$, and $1.1 \times 10^{3}$ kg/m$^3$, respectively, for Mercury, Venus, Earth, Mars, Jupiter, Saturn, Uranus, Neptune, and Pluto
42. 171 km
44. 8.9 m; 9300 tons

### Chapter 2

2. 23 mi/h
4. $1.9 \times 10^{10}$ years
6. 13 m/s, 0.77 s
8. (a) 13.8 s, 388 m; (b) 72 m
10. 6.27 m/s; 0 m/s
12. 32.4 m/s
14. $3.0 \times 10^{3}$ m/s$^2$
16. (a) 4.25 m, 3.0 m/s, −6.0 m/s$^2$; (b) 2.0 m, −6.0 m/s, −6.0 m/s$^2$; (c) −1.5 m/s, −6.0 m/s$^2$
18. (b) 1720 m; (c) approximately 1720 m
20. (b) 1.6 s and also at times earlier or later than this by a multiple of 3.14s, ±2.0 m/s, 0 m/s$^2$; (c) 0 s and also at times earlier or later than this by a multiple of 3.14s, 0 m/s, ±2.0 m/s$^2$
22. 2.4 m/s$^2$
24. 4.0 years; $6.2 \times 10^{8}$ m/s
26. 350 m/s$^2$; yes
28. 7.1 m/s$^2$; 3.7 s
30. (a) 65.2 ft/s$^2$; (b) at constant acceleration the distance would have been 345 yd; (c) 319 mi/h
32. (a) 7.3 m/s; (b) 0.53 m/s$^2$, 15 m/s
34. 76 km/h
36. 0.48 m/s$^2$; 14 s
38. 8.8 s; 86 m/s
40. $6.8 \times 10^{3}$ m
42. 849 m/s
44. 33 m/s; $2.2 \times 10^{3}$ m/s$^2$
46. $1.6 \times 10^{4}$ m/s$^2$
48. $1.88 \times 10^{3}$ m/s; 255 s
50. 802 m/s; 1.89 s
52. (a) $n\sqrt{2h/g}$; (b) $\frac{3}{4}h$; (c) $\frac{2}{3}h$

### Chapter 3

2. 609 m at 38° west of north
4. 11.2 km at 2.3° north of east
6. 3.4 m at 22° west of north
10. (b) $2.12 \times 10^{11}$ m
14. $4\hat{\mathbf{x}} + 5\hat{\mathbf{y}} + 3\hat{\mathbf{z}}$; 7.1 m; 9.4 m
16. 5690 km
18. 9.2 km; 7.7 km
20. (b) $C_x = 2$ cm, $C_y = 5$ cm
22. $A_x = 4.2$, $A_y = 0.5$, $A_z = \pm 4.2$
24. 3.74
26. 5.9; 32°, −60°, 80°
28. $17\hat{\mathbf{x}} + 3\hat{\mathbf{y}} - 16\hat{\mathbf{z}}$
30. $-2.24 \times 10^{6}$ m$^2$
32. −304 m$^2$
34. 2.4; 1.8
36. $0.600\hat{\mathbf{x}} + 0.097\hat{\mathbf{y}} + 0.794\hat{\mathbf{z}}$
38. 54.7°

40. 47.3 $m^2$ down
42. $-12\hat{\mathbf{x}} - 14\hat{\mathbf{y}} - 9\hat{\mathbf{z}}$
44. $0.45\hat{\mathbf{x}} - 0.59\hat{\mathbf{y}} - 0.67\hat{\mathbf{z}}$
48. (a) 2.80 km, 4.15 km; (b) 3.59 km, 3.48 km
52. rotate coordinate system by an angle $\theta = -26.6°$
54. (a) $x'' = x'$, $y'' = y' \cos\phi + z' \sin\phi$, $z'' = -y' \sin\phi + z' \cos\phi$; (b) $x'' = x\cos\theta + y\sin\theta$, $y'' = -x\sin\theta\cos\phi + y\cos\theta\cos\phi + z\sin\phi$, $z'' = x\sin\theta\sin\phi - y\cos\theta\sin\phi + z\cos\phi$

## Chapter 4

2. 29.9 km/s; 19.0 km/s
4. (a) $8\hat{\mathbf{x}} + 10\hat{\mathbf{y}}$, 12.8 m/s; (b) $4\hat{\mathbf{x}} + 6\hat{\mathbf{y}}$, 7.2 $m/s^2$
6. (a) $2\hat{\mathbf{x}} + (5 + 8t)\hat{\mathbf{y}} + (-2 - 6t)\hat{\mathbf{z}}$; (b) $8\hat{\mathbf{y}} - 6\hat{\mathbf{z}}$; 10 $m/s^2$ at 37° down from $y$ axis
8. 0.97 m
10. 135 km/h
12. 65.8 m/s; 93.4 m/s
14. (a) 7.25°; (b) 13 m
16. (a) $2.5 \times 10^4$ m; (b) $5.0 \times 10^4$ m; (c) no
18. 0.11 m
20. (a) 22 m/s; (b) 3.5 rev/s
22. 65 m; 14 liters
24. (a) 71 m/s; (b) 4 m; (c) 8 m
26. 43°; 35 ft/s
28. yes, by 0.5 s
30. (a) 270 m; (b) air resistance
32. (a) 6.0 m; (b) 1.5 m; (c) 1.1 s
34. 7.8 m/s; 63°
36. no; 12.2 m
38. $R + gR^2/u^2 + (1 - g^2R^2/u^4)u^2/2g$; 4.4 m
40. 9.95°; 205 m, or 42 minutes of arc
42. 73 m/s; $5.4 \times 10^4$ $m/s^2$
44. 8.9 $m/s^2$
46. $2.2 \times 10^3$ $m/s^2$
48. 9.45 $m/s^2$, 29.7 $m/s^2$; −0.36 $m/s^2$, 39.5 $m/s^2$
50. 13 m/s at 40° from vertical
52. 27 m/s at 68° from vertical
54. 329 m/s
56. 27 km/h at 43° east of north; 33°
58. (a) 50 km; (b) 33 km, 67 km
60. 15 km/h at 15° east of north
62. $v = 4v_0^2t/(4v_0^2t^2 + h^2)^{1/2}$
64. 528 km/h

## Chapter 5

2. 17 $m/s^2$
4. $6.6 \times 10^3$ N, 12 times weight
6. $1.2 \times 10^4$ N
8. $1.8 \times 10^3$ $m/s^2$; $1.3 \times 10^5$ N
10. $1.46 \times 10^6$ N
12. $2.9 \times 10^3$ N, $1.8 \times 10^3$ N
14. $-2773 + 296t$ measured in N
16. no, since with 1150 children on each side the tension would be 150,000 N
18. $4.7 \times 10^{20}$ N at 25° with the Sun–Moon line
20. 770 N toward dock
22. $1.2 \times 10^3$ $m/s^2$; 23 m/s
24. $6.9 \times 10^5$ N; $2.8 \times 10^3$ N
26. 180 N
28. $8.2 \times 10^3$ N
30. 165 N at 20° downward from horizontal
32. 150 N
34. $1.6 \times 10^4$ kg · m/s; 8.1 km/h
36. 5.7 kg · m/s; 8.7 kg · m/s
38. $7.4 \times 10^2$ N
40. $p_x = 8.5 \times 10^{-23}$ kg · m/s, $p_y = 3.4 \times 10^{-23}$ kg · m/s

## Chapter 6

2. (a) 0.0010; (b) no, since price is based on mass
4. 29 N, 128 N
6. $9.2 \times 10^2$ N; $1.4 \times 10^2$ N
8. 4.1 m/s
10. (b) [weight] = $7.4 \times 10^2$ N, [normal force] = $6.0 \times 10^2$ N, [resultant] = $4.2 \times 10^2$ N; (c) 5.6 $m/s^2$
12. 64 m; 5.1 s
14. 9.8 $m/s^2$
16. 7.0°
18. $a_1 = g(m_1 - 2m_2)/(m_1 - 4m_2)$; $-a_1$; $-2a_1$
20. $a_1 = g(4m_2m_3 - m_1m_2 - m_1m_3)/(4m_2m_3 + m_1m_2 + m_1m_3)$, $a_2 = g(3m_1m_3 - 4m_2m_3 - m_1m_2)/(4m_2m_3 + m_1m_2 + m_1m_3)$, $a_3 = g(3m_1m_2 - 4m_2m_3 - m_1m_3)/(4m_2m_3 + m_1m_2 + m_1m_3)$, $T_1 = 2T_2 = 8m_1m_2m_3g/(4m_2m_3 + m_1m_2 + m_1m_3)$
22. $4.3 \times 10^3$ N
24. 3.9 $m/s^2$
26. 40 m
28. 53 m
30. 0.17 $m/s^2$
32. $1.9 \times 10^2$ N; $7.4 \times 10^3$ N
34. $1.2 \times 10^2$ m; 6.5 s
36. 27°
38. 4.4 $m/s^2$, 1.3 N
40. $T = \mu_k mg/\sqrt{1 + \mu_k^2}$
42. $a_1 = F/m_1 - \mu_1 g$, $a_2 = \mu_1 m_1 g/m_2 - \mu_2 g(m_1 + m_2)/m_2$
44. $6.8 \times 10^2$ N
46. 74 N/m
48. 0.15 m
50. 180 N/m, 360 N/m
54. 3.4 s
56. $4.4 \times 10^2$ N
58. $7.6 \times 10^2$ N; $8.1 \times 10^2$ N
60. 76 km/h
62. outer; inner
64. 3.0 m/s
66. 68°
68. $v = \sqrt{gl \tan\theta \sin\theta}$
70. 28°
72. $T = mv^2/(2\pi r)$
74. $\phi = \tan^{-1}[\tan\theta/(1 - v_E^2/gR_E)] - \theta$ or approximately $(\sin\theta\cos\theta)v_E^2/gR_E$, where $R_E$ and $v_E$ are the equatorial radius and speed of the Earth and $\theta$ is the latitude angle; 0.099°
76. $v_0^2 = (gR^2 \tan\alpha)/(R + L\sin\alpha)$; 0.95 m/s

## Chapter 7

2. $4.93 \times 10^3$ J
4. $8.8 \times 10^3$ J
6. $2.4 \times 10^5$ J; $3.6 \times 10^2$ J/s
8. 6 J
10. $2.6 \times 10^3$ J
12. 24 J
14. 40 J
16. (a) $1.2 \times 10^4$ J; (b) 290 N, $1.2 \times 10^4$ J
18. (a) $7.1 \times 10^3$ N; (b) $2.2 \times 10^5$ J, $8.1 \times 10^3$ N
20. curved ramp: $W = mgR[(1 - \sqrt{2}/2) + \mu_k\sqrt{2}/2]$; straight ramp: $W = mgR(1 - \sqrt{2}/2)[1 + \mu_k \cot(45°/2)]$, which is the same
22. $2.66 \times 10^{33}$ J
24. $1.3 \times 10^3$ J, $5.8 \times 10^3$ J; 22 times
26. (a) $4.0 \times 10^5$ J; (b) $2.5 \times 10^4$ J; (c) $1.2 \times 10^6$ J
28. $9.79 \times 10^6$ J, $5.72 \times 10^6$ J; $4.07 \times 10^6$ J
30. $h/2$; $3h/4$
32. $W = --\frac{1}{2}a(x_2^2 - x_1^2) + \frac{1}{4}b(x_2^4 - x_1^4)$; $\Delta K = W$
34. (a) $1.2 \times 10^4$ N; (b) 39 m; (c) $4.7 \times 10^5$ J; (d) $4.7 \times 10^5$ J
36. $1.9 \times 10^5$ J
38. $2.2 \times 10^9$ J, $6.4 \times 10^9$ J
40. $8.2 \times 10^6$ m$^3$
42. 5.1 m; yes, since the jumper does some extra work with his arms
44. 99 m/s; $9.8 \times 10^{10}$ J; 23 tons
46. (a) $2.5 \times 10^4$ N; (b) $1.2 \times 10^7$ J; (c) 15 m/s
48. 50 J, 17 J
50. 1.6 m/s; 26°
52. 9.9 m/s; 30 m/s
54. 48° from top

## Chapter 8

4. $U = K/(3x^3)$
6. 217 J
8. (a) 5 J; (b) −4 J, no
10. (a) 47 J; (b) 0.89 m
12. 0.26 m
14. 1.9 s; 9.6 m; $2.9 \times 10^5$ J
16. $t = v/\mu_k g$; $x = v^2/2\mu_k g$; $mv^2/2$; $mv^2$
18. $\mathbf{F} = (2K/x^3)\hat{\mathbf{x}}$
20. 0.19 Å, 0.80 Å
22. (a) none and 0.2 m for $E_1$, 3.1 m and 0.3 m for $E_2$, 1.3 m and 0.5 m for $E_3$; (b) absolute maximum at 0.9 m, absolute minimum at turning points, local maximum at 2.2 m, local minimum at 1.7 m; (c) unbound for $E_1$, bound for $E_2$ and $E_3$
24. 11 eV
26. $2.2 \times 10^6$, $1.1 \times 10^7$, $2.7 \times 10^6$, $5.8 \times 10^5$, $3.6 \times 10^6$, and $9.8 \times 10^6$ J; bus; snowmobile
28. 1.65 kcal/kg, 2.81 kcal/kg, 2.82 kcal/kg
30. $1.0 \times 10^6$ eV
32. $9.4 \times 10^8$ eV
34. 540 kcal
36. 18 kW · h
38. $1.1 \times 10^3$ W; 0 W
40. 0.61 hp
42. 746 J
44. 9.1 gal./h
46. $4.2 \times 10^5$ W
48. $2.5 \times 10^3$ km$^2$
50. (a) $1.7 \times 10^{10}$ J; (b) 17 min; (c) 17 min, 4 km
52. $1.2 \times 10^{-4}$ W
54. 32 km/h
56. 52 hp
58. 37%
60. (a) $3.2 \times 10^4$ W; (b) $7.8 \times 10^2$ W; (c) $3.1 \times 10^4$ W
62. (a) $45.36\,(61.14 + 6.52t)(655.9 - 61.14t + 3.26t^2)$ in joules; (b) $9.76 \times 10^6$ J, $5.71 \times 10^6$ J; (c) $1.35 \times 10^6$ W
64. $2.0 \times 10^5$ W
66. (a) $1.08 \times 10^4$ N (weight), $1.05 \times 10^4$ N (lift), $2.43 \times 10^3$ N (friction); (b) $4.85 \times 10^3$ (push), same weight, lift, friction; (c) 300 hp

## Chapter 9

2. $4.7 \times 10^{-35}$ N
4. 8.88 m/s$^2$, 3.71 m/s$^2$, 3.71 m/s$^2$
6. $4.34 \times 10^{20}$ N, $1.99 \times 10^{20}$ N; $6.32 \times 10^{20}$ N toward Sun; $2.35 \times 10^{20}$ N toward Sun
8. $4.0 \times 10^{-4}$ radian/s, or 0.23 rev/h
10. $2 \times 10^8$ years, $3 \times 10^5$ m/s
12. Lincoln, Nebraska, 22.6° west of New York City
14. $m_1/m_2 = 1.6$
16. $3.0 \times 10^{10}$ m
18. (a) $7.5 \times 10^3$ m/s, $8.32 \times 10^3$ m/s; (b) $3.94 \times 10^8$ J, $4.85 \times 10^8$ J
20. $8.21 \times 10^3$ m/s
22. 3.5 days
24. $2.66 \times 10^{33}$ J, $-5.31 \times 10^{33}$ J; $-2.66 \times 10^{33}$ J
26. (a) $6.0 \times 10^{-14}$ m/s$^2$, $9.1 \times 10^{-14}$ m/s$^2$; (b) $1.1 \times 10^5$ m/s, $1.6 \times 10^5$ m/s; (c) $3.4 \times 10^{51}$ J, $5.1 \times 10^{51}$ J, $7.7 \times 10^{51}$ J, yes
28. (a) $1.1 \times 10^4$ m/s; (b) $1.2 \times 10^{11}$ J, 29 tons; (c) $1.2 \times 10^5$ m/s$^2$
30. (a) $-1.1 \times 10^{11}$ J, $-2.2 \times 10^{11}$ J, $-1.1 \times 10^{11}$ J; (b) yes, yes
32. (a) $1.1 \times 10^4$ m/s; (b) $2.4 \times 10^3$ m/s
34. (a) I and III elliptical, II circular
36. 16 km/s
38. (a) yes; (b) yes
40. (a) $4.21 \times 10^4$ m/s; (b) $7.19 \times 10^4$ m/s, $1.23 \times 10^4$ m/s
42. $7.78 \times 10^7$ m
46. (a) $2.9 \times 10^3$ m/s; (b) $2.6 \times 10^3$ m/s; (c) 0.71 year; (d) 44° ahead of Earth; 256° from initial position
48. (c) 23 km/s
50. $9.0 \times 10^{-3}$ m/s$^2$; larger
52. 1.0 rev/min

## Chapter 10

2. 1.36 m/s
4. 24 km/h; 89 km/h, 41 km/h
6. 150 bullets
8. $5.3 \times 10^2$ kg · m/s; 0.18 m/s; no

10. $\sqrt{2}v$, at 135° with respect to the direction of the other fragments
12. (a) 42 km/h at 56° east of north; (b) $3.6 \times 10^5$ J
14. $3.5 \times 10^2$ N
16. (a) 0.10 kg/s; (b) 2.3 kg · m/s, 2.3 N
18. $v_n = \sum_{k=1}^{n} mu/(M - km)$
20. 0.19 m from center of seesaw
22. $2.7 \times 10^{-4}$ m
24. 2.28 Å from H
26. (950, 180, 820)
28. on diagonal of cube, at $\sqrt{3}/3L$ from vertex
30. 0.333 m above bottom stick
32. $L/(64/\pi - 4)$ from center of cube, toward hole
36. 950 m
38. $9.0 \times 10^7$ J
40. $2.4 \times 10^3$ m/s
42. 950 kg
44. 1.0 m
46. $1.8 \times 10^4$ m/s, at 127°
48. (a) $9.2 \times 10^6$ J, 0; (b) $9.2 \times 10^6$ J, $1.7 \times 10^6$ J, extra energy comes from explosive chemical reactions
50. (a) $3.4 \times 10^5$ J, $3.6 \times 10^5$ J; (b) $3.4 \times 10^5$ J, 0
52. 11 tons/s
54. $4.4 \times 10^3$ m/s

## Chapter 11

2. $-7.8 \times 10^8$ N; 1.1 m/s$^2$
4. $1.0 \times 10^3$ m/s$^2$; $6.5 \times 10^4$ N
6. 9.3 kg · m/s, 24° up
8. (a) – 0.27 m/s, 0.53 m/s; (b) 0.019 J, 0.0021 J, 0; 0.017 J
10. 14 m/s
12. 2.6 km/h, 12.6 km/h
14. (a) 18 m/s; (b) 30 m/s
16. 1.1 m/s
18. 39 m/s
20. 0.57 J
22. $4.0 \times 10^{-13}$ J
24. (a) $h/9$, $4h/9$; (b) $h$, 0
26. $v/3$ to left, $2v/9$ to right, $8v/9$ to right
28. 13.5 m/s
30. (a) 3.0 m/s; (b) $2.1 \times 10^2$ m/s$^2$
32. 21 m/s
34. $v/5$ to left, $2\sqrt{3}v/5$ up 30° to right, and $2\sqrt{3}v/5$ down 30° to right
36. $2.6 \times 10^7$ m/s
38. 45° each, $4.0 \times 10^{-13}$ J each
40. $1.9 \times 10^5$ eV, 70°
42. $1.25 \times 10^{-13}$ J
44. $4.3 \times 10^{-11}$ J

## Chapter 12

2. $7.3 \times 10^{-5}$ radian/s; 460 m/s, 350 m/s
4. 160 km/h horizontal; 0; 113 km/h down 45°
6. $5.50 \times 10^{-3}$ m/s; 0.509 m/s; 0.221 m/s
8. (b) $5.4 \times 10^{-3}$ m
10. 12 radian/s$^2$
12. $9.6 \times 10^{-22}$ radian/s$^2$
14. $1.21 \times 10^{-10}$ m
16. $6.50 \times 10^{-46}$ kg · m$^2$
18. 0.44 kg · m$^2$
20. $-1.1 \times 10^{22}$ kg · m$^2$
22. $0.38 M_E R_E^2$
24. $\frac{1}{12} ML^2 \sin^2\theta$
28. $(83/256)MR^2$
30. $\frac{1}{2}MR^2 - (r^2M/R^2)(\frac{1}{2}r^2 + d^2)$
32. $\frac{1}{2}MR^2 - (2r^2M/R^2)(r^2 + \frac{1}{2}R^2)$
34. $0.338 M_E R_E^2$
36. (a) $2.2 \times 10^2$ kg · m$^2$; (b) $4.4 \times 10^3$ J
38. $3.44 \times 10^5$ J, 3.91%
40. 2.5 km
42. $2.74 \times 10^3$ J; 13.1 J; $4.74 \times 10^{-3}$
44. $\frac{1}{2} MR^2$
46. $\frac{1}{6}Ml^2$
48. $\frac{1}{2} M (R^4 - r^4 - 8hr^3/3\pi - r^2h^2/2)/(R^2 - r^2)$
52. $5.3 \times 10^{-13}$ kg · m$^2$/s
54. $2.1 \times 10^{12}$ kg · m$^2$/s in north direction
56. (a) $1.1 \times 10^{14}$ kg · m$^2$/s; (b) $4.3 \times 10^{36}$ kg · m$^2$/s
58. (a) 0 kg · m$^2$/s; (b) $2.7 \times 10^{40}$ kg · m$^2$/s, $5.4 \times 10^{40}$ kg · m$^2$/s, $2.7 \times 10^{40}$ kg · m$^2$/s, no
60. $2\hbar/[(1 - 2\sqrt{2}/3)m_e a_0] = 7.6 \times 10^7$ m/s; $2\hbar/[(1 + 2\sqrt{2}/3)m_e a_0)] = 2.2 \times 10^6$ m/s
62. 0.052 kg · m$^2$/s$^2$
64. $5.6 \times 10^{41}$ J · s; $3.1 \times 10^{43}$ J · s; 1.8%
66. $1.8 \times 10^{22}$ kg · m$^2$/s$^2$

## Chapter 13

2. 400 N
4. $1.2 \times 10^4$ N; $7.2 \times 10^3$ N · m; depress
6. $2.7 \times 10^4$ N · m
8. 9.7 m/s$^2$
10. $8.2 \times 10^2$ N
12. (a) 110 J · s; (b) 34 kg · m$^2$/s$^2$; (c) 34 N · m
14. (a) 76°; (b) 290 N, 250 N
16. $1.6 \times 10^8$ J · s east or west; $1.1 \times 10^4$ J · s/s; $1.1 \times 10^4$ N · m; $9.5 \times 10^3$ N on each side
18. (a) $\omega_1 R_1/R_2$; (b) $(T - T')R_1$, $(T - T')R_2$; (c) $(T - T')R_1\omega_1$, $(T - T')R_2\omega_2$, yes
20. 1.1 kcal/min
22. 27 N · m
24. 4.6 radian/s
26. (a) 1140 N · m; (b) 2.1 m/s$^2$
28. (a) $2.4 \times 10^{19}$ J · s; (b) $3.0 \times 10^{-19}$ radian/s
30. 7.2°
32. (a) 5.0 m/s; (b) $8.7 \times 10^{-3}$ radian/s; (c) 19 times
34. [height of end] $= m^2v^2/[2g(M/3 + m)(M/2 + m)]$
36. $v_{CM} = mv/(m + M)$, $\omega = v_{CM}/[\frac{1}{3}l + \frac{1}{4}l/(1 + m/M)^2 + \frac{1}{4}lm/M]$
40. 9.6 m/s$^2$
42. 100 N in forward direction; 300 N in forward direction
44. $2a/7$
46. $a = g(\sin\theta - \mu_k \cos\theta)$; $\alpha = (2/R)g\mu_k \cos\theta$
48. $a = (g \sin\theta)(m + 4M)/(m + 6M)$
50. 0.37 radian/s
52. 0.85 radian/s

## Chapter 14

2. $1.2 \times 10^4$ N; $4.6 \times 10^4$ N
4. 99 N at 27° right of vertical at the top, and at 27° left of vertical at the bottom
6. 2.3 N
8. $9.0 \times 10^3$ N, $8.9 \times 10^3$ N; the second
10. 0.408 $Mg$
12. $Mg\sqrt{1 - R^2/(R+l)^2}$, $MgR/(R+l)$
14. 0.11 $Mg$
16. (a) $U = 5.3 \times 10^2$ J $\times$ {sin($\theta + \tan^{-1} 2$) + [constant]}; (b) 27°; (c) 56 J
18. 1.04 book lengths; infinite
20. $\sqrt{Rgl/2h} = 18$ m/s
22. 7.4 m/s$^2$
24. (a) 8.8 m/s$^2$; (b) 5.4 m/s$^2$; (c) 3.7 m/s$^2$
26. $T_{AB} = T_{DE} = 0.375$, $T_{AF} = T_{EH} = -0.530$, $T_{BF} = T_{CG} = T_{DH} = 0.250$, $T_{FG} = T_{GH} = -0.375$ in units of $Mg$
28. 0.577 $Mg$ tension in cables; 0.289 $Mg$ compression in top beam, $\frac{3}{8}$ $Mg$ and $\frac{1}{8}$ $Mg$ tension at top and at bottom of side beams
30. 2 $mg$, $mg/\sqrt{3}$, $mg/\sqrt{3}$
32. $\tan^{-1}(1/2\mu_s)$
36. $1.7 \times 10^3$ N; $1.5 \times 10^3$ N
38. $1.9 \times 10^3$ N
40. $7.4 \times 10^2$ N
42. 8.9
44. 393
46. 34.6
48. $2.9 \times 10^4$ N/m
50. $4.0 \times 10^{-6}$ m
52. 5.6%
54. 6.8, 11.9, 8.0, and 12.8 $\times 10^{10}$ N/m$^2$
56. $8.1 \times 10^5$ N/m$^2$
58. $4.3 \times 10^3$ m
60. $6.8 \times 10^{-3}$
62. 130° C
64. $3.8 \times 10^5$ N/m$^2$
66. $2.8 \times 10^{-6}$ m

## Chapter 15

2. $1.7 \times 10^4$ m/s$^2$, 27 m/s; $2.0 \times 10^4$ N
4. (a) $A = 3.0$ m, $\nu = 0.32$ Hz, $\omega = 2.0$ radian/s, $T = 3.1$ s; (b) 0.26 s, 1.05 s; (c) at times larger and smaller by a multiple of one-half period
6. (a) $0.20 \cos 6\pi t + (4.0/6\pi) \sin 6\pi t$; (b) 0.043 s, $-1.0 \times 10^2$ m/s$^2$
8. $1.9 \times 10^4$ N
10. $1.13 \times 10^{14}$ Hz
12. (a) 4.5 Hz; (b) 8.5 m/s, strong vibrations
14. (a) $x_1 = -x_2 = (-v_1/\omega) \cos \omega t$ with $\omega = \sqrt{3k/m}$; (b) $x_1 = x_2 = (-v_1/\omega) \cos \omega t$ with $\omega = \sqrt{k/m}$
16. (a) 6.6 J; (b) $t = 0.30$ s, $t = 0$; (c) $t = 0.15$ s
18. 18 eV; no
22. 0.19 Hz
24. (a) 0.73 min; (b) 1.0 mm
26. $4.189\sqrt{l/g}$
28. (a) $1.3 \times 10^{-3}$ J; (b) 0.14 m/s
30. (a) 9.4 J; (b) $1.6 \times 10^{-5}$ W; (c) $6.0 \times 10^{-3}$ N · m/radian
34. 1.988 s
36. 1.5 s
38. (b) $l = R/\sqrt{2}$
40. $\omega = \sqrt{(\kappa - mgR)/mR^2}$; $m = \kappa/gR$
42. 3.6 radians/s; 3.3 m/s
44. $2\pi\sqrt{4l/5g}$
46. (a) $T = mg[1 - \frac{1}{2}A^2 + \frac{3}{2}A^2 \sin^2(\sqrt{g/l}\, t)]$; (b) $t = \frac{1}{2}\pi/\sqrt{g/l}$ gives $T = mg(1 + A^2)$
48. $\gamma = 2.5 \times 10^{-3}$/s
50. 18

## Chapter 16

2. 0.114 Hz, 0.716 radian/s, 0.0524/m, 13.7 m/s
4. (a) $6.0 \times 10^{-3}$ m, 0.31 m, 20/m, 0.64 Hz, 4.0 radian/s, negative $x$, 0.20 m/s; (b) −0.26 s
6. 7.8 m/s
8. $5.5 \times 10^{-7}$ m; $4.2 \times 10^{-7}$ m
12. 0.69 s
14. 1.6 s
16. 0.026 s
18. $\mu v^2$
22. 1:2.4
24. 2.8 cm
26. $A = 5.0$; $\delta = -\tan^{-1} \frac{4}{3}$
28. 0.007 Hz
30. 1.00/day, 0.92/day; 1.00 day, 1.09 day; first is due to Sun, second due to Moon
32. 392 Hz, 588 Hz, 784 Hz, 980 Hz
34.. 1.6 Hz, 3.2 Hz
36. 71 N
38. $1.25 \times 10^{-7}$ m; $2.5 \times 10^{-7}$ m
40. 3.7 m/s; $6.8 \times 10^3$ m/s$^2$
42. (a) $9.6 \times 10^3$ N; (b) 12 Hz
44. 0.018 J/m $\times$ ($\sin^2 \omega t \cos^2 kx + \cos^2 \omega t \sin^2 kx$), where $\omega = 3.7 \times 10^3$ radian/s and $k = 28$/m; location of maxima depends on time: maxima are at 0.057 m, 0.17 m, 0.28 m when $\sin^2 \omega t < \cos^2 \omega t$, and maxima are at 0 m, 0.11 m, 0.23 m, 0.34 m when $\sin^2 \omega t > \cos^2 \omega t$; 0.018 J/m $\times \sin^2 \omega t$ or 0.018 J/m $\times \cos^2 \omega t$

## Chapter 17

2. 55 Hz, 4187 Hz
4. 1.9, 1.8, 1.7, 1.6, 1.5, 1.5, 1.3, 1.3, 1.2, 1.1, 1.1, and 1.0 cm
6. 123 dB
8. 3.1 m
10. 130 times; 21 dB
12. 9.1 s; use visual signal
14. 150 m
16. (a) $1.2 \times 10^{-4}$ s; (b) on the glass
18. 9 cm, 4000 Hz
20. 337 m/s
24. (a) 0.63 m; (b) 3.5, 3.3, 3.2, 3.0, 2.8, 2.7, 2.5, 2.4, 2.2, 2.1, 2.0, and 1.9 cm

28. 619 Hz
30. 22 m/s, 0.22 Hz
32. 594 Hz, 596 Hz
34. 476 Hz
36. 481 Hz
38. (a) 33.4°; (b) 30.4 s
40. 4.0 m/s, 0.40 Hz; 12 m/s, 0.12 Hz
42. 2.1 m/s²; 4.9 m/s
44. $1.7 \times 10^2$ m
46. (a) 2.3/h; (b) 320 km, 740 km/h
48. (a) 31 m/s; (b) 0.11/h, 9.0 h; (c) probably
50. (b) 1.6 m, 3.2 m
52. $4.0 \times 10^3$ km
54. 7.2
56. (a) 6.3 m/s; (b) 1.0 m

## Chapter 18

2. (a) 11 m; (b) 8.1 cm, 11.7 cm
4. $6.1 \times 10^3$ N; layer of air under the paper exerts an opposite pressure force
6. 130 cm²
8. $2.34 \times 10^5$ N/m²
10. $5.1 \times 10^4$ N
12. 52 atm
14. (a) 360 N; (b) 330 N
16. 0.86 m
18. 10.3 m
20. (b) $5.0 \times 10^{11}$ N/m²
22. (a) $\frac{1}{3}\rho g h \pi\,(2R_2^2 - R_1^2 - R_1R_2)$ up on sides, $\rho g h \pi R_2^2$ down on bottom, $\frac{1}{3}\rho g h \pi\,(R_2^2 + R_1^2 + R_1R_2)$ down; (b) $\frac{1}{3}\rho g h \pi\,(R_2^2 + R_1R_2 - 2R_1^2)$ down on sides, $\rho g h \pi R_1^2$ down on bottom, $\frac{1}{3}\rho g h \pi\,(R_2^2 + R_1^2 + R_1R_2)$ down
24. (a) $4.7 \times 10^7$ m³; (b) $4.8 \times 10^{10}$ kg
26. yes
28. (a) 9.5 m; (b) 28.4 m
30. 0.94 kg/m³; 93 N/m²
32. 2.2 m
34. 0.115%; 0.011%; 0.002%
36. $\rho = M/[\frac{4}{3}\pi R^3 + R'^2\,(l - h)]$
38. 19 cm
40. 28 N/m²
42. (a) 15 m/s; (b) $1.1 \times 10^5$ N/m² (overpressure)
44. (a) $v = \sqrt{2g(h_2 - h_1)}$; (b) $p_{atm} - \rho g h_2$; (c) $p_{atm}/\rho g$
46. 2.3 hp
50. $z = \frac{1}{2}h$; $h$
52. $(A/A')\sqrt{2l/g}$

## Chapter 19

2. 58.0°C = 331.2 K; −88.3°C = 184.9 K
4. $2.69 \times 10^{22}$ molecules
6. 3.4 atm
8. 0.060 kg/m³
10. $1.4 \times 10^{-9}$ N/m²
12. 11.6 kg/m³
14. $3.6 \times 10^6$ N/m²
16. 1.3 cm
18. $4.5 \times 10^7$ N/m²
20. 96.3 g
22. (a) $3.1 \times 10^3$ kg; (b) $2.8 \times 10^3$ m³
24. (a) 1.2 m; (b) 2.5 kg at 2.5 atm
26. 29.0 g
30. $5.7 \times 10^{-21}$ J
32. 428 m/s, 349 m/s; $4.9 \times 10^{-21}$ J
34. 0.43%
36. $1.9 \times 10^5$ J; $\frac{3}{5}$; $\frac{2}{5}$
38. 284 K
40. $2.4 \times 10^{-10}$ m
42. 4.4 cm
44. 17 m
46. 9.8 h

## Chapter 20

2. 97 kW
4. 25°C
6. 97 W
8. 4.3 cal/s
10. 14.5 %
12. 3.5 h
14. 1.4 liters/min
16. 0.34 m
18. ±0.08°C
20. (a) 1.1 cm; (b) 47 cm
26. (a) 709 kg/m³; (b) 41.1¢/kg, 42.3¢/kg
28. 9.4 cm
30. 100.32°C
32. $5.9 \times 10^3$ cal/s; 110 times as large
34. (a) 10 m² · °C · s/cal; (b) 13.6 ft² · °F · h/Btu
36. 0.085 kg/h
38. 1.8 cm
40. 0.45 cm/h
44. (a) $1.0 \times 10^{11}$ kcal; (b) 5 A-bombs
46. 41 kcal/kg; the same
48. 22°C
50. 0.65 kg/h
52. $2.5 \times 10^3$ J; 50 m/s
54. 4.94 cal/K · mole, 6.93 cal/K · mole
56. 50 kcal; 119 kcal
58. $1.5 \times 10^4$ kcal
60. (a) 2.62 min; (b) 2.64 min, no
62. 30.8 atm; 507°C
64. −225°C
66. 1.0°C
68. $1.1 \times 10^4$ J

## Chapter 21

2. $\Delta Q = 1.96$ J, $\Delta W = 1.96$ J, $\Delta E = 0$
4. 518 kcal/kg
6. 0.35, or 35%
8. 0.14, or 14%
10. 0.19 kg/s
14. (a) $2.0 \times 10^4$ J; (b) $3.9 \times 10^4$ J, $1.9 \times 10^4$ J
16. $1 - 3 \times 10^{-9}$, or 99.9999997%
18. $3.9 \times 10^2$ W
20. (a) $3.1 \times 10^2$ MW; (b) $1.0 \times 10^3$ MW
22. 7.1 kW
24. $1.7 \times 10^2$ J
26. $8.2 \times 10^3$ cal/K · s

28. $3.7 \times 10^2$ cal/K · day
30. $-2.9 \times 10^2$ cal/K
32. 3.2 cal/K · s
34. (a) $6.9 \times 10^3$ cal/K; (b) $4.6 \times 10^2$ cal/K
36. (b) 0.19 kcal/K
38. 14.5 cal/K

# PHOTOGRAPH CREDITS

## PRELUDE

page 1 James L. Mairs
page 2 (top) James L. Mairs; (bottom) James L. Mairs
page 3 (top) James L. Mairs; (center) Lockwood, Kessler, and Barlett, Inc.; (bottom) NASA
page 4 (top) U.S. Geological Survey; (center) General Electric Space Systems; (bottom) NASA
page 5 NASA
page 9 (top) Hale Observatories; (bottom) Hale Observatories
page 10 (top) Hale Observatories; (bottom) Hale Observatories
page 11 (top) J. R. Kuhn, Dept. of Physics, Princeton Univ.; (bottom) James L. Mairs
page 12 (top) Zeiss; (center) Dr. Ronald Radius, The Eye Institute, Milwaukee, Wis.; (bottom) From Tissue and Organs: A Test Atlas of Scanning Electron Microscopy by Richard G. Kessel and Randy H. Kardon, W. H. Freeman & Co., San Francisco, © 1979
page 13 (top) From Tissue and Organs: A Test Atlas of Scanning Electron Microscopy by Richard G. Kessel and Randy H. Kardon, W. H. Freeman & Co., San Francisco, © 1979. (bottom) Professor T. T. Tsong, Pennsylvania State Univ.
page 14 A. V. Crewe and M. Utlaut, Univ. of Chicago
page 15 L. S. Bartell, Univ. of Michigan

## CHAPTER 1

Fig. 1.7 BIPM-Picture
Fig. 1.9 U.S. National Bueau of Standards
Fig. 1.10 BIPM-Picture
Fig. 1.11 U.S. National Bureau of Standards
Fig. 1.12 U.S. National Bureau of Standards
Fig. 1.13 Photo courtesy of Hewlett-Packard Company
Fig. 1.14 U.S. National Bureau of Standards
Fig. 1.15 U.S. National Bureau of Standards

## CHAPTER 2

Fig. 2.10 PSSC Physics, 2nd ed., 1965, D. C. Heath & Co. with the Educational Development Center, Newton, Mass.
Fig. 2.11 BIPM-Picture
Fig. 2.15 NASA
page 37 (Galileo Galilei) AIP Niels Bohr Library

## INTERLUDE I

Fig. I.1 Reproduced by gracious permission of Her Majesty Queen Elizabeth II; from the Royal Library, Windsor Castle, Berkshire
Fig. I.2 Duvall Corporation
Fig. I.3 R. Gronsky, National Center for Electron Microscopy, Lawrence Berkeley Laboratory
Fig. I.4 Martin J. Buerger, Institute Professor, MIT
Fig. I.5 Professor T. T. Tsong, Pennsylvania State Univ.
Fig. I.6 R. P. Goehner, General Electric
Fig. I.7 J. M. Karansinski, IBM Watson Research Center
Fig. I.8 David Scharf, 1977, through Stockton Books, Inc.
Fig. I.9 Earth Scenes, by Breck P. Kent
Fig. I.10 W. A. Bentley; reproduced from Snow Crystals by W. A. Bentley and W. J. Humphreys, Dover, New York, 1962
Fig. I.11 Hans C. Ohanian
Fig. I.12 Courtesy H. B. Huntington, RPI
Fig. I.13 © Beeldrecht, Amsterdam/VAGA, New York, Collection Haags Gemeentemuseum, The Hague
Fig. I.14 © Beeldrecht, Amsterdam/VAGA, New York, Collection Haags Gemeentemuseum, The Hague
Fig. I.15 © Beeldrecht, Amsterdam/VAGA, New York, Collection Haags Gemeentemuseum, The Hague
Fig. I.16 © Beeldrecht, Amsterdam/VAGA, New York, Collection Haags Gemeentemuseum, The Hague
Fig. I.26 C. S. Smith, MIT
Fig. I.32 Otsuka Kogeisha & Co., Ltd.
Fig. I.33 C. S. Smith, MIT
Fig. I.34 Metropolitan Museum of Art

## CHAPTER 4

Fig. 4.5 PSSC Physics, 2nd ed., 1965, D. C. Heath & Co. with the Educational Development Center, Newton, Mass.
Fig. 4.7 © Smithsonian Institution
Fig. 4.8 Dr. Harold E. Edgerton, MIT
Fig. 4.11 (a) Photo courtesy of The Schenectady Gazette, Schenectady, N.Y.
Fig. 4.16 NASA
Fig. 4.27 Photo by Jeffrey Wood, PH1

## CHAPTER 5

Fig. 5.2 The Smithsonian Institution
Fig. 5.4 NASA
Fig. 5.5 Road & Track
Fig. 5.20 Honeywell, Inc., Sperry Commercial Flight System Group
Fig. 5.22 The Science Museum, London

page 99 (Sir Isaac Newton) Burndy Library, courtesy AIP Niels Bohr Library
page 102 (Ernst Mach) AIP Niels Bohr Library

## CHAPTER 6
Fig. 6.1 (a) H. E. Edgerton; (b) H. E. Edgerton; (c) The Warder Collection; (d) Photo CERN
Fig. 6.3 NASA
Fig. 6.4 NASA
Fig. 6.11 S. J. Calabrese, RPI
Fig. 6.12 S. J. Calabrese, RPI
Fig. 6.31 Leo De Wys, Inc. © Larry Miller
Fig. 6.35 U.S. National Bureau of Standards
page 123 (Pierre Simon, Marquis de Place) AIP Niels Bohr Library
page 136 (Leonardo da Vinci) NMAH Archives Center, Smithsonian Institution

## CHAPTER 7
Fig. 7.21 Neil Liefer, Sports Illustrated
Fig. 7.23 Six Flags Great Adventure Family Entertainment Center
page 161 (James Prescott Joule) © British Museum
page 174 (Christiaan Huygens) The Bettmann Archive

## CHAPTER 8
Fig. 8.4 Courtesy Calspan Corp., Buffalo, N.Y.
Fig. 8.10 U.S. Air Force Photo
page 186 (Joseph Louis LaGrange) The Granger Collection
page 196 (Hermann von Helmholz) AIP Niels Bohr Library
page 198 (James Watt) © British Museum

## CHAPTER 9
Fig. 9.5 (a) From Cavendish, Philosophical Transactions of the Royal Society 18 (1798): 388
Fig. 9.14 Sovfoto Agency
Fig. 9.15 NASA
Fig. 9.34 NASA
Fig. 9.37 NFB Phototheque, photos by G. Blouin, 1949
Fig. 9.38 © 1987 Esselte Map Service AB Stockholm
Fig. 9.40 NASA
Fig. 9.45 National Optical Astronomy Observatory
page 215 (Henry Cavendish) © British Museum
page 216 (Nicolas Copernicus) National Portrait Gallery, Smithsonian Institution
page 219 (Johannes Kepler) AIP Niels Bohr Library

## INTERLUDE II
Fig. II.1 (a) Palomar Observatory Photograph; (b) Palomar Observatory Photograph
Fig. II.2 Hale Observatories
Fig. II.3 Hale Observatories
Fig. II.4 Palomar Observatory Photograph
Fig. II.5 Palomar Observatory Photograph
Fig. II.6 Palomar Observatory Photograph
Fig. II.7 Palomar Observatory Photograph
Fig. II.8 Palomar Observatory Photograph
Fig. II.9 Palomar Observatory Photograph
Fig. II.11 Palomar Observatory Photograph
Fig. II.12 AT&T Bell Laboratories

## CHAPTER 10
Fig. 10.6 PSSC Physics, 2nd ed., 1965, D. C. Heath & Co., with the Educational Development Center, Newton, Mass.
Fig. 10.12 The Bettmann Archive
Fig. 10.15 (a) NASA
Fig. 10.17 NASA
Fig. 10.18 Duomo Photography, Inc.

## CHAPTER 11
Fig. 11.1 Mercedes-Benz of North America, Inc.
Fig. 11.9 Reprinted from Philosophical Magazine, R. H. Brown et al., vol. 40, 1949
Fig. 11.10 © Fundamental Photographs, 1972
Fig. 11.11 W. B. Hamilton, U.S. Geological Survey
Fig. 11.18 C. F. Powell and G. P. S. Occhialini, Nuclear Physics in Photographs, Oxford Univ. Press, New York

## INTERLUDE III
Fig. III.4 Adapted from L. C. Lundstrom, SAE Transactions 75, 1967; reprinted with permission © 1967 Society of Automotive Engineers, Inc.
Fig. III.5 From Modern Automotive Structural Analysis by J. A. Wolf, Van Nostrand Reinhold, New York, 1982
Fig. III.6 Adapted from K. H. Lin, M. M. Kamal, J. W. Justusson, SAE paper 750117, 1975; reprinted with permission © 1975 Society of Automotive Engineers, Inc.
Fig. III.7 Volvo Cars of North America
Fig. III.12 Humanoid Systems, Humanetics, Inc.
Fig. III.13 Insurance Institute for Highway Safety
Fig. III.16 From The Prevention of Highway Injury, Marvin L. Selzer, Paul W. Gikas, and Donald F. Huelke, eds., Highway Safety Research Institute, 1967
Fig. III.17 Bob Costanzo, Daytona International Speedway
Fig. III.18 Mercedes-Benz of North America, Inc.
Fig. III.19 Courtesy Colonel John P. Stapp
Fig. III.22 Minicars, Inc.
Fig. III.24 Volvo Cars of North America

## CHAPTER 12
Fig. 12.1 Dr. Harold E. Edgerton, MIT

## CHAPTER 13
Fig. 13.14 Sperry Flight Systems
Fig. 13.21 PSSC Physics, 2nd ed., 1965, D. C. Heath & Co. with the Educational Development Center, Newton, Mass.
Fig. 13.27 Wide World Photos
Fig. 13.32 Chicago Historical Society
Fig. 13.41 NASA
Fig. 13.44 Lufthansa Photo

## CHAPTER 14
Fig. 14.9 Golden Gate Bridge, Highway & Transportation District
Fig. 14.26 The Port Authority of NJ & NY
Fig. 14.37 Joseph Daniel, Wildpic

## CHAPTER 15
Fig. 15.6 NASA
Fig. 15.8 Adapted from Energy, A Sequel to IPS, by Uri Haber-Schaim, Prentice-Hall, Englewood Cliffs, N.J., 1983
Fig. 15.14 © Lester Lefkowitz
Fig. 15.17 U.S. National Bureau of Standards
Fig. 15.25 Portescap, La Chaux-de-Fonds, Suisse; Service de Presse, Nicole Rouge

## CHAPTER 16

Fig. 16.22 UPI © Bashford Thompson, Tacoma, Wash.

## CHAPTER 17

Fig. 17.1 Fundamental Photographs
Fig. 17.2 Engineering Applications of Lasers and Holography by Winston E. Kock, Plenum Publishing Co., New York, 1975
Fig. 17.5 William B. Joyce
Fig. 17.8 S. Gilmore, General Electric Research and Development Center
Fig. 17.9 Courtesy of C. F. Quate and L. Lam, Hansen Laboratories, Stanford Univ.
Fig. 17.18 Dr. Harold E. Edgerton, MIT
Fig. 17.23 Mrs. Harry A. Simms, Sr.
Fig. 17.25 California Historical Society, San Francisco; Photographer: Arnold Genthe, New York, FN-27239
Fig. 17.26 R. Such, Lamont-Doherty Geological Observatory of Columbia University
Fig. 17.28 Courtesy of John S. Shelton
Fig. 17.29 PSSC Physics, 2nd ed., 1965, D. C. Heath & Co. with the Educational Development Center, Newton, Mass.
Fig. 17.30 PSSC Physics, 2nd ed., 1965, D. C. Heath & Co. with the Educational Development Center, Newton, Mass.
Fig. 17.31 Educational Development Center, Newton, Mass.
Fig. 17.32 Educational Development Center, Newton, Mass.
page 446 (Christian Doppler) AIP Niels Bohr Library

## INTERLUDE IV

Fig. IV.2 Susan Leavines, Photo Researchers, Inc.
Fig. IV.3 Reprinted from the Proceedings of the Royal Society, P. M. S. Blackett and D. S. Lees, 1932
Fig. IV.8 Reprinted from the Proceedings of the Royal Society, P. M. S. Blackett and D. S. Lees, 1932
Fig. IV.13 Nuclear Medicine Section, Dept. of Radiology, Hospital of the Univ. of Pennsylvania
Fig. IV.14 Michael G. Velchik, M.D., Nuclear Medicine Section, Dept. of Radiology, Hospital of the Univ. of Pennsylvania
Fig. IV.15 Michael G. Velchik, M.D., Nuclear Medicine Section, Dept. of Radiology, Hospital of the Univ. of Pennsylvania
Fig. IV.16 Micheal G. Velchik, M.D., Nuclear Medicine Section, Dept. of Radiology, Hospital of the Univ. of Pennsylvania
Fig. IV.17 M. D. Anderson, Hospital and Tumor Institute, Univ. of Texas, Houston

## CHAPTER 18

Fig. 18.1 Boston Museum of Science photo by Bradford Washburn
Fig. 18.8 A. D. Moore, Univ. of Michigan, from Introduction to Electric Fields by W. E. Rogers, McGraw-Hill Book Co., New York, 1954
Fig. 18.9 D. C. Hazen and R. F. Lehnert, Subsonic Aerodynamics Laboratory, Princeton
Fig. 18.10 D. C. Hazen and R. F. Lehnert, Subsonic Aerodynamics Laboratory, Princeton
Fig. 18.11 Photo courtesy of MCO Properties, Inc.
Fig. 18.14 (a) Naval Photographic Center
Fig. 18.23 The Balloon Works, Statesville, N.C.
Fig. 18.31 Hans C. Ohanian
Fig. 18.33 NASA
Fig. 18.35 U.S. Navy Photo
Fig. 18.41
page 473 (Blaise Pascal) National Portrait Gallery, Smithsonian Institution
page 480 (Daniel Bernoulli) National Portrait Gallery, Smithsonian Institution

## CHAPTER 19

Fig. 19.12 Physical Science Laboratory, National Scientific Balloon Facility
page 495 (Lord Kelvin) Wellcome Institute Library, London
page 496 (Robert Boyle) © British Museum
page 506 (Ludwig Boltzmann) AIP Niels Bohr Library

## CHAPTER 20

Fig. 20.5 The Warder Collection
Fig. 20.6 AP, Wide World Photos
Fig. 20.9 U.S. National Bureau of Standards
Fig. 20.10 (b) U.S. National Bureau of Standards
Fig. 20.12 Photo CERN
Fig. 20.17 Robert Azzi, Woodfin Camp & Associates
Fig. 20.18 VANSCAN® thermogram by Daedalus Enterprises, Inc.
page 515 (Benjamin Rumford) © British Museum

## CHAPTER 21

Fig. 21.22 Vorpal Galleries: New York, San Francisco, Laguna Beach, Calif.
page 550 (Sadi Carnot) AIP Niels Bohr Library
page 555 (Rudolph Clausius) National Portrait Gallery, Smithsonian Institution
page 564 (Walther Nernst) NMAH Archives Center, Smithsonian Institution

## INTERLUDE V

Fig. V.6 UPI, Bettmann Archives
Fig. V.7 Los Alamos National Laboratory
Fig. V.8 Bureau of Reclamation
Fig. V.9 NASA
Fig. V.10 French Embassy Press & Information Division
Fig. V.11 Southern California Edison Co.
Fig. V.12 (b) Advanced Energy Technologies
Fig. V.13 NASA
Fig. V.14 GM Hughes Electronics
Fig. V.15 PG&E Photographs
Fig. V.16 PG&E Photographs
Fig. V.17 French Embassy Press & Information Division
Fig. V.18 New York Power Authority
Fig. V.19 PG&E Photographs
Fig. V.20 Novosti, Gamma
Fig. V.22 Kansas City Power & Light, Kansas Gas & Electric

# INDEX



In the two-volume edition, all pages after 570 are in Volume Two.
Interlude V is the last one in Volume One.

In the two-volume edition, all pages after 570 are in Volume Two.
Interlude V is the last one in Volume One.

In the two-volume edition, all pages after 570 are in Volume Two.
Interlude V is the last one in Volume One.

In the two-volume edition, all pages after 570 are in Volume Two.
Interlude V is the last one in Volume One.

In the two-volume edition, all pages after 570 are in Volume Two.
Interlude V is the last one in Volume One.

In the two-volume edition, all pages after 570 are in Volume Two.
Interlude V is the last one in Volume One.

In the two-volume edition, all pages after 570 are in Volume Two.
Interlude V is the last one in Volume One.

In the two-volume edition, all pages after 570 are in Volume Two.
Interlude V is the last one in Volume One.

In the two-volume edition, all pages after 570 are in Volume Two.
Interlude V is the last one in Volume One.

In the two-volume edition, all pages after 570 are in Volume Two.
Interlude V is the last one in Volume One.

In the two-volume edition, all pages after 570 are in Volume Two.
Interlude V is the last one in Volume One.

In the two-volume edition, all pages after 570 are in Volume Two.
Interlude V is the last one in Volume One.

In the two-volume edition, all pages after 570 are in Volume Two.
Interlude V is the last one in Volume One.

In the two-volume edition, all pages after 570 are in Volume Two.
Interlude V is the last one in Volume One.

In the two-volume edition, all pages after 570 are in Volume Two.
Interlude V is the last one in Volume One.

In the two-volume edition, all pages after 570 are in Volume Two.
Interlude V is the last one in Volume One.

In the two-volume edition, all pages after 570 are in Volume Two.
Interlude V is the last one in Volume One.

In the two-volume edition, all pages after 570 are in Volume Two.
Interlude V is the last one in Volume One.

In the two-volume edition, all pages after 570 are in Volume Two.
Interlude V is the last one in Volume One.

In the two-volume edition, all pages after 570 are in Volume Two.
Interlude V is the last one in Volume One.

In the two-volume edition, all pages after 570 are in Volume Two.
Interlude V is the last one in Volume One.

In the two-volume edition, all pages after 570 are in Volume Two.
Interlude V is the last one in Volume One.